STUDENT RESOURCE MANUAL

Neil Wigley
University of Windsor
Albert Herr

With Sample Exams Prepared By
Henry Smith
Southeastern Louisiana University

To Accompany

CALCULUS

EARLY TRANSCENDENTALS

Seventh Edition

Howard Anton
Drexel University
Irl C. Bivens
Davidson College
Stephen L. Davis
Davidson College

John Wiley & Sons, Inc.

Cover Design: Norm Christensen

To order books or for customer service call 1-800-CALL-WILEY (225-5945).

ISBN 0-471-44172-4

Printed in the United States of America

10 9 8 7 6 5 4 3

Printed and bound by Victor Graphics Inc.

CONTENTS

Solutions

Sample Exams

CHAPTER 1
Functions

EXERCISE SET 1.1

1. **(a)** around 1943 **(b)** 1960; 4200
(c) no; you need the year's population **(d)** war; marketing techniques
(e) news of health risk; social pressure, antismoking campaigns, increased taxation

3. **(a)** $-2.9, -2.0, 2.35, 2.9$ **(b)** none **(c)** $y = 0$
(d) $-1.75 \leq x \leq 2.15$ **(e)** $y_{\max} = 2.8$ at $x = -2.6$; $y_{\min} = -2.2$ at $x = 1.2$

5. **(a)** $x = 2, 4$ **(b)** none **(c)** $x \leq 2$; $4 \leq x$ **(d)** $y_{\min} = -1$; no maximum value

7. **(a)** Breaks could be caused by war, pestilence, flood, earthquakes, for example.
(b) C decreases for eight hours, takes a jump upwards, and then repeats.

9. **(a)** The side adjacent to the building has length x, so $L = x + 2y$. Since $A = xy = 1000$, $L = x + 2000/x$.
(b) $x > 0$ and x must be smaller than the width of the building, which was not given.
(c)

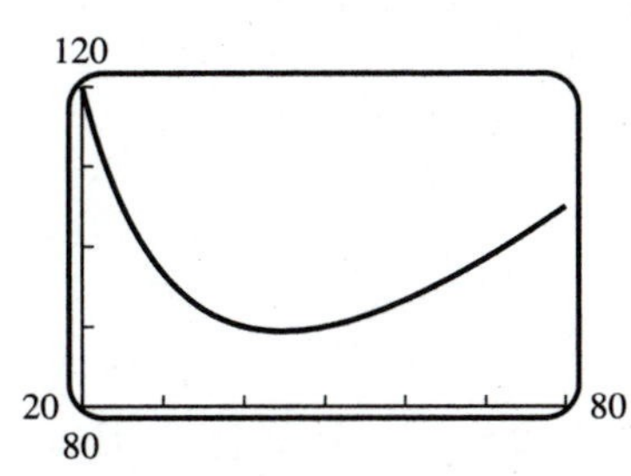

(d) $L_{\min} \approx 89.44$ ft

11. **(a)** $V = 500 = \pi r^2 h$ so $h = \dfrac{500}{\pi r^2}$. Then

$$C = (0.02)(2)\pi r^2 + (0.01)2\pi r h = 0.04\pi r^2 + 0.02\pi r \frac{500}{\pi r^2}$$

$$= 0.04\pi r^2 + \frac{10}{r};\ C_{\min} \approx 4.39 \text{ at } r \approx 3.4,\ h \approx 13.8.$$

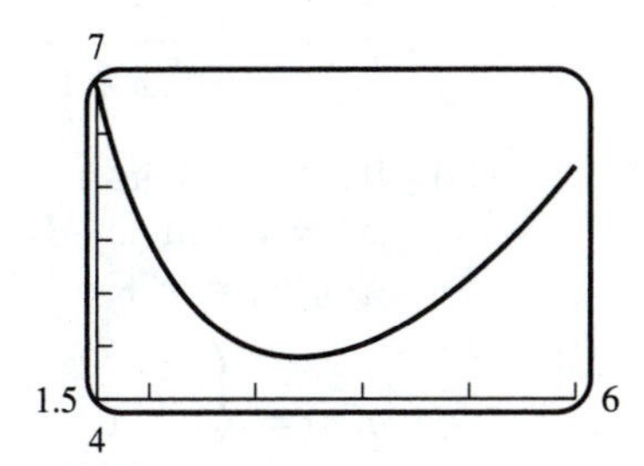

(b) $C = (0.02)(2)(2r)^2 + (0.01)2\pi r h = 0.16r^2 + \dfrac{10}{r}$. Since $0.04\pi < 0.16$, the top and bottom now get more weight. Since they cost more, we diminish their sizes in the solution, and the cans become taller.

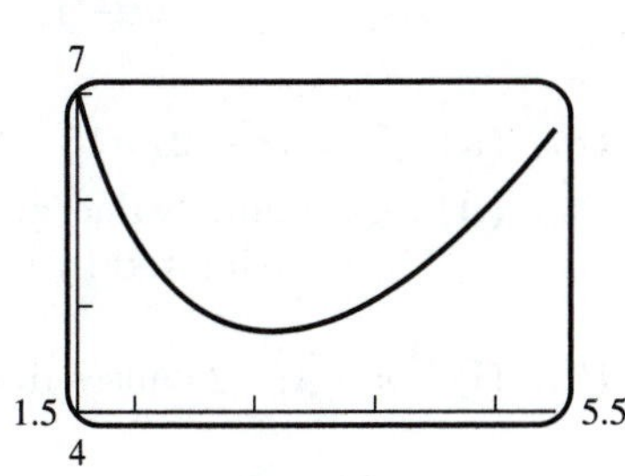

(c) $r \approx 3.1$ cm, $h \approx 16.0$ cm, $C \approx 4.76$ cents

EXERCISE SET 1.2

1. **(a)** $f(0) = 3(0)^2 - 2 = -2$; $f(2) = 3(2)^2 - 2 = 10$; $f(-2) = 3(-2)^2 - 2 = 10$; $f(3) = 3(3)^2 - 2 = 25$; $f(\sqrt{2}) = 3(\sqrt{2})^2 - 2 = 4$; $f(3t) = 3(3t)^2 - 2 = 27t^2 - 2$

(b) $f(0) = 2(0) = 0$; $f(2) = 2(2) = 4$; $f(-2) = 2(-2) = -4$; $f(3) = 2(3) = 6$; $f(\sqrt{2}) = 2\sqrt{2}$; $f(3t) = 1/3t$ for $t > 1$ and $f(3t) = 6t$ for $t \le 1$.

3. **(a)** $x \ne 3$ **(b)** $x \le -\sqrt{3}$ or $x \ge \sqrt{3}$

(c) $x^2 - 2x + 5 = 0$ has no real solutions so $x^2 - 2x + 5$ is always positive or always negative. If $x = 0$, then $x^2 - 2x + 5 = 5 > 0$; domain: $(-\infty, +\infty)$.

(d) $x \ne 0$ **(e)** $\sin x \ne 1$, so $x \ne (2n + \frac{1}{2})\pi$, $n = 0, \pm 1, \pm 2, \ldots$

5. **(a)** $x \le 3$ **(b)** $-2 \le x \le 2$ **(c)** $x \ge 0$ **(d)** all x **(e)** all x

7. **(a)** yes **(b)** yes

(c) no (vertical line test fails) **(d)** no (vertical line test fails)

9. The cosine of θ is $(L - h)/L$ (side adjacent over hypotenuse), so $h = L(1 - \cos\theta)$.

11.

h

t

13. **(a)** If $x < 0$, then $|x| = -x$ so $f(x) = -x + 3x + 1 = 2x + 1$. If $x \ge 0$, then $|x| = x$ so $f(x) = x + 3x + 1 = 4x + 1$;

$$f(x) = \begin{cases} 2x + 1, & x < 0 \\ 4x + 1, & x \ge 0 \end{cases}$$

(b) If $x < 0$, then $|x| = -x$ and $|x - 1| = 1 - x$ so $g(x) = -x + 1 - x = 1 - 2x$. If $0 \le x < 1$, then $|x| = x$ and $|x - 1| = 1 - x$ so $g(x) = x + 1 - x = 1$. If $x \ge 1$, then $|x| = x$ and $|x - 1| = x - 1$ so $g(x) = x + x - 1 = 2x - 1$;

$$g(x) = \begin{cases} 1 - 2x, & x < 0 \\ 1, & 0 \le x < 1 \\ 2x - 1, & x \ge 1 \end{cases}$$

15. **(a)** $V = (8 - 2x)(15 - 2x)x$ **(b)** $-\infty < x < +\infty, -\infty < V < +\infty$ **(c)** $0 < x < 4$

(d) minimum value at $x = 0$ or at $x = 4$; maximum value somewhere in between (can be approximated by zooming with graphing calculator)

17. **(i)** $x = 1, -2$ causes division by zero **(ii)** $g(x) = x + 1$, all x

19. **(a)** 25°F **(b)** 2°F **(c)** −15°F

21. If $v = 8$ then $-10 = \text{WCI} = 91.4 + (91.4 - T)(0.0203(8) - 0.304\sqrt{8} - 0.474)$; thus $T = 91.4 + (10 + 91.4)/(0.0203(8) - 0.304\sqrt{8} - 0.474)$ and $T = 5$°F

23. Let t denote time in minutes after 9:23 AM. Then $D(t) = 1000 - 20t$ ft.

EXERCISE SET 1.3

1. (e) seems best, though only (a) is bad.

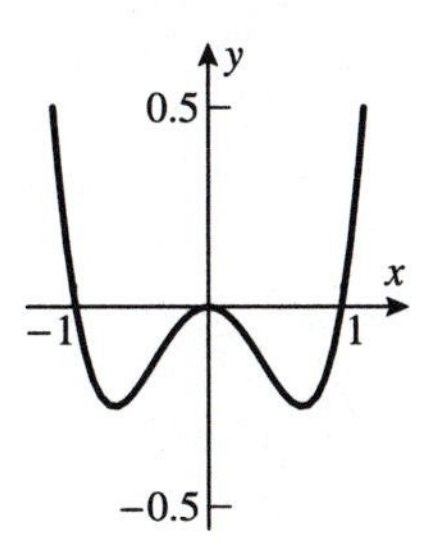

3. (b) and (c) are good; (a) is very bad.

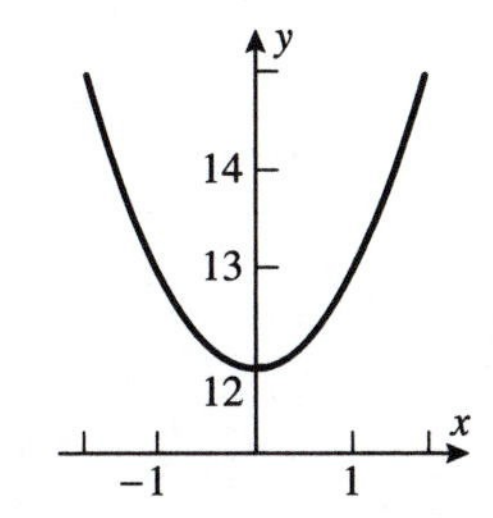

5. $[-3, 3] \times [0, 5]$

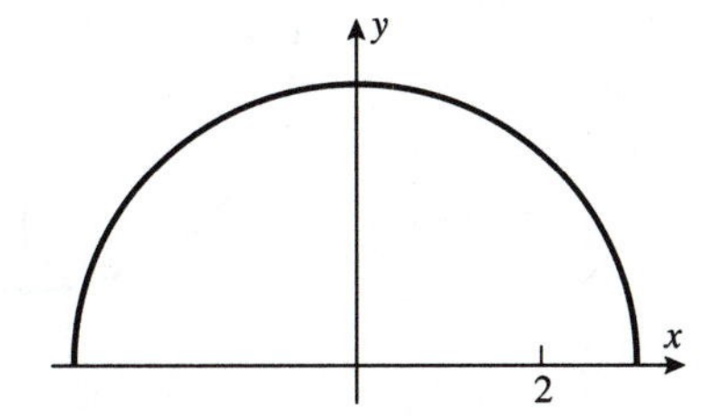

7. **(a)** window too narrow, too short
(b) window wide enough, but too short
(c) good window, good spacing
(d) window too narrow, too short

(e) window too narrow, too short

9. $[-5, 14] \times [-60, 40]$

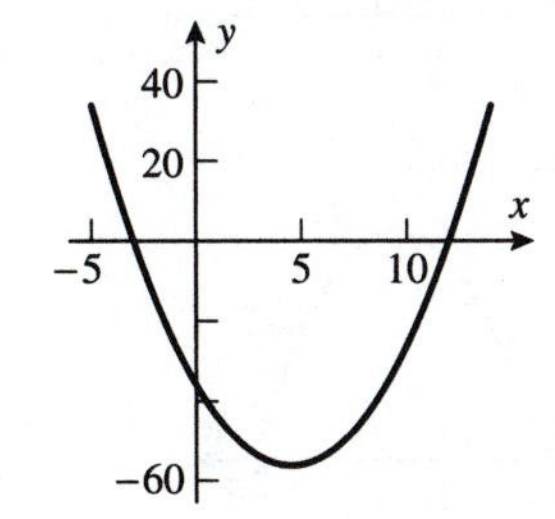

11. $[-0.1, 0.1] \times [-3, 3]$

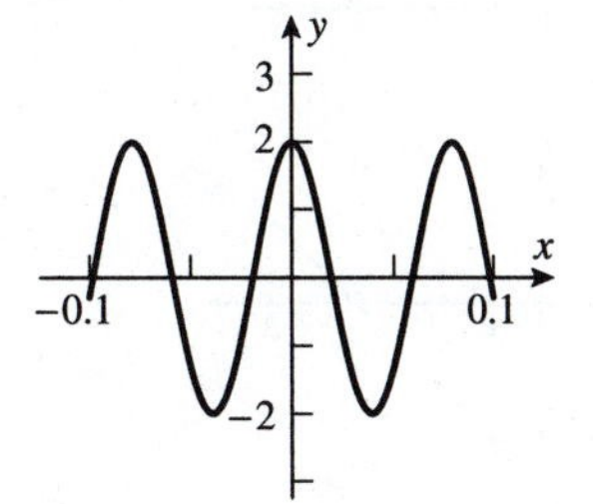

13. $[-250, 1050] \times [-1500000, 600000]$

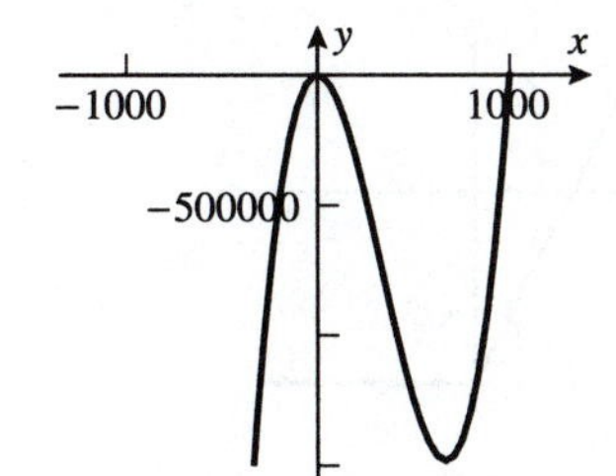

15. $[-2, 2] \times [-20, 20]$

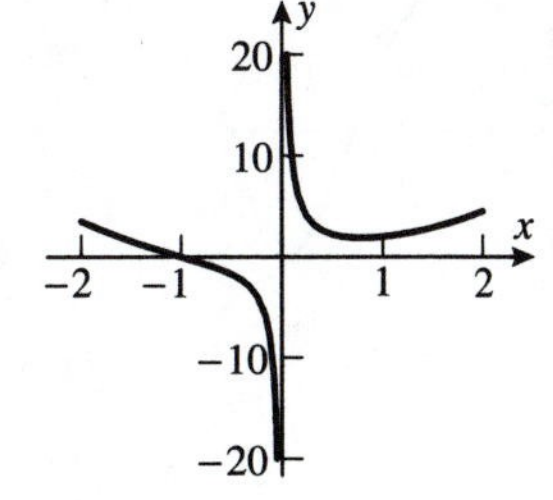

17. depends on graphing utility

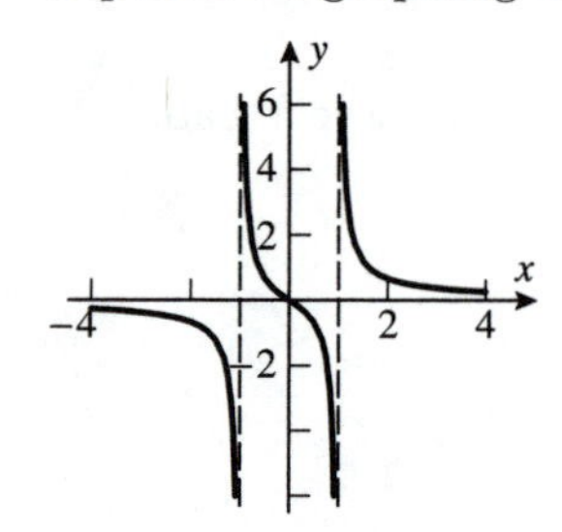

19. **(a)** $f(x) = \sqrt{16 - x^2}$

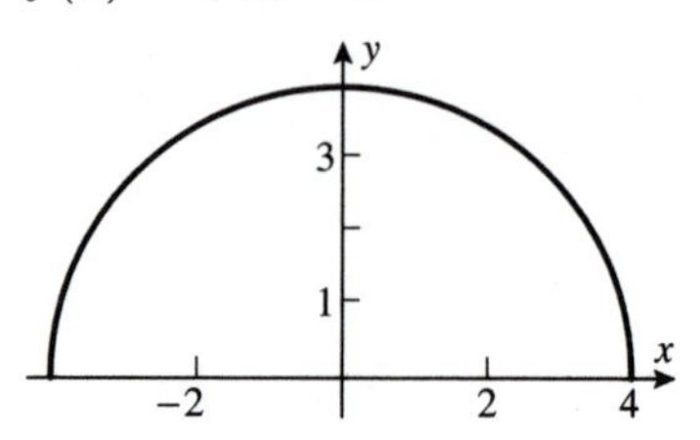

(b) $f(x) = -\sqrt{16 - x^2}$

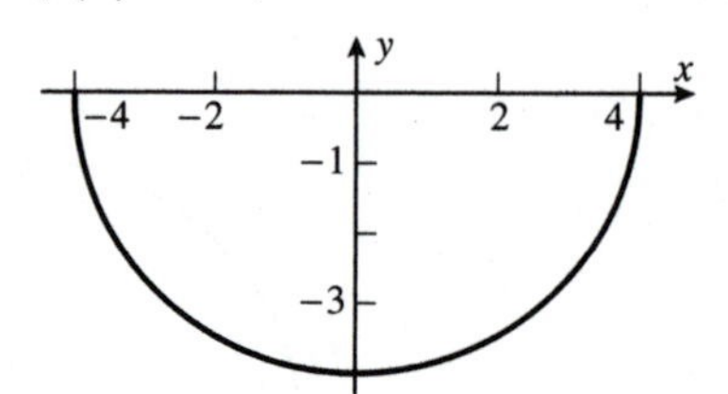

(c)

(d)

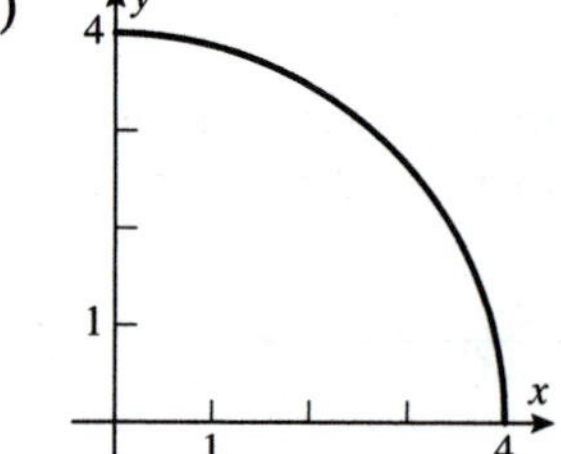

(e) No; the vertical line test fails.

21. **(a)**

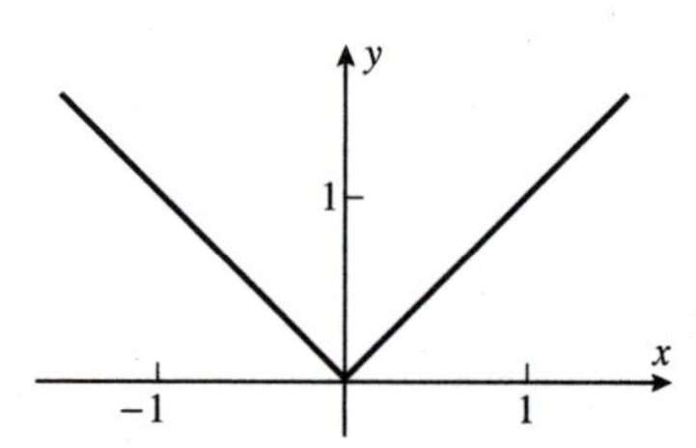

(b)

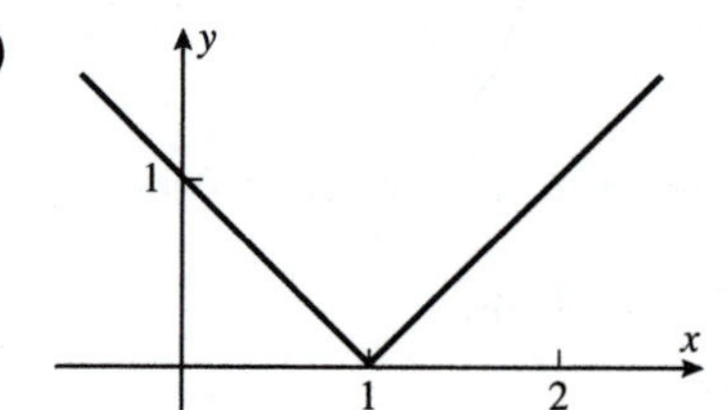

(c)

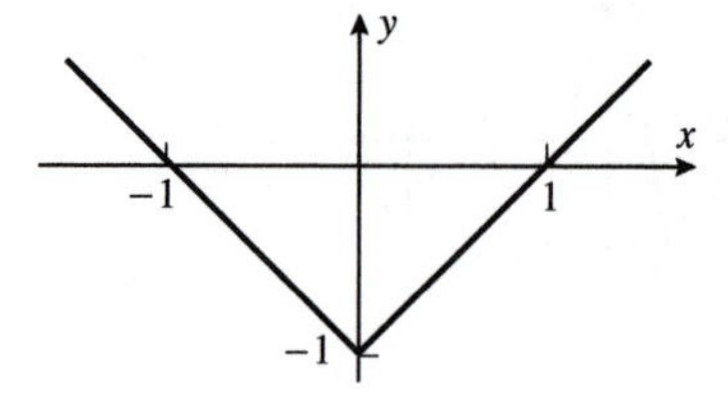

(d)

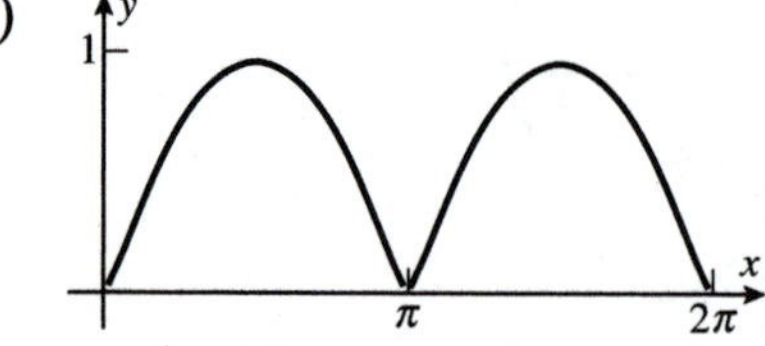

(e)

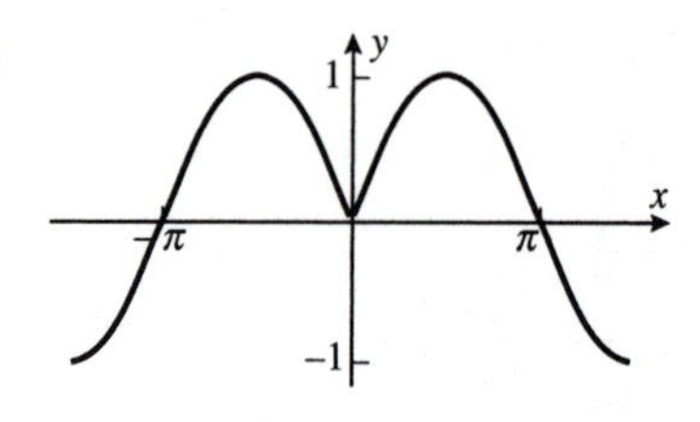

(f)

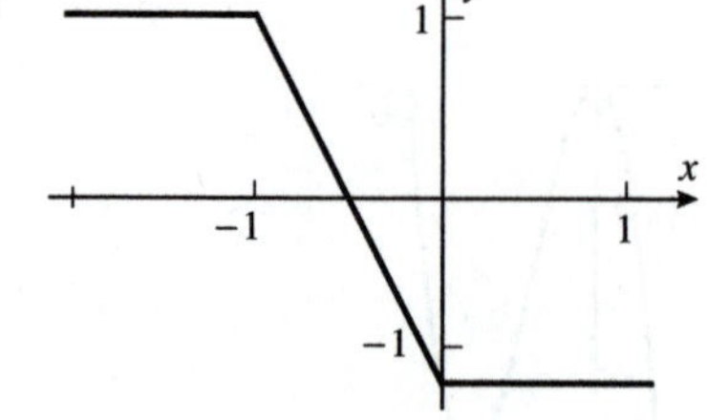

23. The portions of the graph of $y = f(x)$ which lie below the x-axis are reflected over the x-axis to give the graph of $y = |f(x)|$.

25. **(a)** for example, let $a = 1.1$

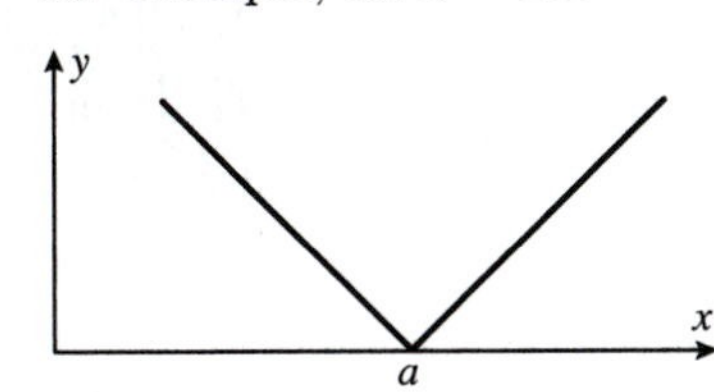

(b)

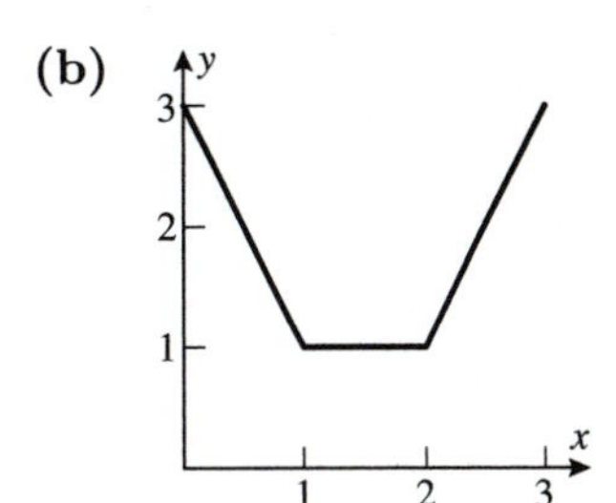

27.

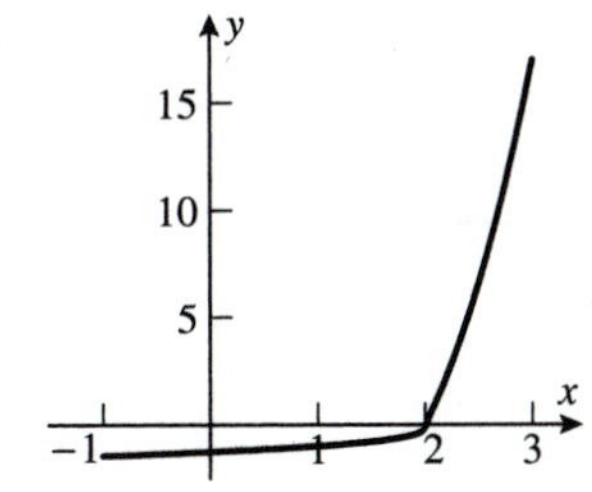

29. **(a)**

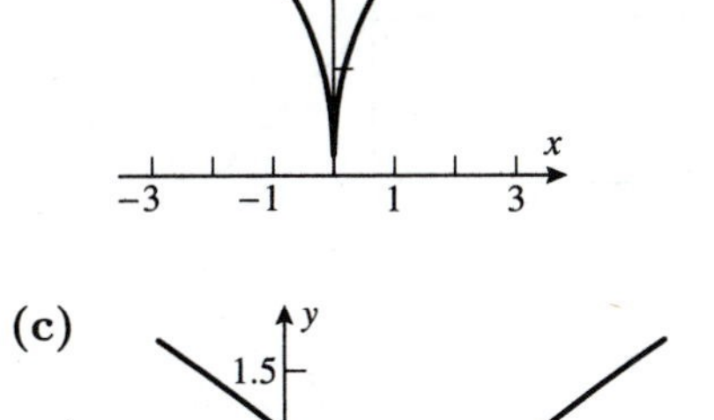

(b)

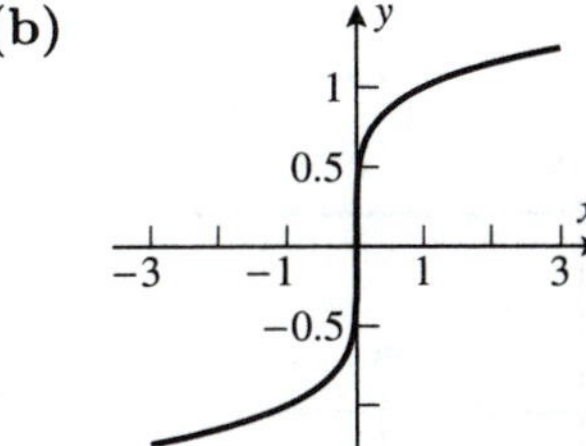

(c)

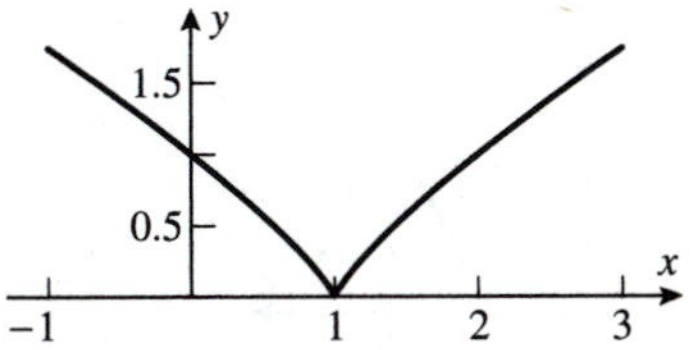

(d)

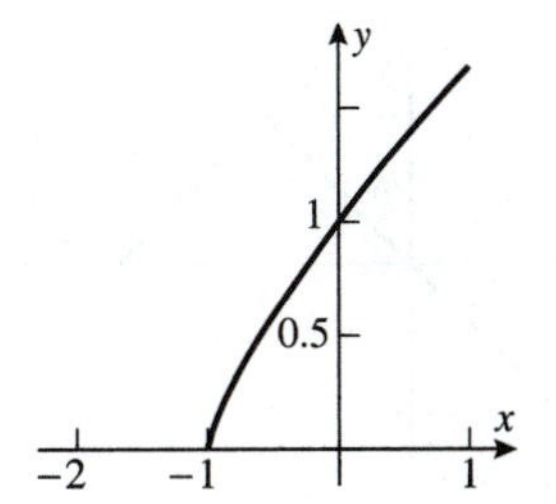

31. **(a)** stretches or shrinks the graph in the y-direction; reflects it over the x-axis if c changes sign

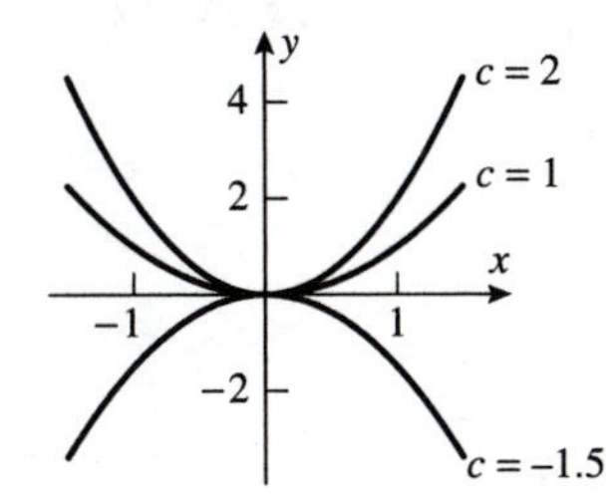

(b) As c increases, the parabola moves down and to the left. If c increases, up and right.

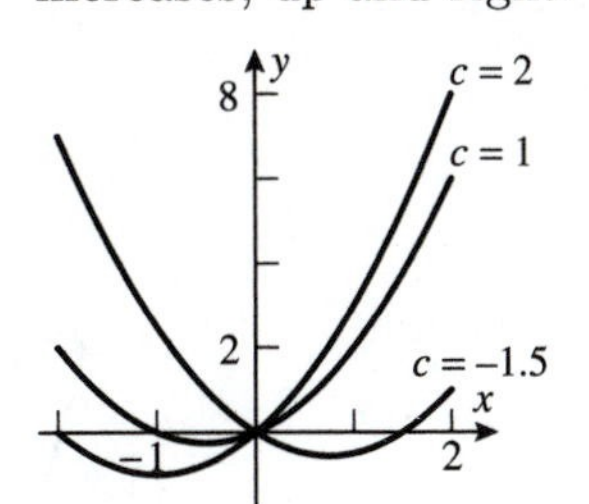

(c) The graph rises or falls in the y-direction with changes in c.

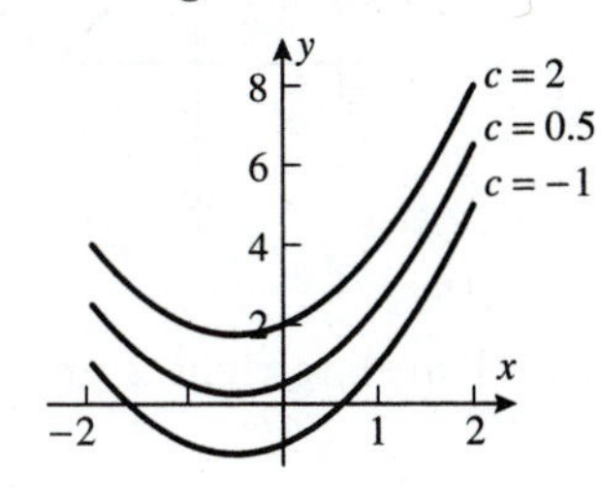

33. The curve oscillates between the lines $y = x$ and $y = -x$ with increasing rapidity as $|x|$ increases.

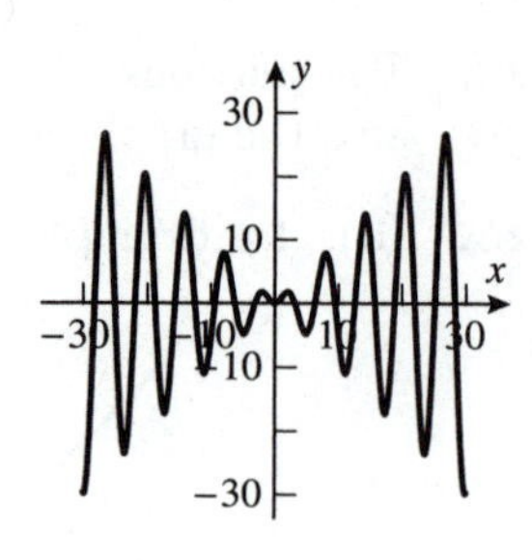

EXERCISE SET 1.4

1. **(a)**

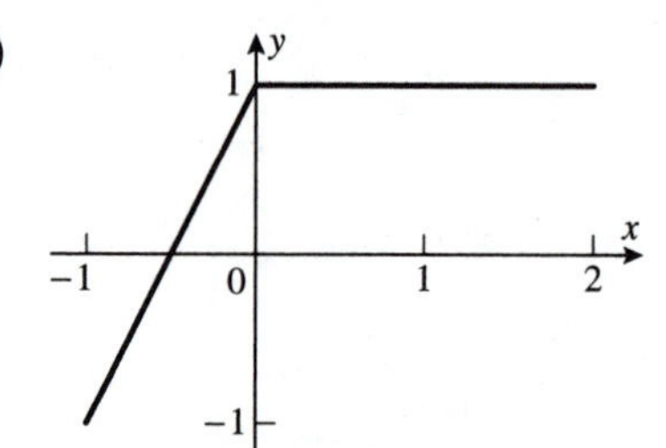

(b)

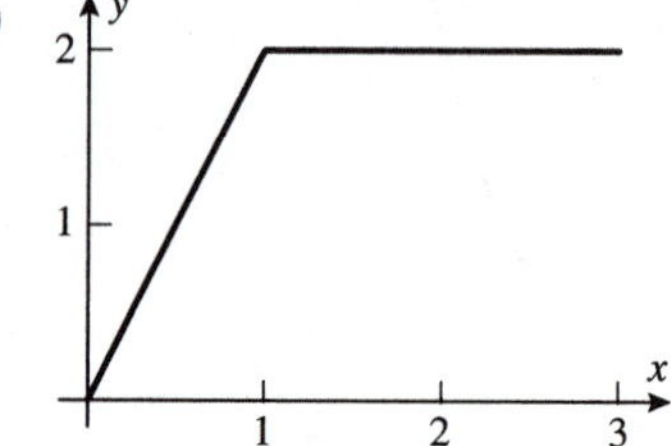

(c)

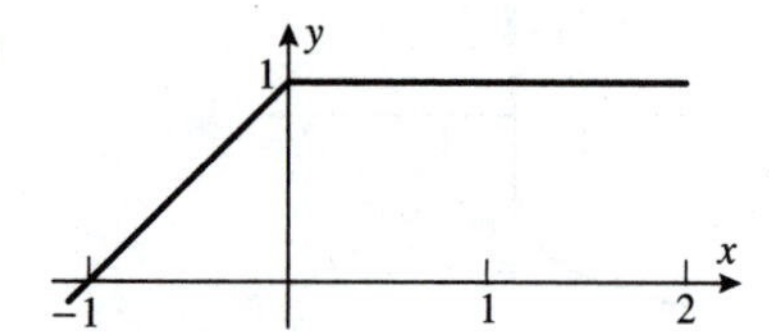

(d)

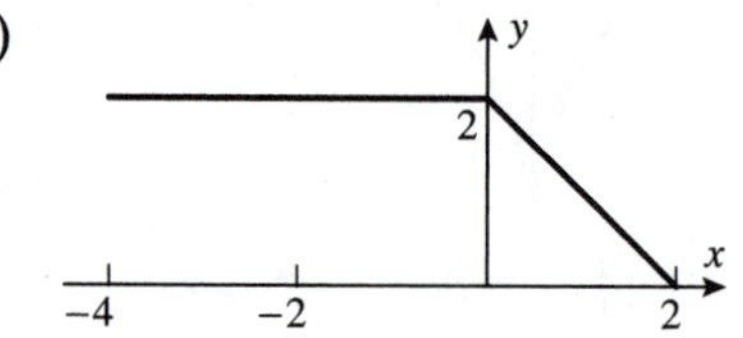

3. **(a)**

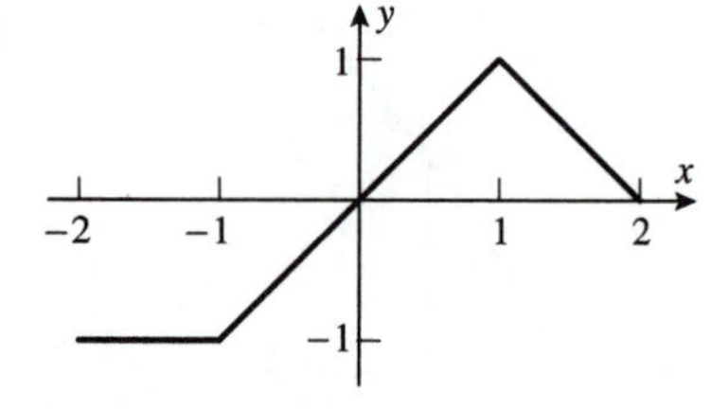

(b)

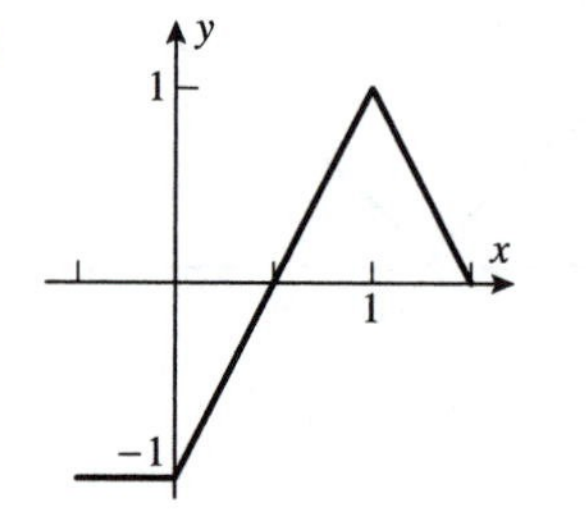

(c)

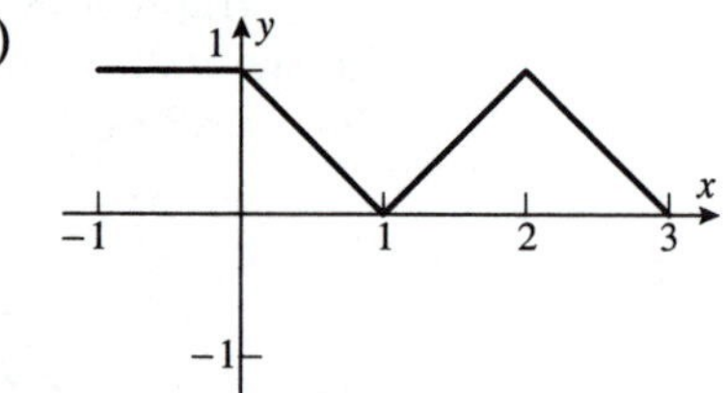

(d)

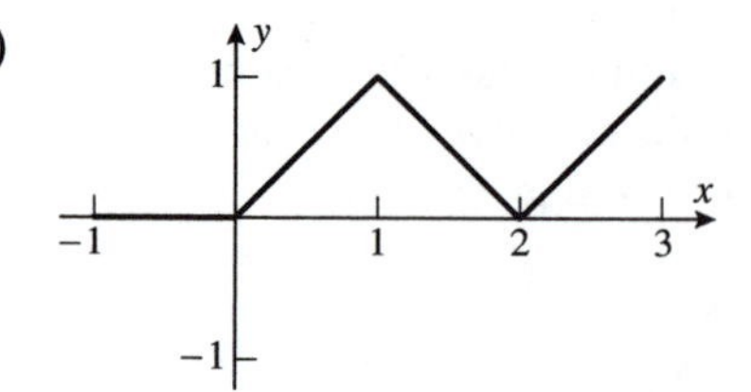

5. Translate right 2 units, and up one unit.

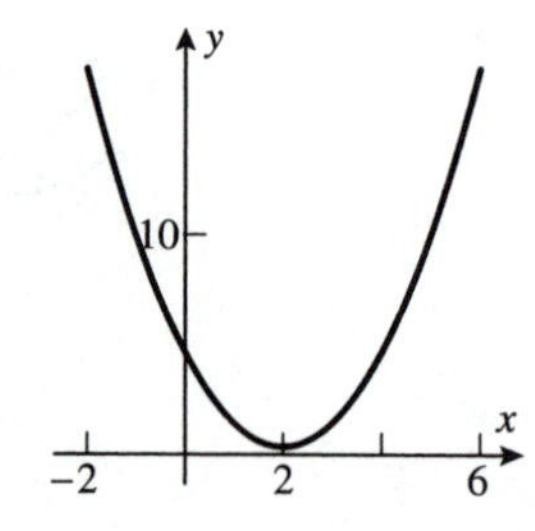

7. Translate left 1 unit, stretch vertically by a factor of 2, reflect over x-axis, translate down 3 units.

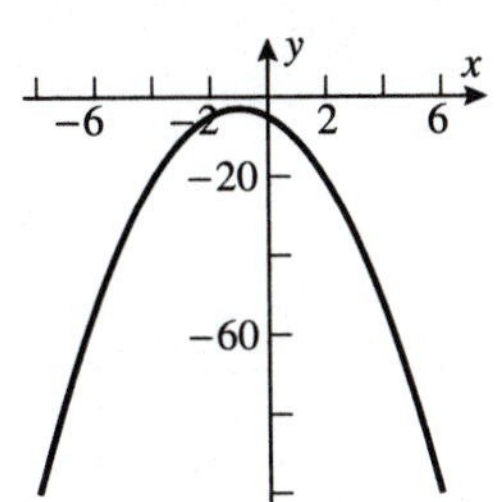

9. $y = (x+3)^2 - 9$; translate left 3 units and down 9 units.

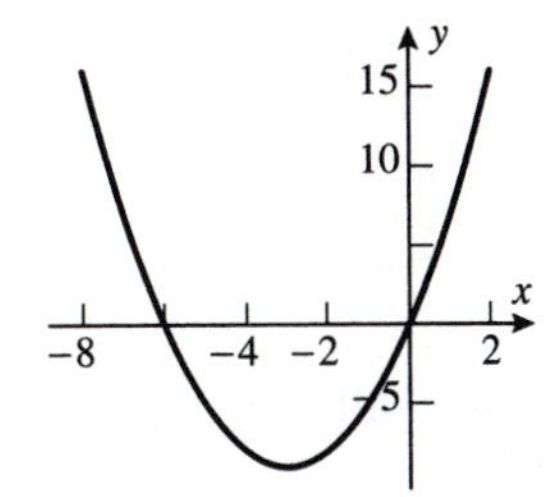

11. $y = -(x-1)^2 + 2$; translate right 1 unit, reflect over x-axis, translate up 2 units.

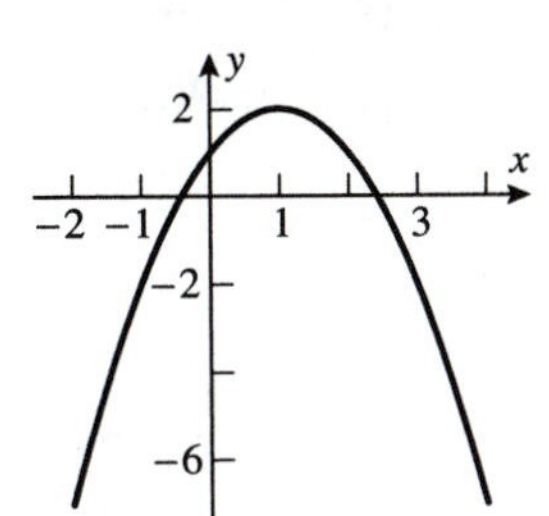

13. Translate left 1 unit, reflect over x-axis, translate up 3 units.

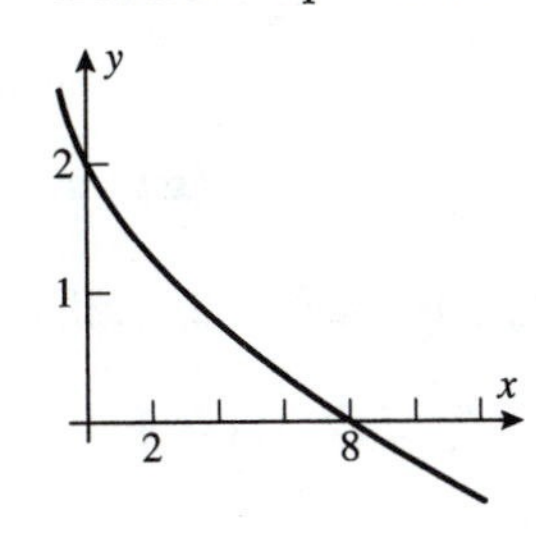

15. Compress vertically by a factor of $\frac{1}{2}$, translate up 1 unit.

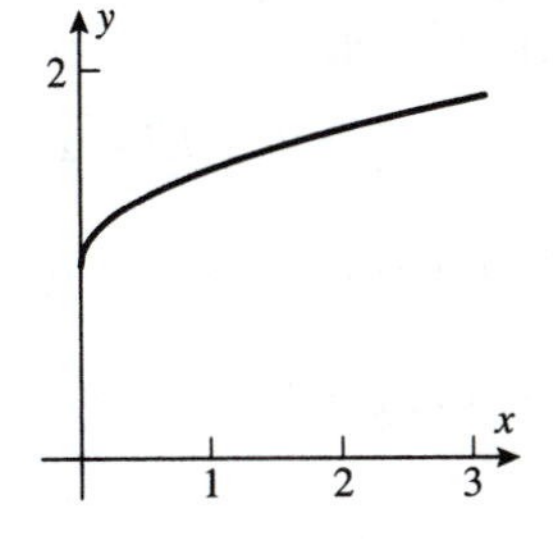

17. Translate right 3 units.

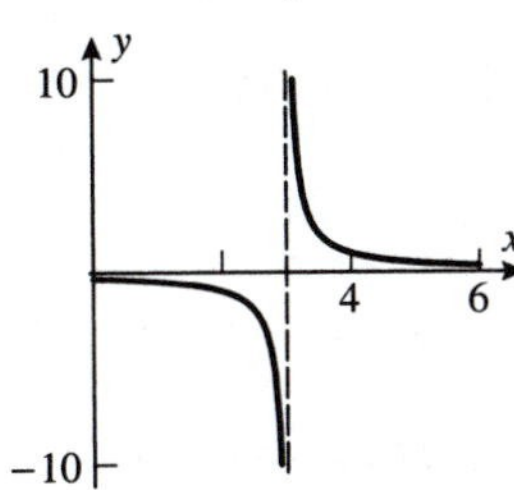

19. Translate left 1 unit, reflect over x-axis, translate up 2 units.

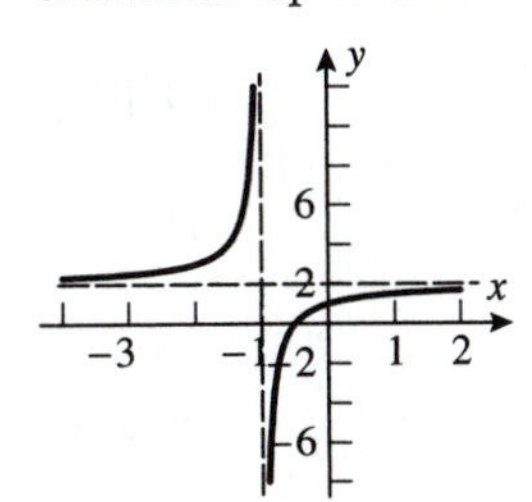

21. Translate left 2 units and down 2 units.

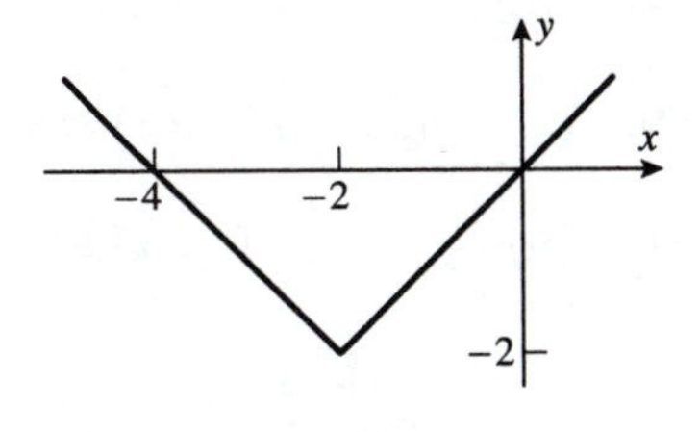

23. Stretch vertically by a factor of 2, translate right 1 unit and up 1 unit.

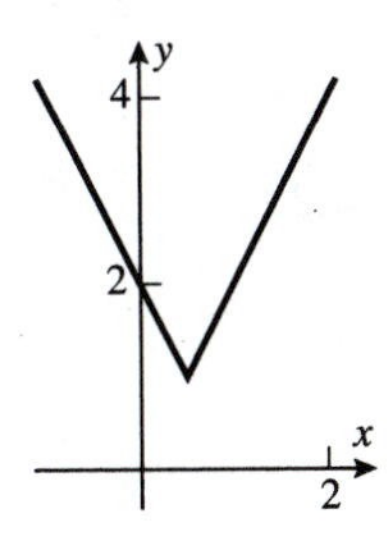

25. Stretch vertically by a factor of 2, reflect over x-axis, translate up 1 unit.

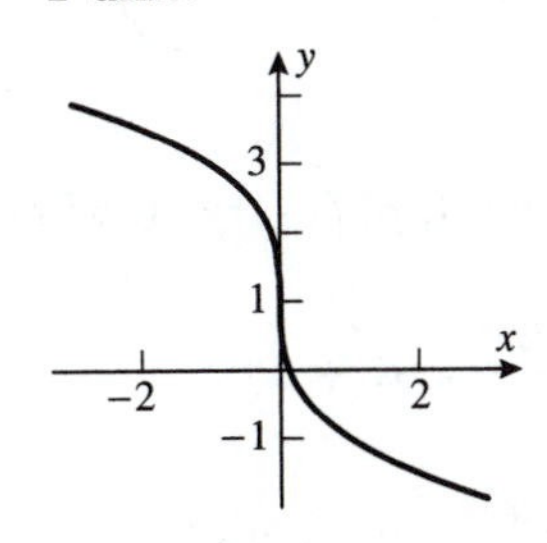

27. Translate left 1 unit and up 2 units.

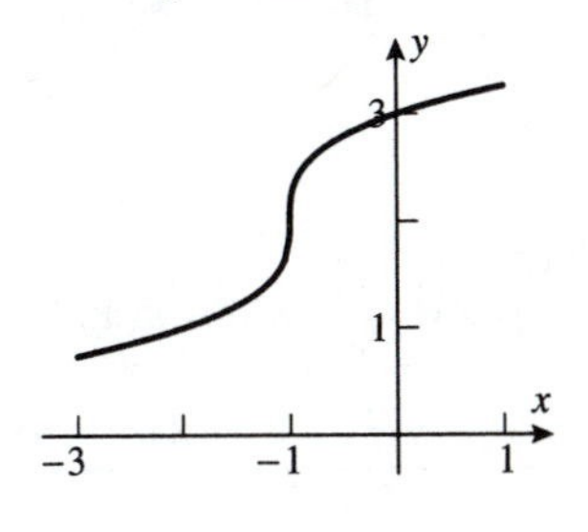

29. (a)

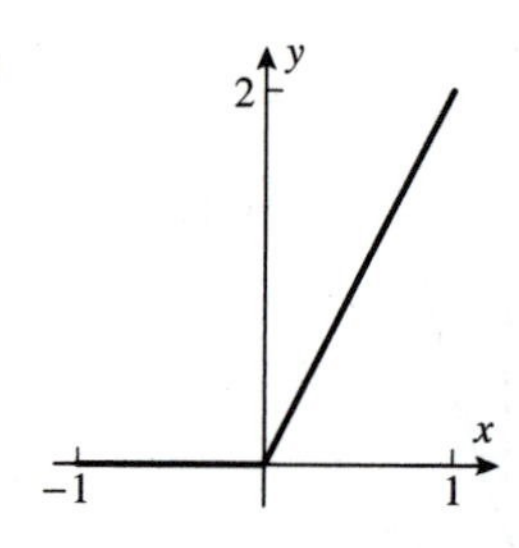

(b) $y = \begin{cases} 0 \text{ if } x \le 0 \\ 2x \text{ if } 0 < x \end{cases}$

31. $(f+g)(x) = x^2 + 2x + 1$, all x; $(f-g)(x) = 2x - x^2 - 1$, all x; $(fg)(x) = 2x^3 + 2x$, all x; $(f/g)(x) = 2x/(x^2+1)$, all x

33. $(f+g)(x) = 3\sqrt{x-1}$, $x \ge 1$; $(f-g)(x) = \sqrt{x-1}$, $x \ge 1$; $(fg)(x) = 2x - 2$, $x \ge 1$; $(f/g)(x) = 2, x > 1$

35. (a) 3 (b) 9 (c) 2 (d) 2

37. (a) $t^4 + 1$ (b) $t^2 + 4t + 5$ (c) $x^2 + 4x + 5$ (d) $\dfrac{1}{x^2} + 1$
(e) $x^2 + 2xh + h^2 + 1$ (f) $x^2 + 1$ (g) $x + 1$ (h) $9x^2 + 1$

39. $(f \circ g)(x) = 2x^2 - 2x + 1$, all x; $(g \circ f)(x) = 4x^2 + 2x$, all x

41. $(f \circ g)(x) = 1 - x$, $x \le 1$; $(g \circ f)(x) = \sqrt{1 - x^2}$, $|x| \le 1$

43. $(f \circ g)(x) = \dfrac{1}{1 - 2x}$, $x \ne \dfrac{1}{2}, 1$; $(g \circ f)(x) = -\dfrac{1}{2x} - \dfrac{1}{2}$, $x \ne 0, 1$

45. $x^{-6} + 1$

47. (a) $g(x) = \sqrt{x}$, $h(x) = x + 2$ (b) $g(x) = |x|$, $h(x) = x^2 - 3x + 5$

49. (a) $g(x) = x^2$, $h(x) = \sin x$ (b) $g(x) = 3/x$, $h(x) = 5 + \cos x$

51. (a) $f(x) = x^3$, $g(x) = 1 + \sin x$, $h(x) = x^2$ (b) $f(x) = \sqrt{x}$, $g(x) = 1 - x$, $h(x) = \sqrt[3]{x}$

53.

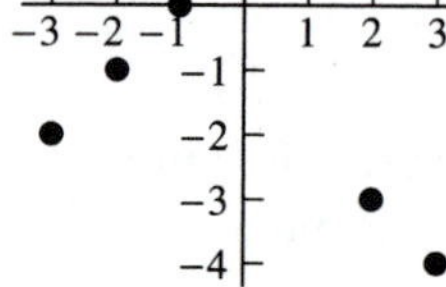

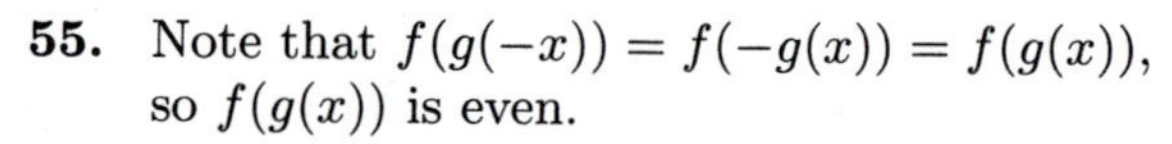

55. Note that $f(g(-x)) = f(-g(x)) = f(g(x))$, so $f(g(x))$ is even.

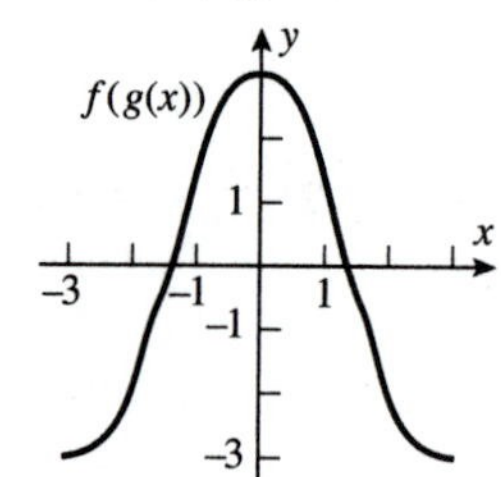

57. $f(g(x)) = 0$ when $g(x) = \pm 2$, so $x = \pm 1.4$; $g(f(x)) = 0$ when $f(x) = 0$, so $x = \pm 2$.

59. $\dfrac{3(x+h)^2 - 5 - (3x^2 - 5)}{h} = \dfrac{6xh + 3h^2}{h} = 6x + 3h$;

$\dfrac{3w^2 - 5 - (3x^2 - 5)}{w - x} = \dfrac{3(w-x)(w+x)}{w - x} = 3w + 3x$

61. $\dfrac{1/(x+h)-1/x}{h}=\dfrac{x-(x+h)}{xh(x+h)}=\dfrac{-1}{x(x+h)};\ \dfrac{1/w-1/x}{w-x}=\dfrac{x-w}{wx(w-x)}=-\dfrac{1}{xw}$

63. **(a)** the origin **(b)** the x-axis **(c)** the y-axis **(d)** none

65. **(a)**

x	-3	-2	-1	0	1	2	3
$f(x)$	1	-5	-1	0	-1	-5	1

(b)

x	-3	-2	-1	0	1	2	3
$f(x)$	1	5	-1	0	1	-5	-1

67. **(a)** even **(b)** odd **(c)** odd **(d)** neither

69. **(a)** $f(-x)=(-x)^2=x^2=f(x)$, even **(b)** $f(-x)=(-x)^3=-x^3=-f(x)$, odd

(c) $f(-x)=|-x|=|x|=f(x)$, even **(d)** $f(-x)=-x+1$, neither

(e) $f(-x)=\dfrac{(-x)^3-(-x)}{1+(-x)^2}=-\dfrac{x^3+x}{1+x^2}=-f(x)$, odd

(f) $f(-x)=2=f(x)$, even

71. **(a)** y-axis, because $(-x)^4=2y^3+y$ gives $x^4=2y^3+y$

(b) origin, because $(-y)=\dfrac{(-x)}{3+(-x)^2}$ gives $y=\dfrac{x}{3+x^2}$

(c) x-axis, y-axis, and origin because $(-y)^2=|x|-5$, $y^2=|-x|-5$, and $(-y)^2=|-x|-5$ all give $y^2=|x|-5$

73.

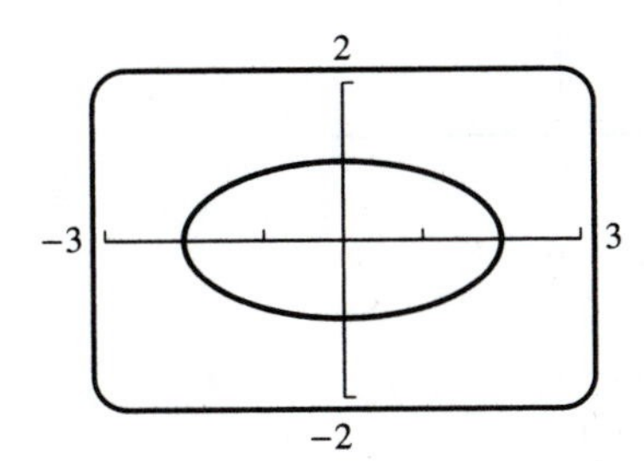

75.

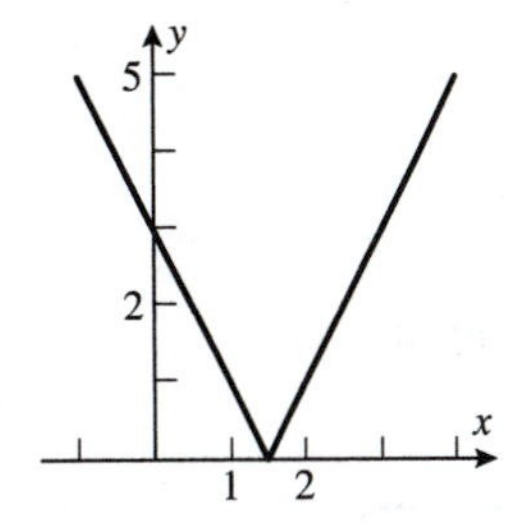

77. **(a)**

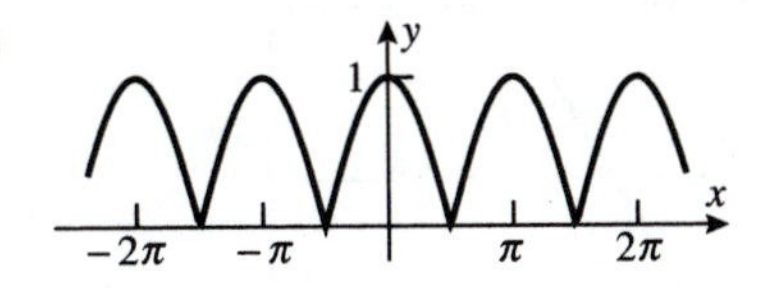

(b)

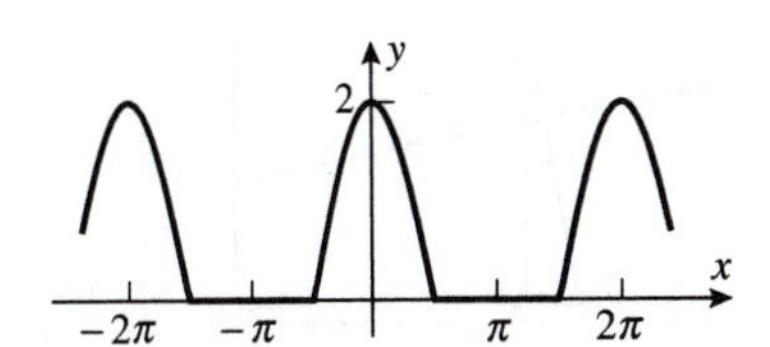

79. Yes, e.g. $f(x)=x^k$ and $g(x)=x^n$ where k and n are integers.

EXERCISE SET 1.5

1. **(a)** $\dfrac{3-0}{0-2}=-\dfrac{3}{2},\dfrac{3-(8/3)}{0-6}=-\dfrac{1}{18},\ \dfrac{0-(8/3)}{2-6}=\dfrac{2}{3}$

(b) Yes; the first and third slopes above are negative reciprocals of each other.

3. III < II < IV < I

5. **(a)** $\dfrac{1-(-5)}{1-(-2)} = 2$, $\dfrac{-5-(-1)}{-2-0} = 2$, $\dfrac{1-(-1)}{1-0} = 2$. Since the slopes connecting all pairs of points are equal, they lie on a line.

(b) $\dfrac{4-2}{-2-0} = -1$, $\dfrac{2-5}{0-1} = 3$, $\dfrac{4-5}{-2-1} = \dfrac{1}{3}$. Since the slopes connecting the pairs of points are not equal, the points do not lie on a line.

7. The slope, $m = 3$, is equal to $\dfrac{y-2}{x-1}$, and thus $y - 2 = 3(x - 1)$.

(a) If $x = 5$ then $y = 14$.

(b) If $y = -2$ then $x = -1/3$.

9. **(a)** The first slope is $\dfrac{2-0}{1-x}$ and the second is $\dfrac{5-0}{4-x}$. Since they are negatives of each other we get $2(4-x) = -5(1-x)$ or $7x = 13$, $x = 13/7$.

11. **(a)** 153° **(b)** 45° **(c)** 117° **(d)** 89°

13. **(a)** $m = \tan\phi = \sqrt{3}$, so $\phi = 60°$

(b) $m = \tan\phi = -2$, so $\phi = 117°$

15. $y = -2x + 4$

17. Parallel means the lines have equal slopes, so $y = 4x + 7$.

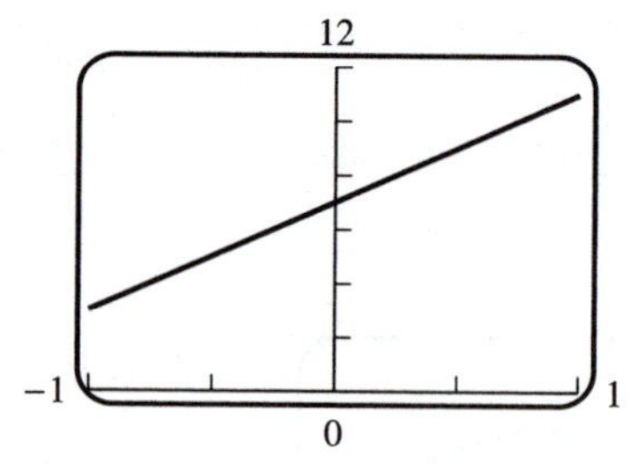

19. The negative reciprocal of 5 is $-1/5$, so $y = -\frac{1}{5}x + 6$.

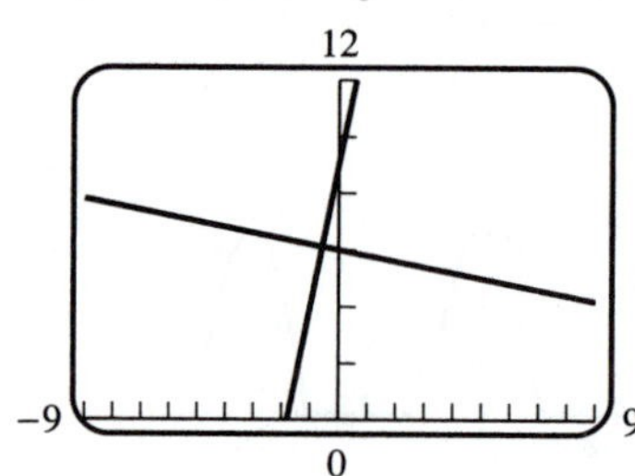

21. $m = \dfrac{4-(4-7)}{2-1} = 11$,
so $y - (-7) = 11(x - 1)$,
or $y = 11x - 18$

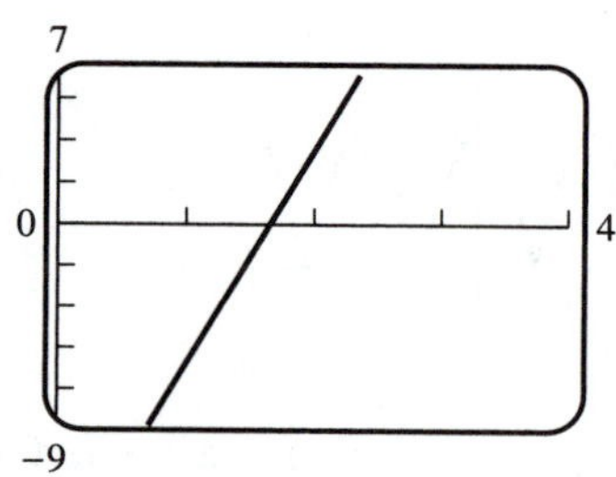

23. **(a)** $m_1 = m_2 = 4$, parallel

(b) $m_1 = 2 = -1/m_2$, perpendicular

(c) $m_1 = m_2 = 5/3$, parallel

(d) If $A \neq 0$ and $B \neq 0$ then $m_1 = -A/B = -1/m_2$, perpendicular; if $A = 0$ or $B = 0$ (not both) then one line is horizontal, the other vertical, so perpendicular.

(e) neither

25. **(a)** $m = (0 - (-3))/(2 - 0)) = 3/2$ so $y = 3x/2 - 3$

(b) $m = (-3 - 0)/(4 - 0) = -3/4$ so $y = -3x/4$

27. **(a)** The velocity is the slope, which is $\dfrac{5-(-4)}{10-0} = 9/10$ ft/s.

(b) $x = -4$

(c) The line has slope 9/10 and passes through $(0, -4)$, so has equation $x = 9t/10 - 4$; at $t = 2$, $x = -2.2$.

(d) $t = 80/9$

29. **(a)** The acceleration is the slope of the velocity, so $a = \dfrac{3-(-1)}{1-4} = -\dfrac{4}{3}$ ft/s^2.

(b) $v - 3 = -\frac{4}{3}(t-1)$, or $v = -\frac{4}{3}t + \frac{13}{3}$

(c) $v = \frac{13}{3}$ ft/s

31. **(a)** It moves (to the left) 6 units with velocity $v = -3$ cm/s, then remains motionless for 5 s, then moves 3 units to the left with velocity $v = -1$ cm/s.

(b) $v_{\text{ave}} = \dfrac{0-9}{10-0} = -\dfrac{9}{10}$ cm/s

(c) Since the motion is in one direction only, the speed is the negative of the velocity, so $s_{\text{ave}} = \frac{9}{10}$ cm/s.

33. **(a)** If x_1 denotes the final position and x_0 the initial position, then $v = (x_1 - x_0)/(t_1 - t_0) = 0$ mi/h, since $x_1 = x_0$.

(b) If the distance traveled in one direction is d, then the outward journey took $t = d/40$ h. Thus $s_{\text{ave}} = \dfrac{\text{total dist}}{\text{total time}} = \dfrac{2d}{t + (2/3)t} = \dfrac{80t}{t + (2/3)t} = 48$ mi/h.

(c) $t + (2/3)t = 5$, so $t = 3$ and $2d = 80t = 240$ mi round trip

35. **(a)**

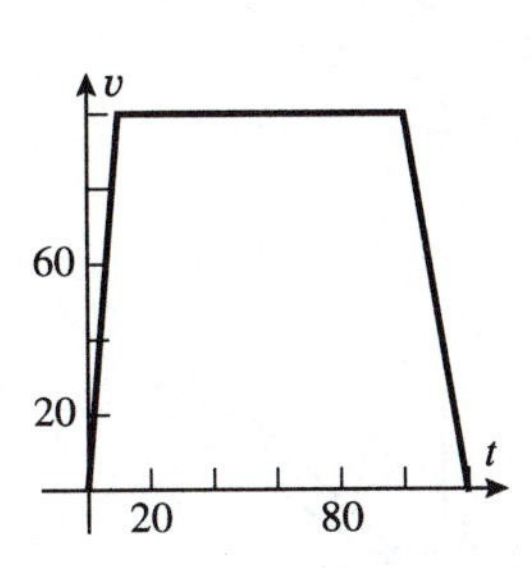

(b) $v = \begin{cases} 10t & \text{if } 0 \le t \le 10 \\ 100 & \text{if } 10 \le t \le 100 \\ 600 - 5t & \text{if } 100 \le t \le 120 \end{cases}$

37. **(a)** $y = 20 - 15 = 5$ when $x = 45$, so $5 = 45k$, $k = 1/9$, $y = x/9$

(b)

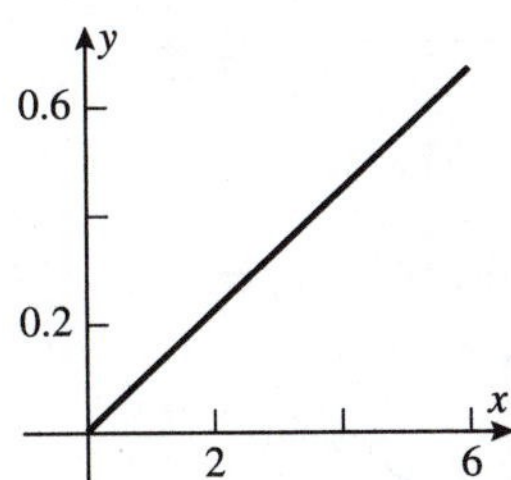

(c) $l = 15 + y = 15 + 100(1/9) = 26.11$ in.

(d) If $y_{\text{max}} = 15$ then solve $15 = kx = x/9$ for $x = 135$ lb.

39. Each increment of 1 in the value of x yields the increment of 1.2 for y, so the relationship is linear. If $y = mx + b$ then $m = 1.2$; from $x = 0$, $y = 2$, follows $b = 2$, so $y = 1.2x + 2$

41. **(a)** With T_F as independent variable, we have $\dfrac{T_C - 100}{T_F - 212} = \dfrac{0 - 100}{32 - 212}$, so $T_C = \dfrac{5}{9}(T_F - 32)$.

(b) 5/9

(c) Set $T_F = T_C = \frac{5}{9}(T_F - 32)$ and solve for T_F: $T_F = T_C = -40°$ (F or C).

(d) 37° C

43. (a) $\dfrac{p-1}{h-0} = \dfrac{5.9-1}{50-0}$, or $p = 0.098h + 1$ (b) when $p = 2$, or $h = 1/0.098 \approx 10.20$ m

45. (a) $\dfrac{r-0.80}{t-0} = \dfrac{0.75-0.80}{4-0}$, so $r = -0.0125t + 0.8$ (b) 64 days

47. (a) For x trips we have $C_1 = 2x$ and $C_2 = 25 + x/4$

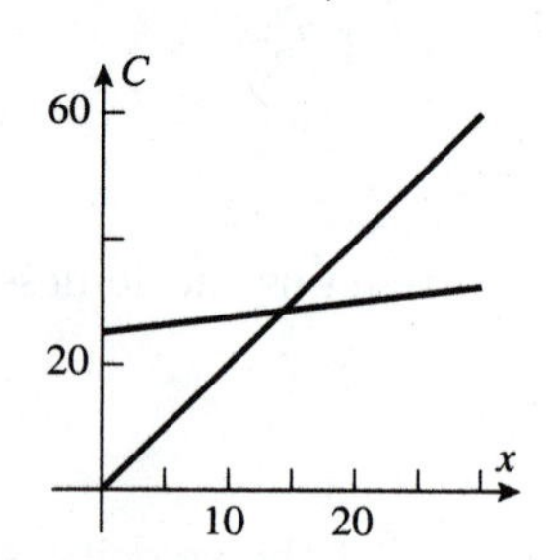

(b) $2x = 25 + x/4$, or $x = 100/7$, so the commuter pass becomes worthwhile at $x = 15$.

EXERCISE SET 1.6

1. (a) $y = 3x + b$ (b) $y = 3x + 6$ (c)

y = 3x + 6
y = 3x + 2
y = 3x − 4
10
−10
−2
2

3. (a) $y = mx + 2$

(b) $m = \tan\phi = \tan 135^\circ = -1$, so $y = -x + 2$ (c)

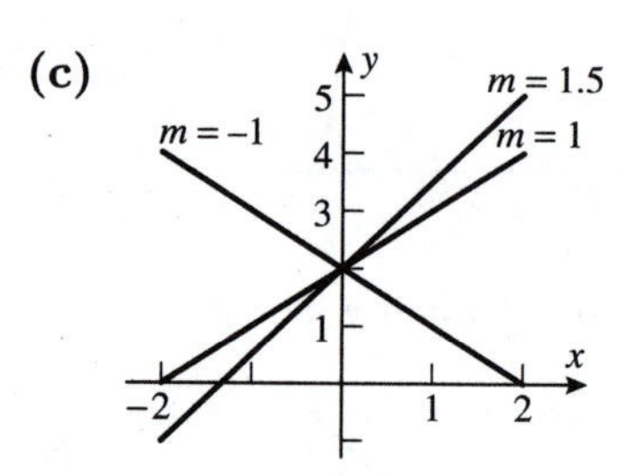

5. (a) The slope is -1.

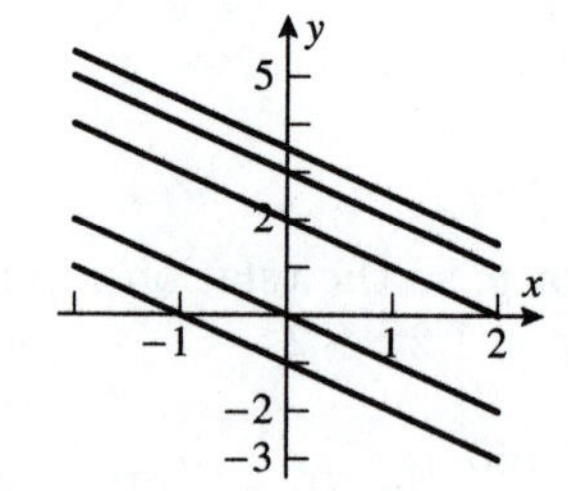

(b) The y-intercept is $y = -1$.

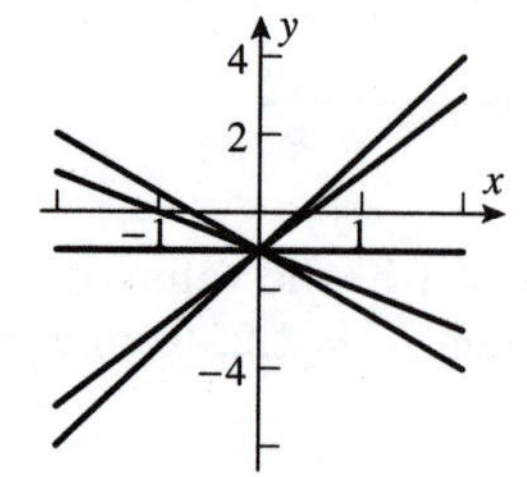

(c) They pass through the point $(-4, 2)$.

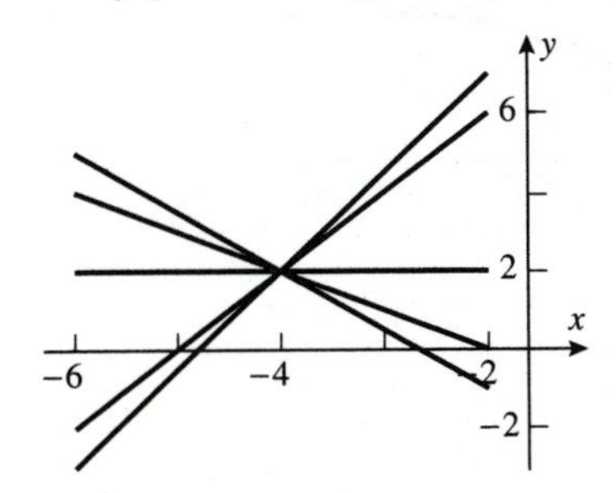

(d) The x-intercept is $x = 1$.

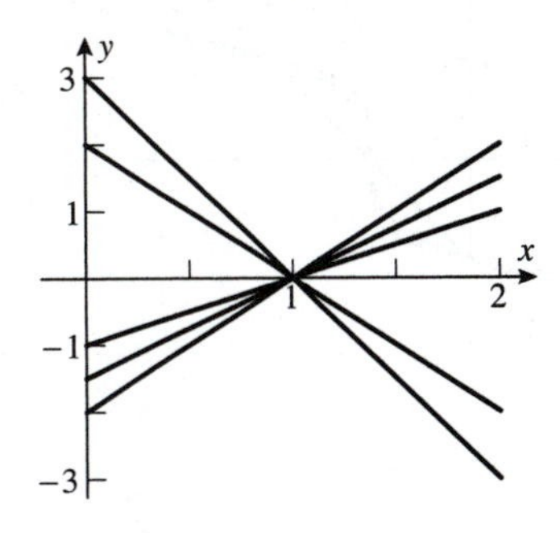

7. Let the line be tangent to the circle at the point (x_0, y_0) where $x_0^2 + y_0^2 = 9$. The slope of the tangent line is the negative reciprocal of y_0/x_0 (why?), so $m = -x_0/y_0$ and $y = -(x_0/y_0)x + b$. Substituting the point (x_0, y_0) as well as $y_0 = \pm\sqrt{9 - x_0^2}$ we get $y = \pm\dfrac{9 - x_0 x}{\sqrt{9 - x_0^2}}$.

9. The x-intercept is $x = 10$ so that with depreciation at 10% per year the final value is always zero, and hence $y = m(x - 10)$. The y-intercept is the original value.

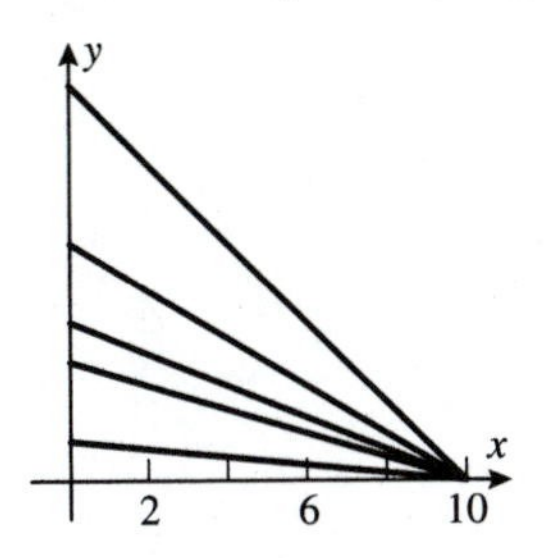

11. **(a)** VI **(b)** IV **(c)** III **(d)** V **(e)** I **(f)** II

13. **(a)**

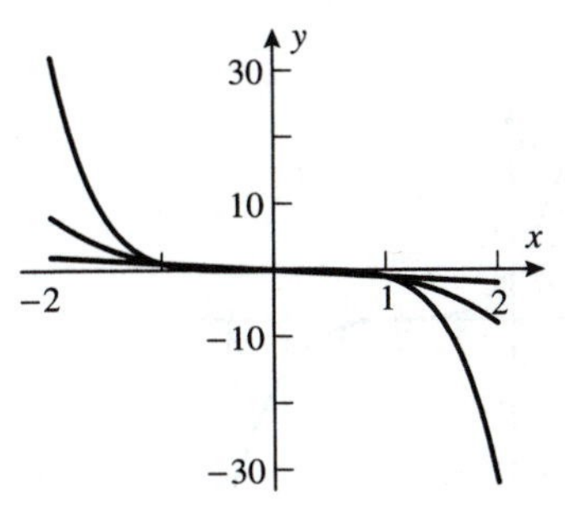

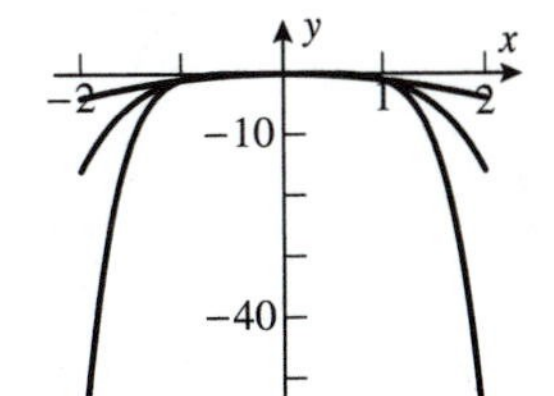

(b)

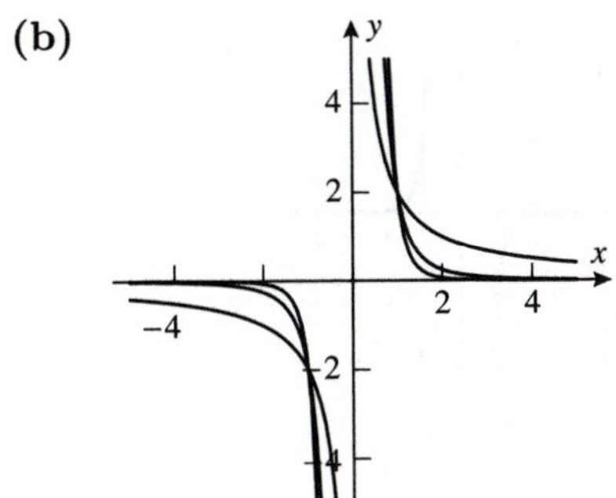

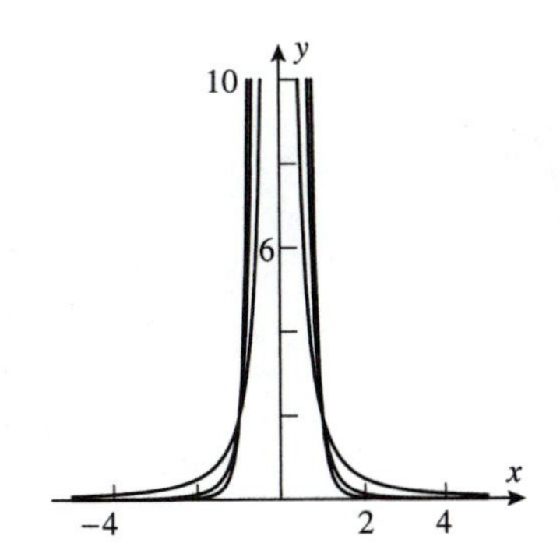

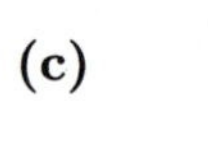

(c)

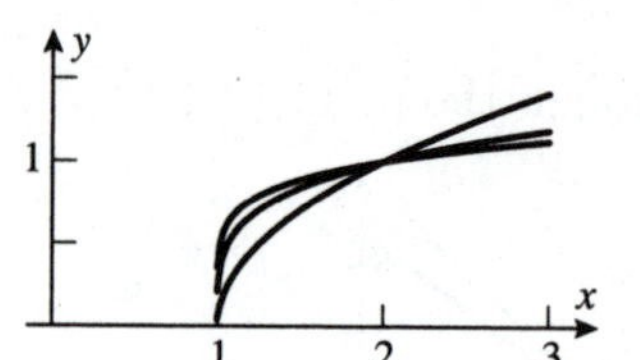

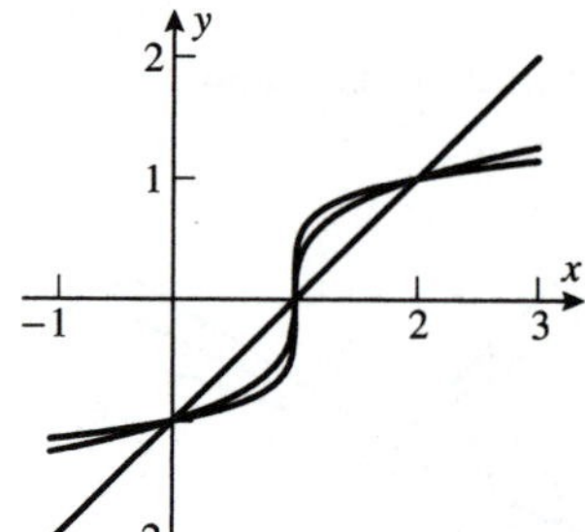

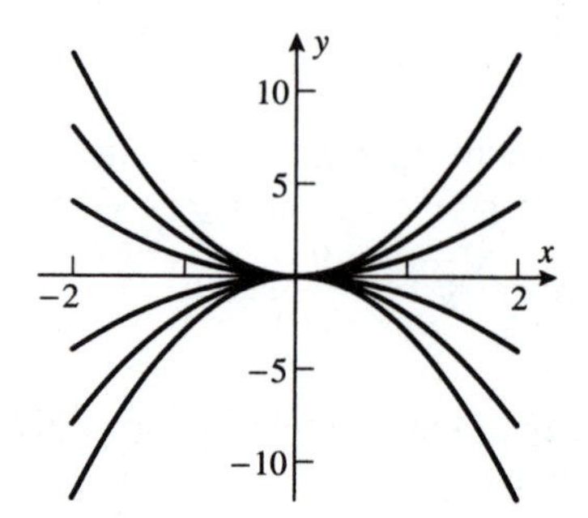

15. (a)

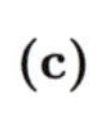

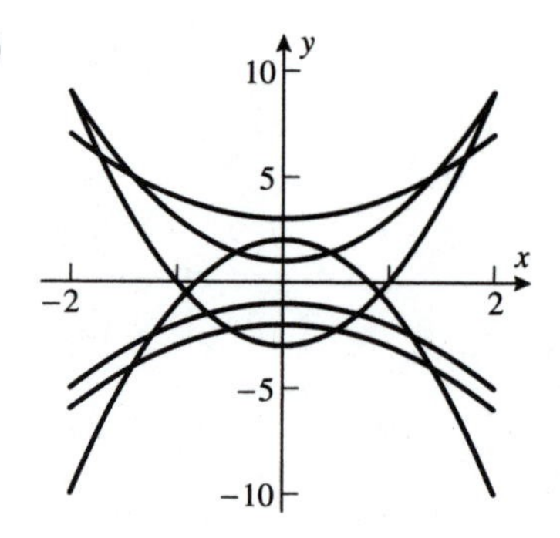

(b)

(c)

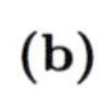

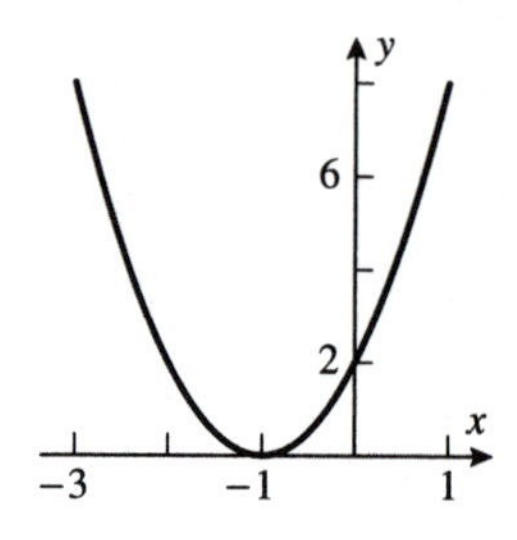

17. (a)

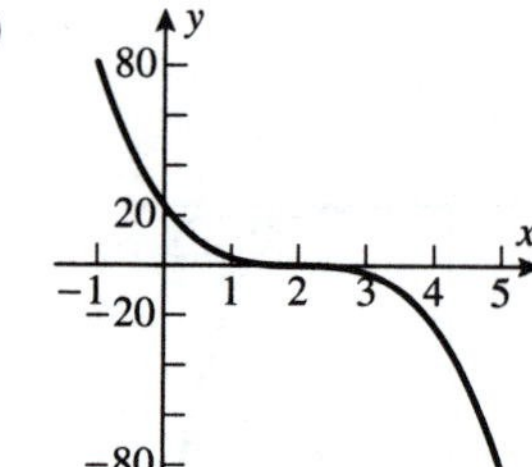

(b)

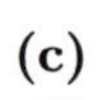

(c)

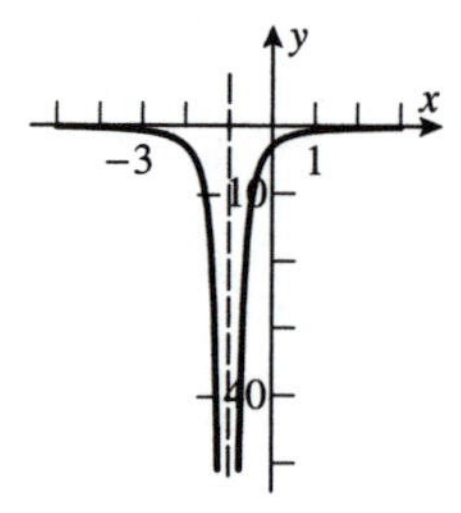

(d)

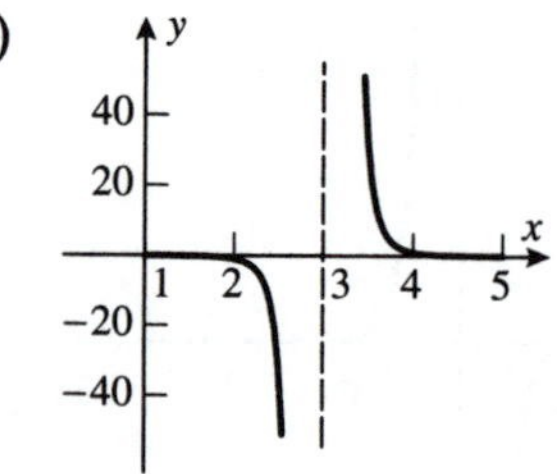

19. (a)

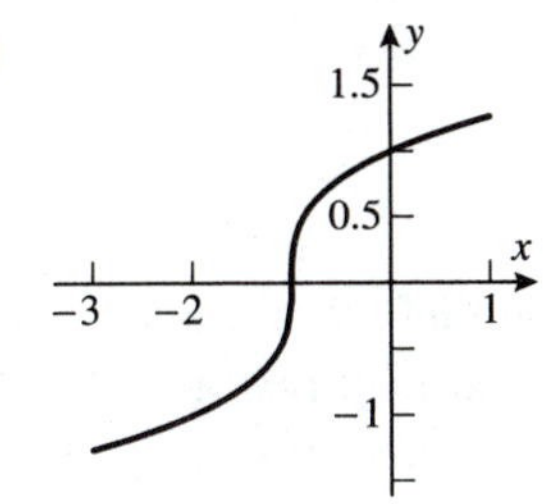

(b)

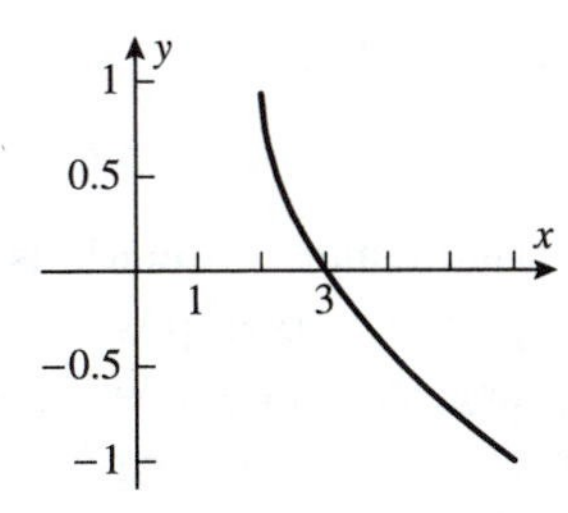

(c)

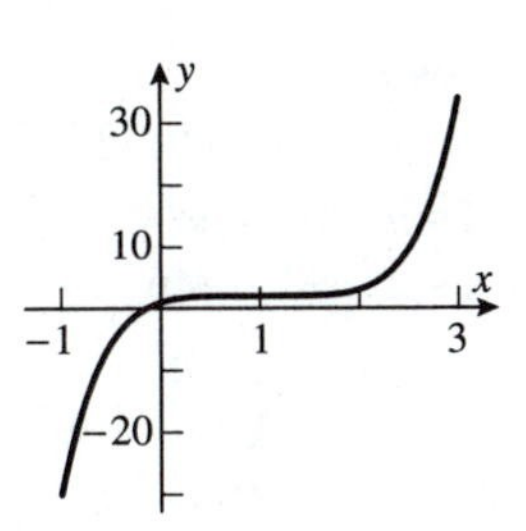

(d)

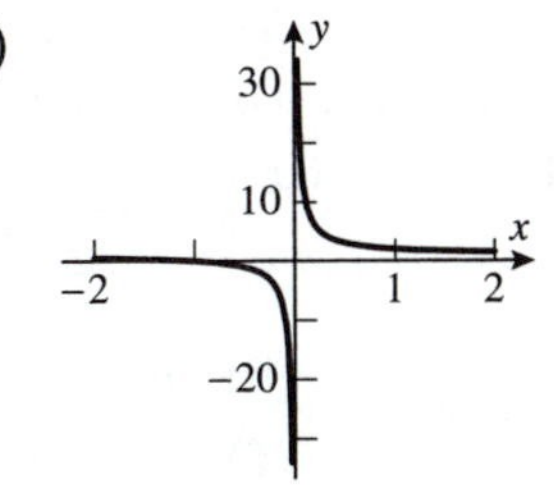

21. $y = x^2 + 2x = (x+1)^2 - 1$

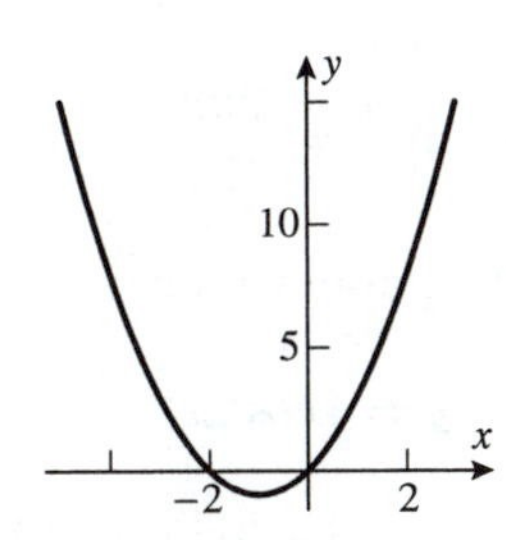

23. (a) N·m

(b) k = 20 N·m

(c)

V(L)	0.25	0.5	1.0	1.5	2.0
P (N/m^2)	80×10^3	40×10^3	20×10^3	13.3×10^3	10×10^3

(d)

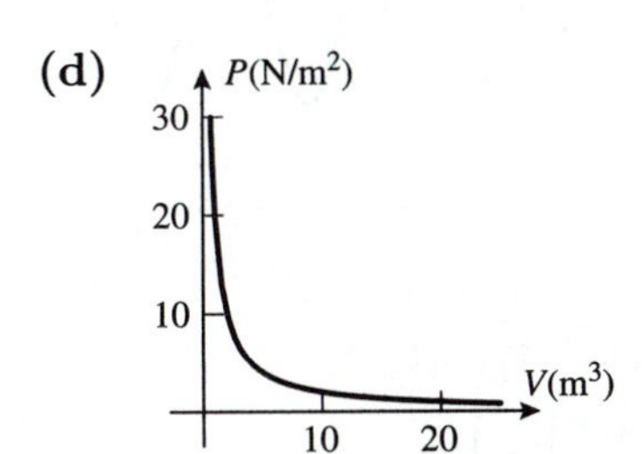

25. (a) $F = k/x^2$ so $0.0005 = k/(0.3)^2$ and $k = 0.000045$ N·m^2.

(b) $F = 0.000005$ N

(c)

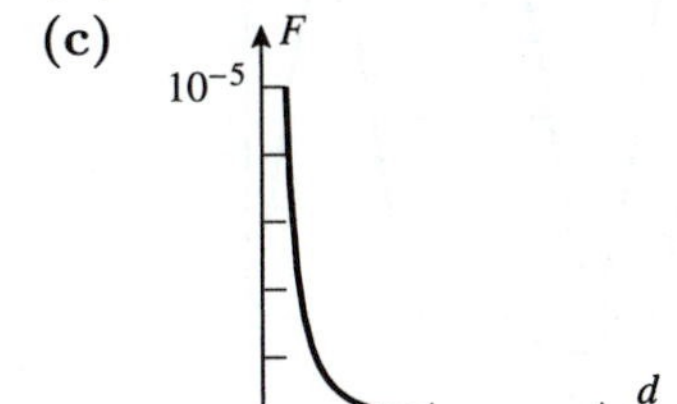

(d) When they approach one another, the force becomes infinite; when they get far apart it tends to zero.

27. **(a)** II; $y=1$, $x=-1,2$ **(b)** I; $y=0$, $x=-2,3$
(c) IV; $y=2$ **(d)** III; $y=0$, $x=-2$

29. Order the six trigonometric functions as sin, cos, tan, cot, sec, csc:
(a) pos, pos, pos, pos, pos, pos **(b)** neg, zero, undef, zero, undef, neg
(c) pos, neg, neg, neg, neg, pos **(d)** neg, pos, neg, neg, pos, neg
(e) neg, neg, pos, pos, neg, neg **(f)** neg, pos, neg, neg, pos, neg

31. **(a)** $\sin(\pi - x) = \sin x$; 0.588 **(b)** $\cos(-x) = \cos x$; 0.924
(c) $\sin(2\pi + x) = \sin x$; 0.588 **(d)** $\cos(\pi - x) = -\cos x$; -0.924
(e) $\cos^2 x = 1 - \sin^2 x; 0.655$ **(f)** $\sin^2 2x = 4\sin^2 x\cos^2 x = 4\sin^2 x(1-\sin^2 x); 0.905$

33. **(a)** $-a$ **(b)** b **(c)** $-c$ **(d)** $\pm\sqrt{1-a^2}$
(e) $-b$ **(f)** $-a$ **(g)** $\pm 2b\sqrt{1-b^2}$ **(h)** $2b^2-1$
(i) $1/b$ **(j)** $-1/a$ **(k)** $1/c$ **(l)** $(1-b)/2$

35. If the arc length is x, then solve the ratio $\dfrac{x}{1} = \dfrac{2\pi r}{27.3}$ to get $x \approx 87,458$ km.

37. The second quarter revolves twice (720°) about its own center.

39. **(a)** $y = 3\sin(x/2)$ **(b)** $y = 4\cos 2x$ **(c)** $y = -5\sin 4x$

41. **(a)** $y = \sin(x + \pi/2)$ **(b)** $y = 3 + 3\sin(2x/9)$ **(c)** $y = 1 + 2\sin(2(x - \pi/4))$

43. **(a)** $3, \pi/2, 0$ **(b)** $2, 2, 0$ **(c)** $1, 4\pi, 0$

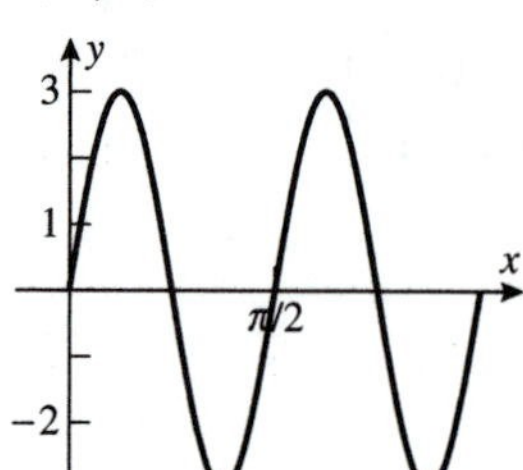

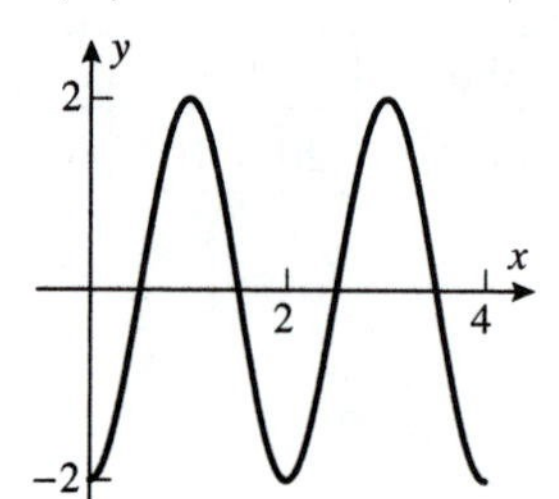

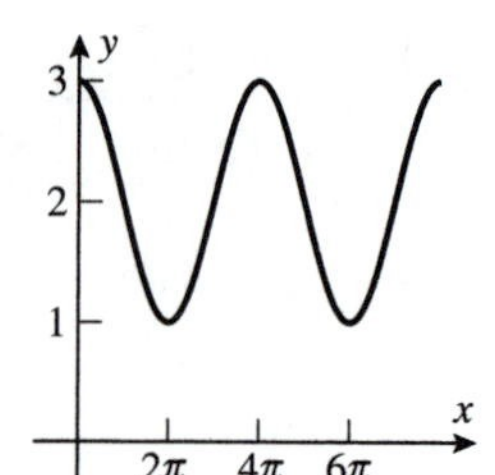

45. **(a)** $A\sin(\omega t + \theta) = A\sin(\omega t)\cos\theta + A\cos(\omega t)\sin\theta = A_1\sin(\omega t) + A_2\cos(\omega t)$

(b) $A_1 = A\cos\theta, A_2 = A\sin\theta$, so $A = \sqrt{A_1^2 + A_2^2}$ and $\theta = \tan^{-1}(A_2/A_1)$.

(c) $A = 5\sqrt{13}/2$, $\theta = \tan^{-1}\dfrac{1}{2\sqrt{3}}$;

$$x = \frac{5\sqrt{13}}{2}\sin\left(2\pi t + \tan^{-1}\frac{1}{2\sqrt{3}}\right)$$

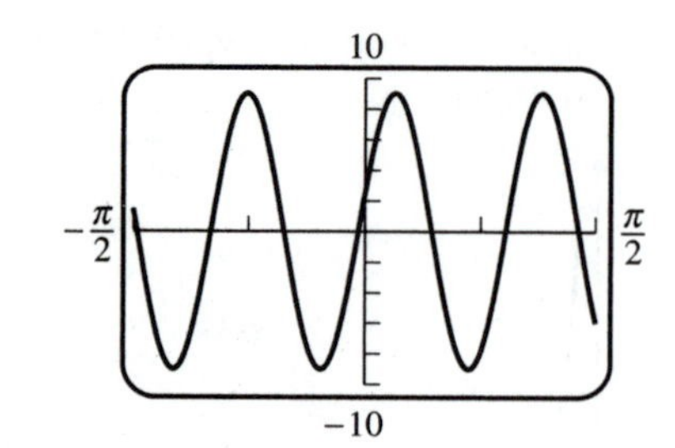

EXERCISE SET 1.7

1. The sum of the squares for the residuals for line I is approximately $1^2 + 1^2 + 1^2 + 0^2 + 2^2 + 1^2 + 1^2 + 1^2 = 10$, and the same for line II is approximately $0^2 + (0.4)^2 + (1.2)^2 + 0^2 + (2.2)^2 + (0.6)^2 + (0.2)^2 + 0^2 = 6.84$; line II is the regression line.

3. Least squares line $S = 1.5388t - 2842.9$, correlation coefficient 0.83409

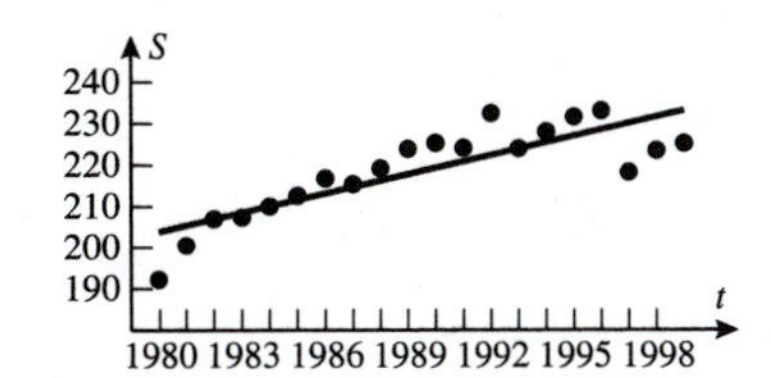

5. **(a)** Least squares line $p = 0.0146T + 3.98$, correlation coefficient 0.9999
(b) $p = 3.25$ atm **(c)** $T = -272°\mathrm{C}$

7. **(a)** $R = 0.00723T + 1.55$ **(b)** $T = -214°\mathrm{C}$

9. **(a)** $S = 0.50179w - 0.00643$ **(b)** $S = 8$, $w = 16$ lb

11. **(a)** Let h denote the height in inches and y the number of rebounds per minute. Then $y = 0.00630h - 0.266$, $r = 0.313$

(b)

(c) No, the data points are too widely scattered.

13. **(a)** $H \approx 20000/110 \approx 181$ km/s/Mly
(b) One light year is 9.408×10^{12} km and
$$t = \frac{d}{v} = \frac{1}{H} = \frac{1}{20\text{km/s/Mly}} = \frac{9.408 \times 10^{18}\text{km}}{20\text{km/s}} = 4.704 \times 10^{17}\ \text{s} = 1.492 \times 10^{10}\ \text{years}.$$
(c) The Universe would be even older.

15. **(a)** $P = 0.322t^2 + 0.0671t + 0.00837$ **(b)** $P = 1.43$ cm

17. As in Example 4, a possible model is of the form $T = D + A\sin\left[B\left(t - \frac{C}{B}\right)\right]$. Since the longest day has 993 minutes and the shortest has 706, take $2A = 993 - 706 = 287$ or $A = 143.5$. The midpoint between the longest and shortest days is 849.5 minutes, so there is a vertical shift of $D = 849.5$. The period is about 365.25 days, so $2\pi/B = 365.25$ or $B = \pi/183$. Note that the sine function takes the value -1 when $t - \frac{C}{B} = -91.8125$, and T is a minimum at about $t = 0$. Thus the phase shift $\frac{C}{B} \approx 91.5$. Hence $T = 849.5 + 143.5\sin\left[\frac{\pi}{183}t - \frac{\pi}{2}\right]$ is a model for the temperature.

19. $t = 0.445\sqrt{d}$

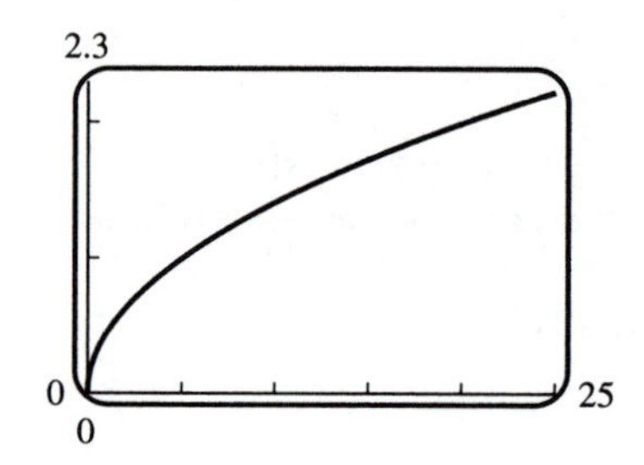

EXERCISE SET 1.8

1. **(a)** $x + 1 = t = y - 1$, $y = x + 2$

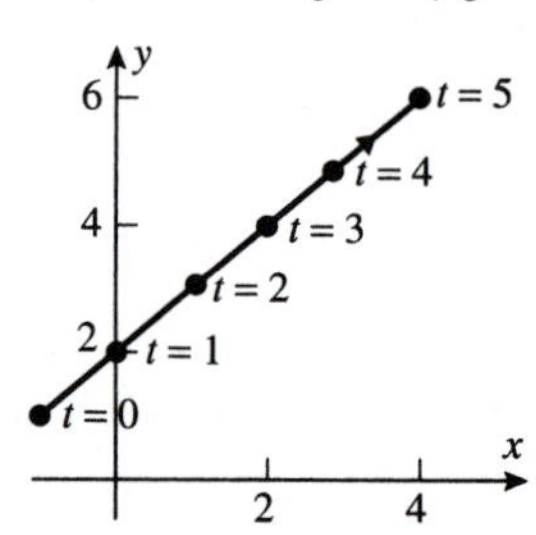

(c)

t	0	1	2	3	4	5
x	−1	0	1	2	3	4
y	1	2	3	4	5	6

3. $t = (x + 4)/3$; $y = 2x + 10$

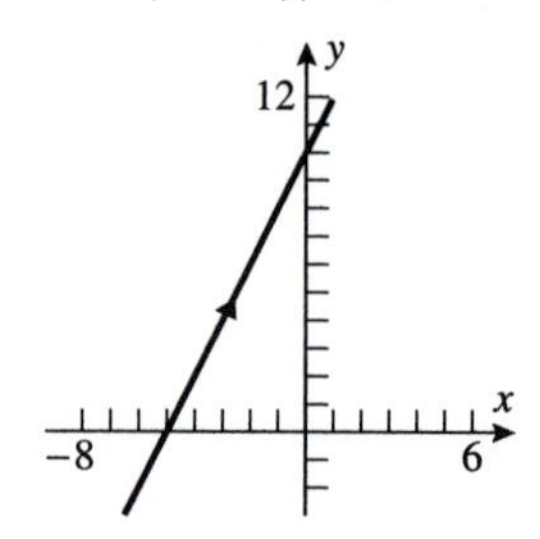

5. $\cos t = x/2$, $\sin t = y/5$;
$x^2/4 + y^2/25 = 1$

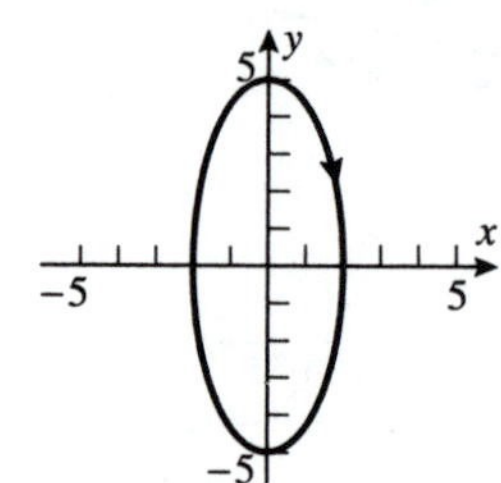

7. $\cos t = (x - 3)/2$, $\sin t = (y - 2)/4$;
$(x - 3)^2/4 + (y - 2)^2/16 = 1$

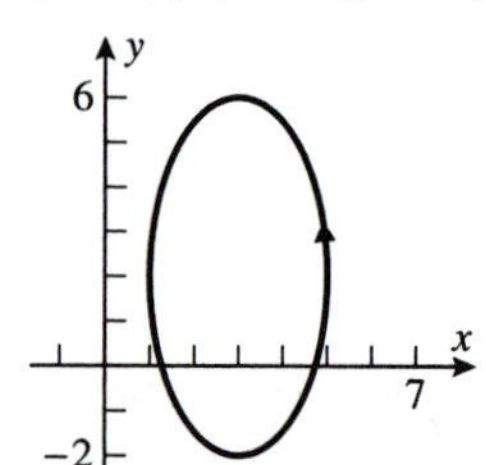

9. $\cos 2t = 1 - 2\sin^2 t$;
$x = 1 - 2y^2$, $-1 \le y \le 1$

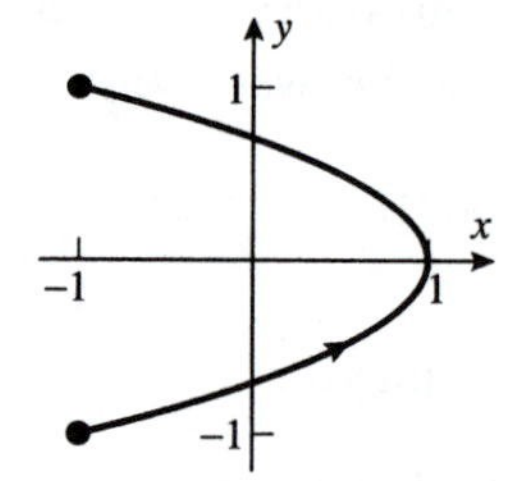

11. $x/2 + y/3 = 1, 0 \leq x \leq 2, 0 \leq y \leq 3$

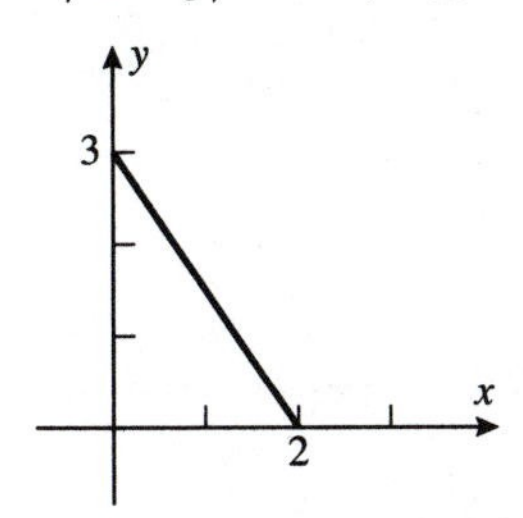

13. $x = 5\cos t$, $y = -5\sin t$, $0 \leq t \leq 2\pi$

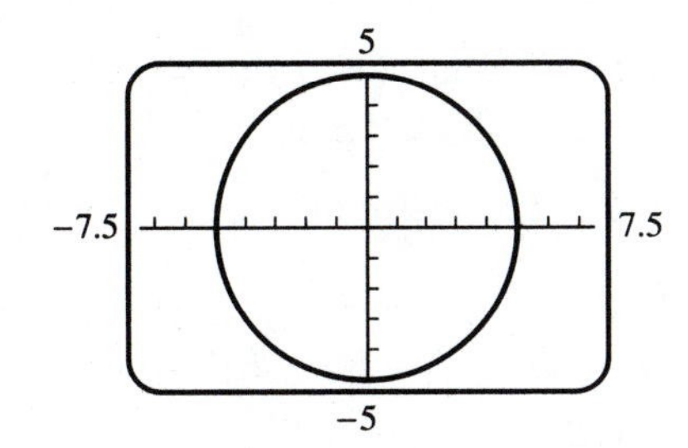

15. $x = 2$, $y = t$

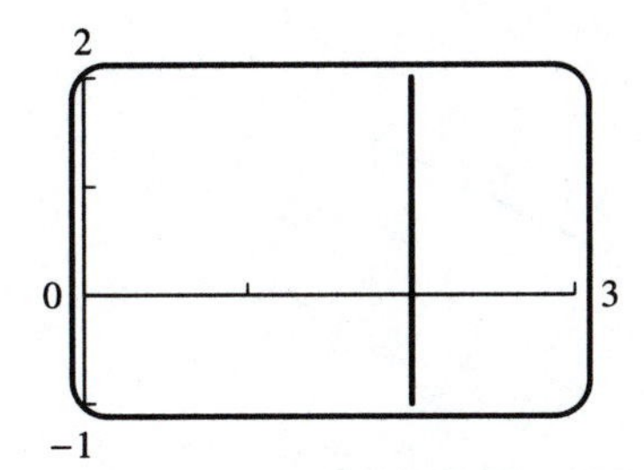

17. $x = t^2$, $y = t$, $-1 \leq t \leq 1$

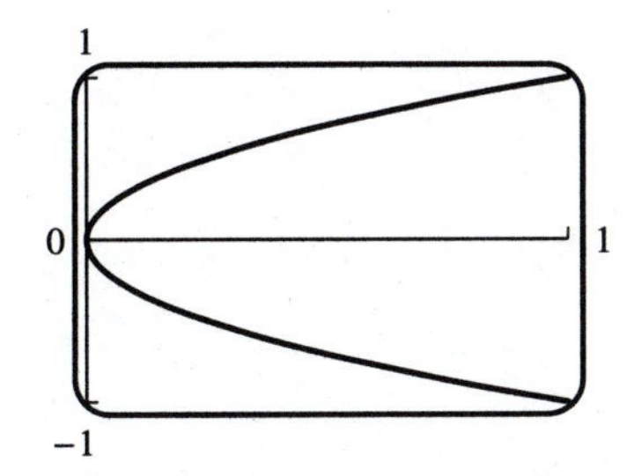

19. **(a)** IV, because x always increases whereas y oscillates.

(b) II, because $(x/2)^2 + (y/3)^2 = 1$, an ellipse.

(c) V, because $x^2 + y^2 = t^2$ increases in magnitude while x and y keep changing sign.

(d) VI; examine the cases $t < -1$ and $t > -1$ and you see the curve lies in the first, second and fourth quadrants only.

(e) III because $y > 0$.

(f) I; since x and y are bounded, the answer must be I or II; but as t runs, say, from 0 to π, x goes directly from 2 to -2, but y goes from 0 to 1 to 0 to -1 and back to 0, which describes I but not II.

21. **(a)**

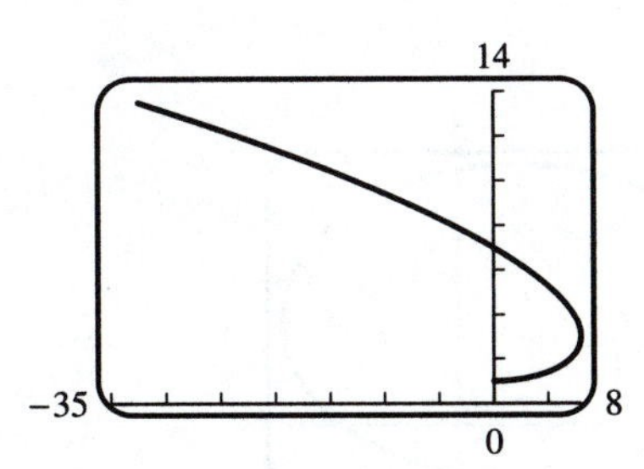

(b)

t	0	1	2	3	4	5
x	0	5.5	8	4.5	-8	-32.5
y	1	1.5	3	5.5	9	13.5

(c) $x = 0$ when $t = 0, 2\sqrt{3}$.

(d) for $0 < t < 2\sqrt{2}$

(e) at $t = 2$

23. **(a)**

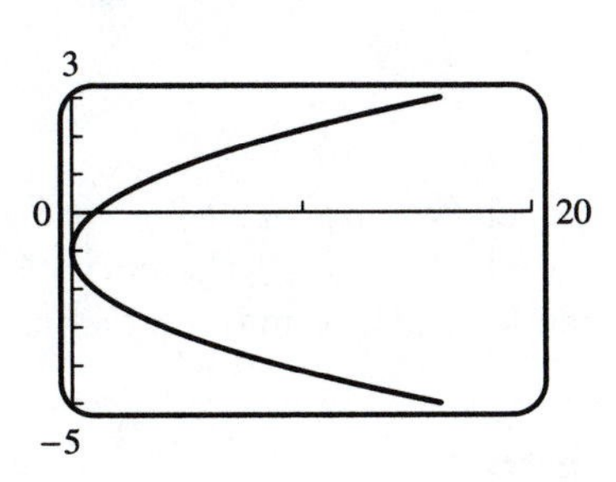

(b)

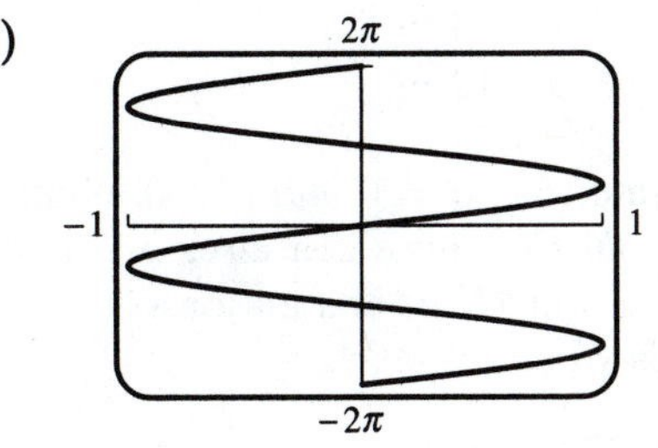

25. **(a)** Eliminate t to get $\dfrac{x - x_0}{x_1 - x_0} = \dfrac{y - y_0}{y_1 - y_0}$

(b) Set $t = 0$ to get (x_0, y_0); $t = 1$ for (x_1, y_1).

(c) $x = 1 + t$, $y = -2 + 6t$

(d) $x = 2 - t$, $y = 4 - 6t$

27. **(a)** $|R-P|^2 = (x-x_0)^2+(y-y_0)^2 = t^2[(x_1-x_0)^2+(y_1-y_0)^2]$ and $|Q-P|^2 = (x_1-x_0)^2+(y_1-y_0)^2$, so $r = |R - P| = |Q - P|t = qt$.

(b) $t = 1/2$

(c) $t = 3/4$

29. The two branches corresponding to $-1 \le t \le 0$ and $0 \le t \le 1$ coincide.

31. **(a)** $\dfrac{x - b}{a} = \dfrac{y - d}{c}$

(b)

33.

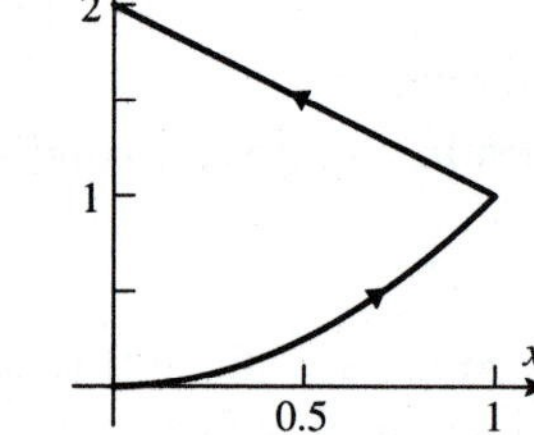

35. **(a)** $x = 4\cos t$, $y = 3\sin t$

(b) $x = -1 + 4\cos t$, $y = 2 + 3\sin t$

(c)

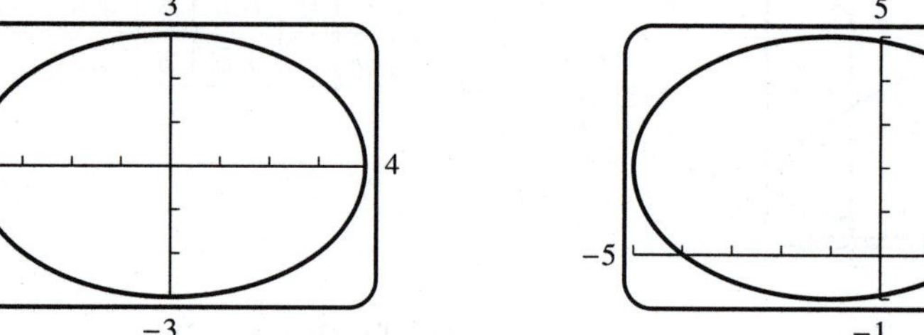

37. **(a)** From Exercise 36, $x = 400\sqrt{2}t$, $y = 400\sqrt{2}t - 4.9t^2$.

(b) 16,326.53 m

(c) 65,306.12 m

39. Assume that $a \neq 0$ and $b \neq 0$; eliminate the parameter to get $(x - h)^2/a^2 + (y - k)^2/b^2 = 1$. If $|a| = |b|$ the curve is a circle with center (h, k) and radius $|a|$; if $|a| \neq |b|$ the curve is an ellipse with center (h, k) and major axis parallel to the x-axis when $|a| > |b|$, or major axis parallel to the y-axis when $|a| < |b|$.

(a) ellipses with a fixed center and varying axes of symmetry

(b) (assume $a \neq 0$ and $b \neq 0$) ellipses with varying center and fixed axes of symmetry

(c) circles of radius 1 with centers on the line $y = x - 1$

41. **(a)**

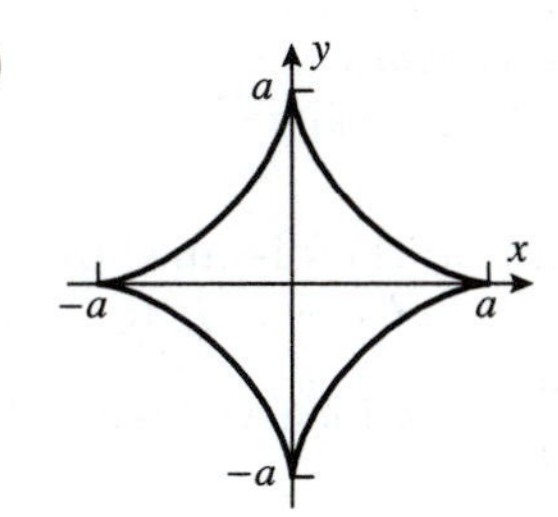

(b) Use $b = a/4$ in the equations of Exercise 40 to get
$x = \frac{3}{4}a\cos\phi + \frac{1}{4}a\cos 3\phi$, $y = \frac{3}{4}a\sin\phi - \frac{1}{4}a\sin 3\phi$;
but trigonometric identities yield $\cos 3\phi = 4\cos^3\phi - 3\cos\phi$, $\sin 3\phi = 3\sin\phi - 4\sin^3\phi$,
so $x = a\cos^3\phi$, $y = a\sin^3\phi$.

(c) $x^{2/3} + y^{2/3} = a^{2/3}(\cos^2\phi + \sin^2\phi) = a^{2/3}$

CHAPTER 1 SUPPLEMENTARY EXERCISES

1. 1940-45; the greatest five-year slope

3.

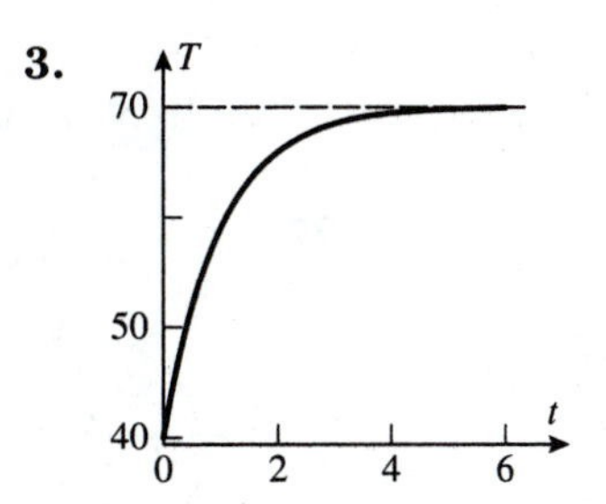

5. If the side has length x and height h, then $V = 8 = x^2h$, so $h = 8/x^2$. Then the cost $C = 5x^2 + 2(4)(xh) = 5x^2 + 64/x$.

7.

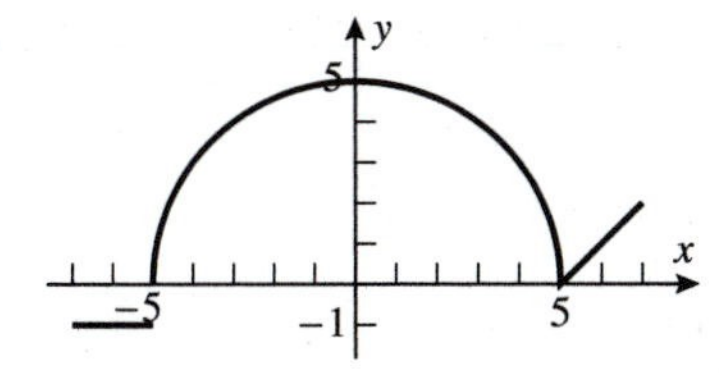

9. **(a)** The base has sides $(10 - 2x)/2$ and $6 - 2x$, and the height is x, so $V = (6 - 2x)(5 - x)x$ ft^3.

(b) From the picture we see that $x < 5$ and $2x < 6$, so $0 < x < 3$.

(c) 3.57 ft ×3.79 ft ×1.21 ft

11. $f(g(x)) = (3x+2)^2+1, g(f(x)) = 3(x^2+1)+2$, so $9x^2+12x+5 = 3x^2+5, 6x^2+12x = 0, x = 0, -2$

13. $1/(2 - x^2)$

15.

x	−4	−3	−2	−1	0	1	2	3	4
$f(x)$	0	−1	2	1	3	−2	−3	4	−4
$g(x)$	3	2	1	−3	−1	−4	4	−2	0
$(f \circ g)(x)$	4	−3	−2	−1	1	0	−4	2	3
$(g \circ f)(x)$	−1	−3	4	−4	−2	1	2	0	3

17. **(a)** even $\times$ odd = odd **(b)** a square is even
(c) even + odd is neither **(d)** odd $\times$ odd = even

19. **(a)** If x denotes the distance from A to the base of the tower, and y the distance from B to the base, then $x^2+d^2 = y^2$. Moreover $h = x\tan\alpha = y\tan\beta$, so $d^2 = y^2-x^2 = h^2(\cot^2\beta-\cot^2\alpha)$, $h^2 = \dfrac{d^2}{\cot^2\beta-\cot^2\alpha} = \dfrac{d^2}{1/\tan^2\beta - 1/\tan^2\alpha} = \dfrac{d^2\tan^2\alpha\tan^2\beta}{\tan^2\alpha-\tan^2\beta}$, which yields the result.

(b) 295.72 ft.

21. When $x = 0$ the value of the green curve is higher than that of the blue curve, therefore the blue curve is given by $y = 1+2\sin x$.
The points A, B, C, D are the points of intersection of the two curves, i.e. where $1+2\sin x = 2\sin(x/2)+2\cos(x/2)$. Let $\sin(x/2) = p, \cos(x/2) = q$. Then $2\sin x = 4\sin(x/2)\cos(x/2)$, so the equation which yields the points of intersection becomes $1+4pq = 2p+2q$, $4pq-2p-2q+1 = 0, (2p-1)(2q-1) = 0$; thus whenever either $\sin(x/2) = 1/2$ or $\cos(x/2) = 1/2$, i.e. when $x/2 = \pi/6, 5\pi/6, \pm\pi/3$. Thus A has coordinates $(-2\pi/3, 1-\sqrt{3})$, B has coordinates $(\pi/3, 1+\sqrt{3})$, C has coordinates $(2\pi/3, 1+\sqrt{3})$, and D has coordinates $(5\pi/3, 1-\sqrt{3})$.

23. **(a)** The circle of radius 1 centered at (a, a^2); therefore, the family of all circles of radius 1 with centers on the parabola $y = x^2$.

(b) All parabolas which open up, have latus rectum equal to 1 and vertex on the line $y = x/2$.

25.

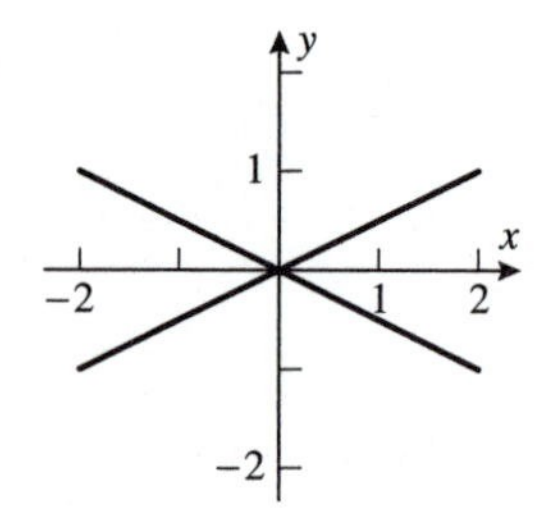

27.

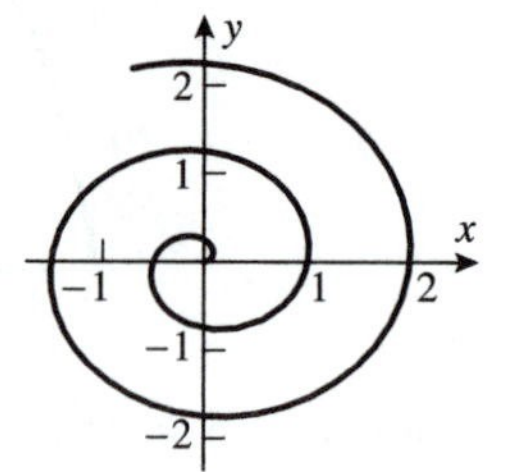

29. $d = \sqrt{(x-1)^2+(\sqrt{x}-2)^2}$;
$d = 9.1$ at $x = 1.358094$

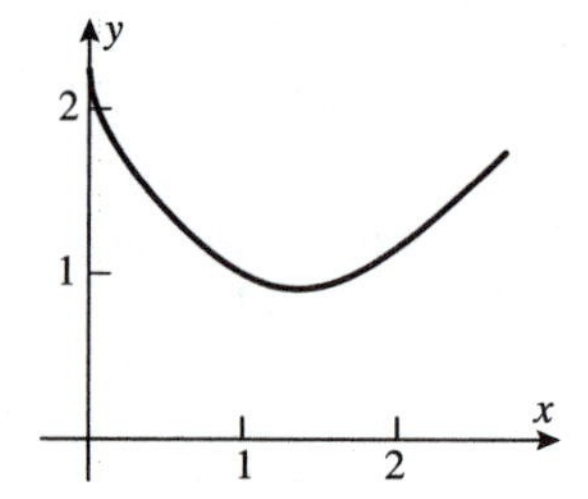

31. $w = 63.9V$, $w = 63.9\pi h^2(5/2 - h/3)$; $h = 0.48$ ft when $w = 108$ lb

33. **(a)**

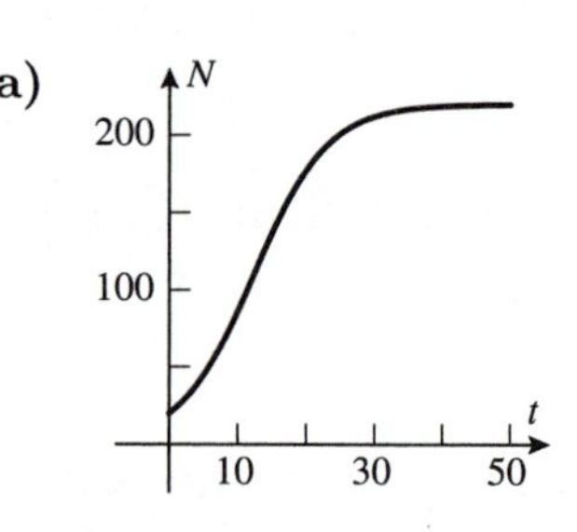

(b) $N = 80$ when $t = 9.35$ yrs
(c) 220 sheep

35. **(a)**

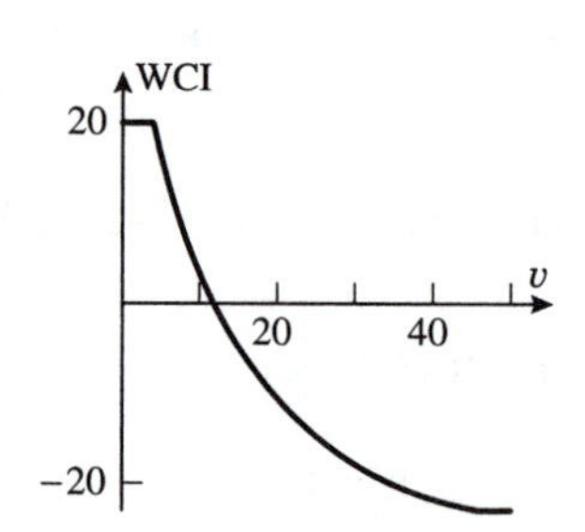

(b) $T = 3°$F, $-11°$F, $-18°$F, $-22°$F
(c) $v = 35, 19, 12, 7$ mi/h

37. The domain is the set $-0.7245 \le x \le 1.2207$, the range is $-1.0551 \le y \le 1.4902$.

39. **(a)**

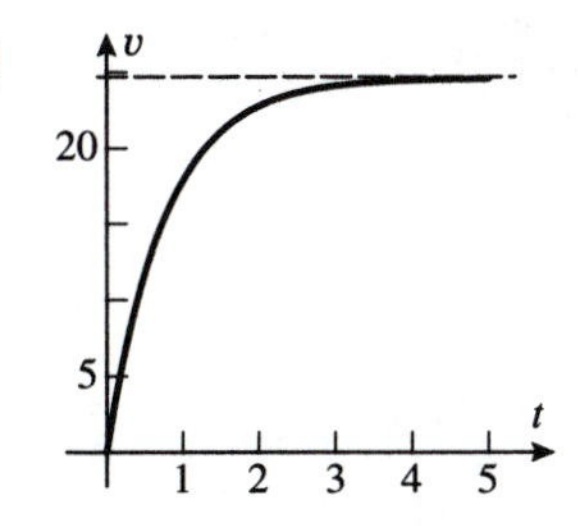

(b) As $t \to \infty$, $(0.273)^t \to 0$, and thus $v \to 24.61$ ft/s.

(c) For large t the velocity approaches c.

(d) No; but it comes very close (arbitrarily close).

(e) 3.013 s

41. **(a)**

1.90	1.92	1.94	1.96	1.98	2.00	2.02	2.04	2.06	2.08	2.10
3.4161	3.4639	3.5100	3.5543	3.5967	3.6372	3.6756	3.7119	3.7459	3.7775	3.8068

(b) $y = 1.9589x - 0.2910$

(c) $y - 3.6372 = 1.9589(x - 2)$, or $y = 1.9589x - 0.2806$

(d) As one zooms in on the point $(2, f(2))$ the two curves seem to converge to one line.

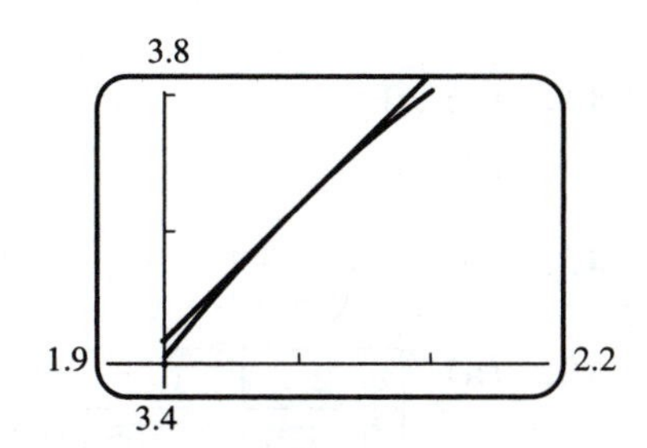

43. The data are periodic, so it is reasonable that a trigonometric function might approximate them. A possible model is of the form $T = D + A\sin\left[B\left(t - \frac{C}{B}\right)\right]$. Since the highest level is 1.032 meters and the lowest is 0.045, take $2A = 1.032 - 0.042 = 0.990$ or $A = 0.495$. The midpoint between the lowest and highest levels is 0.537 meters, so there is a vertical shift of $D = 0.537$. The period is about 12 hours, so $2\pi/B = 12$ or $B = \pi/6$. The phase shift $\frac{C}{B} \approx 6.5$. Hence $T = 0.537 + 0.495\sin\left[\frac{\pi}{6}(t - 6.5)\right]$ is a model for the temperature.

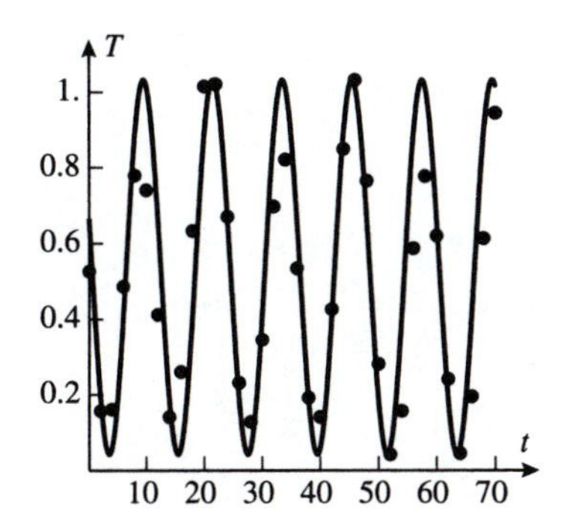

CHAPTER 2
Limits and Continuity

EXERCISE SET 2.1

1. **(a)** -1 **(b)** 3 **(c)** does not exist
(d) 1 **(e)** -1 **(f)** 3

3. **(a)** 1 **(b)** 1 **(c)** 1
(d) 1 **(e)** $-\infty$ **(f)** $+\infty$

5. **(a)** 0 **(b)** 0 **(c)** 0
(d) 3 **(e)** $+\infty$ **(f)** $+\infty$

7. **(a)** $-\infty$ **(b)** $+\infty$ **(c)** does not exist
(d) undef **(e)** 2 **(f)** 0

9. **(a)** $-\infty$ **(b)** $-\infty$ **(c)** $-\infty$
(d) 1 **(e)** 1 **(f)** 2

11. **(a)** 0 **(b)** 0 **(c)** 0
(d) 0 **(e)** does not exist **(f)** does not exist

13. for all $x_0 \neq -4$

19. **(a)**

2	1.5	1.1	1.01	1.001	0	0.5	0.9	0.99	0.999
0.1429	0.2105	0.3021	0.3300	0.3330	1.0000	0.5714	0.3690	0.3367	0.3337

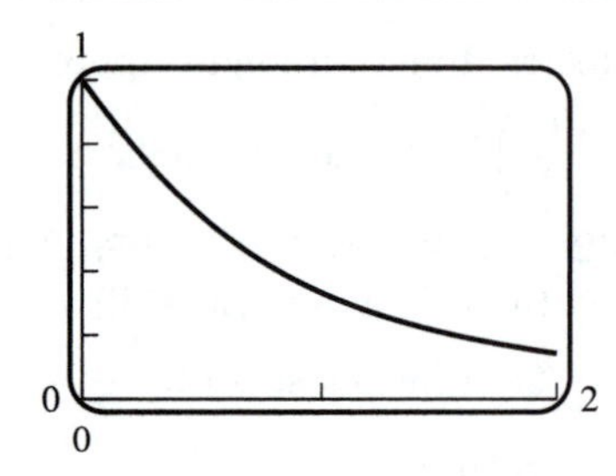

The limit is 1/3.

(b)

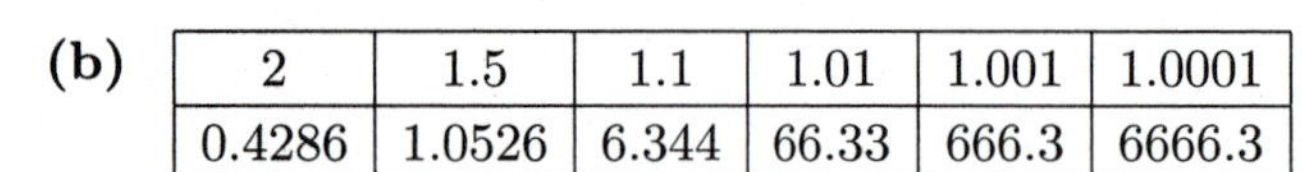

2	1.5	1.1	1.01	1.001	1.0001
0.4286	1.0526	6.344	66.33	666.3	6666.3

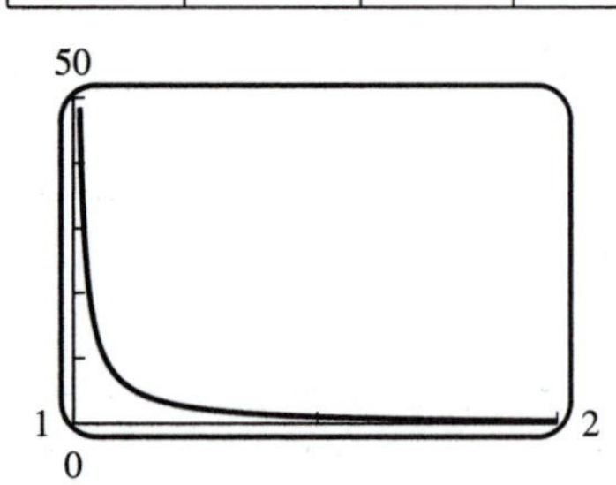

The limit is $+\infty$.

(c)

0	0.5	0.9	0.99	0.999	0.9999
−1	−1.7143	−7.0111	−67.001	−667.0	−6667.0

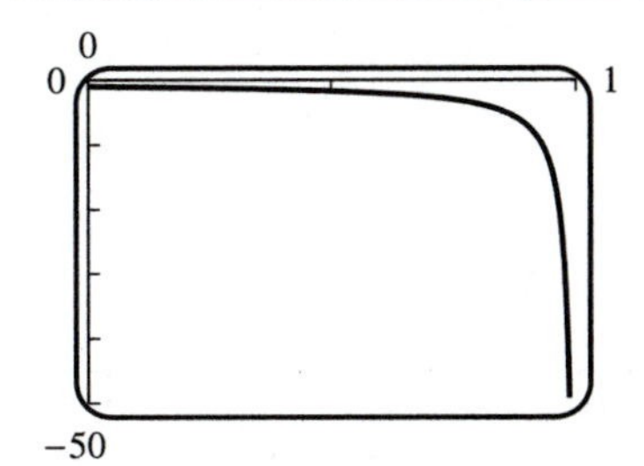

The limit is $-\infty$.

21. (a)

−0.25	−0.1	−0.001	−0.0001	0.0001	0.001	0.1	0.25
2.7266	2.9552	3.0000	3.0000	3.0000	3.0000	2.9552	2.7266

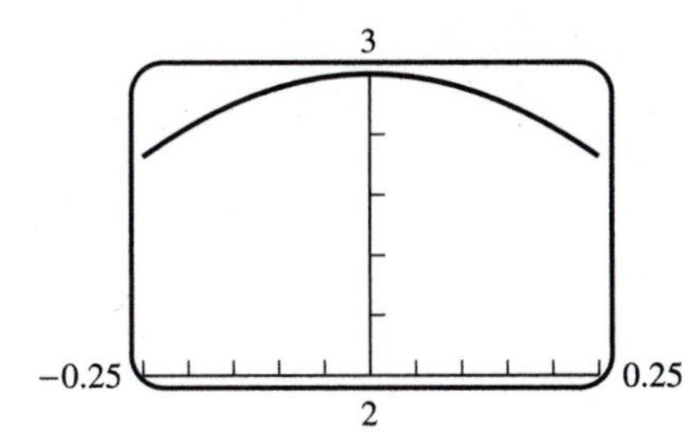

The limit is 3.

(b)

0	−0.5	−0.9	−0.99	−0.999	−1.5	−1.1	−1.01	−1.001
1	1.7552	6.2161	54.87	541.1	−0.1415	−4.536	−53.19	−539.5

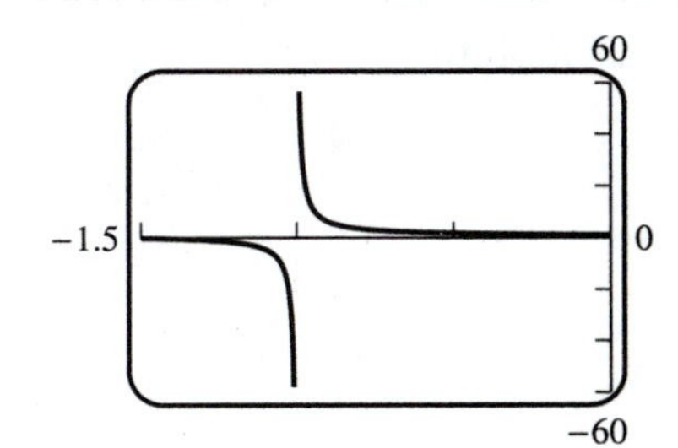

The limit does not exist.

23. The height of the ball at time $t = 0.25 + \Delta t$ is $s(0.25+\Delta t) = -16(0.25+\Delta t)^2 + 29(0.25+\Delta t) + 6$, so the distance traveled over the interval from $t = 0.25 - \Delta t$ to $t = 0.25 + \Delta t$ is
$s(0.25 + \Delta t) - s(0.25 - \Delta t) = -64(0.25)\Delta t + 58\Delta t$.
Thus the average velocity over the same interval is given by
$v_{\text{ave}} = [s(0.25 + \Delta t) - s(0.25 - \Delta t)]/2\Delta t = (-64(0.25)\Delta t + 58\Delta t)/2\Delta t = 21$ ft/s,
and this will also be the instantaneous velocity, since it happens to be independent of Δt.

25. (a)

−100,000,000	−100,000	−1000	−100	−10	10	100	1000
2.0000	2.0001	2.0050	2.0521	2.8333	1.6429	1.9519	1.9950

100,000	100,000,000
2.0000	2.0000

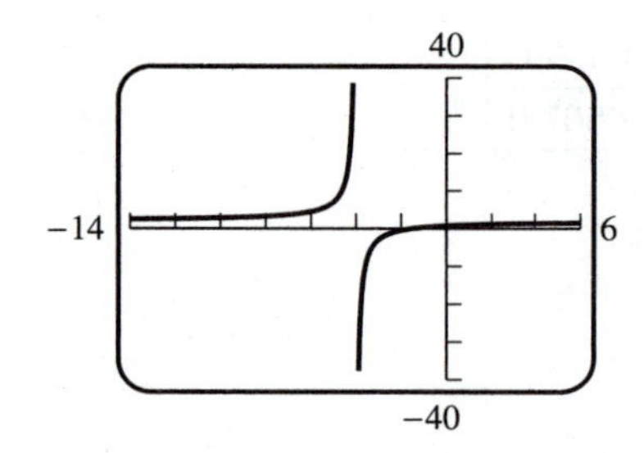

asymptote $y = 2$ as $x \to \pm\infty$

(b)

−100,000,000	−100,000	−1000	−100	−10	10	100	1000	100,000	100,000,000
20.0855	20.0864	20.1763	21.0294	35.4013	13.7858	19.2186	19.9955	20.0846	20.0855

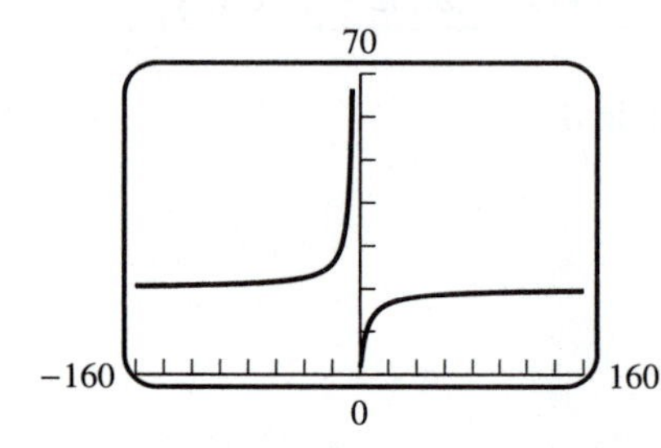

asymptote $y = 20.086$.

(c)

−100,000,000	−100,000	−1000	−100	−10	10	100	1000	100,000	100,000,000
−100,000,001	−100,000	−1001	−101.0	−11.2	9.2	99.0	999.0	99,999	99,999,999

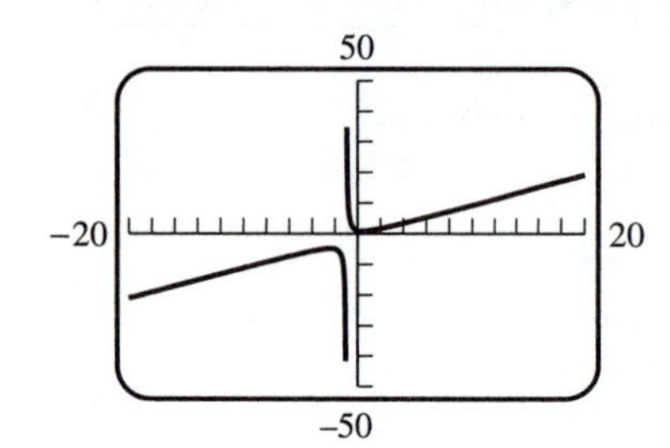

no horizontal asymptote

27. It appears that $\lim_{t \to +\infty} n(t) = +\infty$, and $\lim_{t \to +\infty} e(t) = c$.

29. **(a)** $\lim_{t \to 0^+} \dfrac{\sin t}{t}$ **(b)** $\lim_{t \to 0^+} \dfrac{t-1}{t+1}$ **(c)** $\lim_{t \to 0^-} (1+2t)^{1/t}$

31. $\lim_{x \to -\infty} f(x) = L$ and $\lim_{x \to +\infty} = L$

33. **(a)** The limit appears to be 3.

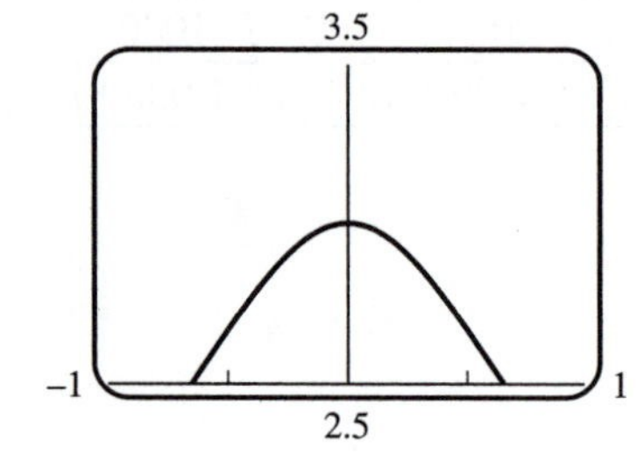

(b) The limit appears to be 3.

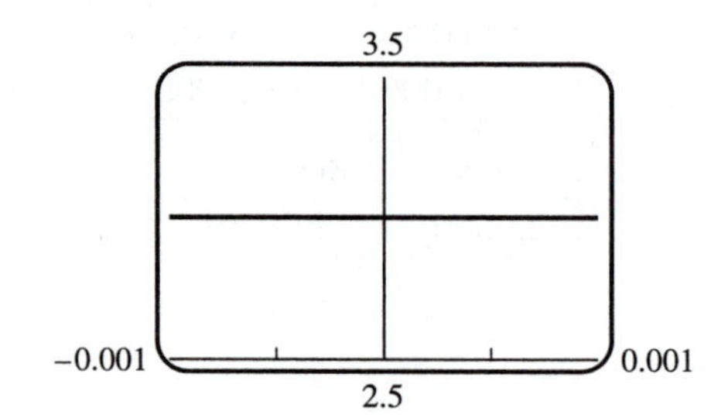

(c) The limit does not exist.

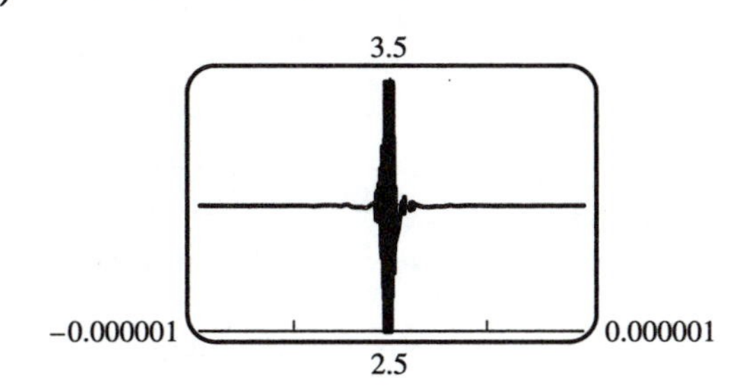

35. **(a)** The plot over the interval $[-a, a]$ becomes subject to catastrophic subtraction if a is small enough (the size depending on the machine).

(c) It does not.

EXERCISE SET 2.2

1. **(a)** 7 **(b)** π **(c)** -6 **(d)** 36

3. **(a)** -6 **(b)** 13 **(c)** -8 **(d)** 16 **(e)** 2 **(f)** $-1/2$

(g) The limit doesn't exist because the denominator tends to zero but the numerator doesn't.

(h) The limit doesn't exist because the denominator tends to zero but the numerator doesn't.

5. 0 **7.** 8 **9.** 4 **11.** $-4/5$

13. $3/2$ **15.** $+\infty$ **17.** does not exist

19. $-\infty$ **21.** $+\infty$ **23.** does not exist

25. $+\infty$ **27.** $+\infty$ **29.** 6

33. **(a)** 2 **(b)** 2 **(c)** 2 **35.** **(a)** 3 **(b)**

37. **(a)** Theorem 2.2.2(a) doesn't apply; moreover one cannot add/subtract infinities.

(b) $\lim_{x\to 0^+}\left(\frac{1}{x}-\frac{1}{x^2}\right) = \lim_{x\to 0^+}\left(\frac{x-1}{x^2}\right) = -\infty$

39. $\lim_{x\to 0} \frac{x}{x\left(\sqrt{x+4}+2\right)} = \frac{1}{4}$

41. The left and/or right limits could be plus or minus infinity; or the limit could exist, or equal any preassigned real number. For example, let $q(x) = x - x_0$ and let $p(x) = a(x - x_0)^n$ where n takes on the values $0, 1, 2$.

EXERCISE SET 2.3

1. **(a)** -3 **(b)** $-\infty$

3. **(a)** -12 **(b)** 21 **(c)** -15 **(d)** 25
(e) 2 **(f)** $-3/5$ **(g)** 0
(h) The limit doesn't exist because the denominator tends to zero but the numerator doesn't.

5. $+\infty$ **7.** $-\infty$ **9.** $+\infty$ **11.** $3/2$

13. 0 **15.** 0 **17.** $-5^{1/3}/2$ **19.** $-\sqrt{5}$

21. $1/\sqrt{6}$ **23.** $\sqrt{3}$ **25.** $-\infty$ **27.** $-1/7$

29. **(a)** $+\infty$ **(b)** -5

31. $\lim_{x\to+\infty} (\sqrt{x^2+3}-x)\dfrac{\sqrt{x^2+3}+x}{\sqrt{x^2+3}+x} = \lim_{x\to+\infty} \dfrac{3}{\sqrt{x^2+3}+x} = 0$

33. $\lim_{x\to+\infty} \left(\sqrt{x^2+ax}-x\right)\dfrac{\sqrt{x^2+ax}+x}{\sqrt{x^2+ax}+x} = \lim_{x\to+\infty} \dfrac{ax}{\sqrt{x^2+ax}+x} = a/2$

35. $\lim_{x\to+\infty} p(x) = (-1)^n\infty$ and $\lim_{x\to-\infty} p(x) = +\infty$

37. If $m > n$ the limits are both zero. If $m = n$ the limits are both equal to a_m, the leading coefficient of p. If $n > m$ the limits are $\pm\infty$ where the sign depends on the sign of a_m and whether n is even or odd.

39. If $m > n$ the limit is 0. If $m = n$ the limit is -3. If $m < n$ and $n - m$ is odd, then the limit is $+\infty$; if $m < n$ and $n - m$ is even, then the limit is $-\infty$.

41. $f(x) = x + 2 + \dfrac{2}{x-2}$, so $\lim_{x\to\pm\infty} (f(x) - (x+2)) = 0$ and $f(x)$ is asymptotic to $y = x + 2$.

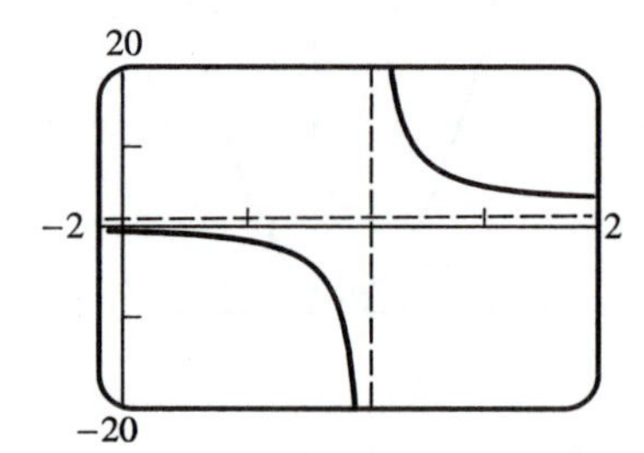

43. $f(x) = -x^2 + 1 + 2/(x-3)$ so $\lim_{x\to\pm\infty} [f(x) - (-x^2+1)] = 0$ and $f(x)$ is asymptotic to $y = -x^2 + 1$.

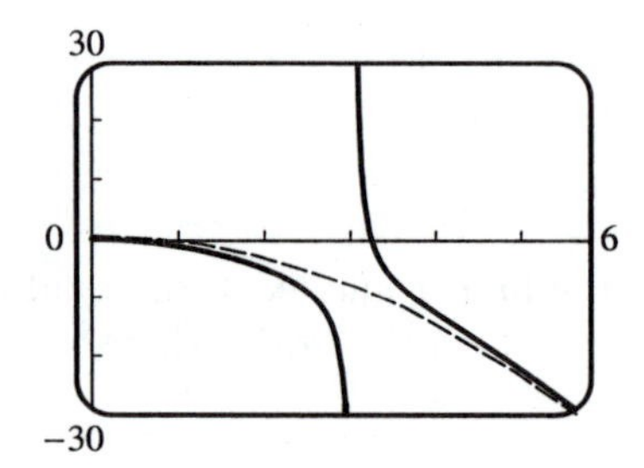

45. $f(x) - \sin x = 0$ and $f(x)$ is asymptotic to $y = \sin x$.

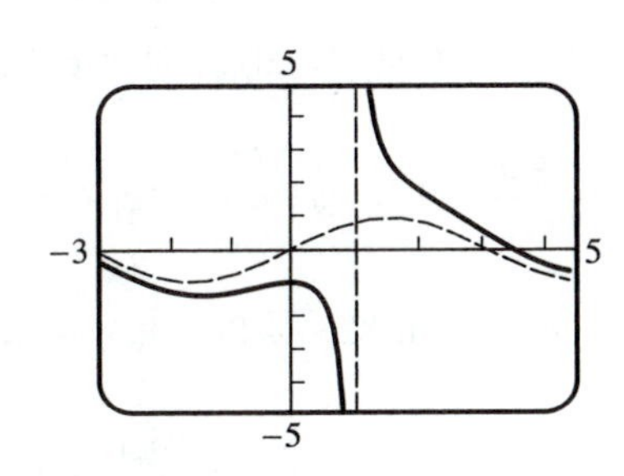

EXERCISE SET 2.4

1. **(a)** $|f(x) - f(0)| = |x + 2 - 2| = |x| < 0.1$ if and only if $|x| < 0.1$

(b) $|f(x) - f(3)| = |(4x - 5) - 7| = 4|x - 3| < 0.1$ if and only if $|x - 3| < (0.1)/4 = 0.0025$

(c) $|f(x) - f(4)| = |x^2 - 16| < \epsilon$ if $|x - 4| < \delta$. We get $f(x) = 16 + \epsilon = 16.001$ at $x = 4.000124998$, which corresponds to $\delta = 0.000124998$; and $f(x) = 16 - \epsilon = 15.999$ at $x = 3.999874998$, for which $\delta = 0.000125002$. Use the smaller δ: thus $|f(x) - 16| < \epsilon$ provided $|x - 4| < 0.000125$ (to six decimals).

3. **(a)** $x_1 = (1.95)^2 = 3.8025, x_2 = (2.05)^2 = 4.2025$

(b) $\delta = \min(\,|4 - 3.8025|, |4 - 4.2025|\,) = 0.1975$

5. $|(x^3 - 4x + 5) - 2| < 0.05$, $-0.05 < (x^3 - 4x + 5) - 2 < 0.05$, $1.95 < x^3 - 4x + 5 < 2.05$; $x^3 - 4x + 5 = 1.95$ at $x = 1.0616$, $x^3 - 4x + 5 = 2.05$ at $x = 0.9558$; $\delta = \min(1.0616 - 1, 1 - 0.9558) = 0.0442$

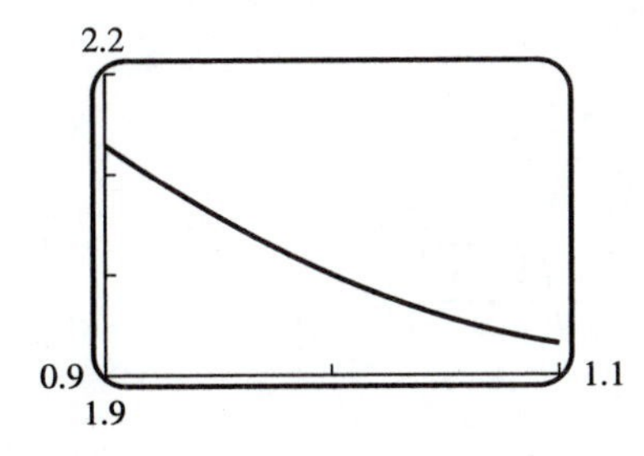

7. With the TRACE feature of a calculator we discover that (to five decimal places) (0.87000, 1.80274) and (1.13000, 2.19301) belong to the graph. Set $x_0 = 0.87$ and $x_1 = 1.13$. Since the graph of $f(x)$ rises from left to right, we see that if $x_0 < x < x_1$ then $1.80274 < f(x) < 2.19301$, and therefore $1.8 < f(x) < 2.2$. So we can take $\delta = 0.13$.

9. $|2x - 8| = 2|x - 4| < 0.1$ if $|x - 4| < 0.05$, $\delta = 0.05$

11. $|7x + 5 - (-2)| = 7|x - (-1)| < 0.01$ if $|x + 1| < \frac{1}{700}$, $\delta = \frac{1}{700}$

13. $\left|\dfrac{x^2 - 4}{x - 2} - 4\right| = \left|\dfrac{x^2 - 4 - 4x + 8}{x - 2}\right| = |x - 2| < 0.05$ if $|x - 2| < 0.05$, $\delta = 0.05$

15. if $\delta < 1$ then $|x^2 - 16| = |x - 4||x + 4| < 9|x - 4| < 0.001$ if $|x - 4| < \frac{1}{9000}$, $\delta = \frac{1}{9000}$

17. if $\delta \le 1$ then $\left|\dfrac{1}{x} - \dfrac{1}{5}\right| = \dfrac{|x - 5|}{5|x|} \le \dfrac{|x - 5|}{20} < 0.05$ if $|x - 5| < 1$, $\delta = 1$

19. $|3x - 15| = 3|x - 5| < \epsilon$ if $|x - 5| < \frac{1}{3}\epsilon$, $\delta = \frac{1}{3}\epsilon$

21. $|2x - 7 - (-3)| = 2|x - 2| < \epsilon$ if $|x - 2| < \frac{1}{2}\epsilon$, $\delta = \frac{1}{2}\epsilon$

23. $\left|\dfrac{x^2 + x}{x} - 1\right| = |x| < \epsilon$ if $|x| < \epsilon$, $\delta = \epsilon$

25. if $\delta < 1$ then $|2x^2 - 2| = 2|x - 1||x + 1| < 6|x - 1| < \epsilon$ if $|x - 1| < \frac{1}{6}\epsilon$, $\delta = \min(1, \frac{1}{6}\epsilon)$

27. if $\delta < \dfrac{1}{6}$ then $\left|\dfrac{1}{x} - 3\right| = \dfrac{3|x - \frac{1}{3}|}{|x|} < 18\left|x - \dfrac{1}{3}\right| < \epsilon$ if $\left|x - \dfrac{1}{3}\right| < \frac{1}{18}\epsilon$, $\delta = \min\left(\dfrac{1}{6}, \dfrac{1}{18}\epsilon\right)$

29. $|\sqrt{x} - 2| = \left|(\sqrt{x} - 2)\dfrac{\sqrt{x} + 2}{\sqrt{x} + 2}\right| = \left|\dfrac{x - 4}{\sqrt{x} + 2}\right| < \dfrac{1}{2}|x - 4| < \epsilon$ if $|x - 4| < 2\epsilon$, $\delta = 2\epsilon$

31. $|f(x) - 3| = |x + 2 - 3| = |x - 1| < \epsilon$ if $0 < |x - 1| < \epsilon$, $\delta = \epsilon$

33. **(a)** $|f(x) - L| = \dfrac{1}{x^2} < 0.1$ if $x > \sqrt{10}$, $N = \sqrt{10}$

(b) $|f(x) - L| = \left|\dfrac{x}{x + 1} - 1\right| = \left|\dfrac{1}{x + 1}\right| < 0.01$ if $x + 1 > 100$, $N = 99$

(c) $|f(x) - L| = \left|\dfrac{1}{x^3}\right| < \dfrac{1}{1000}$ if $|x| > 10$, $x < -10$, $N = -10$

(d) $|f(x) - L| = \left|\dfrac{x}{x + 1} - 1\right| = \left|\dfrac{1}{x + 1}\right| < 0.01$ if $|x + 1| > 100$, $-x - 1 > 100$, $x < -101$, $N = -101$

35. **(a)** $\dfrac{x_1^2}{1 + x_1^2} = 1 - \epsilon$, $x_1 = -\sqrt{\dfrac{1 - \epsilon}{\epsilon}}$; $\dfrac{x_2^2}{1 + x_2^2} = 1 - \epsilon$, $x_2 = \sqrt{\dfrac{1 - \epsilon}{\epsilon}}$

(b) $N = \sqrt{\dfrac{1 - \epsilon}{\epsilon}}$ **(c)** $N = -\sqrt{\dfrac{1 - \epsilon}{\epsilon}}$

37. $\dfrac{1}{x^2} < 0.01$ if $|x| > 10$, $N = 10$

39. $\left|\dfrac{x}{x + 1} - 1\right| = \left|\dfrac{1}{x + 1}\right| < 0.001$ if $|x + 1| > 1000$, $x > 999$, $N = 999$

41. $\left|\dfrac{1}{x + 2} - 0\right| < 0.005$ if $|x + 2| > 200$, $-x - 2 > 200$, $x < -202$, $N = -202$

43. $\left|\dfrac{4x - 1}{2x + 5} - 2\right| = \left|\dfrac{11}{2x + 5}\right| < 0.1$ if $|2x + 5| > 110$, $-2x - 5 > 110$, $2x < -115$, $x < -57.5$, $N = -57.5$

45. $\left|\dfrac{1}{x^2}\right| < \epsilon$ if $|x| > \dfrac{1}{\sqrt{\epsilon}}$, $N = \dfrac{1}{\sqrt{\epsilon}}$

47. $\left|\dfrac{1}{x + 2}\right| < \epsilon$ if $|x + 2| > \dfrac{1}{\epsilon}$, $-x - 2 < \dfrac{1}{\epsilon}$, $x > -2 - \dfrac{1}{\epsilon}$, $N = -2 - \dfrac{1}{\epsilon}$

49. $\left|\dfrac{x}{x + 1} - 1\right| = \left|\dfrac{1}{x + 1}\right| < \epsilon$ if $|x + 1| > \dfrac{1}{\epsilon}$, $x > \dfrac{1}{\epsilon} - 1$, $N = \dfrac{1}{\epsilon} - 1$

51. $\left|\dfrac{4x-1}{2x+5}-2\right|=\left|\dfrac{11}{2x+5}\right|<\epsilon$ if $|2x+5|>\dfrac{11}{\epsilon}$, $-2x-5>\dfrac{11}{\epsilon}$, $2x<-\dfrac{11}{\epsilon}-5$, $x<-\dfrac{11}{2\epsilon}-\dfrac{5}{2}$, $N=-\dfrac{5}{2}-\dfrac{11}{2\epsilon}$

53. **(a)** $\dfrac{1}{x^2}>100$ if $|x|<\dfrac{1}{10}$ **(b)** $\dfrac{1}{|x-1|}>1000$ if $|x-1|<\dfrac{1}{1000}$

(c) $\dfrac{-1}{(x-3)^2}<-1000$ if $|x-3|<\dfrac{1}{10\sqrt{10}}$ **(d)** $-\dfrac{1}{x^4}<-10000$ if $x^4<\dfrac{1}{10000}$, $|x|<\dfrac{1}{10}$

55. if $M>0$ then $\dfrac{1}{(x-3)^2}>M$, $0<(x-3)^2<\dfrac{1}{M}$, $0<|x-3|<\dfrac{1}{\sqrt{M}}$, $\delta=\dfrac{1}{\sqrt{M}}$

57. if $M>0$ then $\dfrac{1}{|x|}>M$, $0<|x|<\dfrac{1}{M}$, $\delta=\dfrac{1}{M}$

59. if $M<0$ then $-\dfrac{1}{x^4}<M$, $0<x^4<-\dfrac{1}{M}$, $|x|<\dfrac{1}{(-M)^{1/4}}$, $\delta=\dfrac{1}{(-M)^{1/4}}$

61. if $x>2$ then $|x+1-3|=|x-2|=x-2<\epsilon$ if $2<x<2+\epsilon$, $\delta=\epsilon$

63. if $x>4$ then $\sqrt{x-4}<\epsilon$ if $x-4<\epsilon^2$, $4<x<4+\epsilon^2$, $\delta=\epsilon^2$

65. if $x>2$ then $|f(x)-2|=|x-2|=x-2<\epsilon$ if $2<x<2+\epsilon$, $\delta=\epsilon$

67. **(a)** if $M<0$ and $x>1$ then $\dfrac{1}{1-x}<M$, $x-1<-\dfrac{1}{M}$, $1<x<1-\dfrac{1}{M}$, $\delta=-\dfrac{1}{M}$

(b) if $M>0$ and $x<1$ then $\dfrac{1}{1-x}>M$, $1-x<\dfrac{1}{M}$, $1-\dfrac{1}{M}<x<1$, $\delta=\dfrac{1}{M}$

69. **(a)** Given any $M>0$ there corresponds $N>0$ such that if $x>N$ then $f(x)>M$, $x+1>M$, $x>M-1$, $N=M-1$.

(b) Given any $M<0$ there corresponds $N<0$ such that if $x<N$ then $f(x)<M$, $x+1<M$, $x<M-1$, $N=M-1$.

71. if $\delta\le 2$ then $|x-3|<2$, $-2<x-3<2$, $1<x<5$, and $|x^2-9|=|x+3||x-3|<8|x-3|<\epsilon$ if $|x-3|<\frac{1}{8}\epsilon$, $\delta=\min\left(2,\frac{1}{8}\epsilon\right)$

EXERCISE SET 2.5

1. **(a)** no, $x=2$ **(b)** no, $x=2$ **(c)** no, $x=2$ **(d)** yes
(e) yes **(f)** yes

3. **(a)** no, $x=1,3$ **(b)** yes **(c)** no, $x=1$ **(d)** yes
(e) no, $x=3$ **(f)** yes

5. **(a)** At $x=3$ the one-sided limits fail to exist.
(b) At $x=-2$ the two-sided limit exists but is not equal to $F(-2)$.
(c) At $x=3$ the limit fails to exist.

7. **(a)** 3 **(b)** 3

9. (a)

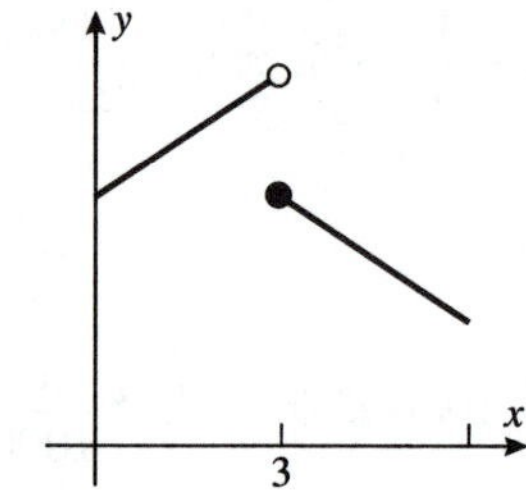

(b)

(c)

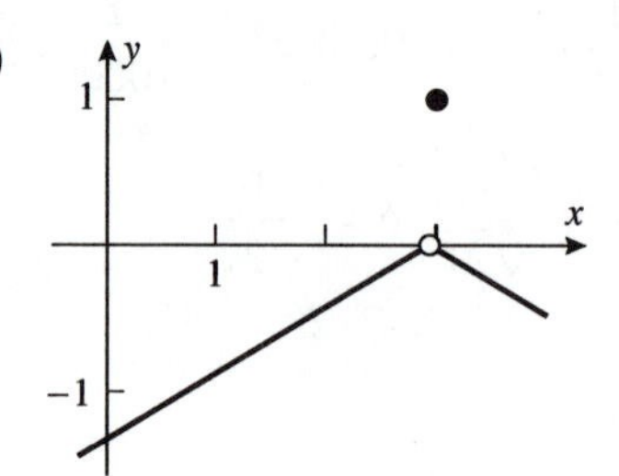

(d)

11. (a)

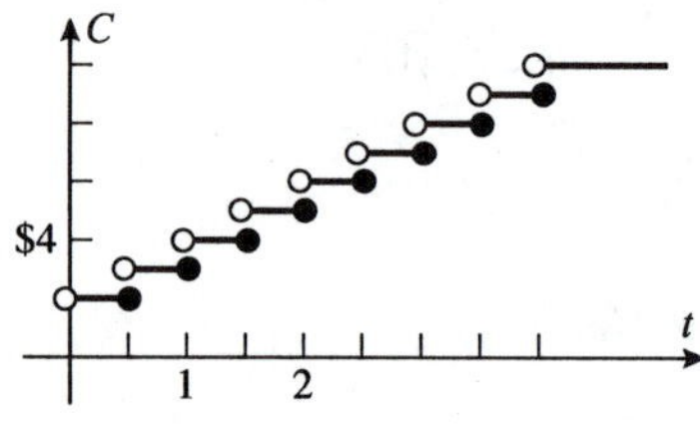
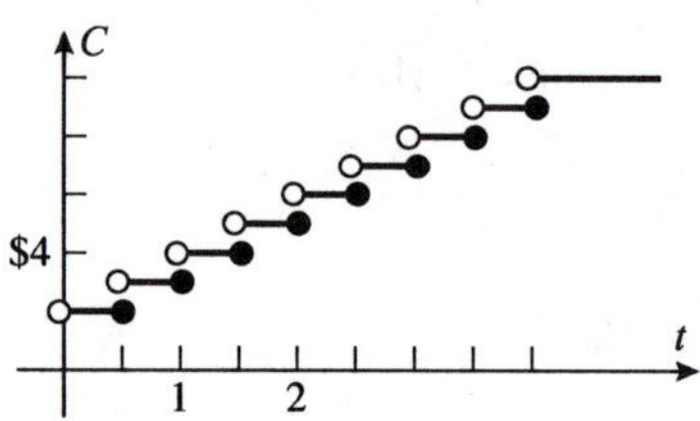

(b) One second could cost you one dollar.

13. none

15. none

17. f is not defined at $x = \pm 4$

19. f is not defined at $x = \pm 3$

21. none

23. none; $f(x) = 2x + 3$ is continuous on $x < 4$ and $f(x) = 7 + \dfrac{16}{x}$ is continuous on $4 < x$; $\lim_{x \to 4^-} f(x) = \lim_{x \to 4^+} f(x) = f(4) = 11$ so f is continuous at $x = 4$

25. (a) f is continuous for $x < 1$, and for $x > 1$; $\lim_{x \to 1^-} f(x) = 5$, $\lim_{x \to 1^+} f(x) = k$, so if $k = 5$ then f is continuous for all x

(b) f is continuous for $x < 2$, and for $x > 2$; $\lim_{x \to 2^-} f(x) = 4k$, $\lim_{x \to 2^+} f(x) = 4 + k$, so if $4k = 4 + k$, $k = 4/3$ then f is continuous for all x

27. (a)

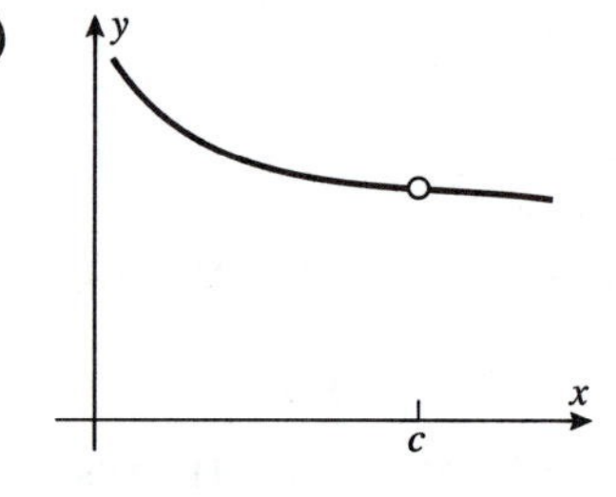

(b)

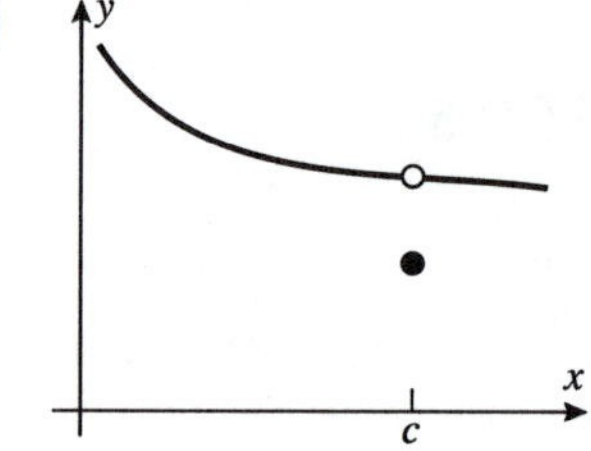

29. (a) $x = 0$, $\lim_{x \to 0^-} f(x) = -1 \neq +1 = \lim_{x \to 0^+} f(x)$ so the discontinuity is not removable

(b) $x = -3$; define $f(-3) = -3 = \lim_{x \to -3} f(x)$, then the discontinuity is removable

(c) f is undefined at $x = \pm 2$; at $x = 2$, $\lim_{x \to 2} f(x) = 1$, so define $f(2) = 1$ and f becomes continuous there; at $x = -2$, $\lim_{x \to -2}$ does not exist, so the discontinuity is not removable

31. **(a)** discontinuity at $x = 1/2$, not removable; at $x = -3$, removable

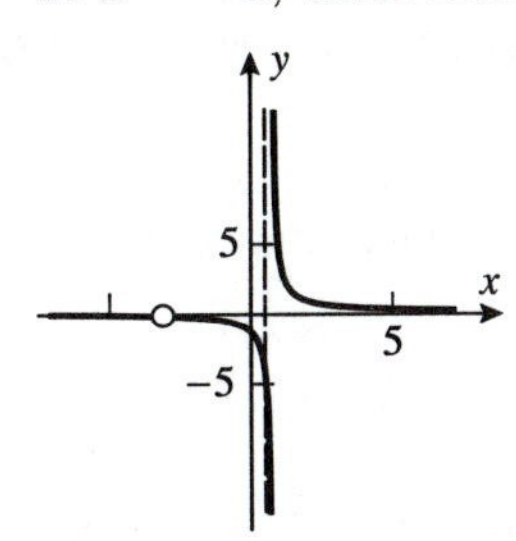

(b) 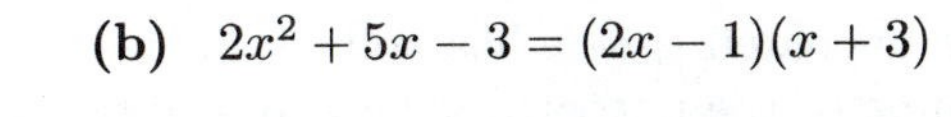
$2x^2 + 5x - 3 = (2x - 1)(x + 3)$

33. For $x > 0$, $f(x) = x^{3/5} = (x^3)^{1/5}$ is the composition (Theorem 2.4.6) of the two continuous functions $g(x) = x^3$ and $h(x) = x^{1/5}$ and is thus continuous. For $x < 0$, $f(x) = f(-x)$ which is the composition of the continuous functions $f(x)$ (for positive x) and the continuous function $y = -x$. Hence $f(-x)$ is continuous for all $x > 0$. At $x = 0$, $f(0) = \lim_{x \to 0} f(x) = 0$.

35. **(a)** Let $f(x) = k$ for $x \neq c$ and $f(c) = 0$; $g(x) = l$ for $x \neq c$ and $g(c) = 0$. If $k = -l$ then $f + g$ is continuous; otherwise it's not.

(b) $f(x) = k$ for $x \neq c$, $f(c) = 1$; $g(x) = l \neq 0$ for $x \neq c$, $g(c) = 1$. If $kl = 1$, then fg is continuous; otherwise it's not.

37. Since f and g are continuous at $x = c$ we know that $\lim_{x \to c} f(x) = f(c)$ and $\lim_{x \to c} g(x) = g(c)$. In the following we use Theorem 2.2.2.

(a) $f(c) + g(c) = \lim_{x \to c} f(x) + \lim_{x \to c} g(x) = \lim_{x \to c}(f(x) + g(x))$ so $f + g$ is continuous at $x = c$.

(b) same as (a) except the + sign becomes a $-$ sign

(c) $\dfrac{f(c)}{g(c)} = \dfrac{\lim_{x \to c} f(x)}{\lim_{x \to c} g(x)} = \lim_{x \to c} \dfrac{f(x)}{g(x)}$ so $\dfrac{f}{g}$ is continuous at $x = c$

39. Of course such a function must be discontinuous. Let $f(x) = 1$ on $0 \le x < 1$, and $f(x) = -1$ on $1 \le x \le 2$.

41. The cone has volume $\pi r^2 h/3$. The function $V(r) = \pi r^2 h$ (for variable r and fixed h) gives the volume of a right circular cylinder of height h and radius r, and satisfies $V(0) < \pi r^2 h/3 < V(r)$. By the Intermediate Value Theorem there is a value c between 0 and r such that $V(c) = \pi r^2 h/3$, so the cylinder of radius c (and height h) has volume equal to that of the cone.

43. If $f(x) = x^3 + x^2 - 2x$ then $f(-1) = 2$, $f(1) = 0$. Use the Intermediate Value Theorem.

45. For the negative root, use intervals on the x-axis as follows: $[-2, -1]$; since $f(-1.3) < 0$ and $f(-1.2) > 0$, the midpoint $x = -1.25$ of $[-1.3, -1.2]$ is the required approximation of the root. For the positive root use the interval $[0, 1]$; since $f(0.7) < 0$ and $f(0.8) > 0$, the midpoint $x = 0.75$ of $[0.7, 0.8]$ is the required approximation.

47. For the negative root, use intervals on the x-axis as follows: $[-2, -1]$; since $f(-1.7) < 0$ and $f(-1.6) > 0$, use the interval $[-1.7, -1.6]$. Since $f(-1.61) < 0$ and $f(-1.60) > 0$ the midpoint $x = -1.605$ of $[-1.61, -1.60]$ is the required approximation of the root. For the positive root use the interval $[1, 2]$; since $f(1.3) > 0$ and $f(1.4) < 0$, use the interval $[1.3, 1.4]$. Since $f(1.37) > 0$ and $f(1.38) < 0$, the midpoint $x = 1.375$ of $[1.37, 1.38]$ is the required approximation.

49. $x = 2.24$

51. The uncoated sphere has volume $4\pi(x-1)^3/3$ and the coated sphere has volume $4\pi x^3/3$. If the volume of the uncoated sphere and of the coating itself are the same, then the coated sphere has twice the volume of the uncoated sphere. Thus $2(4\pi(x-1)^3/3) = 4\pi x^3/3$, or $x^3 - 6x^2 + 6x - 2 = 0$, with the solution $x = 4.847$ cm.

53. We must show $\lim_{x\to c} f(x) = f(c)$. Let $\epsilon > 0$; then there exists $\delta > 0$ such that if $|x-c| < \delta$ then $|f(x) - f(c)| < \epsilon$. But this certainly satisfies Definition 2.4.1.

EXERCISE SET 2.6

1. none

3. $x = n\pi$, $n = 0, \pm 1, \pm 2, \ldots$

5. $x = n\pi$, $n = 0, \pm 1, \pm 2, \ldots$

7. none

9. $2n\pi + \pi/6, 2n\pi + 5\pi/6$, $n = 0, \pm 1, \pm 2, \ldots$

11. **(a)** $\sin x, x^3 + 7x + 1$ **(b)** $|x|, \sin x$ **(c)** $x^3, \cos x, x + 1$
(d) $\sqrt{x}, 3 + x, \sin x, 2x$ **(e)** $\sin x, \sin x$ **(f)** $x^5 - 2x^3 + 1, \cos x$

13. $\cos\left(\lim_{x\to+\infty} \dfrac{1}{x}\right) = \cos 0 = 1$

15. $\sin\left(\lim_{x\to+\infty} \dfrac{\pi x}{2 - 3x}\right) = \sin\left(-\dfrac{\pi}{3}\right) = -\dfrac{\sqrt{3}}{2}$

17. $3 \lim_{\theta\to 0} \dfrac{\sin 3\theta}{3\theta} = 3$

19. $-\lim_{x\to 0^-} \dfrac{\sin x}{x} = -1$

21. $\dfrac{1}{5} \lim_{x\to 0^+} \sqrt{x} \lim_{x\to 0^+} \dfrac{\sin x}{x} = 0$

23. $\dfrac{\tan 7x}{\sin 3x} = \dfrac{7}{3\cos 7x} \dfrac{\sin 7x}{7x} \dfrac{3x}{\sin 3x}$ so $\lim_{x\to 0} \dfrac{\tan 7x}{\sin 3x} = \dfrac{7}{3(1)}(1)(1) = \dfrac{7}{3}$

25. $\left(\lim_{h\to 0} \cos h\right) \lim_{h\to 0} \dfrac{h}{\sin h} = 1$

27. $\dfrac{\theta^2}{1 - \cos\theta} \dfrac{1 + \cos\theta}{1 + \cos\theta} = \dfrac{\theta^2(1 + \cos\theta)}{1 - \cos^2\theta} = \left(\dfrac{\theta}{\sin\theta}\right)^2 (1 + \cos\theta)$ so $\lim_{\theta\to 0} \dfrac{\theta^2}{1 - \cos\theta} = (1)^2 2 = 2$

29. 0

31. $\dfrac{1 - \cos 5h}{\cos 7h - 1} = \dfrac{(1 - \cos 5h)(1 + \cos 5h)(1 + \cos 7h)}{(\cos 7h - 1)(1 + \cos 5h)(1 + \cos 7h)} = -\dfrac{25}{49}\left(\dfrac{\sin 5h}{5h}\right)^2 \left(\dfrac{7h}{\sin 7h}\right)^2 \dfrac{1 + \cos 7h}{1 + \cos 5h}$ so $\lim_{h\to 0} \dfrac{1 - \cos 5h}{\cos 7h - 1} = -\dfrac{25}{49}$

33. $\lim_{x\to 0^+} \cos\left(\dfrac{1}{x}\right) = \lim_{t\to+\infty} \cos t$; limit does not exist

35. $2 + \lim_{x\to 0} \dfrac{\sin x}{x} = 3$

37.

2.1	2.01	2.001	2.0001	2.00001	1.9	1.99	1.999	1.9999	1.99999
0.484559	0.498720	0.499875	0.499987	0.499999	0.509409	0.501220	0.500125	0.500012	0.500001

The limit is 0.5.

39.

-0.9	-0.99	-0.999	-0.9999	-0.99999	-1.1	-1.01	-1.001	-1.0001	-1.00001
0.405086	0.340050	0.334001	0.333400	0.333340	0.271536	0.326717	0.332667	0.333267	0.333327

The limit is $1/3$.

41. $\lim_{x\to 0^-} f(x) = k \lim_{x\to 0} \dfrac{\sin kx}{kx \cos kx} = k$, $\lim_{x\to 0^+} f(x) = 2k^2$, so $k = 2k^2$, $k = \dfrac{1}{2}$

43. (a) $\lim_{t\to 0^+} \dfrac{\sin t}{t} = 1$

(b) $\lim_{t\to 0^-} \dfrac{1-\cos t}{t} = 0$ (Theorem 2.6.3)

(c) $\sin(\pi - t) = \sin t$, so $\lim_{x\to\pi} \dfrac{\pi - x}{\sin x} = \lim_{t\to 0} \dfrac{t}{\sin t} = 1$

45. $t = x - 1$; $\sin(\pi x) = \sin(\pi t + \pi) = -\sin \pi t$; and $\lim_{x\to 1} \dfrac{\sin(\pi x)}{x-1} = -\lim_{t\to 0} \dfrac{\sin \pi t}{t} = -\pi$

47. $-|x| \le x \cos\left(\dfrac{50\pi}{x}\right) \le |x|$

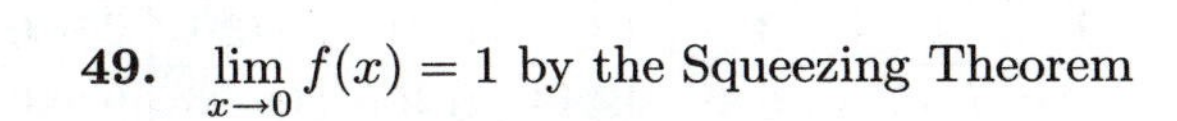

49. $\lim_{x\to 0} f(x) = 1$ by the Squeezing Theorem

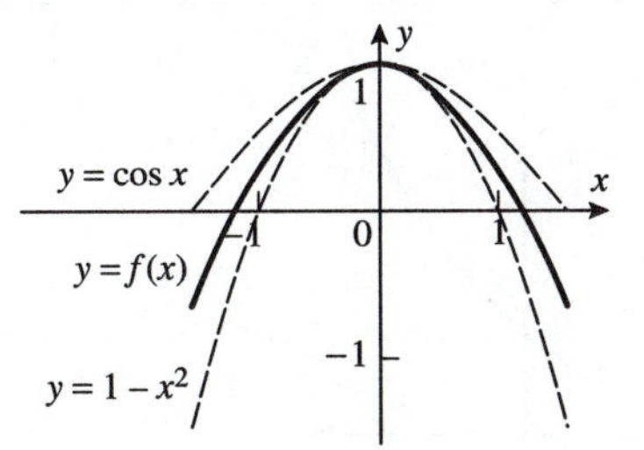

51. Let $g(x) = -\dfrac{1}{x}$ and $h(x) = \dfrac{1}{x}$; thus $\lim_{x\to+\infty} \dfrac{\sin x}{x} = 0$ by the Squeezing Theorem.

53. (a) $\sin x = \sin t$ where x is measured in degrees, t is measured in radians and $t = \dfrac{\pi x}{180}$. Thus $\lim_{x\to 0} \dfrac{\sin x}{x} = \lim_{t\to 0} \dfrac{\sin t}{(180t/\pi)} = \dfrac{\pi}{180}$.

55. (a) $\sin 10^\circ = 0.17365$

(b) $\sin 10^\circ = \sin \dfrac{\pi}{18} \approx \dfrac{\pi}{18} = 0.17453$

57. (a) 0.08749

(b) $\tan 5^\circ \approx \dfrac{\pi}{36} = 0.08727$

59. (a) Let $f(x) = x - \cos x$; $f(0) = -1$, $f(\pi/2) = \pi/2$. By the IVT there must be a solution of $f(x) = 0$.

(b)

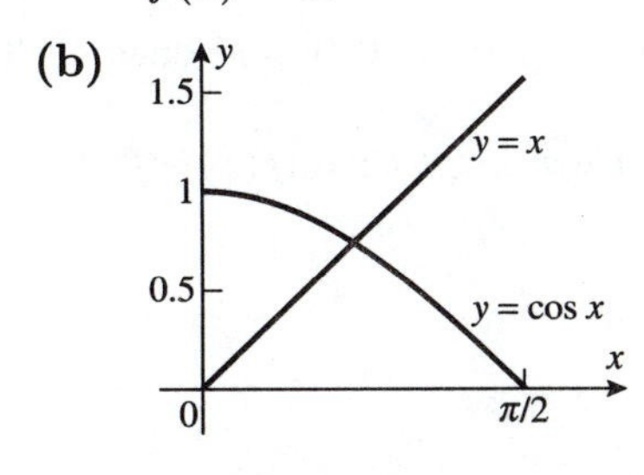

(c) 0.739

61. (a) There is symmetry about the equatorial plane.

(b) Let $g(\phi)$ be the given function. Then $g(38) < 9.8$ and $g(39) > 9.8$, so by the Intermediate Value Theorem there is a value c between 38 and 39 for which $g(c) = 9.8$ exactly.

CHAPTER 2 SUPPLEMENTARY EXERCISES

1. **(a)** 1 **(b)** no limit **(c)** no limit
(d) 1 **(e)** 3 **(f)** 0
(g) 0 **(h)** 2 **(i)** 1/2

5. **(a)** $0.222\ldots, 0.24390, 0.24938, 0.24994, 0.24999, 0.25000$; for $x \neq 2$, $f(x) = \dfrac{1}{x+2}$, so the limit is 1/4.

(b) $1.15782, 4.22793, 4.00213, 4.00002, 4.00000, 4.00000$; to prove, use $\dfrac{\tan 4x}{x} = \dfrac{\sin 4x}{x \cos 4x} = \dfrac{4}{\cos 4x}\dfrac{\sin 4x}{4x}$, the limit is 4.

7. **(a)**

x	1	0.1	0.01	0.001	0.0001	0.00001	0.000001
$f(x)$	1.000	0.443	0.409	0.406	0.406	0.405	0.405

(b)

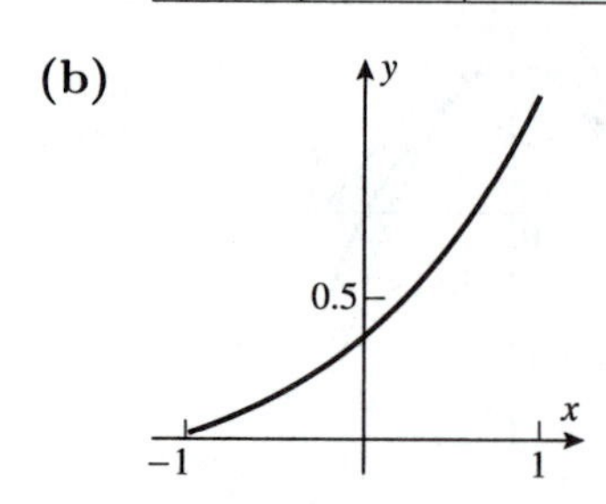

9. **(a)**

(b) Let $g(x) = x - f(x)$. Then $g(1) \geq 0$ and $g(0) \leq 0$; by the Intermediate Value Theorem there is a solution c in $[0, 1]$ of $g(c) = 0$.

11. If, on the contrary, $f(x_0) < 0$ for some x_0 in $[0, 1]$, then by the Intermediate Value Theorem we would have a solution of $f(x) = 0$ in $[0, x_0]$, contrary to the hypothesis.

13. $f(-6) = 185$, $f(0) = -1$, $f(2) = 65$; apply Theorem 2.4.9 twice, once on $[-6, 0]$ and once on $[0, 2]$

15. Let $\epsilon = f(x_0)/2 > 0$; then there corresponds $\delta > 0$ such that if $|x - x_0| < \delta$ then $|f(x) - f(x_0)| < \epsilon$, $-\epsilon < f(x) - f(x_0) < \epsilon$, $f(x) > f(x_0) - \epsilon = f(x_0)/2 > 0$ for $x_0 - \delta < x < x_0 + \delta$.

17. **(a)** $-3.449, 1.449$ **(b)** $x = 0, \pm 1.896$

19. **(a)** $\sqrt{5}$, no limit, $\sqrt{10}$, $\sqrt{10}$, no limit, $+\infty$, no limit
(b) $5, 10, 0, 0, 10, -\infty, +\infty$

21. a/b **23.** does not exist

25. 0

27. $3-k$

31.

x	0.1	0.01	0.001	0.0001	0.00001	0.000001
$f(x)$	2.59	2.70	2.717	2.718	2.7183	2.71828

33.

x	1.1	1.01	1.001	1.0001	1.00001	1.000001
$f(x)$	0.49	0.54	0.540	0.5403	0.54030	0.54030

35.

x	100	1000	10^4	10^5	10^6	10^7
$f(x)$	0.48809	0.49611	0.49876	0.49961	0.49988	0.49996

37. $\delta \approx 0.07747$ (use a graphing utility)

39. **(a)** $x^3 - x - 1 = 0$, $x^3 = x + 1$, $x = \sqrt[3]{x+1}$.

(b)

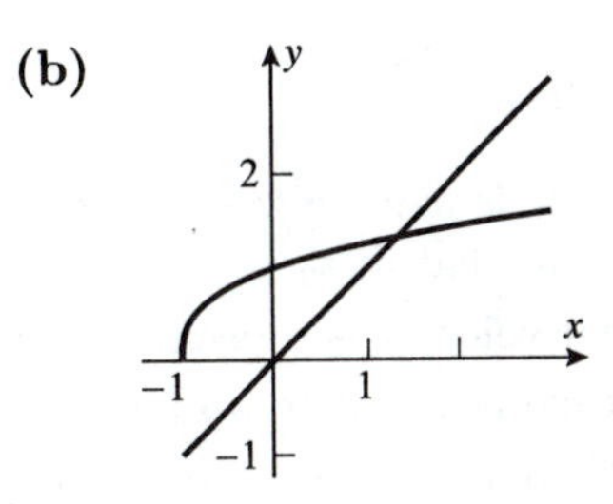

(c)

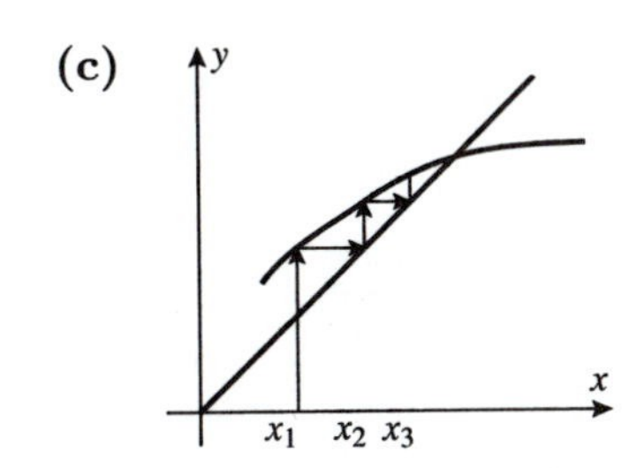

(d) $1, 1.26, 1.31, 1.322, 1.324, 1.3246, 1.3247$

41. $x = \sqrt[5]{x+2}$; 1.267168

CHAPTER 3
The Derivative

EXERCISE SET 3.1

1. **(a)** $m_{\text{tan}} = (50-10)/(15-5)$
$= 40/10$
$= 4$ m/s

(b)

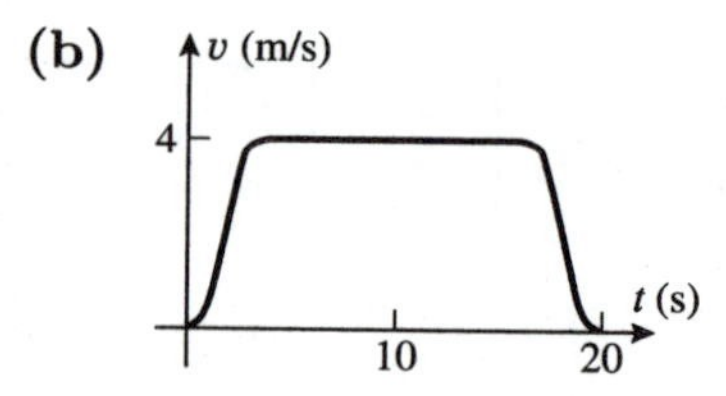

3. From the figure:

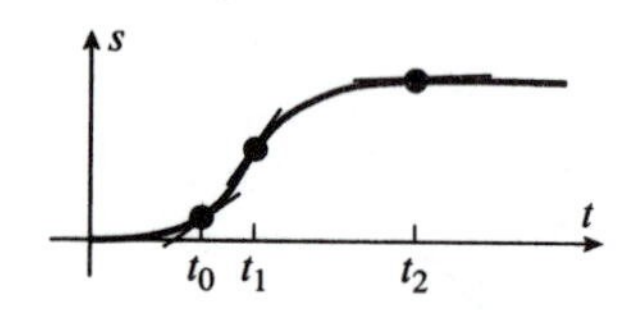

(a) The particle is moving faster at time t_0 because the slope of the tangent to the curve at t_0 is greater than that at t_2.

(b) The initial velocity is 0 because the slope of a horizontal line is 0.

(c) The particle is speeding up because the slope increases as t increases from t_0 to t_1.

(d) The particle is slowing down because the slope decreases as t increases from t_1 to t_2.

5. It is a straight line with slope equal to the velocity.

7. **(a)** $m_{\text{sec}} = \dfrac{f(4)-f(3)}{4-3} = \dfrac{(4)^2/2-(3)^2/2}{1} = \dfrac{7}{2}$

(b)
$$m_{\text{tan}} = \lim_{x_1\to 3}\frac{f(x_1)-f(3)}{x_1-3} = \lim_{x_1\to 3}\frac{x_1^2/2-9/2}{x_1-3}$$
$$= \lim_{x_1\to 3}\frac{x_1^2-9}{2(x_1-3)} = \lim_{x_1\to 3}\frac{(x_1+3)(x_1-3)}{2(x_1-3)} = \lim_{x_1\to 3}\frac{x_1+3}{2} = 3$$

(c)
$$m_{\text{tan}} = \lim_{x_1\to x_0}\frac{f(x_1)-f(x_0)}{x_1-x_0}$$
$$= \lim_{x_1\to x_0}\frac{x_1^2/2-x_0^2/2}{x_1-x_0}$$
$$= \lim_{x_1\to x_0}\frac{x_1^2-x_0^2}{2(x_1-x_0)}$$
$$= \lim_{x_1\to x_0}\frac{x_1+x_0}{2} = x_0$$

(d)

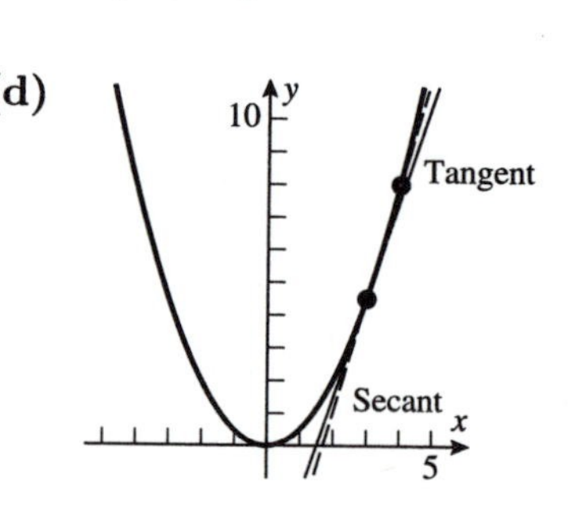

9. **(a)** $m_{\text{sec}} = \dfrac{f(3)-f(2)}{3-2} = \dfrac{1/3-1/2}{1} = -\dfrac{1}{6}$

(b)
$$m_{\text{tan}} = \lim_{x_1\to 2}\frac{f(x_1)-f(2)}{x_1-2} = \lim_{x_1\to 2}\frac{1/x_1-1/2}{x_1-2}$$
$$= \lim_{x_1\to 2}\frac{2-x_1}{2x_1(x_1-2)} = \lim_{x_1\to 2}\frac{-1}{2x_1} = -\frac{1}{4}$$

(c) $$m_{\tan} = \lim_{x_1 \to x_0} \frac{f(x_1) - f(x_0)}{x_1 - x_0} = \lim_{x_1 \to x_0} \frac{1/x_1 - 1/x_0}{x_1 - x_0} = \lim_{x_1 \to x_0} \frac{x_0 - x_1}{x_0 x_1 (x_1 - x_0)} = \lim_{x_1 \to x_0} \frac{-1}{x_0 x_1} = -\frac{1}{x_0^2}$$

(d)

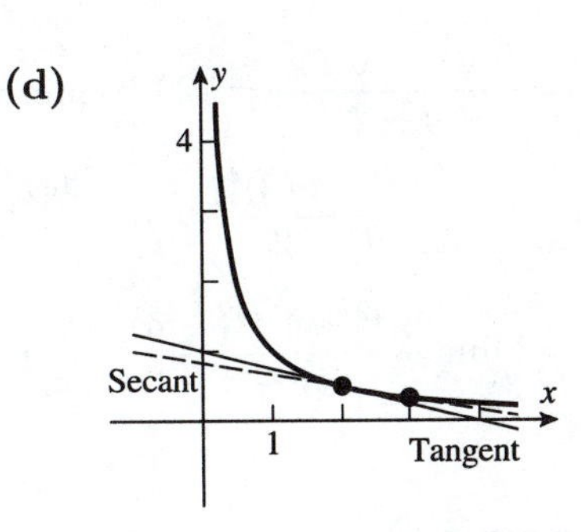

11. (a) $$m_{\tan} = \lim_{x_1 \to x_0} \frac{f(x_1) - f(x_0)}{x_1 - x_0} = \lim_{x_1 \to x_0} \frac{(x_1^2 + 1) - (x_0^2 + 1)}{x_1 - x_0} = \lim_{x_1 \to x_0} \frac{x_1^2 - x_0^2}{x_1 - x_0} = \lim_{x_1 \to x_0} (x_1 + x_0) = 2x_0$$

(b) $m_{\tan} = 2(2) = 4$

13. (a) $$m_{\tan} = \lim_{x_1 \to x_0} \frac{f(x_1) - f(x_0)}{x_1 - x_0} = \lim_{x_1 \to x_0} \frac{\sqrt{x_1} - \sqrt{x_0}}{x_1 - x_0} = \lim_{x_1 \to x_0} \frac{1}{\sqrt{x_1} + \sqrt{x_0}} = \frac{1}{2\sqrt{x_0}}$$

(b) $m_{\tan} = \dfrac{1}{2\sqrt{1}} = \dfrac{1}{2}$

15. (a) 72°F at about 4:30 P.M.

(b) about $(67 - 43)/6 = 4$°F/h

(c) decreasing most rapidly at about 9 P.M.; rate of change of temperature is about -7°F/h (slope of estimated tangent line to curve at 9 P.M.)

17. (a) during the first year after birth

(b) about 6 cm/year (slope of estimated tangent line at age 5)

(c) the growth rate is greatest at about age 14; about 10 cm/year

(d)

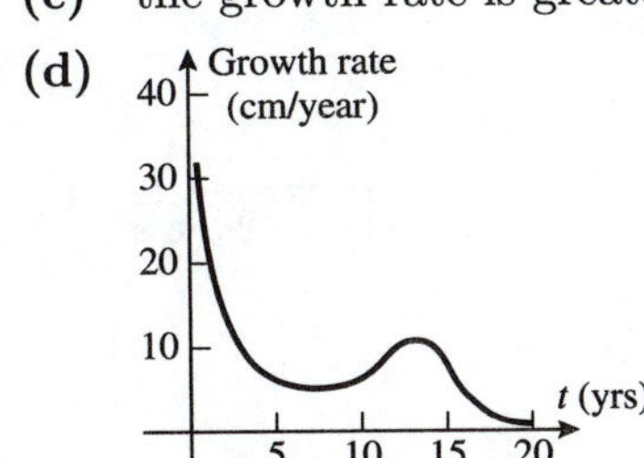

19. (a) $5(40)^3 = 320{,}000$ ft

(b) $v_{\text{ave}} = 320{,}000/40 = 8{,}000$ ft/s

(c) $5t^3 = 135$ when the rocket has gone 135 ft, so $t^3 = 27$, $t = 3$ s; $v_{\text{ave}} = 135/3 = 45$ ft/s.

(d) $$v_{\text{inst}} = \lim_{t_1 \to 40} \frac{5t_1^3 - 5(40)^3}{t_1 - 40} = \lim_{t_1 \to 40} \frac{5(t_1^3 - 40^3)}{t_1 - 40} = \lim_{t_1 \to 40} 5(t_1^2 + 40t_1 + 1600) = 24{,}000 \text{ ft/s}$$

21. **(a)** $v_{\text{ave}} = \dfrac{6(4)^4 - 6(2)^4}{4-2} = 720$ ft/min

(b) $v_{\text{inst}} = \lim\limits_{t_1 \to 2} \dfrac{6t_1^4 - 6(2)^4}{t_1 - 2} = \lim\limits_{t_1 \to 2} \dfrac{6(t_1^4 - 16)}{t_1 - 2}$

$= \lim\limits_{t_1 \to 2} \dfrac{6(t_1^2+4)(t_1^2-4)}{t_1 - 2} = \lim\limits_{t_1 \to 2} 6(t_1^2+4)(t_1+2) = 192$ ft/min

EXERCISE SET 3.2

1. $f'(1) = 2,\ f'(3) = 0,\ f'(5) = -2,\ f'(6) = -1/2$

3. **(b)** $m = f'(2) = 3$ **(c)** the same, $f'(2) = 3$

5.

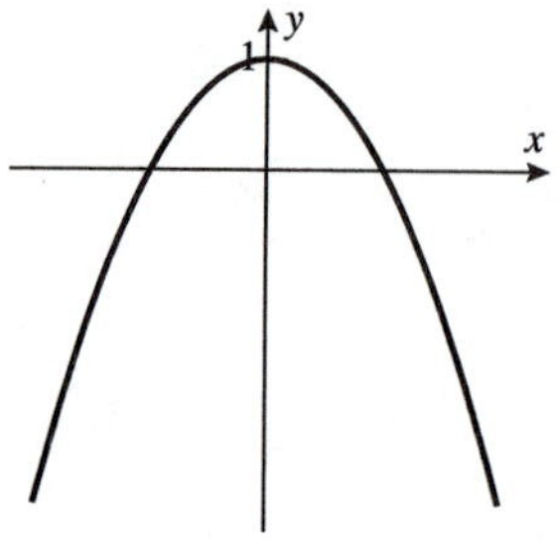

7. $y - (-1) = 5(x-3),\ y = 5x - 16$

9. $f'(x) = \lim\limits_{w \to x} \dfrac{f(w) - f(x)}{w - x} = \lim\limits_{w \to x} \dfrac{3w^2 - 3x^2}{w - x} = \lim\limits_{w \to x} 3(w + x) = 6x;\ f(3) = 3(3)^2 = 27,\ f'(3) = 18$
so $y - 27 = 18(x-3),\ y = 18x - 27$

11. $f'(x) = \lim\limits_{w \to x} \dfrac{f(w) - f(x)}{w - x} = \lim\limits_{w \to x} \dfrac{w^3 - x^3}{w - x} = \lim\limits_{w \to x} (w^2 + wx + x^2) = 3x^2;\ f(0) = 0^3 = 0,$
$f'(0) = 0$ so $y - 0 = (0)(x - 0),\ y = 0$

13. $f'(x) = \lim\limits_{w \to x} \dfrac{f(w) - f(x)}{w - x} = \lim\limits_{w \to x} \dfrac{\sqrt{w+1} - \sqrt{x+1}}{w - x}$

$= \lim\limits_{w \to x} \dfrac{\sqrt{w+1} - \sqrt{x+1}}{w - x} \dfrac{\sqrt{w+1} + \sqrt{x+1}}{\sqrt{w+1} + \sqrt{x+1}} = \lim\limits_{w \to x} \dfrac{1}{(\sqrt{w+1} + \sqrt{x+1})} = \dfrac{1}{2\sqrt{x+1}};$

$f(8) = \sqrt{8+1} = 3,\ f'(8) = \dfrac{1}{6}$ so $y - 3 = \dfrac{1}{6}(x - 8),\ y = \dfrac{1}{6}x + \dfrac{5}{3}$

15. $f'(x) = \lim\limits_{\Delta x \to 0} \dfrac{\dfrac{1}{x + \Delta x} - \dfrac{1}{x}}{\Delta x} = \lim\limits_{\Delta x \to 0} \dfrac{\dfrac{x - (x + \Delta x)}{x(x + \Delta x)}}{\Delta x}$

$= \lim\limits_{\Delta x \to 0} \dfrac{-\Delta x}{x \Delta x (x + \Delta x)} = \lim\limits_{\Delta x \to 0} -\dfrac{1}{x(x + \Delta x)} = -\dfrac{1}{x^2}$

17. $f'(x) = \lim\limits_{\Delta x \to 0} \dfrac{[a(x + \Delta x)^2 + b] - [ax^2 + b]}{\Delta x} = \lim\limits_{\Delta x \to 0} \dfrac{ax^2 + 2ax\Delta x + a(\Delta x)^2 + b - ax^2 - b}{\Delta x}$

$= \lim\limits_{\Delta x \to 0} \dfrac{2ax\Delta x + a(\Delta x)^2}{\Delta x} = \lim\limits_{\Delta x \to 0} (2ax + a\Delta x) = 2ax$

19. $f'(x) = \lim_{\Delta x \to 0} \dfrac{\dfrac{1}{\sqrt{x+\Delta x}} - \dfrac{1}{\sqrt{x}}}{\Delta x} = \lim_{\Delta x \to 0} \dfrac{\sqrt{x} - \sqrt{x+\Delta x}}{\Delta x\sqrt{x}\sqrt{x+\Delta x}}$

$= \lim_{\Delta x \to 0} \dfrac{x - (x+\Delta x)}{\Delta x\sqrt{x}\sqrt{x+\Delta x}(\sqrt{x}+\sqrt{x+\Delta x})} = \lim_{\Delta x \to 0} \dfrac{-1}{\sqrt{x}\sqrt{x+\Delta x}(\sqrt{x}+\sqrt{x+\Delta x})} = -\dfrac{1}{2x^{3/2}}$

21. $f'(t) = \lim_{h \to 0} \dfrac{f(t+h) - f(t)}{h} = \lim_{h \to 0} \dfrac{[4(t+h)^2 + (t+h)] - [4t^2 + t]}{h}$

$= \lim_{h \to 0} \dfrac{4t^2 + 8th + 4h^2 + t + h - 4t^2 - t}{h}$

$= \lim_{h \to 0} \dfrac{8th + 4h^2 + h}{h} = \lim_{h \to 0}(8t + 4h + 1) = 8t + 1$

23. **(a)** D **(b)** F **(c)** B **(d)** C **(e)** A **(f)** E

25. **(a)**

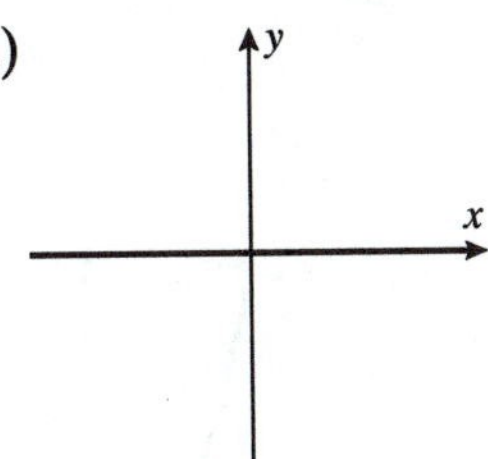

(b)

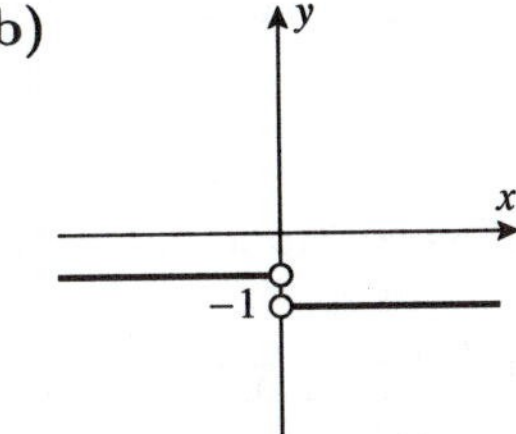

(c)

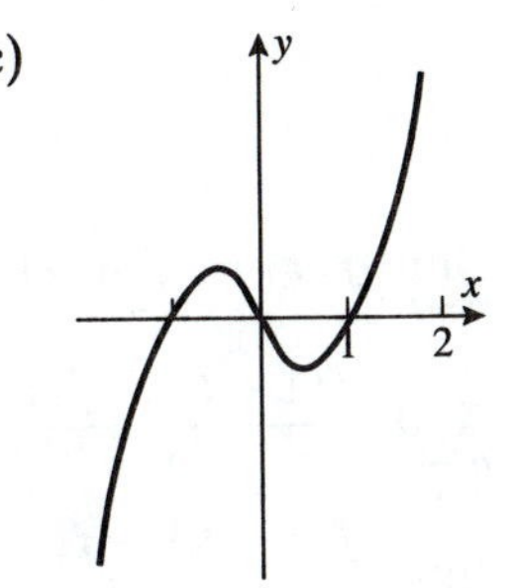

27. **(a)** $f(x) = x^2$ and $a = 3$ **(b)** $f(x) = \sqrt{x}$ and $a = 1$

29. $\dfrac{dy}{dx} = \lim_{h \to 0} \dfrac{[4(x+h)^2 + 1] - [4x^2 + 1]}{h} = \lim_{h \to 0} \dfrac{4x^2 + 8xh + 4h^2 + 1 - 4x^2 - 1}{h} = \lim_{h \to 0}(8x + 4h) = 8x$

$\left.\dfrac{dy}{dx}\right|_{x=1} = 8(1) = 8$

31. $y = -2x + 1$

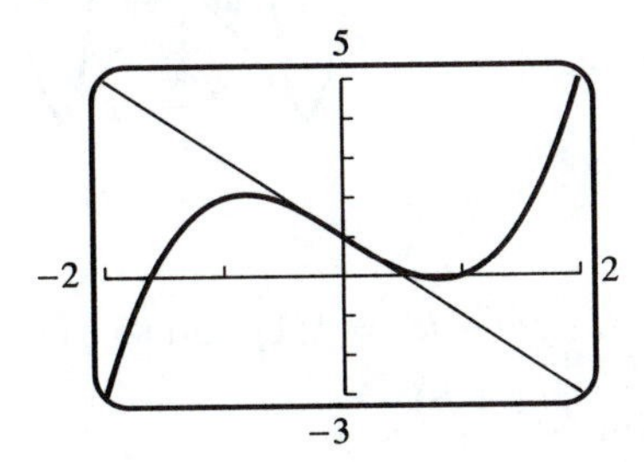

33. **(b)**

h	0.5	0.1	0.01	0.001	0.0001	0.00001
$(f(1+h) - f(1))/h$	1.6569	1.4355	1.3911	1.3868	1.3863	1.3863

35. **(a)** dollars/ft

(b) As you go deeper the price per foot may increase dramatically, so $f'(x)$ is roughly the price per additional foot.

(c) If each additional foot costs extra money (this is to be expected) then $f'(x)$ remains positive.

(d) From the approximation $1000 = f'(300) \approx \dfrac{f(301) - f(300)}{301 - 300}$
we see that $f(301) \approx f(300) + 1000$, so the extra foot will cost around \$1000.

37. **(a)** $F \approx 200$ lb, $dF/d\theta \approx 50$ lb/rad **(b)** $\mu = (dF/d\theta)/F \approx 50/200 = 0.25$

39. **(a)** $T \approx 115°\text{F}$, $dT/dt \approx -3.35°\text{F/min}$
(b) $k = (dT/dt)/(T - T_0) \approx (-3.35)/(115 - 75) = -0.084$

41. $\lim_{x\to 0} f(x) = \lim_{x\to 0} \sqrt[3]{x} = 0 = f(0)$, so f is continuous at $x = 0$.

$$\lim_{h\to 0} \frac{f(0+h) - f(0)}{h} = \lim_{h\to 0} \frac{\sqrt[3]{h} - 0}{h} = \lim_{h\to 0} \frac{1}{h^{2/3}} = +\infty, \text{ so}$$

$f'(0)$ does not exist.

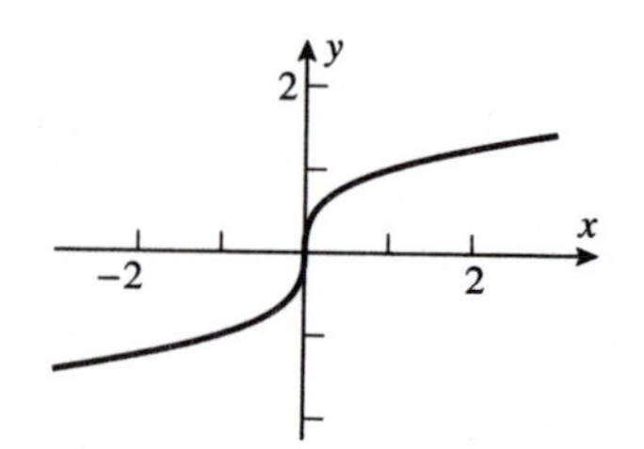

43. $\lim_{x\to 1^-} f(x) = \lim_{x\to 1^+} f(x) = f(1)$, so f is continuous at $x = 1$.

$$\lim_{h\to 0^-} \frac{f(1+h) - f(1)}{h} = \lim_{h\to 0^-} \frac{[(1+h)^2 + 1] - 2}{h} = \lim_{h\to 0^-} (2 + h) = 2;$$

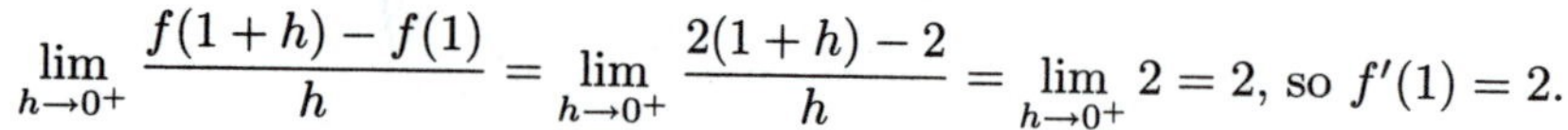

$$\lim_{h\to 0^+} \frac{f(1+h) - f(1)}{h} = \lim_{h\to 0^+} \frac{2(1+h) - 2}{h} = \lim_{h\to 0^+} 2 = 2, \text{ so } f'(1) = 2.$$

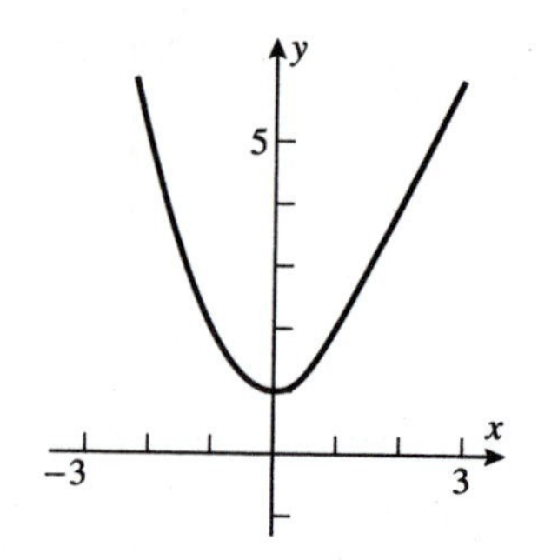

45. Since $-|x| \le x \sin(1/x) \le |x|$ it follows by the Squeezing Theorem (Theorem 2.6.2) that $\lim_{x\to 0} x \sin(1/x) = 0$. The derivative cannot exist: consider $\dfrac{f(x) - f(0)}{x} = \sin(1/x)$. This function oscillates between -1 and $+1$ and does not tend to zero as x tends to zero.

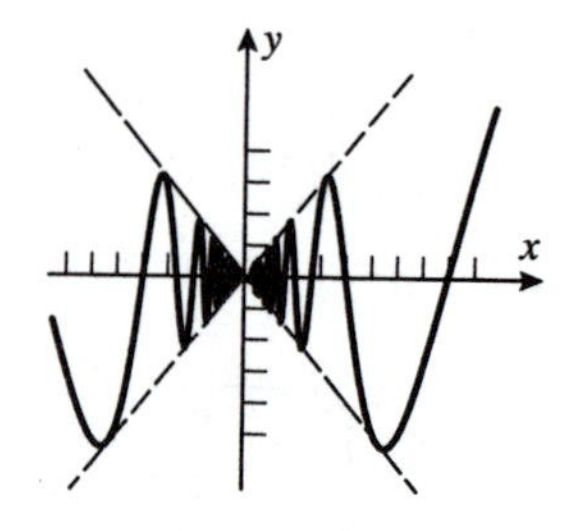

47. f is continuous at $x = 1$ because it is differentiable there, thus $\lim_{h\to 0} f(1+h) = f(1)$ and so $f(1) = 0$ because $\lim_{h\to 0} \dfrac{f(1+h)}{h}$ exists; $f'(1) = \lim_{h\to 0} \dfrac{f(1+h) - f(1)}{h} = \lim_{h\to 0} \dfrac{f(1+h)}{h} = 5$.

49.
$$f'(x) = \lim_{h\to 0} \frac{f(x+h) - f(x)}{h} = \lim_{h\to 0} \frac{f(x)f(h) - f(x)}{h} = \lim_{h\to 0} \frac{f(x)[f(h) - 1]}{h}$$
$$= f(x) \lim_{h\to 0} \frac{f(h) - f(0)}{h} = f(x)f'(0) = f(x)$$

EXERCISE SET 3.3

1. $28x^6$ **3.** $24x^7 + 2$ **5.** 0 **7.** $-\frac{1}{3}(7x^6 + 2)$

9. $3ax^2 + 2bx + c$ **11.** $24x^{-9} + 1/\sqrt{x}$ **13.** $-3x^{-4} - 7x^{-8}$

15. $$f'(x) = (3x^2+6)\frac{d}{dx}\left(2x - \frac{1}{4}\right) + \left(2x - \frac{1}{4}\right)\frac{d}{dx}(3x^2+6) = (3x^2+6)(2) + \left(2x - \frac{1}{4}\right)(6x)$$
$$= 18x^2 - \frac{3}{2}x + 12$$

17. $$f'(x) = (x^3 + 7x^2 - 8)\frac{d}{dx}(2x^{-3} + x^{-4}) + (2x^{-3} + x^{-4})\frac{d}{dx}(x^3 + 7x^2 - 8)$$
$$= (x^3 + 7x^2 - 8)(-6x^{-4} - 4x^{-5}) + (2x^{-3} + x^{-4})(3x^2 + 14x)$$
$$= -15x^{-2} - 14x^{-3} + 48x^{-4} + 32x^{-5}$$

19. $12x(3x^2 + 1)$

21. $$\frac{dy}{dx} = \frac{(5x-3)\frac{d}{dx}(1) - (1)\frac{d}{dx}(5x-3)}{(5x-3)^2} = -\frac{5}{(5x-3)^2};\ y'(1) = -5/4$$

23. $$\frac{dx}{dt} = \frac{(2t+1)\frac{d}{dt}(3t) - (3t)\frac{d}{dt}(2t+1)}{(2t+1)^2} = \frac{(2t+1)(3) - (3t)(2)}{(2t+1)^2} = \frac{3}{(2t+1)^2}$$

25. $$\frac{dy}{dx} = \frac{(x+3)\frac{d}{dx}(2x-1) - (2x-1)\frac{d}{dx}(x+3)}{(x+3)^2}$$
$$= \frac{(x+3)(2) - (2x-1)(1)}{(x+3)^2} = \frac{7}{(x+3)^2};\ \left.\frac{dy}{dx}\right|_{x=1} = \frac{7}{16}$$

27. $$\frac{dy}{dx} = \left(\frac{3x+2}{x}\right)\frac{d}{dx}\left(x^{-5}+1\right) + \left(x^{-5}+1\right)\frac{d}{dx}\left(\frac{3x+2}{x}\right)$$
$$= \left(\frac{3x+2}{x}\right)(-5x^{-6}) + (x^{-5}+1)\left[\frac{x(3) - (3x+2)(1)}{x^2}\right]$$
$$= \left(\frac{3x+2}{x}\right)(-5x^{-6}) + (x^{-5}+1)\left(-\frac{2}{x^2}\right);$$
$$\left.\frac{dy}{dx}\right|_{x=1} = 5(-5) + 2(-2) = -29$$

29. $f'(1) \approx \dfrac{f(1.01) - f(1)}{0.01} = \dfrac{0.999699 - (-1)}{0.01} = 0.0301$, and by differentiation, $f'(1) = 3(1)^2 - 3 = 0$

31. $f'(1) = 0$ **33.** $32t$ **35.** $3\pi r^2$

37. **(a)** $\dfrac{dV}{dr} = 4\pi r^2$ **(b)** $\left.\dfrac{dV}{dr}\right|_{r=5} = 4\pi(5)^2 = 100\pi$

39. **(a)** $g'(x) = \sqrt{x}f'(x) + \dfrac{1}{2\sqrt{x}}f(x)$, $g'(4) = (2)(-5) + \dfrac{1}{4}(3) = -37/4$

(b) $g'(x) = \dfrac{xf'(x) - f(x)}{x^2}$, $g'(4) = \dfrac{(4)(-5) - 3}{16} = -23/16$

41. **(a)** $F'(x) = 5f'(x) + 2g'(x), F'(2) = 5(4) + 2(-5) = 10$
(b) $F'(x) = f'(x) - 3g'(x), F'(2) = 4 - 3(-5) = 19$
(c) $F'(x) = f(x)g'(x) + g(x)f'(x), F'(2) = (-1)(-5) + (1)(4) = 9$
(d) $F'(x) = [g(x)f'(x) - f(x)g'(x)]/g^2(x), F'(2) = [(1)(4) - (-1)(-5)]/(1)^2 = -1$

43. $y - 2 = 5(x + 3)$, $y = 5x + 17$

45. **(a)** $dy/dx = 21x^2 - 10x + 1$, $d^2y/dx^2 = 42x - 10$
(b) $dy/dx = 24x - 2$, $d^2y/dx^2 = 24$
(c) $dy/dx = -1/x^2$, $d^2y/dx^2 = 2/x^3$
(d) $y = 35x^5 - 16x^3 - 3x$, $dy/dx = 175x^4 - 48x^2 - 3$, $d^2y/dx^2 = 700x^3 - 96x$

47. **(a)** $y' = -5x^{-6} + 5x^4$, $y'' = 30x^{-7} + 20x^3$, $y''' = -210x^{-8} + 60x^2$
(b) $y = x^{-1}$, $y' = -x^{-2}$, $y'' = 2x^{-3}$, $y''' = -6x^{-4}$
(c) $y' = 3ax^2 + b$, $y'' = 6ax$, $y''' = 6a$

49. **(a)** $f'(x) = 6x$, $f''(x) = 6$, $f'''(x) = 0$, $f'''(2) = 0$

(b) $\dfrac{dy}{dx} = 30x^4 - 8x$, $\dfrac{d^2y}{dx^2} = 120x^3 - 8$, $\left.\dfrac{d^2y}{dx^2}\right|_{x=1} = 112$

(c) $\dfrac{d}{dx}\left[x^{-3}\right] = -3x^{-4}$, $\dfrac{d^2}{dx^2}\left[x^{-3}\right] = 12x^{-5}$, $\dfrac{d^3}{dx^3}\left[x^{-3}\right] = -60x^{-6}$, $\dfrac{d^4}{dx^4}\left[x^{-3}\right] = 360x^{-7}$,
$\left.\dfrac{d^4}{dx^4}\left[x^{-3}\right]\right|_{x=1} = 360$

51. $y' = 3x^2 + 3$, $y'' = 6x$, and $y''' = 6$ so
$y''' + xy'' - 2y' = 6 + x(6x) - 2(3x^2 + 3) = 6 + 6x^2 - 6x^2 - 6 = 0$

53. $F'(x) = xf'(x) + f(x)$, $F''(x) = xf''(x) + f'(x) + f'(x) = xf''(x) + 2f'(x)$

55. The graph has a horizontal tangent at points where $\dfrac{dy}{dx} = 0$, but $\dfrac{dy}{dx} = x^2 - 3x + 2 = (x - 1)(x - 2) = 0$ if $x = 1, 2$. The corresponding values of y are $5/6$ and $2/3$ so the tangent line is horizontal at $(1, 5/6)$ and $(2, 2/3)$.

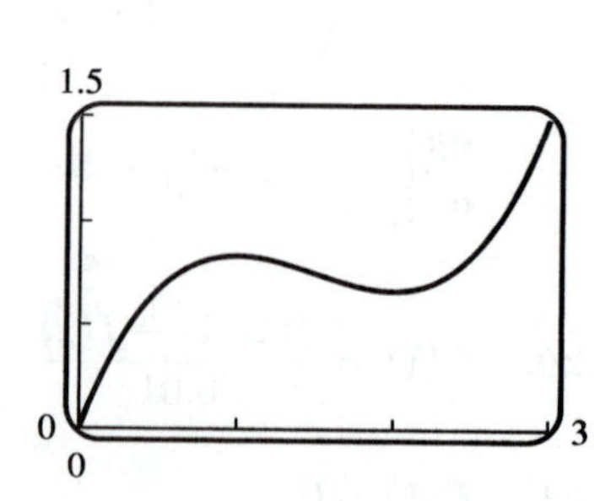

57. The y-intercept is -2 so the point $(0,-2)$ is on the graph; $-2 = a(0)^2 + b(0) + c$, $c = -2$. The x-intercept is 1 so the point (1,0) is on the graph; $0 = a + b - 2$. The slope is $dy/dx = 2ax + b$; at $x = 0$ the slope is b so $b = -1$, thus $a = 3$. The function is $y = 3x^2 - x - 2$.

59. The points $(-1,1)$ and $(2,4)$ are on the secant line so its slope is $(4-1)/(2+1) = 1$. The slope of the tangent line to $y = x^2$ is $y' = 2x$ so $2x = 1$, $x = 1/2$.

61. $y' = -2x$, so at any point (x_0, y_0) on $y = 1 - x^2$ the tangent line is $y - y_0 = -2x_0(x - x_0)$, or $y = -2x_0 x + x_0^2 + 1$. The point $(2,0)$ is to be on the line, so $0 = -4x_0 + x_0^2 + 1$, $x_0^2 - 4x_0 + 1 = 0$. Use the quadratic formula to get $x_0 = \dfrac{4 \pm \sqrt{16-4}}{2} = 2 \pm \sqrt{3}$.

63. $y' = 3ax^2 + b$; the tangent line at $x = x_0$ is $y - y_0 = (3ax_0^2 + b)(x - x_0)$ where $y_0 = ax_0^3 + bx_0$. Solve with $y = ax^3 + bx$ to get

$$
\begin{aligned}
(ax^3 + bx) - (ax_0^3 + bx_0) &= (3ax_0^2 + b)(x - x_0) \\
ax^3 + bx - ax_0^3 - bx_0 &= 3ax_0^2 x - 3ax_0^3 + bx - bx_0 \\
x^3 - 3x_0^2 x + 2x_0^3 &= 0 \\
(x - x_0)(x^2 + xx_0 - 2x_0^2) &= 0 \\
(x - x_0)^2(x + 2x_0) &= 0, \text{ so } x = -2x_0.
\end{aligned}
$$

65. $y' = -\dfrac{1}{x^2}$; the tangent line at $x = x_0$ is $y - y_0 = -\dfrac{1}{x_0^2}(x - x_0)$, or $y = -\dfrac{x}{x_0^2} + \dfrac{2}{x_0}$. The tangent line crosses the x-axis at $2x_0$, the y-axis at $2/x_0$, so that the area of the triangle is $\dfrac{1}{2}(2/x_0)(2x_0) = 2$.

67. $F = GmMr^{-2}$, $\dfrac{dF}{dr} = -2GmMr^{-3} = -\dfrac{2GmM}{r^3}$

69. $f'(x) = 1 + 1/x^2 > 0$ for all $x \neq 0$

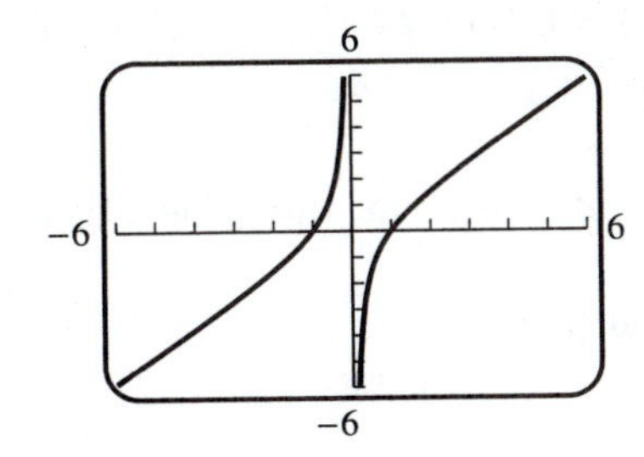

71. $(f \cdot g \cdot h)' = [(f \cdot g) \cdot h]' = (f \cdot g)h' + h(f \cdot g)' = (f \cdot g)h' + h[fg' + f'g] = fgh' + fg'h + f'gh$

73. **(a)** $2(1 + x^{-1})(x^{-3} + 7) + (2x + 1)(-x^{-2})(x^{-3} + 7) + (2x + 1)(1 + x^{-1})(-3x^{-4})$

(b) $(x^7 + 2x - 3)^3 = (x^7 + 2x - 3)(x^7 + 2x - 3)(x^7 + 2x - 3)$ so

$$
\begin{aligned}
\frac{d}{dx}(x^7 + 2x - 3)^3 &= (7x^6 + 2)(x^7 + 2x - 3)(x^7 + 2x - 3) \\
&\quad + (x^7 + 2x - 3)(7x^6 + 2)(x^7 + 2x - 3) \\
&\quad + (x^7 + 2x - 3)(x^7 + 2x - 3)(7x^6 + 2) \\
&= 3(7x^6 + 2)(x^7 + 2x - 3)^2
\end{aligned}
$$

75. f is continuous at 1 because $\lim_{x\to1^-} f(x) = \lim_{x\to1^+} f(x) = f(1)$; also $\lim_{x\to1^-} f'(x) = \lim_{x\to1^-}(2x+1) = 3$ and $\lim_{x\to1^+} f'(x) = \lim_{x\to1^+} 3 = 3$ so f is differentiable at 1.

77. f is continuous at 1 because $\lim_{x\to1^-} f(x) = \lim_{x\to1^+} f(x) = f(1)$, also $\lim_{x\to1^-} f'(x) = \lim_{x\to1^-} 2x = 2$ and $\lim_{x\to1^+} f'(x) = \lim_{x\to1^+} \frac{1}{2\sqrt{x}} = \frac{1}{2}$ so f is not differentiable at 1.

79. **(a)** $f(x) = 3x-2$ if $x \ge 2/3$, $f(x) = -3x+2$ if $x < 2/3$ so f is differentiable everywhere except perhaps at $2/3$. f is continuous at $2/3$, also $\lim_{x\to2/3^-} f'(x) = \lim_{x\to2/3^-}(-3) = -3$ and $\lim_{x\to2/3^+} f'(x) = \lim_{x\to2/3^+}(3) = 3$ so f is not differentiable at $x = 2/3$.

(b) $f(x) = x^2-4$ if $|x| \ge 2$, $f(x) = -x^2+4$ if $|x| < 2$ so f is differentiable everywhere except perhaps at ±2. f is continuous at -2 and 2, also $\lim_{x\to2^-} f'(x) = \lim_{x\to2^-}(-2x) = -4$ and $\lim_{x\to2^+} f'(x) = \lim_{x\to2^+}(2x) = 4$ so f is not differentiable at $x = 2$. Similarly, f is not differentiable at $x = -2$.

81. **(a)** $$\frac{d^2}{dx^2}[cf(x)] = \frac{d}{dx}\left[\frac{d}{dx}[cf(x)]\right] = \frac{d}{dx}\left[c\frac{d}{dx}[f(x)]\right] = c\frac{d}{dx}\left[\frac{d}{dx}[f(x)]\right] = c\frac{d^2}{dx^2}[f(x)]$$

$$\frac{d^2}{dx^2}[f(x)+g(x)] = \frac{d}{dx}\left[\frac{d}{dx}[f(x)+g(x)]\right] = \frac{d}{dx}\left[\frac{d}{dx}[f(x)] + \frac{d}{dx}[g(x)]\right]$$
$$= \frac{d^2}{dx^2}[f(x)] + \frac{d^2}{dx^2}[g(x)]$$

(b) yes, by repeated application of the procedure illustrated in Part (a)

83. **(a)** $f'(x) = nx^{n-1}$, $f''(x) = n(n-1)x^{n-2}$, $f'''(x) = n(n-1)(n-2)x^{n-3}, \ldots,$ $f^{(n)}(x) = n(n-1)(n-2)\cdots1$

(b) from Part (a), $f^{(k)}(x) = k(k-1)(k-2)\cdots1$ so $f^{(k+1)}(x) = 0$ thus $f^{(n)}(x) = 0$ if $n > k$

(c) from Parts (a) and (b), $f^{(n)}(x) = a_n n(n-1)(n-2)\cdots1$

85. **(a)** If a function is differentiable at a point then it is continuous at that point, thus f' is continuous on (a,b) and consequently so is f.

(b) f and all its derivatives up to $f^{(n-1)}(x)$ are continuous on (a,b)

EXERCISE SET 3.4

1. $f'(x) = -2\sin x - 3\cos x$

3. $f'(x) = \dfrac{x(\cos x) - (\sin x)(1)}{x^2} = \dfrac{x\cos x - \sin x}{x^2}$

5. $f'(x) = x^3(\cos x) + (\sin x)(3x^2) - 5(-\sin x) = x^3\cos x + (3x^2+5)\sin x$

7. $f'(x) = \sec x\tan x - \sqrt{2}\sec^2 x$

9. $f'(x) = \sec x(\sec^2 x) + (\tan x)(\sec x\tan x) = \sec^3 x + \sec x\tan^2 x$

11. $f'(x) = (\csc x)(-\csc^2 x) + (\cot x)(-\csc x\cot x) = -\csc^3 x - \csc x\cot^2 x$

13. $f'(x) = \dfrac{(1+\csc x)(-\csc^2 x) - \cot x(0 - \csc x \cot x)}{(1+\csc x)^2} = \dfrac{\csc x(-\csc x - \csc^2 x + \cot^2 x)}{(1+\csc x)^2}$ but

$1 + \cot^2 x = \csc^2 x$ (identity) thus $\cot^2 x - \csc^2 x = -1$ so

$f'(x) = \dfrac{\csc x(-\csc x - 1)}{(1+\csc x)^2} = -\dfrac{\csc x}{1+\csc x}$

15. $f(x) = \sin^2 x + \cos^2 x = 1$ (identity) so $f'(x) = 0$

17. $f(x) = \dfrac{\tan x}{1 + x\tan x}$ (because $\sin x \sec x = (\sin x)(1/\cos x) = \tan x$),

$f'(x) = \dfrac{(1 + x\tan x)(\sec^2 x) - \tan x[x(\sec^2 x) + (\tan x)(1)]}{(1 + x\tan x)^2}$

$= \dfrac{\sec^2 x - \tan^2 x}{(1 + x\tan x)^2} = \dfrac{1}{(1 + x\tan x)^2}$ (because $\sec^2 x - \tan^2 x = 1$)

19. $dy/dx = -x\sin x + \cos x$, $d^2y/dx^2 = -x\cos x - \sin x - \sin x = -x\cos x - 2\sin x$

21. $dy/dx = x(\cos x) + (\sin x)(1) - 3(-\sin x) = x\cos x + 4\sin x$,
$d^2y/dx^2 = x(-\sin x) + (\cos x)(1) + 4\cos x = -x\sin x + 5\cos x$

23. $dy/dx = (\sin x)(-\sin x) + (\cos x)(\cos x) = \cos^2 x - \sin^2 x$,
$d^2y/dx^2 = (\cos x)(-\sin x) + (\cos x)(-\sin x) - [(\sin x)(\cos x) + (\sin x)(\cos x)] = -4\sin x\cos x$

25. Let $f(x) = \tan x$, then $f'(x) = \sec^2 x$.

(a) $f(0) = 0$ and $f'(0) = 1$ so $y - 0 = (1)(x - 0)$, $y = x$.

(b) $f\left(\frac{\pi}{4}\right) = 1$ and $f'\left(\frac{\pi}{4}\right) = 2$ so $y - 1 = 2\left(x - \frac{\pi}{4}\right)$, $y = 2x - \frac{\pi}{2} + 1$.

(c) $f\left(-\frac{\pi}{4}\right) = -1$ and $f'\left(-\frac{\pi}{4}\right) = 2$ so $y + 1 = 2\left(x + \frac{\pi}{4}\right)$, $y = 2x + \frac{\pi}{2} - 1$.

27. **(a)** If $y = x\sin x$ then $y' = \sin x + x\cos x$ and $y'' = 2\cos x - x\sin x$ so $y'' + y = 2\cos x$.

(b) If $y = x\sin x$ then $y' = \sin x + x\cos x$ and $y'' = 2\cos x - x\sin x$ so $y'' + y = 2\cos x$; differentiate twice more to get $y^{(4)} + y'' = -2\cos x$.

29. **(a)** $f'(x) = \cos x = 0$ at $x = \pm\pi/2, \pm 3\pi/2$.

(b) $f'(x) = 1 - \sin x = 0$ at $x = -3\pi/2, \pi/2$.

(c) $f'(x) = \sec^2 x \geq 1$ always, so no horizontal tangent line.

(d) $f'(x) = \sec x \tan x = 0$ when $\sin x = 0$, $x = \pm 2\pi, \pm\pi, 0$

31. $x = 10\sin\theta$, $dx/d\theta = 10\cos\theta$; if $\theta = 60°$, then
$dx/d\theta = 10(1/2) = 5$ ft/rad $= \pi/36$ ft/deg ≈ 0.087 ft/deg

33. $D = 50\tan\theta$, $dD/d\theta = 50\sec^2\theta$; if $\theta = 45°$, then
$dD/d\theta = 50(\sqrt{2})^2 = 100$ m/rad $= 5\pi/9$ m/deg ≈ 1.75 m/deg

35. **(a)** $\dfrac{d^4}{dx^4}\sin x = \sin x$, so $\dfrac{d^{4k}}{dx^{4k}}\sin x = \sin x$; $\dfrac{d^{87}}{dx^{87}}\sin x = \dfrac{d^3}{dx^3}\dfrac{d^{4\cdot 21}}{dx^{4\cdot 21}}\sin x = \dfrac{d^3}{dx^3}\sin x = -\cos x$

(b) $\dfrac{d^{100}}{dx^{100}}\cos x = \dfrac{d^{4k}}{dx^{4k}}\cos x = \cos x$

37. **(a)** all x **(b)** all x
(c) $x \neq \pi/2 + n\pi$, $n = 0, \pm 1, \pm 2, \ldots$ **(d)** $x \neq n\pi$, $n = 0, \pm 1, \pm 2, \ldots$
(e) $x \neq \pi/2 + n\pi$, $n = 0, \pm 1, \pm 2, \ldots$ **(f)** $x \neq n\pi$, $n = 0, \pm 1, \pm 2, \ldots$
(g) $x \neq (2n+1)\pi$, $n = 0, \pm 1, \pm 2, \ldots$ **(h)** $x \neq n\pi/2$, $n = 0, \pm 1, \pm 2, \ldots$
(i) all x

39. $f'(x) = -\sin x$, $f''(x) = -\cos x$, $f'''(x) = \sin x$, and $f^{(4)}(x) = \cos x$ with higher order derivatives repeating this pattern, so $f^{(n)}(x) = \sin x$ for $n = 3, 7, 11, \ldots$

41. $\lim_{x \to 0} \dfrac{\tan(x+y) - \tan y}{x} = \lim_{h \to 0} \dfrac{\tan(y+h) - \tan y}{h} = \dfrac{d}{dy}(\tan y) = \sec^2 y$

43. Let t be the radian measure, then $h = \dfrac{180}{\pi} t$ and $\cos h = \cos t$, $\sin h = \sin t$.

(a) $\lim_{h \to 0} \dfrac{\cos h - 1}{h} = \lim_{t \to 0} \dfrac{\cos t - 1}{180t/\pi} = \dfrac{\pi}{180} \lim_{t \to 0} \dfrac{\cos t - 1}{t} = 0$

(b) $\lim_{h \to 0} \dfrac{\sin h}{h} = \lim_{t \to 0} \dfrac{\sin t}{180t/\pi} = \dfrac{\pi}{180} \lim_{t \to 0} \dfrac{\sin t}{t} = \dfrac{\pi}{180}$

(c) $\dfrac{d}{dx}[\sin x] = \sin x \lim_{h \to 0} \dfrac{\cos h - 1}{h} + \cos x \lim_{h \to 0} \dfrac{\sin h}{h} = (\sin x)(0) + (\cos x)(\pi/180) = \dfrac{\pi}{180} \cos x$

EXERCISE SET 3.5

1. $(f \circ g)'(x) = f'(g(x))g'(x)$ so $(f \circ g)'(0) = f'(g(0))g'(0) = f'(0)(3) = (2)(3) = 6$

3. **(a)** $(f \circ g)(x) = f(g(x)) = (2x-3)^5$ and $(f \circ g)'(x) = f'(g(x)g'(x) = 5(2x-3)^4(2) = 10(2x-3)^4$
(b) $(g \circ f)(x) = g(f(x)) = 2x^5 - 3$ and $(g \circ f)'(x) = g'(f(x))f'(x) = 2(5x^4) = 10x^4$

5. **(a)** $F'(x) = f'(g(x))g'(x) = f'(g(3))g'(3) = -1(7) = -7$
(b) $G'(x) = g'(f(x))f'(x) = g'(f(3))f'(3) = 4(-2) = -8$

7. $f'(x) = 37(x^3 + 2x)^{36} \dfrac{d}{dx}(x^3 + 2x) = 37(x^3 + 2x)^{36}(3x^2 + 2)$

9. $f'(x) = -2\left(x^3 - \dfrac{7}{x}\right)^{-3} \dfrac{d}{dx}\left(x^3 - \dfrac{7}{x}\right) = -2\left(x^3 - \dfrac{7}{x}\right)^{-3}\left(3x^2 + \dfrac{7}{x^2}\right)$

11. $f(x) = 4(3x^2 - 2x + 1)^{-3}$,
$f'(x) = -12(3x^2 - 2x + 1)^{-4} \dfrac{d}{dx}(3x^2 - 2x + 1) = -12(3x^2 - 2x + 1)^{-4}(6x - 2) = \dfrac{24(1 - 3x)}{(3x^2 - 2x + 1)^4}$

13. $f'(x) = \dfrac{1}{2\sqrt{4 + 3\sqrt{x}}} \dfrac{d}{dx}(4 + 3\sqrt{x}) = \dfrac{3}{4\sqrt{x}\sqrt{4 + 3\sqrt{x}}}$

15. $f'(x) = \cos(x^3) \dfrac{d}{dx}(x^3) = 3x^2 \cos(x^3)$

17. $f'(x) = 20 \cos^4 x \dfrac{d}{dx}(\cos x) = 20 \cos^4 x(-\sin x) = -20 \cos^4 x \sin x$

19. $f'(x) = \cos(1/x^2) \dfrac{d}{dx}(1/x^2) = -\dfrac{2}{x^3} \cos(1/x^2)$

21. $f'(x) = 4\sec(x^7)\dfrac{d}{dx}[\sec(x^7)] = 4\sec(x^7)\sec(x^7)\tan(x^7)\dfrac{d}{dx}(x^7) = 28x^6\sec^2(x^7)\tan(x^7)$

23. $f'(x) = \dfrac{1}{2\sqrt{\cos(5x)}}\dfrac{d}{dx}[\cos(5x)] = -\dfrac{5\sin(5x)}{2\sqrt{\cos(5x)}}$

25. $f'(x) = -3\left[x + \csc(x^3+3)\right]^{-4}\dfrac{d}{dx}\left[x + \csc(x^3+3)\right]$

$$= -3\left[x + \csc(x^3+3)\right]^{-4}\left[1 - \csc(x^3+3)\cot(x^3+3)\frac{d}{dx}(x^3+3)\right]$$

$$= -3\left[x + \csc(x^3+3)\right]^{-4}\left[1 - 3x^2\csc(x^3+3)\cot(x^3+3)\right]$$

27. $\dfrac{dy}{dx} = x^3(2\sin 5x)\dfrac{d}{dx}(\sin 5x) + 3x^2\sin^2 5x = 10x^3\sin 5x\cos 5x + 3x^2\sin^2 5x$

29. $\dfrac{dy}{dx} = x^5\sec\left(\dfrac{1}{x}\right)\tan\left(\dfrac{1}{x}\right)\dfrac{d}{dx}\left(\dfrac{1}{x}\right) + \sec\left(\dfrac{1}{x}\right)(5x^4)$

$$= x^5\sec\left(\frac{1}{x}\right)\tan\left(\frac{1}{x}\right)\left(-\frac{1}{x^2}\right) + 5x^4\sec\left(\frac{1}{x}\right)$$

$$= -x^3\sec\left(\frac{1}{x}\right)\tan\left(\frac{1}{x}\right) + 5x^4\sec\left(\frac{1}{x}\right)$$

31. $\dfrac{dy}{dx} = -\sin(\cos x)\dfrac{d}{dx}(\cos x) = -\sin(\cos x)(-\sin x) = \sin(\cos x)\sin x$

33. $\dfrac{dy}{dx} = 3\cos^2(\sin 2x)\dfrac{d}{dx}[\cos(\sin 2x)] = 3\cos^2(\sin 2x)[-\sin(\sin 2x)]\dfrac{d}{dx}(\sin 2x)$

$$= -6\cos^2(\sin 2x)\sin(\sin 2x)\cos 2x$$

35. $\dfrac{dy}{dx} = (5x+8)^{13}12(x^3+7x)^{11}\dfrac{d}{dx}(x^3+7x) + (x^3+7x)^{12}13(5x+8)^{12}\dfrac{d}{dx}(5x+8)$

$$= 12(5x+8)^{13}(x^3+7x)^{11}(3x^2+7) + 65(x^3+7x)^{12}(5x+8)^{12}$$

37. $\dfrac{dy}{dx} = 3\left[\dfrac{x-5}{2x+1}\right]^2\dfrac{d}{dx}\left[\dfrac{x-5}{2x+1}\right] = 3\left[\dfrac{x-5}{2x+1}\right]^2 \cdot \dfrac{11}{(2x+1)^2} = \dfrac{33(x-5)^2}{(2x+1)^4}$

39. $\dfrac{dy}{dx} = \dfrac{(4x^2-1)^8(3)(2x+3)^2(2) - (2x+3)^3(8)(4x^2-1)^7(8x)}{(4x^2-1)^{16}}$

$$= \frac{2(2x+3)^2(4x^2-1)^7[3(4x^2-1) - 32x(2x+3)]}{(4x^2-1)^{16}} = -\frac{2(2x+3)^2(52x^2+96x+3)}{(4x^2-1)^9}$$

41. $\dfrac{dy}{dx} = 5\left[x\sin 2x + \tan^4(x^7)\right]^4\dfrac{d}{dx}\left[x\sin 2x \tan^4(x^7)\right]$

$$= 5\left[x\sin 2x + \tan^4(x^7)\right]^4\left[x\cos 2x\frac{d}{dx}(2x) + \sin 2x + 4\tan^3(x^7)\frac{d}{dx}\tan(x^7)\right]$$

$$= 5\left[x\sin 2x + \tan^4(x^7)\right]^4\left[2x\cos 2x + \sin 2x + 28x^6\tan^3(x^7)\sec^2(x^7)\right]$$

43. $\dfrac{dy}{dx} = \cos 3x - 3x\sin 3x$; if $x = \pi$ then $\dfrac{dy}{dx} = -1$ and $y = -\pi$, so the equation of the tangent line is $y + \pi = -(x - \pi)$, $y = x$

45. $\dfrac{dy}{dx} = -3\sec^3(\pi/2 - x)\tan(\pi/2 - x)$; if $x = -\pi/2$ then $\dfrac{dy}{dx} = 0, y = -1$ so the equation of the tangent line is $y + 1 = 0, y = -1$

47. $\dfrac{dy}{dx} = \sec^2(4x^2)\dfrac{d}{dx}(4x^2) = 8x\sec^2(4x^2)$, $\left.\dfrac{dy}{dx}\right|_{x=\sqrt{\pi}} = 8\sqrt{\pi}\sec^2(4\pi) = 8\sqrt{\pi}$. When $x = \sqrt{\pi}$, $y = \tan(4\pi) = 0$, so the equation of the tangent line is $y = 8\sqrt{\pi}(x - \sqrt{\pi}) = 8\sqrt{\pi}x - 8\pi$.

49. $\dfrac{dy}{dx} = 2x\sqrt{5 - x^2} + \dfrac{x^2}{2\sqrt{5 - x^2}}(-2x)$, $\left.\dfrac{dy}{dx}\right|_{x=1} = 4 - 1/2 = 7/2$. When $x = 1, y = 2$, so the equation of the tangent line is $y - 2 = (7/2)(x - 1)$, or $y = \dfrac{7}{2}x - \dfrac{3}{2}$.

51.
$$\frac{dy}{dx} = x(-\sin(5x))\frac{d}{dx}(5x) + \cos(5x) - 2\sin x\frac{d}{dx}(\sin x)$$
$$= -5x\sin(5x) + \cos(5x) - 2\sin x\cos x = -5x\sin(5x) + \cos(5x) - \sin(2x),$$
$$\frac{d^2y}{dx^2} = -5x\cos(5x)\frac{d}{dx}(5x) - 5\sin(5x) - \sin(5x)\frac{d}{dx}(5x) - \cos(2x)\frac{d}{dx}(2x)$$
$$= -25x\cos(5x) - 10\sin(5x) - 2\cos(2x)$$

53. $\dfrac{dy}{dx} = \dfrac{(1-x)+(1+x)}{(1-x)^2} = \dfrac{2}{(1-x)^2} = 2(1-x)^{-2}$ and $\dfrac{d^2y}{dx^2} = -2(2)(-1)(1-x)^{-3} = 4(1-x)^{-3}$

55. $y = \cot^3(\pi - \theta) = -\cot^3\theta$ so $dy/dx = 3\cot^2\theta\csc^2\theta$

57.
$$\frac{d}{d\omega}[a\cos^2\pi\omega + b\sin^2\pi\omega] = -2\pi a\cos\pi\omega\sin\pi\omega + 2\pi b\sin\pi\omega\cos\pi\omega$$
$$= \pi(b - a)(2\sin\pi\omega\cos\pi\omega) = \pi(b - a)\sin 2\pi\omega$$

59. (a)

(c) $f'(x) = x\dfrac{-x}{\sqrt{4 - x^2}} + \sqrt{4 - x^2} = \dfrac{4 - 2x^2}{\sqrt{4 - x^2}}$

(d) $f(1) = \sqrt{3}$ and $f'(1) = \dfrac{2}{\sqrt{3}}$ so the tangent line has the equation $y - \sqrt{3} = \dfrac{2}{\sqrt{3}}(x - 1)$.

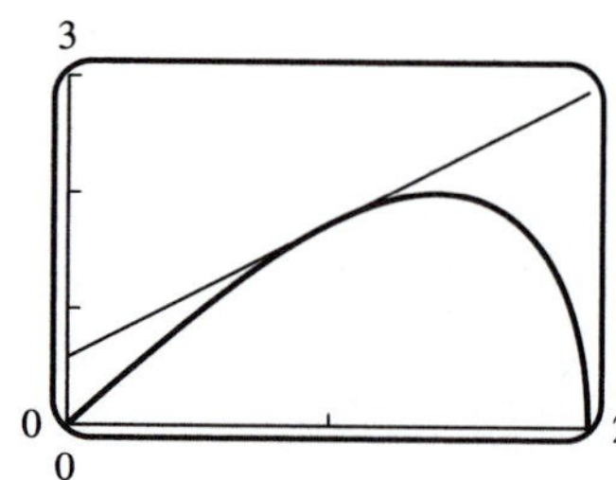

61. **(a)** $dy/dt = -A\omega \sin \omega t, d^2y/dt^2 = -A\omega^2 \cos \omega t = -\omega^2 y$

(b) one complete oscillation occurs when ωt increases over an interval of length 2π, or if t increases over an interval of length $2\pi/\omega$

(c) $f = 1/T$

(d) amplitude $= 0.6$ cm, $T = 2\pi/15$ s/oscillation, $f = 15/(2\pi)$ oscillations/s

63. **(a)** $p \approx 10$ lb/in^2, $dp/dh \approx -2$ lb/in^2/mi

(b) $\dfrac{dp}{dt} = \dfrac{dp}{dh}\dfrac{dh}{dt} \approx (-2)(0.3) = -0.6$ lb/in^2/s

65. With $u = \sin x$, $\dfrac{d}{dx}(|\sin x|) = \dfrac{d}{dx}(|u|) = \dfrac{d}{du}(|u|)\dfrac{du}{dx} = \dfrac{d}{du}(|u|)\cos x = \begin{cases} \cos x, & u > 0 \\ -\cos x, & u < 0 \end{cases}$

$$= \begin{cases} \cos x, & \sin x > 0 \\ -\cos x, & \sin x < 0 \end{cases} = \begin{cases} \cos x, & 0 < x < \pi \\ -\cos x, & -\pi < x < 0 \end{cases}$$

67. **(a)** For $x \neq 0, |f(x)| \leq |x|$, and $\lim\limits_{x\to 0} |x| = 0$, so by the Squeezing Theorem, $\lim\limits_{x\to 0} f(x) = 0$.

(b) If $f'(0)$ were to exist, then the limit $\dfrac{f(x) - f(0)}{x - 0} = \sin(1/x)$ would have to exist, but it doesn't.

(c) for $x \neq 0$, $f'(x) = x\left(\cos\dfrac{1}{x}\right)\left(-\dfrac{1}{x^2}\right) + \sin\dfrac{1}{x} = -\dfrac{1}{x}\cos\dfrac{1}{x} + \sin\dfrac{1}{x}$

(d) $\lim\limits_{x\to 0} \dfrac{f(x) - f(0)}{x - 0} = \lim\limits_{x\to 0} \sin(1/x)$, which does not exist, thus $f'(0)$ does not exist.

69. **(a)** $g'(x) = 3[f(x)]^2 f'(x)$, $g'(2) = 3[f(2)]^2 f'(2) = 3(1)^2(7) = 21$

(b) $h'(x) = f'(x^3)(3x^2)$, $h'(2) = f'(8)(12) = (-3)(12) = -36$

71. $F'(x) = f'(g(x))g'(x) = f'(\sqrt{3x-1})\dfrac{3}{2\sqrt{3x-1}} = \dfrac{\sqrt{3x-1}}{(3x-1)+1}\dfrac{3}{2\sqrt{3x-1}} = \dfrac{1}{2x}$

73. $\dfrac{d}{dx}[f(3x)] = f'(3x)\dfrac{d}{dx}(3x) = 3f'(3x) = 6x$, so $f'(3x) = 2x$. Let $u = 3x$ to get $f'(u) = \dfrac{2}{3}u$; $\dfrac{d}{dx}[f(x)] = f'(x) = \dfrac{2}{3}x$.

75. For an even function, the graph is symmetric about the y-axis; the slope of the tangent line at $(a, f(a))$ is the negative of the slope of the tangent line at $(-a, f(-a))$. For an odd function, the graph is symmetric about the origin; the slope of the tangent line at $(a, f(a))$ is the same as the slope of the tangent line at $(-a, f(-a))$.

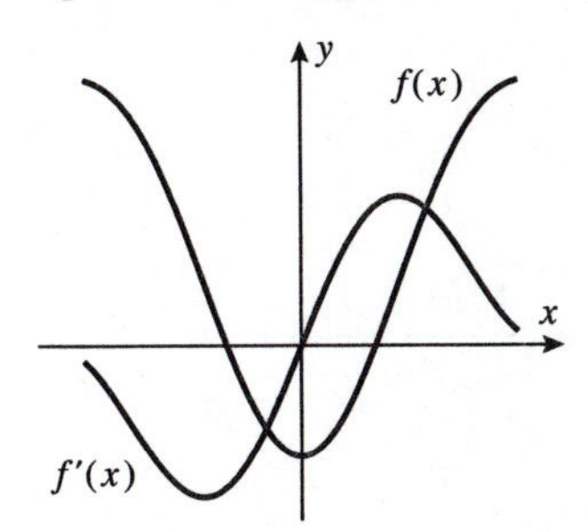

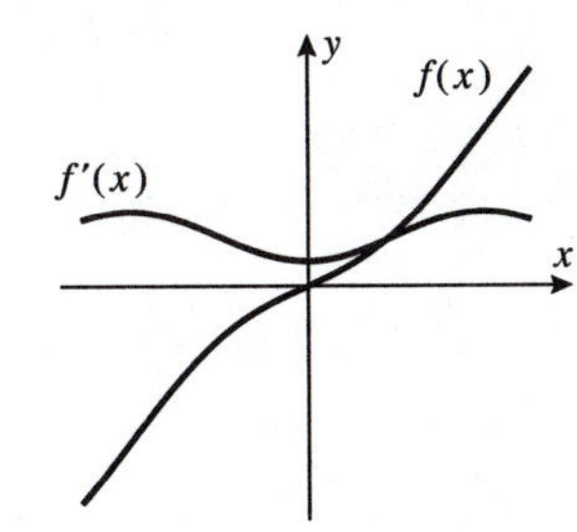

77. $\dfrac{d}{dx}[f(g(h(x)))] = \dfrac{d}{dx}[f(g(u))], \quad u = h(x)$

$$= \frac{d}{du}[f(g(u))]\frac{du}{dx} = f'(g(u))g'(u)\frac{du}{dx} = f'(g(h(x)))g'(h(x))h'(x)$$

EXERCISE SET 3.6

1. $y = (2x-5)^{1/3}$; $dy/dx = \dfrac{2}{3}(2x-5)^{-2/3}$

3. $dy/dx = \dfrac{3}{2}\left[\dfrac{x-1}{x+2}\right]^{1/2}\dfrac{d}{dx}\left[\dfrac{x-1}{x+2}\right] = \dfrac{9}{2(x+2)^2}\left[\dfrac{x-1}{x+2}\right]^{1/2}$

5. $dy/dx = x^3\left(-\dfrac{2}{3}\right)(5x^2+1)^{-5/3}(10x) + 3x^2(5x^2+1)^{-2/3} = \dfrac{1}{3}x^2(5x^2+1)^{-5/3}(25x^2+9)$

7. $dy/dx = \dfrac{5}{2}[\sin(3/x)]^{3/2}[\cos(3/x)](-3/x^2) = -\dfrac{15[\sin(3/x)]^{3/2}\cos(3/x)}{2x^2}$

9. **(a)** $3x^2 + x\dfrac{dy}{dx} + y - 2 = 0$, $\dfrac{dy}{dx} = \dfrac{2-3x^2-y}{x}$

(b) $y = \dfrac{1+2x-x^3}{x} = \dfrac{1}{x} + 2 - x^2$ so $\dfrac{dy}{dx} = -\dfrac{1}{x^2} - 2x$

(c) from Part (a), $\dfrac{dy}{dx} = \dfrac{2-3x^2-y}{x} = \dfrac{2-3x^2-(1/x+2-x^2)}{x} = -2x - \dfrac{1}{x^2}$

11. $2x + 2y\dfrac{dy}{dx} = 0$ so $\dfrac{dy}{dx} = -\dfrac{x}{y}$

13. $x^2\dfrac{dy}{dx} + 2xy + 3x(3y^2)\dfrac{dy}{dx} + 3y^3 - 1 = 0$

$(x^2+9xy^2)\dfrac{dy}{dx} = 1 - 2xy - 3y^3$ so $\dfrac{dy}{dx} = \dfrac{1-2xy-3y^3}{x^2+9xy^2}$

15. $-\dfrac{1}{y^2}\dfrac{dy}{dx} - \dfrac{1}{x^2} = 0$ so $\dfrac{dy}{dx} = -\dfrac{y^2}{x^2}$

17. $\cos(x^2y^2)\left[x^2(2y)\dfrac{dy}{dx} + 2xy^2\right] = 1$, $\dfrac{dy}{dx} = \dfrac{1-2xy^2\cos(x^2y^2)}{2x^2y\cos(x^2y^2)}$

19. $3\tan^2(xy^2+y)\sec^2(xy^2+y)\left(2xy\dfrac{dy}{dx} + y^2 + \dfrac{dy}{dx}\right) = 1$

so $\dfrac{dy}{dx} = \dfrac{1-3y^2\tan^2(xy^2+y)\sec^2(xy^2+y)}{3(2xy+1)\tan^2(xy^2+y)\sec^2(xy^2+y)}$

21. $\dfrac{dy}{dx} = \dfrac{3x}{4y}$, $\dfrac{d^2y}{dx^2} = \dfrac{(4y)(3)-(3x)(4dy/dx)}{16y^2} = \dfrac{12y-12x(3x/(4y))}{16y^2} = \dfrac{12y^2-9x^2}{16y^3} = \dfrac{-3(3x^2-4y^2)}{16y^3}$,

but $3x^2 - 4y^2 = 7$ so $\dfrac{d^2y}{dx^2} = \dfrac{-3(7)}{16y^3} = -\dfrac{21}{16y^3}$

23. $\dfrac{dy}{dx} = -\dfrac{y}{x}$, $\dfrac{d^2y}{dx^2} = -\dfrac{x(dy/dx)-y(1)}{x^2} = -\dfrac{x(-y/x)-y}{x^2} = \dfrac{2y}{x^2}$

25. $\dfrac{dy}{dx} = (1+\cos y)^{-1}$, $\dfrac{d^2y}{dx^2} = -(1+\cos y)^{-2}(-\sin y)\dfrac{dy}{dx} = \dfrac{\sin y}{(1+\cos y)^3}$

27. By implicit differentiation, $2x + 2y(dy/dx) = 0$, $\dfrac{dy}{dx} = -\dfrac{x}{y}$; at $(1/\sqrt{2}, 1/\sqrt{2})$, $\dfrac{dy}{dx} = -1$; at $(1/\sqrt{2}, -1/\sqrt{2})$, $\dfrac{dy}{dx} = +1$. Directly, at the upper point $y = \sqrt{1-x^2}$, $\dfrac{dy}{dx} = \dfrac{-x}{\sqrt{1-x^2}} = -1$ and at the lower point $y = -\sqrt{1-x^2}$, $\dfrac{dy}{dx} = \dfrac{x}{\sqrt{1-x^2}} = +1$.

29. $4x^3 + 4y^3\dfrac{dy}{dx} = 0$, so $\dfrac{dy}{dx} = -\dfrac{x^3}{y^3} = -\dfrac{1}{15^{3/4}} \approx -0.1312$.

31. $4(x^2+y^2)\left(2x + 2y\dfrac{dy}{dx}\right) = 25\left(2x - 2y\dfrac{dy}{dx}\right)$,

$\dfrac{dy}{dx} = \dfrac{x[25 - 4(x^2+y^2)]}{y[25 + 4(x^2+y^2)]}$; at $(3,1)$ $\dfrac{dy}{dx} = -9/13$

35. (a)

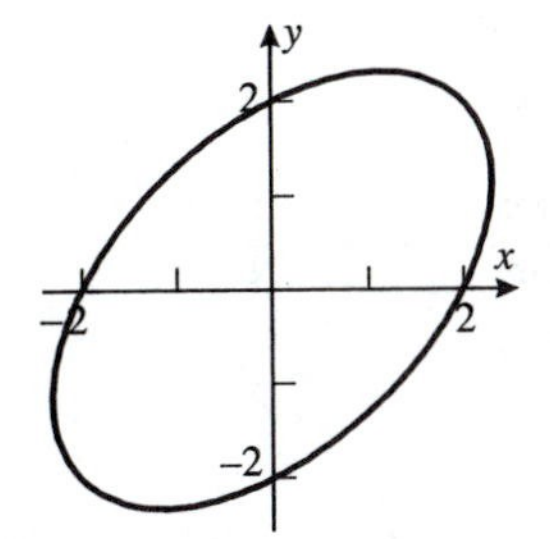

(b) ± 1.1547

(c) Implicit differentiation yields $2x - x\dfrac{dy}{dx} - y + 2y\dfrac{dy}{dx} = 0$. Solve for $\dfrac{dy}{dx} = \dfrac{y-2x}{2y-x}$. If $\dfrac{dy}{dx} = 0$ then $y - 2x = 0$ or $y = 2x$. Thus $4 = x^2 - xy + y^2 = x^2 - 2x^2 + 4x^2 = 3x^2$, $x = \pm\dfrac{2}{\sqrt{3}}$.

37. $4a^3\dfrac{da}{dt} - 4t^3 = 6\left(a^2 + 2at\dfrac{da}{dt}\right)$, solve for $\dfrac{da}{dt}$ to get $\dfrac{da}{dt} = \dfrac{2t^3 + 3a^2}{2a^3 - 6at}$

39. $2a^2\omega\dfrac{d\omega}{d\lambda} + 2b^2\lambda = 0$ so $\dfrac{d\omega}{d\lambda} = -\dfrac{b^2\lambda}{a^2\omega}$

41. The point (1,1) is on the graph, so $1 + a = b$. The slope of the tangent line at (1,1) is $-4/3$; use implicit differentiation to get $\dfrac{dy}{dx} = -\dfrac{2xy}{x^2 + 2ay}$ so at (1,1), $-\dfrac{2}{1+2a} = -\dfrac{4}{3}$, $1 + 2a = 3/2$, $a = 1/4$ and hence $b = 1 + 1/4 = 5/4$.

43. Let $P(x_0, y_0)$ be a point where a line through the origin is tangent to the curve $x^2 - 4x + y^2 + 3 = 0$. Implicit differentiation applied to the equation of the curve gives $dy/dx = (2-x)/y$. At P the slope of the curve must equal the slope of the line so $(2 - x_0)/y_0 = y_0/x_0$, or $y_0^2 = 2x_0 - x_0^2$. But $x_0^2 - 4x_0 + y_0^2 + 3 = 0$ because (x_0, y_0) is on the curve, and elimination of y_0^2 in the latter two equations gives $x_0^2 - 4x_0 + (2x_0 - x_0^2) + 3 = 0$, $x_0 = 3/2$ which when substituted into $y_0^2 = 2x_0 - x_0^2$ yields $y_0^2 = 3/4$, so $y_0 = \pm\sqrt{3}/2$. The slopes of the lines are $(\pm\sqrt{3}/2)/(3/2) = \pm\sqrt{3}/3$ and their equations are $y = (\sqrt{3}/3)x$ and $y = -(\sqrt{3}/3)x$.

45. By the chain rule, $\dfrac{dy}{dx} = \dfrac{dy}{dt}\dfrac{dt}{dx}$. Use implicit differentiation on $2y^3t + t^3y = 1$ to get

$\dfrac{dy}{dt} = -\dfrac{2y^3 + 3t^2y}{6ty^2 + t^3}$, but $\dfrac{dt}{dx} = \dfrac{1}{\cos t}$ so $\dfrac{dy}{dx} = -\dfrac{2y^3 + 3t^2y}{(6ty^2 + t^3)\cos t}$.

47. $2xy\dfrac{dy}{dt} = y^2\dfrac{dx}{dt} = 3(\cos 3x)\dfrac{dx}{dt},\ \dfrac{dy}{dt} = \dfrac{3\cos 3x - y^2}{2xy}\dfrac{dx}{dt}$

49. $y' = rx^{r-1}, y'' = r(r-1)x^{r-2}$ so $3x^2\left[r(r-1)x^{r-2}\right] + 4x\left(rx^{r-1}\right) - 2x^r = 0$,
$3r(r-1)x^r + 4rx^r - 2x^r = 0, (3r^2 + r - 2)x^r = 0$,
$3r^2 + r - 2 = 0,\ (3r-2)(r+1) = 0;\ r = -1, 2/3$

51. We shall find when the curves intersect and check that the slopes are negative reciprocals. For the intersection solve the simultaneous equations $x^2 + (y-c)^2 = c^2$ and $(x-k)^2 + y^2 = k^2$ to obtain $cy = kx = \dfrac{1}{2}(x^2 + y^2)$. Thus $x^2 + y^2 = cy + kx$, or $y^2 - cy = -x^2 + kx$, and $\dfrac{y-c}{x} = -\dfrac{x-k}{y}$. Differentiating the two families yields (black) $\dfrac{dy}{dx} = -\dfrac{x}{y-c}$, and (gray) $\dfrac{dy}{dx} = -\dfrac{x-k}{y}$. But it was proven that these quantities are negative reciprocals of each other.

EXERCISE SET 3.7

1. $\dfrac{dy}{dt} = 3\dfrac{dx}{dt}$

(a) $\dfrac{dy}{dt} = 3(2) = 6$

(b) $-1 = 3\dfrac{dx}{dt},\ \dfrac{dx}{dt} = -\dfrac{1}{3}$

3. $2x\dfrac{dx}{dt} + 2y\dfrac{dy}{dt} = 0$

(a) $2\dfrac{1}{2}(1) + 2\dfrac{\sqrt{3}}{2}\dfrac{dy}{dt} = 0$, so $\dfrac{dy}{dt} = -\dfrac{1}{\sqrt{3}}$.

(b) $2\dfrac{\sqrt{2}}{2}\dfrac{dx}{dt} + 2\dfrac{\sqrt{2}}{2}(-2) = 0$, so $\dfrac{dx}{dt} = 2$.

5. (b) $A = x^2$

(c) $\dfrac{dA}{dt} = 2x\dfrac{dx}{dt}$

Find $\left.\dfrac{dA}{dt}\right|_{x=3}$ given that $\left.\dfrac{dx}{dt}\right|_{x=3} = 2$. From Part (c), $\left.\dfrac{dA}{dt}\right|_{x=3} = 2(3)(2) = 12\ \text{ft}^2/\text{min}$.

7. (a) $V = \pi r^2 h$, so $\dfrac{dV}{dt} = \pi\left(r^2\dfrac{dh}{dt} + 2rh\dfrac{dr}{dt}\right)$.

(b) Find $\left.\dfrac{dV}{dt}\right|_{\substack{h=6,\\ r=10}}$ given that $\left.\dfrac{dh}{dt}\right|_{\substack{h=6,\\ r=10}} = 1$ and $\left.\dfrac{dr}{dt}\right|_{\substack{h=6,\\ r=10}} = -1$. From Part (a),
$\left.\dfrac{dV}{dt}\right|_{\substack{h=6,\\ r=10}} = \pi[10^2(1) + 2(10)(6)(-1)] = -20\pi\ \text{in}^3/\text{s}$; the volume is decreasing.

9. (a) $\tan\theta = \dfrac{y}{x}$, so $\sec^2\theta\dfrac{d\theta}{dt} = \dfrac{x\dfrac{dy}{dt} - y\dfrac{dx}{dt}}{x^2},\ \dfrac{d\theta}{dt} = \dfrac{\cos^2\theta}{x^2}\left(x\dfrac{dy}{dt} - y\dfrac{dx}{dt}\right)$

(b) Find $\left.\dfrac{d\theta}{dt}\right|_{\substack{x=2,\\ y=2}}$ given that $\left.\dfrac{dx}{dt}\right|_{\substack{x=2,\\ y=2}} = 1$ and $\left.\dfrac{dy}{dt}\right|_{\substack{x=2,\\ y=2}} = -\dfrac{1}{4}$.
When $x = 2$ and $y = 2$, $\tan\theta = 2/2 = 1$ so $\theta = \dfrac{\pi}{4}$ and $\cos\theta = \cos\dfrac{\pi}{4} = \dfrac{1}{\sqrt{2}}$. Thus from Part (a), $\left.\dfrac{d\theta}{dt}\right|_{\substack{x=2,\\ y=2}} = \dfrac{(1/\sqrt{2})^2}{2^2}\left[2\left(-\dfrac{1}{4}\right) - 2(1)\right] = -\dfrac{5}{16}$ rad/s; θ is decreasing.

11. Let A be the area swept out, and θ the angle through which the minute hand has rotated. Find $\dfrac{dA}{dt}$ given that $\dfrac{d\theta}{dt} = \dfrac{\pi}{30}$ rad/min; $A = \dfrac{1}{2}r^2\theta = 8\theta$, so $\dfrac{dA}{dt} = 8\dfrac{d\theta}{dt} = \dfrac{4\pi}{15}$ in^2/min.

13. Find $\left.\dfrac{dr}{dt}\right|_{A=9}$ given that $\dfrac{dA}{dt} = 6$. From $A = \pi r^2$ we get $\dfrac{dA}{dt} = 2\pi r\dfrac{dr}{dt}$ so $\dfrac{dr}{dt} = \dfrac{1}{2\pi r}\dfrac{dA}{dt}$. If $A = 9$ then $\pi r^2 = 9$, $r = 3/\sqrt{\pi}$ so $\left.\dfrac{dr}{dt}\right|_{A=9} = \dfrac{1}{2\pi(3/\sqrt{\pi})}(6) = 1/\sqrt{\pi}$ mi/h.

15. Find $\left.\dfrac{dV}{dt}\right|_{r=9}$ given that $\dfrac{dr}{dt} = -15$. From $V = \dfrac{4}{3}\pi r^3$ we get $\dfrac{dV}{dt} = 4\pi r^2\dfrac{dr}{dt}$ so

$\left.\dfrac{dV}{dt}\right|_{r=9} = 4\pi(9)^2(-15) = -4860\pi$. Air must be removed at the rate of 4860π cm^3/min.

17. Find $\left.\dfrac{dx}{dt}\right|_{y=5}$ given that $\dfrac{dy}{dt} = -2$. From $x^2 + y^2 = 13^2$ we get $2x\dfrac{dx}{dt} + 2y\dfrac{dy}{dt} = 0$ so $\dfrac{dx}{dt} = -\dfrac{y}{x}\dfrac{dy}{dt}$. Use $x^2 + y^2 = 169$ to find that $x = 12$ when $y = 5$ so $\left.\dfrac{dx}{dt}\right|_{y=5} = -\dfrac{5}{12}(-2) = \dfrac{5}{6}$ ft/s.

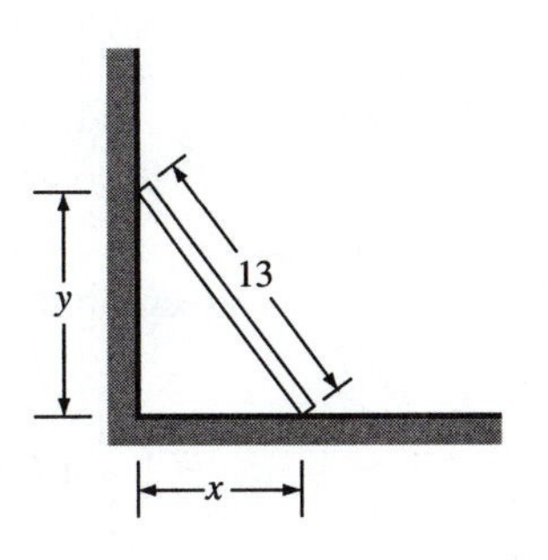

19. Let x denote the distance from first base and y the distance from home plate. Then $x^2 + 60^2 = y^2$ and $2x\dfrac{dx}{dt} = 2y\dfrac{dy}{dt}$. When $x = 50$ then $y = 10\sqrt{61}$ so $\dfrac{dy}{dt} = \dfrac{x}{y}\dfrac{dx}{dt} = \dfrac{50}{10\sqrt{61}}(25) = \dfrac{125}{\sqrt{61}}$ ft/s.

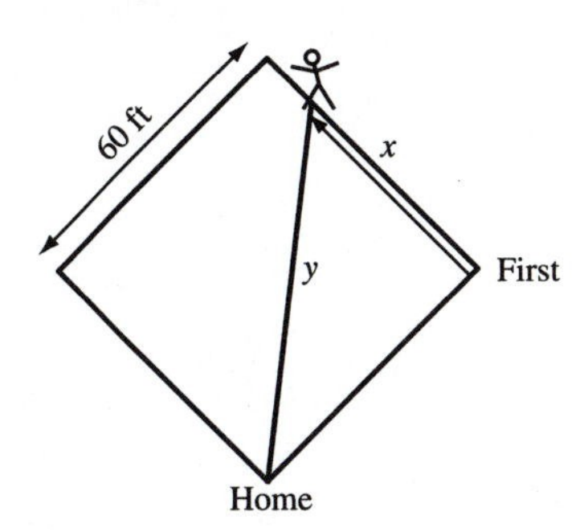

21. Find $\left.\dfrac{dy}{dt}\right|_{x=4000}$ given that $\left.\dfrac{dx}{dt}\right|_{x=4000} = 880$. From $y^2 = x^2 + 3000^2$ we get $2y\dfrac{dy}{dt} = 2x\dfrac{dx}{dt}$ so $\dfrac{dy}{dt} = \dfrac{x}{y}\dfrac{dx}{dt}$. If $x = 4000$, then $y = 5000$ so $\left.\dfrac{dy}{dt}\right|_{x=4000} = \dfrac{4000}{5000}(880) = 704$ ft/s.

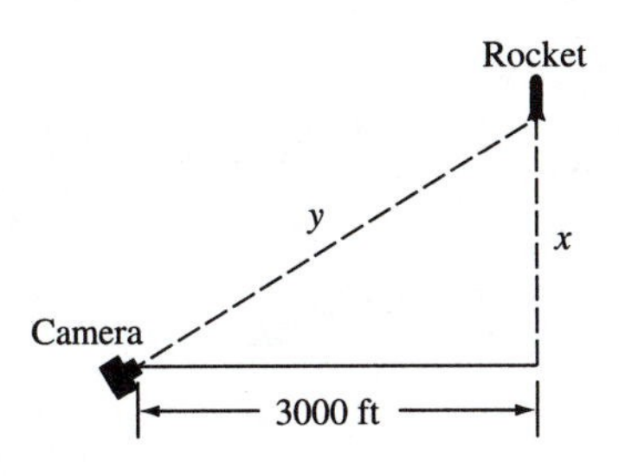

23. **(a)** If x denotes the altitude, then $r - x = 3960$, the radius of the Earth. $\theta = 0$ at perigee, so $r = 4995/1.12 \approx 4460$; the altitude is $x = 4460 - 3960 = 500$ miles. $\theta = \pi$ at apogee, so $r = 4995/0.88 \approx 5676$; the altitude is $x = 5676 - 3960 = 1716$ miles.

(b) If $\theta = 120°$, then $r = 4995/0.94 \approx 5314$; the altitude is $5314 - 3960 = 1354$ miles. The rate of change of the altitude is given by

$$\frac{dx}{dt} = \frac{dr}{dt} = \frac{dr}{d\theta}\frac{d\theta}{dt} = \frac{4995(0.12\sin\theta)}{(1 + 0.12\cos\theta)^2}\frac{d\theta}{dt}.$$

Use $\theta = 120°$ and $d\theta/dt = 2.7°/\text{min} = (2.7)(\pi/180)$ rad/min to get $dr/dt \approx 27.7$ mi/min.

25. Find $\left.\dfrac{dh}{dt}\right|_{h=16}$ given that $\dfrac{dV}{dt} = 20$. The volume of water in the tank at a depth h is $V = \dfrac{1}{3}\pi r^2 h$. Use similar triangles (see figure) to get $\dfrac{r}{h} = \dfrac{10}{24}$ so $r = \dfrac{5}{12}h$ thus $V = \dfrac{1}{3}\pi\left(\dfrac{5}{12}h\right)^2 h = \dfrac{25}{432}\pi h^3$, $\dfrac{dV}{dt} = \dfrac{25}{144}\pi h^2\dfrac{dh}{dt}$; $\dfrac{dh}{dt} = \dfrac{144}{25\pi h^2}\dfrac{dV}{dt}$, $\left.\dfrac{dh}{dt}\right|_{h=16} = \dfrac{144}{25\pi(16)^2}(20) = \dfrac{9}{20\pi}$ ft/min.

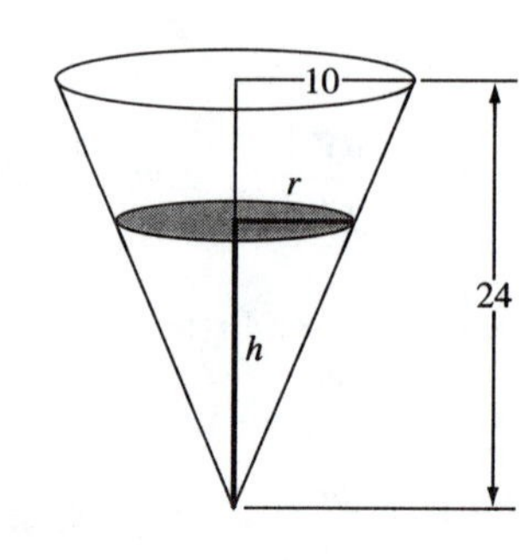

27. Find $\left.\dfrac{dV}{dt}\right|_{h=10}$ given that $\dfrac{dh}{dt} = 5$. $V = \dfrac{1}{3}\pi r^2 h$, but $r = \dfrac{1}{2}h$ so $V = \dfrac{1}{3}\pi\left(\dfrac{h}{2}\right)^2 h = \dfrac{1}{12}\pi h^3$, $\dfrac{dV}{dt} = \dfrac{1}{4}\pi h^2\dfrac{dh}{dt}$, $\left.\dfrac{dV}{dt}\right|_{h=10} = \dfrac{1}{4}\pi(10)^2(5) = 125\pi$ ft^3/min.

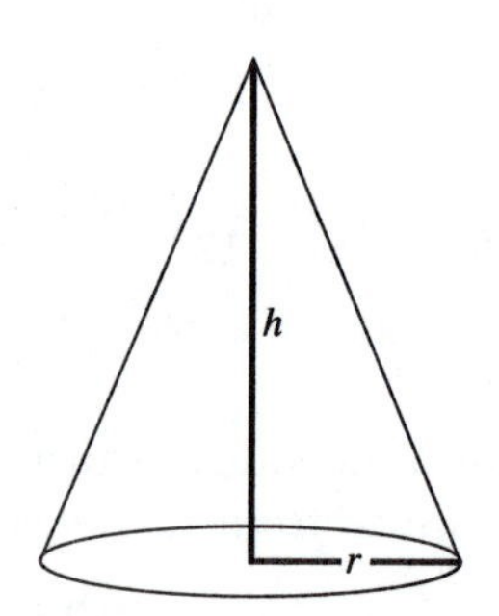

29. With s and h as shown in the figure, we want to find $\dfrac{dh}{dt}$ given that $\dfrac{ds}{dt} = 500$. From the figure, $h = s\sin 30° = \dfrac{1}{2}s$ so $\dfrac{dh}{dt} = \dfrac{1}{2}\dfrac{ds}{dt} = \dfrac{1}{2}(500) = 250$ mi/h.

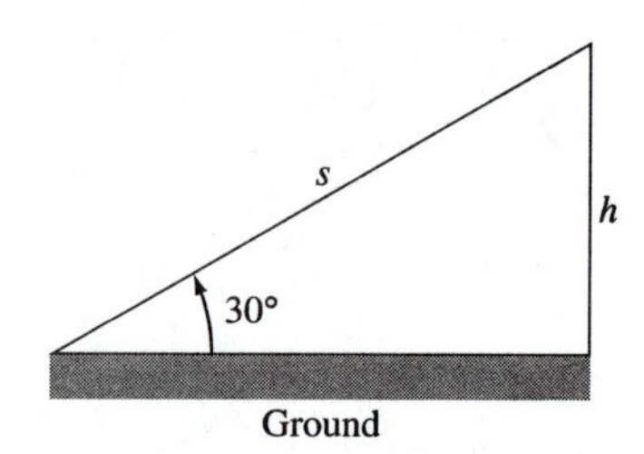

31. Find $\dfrac{dy}{dt}$ given that $\left.\dfrac{dx}{dt}\right|_{y=125} = -12$. From $x^2 + 10^2 = y^2$ we get $2x\dfrac{dx}{dt} = 2y\dfrac{dy}{dt}$ so $\dfrac{dy}{dt} = \dfrac{x}{y}\dfrac{dx}{dt}$. Use $x^2 + 100 = y^2$ to find that $x = \sqrt{15,525} = 15\sqrt{69}$ when $y = 125$ so $\dfrac{dy}{dt} = \dfrac{15\sqrt{69}}{125}(-12) = -\dfrac{36\sqrt{69}}{25}$. The rope must be pulled at the rate of $\dfrac{36\sqrt{69}}{25}$ ft/min.

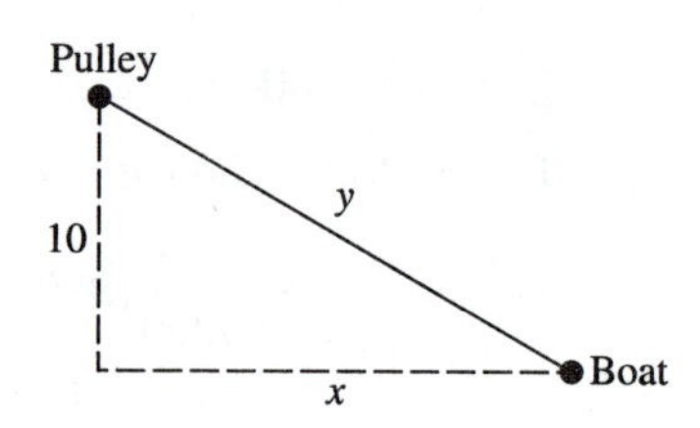

33. Find $\left.\dfrac{dx}{dt}\right|_{\theta=\pi/4}$ given that $\dfrac{d\theta}{dt} = \dfrac{2\pi}{10} = \dfrac{\pi}{5}$ rad/s.

Then $x = 4\tan\theta$ (see figure) so $\dfrac{dx}{dt} = 4\sec^2\theta\dfrac{d\theta}{dt}$,

$\left.\dfrac{dx}{dt}\right|_{\theta=\pi/4} = 4\left(\sec^2\dfrac{\pi}{4}\right)\left(\dfrac{\pi}{5}\right) = 8\pi/5$ km/s.

35. We wish to find $\left.\dfrac{dz}{dt}\right|_{\substack{x=2,\\ y=4}}$ given $\dfrac{dx}{dt} = -600$ and $\left.\dfrac{dy}{dt}\right|_{\substack{x=2,\\ y=4}} = -1200$ (see figure). From the law of cosines,

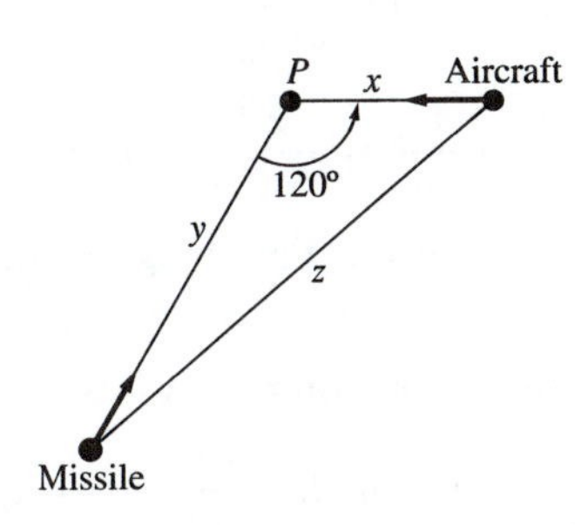

$z^2 = x^2 + y^2 - 2xy\cos 120° = x^2 + y^2 - 2xy(-1/2)$

$= x^2+y^2+xy$, so $2z\dfrac{dz}{dt} = 2x\dfrac{dx}{dt}+2y\dfrac{dy}{dt}+x\dfrac{dy}{dt}+y\dfrac{dx}{dt}$,

$$\frac{dz}{dt} = \frac{1}{2z}\left[(2x+y)\frac{dx}{dt} + (2y+x)\frac{dy}{dt}\right].$$

When $x = 2$ and $y = 4$, $z^2 = 2^2 + 4^2 + (2)(4) = 28$, so $z = \sqrt{28} = 2\sqrt{7}$, thus

$$\left.\frac{dz}{dt}\right|_{\substack{x=2,\\ y=4}} = \frac{1}{2(2\sqrt{7})}[(2(2)+4)(-600) + (2(4)+2)(-1200)] = -\frac{4200}{\sqrt{7}} = -600\sqrt{7} \text{ mi/h};$$

the distance between missile and aircraft is decreasing at the rate of $600\sqrt{7}$ mi/h.

37. **(a)** We want $\left.\dfrac{dy}{dt}\right|_{\substack{x=1,\\ y=2}}$ given that $\left.\dfrac{dx}{dt}\right|_{\substack{x=1,\\ y=2}} = 6$. For convenience, first rewrite the equation as

$xy^3 = \dfrac{8}{5} + \dfrac{8}{5}y^2$ then $3xy^2\dfrac{dy}{dt} + y^3\dfrac{dx}{dt} = \dfrac{16}{5}y\dfrac{dy}{dt}$, $\dfrac{dy}{dt} = \dfrac{y^3}{\dfrac{16}{5}y - 3xy^2}\dfrac{dx}{dt}$ so

$$\left.\frac{dy}{dt}\right|_{\substack{x=1,\\ y=2}} = \frac{2^3}{\dfrac{16}{5}(2) - 3(1)2^2}(6) = -60/7 \text{ units/s}.$$

(b) falling, because $\dfrac{dy}{dt} < 0$

39. The coordinates of P are $(x, 2x)$, so the distance between P and the point $(3, 0)$ is

$D = \sqrt{(x-3)^2 + (2x-0)^2} = \sqrt{5x^2 - 6x + 9}$. Find $\left.\dfrac{dD}{dt}\right|_{x=3}$ given that $\left.\dfrac{dx}{dt}\right|_{x=3} = -2$.

$\dfrac{dD}{dt} = \dfrac{5x-3}{\sqrt{5x^2-6x+9}}\dfrac{dx}{dt}$, so $\left.\dfrac{dD}{dt}\right|_{x=3} = \dfrac{12}{\sqrt{36}}(-2) = -4$ units/s.

41. Solve $\frac{dx}{dt} = 3\frac{dy}{dt}$ given $y = x/(x^2+1)$. Then $y(x^2+1) = x$. Differentiating with respect to x, $(x^2+1)\frac{dy}{dx} + y(2x) = 1$. But $\frac{dy}{dx} = \frac{dy/dt}{dx/dt} = \frac{1}{3}$ so $(x^2+1)\frac{1}{3} + 2xy = 1$, $x^2+1+6xy = 3$, $x^2+1+6x^2/(x^2+1) = 3$, $(x^2+1)^2+6x^2-3x^2-3 = 0$, $x^4+5x^2-3 = 0$. By the binomial theorem applied to x^2 we obtain $x^2 = (-5 \pm \sqrt{25+12})/2$. The minus sign is spurious since x^2 cannot be negative, so $x^2 = (\sqrt{33}-5)/2$, $x \approx \pm 0.6101486081, y = \pm 0.4446235604$.

43. Find $\left.\frac{dS}{dt}\right|_{s=10}$ given that $\left.\frac{ds}{dt}\right|_{s=10} = -2$. From $\frac{1}{s} + \frac{1}{S} = \frac{1}{6}$ we get $-\frac{1}{s^2}\frac{ds}{dt} - \frac{1}{S^2}\frac{dS}{dt} = 0$, so $\frac{dS}{dt} = -\frac{S^2}{s^2}\frac{ds}{dt}$. If $s = 10$, then $\frac{1}{10} + \frac{1}{S} = \frac{1}{6}$ which gives $S = 15$. So $\left.\frac{dS}{dt}\right|_{s=10} = -\frac{225}{100}(-2) = 4.5$ cm/s. The image is moving away from the lens.

45. Let r be the radius, V the volume, and A the surface area of a sphere. Show that $\frac{dr}{dt}$ is a constant given that $\frac{dV}{dt} = -kA$, where k is a positive constant. Because $V = \frac{4}{3}\pi r^3$,

$$\frac{dV}{dt} = 4\pi r^2 \frac{dr}{dt} \qquad (1)$$

But it is given that $\frac{dV}{dt} = -kA$ or, because $A = 4\pi r^2$, $\frac{dV}{dt} = -4\pi r^2 k$ which when substituted into equation (1) gives $-4\pi r^2 k = 4\pi r^2 \frac{dr}{dt}$, $\frac{dr}{dt} = -k$.

47. Extend sides of cup to complete the cone and let V_0 be the volume of the portion added, then (see figure) $V = \frac{1}{3}\pi r^2 h - V_0$ where $\frac{r}{h} = \frac{4}{12} = \frac{1}{3}$ so $r = \frac{1}{3}h$ and $V = \frac{1}{3}\pi\left(\frac{h}{3}\right)^2 h - V_0 = \frac{1}{27}\pi h^3 - V_0$, $\frac{dV}{dt} = \frac{1}{9}\pi h^2 \frac{dh}{dt}$, $\frac{dh}{dt} = \frac{9}{\pi h^2}\frac{dV}{dt}$, $\left.\frac{dh}{dt}\right|_{h=9} = \frac{9}{\pi(9)^2}(20) = \frac{20}{9\pi}$ cm/s.

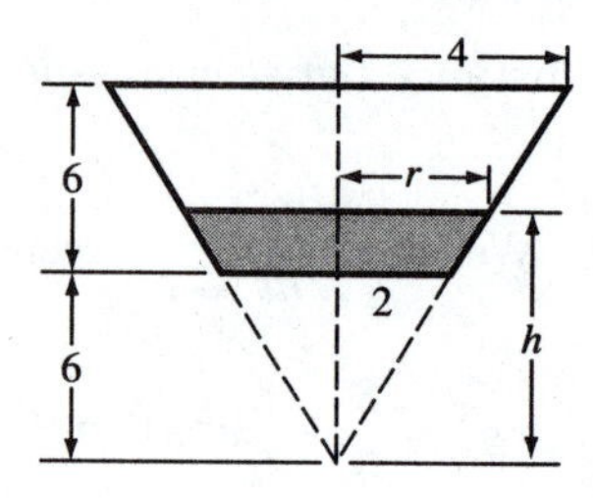

EXERCISE SET 3.8

1. (a) $f(x) \approx f(1) + f'(1)(x-1) = 1 + 3(x-1)$

(b) $f(1+\Delta x) \approx f(1) + f'(1)\Delta x = 1 + 3\Delta x$

(c) From Part (a), $(1.02)^3 \approx 1 + 3(0.02) = 1.06$. From Part (b), $(1.02)^3 \approx 1 + 3(0.02) = 1.06$.

3. (a) $f(x) \approx f(x_0) + f'(x_0)(x - x_0) = 1 + (1/(2\sqrt{1})(x-0) = 1 + (1/2)x$, so with $x_0 = 0$ and $x = -0.1$, we have $\sqrt{0.9} = f(-0.1) \approx 1 + (1/2)(-0.1) = 1 - 0.05 = 0.95$. With $x = 0.1$ we have $\sqrt{1.1} = f(0.1) \approx 1 + (1/2)(0.1) = 1.05$.

(b)

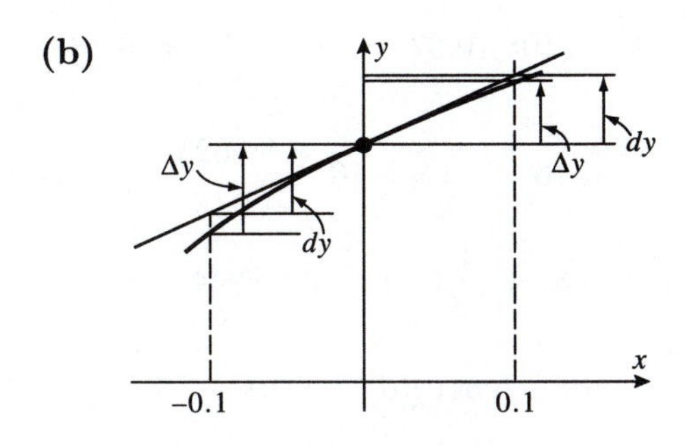

5. $f(x) = (1+x)^{15}$ and $x_0 = 0$. Thus $(1+x)^{15} \approx f(x_0)+f'(x_0)(x-x_0) = 1+15(1)^{14}(x-0) = 1+15x$.

7. $\tan x \approx \tan(0) + \sec^2(0)(x-0) = x$

9. $x^4 \approx (1)^4 + 4(1)^3(x-1)$. Set $\Delta x = x - 1$; then $x = \Delta x + 1$ and $(1+\Delta x)^4 = 1 + 4\Delta x$.

11. $\dfrac{1}{2+x} \approx \dfrac{1}{2+1} - \dfrac{1}{(2+1)^2}(x-1)$, and $2 + x = 3 + \Delta x$, so $\dfrac{1}{3+\Delta x} \approx \dfrac{1}{3} - \dfrac{1}{9}\Delta x$

13. $f(x) = \sqrt{x+3}$ and $x_0 = 0$, so
$\sqrt{x+3} \approx \sqrt{3} + \dfrac{1}{2\sqrt{3}}(x-0) = \sqrt{3} + \dfrac{1}{2\sqrt{3}}x$, and
$\left|f(x) - \left(\sqrt{3} + \dfrac{1}{2\sqrt{3}}x\right)\right| < 0.1$ if $|x| < 1.692$.

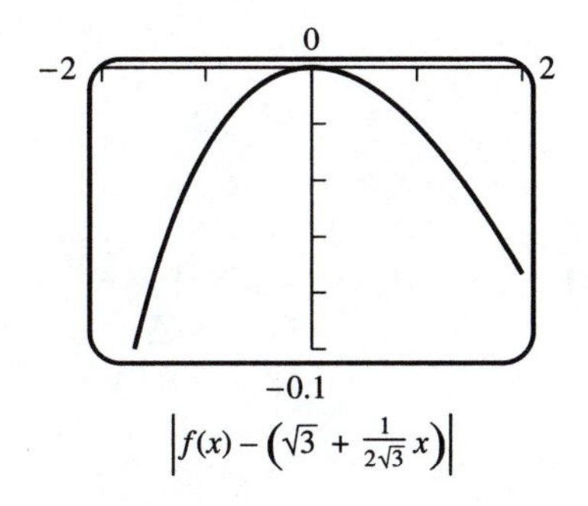

15. $\tan x \approx \tan 0 + (\sec^2 0)(x-0) = x$,
and $|\tan x - x| < 0.1$ if $|x| < 0.6316$

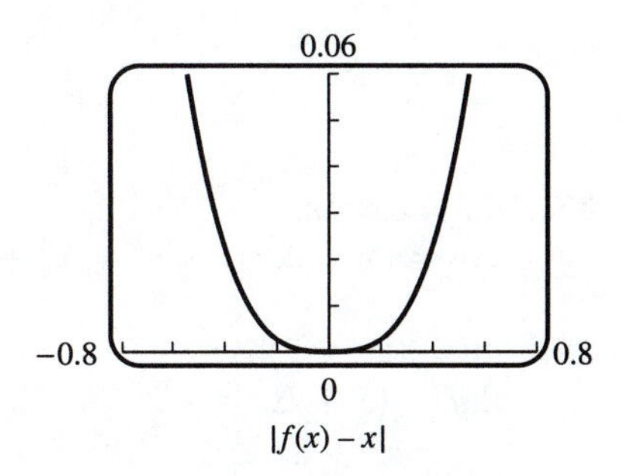

17. **(a)** The local linear approximation $\sin x \approx x$ gives $\sin 1° = \sin(\pi/180) \approx \pi/180 = 0.0174533$ and a calculator gives $\sin 1° = 0.0174524$. The relative error $|\sin(\pi/180) - (\pi/180)|/(\sin \pi/180) = 0.000051$ is very small, so for such a small value of x the approximation is very good.

(b) Use $x_0 = 45°$ (this assumes you know, or can approximate, $\sqrt{2}/2$).

(c) $44° = \dfrac{44\pi}{180}$ radians, and $45° = \dfrac{45\pi}{180} = \dfrac{\pi}{4}$ radians. With $x = \dfrac{44\pi}{180}$ and $x_0 = \dfrac{\pi}{4}$ we obtain $\sin 44° = \sin \dfrac{44\pi}{180} \approx \sin \dfrac{\pi}{4} + \left(\cos \dfrac{\pi}{4}\right)\left(\dfrac{44\pi}{180} - \dfrac{\pi}{4}\right) = \dfrac{\sqrt{2}}{2} + \dfrac{\sqrt{2}}{2}\left(\dfrac{-\pi}{180}\right) = 0.694765$. With a calculator, $\sin 44° = 0.694658$.

19. $f(x) = x^4$, $f'(x) = 4x^3$, $x_0 = 3$, $\Delta x = 0.02$; $(3.02)^4 \approx 3^4 + (108)(0.02) = 81 + 2.16 = 83.16$

21. $f(x) = \sqrt{x}$, $f'(x) = \dfrac{1}{2\sqrt{x}}$, $x_0 = 64$, $\Delta x = 1$; $\sqrt{65} \approx \sqrt{64} + \dfrac{1}{16}(1) = 8 + \dfrac{1}{16} = 8.0625$

23. $f(x) = \sqrt{x}$, $f'(x) = \dfrac{1}{2\sqrt{x}}$, $x_0 = 81$, $\Delta x = -0.1$; $\sqrt{80.9} \approx \sqrt{81} + \dfrac{1}{18}(-0.1) \approx 8.9944$

25. $f(x) = \sin x$, $f'(x) = \cos x$, $x_0 = 0$, $\Delta x = 0.1$; $\sin 0.1 \approx \sin 0 + (\cos 0)(0.1) = 0.1$

27. $f(x) = \cos x$, $f'(x) = -\sin x$, $x_0 = \pi/6$, $\Delta x = \pi/180$;
$\cos 31^\circ \approx \cos 30^\circ + \left(-\dfrac{1}{2}\right)\left(\dfrac{\pi}{180}\right) = \dfrac{\sqrt{3}}{2} - \dfrac{\pi}{360} \approx 0.8573$

29. **(a)** $dy = f'(x)dx = 2x dx = 4(1) = 4$ and
$\Delta y = (x + \Delta x)^2 - x^2 = (2 + 1)^2 - 2^2 = 5$

(b)

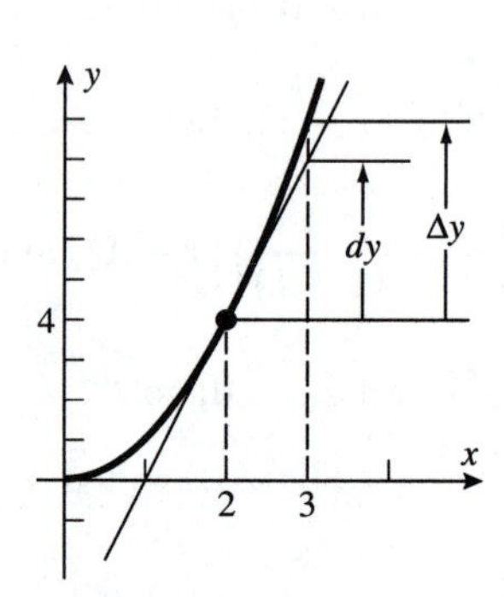

31. **(a)** $dy = (-1/x^2)dx = (-1)(-0.5) = 0.5$ and
$\Delta y = 1/(x + \Delta x) - 1/x$
$= 1/(1 - 0.5) - 1/1 = 2 - 1 = 1$

(b)

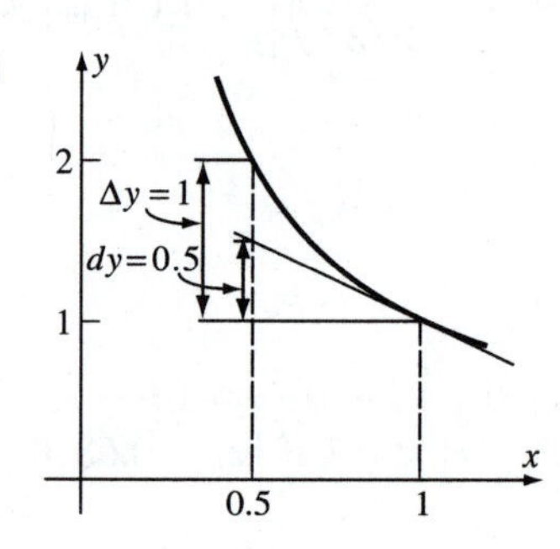

33. $dy = 3x^2 dx$;
$\Delta y = (x + \Delta x)^3 - x^3 = x^3 + 3x^2\Delta x + 3x(\Delta x)^2 + (\Delta x)^3 - x^3 = 3x^2\Delta x + 3x(\Delta x)^2 + (\Delta x)^3$

35. $dy = (2x - 2)dx$;
$\Delta y = [(x + \Delta x)^2 - 2(x + \Delta x) + 1] - [x^2 - 2x + 1]$
$= x^2 + 2x\,\Delta x + (\Delta x)^2 - 2x - 2\Delta x + 1 - x^2 + 2x - 1 = 2x\,\Delta x + (\Delta x)^2 - 2\Delta x$

37. **(a)** $dy = (12x^2 - 14x)dx$
(b) $dy = x\,d(\cos x) + \cos x\,dx = x(-\sin x)dx + \cos x dx = (-x \sin x + \cos x)dx$

39. **(a)** $dy = \left(\sqrt{1 - x} - \dfrac{x}{2\sqrt{1 - x}}\right) dx = \dfrac{2 - 3x}{2\sqrt{1 - x}} dx$
(b) $dy = -17(1 + x)^{-18} dx$

41. $dy = \dfrac{3}{2\sqrt{3x-2}}dx$, $x = 2$, $dx = 0.03$; $\Delta y \approx dy = \dfrac{3}{4}(0.03) = 0.0225$

43. $dy = \dfrac{1-x^2}{(x^2+1)^2}dx$, $x = 2$, $dx = -0.04$; $\Delta y \approx dy = \left(-\dfrac{3}{25}\right)(-0.04) = 0.0048$

45. (a) $A = x^2$ where x is the length of a side; $dA = 2x\,dx = 2(10)(\pm 0.1) = \pm 2$ ft^2.

(b) relative error in x is $\approx \dfrac{dx}{x} = \dfrac{\pm 0.1}{10} = \pm 0.01$ so percentage error in x is $\approx \pm 1\%$; relative error in A is $\approx \dfrac{dA}{A} = \dfrac{2x\,dx}{x^2} = 2\dfrac{dx}{x} = 2(\pm 0.01) = \pm 0.02$ so percentage error in A is $\approx \pm 2\%$

47. (a) $x = 10\sin\theta$, $y = 10\cos\theta$ (see figure),

$dx = 10\cos\theta d\theta = 10\left(\cos\dfrac{\pi}{6}\right)\left(\pm\dfrac{\pi}{180}\right) = 10\left(\dfrac{\sqrt{3}}{2}\right)\left(\pm\dfrac{\pi}{180}\right)$

$\approx \pm 0.151$ in,

$dy = -10(\sin\theta)d\theta = -10\left(\sin\dfrac{\pi}{6}\right)\left(\pm\dfrac{\pi}{180}\right) = -10\left(\dfrac{1}{2}\right)\left(\pm\dfrac{\pi}{180}\right)$

$\approx \pm 0.087$ in

10″
x
θ
y

(b) relative error in x is $\approx \dfrac{dx}{x} = (\cot\theta)d\theta = \left(\cot\dfrac{\pi}{6}\right)\left(\pm\dfrac{\pi}{180}\right) = \sqrt{3}\left(\pm\dfrac{\pi}{180}\right) \approx \pm 0.030$
so percentage error in x is $\approx \pm 3.0\%$;
relative error in y is $\approx \dfrac{dy}{y} = -\tan\theta d\theta = -\left(\tan\dfrac{\pi}{6}\right)\left(\pm\dfrac{\pi}{180}\right) = -\dfrac{1}{\sqrt{3}}\left(\pm\dfrac{\pi}{180}\right) \approx \pm 0.010$
so percentage error in y is $\approx \pm 1.0\%$

49. $\dfrac{dR}{R} = \dfrac{(-2k/r^3)dr}{(k/r^2)} = -2\dfrac{dr}{r}$, but $\dfrac{dr}{r} \approx \pm 0.05$ so $\dfrac{dR}{R} \approx -2(\pm 0.05) = \pm 0.10$; percentage error in R is $\approx \pm 10\%$

51. $A = \dfrac{1}{4}(4)^2\sin 2\theta = 4\sin 2\theta$ thus $dA = 8\cos 2\theta d\theta$ so, with $\theta = 30° = \pi/6$ radians and $d\theta = \pm 15' = \pm 1/4° = \pm\pi/720$ radians, $dA = 8\cos(\pi/3)(\pm\pi/720) = \pm\pi/180 \approx \pm 0.017$ cm^2

53. $V = x^3$ where x is the length of a side; $\dfrac{dV}{V} = \dfrac{3x^2dx}{x^3} = 3\dfrac{dx}{x}$, but $\dfrac{dx}{x} \approx \pm 0.02$
so $\dfrac{dV}{V} \approx 3(\pm 0.02) = \pm 0.06$; percentage error in V is $\approx \pm 6\%$.

55. $A = \dfrac{1}{4}\pi D^2$ where D is the diameter of the circle; $\dfrac{dA}{A} = \dfrac{(\pi D/2)dD}{\pi D^2/4} = 2\dfrac{dD}{D}$, but $\dfrac{dA}{A} \approx \pm 0.01$ so $2\dfrac{dD}{D} \approx \pm 0.01$, $\dfrac{dD}{D} \approx \pm 0.005$; maximum permissible percentage error in D is $\approx \pm 0.5\%$.

57. V = volume of cylindrical rod $= \pi r^2 h = \pi r^2(15) = 15\pi r^2$; approximate ΔV by dV if $r = 2.5$ and $dr = \Delta r = 0.001$. $dV = 30\pi r\,dr = 30\pi(2.5)(0.001) \approx 0.236$ cm^3.

59. (a) $\alpha = \Delta L/(L\Delta T) = 0.006/(40 \times 10) = 1.5 \times 10^{-5}/°\text{C}$

(b) $\Delta L = 2.3 \times 10^{-5}(180)(25) \approx 0.1$ cm, so the pole is about 180.1 cm long.

CHAPTER 3 SUPPLEMENTARY EXERCISES

5. Set $f'(x) = 0$: $f'(x) = 6(2)(2x+7)^5(x-2)^5 + 5(2x+7)^6(x-2)^4 = 0$, so $2x+7 = 0$ or $x-2 = 0$ or, factoring out $(2x+7)^5(x-2)^4$, $12(x-2)+5(2x+7) = 0$. This reduces to $x = -7/2$, $x = 2$, or $22x+11 = 0$, so the tangent line is horizontal at $x = -7/2, 2, -1/2$.

7. Set $f'(x) = \dfrac{3}{2\sqrt{3x+1}}(x-1)^2 + 2\sqrt{3x+1}(x-1) = 0$. If $x = 1$ then $y' = 0$. If $x \neq 1$ then divide out $x-1$ and multiply through by $2\sqrt{3x+1}$ (at points where f is differentiable we must have $\sqrt{3x+1} \neq 0$) to obtain $3(x-1)+4(3x+1) = 0$, or $15x+1 = 0$. So the tangent line is horizontal at $x = 1, -1/15$.

9. **(a)** $x = -2, -1, 1, 3$

(b) $(-\infty, -2)$, $(-1, 1)$, $(3, +\infty)$

(c) $(-2, -1)$, $(1, 3)$

(d) $g''(x) = f''(x)\sin x + 2f'(x)\cos x - f(x)\sin x$; $g''(0) = 2f'(0)\cos 0 = 2(2)(1) = 4$

11. The equation of such a line has the form $y = mx$. The points (x_0, y_0) which lie on both the line and the parabola and for which the slopes of both curves are equal satisfy $y_0 = mx_0 = x_0^3 - 9x_0^2 - 16x_0$, so that $m = x_0^2 - 9x_0 - 16$. By differentiating, the slope is also given by $m = 3x_0^2 - 18x_0 - 16$. Equating, we have $x_0^2 - 9x_0 - 16 = 3x_0^2 - 18x_0 - 16$, or $2x_0^2 - 9x_0 = 0$. The root $x_0 = 0$ corresponds to $m = -16, y_0 = 0$ and the root $x_0 = 9/2$ corresponds to $m = -145/4, y_0 = -1305/8$. So the line $y = -16x$ is tangent to the curve at the point $(0, 0)$, and the line $y = -145x/4$ is tangent to the curve at the point $(9/2, -1305/8)$.

13. The line $y - x = 2$ has slope $m_1 = 1$ so we set $m_2 = \dfrac{d}{dx}(3x - \tan x) = 3 - \sec^2 x = 1$, or $\sec^2 x = 2$, $\sec x = \pm\sqrt{2}$ so $x = n\pi \pm \pi/4$ where $n = 0, \pm 1, \pm 2, \ldots$.

15. The slope of the tangent line is the derivative

$y' = 2x\Big|_{x=\frac{1}{2}(a+b)} = a + b$. The slope of the secant is

$\dfrac{a^2 - b^2}{a - b} = a + b$, so they are equal.

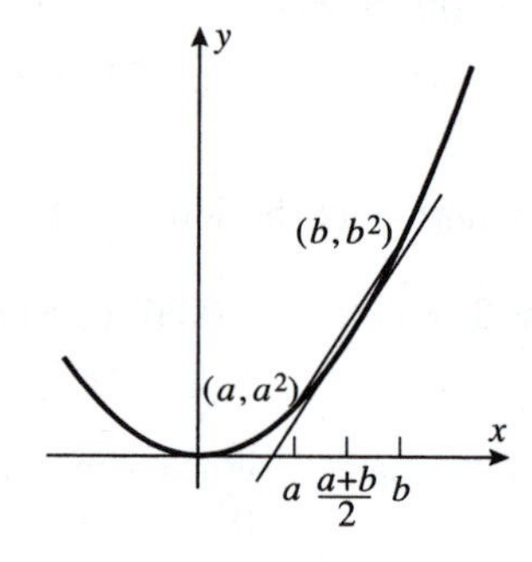

17. **(a)** $\Delta x = 1.5 - 2 = -0.5$; $dy = \dfrac{-1}{(x-1)^2}\Delta x = \dfrac{-1}{(2-1)^2}(-0.5) = 0.5$; and

$\Delta y = \dfrac{1}{(1.5-1)} - \dfrac{1}{(2-1)} = 2 - 1 = 1.$

(b) $\Delta x = 0 - (-\pi/4) = \pi/4$; $dy = \left(\sec^2(-\pi/4)\right)(\pi/4) = \pi/2$; and $\Delta y = \tan 0 - \tan(-\pi/4) = 1$.

(c) $\Delta x = 3 - 0 = 3$; $dy = \dfrac{-x}{\sqrt{25 - x^2}} = \dfrac{-0}{\sqrt{25 - (0)^2}}(3) = 0$; and

$\Delta y = \sqrt{25 - 3^2} - \sqrt{25 - 0^2} = 4 - 5 = -1.$

19. **(a)** $\frac{dW}{dt} = 200(t - 15)$; at $t = 5$, $\frac{dW}{dt} = -2000$; the water is running out at the rate of 2000 gal/min.

(b) $\frac{W(5) - W(0)}{5 - 0} = \frac{10000 - 22500}{5} = -2500$; the average rate of flow out is 2500 gal/min.

21. **(a)** $h = 115 \tan \phi$, $dh = 115 \sec^2 \phi \, d\phi$; with $\phi = 51° = \frac{51}{180}\pi$ radians and

$d\phi = \pm 0.5° = \pm 0.5 \left(\frac{\pi}{180}\right)$ radians, $h \pm dh = 115(1.2349) \pm 2.5340 = 142.0135 \pm 2.5340$, so the height lies between 139.48 m and 144.55 m.

(b) If $|dh| \le 5$ then $|d\phi| \le \frac{5}{115} \cos^2 \frac{51}{180}\pi \approx 0.017$ radians, or $|d\phi| \le 0.98°$.

23. **(a)** $f'(x) = 2x$, $f'(1.8) = 3.6$

(b) $f'(x) = (x^2 - 4x)/(x - 2)^2$, $f'(3.5) \approx -0.777778$

25. $f'(2) \approx 2.772589$; $f'(2) = 4 \ln 2$

27. $v_{\text{inst}} = \lim_{h \to 0} \frac{3(h + 1)^{2.5} + 580h - 3}{10h} = 58 + \frac{1}{10} \frac{d}{dx} 3x^{2.5}\Big|_{x=1} = 58 + \frac{1}{10}(2.5)(3)(1)^{1.5} = 58.75$ ft/s

29. Solve $3x^2 - \cos x = 0$ to get $x = \pm 0.535428$.

31. **(a)** $f'(x) = 5x^4$ **(b)** $f'(x) = -1/x^2$ **(c)** $f'(x) = -1/2x^{3/2}$

(d) $f'(x) = -3/(x - 1)^2$ **(e)** $f'(x) = 3x/\sqrt{3x^2 + 5}$ **(f)** $f'(x) = 3 \cos 3x$

33. $f'(x) = \frac{1 - 2\sqrt{x} \sin 2x}{2\sqrt{x}}$

35. $f'(x) = \frac{(1 + x^2) \sec^2 x - 2x \tan x}{(1 + x^2)^2}$

37. $f'(x) = \frac{-2x^5 \sin x - 2x^4 \cos x + 4x^4 + 6x^2 \sin x + 6x - 3x \cos x - 4x \sin x + 4 \cos x - 8}{2x^2 \sqrt{x^4 - 3 + 2(2 - \cos x)^2}}$

39. Differentiating, $(xy' + y) \cos xy = y'$. With $x = \pi/2$ and $y = 1$ this becomes $y' = 0$, so the equation of the tangent line is $y - 1 = 0(x - \pi/2)$ or $y = 1$.

CHAPTER 4

Exponential, Logarithmic, and Inverse Trigonometric Functions

EXERCISE SET 4.1

1. **(a)** $f(g(x)) = 4(x/4) = x$, $g(f(x)) = (4x)/4 = x$, f and g are inverse functions
(b) $f(g(x)) = 3(3x-1)+1 = 9x-2 \neq x$ so f and g are not inverse functions
(c) $f(g(x)) = \sqrt[3]{(x^3+2)-2} = x$, $g(f(x)) = (x-2)+2 = x$, f and g are inverse functions
(d) $f(g(x)) = (x^{1/4})^4 = x$, $g(f(x)) = (x^4)^{1/4} = |x| \neq x$, f and g are not inverse functions

3. **(a)** yes; all outputs (the elements of row two) are distinct
(b) no; $f(1) = f(6)$

5. **(a)** yes **(b)** yes **(c)** no **(d)** yes **(e)** no **(f)** no

7. **(a)** no, the horizontal line test fails
(b) no, the horizontal line test fails
(c) yes, horizontal line test

9. **(a)** f has an inverse because the graph passes the horizontal line test. To compute $f^{-1}(2)$ start at 2 on the y-axis and go to the curve and then down, so $f^{-1}(2) = 8$; similarly, $f^{-1}(-1) = -1$ and $f^{-1}(0) = 0$.
(b) domain of f^{-1} is $[-2, 2]$, range is $[-8, 8]$ **(c)**

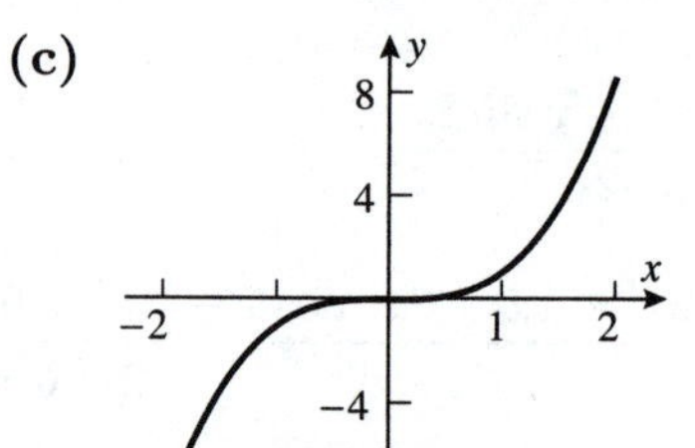

11. **(a)** $f'(x) = 2x + 8$; $f' < 0$ on $(-\infty, -4)$ and $f' > 0$ on $(-4, +\infty)$; not one-to-one
(b) $f'(x) = 10x^4 + 3x^2 + 3 \geq 3 > 0$; $f'(x)$ is positive for all x, so f is one-to-one
(c) $f'(x) = 2 + \cos x \geq 1 > 0$ for all x, so f is one-to-one

13. $y = f^{-1}(x)$, $x = f(y) = y^5$, $y = x^{1/5} = f^{-1}(x)$

15. $y = f^{-1}(x)$, $x = f(y) = 7y - 6$, $y = \dfrac{1}{7}(x+6) = f^{-1}(x)$

17. $y = f^{-1}(x)$, $x = f(y) = 3y^3 - 5$, $y = \sqrt[3]{(x+5)/3} = f^{-1}(x)$

19. $y = f^{-1}(x)$, $x = f(y) = \sqrt[3]{2y-1}$, $y = (x^3+1)/2 = f^{-1}(x)$

21. $y = f^{-1}(x)$, $x = f(y) = 3/y^2$, $y = -\sqrt{3/x} = f^{-1}(x)$

23. $y = f^{-1}(x), x = f(y) = \begin{cases} 5/2 - y, & y < 2 \\ 1/y, & y \geq 2 \end{cases}$, $y = f^{-1}(x) = \begin{cases} 5/2 - x, & x > 1/2 \\ 1/x, & 0 < x \leq 1/2 \end{cases}$

25. $y = f^{-1}(x)$, $x = f(y) = (y+2)^4$ for $y \geq 0$, $y = f^{-1}(x) = x^{1/4} - 2$ for $x \geq 16$

27. $y = f^{-1}(x)$, $x = f(y) = -\sqrt{3-2y}$ for $y \leq 3/2$, $y = f^{-1}(x) = (3 - x^2)/2$ for $x \leq 0$

29. $y = f^{-1}(x)$, $x = f(y) = y - 5y^2$ for $y \geq 1$, $5y^2 - y + x = 0$ for $y \geq 1$,
$y = f^{-1}(x) = (1 + \sqrt{1-20x})/10$ for $x \leq -4$

31. **(a)** $y = f(x) = (6.214 \times 10^{-4})x$ **(b)** $x = f^{-1}(y) = \dfrac{10^4}{6.214}y$
(c) how many meters in y miles

33. **(a)** $f(g(x)) = f(\sqrt{x})$
$= (\sqrt{x})^2 = x, x > 1;$
$g(f(x)) = g(x^2)$
$= \sqrt{x^2} = x, x > 1$

(b)

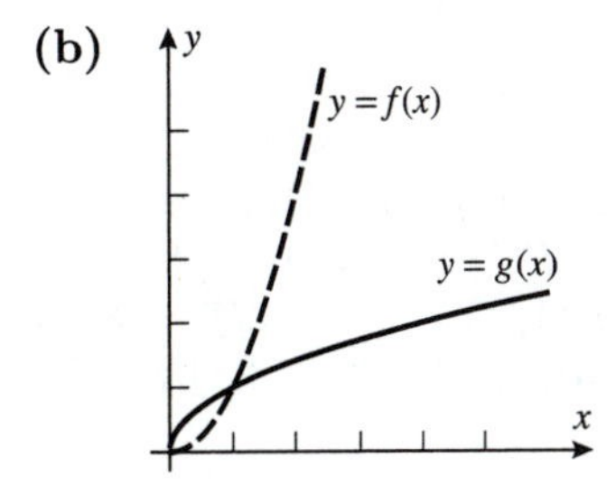

(c) no, because $f(g(x)) = x$ for every x in the domain of g is not satisfied (the domain of g is $x \geq 0$)

35. **(a)** $f(f(x)) = \dfrac{3 - \dfrac{3-x}{1-x}}{1 - \dfrac{3-x}{1-x}} = \dfrac{3 - 3x - 3 + x}{1 - x - 3 + x} = x$ so $f = f^{-1}$

(b) symmetric about the line $y = x$

37. **(a)** $f(x) = x^3 - 3x^2 + 2x = x(x-1)(x-2)$ so $f(0) = f(1) = f(2) = 0$ thus f is not one-to-one.

(b) $f'(x) = 3x^2 - 6x + 2$, $f'(x) = 0$ when $x = \dfrac{6 \pm \sqrt{36-24}}{6} = 1 \pm \sqrt{3}/3$. $f'(x) > 0$ (f is increasing) if $x < 1 - \sqrt{3}/3$, $f'(x) < 0$ (f is decreasing) if $1 - \sqrt{3}/3 < x < 1 + \sqrt{3}/3$, so $f(x)$ takes on values less than $f(1 - \sqrt{3}/3)$ on both sides of $1 - \sqrt{3}/3$ thus $1 - \sqrt{3}/3$ is the largest value of k.

39. if $f^{-1}(x) = 1$, then $x = f(1) = 2(1)^3 + 5(1) + 3 = 10$

41.

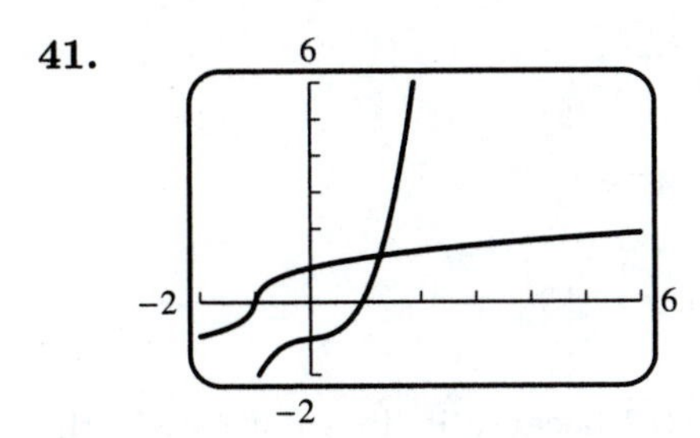

43.

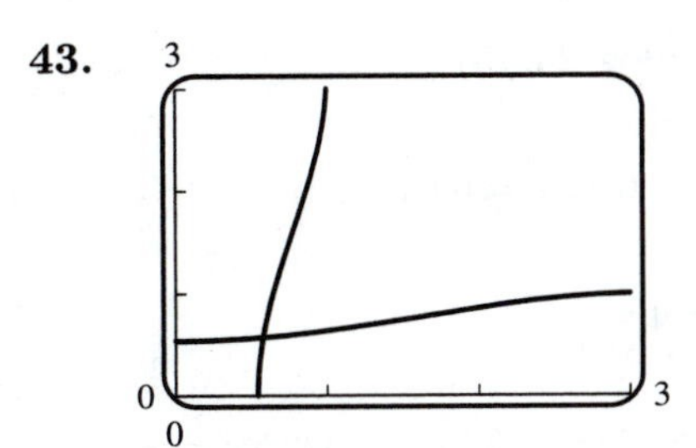

45. $y = f^{-1}(x)$, $x = f(y) = 5y^3 + y - 7$, $\dfrac{dx}{dy} = 15y^2 + 1$, $\dfrac{dy}{dx} = \dfrac{1}{15y^2+1}$;

check: $1 = 15y^2 \dfrac{dy}{dx} + \dfrac{dy}{dx}$, $\dfrac{dy}{dx} = \dfrac{1}{15y^2+1}$

47. $y = f^{-1}(x)$, $x = f(y) = 2y^5 + y^3 + 1$, $\dfrac{dx}{dy} = 10y^4 + 3y^2$, $\dfrac{dy}{dx} = \dfrac{1}{10y^4 + 3y^2}$;
check: $1 = 10y^4 \dfrac{dy}{dx} + 3y^2 \dfrac{dy}{dx}$, $\dfrac{dy}{dx} = \dfrac{1}{10y^4 + 3y^2}$

49. $f(f(x)) = x$ thus $f = f^{-1}$ so the graph is symmetric about $y = x$.

51.

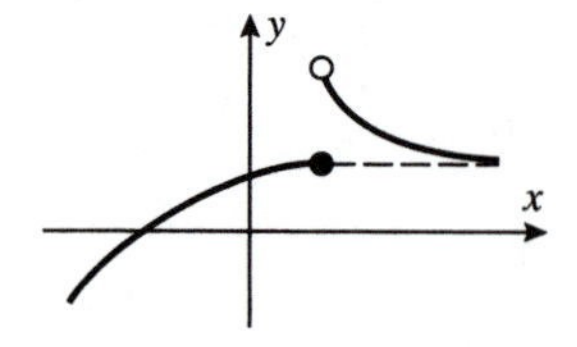

53. $F'(x) = 2f'(2g(x))g'(x)$ so $F'(3) = 2f'(2g(3))g'(3)$. By inspection $f(1) = 3$, so $g(3) = f^{-1}(3) = 1$ and $g'(3) = (f^{-1})'(3) = 1/f'(f^{-1}(3)) = 1/f'(1) = 1/7$ because $f'(x) = 4x^3 + 3x^2$. Thus $F'(3) = 2f'(2)(1/7) = 2(44)(1/7) = 88/7$.
$F(3) = f(2g(3)), g(3) = f^{-1}(3)$; by inspection $f(1) = 3$, so $g(3) = f^{-1}(3) = 1, F(3) = f(2) = 25$.

EXERCISE SET 4.2

1. **(a)** -4 **(b)** 4 **(c)** $1/4$

3. **(a)** 2.9690 **(b)** 0.0341

5. **(a)** $\log_2 16 = \log_2(2^4) = 4$ **(b)** $\log_2\left(\dfrac{1}{32}\right) = \log_2(2^{-5}) = -5$
(c) $\log_4 4 = 1$ **(d)** $\log_9 3 = \log_9(9^{1/2}) = 1/2$

7. **(a)** 1.3655 **(b)** -0.3011

9. **(a)** $2\ln a + \dfrac{1}{2}\ln b + \dfrac{1}{2}\ln c = 2r + s/2 + t/2$ **(b)** $\ln b - 3\ln a - \ln c = s - 3r - t$

11. **(a)** $1 + \log x + \dfrac{1}{2}\log(x-3)$ **(b)** $2\ln|x| + 3\ln\sin x - \dfrac{1}{2}\ln(x^2+1)$

13. $\log \dfrac{2^4(16)}{3} = \log(256/3)$

15. $\ln \dfrac{\sqrt[3]{x}(x+1)^2}{\cos x}$

17. $\sqrt{x} = 10^{-1} = 0.1$, $x = 0.01$

19. $1/x = e^{-2}$, $x = e^2$

21. $2x = 8$, $x = 4$

23. $\log_{10} x = 5$, $x = 10^5$

25. $\ln 2x^2 = \ln 3$, $2x^2 = 3$, $x^2 = 3/2$, $x = \sqrt{3/2}$ (we discard $-\sqrt{3/2}$ because it does not satisfy the original equation)

27. $\ln 5^{-2x} = \ln 3$, $-2x\ln 5 = \ln 3$, $x = -\dfrac{\ln 3}{2\ln 5}$

29. $e^{3x} = 7/2$, $3x = \ln(7/2)$, $x = \dfrac{1}{3}\ln(7/2)$

31. $e^{-x}(x+2)=0$ so $e^{-x}=0$ (impossible) or $x+2=0,\ x=-2$

33. $e^{-2x}-3e^{-x}+2=(e^{-x}-2)(e^{-x}-1)=0$ so $e^{-x}=2,\ x=-\ln 2$ or $e^{-x}=1,\ x=0$

35. **(a)**

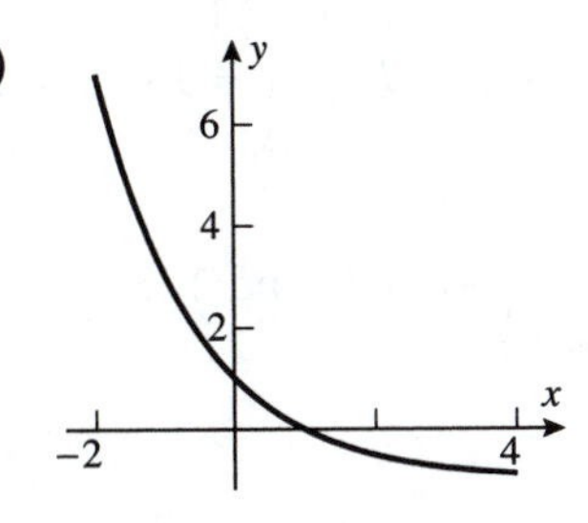

(b)

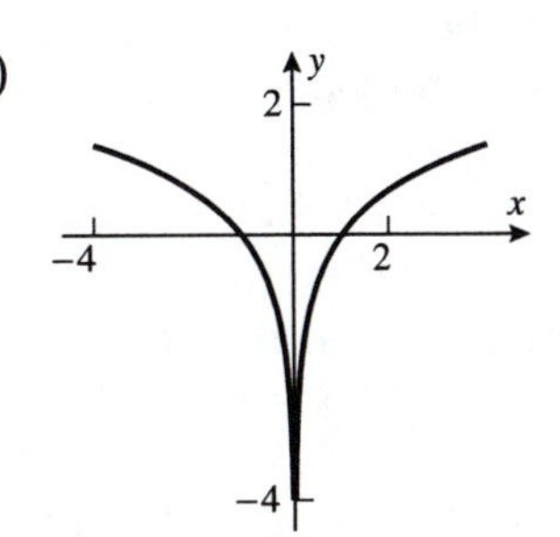

37. $\log_2 7.35=(\log 7.35)/(\log 2)=(\ln 7.35)/(\ln 2)\approx 2.8777$;
$\log_5 0.6=(\log 0.6)/(\log 5)=(\ln 0.6)/(\ln 5)\approx -0.3174$

39.

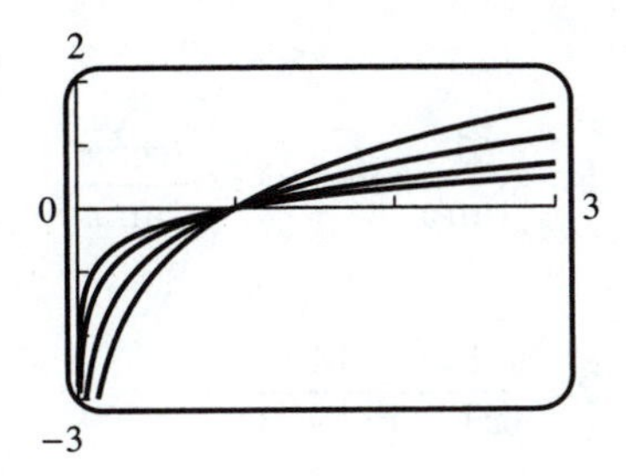

41. **(a)** $x=3.6541, y=1.2958$

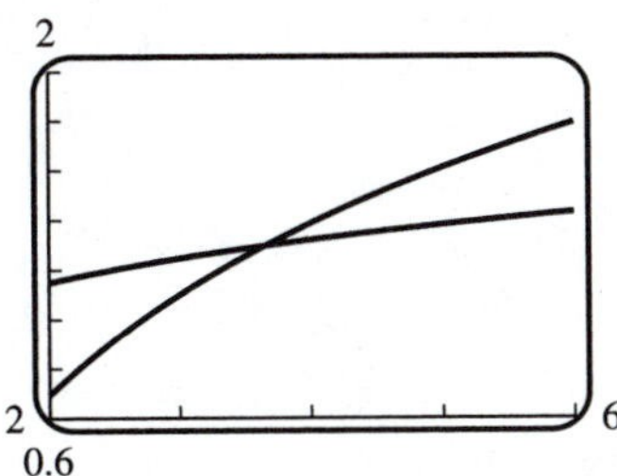

(b) $x\approx 332105.11, y\approx 12.7132$

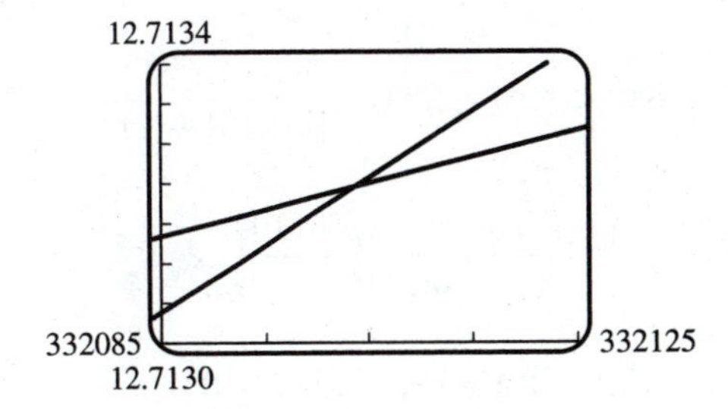

43. **(a)** no, the curve passes through the origin

(b) $y=2^{x/4}$

(c) $y=2^{-x}$

(d) $y=(\sqrt{5})^x$

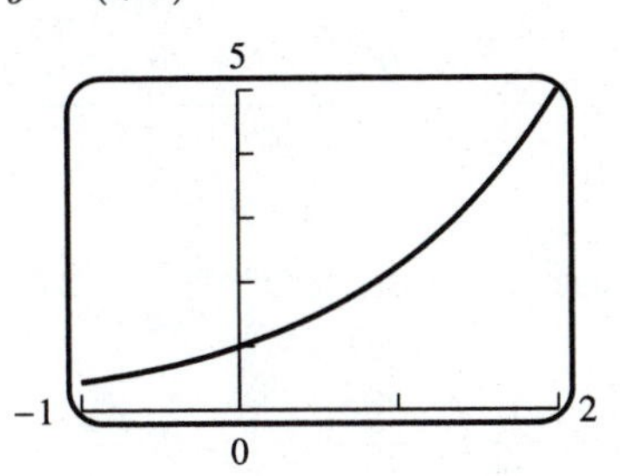

45. $\log(1/2)<0$ so $3\log(1/2)<2\log(1/2)$

47. $75e^{-t/125}=15, t=-125\ln(1/5)=125\ln 5\approx 201$ days.

49. **(a)** 7.4; basic **(b)** 4.2; acidic **(c)** 6.4; acidic **(d)** 5.9; acidic

51. **(a)** 140 dB; damage **(b)** 120 dB; damage
(c) 80 dB; no damage **(d)** 75 dB; no damage

53. Let I_A and I_B be the intensities of the automobile and blender, respectively. Then $\log_{10} I_A/I_0 = 7$ and $\log_{10} I_B/I_0 = 9.3$, $I_A = 10^7 I_0$ and $I_B = 10^{9.3} I_0$, so $I_B/I_A = 10^{2.3} \approx 200$.

55. **(a)** $\log E = 4.4 + 1.5(8.2) = 16.7, E = 10^{16.7} \approx 5 \times 10^{16}$ J

(b) Let M_1 and M_2 be the magnitudes of earthquakes with energies of E and $10E$, respectively. Then $1.5(M_2 - M_1) = \log(10E) - \log E = \log 10 = 1$, $M_2 - M_1 = 1/1.5 = 2/3 \approx 0.67$.

57. If $t = -2x$, then $x = -t/2$ and $\lim_{x\to 0}(1 - 2x)^{1/x} = \lim_{t\to 0}(1 + t)^{-2/t} = \lim_{t\to 0}[(1 + t)^{1/t}]^{-2} = e^{-2}$.

EXERCISE SET 4.3

1. $\dfrac{1}{2x}(2) = 1/x$

3. $2(\ln x)\left(\dfrac{1}{x}\right) = \dfrac{2\ln x}{x}$

5. $\dfrac{1}{\tan x}(\sec^2 x) = \dfrac{\sec^2 x}{\tan x}$

7. $\dfrac{1}{x/(1+x^2)}\left[\dfrac{(1+x^2)(1) - x(2x)}{(1+x^2)^2}\right] = \dfrac{1-x^2}{x(1+x^2)}$

9. $\dfrac{3x^2 - 14x}{x^3 - 7x^2 - 3}$

11. $\dfrac{1}{2}(\ln x)^{-1/2}\left(\dfrac{1}{x}\right) = \dfrac{1}{2x\sqrt{\ln x}}$

13. $-\dfrac{1}{x}\sin(\ln x)$

15. $3x^2 \log_2(3 - 2x) + \dfrac{-2x^3}{(\ln 2)(3 - 2x)}$

17. $\dfrac{2x(1 + \log x) - x/(\ln 10)}{(1 + \log x)^2}$

19. $7e^{7x}$

21. $x^3e^x + 3x^2e^x = x^2e^x(x + 3)$

23.
$$\frac{dy}{dx} = \frac{(e^x + e^{-x})(e^x + e^{-x}) - (e^x - e^{-x})(e^x - e^{-x})}{(e^x + e^{-x})^2}$$
$$= \frac{(e^{2x} + 2 + e^{-2x}) - (e^{2x} - 2 + e^{-2x})}{(e^x + e^{-x})^2} = 4/(e^x + e^{-x})^2$$

25. $(x\sec^2 x + \tan x)e^{x\tan x}$

27. $(1 - 3e^{3x})e^{(x - e^{3x})}$

29. $\dfrac{(x-1)e^{-x}}{1 - xe^{-x}} = \dfrac{x-1}{e^x - x}$

31. $\dfrac{dy}{dx} + \dfrac{1}{xy}\left(x\dfrac{dy}{dx} + y\right) = 0$, $\dfrac{dy}{dx} = -\dfrac{y}{x(y+1)}$

33. $\dfrac{d}{dx}\left[\ln\cos x - \dfrac{1}{2}\ln(4 - 3x^2)\right] = -\tan x + \dfrac{3x}{4 - 3x^2}$

35. $\ln|y| = \ln|x| + \dfrac{1}{3}\ln|1 + x^2|$, $\dfrac{dy}{dx} = x\sqrt[3]{1 + x^2}\left[\dfrac{1}{x} + \dfrac{2x}{3(1 + x^2)}\right]$

37. $\ln|y| = \frac{1}{3}\ln|x^2-8| + \frac{1}{2}\ln|x^3+1| - \ln|x^6-7x+5|$

$$\frac{dy}{dx} = \frac{(x^2-8)^{1/3}\sqrt{x^3+1}}{x^6-7x+5}\left[\frac{2x}{3(x^2-8)} + \frac{3x^2}{2(x^3+1)} - \frac{6x^5-7}{x^6-7x+5}\right]$$

39. $f'(x) = 2^x \ln 2$; $y = 2^x$, $\ln y = x\ln 2$, $\frac{1}{y}y' = \ln 2$, $y' = y\ln 2 = 2^x \ln 2$

41. $f'(x) = \pi^{\sin x}(\ln\pi)\cos x$;

$y = \pi^{\sin x}$, $\ln y = (\sin x)\ln\pi$, $\frac{1}{y}y' = (\ln\pi)\cos x$, $y' = \pi^{\sin x}(\ln\pi)\cos x$

43. $\ln y = (\ln x)\ln(x^3-2x)$, $\frac{1}{y}\frac{dy}{dx} = \frac{3x^2-2}{x^3-2x}\ln x + \frac{1}{x}\ln(x^3-2x)$,

$$\frac{dy}{dx} = (x^3-2x)^{\ln x}\left[\frac{3x^2-2}{x^3-2x}\ln x + \frac{1}{x}\ln(x^3-2x)\right]$$

45. $\ln y = (\tan x)\ln(\ln x)$, $\frac{1}{y}\frac{dy}{dx} = \frac{1}{x\ln x}\tan x + (\sec^2 x)\ln(\ln x)$,

$$\frac{dy}{dx} = (\ln x)^{\tan x}\left[\frac{\tan x}{x\ln x} + (\sec^2 x)\ln(\ln x)\right]$$

47. $f'(x) = ex^{e-1}$

49. **(a)** $\log_x e = \frac{\ln e}{\ln x} = \frac{1}{\ln x}$, $\frac{d}{dx}[\log_x e] = -\frac{1}{x(\ln x)^2}$

(b) $\log_x 2 = \frac{\ln 2}{\ln x}$, $\frac{d}{dx}[\log_x 2] = -\frac{\ln 2}{x(\ln x)^2}$

51. **(a)** $f'(x) = ke^{kx}$, $f''(x) = k^2e^{kx}$, $f'''(x) = k^3e^{kx}, \ldots, f^{(n)}(x) = k^n e^{kx}$

(b) $f'(x) = -ke^{-kx}$, $f''(x) = k^2e^{-kx}$, $f'''(x) = -k^3e^{-kx}, \ldots, f^{(n)}(x) = (-1)^n k^n e^{-kx}$

53.
$$f'(x) = \frac{1}{\sqrt{2\pi}\sigma}\exp\left[-\frac{1}{2}\left(\frac{x-\mu}{\sigma}\right)^2\right]\frac{d}{dx}\left[-\frac{1}{2}\left(\frac{x-\mu}{\sigma}\right)^2\right]$$
$$= \frac{1}{\sqrt{2\pi}\sigma}\exp\left[-\frac{1}{2}\left(\frac{x-\mu}{\sigma}\right)^2\right]\left[-\left(\frac{x-\mu}{\sigma}\right)\left(\frac{1}{\sigma}\right)\right]$$
$$= -\frac{1}{\sqrt{2\pi}\sigma^3}(x-\mu)\exp\left[-\frac{1}{2}\left(\frac{x-\mu}{\sigma}\right)^2\right]$$

55. $y = Ae^{2x} + Be^{-4x}$, $y' = 2Ae^{2x} - 4Be^{-4x}$, $y'' = 4Ae^{2x} + 16Be^{-4x}$ so
$y'' + 2y' - 8y = (4Ae^{2x} + 16Be^{-4x}) + 2(2Ae^{2x} - 4Be^{-4x}) - 8(Ae^{2x} + Be^{-4x}) = 0$

57. **(a)** $f(w) = \ln w$; $f'(1) = \lim_{h\to 0}\frac{\ln(1+h) - \ln 1}{h} = \lim_{h\to 0}\frac{\ln(1+h)}{h} = \left.\frac{1}{w}\right|_{w=1} = 1$

(b) $f(w) = 10^w$; $f'(0) = \lim_{h\to 0}\frac{10^h - 1}{h} = \left.\frac{d}{dw}(10^w)\right|_{w=0} = \left.10^w \ln 10\right|_{w=0} = \ln 10$

EXERCISE SET 4.4

1. **(a)** $-\pi/2$ **(b)** π **(c)** $-\pi/4$ **(d)** 0

3. $\theta = -\pi/3$; $\cos\theta = 1/2$, $\tan\theta = -\sqrt{3}$, $\cot\theta = -1/\sqrt{3}$, $\sec\theta = 2$, $\csc\theta = -2/\sqrt{3}$

5. $\tan\theta = 4/3$, $0 < \theta < \pi/2$; use the triangle shown to get $\sin\theta = 4/5$, $\cos\theta = 3/5$, $\cot\theta = 3/4$, $\sec\theta = 5/3$, $\csc\theta = 5/4$

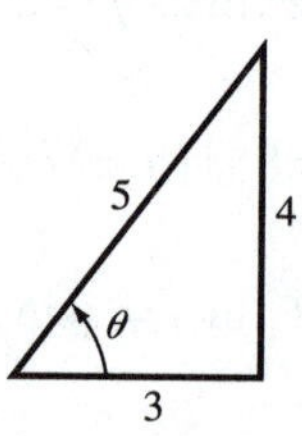

7. **(a)** $\pi/7$
(b) $\sin^{-1}(\sin\pi) = \sin^{-1}(\sin 0) = 0$
(c) $\sin^{-1}(\sin(5\pi/7)) = \sin^{-1}(\sin(2\pi/7)) = 2\pi/7$
(d) Note that $\pi/2 < 630 - 200\pi < \pi$ so
$\sin(630) = \sin(630 - 200\pi) = \sin(\pi - (630 - 200\pi)) = \sin(201\pi - 630)$ where
$0 < 201\pi - 630 < \pi/2$; $\sin^{-1}(\sin 630) = \sin^{-1}(\sin(201\pi - 630)) = 201\pi - 630$.

9. **(a)** $0 \le x \le \pi$ **(b)** $-1 \le x \le 1$
(c) $-\pi/2 < x < \pi/2$ **(d)** $-\infty < x < +\infty$

11. Let $\theta = \cos^{-1}(3/5)$, $\sin 2\theta = 2\sin\theta\cos\theta = 2(4/5)(3/5) = 24/25$

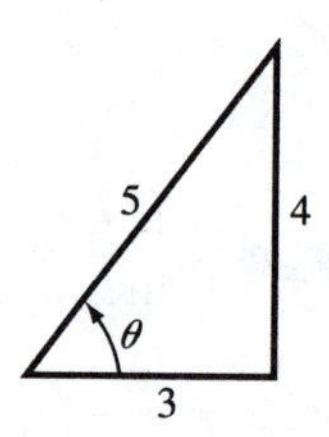

13. **(a)** $\cos(\tan^{-1} x) = \dfrac{1}{\sqrt{1+x^2}}$ **(b)** $\tan(\cos^{-1} x) = \dfrac{\sqrt{1-x^2}}{x}$

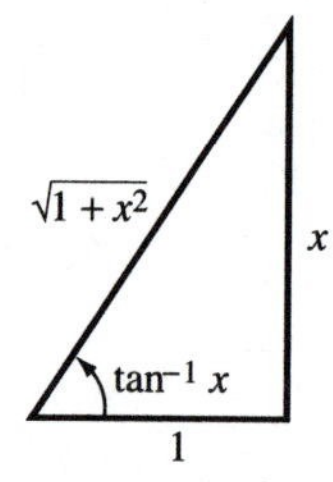

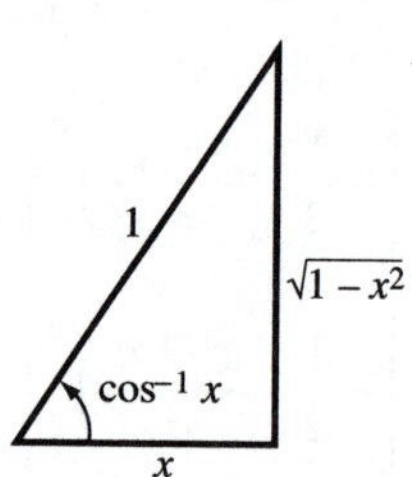

(c) $\sin(\sec^{-1} x) = \dfrac{\sqrt{x^2-1}}{x}$ **(d)** $\cot(\sec^{-1} x) = \dfrac{1}{\sqrt{x^2-1}}$

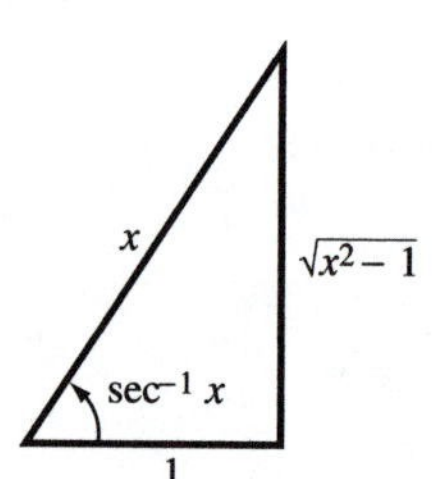

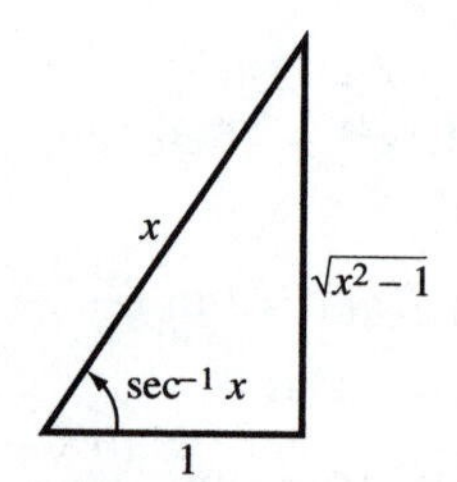

15. (a)

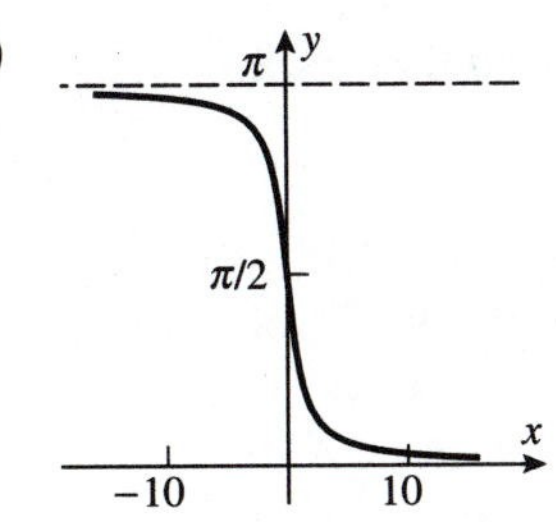

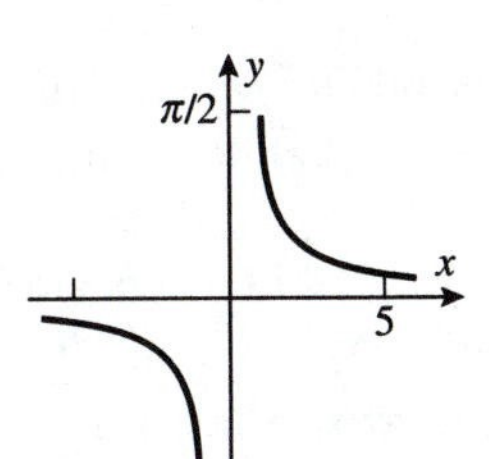

(b) The domain of $\cot^{-1} x$ is $(-\infty, +\infty)$, the range is $(0, \pi)$; the domain of $\csc^{-1} x$ is $(-\infty, -1] \cup [1, +\infty)$, the range is $[-\pi/2, 0) \cup (0, \pi/2]$.

17. (a) 55.0° (b) 33.6° (c) 25.8°

19. (a) By the chain rule, $\dfrac{d}{dx}[\csc^{-1} x] = -\dfrac{1}{x^2}\dfrac{1}{\sqrt{1-(1/x)^2}} = \dfrac{-1}{|x|\sqrt{x^2-1}}$

(b) By the chain rule, $\dfrac{d}{dx}[\csc^{-1} u] = \dfrac{du}{dx}\dfrac{d}{du}[\csc^{-1} u] = \dfrac{-1}{|u|\sqrt{u^2-1}}\dfrac{du}{dx}$

21. (a) $x = \pi + \cos^{-1}(0.85) \approx 3.6964$ rad (b) $\theta = -\cos^{-1}(0.23) \approx -76.7°$

23. (a) $\dfrac{1}{\sqrt{1-x^2/9}}(1/3) = 1/\sqrt{9-x^2}$ (b) $-2/\sqrt{1-(2x+1)^2}$

25. (a) $\dfrac{1}{|x|^7\sqrt{x^{14}-1}}(7x^6) = \dfrac{7}{|x|\sqrt{x^{14}-1}}$ (b) $-1/\sqrt{e^{2x}-1}$

27. (a) $\dfrac{1}{\sqrt{1-1/x^2}}(-1/x^2) = -\dfrac{1}{|x|\sqrt{x^2-1}}$

(b) $\dfrac{\sin x}{\sqrt{1-\cos^2 x}} = \dfrac{\sin x}{|\sin x|} = \begin{cases} 1, & \sin x > 0 \\ -1, & \sin x < 0 \end{cases}$

29. (a) $\dfrac{e^x}{|x|\sqrt{x^2-1}} + e^x \sec^{-1} x$ (b) $\dfrac{3x^2(\sin^{-1} x)^2}{\sqrt{1-x^2}} + 2x(\sin^{-1} x)^3$

31. $x^3 + x\tan^{-1} y = e^y$, $3x^2 + \dfrac{x}{1+y^2}y' + \tan^{-1} y = e^y y'$, $y' = \dfrac{(3x^2+\tan^{-1} y)(1+y^2)}{(1+y^2)e^y - x}$

33. (a)

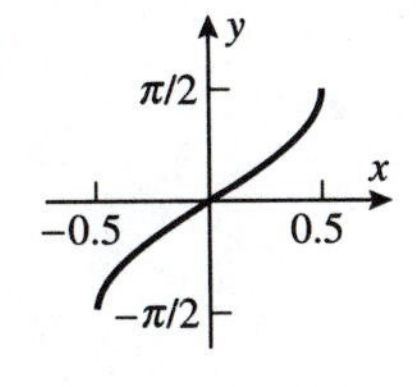

(b)

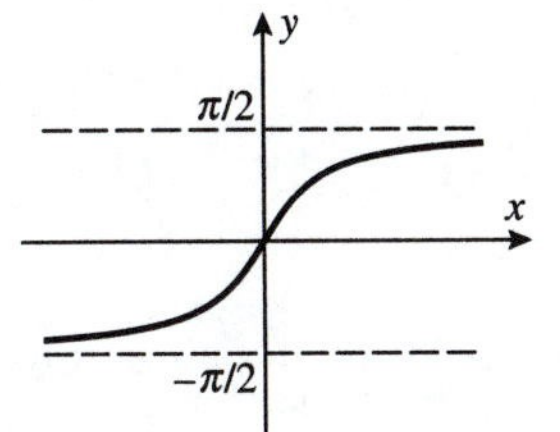

35. (b) $\theta = \sin^{-1}\dfrac{R}{R+h} = \sin^{-1}\dfrac{6378}{16,378} \approx 23°$

37. $\sin 2\theta = gR/v^2 = (9.8)(18)/(14)^2 = 0.9$, $2\theta = \sin^{-1}(0.9)$ or $2\theta = 180° - \sin^{-1}(0.9)$ so $\theta = \frac{1}{2}\sin^{-1}(0.9) \approx 32°$ or $\theta = 90° - \frac{1}{2}\sin^{-1}(0.9) \approx 58°$. The ball will have a lower parabolic trajectory for $\theta = 32°$ and hence will result in the shorter time of flight.

39. $y = 0$ when $x^2 = 6000v^2/g$, $x = 10v\sqrt{60/g} = 1000\sqrt{30}$ for $v = 400$ and $g = 32$;
$\tan\theta = 3000/x = 3/\sqrt{30}$, $\theta = \tan^{-1}(3/\sqrt{30}) \approx 29°$.

41. **(a)** Let $\theta = \sin^{-1}(-x)$ then $\sin\theta = -x$, $-\pi/2 \le \theta \le \pi/2$. But $\sin(-\theta) = -\sin\theta$ and $-\pi/2 \le -\theta \le \pi/2$ so $\sin(-\theta) = -(-x) = x$, $-\theta = \sin^{-1} x$, $\theta = -\sin^{-1} x$.

(b) proof is similar to that in Part (a)

43. **(a)** $\sin^{-1} x = \tan^{-1} \dfrac{x}{\sqrt{1-x^2}}$ (see figure)

(b) $\sin^{-1} x + \cos^{-1} x = \pi/2$; $\cos^{-1} x = \pi/2 - \sin^{-1} x = \pi/2 - \tan^{-1} \dfrac{x}{\sqrt{1-x^2}}$

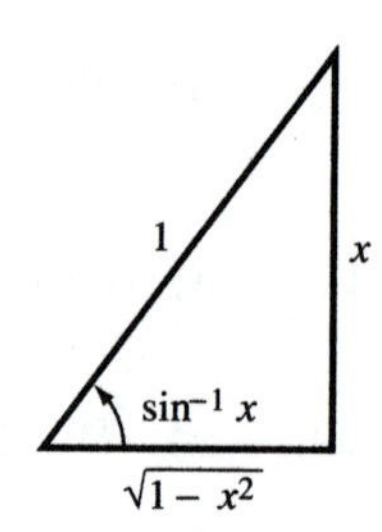

45. **(a)** $\tan^{-1}\dfrac{1}{2} + \tan^{-1}\dfrac{1}{3} = \tan^{-1}\dfrac{1/2 + 1/3}{1 - (1/2)(1/3)} = \tan^{-1} 1 = \pi/4$

(b) $2\tan^{-1}\dfrac{1}{3} = \tan^{-1}\dfrac{1}{3} + \tan^{-1}\dfrac{1}{3} = \tan^{-1}\dfrac{1/3 + 1/3}{1 - (1/3)(1/3)} = \tan^{-1}\dfrac{3}{4}$,

$2\tan^{-1}\dfrac{1}{3} + \tan^{-1}\dfrac{1}{7} = \tan^{-1}\dfrac{3}{4} + \tan^{-1}\dfrac{1}{7} = \tan^{-1}\dfrac{3/4 + 1/7}{1 - (3/4)(1/7)} = \tan^{-1} 1 = \pi/4$

EXERCISE SET 4.5

1. **(a)** $\lim_{x\to 2} \dfrac{x^2 - 4}{x^2 + 2x - 8} = \lim_{x\to 2} \dfrac{(x-2)(x+2)}{(x+4)(x-2)} = \lim_{x\to 2} \dfrac{x+2}{x+4} = \dfrac{2}{3}$

(b) $\lim_{x\to+\infty} \dfrac{2x-5}{3x+7} = \dfrac{2 - \lim_{x\to+\infty} \frac{5}{x}}{3 + \lim_{x\to+\infty} \frac{7}{x}} = \dfrac{2}{3}$

3. $\lim_{x\to 1} \dfrac{1/x}{1} = 1$

5. $\lim_{x\to 0} \dfrac{e^x}{\cos x} = 1$

7. $\lim_{\theta\to 0} \dfrac{\sec^2\theta}{1} = 1$

9. $\lim_{x\to\pi^+} \dfrac{\cos x}{1} = -1$

11. $\lim_{x\to+\infty} \dfrac{1/x}{1} = 0$

13. $\lim_{x\to 0^+} \dfrac{-\csc^2 x}{1/x} = \lim_{x\to 0^+} \dfrac{-x}{\sin^2 x} = \lim_{x\to 0^+} \dfrac{-1}{2\sin x\cos x} = -\infty$

15. $\lim_{x\to+\infty} \dfrac{100x^{99}}{e^x} = \lim_{x\to+\infty} \dfrac{(100)(99)x^{98}}{e^x} = \cdots = \lim_{x\to+\infty} \dfrac{(100)(99)(98)\cdots(1)}{e^x} = 0$

17. $\lim_{x\to 0} \dfrac{2/\sqrt{1-4x^2}}{1} = 2$

19. $\lim_{x\to+\infty} xe^{-x} = \lim_{x\to+\infty} \dfrac{x}{e^x} = \lim_{x\to+\infty} \dfrac{1}{e^x} = 0$

21. $\lim_{x\to+\infty} x\sin(\pi/x) = \lim_{x\to+\infty}\dfrac{\sin(\pi/x)}{1/x} = \lim_{x\to+\infty}\dfrac{(-\pi/x^2)\cos(\pi/x)}{-1/x^2} = \lim_{x\to+\infty}\pi\cos(\pi/x) = \pi$

23. $\lim_{x\to(\pi/2)^-}\sec 3x\cos 5x = \lim_{x\to(\pi/2)^-}\dfrac{\cos 5x}{\cos 3x} = \lim_{x\to(\pi/2)^-}\dfrac{-5\sin 5x}{-3\sin 3x} = \dfrac{-5(+1)}{(-3)(-1)} = -\dfrac{5}{3}$

25. $y = (1-3/x)^x$, $\lim_{x\to+\infty}\ln y = \lim_{x\to+\infty}\dfrac{\ln(1-3/x)}{1/x} = \lim_{x\to+\infty}\dfrac{-3}{1-3/x} = -3$, $\lim_{x\to+\infty} y = e^{-3}$

27. $y = (e^x + x)^{1/x}$, $\lim_{x\to0}\ln y = \lim_{x\to0}\dfrac{\ln(e^x+x)}{x} = \lim_{x\to0}\dfrac{e^x+1}{e^x+x} = 2$, $\lim_{x\to0} y = e^2$

29. $y = (2-x)^{\tan(\pi x/2)}$, $\lim_{x\to1}\ln y = \lim_{x\to1}\dfrac{\ln(2-x)}{\cot(\pi x/2)} = \lim_{x\to1}\dfrac{2\sin^2(\pi x/2)}{\pi(2-x)} = 2/\pi$, $\lim_{x\to1} y = e^{2/\pi}$

31. $\lim_{x\to0}\left(\dfrac{1}{\sin x} - \dfrac{1}{x}\right) = \lim_{x\to0}\dfrac{x-\sin x}{x\sin x} = \lim_{x\to0}\dfrac{1-\cos x}{x\cos x + \sin x} = \lim_{x\to0}\dfrac{\sin x}{2\cos x - x\sin x} = 0$

33. $\lim_{x\to+\infty}\dfrac{(x^2+x)-x^2}{\sqrt{x^2+x}+x} = \lim_{x\to+\infty}\dfrac{x}{\sqrt{x^2+x}+x} = \lim_{x\to+\infty}\dfrac{1}{\sqrt{1+1/x}+1} = 1/2$

35. $\lim_{x\to+\infty}[x - \ln(x^2+1)] = \lim_{x\to+\infty}[\ln e^x - \ln(x^2+1)] = \lim_{x\to+\infty}\ln\dfrac{e^x}{x^2+1}$,

$\lim_{x\to+\infty}\dfrac{e^x}{x^2+1} = \lim_{x\to+\infty}\dfrac{e^x}{2x} = \lim_{x\to+\infty}\dfrac{e^x}{2} = +\infty$ so $\lim_{x\to+\infty}[x-\ln(x^2+1)] = +\infty$

39. **(a)** L'Hôpital's Rule does not apply to the problem $\lim_{x\to1}\dfrac{3x^2-2x+1}{3x^2-2x}$ because it is not a $\dfrac{0}{0}$ form.

(b) $\lim_{x\to1}\dfrac{3x^2-2x+1}{3x^2-2x} = 2$

41. $\lim_{x\to+\infty}\dfrac{1/(x\ln x)}{1/(2\sqrt{x})} = \lim_{x\to+\infty}\dfrac{2}{\sqrt{x}\ln x} = 0$

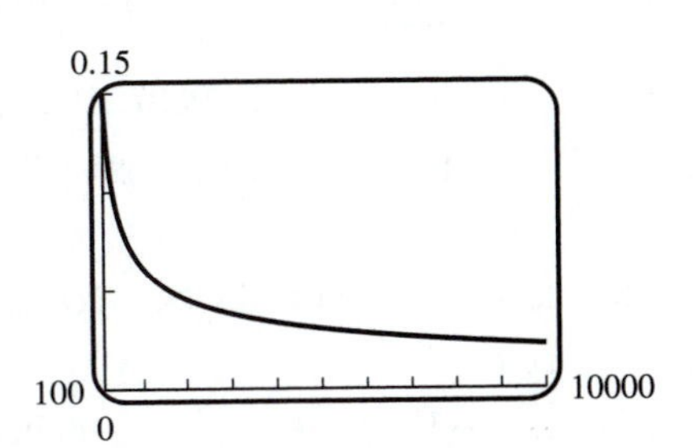

43. $y = (\sin x)^{3/\ln x}$,

$\lim_{x\to0^+}\ln y = \lim_{x\to0^+}\dfrac{3\ln\sin x}{\ln x} = \lim_{x\to0^+}(3\cos x)\dfrac{x}{\sin x} = 3$,

$\lim_{x\to0^+} y = e^3$

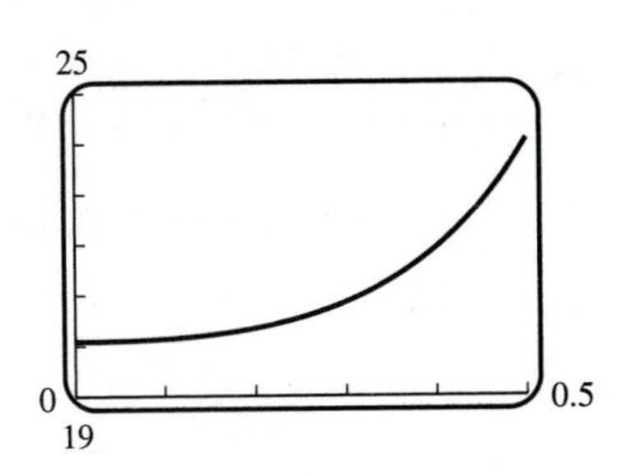

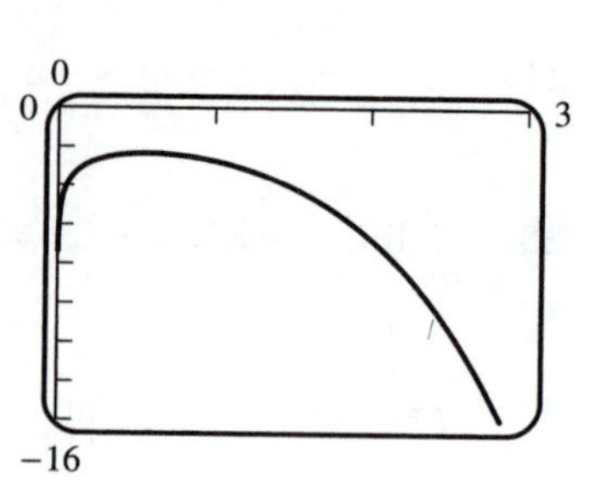

45. $\ln x - e^x = \ln x - \dfrac{1}{e^{-x}} = \dfrac{e^{-x}\ln x - 1}{e^{-x}}$;

$\lim\limits_{x\to+\infty} e^{-x}\ln x = \lim\limits_{x\to+\infty} \dfrac{\ln x}{e^x} = \lim\limits_{x\to+\infty} \dfrac{1/x}{e^x} = 0$ by L'Hôpital's Rule,

so $\lim\limits_{x\to+\infty} [\ln x - e^x] = \lim\limits_{x\to+\infty} \dfrac{e^{-x}\ln x - 1}{e^{-x}} = -\infty$

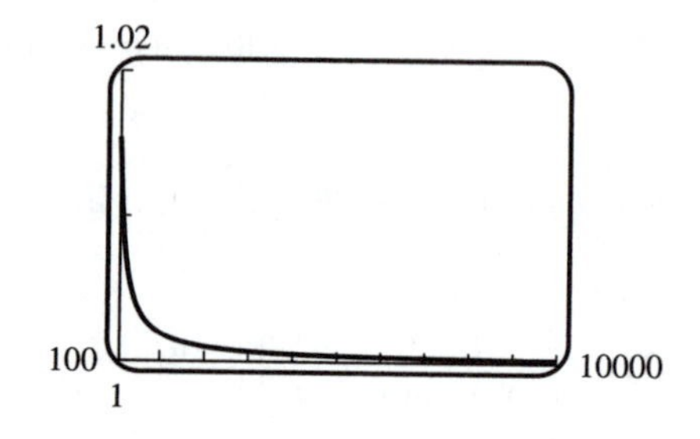

47. $y = (\ln x)^{1/x}$,

$\lim\limits_{x\to+\infty} \ln y = \lim\limits_{x\to+\infty} \dfrac{\ln(\ln x)}{x} = \lim\limits_{x\to+\infty} \dfrac{1}{x\ln x} = 0$;

$\lim\limits_{x\to+\infty} y = 1$, $y = 1$ is the horizontal asymptote

49. (a) 0 (b) $+\infty$ (c) 0 (d) $-\infty$ (e) $+\infty$ (f) $-\infty$

51. $\lim\limits_{x\to+\infty} \dfrac{1 + 2\cos 2x}{1}$ does not exist, nor is it $\pm\infty$; $\lim\limits_{x\to+\infty} \dfrac{x + \sin 2x}{x} = \lim\limits_{x\to+\infty} \left(1 + \dfrac{\sin 2x}{x}\right) = 1$

53. $\lim\limits_{x\to+\infty} (2 + x\cos 2x + \sin 2x)$ does not exist, nor is it $\pm\infty$; $\lim\limits_{x\to+\infty} \dfrac{x(2 + \sin 2x)}{x+1} = \lim\limits_{x\to+\infty} \dfrac{2 + \sin 2x}{1 + 1/x}$, which does not exist because $\sin 2x$ oscillates between -1 and 1 as $x \to +\infty$

55. $\lim\limits_{R\to 0^+} \dfrac{\frac{Vt}{L}e^{-Rt/L}}{1} = \dfrac{Vt}{L}$

57. (b) $\lim\limits_{x\to+\infty} x(k^{1/x} - 1) = \lim\limits_{t\to 0^+} \dfrac{k^t - 1}{t} = \lim\limits_{t\to 0^+} \dfrac{(\ln k)k^t}{1} = \ln k$

(c) $\ln 0.3 = -1.20397$, $1024\left(\sqrt[1024]{0.3} - 1\right) = -1.20327$;
$\ln 2 = 0.69315$, $1024\left(\sqrt[1024]{2} - 1\right) = 0.69338$

59. If $k \neq -1$ then $\lim\limits_{x\to 0}(k + \cos \ell x) = k + 1 \neq 0$, so $\lim\limits_{x\to 0} \dfrac{k + \cos \ell x}{x^2} = \pm\infty$. Hence $k = -1$, and by the rule

$\lim\limits_{x\to 0} \dfrac{-1 + \cos \ell x}{x^2} = \lim\limits_{x\to 0} \dfrac{-\ell \sin \ell x}{2x} = \lim\limits_{x\to 0} \dfrac{-\ell^2 \cos \ell x}{2} = -\dfrac{\ell^2}{2} = -4$ if $\ell = \pm 2\sqrt{2}$.

61. $\lim\limits_{x\to 0^+} \dfrac{\sin(1/x)}{(\sin x)/x}$, $\lim\limits_{x\to 0^+} \dfrac{\sin x}{x} = 1$ but $\lim\limits_{x\to 0^+} \sin(1/x)$ does not exist because $\sin(1/x)$ oscillates between -1 and 1 as $x \to +\infty$, so $\lim\limits_{x\to 0^+} \dfrac{x\sin(1/x)}{\sin x}$ does not exist.

CHAPTER 4 SUPPLEMENTARY EXERCISES

1. **(a)** $f(g(x)) = x$ for all x in the domain of g, and $g(f(x)) = x$ for all x in the domain of f.

(b) They are reflections of each other through the line $y = x$.

(c) The domain of one is the range of the other and vice versa.

(d) The equation $y = f(x)$ can always be solved for x as a function of y. Functions with no inverses include $y = x^2$, $y = \sin x$.

(e) Yes, g is continuous; this is evident from the statement about the graphs in Part (b) above.

(f) Yes, g must be differentiable (where $f' \neq 0$); this can be inferred from the graphs. Note that if $f' = 0$ at a point then g' cannot exist (infinite slope).

3. **(a)** $x = f(y) = 8y^3 - 1$; $y = f^{-1}(x) = \left(\dfrac{x+1}{8}\right)^{1/3} = \dfrac{1}{2}(x+1)^{1/3}$

(b) $f(x) = (x-1)^2$; f does not have an inverse because f is not one-to-one, for example $f(0) = f(2) = 1$.

(c) $x = f(y) = (e^y)^2 + 1$; $y = f^{-1}(x) = \ln\sqrt{x-1} = \frac{1}{2}\ln(x-1)$

(d) $x = f(y) = \dfrac{y+2}{y-1}$; $y = f^{-1}(x) = \dfrac{x+2}{x-1}$

5. $3\ln\left(e^{2x}(e^x)^3\right) + 2\exp(\ln 1) = 3\ln e^{2x} + 3\ln(e^x)^3 + 2\cdot 1 = 3(2x) + (3\cdot 3)x + 2 = 15x + 2$

7. **(a)** $f'(x) = -3/(x+1)^2$. If $x = f(y) = 3/(y+1)$ then $y = f^{-1}(x) = (3/x) - 1$, so $\dfrac{d}{dx}f^{-1}(x) = -\dfrac{3}{x^2}$; and $\dfrac{1}{f'(f^{-1}(x))} = -\dfrac{(f^{-1}(x)+1)^2}{3} = -\dfrac{(3/x)^2}{3} = -\dfrac{3}{x^2}$.

(b) $f(x) = e^{x/2}$, $f'(x) = \frac{1}{2}e^{x/2}$. If $x = f(y) = e^{y/2}$ then $y = f^{-1}(x) = 2\ln x$, so $\dfrac{d}{dx}f^{-1}(x) = \dfrac{2}{x}$; and $\dfrac{1}{f'(f^{-1}(x))} = 2e^{-f^{-1}(x)/2} = 2e^{-\ln x} = 2x^{-1} = \dfrac{2}{x}$

9. **(a)**

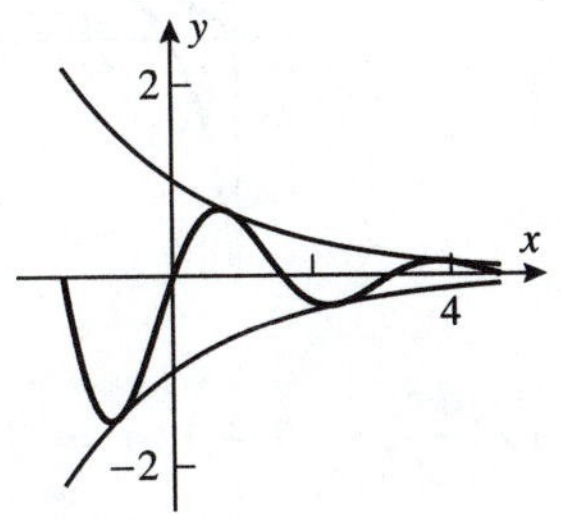

(b) The curve $y = e^{-x/2}\sin 2x$ has x-intercepts at $x = -\pi/2, 0, \pi/2, \pi, 3\pi/2$. It intersects the curve $y = e^{-x/2}$ at $x = \pi/4, 5\pi/4$, and it intersects the curve $y = -e^{-x/2}$ at $x = -\pi/4, 3\pi/4$.

11. **(a)** The function $\ln x - x^{0.2}$ is negative at $x = 1$ and positive at $x = 4$, so it must be zero in between (IVT).

(b) $x = 3.654$

13. $\ln y = \ln 5000 + 1.07x$; $\dfrac{dy/dx}{y} = 1.07$, or $\dfrac{dy}{dx} = 1.07y$

15. **(a)** $y = x^3 + 1$ so $y' = 3x^2$.

(b) $y' = \dfrac{abe^{-x}}{(1+be^{-x})^2}$

(c) $y = \dfrac{1}{2}\ln x + \dfrac{1}{3}\ln(x+1) - \ln\sin x + \ln\cos x$, so

$$y' = \frac{1}{2x} + \frac{1}{3(x+1)} - \frac{\cos x}{\sin x} - \frac{\sin x}{\cos x} = \frac{5x+3}{6x(x+1)} - \cot x - \tan x.$$

(d) $\ln y = \dfrac{\ln(1+x)}{x}$, $\dfrac{y'}{y} = \dfrac{x/(1+x) - \ln(1+x)}{x^2} = \dfrac{1}{x(1+x)} - \dfrac{\ln(1+x)}{x^2}$,

$$\frac{dy}{dx} = \frac{1}{x}(1+x)^{(1/x)-1} - \frac{(1+x)^{(1/x)}}{x^2}\ln(1+x)$$

(e) $\ln y = e^x \ln x$, $\dfrac{y'}{y} = e^x\left(\dfrac{1}{x} + \ln x\right)$, $\dfrac{dy}{dx} = x^{e^x}e^x\left(\dfrac{1}{x} + \ln x\right) = e^x\left[x^{e^x-1} + x^{e^x}\ln x\right]$

(f) $y = \ln\dfrac{(1+e^x+e^{2x})}{(1-e^x)(1+e^x+e^{2x})} = -\ln(1-e^x)$, $\dfrac{dy}{dx} = \dfrac{e^x}{1-e^x}$

17. $\sin(\tan^{-1} x) = x/\sqrt{1+x^2}$ and $\cos(\tan^{-1} x) = 1/\sqrt{1+x^2}$, and $y' = \dfrac{1}{1+x^2}$, $y'' = \dfrac{-2x}{(1+x^2)^2}$, hence

$$y'' + 2\sin y\cos^3 y = \frac{-2x}{(1+x^2)^2} + 2\frac{x}{\sqrt{1+x^2}}\frac{1}{(1+x^2)^{3/2}} = 0.$$

19. Set $y = \log_b x$ and solve $y' = 1$: $y' = \dfrac{1}{x\ln b} = 1$ so $x = \dfrac{1}{\ln b}$. The curves intersect when (x, x) lies on the graph of $y = \log_b x$, so $x = \log_b x$. From Formula (9), Section 4.2, $\log_b x = \dfrac{\ln x}{\ln b}$ from which $\ln x = 1$, $x = e$, $\ln b = 1/e$, $b = e^{1/e} \approx 1.4447$.

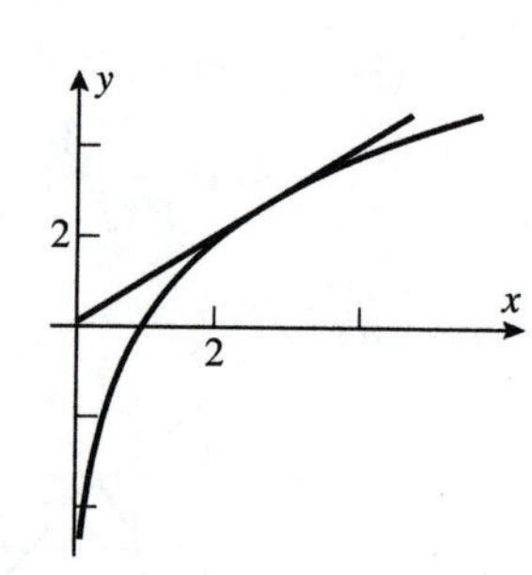

21. Solve $\dfrac{dy}{dt} = 3\dfrac{dx}{dt}$ given $y = x\ln x$. Then $\dfrac{dy}{dt} = \dfrac{dy}{dx}\dfrac{dx}{dt} = (1+\ln x)\dfrac{dx}{dt}$, so $1 + \ln x = 3$, $\ln x = 2$, $x = e^2$.

23. $\dfrac{dk}{dT} = k_0 \exp\left[-\dfrac{q(T-T_0)}{2T_0T}\right]\left(-\dfrac{q}{2T^2}\right) = -\dfrac{qk_0}{2T^2}\exp\left[-\dfrac{q(T-T_0)}{2T_0T}\right]$

25. (a)

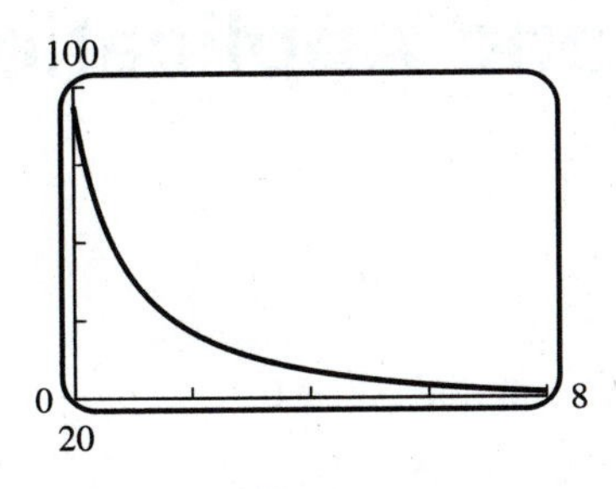

(b) as t tends to $+\infty$, the population tends to 19
$$\lim_{t\to+\infty} P(t) = \lim_{t\to+\infty} \frac{95}{5-4e^{-t/4}} = \frac{95}{5-4\lim\limits_{t\to+\infty} e^{-t/4}} = \frac{95}{5} = 19$$

(c) the rate of population growth tends to zero

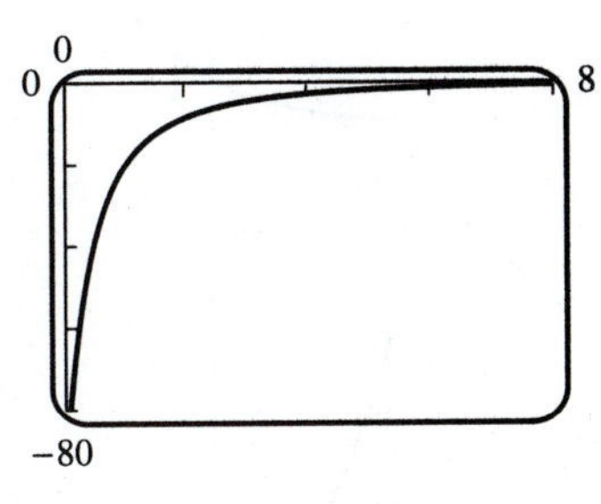

27. (b)

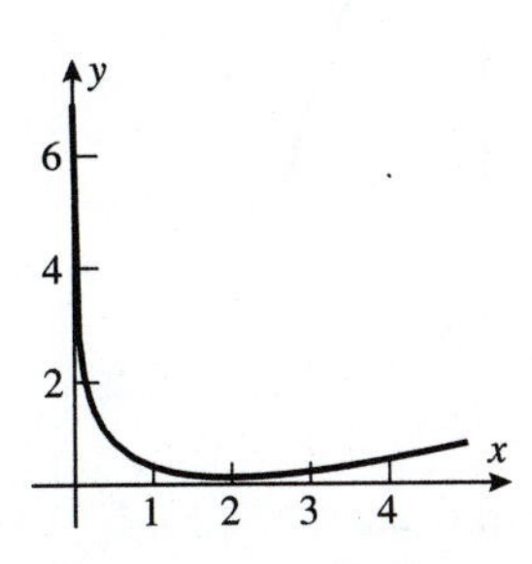

(c) $\dfrac{dy}{dx} = \dfrac{1}{2} - \dfrac{1}{x}$ so $\dfrac{dy}{dx} < 0$ at $x = 1$ and $\dfrac{dy}{dx} > 0$ at $x = e$

(d) The slope is a continuous function which goes from a negative value to a positive value; therefore it must take the value zero in between, by the Intermediate Value Theorem.

(e) $\dfrac{dy}{dx} = 0$ when $x = 2$

29. (a) when the limit takes the form $0/0$ or ∞/∞

(b) Not necessarily; only if $\lim\limits_{x\to a} f(x) = 0$. Consider $g(x) = x$; $\lim\limits_{x\to 0} g(x) = 0$. For $f(x)$ choose $\cos x$, x^2, and $|x|^{1/2}$. Then: $\lim\limits_{x\to 0} \dfrac{\cos x}{x}$ does not exist, $\lim\limits_{x\to 0} \dfrac{x^2}{x} = 0$, and $\lim\limits_{x\to 0} \dfrac{|x|^{1/2}}{x^2} = +\infty$.

CHAPTER 5
The Derivative in Graphing and Applications

EXERCISE SET 5.1

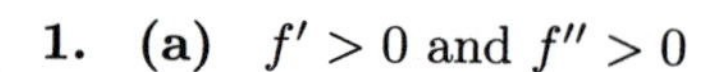

1. **(a)** $f' > 0$ and $f'' > 0$

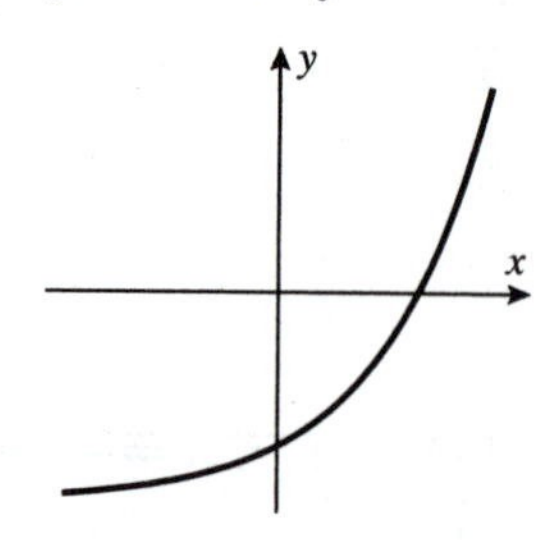

(b) $f' > 0$ and $f'' < 0$

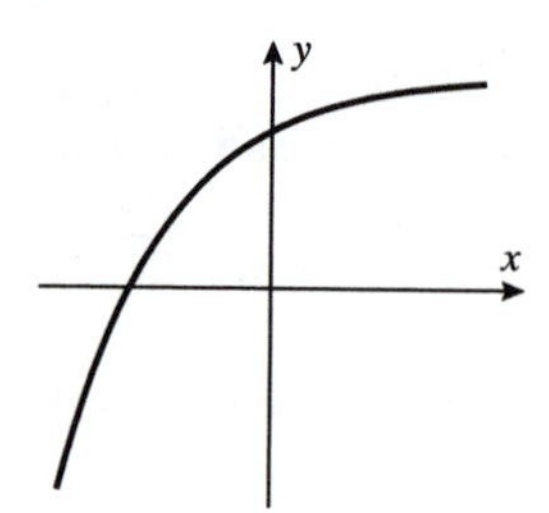

(c) $f' < 0$ and $f'' > 0$

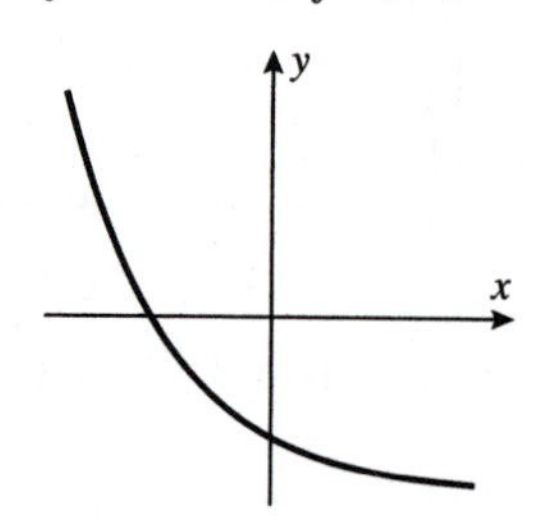

(d) $f' < 0$ and $f'' < 0$

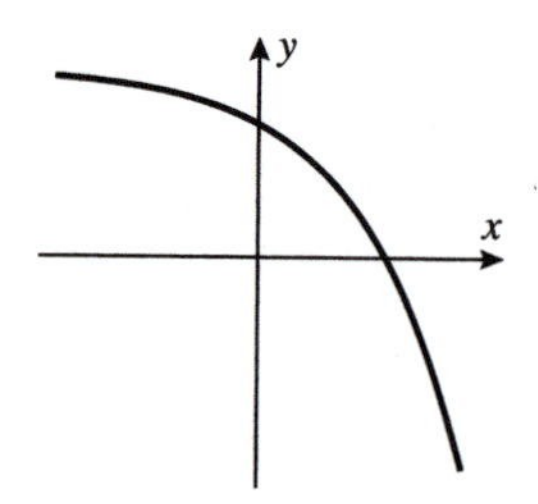

3. A: $dy/dx < 0$, $d^2y/dx^2 > 0$
B: $dy/dx > 0$, $d^2y/dx^2 < 0$
C: $dy/dx < 0$, $d^2y/dx^2 < 0$

5. An inflection point occurs when f'' changes sign: at $x = -1, 0, 1$ and 2.

7. **(a)** $[4, 6]$ **(b)** $[1, 4]$ and $[6, 7]$ **(c)** $(1, 2)$ and $(3, 5)$
(d) $(2, 3)$ and $(5, 7)$ **(e)** $x = 2, 3, 5$

9. **(a)** f is increasing on $[1, 3]$ **(b)** f is decreasing on $(-\infty, 1], [3, +\infty]$
(c) f is concave up on $(-\infty, 2), (4, +\infty)$ **(d)** f is concave down on $(2, 4)$
(e) points of inflection at $x = 2, 4$

11. $f'(x) = 2x - 5$
$f''(x) = 2$
(a) $[5/2, +\infty)$ **(b)** $(-\infty, 5/2]$
(c) $(-\infty, +\infty)$ **(d)** none
(e) none

13. $f'(x) = 3(x + 2)^2$
$f''(x) = 6(x + 2)$
(a) $(-\infty, +\infty)$ **(b)** none
(c) $(-2, +\infty)$ **(d)** $(-\infty, -2)$
(e) -2

15. $f'(x) = 12x^2(x - 1)$
$f''(x) = 36x(x - 2/3)$
(a) $[1, +\infty)$ **(b)** $(-\infty, 1]$
(c) $(-\infty, 0)$, $(2/3, +\infty)$ **(d)** $(0, 2/3)$
(e) $0, 2/3$

17. $f'(x) = \dfrac{4x}{(x^2 + 2)^2}$ $\quad f''(x) = -4\dfrac{3x^2 - 2}{(x^2 + 2)^3}$

(a) $[0, +\infty)$ **(b)** $(-\infty, 0]$ **(c)** $(-\sqrt{2/3}, \sqrt{2/3})$
(d) $(-\infty, -\sqrt{2/3})$, $(\sqrt{2/3}, +\infty)$ **(e)** $-\sqrt{2/3}, \sqrt{2/3}$

19. $f'(x) = \frac{1}{3}(x+2)^{-2/3}$
$f''(x) = -\frac{2}{9}(x+2)^{-5/3}$

(a) $(-\infty, +\infty)$ **(b)** none
(c) $(-\infty, -2)$ **(d)** $(-2, +\infty)$
(e) -2

21. $f'(x) = \dfrac{4(x+1)}{3x^{2/3}}$
$f''(x) = \dfrac{4(x-2)}{9x^{5/3}}$

(a) $[-1, +\infty)$ **(b)** $(-\infty, -1]$
(c) $(-\infty, 0)$, $(2, +\infty)$ **(d)** $(0, 2)$
(e) $0, 2$

23. $f'(x) = -xe^{-x^2/2}$
$f''(x) = (-1 + x^2)e^{-x^2/2}$

(a) $(-\infty, 0]$ **(b)** $[0, +\infty)$
(c) $(-\infty, -1)$, $(1, +\infty)$ **(d)** $(-1, 1)$
(e) $-1, 1$

25. $f'(x) = \dfrac{2x}{1+x^2}$
$f''(x) = 2\dfrac{1-x^2}{(1+x^2)^2}$

(a) $[0, +\infty)$ **(b)** $(-\infty, 0]$
(c) $(-1, 1)$ **(d)** $(-\infty, -1), (1, +\infty)$
(e) $-1, 1$

27. $f'(x) = -\sin x$
$f''(x) = -\cos x$

(a) $[\pi, 2\pi]$ **(b)** $[0, \pi]$
(c) $(\pi/2, 3\pi/2)$ **(d)** $(0, \pi/2)$, $(3\pi/2, 2\pi)$
(e) $\pi/2$, $3\pi/2$

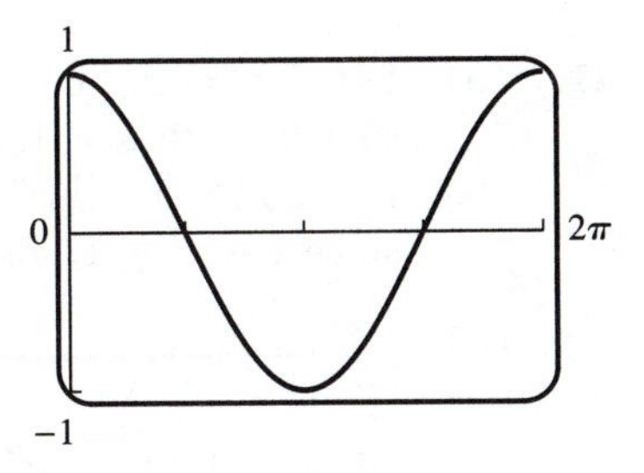

29. $f'(x) = \sec^2 x$
$f''(x) = 2\sec^2 x \tan x$

(a) $(-\pi/2, \pi/2)$ **(b)** none
(c) $(0, \pi/2)$ **(d)** $(-\pi/2, 0)$
(e) 0

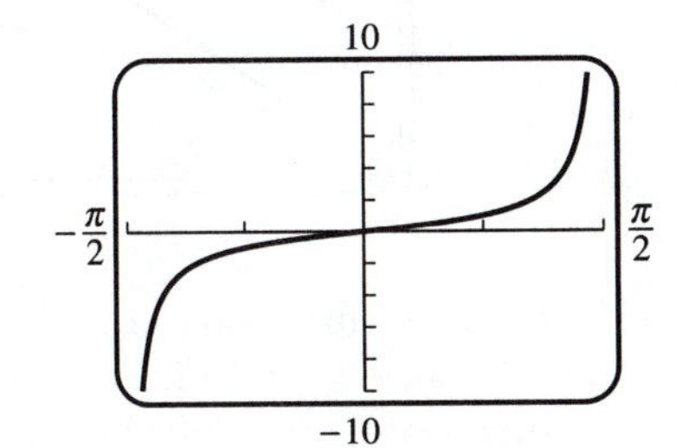

31. $f'(x) = \cos 2x$
$f''(x) = -2\sin 2x$

(a) $[0, \pi/4]$, $[3\pi/4, \pi]$ **(b)** $[\pi/4, 3\pi/4]$
(c) $(\pi/2, \pi)$ **(d)** $(0, \pi/2)$
(e) $\pi/2$

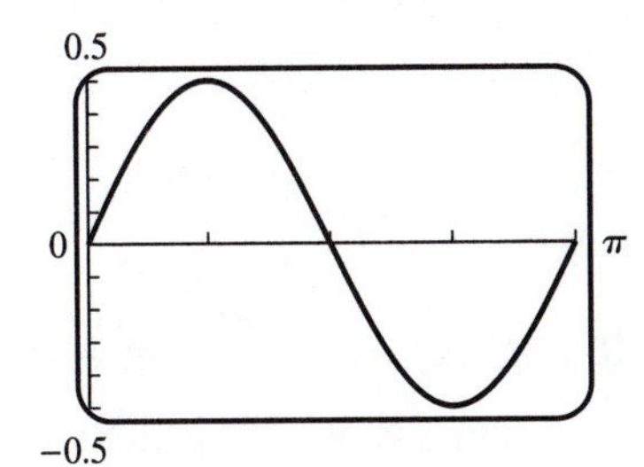

33.  **(a)**

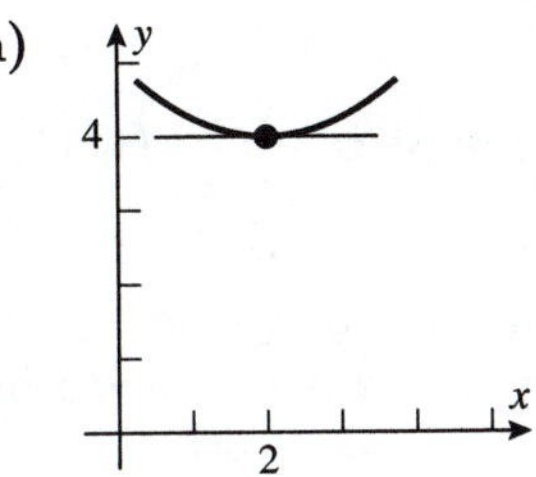

(b)

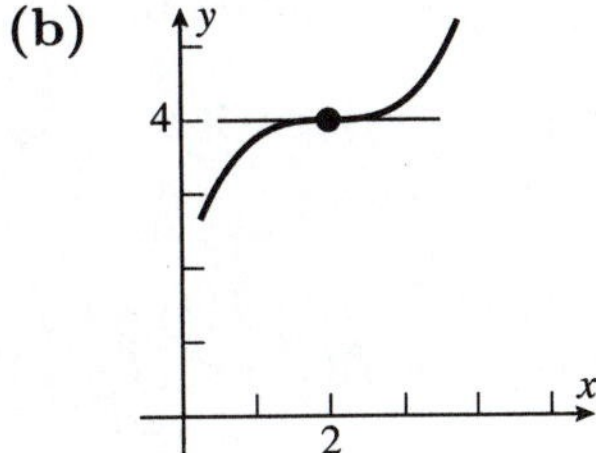

(c)

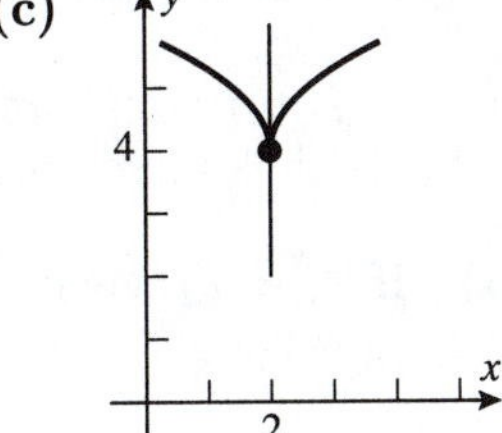

35. **(a)** $f'(x) = 3(x-a)^2$, $f''(x) = 6(x-a)$; inflection point is $(a, 0)$
(b) $f'(x) = 4(x-a)^3$, $f''(x) = 12(x-a)^2$; no inflection points

37. $f'(x) = 1/3 - 1/[3(1+x)^{2/3}]$ so f is increasing on $[0, +\infty)$ thus if $x > 0$, then $f(x) > f(0) = 0$, $1 + x/3 - \sqrt[3]{1+x} > 0$, $\sqrt[3]{1+x} < 1 + x/3$.

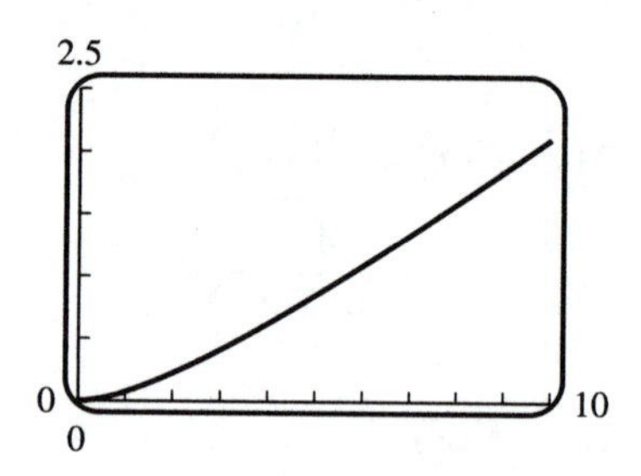

39. $x \geq \sin x$ on $[0, +\infty)$: let $f(x) = x - \sin x$. Then $f(0) = 0$ and $f'(x) = 1 - \cos x \geq 0$, so $f(x)$ is increasing on $[0, +\infty)$.

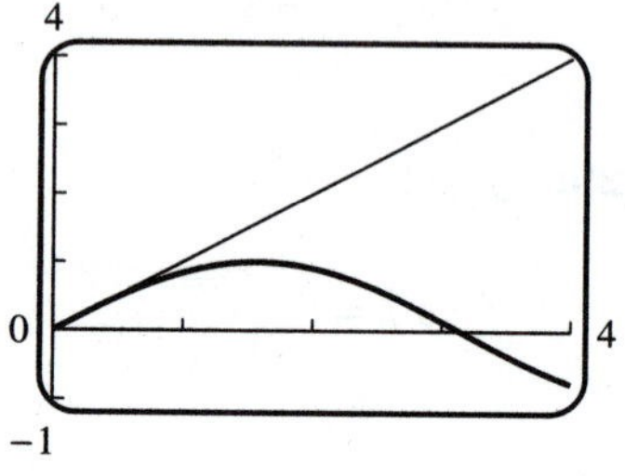

41. **(a)** Let $f(x) = x - \ln(x+1)$ for $x \geq 0$. Then $f(0) = 0$ and $f'(x) = 1 - 1/(x+1) \geq 0$ for $x \geq 0$, so f is increasing for $x \geq 0$ and thus $\ln(x+1) \leq x$ for $x \geq 0$.
(b) Let $g(x) = x - \frac{1}{2}x^2 - \ln(x+1)$. Then $g(0) = 0$ and $g'(x) = 1 - x - 1/(x+1) \leq 0$ for $x \geq 0$ since $1 - x^2 \leq 1$. Thus g is decreasing and thus $\ln(x+1) \geq x - \frac{1}{2}x^2$ for $x \geq 0$.
(c)

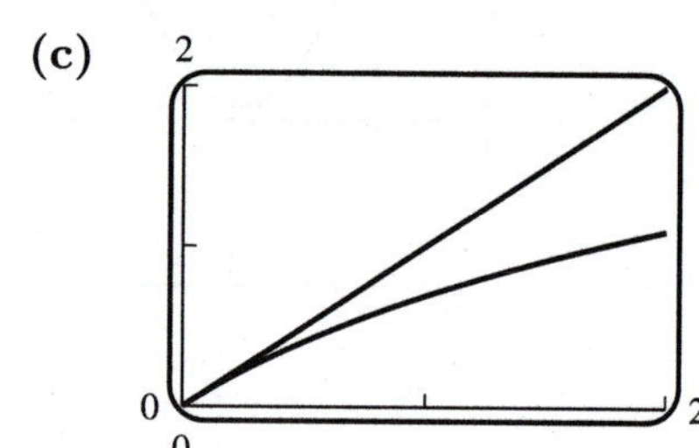

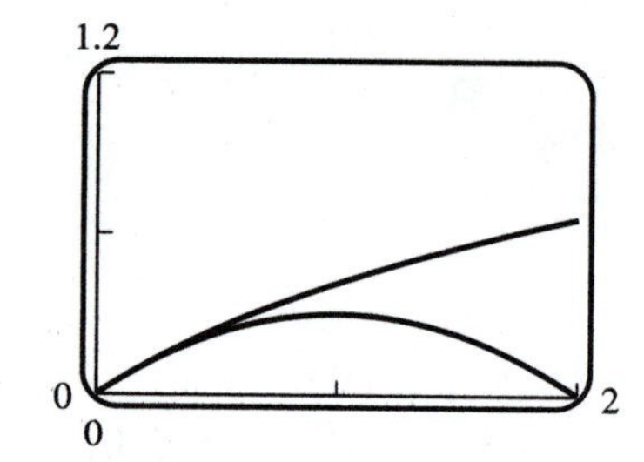

43. Points of inflection at $x = -2, +2$. Concave up on $(-5, -2)$ and $(2, 5)$; concave down on $(-2, 2)$. Increasing on $[-3.5829, 0.2513]$ and $[3.3316, 5]$, and decreasing on $[-5, -3.5829]$ and $[0.2513, 3.3316]$.

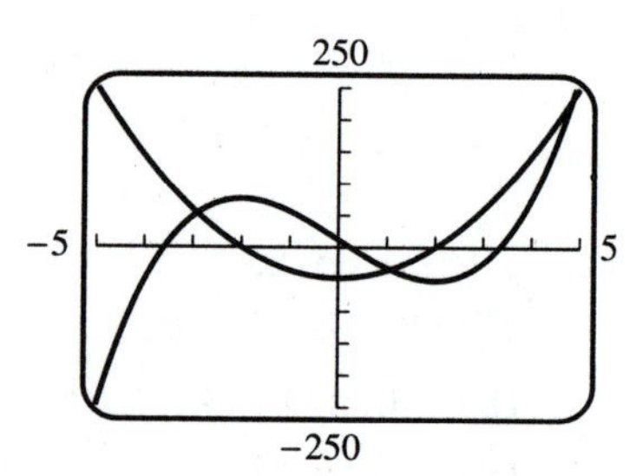

45. $f''(x) = 2\dfrac{90x^3 - 81x^2 - 585x + 397}{(3x^2 - 5x + 8)^3}$. The denominator has complex roots, so is always positive; hence the x-coordinates of the points of inflection of $f(x)$ are the roots of the numerator (if it changes sign). A plot of the numerator over $[-5, 5]$ shows roots lying in $[-3, -2]$, $[0, 1]$, and $[2, 3]$. To six decimal places the roots are $x = -2.464202, 0.662597, 2.701605$.

47. $f(x_1) - f(x_2) = x_1^2 - x_2^2 = (x_1 + x_2)(x_1 - x_2) < 0$ if $x_1 < x_2$ for x_1, x_2 in $[0, +\infty)$, so $f(x_1) < f(x_2)$ and f is thus increasing.

49. **(a)** If $x_1 < x_2$ where x_1 and x_2 are in I, then $f(x_1) < f(x_2)$ and $g(x_1) < g(x_2)$, so $f(x_1) + g(x_1) < f(x_2) + g(x_2)$, $(f+g)(x_1) < (f+g)(x_2)$. Thus $f + g$ is increasing on I.

(b) Case I: If f and g are ≥ 0 on I, and if $x_1 < x_2$ where x_1 and x_2 are in I, then $0 < f(x_1) < f(x_2)$ and $0 < g(x_1) < g(x_2)$, so $f(x_1)g(x_1) < f(x_2)g(x_2)$, $(f \cdot g)(x_1) < (f \cdot g)(x_2)$. Thus $f \cdot g$ is increasing on I.

Case II: If f and g are not necessarily positive on I then no conclusion can be drawn: for example, $f(x) = g(x) = x$ are both increasing on $(-\infty, 0)$, but $(f \cdot g)(x) = x^2$ is decreasing there.

51. **(a)** $f''(x) = 6ax + 2b = 6a\left(x + \dfrac{b}{3a}\right)$, $f''(x) = 0$ when $x = -\dfrac{b}{3a}$. f changes its direction of concavity at $x = -\dfrac{b}{3a}$ so $-\dfrac{b}{3a}$ is an inflection point.

(b) If $f(x) = ax^3 + bx^2 + cx + d$ has three x-intercepts, then it has three roots, say x_1, x_2 and x_3, so we can write $f(x) = a(x - x_1)(x - x_2)(x - x_3) = ax^3 + bx^2 + cx + d$, from which it follows that $b = -a(x_1 + x_2 + x_3)$. Thus $-\dfrac{b}{3a} = \dfrac{1}{3}(x_1 + x_2 + x_3)$, which is the average.

(c) $f(x) = x(x^2 - 3x^2 + 2) = x(x-1)(x-2)$ so the intercepts are 0, 1, and 2 and the average is 1. $f''(x) = 6x - 6 = 6(x-1)$ changes sign at $x = 1$.

53. **(a)** Let $x_1 < x_2$ belong to (a, b). If both belong to $(a, c]$ or both belong to $[c, b)$ then we have $f(x_1) < f(x_2)$ by hypothesis. So assume $x_1 < c < x_2$. We know by hypothesis that $f(x_1) < f(c)$, and $f(c) < f(x_2)$. We conclude that $f(x_1) < f(x_2)$.

(b) Use the same argument as in Part (a), but with inequalities reversed.

55. By Theorem 5.1.2, $f$is decreasing on any interval $[(2n\pi + \pi/2, 2(n+1)\pi + \pi/2]$ $(n = 0, \pm 1, \pm 2, \cdots)$, because $f'(x) = -\sin x + 1 < 0$ on $((2n\pi + \pi/2, 2(n+1)\pi + \pi/2)$. By Exercise 53 (b) we can piece these intervals together to show that $f(x)$ is decreasing on $(-\infty, +\infty)$.

57.

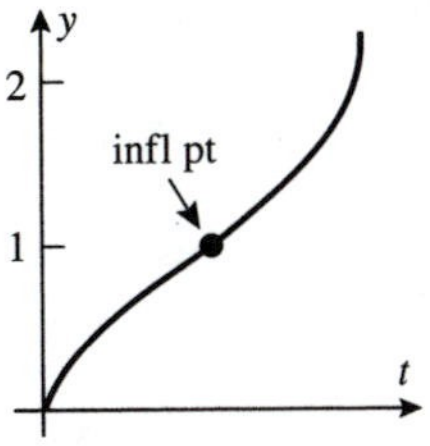

59. **(a)** $y'(t) = \dfrac{LAke^{-kt}}{(1 + Ae^{-kt})^2}S$, so $y'(0) = \dfrac{LAk}{(1 + A)^2}$

(b) The rate of growth increases to its maximum, which occurs when y is halfway between 0 and L, or when $t = \dfrac{1}{k}\ln A$; it then decreases back towards zero.

(c) From (2) one sees that $\dfrac{dy}{dt}$ is maximized when y lies half way between 0 and L, i.e. $y = L/2$. This follows since the right side of (2) is a parabola (with y as independent variable) with y-intercepts $y = 0, L$. The value $y = L/2$ corresponds to $t = \dfrac{1}{k}\ln A$, from (4).

61. $t = 7.67$

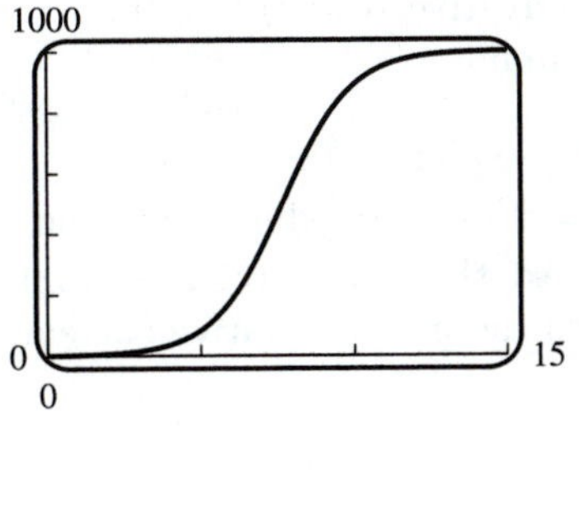

63. **(a)** $g(x)$ has no zeros:
There can be no zero of $g(x)$ on the interval $-\infty < x < 0$ because if there were, say $g(x_0) = 0$ where $x_0 < 0$, then $g'(x)$ would have to be positive between $x = x_0$ and $x = 0$, say $g'(x_1) > 0$ where $x_0 < x_1 < 0$. But then $g'(x)$ cannot be concave up on the interval $(x_1, 0)$, a contradiction.

There can be no zero of $g(x)$ on $0 < x < 4$ because $g(x)$ is concave up for $0 < x < 4$ and thus the graph of $g(x)$, for $0 < x < 4$, must lie above the line $y = -\frac{2}{3}x + 2$, which is the tangent line to the curve at $(0, 2)$, and above the line $y = 3(x - 4) + 3 = 3x - 9$ also for $0 < x < 4$ (see figure). The first condition says that $g(x)$ could only be zero for $x > 3$ and the second condition says that $g(x)$ could only be zero for $x < 3$, thus $g(x)$ has no zeros for $0 < x < 4$.

Finally, if $4 < x < +\infty$, $g(x)$ could only have a zero if $g'(x)$ were negative somewhere for $x > 4$, and since $g'(x)$ is decreasing there we would ultimately have $g(x) < -10$, a contradiction.

(b) one, between 0 and 4

(c) We must have $\lim_{x\to+\infty} g'(x) = 0$; if the limit were -5 then $g(x)$ would at some time cross the line $x = -10$; if the limit were 5 then, since g is concave down for $x > 4$ and $g'(4) = 3$, g' must decrease for $x > 4$ and thus the limit would be < 4.

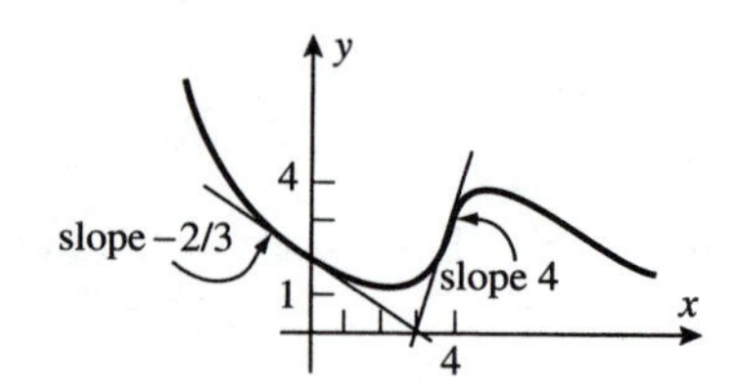

EXERCISE SET 5.2

1. **(a)**

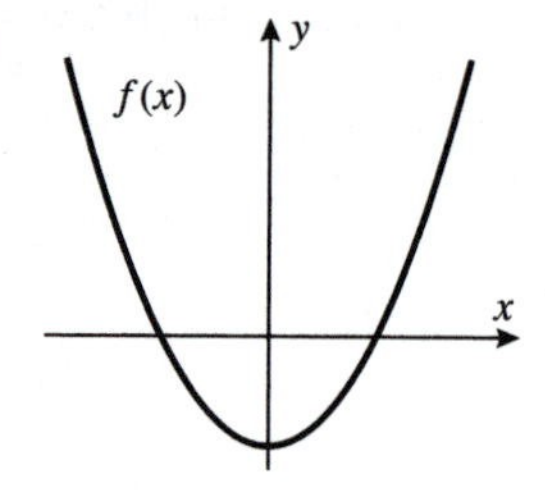

(b)

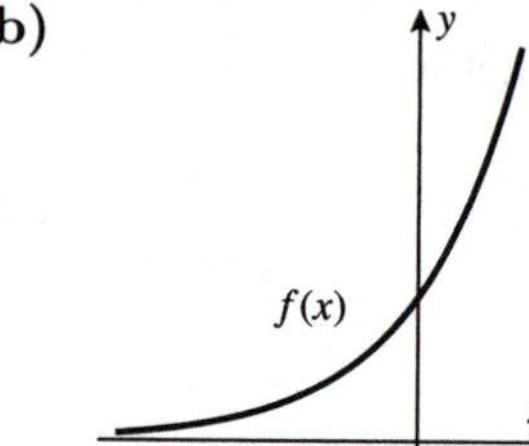

(c)

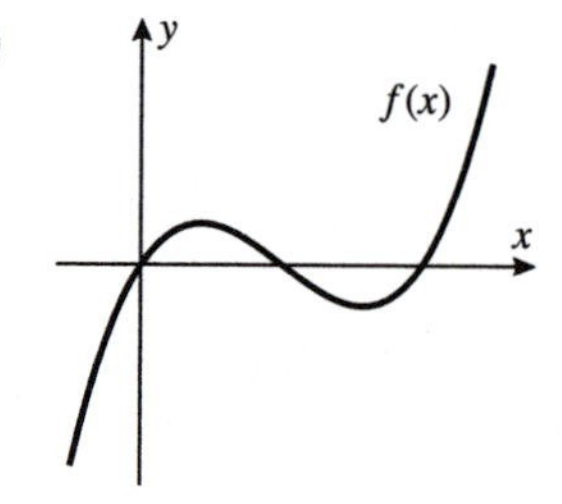

(d)

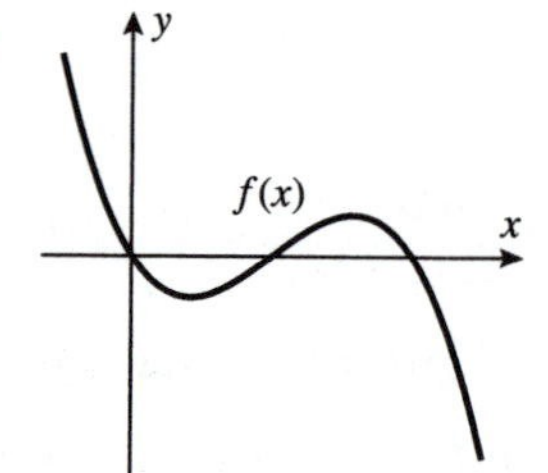

3. **(a)** $f'(x) = 6x - 6$ and $f''(x) = 6$, with $f'(1) = 0$. For the first derivative test, $f' < 0$ for $x < 1$ and $f' > 0$ for $x > 1$. For the second derivative test, $f''(1) > 0$.

(b) $f'(x) = 3x^2 - 3$ and $f''(x) = 6x$. $f'(x) = 0$ at $x = \pm 1$. First derivative test: $f' > 0$ for $x < -1$ and $x > 1$, and $f' < 0$ for $-1 < x < 1$, so there is a relative maximum at $x = -1$, and a relative minimum at $x = 1$. Second derivative test: $f'' < 0$ at $x = -1$, a relative maximum; and $f'' > 0$ at $x = 1$, a relative minimum.

5. **(a)** $f'(x) = 4(x - 1)^3$, $g'(x) = 3x^2 - 6x + 3$ so $f'(1) = g'(1) = 0$.

(b) $f''(x) = 12(x - 1)^2$, $g''(x) = 6x - 6$, so $f''(1) = g''(1) = 0$, which yields no information.

(c) $f' < 0$ for $x < 1$ and $f' > 0$ for $x > 1$, so there is a relative minimum at $x = 1$; $g'(x) = 3(x - 1)^2 > 0$ on both sides of $x = 1$, so there is no relative extremum at $x = 1$.

7. **(a)** $f'(x) = 3x^2 + 6x - 9 = 3(x+3)(x-1)$, $f'(x) = 0$ when $x = -3, 1$ (stationary points).
(b) $f'(x) = 4x(x^2 - 3)$, $f'(x) = 0$ when $x = 0$, $\pm\sqrt{3}$ (stationary points).

9. **(a)** $f'(x) = (2 - x^2)/(x^2 + 2)^2$, $f'(x) = 0$ when $x = \pm\sqrt{2}$ (stationary points).
(b) $f'(x) = \frac{2}{3}x^{-1/3} = 2/(3x^{1/3})$, $f'(x)$ does not exist when $x = 0$.

11. **(a)** $f'(x) = \dfrac{4(x+1)}{3x^{2/3}}$, $f'(x) = 0$ when $x = -1$ (stationary point), $f'(x)$ does not exist when $x = 0$.
(b) $f'(x) = -3\sin 3x$, $f'(x) = 0$ when $\sin 3x = 0, 3x = n\pi, n = 0, \pm 1, \pm 2, \cdots$
$x = n\pi/3, n = 0, \pm 1, \pm 2, \cdots$ (stationary points)

13. **(a)** none
(b) $x = 1$ because f' changes sign from $+$ to $-$ there
(c) none because $f'' = 0$ (never changes sign)

15. **(a)** $x = 2$ because $f'(x)$ changes sign from $-$ to $+$ there.
(b) $x = 0$ because $f'(x)$ changes sign from $+$ to $-$ there.
(c) $x = 1, 3$ because $f''(x)$ changes sign at these points.

17. **(a)** critical numbers $x = 0, \pm\sqrt{5}$; f':

– – 0 + + 0 – – 0 + +
$-\sqrt{5}$ 0 $\sqrt{5}$

$x = 0$: relative maximum; $x = \pm\sqrt{5}$: relative minimum

(b) critical number $x = 1, -1$; f':

+ + + 0 – – – 0 + + +
-1 1

$x = -1$: relative maximum; $x = 1$: relative minimum

19. **(a)** critical point $x = 0$; f':
$x = 0$: relative minimum

– – – 0 + + +
0

(b) critical point $x = \ln 2$: f':
$x = \ln 2$: relative minimum

– – – 0 + + +
ln 2

21. $f'(x) = -2(x+2)$; critical number $x = -2$; $f'(x)$:

+ + + 0 – – –
-2

$f''(x) = -2$; $f''(-2) < 0$, $f(-2) = 5$; relative maximum of 5 at $x = -2$

23. $f'(x) = 2\sin x \cos x = \sin 2x$;
critical numbers $x = \pi/2$, π, $3\pi/2$; $f'(x)$:

+ + 0 – – 0 + + 0 – –
$\frac{\pi}{2}$ π $\frac{3\pi}{2}$

$f''(x) = 2\cos 2x$; $f''(\pi/2) < 0$, $f''(\pi) > 0$, $f''(3\pi/2) < 0$, $f(\pi/2) = f(3\pi/2) = 1$,
$f(\pi) = 0$; relative minimum of 0 at $x = \pi$, relative maximum of 1 at $x = \pi/2$, $3\pi/2$

25. $f'(x) = 3x^2 + 5$; no relative extrema because there are no critical numbers.

27. $f'(x) = (x-1)(3x-1)$; critical numbers $x = 1, 1/3$
$f''(x) = 6x - 4$; $f''(1) > 0$, $f''(1/3) < 0$
relative minimum of 0 at $x = 1$, relative maximum of 4/27 at $x = 1/3$

29. $f'(x) = 4x(1 - x^2)$; critical numbers $x = 0, 1, -1$
$f''(x) = 4 - 12x^2$; $f''(0) > 0$, $f''(1) < 0$, $f''(-1) < 0$
relative minimum of 0 at $x = 0$, relative maximum of 1 at $x = 1, -1$

31. $f'(x) = \frac{4}{5}x^{-1/5}$; critical number $x = 0$; relative minimum of 0 at $x = 0$ (first derivative test)

33. $f'(x) = 2x/(x^2 + 1)^2$; critical number $x = 0$; relative minimum of 0 at $x = 0$

35. $f'(x) = 2x/(1 + x^2)$; critical point at $x = 0$; relative minimum of 0 at $x = 0$ (first derivative test)

37. $f'(x) = 2x$ if $|x| > 2$, $f'(x) = -2x$ if $|x| < 2$,
$f'(x)$ does not exist when $x = \pm 2$; critical numbers $x = 0, 2, -2$
relative minimum of 0 at $x = 2, -2$, relative maximum of 4 at $x = 0$

39. $f'(x) = 2\cos 2x$ if $\sin 2x > 0$,
$f'(x) = -2\cos 2x$ if $\sin 2x < 0$,
$f'(x)$ does not exist when $x = \pi/2, \pi, 3\pi/2$;
critical numbers $x = \pi/4, 3\pi/4, 5\pi/4, 7\pi/4, \pi/2, \pi, 3\pi/2$
relative minimum of 0 at $x = \pi/2, \pi, 3\pi/2$;
relative maximum of 1
at $x = \pi/4, 3\pi/4, 5\pi/4, 7\pi/4$

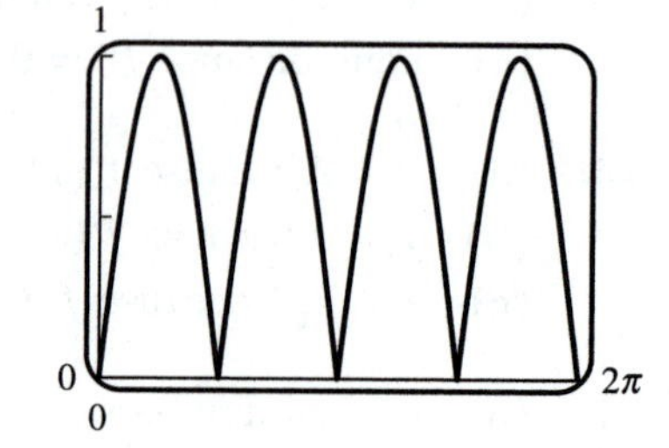

41. $f'(x) = -\sin 2x$;
critical numbers $x = \pi/2, \pi, 3\pi/2$
relative minimum of 0 at $x = \pi/2, 3\pi/2$;
relative maximum of 1 at $x = \pi$

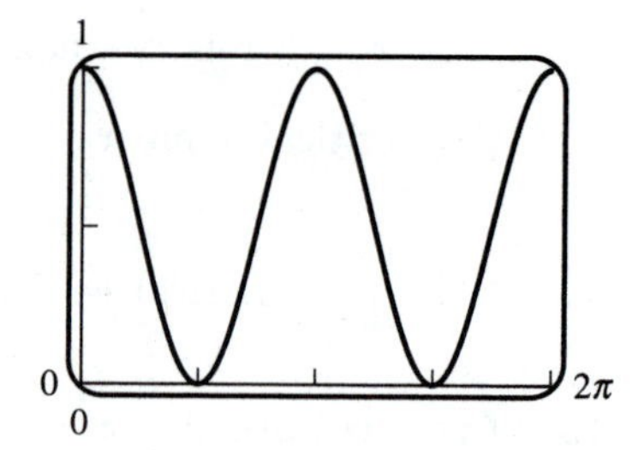

43. $f'(x) = \ln x + 1$, $f''(x) = 1/x$; $f'(1/e) = 0$, $f''(1/e) > 0$;
relative minimum of $-1/e$ at $x = 1/e$

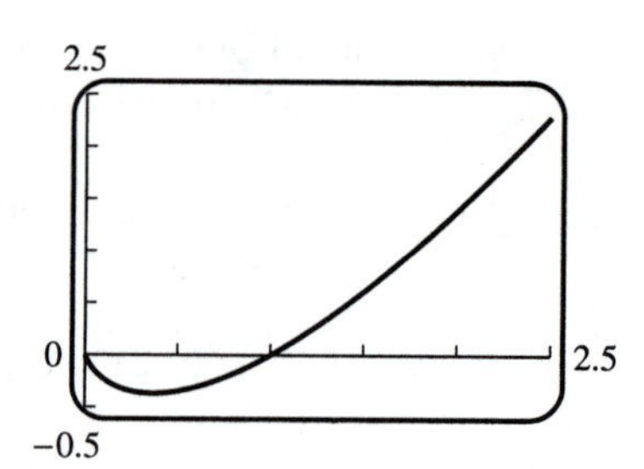

45. $f'(x) = 2x(1 - x)e^{-2x} = 0$ at $x = 0, 1$.
$f''(x) = (4x^2 - 8x + 2)e^{-2x}$;
$f''(0) > 0$ and $f''(1) < 0$, so a
relative minimum of 0 at $x = 0$ and a
relative maximum of $1/e^2$ at $x = 1$.

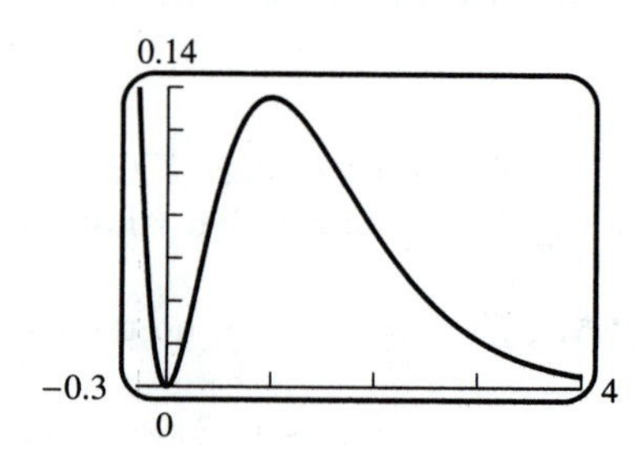

47. Relative minima at $x = -3.58, 3.33$; relative maximum at $x = 0.25$

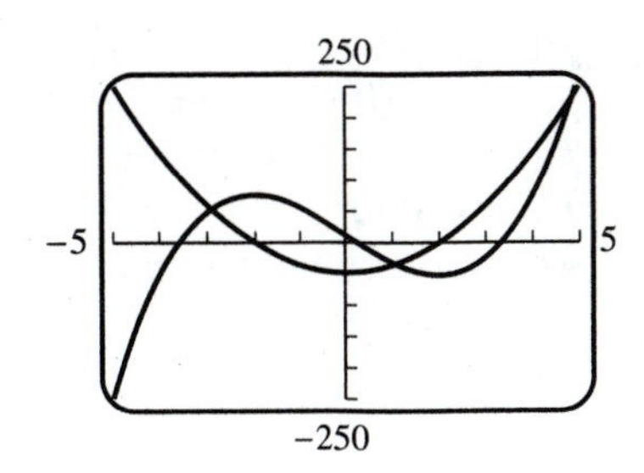

49. Relative minimum at $x = -1.20$ and a relative maximum at $x = 1.80$

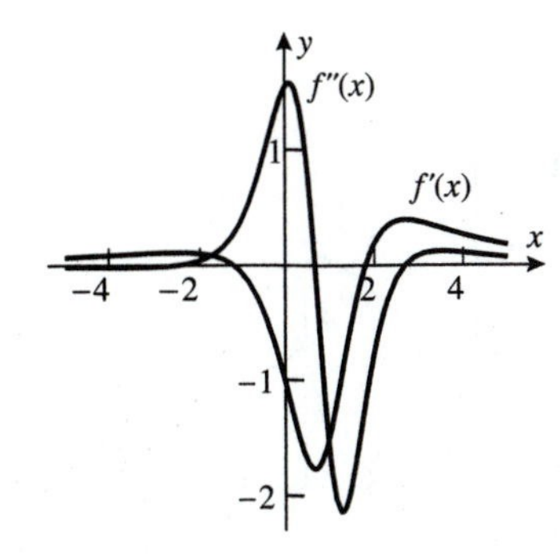

51. $f'(x) = \dfrac{x^4 + 3x^2 - 2x}{(x^2+1)^2}$

$f''(x) = -2\dfrac{x^3 - 3x^2 - 3x + 1}{(x^2+1)^3}$

Relative maximum at $x = 0$, relative minimum at $x \approx 0.59$

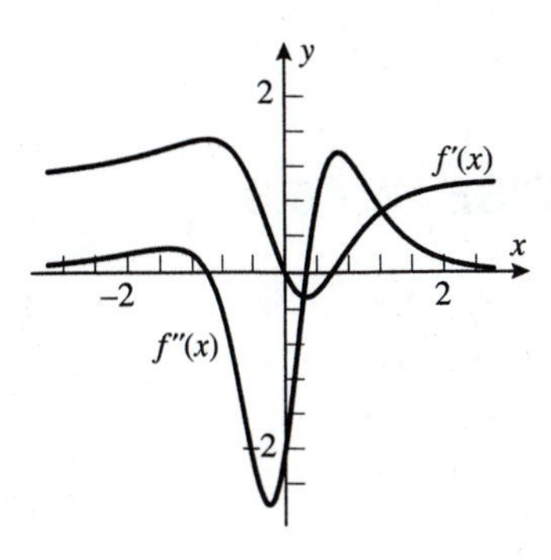

53. **(a)** Let $f(x) = x^2 + \dfrac{k}{x}$, then $f'(x) = 2x - \dfrac{k}{x^2} = \dfrac{2x^3 - k}{x^2}$. f has a relative extremum when $2x^3 - k = 0$, so $k = 2x^3 = 2(3)^3 = 54$.

(b) Let $f(x) = \dfrac{x}{x^2 + k}$, then $f'(x) = \dfrac{k - x^2}{(x^2+k)^2}$. f has a relative extremum when $k - x^2 = 0$, so $k = x^2 = 3^2 = 9$.

55. **(a)** $(-2.2, 4), (2, 1.2), (4.2, 3)$

(b) f' exists everywhere, so the critical numbers are when $f' = 0$, i.e. when $x = \pm 2$ or $r(x) = 0$, so $x \approx -5.1, -2, 0.2, 2$. At $x = -5.1$ f' changes sign from $-$ to $+$, so minimum; at $x = -2$ f' changes sign from $-$ to $+$, so minimum; at $x = 0.2$ f' doesn't change sign, so neither; at $x = 2$ f' changes sign from $+$ to $-$, so maximum.
Finally, $f''(1) = (1^2 - 4)r'(1) + 2r(1) \approx -3(0.6) + 2(0.3) = -1.2$.

57. $f'(x) = 3ax^2 + 2bx + c$ and $f'(x)$ has roots at $x = 0, 1$, so $f'(x)$ must be of the form $f'(x) = 3ax(x-1)$; thus $c = 0$ and $2b = -3a$, $b = -3a/2$. $f''(x) = 6ax + 2b = 6ax - 3a$, so $f''(0) > 0$ and $f''(1) < 0$ provided $a < 0$. Finally $f(0) = d$, so $d = 0$; and $f(1) = a + b + c + d = a + b = -a/2$ so $a = -2$. Thus $f(x) = -2x^3 + 3x^2$.

59. (a) $f'(x) = -xf(x)$. Since $f(x)$ is always positive, $f'(x) = 0$ at $x = 0$, $f'(x) > 0$ for $x < 0$ and $f'(x) < 0$ for $x > 0$, so $x = 0$ is a maximum.

(b)

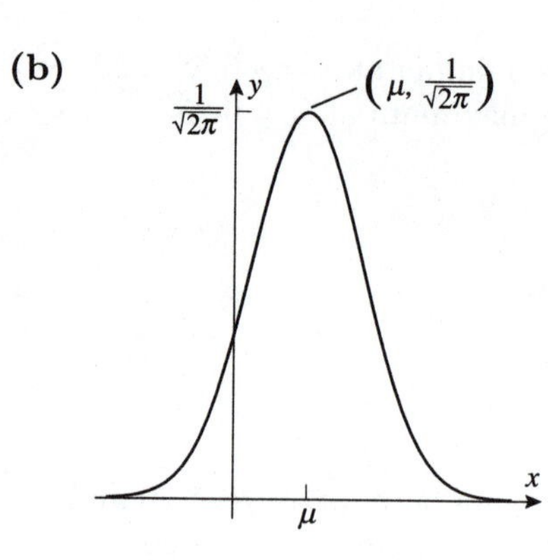

61. (a)

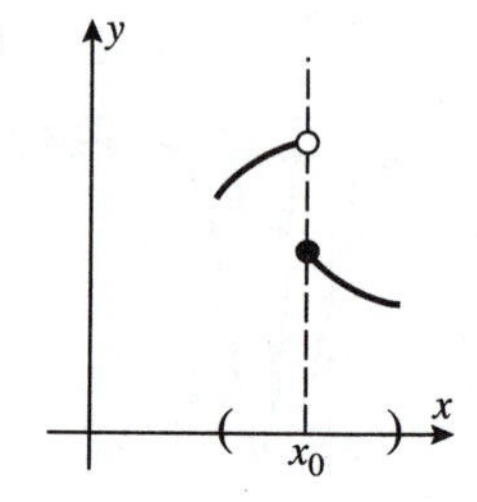

$f(x_0)$ is not an extreme value.

(b)

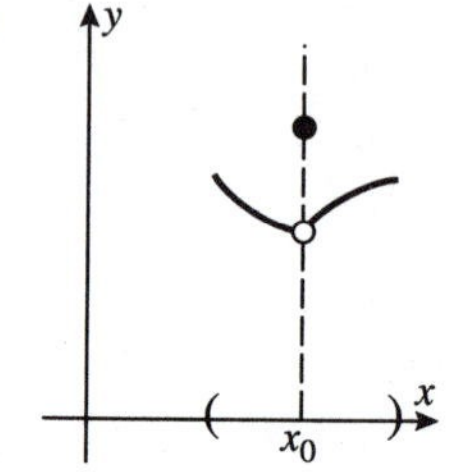

$f(x_0)$ is a relative maximum.

(c)

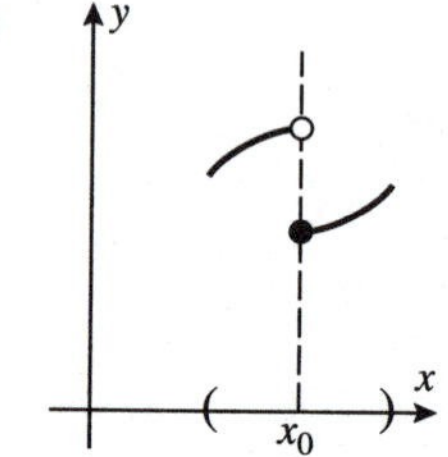

$f(x_0)$ is a relative minimum.

EXERCISE SET 5.3

1. $y = x^2 - 2x - 3$;
$y' = 2(x - 1)$;
$y'' = 2$

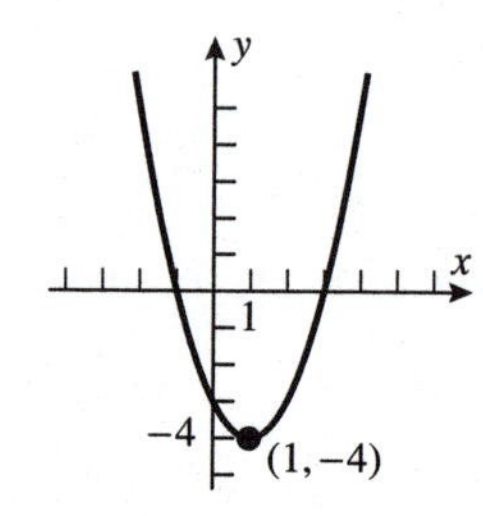

3. $y = x^3 - 3x + 1$;
$y' = 3(x^2 - 1)$;
$y'' = 6x$

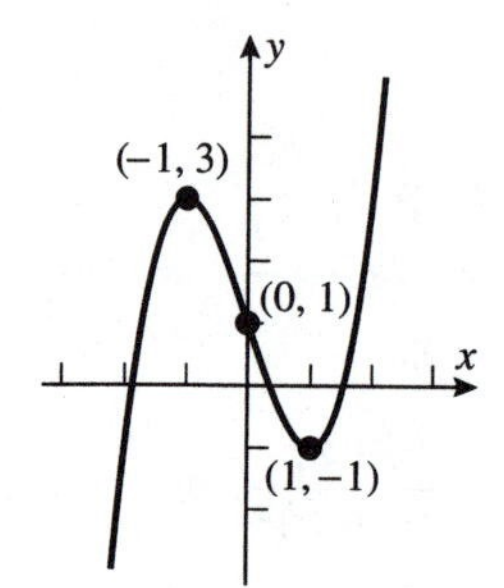

5. $y = x^4 + 2x^3 - 1$;
$y' = 4x^2(x + 3/2)$;
$y'' = 12x(x + 1)$

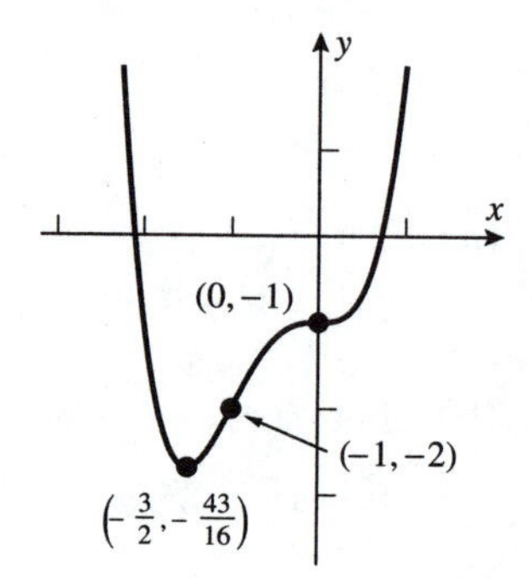

7. $y = x^3(3x^2 - 5)$;
$y' = 15x^2(x^2 - 1)$;
$y'' = 30x(2x^2 - 1)$

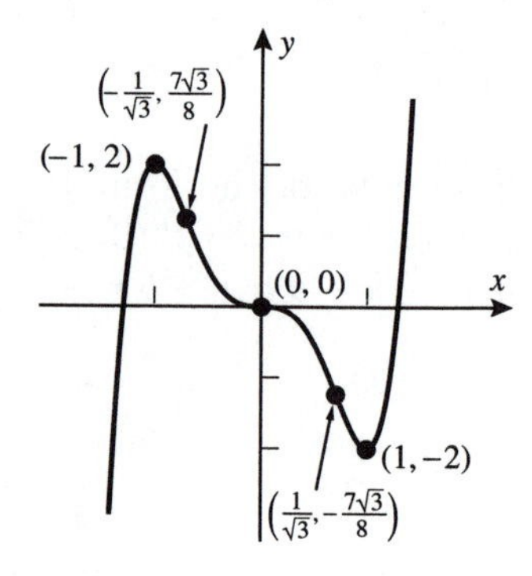

9. $y = x(x - 1)^3$;
$y' = (4x - 1)(x - 1)^2$;
$y'' = 6(2x - 1)(x - 1)$

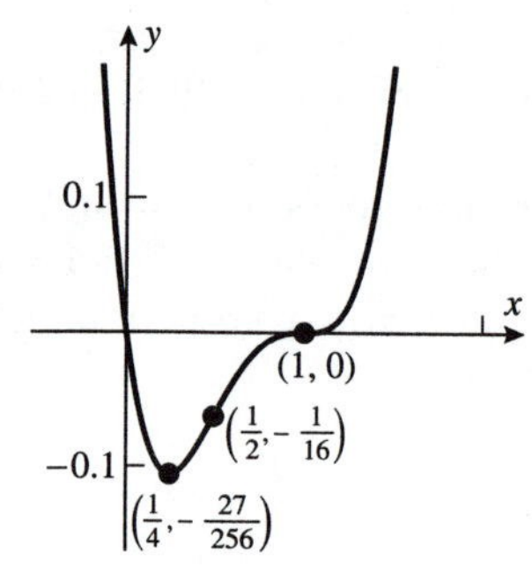

11. $y = 2x/(x - 3)$;
$y' = -6/(x - 3)^2$;
$y'' = 12/(x - 3)^3$

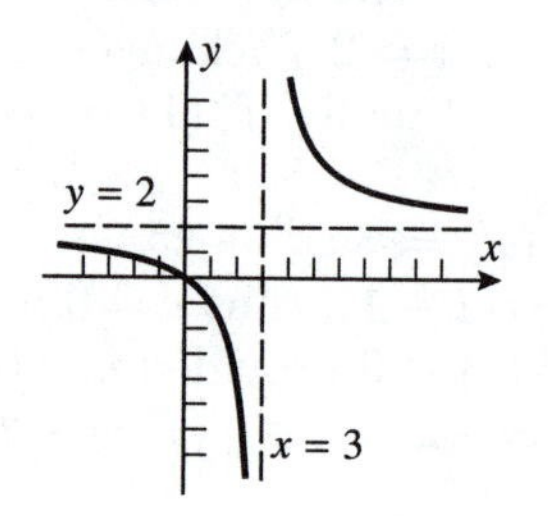

13. $y = \dfrac{x^2}{x^2-1};$

$y' = -\dfrac{2x}{(x^2-1)^2};$

$y'' = \dfrac{2(3x^2+1)}{(x^2-1)^3}$

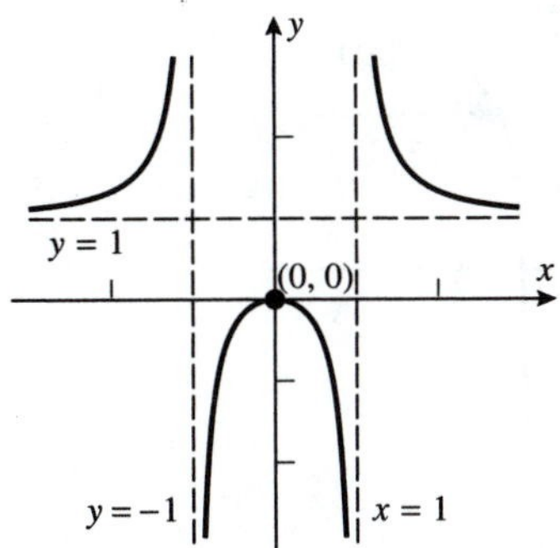

15. $y = x^2 - \dfrac{1}{x} = \dfrac{x^3-1}{x};$

$y' = \dfrac{2x^3+1}{x^2},$

$y' = 0$ when $x = -\sqrt[3]{\dfrac{1}{2}} \approx -0.8;$

$y'' = \dfrac{2(x^3-1)}{x^3}$

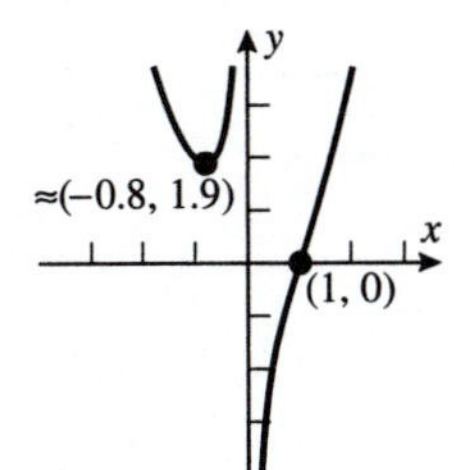

17. $y = \dfrac{x^3-1}{x^3+1};$

$y' = \dfrac{6x^2}{(x^3+1)^2};$

$y'' = \dfrac{12x(1-2x^3)}{(x^3+1)^3}$

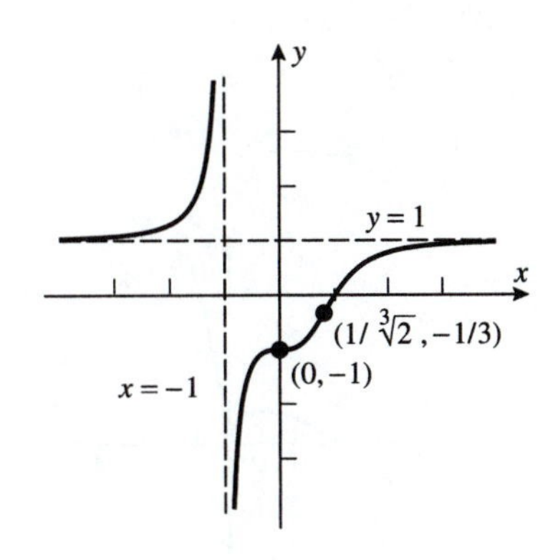

19. $y = \dfrac{x-1}{x^2-4};$

$y' = -\dfrac{x^2-2x+4}{(x^2-4)^2}$

$y = 2\dfrac{x^3-3x^2+12x-4}{(x^2-4)^3}$

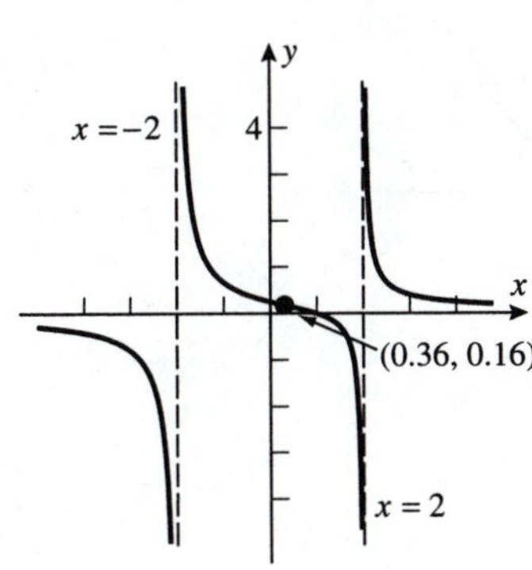

21. $y = \dfrac{1}{x-2};$

$y' = \dfrac{-1}{(x-2)^2}$

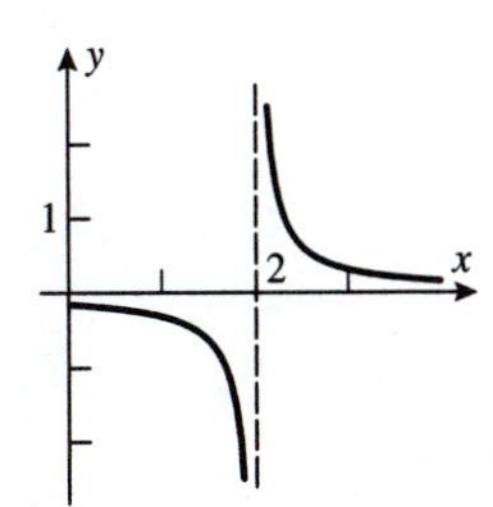

23. $y = \dfrac{(x-1)^2}{x^2}$;

$y' = \dfrac{2(x-1)}{x^3}$;

$y'' = \dfrac{2(3-2x)}{x^4}$

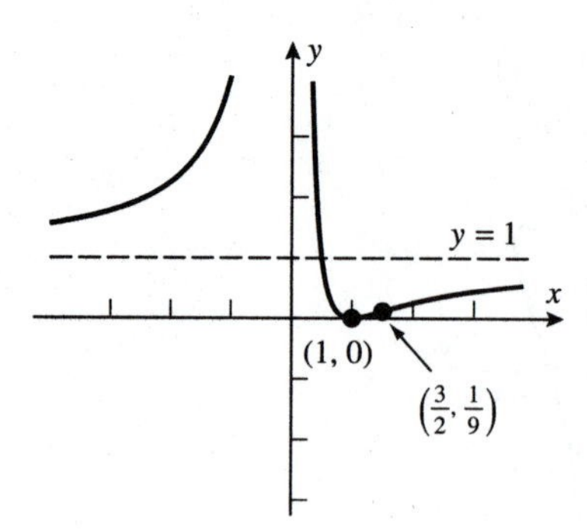

25. $y = 4 + \dfrac{x-1}{x^4}$;

$y' = -\dfrac{3x-4}{x^5}$;

$y'' = 4\dfrac{3x-5}{x^6}$

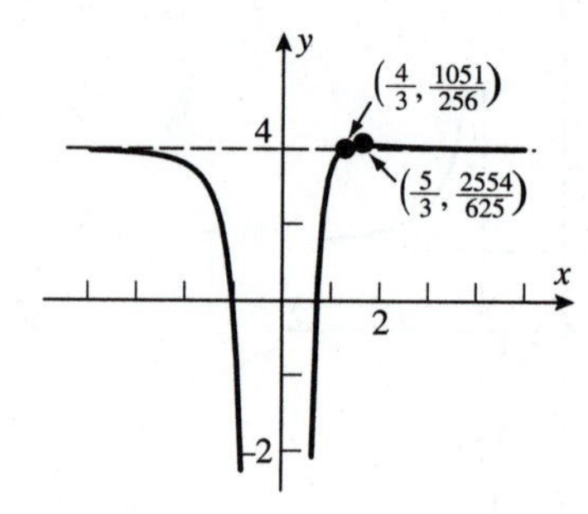

27. **(a)** VI **(b)** I **(c)** III **(d)** V **(e)** IV **(f)** II

29. $y = \sqrt{x^2-1}$;

$y' = \dfrac{x}{\sqrt{x^2-1}}$;

$y'' = -\dfrac{1}{(x^2-1)^{3/2}}$

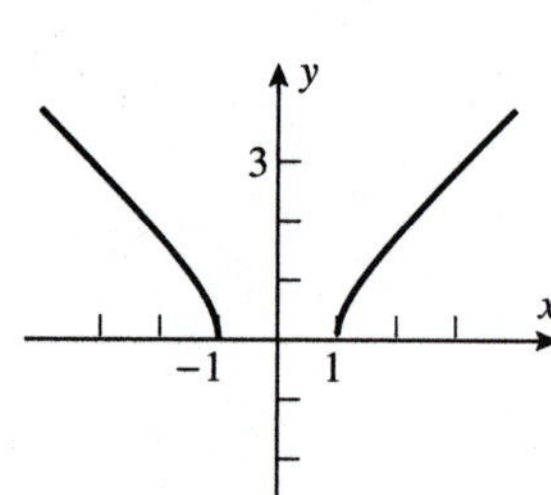

31. $y = 2x + 3x^{2/3}$;

$y' = 2 + 2x^{-1/3}$;

$y'' = -\dfrac{2}{3}x^{-4/3}$

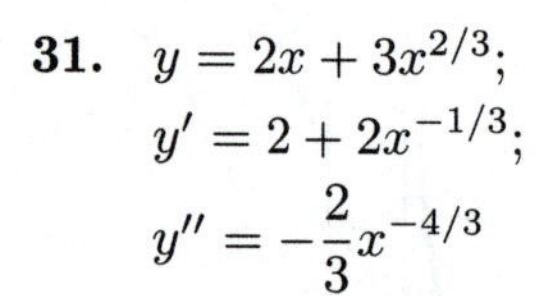

33. $y = x(3-x)^{1/2}$;

$y' = \dfrac{3(2-x)}{2\sqrt{3-x}}$;

$y'' = \dfrac{3(x-4)}{4(3-x)^{3/2}}$

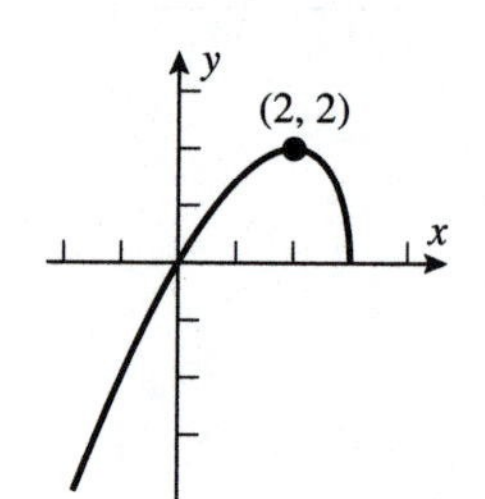

35. $y = \dfrac{8(\sqrt{x}-1)}{x}$;

$y' = \dfrac{4(2-\sqrt{x})}{x^2}$;

$y'' = \dfrac{2(3\sqrt{x}-8)}{x^3}$

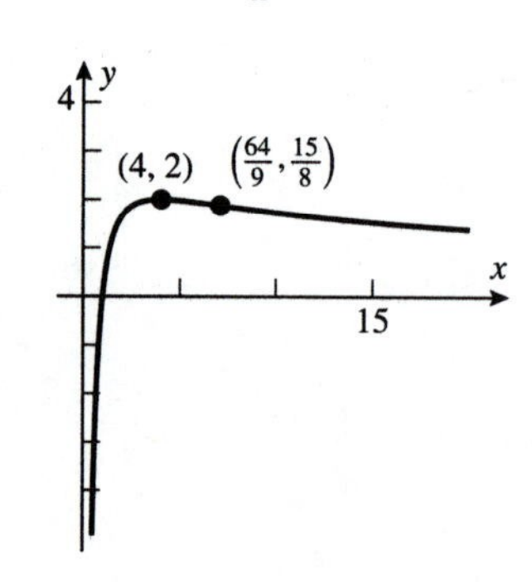

37. $y = x + \sin x$;

$y' = 1 + \cos x$, $y' = 0$ when $x = \pi + 2n\pi$;

$y'' = -\sin x$; $y'' = 0$ when $x = n\pi$

$n = 0, \pm 1, \pm 2, \ldots$

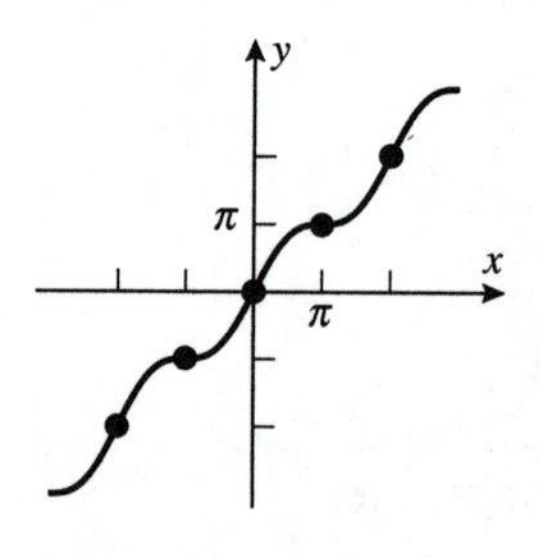

39. $y = \sin x + \cos x$;
$y' = \cos x - \sin x$;
$y' = 0$ when $x = \pi/4 + n\pi$;
$y'' = -\sin x - \cos x$;
$y'' = 0$ when $x = 3\pi/4 + n\pi$

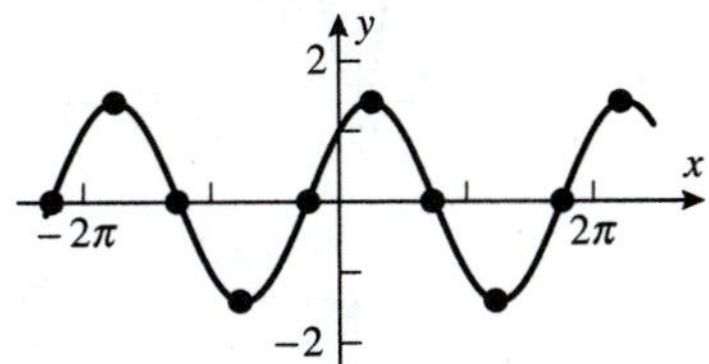

41. $y = \sin^2 x,\ 0 \le x \le 2\pi$;
$y' = 2\sin x \cos x = \sin 2x$;
$y'' = 2\cos 2x$

43. **(a)** 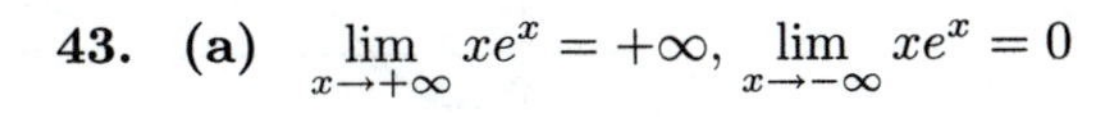
$\lim_{x\to+\infty} xe^x = +\infty,\ \lim_{x\to-\infty} xe^x = 0$

(b) $y = xe^x$;
$y' = (x+1)e^x$;
$y'' = (x+2)e^x$

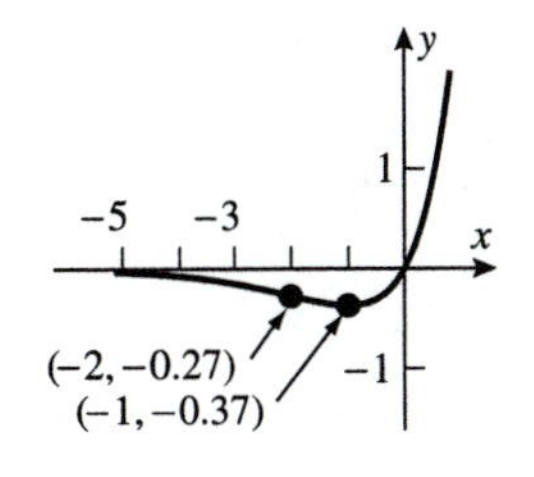

45. **(a)** $\lim_{x\to+\infty} \dfrac{x^2}{e^{2x}} = 0,\ \lim_{x\to-\infty} \dfrac{x^2}{e^{2x}} = +\infty$

(b) $y = x^2/e^{2x} = x^2e^{-2x}$;
$y' = 2x(1-x)e^{-2x}$;
$y'' = 2(2x^2 - 4x + 1)e^{-2x}$;
$y'' = 0$ if $2x^2 - 4x + 1 = 0$, when
$x = \dfrac{4 \pm \sqrt{16-8}}{4} = 1 \pm \sqrt{2}/2 \approx 0.29, 1.71$

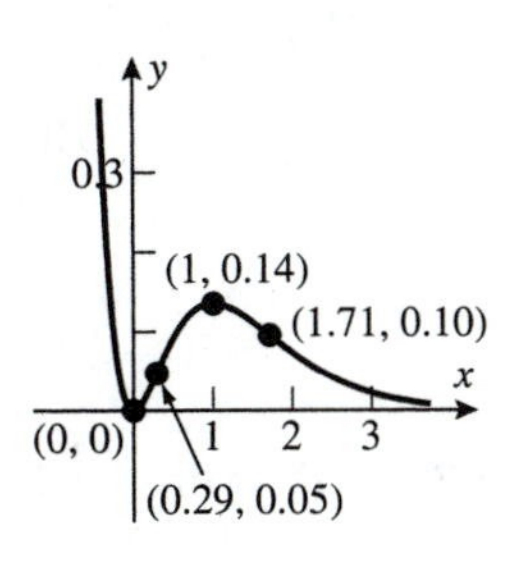

47. **(a)** $\lim_{x\to+\infty} f(x) = +\infty,\ \lim_{x\to-\infty} f(x) = -\infty$

(b) $y = xe^{x^2}$;
$y' = (1 + 2x^2)e^{x^2}$;
$y'' = 2x(3 + 2x^2)e^{x^2}$
no relative extrema, inflection point at $(0, 0)$

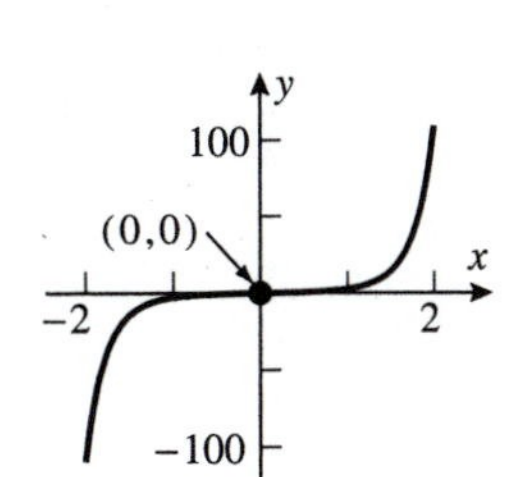

49. $\lim_{x\to+\infty} f(x) = +\infty,\ \lim_{x\to-\infty} f(x) = 0$

$f'(x) = e^x \dfrac{x-1}{x^2},\ f''(x) = e^x \dfrac{x^2 - 2x + 2}{x^3}$

critical point at $x = 1$;
relative minimum at $x = 1$
no points of inflection
vertical asymptote $x = 0$,
horizontal asymptote $y = 0$ for $x \to -\infty$

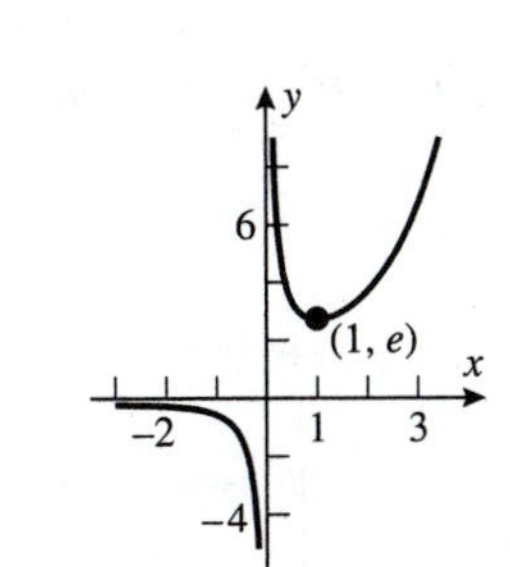

51. $\lim_{x\to+\infty} f(x) = 0$, $\lim_{x\to-\infty} f(x) = +\infty$
$f'(x) = x(2-x)e^{1-x}$, $f''(x) = (x^2 - 4x + 2)e^{1-x}$
critical points at $x = 0, 2$;
relative minimum at $x = 0$,
relative maximum at $x = 2$
points of inflection at $x = 2 \pm \sqrt{2}$
horizontal asymptote $y = 0$ as $x \to +\infty$

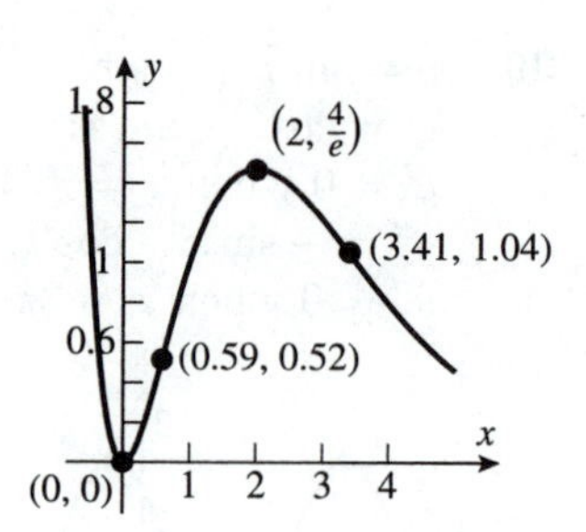

53. **(a)** $\lim_{x\to0^+} y = \lim_{x\to0^+} x\ln x = \lim_{x\to0^+} \frac{\ln x}{1/x} = \lim_{x\to0^+} \frac{1/x}{-1/x^2} = 0$;
$\lim_{x\to+\infty} y = +\infty$

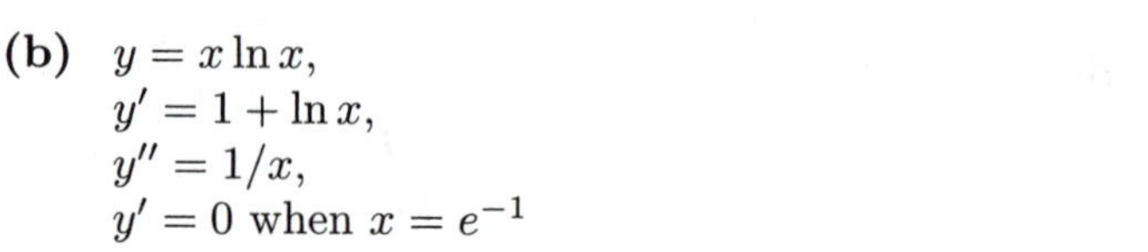

(b) $y = x\ln x$,
$y' = 1 + \ln x$,
$y'' = 1/x$,
$y' = 0$ when $x = e^{-1}$

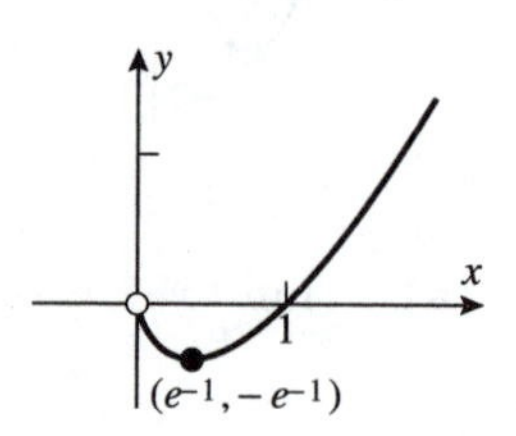

55. **(a)** $\lim_{x\to0^+} y = \lim_{x\to0^+} \frac{\ln x}{x^2} = -\infty$;
$\lim_{x\to+\infty} y = \lim_{x\to+\infty} \frac{\ln x}{x^2} = \lim_{x\to+\infty} \frac{1/x}{2x} = 0$

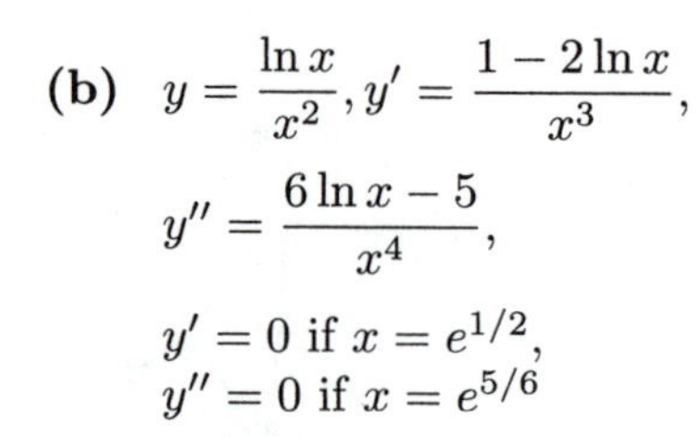

(b) $y = \frac{\ln x}{x^2}, y' = \frac{1 - 2\ln x}{x^3}$,
$y'' = \frac{6\ln x - 5}{x^4}$,
$y' = 0$ if $x = e^{1/2}$,
$y'' = 0$ if $x = e^{5/6}$

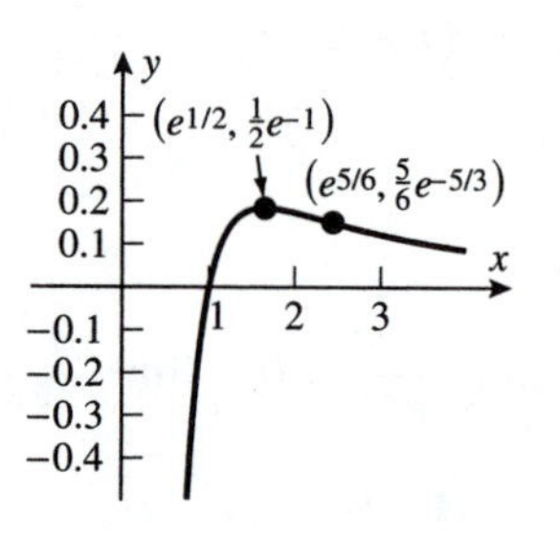

57. **(a)** $\lim_{x\to0^+} x^2\ln x = 0$ by the rule given, $\lim_{x\to+\infty} x^2 \ln x = +\infty$ by inspection, and $f(x)$ not defined for $x < 0$

(b) $y = x^2\ln 2x$, $y' = 2x\ln 2x + x$
$y'' = 2\ln 2x + 3$
$y' = 0$ if $x = 1/(2\sqrt{e})$,
$y'' = 0$ if $x = 1/(2e^{3/2})$

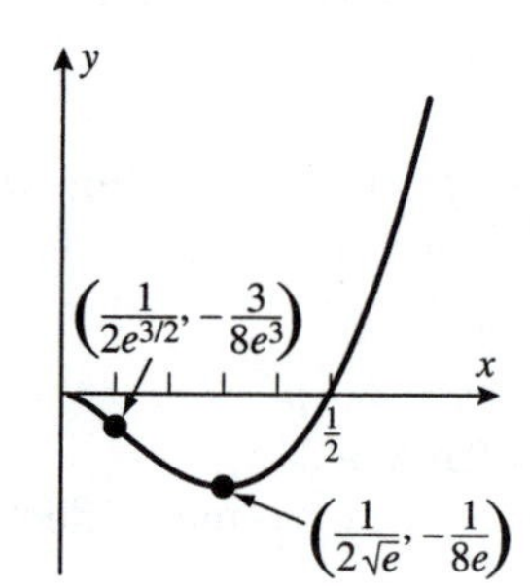

59. **(a)** $\lim_{x\to-\infty} y = -\infty$, $\lim_{x\to+\infty} y = +\infty$;
curve crosses x-axis at $x = 0, 1, -1$

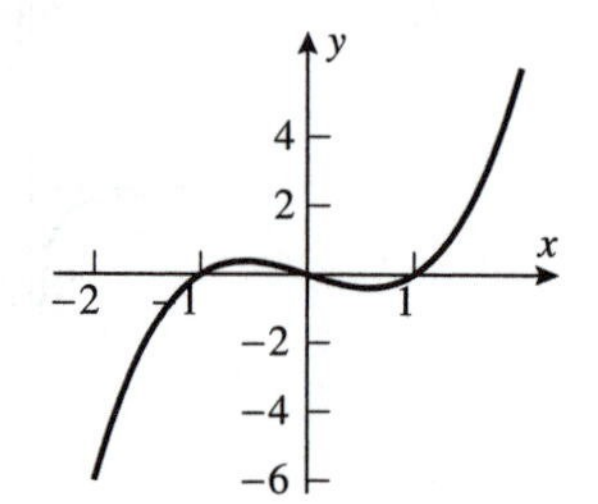

(b) $\lim_{x\to\pm\infty} y = +\infty$;
curve never crosses x-axis

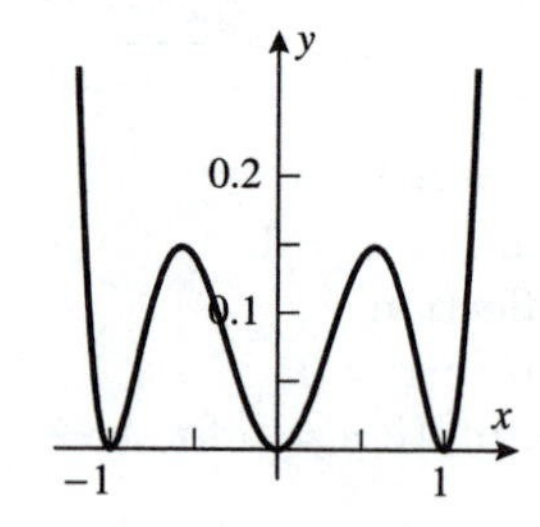

(c) $\lim_{x\to-\infty} y = -\infty$, $\lim_{x\to+\infty} y = +\infty$;
curve crosses x-axis at $x = -1$

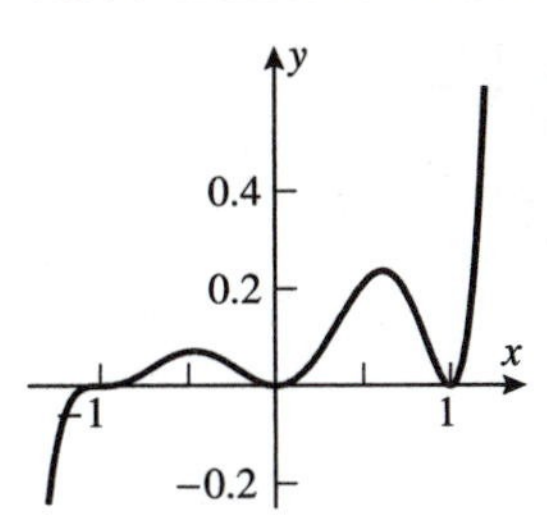

(d) $\lim_{x\to\pm\infty} y = +\infty$;
curve crosses x-axis at $x = 0, 1$

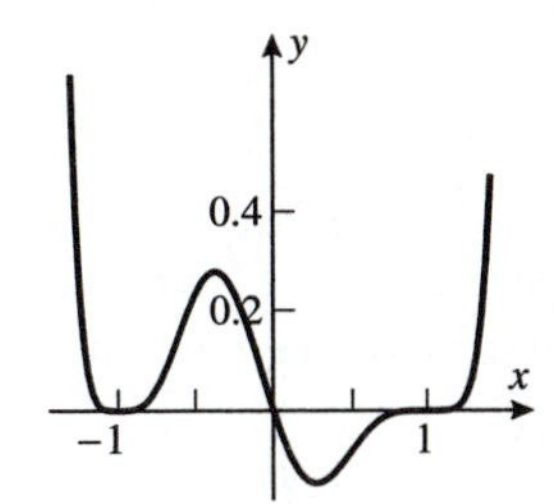

61. (a) horizontal asymptote $y = 3$ as $x \to \pm\infty$, vertical asymptotes of $x = \pm 2$

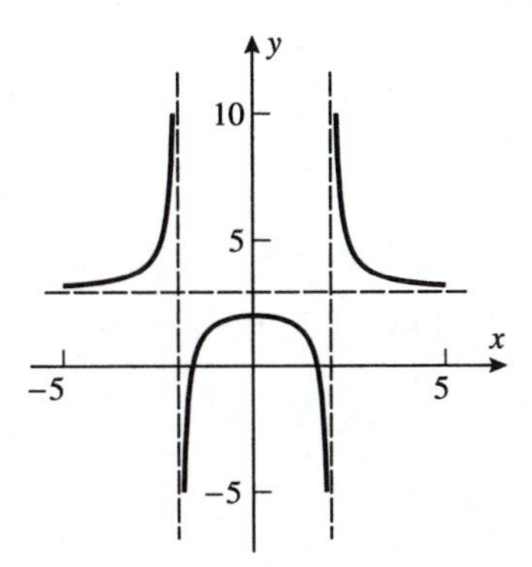

(b) horizontal asymptote of $y = 1$ as $x \to \pm\infty$, vertical asymptotes at $x = \pm 1$

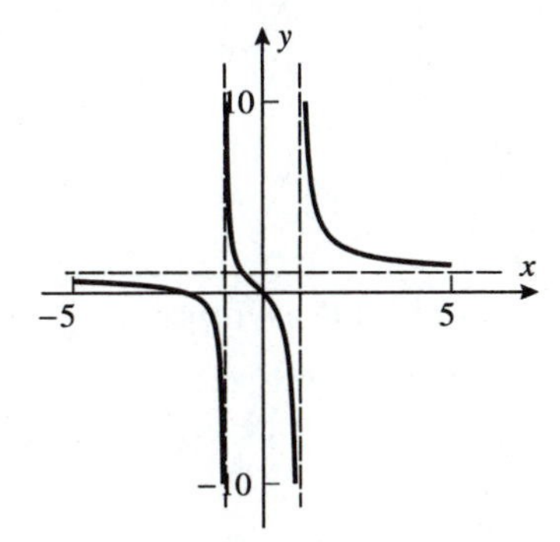

(c) horizontal asymptote of $y = -1$ as $x \to \pm\infty$, vertical asymptotes at $x = -2, 1$

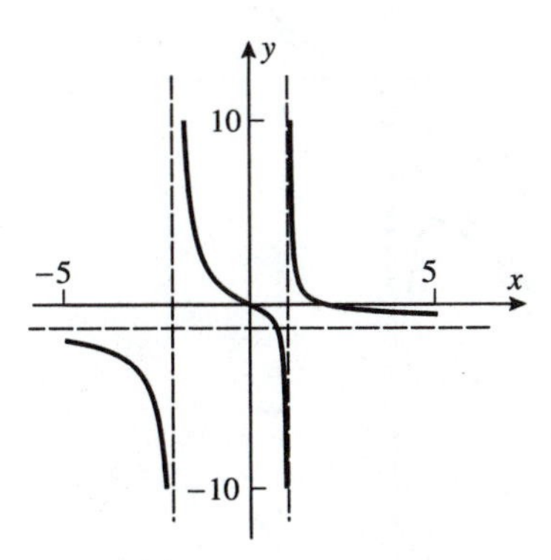

(d) horizontal asymptote of $y = 1$ as $x \to \pm\infty$, vertical asymptote at $x = -1, 2$

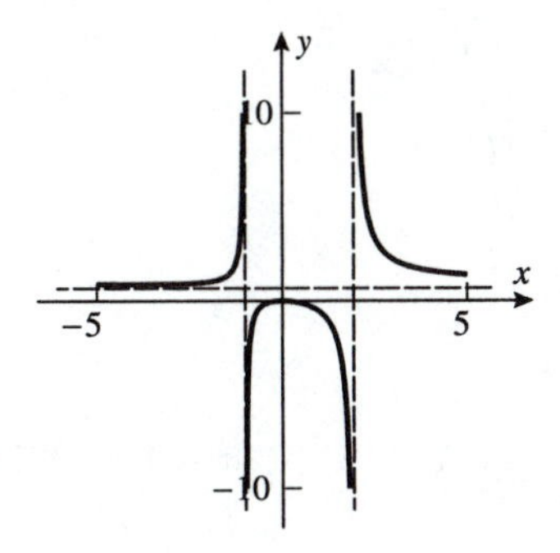

63. (a)

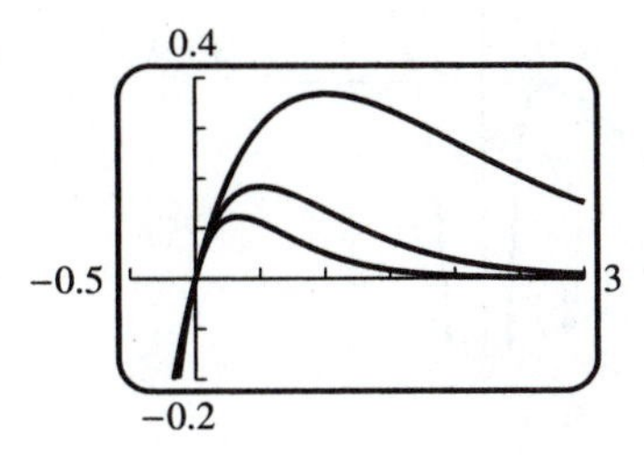

(b) $y' = (1 - bx)e^{-bx}$, $y'' = b^2(x - 2/b)e^{-bx}$; relative maximum at $x = 1/b$, $y = 1/be$; point of inflection at $x = 2/b$, $y = 2/be^2$. Increasing b moves the relative maximum and the point of inflection to the left and down, i.e. towards the origin.

65. (a) The oscillations of $e^x \cos x$ about zero increase as $x \to \pm\infty$ so the limit does not exist.

(b)

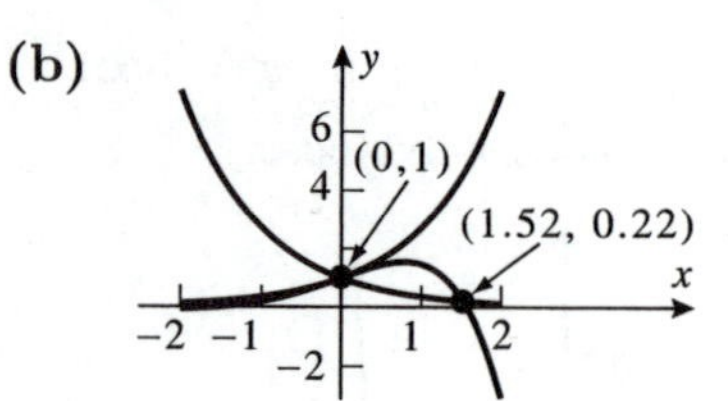

(c) The curve $y = e^{ax}\cos bx$ oscillates between $y = e^{ax}$ and $y = -e^{ax}$. The frequency of oscillation increases when b increases.

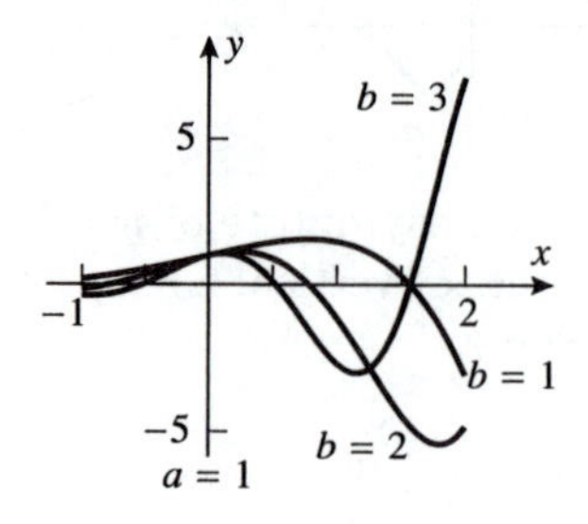

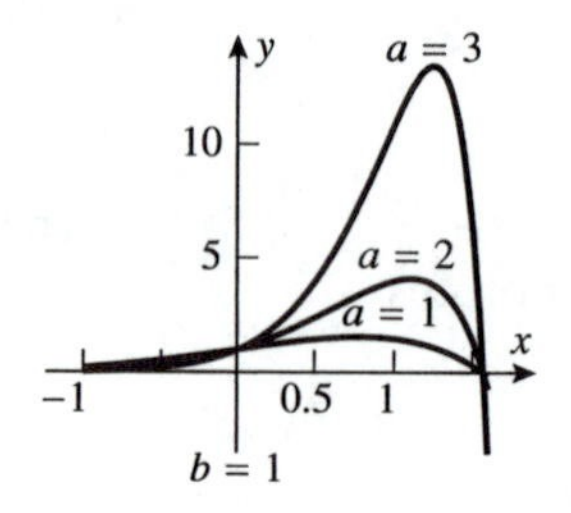

67. $y = \dfrac{x^2-2}{x} = x - \dfrac{2}{x}$ so

$y = x$ is an oblique asymptote;

$y' = \dfrac{x^2+2}{x^2}$,

$y'' = -\dfrac{4}{x^3}$

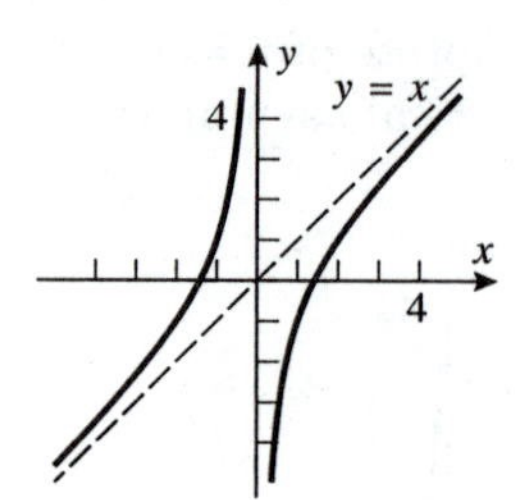

69. $y = \dfrac{(x-2)^3}{x^2} = x - 6 + \dfrac{12x-8}{x^2}$ so

$y = x - 6$ is an oblique asymptote;

$y' = \dfrac{(x-2)^2(x+4)}{x^3}$,

$y'' = \dfrac{24(x-2)}{x^4}$

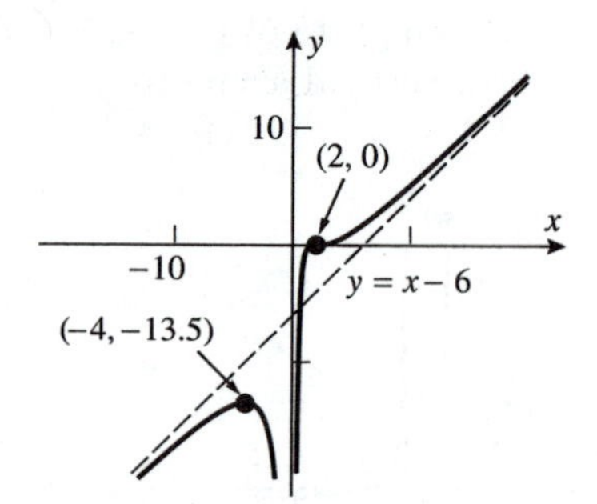

71. $y = x + 1 - \dfrac{1}{x} - \dfrac{1}{x^2} = \dfrac{(x-1)(x+1)^2}{x^2}$,

$y = x + 1$ is an oblique asymptote;

$y' = \dfrac{(x+1)(x^2-x+2)}{x^3}$,

$y'' = -\dfrac{2(x+3)}{x^4}$

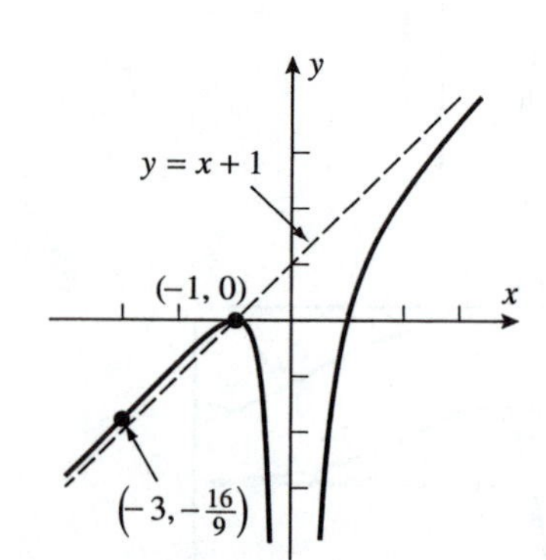

73. $\lim_{x\to\pm\infty}[f(x)-x^2]=\lim_{x\to\pm\infty}(1/x)=0$

$$y=x^2+\frac{1}{x}=\frac{x^3+1}{x},\ y'=2x-\frac{1}{x^2}=\frac{2x^3-1}{x^2},$$

$$y''=2+\frac{2}{x^3}=\frac{2(x^3+1)}{x^3},\ y'=0 \text{ when } x=1/\sqrt[3]{2}\approx 0.8,$$

$$y=3\sqrt[3]{2}/2\approx 1.9;\ y''=0 \text{ when } x=-1, y=0$$

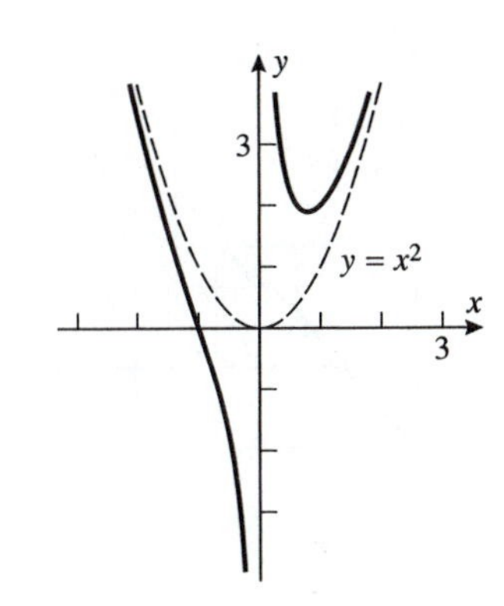

75. Let y be the length of the other side of the rectangle, then $L=2x+2y$ and $xy=400$ so $y=400/x$ and hence $L=2x+800/x$. $L=2x$ is an oblique asymptote (see Exercise 66)

$$L=2x+\frac{800}{x}=\frac{2(x^2+400)}{x},$$

$$L'=2-\frac{800}{x^2}=\frac{2(x^2-400)}{x^2},$$

$$L''=\frac{1600}{x^3},$$

$L'=0$ when $x=20, L=80$

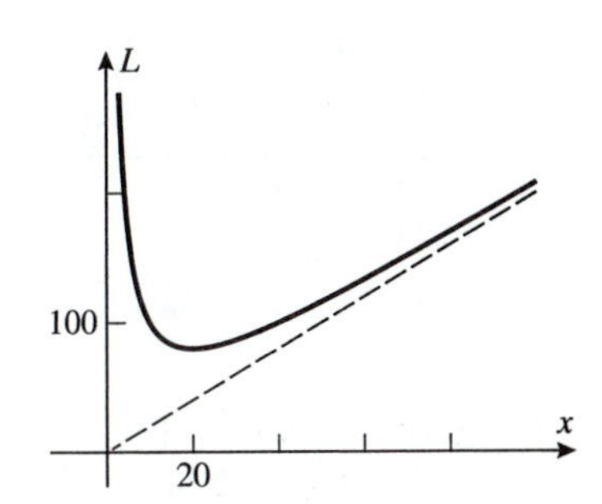

77. $y'=0.1x^4(6x-5)$;
critical numbers: $x=0$, $x=5/6$;
relative minimum at $x=5/6$,
$y\approx -6.7\times 10^{-3}$

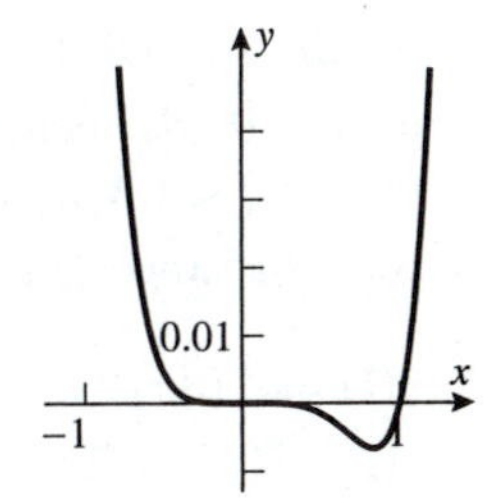

EXERCISE SET 5.4

1. **(a)** positive, negative, slowing down
(b) positive, positive, speeding up
(c) negative, positive, slowing down

3. **(a)** left because $v=ds/dt<0$ at t_0
(b) negative because $a=d^2s/dt^2$ and the curve is concave down at $t_0(d^2s/dt^2<0)$
(c) speeding up because v and a have the same sign
(d) $v<0$ and $a>0$ at t_1 so the particle is slowing down because v and a have opposite signs.

5. 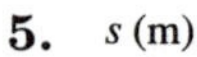

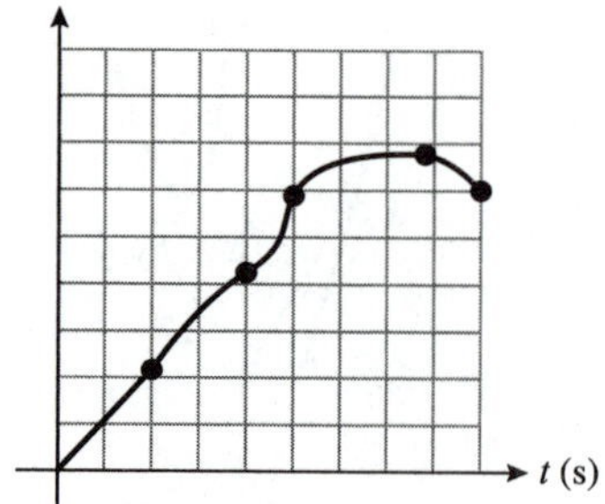

7. 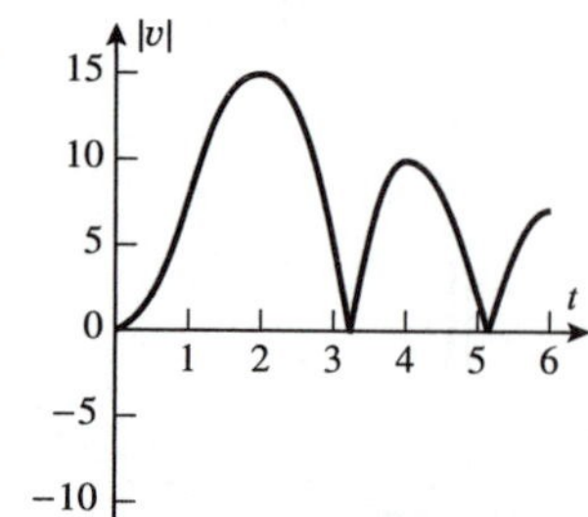

9. **(a)** At 60 mi/h the slope of the estimated tangent line is about 4.6 mi/h/s. Use 1 mi $= 5,280$ ft and 1 h $= 3600$ s to get $a = dv/dt \approx 4.6(5{,}280)/(3600) \approx 6.7$ ft/s^2.

(b) The slope of the tangent to the curve is maximum at $t = 0$ s.

11. **(a)** $v(t) = 3t^2 - 12t$, $a(t) = 6t - 12$

(b) $s(1) = -5$ ft, $v(1) = -9$ ft/s, speed $= 9$ ft/s, $a(1) = -6$ ft/s^2

(c) $v = 0$ at $t = 0, 4$

(d) for $t \geq 0$, $v(t)$ changes sign at $t = 4$, and $a(t)$ changes sign at $t = 2$; so the particle is speeding up for $0 < t < 2$ and $4 < t$ and is slowing down for $2 < t < 4$

(e) total distance $= |s(4) - s(0)| + |s(5) - s(4)| = |-32 - 0| + |-25 - (-32)| = 39$ ft

13. **(a)** $v(t) = -(3\pi/2)\sin(\pi t/2)$, $a(t) = -(3\pi^2/4)\cos(\pi t/2)$

(b) $s(1) = 0$ ft, $v(1) = -3\pi/2$ ft/s, speed $= 3\pi/2$ ft/s, $a(1) = 0$ ft/s^2

(c) $v = 0$ at $t = 0, 2, 4$

(d) v changes sign at $t = 0, 2, 4$ and a changes sign at $t = 1, 3, 5$, so the particle is speeding up for $0 < t < 1$, $2 < t < 3$ and $4 < t < 5$, and it is slowing down for $1 < t < 2$ and $3 < t < 4$

(e)
$$\begin{aligned}\text{total distance} &= |s(2) - s(0)| + |s(4) - s(2)| + |s(5) - s(4)| \\ &= |-3 - 3| + |3 - (-3)| + |0 - 3| = 15 \text{ ft}\end{aligned}$$

15. $v(t) = \dfrac{5 - t^2}{(t^2 + 5)^2}$, $a(t) = \dfrac{2t(t^2 - 15)}{(t^2 + 5)^3}$

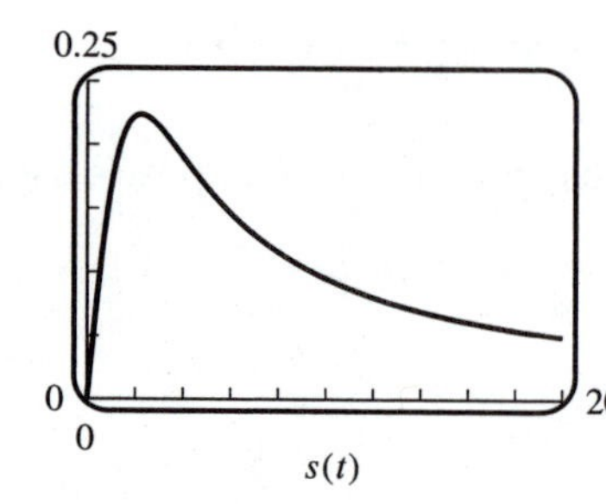

$s(t)$

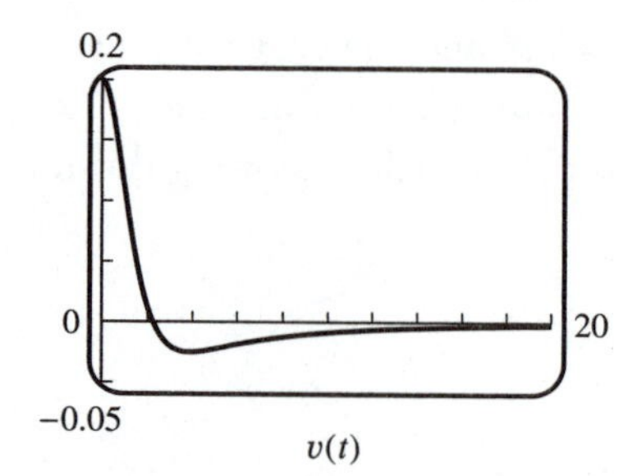

$v(t)$

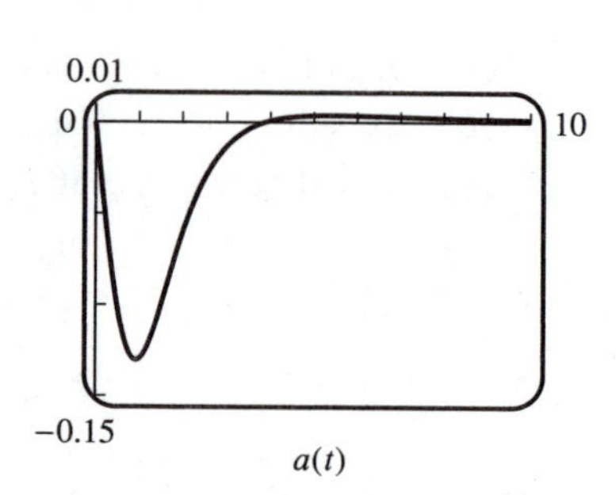

$a(t)$

(a) $v = 0$ at $t = \sqrt{5}$ **(b)** $s = \sqrt{5}/10$ at $t = \sqrt{5}$

(c) a changes sign at $t = \sqrt{15}$, so the particle is speeding up for $\sqrt{5} < t < \sqrt{15}$ and slowing down for $0 < t < \sqrt{5}$ and $\sqrt{15} < t$

17. 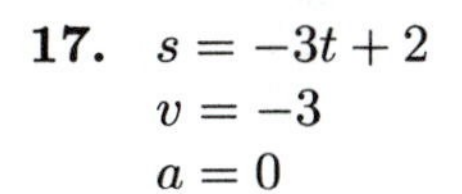
$s = -3t + 2$
$v = -3$
$a = 0$

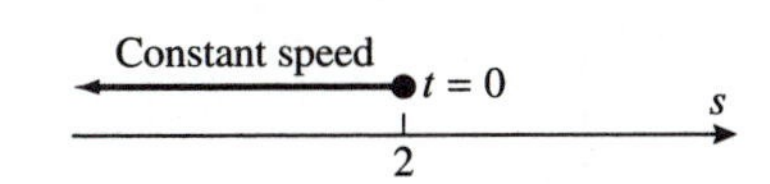

19. $s = t^3 - 9t^2 + 24t$
$v = 3(t - 2)(t - 4)$
$a = 6(t - 3)$

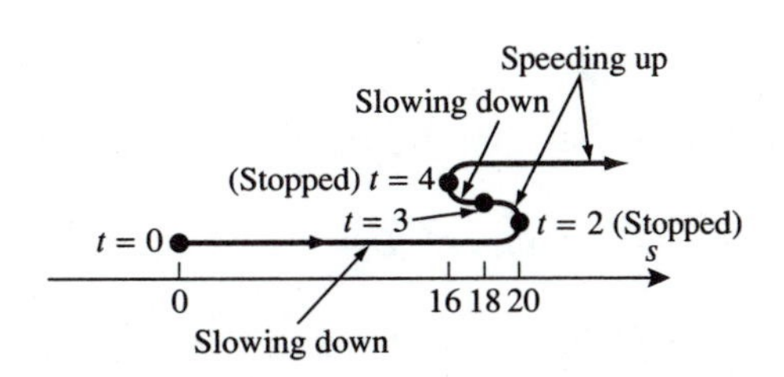

21. $s = \begin{cases} \cos t, & 0 \le t \le 2\pi \\ 1, & t > 2\pi \end{cases}$

$v = \begin{cases} -\sin t, & 0 \le t \le 2\pi \\ 0, & t > 2\pi \end{cases}$

$a = \begin{cases} -\cos t, & 0 \le t < 2\pi \\ 0, & t > 2\pi \end{cases}$

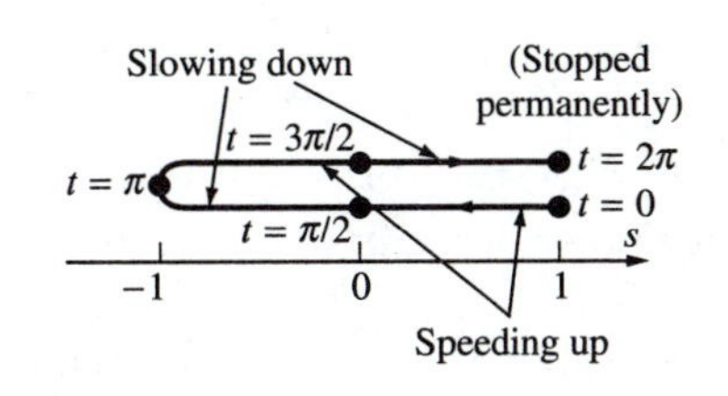

23. **(a)** $v = 10t - 22$, speed $= |v| = |10t - 22|$. $d|v|/dt$ does not exist at $t = 2.2$ which is the only critical point. If $t = 1, 2.2, 3$ then $|v| = 12, 0, 8$. The maximum speed is 12 ft/s.

(b) the distance from the origin is $|s| = |5t^2 - 22t| = |t(5t - 22)|$, but $t(5t - 22) < 0$ for $1 \le t \le 3$ so $|s| = -(5t^2 - 22t) = 22t - 5t^2$, $d|s|/dt = 22 - 10t$, thus the only critical point is $t = 2.2$. $d^2|s|/dt^2 < 0$ so the particle is farthest from the origin when $t = 2.2$. Its position is $s = 5(2.2)^2 - 22(2.2) = -24.2$.

25. $s(t) = s_0 - \frac{1}{2}gt^2 = s_0 - 4.9t^2$ m, $v = -9.8t$ m/s, $a = -9.8$ m/s^2

(a) $|s(1.5) - s(0)| = 11.025$ m

(b) $v(1.5) = -14.7$ m/s

(c) $|v(t)| = 12$ when $t = 12/9.8 = 1.2245$ s

(d) $s(t) - s_0 = -100$ when $4.9t^2 = 100$, $t = 4.5175$ s

27. $s(t) = s_0 + v_0 t - \frac{1}{2}gt^2 = 60t - 4.9t^2$ m and $v(t) = v_0 - gt = 60 - 9.8t$ m/s

(a) $v(t) = 0$ when $t = 60/9.8 \approx 6.12$ s

(b) $s(60/9.8) \approx 183.67$ m

(c) another 6.12 s; solve for t in $s(t) = 0$ to get this result, or use the symmetry of the parabola $s = 60t - 4.9t^2$ about the line $t = 6.12$ in the t-s plane

(d) also 60 m/s, as seen from the symmetry of the parabola (or compute $v(6.12)$)

29. If g $= 32$ ft/s^2, $s_0 = 7$ and v_0 is unknown, then $s(t) = 7 + v_0 t - 16t^2$ and $v(t) = v_0 - 32t$; $s = s_{\max}$ when $v = 0$, or $t = v_0/32$; and $s_{\max} = 208$ yields
$208 = s(v_0/32) = 7 + v_0(v_0/32) - 16(v_0/32)^2 = 7 + v_0^2/64$, so $v_0 = 8\sqrt{201} \approx 113.42$ ft/s.

31. $v_0 = 0$ and $g = 9.8$, so $v^2 = -19.6(s - s_0)$; since $v = 24$ when $s = 0$ it follows that $19.6s_0 = 24^2$ or $s_0 = 29.39$ m.

33. **(a)** $s = s_{\max}$ when $v = 0$, so $0 = v_0^2 - 2g(s_{\max} - s_0)$, $s_{\max} = v_0^2/2g + s_0$.

(b) $s_0 = 7$, $s_{\max} = 208$, $g = 32$ and v_0 is unknown, so from Part (a) $v_0^2 = 2g(208 - 7) = 64 \cdot 201$, $v_0 = 8\sqrt{201} \approx 113.42$ ft/s.

35. (a)

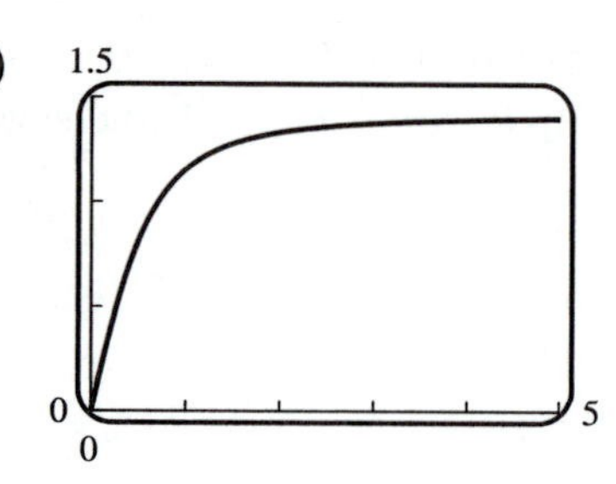

(b) $v = \dfrac{2t}{\sqrt{2t^2+1}}$, $\lim\limits_{t\to+\infty} v = \dfrac{2}{\sqrt{2}} = \sqrt{2}$

37. (a) $s_1 = s_2$ if they collide, so $\frac{1}{2}t^2 - t + 3 = -\frac{1}{4}t^2 + t + 1$, $\frac{3}{4}t^2 - 2t + 2 = 0$ which has no real solution.

(b) Find the minimum value of $D = |s_1 - s_2| = \left|\frac{3}{4}t^2 - 2t + 2\right|$. From Part (a), $\frac{3}{4}t^2 - 2t + 2$ is never zero, and for $t = 0$ it is positive, hence it is always positive, so $D = \frac{3}{4}t^2 - 2t + 2$.

$\dfrac{dD}{dt} = \frac{3}{2}t - 2 = 0$ when $t = \frac{4}{3}$. $\dfrac{d^2D}{dt^2} > 0$ so D is minimum when $t = \frac{4}{3}$, $D = \frac{2}{3}$.

(c) $v_1 = t - 1$, $v_2 = -\dfrac{1}{2}t + 1$. $v_1 < 0$ if $0 \le t < 1$, $v_1 > 0$ if $t > 1$; $v_2 < 0$ if $t > 2$, $v_2 > 0$ if $0 \le t < 2$. They are moving in opposite directions during the intervals $0 \le t < 1$ and $t > 2$.

39. $r(t) = \sqrt{v^2(t)}$, $r'(t) = 2v(t)v'(t)/[2\sqrt{v(t)}] = v(t)a(t)/|v(t)|$ so $r'(t) > 0$ (speed is increasing) if v and a have the same sign, and $r'(t) < 0$ (speed is decreasing) if v and a have opposite signs.
If $v(t) > 0$ then $r(t) = v(t)$ and $r'(t) = a(t)$, so if $a(t) > 0$ then the particle is speeding up and a and v have the same sign; if $a(t) < 0$, then the particle is slowing down, and a and v have opposite signs.
If $v(t) < 0$ then $r(t) = -v(t)$, $r'(t) = -a(t)$, and if $a(t) > 0$ then the particle is speeding up and a and v have opposite signs; if $a(t) < 0$ then the particle is slowing down and a and v have the same sign.

EXERCISE SET 5.5

1. relative maxima at $x = 2, 6$; absolute maximum at $x = 6$; relative and absolute minimum at $x = 4$

3. (a)

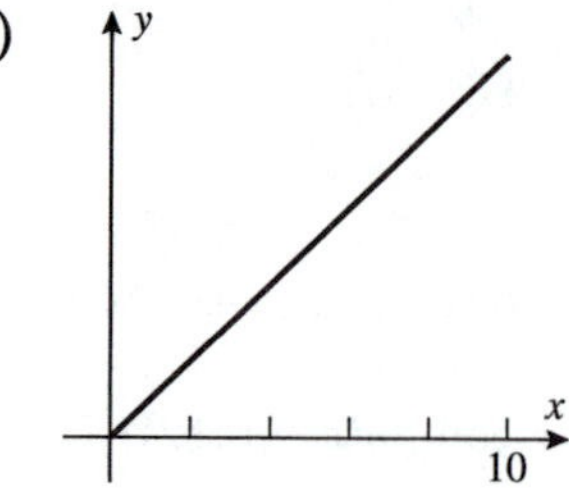

(b)

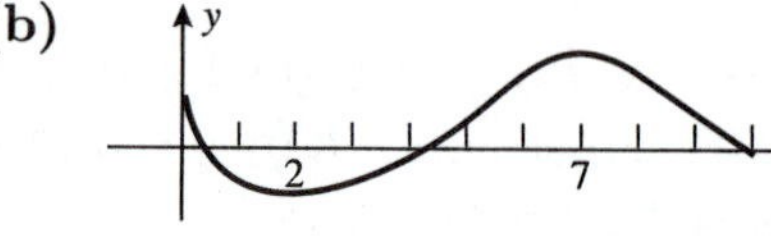

(c) y, x; 3, 5, 7

5. $f'(x) = 8x - 4$, $f'(x) = 0$ when $x = 1/2$; $f(0) = 1$, $f(1/2) = 0$, $f(1) = 1$ so the maximum value is 1 at $x = 0, 1$ and the minimum value is 0 at $x = 1/2$.

7. $f'(x) = 3(x-1)^2$, $f'(x) = 0$ when $x = 1$; $f(0) = -1$, $f(1) = 0$, $f(4) = 27$ so the maximum value is 27 at $x = 4$ and the minimum value is -1 at $x = 0$.

9. $f'(x) = 3/(4x^2+1)^{3/2}$, no critical points; $f(-1) = -3/\sqrt{5}$, $f(1) = 3/\sqrt{5}$ so the maximum value is $3/\sqrt{5}$ at $x = 1$ and the minimum value is $-3/\sqrt{5}$ at $x = -1$.

11. $f'(x) = 1 - \sec^2 x$, $f'(x) = 0$ for x in $(-\pi/4, \pi/4)$ when $x = 0$; $f(-\pi/4) = 1 - \pi/4$, $f(0) = 0$, $f(\pi/4) = \pi/4 - 1$ so the maximum value is $1 - \pi/4$ at $x = -\pi/4$ and the minimum value is $\pi/4 - 1$ at $x = \pi/4$.

13. $f(x) = 1 + |9 - x^2| = \begin{cases} 10 - x^2, & |x| \le 3 \\ -8 + x^2, & |x| > 3 \end{cases}$, $f'(x) = \begin{cases} -2x, & |x| < 3 \\ 2x, & |x| > 3 \end{cases}$ thus $f'(x) = 0$ when $x = 0$, $f'(x)$ does not exist for x in $(-5, 1)$ when $x = -3$ because $\lim_{x \to -3^-} f'(x) \neq \lim_{x \to -3^+} f'(x)$ (see Theorem preceding Exercise 75, Section 3.3); $f(-5) = 17$, $f(-3) = 1$, $f(0) = 10$, $f(1) = 9$ so the maximum value is 17 at $x = -5$ and the minimum value is 1 at $x = -3$.

15. $f'(x) = 2x - 3$; critical point $x = 3/2$. Minimum value $f(3/2) = -13/4$, no maximum.

17. $f'(x) = 12x^2(1 - x)$; critical points $x = 0, 1$. Maximum value $f(1) = 1$, no minimum because $\lim_{x \to +\infty} f(x) = -\infty$.

19. No maximum or minimum because $\lim_{x \to +\infty} f(x) = +\infty$ and $\lim_{x \to -\infty} f(x) = -\infty$.

21. $f'(x) = x(x+2)/(x+1)^2$; critical point $x = -2$ in $(-5, -1)$. Maximum value $f(-2) = -4$, no minimum.

23. $(x^2 - 1)^2$ can never be less than zero because it is the square of $x^2 - 1$; the minimum value is 0 for $x = \pm 1$, no maximum because $\lim_{x \to +\infty} f(x) = +\infty$.

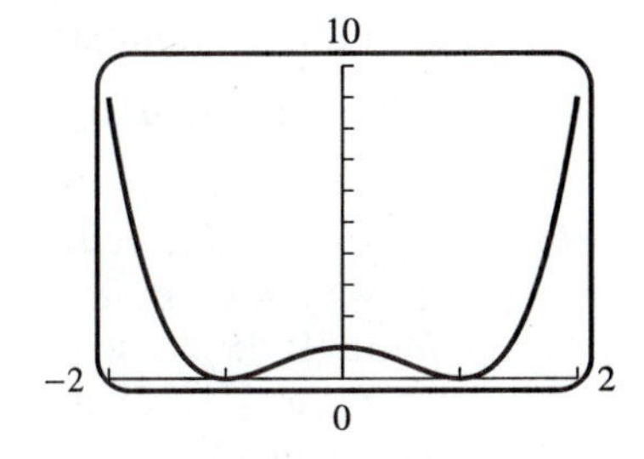

25. $f'(x) = \dfrac{5(8 - x)}{3x^{1/3}}$, $f'(x) = 0$ when $x = 8$ and $f'(x)$ does not exist when $x = 0$; $f(-1) = 21$, $f(0) = 0$, $f(8) = 48$, $f(20) = 0$ so the maximum value is 48 at $x = 8$ and the minimum value is 0 at $x = 0, 20$.

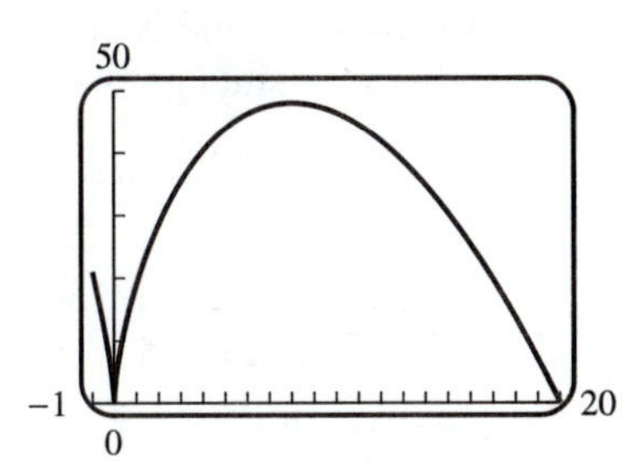

27. $f'(x) = -1/x^2$; no maximum or minimum because there are no critical points in $(0, +\infty)$.

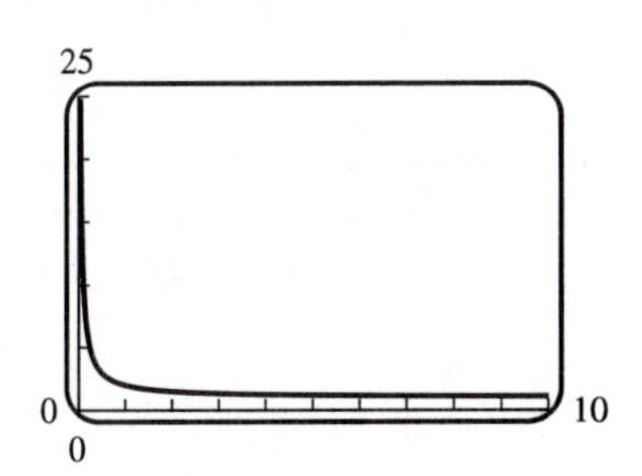

29. $f'(x) = 2\sec x \tan x - \sec^2 x = (2\sin x - 1)/\cos^2 x$, $f'(x) = 0$ for x in $(0, \pi/4)$ when $x = \pi/6$; $f(0) = 2$, $f(\pi/6) = \sqrt{3}$, $f(\pi/4) = 2\sqrt{2}-1$ so the maximum value is 2 at $x = 0$ and the minimum value is $\sqrt{3}$ at $x = \pi/6$.

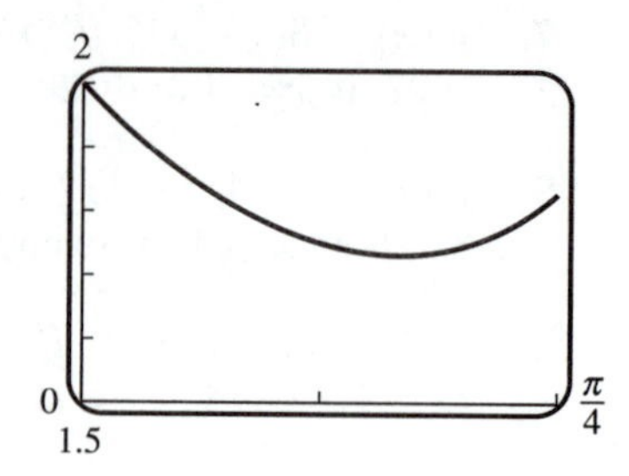

31. $f'(x) = x^2(2x-3)e^{-2x}$, $f'(x) = 0$ for x in $[1, 4]$ when $x = 3/2$;
if $x = 1, 3/2, 4$, then $f(x) = e^{-2}, \dfrac{27}{8}e^{-3}, 64e^{-8}$;
critical point at $x = 3/2$; absolute maximum of $\dfrac{27}{8}e^{-3}$ at $x = 3/2$,
absolute minimum of $64e^{-8}$ at $x = 4$

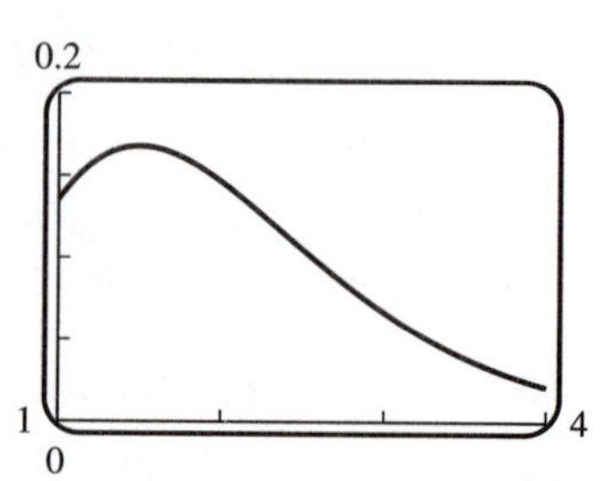

33. $f'(x) = 1/2 + 2x/(x^2+1)$,
$f'(x) = 0$ on $[-4, 0]$ for $x = -2 \pm \sqrt{3}$
if $x = -2 - \sqrt{3}, -2 + \sqrt{3}$ then
$f(x) = -1 - \sqrt{3}/2 + \ln 4 + \ln(2+\sqrt{3}) \approx 0.84$,
$-1 + \sqrt{3}/2 + \ln 4 + \ln(2 - \sqrt{3}) \approx -0.06$,
absolute maximum at $x = -2 - \sqrt{3}$,
absolute minimum at $x = -2 + \sqrt{3}$

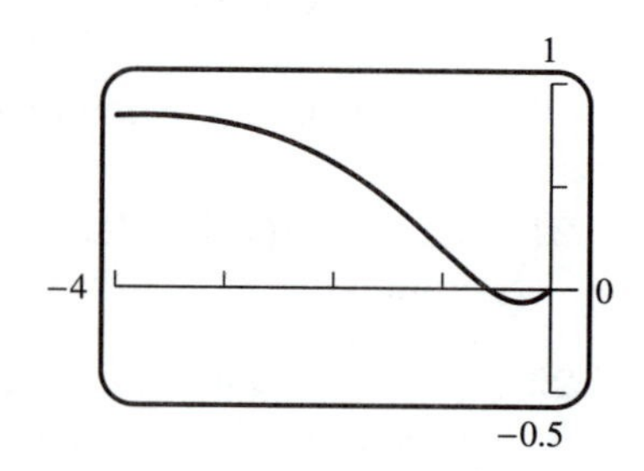

35. $f'(x) = -[\cos(\cos x)]\sin x$; $f'(x) = 0$ if $\sin x = 0$ or if $\cos(\cos x) = 0$. If $\sin x = 0$, then $x = \pi$ is the critical point in $(0, 2\pi)$; $\cos(\cos x) = 0$ has no solutions because $-1 \le \cos x \le 1$. Thus $f(0) = \sin(1)$, $f(\pi) = \sin(-1) = -\sin(1)$, and $f(2\pi) = \sin(1)$ so the maximum value is $\sin(1) \approx 0.84147$ and the minimum value is $-\sin(1) \approx -0.84147$.

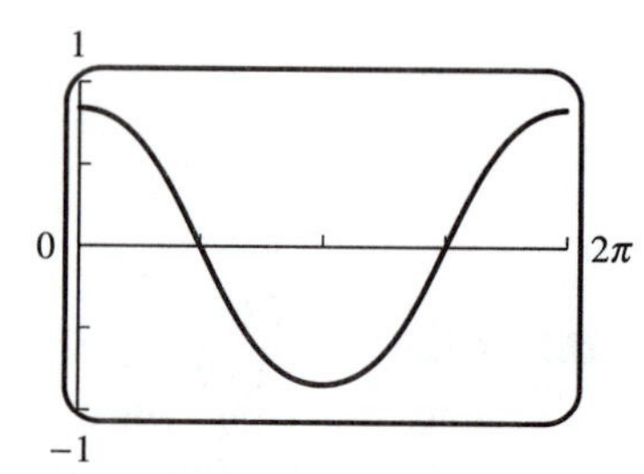

37. $f'(x) = \begin{cases} 4, & x < 1 \\ 2x - 5, & x > 1 \end{cases}$ so $f'(x) = 0$ when $x = 5/2$, and $f'(x)$ does not exist when $x = 1$ because $\lim\limits_{x\to 1^-} f'(x) \ne \lim\limits_{x \to 1^+} f'(x)$ (see Theorem preceding Exercise 75, Section 3.3); $f(1/2) = 0$, $f(1) = 2$, $f(5/2) = -1/4$, $f(7/2) = 3/4$ so the maximum value is 2 and the minimum value is $-1/4$.

39. $\sin 2x$ has a period of π, and $\sin 4x$ a period of $\pi/2$ so $f(x)$ is periodic with period π. Consider the interval $[0, \pi]$. $f'(x) = 4\cos 2x + 4\cos 4x$, $f'(x) = 0$ when $\cos 2x + \cos 4x = 0$, but $\cos 4x = 2\cos^2 2x - 1$ (trig identity) so

$$2\cos^2 2x + \cos 2x - 1 = 0$$
$$(2\cos 2x - 1)(\cos 2x + 1) = 0$$
$$\cos 2x = 1/2 \quad \text{or} \quad \cos 2x = -1.$$

From $\cos 2x = 1/2$, $2x = \pi/3$ or $5\pi/3$ so $x = \pi/6$ or $5\pi/6$. From $\cos 2x = -1$, $2x = \pi$ so $x = \pi/2$. $f(0) = 0$, $f(\pi/6) = 3\sqrt{3}/2$, $f(\pi/2) = 0$, $f(5\pi/6) = -3\sqrt{3}/2$, $f(\pi) = 0$. The maximum value is $3\sqrt{3}/2$ at $x = \pi/6 + n\pi$ and the minimum value is $-3\sqrt{3}/2$ at $x = 5\pi/6 + n\pi$, $n = 0, \pm 1, \pm 2, \cdots$.

41. Let $f(x) = x - \sin x$, then $f'(x) = 1 - \cos x$ and so $f'(x) = 0$ when $\cos x = 1$ which has no solution for $0 < x < 2\pi$ thus the minimum value of f must occur at 0 or 2π. $f(0) = 0$, $f(2\pi) = 2\pi$ so 0 is the minimum value on $[0, 2\pi]$ thus $x - \sin x \geq 0$, $\sin x \leq x$ for all x in $[0, 2\pi]$.

43. Let $m =$ slope at x, then $m = f'(x) = 3x^2 - 6x + 5$, $dm/dx = 6x - 6$; critical point for m is $x = 1$, minimum value of m is $f'(1) = 2$

45. $f'(x) = \dfrac{2x(x^3 - 24x^2 + 192x - 640)}{(x-8)^3}$; real root of $x^3 - 24x^2 + 192x - 640$ at $x = 4(2 + \sqrt[3]{2})$. Since $\lim_{x \to 8^+} f(x) = \lim_{x \to +\infty} f(x) = +\infty$ and there is only one relative extremum, it must be a minimum.

47. The slope of the line is -1, and the slope of the tangent to $y = -x^2$ is $-2x$ so $-2x = -1$, $x = 1/2$. The line lies above the curve so the vertical distance is given by $F(x) = 2 - x + x^2$; $F(-1) = 4$, $F(1/2) = 7/4$, $F(3/2) = 11/4$. The point $(1/2, -1/4)$ is closest, the point $(-1, -1)$ farthest.

49. The absolute extrema of $y(t)$ can occur at the endpoints $t = 0, 12$ or when $dy/dt = 2\sin t = 0$, i.e. $t = 0, 12, k\pi$, $k = 1, 2, 3$; the absolute maximum is $y = 4$ at $t = \pi, 3\pi$; the absolute minimum is $y = 0$ at $t = 0, 2\pi$.

51. $f'(x) = 2ax + b$; critical point is $x = -\dfrac{b}{2a}$

$f''(x) = 2a > 0$ so $f\left(-\dfrac{b}{2a}\right)$ is the minimum value of f, but

$f\left(-\dfrac{b}{2a}\right) = a\left(-\dfrac{b}{2a}\right)^2 + b\left(-\dfrac{b}{2a}\right) + c = \dfrac{-b^2 + 4ac}{4a}$ thus $f(x) \geq 0$ if and only if

$f\left(-\dfrac{b}{2a}\right) \geq 0$, $\dfrac{-b^2 + 4ac}{4a} \geq 0$, $-b^2 + 4ac \geq 0$, $b^2 - 4ac \leq 0$

EXERCISE SET 5.6

1. Let $x =$ one number, $y =$ the other number, and $P = xy$ where $x + y = 10$. Thus $y = 10 - x$ so $P = x(10 - x) = 10x - x^2$ for x in $[0, 10]$. $dP/dx = 10 - 2x$, $dP/dx = 0$ when $x = 5$. If $x = 0, 5, 10$ then $P = 0, 25, 0$ so P is maximum when $x = 5$ and, from $y = 10 - x$, when $y = 5$.

3. If $y = x + 1/x$ for $1/2 \leq x \leq 3/2$ then $dy/dx = 1 - 1/x^2 = (x^2 - 1)/x^2$, $dy/dx = 0$ when $x = 1$. If $x = 1/2, 1, 3/2$ then $y = 5/2, 2, 13/6$ so

(a) y is as small as possible when $x = 1$.

(b) y is as large as possible when $x = 1/2$.

5. Let x and y be the dimensions shown in the figure and A the area, then $A = xy$ subject to the cost condition $3(2x) + 2(2y) = 6000$, or $y = 1500 - 3x/2$. Thus $A = x(1500 - 3x/2) = 1500x - 3x^2/2$ for x in $[0, 1000]$. $dA/dx = 1500 - 3x$, $dA/dx = 0$ when $x = 500$. If $x = 0$ or 1000 then $A = 0$, if $x = 500$ then $A = 375{,}000$ so the area is greatest when $x = 500$ ft and (from $y = 1500 - 3x/2$) when $y = 750$ ft.

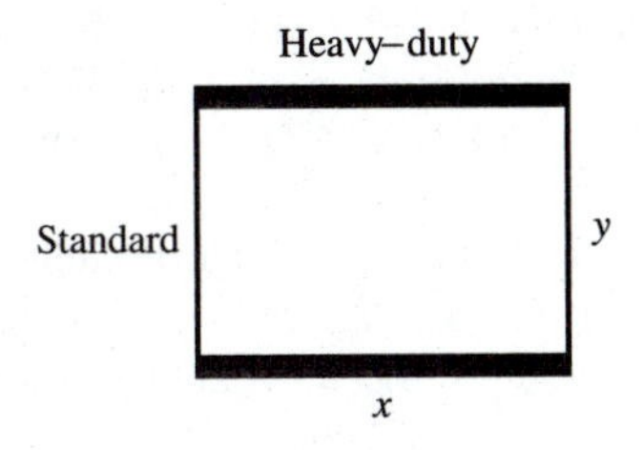

7. Let x, y, and z be as shown in the figure and A the area of the rectangle, then $A = xy$ and, by similar triangles, $z/10 = y/6$, $z = 5y/3$; also $x/10 = (8 - z)/8 = (8 - 5y/3)/8$ thus $y = 24/5 - 12x/25$ so $A = x(24/5 - 12x/25) = 24x/5 - 12x^2/25$ for x in $[0, 10]$. $dA/dx = 24/5 - 24x/25$, $dA/dx = 0$ when $x = 5$. If $x = 0, 5, 10$ then $A = 0, 12, 0$ so the area is greatest when $x = 5$ in. and $y = 12/5$ in.

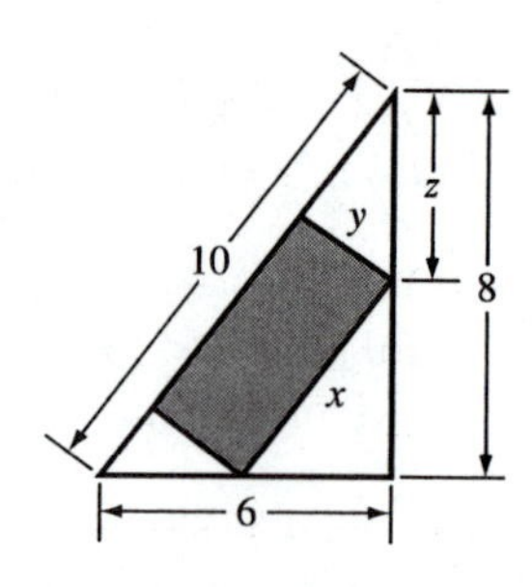

9. $A = xy$ where $x^2 + y^2 = 20^2 = 400$ so $y = \sqrt{400 - x^2}$ and $A = x\sqrt{400 - x^2}$ for $0 \le x \le 20$;
$dA/dx = 2(200 - x^2)/\sqrt{400 - x^2}$, $dA/dx = 0$ when $x = \sqrt{200} = 10\sqrt{2}$. If $x = 0, 10\sqrt{2}, 20$ then $A = 0, 200, 0$ so the area is maximum when $x = 10\sqrt{2}$ and $y = \sqrt{400 - 200} = 10\sqrt{2}$.

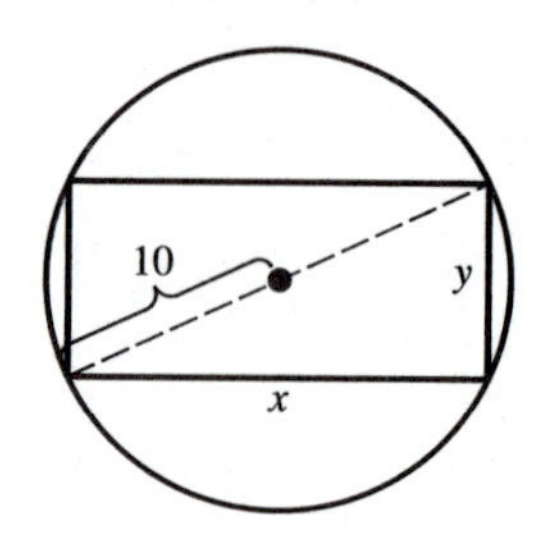

11. Let $x =$ length of each side that uses the \$1 per foot fencing,
$y =$ length of each side that uses the \$2 per foot fencing.
The cost is $C = (1)(2x) + (2)(2y) = 2x + 4y$, but $A = xy = 3200$ thus $y = 3200/x$ so

$$C = 2x + 12800/x \text{ for } x > 0,$$
$$dC/dx = 2 - 12800/x^2, \ dC/dx = 0 \text{ when } x = 80, \ d^2C/dx^2 > 0 \text{ so}$$

C is least when $x = 80$, $y = 40$.

13. Let x and y be the dimensions of a rectangle; the perimeter is $p = 2x + 2y$. But $A = xy$ thus $y = A/x$ so $p = 2x + 2A/x$ for $x > 0$, $dp/dx = 2 - 2A/x^2 = 2(x^2 - A)/x^2$, $dp/dx = 0$ when $x = \sqrt{A}$, $d^2p/dx^2 = 4A/x^3 > 0$ if $x > 0$ so p is a minimum when $x = \sqrt{A}$ and $y = \sqrt{A}$ and thus the rectangle is a square.

15. (a) $\dfrac{dN}{dt} = 250(20 - t)e^{-t/20} = 0$ at $t = 20$, $N(0) = 125{,}000$, $N(20) \approx 161{,}788$, and $N(100) \approx 128{,}369$; the absolute maximum is $N = 161788$ at $t = 20$, the absolute minimum is $N = 125{,}000$ at $t = 0$.

(b) The absolute minimum of $\dfrac{dN}{dt}$ occurs when $\dfrac{d^2N}{dt^2} = 12.5(t - 40)e^{-t/20} = 0$, $t = 40$.

17. $V = x(12 - 2x)^2$ for $0 \le x \le 6$;
$dV/dx = 12(x - 2)(x - 6)$, $dV/dx = 0$ when $x = 2$ for $0 < x < 6$. If $x = 0, 2, 6$ then $V = 0, 128, 0$ so the volume is largest when $x = 2$ in.

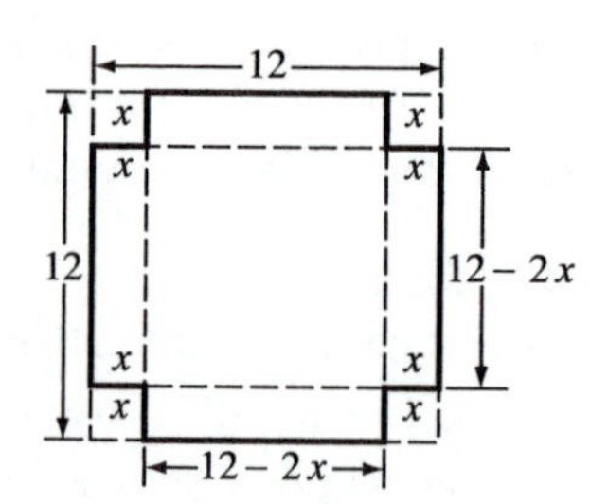

19. Let x be the length of each side of a square, then $V = x(3 - 2x)(8 - 2x) = 4x^3 - 22x^2 + 24x$ for $0 \le x \le 3/2$; $dV/dx = 12x^2 - 44x + 24 = 4(3x - 2)(x - 3)$, $dV/dx = 0$ when $x = 2/3$ for $0 < x < 3/2$. If $x = 0, 2/3, 3/2$ then $V = 0, 200/27, 0$ so the maximum volume is $200/27$ ft^3.

21. Let $x =$ length of each edge of base, $y =$ height, $k = \$/\text{cm}^2$ for the sides. The cost is $C = (2k)(2x^2) + (k)(4xy) = 4k(x^2 + xy)$, but $V = x^2y = 2000$ thus $y = 2000/x^2$ so $C = 4k(x^2 + 2000/x)$ for $x > 0$ $dC/dx = 4k(2x - 2000/x^2)$, $dC/dx = 0$ when $x = \sqrt[3]{1000} = 10$, $d^2C/dx^2 > 0$ so C is least when $x = 10$, $y = 20$.

23. Let x = height and width, y = length. The surface area is $S = 2x^2 + 3xy$ where $x^2y = V$, so $y = V/x^2$ and $S = 2x^2 + 3V/x$ for $x > 0$; $dS/dx = 4x - 3V/x^2$, $dS/dx = 0$ when $x = \sqrt[3]{3V/4}$, $d^2S/dx^2 > 0$ so S is minimum when $x = \sqrt[3]{\dfrac{3V}{4}}$, $y = \dfrac{4}{3}\sqrt[3]{\dfrac{3V}{4}}$.

25. Let r and h be the dimensions shown in the figure, then the surface area is $S = 2\pi rh + 2\pi r^2$.

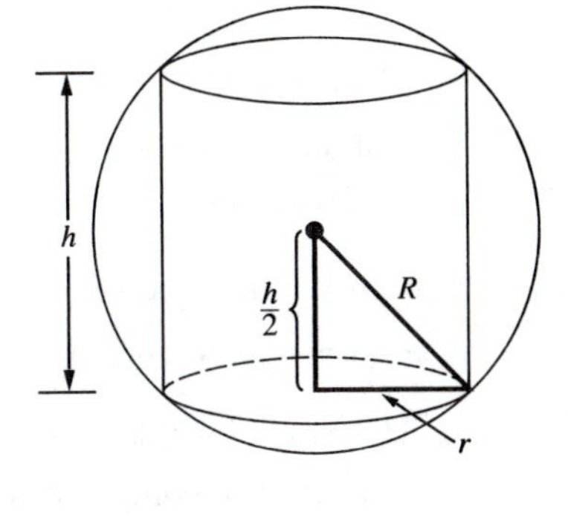

But $r^2 + \left(\dfrac{h}{2}\right)^2 = R^2$ thus $h = 2\sqrt{R^2 - r^2}$ so

$S = 4\pi r\sqrt{R^2 - r^2} + 2\pi r^2$ for $0 \le r \le R$,

$$\frac{dS}{dr} = \frac{4\pi(R^2 - 2r^2)}{\sqrt{R^2 - r^2}} + 4\pi r;\ \frac{dS}{dr} = 0 \text{ when}$$

$$\frac{R^2 - 2r^2}{\sqrt{R^2 - r^2}} = -r \qquad (1)$$

$$R^2 - 2r^2 = -r\sqrt{R^2 - r^2}$$

$$R^4 - 4R^2r^2 + 4r^4 = r^2(R^2 - r^2)$$
$$5r^4 - 5R^2r^2 + R^4 = 0$$

and using the quadratic formula $r^2 = \dfrac{5R^2 \pm \sqrt{25R^4 - 20R^4}}{10} = \dfrac{5 \pm \sqrt{5}}{10}R^2$, $r = \sqrt{\dfrac{5 \pm \sqrt{5}}{10}}R$, of which only $r = \sqrt{\dfrac{5+\sqrt{5}}{10}}R$ satisfies (1). If $r = 0, \sqrt{\dfrac{5+\sqrt{5}}{10}}R, 0$ then $S = 0, (5+\sqrt{5})\pi R^2, 2\pi R^2$ so the surface area is greatest when $r = \sqrt{\dfrac{5+\sqrt{5}}{10}}R$ and, from $h = 2\sqrt{R^2 - r^2}$, $h = 2\sqrt{\dfrac{5-\sqrt{5}}{10}}R$.

27. From (13), $S = 2\pi r^2 + 2\pi rh$. But $V = \pi r^2 h$ thus $h = V/(\pi r^2)$ and so $S = 2\pi r^2 + 2V/r$ for $r > 0$. $dS/dr = 4\pi r - 2V/r^2, dS/dr = 0$ if $r = \sqrt[3]{V/(2\pi)}$. Since $d^2S/dr^2 = 4\pi + 4V/r^3 > 0$, the minimum surface area is achieved when $r = \sqrt[3]{V/2\pi}$ and so $h = V/(\pi r^2) = [V/(\pi r^3)]r = 2r$.

29. The surface area is $S = \pi r^2 + 2\pi rh$
where $V = \pi r^2 h = 500$ so $h = 500/(\pi r^2)$
and $S = \pi r^2 + 1000/r$ for $r > 0$;
$dS/dr = 2\pi r - 1000/r^2 = (2\pi r^3 - 1000)/r^2$,
$dS/dr = 0$ when $r = \sqrt[3]{500/\pi}$, $d^2S/dr^2 > 0$
for $r > 0$ so S is minimum when $r = \sqrt[3]{500/\pi}$ cm and

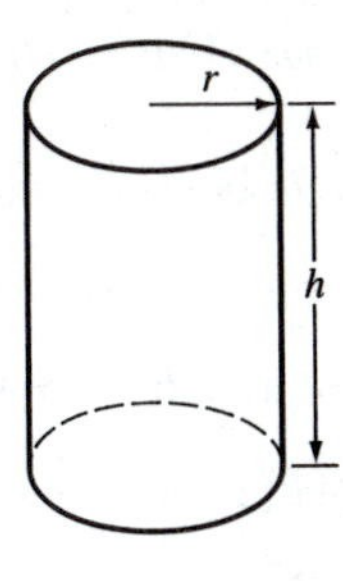

$$h = \frac{500}{\pi r^2} = \frac{500}{\pi}\left(\frac{\pi}{500}\right)^{2/3}$$
$$= \sqrt[3]{500/\pi} \text{ cm}$$

37. Let x be the length of each side of the squares and y the height of the frame, then the volume is $V = x^2y$. The total length of the wire is L thus $8x + 4y = L$, $y = (L - 8x)/4$ so $V = x^2(L - 8x)/4 = (Lx^2 - 8x^3)/4$ for $0 \le x \le L/8$. $dV/dx = (2Lx - 24x^2)/4$, $dV/dx = 0$ for $0 < x < L/8$ when $x = L/12$. If $x = 0, L/12, L/8$ then $V = 0, L^3/1728, 0$ so the volume is greatest when $x = L/12$ and $y = L/12$.

33. Let h and r be the dimensions shown in the figure, then the volume is $V = \frac{1}{3}\pi r^2 h$. But $r^2 + h^2 = L^2$ thus $r^2 = L^2 - h^2$ so $V = \frac{1}{3}\pi(L^2 - h^2)h = \frac{1}{3}\pi(L^2 h - h^3)$ for $0 \le h \le L$. $\frac{dV}{dh} = \frac{1}{3}\pi(L^2 - 3h^2)$. $\frac{dV}{dh} = 0$ when $h = L/\sqrt{3}$. If $h = 0, L/\sqrt{3}, 0$ then $V = 0, \frac{2\pi}{9\sqrt{3}}L^3, 0$ so the volume is as large as possible when $h = L/\sqrt{3}$ and $r = \sqrt{2/3}L$.

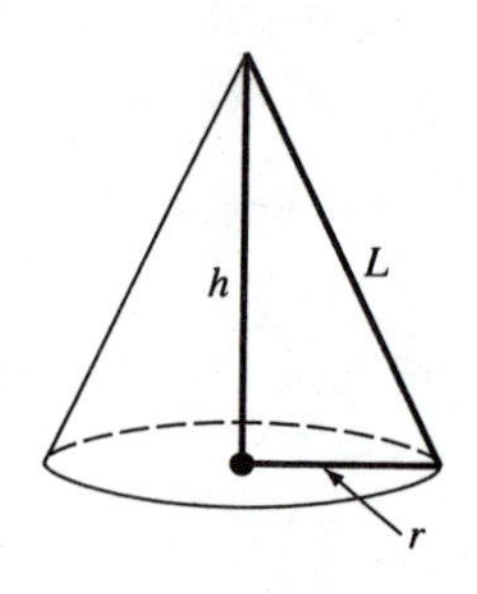

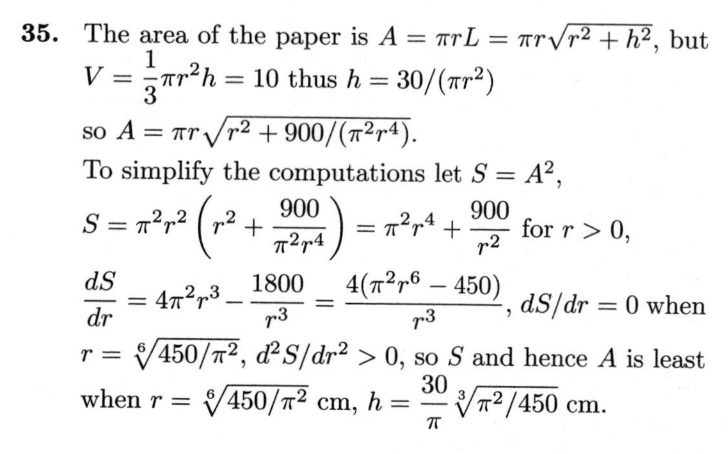

35. The area of the paper is $A = \pi r L = \pi r\sqrt{r^2 + h^2}$, but $V = \frac{1}{3}\pi r^2 h = 10$ thus $h = 30/(\pi r^2)$
so $A = \pi r\sqrt{r^2 + 900/(\pi^2 r^4)}$.
To simplify the computations let $S = A^2$,
$S = \pi^2 r^2\left(r^2 + \frac{900}{\pi^2 r^4}\right) = \pi^2 r^4 + \frac{900}{r^2}$ for $r > 0$,
$\frac{dS}{dr} = 4\pi^2 r^3 - \frac{1800}{r^3} = \frac{4(\pi^2 r^6 - 450)}{r^3}$, $dS/dr = 0$ when $r = \sqrt[6]{450/\pi^2}$, $d^2S/dr^2 > 0$, so S and hence A is least when $r = \sqrt[6]{450/\pi^2}$ cm, $h = \frac{30}{\pi}\sqrt[3]{\pi^2/450}$ cm.

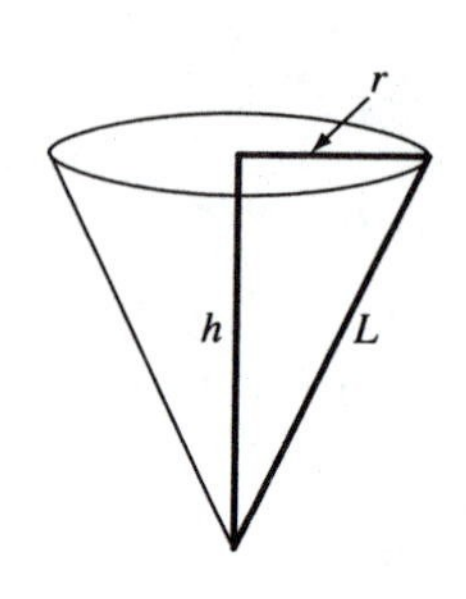

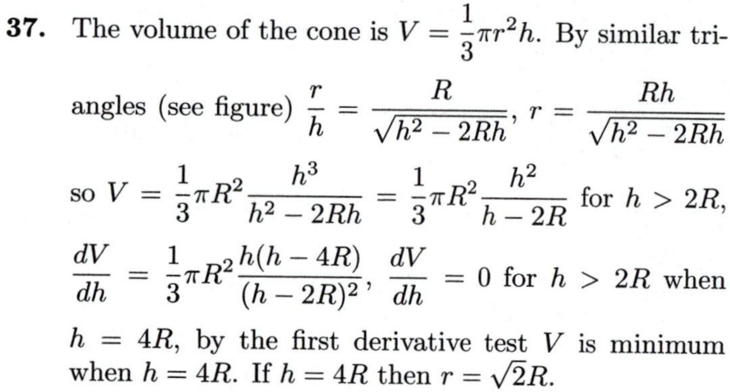

37. The volume of the cone is $V = \frac{1}{3}\pi r^2 h$. By similar triangles (see figure) $\frac{r}{h} = \frac{R}{\sqrt{h^2 - 2Rh}}$, $r = \frac{Rh}{\sqrt{h^2 - 2Rh}}$
so $V = \frac{1}{3}\pi R^2 \frac{h^3}{h^2 - 2Rh} = \frac{1}{3}\pi R^2 \frac{h^2}{h - 2R}$ for $h > 2R$,
$\frac{dV}{dh} = \frac{1}{3}\pi R^2 \frac{h(h - 4R)}{(h - 2R)^2}$, $\frac{dV}{dh} = 0$ for $h > 2R$ when $h = 4R$, by the first derivative test V is minimum when $h = 4R$. If $h = 4R$ then $r = \sqrt{2}R$.

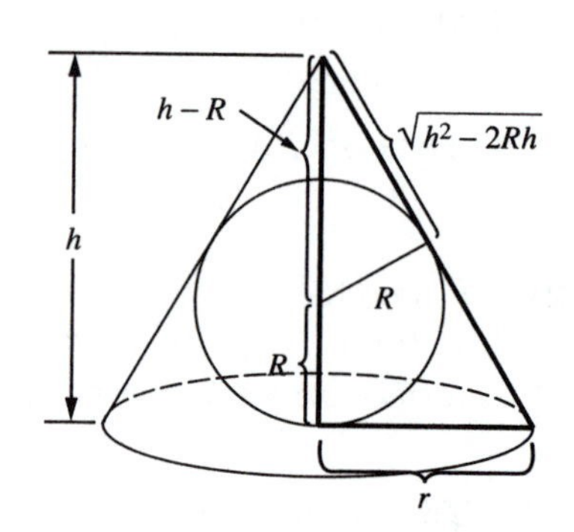

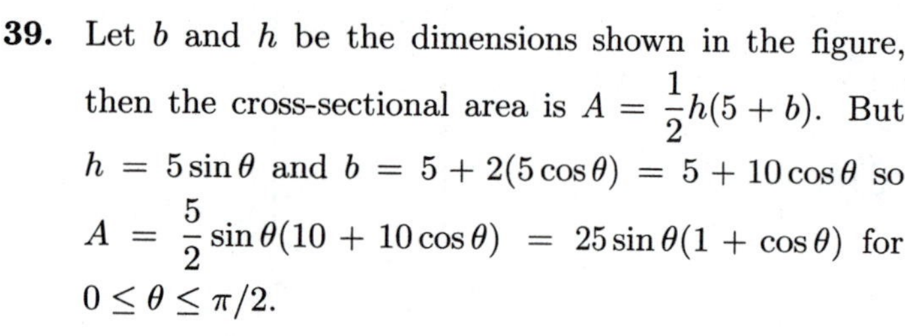

39. Let b and h be the dimensions shown in the figure, then the cross-sectional area is $A = \frac{1}{2}h(5 + b)$. But $h = 5\sin\theta$ and $b = 5 + 2(5\cos\theta) = 5 + 10\cos\theta$ so $A = \frac{5}{2}\sin\theta(10 + 10\cos\theta) = 25\sin\theta(1 + \cos\theta)$ for $0 \le \theta \le \pi/2$.

$$\begin{aligned} dA/d\theta &= -25\sin^2\theta + 25\cos\theta(1 + \cos\theta) \\ &= 25(-\sin^2\theta + \cos\theta + \cos^2\theta) \\ &= 25(-1 + \cos^2\theta + \cos\theta + \cos^2\theta) \\ &= 25(2\cos^2\theta + \cos\theta - 1) = 25(2\cos\theta - 1)(\cos\theta + 1). \end{aligned}$$

$dA/d\theta = 0$ for $0 < \theta < \pi/2$ when $\cos\theta = 1/2$, $\theta = \pi/3$.
If $\theta = 0, \pi/3, \pi/2$ then $A = 0, 75\sqrt{3}/4, 25$ so the cross-sectional area is greatest when $\theta = \pi/3$.

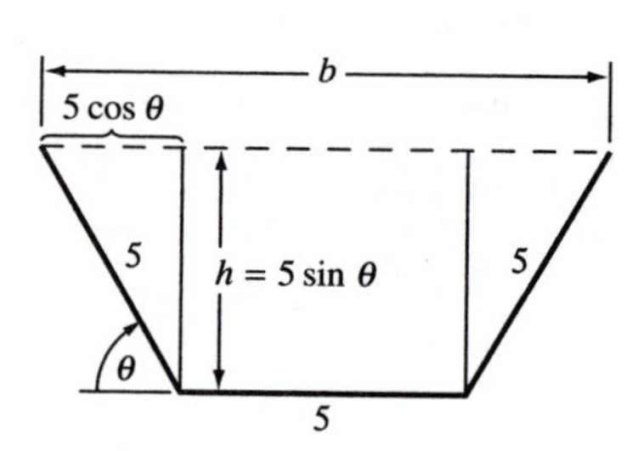

41. Let L, L_1, and L_2 be as shown in the figure, then $L = L_1 + L_2 = 8\csc\theta + \sec\theta$,

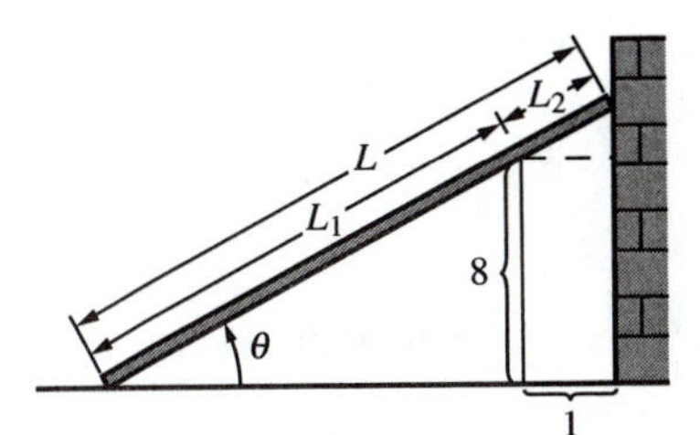

$$\frac{dL}{d\theta} = -8\csc\theta\cot\theta + \sec\theta\tan\theta,\ 0 < \theta < \pi/2$$

$$= -\frac{8\cos\theta}{\sin^2\theta} + \frac{\sin\theta}{\cos^2\theta} = \frac{-8\cos^3\theta + \sin^3\theta}{\sin^2\theta\cos^2\theta};$$

$\dfrac{dL}{d\theta} = 0$ if $\sin^3\theta = 8\cos^3\theta$, $\tan^3\theta = 8$, $\tan\theta = 2$ which gives the absolute minimum for L because $\lim_{\theta\to 0^+} L = \lim_{\theta\to\pi/2^-} L = +\infty$.

If $\tan\theta = 2$, then $\csc\theta = \sqrt{5}/2$ and $\sec\theta = \sqrt{5}$ so $L = 8(\sqrt{5}/2) + \sqrt{5} = 5\sqrt{5}$ ft.

43. (a) The daily profit is

$$\begin{aligned} P &= \text{(revenue)} - \text{(production cost)} = 100x - (100,000 + 50x + 0.0025x^2) \\ &= -100,000 + 50x - 0.0025x^2 \end{aligned}$$

for $0 \le x \le 7000$, so $dP/dx = 50 - 0.005x$ and $dP/dx = 0$ when $x = 10,000$. Because 10,000 is not in the interval $[0, 7000]$, the maximum profit must occur at an endpoint. When $x = 0$, $P = -100,000$; when $x = 7000$, $P = 127,500$ so 7000 units should be manufactured and sold daily.

(b) Yes, because $dP/dx > 0$ when $x = 7000$ so profit is increasing at this production level.

(c) $dP/dx = 15$ when $x = 7000$, so $P(7001) - P(7000) \approx 15$, and the marginal profit is \$15.

45. The profit is

$$P = \text{(profit on nondefective)} - \text{(loss on defective)} = 100(x - y) - 20y = 100x - 120y$$

but $y = 0.01x + 0.00003x^2$ so $P = 100x - 120(0.01x + 0.00003x^2) = 98.8x - 0.0036x^2$ for $x > 0$, $dP/dx = 98.8 - 0.0072x$, $dP/dx = 0$ when $x = 98.8/0.0072 \approx 13,722$, $d^2P/dx^2 < 0$ so the profit is maximum at a production level of about 13,722 pounds.

47. The distance between the particles is $D = \sqrt{(1 - t - t)^2 + (t - 2t)^2} = \sqrt{5t^2 - 4t + 1}$ for $t \ge 0$. For convenience, we minimize D^2 instead, so $D^2 = 5t^2 - 4t + 1$, $dD^2/dt = 10t - 4$, which is 0 when $t = 2/5$. $d^2D^2/dt^2 > 0$ so D^2 and hence D is minimum when $t = 2/5$. The minimum distance is $D = 1/\sqrt{5}$.

49. Let $P(x, y)$ be a point on the curve $x^2 + y^2 = 1$. The distance between $P(x, y)$ and $P_0(2, 0)$ is $D = \sqrt{(x - 2)^2 + y^2}$, but $y^2 = 1 - x^2$ so $D = \sqrt{(x - 2)^2 + 1 - x^2} = \sqrt{5 - 4x}$ for $-1 \le x \le 1$, $\dfrac{dD}{dx} = -\dfrac{2}{\sqrt{5 - 4x}}$ which has no critical points for $-1 < x < 1$. If $x = -1, 1$ then $D = 3, 1$ so the closest point occurs when $x = 1$ and $y = 0$.

51. Let (x, y) be a point on the curve, then the square of the distance between (x, y) and $(0, 2)$ is $S = x^2 + (y - 2)^2$ where $x^2 - y^2 = 1$, $x^2 = y^2 + 1$ so
$S = (y^2 + 1) + (y - 2)^2 = 2y^2 - 4y + 5$ for any y, $dS/dy = 4y - 4$, $dS/dy = 0$ when $y = 1$, $d^2S/dy^2 > 0$ so S is least when $y = 1$ and $x = \pm\sqrt{2}$.

53. If $P(x_0, y_0)$ is on the curve $y = 1/x^2$, then $y_0 = 1/x_0^2$. At P the slope of the tangent line is $-2/x_0^3$ so its equation is $y - \dfrac{1}{x_0^2} = -\dfrac{2}{x_0^3}(x - x_0)$, or $y = -\dfrac{2}{x_0^3}x + \dfrac{3}{x_0^2}$. The tangent line crosses the y-axis at $\dfrac{3}{x_0^2}$, and the x-axis at $\dfrac{3}{2}x_0$. The length of the segment then is $L = \sqrt{\dfrac{9}{x_0^4} + \dfrac{9}{4}x_0^2}$ for $x_0 > 0$. For convenience, we minimize L^2 instead, so $L^2 = \dfrac{9}{x_0^4} + \dfrac{9}{4}x_0^2$, $\dfrac{dL^2}{dx_0} = -\dfrac{36}{x_0^5} + \dfrac{9}{2}x_0 = \dfrac{9(x_0^6 - 8)}{2x_0^5}$, which is 0 when $x_0^6 = 8$, $x_0 = \sqrt{2}$. $\dfrac{d^2L^2}{dx_0^2} > 0$ so L^2 and hence L is minimum when $x_0 = \sqrt{2}$, $y_0 = 1/2$.

55. At each point (x, y) on the curve the slope of the tangent line is $m = \dfrac{dy}{dx} = -\dfrac{2x}{(1+x^2)^2}$ for any x, $\dfrac{dm}{dx} = \dfrac{2(3x^2-1)}{(1+x^2)^3}$, $\dfrac{dm}{dx} = 0$ when $x = \pm 1/\sqrt{3}$, by the first derivative test the only relative maximum occurs at $x = -1/\sqrt{3}$, which is the absolute maximum because $\lim_{x \to \pm\infty} m = 0$. The tangent line has greatest slope at the point $(-1/\sqrt{3}, 3/4)$.

57. With x and y as shown in the figure, the maximum length of pipe will be the smallest value of $L = x + y$. By similar triangles

$$\frac{y}{8} = \frac{x}{\sqrt{x^2-16}}, \quad y = \frac{8x}{\sqrt{x^2-16}} \text{ so}$$

$$L = x + \frac{8x}{\sqrt{x^2-16}} \text{ for } x > 4, \quad \frac{dL}{dx} = 1 - \frac{128}{(x^2-16)^{3/2}},$$

$\dfrac{dL}{dx} = 0$ when

$$\begin{aligned}(x^2-16)^{3/2} &= 128\\ x^2 - 16 &= 128^{2/3} = 16(2^{2/3})\\ x^2 &= 16(1+2^{2/3})\\ x &= 4(1+2^{2/3})^{1/2},\end{aligned}$$

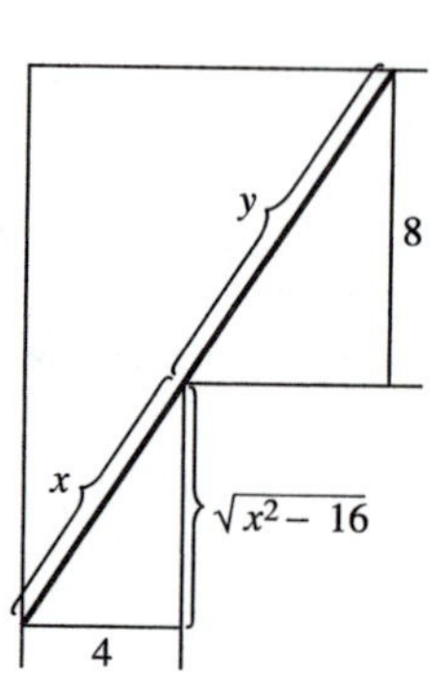

$d^2L/dx^2 = 384x/(x^2-16)^{5/2} > 0$ if $x > 4$ so L is smallest when $x = 4(1+2^{2/3})^{1/2}$. For this value of x, $L = 4(1+2^{2/3})^{3/2}$ ft.

59. Let x = distance from the weaker light source, I = the intensity at that point, and k the constant of proportionality. Then

$$I = \frac{kS}{x^2} + \frac{8kS}{(90-x)^2} \text{ if } 0 < x < 90;$$

$$\frac{dI}{dx} = -\frac{2kS}{x^3} + \frac{16kS}{(90-x)^3} = \frac{2kS[8x^3 - (90-x)^3]}{x^3(90-x)^3} = 18\frac{kS(x-30)(x^2+2700)}{x^3(x-90)^3},$$

which is 0 when $x = 30$; $\dfrac{dI}{dx} < 0$ if $x < 30$, and $\dfrac{dI}{dx} > 0$ if $x > 30$, so the intensity is minimum at a distance of 30 cm from the weaker source.

61.

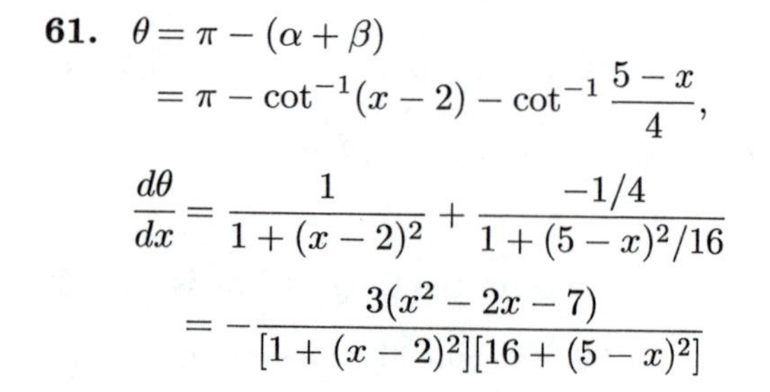

$$\begin{aligned}\theta &= \pi - (\alpha + \beta)\\ &= \pi - \cot^{-1}(x-2) - \cot^{-1}\frac{5-x}{4},\end{aligned}$$

$$\begin{aligned}\frac{d\theta}{dx} &= \frac{1}{1+(x-2)^2} + \frac{-1/4}{1+(5-x)^2/16}\\ &= -\frac{3(x^2-2x-7)}{[1+(x-2)^2][16+(5-x)^2]}\end{aligned}$$

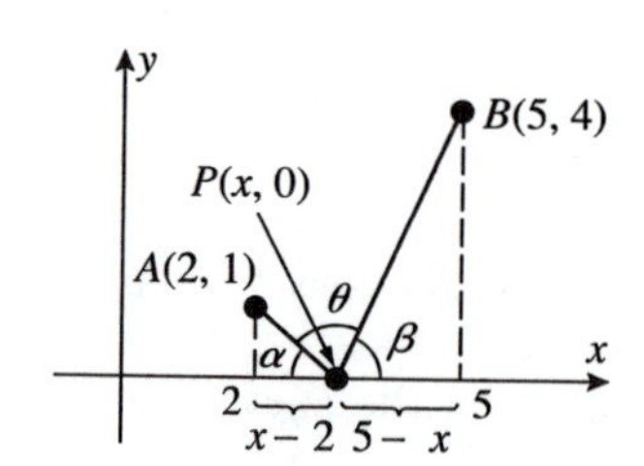

$$d\theta/dx = 0 \text{ when } x = \frac{2 \pm \sqrt{4+28}}{2} = 1 \pm 2\sqrt{2},$$

only $1+2\sqrt{2}$ is in $[2,5]$; $d\theta/dx > 0$ for x in $[2, 1+2\sqrt{2})$,
$d\theta/dx < 0$ for x in $(1+2\sqrt{2}, 5]$, θ is maximum when $x = 1+2\sqrt{2}$.

63. Let v = speed of light in the medium. The total time required for the light to travel from A to P to B is

$$t = \text{(total distance from } A \text{ to } P \text{ to } B)/v = \frac{1}{v}\left(\sqrt{(c-x)^2+a^2} + \sqrt{x^2+b^2}\right),$$

$$\frac{dt}{dx} = \frac{1}{v}\left[-\frac{c-x}{\sqrt{(c-x)^2+a^2}} + \frac{x}{\sqrt{x^2+b^2}}\right]$$

and $\dfrac{dt}{dx} = 0$ when $\dfrac{x}{\sqrt{x^2+b^2}} = \dfrac{c-x}{\sqrt{(c-x)^2+a^2}}$. But $x/\sqrt{x^2+b^2} = \sin\theta_2$ and
$(c-x)/\sqrt{(c-x)^2+a^2} = \sin\theta_1$ thus $dt/dx = 0$ when $\sin\theta_2 = \sin\theta_1$ so $\theta_2 = \theta_1$.

65. **(a)** The rate at which the farmer walks is analogous to the speed of light in Fermat's principle.

(b) the best path occurs when $\theta_1 = \theta_2$ (see figure).

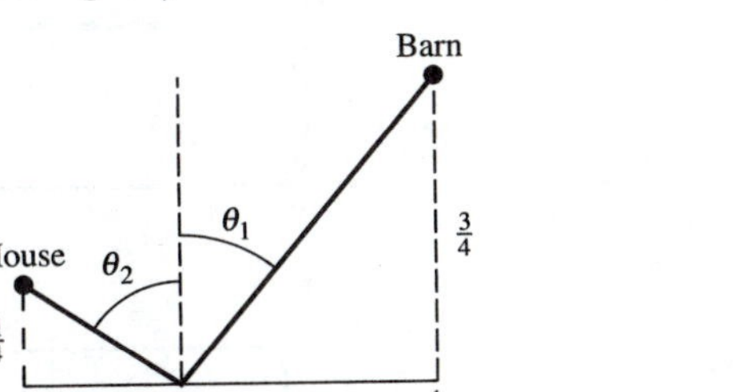

(c) by similar triangles,
$$x/(1/4) = (1-x)/(3/4)$$
$$3x = 1-x$$
$$4x = 1$$
$$x = 1/4 \text{ mi.}$$

EXERCISE SET 5.7

1. $f(x) = x^2 - 2$, $f'(x) = 2x$, $x_{n+1} = x_n - \dfrac{x_n^2 - 2}{2x_n}$

$x_1 = 1$, $x_2 = 1.5$, $x_3 = 1.416666667, \ldots$, $x_5 = x_6 = 1.414213562$

3. $f(x) = x^3 - 6$, $f'(x) = 3x^2$, $x_{n+1} = x_n - \dfrac{x_n^3 - 6}{3x_n^2}$

$x_1 = 2$, $x_2 = 1.833333333$, $x_3 = 1.817263545, \ldots$, $x_5 = x_6 = 1.817120593$

5. $f(x) = x^3 - x + 3$, $f'(x) = 3x^2 - 1$, $x_{n+1} = x_n - \dfrac{x_n^3 - x_n + 3}{3x_n^2 - 1}$

$x_1 = -2$, $x_2 = -1.727272727$, $x_3 = -1.673691174, \ldots, x_5 = x_6 = -1.671699882$

7. $f(x) = x^5 + x^4 - 5$, $f'(x) = 5x^4 + 4x^3$, $x_{n+1} = x_n - \dfrac{x_n^5 + x_n^4 - 5}{5x_n^4 + 4x_n^3}$

$x_1 = 1$, $x_2 = 1.333333333$, $x_3 = 1.239420573, \ldots, x_6 = x_7 = 1.224439550$

9. $f(x) = x^4 + x - 3$, $f'(x) = 4x^3 + 1$,

$$x_{n+1} = x_n - \frac{x_n^4 + x_n - 3}{4x_n^3 + 1}$$

$x_1 = -2$, $x_2 = -1.645161290$,
$x_3 = -1.485723955, \ldots, x_6 = x_7 = -1.452626879$

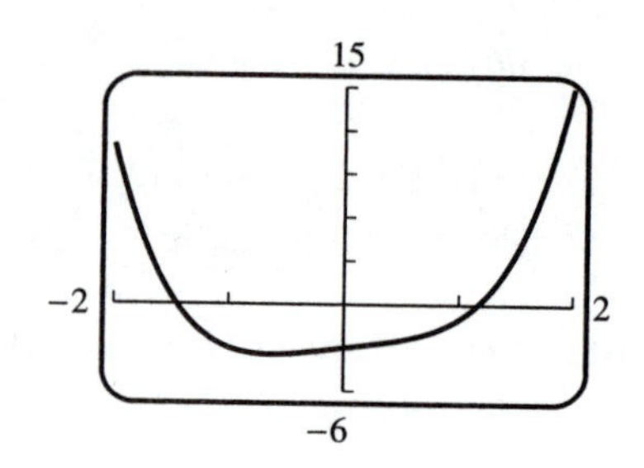

11. $f(x) = 2\sin x - x$, $f'(x) = 2\cos x - 1$,

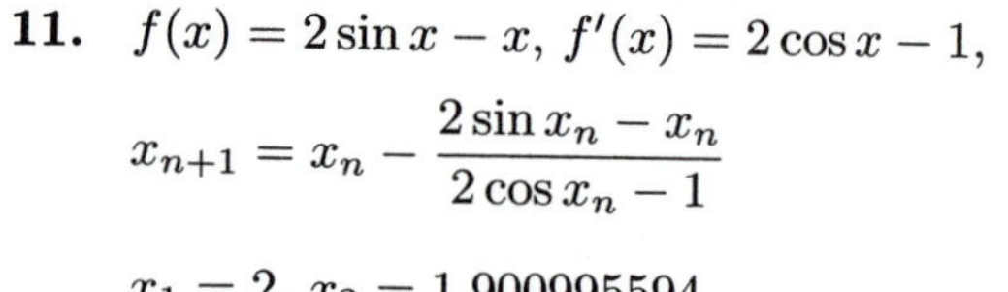

$$x_{n+1} = x_n - \frac{2\sin x_n - x_n}{2\cos x_n - 1}$$

$x_1 = 2$, $x_2 = 1.900995594$,
$x_3 = 1.895511645$, $x_4 = x_5 = 1.895494267$

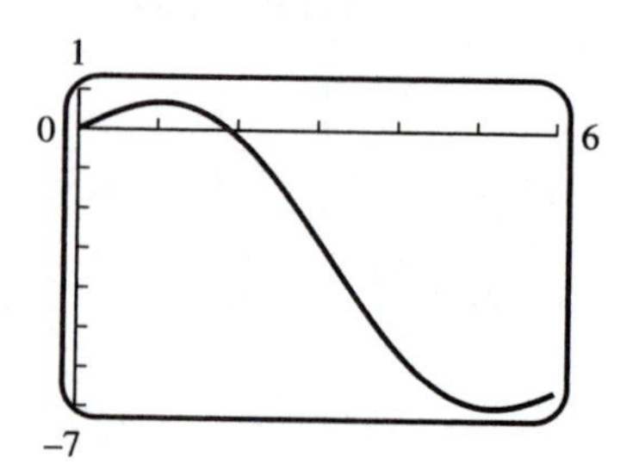

13. $f(x) = x - \tan x$,

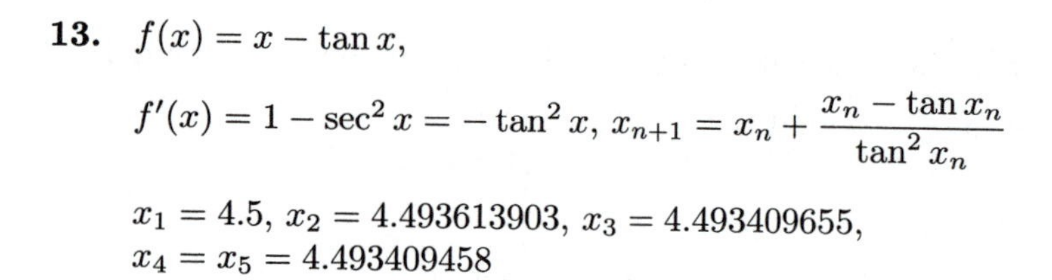

$f'(x) = 1 - \sec^2 x = -\tan^2 x$, $x_{n+1} = x_n + \dfrac{x_n - \tan x_n}{\tan^2 x_n}$

$x_1 = 4.5$, $x_2 = 4.493613903$, $x_3 = 4.493409655$,
$x_4 = x_5 = 4.493409458$

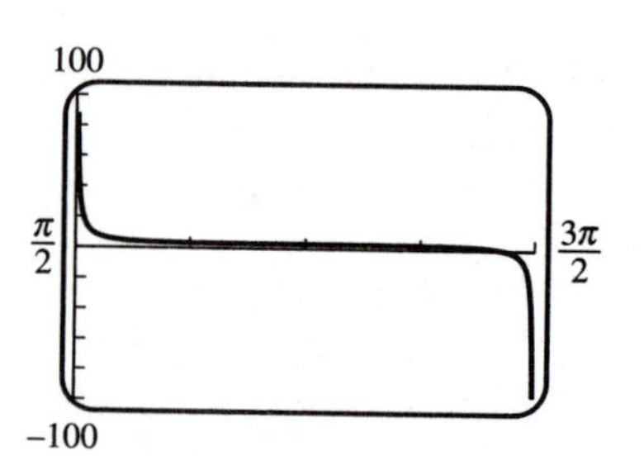

15. At the point of intersection, $x^3 = 0.5x - 1$, $x^3 - 0.5x + 1 = 0$. Let $f(x) = x^3 - 0.5x + 1$. By graphing $y = x^3$ and $y = 0.5x - 1$ it is evident that there is only one point of intersection and it occurs in the interval $[-2, -1]$; note that $f(-2) < 0$ and $f(-1) > 0$. $f'(x) = 3x^2 - 0.5$ so

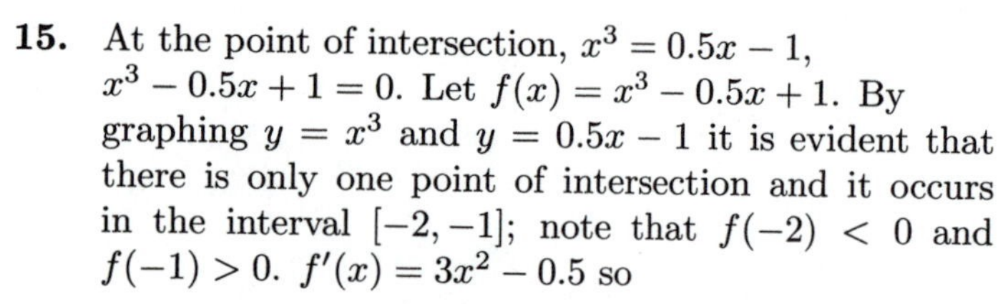

$$x_{n+1} = x_n - \frac{x_n^3 - 0.5x + 1}{3x_n^2 - 0.5};$$

$x_1 = -1$, $x_2 = -1.2$,
$x_3 = -1.166492147, \ldots,$
$x_5 = x_6 = -1.165373043$

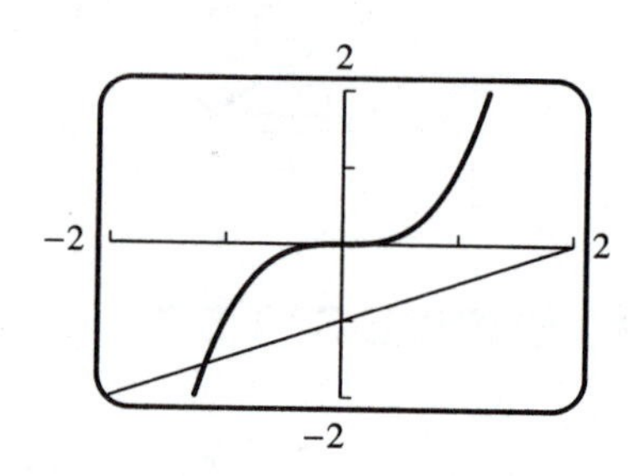

17. The graphs of $y = x^2$ and $y = \sqrt{2x+1}$ intersect at points near $x = -0.5$ and $x = 1$; $x^2 = \sqrt{2x+1}$, $x^4 - 2x - 1 = 0$. Let $f(x) = x^4 - 2x - 1$, then $f'(x) = 4x^3 - 2$ so

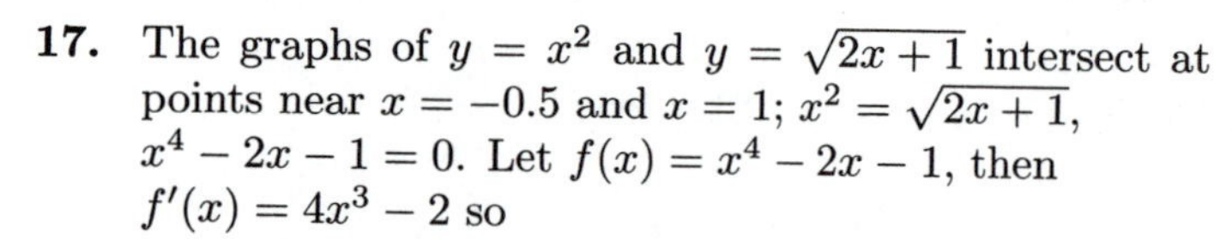

$$x_{n+1} = x_n - \frac{x_n^4 - 2x_n - 1}{4x_n^3 - 2}.$$

If $x_1 = -0.5$, then $x_2 = -0.475$,
$x_3 = -0.474626695$,
$x_4 = x_5 = -0.474626618$; if
$x_1 = 1$, then $x_2 = 2$,
$x_3 = 1.633333333, \ldots, x_8 = x_9 = 1.395336994$.

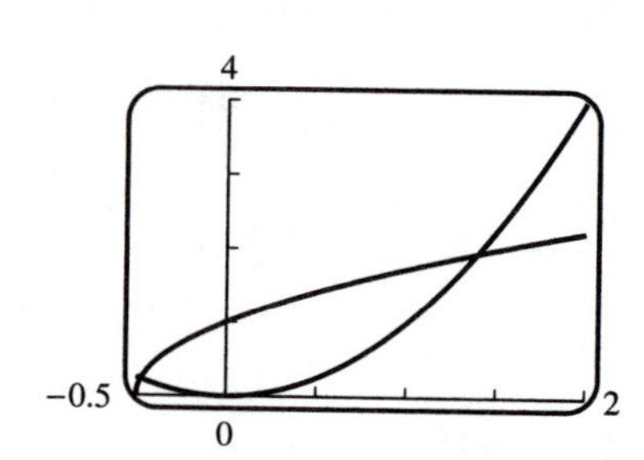

19. $x = 0$; also set
$f(x) = 1 - e^x \cos x$, $f'(x) = e^x(\sin x - \cos x)$,

$$x_{n+1} = x_n - \frac{1 - e^x \cos x}{e^x(\sin x - \cos x)}$$

$x_1 = 1$, $x_2 = 1.572512605$,
$x_3 = 1.363631415$, $x_7 = x_8 = 1.292695719$

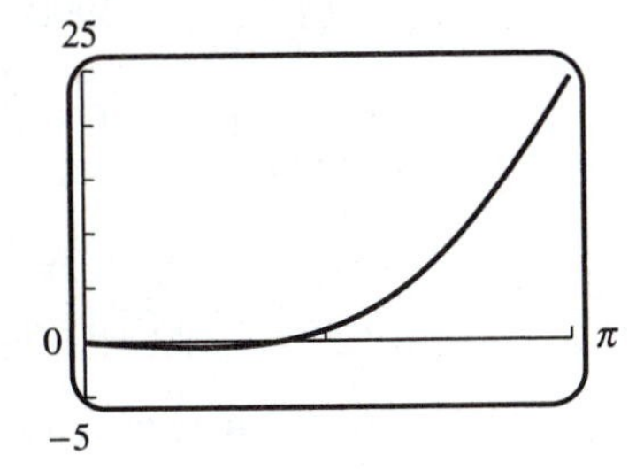

21. **(a)** $f(x) = x^2 - a$, $f'(x) = 2x$, $x_{n+1} = x_n - \dfrac{x_n^2 - a}{2x_n} = \dfrac{1}{2}\left(x_n + \dfrac{a}{x_n}\right)$

(b) $a = 10$; $x_1 = 3$, $x_2 = 3.166666667$, $x_3 = 3.162280702$, $x_4 = x_5 = 3.162277660$

23. $f'(x) = x^3 + 2x + 5$; solve $f'(x) = 0$ to find the critical points. Graph $y = x^3$ and $y = -2x - 5$ to see that they intersect at a point near $x = -1$; $f''(x) = 3x^2 + 2$ so $x_{n+1} = x_n - \dfrac{x_n^3 + 2x_n + 5}{3x_n^2 + 2}$.
$x_1 = -1$, $x_2 = -1.4$, $x_3 = -1.330964467, \cdots, x_5 = x_6 = -1.328268856$ so the minimum value of $f(x)$ occurs at $x \approx -1.328268856$ because $f''(x) > 0$; its value is approximately -4.098859132.

25. A graphing utility shows that there are two inflection points at $x \approx -0.25, 1.25$. These points are the zeros of $f''(x) = (x^4 - 4x^3 + 8x^2 - 4x - 1)\dfrac{e^x}{(x^2+1)^3}$. It is equivalent to find the zeros of $g(x) = x^4 - 4x^3 + 8x^2 - 4x - 1$. One root is $x = 1$ by inspection. Since $g'(x) = 4x^3 - 12x^2 + 16x - 4$, Newton's Method becomes

$$x_{n+1} = x_n - \frac{x_n^4 - 4x_n^3 + 8x_n^2 - 4x_n - 1}{4x_n^3 - 12x_n^2 + 16x_n - 4}$$

With $x_0 = -0.25$, $x_1 = -0.18572695, x_2 = -0.179563312, x_3 = -0.179509029$,
$x_4 = x_5 = -0.179509025$. So the points of inflection are at $x \approx -0.18, x = 1$.

27. Let $f(x)$ be the square of the distance between $(1, 0)$ and any point (x, x^2) on the parabola, then $f(x) = (x-1)^2 + (x^2 - 0)^2 = x^4 + x^2 - 2x + 1$ and $f'(x) = 4x^3 + 2x - 2$. Solve $f'(x) = 0$ to find the critical points; $f''(x) = 12x^2 + 2$ so $x_{n+1} = x_n - \dfrac{4x_n^3 + 2x_n - 2}{12x_n^2 + 2} = x_n - \dfrac{2x_n^3 + x_n - 1}{6x_n^2 + 1}$.
$x_1 = 1$, $x_2 = 0.714285714$, $x_3 = 0.605168701, \ldots, x_6 = x_7 = 0.589754512$; the coordinates are approximately $(0.589754512, 0.347810385)$.

29. **(a)** Let s be the arc length, and L the length of the chord, then $s = 1.5L$. But $s = r\theta$ and $L = 2r\sin(\theta/2)$ so $r\theta = 3r\sin(\theta/2)$, $\theta - 3\sin(\theta/2) = 0$.

(b) Let $f(\theta) = \theta - 3\sin(\theta/2)$, then $f'(\theta) = 1 - 1.5\cos(\theta/2)$ so $\theta_{n+1} = \theta_n - \dfrac{\theta_n - 3\sin(\theta_n/2)}{1 - 1.5\cos(\theta_n/2)}$.

$\theta_1 = 3$, $\theta_2 = 2.991592920$, $\theta_3 = 2.991563137$, $\theta_4 = \theta_5 = 2.991563136$ rad so $\theta \approx 171°$.

31. If $x = 1$, then $y^4 + y = 1$, $y^4 + y - 1 = 0$. Graph $z = y^4$ and $z = 1 - y$ to see that they intersect near $y = -1$ and $y = 1$. Let $f(y) = y^4 + y - 1$, then $f'(y) = 4y^3 + 1$ so $y_{n+1} = y_n - \dfrac{y_n^4 + y_n - 1}{4y_n^3 + 1}$.

If $y_1 = -1$, then $y_2 = -1.333333333$, $y_3 = -1.235807860, \ldots, y_6 = y_7 = -1.220744085$;
if $y_1 = 1$, then $y_2 = 0.8$, $y_3 = 0.731233596, \ldots, y_6 = y_7 = 0.724491959$.

33. $S(25) = 250{,}000 = \dfrac{5000}{i}\left[(1+i)^{25} - 1\right]$; set $f(i) = 50i - (1+i)^{25} + 1$, $f'(i) = 50 - 25(1+i)^{24}$; solve $f(i) = 0$. Set $i_0 = .06$ and $i_{k+1} = i_k - \left[50i - (1+i)^{25} + 1\right] / \left[50 - 25(1+i)^{24}\right]$. Then $i_1 = 0.05430$, $i_2 = 0.05338$, $i_3 = 0.05336$, $\ldots$, $i = 0.053362$.

35. **(a)**

x_1	x_2	x_3	x_4	x_5	x_6	x_7	x_8	x_9	x_{10}
0.5000	−0.7500	0.2917	−1.5685	−0.4654	0.8415	−0.1734	2.7970	1.2197	0.1999

(b) The sequence x_n must diverge, since if it did converge then $f(x) = x^2 + 1 = 0$ would have a solution. It seems the x_n are oscillating back and forth in a quasi-cyclical fashion.

EXERCISE SET 5.8

1. $f(0) = f(4) = 0$; $f'(3) = 0$; $[0, 4]$, $c = 3$

3. $f(2) = f(4) = 0$, $f'(x) = 2x - 6$, $2c - 6 = 0$, $c = 3$

5. $f(\pi/2) = f(3\pi/2) = 0$, $f'(x) = -\sin x$, $-\sin c = 0$, $c = \pi$

7. $f(0) = f(4) = 0$, $f'(x) = \dfrac{1}{2} - \dfrac{1}{2\sqrt{x}}$, $\dfrac{1}{2} - \dfrac{1}{2\sqrt{c}} = 0$, $c = 1$

9. $\dfrac{f(8) - f(0)}{8 - 0} = \dfrac{6}{8} = \dfrac{3}{4} = f'(1.54)$; $c = 1.54$

11. $f(-4) = 12$, $f(6) = 42$, $f'(x) = 2x + 1$, $2c + 1 = \dfrac{42 - 12}{6 - (-4)} = 3$, $c = 1$

13. $f(0) = 1$, $f(3) = 2$, $f'(x) = \dfrac{1}{2\sqrt{x+1}}$, $\dfrac{1}{2\sqrt{c+1}} = \dfrac{2-1}{3-0} = \dfrac{1}{3}$, $\sqrt{c+1} = 3/2$, $c + 1 = 9/4$, $c = 5/4$

15. $f(-5) = 0$, $f(3) = 4$, $f'(x) = -\dfrac{x}{\sqrt{25 - x^2}}$, $-\dfrac{c}{\sqrt{25 - c^2}} = \dfrac{4 - 0}{3 - (-5)} = \dfrac{1}{2}$, $-2c = \sqrt{25 - c^2}$, $4c^2 = 25 - c^2$, $c^2 = 5$, $c = -\sqrt{5}$
(we reject $c = \sqrt{5}$ because it does not satisfy the equation $-2c = \sqrt{25 - c^2}$)

17. **(a)** $f(-2) = f(1) = 0$
The interval is $[-2, 1]$

(b) $c = -1.29$

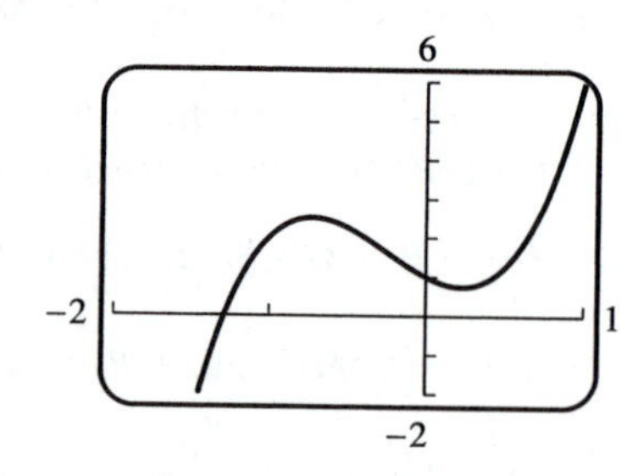

(c) $x_0 = -1$, $x_1 = -1.5$, $x_2 = -1.32$, $x_3 = -1.290$, $x_4 = -1.2885843$

19. **(a)** $f'(x) = \sec^2 x$, $\sec^2 c = 0$ has no solution **(b)** $\tan x$ is not continuous on $[0, \pi]$

21. **(a)** Two x-intercepts of f determine two solutions a and b of $f(x) = 0$; by Rolle's Theorem there exists a point c between a and b such that $f'(c) = 0$, i.e. c is an x-intercept for f'.

(b) $f(x) = \sin x = 0$ at $x = n\pi$, and $f'(x) = \cos x = 0$ at $x = n\pi + \pi/2$, which lies between $n\pi$ and $(n+1)\pi$, $(n = 0, \pm 1, \pm 2, \ldots)$

23. Let $s(t)$ be the position function of the automobile for $0 \le t \le 5$, then by the Mean-Value Theorem there is at least one point c in $(0, 5)$ where
$s'(c) = v(c) = [s(5) - s(0)]/(5-0) = 4/5 = 0.8$ mi/min $= 48$ mi/h.

25. Let $f(t)$ and $g(t)$ denote the distances from the first and second runners to the starting point, and let $h(t) = f(t) - g(t)$. Since they start (at $t = 0$) and finish (at $t = t_1$) at the same time, $h(0) = h(t_1) = 0$, so by Rolle's Theorem there is a time t_2 for which $h'(t_2) = 0$, i.e. $f'(t_2) = g'(t_2)$; so they have the same velocity at time t_2.

27. (a) By the Constant Difference Theorem $f(x) - g(x) = k$ for some k; since $f(x_0) = g(x_0)$, $k = 0$, so $f(x) = g(x)$ for all x.

(b) Set $f(x) = \sin^2 x + \cos^2 x$, $g(x) = 1$; then $f'(x) = 2\sin x\cos x - 2\cos x\sin x = 0 = g'(x)$. Since $f(0) = 1 = g(0)$, $f(x) = g(x)$ for all x.

29. If $f'(x) = g'(x)$, then $f(x) = g(x) + k$. Let $x = 1$,
$f(1) = g(1) + k = (1)^3 - 4(1) + 6 + k = 3 + k = 2$, so $k = -1$. $f(x) = x^3 - 4x + 5$.

31. (a) If x, y belong to I and $x < y$ then for some c in I, $\dfrac{f(y) - f(x)}{y - x} = f'(c)$,
so $|f(x) - f(y)| = |f'(c)||x - y| \le M|x - y|$; if $x > y$ exchange x and y; if $x = y$ the inequality also holds.

(b) $f(x) = \sin x$, $f'(x) = \cos x$, $|f'(x)| \le 1 = M$, so $|f(x) - f(y)| \le |x - y|$ or $|\sin x - \sin y| \le |x - y|$.

33. (a) Let $f(x) = \sqrt{x}$. By the Mean-Value Theorem there is a number c between x and y such that $\dfrac{\sqrt{y} - \sqrt{x}}{y - x} = \dfrac{1}{2\sqrt{c}} < \dfrac{1}{2\sqrt{x}}$ for c in (x, y), thus $\sqrt{y} - \sqrt{x} < \dfrac{y - x}{2\sqrt{x}}$

(b) multiply through and rearrange to get $\sqrt{xy} < \dfrac{1}{2}(x + y)$.

35. (a) If $f(x) = x^3 + 4x - 1$ then $f'(x) = 3x^2 + 4$ is never zero, so by Exercise 34 f has at most one real root; since f is a cubic polynomial it has at least one real root, so it has exactly one real root.

(b) Let $f(x) = ax^3 + bx^2 + cx + d$. If $f(x) = 0$ has at least two distinct real solutions r_1 and r_2, then $f(r_1) = f(r_2) = 0$ and by Rolle's Theorem there is at least one number between r_1 and r_2 where $f'(x) = 0$. But $f'(x) = 3ax^2 + 2bx + c = 0$ for
$x = (-2b \pm \sqrt{4b^2 - 12ac})/(6a) = (-b \pm \sqrt{b^2 - 3ac})/(3a)$, which are not real if $b^2 - 3ac < 0$ so $f(x) = 0$ must have fewer than two distinct real solutions.

37. By the Mean-Value Theorem on the interval $[0, x]$,

$\dfrac{\tan^{-1} x - \tan^{-1} 0}{x - 0} = \dfrac{\tan^{-1} x}{x} = \dfrac{1}{1 + c^2}$ for c in $(0, x)$, but

$\dfrac{1}{1 + x^2} < \dfrac{1}{1 + c^2} < 1$ for c in $(0, x)$ so $\dfrac{1}{1 + x^2} < \dfrac{\tan^{-1} x}{x} < 1$, $\dfrac{x}{1 + x^2} < \tan^{-1} x < x$.

39. (a) $\dfrac{d}{dx}[f^2(x) + g^2(x)] = 2f(x)f'(x) + 2g(x)g'(x) = 2f(x)g(x) + 2g(x)[-f(x)] = 0$,
so $f^2(x) + g^2(x)$ is constant.

(b) $f(x) = \sin x$ and $g(x) = \cos x$

41.

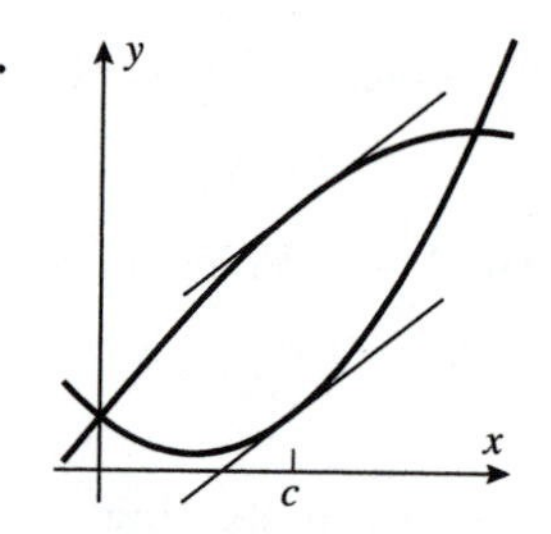

43. **(a)** similar to the proof of Part (a) with $f'(c) < 0$

(b) similar to the proof of Part (a) with $f'(c) = 0$

45. If f is differentiable at $x = 1$, then f is continuous there;
$\lim_{x\to 1^+} f(x) = \lim_{x\to 1^-} f(x) = f(1) = 3$, $a + b = 3$; $\lim_{x\to 1^+} f'(x) = a$ and $\lim_{x\to 1^-} f'(x) = 6$ so $a = 6$ and $b = 3 - 6 = -3$.

47. From Section 3.2 a function has a vertical tangent line at a point of its graph if the slopes of secant lines through the point approach $+\infty$ or $-\infty$. Suppose f is continuous at $x = x_0$ and $\lim_{x\to x_0^+} f(x) = +\infty$. Then a secant line through $(x_1, f(x_1))$ and $(x_0, f(x_0))$, assuming $x_1 > x_0$, will have slope $\dfrac{f(x_1) - f(x_0)}{x_1 - x_0}$. By the Mean Value Theorem, this quotient is equal to $f'(c)$ for some c between x_0 and x_1. But as x_1 approaches x_0, c must also approach x_0, and it is given that $\lim_{c\to x_0^+} f'(c) = +\infty$, so the slope of the secant line approaches $+\infty$. The argument can be altered appropriately for $x_1 < x_0$, and/or for $f'(c)$ approaching $-\infty$.

SUPPLEMENTARY EXERCISES FOR CHAPTER 5

7. **(a)** $f'(x) = \dfrac{7(x-7)(x-1)}{3x^{2/3}}$; critical numbers at $x = 0, 1, 7$;
neither at $x = 0$, relative maximum at $x = 1$, relative minimum at $x = 7$ (First Derivative Test)

(b) $f'(x) = 2\cos x(1 + 2\sin x)$; critical numbers at $x = \pi/2, 3\pi/2, 7\pi/6, 11\pi/6$;
relative maximum at $x = \pi/2, 3\pi/2$, relative minimum at $x = 7\pi/6, 11\pi/6$

(c) $f'(x) = 3 - \dfrac{3\sqrt{x-1}}{2}$; critical numbers at $x = 5$; relative maximum at $x = 5$

9. $\lim_{x\to -\infty} f(x) = +\infty$, $\lim_{x\to +\infty} f(x) = +\infty$
$f'(x) = x(4x^2 - 9x + 6)$, $f''(x) = 6(2x-1)(x-1)$
relative minimum at $x = 0$,
points of inflection when $x = 1/2, 1$,
no asymptotes

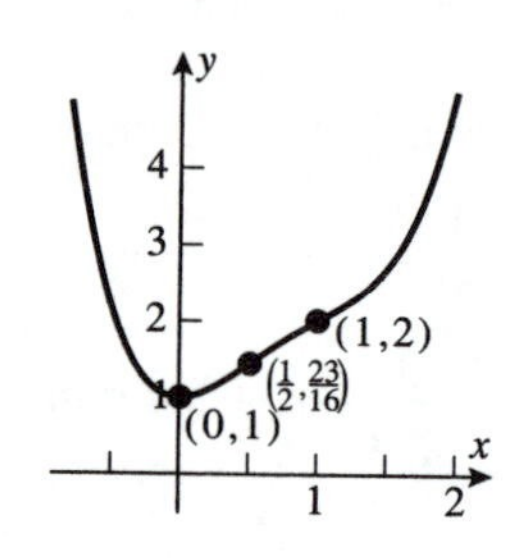

11. $\lim_{x\to\pm\infty} f(x)$ doesn't exist

$f'(x) = 2x\sec^2(x^2+1)$,

$f''(x) = 2\sec^2(x^2+1)\left[1+4x^2\tan(x^2+1)\right]$

critical number at $x = 0$;

relative minimum at $x = 0$

point of inflection when $1+4x^2\tan(x^2+1) = 0$

vertical asymptotes at $x = \pm\sqrt{\pi(n+\frac{1}{2})-1}$, $n = 0, 1, 2, \ldots$

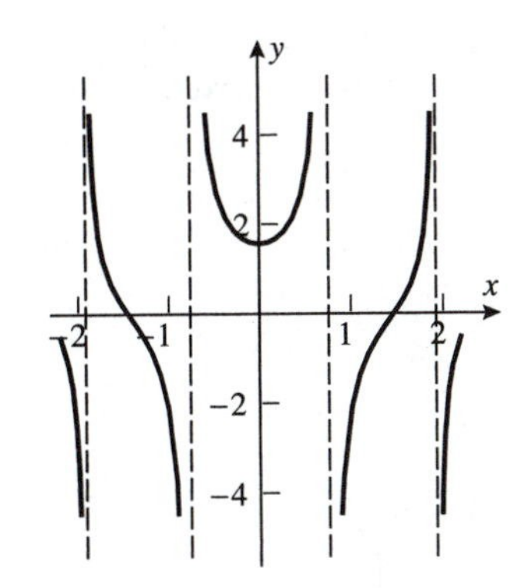

13. $f'(x) = 2\dfrac{x(x+5)}{(x^2+2x+5)^2}$, $f''(x) = -2\dfrac{2x^3+15x^2-25}{(x^2+2x+5)^3}$

critical numbers at $x = -5, 0$;

relative maximum at $x = -5$,

relative minimum at $x = 0$

points of inflection at $x = -7.26, -1.44, 1.20$

horizontal asymptote $y = 1$ as $x \to \pm\infty$

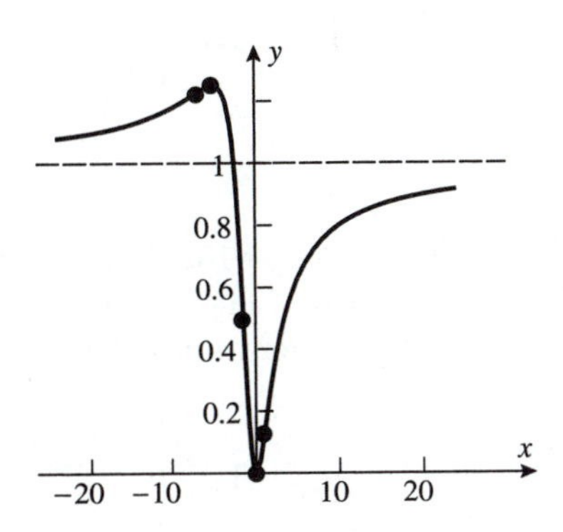

15. $\lim_{x\to-\infty} f(x) = +\infty$, $\lim_{x\to+\infty} f(x) = -\infty$

$f'(x) = \begin{cases} x \\ -2x \end{cases}$ if $\begin{cases} x \le 0 \\ x > 0 \end{cases}$

critical number at $x = 0$, no extrema

inflection point at $x = 0$ (f changes concavity)

no asymptotes

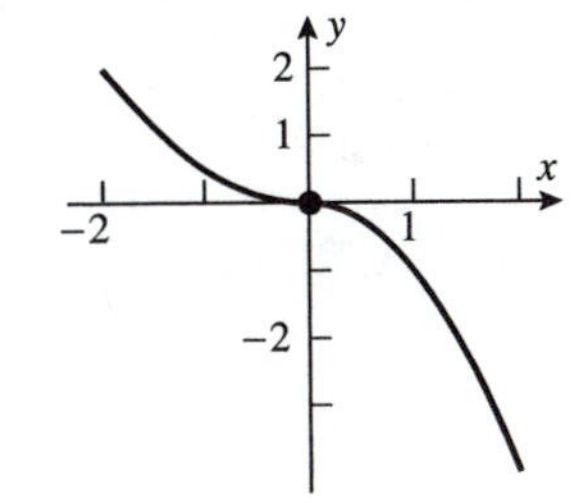

17. **(a)** 40, -5, 5, -40

(b) $f'(x) = x^2 - \dfrac{1}{400}$, $f''(x) = 2x$

critical points at $x = \pm\dfrac{1}{20}$;

relative maximum at $x = -\dfrac{1}{20}$,

relative minimum at $x = \dfrac{1}{20}$

(c) The finer details can be seen when graphing over a much smaller x-window.

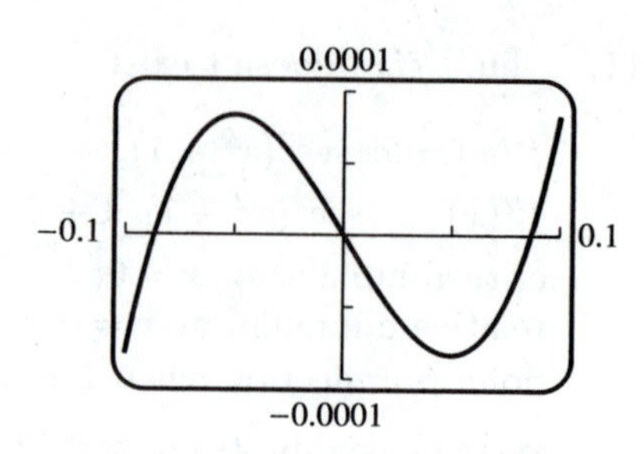

19. **(a)**

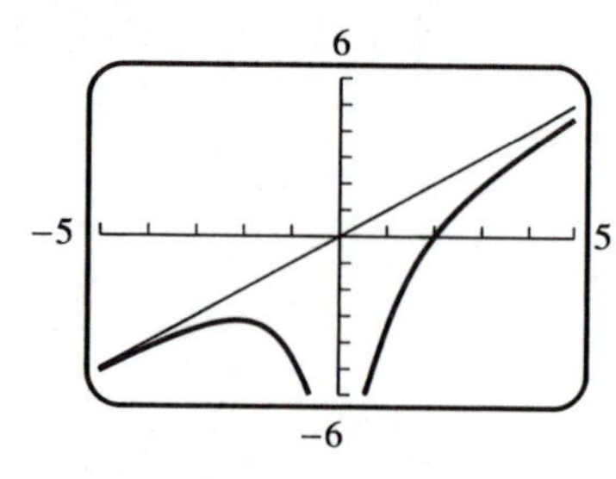

(b) Divide $y = x^2 + 1$ into $y = x^3 - 8$ to get the asymptote $ax + b = x$

21. $f'(x) = 4x^3 - 18x^2 + 24x - 8$, $f''(x) = 12(x-1)(x-2)$
$f''(1) = 0$, $f'(1) = 2$, $f(1) = 2$; $f''(2) = 0$, $f'(2) = 0$, $f(2) = 3$,
so the tangent lines at the inflection points are $y = 2x$ and $y = 3$.

23. $f(x) = \dfrac{(2x-1)(x^2+x-7)}{(2x-1)(3x^2+x-1)} = \dfrac{x^2+x-7}{3x^2+x-1}, \quad x \neq 1/2$

horizontal asymptote: $y = 1/3$,
vertical asymptotes: $x = (-1 \pm \sqrt{13})/6$

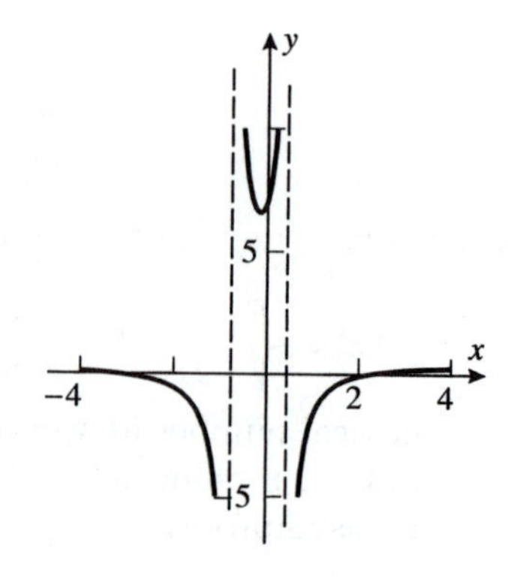

25. $f'(x) = 2ax + b$; $f'(x) > 0$ or $f'(x) < 0$ on $[0, +\infty)$ if $f'(x) = 0$ has no positive solution, so the polynomial is always increasing or always decreasing on $[0, +\infty)$ provided $-b/2a \leq 0$.

27. **(a)** relative minimum -0.232466 at $x = 0.450184$

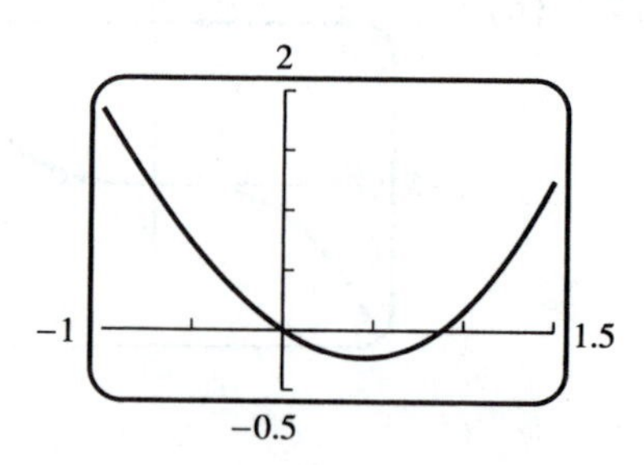

(b) relative maximum 0 at $x = 0$;
relative minimum -0.107587 at $x = \pm 0.674841$

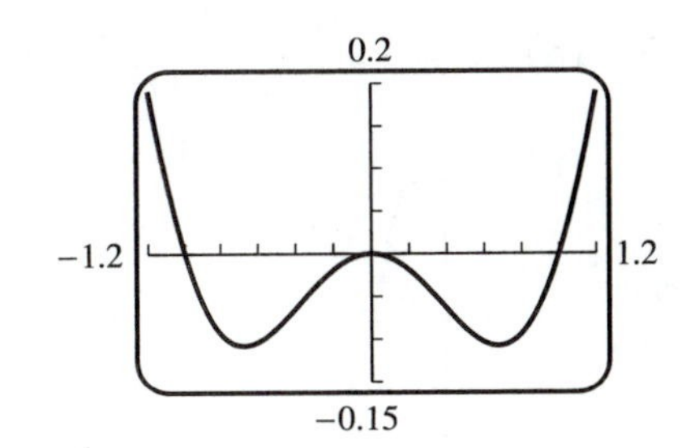

(c) relative maximum 0.876839; at $x = 0.886352$;
relative minimum -0.355977 at $x = -1.244155$

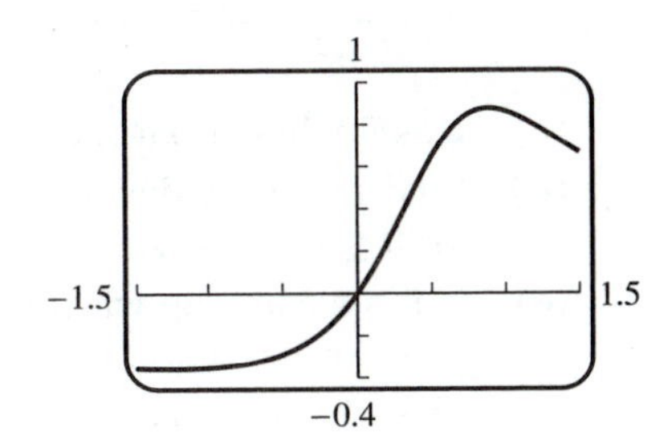

29. **(a)** If $a = k$, a constant, then $v = kt + b$ where b is constant; so the velocity changes sign at $t = -b/k$.

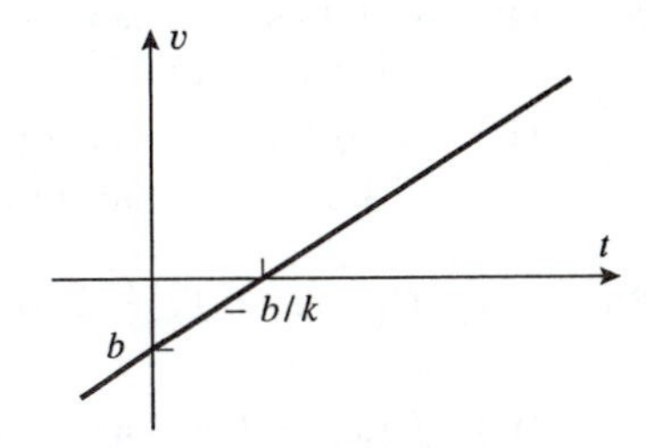

(b) Consider the equation $s = 5 - t^3/6$, $v = -t^2/2$, $a = -t$. Then for $t > 0$, a is decreasing and $av > 0$, so the particle is speeding up.

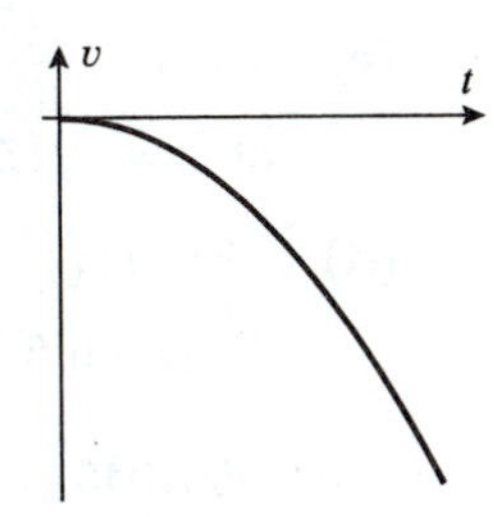

31. **(a)** $s(t) = s_0 + v_0 t - \frac{1}{2}gt^2 = v_0 t - 4.9t^2$, $v(t) = v_0 - 9.8t$; $s_{\max}$ occurs when $v = 0$, i.e. $t = v_0/9.8$, and then $0.76 = s_{\max} = v_0(v_0/9.8) - 4.9(v_0/9.8)^2 = v_0^2/19.6$, so $v_0 = \sqrt{0.76 \cdot 19.6} = 3.86$ m/s and $s(t) = 3.86t - 4.9t^2$. Then $s(t) = 0$ when $t = 0, 0.7878$, $s(t) = 0.15$ when $t = 0.0410, 0.7468$, and $s(t) = 0.76 - 0.15 = 0.61$ when $t = 0.2188, 0.5689$, so the player spends $0.5689 - 0.2188 = 0.3501$ s in the top 15.0 cm of the jump and $0.0410 + (0.7878 - 0.7468) = 0.0820$ s in the bottom 15.0 cm.

(b) The height *vs* time plot is a parabola that opens down, and the slope is smallest near the top of the parabola, so a given change Δh in height corresponds to a large time change Δt near the top of the parabola and a narrower time change at points farther away from the top.

33. **(a)** $v = -2\dfrac{t(t^4 + 2t^2 - 1)}{(t^4 + 1)^2}$, $a = 2\dfrac{3t^8 + 10t^6 - 12t^4 - 6t^2 + 1}{(t^4 + 1)^3}$

(b)

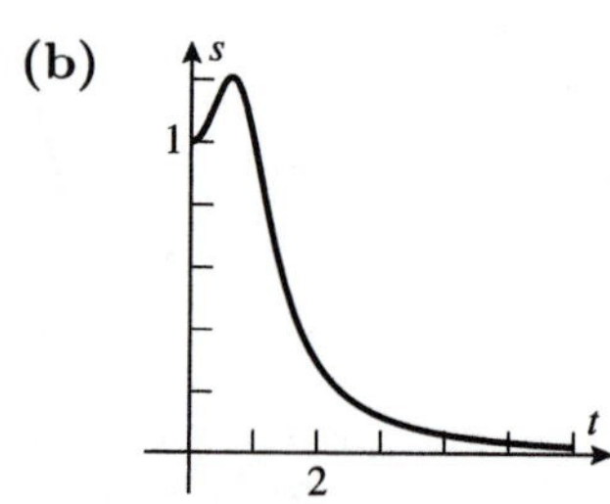

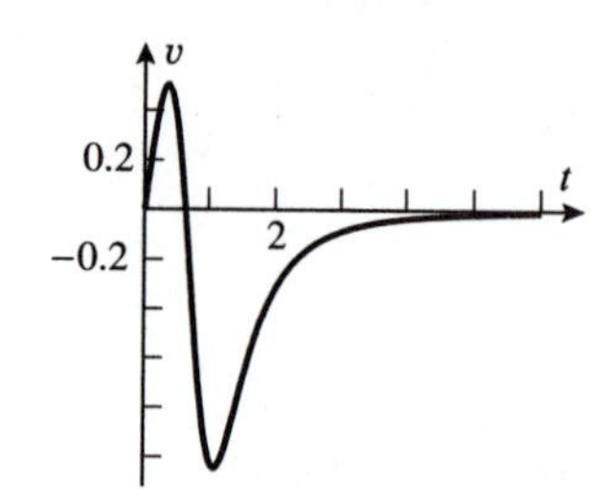

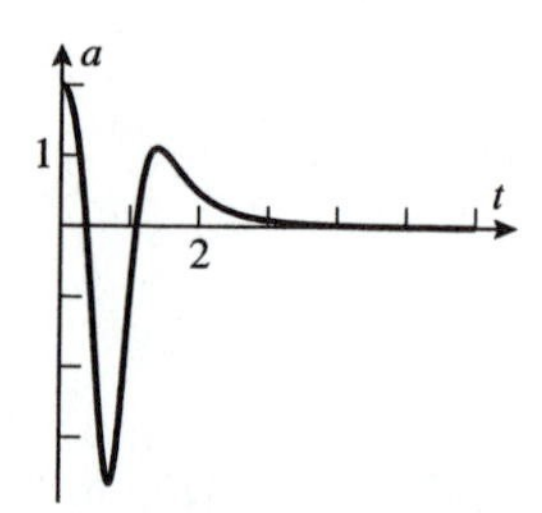

(c) It is farthest from the origin at approximately $t = 0.64$ (when $v = 0$) and $s = 1.2$

(d) Find t so that the velocity $v = ds/dt > 0$. The particle is moving in the positive direction for $0 \le t \le 0.64$ s.

(e) It is speeding up when $a, v > 0$ or $a, v < 0$, so for $0 \le t < 0.36$ and $0.64 < t < 1.1$, otherwise it is slowing down.

(f) Find the maximum value of $|v|$ to obtain: maximum speed $= 1.05$ m/s when $t = 1.10$ s.

37. **(a)** If f has an absolute extremum at a point of (a, b) then it must, by Theorem 5.5.4, be at a critical point of f; since f is differentiable on (a, b) the critical point is a stationary point.

(b) It could occur at a critical point which is not a stationary point: for example, $f(x) = |x|$ on $[-1, 1]$ has an absolute minimum at $x = 0$ but is not differentiable there.

39. **(a)** $f'(x) = -1/x^2 \neq 0$, no critical points; by inspection $M = -1/2$ at $x = -2$; $m = -1$ at $x = -1$

(b) $f'(x) = 3x^2 - 4x^3 = 0$ at $x = 0, 3/4$; $f(-1) = -2, f(0) = 0, f(3/4) = 27/256$, $f(3/2) = -27/16$, so $m = -2$ at $x = -1$, $M = 27/256$ at $x = 3/4$

(c) $f'(x) = \dfrac{x(7x - 12)}{3(x - 2)^{2/3}}$, critical points at $x = 12/7, 2$; $m = f(12/7) = \dfrac{144}{49}\left(-\dfrac{2}{7}\right)^{1/3} \approx -1.9356$ at $x = 12/7$, $M = 9$ at $x = 3$

(d) $\lim_{x\to 0^+} f(x) = \lim_{x\to +\infty} f(x) = +\infty$ and $f'(x) = \dfrac{e^x(x-2)}{x^3}$, stationary point at $x = 2$; by Theorem 5.5.5 $f(x)$ has an absolute minimum at $x = 2$, and $m = e^2/4$.

43. $x = 2.3561945$

45. **(a)** yes; $f'(0) = 0$

(b) no, f is not differentiable on $(-1, 1)$

(c) yes, $f'(\sqrt{\pi/2}) = 0$

47. Let k be the amount of light admitted per unit area of clear glass. The total amount of light admitted by the entire window is

$T = k \cdot$ (area of clear glass) $+ \frac{1}{2}k \cdot$ (area of blue glass) $= 2krh + \frac{1}{4}\pi kr^2$.

But $P = 2h + 2r + \pi r$ which gives $2h = P - 2r - \pi r$ so

$$T = kr(P - 2r - \pi r) + \frac{1}{4}\pi kr^2 = k\left[Pr - \left(2 + \pi - \frac{\pi}{4}\right)r^2\right]$$

$$= k\left[Pr - \frac{8+3\pi}{4}r^2\right] \text{ for } 0 < r < \frac{P}{2+\pi},$$

$$\frac{dT}{dr} = k\left(P - \frac{8+3\pi}{2}r\right), \frac{dT}{dr} = 0 \text{ when } r = \frac{2P}{8+3\pi}.$$

This is the only critical point and $d^2T/dr^2 < 0$ there so the most light is admitted when $r = 2P/(8+3\pi)$ ft.

49. **(a)**

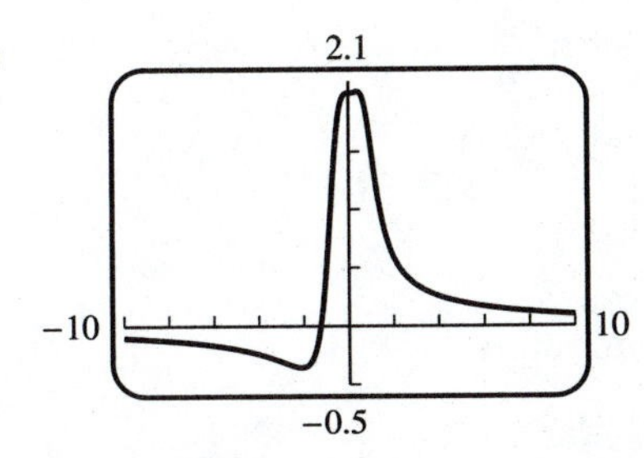

(b) minimum: $(-2.111985, -0.355116)$
maximum: $(0.372591, 2.012931)$

51. Solve $\phi - 0.0167\sin\phi = 2\pi(90)/365$ to get $\phi = 1.565978$ so $r = 150 \times 10^6(1 - 0.0167\cos\phi) = 149.988 \times 10^6$ km.

CHAPTER 6
Integration

EXERCISE SET 6.1

1. Endpoints $0, \frac{1}{n}, \frac{2}{n}, \ldots, \frac{n-1}{n}, 1$; using right endpoints,

$$A_n = \left[\sqrt{\frac{1}{n}} + \sqrt{\frac{2}{n}} + \cdots + \sqrt{\frac{n-1}{n}} + 1\right]\frac{1}{n}$$

n	2	5	10	50	100
A_n	0.853553	0.749739	0.710509	0.676095	0.671463

3. Endpoints $0, \frac{\pi}{n}, \frac{2\pi}{n}, \ldots, \frac{(n-1)\pi}{n}, \pi$; using right endpoints,

$$A_n = [\sin(\pi/n) + \sin(2\pi/n) + \cdots + \sin(\pi(n-1)/n) + \sin\pi]\,\frac{\pi}{n}$$

n	2	5	10	50	100
A_n	1.57080	1.93376	1.98352	1.99935	1.99984

5. Endpoints $1, \frac{n+1}{n}, \frac{n+2}{n}, \ldots, \frac{2n-1}{n}, 2$; using right endpoints,

$$A_n = \left[\frac{n}{n+1} + \frac{n}{n+2} + \cdots + \frac{n}{2n-1} + \frac{1}{2}\right]\frac{1}{n}$$

n	2	5	10	50	100
A_n	0.583333	0.645635	0.668771	0.688172	0.690653

7. Endpoints $0, \frac{1}{n}, \frac{2}{n}, \ldots, \frac{n-1}{n}, 1$; using right endpoints,

$$A_n = \left[\sqrt{1-\left(\frac{1}{n}\right)^2} + \sqrt{1-\left(\frac{2}{n}\right)^2} + \cdots + \sqrt{1-\left(\frac{n-1}{n}\right)^2} + 0\right]\frac{1}{n}$$

n	2	5	10	50	100
A_n	0.433013	0.659262	0.726130	0.774567	0.780106

9. $3(x-1)$

11. $x(x+2)$

13. $(x+3)(x-1)$

15. The area in Exercise 13 is always 3 less than the area in Exercise 11. The regions are identical except that the area in Exercise 11 has the extra trapezoid with vertices at $(0,0), (1,0), (0,2), (1,4)$ (with area 3).

17. B is also the area between the graph of $f(x) = \sqrt{x}$ and the interval $[0,1]$ on the y–axis, so $A+B$ is the area of the square.

EXERCISE SET 6.2

1. **(a)** $\displaystyle\int \frac{x}{\sqrt{1+x^2}}dx = \sqrt{1+x^2} + C$

(b) $\displaystyle\int (x+1)e^x dx = xe^x + C$

3. $\dfrac{d}{dx}\left[\sqrt{x^3+5}\right] = \dfrac{3x^2}{2\sqrt{x^3+5}}$ so $\displaystyle\int \frac{3x^2}{2\sqrt{x^3+5}}dx = \sqrt{x^3+5}+C$

5. $\dfrac{d}{dx}\left[\sin\left(2\sqrt{x}\right)\right] = \dfrac{\cos\left(2\sqrt{x}\right)}{\sqrt{x}}$ so $\displaystyle\int \frac{\cos\left(2\sqrt{x}\right)}{\sqrt{x}}dx = \sin\left(2\sqrt{x}\right)+C$

7. (a) $x^9/9+C$ (b) $\dfrac{7}{12}x^{12/7}+C$ (c) $\dfrac{2}{9}x^{9/2}+C$

9. (a) $\dfrac{1}{2}\displaystyle\int x^{-3}dx = -\frac{1}{4}x^{-2}+C$ (b) $u^4/4-u^2+7u+C$

11. $\displaystyle\int (x^{-3}+x^{1/2}-3x^{1/4}+x^2)dx = -\frac{1}{2}x^{-2}+\frac{2}{3}x^{3/2}-\frac{12}{5}x^{5/4}+\frac{1}{3}x^3+C$

13. $\displaystyle\int (x+x^4)dx = x^2/2+x^5/5+C$

15. $\displaystyle\int x^{1/3}(4-4x+x^2)dx = \int (4x^{1/3}-4x^{4/3}+x^{7/3})dx = 3x^{4/3}-\frac{12}{7}x^{7/3}+\frac{3}{10}x^{10/3}+C$

17. $\displaystyle\int (x+2x^{-2}-x^{-4})dx = x^2/2-2/x+1/(3x^3)+C$

19. $\displaystyle\int \left[\frac{2}{x}+3e^x\right]dx = 2\ln|x|+3e^x+C$

21. $-4\cos x+2\sin x+C$

23. $\displaystyle\int (\sec^2 x+\sec x\tan x)dx = \tan x+\sec x+C$

25. $\displaystyle\int \frac{\sec\theta}{\cos\theta}\,d\theta = \int \sec^2\theta\,d\theta = \tan\theta+C$

27. $\displaystyle\int \sec x\tan x\,dx = \sec x+C$

29. $\displaystyle\int (1+\sin\theta)d\theta = \theta-\cos\theta+C$

31. $\displaystyle\int \left[\frac{1}{2\sqrt{1-x^2}}-\frac{3}{1+x^2}\right]dx = \frac{1}{2}\sin^{-1}x-3\tan^{-1}x+C$

33. $\displaystyle\int \frac{1-\sin x}{1-\sin^2 x}dx = \int \frac{1-\sin x}{\cos^2 x}\,dx = \int \left(\sec^2 x-\sec x\tan x\right)dx = \tan x-\sec x+C$

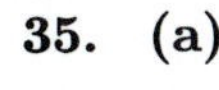

35. (a)

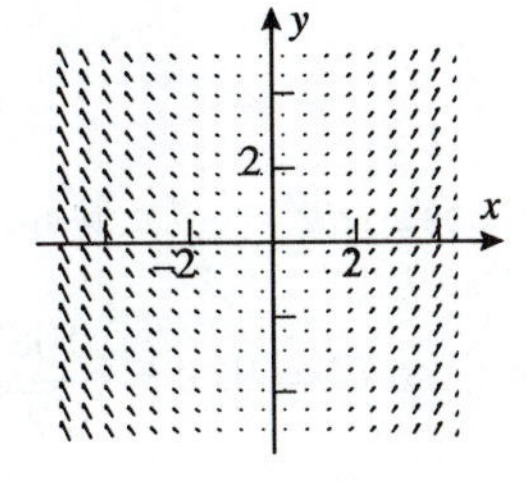

(b)

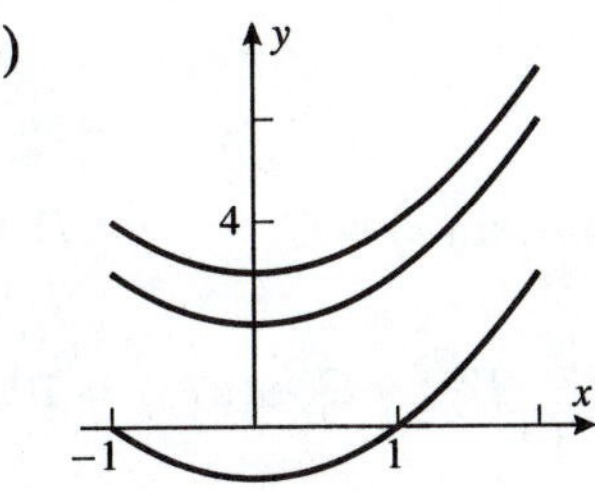

(c) $f(x)=x^2/2-1$

37.

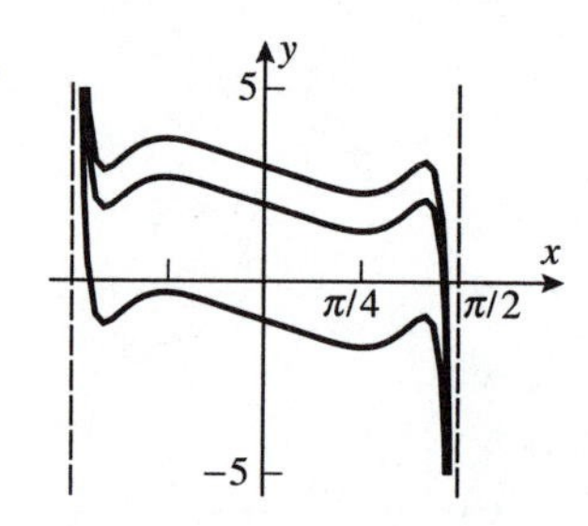

39. $f'(x) = m = -\sin x$ so $f(x) = \int(-\sin x)dx = \cos x + C$; $f(0) = 2 = 1 + C$ so $C = 1$, $f(x) = \cos\ x\ +\ 1$

41. **(a)** $y(x) = \int x^{1/3}dx = \frac{3}{4}x^{4/3} + C$, $y(1) = \frac{3}{4} + C = 2, C = \frac{5}{4}$; $y(x) = \frac{3}{4}x^{4/3} + \frac{5}{4}$

(b) $y(t) = \int(\sin t + 1)\,dt = -\cos t + t + C$, $y\left(\frac{\pi}{3}\right) = -\frac{1}{2} + \frac{\pi}{3} + C = 1/2$, $C = 1 - \frac{\pi}{3}$; $y(t) = -\cos t + t + 1 - \frac{\pi}{3}$

(c) $y(x) = \int(x^{1/2} + x^{-1/2})dx = \frac{2}{3}x^{3/2} + 2x^{1/2} + C$, $y(1) = 0 = \frac{8}{3} + C$, $C = -\frac{8}{3}$, $y(x) = \frac{2}{3}x^{3/2} + 2x^{1/2} - \frac{8}{3}$

43. **(a)** $y = \int 4e^x\,dx = 4e^x + C, 1 = y(0) = 4 + C, C = -3, y = 4e^x - 3$

(b) $y(t) = \int t^{-1}dt = \ln|t| + C$, $y(-1) = C = 5$, $C = 5$; $y(t) = \ln|t| + 5$

45. $f'(x) = \frac{2}{3}x^{3/2} + C_1$; $f(x) = \frac{4}{15}x^{5/2} + C_1x + C_2$

47. $dy/dx = 2x + 1, y = \int(2x + 1)dx = x^2 + x + C$; $y = 0$ when $x = -3$ so $(-3)^2 + (-3) + C = 0, C = -6$ thus $y = x^2 + x - 6$

49. $dy/dx = \int 6x dx = 3x^2 + C_1$. The slope of the tangent line is -3 so $dy/dx = -3$ when $x = 1$. Thus $3(1)^2 + C_1 = -3$, $C_1 = -6$ so $dy/dx = 3x^2 - 6$, $y = \int(3x^2 - 6)dx = x^3 - 6x + C_2$. If $x = 1$, then $y = 5 - 3(1) = 2$ so $(1)^2 - 6(1) + C_2 = 2, C_2 = 7$ thus $y = x^3 - 6x + 7$.

51. **(a)** $F'(x) = G'(x) = 3x + 4$

(b) $F(0) = 16/6 = 8/3$, $G(0) = 0$, so $F(0) - G(0) = 8/3$

(c) $F(x) = (9x^2 + 24x + 16)/6 = 3x^2/2 + 4x + 8/3 = G(x) + 8/3$

53. $\int(\sec^2 x - 1)dx = \tan x - x + C$

55. **(a)** $\frac{1}{2}\int(1 - \cos x)dx = \frac{1}{2}(x - \sin x) + C$ **(b)** $\frac{1}{2}\int(1 + \cos x)\,dx = \frac{1}{2}(x + \sin x) + C$

57. $v = \frac{1087}{2\sqrt{273}}\int T^{-1/2}\,dT = \frac{1087}{\sqrt{273}}T^{1/2} + C$, $v(273) = 1087 = 1087 + C$ so $C = 0$, $v = \frac{1087}{\sqrt{273}}T^{1/2}$ ft/s

EXERCISE SET 6.3

1. (a) $\int u^{23}du = u^{24}/24 + C = (x^2+1)^{24}/24 + C$

(b) $-\int u^3 du = -u^4/4 + C = -(\cos^4 x)/4 + C$

(c) $2\int \sin u\, du = -2\cos u + C = -2\cos\sqrt{x} + C$

(d) $\frac{3}{8}\int u^{-1/2}du = \frac{3}{4}u^{1/2} + C = \frac{3}{4}\sqrt{4x^2+5} + C$

3. (a) $-\int u\, du = -\frac{1}{2}u^2 + C = -\frac{1}{2}\cot^2 x + C$

(b) $\int u^9 du = \frac{1}{10}u^{10} + C = \frac{1}{10}(1+\sin t)^{10} + C$

(c) $\frac{1}{2}\int \cos u\, du = \frac{1}{2}\sin u + C = \frac{1}{2}\sin 2x + C$

(d) $\frac{1}{2}\int \sec^2 u\, du = \frac{1}{2}\tan u + C = \frac{1}{2}\tan x^2 + C$

5. (a) $\int \frac{1}{u}du = \ln|u| + C = \ln|\ln x| + C$

(b) $-\frac{1}{5}\int e^u\, du = -\frac{1}{5}e^u + C = -\frac{1}{5}e^{-5x} + C$

(c) $-\frac{1}{3}\int \frac{1}{u}du = -\frac{1}{3}\ln|u| + C = -\frac{1}{3}\ln|1+\cos 3\theta| + C$

(d) $\int \frac{du}{u} = \ln u + C = \ln(1+e^x) + C$

7. $u = 2 - x^2$, $du = -2x\,dx$; $-\frac{1}{2}\int u^3 du = -u^4/8 + C = -(2-x^2)^4/8 + C$

9. $u = 8x$, $du = 8dx$; $\frac{1}{8}\int \cos u\, du = \frac{1}{8}\sin u + C = \frac{1}{8}\sin 8x + C$

11. $u = 4x$, $du = 4dx$; $\frac{1}{4}\int \sec u \tan u\, du = \frac{1}{4}\sec u + C = \frac{1}{4}\sec 4x + C$

13. $u = 2x$, $du = 2dx$; $\frac{1}{2}\int e^u\, du = \frac{1}{2}e^u + C = \frac{1}{2}e^{2x} + C$

15. $u = 2x$, $\frac{1}{2}\int \frac{1}{\sqrt{1-u^2}}du = \frac{1}{2}\sin^{-1}(2x) + C$

17. $u = 7t^2 + 12$, $du = 14t\,dt$; $\frac{1}{14}\int u^{1/2}du = \frac{1}{21}u^{3/2} + C = \frac{1}{21}(7t^2+12)^{3/2} + C$

19. $u = x^3 + 1$, $du = 3x^2dx$; $\frac{1}{3}\int u^{-1/2}du = \frac{2}{3}u^{1/2} + C = \frac{2}{3}\sqrt{x^3+1} + C$

21. $u = 4x^2 + 1$, $du = 8x\,dx$; $\frac{1}{8}\int u^{-3}du = -\frac{1}{16}u^{-2} + C = -\frac{1}{16}(4x^2+1)^{-2} + C$

23. $u = \sin x$, $du = \cos x\, dx$; $\displaystyle\int e^u\, du = e^u + C = e^{\sin x} + C$

25. $u = -2x^3$, $du = -6x^2$, $\displaystyle -\frac{1}{6}\int e^u du = -\frac{1}{6}e^u + C = -\frac{1}{6}e^{-2x^3} + C$

27. $u = e^x$, $\displaystyle\int \frac{1}{1+u^2}du = \tan^{-1}(e^x) + C$

29. $u = 5/x$, $du = -(5/x^2)dx$; $\displaystyle -\frac{1}{5}\int \sin u\, du = \frac{1}{5}\cos u + C = \frac{1}{5}\cos(5/x) + C$

31. $u = x^3$, $du = 3x^2 dx$; $\displaystyle\frac{1}{3}\int \sec^2 u\, du = \frac{1}{3}\tan u + C = \frac{1}{3}\tan(x^3) + C$

33. $u = \sin 3t$, $du = 3\cos 3t\, dt$; $\displaystyle\frac{1}{3}\int u^5 du = \frac{1}{18}u^6 + C = \frac{1}{18}\sin^6 3t + C$

35. $u = 2 - \sin 4\theta$, $du = -4\cos 4\theta\, d\theta$; $\displaystyle -\frac{1}{4}\int u^{1/2}du = -\frac{1}{6}u^{3/2} + C = -\frac{1}{6}(2 - \sin 4\theta)^{3/2} + C$

37. $u = \tan x$, $\displaystyle\int \frac{1}{\sqrt{1-u^2}}\, du = \sin^{-1}(\tan x) + C$

39. $u = \sec 2x$, $du = 2\sec 2x \tan 2x\, dx$; $\displaystyle\frac{1}{2}\int u^2 du = \frac{1}{6}u^3 + C = \frac{1}{6}\sec^3 2x + C$

41. $\displaystyle\int e^{-x}dx$; $u = -x$, $du = -dx$; $\displaystyle -\int e^u du = -e^u + C = -e^{-x} + C$

43. $u = \sqrt{y+1}$, $\displaystyle du = \frac{1}{2\sqrt{y+1}}dy$, $\displaystyle 2\int e^u du = 2e^u + C = 2e^{\sqrt{y+1}} + C$

45. $u = x - 3$, $x = u + 3$, $dx = du$

$$\int (u+3)u^{1/2}du = \int (u^{3/2} + 3u^{1/2})du = \frac{2}{5}u^{5/2} + 2u^{3/2} + C = \frac{2}{5}(x-3)^{5/2} + 2(x-3)^{3/2} + C$$

47. $\displaystyle\int \sin^2 2\theta \sin 2\theta\, d\theta = \int (1 - \cos^2 2\theta)\sin 2\theta\, d\theta$; $u = \cos 2\theta$, $du = -2\sin 2\theta\, d\theta$,

$$-\frac{1}{2}\int (1-u^2)du = -\frac{1}{2}u + \frac{1}{6}u^3 + C = -\frac{1}{2}\cos 2\theta + \frac{1}{6}\cos^3 2\theta + C$$

49. $\displaystyle\int \left(1 + \frac{1}{t}\right) dt = t + \ln|t| + C$

51. $\ln(e^x) + \ln(e^{-x}) = \ln(e^x e^{-x}) = \ln 1 = 0$ so $\displaystyle\int [\ln(e^x) + \ln(e^{-x})]dx = C$

53. **(a)** $\sin^{-1}(x/3) + C$ **(b)** $(1/\sqrt{5})\tan^{-1}(x/\sqrt{5}) + C$
(c) $(1/\sqrt{\pi})\sec^{-1}(x/\sqrt{\pi}) + C$

55. $u = a + bx$, $du = bdx$,

$$\int (a+bx)^n dx = \frac{1}{b}\int u^n du = \frac{(a+bx)^{n+1}}{b(n+1)} + C$$

57. $u = \sin(a+bx)$, $du = b\cos(a+bx)dx$

$$\frac{1}{b}\int u^n du = \frac{1}{b(n+1)}u^{n+1} + C = \frac{1}{b(n+1)}\sin^{n+1}(a+bx) + C$$

59. **(a)** with $u = \sin x$, $du = \cos x\, dx$; $\int u\, du = \frac{1}{2}u^2 + C_1 = \frac{1}{2}\sin^2 x + C_1$;

with $u = \cos x$, $du = -\sin x\, dx$; $-\int u\, du = -\frac{1}{2}u^2 + C_2 = -\frac{1}{2}\cos^2 x + C_2$

(b) because they differ by a constant:

$$\left(\frac{1}{2}\sin^2 x + C_1\right) - \left(-\frac{1}{2}\cos^2 x + C_2\right) = \frac{1}{2}(\sin^2 x + \cos^2 x) + C_1 - C_2 = 1/2 + C_1 - C_2$$

61. $y(x) = \int \sqrt{3x+1}dx = \frac{2}{9}(3x+1)^{3/2} + C,$

$y(1) = \frac{16}{9} + C = 5$, $C = \frac{29}{9}$ so $y(x) = \frac{2}{9}(3x+1)^{3/2} + \frac{29}{9}$

63. $y(t) = \int 2e^{-t}\, dt = -2e^{-t} + C$, $y(1) = -\frac{2}{e} + C = 3 - \frac{2}{e}, C = 3$; $y(t) = -2e^{-t} + 3$

65.

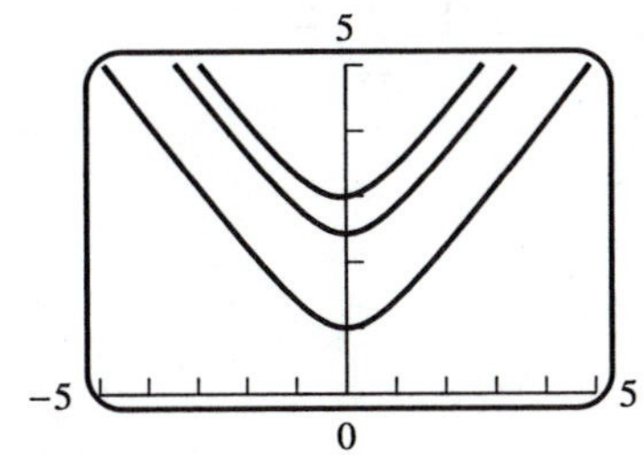

67. $f'(x) = m = \sqrt{3x+1}$, $f(x) = \int (3x+1)^{1/2}dx = \frac{2}{9}(3x+1)^{3/2} + C$

$f(0) = 1 = \frac{2}{9} + C$, $C = \frac{7}{9}$, so $f(x) = \frac{2}{9}(3x+1)^{3/2} + \frac{7}{9}$

69. $u = a\sin\theta, du = a\cos\theta\, d\theta$; $\int \frac{du}{\sqrt{a^2-u^2}} = a\theta + C = \sin^{-1}\frac{u}{a} + C$

EXERCISE SET 6.4

1. **(a)** $1+8+27 = 36$ **(b)** $5+8+11+14+17 = 55$
(c) $20+12+6+2+0+0 = 40$ **(d)** $1+1+1+1+1+1 = 6$
(e) $1-2+4-8+16 = 11$ **(f)** $0+0+0+0+0+0 = 0$

3. $\sum_{k=1}^{10} k$ **5.** $\sum_{k=1}^{10} 2k$ **7.** $\sum_{k=1}^{6}(-1)^{k+1}(2k-1)$

9. **(a)** $\sum_{k=1}^{50} 2k$ **(b)** $\sum_{k=1}^{50}(2k-1)$

11. $\frac{1}{2}(100)(100+1) = 5050$

13. $\frac{1}{6}(20)(21)(41) = 2870$

15. $\sum_{k=1}^{30} k(k^2-4) = \sum_{k=1}^{30}(k^3-4k) = \sum_{k=1}^{30} k^3 - 4\sum_{k=1}^{30} k = \frac{1}{4}(30)^2(31)^2 - 4\cdot\frac{1}{2}(30)(31) = 214{,}365$

17. $\sum_{k=1}^{n} \frac{3k}{n} = \frac{3}{n}\sum_{k=1}^{n} k = \frac{3}{n}\cdot\frac{1}{2}n(n+1) = \frac{3}{2}(n+1)$

19. $\sum_{k=1}^{n-1} \frac{k^3}{n^2} = \frac{1}{n^2}\sum_{k=1}^{n-1} k^3 = \frac{1}{n^2}\cdot\frac{1}{4}(n-1)^2n^2 = \frac{1}{4}(n-1)^2$

23. $\frac{1+2+3+\cdots+n}{n^2} = \sum_{k=1}^{n}\frac{k}{n^2} = \frac{1}{n^2}\sum_{k=1}^{n} k = \frac{1}{n^2}\cdot\frac{1}{2}n(n+1) = \frac{n+1}{2n}$; $\lim_{n\to+\infty}\frac{n+1}{2n} = \frac{1}{2}$

25. $\sum_{k=1}^{n}\frac{5k}{n^2} = \frac{5}{n^2}\sum_{k=1}^{n} k = \frac{5}{n^2}\cdot\frac{1}{2}n(n+1) = \frac{5(n+1)}{2n}$; $\lim_{n\to+\infty}\frac{5(n+1)}{2n} = \frac{5}{2}$

27. (a) $\sum_{j=0}^{5} 2^j$ (b) $\sum_{j=1}^{6} 2^{j-1}$ (c) $\sum_{j=2}^{7} 2^{j-2}$

29. Endpoints $2, 3, 4, 5, 6$; $\Delta x = 1$;

(a) Left endpoints: $\sum_{k=1}^{4} f(x_k^*)\Delta x = 7 + 10 + 13 + 16 = 46$

(b) Midpoints: $\sum_{k=1}^{4} f(x_k^*)\Delta x = 8.5 + 11.5 + 14.5 + 17.5 = 52$

(c) Right endpoints: $\sum_{k=1}^{4} f(x_k^*)\Delta x = 10 + 13 + 16 + 19 = 58$

31. Endpoints: $0, \pi/4, \pi/2, 3\pi/4, \pi$; $\Delta x = \pi/4$

(a) Left endpoints: $\sum_{k=1}^{4} f(x_k^*)\Delta x = \left(1 + \sqrt{2}/2 + 0 - \sqrt{2}/2\right)(\pi/4) = \pi/4$

(b) Midpoints: $\sum_{k=1}^{4} f(x_k^*)\Delta x = [\cos(\pi/8) + \cos(3\pi/8) + \cos(5\pi/8) + \cos(7\pi/8)](\pi/4)$

$= [\cos(\pi/8) + \cos(3\pi/8) - \cos(3\pi/8) - \cos(\pi/8)](\pi/4) = 0$

(c) Right endpoints: $\sum_{k=1}^{4} f(x_k^*)\Delta x = \left(\sqrt{2}/2 + 0 - \sqrt{2}/2 - 1\right)(\pi/4) = -\pi/4$

33. (a) 0.718771403, 0.705803382, 0.698172179
(b) 0.668771403, 0.680803382, 0.688172179
(c) 0.692835360, 0.693069098, 0.693134682

35. (a) 4.884074734, 5.115572731, 5.248762738
(b) 5.684074734, 5.515572731, 5.408762738
(c) 5.34707029, 5.338362719, 5.334644416

37. $\Delta x = \dfrac{3}{n}$, $x_k^* = 1 + \dfrac{3}{n}k$; $f(x_k^*)\Delta x = \dfrac{1}{2}x_k^*\Delta x = \dfrac{1}{2}\left(1 + \dfrac{3}{n}k\right)\dfrac{3}{n} = \dfrac{3}{2}\left[\dfrac{1}{n} + \dfrac{3}{n^2}k\right]$

$$\sum_{k=1}^{n} f(x_k^*)\Delta x = \frac{3}{2}\left[\sum_{k=1}^{n}\frac{1}{n} + \sum_{k=1}^{n}\frac{3}{n^2}k\right] = \frac{3}{2}\left[1 + \frac{3}{n^2}\cdot\frac{1}{2}n(n+1)\right] = \frac{3}{2}\left[1 + \frac{3}{2}\frac{n+1}{n}\right]$$

$$A = \lim_{n\to+\infty}\frac{3}{2}\left[1 + \frac{3}{2}\left(1 + \frac{1}{n}\right)\right] = \frac{3}{2}\left(1 + \frac{3}{2}\right) = \frac{15}{4}$$

39. $\Delta x = \dfrac{3}{n}, x_k^* = 0 + k\dfrac{3}{n}; f(x_k^*)\Delta x = (9 - 9\dfrac{k^2}{n^2})\dfrac{3}{n}$

$$\sum_{k=1}^{n} f(x_k^*)\Delta x = \sum_{k=1}^{n}(9 - 9\frac{k^2}{n^2})\frac{3}{n} = \frac{27}{n}\sum_{k=1}^{n}\left(1 - \frac{k^2}{n^2}\right) = 27 - \frac{27}{n^3}\sum_{k=1}^{n}k^2$$

$$A = \lim_{n\to+\infty}\left[27 - \frac{27}{n^3}\sum_{k=1}^{n}k^2\right] = 27 - 27\left(\frac{1}{3}\right) = 18$$

41. $\Delta x = \dfrac{4}{n}$, $x_k^* = 2 + k\dfrac{4}{n}$

$$f(x_k^*)\Delta x = (x_k^*)^3\Delta x = \left[2 + \frac{4}{n}k\right]^3\frac{4}{n} = \frac{32}{n}\left[1 + \frac{2}{n}k\right]^3 = \frac{32}{n}\left[1 + \frac{6}{n}k + \frac{12}{n^2}k^2 + \frac{8}{n^3}k^3\right]$$

$$\sum_{k=1}^{n} f(x_k^*)\Delta x = \frac{32}{n}\left[\sum_{k=1}^{n}1 + \frac{6}{n}\sum_{k=1}^{n}k + \frac{12}{n^2}\sum_{k=1}^{n}k^2 + \frac{8}{n^3}\sum_{k=1}^{n}k^3\right]$$

$$= \frac{32}{n}\left[n + \frac{6}{n}\cdot\frac{1}{2}n(n+1) + \frac{12}{n^2}\cdot\frac{1}{6}n(n+1)(2n+1) + \frac{8}{n^3}\cdot\frac{1}{4}n^2(n+1)^2\right]$$

$$= 32\left[1 + 3\frac{n+1}{n} + 2\frac{(n+1)(2n+1)}{n^2} + 2\frac{(n+1)^2}{n^2}\right]$$

$$A = \lim_{n\to+\infty} 32\left[1 + 3\left(1 + \frac{1}{n}\right) + 2\left(1 + \frac{1}{n}\right)\left(2 + \frac{1}{n}\right) + 2\left(1 + \frac{1}{n}\right)^2\right]$$

$$= 32[1 + 3(1) + 2(1)(2) + 2(1)^2] = 320$$

43. $\Delta x = \dfrac{3}{n}$, $x_k^* = 1 + (k-1)\dfrac{3}{n}$

$$f(x_k^*)\Delta x = \frac{1}{2}x_k^*\Delta x = \frac{1}{2}\left[1 + (k-1)\frac{3}{n}\right]\frac{3}{n} = \frac{1}{2}\left[\frac{3}{n} + (k-1)\frac{9}{n^2}\right]$$

$$\sum_{k=1}^{n} f(x_k^*)\Delta x = \frac{1}{2}\left[\sum_{k=1}^{n}\frac{3}{n} + \frac{9}{n^2}\sum_{k=1}^{n}(k-1)\right] = \frac{1}{2}\left[3 + \frac{9}{n^2}\cdot\frac{1}{2}(n-1)n\right] = \frac{3}{2} + \frac{9}{4}\frac{n-1}{n}$$

$$A = \lim_{n\to+\infty}\left[\frac{3}{2} + \frac{9}{4}\left(1 - \frac{1}{n}\right)\right] = \frac{3}{2} + \frac{9}{4} = \frac{15}{4}$$

45. $\Delta x = \dfrac{3}{n}, x_k^* = 0 + (k-1)\dfrac{3}{n}; f(x_k^*)\Delta x = (9 - 9\dfrac{(k-1)^2}{n^2})\dfrac{3}{n}$

$$\sum_{k=1}^{n} f(x_k^*)\Delta x = \sum_{k=1}^{n}\left[9 - 9\frac{(k-1)^2}{n^2}\right]\frac{3}{n} = \frac{27}{n}\sum_{k=1}^{n}\left(1 - \frac{(k-1)^2}{n^2}\right) = 27 - \frac{27}{n^3}\sum_{k=1}^{n}k^2 + \frac{54}{n^3}\sum_{k=1}^{n}k - \frac{27}{n^2}$$

$$A = \lim_{n\to+\infty} = 27 - 27\left(\frac{1}{3}\right) + 0 + 0 = 18$$

47. $\Delta x = \dfrac{1}{n}, x_k^* = \dfrac{2k-1}{2n}$

$$f(x_k^*)\Delta x = \frac{(2k-1)^2}{(2n)^2}\frac{1}{n} = \frac{k^2}{n^3} - \frac{k}{n^3} + \frac{1}{4n^3}$$

$$\sum_{k=1}^{n} f(x_k^*)\Delta x = \frac{1}{n^3}\sum_{k=1}^{n} k^2 - \frac{1}{n^3}\sum_{k=1}^{n} k + \frac{1}{4n^3}\sum_{k=1}^{n} 1$$

Using Theorem 6.4.4,

$$A = \lim_{n\to+\infty}\sum_{k=1}^{n} f(x_k^*)\Delta x = \frac{1}{3} + 0 + 0 = \frac{1}{3}$$

49. $\Delta x = \dfrac{2}{n}, x_k^* = -1 + \dfrac{2k}{n}$

$$f(x_k^*)\Delta x = \left(-1 + \frac{2k}{n}\right)\frac{2}{n} = -\frac{2}{n} + 4\frac{k}{n^2}$$

$$\sum_{k=1}^{n} f(x_k^*)\Delta x = -2 + \frac{4}{n^2}\sum_{k=1}^{n} k = -2 + \frac{4}{n^2}\frac{n(n+1)}{2} = -2 + 2 + \frac{2}{n}$$

$$A = \lim_{n\to+\infty}\sum_{k=1}^{n} f(x_k^*)\Delta x = 0$$

The area below the x-axis cancels the area above the x-axis.

51. $\Delta x = \dfrac{2}{n}, x_k^* = \dfrac{2k}{n}$

$$f(x_k^*) = \left[\left(\frac{2k}{n}\right)^2 - 1\right]\frac{2}{n} = \frac{8k^2}{n^3} - \frac{2}{n}$$

$$\sum_{k=1}^{n} f(x_k^*)\Delta x = \frac{8}{n^3}\sum_{k=1}^{n} k^2 - \frac{2}{n}\sum_{k=1}^{n} 1 = \frac{8}{n^3}\frac{n(n+1)(2n+1)}{6} - 2$$

$$A = \lim_{n\to+\infty}\sum_{k=1}^{n} f(x_k^*)\Delta x = \frac{16}{6} - 2 = \frac{2}{3}$$

53. $\Delta x = \dfrac{b-a}{n},\ x_k^* = a + \dfrac{b-a}{n}(k-1)$

$$f(x_k^*)\Delta x = mx_k^*\Delta x = m\left[a + \frac{b-a}{n}(k-1)\right]\frac{b-a}{n} = m(b-a)\left[\frac{a}{n} + \frac{b-a}{n^2}(k-1)\right]$$

$$\sum_{k=1}^{n} f(x_k^*)\Delta x = m(b-a)\left[a + \frac{b-a}{2}\cdot\frac{n-1}{n}\right]$$

$$A = \lim_{n\to+\infty} m(b-a)\left[a + \frac{b-a}{2}\left(1 - \frac{1}{n}\right)\right] = m(b-a)\frac{b+a}{2} = \frac{1}{2}m(b^2 - a^2)$$

55. **(a)** With x_k^* as the right endpoint, $\Delta x = \dfrac{b}{n},\ x_k^* = \dfrac{b}{n}k$

$$f(x_k^*)\Delta x = (x_k^*)^3\Delta x = \frac{b^4}{n^4}k^3,\ \sum_{k=1}^{n} f(x_k^*)\Delta x = \frac{b^4}{n^4}\sum_{k=1}^{n} k^3 = \frac{b^4}{4}\frac{(n+1)^2}{n^2}$$

$$A = \lim_{n\to+\infty}\frac{b^4}{4}\left(1 + \frac{1}{n}\right)^2 = b^4/4$$

(b) $\Delta x = \dfrac{b-a}{n}$, $x_k^* = a + \dfrac{b-a}{n}k$

$$f(x_k^*)\Delta x = (x_k^*)^3\Delta x = \left[a + \frac{b-a}{n}k\right]^3 \frac{b-a}{n}$$

$$= \frac{b-a}{n}\left[a^3 + \frac{3a^2(b-a)}{n}k + \frac{3a(b-a)^2}{n^2}k^2 + \frac{(b-a)^3}{n^3}k^3\right]$$

$$\sum_{k=1}^{n} f(x_k^*)\Delta x = (b-a)\left[a^3 + \frac{3}{2}a^2(b-a)\frac{n+1}{n} + \frac{1}{2}a(b-a)^2\frac{(n+1)(2n+1)}{n^2} + \frac{1}{4}(b-a)^3\frac{(n+1)^2}{n^2}\right]$$

$$A = \lim_{n\to+\infty}\sum_{k=1}^{n} f(x_k^*)\Delta x$$

$$= (b-a)\left[a^3 + \frac{3}{2}a^2(b-a) + a(b-a)^2 + \frac{1}{4}(b-a)^3\right] = \frac{1}{4}(b^4 - a^4).$$

57. If $n = 2m$ then $2m + 2(m-1) + \cdots + 2\cdot 2 + 2 = 2\sum_{k=1}^{m} k = 2\cdot\dfrac{m(m+1)}{2} = m(m+1) = \dfrac{n^2+2n}{4}$;

if $n = 2m+1$ then $(2m+1) + (2m-1) + \cdots + 5 + 3 + 1 = \sum_{k=1}^{m+1}(2k-1)$

$$= 2\sum_{k=1}^{m+1} k - \sum_{k=1}^{m+1} 1 = 2\cdot\frac{(m+1)(m+2)}{2} - (m+1) = (m+1)^2 = \frac{n^2+2n+1}{4}$$

59. both are valid

61. $$\sum_{k=1}^{n}(a_k - b_k) = (a_1 - b_1) + (a_2 - b_2) + \cdots + (a_n - b_n)$$

$$= (a_1 + a_2 + \cdots + a_n) - (b_1 + b_2 + \cdots + b_n) = \sum_{k=1}^{n} a_k - \sum_{k=1}^{n} b_k$$

63. **(a)** $\sum_{k=1}^{n} 1$ means add 1 to itself n times, which gives the result.

(b) $\dfrac{1}{n^2}\sum_{k=1}^{n} k = \dfrac{1}{n^2}\dfrac{n(n+1)}{2} = \dfrac{1}{2} + \dfrac{1}{2n}$, so $\lim_{n\to+\infty}\dfrac{1}{n^2}\sum_{k=1}^{n} k = \dfrac{1}{2}$

(c) $\dfrac{1}{n^3}\sum_{k=1}^{n} k^2 = \dfrac{1}{n^3}\dfrac{n(n+1)(2n+1)}{6} = \dfrac{2}{6} + \dfrac{3}{6n} + \dfrac{1}{6n^2}$, so $\lim_{n\to+\infty}\dfrac{1}{n^3}\sum_{k=1}^{n} k^2 = \dfrac{1}{3}$

(d) $\dfrac{1}{n^4}\sum_{k=1}^{n} k^3 = \dfrac{1}{n^4}\left(\dfrac{n(n+1)}{2}\right)^2 = \dfrac{1}{4} + \dfrac{1}{2n} + \dfrac{1}{4n^2}$, so $\lim_{n\to+\infty}\dfrac{1}{n^4}\sum_{k=1}^{n} k^3 = \dfrac{1}{4}$

EXERCISE SET 6.5

1. **(a)** $(4/3)(1) + (5/2)(1) + (4)(2) = 71/6$ **(b)** 2

3. **(a)** $(-9/4)(1) + (3)(2) + (63/16)(1) + (-5)(3) = -117/16$
(b) 3

5. $\int_{-1}^{2} x^2\,dx$

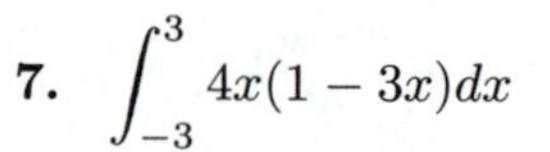

7. $\int_{-3}^{3} 4x(1-3x)dx$

9. **(a)** $\lim_{\max \Delta x_k \to 0} \sum_{k=1}^{n} 2x_k^* \Delta x_k$; $a = 1$, $b = 2$

(b) $\lim_{\max \Delta x_k \to 0} \sum_{k=1}^{n} \frac{x_k^*}{x_k^* + 1} \Delta x_k$; $a = 0$, $b = 1$

11. **(a)** $A = \frac{1}{2}(3)(3) = 9/2$

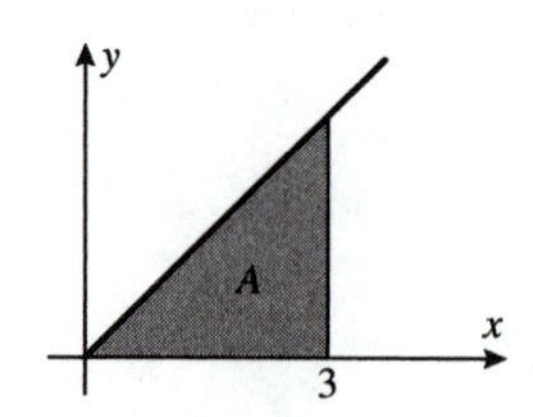

(b) $-A = -\frac{1}{2}(1)(1+2) = -3/2$

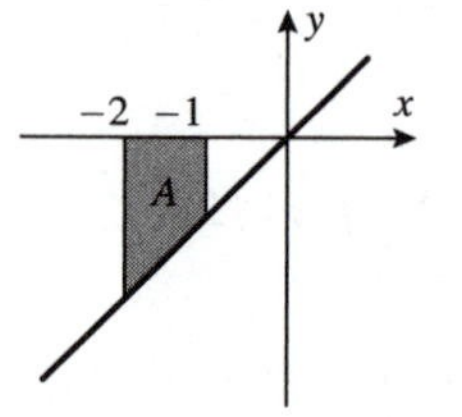

(c) $-A_1 + A_2 = -\frac{1}{2} + 8 = 15/2$

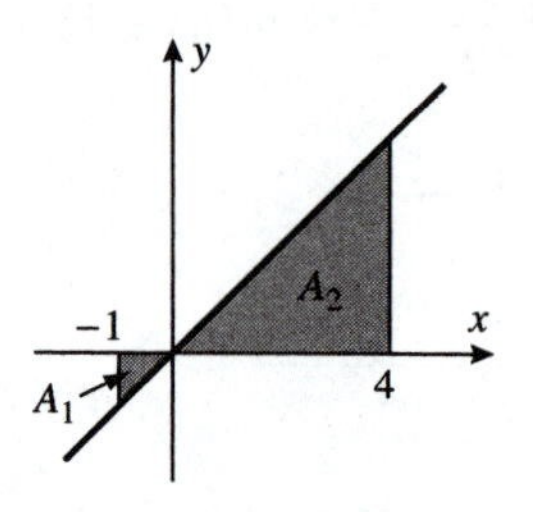

(d) $-A_1 + A_2 = 0$

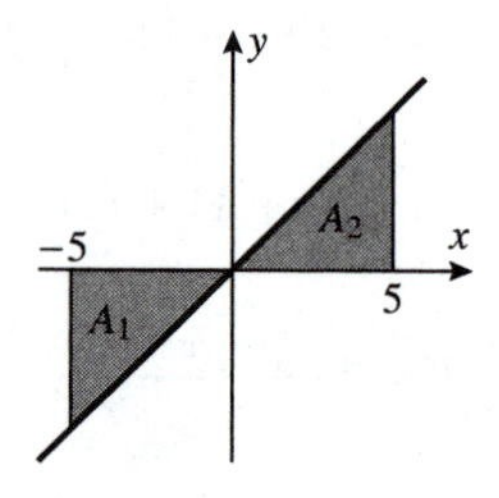

13. **(a)** $A = 2(5) = 10$

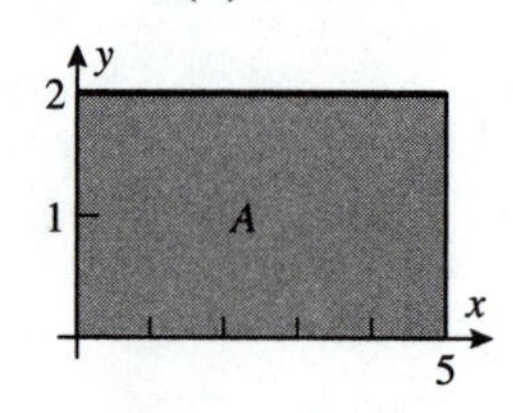

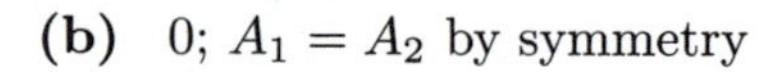

(b) 0; $A_1 = A_2$ by symmetry

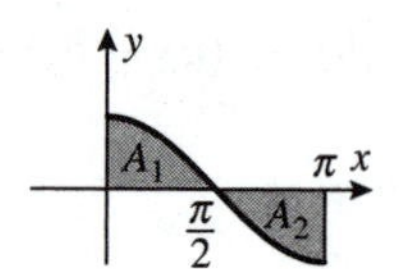

(c) $A_1 + A_2 = \frac{1}{2}(5)(5/2) + \frac{1}{2}(1)(1/2)$

$= 13/2$

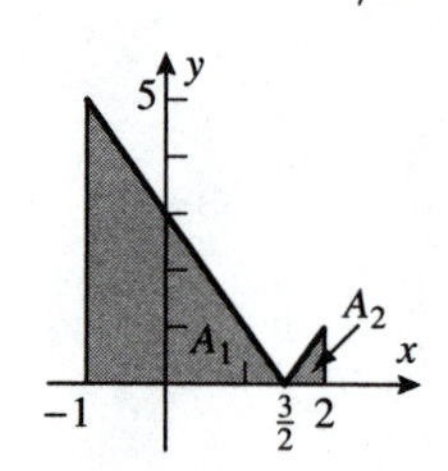

(d) $\frac{1}{2}[\pi(1)^2] = \pi/2$

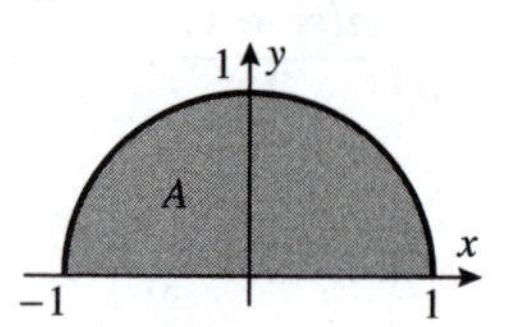

15. **(a)** 0.8 **(b)** −2.6 **(c)** −1.8 **(d)** −0.3

17. $\int_{-1}^{2} f(x)dx + 2\int_{-1}^{2} g(x)dx = 5 + 2(-3) = -1$

19. $\int_{1}^{5} f(x)dx = \int_{0}^{5} f(x)dx - \int_{0}^{1} f(x)dx = 1 - (-2) = 3$

21. **(a)** $\int_0^1 x dx + 2\int_0^1 \sqrt{1-x^2}dx = 1/2 + 2(\pi/4) = (1+\pi)/2$

(b) $4\int_{-1}^3 dx - 5\int_{-1}^3 x dx = 4\cdot 4 - 5(-1/2 + (3\cdot 3)/2) = -4$

23. **(a)** $\sqrt{x} > 0,\ 1-x < 0$ on $[2,3]$ so the integral is negative

(b) $x^2 > 0,\ 3 - \cos x > 0$ for all x so the integral is positive

25. $\int_0^{10} \sqrt{25-(x-5)^2}dx = \pi(5)^2/2 = 25\pi/2$

27. $\int_0^1 (3x+1)dx = 5/2$

29. **(a)** f is continuous on $[-1,1]$ so f is integrable there by Part (a) of Theorem 6.5.8

(b) $|f(x)| \le 1$ so f is bounded on $[-1,1]$, and f has one point of discontinuity, so by Part (b) of Theorem 6.5.8 f is integrable on $[-1,1]$

(c) f is not bounded on [-1,1] because $\lim_{x\to 0} f(x) = +\infty$, so f is not integrable on [0,1]

(d) $f(x)$ is discontinuous at the point $x=0$ because $\lim_{x\to 0}\sin\frac{1}{x}$ does not exist. f is continuous elsewhere. $-1 \le f(x) \le 1$ for x in $[-1,1]$ so f is bounded there. By Part (b), Theorem 6.5.8, f is integrable on $[-1,1]$.

31. **(a)** Let $S_n = \sum_{k=1}^{n} f(x_k^*)\Delta x_k$ and $S = \int_a^b f(x)dx$ then $\sum_{k=1}^{n} cf(x_k^*)\Delta x_k = cS_n$ and we want to prove that $\lim_{\max \Delta x_k \to 0} cS_n = cS$. If $c=0$ the result follows immediately, so suppose that $c \ne 0$ then for any $\epsilon > 0$, $|cS_n - cS| = |c||S_n - S| < \epsilon$ if $|S_n - S| < \epsilon/|c|$. But because f is integrable on $[a,b]$, there is a number $\delta > 0$ such that $|S_n - S| < \epsilon/|c|$ whenever $\max \Delta x_k < \delta$ so $|cS_n - cS| < \epsilon$ and hence $\lim_{\max \Delta x_k \to 0} cS_n = cS$.

(b) Let $R_n = \sum_{k=1}^{n} f(x_k^*)\Delta x_k$, $S_n = \sum_{k=1}^{n} g(x_k^*)\Delta x_k$, $T_n = \sum_{k=1}^{n}[f(x_k^*) + g(x_k^*)]\Delta x_k$, $R = \int_a^b f(x)dx$, and $S = \int_a^b g(x)dx$ then $T_n = R_n + S_n$ and we want to prove that $\lim_{\max \Delta x_k \to 0} T_n = R + S$.
$|T_n - (R+S)| = |(R_n - R) + (S_n - S)| \le |R_n - R| + |S_n - S|$
so for any $\epsilon > 0$ $|T_n - (R+S)| < \epsilon$ if $|R_n - R| + |S_n - S| < \epsilon$.
Because f and g are integrable on $[a,b]$, there are numbers δ_1 and δ_2 such that $|R_n - R| < \epsilon/2$ for max $\Delta x_k < \delta_1$ and $|S_n - S| < \epsilon/2$ for max $\Delta x_k < \delta_2$.
If $\delta = \min(\delta_1, \delta_2)$ then $|R_n - R| < \epsilon/2$ and $|S_n - S| < \epsilon/2$ for $\max \Delta x_k < \delta$ thus $|R_n - R| + |S_n - S| < \epsilon$ and so $|T_n - (R+S)| < \epsilon$ for $\max \Delta x_k < \delta$ which shows that $\lim_{\max \Delta x_k \to 0} T_n = R + S$.

33. $\Delta x_k = \frac{4k^2}{n^2} - \frac{4(k-1)^2}{n^2} = \frac{4}{n^2}(2k-1)$, $x_k^* = \frac{4k^2}{n^2}$,

$f(x_k^*) = \frac{2k}{n}$, $f(x_k^*)\Delta x_k = \frac{8k}{n^3}(2k-1) = \frac{8}{n^3}(2k^2 - k)$,

$$\sum_{k=1}^{n} f(x_k^*)\Delta x_k = \frac{8}{n^3}\sum_{k=1}^{n}(2k^2 - k) = \frac{8}{n^3}\left[\frac{1}{3}n(n+1)(2n+1) - \frac{1}{2}n(n+1)\right] = \frac{4}{3}\frac{(n+1)(4n-1)}{n^2},$$

$$\lim_{n\to+\infty}\sum_{k=1}^{n} f(x_k^*)\Delta x_k = \lim_{n\to+\infty}\frac{4}{3}\left(1+\frac{1}{n}\right)\left(4-\frac{1}{n}\right) = \frac{16}{3}.$$

35. With $f(x) = g(x)$ then $f(x) - g(x) = 0$ for $a < x \le b$. By Theorem 6.5.4(b)

$$\int_a^b f(x)\,dx = \int_a^b [(f(x) - g(x) + g(x)]dx = \int_a^b [f(x) - g(x)]dx + \int_a^b g(x)dx.$$

But the first term on the right hand side is zero (from Exercise 34), so

$$\int_a^b f(x)\,dx = \int_a^b g(x)\,dx$$

EXERCISE SET 6.6

1. (a) $\displaystyle\int_0^2 (2-x)dx = (2x - x^2/2)\Big]_0^2 = 4 - 4/2 = 2$

(b) $\displaystyle\int_{-1}^1 2dx = 2x\Big]_{-1}^1 = 2(1) - 2(-1) = 4$

(c) $\displaystyle\int_1^3 (x+1)dx = (x^2/2 + x)\Big]_1^3 = 9/2 + 3 - (1/2 + 1) = 6$

3. $\displaystyle\int_2^3 x^3 dx = x^4/4\Big]_2^3 = 81/4 - 16/4 = 65/4$

5. $\displaystyle\int_1^9 \sqrt{x}dx = \frac{2}{3}x^{3/2}\Big]_1^9 = \frac{2}{3}(27 - 1) = 52/3$

7. $\displaystyle\int_1^3 e^x dx = e^x\Big]_1^3 = e^3 - e$

9. $\displaystyle\left(\frac{1}{3}x^3 - 2x^2 + 7x\right)\Big]_{-3}^0 = 48$

11. $\displaystyle\int_1^3 x^{-2}dx = -\frac{1}{x}\Big]_1^3 = 2/3$

13. $\displaystyle\frac{4}{5}x^{5/2}\Big]_4^9 = 844/5$

15. $\displaystyle -\cos\theta\Big]_{-\pi/2}^{\pi/2} = 0$

17. $\displaystyle \sin x\Big]_{-\pi/4}^{\pi/4} = \sqrt{2}$

19. $\displaystyle 5e^x\Big]_{\ln 2}^3 = 5e^3 - 5(2) = 5e^3 - 10$

21. $\displaystyle \sin^{-1} x\Big]_0^{1/\sqrt{2}} = \sin^{-1}(1/\sqrt{2}) - \sin^{-1} 0 = \pi/4$

23. $\displaystyle \sec^{-1} x\Big]_{\sqrt{2}}^2 = \sec^{-1} 2 - \sec^{-1}\sqrt{2} = \pi/3 - \pi/4 = \pi/12$

25. $\displaystyle \left(6\sqrt{t} - \frac{10}{3}t^{3/2} + \frac{2}{\sqrt{t}}\right)\Big]_1^4 = -55/3$

27. $\displaystyle \left(\frac{1}{2}x^2 - 2\cot x\right)\Big]_{\pi/6}^{\pi/2} = \pi^2/9 + 2\sqrt{3}$

29. (a) $\displaystyle\int_0^{3/2} (3-2x)dx + \int_{3/2}^2 (2x-3)dx = (3x - x^2)\Big]_0^{3/2} + (x^2 - 3x)\Big]_{3/2}^2 = 9/4 + 1/4 = 5/2$

(b) $\displaystyle\int_0^{\pi/2} \cos x\,dx + \int_{\pi/2}^{3\pi/4} (-\cos x)dx = \sin x\Big]_0^{\pi/2} - \sin x\Big]_{\pi/2}^{3\pi/4} = 2 - \sqrt{2}/2$

31. (a) $\int_{-1}^{0}(1-e^x)dx+\int_0^1(e^x-1)dx = (x-e^x)\Big]_{-1}^{0}+(e^x-x)\Big]_0^1 = -1-(-1-e^{-1})+e-1-1 = e+1/e-2$

(b) $\int_1^2 \frac{2-x}{x}\,dx + \int_2^4 \frac{x-2}{x}\,dx = 2\ln x\Big]_1^2 - 1 + 2 - 2\ln x\Big]_2^4 = 2\ln 2 + 1 - 2\ln 4 + 2\ln 2 = 1$

33. (a) 17/6 (b) $F(x) = \begin{cases} \frac{1}{2}x^2, & x \le 1 \\ \frac{1}{3}x^3 + \frac{1}{6}, & x > 1 \end{cases}$

35. 0.665867079; $\int_1^3 \frac{1}{x^2}\,dx = -\frac{1}{x}\Big]_1^3 = 2/3$

37. 3.106017890; $\int_{-1}^{1} \sec^2 x\,dx = \tan x\Big|_{-1}^{1} = 2\tan 1 \approx 3.114815450$

39. $A = \int_0^3 (x^2+1)dx = \left(\frac{1}{3}x^3 + x\right)\Big]_0^3 = 12$

41. $A = \int_0^{2\pi/3} 3\sin x\,dx = -3\cos x\Big]_0^{2\pi/3} = 9/2$

43. (a) $A = \int_0^{0.8} \frac{1}{\sqrt{1-x^2}}dx = \sin^{-1} x\Big]_0^{0.8} = \sin^{-1}(0.8)$

(b) The calculator was in degree mode instead of radian mode; the correct answer is 0.93.

45. (a) the area between the curve and the x-axis breaks into equal parts, one above and one below the x-axis, so the integral is zero

(b) $\int_{-1}^{1} x^3dx = \frac{1}{4}x^4\Big]_{-1}^{1} = \frac{1}{4}(1^4 - (-1)^4) = 0;$

$\int_{-\pi/2}^{\pi/2} \sin x dx = -\cos x\Big]_{-\pi/2}^{\pi/2} = -\cos(\pi/2) + \cos(-\pi/2) = 0 + 0 = 0$

(c) The area on the left side of the y-axis is equal to the area on the right side, so $\int_{-a}^{a} f(x)dx = 2\int_0^a f(x)dx$

(d) $\int_{-1}^{1} x^2dx = \frac{1}{3}x^3\Big]_{-1}^{1} = \frac{1}{3}(1^3 - (-1)^3) = \frac{2}{3} = 2\int_0^1 x^2dx;$

$\int_{-\pi/2}^{\pi/2} \cos x dx = \sin x\Big]_{-\pi/2}^{\pi/2} = \sin(\pi/2) - \sin(-\pi/2) = 1 + 1 = 2 = 2\int_0^{\pi/2} \cos x dx$

47. (a) $x^3 + 1$ (b) $F(x) = \left(\frac{1}{4}t^4 + t\right)\Big]_1^x = \frac{1}{4}x^4 + x - \frac{5}{4};\ F'(x) = x^3 + 1$

49. (a) $\sin\sqrt{x}$ (b) e^{x^2}

51. $-\frac{x}{\cos x}$

53. $F'(x) = \sqrt{3x^2+1}$, $F''(x) = \dfrac{3x}{\sqrt{3x^2+1}}$

(a) 0 **(b)** $\sqrt{13}$ **(c)** $6/\sqrt{13}$

55. **(a)** $F'(x) = \dfrac{x-3}{x^2+7} = 0$ when $x = 3$, which is a relative minimum, and hence the absolute minimum, by the first derivative test.

(b) increasing on $[3, +\infty)$, decreasing on $(-\infty, 3]$

(c) $F''(x) = \dfrac{7+6x-x^2}{(x^2+7)^2} = \dfrac{(7-x)(1+x)}{(x^2+7)^2}$; concave up on $(-1, 7)$, concave down on $(-\infty, -1)$ and on $(7, +\infty)$

57. **(a)** $(0, +\infty)$ because f is continuous there and 1 is in $(0, +\infty)$

(b) at $x = 1$ because $F(1) = 0$

59. **(a)** $f_{\text{ave}} = \dfrac{1}{9}\displaystyle\int_0^9 x^{1/2}dx = 2$; $\sqrt{x^*} = 2$, $x^* = 4$

(b) $f_{\text{ave}} = \dfrac{1}{3}\displaystyle\int_{-1}^2 (3x^2+2x+1)\,dx = \dfrac{1}{3}(x^3+x^2+x)\Big]_{-1}^2 = 5$; $3x^{*2} + 2x^* + 1 = 5$, with solutions $x^* = -(1/3)(1 \pm \sqrt{13})$, but only $x^* = -(1/3)(1-\sqrt{13})$ lies in the interval $[-1, 2]$.

61. $\sqrt{2} \le \sqrt{x^3+2} \le \sqrt{29}$, so $3\sqrt{2} \le \displaystyle\int_0^3 \sqrt{x^3+2}dx \le 3\sqrt{29}$

63. **(a)** $\left[cF(x)\right]_a^b = cF(b) - cF(a) = c[F(b) - F(a)] = c\left[F(x)\right]_a^b$

(b)
$$\begin{aligned}\left[F(x) + G(x)\right]_a^b &= [F(b) + G(b)] - [F(a) + G(a)] \\ &= [F(b) - F(a)] + [G(b) - G(a)] = F(x)\Big]_a^b + G(x)\Big]_a^b\end{aligned}$$

(c)
$$\begin{aligned}\left[F(x) - G(x)\right]_a^b &= [F(b) - G(b)] - [F(a) - G(a)] \\ &= [F(b) - F(a)] - [G(b) - G(a)] = F(x)\Big]_a^b - G(x)\Big]_a^b\end{aligned}$$

65. $\displaystyle\sum_{k=1}^n \frac{\pi}{4n}\sec^2\left(\frac{\pi k}{4n}\right) = \sum_{k=1}^n f(x_k^*)\Delta x$ where $f(x) = \sec^2 x$, $x_k^* = \dfrac{\pi k}{4n}$ and $\Delta x = \dfrac{\pi}{4n}$ for $0 \le x \le \dfrac{\pi}{4}$.

Thus $\displaystyle\lim_{n\to+\infty}\sum_{k=1}^n \frac{\pi}{4n}\sec^2\left(\frac{\pi k}{4n}\right)\lim_{n\to+\infty}\sum_{k=1}^n f(x_k^*)\Delta x = \int_0^{\pi/4}\sec^2 x\,dx = \tan x\Big]_0^{\pi/4} = 1$

EXERCISE SET 6.7

1. **(a)** the increase in height in inches, during the first ten years

(b) the change in the radius in centimeters, during the time interval $t = 1$ to $t = 2$ seconds

(c) the change in the speed of sound in ft/s, during an increase in temperature from $t = 32°\text{F}$ to $t = 100°\text{F}$

(d) the displacement of the particle in cm, during the time interval $t = t_1$ to $t = t_2$ seconds

3. **(a)** $\text{displ} = s(3) - s(0)$

$$= \int_0^3 v(t)dt = \int_0^2 (1-t)dt + \int_2^3 (t-3)dt = (t - t^2/2)\Big]_0^2 + (t^2/2 - 3t)\Big]_2^3 = -1/2;$$

$$\text{dist} = \int_0^3 |v(t)|dt = (t - t^2/2)\Big]_0^1 + (t^2/2 - t)\Big]_1^2 - (t^2/2 - 3t)\Big]_2^3 = 3/2$$

(b) $\text{displ} = s(3) - s(0)$

$$= \int_0^3 v(t)dt = \int_0^1 tdt + \int_1^2 dt + \int_2^3 (5-2t)dt = t^2/2\Big]_0^1 + t\Big]_1^2 + (5t - t^2)\Big]_2^3 = 3/2;$$

$$\text{dist} = \int_0^1 tdt + \int_1^2 dt + \int_2^{5/2} (5-2t)dt + \int_{5/2}^3 (2t-5)dt$$

$$= t^2/2\Big]_0^1 + t\Big]_1^2 + (5t - t^2)\Big]_2^{5/2} + (t^2 - 5t)\Big]_{5/2}^3 = 2$$

5. **(a)** $v(t) = 20 + \int_0^t a(u)du$; add areas of the small blocks to get

$v(4) \approx 20 + 1.4 + 3.0 + 4.7 + 6.2 = 35.3$ m/s

(b) $v(6) = v(4) + \int_4^6 a(u)du \approx 35.3 + 7.5 + 8.6 = 51.4$ m/s

7. **(a)** $s(t) = \int (t^3 - 2t^2 + 1)dt = \frac{1}{4}t^4 - \frac{2}{3}t^3 + t + C,$

$s(0) = \frac{1}{4}(0)^4 - \frac{2}{3}(0)^3 + 0 + C = 1,\ C = 1,\ s(t) = \frac{1}{4}t^4 - \frac{2}{3}t^3 + t + 1$

(b) $v(t) = \int 4\cos 2t\, dt = 2\sin 2t + C_1,\ v(0) = 2\sin 0 + C_1 = -1,\ C_1 = -1,$

$v(t) = 2\sin 2t - 1,\ s(t) = \int (2\sin 2t - 1)dt = -\cos 2t - t + C_2,$

$s(0) = -\cos 0 - 0 + C_2 = -3,\ C_2 = -2,\ s(t) = -\cos 2t - t - 2$

9. **(a)** $s(t) = \int (2t - 3)dt = t^2 - 3t + C,\ s(1) = (1)^2 - 3(1) + C = 5,\ C = 7,\ s(t) = t^2 - 3t + 7$

(b) $v(t) = \int \cos t dt = \sin t + C_1,\ v(\pi/2) = 2 = 1 + C_1,\ C_1 = 1,\ v(t) = \sin t + 1,$

$s(t) = \int (\sin t + 1)dt = -\cos t + t + C_2,\ s(\pi/2) = 0 = \pi/2 + C_2,\ C_2 = -\pi/2,$

$s(t) = -\cos t + t - \pi/2$

11. **(a)** $\text{displacement} = s(\pi/2) - s(0) = \int_0^{\pi/2} \sin t dt = -\cos t\Big]_0^{\pi/2} = 1$ m

$\text{distance} = \int_0^{\pi/2} |\sin t| dt = 1$ m

(b) $\text{displacement} = s(2\pi) - s(\pi/2) = \int_{\pi/2}^{2\pi} \cos t dt = \sin t\Big]_{\pi/2}^{2\pi} = -1$ m

$\text{distance} = \int_{\pi/2}^{2\pi} |\cos t| dt = -\int_{\pi/2}^{3\pi/2} \cos t dt + \int_{3\pi/2}^{2\pi} \cos t dt = 3$ m

13. (a) $v(t) = t^3 - 3t^2 + 2t = t(t-1)(t-2)$

$$\text{displacement} = \int_0^3 (t^3 - 3t^2 + 2t)dt = 9/4 \text{ m}$$

$$\text{distance} = \int_0^3 |v(t)|dt = \int_0^1 v(t)dt + \int_1^2 -v(t)dt + \int_2^3 v(t)dt = 11/4 \text{ m}$$

(b) $$\text{displacement} = \int_0^3 (\sqrt{t} - 2)dt = 2\sqrt{3} - 6 \text{ m}$$

$$\text{distance} = \int_0^3 |v(t)|dt = -\int_0^3 v(t)dt = 6 - 2\sqrt{3} \text{ m}$$

15. $v(t) = -2t + 3$

$$\text{displacement} = \int_1^4 (-2t+3)dt = -6 \text{ m}$$

$$\text{distance} = \int_1^4 |-2t+3|dt = \int_1^{3/2} (-2t+3)dt + \int_{3/2}^4 (2t-3)dt = 13/2 \text{ m}$$

17. $v(t) = \frac{2}{5}\sqrt{5t+1} + \frac{8}{5}$

$$\text{displacement} = \int_0^3 \left(\frac{2}{5}\sqrt{5t+1} + \frac{8}{5}\right) dt = \frac{4}{75}(5t+1)^{3/2} + \frac{8}{5}t\Big]_0^3 = 204/25 \text{ m}$$

$$\text{distance} = \int_0^3 |v(t)|dt = \int_0^3 v(t)dt = 204/25 \text{ m}$$

19. (a) $s = \int \sin \frac{1}{2}\pi t\, dt = -\frac{2}{\pi} \cos \frac{1}{2}\pi t + C$

$s = 0$ when $t = 0$ which gives $C = \frac{2}{\pi}$ so $s = -\frac{2}{\pi} \cos \frac{1}{2}\pi t + \frac{2}{\pi}$.

$a = \frac{dv}{dt} = \frac{\pi}{2} \cos \frac{1}{2}\pi t$. When $t = 1 : s = 2/\pi,\ v = 1,\ |v| = 1,\ a = 0$.

(b) $v = -3\int t\, dt = -\frac{3}{2}t^2 + C_1$, $v = 0$ when $t = 0$ which gives $C_1 = 0$ so $v = -\frac{3}{2}t^2$

$s = -\frac{3}{2}\int t^2 dt = -\frac{1}{2}t^3 + C_2$, $s = 1$ when $t = 0$ which gives $C_2 = 1$ so $s = -\frac{1}{2}t^3 + 1$.

When $t = 1 : s = 1/2,\ v = -3/2,\ |v| = 3/2,\ a = -3$.

21. $$A = A_1 + A_2 = \int_0^1 (1 - x^2)dx + \int_1^3 (x^2 - 1)dx = 2/3 + 20/3 = 22/3$$

23. $$A = A_1 + A_2 = \int_{-1}^0 \left[1 - \sqrt{x+1}\right] dx + \int_0^1 [\sqrt{x+1} - 1]\, dx$$

$$= \left(x - \frac{2}{3}(x+1)^{3/2}\right)\Big]_{-1}^0 + \left(\frac{2}{3}(x+1)^{3/2} - x\right)\Big]_0^1 = -\frac{2}{3} + 1 + \frac{4\sqrt{2}}{3} - 1 - \frac{2}{3} = 4\frac{\sqrt{2}-1}{3}$$

25. $$A = A_1 + A_2 = \int_{-1}^0 (1 - e^x)dx + \int_0^1 (e^x - 1)dx = 1/e + e - 2$$

27. By inspection the velocity is positive for $t > 0$, and during the first second the particle is at most 5/2 cm from the starting position. For $T > 1$ the displacement of the particle during the time interval $[0, T]$ is given by

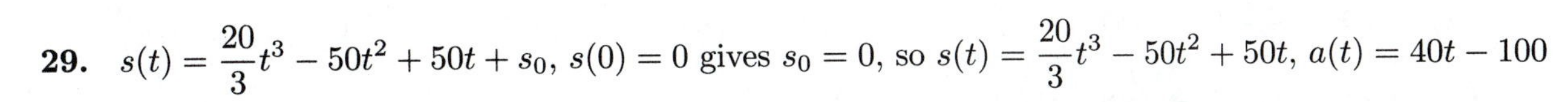

$$\int_0^T v(t)\,dt = 5/2 + \int_1^T (6\sqrt{t} - 1/t)\,dt = 5/2 + (4t^{3/2} - \ln t)\Big]_1^T = -3/2 + 4T^{3/2} - \ln T,$$

and the displacement equals 4 cm if $4T^{3/2} - \ln T = 11/2, T \approx 1.272$ s

29. $s(t) = \dfrac{20}{3}t^3 - 50t^2 + 50t + s_0$, $s(0) = 0$ gives $s_0 = 0$, so $s(t) = \dfrac{20}{3}t^3 - 50t^2 + 50t$, $a(t) = 40t - 100$

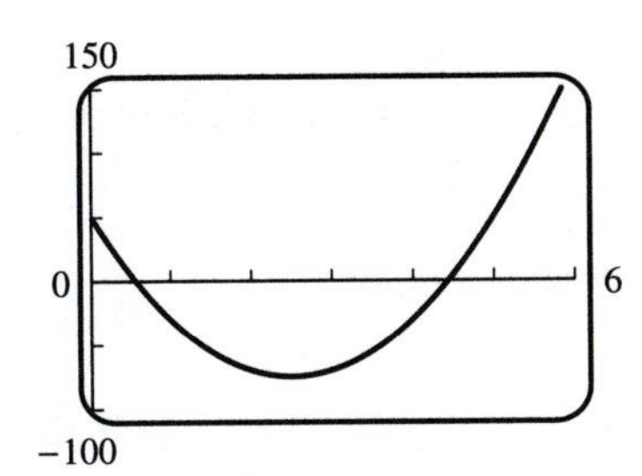

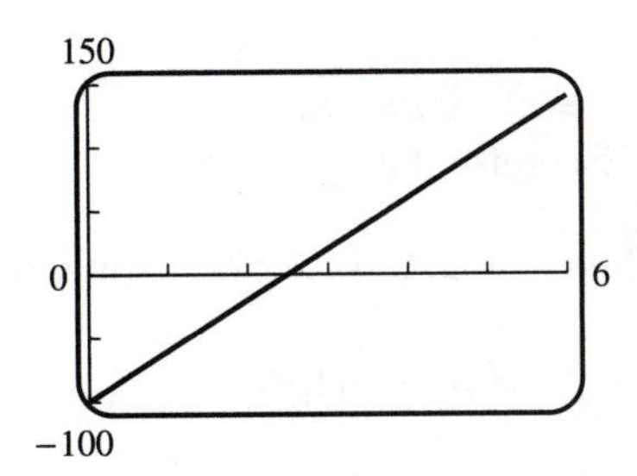

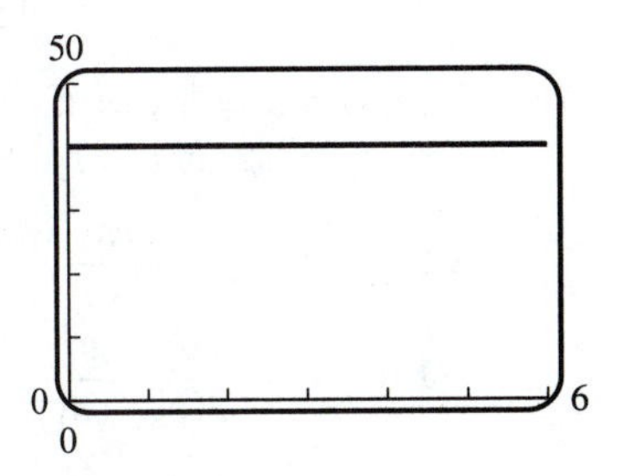

31. (a) From the graph the velocity is at first positive, but then turns negative, then positive again. The displacement, which is the cumulative area from $x = 0$ to $x = 5$, starts positive, turns negative, and then turns positive again.

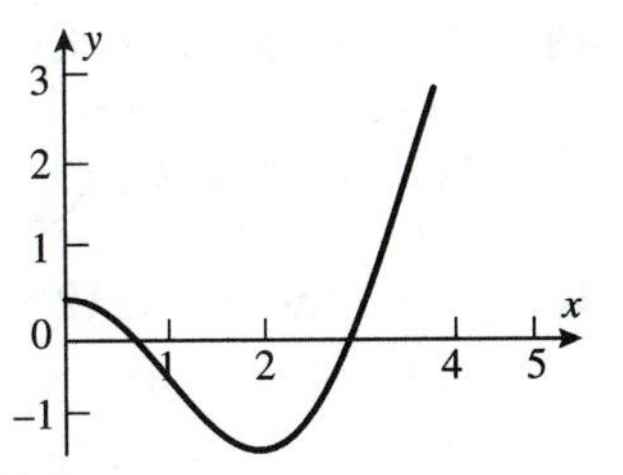

(b) displ $= 5/2 - \sin 5 + 5\cos 5$

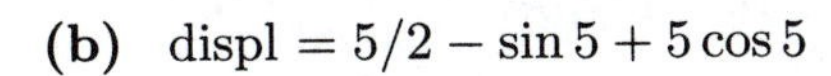

33. (a) From the graph the velocity is positive, so the displacement is always increasing and is therefore positive.

(b) $s(t) = t/2 + (t + 1)e^{-t}$

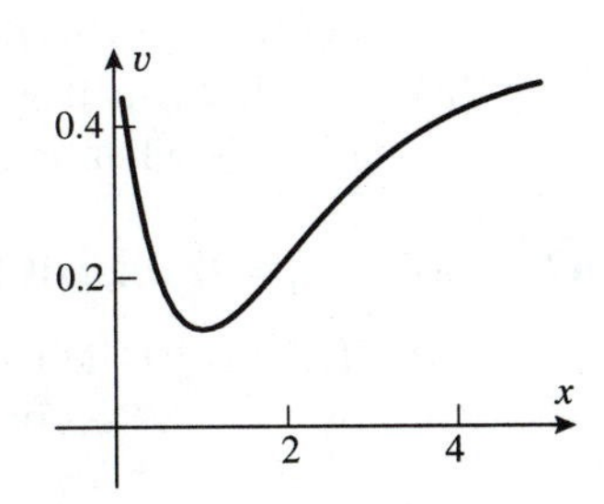

35. (a) $a(t) = \begin{cases} 0, & t < 4 \\ -10, & t > 4 \end{cases}$

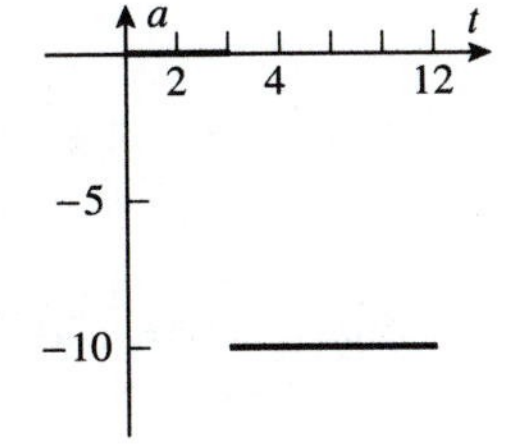

(b) $v(t) = \begin{cases} 25, & t < 4 \\ 65 - 10t, & t > 4 \end{cases}$

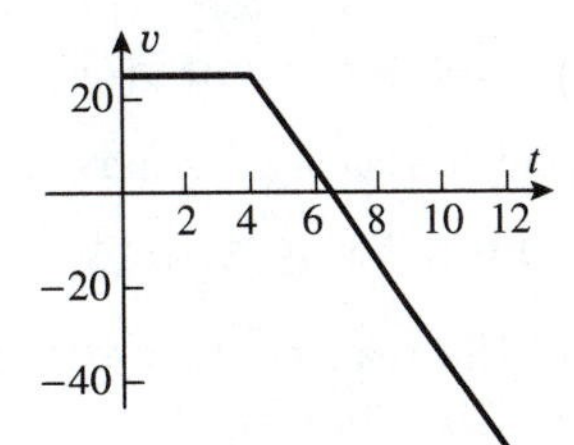

(c) $x(t) = \begin{cases} 25t, & t < 4 \\ 65t - 5t^2 - 80, & t > 4 \end{cases}$, so $x(8) = 120$, $x(12) = -20$

(d) $x(6.5) = 131.25$

37. **(a)** $a = -1$ mi/h/s $= -22/15$ ft/s^2 **(b)** $a = 30$ km/h/min $= 1/7200$ km/s^2

39. $a = a_0$ ft/s^2, $v = a_0 t + v_0 = a_0 t + 132$ ft/s, $s = a_0 t^2/2 + 132t + s_0 = a_0 t^2/2 + 132t$ ft; $s = 200$ ft when $v = 88$ ft/s. Solve $88 = a_0 t + 132$ and $200 = a_0 t^2/2 + 132t$ to get $a_0 = -\dfrac{121}{5}$ when $t = \dfrac{20}{11}$, so $s = -12.1t^2 + 132t$, $v = -\dfrac{121}{5}t + 132$.

(a) $a_0 = -\dfrac{121}{5}$ ft/s^2 **(b)** $v = 55$ mi/h $= \dfrac{242}{3}$ ft/s when $t = \dfrac{70}{33}$ s

(c) $v = 0$ when $t = \dfrac{60}{11}$ s

41. Suppose $s = s_0 = 0$, $v = v_0 = 0$ at $t = t_0 = 0$; $s = s_1 = 120$, $v = v_1$ at $t = t_1$; and $s = s_2$, $v = v_2 = 12$ at $t = t_2$. From Exercise 36(a),

$2.6 = a = \dfrac{v_1^2 - v_0^2}{2(s_1 - s_0)}$, $v_1^2 = 2as_1 = 5.2(120) = 624$. Applying the formula again,

$-1.5 = a = \dfrac{v_2^2 - v_1^2}{2(s_2 - s_1)}$, $v_2^2 = v_1^2 - 3(s_2 - s_1)$, so

$s_2 = s_1 - (v_2^2 - v_1^2)/3 = 120 - (144 - 624)/3 = 280$ m.

43. The truck's velocity is $v_T = 50$ and its position is $s_T = 50t + 5000$. The car's acceleration is $a_C = 2$, so $v_C = 2t$, $s_C = t^2$ (initial position and initial velocity of the car are both zero). $s_T = s_C$ when $50t + 5000 = t^2$, $t^2 - 50t - 5000 = (t+50)(t-100) = 0$, $t = 100$ s and $s_C = s_T = t^2 = 10{,}000$ ft.

45. $s = 0$ and $v = 112$ when $t = 0$ so $v(t) = -32t + 112$, $s(t) = -16t^2 + 112t$

(a) $v(3) = 16$ ft/s, $v(5) = -48$ ft/s

(b) $v = 0$ when the projectile is at its maximum height so $-32t + 112 = 0$, $t = 7/2$ s, $s(7/2) = -16(7/2)^2 + 112(7/2) = 196$ ft.

(c) $s = 0$ when it reaches the ground so $-16t^2 + 112t = 0$, $-16t(t-7) = 0$, $t = 0, 7$ of which $t = 7$ is when it is at ground level on its way down. $v(7) = -112$, $|v| = 112$ ft/s.

47. **(a)** $s(t) = 0$ when it hits the ground, $s(t) = -16t^2 + 16t = -16t(t-1) = 0$ when $t = 1$ s.

(b) The projectile moves upward until it gets to its highest point where $v(t) = 0$, $v(t) = -32t + 16 = 0$ when $t = 1/2$ s.

49. **(a)** $s(t) = 0$ when the package hits the ground, $s(t) = -16t^2 + 20t + 200 = 0$ when $t = (5 + 5\sqrt{33})/8$ s

(b) $v(t) = -32t + 20$, $v[(5 + 5\sqrt{33})/8] = -20\sqrt{33}$, the speed at impact is $20\sqrt{33}$ ft/s

51. $s(t) = -4.9t^2 + 49t + 150$ and $v(t) = -9.8t + 49$

(a) the projectile reaches its maximum height when $v(t) = 0$, $-9.8t + 49 = 0$, $t = 5$ s

(b) $s(5) = -4.9(5)^2 + 49(5) + 150 = 272.5$ m

(c) the projectile reaches its starting point when $s(t) = 150$, $-4.9t^2 + 49t + 150 = 150$, $-4.9t(t-10) = 0$, $t = 10$ s

(d) $v(10) = -9.8(10) + 49 = -49$ m/s

(e) $s(t) = 0$ when the projectile hits the ground, $-4.9t^2 + 49t + 150 = 0$ when (use the quadratic formula) $t \approx 12.46$ s

(f) $v(12.46) = -9.8(12.46) + 49 \approx -73.1$, the speed at impact is about 73.1 m/s

53. $g = 9.8/6 = 4.9/3$ m/s^2, so $v = -(4.9/3)t$, $s = -(4.9/6)t^2 + 5$, $s = 0$ when $t = \sqrt{30/4.9}$ and $v = -(4.9/3)\sqrt{30/4.9} \approx -4.04$, so the speed of the module upon landing is 4.04 m/s

55. $f_{\text{ave}} = \dfrac{1}{3-1}\displaystyle\int_1^3 3x\,dx = \dfrac{3}{4}x^2\Big]_1^3 = 6$

57. $f_{\text{ave}} = \dfrac{1}{\pi - 0}\displaystyle\int_0^\pi \sin x\,dx = -\dfrac{1}{\pi}\cos x\Big]_0^\pi = 2/\pi$

59. $f_{\text{ave}} = \dfrac{1}{e-1}\displaystyle\int_1^e \dfrac{1}{x}\,dx = \dfrac{1}{e-1}(\ln e - \ln 1) = \dfrac{1}{e-1}$

61. **(a)** $f_{\text{ave}} = \dfrac{1}{2-0}\displaystyle\int_0^2 x^2 dx = 4/3$

(b) $(x^*)^2 = 4/3, x^* = \pm 2/\sqrt{3}$, but only $2/\sqrt{3}$ is in $[0, 2]$

(c)

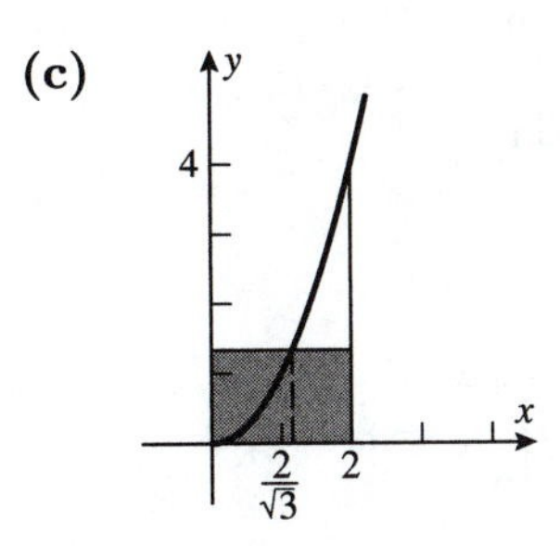

63. **(a)** $v_{\text{ave}} = \dfrac{1}{4-1}\displaystyle\int_1^4 (3t^3 + 2)dt = \dfrac{1}{3}\dfrac{789}{4} = \dfrac{263}{4}$

(b) $v_{\text{ave}} = \dfrac{s(4) - s(1)}{4-1} = \dfrac{100 - 7}{3} = 31$

65. time to fill tank = (volume of tank)/(rate of filling) = $[\pi(3)^2 5]/(1) = 45\pi$, weight of water in tank at time $t = (62.4)$ (rate of filling)(time) $= 62.4t$,

$\text{weight}_{\text{ave}} = \dfrac{1}{45\pi}\displaystyle\int_0^{45\pi} 62.4t\,dt = 1404\pi$ lb

67. **(a)** amount of water = (rate of flow)(time) = $4t$ gal, total amount = $4(30) = 120$ gal

(b) amount of water $= \displaystyle\int_0^{60} (4 + t/10)dt = 420$ gal

(c) amount of water $= \displaystyle\int_0^{120} (10 + \sqrt{t})dt = 1200 + 160\sqrt{30} \approx 2076.36$ gal

69. **(a)** $\displaystyle\int_a^b [f(x) - f_{\text{ave}}]\,dx = \int_a^b f(x)dx - \int_a^b f_{\text{ave}}dx = \int_a^b f(x)dx - f_{\text{ave}}(b-a) = 0$

because $f_{\text{ave}}(b-a) = \displaystyle\int_a^b f(x)dx$

(b) no, because if $\displaystyle\int_a^b [f(x) - c]dx = 0$ then $\displaystyle\int_a^b f(x)dx - c(b-a) = 0$ so

$c = \dfrac{1}{b-a}\displaystyle\int_a^b f(x)dx = f_{\text{ave}}$ is the only value

EXERCISE SET 6.8

1. **(a)** $\int_1^3 u^7 du$ **(b)** $-\frac{1}{2}\int_7^4 u^{1/2} du$ **(c)** $\frac{1}{\pi}\int_{-\pi}^{\pi} \sin u\, du$ **(d)** $\int_{-3}^{0} (u+5)u^{20} du$

3. **(a)** $\frac{1}{2}\int_{-1}^{1} e^u\, du$ **(b)** $\int_1^2 u\, du$

5. $u = 2x+1$, $\frac{1}{2}\int_1^3 u^4 du = \frac{1}{10}u^5\Big]_1^3 = 121/5$, or $\frac{1}{10}(2x+1)^5\Big]_0^1 = 121/5$

7. $u = 1-2x$, $-\frac{1}{2}\int_3^1 u^3 du = -\frac{1}{8}u^4\Big]_3^1 = 10$, or $-\frac{1}{8}(1-2x)^4\Big]_{-1}^0 = 10$

9. $u = 1+x$, $\int_1^9 (u-1)u^{1/2} du = \int_1^9 (u^{3/2} - u^{1/2}) du = \frac{2}{5}u^{5/2} - \frac{2}{3}u^{3/2}\Big]_1^9 = 1192/15$,

or $\frac{2}{5}(1+x)^{5/2} - \frac{2}{3}(1+x)^{3/2}\Big]_0^8 = 1192/15$

11. $u = x/2$, $8\int_0^{\pi/4} \sin u\, du = -8\cos u\Big]_0^{\pi/4} = 8 - 4\sqrt{2}$, or $-8\cos(x/2)\Big]_0^{\pi/2} = 8 - 4\sqrt{2}$

13. $u = x^2+2$, $\frac{1}{2}\int_6^3 u^{-3} du = -\frac{1}{4u^2}\Big]_6^3 = -1/48$, or $-\frac{1}{4}\frac{1}{(x^2+2)^2}\Big]_{-2}^{-1} = -1/48$

15. $u = e^x + 4$, $du = e^x dx$, $u = e^{-\ln 3} + 4 = \frac{1}{3} + 4 = \frac{13}{3}$ when $x = -\ln 3$,

$u = e^{\ln 3} + 4 = 3 + 4 = 7$ when $x = \ln 3$, $\int_{13/3}^{7} \frac{1}{u} du = \ln u\Big]_{13/3}^{7} = \ln(7) - \ln(13/3) = \ln(21/13)$

17. $u = \sqrt{x}$, $2\int_1^{\sqrt{3}} \frac{1}{u^2+1} du = 2\tan^{-1} u\Big]_1^{\sqrt{3}} = 2(\tan^{-1}\sqrt{3} - \tan^{-1} 1) = 2(\pi/3 - \pi/4) = \pi/6$

19. $\frac{1}{3}\int_0^5 \sqrt{25-u^2}\, du = \frac{1}{3}\left[\frac{1}{4}\pi(5)^2\right] = \frac{25}{12}\pi$

21. $-\frac{1}{2}\int_1^0 \sqrt{1-u^2}\, du = \frac{1}{2}\int_0^1 \sqrt{1-u^2}\, du = \frac{1}{2}\cdot\frac{1}{4}[\pi(1)^2] = \pi/8$

23. $\int_0^1 \sin \pi x dx = -\frac{1}{\pi}\cos \pi x\Big]_0^1 = -\frac{1}{\pi}(-1-1) = 2/\pi$

25. $\int_3^7 (x+5)^{-2} dx = -(x+5)^{-1}\Big]_3^7 = -\frac{1}{12} + \frac{1}{8} = \frac{1}{24}$

27. $A = \int_0^{1/6} \frac{1}{\sqrt{1-9x^2}} dx = \frac{1}{3}\int_0^{1/2} \frac{1}{\sqrt{1-u^2}} du = \frac{1}{3}\sin^{-1} u\Big]_0^{1/2} = \pi/18$

29. $\dfrac{1}{2-0}\displaystyle\int_0^2 \frac{x}{(5x^2+1)^2}\,dx = -\frac{1}{2}\frac{1}{10}\frac{1}{5x^2+1}\bigg|_0^2 = \frac{1}{21}$

31. $f_{\text{ave}} = \dfrac{1}{4}\displaystyle\int_0^4 e^{-2x}\,dx = -\frac{1}{8}e^{-2x}\bigg]_0^4 = \frac{1}{8}(1-e^{-8})$

33. $\dfrac{2}{3}(3x+1)^{1/2}\bigg]_0^1 = 2/3$

35. $\dfrac{2}{3}(x^3+9)^{1/2}\bigg]_{-1}^1 = \dfrac{2}{3}(\sqrt{10}-2\sqrt{2})$

37. $u = x^2+4x+7$, $\dfrac{1}{2}\displaystyle\int_{12}^{28} u^{-1/2}du = u^{1/2}\bigg]_{12}^{28} = \sqrt{28}-\sqrt{12} = 2(\sqrt{7}-\sqrt{3})$

39. $\dfrac{1}{2}\sin^2 x\bigg]_{-3\pi/4}^{\pi/4} = 0$

41. $\dfrac{5}{2}\sin(x^2)\bigg]_0^{\sqrt{\pi}} = 0$

43. $u = 3\theta$, $\dfrac{1}{3}\displaystyle\int_{\pi/4}^{\pi/3} \sec^2 u\,du = \frac{1}{3}\tan u\bigg]_{\pi/4}^{\pi/3} = (\sqrt{3}-1)/3$

45. $u = 4-3y$, $y = \dfrac{1}{3}(4-u)$, $dy = -\dfrac{1}{3}du$

$$-\frac{1}{27}\int_4^1 \frac{16-8u+u^2}{u^{1/2}}du = \frac{1}{27}\int_1^4 (16u^{-1/2}-8u^{1/2}+u^{3/2})du$$

$$= \frac{1}{27}\left[32u^{1/2}-\frac{16}{3}u^{3/2}+\frac{2}{5}u^{5/2}\right]_1^4 = 106/405$$

47. $\ln(x+e)\bigg]_0^e = \ln(2e)-\ln e = \ln 2$

49. $u = \sqrt{3}x^2$, $\dfrac{1}{2\sqrt{3}}\displaystyle\int_0^{\sqrt{3}} \frac{1}{\sqrt{4-u^2}}du = \frac{1}{2\sqrt{3}}\sin^{-1}\frac{u}{2}\bigg]_0^{\sqrt{3}} = \frac{1}{2\sqrt{3}}\left(\frac{\pi}{3}\right) = \frac{\pi}{6\sqrt{3}}$

51. $u = 3x$, $\dfrac{1}{3}\displaystyle\int_0^{2\sqrt{3}} \frac{1}{4+u^2}du = \frac{1}{6}\tan^{-1}\frac{u}{2}\bigg]_0^{2\sqrt{3}} = \frac{1}{6}\frac{\pi}{3} = \frac{\pi}{18}$

53. **(b)** $\displaystyle\int_0^{\pi/6} \sin^4 x(1-\sin^2 x)\cos x\,dx = \left(\frac{1}{5}\sin^5 x - \frac{1}{7}\sin^7 x\right)\bigg|_0^{\pi/6} = \frac{1}{160}-\frac{1}{896} = \frac{23}{4480}$

55. **(a)** $u = 3x+1$, $\dfrac{1}{3}\displaystyle\int_1^4 f(u)du = 5/3$

(b) $u = 3x$, $\dfrac{1}{3}\displaystyle\int_0^9 f(u)du = 5/3$

(c) $u = x^2$, $1/2\displaystyle\int_4^0 f(u)du = -1/2\int_0^4 f(u)du = -1/2$

57. $\sin x = \cos(\pi/2 - x)$,

$$\int_0^{\pi/2} \sin^n x\,dx = \int_0^{\pi/2} \cos^n(\pi/2 - x)dx = -\int_{\pi/2}^{0} \cos^n u\,du \quad (u = \pi/2 - x)$$

$$= \int_0^{\pi/2} \cos^n u\,du = \int_0^{\pi/2} \cos^n x\,dx \quad \text{(by replacing } u \text{ by } x\text{)}$$

59. $y(t) = (802.137)\int e^{1.528t}dt = 524.959e^{1.528t} + C$; $y(0) = 750 = 524.959 + C$, $C = 225.041$,

$y(t) = 524.959e^{1.528t} + 225.041$, $y(12) = 48,233,500,000$

61. $s(t) = \int (25 + 10e^{-0.05t})dt = 25t - 200e^{-0.05t} + C$

(a) $s(10) - s(0) = 250 - 200(e^{-0.5} - 1) = 450 - 200/\sqrt{e} \approx 328.69$ ft

(b) yes; without it the distance would have been 250 ft

63. The area is given by $\int_0^2 1/(1 + kx^2)dx = (1/\sqrt{k})\tan^{-1}(2\sqrt{k}) = 0.6$; solve for k to get $k = 5.081435$.

65. **(a)** $V_{\text{rms}}^2 = \dfrac{1}{1/f - 0}\displaystyle\int_0^{1/f} V_p^2 \sin^2(2\pi f t)dt = \frac{1}{2} f V_p^2 \int_0^{1/f} [1 - \cos(4\pi f t)]dt$

$$= \frac{1}{2} f V_p^2 \left[t - \frac{1}{4\pi f}\sin(4\pi f t)\right]_0^{1/f} = \frac{1}{2}V_p^2, \text{ so } V_{\text{rms}} = V_p/\sqrt{2}$$

(b) $V_p/\sqrt{2} = 120$, $V_p = 120\sqrt{2} \approx 169.7$ V

67. **(a)** $I = -\displaystyle\int_a^0 \frac{f(a-u)}{f(a-u) + f(u)}du = \int_0^a \frac{f(a-u) + f(u) - f(u)}{f(a-u) + f(u)}du$

$$= \int_0^a du - \int_0^a \frac{f(u)}{f(a-u) + f(u)}du, I = a - I \text{ so } 2I = a, I = a/2$$

(b) $3/2$

(c) $\pi/4$

69. **(a)** Let $u = -x$ then

$$\int_{-a}^{a} f(x)dx = -\int_{a}^{-a} f(-u)du = \int_{-a}^{a} f(-u)du = -\int_{-a}^{a} f(u)du$$

so, replacing u by x in the latter integral,

$$\int_{-a}^{a} f(x)dx = -\int_{-a}^{a} f(x)dx, \; 2\int_{-a}^{a} f(x)dx = 0, \; \int_{-a}^{a} f(x)dx = 0$$

The graph of f is symmetric about the origin so $\int_{-a}^{0} f(x)dx$ is the negative of $\int_0^a f(x)dx$

thus $\int_{-a}^{a} f(x)dx = \int_{-a}^{0} f(x)\,dx + \int_0^a f(x)dx = 0$

(b) $\int_{-a}^{a} f(x)dx = \int_{-a}^{0} f(x)dx + \int_0^a f(x)dx$, let $u = -x$ in $\int_{-a}^{0} f(x)dx$ to get

$$\int_{-a}^{0} f(x)dx = -\int_{a}^{0} f(-u)du = \int_0^a f(-u)du = \int_0^a f(u)du = \int_0^a f(x)dx$$

so $\int_{-a}^{a} f(x)dx = \int_0^a f(x)dx + \int_0^a f(x)dx = 2\int_0^a f(x)dx$

The graph of $f(x)$ is symmetric about the y-axis so there is as much signed area to the left of the y-axis as there is to the right.

EXERCISE SET 6.9

1. **(a)**

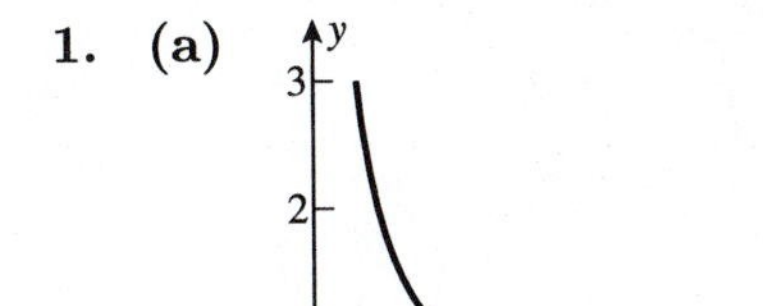

(b)

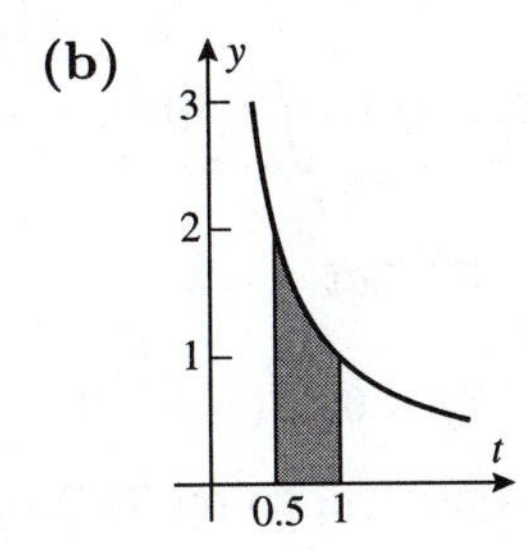

(c)

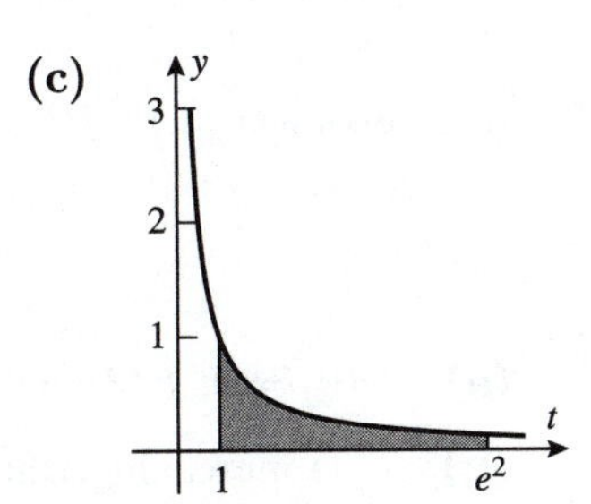

3. **(a)** $\ln t\Big]_1^{ac} = \ln(ac) = \ln a + \ln c = 7$ **(b)** $\ln t\Big]_1^{1/c} = \ln(1/c) = -5$

(c) $\ln t\Big]_1^{a/c} = \ln(a/c) = 2 - 5 = -3$ **(d)** $\ln t\Big]_1^{a^3} = \ln a^3 = 3\ln a = 6$

5. $\ln 5 \approx 1.603210678$; $\ln 5 = 1.609437912$; magnitude of error is < 0.0063

7. **(a)** x^{-1}, $x > 0$ **(b)** x^2, $x \neq 0$

(c) $-x^2$, $-\infty < x < +\infty$ **(d)** $-x$, $-\infty < x < +\infty$

(e) x^3, $x > 0$ **(f)** $\ln x + x$, $x > 0$

(g) $x - \sqrt[3]{x}$, $-\infty < x < +\infty$ **(h)** $\dfrac{e^x}{x}$, $x > 0$

9. **(a)** $3^\pi = e^{\pi \ln 3}$ **(b)** $2^{\sqrt{2}} = e^{\sqrt{2}\ln 2}$

11. **(a)** $\lim_{x\to+\infty}\left[\left(1+\frac{1}{x}\right)^x\right]^2 = \left[\lim_{x\to+\infty}\left(1+\frac{1}{x}\right)^x\right]^2 = e^2$

(b) $y = 2x$, $\lim_{y\to 0}(1+y)^{2/y} = \lim_{y\to 0}\left[(1+y)^{1/y}\right]^2 = e^2$

13. $g'(x) = x^2 - x$

15. **(a)** $\dfrac{1}{x^3}(3x^2) = \dfrac{3}{x}$ **(b)** $e^{\ln x}\dfrac{1}{x} = 1$

17. $F'(x) = \dfrac{\cos x}{x^2+3}$, $F''(x) = \dfrac{-(x^2+3)\sin x - 2x\cos x}{(x^2+3)^2}$

(a) 0 **(b)** 1/3 **(c)** 0

19. **(a)** $\dfrac{d}{dx}\int_1^{x^2} t\sqrt{1+t}\,dt = x^2\sqrt{1+x^2}(2x) = 2x^3\sqrt{1+x^2}$

(b) $\int_1^{x^2} t\sqrt{1+t}\,dt = -\frac{2}{3}(x^2+1)^{3/2} + \frac{2}{5}(x^2+1)^{5/2} - \frac{4\sqrt{2}}{15}$

21. **(a)** $-\sin x^2$ **(b)** $-\dfrac{\tan^2 x}{1+\tan^2 x}\sec^2 x = -\tan^2 x$

23. $-3\dfrac{3x-1}{9x^2+1} + 2x\dfrac{x^2-1}{x^4+1}$

25. **(a)** $\sin^2(x^3)(3x^2) - \sin^2(x^2)(2x) = 3x^2\sin^2(x^3) - 2x\sin^2(x^2)$

(b) $\dfrac{1}{1+x}(1) - \dfrac{1}{1-x}(-1) = \dfrac{2}{1-x^2}$

27. from geometry, $\int_0^3 f(t)dt = 0$, $\int_3^5 f(t)dt = 6$, $\int_5^7 f(t)dt = 0$; and $\int_7^{10} f(t)dt$

$$= \int_7^{10} (4t-37)/3dt = -3$$

(a) $F(0)=0,\ F(3)=0,\ F(5)=6,\ F(7)=6,\ F(10)=3$

(b) F is increasing where $F' = f$ is positive, so on $[3/2,6]$ and $[37/4,10]$, decreasing on $[0,3/2]$ and $[6,37/4]$

(c) critical points when $F'(x) = f(x) = 0$, so $x = 3/2, 6, 37/4$; maximum $15/2$ at $x=6$, minimum $-9/4$ at $x = 3/2$

(d)

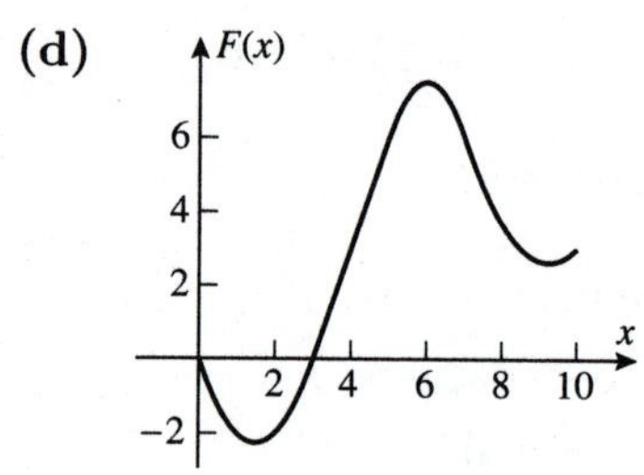

29. $x<0: F(x) = \int_{-1}^{x}(-t)dt = -\dfrac{1}{2}t^2\Big]_{-1}^{x} = \dfrac{1}{2}(1-x^2)$,

$x \ge 0: F(x) = \int_{-1}^{0}(-t)dt + \int_0^x t\,dt = \dfrac{1}{2} + \dfrac{1}{2}x^2$; $F(x) = \begin{cases} (1-x^2)/2, & x<0 \\ (1+x^2)/2, & x \ge 0 \end{cases}$

31. $y(x) = 2 + \int_1^x t^{1/3}dt = 2 + \dfrac{3}{4}t^{4/3}\Big]_1^x = \dfrac{5}{4} + \dfrac{3}{4}x^{4/3}$

33. $y(x) = 1 + \int_{\pi/4}^{x}(\sec^2 t - \sin t)dt = \tan x + \cos x - \sqrt{2}/2$

35. $P(x) = P_0 + \int_0^x r(t)dt$ individuals

37. II has a minimum at $x = 12$, and I has a zero there, so I could be the derivative of II; on the other hand I has a minimum near $x = 1/3$, but II is not zero there, so II could not be the derivative of I, so I is the graph of $f(x)$ and II is the graph of $\int_0^x f(t)\,dt$.

39. **(a)** where $f(t)=0$; by the First Derivative Test, at $t=3$

(b) where $f(t)=0$; by the First Derivative Test, at $t=1,5$

(c) at $t = 0, 1$ or 5; from the graph it is evident that it is at $t=5$

(d) at $t = 0, 3$ or 5; from the graph it is evident that it is at $t = 3$

(e) F is concave up when $F'' = f'$ is positive, i.e. where f is increasing, so on $(0, 1/2)$ and $(2, 4)$; it is concave down on $(1/2, 2)$ and $(4, 5)$

(f)

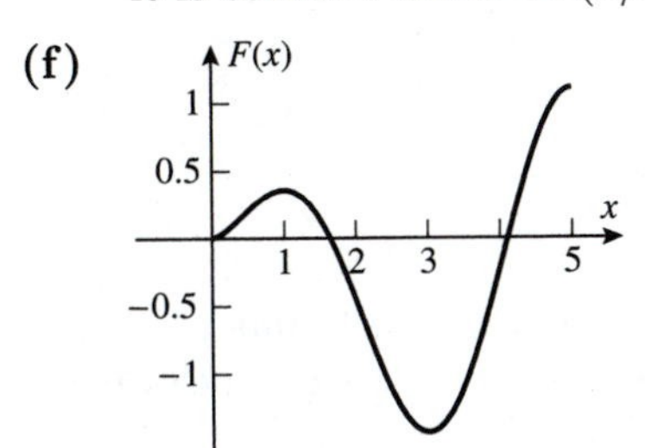

41. $C'(x) = \cos(\pi x^2/2)$, $C''(x) = -\pi x \sin(\pi x^2/2)$

(a) $\cos t$ goes from negative to positive at $2k\pi - \pi/2$, and from positive to negative at $t = 2k\pi + \pi/2$, so $C(x)$ has relative minima when $\pi x^2/2 = 2k\pi - \pi/2$, $x = \pm\sqrt{4k-1}$, $k = 1, 2, \ldots$, and $C(x)$ has relative maxima when $\pi x^2/2 = (4k+1)\pi/2$, $x = \pm\sqrt{4k+1}$, $k = 0, 1, \ldots$.

(b) $\sin t$ changes sign at $t = k\pi$, so $C(x)$ has inflection points at $\pi x^2/2 = k\pi$, $x = \pm\sqrt{2k}$, $k = 1, 2, \ldots$; the case $k = 0$ is distinct due to the factor of x in $C''(x)$, but x changes sign at $x = 0$ and $\sin(\pi x^2/2)$ does not, so there is also a point of inflection at $x = 0$

43. Differentiate: $f(x) = 3e^{3x}$, so $2 + \int_a^x f(t)dt = 2 + \int_a^x 3e^{3t}dt = 2 + e^{3t}\Big]_a^x = 2 + e^{3x} - e^{3a} = e^{3x}$ provided $e^{3a} = 2$, $a = (\ln 2)/3$.

45. From Exercise 44(d) $\left| e - \left(1 + \frac{1}{50}\right)^{50} \right| < y(50)$, and from the graph $y(50) < 0.06$

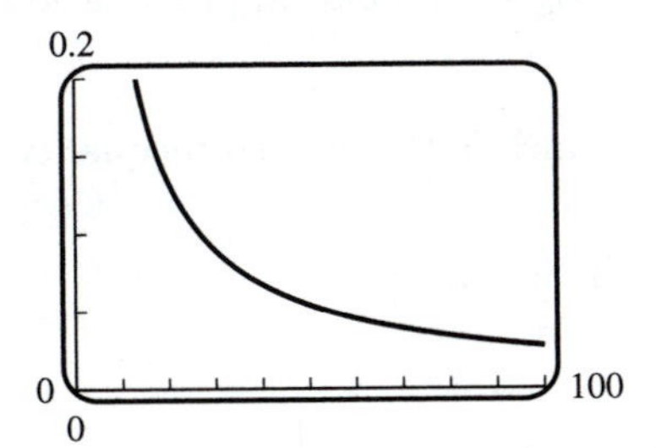

CHAPTER 6 SUPPLEMENTARY EXERCISES

5. If the acceleration $a = \text{const}$, then $v(t) = at + v_0$, $s(t) = \frac{1}{2}at^2 + v_0t + s_0$.

7. **(a)** $\frac{1}{2} + \frac{1}{4} = \frac{3}{4}$

(b) $-1 - \frac{1}{2} = -\frac{3}{2}$

(c) $5\left(-1 - \frac{3}{4}\right) = -\frac{35}{4}$

(d) -2

(e) not enough information

(f) not enough information

9. **(a)** $\int_{-1}^{1} dx + \int_{-1}^{1} \sqrt{1-x^2}\,dx = 2(1) + \pi(1)^2/2 = 2 + \pi/2$

(b) $\frac{1}{3}(x^2+1)^{3/2}\Big]_0^3 - \pi(3)^2/4 = \frac{1}{3}(10^{3/2}-1) - 9\pi/4$

(c) $u = x^2$, $du = 2xdx$; $\frac{1}{2}\int_0^1 \sqrt{1-u^2}du = \frac{1}{2}\pi(1)^2/4 = \pi/8$

11. The rectangle with vertices $(0,0)$, $(\pi,0)$, $(\pi,1)$ and $(0,1)$ has area π and is much too large; so is the triangle with vertices $(0,0)$, $(\pi,0)$ and $(\pi,1)$ which has area $\pi/2$; $1-\pi$ is negative; so the answer is $35\pi/128$.

13. Since $y = e^x$ and $y = \ln x$ are inverse functions, their graphs are symmetric with respect to the line $y = x$; consequently the areas A_1 and A_3 are equal (see figure). But $A_1 + A_2 = e$, so

$$\int_1^e \ln x dx + \int_0^1 e^x dx = A_2 + A_3 = A_2 + A_1 = e$$

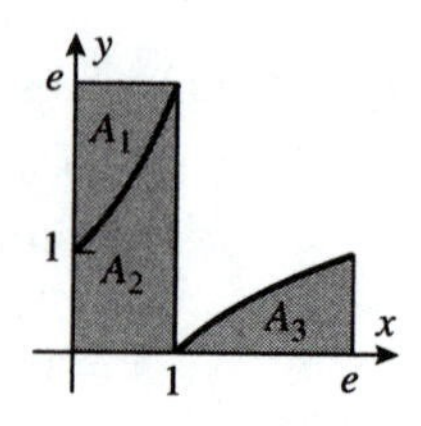

15. Since $f(x) = \frac{1}{x}$ is positive and increasing on the interval $[1,2]$, the left endpoint approximation overestimates the integral of $\frac{1}{x}$ and the right endpoint approximation underestimates it.

(a) For $n = 5$ this becomes

$$0.2\left[\frac{1}{1.2}+\frac{1}{1.4}+\frac{1}{1.6}+\frac{1}{1.8}+\frac{1}{2.0}\right] < \int_1^2 \frac{1}{x}dx < 0.2\left[\frac{1}{1.0}+\frac{1}{1.2}+\frac{1}{1.4}+\frac{1}{1.6}+\frac{1}{1.8}\right]$$

(b) For general n the left endpoint approximation to $\int_1^2 \frac{1}{x}dx = \ln 2$ is

$\frac{1}{n}\sum_{k=1}^{n}\frac{1}{1+(k-1)/n} = \sum_{k=1}^{n}\frac{1}{n+k-1} = \sum_{k=0}^{n-1}\frac{1}{n+k}$ and the right endpoint approximation is $\sum_{k=1}^{n}\frac{1}{n+k}$. This yields $\sum_{k=1}^{n}\frac{1}{n+k} < \int_1^2 \frac{1}{x}dx < \sum_{k=0}^{n-1}\frac{1}{n+k}$ which is the desired inequality.

(c) By telescoping, the difference is $\frac{1}{n} - \frac{1}{2n} = \frac{1}{2n}$ so $\frac{1}{2n} \le 0.1$, $n \ge 5$

(d) $n \ge 1{,}000$

17. **(a)** $1\cdot 2 + 2\cdot 3 + \cdots + n(n+1) = \sum_{k=1}^{n} k(k+1) = \sum_{k=1}^{n} k^2 + \sum_{k=1}^{n} k$

$$= \frac{1}{6}n(n+1)(2n+1) + \frac{1}{2}n(n+1) = \frac{1}{3}n(n+1)(n+2)$$

(b) $\sum_{k=1}^{n-1}\left(\frac{9}{n} - \frac{k}{n^2}\right) = \frac{9}{n}\sum_{k=1}^{n-1} 1 - \frac{1}{n^2}\sum_{k=1}^{n-1} k = \frac{9}{n}(n-1) - \frac{1}{n^2}\cdot\frac{1}{2}(n-1)(n) = \frac{17}{2}\left(\frac{n-1}{n}\right)$;

$\lim_{n\to+\infty} \frac{17}{2}\left(\frac{n-1}{n}\right) = \frac{17}{2}$

(c) $\sum_{i=1}^{3}\left[\sum_{j=1}^{2} i + \sum_{j=1}^{2} j\right] = \sum_{i=1}^{3}\left[2i + \frac{1}{2}(2)(3)\right] = 2\sum_{i=1}^{3} i + \sum_{i=1}^{3} 3 = 2\cdot\frac{1}{2}(3)(4) + (3)(3) = 21$

19. For $1 \le k \le n$ the k-th L-shaped strip consists of the corner square, a strip above and a strip to the right for a combined area of $1+(k-1)+(k-1) = 2k-1$, so the total area is $\sum_{k=1}^{n}(2k-1) = n^2$.

21. $(3^5 - 3^4) + (3^6 - 3^5) + \cdots + (3^{17} - 3^{16}) = 3^{17} - 3^4$

23. $\left(\frac{1}{2^2} - \frac{1}{1^2}\right) + \left(\frac{1}{3^2} - \frac{1}{2^2}\right) + \cdots + \left(\frac{1}{20^2} - \frac{1}{19^2}\right) = \frac{1}{20^2} - 1 = -\frac{399}{400}$

25. **(a)**
$$\sum_{k=1}^{n} \frac{1}{(2k-1)(2k+1)} = \frac{1}{2}\sum_{k=1}^{n}\left(\frac{1}{2k-1} - \frac{1}{2k+1}\right)$$
$$= \frac{1}{2}\left[\left(1-\frac{1}{3}\right) + \left(\frac{1}{3}-\frac{1}{5}\right) + \left(\frac{1}{5}-\frac{1}{7}\right) + \cdots + \left(\frac{1}{2n-1} - \frac{1}{2n+1}\right)\right]$$
$$= \frac{1}{2}\left[1 - \frac{1}{2n+1}\right] = \frac{n}{2n+1}$$

(b) $\lim_{n\to+\infty} \frac{n}{2n+1} = \frac{1}{2}$

27. $\sum_{i=1}^{n}(x_i - \bar{x}) = \sum_{i=1}^{n} x_i - \sum_{i=1}^{n} \bar{x} = \sum_{i=1}^{n} x_i - n\bar{x}$ but $\bar{x} = \frac{1}{n}\sum_{i=1}^{n} x_i$ thus $\sum_{i=1}^{n} x_i = n\bar{x}$ so $\sum_{i=1}^{n}(x_i - \bar{x}) = n\bar{x} - n\bar{x} = 0$

29. **(a)** $\sum_{k=0}^{19} 3^{k+1} = \sum_{k=0}^{19} 3(3^k) = \frac{3(1-3^{20})}{1-3} = \frac{3}{2}(3^{20}-1)$

(b) $\sum_{k=0}^{25} 2^{k+5} = \sum_{k=0}^{25} 2^5 2^k = \frac{2^5(1-2^{26})}{1-2} = 2^{31} - 2^5$

(c) $\sum_{k=0}^{100} (-1)\left(\frac{-1}{2}\right)^k = \frac{(-1)(1-(-1/2)^{101})}{1-(-1/2)} = -\frac{2}{3}(1 + 1/2^{101})$

31. **(a)** If $u = \sec x$, $du = \sec x \tan x dx$, $\int \sec^2 x \tan x dx = \int u du = u^2/2 + C_1 = (\sec^2 x)/2 + C_1$;

if $u = \tan x$, $du = \sec^2 x dx$, $\int \sec^2 x \tan x dx = \int u du = u^2/2 + C_2 = (\tan^2 x)/2 + C_2$.

(b) They are equal only if $\sec^2 x$ and $\tan^2 x$ differ by a constant, which is true.

33. $\int \sqrt{1+x^{-2/3}}dx = \int x^{-1/3}\sqrt{x^{2/3}+1}dx$; $u = x^{2/3}+1$, $du = \frac{2}{3}x^{-1/3}dx$

$\frac{3}{2}\int u^{1/2}du = u^{3/2} + C = (x^{2/3}+1)^{3/2} + C$

35. left endpoints: $x_k^* = 1, 2, 3, 4$; $\sum_{k=1}^{4} f(x_k^*)\Delta x = (2+3+2+1)(1) = 8$

right endpoints: $x_k^* = 2, 3, 4, 5$; $\sum_{k=1}^{4} f(x_k^*)\Delta x = (3+2+1+2)(1) = 8$

37. $f\text{ave} = \dfrac{1}{e-1}\displaystyle\int_1^e \dfrac{1}{x}dx = \dfrac{1}{e-1}\ln x\Big]_1^e = \dfrac{1}{e-1}$; $\dfrac{1}{x^*} = \dfrac{1}{e-1}$, $x^* = e-1$

39. $0.351220577, 0.420535296, 0.386502483$

41. $f(x) = e^x, [a,b] = [0,1], \Delta x = \dfrac{1}{n}$; $\displaystyle\lim_{n\to+\infty}\sum_{k=1}^{n} f(x_k^*) = \int_0^1 e^x\,dx = e-1$

43. **(a)** $\displaystyle\int_1^x \frac{1}{1+e^t}dt$ **(b)** $\displaystyle\int_{-\ln(e^2+e-1)}^{x} \frac{1}{1+e^t}dt$

45. $F'(x) = \dfrac{1}{1+x^2} + \dfrac{1}{1+(1/x)^2}(-1/x^2) = 0$ so F is constant on $(0,+\infty)$.

47. **(a)** The domain is $(-\infty,+\infty)$; $F(x)$ is 0 if $x=1$, positive if $x>1$, and negative if $x<1$, because the integrand is positive, so the sign of the integral depends on the orientation (forwards or backwards).

(b) The domain is $[-2,2]$; $F(x)$ is 0 if $x=-1$, positive if $-1<x\le 2$, and negative if $-2\le x<-1$; same reasons as in Part (*a*).

49. **(a)** no, since the velocity curve is not a straight line

(b) $25 < t < 40$ **(c)** 3.54 ft/s **(d)** 141.5 ft

(e) no since the velocity is positive and the acceleration is never negative

(f) need the position at any one given time (e.g. s_0)

51. $u = 5+2\sin 3x$, $du = 6\cos 3x dx$; $\displaystyle\int \frac{1}{6\sqrt{u}}du = \frac{1}{3}u^{1/2} + C = \frac{1}{3}\sqrt{5+2\sin 3x} + C$

53. $u = ax^3+b$, $du = 3ax^2dx$; $\displaystyle\int \frac{1}{3au^2}du = -\frac{1}{3au} + C = -\frac{1}{3a^2x^3+3ab} + C$

55. $\left(-\dfrac{1}{3u^3} - \dfrac{3}{u} + \dfrac{1}{4u^4}\right)\Big]_{-2}^{-1} = 389/192$

57. $u = \ln x$, $du = (1/x)dx$; $\displaystyle\int_1^2 \frac{1}{u}du = \ln u\Big]_1^2 = \ln 2$

59. $u = e^{-2x}$, $du = -2e^{-2x}dx$; $-\dfrac{1}{2}\displaystyle\int_1^{1/4}(1+\cos u)du = \frac{3}{8} + \frac{1}{2}\left(\sin 1 - \sin\frac{1}{4}\right)$

61. With $b = 1.618034$, area $= \displaystyle\int_0^b (x+x^2-x^3)dx = 1.007514$.

63. **(a)** Solve $\dfrac{1}{4}k^4 - k - k^2 + \dfrac{7}{4} = 0$ to get $k = 2.073948$.

(b) Solve $-\dfrac{1}{2}\cos 2k + \dfrac{1}{3}k^3 + \dfrac{1}{2} = 3$ to get $k = 1.837992$.

65. **(a)** $y = J_0(x)$ **(b)** 0.7651976866 **(c)** $J_0(x) = 0$ if $x = 2.404826$

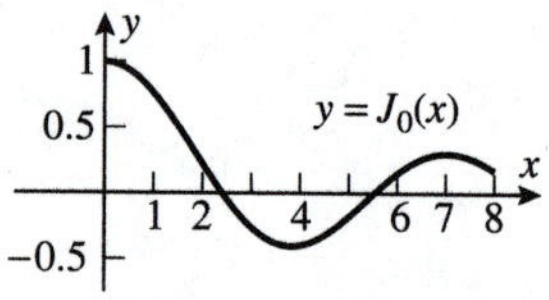

CHAPTER 7
Applications of the Definite Integral in Geometry, Science, and Engineering

EXERCISE SET 7.1

1. $A = \int_{-1}^{2} (x^2 + 1 - x)dx = (x^3/3 + x - x^2/2)\Big]_{-1}^{2} = 9/2$

3. $A = \int_{1}^{2} (y - 1/y^2)dy = (y^2/2 + 1/y)\Big]_{1}^{2} = 1$

5. **(a)** $A = \int_{0}^{4} (4x - x^2)dx = 32/3$ **(b)** $A = \int_{0}^{16} (\sqrt{y} - y/4)dy = 32/3$

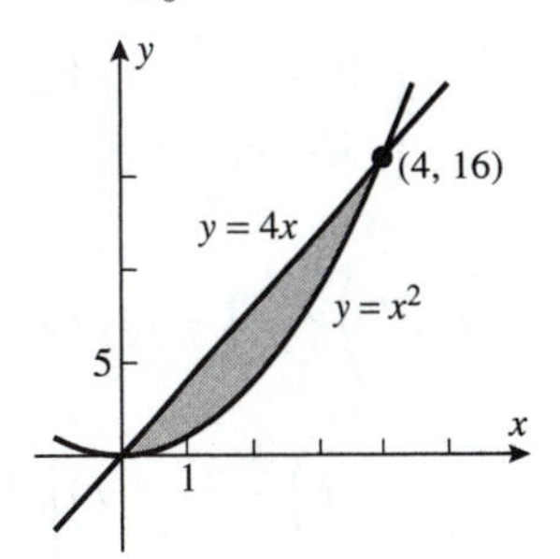

7. $A = \int_{1/4}^{1} (\sqrt{x} - x^2)dx = 49/192$

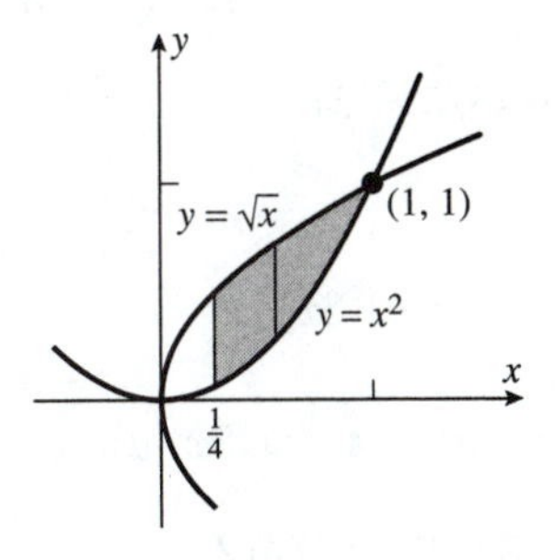

9. $A = \int_{\pi/4}^{\pi/2} (0 - \cos 2x)dx$

$= -\int_{\pi/4}^{\pi/2} \cos 2x\, dx = 1/2$

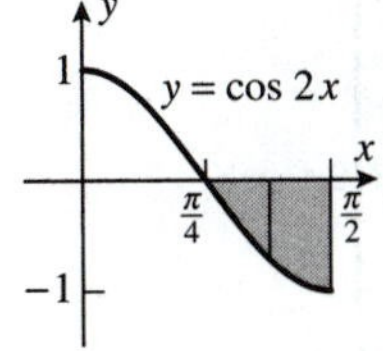

11. $A = \int_{\pi/4}^{3\pi/4} \sin y\, dy = \sqrt{2}$

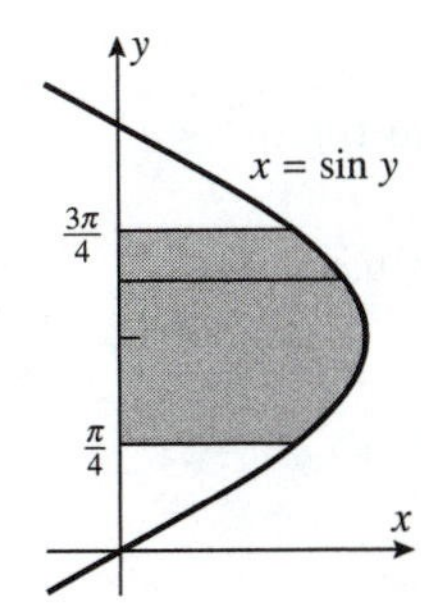

13. $A = \int_0^{\ln 2} (e^{2x} - e^x)\, dx$

$= \left(\frac{1}{2}e^{2x} - e^x\right)\Big]_0^{\ln 2} = 1/2$

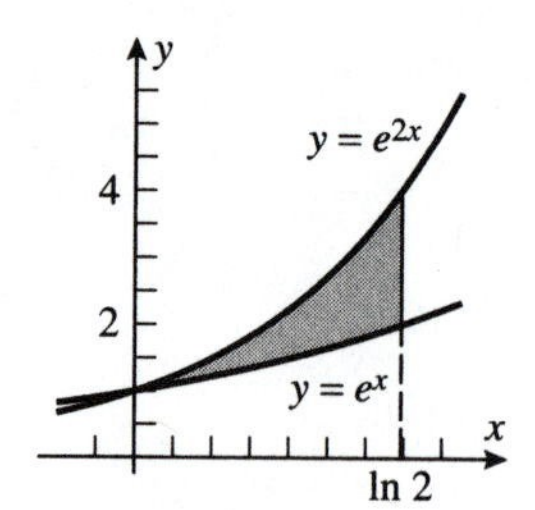

15. $A = \int_{-1}^{1} \left(\frac{2}{1+x^2} - |x|\right) dx$

$= 2\int_0^1 \left(\frac{2}{1+x^2} - x\right) dx$

$= 4\tan^{-1} x - x^2\Big]_0^1 = \pi - 1$

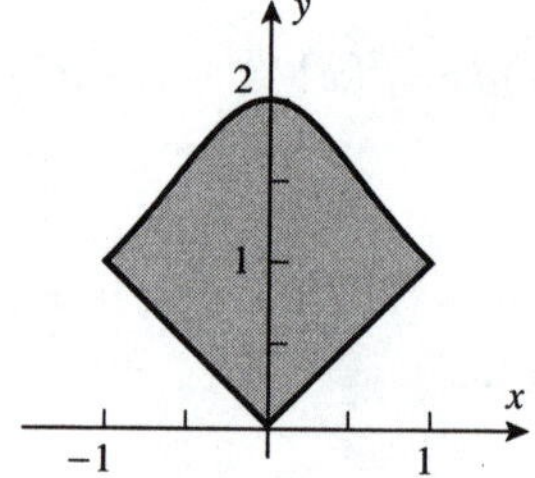

17. $y = 2 + |x-1| = \begin{cases} 3 - x, & x \le 1 \\ 1 + x, & x \ge 1 \end{cases}$,

$A = \int_{-5}^{1} \left[\left(-\frac{1}{5}x + 7\right) - (3 - x)\right] dx$

$+ \int_1^5 \left[\left(-\frac{1}{5}x + 7\right) - (1 + x)\right] dx$

$= \int_{-5}^{1} \left(\frac{4}{5}x + 4\right) dx + \int_1^5 \left(6 - \frac{6}{5}x\right) dx$

$= 72/5 + 48/5 = 24$

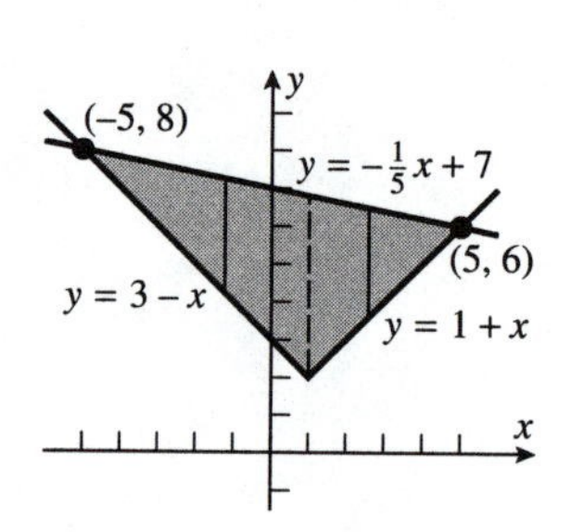

19. $A = \int_0^1 (x^3 - 4x^2 + 3x)dx$

$+ \int_1^3 [-(x^3 - 4x^2 + 3x)]dx$

$= 5/12 + 32/12 = 37/12$

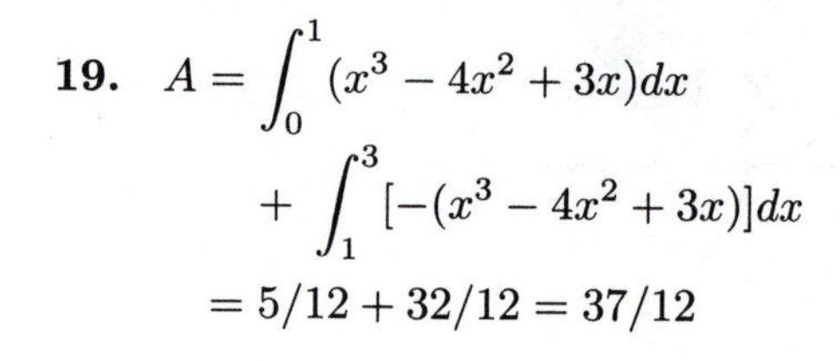

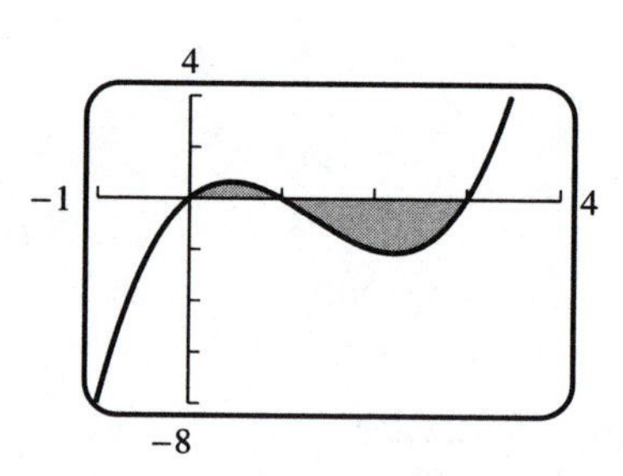

21. From the symmetry of the region

$A = 2\int_{\pi/4}^{5\pi/4} (\sin x - \cos x)dx = 4\sqrt{2}$

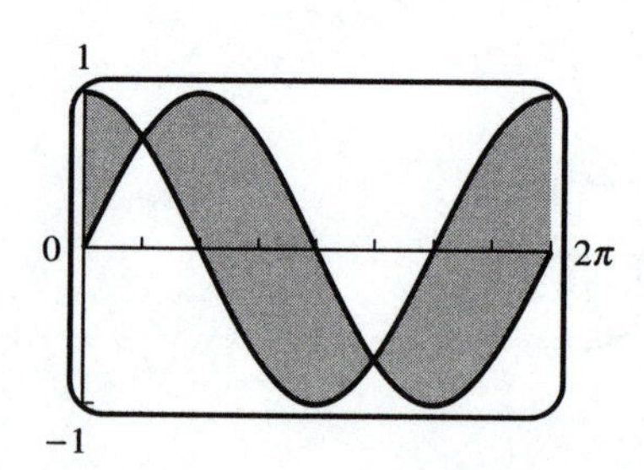

23. $A = \int_{-1}^{0} (y^3 - y)dy + \int_{0}^{1} -(y^3 - y)dy$
$= 1/2$

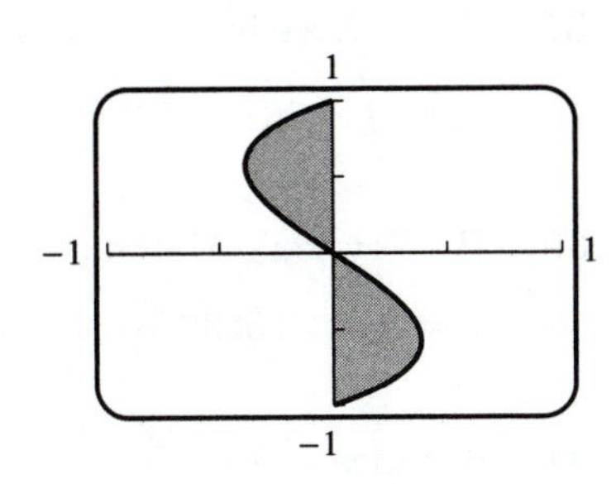

25. The curves meet when $x = \sqrt{\ln 2}$, so

$$A = \int_0^{\sqrt{\ln 2}} (2x - xe^{x^2})\,dx = \left(x^2 - \frac{1}{2}e^{x^2}\right)\Bigg]_0^{\sqrt{\ln 2}} = \ln 2 - \frac{1}{2}$$

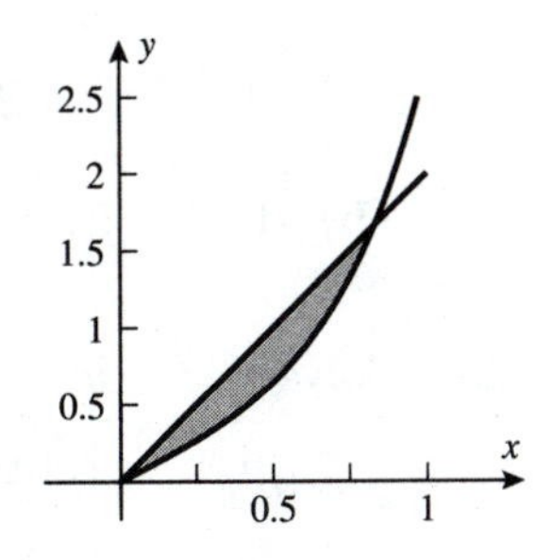

27. The area is given by $\int_0^k (1/\sqrt{1-x^2} - x)dx = \sin^{-1} k - k^2/2 = 1$; solve for k to get $k = 0.997301$.

29. Solve $3-2x = x^6+2x^5-3x^4+x^2$ to find the real roots $x = -3, 1$; from a plot it is seen that the line is above the polynomial when $-3 < x < 1$, so $A = \int_{-3}^{1} (3-2x-(x^6+2x^5-3x^4+x^2))\,dx = 9152/105$

31.
$$\int_0^k 2\sqrt{y}dy = \int_k^9 2\sqrt{y}dy$$
$$\int_0^k y^{1/2}dy = \int_k^9 y^{1/2}dy$$
$$\frac{2}{3}k^{3/2} = \frac{2}{3}(27 - k^{3/2})$$
$$k^{3/2} = 27/2$$
$$k = (27/2)^{2/3} = 9/\sqrt[3]{4}$$

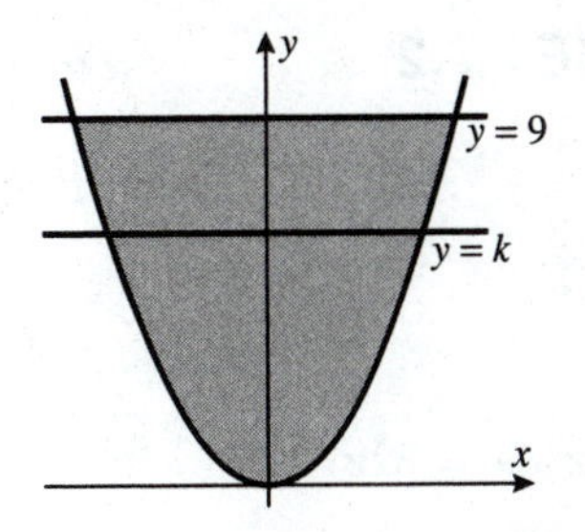

33. **(a)** $A = \int_0^2 (2x - x^2)dx = 4/3$

(b) $y = mx$ intersects $y = 2x - x^2$ where $mx = 2x - x^2, x^2 + (m-2)x = 0, x(x + m - 2) = 0$ so $x = 0$ or $x = 2 - m$. The area below the curve and above the line is

$$\int_0^{2-m} (2x - x^2 - mx)dx = \int_0^{2-m} [(2-m)x - x^2]dx = \left[\frac{1}{2}(2-m)x^2 - \frac{1}{3}x^3\right]_0^{2-m} = \frac{1}{6}(2-m)^3$$

so $(2-m)^3/6 = (1/2)(4/3) = 2/3, (2-m)^3 = 4, m = 2 - \sqrt[3]{4}$.

35. **(a)** It gives the area of the region that is between f and g when $f(x) > g(x)$ minus the area of the region between f and g when $f(x) < g(x)$, for $a \le x \le b$.

(b) It gives the area of the region that is between f and g for $a \le x \le b$.

37. The curves intersect at $x = 0$ and, by Newton's Method, at $x \approx 2.595739080 = b$, so
$A \approx \int_0^b (\sin x - 0.2x)dx = -\left[\cos x + 0.1x^2\right]_0^b \approx 1.180898334$

39. By Newton's Method the points of intersection are $x = x_1 \approx 0.4814008713$ and
$x = x_2 \approx 2.363938870$, and $A \approx \int_{x_1}^{x_2} \left(\frac{\ln x}{x} - (x-2)\right) dx \approx 1.189708441$.

41. distance $= \int |v|\, dt$, so

(a) distance $= \int_0^{60} (3t - t^2/20)\, dt = 1800$ ft.

(b) If $T \le 60$ then distance $= \int_0^T (3t - t^2/20)\, dt = \frac{3}{2}T^2 - \frac{1}{60}T^3$ ft.

43. Solve $x^{1/2} + y^{1/2} = a^{1/2}$ for y to get

$y = (a^{1/2} - x^{1/2})^2 = a - 2a^{1/2}x^{1/2} + x$

$A = \int_0^a (a - 2a^{1/2}x^{1/2} + x)dx = a^2/6$

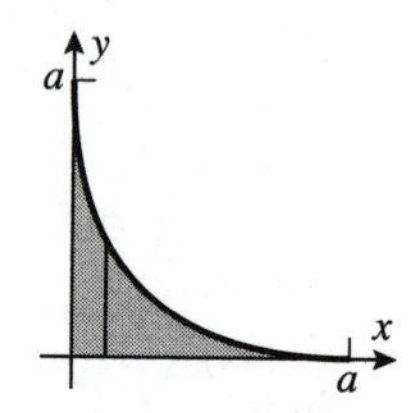

45. Let A be the area between the curve and the x-axis and A_R the area of the rectangle, then
$A = \int_0^b kx^m dx = \frac{k}{m+1}x^{m+1}\Big]_0^b = \frac{kb^{m+1}}{m+1}$, $A_R = b(kb^m) = kb^{m+1}$, so $A/A_R = 1/(m+1)$.

EXERCISE SET 7.2

1. $V = \pi \int_{-1}^{3} (3 - x)dx = 8\pi$

3. $V = \pi \int_0^2 \frac{1}{4}(3 - y)^2 dy = 13\pi/6$

5. $V = \pi \int_0^2 x^4 dx = 32\pi/5$

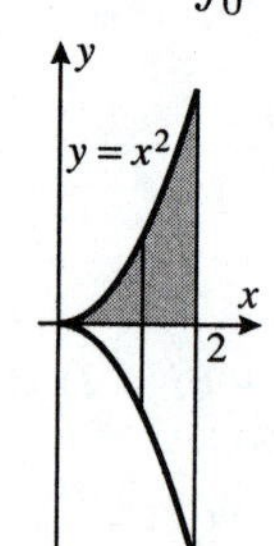

7. $V = \pi \int_{\pi/4}^{\pi/2} \cos x\, dx = (1 - \sqrt{2}/2)\pi$

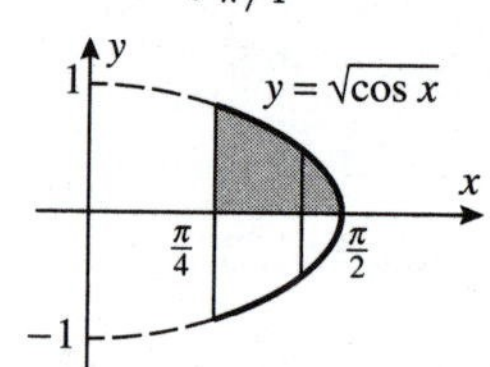

9. $V = \pi \int_{-4}^{4} [(25 - x^2) - 9]dx$

$= 2\pi \int_{0}^{4} (16 - x^2)dx = 256\pi/3$

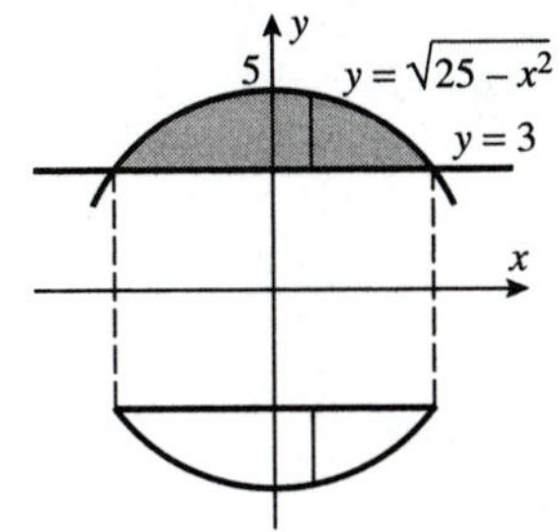

11. $V = \pi \int_{0}^{4} [(4x)^2 - (x^2)^2]dx$

$= \pi \int_{0}^{4} (16x^2 - x^4)dx = 2048\pi/15$

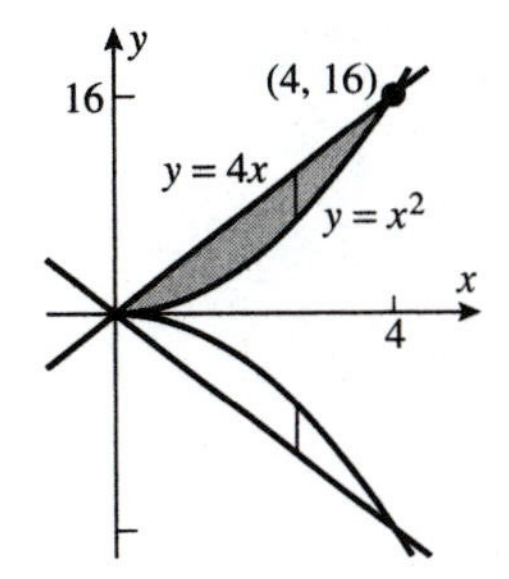

13. $V = \pi \int_{0}^{\ln 3} e^{2x} dx = \frac{\pi}{2} e^{2x}\Big]_{0}^{\ln 3} = 4\pi$

15. $V = \int_{-2}^{2} \pi \frac{1}{4 + x^2} dx = \frac{\pi}{2} \tan^{-1}(x/2)\Big]_{-2}^{2} = \pi^2/4$

17. $V = \pi \int_{0}^{1} y^{2/3} dy = 3\pi/5$

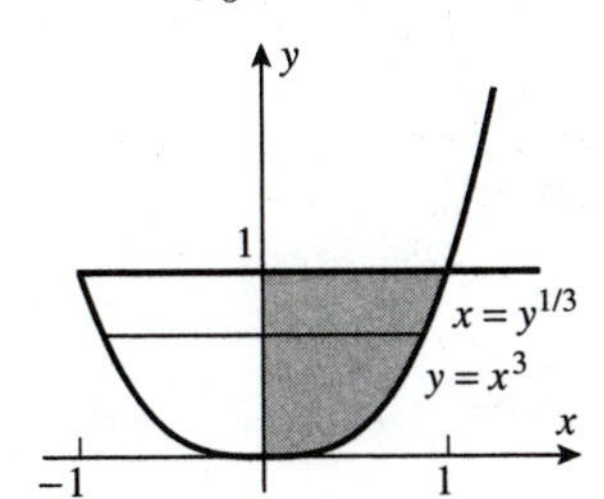

19. $V = \pi \int_{-1}^{3} (1 + y)dy = 8\pi$

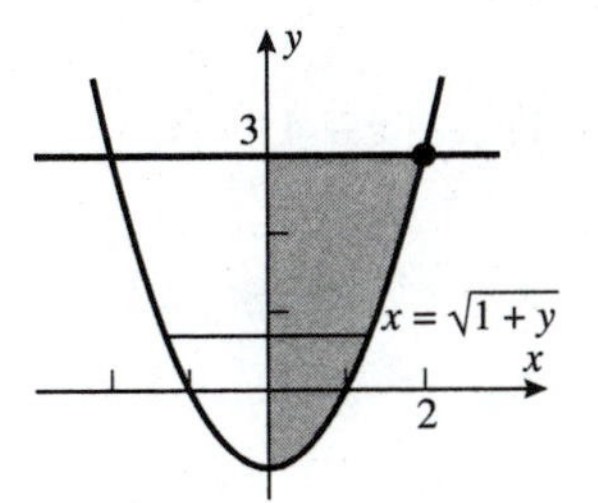

21. $V = \pi \int_{\pi/4}^{3\pi/4} \csc^2 y \, dy = 2\pi$

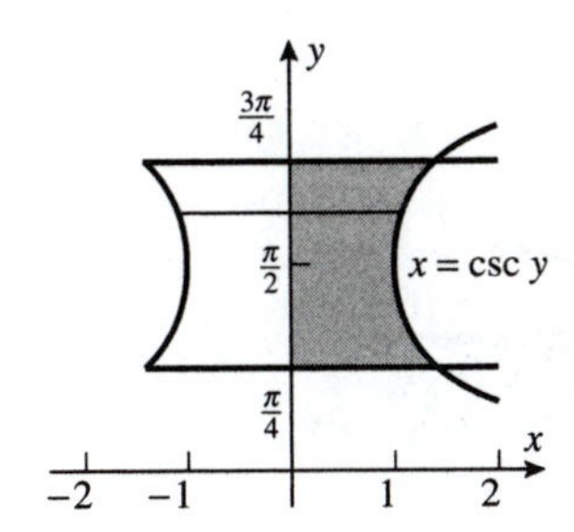

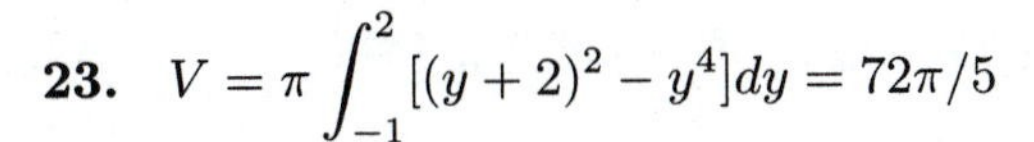

23. $V = \pi \int_{-1}^{2} [(y + 2)^2 - y^4]dy = 72\pi/5$

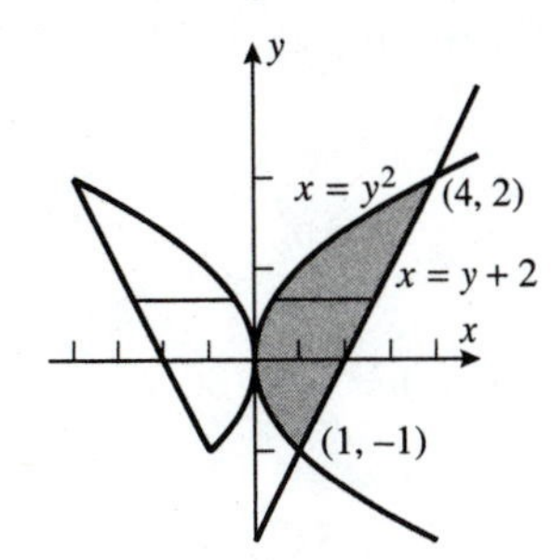

25. $V = \int_0^1 \pi e^{2y}\, dy = \dfrac{\pi}{2}\,(e^2 - 1)$

27. $V = \pi \int_{-a}^{a} \dfrac{b^2}{a^2}(a^2 - x^2)dx = 4\pi ab^2/3$

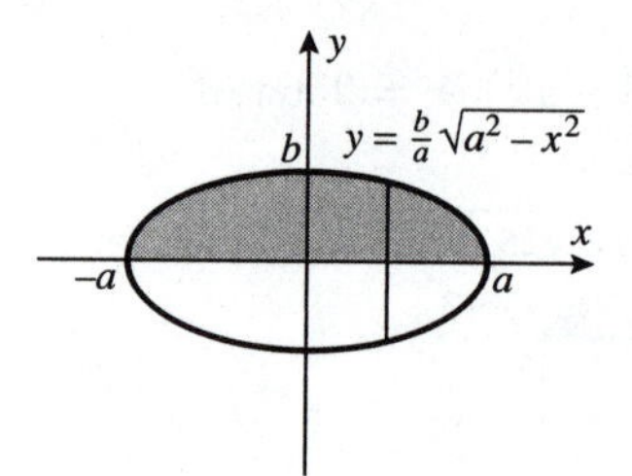

29. $V = \pi \int_{-1}^{0} (x+1)dx$

$+\, \pi \int_0^1 [(x+1) - 2x]dx$

$= \pi/2 + \pi/2 = \pi$

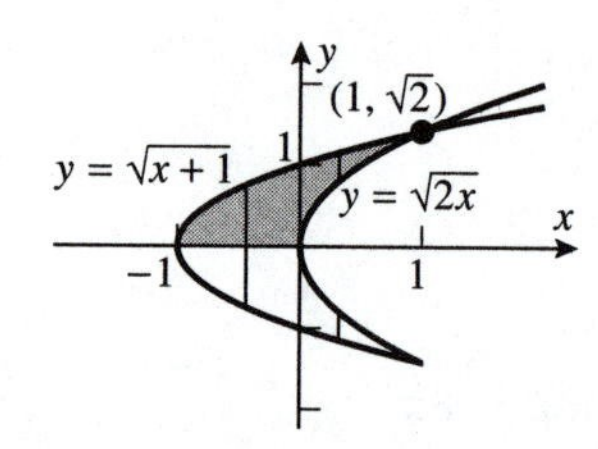

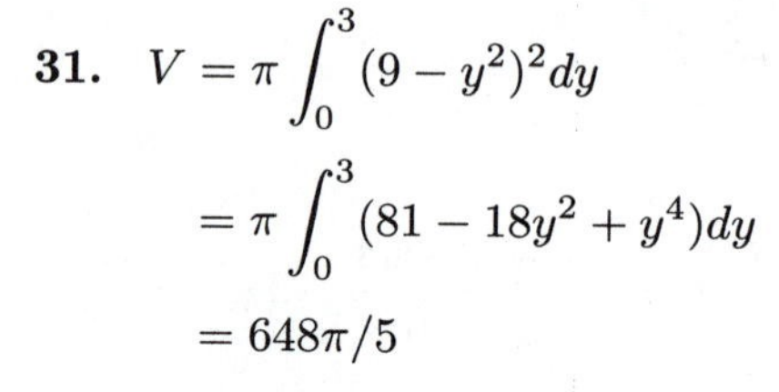

31. $V = \pi \int_0^3 (9 - y^2)^2 dy$

$= \pi \int_0^3 (81 - 18y^2 + y^4)dy$

$= 648\pi/5$

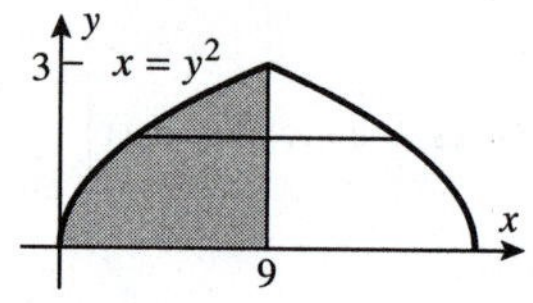

33. $V = \pi \int_0^1 [(\sqrt{x} + 1)^2 - (x+1)^2]dx$

$= \pi \int_0^1 (2\sqrt{x} - x - x^2)dx = \pi/2$

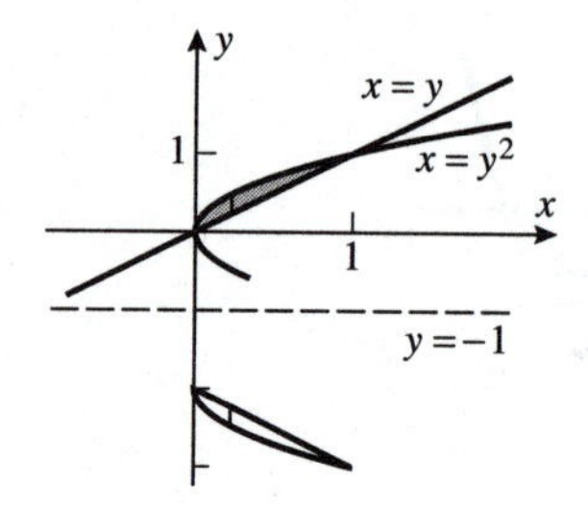

35. $A(x) = \pi(x^2/4)^2 = \pi x^4/16,$

$V = \int_0^{20} (\pi x^4/16)dx = 40{,}000\pi \text{ ft}^3$

37. $V = \int_0^1 (x - x^2)^2 dx$

$= \int_0^1 (x^2 - 2x^3 + x^4)dx = 1/30$

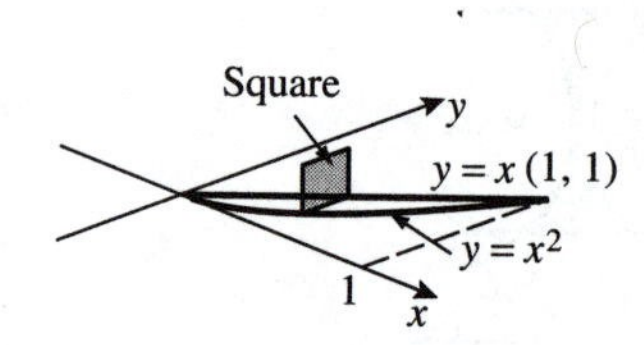

39. On the upper half of the circle, $y = \sqrt{1 - x^2}$, so:

(a) $A(x)$ is the area of a semicircle of radius y, so

$$A(x) = \pi y^2/2 = \pi(1 - x^2)/2; \; V = \frac{\pi}{2}\int_{-1}^{1} (1 - x^2)\, dx = \pi \int_0^1 (1 - x^2)\, dx = 2\pi/3$$

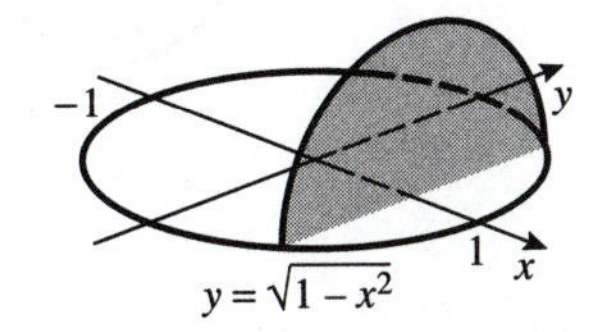

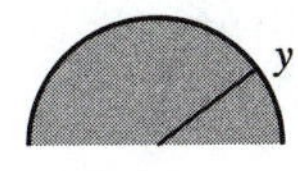

(b) $A(x)$ is the area of a square of side $2y$, so

$$A(x) = 4y^2 = 4(1 - x^2); \; V = 4\int_{-1}^{1} (1 - x^2)\, dx = 8\int_0^1 (1 - x^2)\, dx = 16/3$$

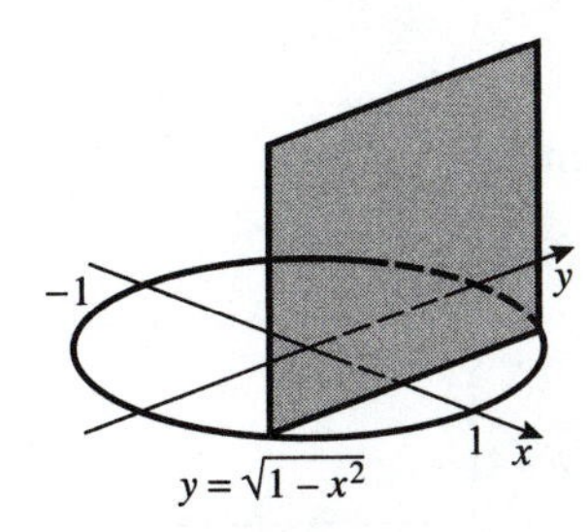

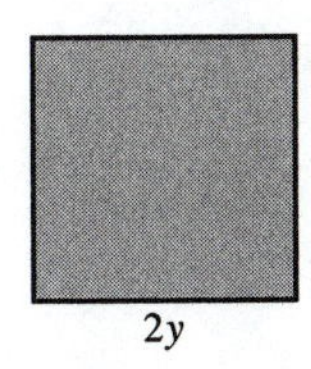

(c) $A(x)$ is the area of an equilateral triangle with sides $2y$, so

$$A(x) = \frac{\sqrt{3}}{4}(2y)^2 = \sqrt{3}y^2 = \sqrt{3}(1 - x^2);$$

$$V = \int_{-1}^{1} \sqrt{3}(1 - x^2)\, dx = 2\sqrt{3}\int_0^1 (1 - x^2)\, dx = 4\sqrt{3}/3$$

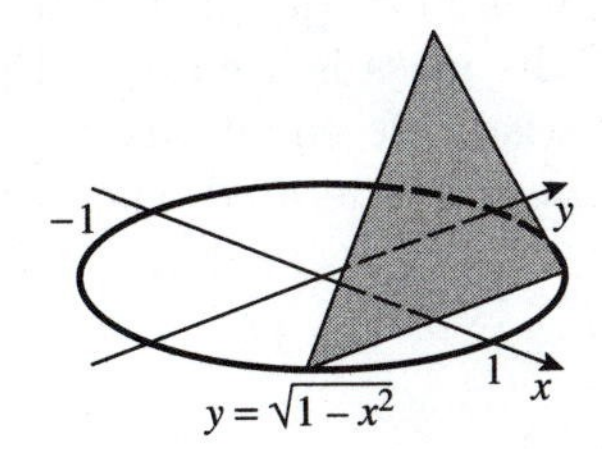

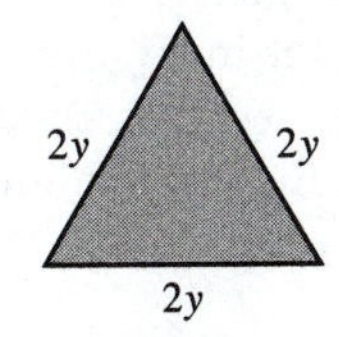

41. The two curves cross at $x = b \approx 1.403288534$, so

$$V = \pi \int_0^b ((2x/\pi)^2 - \sin^{16} x)\, dx + \pi \int_b^{\pi/2} (\sin^{16} x - (2x/\pi)^2)\, dx \approx 0.710172176.$$

43. $V = \pi \int_1^e (1 - (\ln y)^2)\, dy = \pi$

45. (a) $V = \pi \int_{r-h}^{r} (r^2 - y^2)\, dy = \pi(rh^2 - h^3/3) = \frac{1}{3}\pi h^2(3r - h)$

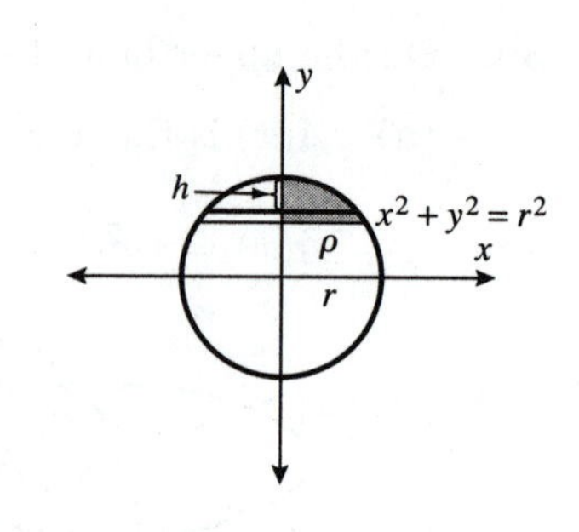

(b) By the Pythagorean Theorem,

$r^2 = (r - h)^2 + \rho^2$, $2hr = h^2 + \rho^2$; from Part (a),

$$V = \frac{\pi h}{3}(3hr - h^2) = \frac{\pi h}{3}\left(\frac{3}{2}(h^2 + \rho^2) - h^2\right)$$
$$= \frac{1}{6}\pi h(h^2 + 3\rho^2)$$

47. (b) $\Delta x = \frac{5}{10} = 0.5$; $\{y_0, y_1, \cdots, y_{10}\} = \{0, 2.00, 2.45, 2.45, 2.00, 1.46, 1.26, 1.25, 1.25, 1.25, 1.25\}$;

$$\text{left} = \pi \sum_{i=0}^{9} \left(\frac{y_i}{2}\right)^2 \Delta x \approx 11.157;$$

$$\text{right} = \pi \sum_{i=1}^{10} \left(\frac{y_i}{2}\right)^2 \Delta x \approx 11.771; \ V \approx \text{average} = 11.464 \text{ cm}^3$$

49. **(a)** **(b)**

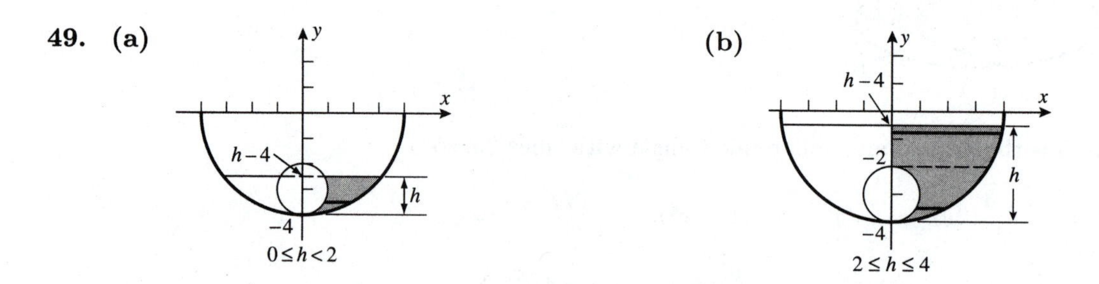

If the cherry is partially submerged then $0 \le h < 2$ as shown in Figure (a); if it is totally submerged then $2 \le h \le 4$ as shown in Figure (b). The radius of the glass is 4 cm and that of the cherry is 1 cm so points on the sections shown in the figures satisfy the equations $x^2 + y^2 = 16$ and $x^2 + (y + 3)^2 = 1$. We will find the volumes of the solids that are generated when the shaded regions are revolved about the y-axis.

For $0 \le h < 2$,

$$V = \pi \int_{-4}^{h-4} [(16 - y^2) - (1 - (y + 3)^2)]dy = 6\pi \int_{-4}^{h-4} (y + 4)dy = 3\pi h^2;$$

for $2 \le h \le 4$,

$$V = \pi \int_{-4}^{-2} [(16 - y^2) - (1 - (y + 3)^2)]dy + \pi \int_{-2}^{h-4} (16 - y^2)dy$$
$$= 6\pi \int_{-4}^{-2} (y + 4)dy + \pi \int_{-2}^{h-4} (16 - y^2)dy = 12\pi + \frac{1}{3}\pi(12h^2 - h^3 - 40)$$
$$= \frac{1}{3}\pi(12h^2 - h^3 - 4)$$

so

$$V = \begin{cases} 3\pi h^2 & \text{if } 0 \le h < 2 \\ \frac{1}{3}\pi(12h^2 - h^3 - 4) & \text{if } 2 \le h \le 4 \end{cases}$$

51. $\tan\theta = h/x$ so $h = x\tan\theta$,

$$A(y) = \frac{1}{2}hx = \frac{1}{2}x^2\tan\theta = \frac{1}{2}(r^2 - y^2)\tan\theta$$

because $x^2 = r^2 - y^2$,

$$V = \frac{1}{2}\tan\theta\int_{-r}^{r}(r^2 - y^2)dy$$

$$= \tan\theta\int_{0}^{r}(r^2 - y^2)dy = \frac{2}{3}r^3\tan\theta$$

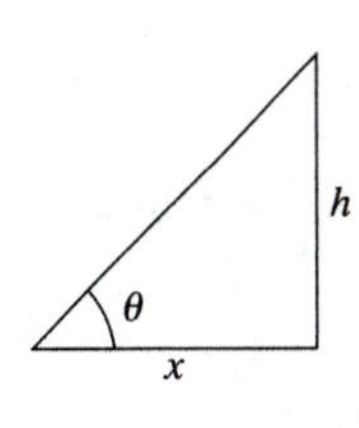

53. Each cross section perpendicular to the y-axis is a square so

$A(y) = x^2 = r^2 - y^2$,

$$\frac{1}{8}V = \int_0^r (r^2 - y^2)dy$$

$$V = 8(2r^3/3) = 16r^3/3$$

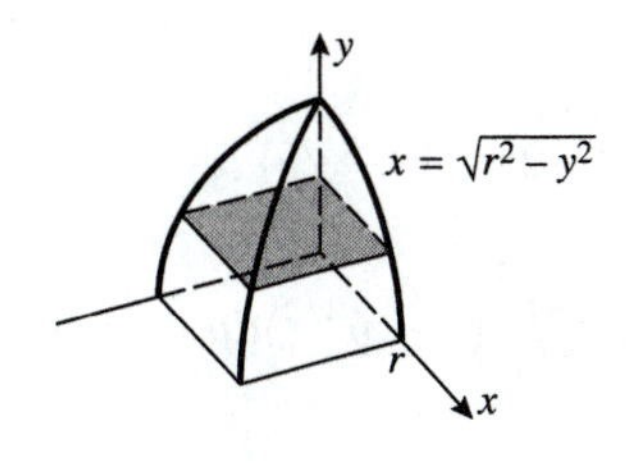

EXERCISE SET 7.3

1. $V = \int_1^2 2\pi x(x^2)dx = 2\pi\int_1^2 x^3 dx = 15\pi/2$

3. $V = \int_0^1 2\pi y(2y - 2y^2)dy = 4\pi\int_0^1 (y^2 - y^3)dy = \pi/3$

5. $V = \int_0^1 2\pi(x)(x^3)dx$

$$= 2\pi\int_0^1 x^4 dx = 2\pi/5$$

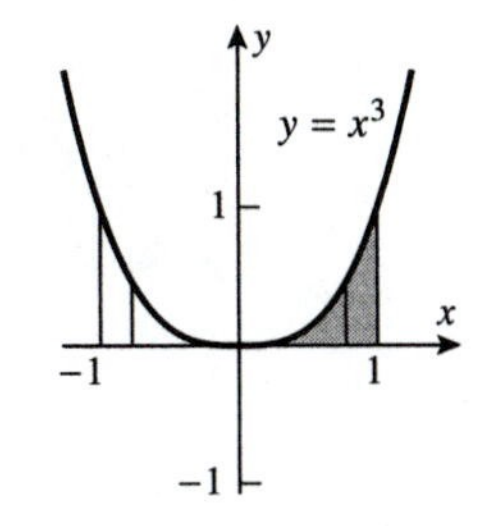

7. $V = \int_1^3 2\pi x(1/x)dx = 2\pi\int_1^3 dx = 4\pi$

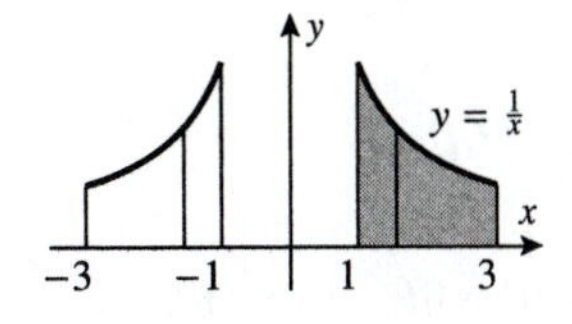

9. $V = \int_1^2 2\pi x[(2x - 1) - (-2x + 3)]dx$

$$= 8\pi\int_1^2 (x^2 - x)dx = 20\pi/3$$

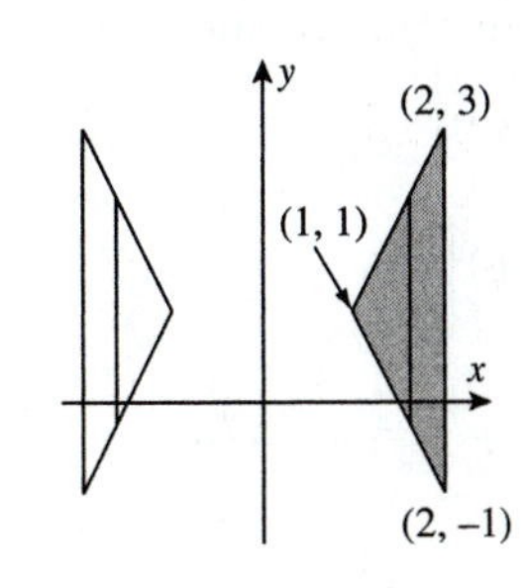

11. $V = 2\pi \int_0^1 \frac{x}{x^2+1}\,dx$

$= \pi \ln(x^2+1)\Big]_0^1 = \pi \ln 2$

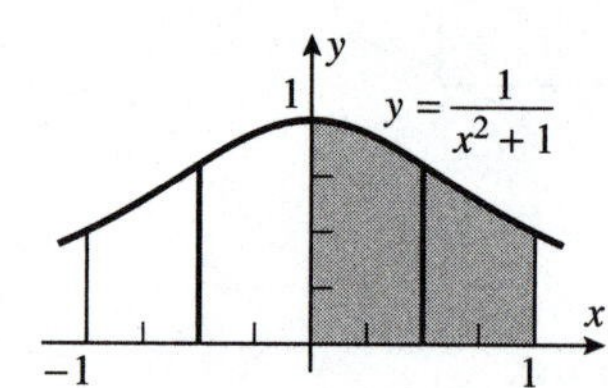

13. $V = \int_0^1 2\pi y^3 dy = \pi/2$

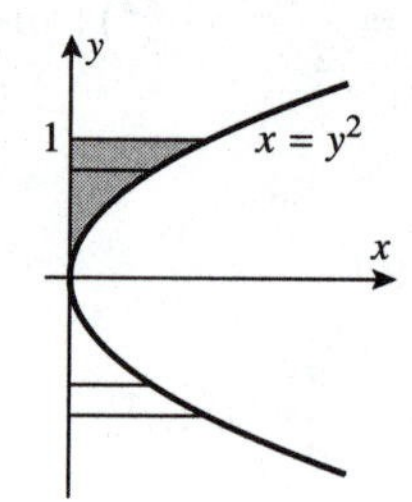

15. $V = \int_0^1 2\pi y(1-\sqrt{y})dy$

$= 2\pi \int_0^1 (y - y^{3/2})dy = \pi/5$

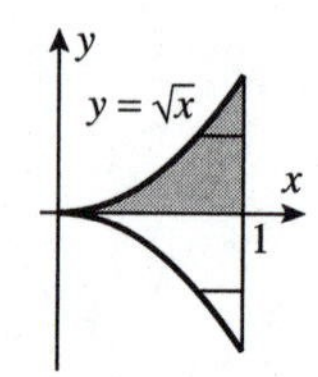

17. $V = 2\pi \int_0^{\pi} x \sin x dx = 2\pi^2$

19. **(a)** $V = \int_0^1 2\pi x(x^3 - 3x^2 + 2x)dx = 7\pi/30$

(b) much easier; the method of slicing would require that x be expressed in terms of y.

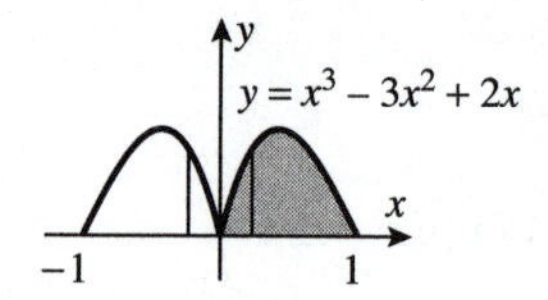

21. $V = \int_0^1 2\pi(1-y)y^{1/3}dy$

$= 2\pi \int_0^1 (y^{1/3} - y^{4/3})dy = 9\pi/14$

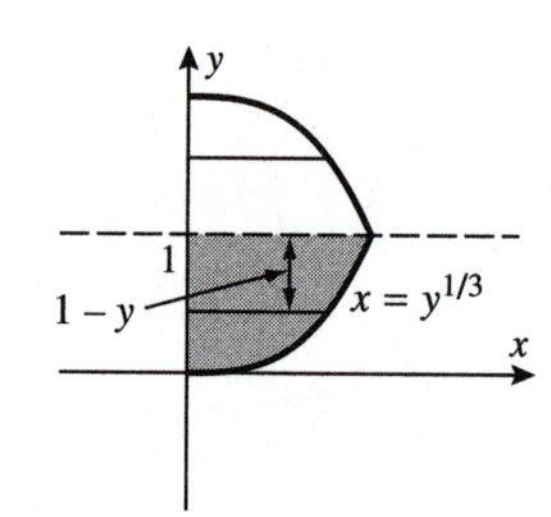

23. $x = \frac{h}{r}(r-y)$ is an equation of the line through $(0, r)$ and $(h, 0)$ so

$V = \int_0^r 2\pi y \left[\frac{h}{r}(r-y)\right] dy$

$= \frac{2\pi h}{r} \int_0^r (ry - y^2)dy = \pi r^2 h/3$

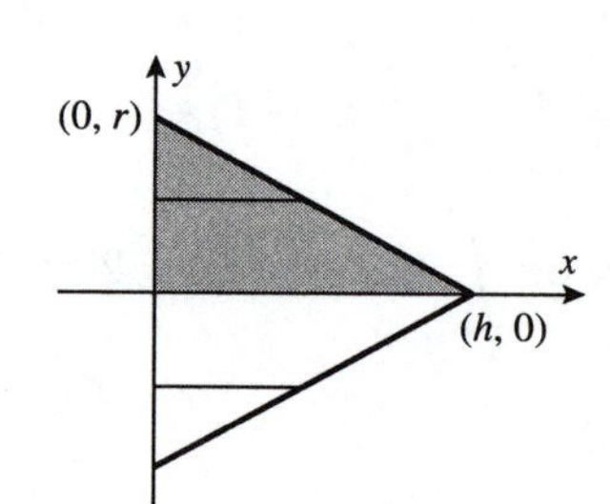

25. $V = \int_0^a 2\pi x(2\sqrt{r^2 - x^2})dx = 4\pi \int_0^a x(r^2 - x^2)^{1/2}dx$

$= -\frac{4\pi}{3}(r^2 - x^2)^{3/2}\Big]_0^a = \frac{4\pi}{3}\left[r^3 - (r^2 - a^2)^{3/2}\right]$

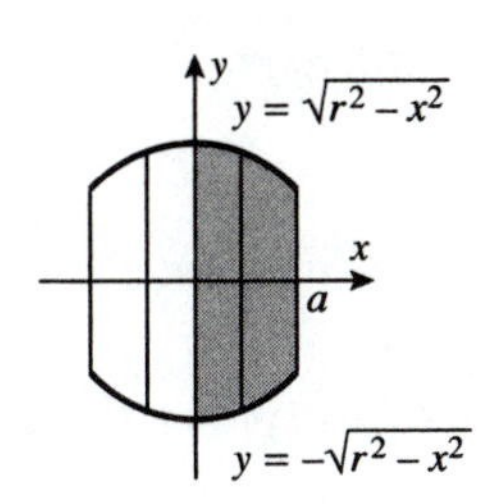

27. $V_x = \pi \int_{1/2}^b \frac{1}{x^2}dx = \pi(2 - 1/b)$, $V_y = 2\pi \int_{1/2}^b dx = \pi(2b - 1)$;

$V_x = V_y$ if $2 - 1/b = 2b - 1$, $2b^2 - 3b + 1 = 0$, solve to get $b = 1/2$ (reject) or $b = 1$.

EXERCISE SET 7.4

1. (a) $\frac{dy}{dx} = 2$, $L = \int_1^2 \sqrt{1+4}dx = \sqrt{5}$

(b) $\frac{dx}{dy} = \frac{1}{2}$, $L = \int_2^4 \sqrt{1 + 1/4}\,dy = 2\sqrt{5}/2 = \sqrt{5}$

3. $f'(x) = \frac{9}{2}x^{1/2}$, $1 + [f'(x)]^2 = 1 + \frac{81}{4}x$,

$L = \int_0^1 \sqrt{1 + 81x/4}\,dx = \frac{8}{243}\left(1 + \frac{81}{4}x\right)^{3/2}\Big]_0^1 = (85\sqrt{85} - 8)/243$

5. $\frac{dy}{dx} = \frac{2}{3}x^{-1/3}$, $1 + \left(\frac{dy}{dx}\right)^2 = 1 + \frac{4}{9}x^{-2/3} = \frac{9x^{2/3} + 4}{9x^{2/3}}$,

$L = \int_1^8 \frac{\sqrt{9x^{2/3} + 4}}{3x^{1/3}}dx = \frac{1}{18}\int_{13}^{40} u^{1/2}du$, $u = 9x^{2/3} + 4$

$= \frac{1}{27}u^{3/2}\Big]_{13}^{40} = \frac{1}{27}(40\sqrt{40} - 13\sqrt{13}) = \frac{1}{27}(80\sqrt{10} - 13\sqrt{13})$

or (alternate solution)

$$x = y^{3/2},\ \frac{dx}{dy} = \frac{3}{2}y^{1/2},\ 1 + \left(\frac{dx}{dy}\right)^2 = 1 + \frac{9}{4}y = \frac{4 + 9y}{4},$$

$$L = \frac{1}{2}\int_1^4 \sqrt{4 + 9y}\,dy = \frac{1}{18}\int_{13}^{40} u^{1/2}du = \frac{1}{27}(80\sqrt{10} - 13\sqrt{13})$$

7. $x = g(y) = \frac{1}{24}y^3 + 2y^{-1}$, $g'(y) = \frac{1}{8}y^2 - 2y^{-2}$,

$1 + [g'(y)]^2 = 1 + \left(\frac{1}{64}y^4 - \frac{1}{2} + 4y^{-4}\right) = \frac{1}{64}y^4 + \frac{1}{2} + 4y^{-4} = \left(\frac{1}{8}y^2 + 2y^{-2}\right)^2$,

$L = \int_2^4 \left(\frac{1}{8}y^2 + 2y^{-2}\right)dy = 17/6$

9. $(dx/dt)^2 + (dy/dt)^2 = (t^2)^2 + (t)^2 = t^2(t^2+1),\ L = \int_0^1 t(t^2+1)^{1/2}dt = (2\sqrt{2}-1)/3$

11. $(dx/dt)^2 + (dy/dt)^2 = (-2\sin 2t)^2 + (2\cos 2t)^2 = 4,\ L = \int_0^{\pi/2} 2\,dt = \pi$

13. $(dx/dt)^2 + (dy/dt)^2 = [e^t(\cos t - \sin t)]^2 + [e^t(\cos t + \sin t)]^2 = 2e^{2t}$,

$$L = \int_0^{\pi/2} \sqrt{2}e^t dt = \sqrt{2}(e^{\pi/2}-1)$$

15. $dy/dx = \dfrac{\sec x \tan x}{\sec x} = \tan x,\ \sqrt{1+(y')^2} = \sqrt{1+\tan^2 x} = \sec x$ when $0 < x < \pi/4$, so

$$L = \int_0^{\pi/4} \sec x\,dx = \ln(1+\sqrt{2})$$

17. **(a)** $(dx/d\theta)^2 + (dy/d\theta)^2 = (a(1-\cos\theta))^2 + (a\sin\theta)^2 = a^2(2-2\cos\theta)$, so

$$L = \int_0^{2\pi} \sqrt{(dx/d\theta)^2 + (dy/d\theta)^2}\,d\theta = a\int_0^{2\pi} \sqrt{2(1-\cos\theta)}\,d\theta$$

19. **(a)**

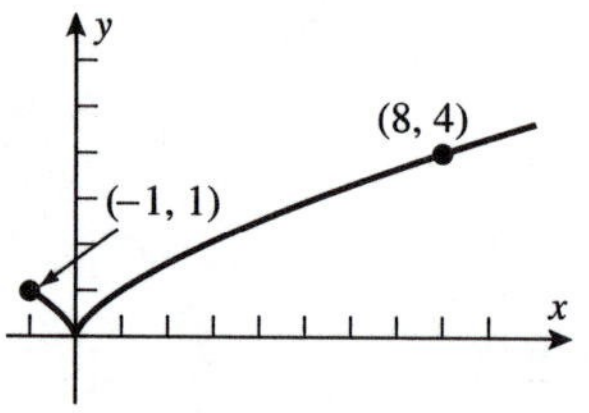

(b) dy/dx does not exist at $x = 0$.

(c) $x = g(y) = y^{3/2},\ g'(y) = \dfrac{3}{2}y^{1/2}$,

$$L = \int_0^1 \sqrt{1+9y/4}\,dy \quad (\text{portion for } -1 \le x \le 0)$$

$$+ \int_0^4 \sqrt{1+9y/4}\,dy \quad (\text{portion for } 0 \le x \le 8)$$

$$= \frac{8}{27}\left(\frac{13}{8}\sqrt{13}-1\right) + \frac{8}{27}(10\sqrt{10}-1) = (13\sqrt{13}+80\sqrt{10}-16)/27$$

21. $L = \int_0^2 \sqrt{1+4x^2}\,dx \approx 4.645975301$

23. Numerical integration yields: in Exercise 21, $L \approx 4.646783762$; in Exercise 22, $L \approx 3.820197788$.

25. $f'(x) = \cos x,\ \sqrt{2}/2 \le \cos x \le 1$ for $0 \le x \le \pi/4$ so

$(\pi/4)\sqrt{1+1/2} \le L \le (\pi/4)\sqrt{1+1},\ \dfrac{\pi}{4}\sqrt{3/2} \le L \le \dfrac{\pi}{4}\sqrt{2}$.

27. **(a)** $(dx/dt)^2 + (dy/dt)^2 = 4\sin^2 t + \cos^2 t = 4\sin^2 t + (1-\sin^2 t) = 1 + 3\sin^2 t$,

$$L = \int_0^{2\pi} \sqrt{1+3\sin^2 t}\,dt = 4\int_0^{\pi/2} \sqrt{1+3\sin^2 t}\,dt$$

(b) 9.69

(c) distance traveled $= \int_{1.5}^{4.8} \sqrt{1+3\sin^2 t}\, dt \approx 5.16$ cm

29. $L = \int_0^{\pi} \sqrt{1+(k\cos x)^2}\, dx$

k	1	2	1.84	1.83	1.832
L	3.8202	5.2704	5.0135	4.9977	5.0008

Experimentation yields the values in the table, which by the Intermediate-Value Theorem show that the true solution k to $L = 5$ lies between $k = 1.83$ and $k = 1.832$, so $k = 1.83$ to two decimal places.

EXERCISE SET 7.5

1. $S = \int_0^1 2\pi(7x)\sqrt{1+49}dx = 70\pi\sqrt{2}\int_0^1 x\, dx = 35\pi\sqrt{2}$

3. $f'(x) = -x/\sqrt{4-x^2},\ 1+[f'(x)]^2 = 1+\dfrac{x^2}{4-x^2} = \dfrac{4}{4-x^2},$

$$S = \int_{-1}^{1} 2\pi\sqrt{4-x^2}(2/\sqrt{4-x^2})dx = 4\pi\int_{-1}^{1} dx = 8\pi$$

5. $S = \int_0^2 2\pi(9y+1)\sqrt{82}dy = 2\pi\sqrt{82}\int_0^2 (9y+1)dy = 40\pi\sqrt{82}$

7. $g'(y) = -y/\sqrt{9-y^2},\ 1+[g'(y)]^2 = \dfrac{9}{9-y^2},\ S = \int_{-2}^{2} 2\pi\sqrt{9-y^2}\cdot\dfrac{3}{\sqrt{9-y^2}}dy = 6\pi\int_{-2}^{2} dy = 24\pi$

9. $f'(x) = \dfrac{1}{2}x^{-1/2} - \dfrac{1}{2}x^{1/2},\ 1+[f'(x)]^2 = 1+\dfrac{1}{4}x^{-1} - \dfrac{1}{2} + \dfrac{1}{4}x = \left(\dfrac{1}{2}x^{-1/2} + \dfrac{1}{2}x^{1/2}\right)^2,$

$$S = \int_1^3 2\pi\left(x^{1/2} - \frac{1}{3}x^{3/2}\right)\left(\frac{1}{2}x^{-1/2} + \frac{1}{2}x^{1/2}\right)dx = \frac{\pi}{3}\int_1^3 (3+2x-x^2)dx = 16\pi/9$$

11. $x = g(y) = \dfrac{1}{4}y^4 + \dfrac{1}{8}y^{-2},\ g'(y) = y^3 - \dfrac{1}{4}y^{-3},$

$$1+[g'(y)]^2 = 1+\left(y^6 - \frac{1}{2} + \frac{1}{16}y^{-6}\right) = \left(y^3 + \frac{1}{4}y^{-3}\right)^2,$$

$$S = \int_1^2 2\pi\left(\frac{1}{4}y^4 + \frac{1}{8}y^{-2}\right)\left(y^3 + \frac{1}{4}y^{-3}\right)dy = \frac{\pi}{16}\int_1^2 (8y^7 + 6y + y^{-5})dy = 16{,}911\pi/1024$$

13. $f'(x) = \cos x,\ 1+[f'(x)]^2 = 1+\cos^2 x,\ S = \int_0^{\pi} 2\pi\sin x\sqrt{1+\cos^2 x}\, dx = 2\pi(\sqrt{2} + \ln(\sqrt{2}+1))$

15. $f'(x) = e^x,\quad 1+[f'(x)]^2 = 1+e^{2x},\ S = \int_0^1 2\pi e^x\sqrt{1+e^{2x}}\, dx \approx 22.94$

17. Revolve the line segment joining the points $(0,0)$ and (h,r) about the x-axis. An equation of the line segment is $y = (r/h)x$ for $0 \le x \le h$ so

$$S = \int_0^h 2\pi(r/h)x\sqrt{1+r^2/h^2}dx = \frac{2\pi r}{h^2}\sqrt{r^2+h^2}\int_0^h x\, dx = \pi r\sqrt{r^2+h^2}$$

19. $g(y)=\sqrt{r^2-y^2}$, $g'(y)=-y/\sqrt{r^2-y^2}$, $1+[g'(y)]^2=r^2/(r^2-y^2)$,

(a) $S=\int_{r-h}^{r} 2\pi\sqrt{r^2-y^2}\sqrt{r^2/(r^2-y^2)}\,dy=2\pi r\int_{r-h}^{r} dy=2\pi rh$

(b) From Part (a), the surface area common to two polar caps of height $h_1>h_2$ is $2\pi rh_1-2\pi rh_2=2\pi r(h_1-h_2)$.

21. $x'=2t, y'=2, (x')^2+(y')^2=4t^2+4$

$$S=2\pi\int_0^4 (2t)\sqrt{4t^2+4}dt=8\pi\int_0^4 t\sqrt{t^2+1}dt=\frac{8\pi}{3}(17\sqrt{17}-1)$$

23. $x'=1$, $y'=4t$, $(x')^2+(y')^2=1+16t^2$, $S=2\pi\int_0^1 t\sqrt{1+16t^2}\,dt=\frac{\pi}{24}(17\sqrt{17}-1)$

25. $x'=-r\sin t$, $y'=r\cos t$, $(x')^2+(y')^2=r^2$,

$$S=2\pi\int_0^\pi r\sin t\sqrt{r^2}\,dt=2\pi r^2\int_0^\pi \sin t\,dt=4\pi r^2$$

27. **(a)** length of arc of sector = circumference of base of cone,

$\ell\theta=2\pi r, \theta=2\pi r/\ell;\ S=$ area of sector $=\frac{1}{2}\ell^2(2\pi r/\ell)=\pi r\ell$

(b) $S=\pi r_2\ell_2-\pi r_1\ell_1=\pi r_2(\ell_1+\ell)-\pi r_1\ell_1=\pi[(r_2-r_1)\ell_1+r_2\ell]$;
Using similar triangles $\ell_2/r_2=\ell_1/r_1, r_1\ell_2=r_2\ell_1, r_1(\ell_1+\ell)=r_2\ell_1, (r_2-r_1)\ell_1=r_1\ell$
so $S=\pi\left(r_1\ell+r_2\ell\right)=\pi\left(r_1+r_2\right)\ell$.

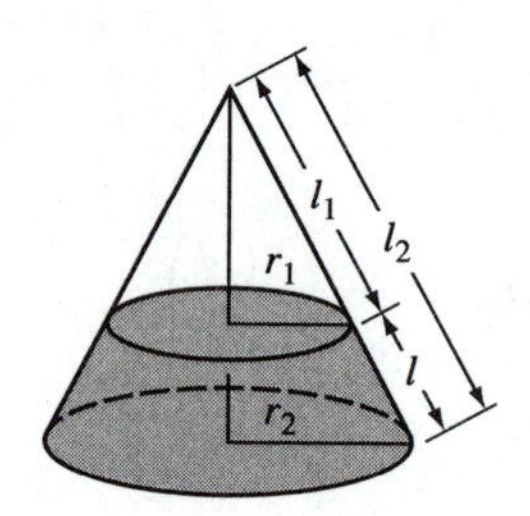

29. $2\pi k\sqrt{1+[f'(x)]^2}\le 2\pi f(x)\sqrt{1+[f'(x)]^2}\le 2\pi K\sqrt{1+[f'(x)]^2}$, so

$$\int_a^b 2\pi k\sqrt{1+[f'(x)]^2}dx\le\int_a^b 2\pi f(x)\sqrt{1+[f'(x)]^2}dx\le\int_a^b 2\pi K\sqrt{1+[f'(x)]^2}dx,$$

$$2\pi k\int_a^b\sqrt{1+[f'(x)]^2}dx\le S\le 2\pi K\int_a^b\sqrt{1+[f'(x)]^2}dx,\ 2\pi kL\le S\le 2\pi KL$$

EXERCISE SET 7.6

1. **(a)** $W=F\cdot d=30(7)=210$ ft·lb

(b) $W=\int_1^6 F(x)\,dx=\int_1^6 x^{-2}\,dx=-\frac{1}{x}\Big]_1^6=5/6$ ft·lb

3. distance traveled $=\int_0^5 v(t)\,dt=\int_0^5\frac{4t}{5}\,dt=\frac{2}{5}t^2\Big]_0^5=10$ ft. The force is a constant 10 lb, so the work done is $10\cdot 10=100$ ft·lb.

5. $F(x) = kx$, $F(0.2) = 0.2k = 100$, $k = 500$ N/m, $W = \int_0^{0.8} 500x\,dx = 160\,\text{J}$

7. $W = \int_0^1 kx\,dx = k/2 = 10$, $k = 20\,\text{lb/ft}$

9. $W = \int_0^6 (9 - x)\rho(25\pi)dx = 900\pi\rho$ ft·lb

11. $w/4 = x/3, w = 4x/3$,

$$W = \int_0^2 (3 - x)(9810)(4x/3)(6)dx$$

$$= 78480 \int_0^2 (3x - x^2)dx$$

$$= 261,600\,\text{J}$$

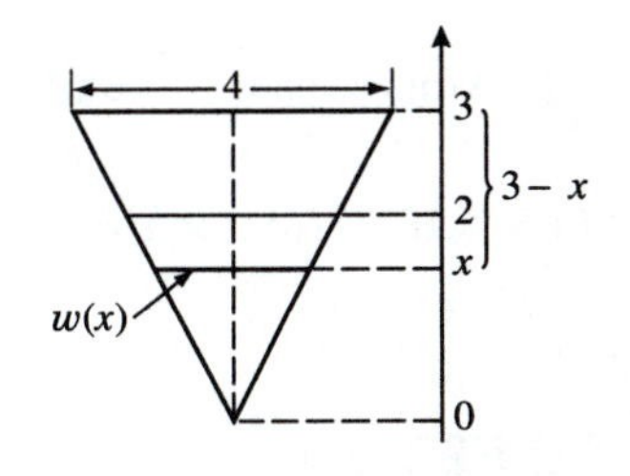

13. **(a)** $W = \int_0^9 (10 - x)62.4(300)dx$

$$= 18{,}720 \int_0^9 (10 - x)dx$$

$$= 926{,}640 \text{ ft·lb}$$

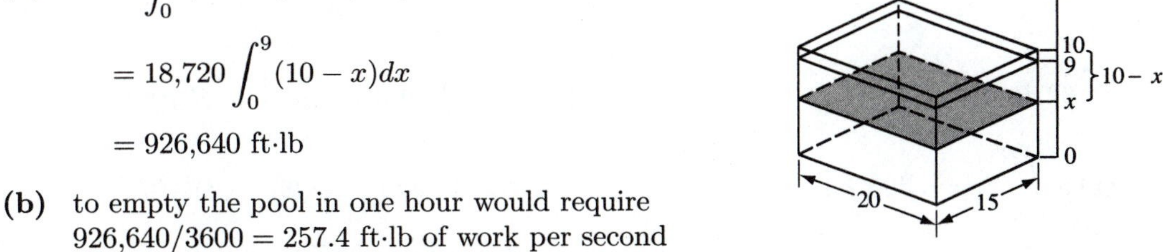

(b) to empty the pool in one hour would require 926,640/3600 = 257.4 ft·lb of work per second so hp of motor = 257.4/550 = 0.468

15. $W = \int_0^{100} 15(100 - x)dx$

$$= 75,000 \text{ ft·lb}$$

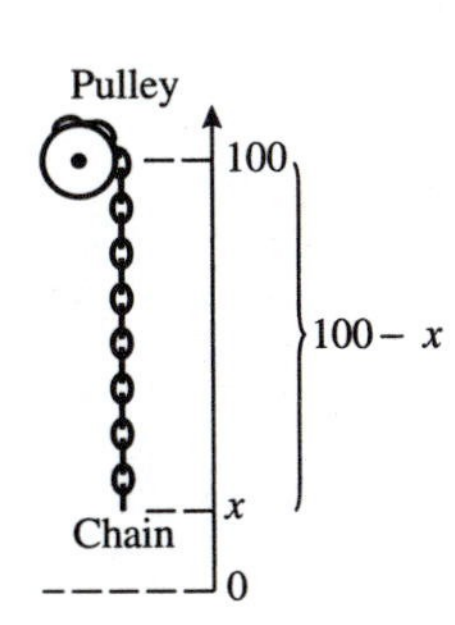

17. When the rocket is x ft above the ground

total weight = weight of rocket + weight of fuel

$$= 3 + [40 - 2(x/1000)]$$

$$= 43 - x/500 \text{ tons},$$

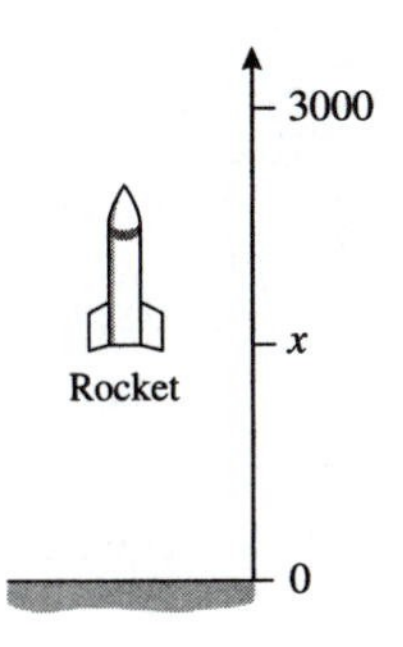

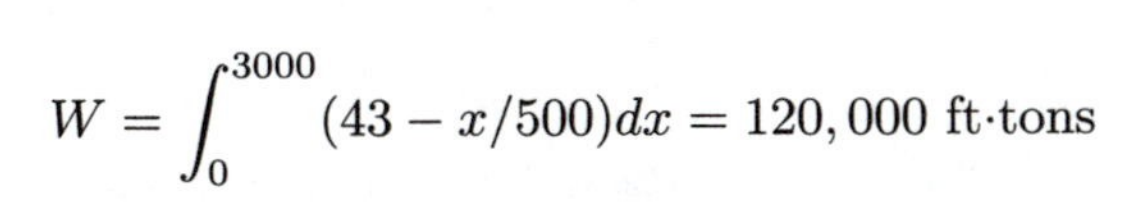

$$W = \int_0^{3000} (43 - x/500)dx = 120,000 \text{ ft·tons}$$

19. **(a)** $150 = k/(4000)^2$, $k = 2.4 \times 10^9$, $w(x) = k/x^2 = 2{,}400{,}000{,}000/x^2$ lb

(b) $6000 = k/(4000)^2$, $k = 9.6 \times 10^{10}$, $w(x) = \left(9.6 \times 10^{10}\right)/(x + 4000)^2$ lb

(c) $W = \int_{4000}^{5000} 9.6(10^{10})x^{-2}dx = 4{,}800{,}000$ mi·lb $= 2.5344 \times 10^{10}$ ft·lb

21. $W = F \cdot d = (6.40 \times 10^5)(3.00 \times 10^3) = 1.92 \times 10^9$ J; from the Work-Energy Relationship (5), $v_f^2 = 2W/m + v_i^2 = 2(1.92 \cdot 10^9)/(4 \cdot 10^5) + 20^2 = 10{,}000$, $v_f = 100$ m/s

23. **(a)** The kinetic energy would have decreased by $\frac{1}{2}mv^2 = \frac{1}{2}4 \cdot 10^6(15000)^2 = 4.5 \times 10^{14}$ J

(b) $(4.5 \times 10^{14})/(4.2 \times 10^{15}) \approx 0.107$

(c) $\frac{1000}{13}(0.107) \approx 8.24$ bombs

EXERCISE SET 7.7

1. **(a)** $F = \rho h A = 62.4(5)(100) = 31{,}200$ lb
$P = \rho h = 62.4(5) = 312$ lb/ft^2

(b) $F = \rho h A = 9810(10)(25) = 2{,}452{,}500$ N
$P = \rho h = 9810(10) = 98.1$ kPa

3. $F = \int_0^2 62.4x(4)dx$

$= 249.6 \int_0^2 x\,dx = 499.2$ lb

5. $F = \int_0^5 9810x(2\sqrt{25 - x^2})dx$

$= 19{,}620 \int_0^5 x(25 - x^2)^{1/2}dx$

$= 8.175 \times 10^5$ N

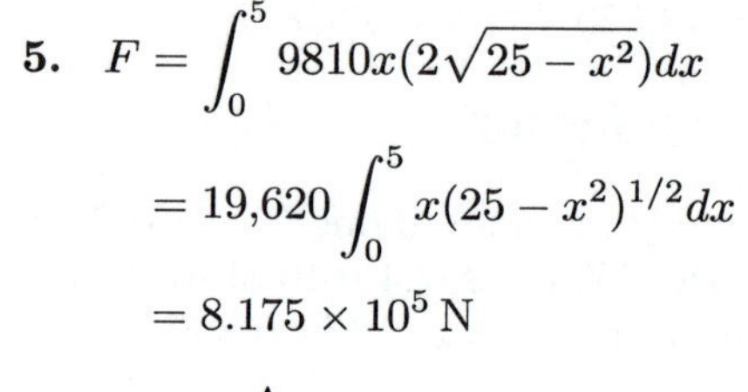

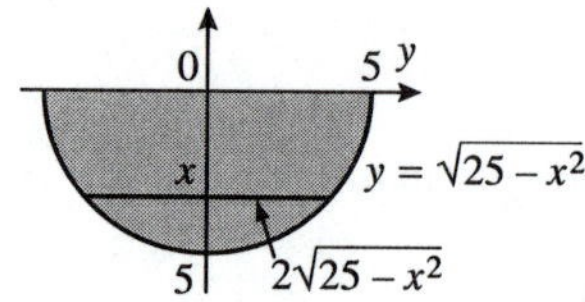

7. By similar triangles

$\frac{w(x)}{6} = \frac{10 - x}{8}$

$w(x) = \frac{3}{4}(10 - x)$,

$F = \int_2^{10} 9810x\left[\frac{3}{4}(10 - x)\right]dx$

$= 7357.5 \int_2^{10} (10x - x^2)dx = 1{,}098{,}720$ N

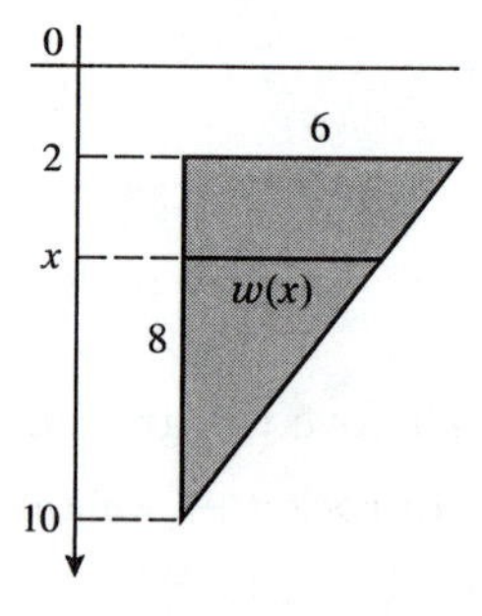

9. Yes: if $\rho_2 = 2\rho_1$ then $F_2 = \int_a^b \rho_2 h(x)w(x)\,dx = \int_a^b 2\rho_1 h(x)w(x)\,dx = 2\int_a^b \rho_1 h(x)w(x)\,dx = 2F_1$.

11. Find the forces on the upper and lower halves and add them:

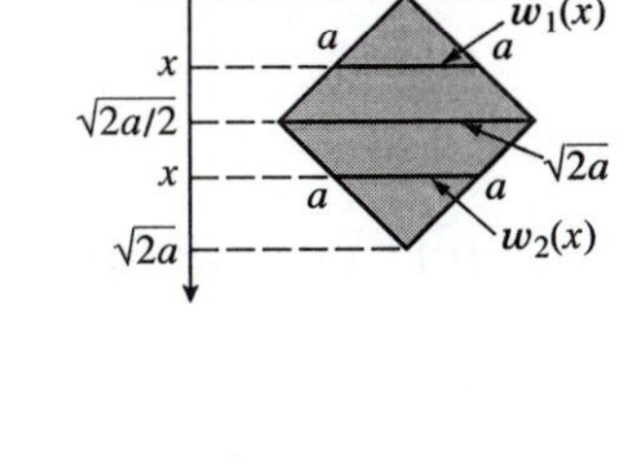

$$\frac{w_1(x)}{\sqrt{2}a} = \frac{x}{\sqrt{2}a/2},\ w_1(x) = 2x$$

$$F_1 = \int_0^{\sqrt{2}a/2} \rho x(2x)dx = 2\rho \int_0^{\sqrt{2}a/2} x^2 dx = \sqrt{2}\rho a^3/6,$$

$$\frac{w_2(x)}{\sqrt{2}a} = \frac{\sqrt{2}a - x}{\sqrt{2}a/2},\ w_2(x) = 2(\sqrt{2}a - x)$$

$$F_2 = \int_{\sqrt{2}a/2}^{\sqrt{2}a} \rho x[2(\sqrt{2}a - x)]dx = 2\rho \int_{\sqrt{2}a/2}^{\sqrt{2}a} (\sqrt{2}ax - x^2)dx = \sqrt{2}\rho a^3/3,$$

$$F = F_1 + F_2 = \sqrt{2}\rho a^3/6 + \sqrt{2}\rho a^3/3 = \rho a^3/\sqrt{2}\ \text{lb}$$

13. $\sqrt{16^2 + 4^2} = \sqrt{272} = 4\sqrt{17}$ is the other dimension of the bottom.
$(h(x) - 4)/4 = x/(4\sqrt{17})$
$h(x) = x/\sqrt{17} + 4,$
$\sec\theta = 4\sqrt{17}/16 = \sqrt{17}/4$

$$F = \int_0^{4\sqrt{17}} 62.4(x/\sqrt{17} + 4)10(\sqrt{17}/4)\,dx$$

$$= 156\sqrt{17} \int_0^{4\sqrt{17}} (x/\sqrt{17} + 4)dx$$

$$= 63{,}648\ \text{lb}$$

15. $h(x) = x\sin 60° = \sqrt{3}x/2,$
$\theta = 30°,\ \sec\theta = 2/\sqrt{3},$

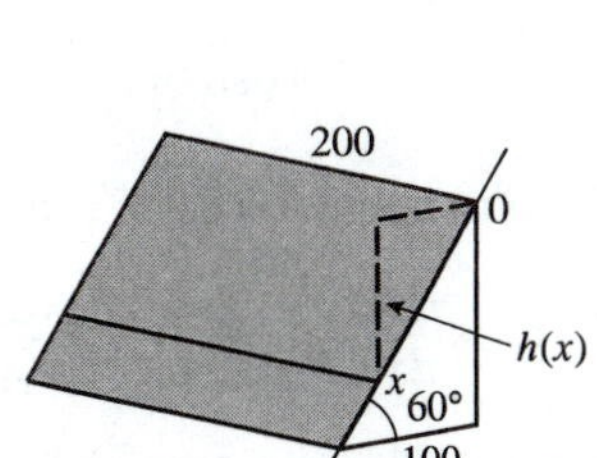

$$F = \int_0^{100} 9810(\sqrt{3}x/2)(200)(2/\sqrt{3})\,dx$$

$$= 200 \cdot 9810 \int_0^{100} x\,dx$$

$$= 9810 \cdot 100^3 = 9.81 \times 10^9\ \text{N}$$

17. **(a)** From Exercise 16, $F = 4\rho_0(h + 1)$ so (assuming that ρ_0 is constant) $dF/dt = 4\rho_0(dh/dt)$ which is a positive constant if dh/dt is a positive constant.
(b) If $dh/dt = 20$ then $dF/dt = 80\rho_0$ lb/min from Part (a).

EXERCISE SET 7.8

1. **(a)** $\sinh 3 \approx 10.0179$
(b) $\cosh(-2) \approx 3.7622$
(c) $\tanh(\ln 4) = 15/17 \approx 0.8824$
(d) $\sinh^{-1}(-2) \approx -1.4436$
(e) $\cosh^{-1} 3 \approx 1.7627$
(f) $\tanh^{-1}\frac{3}{4} \approx 0.9730$

3. **(a)** $\sinh(\ln 3) = \frac{1}{2}(e^{\ln 3} - e^{-\ln 3}) = \frac{1}{2}\left(3 - \frac{1}{3}\right) = \frac{4}{3}$

(b) $\cosh(-\ln 2) = \frac{1}{2}(e^{-\ln 2} + e^{\ln 2}) = \frac{1}{2}\left(\frac{1}{2} + 2\right) = \frac{5}{4}$

(c) $\tanh(2\ln 5) = \dfrac{e^{2\ln 5} - e^{-2\ln 5}}{e^{2\ln 5} + e^{-2\ln 5}} = \dfrac{25 - 1/25}{25 + 1/25} = \dfrac{312}{313}$

(d) $\sinh(-3\ln 2) = \frac{1}{2}(e^{-3\ln 2} - e^{3\ln 2}) = \frac{1}{2}\left(\frac{1}{8} - 8\right) = -\frac{63}{16}$

5.

	$\sinh x_0$	$\cosh x_0$	$\tanh x_0$	$\coth x_0$	$\operatorname{sech} x_0$	$\operatorname{csch} x_0$
(a)	2	$\sqrt{5}$	$2/\sqrt{5}$	$\sqrt{5}/2$	$1/\sqrt{5}$	1/2
(b)	3/4	5/4	3/5	5/3	4/5	4/3
(c)	4/3	5/3	4/5	5/4	3/5	3/4

(a) $\cosh^2 x_0 = 1 + \sinh^2 x_0 = 1 + (2)^2 = 5$, $\cosh x_0 = \sqrt{5}$

(b) $\sinh^2 x_0 = \cosh^2 x_0 - 1 = \frac{25}{16} - 1 = \frac{9}{16}$, $\sinh x_0 = \frac{3}{4}$ (because $x_0 > 0$)

(c) $\operatorname{sech}^2 x_0 = 1 - \tanh^2 x_0 = 1 - \left(\frac{4}{5}\right)^2 = 1 - \frac{16}{25} = \frac{9}{25}$, $\operatorname{sech} x_0 = \frac{3}{5}$,

$\cosh x_0 = \dfrac{1}{\operatorname{sech} x_0} = \dfrac{5}{3}$, from $\dfrac{\sinh x_0}{\cosh x_0} = \tanh x_0$ we get $\sinh x_0 = \left(\frac{5}{3}\right)\left(\frac{4}{5}\right) = \frac{4}{3}$

7. **(a)** $y = \sinh^{-1} x$ if and only if $x = \sinh y$; $1 = \dfrac{dy}{dx}\dfrac{dx}{dy} = \dfrac{dy}{dx}\cosh y$; so

$\dfrac{d}{dx}[\sinh^{-1} x] = \dfrac{dy}{dx} = \dfrac{1}{\cosh y} = \dfrac{1}{\sqrt{1 + \sinh^2 y}} = \dfrac{1}{\sqrt{1 + x^2}}$ for all x.

(b) Let $x \geq 1$. Then $y = \cosh^{-1} x$ if and only if $x = \cosh y$; $1 = \dfrac{dy}{dx}\dfrac{dx}{dy} = \dfrac{dy}{dx}\sinh y$, so

$\dfrac{d}{dx}[\cosh^{-1} x] = \dfrac{dy}{dx} = \dfrac{1}{\sinh y} = \dfrac{1}{\sqrt{\cosh^2 y - 1}} = \dfrac{1}{x^2 - 1}$ for $x \geq 1$.

(c) Let $-1 < x < 1$. Then $y = \tanh^{-1} x$ if and only if $x = \tanh y$; thus

$1 = \dfrac{dy}{dx}\dfrac{dx}{dy} = \dfrac{dy}{dx}\operatorname{sech}^2 y = \dfrac{dy}{dx}(1 - \tanh^2 y) = 1 - x^2$, so $\dfrac{d}{dx}[\tanh^{-1} x] = \dfrac{dy}{dx} = \dfrac{1}{1 - x^2}$.

9. $4\cosh(4x - 8)$

11. $-\dfrac{1}{x}\operatorname{csch}^2(\ln x)$

13. $\dfrac{1}{x^2}\operatorname{csch}(1/x)\coth(1/x)$

15. $\dfrac{2 + 5\cosh(5x)\sinh(5x)}{\sqrt{4x + \cosh^2(5x)}}$

17. $x^{5/2}\tanh(\sqrt{x})\operatorname{sech}^2(\sqrt{x}) + 3x^2\tanh^2(\sqrt{x})$

19. $\dfrac{1}{\sqrt{1 + x^2/9}}\left(\dfrac{1}{3}\right) = 1/\sqrt{9 + x^2}$

21. $1/\left[(\cosh^{-1} x)\sqrt{x^2 - 1}\right]$

23. $-(\tanh^{-1} x)^{-2}/(1 - x^2)$

25. $\dfrac{\sinh x}{\sqrt{\cosh^2 x - 1}} = \dfrac{\sinh x}{|\sinh x|} = \begin{cases} 1, & x > 0 \\ -1, & x < 0 \end{cases}$

27. $-\dfrac{e^x}{2x\sqrt{1-x}}+e^x\,\text{sech}^{-1}x$

31. $\dfrac{1}{7}\sinh^7 x+C$

33. $\dfrac{2}{3}(\tanh x)^{3/2}+C$

35. $\ln(\cosh x)+C$

37. $-\dfrac{1}{3}\text{sech}^3 x\Big]_{\ln 2}^{\ln 3}=37/375$

39. $u=3x$, $\dfrac{1}{3}\displaystyle\int\frac{1}{\sqrt{1+u^2}}du=\frac{1}{3}\sinh^{-1}3x+C$

41. $u=e^x$, $\displaystyle\int\frac{1}{u\sqrt{1-u^2}}du=-\,\text{sech}^{-1}(e^x)+C$

43. $u=2x$, $\displaystyle\int\frac{du}{u\sqrt{1+u^2}}=-\text{csch}^{-1}|u|+C=-\text{csch}^{-1}|2x|+C$

45. $\tanh^{-1}x\Big]_0^{1/2}=\tanh^{-1}(1/2)-\tanh^{-1}(0)=\dfrac{1}{2}\ln\dfrac{1+1/2}{1-1/2}=\dfrac{1}{2}\ln 3$

49. $A=\displaystyle\int_0^{\ln 3}\sinh 2x\,dx=\frac{1}{2}\cosh 2x\Big]_0^{\ln 3}=\frac{1}{2}[\cosh(2\ln 3)-1]$,

but $\cosh(2\ln 3)=\cosh(\ln 9)=\dfrac{1}{2}(e^{\ln 9}+e^{-\ln 9})=\dfrac{1}{2}(9+1/9)=41/9$ so $A=\dfrac{1}{2}[41/9-1]=16/9$.

51. $V=\pi\displaystyle\int_0^5(\cosh^2 2x-\sinh^2 2x)dx=\pi\int_0^5 dx=5\pi$

53. $y'=\sinh x$, $1+(y')^2=1+\sinh^2 x=\cosh^2 x$

$$L=\int_0^{\ln 2}\cosh x\,dx=\sinh x\Big]_0^{\ln 2}=\sinh(\ln 2)=\frac{1}{2}(e^{\ln 2}-e^{-\ln 2})=\frac{1}{2}\left(2-\frac{1}{2}\right)=\frac{3}{4}$$

55. $\sinh(-x)=\dfrac{1}{2}(e^{-x}-e^{x})=-\dfrac{1}{2}(e^x-e^{-x})=-\sinh x$

$\cosh(-x)=\dfrac{1}{2}(e^{-x}+e^{x})=\dfrac{1}{2}(e^x+e^{-x})=\cosh x$

57. **(a)** Divide $\cosh^2 x-\sinh^2 x=1$ by $\cosh^2 x$.

(b) $\tanh(x+y)=\dfrac{\sinh x\cosh y+\cosh x\sinh y}{\cosh x\cosh y+\sinh x\sinh y}=\dfrac{\dfrac{\sinh x}{\cosh x}+\dfrac{\sinh y}{\cosh y}}{1+\dfrac{\sinh x\sinh y}{\cosh x\cosh y}}=\dfrac{\tanh x+\tanh y}{1+\tanh x\tanh y}$

(c) Let $y=x$ in Part (b).

59. **(a)** $\dfrac{d}{dx}(\cosh^{-1}x)=\dfrac{1+x/\sqrt{x^2-1}}{x+\sqrt{x^2-1}}=1/\sqrt{x^2-1}$

(b) $\dfrac{d}{dx}(\tanh^{-1}x)=\dfrac{d}{dx}\left[\dfrac{1}{2}(\ln(1+x)-\ln(1-x))\right]=\dfrac{1}{2}\left(\dfrac{1}{1+x}+\dfrac{1}{1-x}\right)=1/(1-x^2)$

61. If $|u|<1$ then, by Theorem 7.8.6, $\displaystyle\int\frac{du}{1-u^2}=\tanh^{-1}u+C$.

For $|u|>1$, $\displaystyle\int\frac{du}{1-u^2}=\coth^{-1}u+C=\tanh^{-1}(1/u)+C$.

63. (a) $\lim_{x\to+\infty} \sinh x = \lim_{x\to+\infty} \frac{1}{2}(e^x - e^{-x}) = +\infty - 0 = +\infty$

(b) $\lim_{x\to-\infty} \sinh x = \lim_{x\to-\infty} \frac{1}{2}(e^x - e^{-x}) = 0 - \infty = -\infty$

(c) $\lim_{x\to+\infty} \tanh x = \lim_{x\to+\infty} \frac{e^x - e^{-x}}{e^x + e^{-x}} = 1$

(d) $\lim_{x\to-\infty} \tanh x = \lim_{x\to-\infty} \frac{e^x - e^{-x}}{e^x + e^{-x}} = -1$

(e) $\lim_{x\to+\infty} \sinh^{-1} x = \lim_{x\to+\infty} \ln(x + \sqrt{x^2+1}) = +\infty$

(f) $\lim_{x\to1^-} \tanh^{-1} x = \lim_{x\to1^-} \frac{1}{2}[\ln(1+x) - \ln(1-x)] = +\infty$

65. For $|x| < 1$, $y = \tanh^{-1} x$ is defined and $dy/dx = 1/(1-x^2) > 0$; $y'' = 2x/(1-x^2)^2$ changes sign at $x = 0$, so there is a point of inflection there.

67. Using $\sinh x + \cosh x = e^x$ (Exercise 56a), $(\sinh x + \cosh x)^n = (e^x)^n = e^{nx} = \sinh nx + \cosh nx$.

69. (a) $y' = \sinh(x/a), 1 + (y')^2 = 1 + \sinh^2(x/a) = \cosh^2(x/a)$

$$L = 2\int_0^b \cosh(x/a)\,dx = 2a\sinh(x/a)\Big]_0^b = 2a\sinh(b/a)$$

(b) The highest point is at $x = b$, the lowest at $x = 0$, so $S = a\cosh(b/a) - a\cosh(0) = a\cosh(b/a) - a$.

71. From Part (b) of Exercise 69, $S = a\cosh(b/a) - a$ so $30 = a\cosh(200/a) - a$. Let $u = 200/a$, then $a = 200/u$ so $30 = (200/u)[\cosh u - 1], \cosh u - 1 = 0.15u$. If $f(u) = \cosh u - 0.15u - 1$, then $u_{n+1} = u_n - \frac{\cosh u_n - 0.15u_n - 1}{\sinh u_n - 0.15}$; $u_1 = 0.3, \ldots, u_4 = u_5 = 0.297792782 \approx 200/a$ so $a \approx 671.6079505$. From Part (a), $L = 2a\sinh(b/a) \approx 2(671.6079505)\sinh(0.297792782) \approx 405.9$ ft.

CHAPTER 7 SUPPLEMENTARY EXERCISES

7. (a) $A = \int_a^b (f(x) - g(x))\,dx + \int_b^c (g(x) - f(x))\,dx + \int_c^d (f(x) - g(x))\,dx$

(b) $A = \int_{-1}^0 (x^3 - x)\,dx + \int_0^1 (x - x^3)\,dx + \int_1^2 (x^3 - x)\,dx = \frac{1}{4} + \frac{1}{4} + \frac{9}{4} = \frac{11}{4}$

9. By implicit differentiation $\frac{dy}{dx} = -\left(\frac{y}{x}\right)^{1/3}$, so $1 + \left(\frac{dy}{dx}\right)^2 = 1 + \left(\frac{y}{x}\right)^{2/3} = \frac{x^{2/3} + y^{2/3}}{x^{2/3}} = \frac{a^{2/3}}{x^{2/3}}$,

$$L = \int_{-a}^{-a/8} \frac{a^{1/3}}{(-x^{1/3})}dx = -a^{1/3}\int_{-a}^{-a/8} x^{-1/3}dx = 9a/8.$$

11. Let the sphere have radius R, the hole radius r. By the Pythagorean Theorem, $r^2 + (L/2)^2 = R^2$. Use cylindrical shells to calculate the volume of the solid obtained by rotating about the y-axis the region $r < x < R$, $-\sqrt{R^2 - x^2} < y < \sqrt{R^2 - x^2}$:

$$V = \int_r^R (2\pi x)2\sqrt{R^2 - x^2}\,dx = -\frac{4}{3}\pi(R^2 - x^2)^{3/2}\Big]_r^R = \frac{4}{3}\pi(L/2)^3,$$

so the volume is independent of R.

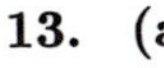

13. (a)

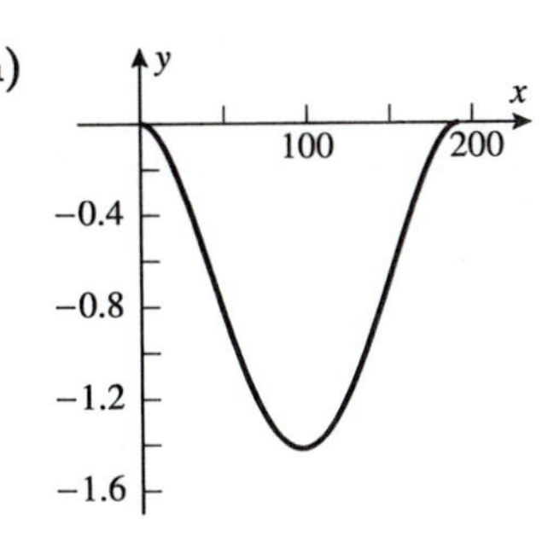

(b) The maximum deflection occurs at $x = 96$ inches (the midpoint of the beam) and is about 1.42 in.

(c) The length of the centerline is $\displaystyle\int_0^{192} \sqrt{1 + (dy/dx)^2}\, dx = 192.026$ in.

15. $x' = e^t(\cos t - \sin t), y' = e^t(\cos t + \sin t), (x')^2 + (y')^2 = 2e^{2t}$

$$S = 2\pi \int_0^{\pi/2} (e^t \sin t)\sqrt{2e^{2t}}dt = 2\sqrt{2}\pi \int_0^{\pi/2} e^{2t} \sin t\, dt$$

$$= 2\sqrt{2}\pi \left[\frac{1}{5}e^{2t}(2\sin t - \cos t)\right]_0^{\pi/2} = \frac{2\sqrt{2}}{5}\pi(2e^{\pi} + 1)$$

17. (a) $F = kx,\ \dfrac{1}{2} = k\dfrac{1}{4},\ k = 2,\ W = \displaystyle\int_0^{1/4} kx\, dx = 1/16$ J

(b) $25 = \displaystyle\int_0^{L} kx\, dx = kL^2/2,\ L = 5$ m

19. (a) $F = \displaystyle\int_0^{1} \rho x 3\, dx$ N

(b) By similar triangles $\dfrac{w(x)}{4} = \dfrac{x}{2},\ w(x) = 2x$, so

$$F = \int_1^{4} \rho(1 + x)2x\, dx \text{ lb/ft}^2.$$

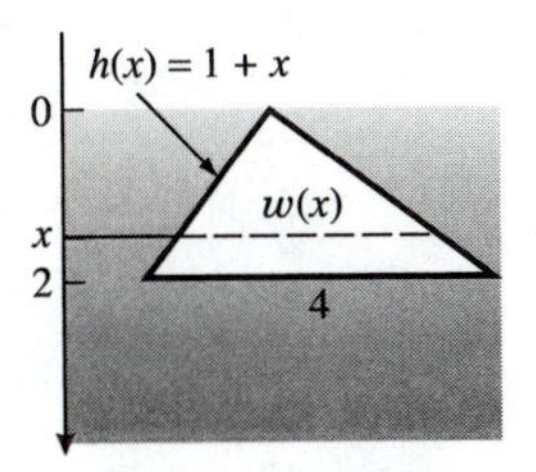

(c) A formula for the parabola is $y = \dfrac{8}{125}x^2 - 10$, so $F = \displaystyle\int_{-10}^{0} 9810|y|2\sqrt{\frac{125}{8}(y + 10)}\, dy$ N.

21. (a)

$$\begin{aligned}\cosh 3x &= \cosh(2x + x) = \cosh 2x \cosh x + \sinh 2x \sinh x\\ &= (2\cosh^2 x - 1)\cosh x + (2\sinh x \cosh x)\sinh x\\ &= 2\cosh^3 x - \cosh x + 2\sinh^2 x \cosh x\\ &= 2\cosh^3 x - \cosh x + 2(\cosh^2 x - 1)\cosh x = 4\cosh^3 x - 3\cosh x\end{aligned}$$

(b) from Theorem 7.8.2 with x replaced by $\dfrac{x}{2}$: $\cosh x = 2\cosh^2 \dfrac{x}{2} - 1$,

$2\cosh^2 \dfrac{x}{2} = \cosh x + 1,\ \cosh^2 \dfrac{x}{2} = \dfrac{1}{2}(\cosh x + 1)$,

$\cosh \dfrac{x}{2} = \sqrt{\dfrac{1}{2}(\cosh x + 1)}$ (because $\cosh \dfrac{x}{2} > 0$)

(c) from Theorem 7.8.2 with x replaced by $\dfrac{x}{2}$: $\cosh x = 2\sinh^2 \dfrac{x}{2} + 1$,

$2\sinh^2 \dfrac{x}{2} = \cosh x - 1,\ \sinh^2 \dfrac{x}{2} = \dfrac{1}{2}(\cosh x - 1),\ \sinh \dfrac{x}{2} = \pm\sqrt{\dfrac{1}{2}(\cosh x - 1)}$

23. Set $a = 68.7672$, $b = 0.0100333$, $c = 693.8597$, $d = 299.2239$.

(a)

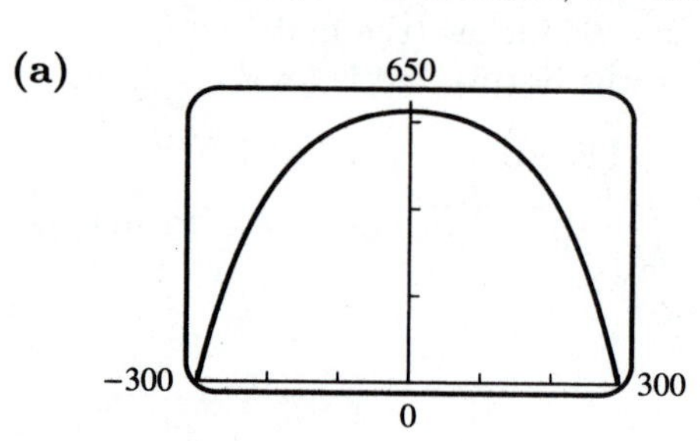

(b) $L = 2\int_0^d \sqrt{1 + a^2b^2\sinh^2 bx}\,dx$

$= 1480.2798$ ft

(c) $x = 283.6249$ ft

(d) $82°$

25. Let (a, k), where $\pi/2 < a < \pi$, be the coordinates of the point of intersection of $y = k$ with $y = \sin x$. Thus $k = \sin a$ and if the shaded areas are equal,

$$\int_0^a (k - \sin x)dx = \int_0^a (\sin a - \sin x)\,dx = a\sin a + \cos a - 1 = 0$$

Solve for a to get $a \approx 2.331122$, so $k = \sin a \approx 0.724611$.

CHAPTER 8

Principles of Integral Evaluation

EXERCISE SET 8.1

1. $u = 3 - 2x, du = -2dx, \quad -\frac{1}{2}\int u^3\,du = -\frac{1}{8}u^4 + C = -\frac{1}{8}(3-2x)^4 + C$

3. $u = x^2, du = 2xdx, \quad \frac{1}{2}\int \sec^2 u\,du = \frac{1}{2}\tan u + C = \frac{1}{2}\tan(x^2) + C$

5. $u = 2 + \cos 3x, du = -3\sin 3xdx, \quad -\frac{1}{3}\int \frac{du}{u} = -\frac{1}{3}\ln|u| + C = -\frac{1}{3}\ln(2+\cos 3x) + C$

7. $u = e^x, du = e^x dx, \quad \int \sinh u\,du = \cosh u + C = \cosh e^x + C$

9. $u = \cot x, du = -\csc^2 xdx, \quad -\int e^u\,du = -e^u + C = -e^{\cot x} + C$

11. $u = \cos 7x, du = -7\sin 7xdx, \quad -\frac{1}{7}\int u^5\,du = -\frac{1}{42}u^6 + C = -\frac{1}{42}\cos^6 7x + C$

13. $u = e^x, du = e^x dx, \quad \int \frac{du}{\sqrt{4+u^2}} = \ln\left(u + \sqrt{u^2+4}\right) + C = \ln\left(e^x + \sqrt{e^{2x}+4}\right) + C$

15. $u = \sqrt{x-2}, du = \frac{1}{2\sqrt{x-2}}\,dx, \quad 2\int e^u\,du = 2e^u + C = 2e^{\sqrt{x-2}} + C$

17. $u = \sqrt{x}, du = \frac{1}{2\sqrt{x}}\,dx, \quad \int 2\cosh u\,du = 2\sinh u + C = 2\sinh\sqrt{x} + C$

19. $u = \sqrt{x}, du = \frac{1}{2\sqrt{x}}\,dx, \quad \int \frac{2\,du}{3^u} = 2\int e^{-u\ln 3}\,du = -\frac{2}{\ln 3}e^{-u\ln 3} + C = -\frac{2}{\ln 3}3^{-\sqrt{x}} + C$

21. $u = \frac{2}{x}, du = -\frac{2}{x^2}\,dx, \quad -\frac{1}{2}\int \operatorname{csch}^2 u\,du = \frac{1}{2}\coth u + C = \frac{1}{2}\coth\frac{2}{x} + C$

23. $u = e^{-x}, du = -e^{-x}dx, \quad -\int \frac{du}{4-u^2} = -\frac{1}{4}\ln\left|\frac{2+u}{2-u}\right| + C = -\frac{1}{4}\ln\left|\frac{2+e^{-x}}{2-e^{-x}}\right| + C$

25. $u = e^x, du = e^x dx, \quad \int \frac{e^x\,dx}{\sqrt{1-e^{2x}}} = \int \frac{du}{\sqrt{1-u^2}} = \sin^{-1} u + C = \sin^{-1} e^x + C$

27. $u = x^2, du = 2xdx, \quad \frac{1}{2}\int \frac{du}{\sec u} = \frac{1}{2}\int \cos u\,du = \frac{1}{2}\sin u + C = \frac{1}{2}\sin(x^2) + C$

29. $4^{-x^2} = e^{-x^2\ln 4}, u = -x^2\ln 4, du = -2x\ln 4\,dx = -x\ln 16\,dx,$

$$-\frac{1}{\ln 16}\int e^u\,du = -\frac{1}{\ln 16}e^u + C = -\frac{1}{\ln 16}e^{-x^2\ln 4} + C = -\frac{1}{\ln 16}4^{-x^2} + C$$

EXERCISE SET 8.2

1. $u = x$, $dv = e^{-x}dx$, $du = dx$, $v = -e^{-x}$; $\int xe^{-x}dx = -xe^{-x} + \int e^{-x}dx = -xe^{-x} - e^{-x} + C$

3. $u = x^2$, $dv = e^x dx$, $du = 2x\,dx$, $v = e^x$; $\int x^2e^x dx = x^2e^x - 2\int xe^x dx$.

For $\int xe^x dx$ use $u = x$, $dv = e^x dx$, $du = dx$, $v = e^x$ to get

$\int xe^x dx = xe^x - e^x + C_1$ so $\int x^2e^x dx = x^2e^x - 2xe^x + 2e^x + C$

5. $u = x$, $dv = \sin 2x\,dx$, $du = dx$, $v = -\frac{1}{2}\cos 2x$;

$\int x\sin 2x\,dx = -\frac{1}{2}x\cos 2x + \frac{1}{2}\int \cos 2x\,dx = -\frac{1}{2}x\cos 2x + \frac{1}{4}\sin 2x + C$

7. $u = x^2$, $dv = \cos x\,dx$, $du = 2x\,dx$, $v = \sin x$; $\int x^2\cos x\,dx = x^2\sin x - 2\int x\sin x\,dx$

For $\int x\sin x\,dx$ use $u = x$, $dv = \sin x\,dx$ to get

$\int x\sin x\,dx = -x\cos x + \sin x + C_1$ so $\int x^2\cos x\,dx = x^2\sin x + 2x\cos x - 2\sin x + C$

9. $u = \ln x$, $dv = \sqrt{x}\,dx$, $du = \frac{1}{x}dx$, $v = \frac{2}{3}x^{3/2}$;

$\int \sqrt{x}\ln x\,dx = \frac{2}{3}x^{3/2}\ln x - \frac{2}{3}\int x^{1/2}dx = \frac{2}{3}x^{3/2}\ln x - \frac{4}{9}x^{3/2} + C$

11. $u = (\ln x)^2$, $dv = dx$, $du = 2\frac{\ln x}{x}dx$, $v = x$; $\int (\ln x)^2 dx = x(\ln x)^2 - 2\int \ln x\,dx$.

Use $u = \ln x$, $dv = dx$ to get $\int \ln x\,dx = x\ln x - \int dx = x\ln x - x + C_1$ so

$\int (\ln x)^2 dx = x(\ln x)^2 - 2x\ln x + 2x + C$

13. $u = \ln(2x+3)$, $dv = dx$, $du = \frac{2}{2x+3}dx$, $v = x$; $\int \ln(2x+3)dx = x\ln(2x+3) - \int \frac{2x}{2x+3}dx$

but $\int \frac{2x}{2x+3}dx = \int \left(1 - \frac{3}{2x+3}\right)dx = x - \frac{3}{2}\ln(2x+3) + C_1$ so

$\int \ln(2x+3)dx = x\ln(2x+3) - x + \frac{3}{2}\ln(2x+3) + C$

15. $u = \sin^{-1}x$, $dv = dx$, $du = 1/\sqrt{1-x^2}dx$, $v = x$;

$\int \sin^{-1}x\,dx = x\sin^{-1}x - \int x/\sqrt{1-x^2}dx = x\sin^{-1}x + \sqrt{1-x^2} + C$

17. $u = \tan^{-1}(2x)$, $dv = dx$, $du = \frac{2}{1+4x^2}dx$, $v = x$;

$\int \tan^{-1}(2x)dx = x\tan^{-1}(2x) - \int \frac{2x}{1+4x^2}dx = x\tan^{-1}(2x) - \frac{1}{4}\ln(1+4x^2) + C$

19. $u = e^x$, $dv = \sin x\,dx$, $du = e^x dx$, $v = -\cos x$; $\int e^x \sin x\,dx = -e^x \cos x + \int e^x \cos x\,dx$.

For $\int e^x \cos x\,dx$ use $u = e^x$, $dv = \cos x\,dx$ to get $\int e^x \cos x = e^x \sin x - \int e^x \sin x\,dx$ so

$$\int e^x \sin x\,dx = -e^x \cos x + e^x \sin x - \int e^x \sin x\,dx,$$

$$2\int e^x \sin x\,dx = e^x(\sin x - \cos x) + C_1,\ \int e^x \sin x\,dx = \frac{1}{2}e^x(\sin x - \cos x) + C$$

21. $u = e^{ax}$, $dv = \sin bx\,dx$, $du = ae^{ax}dx$, $v = -\frac{1}{b}\cos bx \quad (b \neq 0)$;

$$\int e^{ax} \sin bx\,dx = -\frac{1}{b}e^{ax}\cos bx + \frac{a}{b}\int e^{ax}\cos bx\,dx.$$ Use $u = e^{ax}$, $dv = \cos bx\,dx$ to get

$$\int e^{ax}\cos bx\,dx = \frac{1}{b}e^{ax}\sin bx - \frac{a}{b}\int e^{ax}\sin bx\,dx$$ so

$$\int e^{ax}\sin bx\,dx = -\frac{1}{b}e^{ax}\cos bx + \frac{a}{b^2}e^{ax}\sin bx - \frac{a^2}{b^2}\int e^{ax}\sin bx\,dx,$$

$$\int e^{ax}\sin bx\,dx = \frac{e^{ax}}{a^2+b^2}(a\sin bx - b\cos bx) + C$$

23. $u = \sin(\ln x)$, $dv = dx$, $du = \frac{\cos(\ln x)}{x}dx$, $v = x$;

$$\int \sin(\ln x)dx = x\sin(\ln x) - \int \cos(\ln x)dx.$$ Use $u = \cos(\ln x)$, $dv = dx$ to get

$$\int \cos(\ln x)dx = x\cos(\ln x) + \int \sin(\ln x)dx$$ so

$$\int \sin(\ln x)dx = x\sin(\ln x) - x\cos(\ln x) - \int \sin(\ln x)dx,$$

$$\int \sin(\ln x)dx = \frac{1}{2}x[\sin(\ln x) - \cos(\ln x)] + C$$

25. $u = x$, $dv = \sec^2 x\,dx$, $du = dx$, $v = \tan x$;

$$\int x\sec^2 x\,dx = x\tan x - \int \tan x\,dx = x\tan x - \int \frac{\sin x}{\cos x}dx = x\tan x + \ln|\cos x| + C$$

27. $u = x^2$, $dv = xe^{x^2}dx$, $du = 2x\,dx$, $v = \frac{1}{2}e^{x^2}$;

$$\int x^3 e^{x^2}dx = \frac{1}{2}x^2 e^{x^2} - \int xe^{x^2}dx = \frac{1}{2}x^2 e^{x^2} - \frac{1}{2}e^{x^2} + C$$

29. $u = x$, $dv = e^{-5x}dx$, $du = dx$, $v = -\frac{1}{5}e^{-5x}$;

$$\int_0^1 xe^{-5x}dx = -\frac{1}{5}xe^{-5x}\Big]_0^1 + \frac{1}{5}\int_0^1 e^{-5x}dx$$

$$= -\frac{1}{5}e^{-5} - \frac{1}{25}e^{-5x}\Big]_0^1 = -\frac{1}{5}e^{-5} - \frac{1}{25}(e^{-5} - 1) = (1 - 6e^{-5})/25$$

31. $u = \ln x,\ dv = x^2dx,\ du = \frac{1}{x}dx,\ v = \frac{1}{3}x^3;$

$$\int_1^e x^2 \ln x\,dx = \frac{1}{3}x^3 \ln x\Big]_1^e - \frac{1}{3}\int_1^e x^2dx = \frac{1}{3}e^3 - \frac{1}{9}x^3\Big]_1^e = \frac{1}{3}e^3 - \frac{1}{9}(e^3-1) = (2e^3+1)/9$$

33. $u = \ln(x+3),\ dv = dx,\ du = \frac{1}{x+3}dx,\ v = x;$

$$\int_{-2}^2 \ln(x+3)dx = x\ln(x+3)\Big]_{-2}^2 - \int_{-2}^2 \frac{x}{x+3}dx = 2\ln 5 + 2\ln 1 - \int_{-2}^2\left[1 - \frac{3}{x+3}\right]dx$$

$$= 2\ln 5 - [x - 3\ln(x+3)]\Big]_{-2}^2 = 2\ln 5 - (2 - 3\ln 5) + (-2 - 3\ln 1) = 5\ln 5 - 4$$

35. $u = \sec^{-1}\sqrt{\theta},\ dv = d\theta,\ du = \frac{1}{2\theta\sqrt{\theta-1}}d\theta,\ v = \theta;$

$$\int_2^4 \sec^{-1}\sqrt{\theta}d\theta = \theta\sec^{-1}\sqrt{\theta}\Big]_2^4 - \frac{1}{2}\int_2^4 \frac{1}{\sqrt{\theta-1}}d\theta = 4\sec^{-1}2 - 2\sec^{-1}\sqrt{2} - \sqrt{\theta-1}\Big]_2^4$$

$$= 4\left(\frac{\pi}{3}\right) - 2\left(\frac{\pi}{4}\right) - \sqrt{3} + 1 = \frac{5\pi}{6} - \sqrt{3} + 1$$

37. $u = x,\ dv = \sin 4x\,dx,\ du = dx,\ v = -\frac{1}{4}\cos 4x;$

$$\int_0^{\pi/2} x\sin 4x\,dx = -\frac{1}{4}x\cos 4x\Big]_0^{\pi/2} + \frac{1}{4}\int_0^{\pi/2}\cos 4x\,dx = -\pi/8 + \frac{1}{16}\sin 4x\Big]_0^{\pi/2} = -\pi/8$$

39. $u = \tan^{-1}\sqrt{x},\ dv = \sqrt{x}dx,\ du = \frac{1}{2\sqrt{x}(1+x)}dx,\ v = \frac{2}{3}x^{3/2};$

$$\int_1^3 \sqrt{x}\tan^{-1}\sqrt{x}dx = \frac{2}{3}x^{3/2}\tan^{-1}\sqrt{x}\Big]_1^3 - \frac{1}{3}\int_1^3 \frac{x}{1+x}dx$$

$$= \frac{2}{3}x^{3/2}\tan^{-1}\sqrt{x}\Big]_1^3 - \frac{1}{3}\int_1^3\left[1 - \frac{1}{1+x}\right]dx$$

$$= \left[\frac{2}{3}x^{3/2}\tan^{-1}\sqrt{x} - \frac{1}{3}x + \frac{1}{3}\ln|1+x|\right]_1^3 = (2\sqrt{3}\pi - \pi/2 - 2 + \ln 2)/3$$

41. $t = \sqrt{x},\ t^2 = x,\ dx = 2t\,dt$

(a) $\int e^{\sqrt{x}}dx = 2\int te^t\,dt;\ u = t, dv = e^tdt, du = dt, v = e^t,$

$$\int e^{\sqrt{x}}dx = 2te^t - 2\int e^t\,dt = 2(t-1)e^t + C = 2(\sqrt{x}-1)e^{\sqrt{x}} + C$$

(b) $\int \cos\sqrt{x}\,dx = 2\int t\cos t\,dt;\ u = t, dv = \cos t dt, du = dt, v = \sin t,$

$$\int \cos\sqrt{x}\,dx = 2t\sin t - 2\int \sin t dt = 2t\sin t + 2\cos t + C = 2\sqrt{x}\sin\sqrt{x} + 2\cos\sqrt{x} + C$$

43.

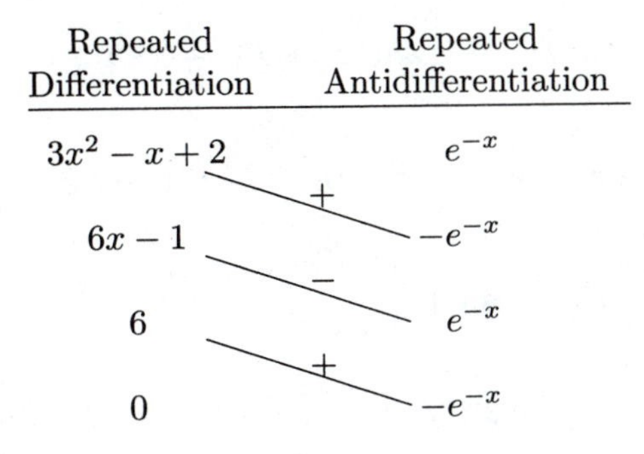

Repeated Differentiation
Repeated Antidifferentiation
$3x^2 - x + 2$
e^{-x}
+
$6x - 1$
$-e^{-x}$
−
6
e^{-x}
+
0
$-e^{-x}$

$$\int (3x^2 - x + 2)e^{-x} = -(3x^2 - x + 2)e^{-x} - (6x - 1)e^{-x} - 6e^{-x} + C = -e^{-x}[3x^2 + 5x + 7] + C$$

45.

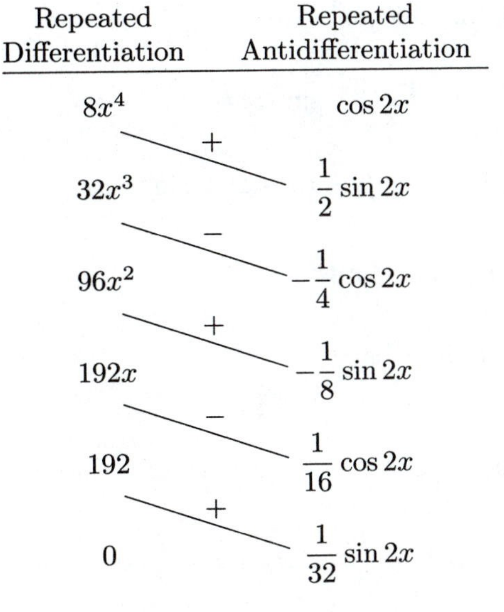

Repeated Differentiation
Repeated Antidifferentiation
$8x^4$
$\cos 2x$
+
$32x^3$
$\frac{1}{2}\sin 2x$
−
$96x^2$
$-\frac{1}{4}\cos 2x$
+
$192x$
$-\frac{1}{8}\sin 2x$
−
192
$\frac{1}{16}\cos 2x$
+
0
$\frac{1}{32}\sin 2x$

$$\int 8x^4 \cos 2x \, dx = (4x^4 - 12x^2 + 6)\sin 2x + (8x^3 - 12x)\cos 2x + C$$

47. **(a)** $A = \int_1^e \ln x \, dx = (x \ln x - x)\Big]_1^e = 1$

(b) $V = \pi \int_1^e (\ln x)^2 dx = \pi\Big[(x(\ln x)^2 - 2x \ln x + 2x)\Big]_1^e = \pi(e - 2)$

49. $V = 2\pi \int_0^\pi x \sin x \, dx = 2\pi(-x\cos x + \sin x)\Big]_0^\pi = 2\pi^2$

51. distance $= \int_0^5 t^2 e^{-t} dt; u = t^2, dv = e^{-t}dt, du = 2tdt, v = -e^{-t},$

distance $= -t^2 e^{-t}\Big]_0^5 + 2\int_0^5 te^{-t}dt; u = 2t, dv = e^{-t}dt, du = 2dt, v = -e^{-t},$

distance $= -25e^{-5} - 2te^{-t}\Big]_0^5 + 2\int_0^5 e^{-t}dt = -25e^{-5} - 10e^{-5} - 2e^{-t}\Big]_0^5$

$= -25e^{-5} - 10e^{-5} - 2e^{-5} + 2 = -37e^{-5} + 2$

53. (a) $\int \sin^3 x\,dx = -\frac{1}{3}\sin^2 x\cos x + \frac{2}{3}\int \sin x\,dx = -\frac{1}{3}\sin^2 x\cos x - \frac{2}{3}\cos x + C$

(b) $\int \sin^4 x\,dx = -\frac{1}{4}\sin^3 x\cos x + \frac{3}{4}\int \sin^2 x\,dx$, $\int \sin^2 x\,dx = -\frac{1}{2}\sin x\cos x + \frac{1}{2}x + C_1$ so

$$\int_0^{\pi/4} \sin^4 x\,dx = \left[-\frac{1}{4}\sin^3 x\cos x - \frac{3}{8}\sin x\cos x + \frac{3}{8}x\right]_0^{\pi/4}$$

$$= -\frac{1}{4}(1/\sqrt{2})^3(1/\sqrt{2}) - \frac{3}{8}(1/\sqrt{2})(1/\sqrt{2}) + 3\pi/32 = 3\pi/32 - 1/4$$

55. $u = \sin^{n-1} x$, $dv = \sin x\,dx$, $du = (n-1)\sin^{n-2} x\cos x\,dx$, $v = -\cos x$;

$$\int \sin^n x\,dx = -\sin^{n-1} x\cos x + (n-1)\int \sin^{n-2} x\cos^2 x\,dx$$

$$= -\sin^{n-1} x\cos x + (n-1)\int \sin^{n-2} x\,(1-\sin^2 x)dx$$

$$= -\sin^{n-1} x\cos x + (n-1)\int \sin^{n-2} x\,dx - (n-1)\int \sin^n x\,dx,$$

$$n\int \sin^n x\,dx = -\sin^{n-1} x\cos x + (n-1)\int \sin^{n-2} x\,dx,$$

$$\int \sin^n x\,dx = -\frac{1}{n}\sin^{n-1} x\cos x + \frac{n-1}{n}\int \sin^{n-2} x\,dx$$

57. (a) $\int \tan^4 x\,dx = \frac{1}{3}\tan^3 x - \int \tan^2 x\,dx = \frac{1}{3}\tan^3 x - \tan x + \int dx = \frac{1}{3}\tan^3 x - \tan x + x + C$

(b) $\int \sec^4 x\,dx = \frac{1}{3}\sec^2 x\tan x + \frac{2}{3}\int \sec^2 x\,dx = \frac{1}{3}\sec^2 x\tan x + \frac{2}{3}\tan x + C$

(c) $\int x^3 e^x dx = x^3 e^x - 3\int x^2 e^x dx = x^3 e^x - 3\left[x^2 e^x - 2\int x e^x dx\right]$

$$= x^3 e^x - 3x^2 e^x + 6\left[x e^x - \int e^x dx\right] = x^3 e^x - 3x^2 e^x + 6x e^x - 6e^x + C$$

59. $u = x$, $dv = f''(x)dx$, $du = dx$, $v = f'(x)$;

$$\int_{-1}^{1} x f''(x)dx = x f'(x)\Big]_{-1}^{1} - \int_{-1}^{1} f'(x)dx$$

$$= f'(1) + f'(-1) - f(x)\Big]_{-1}^{1} = f'(1) + f'(-1) - f(1) + f(-1)$$

61. (a) Use Exercise 60(c);

$$\int_0^{1/2} \sin^{-1} x\,dx = \frac{1}{2}\sin^{-1}\left(\frac{1}{2}\right) - 0\cdot\sin^{-1} 0 - \int_{\sin^{-1}(0)}^{\sin^{-1}(1/2)} \sin x\,dx = \frac{1}{2}\sin^{-1}\left(\frac{1}{2}\right) - \int_0^{\pi/6} \sin x\,dx$$

(b) Use Exercise 60(b);

$$\int_e^{e^2} \ln x\,dx = e^2\ln e^2 - e\ln e - \int_{\ln e}^{\ln e^2} f^{-1}(y)\,dy = 2e^2 - e - \int_1^2 e^y\,dy = 2e^2 - e - \int_1^2 e^x\,dx$$

63. $u = \ln(x+1), dv = dx, du = \dfrac{dx}{x+1}, v = x+1;$

$$\int \ln(x+1)\,dx = \int u\,dv = uv - \int v\,du = (x+1)\ln(x+1) - \int dx = (x+1)\ln(x+1) - x + C$$

65. $u = \tan^{-1} x, dv = x\,dx, du = \dfrac{1}{1+x^2}\,dx, v = \dfrac{1}{2}(x^2+1)$

$$\int x\tan^{-1} x\,dx = \int u\,dv = uv - \int v\,du = \frac{1}{2}(x^2+1)\tan^{-1} x - \frac{1}{2}\int dx$$
$$= \frac{1}{2}(x^2+1)\tan^{-1} x - \frac{1}{2}x + C$$

EXERCISE SET 8.3

1. $u = \cos x, \ -\displaystyle\int u^5 du = -\frac{1}{6}\cos^6 x + C$

3. $u = \sin ax, \ \dfrac{1}{a}\displaystyle\int u\,du = \frac{1}{2a}\sin^2 ax + C, \quad a \neq 0$

5. $\displaystyle\int \sin^2 5\theta\,d\theta = \frac{1}{2}\int (1 - \cos 10\theta)d\theta = \frac{1}{2}\theta - \frac{1}{20}\sin 10\theta + C$

7. $\displaystyle\int \cos^5 \theta d\theta = \int (1 - \sin^2\theta)^2 \cos\theta d\theta = \int (1 - 2\sin^2\theta + \sin^4\theta)\cos\theta d\theta$

$$= \sin\theta - \frac{2}{3}\sin^3\theta + \frac{1}{5}\sin^5\theta + C$$

9. $\displaystyle\int \sin^2 2t\cos^3 2t\,dt = \int \sin^2 2t(1 - \sin^2 2t)\cos 2t\,dt = \int (\sin^2 2t - \sin^4 2t)\cos 2t\,dt$

$$= \frac{1}{6}\sin^3 2t - \frac{1}{10}\sin^5 2t + C$$

11. $\displaystyle\int \sin^2 x\cos^2 x\,dx = \frac{1}{4}\int \sin^2 2x\,dx = \frac{1}{8}\int (1 - \cos 4x)dx = \frac{1}{8}x - \frac{1}{32}\sin 4x + C$

13. $\displaystyle\int \sin x\cos 2x\,dx = \frac{1}{2}\int (\sin 3x - \sin x)dx = -\frac{1}{6}\cos 3x + \frac{1}{2}\cos x + C$

15. $\displaystyle\int \sin x\cos(x/2)dx = \frac{1}{2}\int [\sin(3x/2) + \sin(x/2)]dx = -\frac{1}{3}\cos(3x/2) - \cos(x/2) + C$

17. $\displaystyle\int_0^{\pi/4} \cos^3 x\,dx = \int_0^{\pi/4} (1 - \sin^2 x)\cos x\,dx$

$$= \left[\sin x - \frac{1}{3}\sin^3 x\right]_0^{\pi/4} = (\sqrt{2}/2) - \frac{1}{3}(\sqrt{2}/2)^3 = 5\sqrt{2}/12$$

19. $\displaystyle\int_0^{\pi/3} \sin^4 3x\cos^3 3x\,dx = \int_0^{\pi/3} \sin^4 3x(1 - \sin^2 3x)\cos 3x\,dx = \left[\frac{1}{15}\sin^5 3x - \frac{1}{21}\sin^7 3x\right]_0^{\pi/3} = 0$

21. $\displaystyle\int_0^{\pi/6} \sin 2x \cos 4x\, dx = \frac{1}{2}\int_0^{\pi/6}(\sin 6x - \sin 2x)dx = \left[-\frac{1}{12}\cos 6x + \frac{1}{4}\cos 2x\right]_0^{\pi/6}$
$= [(-1/12)(-1) + (1/4)(1/2)] - [-1/12 + 1/4] = 1/24$

23. $\displaystyle\frac{1}{3}\tan(3x+1) + C$

25. $u = e^{-2x}$, $du = -2e^{-2x}\,dx$; $\displaystyle -\frac{1}{2}\int \tan u\, du = \frac{1}{2}\ln|\cos u| + C = \frac{1}{2}\ln|\cos(e^{-2x})| + C$

27. $\displaystyle\frac{1}{2}\ln|\sec 2x + \tan 2x| + C$

29. $u = \tan x$, $\displaystyle\int u^2 du = \frac{1}{3}\tan^3 x + C$

31. $\displaystyle\int \tan^3 4x(1+\tan^2 4x)\sec^2 4x\, dx = \int(\tan^3 4x + \tan^5 4x)\sec^2 4x\, dx = \frac{1}{16}\tan^4 4x + \frac{1}{24}\tan^6 4x + C$

33. $\displaystyle\int \sec^4 x(\sec^2 x - 1)\sec x \tan x\, dx = \int(\sec^6 x - \sec^4 x)\sec x \tan x\, dx = \frac{1}{7}\sec^7 x - \frac{1}{5}\sec^5 x + C$

35. $\displaystyle\int(\sec^2 x - 1)^2 \sec x\, dx = \int(\sec^5 x - 2\sec^3 x + \sec x)dx = \int \sec^5 x\, dx - 2\int \sec^3 x\, dx + \int \sec x\, dx$

$\displaystyle= \frac{1}{4}\sec^3 x \tan x + \frac{3}{4}\int \sec^3 x\, dx - 2\int \sec^3 x\, dx + \ln|\sec x + \tan x|$

$\displaystyle= \frac{1}{4}\sec^3 x \tan x - \frac{5}{4}\left[\frac{1}{2}\sec x \tan x + \frac{1}{2}\ln|\sec x + \tan x|\right] + \ln|\sec x + \tan x| + C$

$\displaystyle= \frac{1}{4}\sec^3 x \tan x - \frac{5}{8}\sec x \tan x + \frac{3}{8}\ln|\sec x + \tan x| + C$

37. $\displaystyle\int \sec^2 2t(\sec 2t \tan 2t)dt = \frac{1}{6}\sec^3 2t + C$

39. $\displaystyle\int \sec^4 x\, dx = \int(1+\tan^2 x)\sec^2 x\, dx = \int(\sec^2 x + \tan^2 x \sec^2 x)dx = \tan x + \frac{1}{3}\tan^3 x + C$

41. Use equation (19) to get $\displaystyle\int \tan^4 x\, dx = \frac{1}{3}\tan^3 x - \tan x + x + C$

43. $\displaystyle\int \sqrt{\tan x}(1+\tan^2 x)\sec^2 x\, dx = \frac{2}{3}\tan^{3/2} x + \frac{2}{7}\tan^{7/2} x + C$

45. $\displaystyle\int_0^{\pi/6}(\sec^2 2x - 1)dx = \left[\frac{1}{2}\tan 2x - x\right]_0^{\pi/6} = \sqrt{3}/2 - \pi/6$

47. $u = x/2$,

$\displaystyle 2\int_0^{\pi/4} \tan^5 u\, du = \left[\frac{1}{2}\tan^4 u - \tan^2 u - 2\ln|\cos u|\right]_0^{\pi/4} = 1/2 - 1 - 2\ln(1/\sqrt{2}) = -1/2 + \ln 2$

49. $\displaystyle\int(\csc^2 x - 1)\csc^2 x(\csc x \cot x)dx = \int(\csc^4 x - \csc^2 x)(\csc x \cot x)dx = -\frac{1}{5}\csc^5 x + \frac{1}{3}\csc^3 x + C$

51. $\displaystyle\int(\csc^2 x - 1)\cot x\, dx = \int \csc x(\csc x \cot x)dx - \int \frac{\cos x}{\sin x}dx = -\frac{1}{2}\csc^2 x - \ln|\sin x| + C$

53. (a) $\displaystyle\int_0^{2\pi} \sin mx \cos nx\,dx = \frac{1}{2}\int_0^{2\pi}[\sin(m+n)x+\sin(m-n)x]dx = \left[-\frac{\cos(m+n)x}{2(m+n)} - \frac{\cos(m-n)x}{2(m-n)}\right]_0^{2\pi}$

but $\displaystyle\cos(m+n)x\Big]_0^{2\pi} = 0, \cos(m-n)x\Big]_0^{2\pi} = 0.$

(b) $\displaystyle\int_0^{2\pi} \cos mx \cos nx\,dx = \frac{1}{2}\int_0^{2\pi}[\cos(m+n)x + \cos(m-n)x]dx;$

since $m \neq n$, evaluate sin at integer multiples of 2π to get 0.

(c) $\displaystyle\int_0^{2\pi} \sin mx \sin nx\,dx = \frac{1}{2}\int_0^{2\pi}[\cos(m-n)x - \cos(m+n)x]\,dx;$

since $m \neq n$, evaluate sin at integer multiples of 2π to get 0.

55. $y' = \tan x$, $1 + (y')^2 = 1 + \tan^2 x = \sec^2 x$,

$$L = \int_0^{\pi/4}\sqrt{\sec^2 x}\,dx = \int_0^{\pi/4}\sec x\,dx = \ln|\sec x + \tan x|\Big]_0^{\pi/4} = \ln(\sqrt{2}+1)$$

57. $\displaystyle V = \pi\int_0^{\pi/4}(\cos^2 x - \sin^2 x)dx = \pi\int_0^{\pi/4}\cos 2x\,dx = \frac{1}{2}\pi\sin 2x\Big]_0^{\pi/4} = \pi/2$

59. With $0 < \alpha < \beta, D = D_\beta - D_\alpha = \displaystyle\frac{L}{2\pi}\int_\alpha^\beta \sec x\,dx = \frac{L}{2\pi}\ln|\sec x + \tan x|\Big]_\alpha^\beta = \frac{L}{2\pi}\ln\left|\frac{\sec\beta + \tan\beta}{\sec\alpha + \tan\alpha}\right|$

61. (a) $\displaystyle\int \csc x\,dx = \int \sec(\pi/2 - x)dx = -\ln|\sec(\pi/2 - x) + \tan(\pi/2 - x)| + C$

$$= -\ln|\csc x + \cot x| + C$$

(b) $\displaystyle -\ln|\csc x + \cot x| = \ln\frac{1}{|\csc x + \cot x|} = \ln\frac{|\csc x - \cot x|}{|\csc^2 x - \cot^2 x|} = \ln|\csc x - \cot x|,$

$$-\ln|\csc x + \cot x| = -\ln\left|\frac{1}{\sin x} + \frac{\cos x}{\sin x}\right| = \ln\left|\frac{\sin x}{1+\cos x}\right|$$

$$= \ln\left|\frac{2\sin(x/2)\cos(x/2)}{2\cos^2(x/2)}\right| = \ln|\tan(x/2)|$$

63. $\displaystyle a\sin x + b\cos x = \sqrt{a^2+b^2}\left[\frac{a}{\sqrt{a^2+b^2}}\sin x + \frac{b}{\sqrt{a^2+b^2}}\cos x\right] = \sqrt{a^2+b^2}(\sin x\cos\theta + \cos x\sin\theta)$

where $\cos\theta = a/\sqrt{a^2+b^2}$ and $\sin\theta = b/\sqrt{a^2+b^2}$ so $a\sin x + b\cos x = \sqrt{a^2+b^2}\sin(x+\theta)$

and $\displaystyle\int\frac{dx}{a\sin x + b\cos x} = \frac{1}{\sqrt{a^2+b^2}}\int\csc(x+\theta)dx = -\frac{1}{\sqrt{a^2+b^2}}\ln|\csc(x+\theta) + \cot(x+\theta)| + C$

$$= -\frac{1}{\sqrt{a^2+b^2}}\ln\left|\frac{\sqrt{a^2+b^2} + a\cos x - b\sin x}{a\sin x + b\cos x}\right| + C$$

65. (a) $\displaystyle\int_0^{\pi/2}\sin^3 x\,dx = \frac{2}{3}$

(b) $\displaystyle\int_0^{\pi/2}\sin^4 x\,dx = \frac{1\cdot 3}{2\cdot 4}\cdot\frac{\pi}{2} = 3\pi/16$

(c) $\displaystyle\int_0^{\pi/2}\sin^5 x\,dx = \frac{2\cdot 4}{3\cdot 5} = 8/15$

(d) $\displaystyle\int_0^{\pi/2}\sin^6 x\,dx = \frac{1\cdot 3\cdot 5}{2\cdot 4\cdot 6}\cdot\frac{\pi}{2} = 5\pi/32$

EXERCISE SET 8.4

1. $x = 2\sin\theta,\ dx = 2\cos\theta\, d\theta,$

$$4\int \cos^2\theta\, d\theta = 2\int (1+\cos 2\theta)d\theta = 2\theta + \sin 2\theta + C$$

$$= 2\theta + 2\sin\theta\cos\theta + C = 2\sin^{-1}(x/2) + \frac{1}{2}x\sqrt{4-x^2} + C$$

3. $x = 3\sin\theta,\ dx = 3\cos\theta\, d\theta,$

$$9\int \sin^2\theta\, d\theta = \frac{9}{2}\int (1-\cos 2\theta)d\theta = \frac{9}{2}\theta - \frac{9}{4}\sin 2\theta + C = \frac{9}{2}\theta - \frac{9}{2}\sin\theta\cos\theta + C$$

$$= \frac{9}{2}\sin^{-1}(x/3) - \frac{1}{2}x\sqrt{9-x^2} + C$$

5. $x = 2\tan\theta,\ dx = 2\sec^2\theta\, d\theta,$

$$\frac{1}{8}\int \frac{1}{\sec^2\theta}d\theta = \frac{1}{8}\int \cos^2\theta\, d\theta = \frac{1}{16}\int (1+\cos 2\theta)d\theta = \frac{1}{16}\theta + \frac{1}{32}\sin 2\theta + C$$

$$= \frac{1}{16}\theta + \frac{1}{16}\sin\theta\cos\theta + C = \frac{1}{16}\tan^{-1}\frac{x}{2} + \frac{x}{8(4+x^2)} + C$$

7. $x = 3\sec\theta,\ dx = 3\sec\theta\tan\theta\, d\theta,$

$$3\int \tan^2\theta\, d\theta = 3\int (\sec^2\theta - 1)d\theta = 3\tan\theta - 3\theta + C = \sqrt{x^2-9} - 3\sec^{-1}\frac{x}{3} + C$$

9. $x = \sqrt{2}\sin\theta,\ dx = \sqrt{2}\cos\theta\, d\theta,$

$$2\sqrt{2}\int \sin^3\theta\, d\theta = 2\sqrt{2}\int \left[1-\cos^2\theta\right]\sin\theta\, d\theta$$

$$= 2\sqrt{2}\left(-\cos\theta + \frac{1}{3}\cos^3\theta\right) + C = -2\sqrt{2-x^2} + \frac{1}{3}(2-x^2)^{3/2} + C$$

11. $x = \frac{3}{2}\sec\theta,\ dx = \frac{3}{2}\sec\theta\tan\theta\, d\theta,\ \frac{2}{9}\int \frac{1}{\sec\theta}d\theta = \frac{2}{9}\int \cos\theta\, d\theta = \frac{2}{9}\sin\theta + C = \frac{\sqrt{4x^2-9}}{9x} + C$

13. $x = \sin\theta,\ dx = \cos\theta\, d\theta,\ \int \frac{1}{\cos^2\theta}d\theta = \int \sec^2\theta\, d\theta = \tan\theta + C = x/\sqrt{1-x^2} + C$

15. $x = \sec\theta,\ dx = \sec\theta\tan\theta\, d\theta,\ \int \sec\theta\, d\theta = \ln|\sec\theta + \tan\theta| + C = \ln\left|x + \sqrt{x^2-1}\right| + C$

17. $x = \frac{1}{3}\sec\theta,\ dx = \frac{1}{3}\sec\theta\tan\theta\, d\theta,$

$$\frac{1}{3}\int \frac{\sec\theta}{\tan^2\theta}d\theta = \frac{1}{3}\int \csc\theta\cot\theta\, d\theta = -\frac{1}{3}\csc\theta + C = -x/\sqrt{9x^2-1} + C$$

19. $e^x = \sin\theta,\ e^x dx = \cos\theta\, d\theta,$

$$\int \cos^2\theta\, d\theta = \frac{1}{2}\int (1+\cos 2\theta)d\theta = \frac{1}{2}\theta + \frac{1}{4}\sin 2\theta + C = \frac{1}{2}\sin^{-1}(e^x) + \frac{1}{2}e^x\sqrt{1-e^{2x}} + C$$

21. $x = 4\sin\theta$, $dx = 4\cos\theta\, d\theta$,

$$1024\int_0^{\pi/2} \sin^3\theta\cos^2\theta\, d\theta = 1024\left[-\frac{1}{3}\cos^3\theta + \frac{1}{5}\cos^5\theta\right]_0^{\pi/2} = 1024(1/3 - 1/5) = 2048/15$$

23. $x = \sec\theta$, $dx = \sec\theta\tan\theta\, d\theta$, $\displaystyle\int_{\pi/4}^{\pi/3} \frac{1}{\sec\theta}d\theta = \int_{\pi/4}^{\pi/3} \cos\theta\, d\theta = \sin\theta\Big]_{\pi/4}^{\pi/3} = (\sqrt{3} - \sqrt{2})/2$

25. $x = \sqrt{3}\tan\theta$, $dx = \sqrt{3}\sec^2\theta\, d\theta$,

$$\frac{1}{9}\int_{\pi/6}^{\pi/3} \frac{\sec\theta}{\tan^4\theta}d\theta = \frac{1}{9}\int_{\pi/6}^{\pi/3} \frac{\cos^3\theta}{\sin^4\theta}d\theta = \frac{1}{9}\int_{\pi/6}^{\pi/3} \frac{1-\sin^2\theta}{\sin^4\theta}\cos\theta\, d\theta = \frac{1}{9}\int_{1/2}^{\sqrt{3}/2} \frac{1-u^2}{u^4}du \ (u = \sin\theta)$$

$$= \frac{1}{9}\int_{1/2}^{\sqrt{3}/2} (u^{-4} - u^{-2})du = \frac{1}{9}\left[-\frac{1}{3u^3} + \frac{1}{u}\right]_{1/2}^{\sqrt{3}/2} = \frac{10\sqrt{3}+18}{243}$$

27. $u = x^2 + 4$, $du = 2x\, dx$,

$\displaystyle\frac{1}{2}\int \frac{1}{u}du = \frac{1}{2}\ln|u| + C = \frac{1}{2}\ln(x^2+4) + C$; or $x = 2\tan\theta$, $dx = 2\sec^2\theta\, d\theta$,

$$\int \tan\theta\, d\theta = \ln|\sec\theta| + C_1 = \ln\frac{\sqrt{x^2+4}}{2} + C_1 = \ln(x^2+4)^{1/2} - \ln 2 + C_1$$

$$= \frac{1}{2}\ln(x^2+4) + C \text{ with } C = C_1 - \ln 2$$

29. $y' = \dfrac{1}{x}$, $1 + (y')^2 = 1 + \dfrac{1}{x^2} = \dfrac{x^2+1}{x^2}$,

$$L = \int_1^2 \sqrt{\frac{x^2+1}{x^2}}dx = \int_1^2 \frac{\sqrt{x^2+1}}{x}dx; \ x = \tan\theta,\ dx = \sec^2\theta\, d\theta,$$

$$L = \int_{\pi/4}^{\tan^{-1}(2)} \frac{\sec^3\theta}{\tan\theta}d\theta = \int_{\pi/4}^{\tan^{-1}(2)} \frac{\tan^2\theta+1}{\tan\theta}\sec\theta\, d\theta = \int_{\pi/4}^{\tan^{-1}(2)} (\sec\theta\tan\theta + \csc\theta)d\theta$$

$$= \Big[\sec\theta + \ln|\csc\theta - \cot\theta|\Big]_{\pi/4}^{\tan^{-1}(2)} = \sqrt{5} + \ln\left(\frac{\sqrt{5}}{2} - \frac{1}{2}\right) - \left[\sqrt{2} + \ln|\sqrt{2}-1|\right]$$

$$= \sqrt{5} - \sqrt{2} + \ln\frac{2+2\sqrt{2}}{1+\sqrt{5}}$$

31. $y' = 2x$, $1 + (y')^2 = 1 + 4x^2$,

$$S = 2\pi\int_0^1 x^2\sqrt{1+4x^2}dx; \ x = \frac{1}{2}\tan\theta,\ dx = \frac{1}{2}\sec^2\theta\, d\theta,$$

$$S = \frac{\pi}{4}\int_0^{\tan^{-1}2} \tan^2\theta\sec^3\theta\, d\theta = \frac{\pi}{4}\int_0^{\tan^{-1}2} (\sec^2\theta - 1)\sec^3\theta\, d\theta = \frac{\pi}{4}\int_0^{\tan^{-1}2} (\sec^5\theta - \sec^3\theta)d\theta$$

$$= \frac{\pi}{4}\left[\frac{1}{4}\sec^3\theta\tan\theta - \frac{1}{8}\sec\theta\tan\theta - \frac{1}{8}\ln|\sec\theta + \tan\theta|\right]_0^{\tan^{-1}2} = \frac{\pi}{32}[18\sqrt{5} - \ln(2+\sqrt{5})]$$

33. (a) $x = 3\sinh u$, $dx = 3\cosh u\, du$, $\displaystyle\int du = u + C = \sinh^{-1}(x/3) + C$

(b) $x = 3\tan\theta$, $dx = 3\sec^2\theta\, d\theta$,

$$\int \sec\theta\, d\theta = \ln|\sec\theta + \tan\theta| + C = \ln\left(\sqrt{x^2+9}/3 + x/3\right) + C$$

but $\sinh^{-1}(x/3) = \ln\left(x/3 + \sqrt{x^2/9+1}\right) = \ln\left(x/3 + \sqrt{x^2+9}/3\right)$ so the results agree.

(c) $x = \cosh u$, $dx = \sinh u\, du$,

$$\int \sinh^2 u\, du = \frac{1}{2}\int (\cosh 2u - 1)du = \frac{1}{4}\sinh 2u - \frac{1}{2}u + C$$

$$= \frac{1}{2}\sinh u \cosh u - \frac{1}{2}u + C = \frac{1}{2}x\sqrt{x^2-1} - \frac{1}{2}\cosh^{-1} x + C$$

because $\cosh u = x$, and $\sinh u = \sqrt{\cosh^2 u - 1} = \sqrt{x^2-1}$

35. $\displaystyle\int \frac{1}{(x-2)^2+9}dx = \frac{1}{3}\tan^{-1}\left(\frac{x-2}{3}\right) + C$

37. $\displaystyle\int \frac{1}{\sqrt{9-(x-1)^2}}dx = \sin^{-1}\left(\frac{x-1}{3}\right) + C$

39. $\displaystyle\int \frac{1}{\sqrt{(x-3)^2+1}}dx = \ln\left(x-3+\sqrt{(x-3)^2+1}\right) + C$

41. $\displaystyle\int \sqrt{4-(x+1)^2}dx$, let $x+1 = 2\sin\theta$,

$$4\int \cos^2\theta\, d\theta = 2\theta + \sin 2\theta + C = 2\theta + 2\sin\theta\cos\theta + C$$

$$= 2\sin^{-1}\left(\frac{x+1}{2}\right) + \frac{1}{2}(x+1)\sqrt{3-2x-x^2} + C$$

43. $\displaystyle\int \frac{1}{2(x+1)^2+5}dx = \frac{1}{2}\int \frac{1}{(x+1)^2+5/2}dx = \frac{1}{\sqrt{10}}\tan^{-1}\sqrt{2/5}(x+1) + C$

45. $\displaystyle\int_1^2 \frac{1}{\sqrt{4x-x^2}}\,dx = \int_1^2 \frac{1}{\sqrt{4-(x-2)^2}}\,dx = \sin^{-1}\frac{x-2}{2}\bigg]_1^2 = \pi/6$

47. $u = \sin^2 x$, $du = 2\sin x\cos x\, dx$;

$$\frac{1}{2}\int \sqrt{1-u^2}\, du = \frac{1}{4}\left[u\sqrt{1-u^2} + \sin^{-1} u\right] + C = \frac{1}{4}\left[\sin^2 x\sqrt{1-\sin^4 x} + \sin^{-1}(\sin^2 x)\right] + C$$

EXERCISE SET 8.5

1. $\displaystyle\frac{A}{(x-2)} + \frac{B}{(x+5)}$

3. $\displaystyle\frac{2x-3}{x^2(x-1)} = \frac{A}{x} + \frac{B}{x^2} + \frac{C}{x-1}$

5. $\displaystyle\frac{A}{x} + \frac{B}{x^2} + \frac{C}{x^3} + \frac{Dx+E}{x^2+1}$

7. $\displaystyle\frac{Ax+B}{x^2+5} + \frac{Cx+D}{(x^2+5)^2}$

9. $\dfrac{1}{(x+4)(x-1)} = \dfrac{A}{x+4} + \dfrac{B}{x-1}$; $A = -\dfrac{1}{5}$, $B = \dfrac{1}{5}$ so

$$-\frac{1}{5}\int \frac{1}{x+4}dx + \frac{1}{5}\int \frac{1}{x-1}dx = -\frac{1}{5}\ln|x+4| + \frac{1}{5}\ln|x-1| + C = \frac{1}{5}\ln\left|\frac{x-1}{x+4}\right| + C$$

11. $\dfrac{11x+17}{(2x-1)(x+4)} = \dfrac{A}{2x-1} + \dfrac{B}{x+4}$; $A = 5$, $B = 3$ so

$$5\int \frac{1}{2x-1}dx + 3\int \frac{1}{x+4}dx = \frac{5}{2}\ln|2x-1| + 3\ln|x+4| + C$$

13. $\dfrac{2x^2-9x-9}{x(x+3)(x-3)} = \dfrac{A}{x} + \dfrac{B}{x+3} + \dfrac{C}{x-3}$; $A = 1$, $B = 2$, $C = -1$ so

$$\int \frac{1}{x}\,dx + 2\int \frac{1}{x+3}\,dx - \int \frac{1}{x-3}\,dx = \ln|x| + 2\ln|x+3| - \ln|x-3| + C = \ln\left|\frac{x(x+3)^2}{x-3}\right| + C$$

Note that the symbol C has been recycled; to save space this recycling is usually not mentioned.

15. $\dfrac{x^2+2}{x+2} = x - 2 + \dfrac{6}{x+2}$, $\displaystyle\int \left(x - 2 + \frac{6}{x+2}\right) dx = \frac{1}{2}x^2 - 2x + 6\ \ln|x+2| + C$

17. $\dfrac{3x^2-10}{x^2-4x+4} = 3 + \dfrac{12x-22}{x^2-4x+4}$, $\dfrac{12x-22}{(x-2)^2} = \dfrac{A}{x-2} + \dfrac{B}{(x-2)^2}$; $A = 12$, $B = 2$ so

$$\int 3dx + 12\int \frac{1}{x-2}dx + 2\int \frac{1}{(x-2)^2}dx = 3x + 12\ln|x-2| - 2/(x-2) + C$$

19. $\dfrac{x^5+2x^2+1}{x^3-x} = x^2 + 1 + \dfrac{2x^2+x+1}{x^3-x}$,

$\dfrac{2x^2+x+1}{x(x+1)(x-1)} = \dfrac{A}{x} + \dfrac{B}{x+1} + \dfrac{C}{x-1}$; $A = -1$, $B = 1$, $C = 2$ so

$$\int (x^2+1)dx - \int \frac{1}{x}dx + \int \frac{1}{x+1}dx + 2\int \frac{1}{x-1}dx$$

$$= \frac{1}{3}x^3 + x - \ln|x| + \ln|x+1| + 2\ln|x-1| + C = \frac{1}{3}x^3 + x + \ln\left|\frac{(x+1)(x-1)^2}{x}\right| + C$$

21. $\dfrac{2x^2+3}{x(x-1)^2} = \dfrac{A}{x} + \dfrac{B}{x-1} + \dfrac{C}{(x-1)^2}$; $A = 3$, $B = -1$, $C = 5$ so

$$3\int \frac{1}{x}\,dx - \int \frac{1}{x-1}\,dx + 5\int \frac{1}{(x-1)^2}dx = 3\ln|x| - \ln|x-1| - 5/(x-1) + C$$

23. $\dfrac{x^2+x-16}{(x+1)(x-3)^2} = \dfrac{A}{x+1} + \dfrac{B}{x-3} + \dfrac{C}{(x-3)^2}$; $A = -1$, $B = 2$, $C = -1$ so

$$-\int \frac{1}{x+1}\,dx + 2\int \frac{1}{x-3}\,dx - \int \frac{1}{(x-3)^2}\,dx$$

$$= -\ln|x+1| + 2\ln|x-3| + \frac{1}{x-3} + C = \ln\frac{(x-3)^2}{|x+1|} + \frac{1}{x-3} + C$$

25. $\dfrac{x^2}{(x+2)^3} = \dfrac{A}{x+2} + \dfrac{B}{(x+2)^2} + \dfrac{C}{(x+2)^3}$; $A = 1$, $B = -4$, $C = 4$ so

$$\int \frac{1}{x+2}\,dx - 4\int \frac{1}{(x+2)^2}\,dx + 4\int \frac{1}{(x+2)^3}\,dx = \ln|x+2| + \frac{4}{x+2} - \frac{2}{(x+2)^2} + C$$

27. $\dfrac{2x^2-1}{(4x-1)(x^2+1)} = \dfrac{A}{4x-1} + \dfrac{Bx+C}{x^2+1}$; $A = -14/17$, $B = 12/17$, $C = 3/17$ so

$$\int \frac{2x^2-1}{(4x-1)(x^2+1)}dx = -\frac{7}{34}\ln|4x-1| + \frac{6}{17}\ln(x^2+1) + \frac{3}{17}\tan^{-1}x + C$$

29. $\dfrac{x^3+3x^2+x+9}{(x^2+1)(x^2+3)} = \dfrac{Ax+B}{x^2+1} + \dfrac{Cx+D}{x^2+3}$; $A = 0$, $B = 3$, $C = 1$, $D = 0$ so

$$\int \frac{x^3+3x^2+x+9}{(x^2+1)(x^2+3)}\,dx = 3\tan^{-1}x + \frac{1}{2}\ln(x^2+3) + C$$

31. $\dfrac{x^3-3x^2+2x-3}{x^2+1} = x - 3 + \dfrac{x}{x^2+1}$,

$$\int \frac{x^3-3x^2+2x-3}{x^2+1}\,dx = \frac{1}{2}x^2 - 3x + \frac{1}{2}\ln(x^2+1) + C$$

33. Let $x = \sin\theta$ to get $\displaystyle\int \frac{1}{x^2+4x-5}\,dx$, and $\dfrac{1}{(x+5)(x-1)} = \dfrac{A}{x+5} + \dfrac{B}{x-1}$; $A = -1/6$, $B = 1/6$ so we get $\displaystyle -\frac{1}{6}\int \frac{1}{x+5}\,dx + \frac{1}{6}\int \frac{1}{x-1}\,dx = \frac{1}{6}\ln\left|\frac{x-1}{x+5}\right| + C = \frac{1}{6}\ln\left(\frac{1-\sin\theta}{5+\sin\theta}\right) + C.$

35. $V = \pi\displaystyle\int_0^2 \frac{x^4}{(9-x^2)^2}\,dx$, $\dfrac{x^4}{x^4-18x^2+81} = 1 + \dfrac{18x^2-81}{x^4-18x^2+81}$,

$$\frac{18x^2-81}{(9-x^2)^2} = \frac{18x^2-81}{(x+3)^2(x-3)^2} = \frac{A}{x+3} + \frac{B}{(x+3)^2} + \frac{C}{x-3} + \frac{D}{(x-3)^2};$$

$A = -\dfrac{9}{4}$, $B = \dfrac{9}{4}$, $C = \dfrac{9}{4}$, $D = \dfrac{9}{4}$ so

$$V = \pi\left[x - \frac{9}{4}\ln|x+3| - \frac{9/4}{x+3} + \frac{9}{4}\ln|x-3| - \frac{9/4}{x-3}\right]_0^2 = \pi\left(\frac{19}{5} - \frac{9}{4}\ln 5\right)$$

37. $\dfrac{x^2+1}{(x^2+2x+3)^2} = \dfrac{Ax+B}{x^2+2x+3} + \dfrac{Cx+D}{(x^2+2x+3)^2}$; $A = 0$, $B = 1$, $C = D = -2$ so

$$\int \frac{x^2+1}{(x^2+2x+3)^2}dx = \int \frac{1}{(x+1)^2+2}dx - \int \frac{2x+2}{(x^2+2x+3)^2}dx$$

$$= \frac{1}{\sqrt{2}}\tan^{-1}\frac{x+1}{\sqrt{2}} + 1/(x^2+2x+3) + C$$

39. $x^4 - 3x^3 - 7x^2 + 27x - 18 = (x-1)(x-2)(x-3)(x+3)$,

$$\frac{1}{(x-1)(x-2)(x-3)(x+3)} = \frac{A}{x-1} + \frac{B}{x-2} + \frac{C}{x-3} + \frac{D}{x+3};$$

$A = 1/8$, $B = -1/5$, $C = 1/12$, $D = -1/120$ so

$$\int \frac{dx}{x^4 - 3x^3 - 7x^2 + 27x - 18} = \frac{1}{8}\ln|x-1| - \frac{1}{5}\ln|x-2| + \frac{1}{12}\ln|x-3| - \frac{1}{120}\ln|x+3| + C$$

41. **(a)** $x^4 + 1 = (x^4 + 2x^2 + 1) - 2x^2 = (x^2+1)^2 - 2x^2$

$$= [(x^2+1) + \sqrt{2}x][(x^2+1) - \sqrt{2}x]$$

$$= (x^2 + \sqrt{2}x + 1)(x^2 - \sqrt{2}x + 1); a = \sqrt{2}, b = -\sqrt{2}$$

(b) $$\frac{x}{(x^2+\sqrt{2}x+1)(x^2-\sqrt{2}x+1)} = \frac{Ax+B}{x^2+\sqrt{2}x+1} + \frac{Cx+D}{x^2-\sqrt{2}x+1};$$

$A = 0$, $B = -\dfrac{\sqrt{2}}{4}$, $C = 0$, $D = \dfrac{\sqrt{2}}{4}$ so

$$\int_0^1 \frac{x}{x^4+1}dx = -\frac{\sqrt{2}}{4}\int_0^1 \frac{1}{x^2+\sqrt{2}x+1}dx + \frac{\sqrt{2}}{4}\int_0^1 \frac{1}{x^2-\sqrt{2}x+1}dx$$

$$= -\frac{\sqrt{2}}{4}\int_0^1 \frac{1}{(x+\sqrt{2}/2)^2+1/2}dx + \frac{\sqrt{2}}{4}\int_0^1 \frac{1}{(x-\sqrt{2}/2)^2+1/2}dx$$

$$= -\frac{\sqrt{2}}{4}\int_{\sqrt{2}/2}^{1+\sqrt{2}/2} \frac{1}{u^2+1/2}du + \frac{\sqrt{2}}{4}\int_{-\sqrt{2}/2}^{1-\sqrt{2}/2} \frac{1}{u^2+1/2}du$$

$$= -\frac{1}{2}\tan^{-1}\sqrt{2}u\Big]_{\sqrt{2}/2}^{1+\sqrt{2}/2} + \frac{1}{2}\tan^{-1}\sqrt{2}u\Big]_{-\sqrt{2}/2}^{1-\sqrt{2}/2}$$

$$= -\frac{1}{2}\tan^{-1}(\sqrt{2}+1) + \frac{1}{2}\left(\frac{\pi}{4}\right) + \frac{1}{2}\tan^{-1}(\sqrt{2}-1) - \frac{1}{2}\left(-\frac{\pi}{4}\right)$$

$$= \frac{\pi}{4} - \frac{1}{2}[\tan^{-1}(\sqrt{2}+1) - \tan^{-1}(\sqrt{2}-1)]$$

$$= \frac{\pi}{4} - \frac{1}{2}[\tan^{-1}(1+\sqrt{2}) + \tan^{-1}(1-\sqrt{2})]$$

$$= \frac{\pi}{4} - \frac{1}{2}\tan^{-1}\left[\frac{(1+\sqrt{2})+(1-\sqrt{2})}{1-(1+\sqrt{2})(1-\sqrt{2})}\right] \quad \text{(Exercise 78, Section 7.6)}$$

$$= \frac{\pi}{4} - \frac{1}{2}\tan^{-1}1 = \frac{\pi}{4} - \frac{1}{2}\left(\frac{\pi}{4}\right) = \frac{\pi}{8}$$

EXERCISE SET 8.6

1. Formula (60): $\frac{3}{16}\left[4x + \ln|-1+4x|\right] + C$

3. Formula (65): $\frac{1}{5}\ln\left|\frac{x}{5+2x}\right| + C$

5. Formula (102): $\frac{1}{5}(x+1)(-3+2x)^{3/2} + C$

7. Formula (108): $\frac{1}{2}\ln\left|\frac{\sqrt{4-3x}-2}{\sqrt{4-3x}+2}\right| + C$

9. Formula (69): $\frac{1}{2\sqrt{5}}\ln\left|\frac{x+\sqrt{5}}{x-\sqrt{5}}\right| + C$

11. Formula (73): $\frac{x}{2}\sqrt{x^2-3} - \frac{3}{2}\ln\left|x + \sqrt{x^2-3}\right| + C$

13. Formula (95): $\frac{x}{2}\sqrt{x^2+4} - 2\ln(x + \sqrt{x^2+4}) + C$

15. Formula (74): $\frac{x}{2}\sqrt{9-x^2} + \frac{9}{2}\sin^{-1}\frac{x}{3} + C$

17. Formula (79): $\sqrt{3-x^2} - \sqrt{3}\ln\left|\frac{\sqrt{3}+\sqrt{9-x^2}}{x}\right| + C$

19. Formula (38): $-\frac{1}{10}\sin(5x) + \frac{1}{2}\sin x + C$

21. Formula (50): $\frac{x^4}{16}\left[4\ln x - 1\right] + C$

23. Formula (42): $\frac{e^{-2x}}{13}(-2\sin(3x) - 3\cos(3x)) + C$

25. $u = e^{2x}, du = 2e^{2x}dx$, Formula (62): $\frac{1}{2}\int \frac{u\,du}{(4-3u)^2} = \frac{1}{18}\left[\frac{4}{4-3e^{2x}} + \ln\left|4-3e^{2x}\right|\right] + C$

27. $u = 3\sqrt{x}, du = \frac{3}{2\sqrt{x}}dx$, Formula (68): $\frac{2}{3}\int \frac{du}{u^2+4} = \frac{1}{3}\tan^{-1}\frac{3\sqrt{x}}{2} + C$

29. $u = 3x, du = 3dx$, Formula (76): $\frac{1}{3}\int \frac{du}{\sqrt{u^2-4}} = \frac{1}{3}\ln\left|3x + \sqrt{9x^2-4}\right| + C$

31. $u = 3x^2, du = 6x dx, u^2 du = 54x^5 dx$, Formula (81):

$$\frac{1}{54}\int \frac{u^2\,du}{\sqrt{5-u^2}} = -\frac{x^2}{36}\sqrt{5-9x^4} + \frac{5}{108}\sin^{-1}\frac{3x^2}{\sqrt{5}} + C$$

33. $u = \ln x, du = dx/x$, Formula (26): $\int \sin^2 u\,du = \frac{1}{2}\ln x + \frac{1}{4}\sin(2\ln x) + C$

35. $u = -2x, du = -2dx$, Formula (51): $\frac{1}{4}\int ue^u\,du = \frac{1}{4}(-2x-1)e^{-2x} + C$

37. $u = \cos 3x, du = -3\sin 3x$, Formula (67): $-\int \frac{du}{u(u+1)^2} = -\frac{1}{3}\left[\frac{1}{1+\cos 3x} + \ln\left|\frac{\cos 3x}{1+\cos 3x}\right|\right] + C$

39. $u = 4x^2, du = 8xdx$, Formula (70): $\dfrac{1}{8}\displaystyle\int \frac{du}{u^2-1} = \frac{1}{16}\ln\left|\frac{4x^2-1}{4x^2+1}\right| + C$

41. $u = 2e^x, du = 2e^x dx$, Formula (74):

$$\frac{1}{2}\int \sqrt{3-u^2}\,du = \frac{1}{4}u\sqrt{3-u^2} + \frac{3}{4}\sin^{-1}(u/\sqrt{3}) + C = \frac{1}{2}e^x\sqrt{3-4e^{2x}} + \frac{3}{4}\sin^{-1}(2e^x/\sqrt{3}) + C$$

43. $u = 3x, du = 3dx$, Formula (112):

$$\frac{1}{3}\int\sqrt{\frac{5}{3}u - u^2}\,du = \frac{1}{6}\left(u - \frac{5}{6}\right)\sqrt{\frac{5}{3}u - u^2} + \frac{25}{216}\sin^{-1}\left(\frac{u-5}{5}\right) + C$$

$$= \frac{18x-5}{36}\sqrt{5x-9x^2} + \frac{25}{216}\sin^{-1}\left(\frac{18x-5}{5}\right) + C$$

45. $u = 3x, du = 3dx$, Formula (44):

$$\frac{1}{9}\int u\sin u\,du = \frac{1}{9}(\sin u - u\cos u) + C = \frac{1}{9}(\sin 3x - 3x\cos 3x) + C$$

47. $u = -\sqrt{x}, u^2 = x, 2udu = dx$, Formula (51): $2\displaystyle\int ue^u du = -2(\sqrt{x}+1)e^{-\sqrt{x}} + C$

49. $x^2 + 4x - 5 = (x+2)^2 - 9; u = x+2, du = dx$, Formula (70):

$$\int\frac{du}{u^2-9} = \frac{1}{6}\ln\left|\frac{u-3}{u+3}\right| + C = \frac{1}{6}\ln\left|\frac{x-1}{x+5}\right| + C$$

51. $x^2 - 4x - 5 = (x-2)^2 - 9, u = x-2, du = dx$, Formula (77):

$$\int\frac{u+2}{\sqrt{9-u^2}}\,du = \int\frac{u\,du}{\sqrt{9-u^2}} + 2\int\frac{du}{\sqrt{9-u^2}} = -\sqrt{9-u^2} + 2\sin^{-1}\frac{u}{3} + C$$

$$= -\sqrt{5+4x-x^2} + 2\sin^{-1}\left(\frac{x-2}{3}\right) + C$$

53. $u = \sqrt{x-2},\ x = u^2+2,\ dx = 2u\,du;$

$$\int 2u^2(u^2+2)du = 2\int(u^4+2u^2)du = \frac{2}{5}u^5 + \frac{4}{3}u^3 + C = \frac{2}{5}(x-2)^{5/2} + \frac{4}{3}(x-2)^{3/2} + C$$

55. $u = \sqrt{x^3+1},\ x^3 = u^2 - 1,\ 3x^2dx = 2u\,du;$

$$\frac{2}{3}\int u^2(u^2-1)du = \frac{2}{3}\int(u^4-u^2)du = \frac{2}{15}u^5 - \frac{2}{9}u^3 + C = \frac{2}{15}(x^3+1)^{5/2} - \frac{2}{9}(x^3+1)^{3/2} + C$$

57. $u = x^{1/6},\ x = u^6,\ dx = 6u^5du;$

$$\int\frac{6u^5}{u^3+u^2}du = 6\int\frac{u^3}{u+1}du = 6\int\left[u^2 - u + 1 - \frac{1}{u+1}\right]du$$

$$= 2x^{1/2} - 3x^{1/3} + 6x^{1/6} - 6\ \ln(x^{1/6}+1) + C$$

59. $u = x^{1/4},\ x = u^4,\ dx = 4u^3du;$ $4\displaystyle\int\frac{1}{u(1-u)}du = 4\int\left[\frac{1}{u} + \frac{1}{1-u}\right]du = 4\ln\frac{x^{1/4}}{|1-x^{1/4}|} + C$

61. $u = x^{1/6}$, $x = u^6$, $dx = 6u^5du$;

$$6\int \frac{u^3}{u-1}du = 6\int \left[u^2 + u + 1 + \frac{1}{u-1}\right] du = 2x^{1/2} + 3x^{1/3} + 6x^{1/6} + 6\ \ln|x^{1/6} - 1| + C$$

63. $u = \sqrt{1+x^2}$, $x^2 = u^2 - 1$, $2x\,dx = 2u\,du$, $x\,dx = u\,du$;

$$\int (u^2 - 1)du = \frac{1}{3}(1+x^2)^{3/2} - (1+x^2)^{1/2} + C$$

65. $u = \sqrt{x}$, $x = u^2$, $dx = 2u\,du$, Formula (44): $2\int u \sin u\,du = 2\sin\sqrt{x} - 2\sqrt{x}\cos\sqrt{x} + C$

67. $$\int \frac{1}{1 + \dfrac{2u}{1+u^2} + \dfrac{1-u^2}{1+u^2}} \frac{2}{1+u^2}du = \int \frac{1}{u+1}du = \ln|\tan(x/2) + 1| + C$$

69. $u = \tan(\theta/2)$, $\displaystyle\int \frac{d\theta}{1 - \cos\theta} = \int \frac{1}{u^2}du = -\frac{1}{u} + C = -\cot(\theta/2) + C$

71. $u = \tan(x/2)$, $\displaystyle 2\int \frac{1-u^2}{(3u^2+1)(u^2+1)}du$;

$$\frac{1-u^2}{(3u^2+1)(u^2+1)} = \frac{(0)u+2}{3u^2+1} + \frac{(0)u-1}{u^2+1} = \frac{2}{3u^2+1} - \frac{1}{u^2+1} \text{ so}$$

$$\int \frac{\cos x}{2 - \cos x}dx = \frac{4}{\sqrt{3}}\tan^{-1}\left[\sqrt{3}\tan(x/2)\right] - x + C$$

73. $$\int_2^x \frac{1}{t(4-t)}\,dt = \frac{1}{4}\ln\frac{t}{4-t}\bigg]_2^x \quad \text{(Formula (65), } a = 4, b = -1\text{)}$$

$$= \frac{1}{4}\left[\ln\frac{x}{4-x} - \ln 1\right] = \frac{1}{4}\ln\frac{x}{4-x}, \frac{1}{4}\ln\frac{x}{4-x} = 0.5, \ln\frac{x}{4-x} = 2,$$

$$\frac{x}{4-x} = e^2,\ x = 4e^2 - e^2x,\ x(1+e^2) = 4e^2,\ x = 4e^2/(1+e^2) \approx 3.523188312$$

75. $$A = \int_0^4 \sqrt{25 - x^2}\,dx = \left(\frac{1}{2}x\sqrt{25-x^2} + \frac{25}{2}\sin^{-1}\frac{x}{5}\right)\bigg]_0^4 \quad \text{(Formula (74), } a = 5\text{)}$$

$$= 6 + \frac{25}{2}\sin^{-1}\frac{4}{5} \approx 17.59119023$$

77. $\displaystyle A = \int_0^1 \frac{1}{25 - 16x^2}\,dx$; $u = 4x$,

$$A = \frac{1}{4}\int_0^4 \frac{1}{25 - u^2}\,du = \frac{1}{40}\ln\left|\frac{u+5}{u-5}\right|\bigg]_0^4 = \frac{1}{40}\ln 9 \approx 0.054930614 \text{ (Formula (69), } a = 5\text{)}$$

79. $$V = 2\pi\int_0^{\pi/2} x\cos x\,dx = 2\pi(\cos x + x\sin x)\bigg]_0^{\pi/2} = \pi(\pi - 2) \approx 3.586419094 \quad \text{(Formula (45))}$$

81. $V = 2\pi \int_0^3 xe^{-x}dx;\ u = -x,$

$$V = 2\pi \int_0^{-3} ue^u du = 2\pi e^u(u-1)\Big]_0^{-3} = 2\pi(1 - 4e^{-3}) \approx 5.031899801 \quad \text{(Formula (51))}$$

83. $L = \int_0^2 \sqrt{1+16x^2}\,dx;\ u = 4x,$

$$L = \frac{1}{4}\int_0^8 \sqrt{1+u^2}\,du = \frac{1}{4}\left(\frac{u}{2}\sqrt{1+u^2} + \frac{1}{2}\ln\left(u+\sqrt{1+u^2}\right)\right)\Big]_0^8 \quad \text{(Formula (72), } a^2 = 1)$$

$$= \sqrt{65} + \frac{1}{8}\ln(8+\sqrt{65}) \approx 8.409316783$$

85. $S = 2\pi \int_0^{\pi} (\sin x)\sqrt{1+\cos^2 x}\,dx;\ u = \cos x,$

$$S = -2\pi \int_1^{-1} \sqrt{1+u^2}\,du = 4\pi \int_0^1 \sqrt{1+u^2}\,du = 4\pi\left(\frac{u}{2}\sqrt{1+u^2} + \frac{1}{2}\ln\left(u+\sqrt{1+u^2}\right)\right)\Big]_0^1 a^2 = 1$$

$$= 2\pi\left[\sqrt{2} + \ln(1+\sqrt{2})\right] \approx 14.42359945 \quad \text{(Formula (72))}$$

87. (a) $s(t) = 2 + \int_0^t 20\cos^6 u \sin^3 u\,du$

$$= -\frac{20}{9}\sin^2 t\cos^7 t - \frac{40}{63}\cos^7 t + \frac{166}{63}$$

(b)

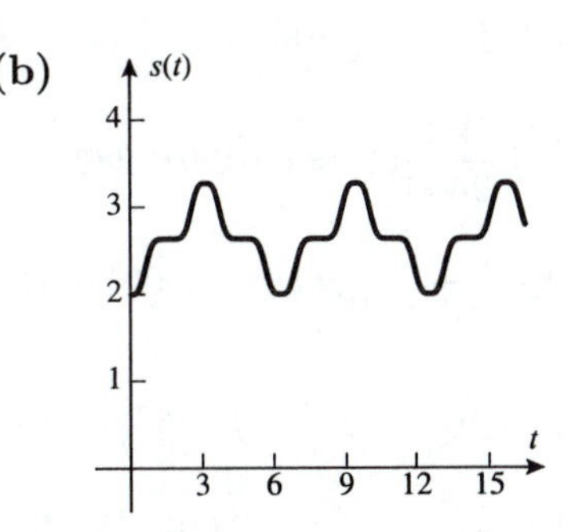

89. (a) $\int \sec x\,dx = \int \frac{1}{\cos x}dx = \int \frac{2}{1-u^2}du = \ln\left|\frac{1+u}{1-u}\right| + C = \ln\left|\frac{1+\tan(x/2)}{1-\tan(x/2)}\right| + C$

$$= \ln\left\{\left|\frac{\cos(x/2)+\sin(x/2)}{\cos(x/2)-\sin(x/2)}\right|\left|\frac{\cos(x/2)+\sin(x/2)}{\cos(x/2)+\sin(x/2)}\right|\right\} + C = \ln\left|\frac{1+\sin x}{\cos x}\right| + C$$

$$= \ln|\sec x + \tan x| + C$$

(b) $\tan\left(\frac{\pi}{4} + \frac{x}{2}\right) = \dfrac{\tan\frac{\pi}{4} + \tan\frac{x}{2}}{1 - \tan\frac{\pi}{4}\tan\frac{x}{2}} = \dfrac{1+\tan\frac{x}{2}}{1-\tan\frac{x}{2}}$

91. Let $u = \tanh(x/2)$ then $\cosh(x/2) = 1/\operatorname{sech}(x/2) = 1/\sqrt{1-\tanh^2(x/2)} = 1/\sqrt{1-u^2}$,

$\sinh(x/2) = \tanh(x/2)\cosh(x/2) = u/\sqrt{1-u^2}$, so $\sinh x = 2\sinh(x/2)\cosh(x/2) = 2u/(1-u^2)$,

$\cosh x = \cosh^2(x/2) + \sinh^2(x/2) = (1+u^2)/(1-u^2)$, $x = 2\tanh^{-1} u$, $dx = [2/(1-u^2)]du$;

$$\int \frac{dx}{2\cosh x + \sinh x} = \int \frac{1}{u^2+u+1}du = \frac{2}{\sqrt{3}}\tan^{-1}\frac{2u+1}{\sqrt{3}} + C = \frac{2}{\sqrt{3}}\tan^{-1}\frac{2\tanh(x/2)+1}{\sqrt{3}} + C.$$

EXERCISE SET 8.7

1. exact value $= 14/3 \approx 4.666666667$

(a) 4.667600663, $|E_M| \approx 0.000933996$
(b) 4.664795679, $|E_T| \approx 0.001870988$
(c) 4.666651630, $|E_S| \approx 0.000015037$

3. exact value $= 2$

(a) 2.008248408, $|E_M| \approx 0.008248408$
(b) 1.983523538, $|E_T| \approx 0.016476462$
(c) 2.000109517, $|E_S| \approx 0.000109517$

5. exact value $= e^{-1} - e^{-3} \approx 0.318092373$

(a) 0.317562837, $|E_M| \approx 0.000529536$
(b) 0.319151975, $|E_T| \approx 0.001059602$
(c) 0.318095187, $|E_S| \approx 0.000002814$

7. $f(x) = \sqrt{x+1}$, $f''(x) = -\frac{1}{4}(x+1)^{-3/2}$, $f^{(4)}(x) = -\frac{15}{16}(x+1)^{-7/2}$; $K_2 = 1/4$, $K_4 = 15/16$

(a) $|E_M| \le \frac{27}{2400}(1/4) = 0.002812500$
(b) $|E_T| \le \frac{27}{1200}(1/4) = 0.005625000$
(c) $|E_S| \le \frac{243}{180 \times 10^4}(15/16) \approx 0.000126563$

9. $f(x) = \sin x$, $f''(x) = -\sin x$, $f^{(4)}(x) = \sin x$; $K_2 = K_4 = 1$

(a) $|E_M| \le \frac{\pi^3}{2400}(1) \approx 0.012919282$
(b) $|E_T| \le \frac{\pi^3}{1200}(1) \approx 0.025838564$
(c) $|E_S| \le \frac{\pi^5}{180 \times 10^4}(1) \approx 0.000170011$

11. $f(x) = e^{-x}$, $f''(x) = f^{(4)}(x) = e^{-x}$; $K_2 = K_4 = e^{-1}$

(a) $|E_M| \le \frac{8}{2400}(e^{-1}) \approx 0.001226265$
(b) $|E_T| \le \frac{8}{1200}(e^{-1}) \approx 0.002452530$
(c) $|E_S| \le \frac{32}{180 \times 10^4}(e^{-1}) \approx 0.000006540$

13. **(a)** $n > \left[\frac{(27)(1/4)}{(24)(5 \times 10^{-4})}\right]^{1/2} \approx 23.7$; $n = 24$
(b) $n > \left[\frac{(27)(1/4)}{(12)(5 \times 10^{-4})}\right]^{1/2} \approx 33.5$; $n=34$
(c) $n > \left[\frac{(243)(15/16)}{(180)(5 \times 10^{-4})}\right]^{1/4} \approx 7.1$; $n = 8$

15. **(a)** $n > \left[\frac{(\pi^3)(1)}{(24)(10^{-3})}\right]^{1/2} \approx 35.9$; $n = 36$
(b) $n > \left[\frac{(\pi^3)(1)}{(12)(10^{-3})}\right]^{1/2} \approx 50.8$; $n = 51$
(c) $n > \left[\frac{(\pi^5)(1)}{(180)(10^{-3})}\right]^{1/4} \approx 6.4$; $n = 8$

17. **(a)** $n > \left[\frac{(8)(e^{-1})}{(24)(10^{-6})}\right]^{1/2} \approx 350.2$; $n = 351$
(b) $n > \left[\frac{(8)(e^{-1})}{(12)(10^{-6})}\right]^{1/2} \approx 495.2$; $n = 496$
(c) $n > \left[\frac{(32)(e^{-1})}{(180)(10^{-6})}\right]^{1/4} \approx 15.99$; $n = 16$

19. $g(X_0) = aX_0^2 + bX_0 + c = 4a + 2b + c = f(X_0) = 1/X_0 = 1/2$; similarly $9a + 3b + c = 1/3, 16a + 4b + c = 1/4$. Three equations in three unknowns, with solution $a = 1/24, b = -3/8, c = 13/12, g(x) = x^2/24 - 3x/8 + 13/12$.

$$\int_0^4 g(x)\,dx = \int \left(\frac{x^2}{24} - \frac{3x}{8} + \frac{13}{12}\right) dx = \frac{25}{36}$$

$$\frac{\Delta x}{3}[f(X_0) + 4f(X_1) + f(X_2)] = \frac{1}{3}\left[\frac{1}{2} + \frac{4}{3} + \frac{1}{4}\right] = \frac{25}{36}$$

21. 0.746824948, 0.746824133

23. 2.129861595, 2.129861293

25. 0.805376152, 0.804776489

27. **(a)** 3.142425985, $|E_M| \approx 0.000833331$
(b) 3.139925989, $|E_T| \approx 0.001666665$
(c) 3.141592614, $|E_S| \approx 0.000000040$

29. $S_{14} = 0.693147984$, $|E_S| \approx 0.000000803 = 8.03 \times 10^{-7}$; the method used in Example 6 results in a value of n which ensures that the magnitude of the error will be less than 10^{-6}, this is not necessarily the *smallest* value of n.

31. $f(x) = x\sin x$, $f''(x) = 2\cos x - x\sin x$, $|f''(x)| \le 2|\cos x| + |x|\,|\sin x| \le 2 + 2 = 4$ so $K_2 \le 4$,

$$n > \left[\frac{(8)(4)}{(24)(10^{-4})}\right]^{1/2} \approx 115.5;\ n = 116 \text{ (a smaller } n \text{ might suffice)}$$

33. $f(x) = \sqrt{x}$, $f''(x) = -\dfrac{1}{4x^{3/2}}$, $\lim_{x\to 0^+} |f''(x)| = +\infty$

35. $L = \displaystyle\int_0^{\pi} \sqrt{1 + \cos^2 x}\,dx \approx 3.820187623$

37.

t (s)	0	5	10	15	20
v (mi/hr)	0	40	60	73	84
v (ft/s)	0	58.67	88	107.07	123.2

$$\int_0^{20} v\,dt \approx \frac{20}{(3)(4)}[0 + 4(58.67) + 2(88) + 4(107.07) + 123.2] \approx 1604 \text{ ft}$$

39. $$\int_0^{180} v\,dt \approx \frac{180}{(3)(6)}[0.00 + 4(0.03) + 2(0.08) + 4(0.16) + 2(0.27) + 4(0.42) + 0.65] = 37.9 \text{ mi}$$

41. $$V = \int_0^{16} \pi r^2 dy = \pi \int_0^{16} r^2 dy \approx \pi\frac{16}{(3)(4)}[(8.5)^2 + 4(11.5)^2 + 2(13.8)^2 + 4(15.4)^2 + (16.8)^2]$$
$$\approx 9270 \text{ cm}^3 \approx 9.3 \text{ L}$$

43. $$\int_a^b f(x)\,dx \approx A_1 + A_2 + \cdots + A_n = \frac{b-a}{n}\left[\frac{1}{2}(y_0 + y_1) + \frac{1}{2}(y_1 + y_2) + \cdots + \frac{1}{2}(y_{n-1} + y_n)\right]$$
$$= \frac{b-a}{2n}[y_0 + 2y_1 + 2y_2 + \cdots + 2y_{n-1} + y_n]$$

45. **(a)** The maximum value of $|f''(x)|$ is approximately 3.844880.
(b) $n = 18$
(c) 0.904741

47. **(a)** The maximum value of $|f^{(4)}(x)|$ is approximately 42.551816.
(b) $n = 8$
(c) 0.904524

EXERCISE SET 8.8

1. **(a)** improper; infinite discontinuity at $x = 3$
(b) continuous integrand, not improper
(c) improper; infinite discontinuity at $x = 0$
(d) improper; infinite interval of integration
(e) improper; infinite interval of integration and infinite discontinuity at $x = 1$
(f) continuous integrand, not improper

3. $\lim_{\ell\to+\infty} (-e^{-x})\Big]_0^\ell = \lim_{\ell\to+\infty} (-e^{-\ell} + 1) = 1$

5. $\lim_{\ell\to+\infty} \ln \dfrac{x-1}{x+1}\Big]_4^\ell = \lim_{\ell\to+\infty} \left(\ln \dfrac{\ell-1}{\ell+1} - \ln \dfrac{3}{5}\right) = -\ln \dfrac{3}{5} = \ln \dfrac{5}{3}$

7. $\lim_{\ell\to+\infty} -\dfrac{1}{2\ln^2 x}\Big]_e^\ell = \lim_{\ell\to+\infty} \left[-\dfrac{1}{2\ln^2 \ell} + \dfrac{1}{2}\right] = \dfrac{1}{2}$

9. $\lim_{\ell\to-\infty} -\dfrac{1}{4(2x-1)^2}\Big]_\ell^0 = \lim_{\ell\to-\infty} \dfrac{1}{4}[-1 + 1/(2\ell-1)^2] = -1/4$

11. $\lim_{\ell\to-\infty} \dfrac{1}{3}e^{3x}\Big]_\ell^0 = \lim_{\ell\to-\infty} \left[\dfrac{1}{3} - \dfrac{1}{3}e^{3\ell}\right] = \dfrac{1}{3}$

13. $\int_{-\infty}^{+\infty} x^3 dx$ converges if $\int_{-\infty}^{0} x^3 dx$ and $\int_{0}^{+\infty} x^3 dx$ both converge; it diverges if either (or both) diverges. $\int_{0}^{+\infty} x^3 dx = \lim_{\ell\to+\infty} \dfrac{1}{4}x^4\Big]_0^\ell = \lim_{\ell\to+\infty} \dfrac{1}{4}\ell^4 = +\infty$ so $\int_{-\infty}^{+\infty} x^3 dx$ is divergent.

15. $\int_0^{+\infty} \dfrac{x}{(x^2+3)^2}dx = \lim_{\ell\to+\infty} -\dfrac{1}{2(x^2+3)}\Big]_0^\ell = \lim_{\ell\to+\infty} \dfrac{1}{2}[-1/(\ell^2+3) + 1/3] = \dfrac{1}{6}$,

similarly $\int_{-\infty}^{0} \dfrac{x}{(x^2+3)^2}dx = -1/6$ so $\int_{-\infty}^{\infty} \dfrac{x}{(x^2+3)^2}dx = 1/6 + (-1/6) = 0$

17. $\lim_{\ell\to3^+} -\dfrac{1}{x-3}\Big]_\ell^4 = \lim_{\ell\to3^+} \left[-1 + \dfrac{1}{\ell-3}\right] = +\infty$, divergent

19. $\lim_{\ell\to\pi/2^-} -\ln(\cos x)\Big]_0^\ell = \lim_{\ell\to\pi/2^-} -\ln(\cos\ell) = +\infty$, divergent

21. $\lim_{\ell\to 1^-} \sin^{-1} x\Big]_0^\ell = \lim_{\ell\to 1^-} \sin^{-1}\ell = \pi/2$

23. $\lim_{\ell\to\pi/6^-} -\sqrt{1-2\sin x}\Big]_0^\ell = \lim_{\ell\to\pi/6^-} (-\sqrt{1-2\sin\ell}+1) = 1$

25. $\int_0^2 \frac{dx}{x-2} = \lim_{\ell\to 2^-} \ln|x-2|\Big]_0^\ell = \lim_{\ell\to 2^-} (\ln|\ell-2| - \ln 2) = -\infty$, divergent

27. $\int_0^8 x^{-1/3}dx = \lim_{\ell\to 0^+} \frac{3}{2}x^{2/3}\Big]_\ell^8 = \lim_{\ell\to 0^+} \frac{3}{2}(4-\ell^{2/3}) = 6,$

$\int_{-1}^0 x^{-1/3}dx = \lim_{\ell\to 0^-} \frac{3}{2}x^{2/3}\Big]_{-1}^\ell = \lim_{\ell\to 0^-} \frac{3}{2}(\ell^{2/3}-1) = -3/2$

so $\int_{-1}^8 x^{-1/3}dx = 6 + (-3/2) = 9/2$

29. Define $\int_0^{+\infty} \frac{1}{x^2}dx = \int_0^a \frac{1}{x^2}dx + \int_a^{+\infty} \frac{1}{x^2}dx$ where $a > 0$; take $a = 1$ for convenience,

$\int_0^1 \frac{1}{x^2}dx = \lim_{\ell\to 0^+} (-1/x)\Big]_\ell^1 = \lim_{\ell\to 0^+} (1/\ell - 1) = +\infty$ so $\int_0^{+\infty} \frac{1}{x^2}dx$ is divergent.

31. $\int_0^{+\infty} \frac{e^{-\sqrt{x}}}{\sqrt{x}}dx = 2\int_0^{+\infty} e^{-u}du = 2\lim_{\ell\to+\infty} (-e^{-u})\Big]_0^\ell = 2\lim_{\ell\to+\infty} (1-e^{-\ell}) = 2$

33. $\int_0^{+\infty} \frac{e^{-x}}{\sqrt{1-e^{-x}}}dx = \int_0^1 \frac{du}{\sqrt{u}} = \lim_{\ell\to 0^+} 2\sqrt{u}\Big]_\ell^1 = \lim_{\ell\to 0^+} 2(1-\sqrt{\ell}) = 2$

35. $\lim_{\ell\to+\infty} \int_0^\ell e^{-x}\cos x\,dx = \lim_{\ell\to+\infty} \frac{1}{2}e^{-x}(\sin x - \cos x)\Big]_0^\ell = 1/2$

37. **(a)** 2.726585 **(b)** 2.804364 **(c)** 0.219384 **(d)** 0.504067

39. $1 + \left(\frac{dy}{dx}\right)^2 = 1 + \frac{4-x^{2/3}}{x^{2/3}} = \frac{4}{x^{2/3}}$; the arc length is $\int_0^8 \frac{2}{x^{1/3}}\,dx = 3x^{2/3}\Big|_0^8 = 12$

41. $\int \ln x\,dx = x\ln x - x + C,$

$\int_0^1 \ln x\,dx = \lim_{\ell\to 0^+} \int_\ell^1 \ln x\,dx = \lim_{\ell\to 0^+} (x\ln x - x)\Big]_\ell^1 = \lim_{\ell\to 0^+} (-1-\ell\ln\ell+\ell),$

but $\lim_{\ell\to 0^+} \ell\ln\ell = \lim_{\ell\to 0^+} \frac{\ln\ell}{1/\ell} = \lim_{\ell\to 0^+} (-\ell) = 0$ so $\int_0^1 \ln x\,dx = -1$

43. $\int_0^{+\infty} e^{-3x}dx = \lim_{\ell\to+\infty} \int_0^\ell e^{-3x}dx = \lim_{\ell\to+\infty} \left(-\frac{1}{3}e^{-3x}\right)\Big]_0^\ell = \lim_{\ell\to+\infty} \left(-\frac{1}{3}e^{-3\ell} + \frac{1}{9}\right) = \frac{1}{3}$

45. (a) $V = \pi \int_0^{+\infty} e^{-2x}dx = -\frac{\pi}{2} \lim_{\ell\to+\infty} e^{-2x}\Big]_0^\ell = \pi/2$

(b) $S = 2\pi \int_0^{+\infty} e^{-x}\sqrt{1+e^{-2x}}dx$, let $u = e^{-x}$ to get

$S = -2\pi \int_1^0 \sqrt{1+u^2}du = 2\pi \left[\frac{u}{2}\sqrt{1+u^2} + \frac{1}{2}\ln\left|u+\sqrt{1+u^2}\right|\right]_0^1 = \pi\left[\sqrt{2}+\ln(1+\sqrt{2})\right]$

47. (a) For $x \geq 1, x^2 \geq x, e^{-x^2} \leq e^{-x}$

(b) $\int_1^{+\infty} e^{-x}\,dx = \lim_{\ell\to+\infty} \int_1^\ell e^{-x}\,dx = \lim_{\ell\to+\infty} -e^{-x}\Big]_1^\ell = \lim_{\ell\to+\infty}(e^{-1}-e^{-\ell}) = 1/e$

(c) By Parts (a) and (b) and Exercise 46(b), $\int_1^{+\infty} e^{-x^2}\,dx$ is convergent and is $\leq 1/e$.

49. $V = \lim_{\ell\to+\infty} \int_1^\ell (\pi/x^2)\,dx = \lim_{\ell\to+\infty} -(\pi/x)\Big]_1^\ell = \lim_{\ell\to+\infty}(\pi - \pi/\ell) = \pi$

$A = \lim_{\ell\to+\infty} \int_1^\ell 2\pi(1/x)\sqrt{1+1/x^4}\,dx$; use Exercise 46(a) with $f(x) = 2\pi/x$, $g(x) = (2\pi/x)\sqrt{1+1/x^4}$ and $a = 1$ to see that the area is infinite.

51. $\int_0^{2x} \sqrt{1+t^3}dt \geq \int_0^{2x} t^{3/2}dt = \frac{2}{5}t^{5/2}\Big]_0^{2x} = \frac{2}{5}(2x)^{5/2}$,

$\lim_{x\to+\infty} \int_0^{2x} t^{3/2}dt = \lim_{x\to+\infty} \frac{2}{5}(2x)^{5/2} = +\infty$ so $\int_0^{+\infty}\sqrt{1+t^3}dt = +\infty$; by L'Hôpital's Rule

$$\lim_{x\to+\infty} \frac{\int_0^{2x}\sqrt{1+t^3}dt}{x^{5/2}} = \lim_{x\to+\infty} \frac{2\sqrt{1+(2x)^3}}{(5/2)x^{3/2}} = \lim_{x\to+\infty} \frac{2\sqrt{1/x^3+8}}{5/2} = 8\sqrt{2}/5$$

53. Let $x = r\tan\theta$ to get $\int \frac{dx}{(r^2+x^2)^{3/2}} = \frac{1}{r^2}\int \cos\theta\,d\theta = \frac{1}{r^2}\sin\theta + C = \frac{x}{r^2\sqrt{r^2+x^2}} + C$

so $u = \frac{2\pi NIr}{k} \lim_{\ell\to+\infty} \frac{x}{r^2\sqrt{r^2+x^2}}\Big]_a^\ell = \frac{2\pi NI}{kr} \lim_{\ell\to+\infty}(\ell/\sqrt{r^2+\ell^2} - a/\sqrt{r^2+a^2})$

$= \frac{2\pi NI}{kr}(1 - a/\sqrt{r^2+a^2})$.

55. (a) Satellite's weight $= w(x) = k/x^2$ lb when $x =$ distance from center of Earth; $w(4000) = 6000$ so $k = 9.6 \times 10^{10}$ and $W = \int_{4000}^{4000+\ell} 9.6 \times 10^{10}x^{-2}dx$ mi·lb.

(b) $\int_{4000}^{+\infty} 9.6 \times 10^{10}x^{-2}dx = \lim_{\ell\to+\infty} -9.6\times 10^{10}/x\Big]_{4000}^\ell = 2.4 \times 10^7$ mi·lb

57. (a) $\mathcal{L}\{f(t)\} = \int_0^{+\infty} te^{-st}\,dt = \lim_{\ell\to+\infty} -(t/s+1/s^2)e^{-st}\Big]_0^\ell = \frac{1}{s^2}$

(b) $\mathcal{L}\{f(t)\} = \int_0^{+\infty} t^2e^{-st}\,dt = \lim_{\ell\to+\infty} -(t^2/s+2t/s^2+2/s^3)e^{-st}\Big]_0^\ell = \frac{2}{s^3}$

(c) $\mathcal{L}\{f(t)\} = \int_3^{+\infty} e^{-st}dt = \lim_{\ell\to+\infty} -\frac{1}{s}e^{-st}\Big]_3^\ell = \frac{e^{-3s}}{s}$

59. (a) $u = \sqrt{a}x, du = \sqrt{a}\,dx, 2\int_0^{+\infty} e^{-ax^2}\,dx = \frac{2}{\sqrt{a}}\int_0^{+\infty} e^{-u^2}\,du = \sqrt{\pi/a}$

(b) $x = \sqrt{2}\sigma u, dx = \sqrt{2}\sigma\,du, \frac{2}{\sqrt{2\pi}\sigma}\int_0^{+\infty} e^{-x^2/2\sigma^2}\,dx = \frac{2}{\sqrt{\pi}}\int_0^{+\infty} e^{-u^2}\,du = 1$

61. (a) $\int_0^4 \frac{1}{x^6+1}dx \approx 1.047;\ \pi/3 \approx 1.047$

(b) $\int_0^{+\infty} \frac{1}{x^6+1}dx = \int_0^4 \frac{1}{x^6+1}dx + \int_4^{+\infty} \frac{1}{x^6+1}dx$ so

$$E = \int_4^{+\infty} \frac{1}{x^6+1}dx < \int_4^{+\infty} \frac{1}{x^6}dx = \frac{1}{5(4)^5} < 2\times 10^{-4}$$

63. If $p = 1$, then $\int_0^1 \frac{dx}{x} = \lim_{\ell\to 0^+} \ln x\Big]_\ell^1 = +\infty$;

if $p \neq 1$, then $\int_0^1 \frac{dx}{x^p} = \lim_{\ell\to 0^+} \frac{x^{1-p}}{1-p}\Big]_\ell^1 = \lim_{\ell\to 0^+}[(1-\ell^{1-p})/(1-p)] = \begin{cases} 1/(1-p), & p<1 \\ +\infty, & p>1 \end{cases}$.

65. $2\int_0^1 \cos(u^2)du \approx 1.809$

CHAPTER 8 SUPPLEMENTARY EXERCISES

1. (a) integration by parts, $u = x$, $dv = \sin x\,dx$
(b) u-substitution: $u = \sin x$
(c) reduction formula
(d) u-substitution: $u = \tan x$
(e) u-substitution: $u = x^3 + 1$
(f) u-substitution: $u = x + 1$
(g) integration by parts: $dv = dx, u = \tan^{-1} x$
(h) trigonometric substitution: $x = 2\sin\theta$
(i) u-substitution: $u = 4 - x^2$

5. (a) #40 (b) #57 (c) #113
(d) #108 (e) #52 (f) #71

7. (a) $u = 2x$,

$$\int \sin^4 2x\,dx = \frac{1}{2}\int \sin^4 u\,du = \frac{1}{2}\left[-\frac{1}{4}\sin^3 u\cos u + \frac{3}{4}\int \sin^2 u\,du\right]$$

$$= -\frac{1}{8}\sin^3 u\cos u + \frac{3}{8}\left[-\frac{1}{2}\sin u\cos u + \frac{1}{2}\int du\right]$$

$$= -\frac{1}{8}\sin^3 u\cos u - \frac{3}{16}\sin u\cos u + \frac{3}{16}u + C$$

$$= -\frac{1}{8}\sin^3 2x\cos 2x - \frac{3}{16}\sin 2x\cos 2x + \frac{3}{8}x + C$$

(b) $u = x^2$,

$$\int x\cos^5(x^2)dx = \frac{1}{2}\int \cos^5 u\,du = \frac{1}{2}\int (\cos u)(1-\sin^2 u)^2\,du$$
$$= \frac{1}{2}\int \cos u\,du - \int \cos u \sin^2 u\,du + \frac{1}{2}\int \cos u \sin^4 u\,du$$
$$= \frac{1}{2}\sin u - \frac{1}{3}\sin^3 u + \frac{1}{10}\sin^5 u + C$$
$$= \frac{1}{2}\sin(x^2) - \frac{1}{3}\sin^3(x^2) + \frac{1}{10}\sin^5(x^2) + C$$

9. (a) With $u = \sqrt{x}$:

$$\int \frac{1}{\sqrt{x}\,\sqrt{2-x}}\,dx = 2\int \frac{1}{\sqrt{2-u^2}}\,du = 2\sin^{-1}(u/\sqrt{2}) + C = 2\sin^{-1}(\sqrt{x/2}) + C;$$

with $u = \sqrt{2-x}$:

$$\int \frac{1}{\sqrt{x}\,\sqrt{2-x}}\,dx = -2\int \frac{1}{\sqrt{2-u^2}}\,du = -2\sin^{-1}(u/\sqrt{2}) + C = -2\sin^{-1}(\sqrt{2-x}/\sqrt{2}) + C;$$

completing the square:

$$\int \frac{1}{\sqrt{1-(x-1)^2}}\,dx = \sin^{-1}(x-1) + C.$$

(b) In the three results in Part (a) the antiderivatives differ by a constant, in particular $2\sin^{-1}(\sqrt{x/2}) = \pi - 2\sin^{-1}(\sqrt{2-x}/\sqrt{2}) = \pi/2 + \sin^{-1}(x-1)$.

11. Solve $y = 1/(1+x^2)$ for x to get $x = \sqrt{\dfrac{1-y}{y}}$ and integrate with respect to y to get $A = \displaystyle\int_0^1 \sqrt{\frac{1-y}{y}}\,dy$ (see figure)

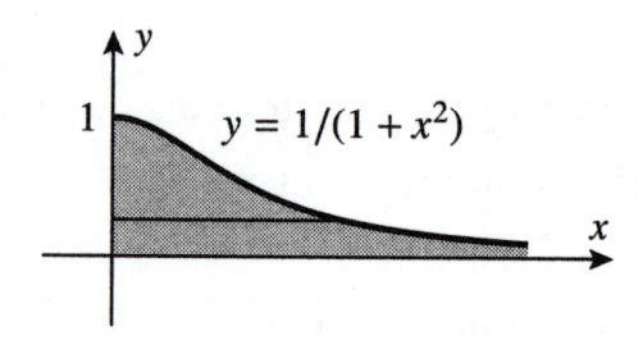

13. $V = 2\pi \displaystyle\int_0^{+\infty} xe^{-x}dx = 2\pi \lim_{\ell\to+\infty} -e^{-x}(x+1)\Big]_0^{\ell} = 2\pi \lim_{\ell\to+\infty}\left[1 - e^{-\ell}(\ell+1)\right]$

but $\displaystyle\lim_{\ell\to+\infty} e^{-\ell}(\ell+1) = \lim_{\ell\to+\infty}\frac{\ell+1}{e^{\ell}} = \lim_{\ell\to+\infty}\frac{1}{e^{\ell}} = 0$ so $V = 2\pi$

15. $u = \cos\theta$, $-\displaystyle\int u^{1/2}du = -\frac{2}{3}\cos^{3/2}\theta + C$

17. $u = \tan(x^2)$, $\dfrac{1}{2}\displaystyle\int u^2du = \frac{1}{6}\tan^3(x^2) + C$

19. $x = \sqrt{3}\tan\theta$, $dx = \sqrt{3}\sec^2\theta\,d\theta$,

$$\frac{1}{3}\int \frac{1}{\sec\theta}d\theta = \frac{1}{3}\int \cos\theta\,d\theta = \frac{1}{3}\sin\theta + C = \frac{x}{3\sqrt{3+x^2}} + C$$

21. $\displaystyle\int \frac{x+3}{\sqrt{(x+1)^2+1}}dx$, let $u = x+1$,

$$\int \frac{u+2}{\sqrt{u^2+1}}du = \int \left[u(u^2+1)^{-1/2} + \frac{2}{\sqrt{u^2+1}}\right]du = \sqrt{u^2+1} + 2\sinh^{-1}u + C$$
$$= \sqrt{x^2+2x+2} + 2\sinh^{-1}(x+1) + C$$

Alternate solution: let $x+1=\tan\theta$,

$$\int(\tan\theta+2)\sec\theta\,d\theta=\int\sec\theta\tan\theta\,d\theta+2\int\sec\theta\,d\theta=\sec\theta+2\ln|\sec\theta+\tan\theta|+C$$
$$=\sqrt{x^2+2x+2}+2\ln(\sqrt{x^2+2x+2}+x+1)+C.$$

23. $\dfrac{1}{(x-1)(x+2)(x-3)}=\dfrac{A}{x-1}+\dfrac{B}{x+2}+\dfrac{C}{x-3}$; $A=-\dfrac{1}{6}$, $B=\dfrac{1}{15}$, $C=\dfrac{1}{10}$ so

$$-\frac{1}{6}\int\frac{1}{x-1}dx+\frac{1}{15}\int\frac{1}{x+2}dx+\frac{1}{10}\int\frac{1}{x-3}dx$$
$$=-\frac{1}{6}\ln|x-1|+\frac{1}{15}\ln|x+2|+\frac{1}{10}\ln|x-3|+C$$

25. $u=\sqrt{x-4}$, $x=u^2+4$, $dx=2u\,du$,

$$\int_0^2\frac{2u^2}{u^2+4}du=2\int_0^2\left[1-\frac{4}{u^2+4}\right]du=\left[2u-4\tan^{-1}(u/2)\right]_0^2=4-\pi$$

27. $u=\sqrt{e^x+1}$, $e^x=u^2-1$, $x=\ln(u^2-1)$, $dx=\dfrac{2u}{u^2-1}du$,

$$\int\frac{2}{u^2-1}du=\int\left[\frac{1}{u-1}-\frac{1}{u+1}\right]du=\ln|u-1|-\ln|u+1|+C=\ln\frac{\sqrt{e^x+1}-1}{\sqrt{e^x+1}+1}+C$$

29. $\lim\limits_{\ell\to+\infty}-\dfrac{1}{2(x^2+1)}\Big]_a^\ell=\lim\limits_{\ell\to+\infty}\left[-\dfrac{1}{2(\ell^2+1)}+\dfrac{1}{2(a^2+1)}\right]=\dfrac{1}{2(a^2+1)}$

31. Let $u=x^4$ to get $\dfrac{1}{4}\displaystyle\int\frac{1}{\sqrt{1-u^2}}du=\frac{1}{4}\sin^{-1}u+C=\frac{1}{4}\sin^{-1}(x^4)+C.$

33. $\displaystyle\int\sqrt{x-\sqrt{x^2-4}}dx=\frac{1}{\sqrt{2}}\int(\sqrt{x+2}-\sqrt{x-2})dx=\frac{\sqrt{2}}{3}[(x+2)^{3/2}-(x-2)^{3/2}]+C$

35. **(a)** $(x+4)(x-5)(x^2+1)^2$; $\dfrac{A}{x+4}+\dfrac{B}{x-5}+\dfrac{Cx+D}{x^2+1}+\dfrac{Ex+F}{(x^2+1)^2}$

(b) $-\dfrac{3}{x+4}+\dfrac{2}{x-5}-\dfrac{x-2}{x^2+1}-\dfrac{3}{(x^2+1)^2}$

(c) $-3\ln|x+4|+2\ln|x-5|+2\tan^{-1}x-\dfrac{1}{2}\ln(x^2+1)-\dfrac{3}{2}\left(\dfrac{x}{x^2+1}+\tan^{-1}x\right)+C$

37. **(a)** $t=-\ln x, x=e^{-t}, dx=-e^{-t}dt$,

$$\int_0^1(\ln x)^n dx=-\int_{+\infty}^0(-t)^n e^{-t}dt=(-1)^n\int_0^{+\infty}t^n e^{-t}dt=(-1)^n\Gamma(n+1)$$

(b) $t=x^n, x=t^{1/n}, dx=(1/n)t^{1/n-1}dt$,

$$\int_0^{+\infty}e^{-x^n}dx=(1/n)\int_0^{+\infty}t^{1/n-1}e^{-t}dt=(1/n)\Gamma(1/n)=\Gamma(1/n+1)$$

CHAPTER 9

Mathematical Modeling with Differential Equations

EXERCISE SET 9.1

1. $y' = 2x^2 e^{x^3/3} = x^2 y$ and $y(0) = 2$ by inspection.

3. **(a)** first order; $\dfrac{dy}{dx} = c;\, (1+x)\dfrac{dy}{dx} = (1+x)c = y$

(b) second order; $y' = c_1 \cos t - c_2 \sin t,\ y'' + y = -c_1 \sin t - c_2 \cos t + (c_1 \sin t + c_2 \cos t) = 0$

5. $\dfrac{1}{y}\dfrac{dy}{dx} = x\dfrac{dy}{dx} + y,\ \dfrac{dy}{dx}(1 - xy) = y^2,\ \dfrac{dy}{dx} = \dfrac{y^2}{1 - xy}$

7. **(a)** IF: $\mu = e^{3\int dx} = e^{3x},\ \dfrac{d}{dx}\left[ye^{3x}\right] = 0, ye^{3x} = C, y = Ce^{-3x}$

separation of variables: $\dfrac{dy}{y} = -3dx, \ln|y| = -3x + C_1, y = \pm e^{-3x}e^{C_1} = Ce^{-3x}$
including $C = 0$ by inspection

(b) IF: $\mu = e^{-2\int dt} = e^{-2t},\ \dfrac{d}{dt}[ye^{-2t}] = 0, ye^{-2t} = C, y = Ce^{2t}$

separation of variables: $\dfrac{dy}{y} = 2dt,\ \ln|y| = 2t + C_1,\ y = \pm e^{C_1}e^{2t} = Ce^{2t}$
including $C = 0$ by inspection

9. $\mu = e^{\int 3dx} = e^{3x},\ e^{3x}y = \displaystyle\int e^x\, dx = e^x + C,\ y = e^{-2x} + Ce^{-3x}$

11. $\mu = e^{\int dx} = e^x,\ e^x y = \displaystyle\int e^x \cos(e^x) dx = \sin(e^x) + C,\ y = e^{-x}\sin(e^x) + Ce^{-x}$

13. $\dfrac{dy}{dx} + \dfrac{x}{x^2+1}y = 0, \mu = e^{\int (x/(x^2+1))dx} = e^{\frac{1}{2}\ln(x^2+1)} = \sqrt{x^2+1},$

$\dfrac{d}{dx}\left[y\sqrt{x^2+1}\right] = 0,\ y\sqrt{x^2+1} = C,\ y = \dfrac{C}{\sqrt{x^2+1}}$

15. $\dfrac{1}{y}dy = \dfrac{1}{x}dx,\ \ln|y| = \ln|x| + C_1,\ \ln\left|\dfrac{y}{x}\right| = C_1,\ \dfrac{y}{x} = \pm e^{C_1} = C,\ y = Cx$
including $C = 0$ by inspection

17. $\dfrac{dy}{1+y} = -\dfrac{x}{\sqrt{1+x^2}}dx, \ln|1+y| = -\sqrt{1+x^2} + C_1, 1 + y = \pm e^{-\sqrt{1+x^2}}e^{C_1} = Ce^{-\sqrt{1+x^2}},$
$y = Ce^{-\sqrt{1+x^2}} - 1, C \neq 0$

19. $\left(\dfrac{1}{y} + y\right)\, dy = e^x dx,\ \ln|y| + y^2/2 = e^x + C$; by inspection, $y = 0$ is also a solution

21. $e^y dy = \dfrac{\sin x}{\cos^2 x}dx = \sec x \tan x\, dx,\ e^y = \sec x + C,\ y = \ln(\sec x + C)$

23. $\dfrac{dy}{y^2-y}=\dfrac{dx}{\sin x},\ \displaystyle\int\left[-\frac{1}{y}+\frac{1}{y-1}\right]dy=\int\csc x\,dx,\ \ln\left|\frac{y-1}{y}\right|=\ln|\csc x-\cot x|+C_1,$

$\dfrac{y-1}{y}=\pm e^{C_1}(\csc x-\cot x)=C(\csc x-\cot x),\ y=\dfrac{1}{1-C(\csc x-\cot x)},C\neq 0;$

by inspection, $y=0$ is also a solution, as is $y=1$.

25. $\dfrac{dy}{dx}+\dfrac{1}{x}y=1,\ \mu=e^{\int(1/x)dx}=e^{\ln x}=x,\ \dfrac{d}{dx}[xy]=x,\ xy=\dfrac{1}{2}x^2+C,\ y=x/2+C/x$

(a) $2=y(1)=\dfrac{1}{2}+C,C=\dfrac{3}{2},y=x/2+3/(2x)$

(b) $2=y(-1)=-1/2-C,C=-5/2,y=x/2-5/(2x)$

27. $\mu=e^{-\int x\,dx}=e^{-x^2/2},\ e^{-x^2/2}y=\displaystyle\int xe^{-x^2/2}dx=-e^{-x^2/2}+C,$

$y=-1+Ce^{x^2/2},\ 3=-1+C,\ C=4,\ y=-1+4e^{x^2/2}$

29. $(y+\cos y)\,dy=4x^2\,dx,\ \dfrac{y^2}{2}+\sin y=\dfrac{4}{3}x^3+C,\ \dfrac{\pi^2}{2}+\sin\pi=\dfrac{4}{3}(1)^3+C,\ \dfrac{\pi^2}{2}=\dfrac{4}{3}+C,$

$C=\dfrac{\pi^2}{2}-\dfrac{4}{3},3y^2+6\sin y=8x^3+3\pi^2-8$

31. $2(y-1)\,dy=(2t+1)\,dt,y^2-2y=t^2+t+C,1+2=C,C=3,y^2-2y=t^2+t+3$

33. (a) $\dfrac{dy}{y}=\dfrac{dx}{2x},\ \ln|y|=\dfrac{1}{2}\ln|x|+C_1,$

$|y|=C|x|^{1/2},\ y^2=Cx;$

by inspection $y=0$ is also a solution.

(b) $1=C(2)^2,C=1/4,y^2=x/4$

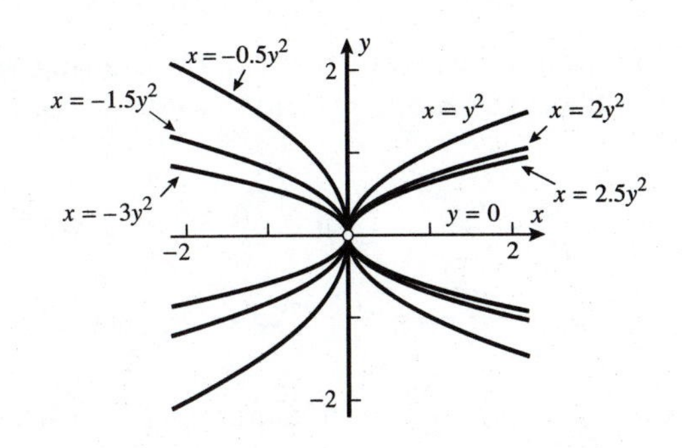

35. $\dfrac{dy}{y}=-\dfrac{x\,dx}{x^2+4},$

$\ln|y|=-\dfrac{1}{2}\ln(x^2+4)+C_1,$

$y=\dfrac{C}{\sqrt{x^2+4}}$

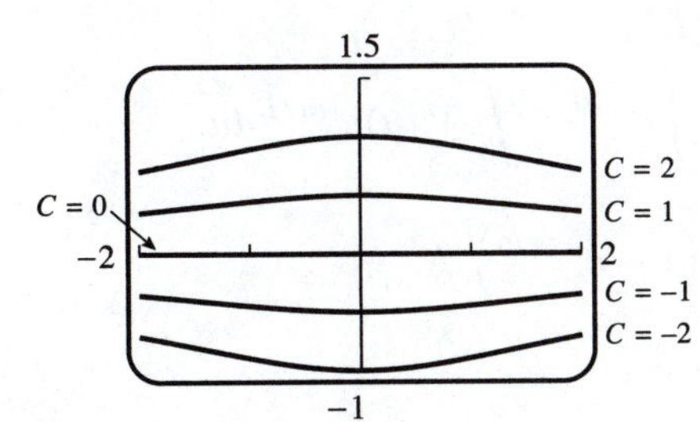

37. $(1-y^2)\,dy=x^2\,dx,$

$y-\dfrac{y^3}{3}=\dfrac{x^3}{3}+C_1,x^3+y^3-3y=C$

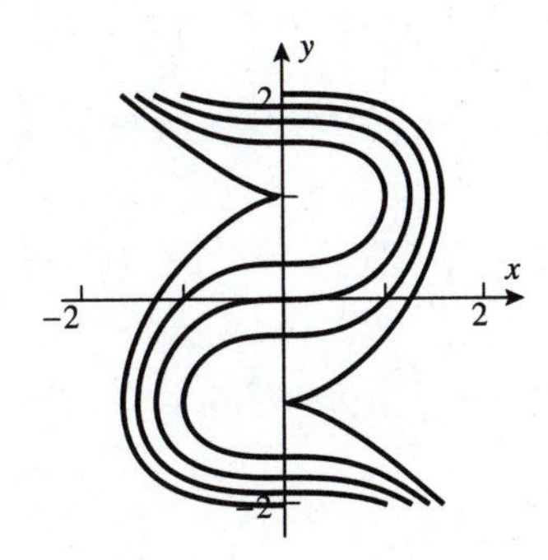

39. Of the solutions $y = \dfrac{1}{2x^2 - C}$, all pass through the point $\left(0, -\dfrac{1}{C}\right)$ and thus never through $(0,0)$. A solution of the initial value problem with $y(0) = 0$ is (by inspection) $y = 0$. The methods of Example 4 fail because the integrals there become divergent when the point $x = 0$ is included in the integral.

41. $\dfrac{dy}{dx} = xe^y, e^{-y}\,dy = x\,dx, -e^{-y} = \dfrac{x^2}{2} + C, x = 2$ when $y = 0$ so $-1 = 2 + C, C = -3, x^2 + 2e^{-y} = 6$

43. $\dfrac{dy}{dt} =$ rate in $-$ rate out, where y is the amount of salt at time t,

$\dfrac{dy}{dt} = (4)(2) - \left(\dfrac{y}{50}\right)(2) = 8 - \dfrac{1}{25}y$, so $\dfrac{dy}{dt} + \dfrac{1}{25}y = 8$ and $y(0) = 25$.

$\mu = e^{\int(1/25)dt} = e^{t/25}$, $e^{t/25}y = \displaystyle\int 8e^{t/25}dt = 200e^{t/25} + C$,

$y = 200 + Ce^{-t/25}$, $25 = 200 + C$, $C = -175$,

(a) $y = 200 - 175e^{-t/25}$ oz

(b) when $t = 25$, $y = 200 - 175e^{-1} \approx 136$ oz

45. The volume V of the (polluted) water is $V(t) = 500 + (20 - 10)t = 500 + 10t$; if $y(t)$ is the number of pounds of particulate matter in the water,

then $y(0) = 50$, and $\dfrac{dy}{dt} = 0 - 10\dfrac{y}{V} = -\dfrac{1}{50+t}y$, $\dfrac{dy}{dt} + \dfrac{1}{50+t}y = 0$; $\mu = e^{\int \frac{dt}{50+t}} = 50 + t$;

$\dfrac{d}{dt}[(50+t)y] = 0$, $(50+t)y = C$, $2500 = 50y(0) = C$, $y(t) = 2500/(50+t)$.

The tank reaches the point of overflowing when $V = 500 + 10t = 1000$, $t = 50$ min, so $y = 2500/(50+50) = 25$ lb.

47. **(a)** $\dfrac{dv}{dt} + \dfrac{c}{m}v = -g, \mu = e^{(c/m)\int dt} = e^{ct/m}, \dfrac{d}{dt}\left[ve^{ct/m}\right] = -ge^{ct/m}, ve^{ct/m} = -\dfrac{gm}{c}e^{ct/m} + C,$

$v = -\dfrac{gm}{c} + Ce^{-ct/m}$, but $v_0 = v(0) = -\dfrac{gm}{c} + C, C = v_0 + \dfrac{gm}{c}, v = -\dfrac{gm}{c} + \left(v_0 + \dfrac{gm}{c}\right)e^{-ct/m}$

(b) Replace $\dfrac{mg}{c}$ with v_τ and $-ct/m$ with $-gt/v_\tau$ in (23).

(c) From Part (b), $s(t) = C - v_\tau t - (v_0 + v_\tau)\dfrac{v_\tau}{g}e^{-gt/v_\tau}$;

$s_0 = s(0) = C - (v_0 + v_\tau)\dfrac{v_\tau}{g}$, $C = s_0 + (v_0 + v_\tau)\dfrac{v_\tau}{g}$, $s(t) = s_0 - v_\tau t + \dfrac{v_\tau}{g}(v_0 + v_\tau)\left(1 - e^{-gt/v_\tau}\right)$

49. $\dfrac{dI}{dt} + \dfrac{R}{L}I = \dfrac{V(t)}{L}, \mu = e^{(R/L)\int dt} = e^{Rt/L}, \dfrac{d}{dt}(e^{Rt/L}I) = \dfrac{V(t)}{L}e^{Rt/L},$

$Ie^{Rt/L} = I(0) + \dfrac{1}{L}\displaystyle\int_0^t V(u)e^{Ru/L}du, I(t) = I(0)e^{-Rt/L} + \dfrac{1}{L}e^{-Rt/L}\int_0^t V(u)e^{Ru/L}du.$

(a) $I(t) = \dfrac{1}{4}e^{-5t/2}\displaystyle\int_0^t 12e^{5u/2}du = \dfrac{6}{5}e^{-5t/2}e^{5u/2}\Big]_0^t = \dfrac{6}{5}\left(1 - e^{-5t/2}\right)$ A.

(b) $\displaystyle\lim_{t\to+\infty} I(t) = \dfrac{6}{5}$ A

51. **(a)** $\dfrac{dv}{dt} = \dfrac{ck}{m_0 - kt} - g, v = -c\ln(m_0 - kt) - gt + C; v = 0$ when $t = 0$ so $0 = -c\ln m_0 + C$,

$C = c\ln m_0, v = c\ln m_0 - c\ln(m_0 - kt) - gt = c\ln\dfrac{m_0}{m_0 - kt} - gt.$

(b) $m_0 - kt = 0.2m_0$ when $t = 100$ so

$v = 2500\ln\dfrac{m_0}{0.2m_0} - 9.8(100) = 2500\ln 5 - 980 \approx 3044\,\text{m/s}.$

53. **(a)** $A(h) = \pi(1)^2 = \pi, \pi\dfrac{dh}{dt} = -0.025\sqrt{h}, \dfrac{\pi}{\sqrt{h}}dh = -0.025dt, 2\pi\sqrt{h} = -0.025t + C; h = 4$ when $t = 0$, so $4\pi = C, 2\pi\sqrt{h} = -0.025t + 4\pi, \sqrt{h} = 2 - \dfrac{0.025}{2\pi}t, h \approx (2 - 0.003979\,t)^2.$

(b) $h = 0$ when $t \approx 2/0.003979 \approx 502.6$ s ≈ 8.4 min.

55. $\dfrac{dv}{dt} = -0.04v^2, \dfrac{1}{v^2}dv = -0.04dt, -\dfrac{1}{v} = -0.04t + C; v = 50$ when $t = 0$ so $-\dfrac{1}{50} = C$,

$-\dfrac{1}{v} = -0.04t - \dfrac{1}{50}, v = \dfrac{50}{2t+1}$ cm/s. But $v = \dfrac{dx}{dt}$ so $\dfrac{dx}{dt} = \dfrac{50}{2t+1}, x = 25\ln(2t+1) + C_1;$

$x = 0$ when $t = 0$ so $C_1 = 0, x = 25\ln(2t+1)$ cm.

57. Differentiate to get $\dfrac{dy}{dx} = -\sin x + e^{-x^2}, y(0) = 1.$

59. Suppose that $H(y) = G(x) + C$. Then $\dfrac{dH}{dy}\dfrac{dy}{dx} = G'(x)$. But $\dfrac{dH}{dy} = h(y)$ and $\dfrac{dG}{dx} = g(x)$, hence $y(x)$ is a solution of (10).

61. Suppose $I_1 \subset I$ is an interval with $I_1 \neq I$,and suppose $Y(x)$ is defined on I_1 and is a solution of (5) there. Let x_0 be a point of I_1. Solve the initial value problem on I with initial value $y(x_0) = Y(x_0)$. Then $y(x)$ is an extension of $Y(x)$ to the interval I, and by Exercise 58(b) applied to the interval I_1, it follows that $y(x) = Y(x)$ for x in I_1.

EXERCISE SET 9.2

1.

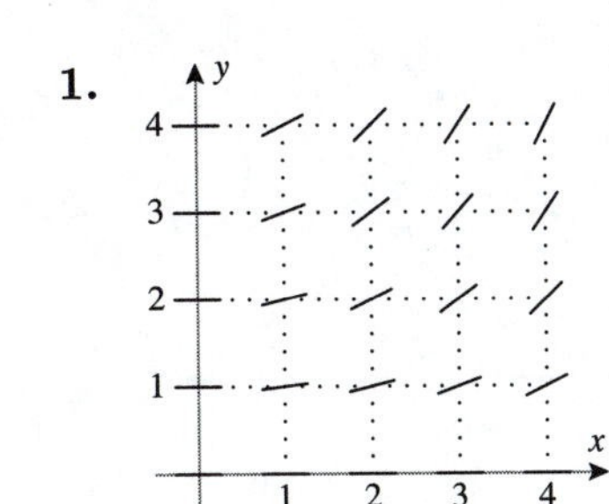

3.

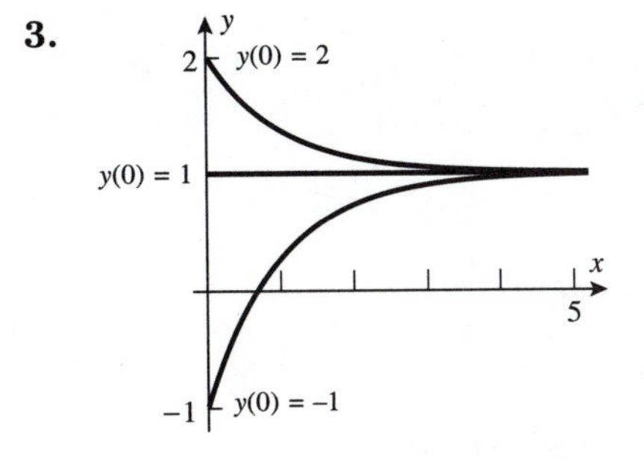

5.

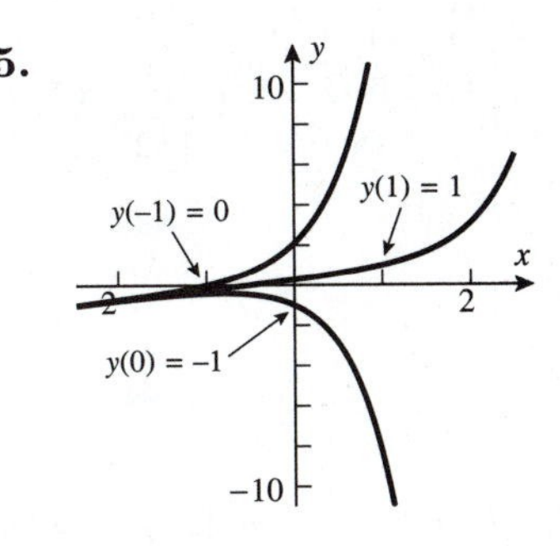

7. $\lim\limits_{x\to+\infty} y = 1$

9. **(a)** IV, since the slope is positive for $x > 0$ and negative for $x < 0$.

(b) VI, since the slope is positive for $y > 0$ and negative for $y < 0$.

(c) V, since the slope is always positive.

(d) II, since the slope changes sign when crossing the lines $y = \pm 1$.

(e) I, since the slope can be positive or negative in each quadrant but is not periodic.

(f) III, since the slope is periodic in both x and y.

11. (a) $y_0 = 1$,
$y_{n+1} = y_n + (x_n + y_n)(0.2) = (x_n + 6y_n)/5$

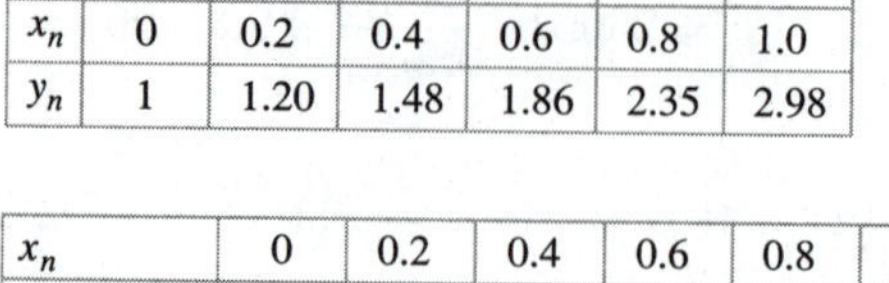

n	0	1	2	3	4	5
x_n	0	0.2	0.4	0.6	0.8	1.0
y_n	1	1.20	1.48	1.86	2.35	2.98

(b) $y' - y = x$, $\mu = e^{-x}$, $\dfrac{d}{dx}\left[ye^{-x}\right] = xe^{-x}$,
$ye^{-x} = -(x+1)e^{-x} + C$, $1 = -1 + C$,
$C = 2$, $y = -(x+1) + 2e^x$

x_n	0	0.2	0.4	0.6	0.8	1.0
$y(x_n)$	1	1.24	1.58	2.04	2.65	3.44
abs. error	0	0.04	0.10	0.19	0.30	0.46
perc. error	0	3	6	9	11	13

(c)

13. $y_0 = 1, y_{n+1} = y_n + \sqrt{y_n}/2$

n	0	1	2	3	4	5	6	7	8
x_n	0	0.5	1	1.5	2	2.5	3	3.5	4
y_n	1	1.50	2.11	2.84	3.68	4.64	5.72	6.91	8.23

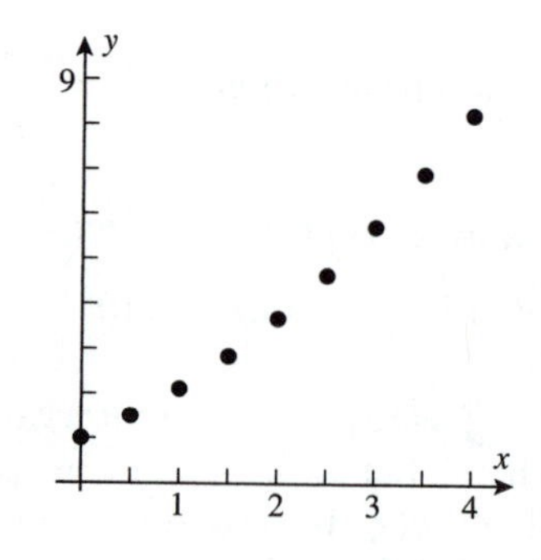

15. $y_0 = 1, y_{n+1} = y_n + \dfrac{1}{2}\sin y_n$

n	0	1	2	3	4
t_n	0	0.5	1	1.5	2
y_n	1	1.42	1.92	2.39	2.73

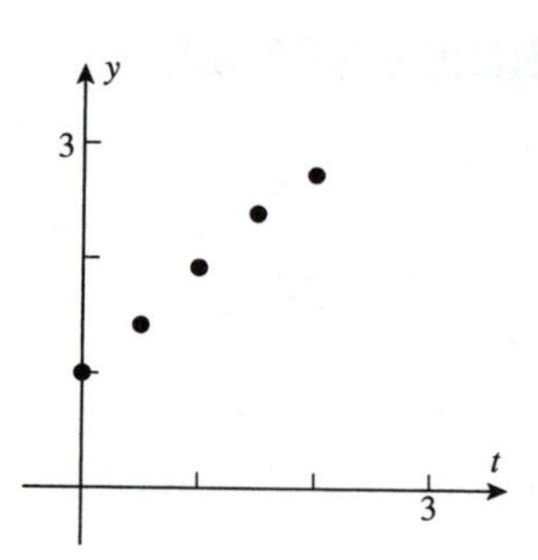

17. $h = 1/5$, $y_0 = 1$, $y_{n+1} = y_n + \dfrac{1}{5}\cos(2\pi n/5)$

n	0	1	2	3	4	5
t_n	0	0.2	0.4	0.6	0.8	1.0
y_n	1.00	1.06	0.90	0.74	0.80	1.00

19. (b) $y\,dy = -x\,dx$, $y^2/2 = -x^2/2 + C_1$, $x^2 + y^2 = C$; if $y(0) = 1$ then $C = 1$ so $y(1/2) = \sqrt{3}/2$.

EXERCISE SET 9.3

1. (a) $\frac{dy}{dt} = ky^2,\ y(0) = y_0, k > 0$

(b) $\frac{dy}{dt} = -ky^2,\ y(0) = y_0, k > 0$

3. (a) $\frac{ds}{dt} = \frac{1}{2}s$

(b) $\frac{d^2s}{dt^2} = 2\frac{ds}{dt}$

5. (a) $\frac{dy}{dt} = 0.01y,\ y_0 = 10{,}000$

(b) $y = 10{,}000e^{t/100}$

(c) $T = \frac{1}{k}\ln 2 = \frac{1}{0.01}\ln 2 \approx 69.31$ h

(d) $45{,}000 = 10{,}000e^{t/100}$,
$t = 100\ln\frac{45{,}000}{10{,}000} \approx 150.41$ h

7. (a) $\frac{dy}{dt} = -ky,\ y(0) = 5.0\times 10^7;\ 3.83 = T = \frac{1}{k}\ln 2$, so $k = \frac{\ln 2}{3.83} \approx 0.1810$

(b) $y = 5.0\times 10^7 e^{-0.181t}$

(c) $y(30) = 5.0\times 10^7 e^{-0.1810(30)} \approx 219{,}000$

(d) $y(t) = (0.1)y_0 = y_0 e^{-kt},\ -kt = \ln 0.1,\ t = -\frac{\ln 0.1}{0.1810} = 12.72$ days

9. $100e^{0.02t} = 5000,\ e^{0.02t} = 50,\ t = \frac{1}{0.02}\ln 50 \approx 196$ days

11. $y(t) = y_0e^{-kt} = 10.0e^{-kt},\ 3.5 = 10.0e^{-k(5)},\ k = -\frac{1}{5}\ln\frac{3.5}{10.0} \approx 0.2100,\ T = \frac{1}{k}\ln 2 \approx 3.30$ days

13. (a) $k = \frac{\ln 2}{5} \approx 0.1386; y \approx 2e^{0.1386t}$

(b) $y(t) = 5e^{0.015t}$

(c) $y = y_0e^{kt},\ 1 = y_0e^k,\ 100 = y_0e^{10k}$. Divide: $100 = e^{9k},\ k = \frac{1}{9}\ln 100 \approx 0.5117$,
$y \approx y_0e^{0.5117t}$; also $y(1) = 1$, so $y_0 = e^{-0.5117} \approx 0.5995, y \approx 0.5995e^{0.5117t}$.

(d) $k = \frac{\ln 2}{T} \approx 0.1386,\ 1 = y(1) \approx y_0e^{0.1386},\ y_0 \approx e^{-0.1386} \approx 0.8706,\ y \approx 0.8706e^{0.1386t}$

17. (a) $T = \frac{\ln 2}{k}$; and $\ln 2 \approx 0.6931$. If k is measured in percent, $k' = 100k$,
then $T = \frac{\ln 2}{k} \approx \frac{69.31}{k'} \approx \frac{70}{k'}$.

(b) 70 yr

(c) 20 yr

(d) 7%

19. From (12), $y(t) = y_0e^{-0.000121t}$. If $0.27 = \frac{y(t)}{y_0} = e^{-0.000121t}$ then $t = -\frac{\ln 0.27}{0.000121} \approx 10{,}820$ yr, and if $0.30 = \frac{y(t)}{y_0}$ then $t = -\frac{\ln 0.30}{0.000121} \approx 9950$, or roughly between 9000 B.C. and 8000 B.C.

21. $y_0 \approx 2,\ L \approx 8$; since the curve $y = \frac{2\cdot 8}{2 + 6e^{-kt}}$ passes through the point $(2, 4)$, $4 = \frac{16}{2 + 6e^{-2k}}$,
$6e^{-2k} = 2,\ k = \frac{1}{2}\ln 3 \approx 0.5493$.

23. **(a)** $y_0 = 5$ **(b)** $L = 12$ **(c)** $k = 1$

(d) $L/2 = 6 = \dfrac{60}{5 + 7e^{-t}}$, $5 + 7e^{-t} = 10$, $t = -\ln(5/7) \approx 0.3365$

(e) $\dfrac{dy}{dt} = \dfrac{1}{12}y(12 - y)$, $y(0) = 5$

25. See (13):

(a) $L = 10$ **(b)** $k = 10$

(c) $\dfrac{dy}{dt} = 10(1 - 0.1y)y = 25 - (y - 5)^2$ is maximized when $y = 5$.

27. Assume $y(t)$ students have had the flu t days after semester break. Then $y(0) = 20$, $y(5) = 35$.

(a) $\dfrac{dy}{dt} = ky(L - y) = ky(1000 - y)$, $y_0 = 20$

(b) Part (a) has solution $y = \dfrac{20000}{20 + 980e^{-kt}} = \dfrac{1000}{1 + 49e^{-kt}}$;

$35 = \dfrac{1000}{1 + 49e^{-5k}}$, $k = 0.115$, $y \approx \dfrac{1000}{1 + 49e^{-0.115t}}$.

(c)

t	0	1	2	3	4	5	6	7	8	9	10	11	12	13	14
$y(t)$	20	22	25	28	31	35	39	44	49	54	61	67	75	83	93

(d)

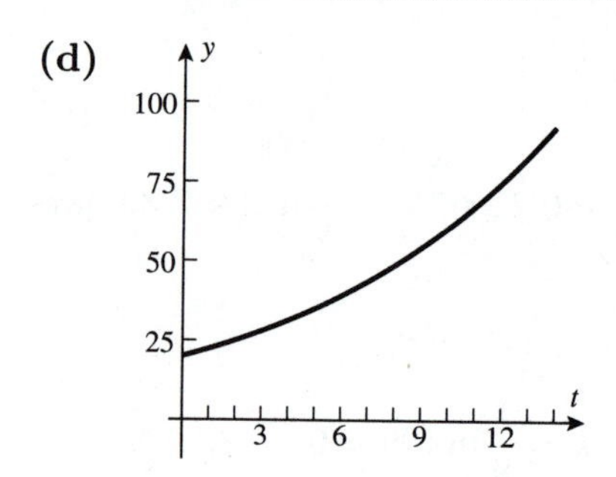

29. **(a)** $\dfrac{dT}{dt} = -k(T - 21)$, $T(0) = 95$, $\dfrac{dT}{T - 21} = -k\,dt$, $\ln(T - 21) = -kt + C_1$,

$T = 21 + e^{C_1}e^{-kt} = 21 + Ce^{-kt}$, $95 = T(0) = 21 + C, C = 74$, $T = 21 + 74e^{-kt}$

(b) $85 = T(1) = 21 + 74e^{-k}$, $k = -\ln\dfrac{64}{74} = -\ln\dfrac{32}{37}$, $T = 21 + 74e^{t\ln(32/37)} = 21 + 74\left(\dfrac{32}{37}\right)^t$,

$T = 51$ when $\dfrac{30}{74} = \left(\dfrac{32}{37}\right)^t$, $t = \dfrac{\ln(30/74)}{\ln(32/37)} \approx 6.22$ min

31. Let T denote the body temperature of McHam's body at time t, the number of hours elapsed after 10:06 P.M.; then $\dfrac{dT}{dt} = -k(T - 72)$, $\dfrac{dT}{T - 72} = -kdt$, $\ln(T - 72) = -kt + C, T = 72 + e^C e^{-kt}$,

$77.9 = 72 + e^C$, $e^C = 5.9$, $T = 72 + 5.9e^{-kt}$, $75.6 = 72 + 5.9e^{-k}$, $k = -\ln\dfrac{3.6}{5.9} \approx 0.4940$,

$T = 72 + 5.9e^{-0.4940t}$. McHam's body temperature was last 98.6° when $t = -\dfrac{\ln(26.6/5.9)}{0.4940} \approx -3.05$, so around 3 hours and 3 minutes before 10:06; the death took place at approximately 7:03 P.M., while Moore was on stage.

33. **(a)** $y = y_0 b^t = y_0 e^{t \ln b} = y_0 e^{kt}$ with $k = \ln b > 0$ since $b > 1$.

(b) $y = y_0 b^t = y_0 e^{t \ln b} = y_0 e^{-kt}$ with $k = -\ln b > 0$ since $0 < b < 1$.

(c) $y = 4(2^t) = 4e^{t \ln 2}$

(d) $y = 4(0.5^t) = 4e^{t \ln 0.5} = 4e^{-t \ln 2}$

EXERCISE SET 9.4

1. **(a)** $y = e^{2x}$, $y' = 2e^{2x}$, $y'' = 4e^{2x}$; $y'' - y' - 2y = 0$
$y = e^{-x}$, $y' = -e^{-x}$, $y'' = e^{-x}$; $y'' - y' - 2y = 0$.

(b) $y = c_1 e^{2x} + c_2 e^{-x}$, $y' = 2c_1 e^{2x} - c_2 e^{-x}$, $y'' = 4c_1 e^{2x} + c_2 e^{-x}$; $y'' - y' - 2y = 0$

3. $m^2 + 3m - 4 = 0$, $(m-1)(m+4) = 0$; $m = 1, -4$ so $y = c_1 e^x + c_2 e^{-4x}$.

5. $m^2 - 2m + 1 = 0$, $(m-1)^2 = 0$; $m = 1$, so $y = c_1 e^x + c_2 x e^x$.

7. $m^2 + 5 = 0$, $m = \pm\sqrt{5}\,i$ so $y = c_1 \cos\sqrt{5}\,x + c_2 \sin\sqrt{5}\,x$.

9. $m^2 - m = 0$, $m(m-1) = 0$; $m = 0, 1$ so $y = c_1 + c_2 e^x$.

11. $m^2 + 4m + 4 = 0$, $(m+2)^2 = 0$; $m = -2$ so $y = c_1 e^{-2t} + c_2 t e^{-2t}$.

13. $m^2 - 4m + 13 = 0$, $m = 2 \pm 3i$ so $y = e^{2x}(c_1 \cos 3x + c_2 \sin 3x)$.

15. $8m^2 - 2m - 1 = 0$, $(4m+1)(2m-1) = 0$; $m = -1/4, 1/2$ so $y = c_1 e^{-x/4} + c_2 e^{x/2}$.

17. $m^2 + 2m - 3 = 0$, $(m+3)(m-1) = 0$; $m = -3, 1$ so $y = c_1 e^{-3x} + c_2 e^x$ and $y' = -3c_1 e^{-3x} + c_2 e^x$. Solve the system $c_1 + c_2 = 1$, $-3c_1 + c_2 = 5$ to get $c_1 = -1$, $c_2 = 2$ so $y = -e^{-3x} + 2e^x$.

19. $m^2 - 6m + 9 = 0$, $(m-3)^2 = 0$; $m = 3$ so $y = (c_1 + c_2 x)e^{3x}$ and $y' = (3c_1 + c_2 + 3c_2 x)e^{3x}$. Solve the system $c_1 = 2$, $3c_1 + c_2 = 1$ to get $c_1 = 2$, $c_2 = -5$ so $y = (2 - 5x)e^{3x}$.

21. $m^2 + 4m + 5 = 0$, $m = -2 \pm i$ so $y = e^{-2x}(c_1 \cos x + c_2 \sin x)$,
$y' = e^{-2x}[(c_2 - 2c_1)\cos x - (c_1 + 2c_2)\sin x]$. Solve the system $c_1 = -3$, $c_2 - 2c_1 = 0$ to get $c_1 = -3$, $c_2 = -6$ so $y = -e^{-2x}(3\cos x + 6\sin x)$.

23. **(a)** $m = 5, -2$ so $(m-5)(m+2) = 0$, $m^2 - 3m - 10 = 0$; $y'' - 3y' - 10y = 0$.

(b) $m = 4, 4$ so $(m-4)^2 = 0$, $m^2 - 8m + 16 = 0$; $y'' - 8y' + 16y = 0$.

(c) $m = -1 \pm 4i$ so $(m + 1 - 4i)(m + 1 + 4i) = 0$, $m^2 + 2m + 17 = 0$; $y'' + 2y' + 17y = 0$.

25. $m^2 + km + k = 0$, $m = \left(-k \pm \sqrt{k^2 - 4k}\right)/2$

(a) $k^2 - 4k > 0$, $k(k-4) > 0$; $k < 0$ or $k > 4$

(b) $k^2 - 4k = 0$; $k = 0, 4$

(c) $k^2 - 4k < 0$, $k(k-4) < 0$; $0 < k < 4$

27. **(a)** $\dfrac{d^2y}{dz^2} + 2\dfrac{dy}{dz} + 2y = 0$, $m^2 + 2m + 2 = 0$; $m = -1 \pm i$ so
$y = e^{-z}(c_1 \cos z + c_2 \sin z) = \dfrac{1}{x}[c_1 \cos(\ln x) + c_2 \sin(\ln x)]$.

(b) $\dfrac{d^2y}{dz^2} - 2\dfrac{dy}{dz} - 2y = 0$, $m^2 - 2m - 2 = 0$; $m = 1 \pm \sqrt{3}$ so
$y = c_1 e^{(1+\sqrt{3})z} + c_2 e^{(1-\sqrt{3})z} = c_1 x^{1+\sqrt{3}} + c_2 x^{1-\sqrt{3}}$

29. **(a)** Neither is a constant multiple of the other, since, e.g. if $y_1 = ky_2$ then $e^{m_1x} = ke^{m_2x}$, $e^{(m_1-m_2)x} = k$. But the right hand side is constant, and the left hand side is constant only if $m_1 = m_2$, which is false.

(b) If $y_1 = ky_2$ then $e^{mx} = kxe^{mx}, kx = 1$ which is impossible. If $y_2 = y_1$ then $xe^{mx} = ke^{mx}$, $x = k$ which is impossible.

31. **(a)** The general solution is $c_1e^{\mu x} + c_2e^{mx}$; let $c_1 = 1/(\mu - m)$, $c_2 = -1/(\mu - m)$.

(b) $\lim\limits_{\mu \to m} \dfrac{e^{\mu x} - e^{mx}}{\mu - m} = \lim\limits_{\mu \to m} xe^{\mu x} = xe^{mx}$.

33. $k/M = 0.25/1 = 0.25$

(a) From (20), $y = 0.3\cos(t/2)$

(b) $T = 2\pi \cdot 2 = 4\pi$ s, $f = 1/T = 1/(4\pi)$ Hz

(c)

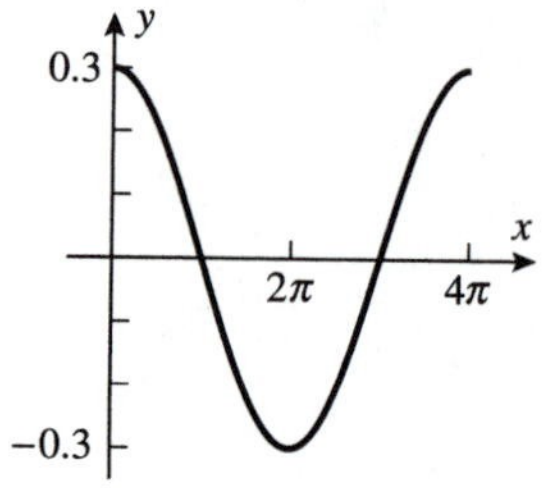

(d) $y = 0$ at the equilibrium position, so $t/2 = \pi/2, t = \pi$ s.

(e) $t/2 = \pi$ at the maximum position below the equlibrium position, so $t = 2\pi$ s.

35. $l = 0.05$, $k/M = g/l = 9.8/0.05 = 196\ \text{s}^{-2}$

(a) From (20), $y = -0.12\cos 14t$.

(b) $T = 2\pi\sqrt{M/k} = 2\pi/14 = \pi/7$ s,
$f = 7/\pi$ Hz

(c)

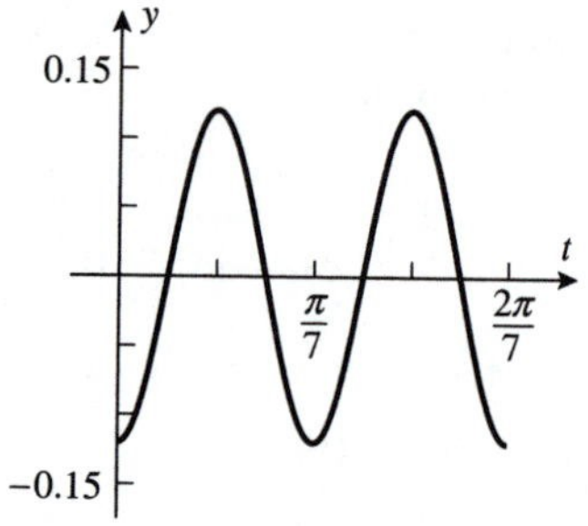

(d) $14t = \pi/2$, $t = \pi/28$ s

(e) $14t = \pi$, $t = \pi/14$ s

37. Assume $y = y_0 \cos\sqrt{\dfrac{k}{M}}\,t$, so $v = \dfrac{dy}{dt} = -y_0\sqrt{\dfrac{k}{M}}\sin\sqrt{\dfrac{k}{M}}\,t$

(a) The maximum speed occurs when $\sin\sqrt{\dfrac{k}{M}}\,t = \pm 1$, $\sqrt{\dfrac{k}{M}}\,t = n\pi + \pi/2$,

so $\cos\sqrt{\dfrac{k}{M}}\,t = 0$, $y = 0$.

(b) The minimum speed occurs when $\sin\sqrt{\dfrac{k}{M}}\,t = 0$, $\sqrt{\dfrac{k}{M}}\,t = n\pi$, so $\cos\sqrt{\dfrac{k}{M}}\,t = \pm 1$, $y = \pm y_0$.

39. By Hooke's Law, $F(t) = -kx(t)$, since the only force is the restoring force of the spring. Newton's Second Law gives $F(t) = Mx''(t)$, so $Mx''(t) + kx(t) = 0$, $x(0) = x_0, x'(0) = 0$.

41. **(a)** $m^2 + 2.4m + 1.44 = 0, (m + 1.2)^2 = 0, m = -1.2,\ y = C_1e^{-6t/5} + C_2te^{-6t/5}$,

$C_1 = 1,\ 2 = y'(0) = -\frac{6}{5}C_1 + C_2, C_2 = \frac{16}{5},\ y = e^{-6t/5} + \frac{16}{5}te^{-6t/5}$

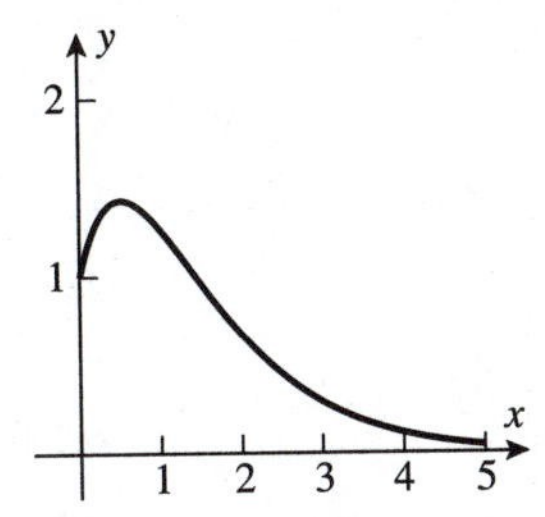

(b) $y'(t) = 0$ when $t = t_1 = 25/48 \approx 0.520833,\ y(t_1) = 1.427364$ cm

(c) $y = \frac{16}{5}\ e^{-6t/5}(t + 5/16) = 0$ only if $t = -5/16$, so $y \neq 0$ for $t \geq 0$.

43. **(a)** $m^2 + m + 5 = 0, m = -1/2 \pm (\sqrt{19}/2)i,\ y = e^{-t/2}\left[C_1\cos(\sqrt{19}t/2) + C_2\sin(\sqrt{19}t/2)\right]$,

$1 = y(0) = C_1, -3.5 = y'(0) = -(1/2)C_1 + (\sqrt{19}/2)C_2,\ C_2 = -6/\sqrt{19}$,

$y = e^{-t/2}\cos(\sqrt{19}\,t/2) - (6/\sqrt{19}\,)e^{-t/2}\sin(\sqrt{19}\,t/2)$

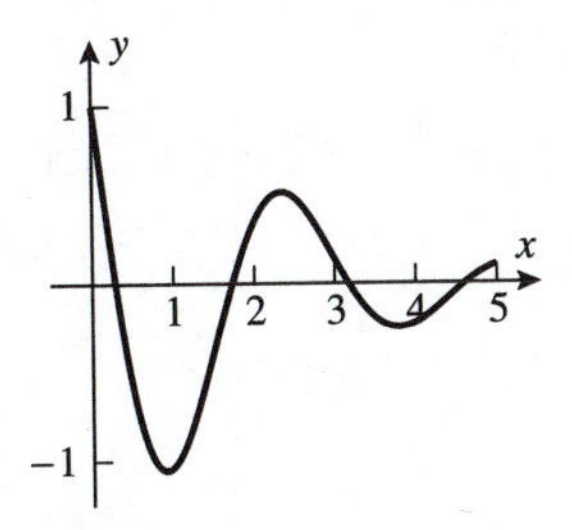

(b) $y'(t) = 0$ for the first time when $t = t_1 = 0.905533,\ y(t_1) = -1.054466$ cm so the maximum distance below the equilibrium position is 1.054466 cm.

(c) $y(t) = 0$ for the first time when $t = t_2 = 0.288274,\ y'(t_2) = -3.210357$ cm/s.

(d) The acceleration is $y''(t)$ so from the differential equation $y'' = -y' - 5y$. But $y = 0$ when the object passes through the equilibrium position, thus $y'' = -y' = 3.210357$ cm/s^2.

45. **(a)** $m^2 + 3.5m + 3 = (m + 1.5)(m + 2), y = C_1e^{-3t/2} + C_2e^{-2t}$,

$1 = y(0) = C_1 + C_2, v_0 = y'(0) = -(3/2)C_1 - 2C_2,\ C_1 = 4 + 2v_0, C_2 = -3 - 2v_0$,

$y(t) = (4 + 2v_0)e^{-3t/2} - (3 + 2v_0)e^{-2t}$

(b) $v_0 = 2, y(t) = 8e^{-3t/2} - 7e^{-2t},\ v_0 = -1, y(t) = 2e^{-3t/2} - e^{-2t}$,

$v_0 = -4, y(t) = -4e^{-3t/2} + 5e^{-2t}$

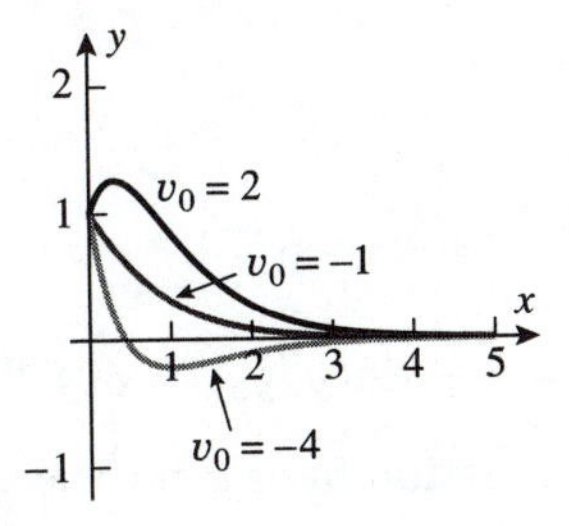

CHAPTER 9 SUPPLEMENTARY EXERCISES

5. (a) linear (b) linear and separable (c) separable (d) neither

7. The parabola $ky(L-y)$ opens down and has its maximum midway between the y-intercepts, that is, at the point $y = \frac{1}{2}(0+L) = L/2$, where $\frac{dy}{dt} = k(L/2)^2 = kL^2/4$.

9. $\frac{dV}{dt} = -kS$; but $V = \frac{4\pi}{3}r^3$, $\frac{dV}{dt} = 4\pi r^2 \frac{dr}{dt}$, $S = 4\pi r^2$, so $dr/dt = -k, r = -kt + C, 4 = C$, $r = -kt + 4, 3 = -k + 4, k = 1, r = 4 - t$ m.

11. (a) Assume the air contains $y(t)$ ft^3 of carbon monoxide at time t. Then $y_0 = 0$ and for $t > 0$, $\frac{dy}{dt} = 0.04(0.1) - \frac{y}{1200}(0.1) = 1/250 - y/12000$, $\frac{d}{dt}\left[ye^{t/12000}\right] = \frac{1}{250}e^{t/12000}$, $ye^{t/12000} = 48e^{t/12000} + C$, $y(0) = 0$, $C = -48$; $y = 48(1 - e^{-t/12000})$. Thus the percentage of carbon monoxide is $P = \frac{y}{1200}100 = 4(1 - e^{-t/12000})$ percent.

(b) $0.012 = 4(1 - e^{-t/12000})$, $t = 36.05$ min

13. $\left(\frac{1}{y^5} + \frac{1}{y}\right) dy = \frac{dx}{x}$, $-\frac{1}{4}y^{-4} + \ln|y| = \ln|x| + C$; $-\frac{1}{4} = C$, $y^{-4} + 4\ln(x/y) = 1$

15. $\frac{dy}{y^2} = 4\sec^2 2x\,dx$, $-\frac{1}{y} = 2\tan 2x + C$, $-1 = 2\tan\left(2\frac{\pi}{8}\right) + C = 2\tan\frac{\pi}{4} + C = 2 + C$, $C = -3$, $y = \frac{1}{3 - 2\tan 2x}$

17. (a) $\mu = e^{-\int dx} = e^{-x}$, $\frac{d}{dx}\left[ye^{-x}\right] = xe^{-x}\sin 3x$,

$$ye^{-x} = \int xe^{-x}\sin 3x\,dx = \left(-\frac{3}{10}x - \frac{3}{50}\right)e^{-x}\cos 3x + \left(-\frac{1}{10}x + \frac{2}{25}\right)e^{-x}\sin 3x + C;$$

$$1 = y(0) = -\frac{3}{50} + C,\ C = \frac{53}{50},\ y = \left(-\frac{3}{10}x - \frac{3}{50}\right)\cos 3x + \left(-\frac{1}{10}x + \frac{2}{25}\right)\sin 3x + \frac{53}{50}e^x$$

(c)

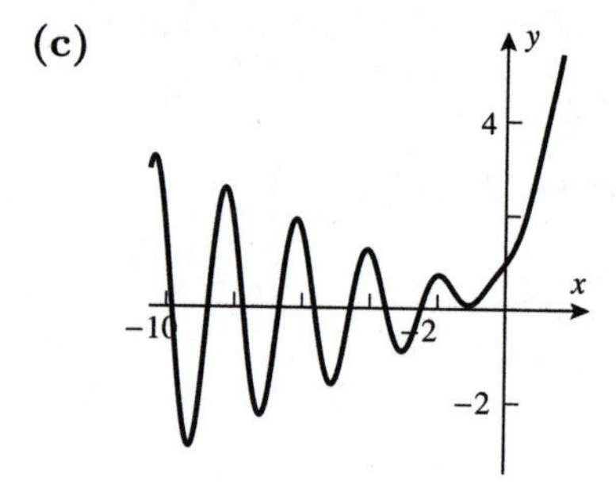

19. (a) Let $T_1 = 5730 - 40 = 5690$, $k_1 = \frac{\ln 2}{T_1} \approx 0.00012182$; $T_2 = 5730 + 40 = 5770$, $k_2 \approx 0.00012013$. With $y/y_0 = 0.92, 0.93$, $t_1 = -\frac{1}{k_1}\ln\frac{y}{y_0} = 684.5, 595.7$; $t_2 = -\frac{1}{k_2}\ln(y/y_0) = 694.1, 604.1$; in 1988 the shroud was at most 695 years old, which places its creation in or after the year 1293.

(b) Suppose T is the true half-life of carbon-14 and $T_1 = T(1+r/100)$ is the false half-life. Then with $k = \dfrac{\ln 2}{T}$, $k_1 = \dfrac{\ln 2}{T_1}$ we have the formulae $y(t) = y_0 e^{-kt}$, $y_1(t) = y_0 e^{-k_1 t}$. At a certain point in time a reading of the carbon-14 is taken resulting in a certain value y, which in the case of the true formula is given by $y = y(t)$ for some t, and in the case of the false formula is given by $y = y_1(t_1)$ for some t_1.

If the true formula is used then the time t since the beginning is given by $t = -\dfrac{1}{k}\ln\dfrac{y}{y_0}$. If the false formula is used we get a false value $t_1 = -\dfrac{1}{k_1}\ln\dfrac{y}{y_0}$; note that in both cases the value y/y_0 is the same. Thus $t_1/t = k/k_1 = T_1/T = 1 + r/100$, so the percentage error in the time to be measured is the same as the percentage error in the half-life.

21. (a) $y = C_1 e^x + C_2 e^{2x}$ (b) $y = C_1 e^{x/2} + C_2 x e^{x/2}$

(c) $y = e^{-x/2}\left[C_1 \cos\dfrac{\sqrt{7}}{2}x + C_2 \sin\dfrac{\sqrt{7}}{2}x\right]$

23. (a) Use (15) in Section 9.3 with $y_0 = 19, L = 95$: $y(t) = \dfrac{1805}{19 + 76e^{-kt}}$, $25 = y(1) = \dfrac{1805}{19 + 76e^{-k}}$, $k \approx 0.3567$; when $0.8L = y(t) = \dfrac{y_0 L}{19 + 76e^{-kt}}$, $19 + 76e^{-kt} = \dfrac{5}{4}y_0 = \dfrac{95}{4}$, $t \approx 7.77$ yr.

(b) From (13), $\dfrac{dy}{dt} = k\left(1 - \dfrac{y}{95}\right)y$, $y(0) = 19$.

25. $y = y_0 \cos\sqrt{\dfrac{k}{M}}t$, $T = 2\pi\sqrt{\dfrac{M}{k}}$, $y = y_0 \cos\dfrac{2\pi t}{T}$

(a) $v = y'(t) = -\dfrac{2\pi}{T}y_0 \sin\dfrac{2\pi t}{T}$ has maximum magnitude $2\pi|y_0|/T$ and occurs when $2\pi t/T = n\pi + \pi/2$, $y = y_0 \cos(n\pi + \pi/2) = 0$.

(b) $a = y''(t) = -\dfrac{4\pi^2}{T^2}y_0 \cos\dfrac{2\pi t}{T}$ has maximum magnitude $4\pi^2|y_0|/T^2$ and occurs when $2\pi t/T = j\pi$, $y = y_0 \cos j\pi = \pm y_0$.

27. (a) $A = 1000e^{(0.08)(5)} = 1000e^{0.4} \approx \$1,491.82$

(b) $Pe^{(0.08)(10)} = 10,000$, $Pe^{0.8} = 10,000$, $P = 10,000e^{-0.8} \approx \$4,493.29$

(c) From (11) of Section 9.3 with $k = r = 0.08$, $T = (\ln 2)/0.08 \approx 8.7$ years.

29. $\dfrac{d}{dt}\left[\dfrac{1}{2}k[y(t)]^2 + \dfrac{1}{2}M(y'(t))^2\right] = ky(t)y'(t) + My'(t)y''(t) = My'(t)[\dfrac{k}{M}y(t) + y''(t)] = 0$, as required.

CHAPTER 10
Infinite Series

EXERCISE SET 10.1

1. **(a)** $f^{(k)}(x) = (-1)^k e^{-x}$, $f^{(k)}(0) = (-1)^k$; $e^{-x} \approx 1 - x + x^2/2$ (quadratic), $e^{-x} \approx 1 - x$ (linear)

(b) $f'(x) = -\sin x, f''(x) = -\cos x, f(0) = 1, f'(0) = 0, f''(0) = -1,$
$\cos x \approx 1 - x^2/2$ (quadratic), $\cos x \approx 1$ (linear)

(c) $f'(x) = \cos x, f''(x) = -\sin x, f(\pi/2) = 1, f'(\pi/2) = 0, f''(\pi/2) = -1,$
$\sin x \approx 1 - (x - \pi/2)^2/2$ (quadratic), $\sin x \approx 1$ (linear)

(d) $f(1) = 1, f'(1) = 1/2, f''(1) = -1/4;$
$\sqrt{x} = 1 + \frac{1}{2}(x-1) - \frac{1}{8}(x-1)^2$ (quadratic), $\sqrt{x} \approx 1 + \frac{1}{2}(x-1)$ (linear)

3. **(a)** $f'(x) = \frac{1}{2}x^{-1/2}$, $f''(x) = -\frac{1}{4}x^{-3/2}$; $f(1) = 1, f'(1) = \frac{1}{2}, f''(1) = -\frac{1}{4}$;
$\sqrt{x} \approx 1 + \frac{1}{2}(x-1) - \frac{1}{8}(x-1)^2$

(b) $x = 1.1, x_0 = 1, \sqrt{1.1} \approx 1 + \frac{1}{2}(0.1) - \frac{1}{8}(0.1)^2 = 1.04875$, calculator value ≈ 1.0488088

5. $f(x) = \tan x$, $61° = \pi/3 + \pi/180$ rad; $x_0 = \pi/3$, $f'(x) = \sec^2 x$, $f''(x) = 2\sec^2 x \tan x$;
$f(\pi/3) = \sqrt{3}, f'(\pi/3) = 4, f''(x) = 8\sqrt{3}$; $\tan x \approx \sqrt{3} + 4(x - \pi/3) + 4\sqrt{3}(x - \pi/3)^2$,
$\tan 61° = \tan(\pi/3 + \pi/180) \approx \sqrt{3} + 4\pi/180 + 4\sqrt{3}(\pi/180)^2 \approx 1.80397443$,
calculator value ≈ 1.80404776

7. $f^{(k)}(x) = (-1)^k e^{-x}$, $f^{(k)}(0) = (-1)^k; p_0(x) = 1$, $p_1(x) = 1 - x$, $p_2(x) = 1 - x + \frac{1}{2}x^2$,
$p_3(x) = 1 - x + \frac{1}{2}x^2 - \frac{1}{3!}x^3$, $p_4(x) = 1 - x + \frac{1}{2}x^2 - \frac{1}{3!}x^3 + \frac{1}{4!}x^4$; $\sum_{k=0}^{n} \frac{(-1)^k}{k!}x^k$

9. $f^{(k)}(0) = 0$ if k is odd, $f^{(k)}(0)$ is alternately π^k and $-\pi^k$ if k is even; $p_0(x) = 1$, $p_1(x) = 1$,
$p_2(x) = 1 - \frac{\pi^2}{2!}x^2$; $p_3(x) = 1 - \frac{\pi^2}{2!}x^2$, $p_4(x) = 1 - \frac{\pi^2}{2!}x^2 + \frac{\pi^4}{4!}x^4$; $\sum_{k=0}^{[\frac{n}{2}]} \frac{(-1)^k \pi^{2k}}{(2k)!}x^{2k}$
NB: The function $[x]$ defined for real x indicates the greatest integer which is $\leq x$.

11. $f^{(0)}(0) = 0$; for $k \geq 1$, $f^{(k)}(x) = \frac{(-1)^{k+1}(k-1)!}{(1+x)^k}$, $f^{(k)}(0) = (-1)^{k+1}(k-1)!$; $p_0(x) = 0$,
$p_1(x) = x$, $p_2(x) = x - \frac{1}{2}x^2$, $p_3(x) = x - \frac{1}{2}x^2 + \frac{1}{3}x^3$, $p_4(x) = x - \frac{1}{2}x^2 + \frac{1}{3}x^3 - \frac{1}{4}x^4$; $\sum_{k=1}^{n} \frac{(-1)^{k+1}}{k}x^k$

13. $f^{(k)}(0) = 0$ if k is odd, $f^{(k)}(0) = 1$ if k is even; $p_0(x) = 1, p_1(x) = 1$,
$p_2(x) = 1 + x^2/2$, $p_3(x) = 1 + x^2/2$, $p_4(x) = 1 + x^2/2 + x^4/4!$; $\sum_{k=0}^{[\frac{n}{2}]} \frac{1}{(2k)!}x^{2k}$

15. $f^{(k)}(x) = \begin{cases} (-1)^{k/2}(x\sin x - k\cos x) & k \text{ even} \\ (-1)^{(k-1)/2}(x\cos x + k\sin x) & k \text{ odd} \end{cases}$, $\quad f^{(k)}(0) = \begin{cases} (-1)^{1+k/2}k & k \text{ even} \\ 0 & k \text{ odd} \end{cases}$

$p_0(x) = 0,\ p_1(x) = 0,\ p_2(x) = x^2, p_3(x) = x^2,\ p_4(x) = x^2 - \frac{1}{6}x^4;\ \sum_{k=0}^{[\frac{n}{2}]-1} \frac{(-1)^k}{(2k+1)!}x^{2k+2}$

17. $f^{(k)}(x_0) = e;\ p_0(x) = e,\ p_1(x) = e + e(x-1),$

$p_2(x) = e + e(x-1) + \frac{e}{2}(x-1)^2,\ p_3(x) = e + e(x-1) + \frac{e}{2}(x-1)^2 + \frac{e}{3!}(x-1)^3,$

$p_4(x) = e + e(x-1) + \frac{e}{2}(x-1)^2 + \frac{e}{3!}(x-1)^3 + \frac{e}{4!}(x-1)^4;\ \sum_{k=0}^{n} \frac{e}{k!}(x-1)^k$

19. $f^{(k)}(x) = \frac{(-1)^k k!}{x^{k+1}},\ f^{(k)}(-1) = -k!;\ p_0(x) = -1;\ p_1(x) = -1 - (x+1);$

$p_2(x) = -1 - (x+1) - (x+1)^2;\ p_3(x) = -1 - (x+1) - (x+1)^2 - (x+1)^3;$

$p_4(x) = -1 - (x+1) - (x+1)^2 - (x+1)^3 - (x+1)^4;\ \sum_{k=0}^{n}(-1)(x+1)^k$

21. $f^{(k)}(1/2) = 0$ if k is odd, $f^{(k)}(1/2)$ is alternately π^k and $-\pi^k$ if k is even;

$p_0(x) = p_1(x) = 1, p_2(x) = p_3(x) = 1 - \frac{\pi^2}{2}(x-1/2)^2,$

$p_4(x) = 1 - \frac{\pi^2}{2}(x-1/2)^2 + \frac{\pi^4}{4!}(x-1/2)^4;\ \sum_{k=0}^{[\frac{n}{2}]} \frac{(-1)^k \pi^{2k}}{(2k)!}(x-1/2)^{2k}$

23. $f(1) = 0$, for $k \geq 1, f^{(k)}(x) = \frac{(-1)^{k-1}(k-1)!}{x^k};\ f^{(k)}(1) = (-1)^{k-1}(k-1)!;$

$p_0(x) = 0,\ p_1(x) = (x-1);\ p_2(x) = (x-1) - \frac{1}{2}(x-1)^2;\ p_3(x) = (x-1) - \frac{1}{2}(x-1)^2 + \frac{1}{3}(x-1)^3,$

$p_4(x) = (x-1) - \frac{1}{2}(x-1)^2 + \frac{1}{3}(x-1)^3 - \frac{1}{4}(x-1)^4;\ \sum_{k=1}^{n} \frac{(-1)^{k-1}}{k}(x-1)^k$

25. **(a)** $f(0) = 1, f'(0) = 2, f''(0) = -2, f'''(0) = 6$, the third MacLaurin polynomial for $f(x)$ is $f(x)$.

(b) $f(1) = 1, f'(1) = 2, f''(1) = -2, f'''(1) = 6$, the third Taylor polynomial for $f(x)$ is $f(x)$.

27. $f^{(k)}(0) = (-2)^k;\ p_0(x) = 1,\ p_1(x) = 1 - 2x,$

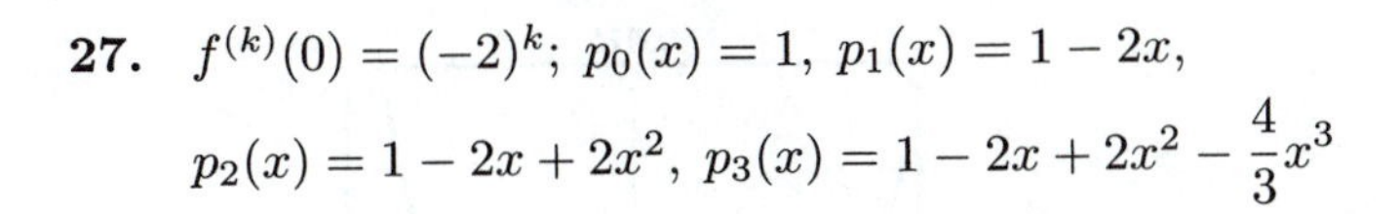

$p_2(x) = 1 - 2x + 2x^2,\ p_3(x) = 1 - 2x + 2x^2 - \frac{4}{3}x^3$

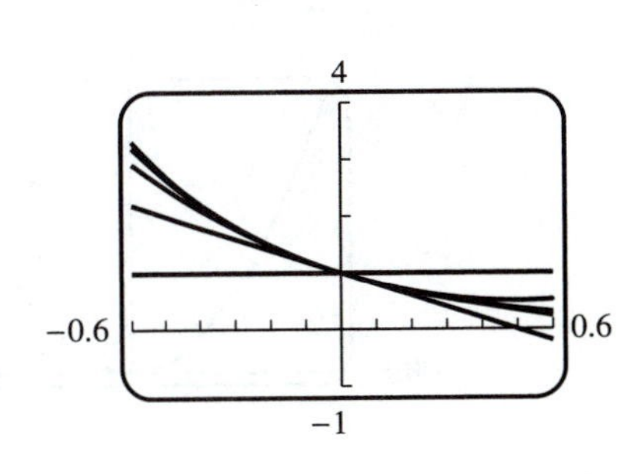

29. $f^{(k)}(\pi) = 0$ if k is odd, $f^{(k)}(\pi)$ is alternately -1 and 1 if k is even; $p_0(x) = -1$, $p_2(x) = -1 + \frac{1}{2}(x-\pi)^2$,

$p_4(x) = -1 + \frac{1}{2}(x-\pi)^2 - \frac{1}{24}(x-\pi)^4$,

$p_6(x) = -1 + \frac{1}{2}(x-\pi)^2 - \frac{1}{24}(x-\pi)^4 + \frac{1}{720}(x-\pi)^6$

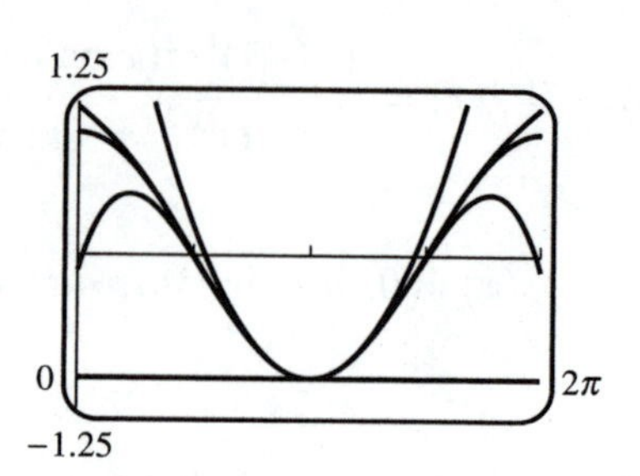

31. $f^{(k)}(x) = e^x, |f^{(k)}(x)| \le e^{1/2} < 2$ on $[0, 1/2]$, let $M = 2$,

$e^{1/2} = 1 + \frac{1}{2} + \frac{1}{8} + \frac{1}{48} + \frac{1}{24 \cdot 16} + \cdots + \frac{1}{n!2^n} + R_n(1/2)$;

$|R_n(1/2)| \le \frac{M}{(n+1)!}(1/2)^{n+1} \le \frac{2}{(n+1)!}(1/2)^{n+1} \le 0.00005$ for $n = 5$;

$e^{1/2} \approx 1 + \frac{1}{2} + \frac{1}{8} + \frac{1}{48} + \frac{1}{24 \cdot 16} + \frac{1}{120 \cdot 32} \approx 1.64870$, calculator value 1.64872

33. $p(0) = 1$, $p(x)$ has slope -1 at $x = 0$, and $p(x)$ is concave up at $x = 0$, eliminating I, II and III respectively and leaving IV.

35. $f^{(k)}(\ln 4) = 15/8$ for k even, $f^{(k)}(\ln 4) = 17/8$ for k odd, which can be written as

$f^{(k)}(\ln 4) = \frac{16 - (-1)^k}{8}$; $\sum_{k=0}^{n} \frac{16 - (-1)^k}{8k!}(x - \ln 4)^k$

37. From Exercise 2(a), $p_1(x) = 1 + x$, $p_2(x) = 1 + x + x^2/2$

(a)

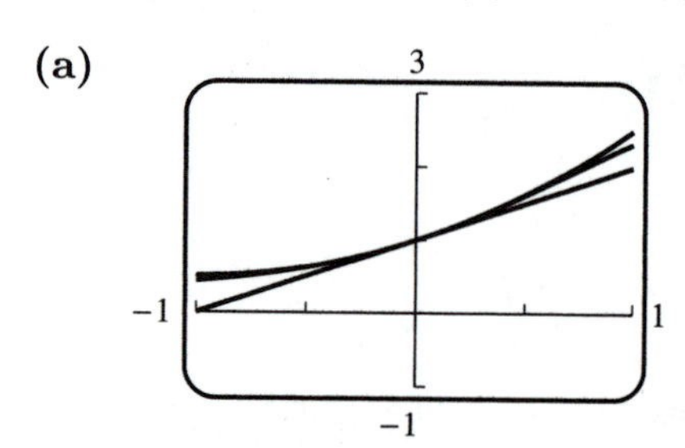

(b)

x	−1.000	−0.750	−0.500	−0.250	0.000	0.250	0.500	0.750	1.000
$f(x)$	0.431	0.506	0.619	0.781	1.000	1.281	1.615	1.977	2.320
$p_1(x)$	0.000	0.250	0.500	0.750	1.000	1.250	1.500	1.750	2.000
$p_2(x)$	0.500	0.531	0.625	0.781	1.000	1.281	1.625	2.031	2.500

(c) 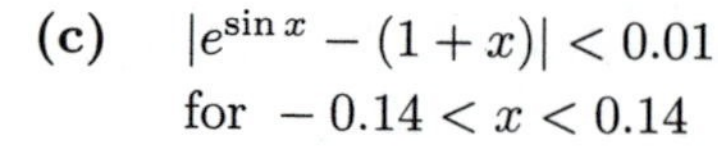
$|e^{\sin x} - (1 + x)| < 0.01$
for $-0.14 < x < 0.14$

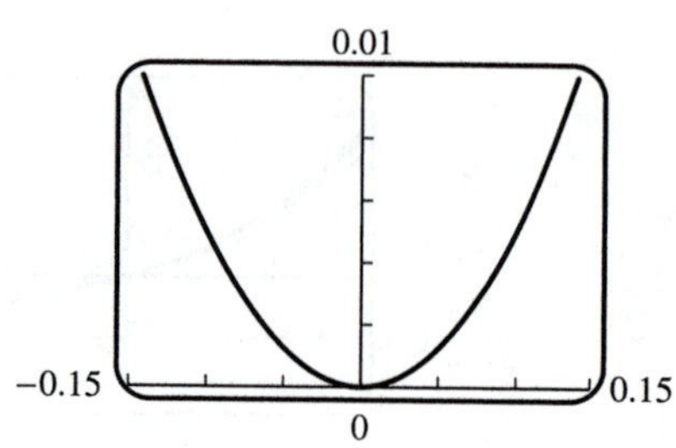

(d) $|e^{\sin x} - (1 + x + x^2/2)| < 0.01$
for $-0.50 < x < 0.50$

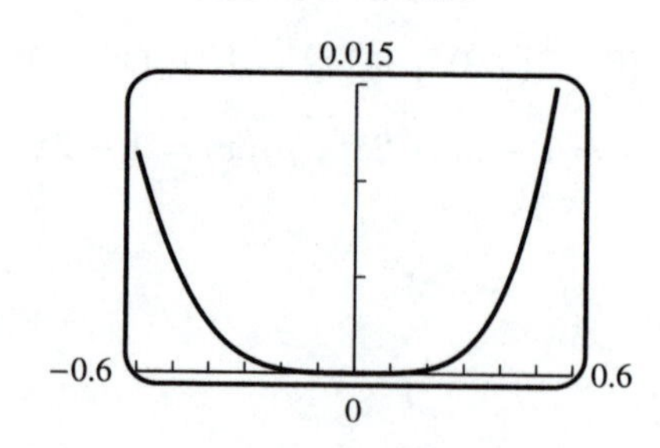

39. **(a)** $\sin x = x - \frac{x^3}{3!} + 0 \cdot x^4 + R_4(x)$,

$|R_4(x)| \le \frac{|x|^5}{5!} < 0.5 \times 10^{-3}$ if $|x|^5 < 0.06$,

$|x| < (0.06)^{1/5} \approx 0.569, (-0.569, 0.569)$

(b)

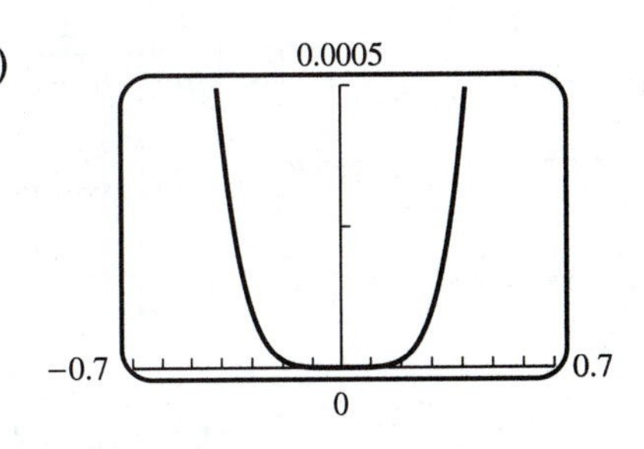

EXERCISE SET 10.2

1. **(a)** $\frac{1}{3^{n-1}}$ **(b)** $\frac{(-1)^{n-1}}{3^{n-1}}$ **(c)** $\frac{2n-1}{2n}$ **(d)** $\frac{n^2}{\pi^{1/(n+1)}}$

3. **(a)** $2, 0, 2, 0$ **(b)** $1, -1, 1, -1$ **(c)** $2(1 + (-1)^n)$; $2 + 2\cos n\pi$

5. $1/3, 2/4, 3/5, 4/6, 5/7, \ldots$; $\lim_{n\to+\infty} \frac{n}{n+2} = 1$, converges

7. $2, 2, 2, 2, 2, \ldots$; $\lim_{n\to+\infty} 2 = 2$, converges

9. $\frac{\ln 1}{1}, \frac{\ln 2}{2}, \frac{\ln 3}{3}, \frac{\ln 4}{4}, \frac{\ln 5}{5}, \ldots$;

$\lim_{n\to+\infty} \frac{\ln n}{n} = \lim_{n\to+\infty} \frac{1}{n} = 0$ $\left(\text{apply L'Hôpital's Rule to } \frac{\ln x}{x}\right)$, converges

11. $0, 2, 0, 2, 0, \ldots$; diverges

13. $-1, 16/9, -54/28, 128/65, -250/126, \ldots$; diverges because odd-numbered terms approach -2, even-numbered terms approach 2.

15. $6/2, 12/8, 20/18, 30/32, 42/50, \ldots$; $\lim_{n\to+\infty} \frac{1}{2}(1 + 1/n)(1 + 2/n) = 1/2$, converges

17. $\cos(3), \cos(3/2), \cos(1), \cos(3/4), \cos(3/5), \ldots$; $\lim_{n\to+\infty} \cos(3/n) = 1$, converges

19. $e^{-1}, 4e^{-2}, 9e^{-3}, 16e^{-4}, 25e^{-5}, \ldots$; $\lim_{x\to+\infty} x^2 e^{-x} = \lim_{x\to+\infty} \frac{x^2}{e^x} = 0$, so $\lim_{n\to+\infty} n^2 e^{-n} = 0$, converges

21. $2, (5/3)^2, (6/4)^3, (7/5)^4, (8/6)^5, \ldots$; let $y = \left[\frac{x+3}{x+1}\right]^x$, converges because

$$\lim_{x\to+\infty} \ln y = \lim_{x\to+\infty} \frac{\ln \frac{x+3}{x+1}}{1/x} = \lim_{x\to+\infty} \frac{2x^2}{(x+1)(x+3)} = 2, \text{ so } \lim_{n\to+\infty} \left[\frac{n+3}{n+1}\right]^n = e^2$$

23. $\left\{\frac{2n-1}{2n}\right\}_{n=1}^{+\infty}$; $\lim_{n\to+\infty} \frac{2n-1}{2n} = 1$, converges

25. $\left\{\frac{1}{3^n}\right\}_{n=1}^{+\infty}$; $\lim_{n\to+\infty} \frac{1}{3^n} = 0$, converges

27. $\left\{\frac{1}{n}-\frac{1}{n+1}\right\}_{n=1}^{+\infty}$; $\lim_{n\to+\infty}\left(\frac{1}{n}-\frac{1}{n+1}\right)=0$, converges

29. $\left\{\sqrt{n+1}-\sqrt{n+2}\right\}_{n=1}^{+\infty}$; converges because

$$\lim_{n\to+\infty}(\sqrt{n+1}-\sqrt{n+2})=\lim_{n\to+\infty}\frac{(n+1)-(n+2)}{\sqrt{n+1}+\sqrt{n+2}}=\lim_{n\to+\infty}\frac{-1}{\sqrt{n+1}+\sqrt{n+2}}=0$$

31. **(a)** $1, 2, 1, 4, 1, 6$ **(b)** $a_n=\begin{cases} n, & n \text{ odd} \\ 1/2^n, & n \text{ even}\end{cases}$ **(c)** $a_n=\begin{cases} 1/n, & n \text{ odd} \\ 1/(n+1), & n \text{ even}\end{cases}$

(d) In Part (a) the sequence diverges, since the even terms diverge to $+\infty$ and the odd terms equal 1; in Part (b) the sequence diverges, since the odd terms diverge to $+\infty$ and the even terms tend to zero; in Part (c) $\lim_{n\to+\infty} a_n = 0$.

33. $\lim_{n\to+\infty}\sqrt[n]{n}=1$, so $\lim_{n\to+\infty}\sqrt[n]{n^3}=1^3=1$

35. $\lim_{n\to+\infty} x_{n+1}=\frac{1}{2}\lim_{n\to+\infty}\left(x_n+\frac{a}{x_n}\right)$ or $L=\frac{1}{2}\left(L+\frac{a}{L}\right)$, $2L^2-L^2-a=0$, $L=\sqrt{a}$ (we reject $-\sqrt{a}$ because $x_n>0$, thus $L\ge 0$.

37. **(a)** $1, \frac{1}{4}+\frac{2}{4}, \frac{1}{9}+\frac{2}{9}+\frac{3}{9}, \frac{1}{16}+\frac{2}{16}+\frac{3}{16}+\frac{4}{16}=1, \frac{3}{4}, \frac{2}{3}, \frac{5}{8}$

(c) $a_n=\frac{1}{n^2}(1+2+\cdots+n)=\frac{1}{n^2}\frac{1}{2}n(n+1)=\frac{1}{2}\frac{n+1}{n}$, $\lim_{n\to+\infty} a_n=1/2$

39. Let $a_n=0, b_n=\frac{\sin^2 n}{n}, c_n=\frac{1}{n}$; then $a_n\le b_n\le c_n$, $\lim_{n\to+\infty}a_n=\lim_{n\to+\infty}c_n=0$, so $\lim_{n\to+\infty}b_n=0$.

41. **(a)** $a_1=(0.5)^2, a_2=a_1^2=(0.5)^4, \ldots, a_n=(0.5)^{2^n}$

(c) $\lim_{n\to+\infty}a_n=\lim_{n\to+\infty}e^{2^n\ln(0.5)}=0$, since $\ln(0.5)<0$.

(d) Replace 0.5 in Part (a) with a_0; then the sequence converges for $-1\le a_0\le 1$, because if $a_0=\pm1$, then $a_n=1$ for $n\ge1$; if $a_0=0$ then $a_n=0$ for $n\ge 1$; and if $0<|a_0|<1$ then $a_1=a_0^2>0$ and $\lim_{n\to+\infty}a_n=\lim_{n\to+\infty}e^{2^{n-1}\ln a_1}=0$ since $0<a_1<1$. This same argument proves divergence to $+\infty$ for $|a|>1$ since then $\ln a_1>0$.

43. **(a)**

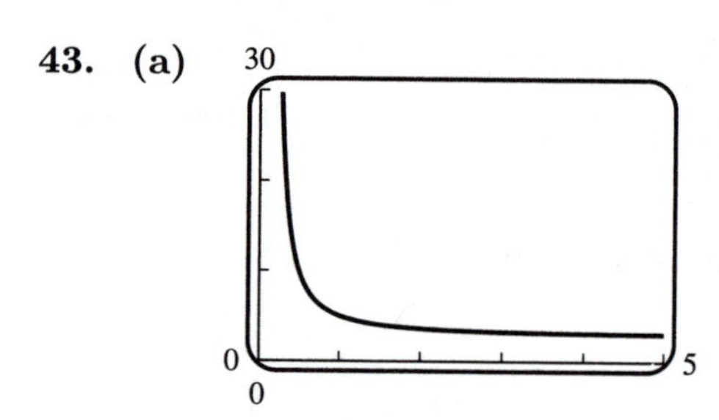

(b) Let $y=(2^x+3^x)^{1/x}$, $\lim_{x\to+\infty}\ln y=\lim_{x\to+\infty}\frac{\ln(2^x+3^x)}{x}=\lim_{x\to+\infty}\frac{2^x\ln 2+3^x\ln 3}{2^x+3^x}$

$=\lim_{x\to+\infty}\frac{(2/3)^x\ln 2+\ln 3}{(2/3)^x+1}=\ln 3$, so $\lim_{n\to+\infty}(2^n+3^n)^{1/n}=e^{\ln 3}=3$

Alternate proof: $3=(3^n)^{1/n}<(2^n+3^n)^{1/n}<(2\cdot 3^n)^{1/n}=3\cdot 2^{1/n}$. Then apply the Squeezing Theorem.

45. $a_n = \dfrac{1}{n-1}\displaystyle\int_1^n \frac{1}{x}\,dx = \frac{\ln n}{n-1},\ \lim_{n\to+\infty} a_n = \lim_{n\to+\infty}\frac{\ln n}{n-1} = \lim_{n\to+\infty}\frac{1}{n} = 0,$

$\left(\text{apply L'Hôpital's Rule to } \dfrac{\ln n}{n-1}\right)$, converges

47. $\left|\dfrac{1}{n} - 0\right| = \dfrac{1}{n} < \epsilon$ if $n > 1/\epsilon$

(a) $1/\epsilon = 1/0.5 = 2$, $N = 3$

(b) $1/\epsilon = 1/0.1 = 10$, $N = 11$

(c) $1/\epsilon = 1/0.001 = 1000$, $N = 1001$

49. **(a)** $\left|\dfrac{1}{n} - 0\right| = \dfrac{1}{n} < \epsilon$ if $n > 1/\epsilon$, choose any $N > 1/\epsilon$.

(b) $\left|\dfrac{n}{n+1} - 1\right| = \dfrac{1}{n+1} < \epsilon$ if $n > 1/\epsilon - 1$, choose any $N > 1/\epsilon - 1$.

EXERCISE SET 10.3

1. $a_{n+1} - a_n = \dfrac{1}{n+1} - \dfrac{1}{n} = -\dfrac{1}{n(n+1)} < 0$ for $n \geq 1$, so strictly decreasing.

3. $a_{n+1} - a_n = \dfrac{n+1}{2n+3} - \dfrac{n}{2n+1} = \dfrac{1}{(2n+1)(2n+3)} > 0$ for $n \geq 1$, so strictly increasing.

5. $a_{n+1} - a_n = (n+1-2^{n+1}) - (n-2^n) = 1 - 2^n < 0$ for $n \geq 1$, so strictly decreasing.

7. $\dfrac{a_{n+1}}{a_n} = \dfrac{(n+1)/(2n+3)}{n/(2n+1)} = \dfrac{(n+1)(2n+1)}{n(2n+3)} = \dfrac{2n^2+3n+1}{2n^2+3n} > 1$ for $n \geq 1$, so strictly increasing.

9. $\dfrac{a_{n+1}}{a_n} = \dfrac{(n+1)e^{-(n+1)}}{ne^{-n}} = (1+1/n)e^{-1} < 1$ for $n \geq 1$, so strictly decreasing.

11. $\dfrac{a_{n+1}}{a_n} = \dfrac{(n+1)^{n+1}}{(n+1)!}\cdot\dfrac{n!}{n^n} = \dfrac{(n+1)^n}{n^n} = (1+1/n)^n > 1$ for $n \geq 1$, so strictly increasing.

13. $f(x) = x/(2x+1)$, $f'(x) = 1/(2x+1)^2 > 0$ for $x \geq 1$, so strictly increasing.

15. $f(x) = 1/(x+\ln x)$, $f'(x) = -\dfrac{1+1/x}{(x+\ln x)^2} < 0$ for $x \geq 1$, so strictly decreasing.

17. $f(x) = \dfrac{\ln(x+2)}{x+2}$, $f'(x) = \dfrac{1-\ln(x+2)}{(x+2)^2} < 0$ for $x \geq 1$, so strictly decreasing.

19. $f(x) = 2x^2 - 7x$, $f'(x) = 4x - 7 > 0$ for $x \geq 2$, so eventually strictly increasing.

21. $f(x) = \dfrac{x}{x^2+10}$, $f'(x) = \dfrac{10-x^2}{(x^2+10)^2} < 0$ for $x \geq 4$, so eventually strictly decreasing.

23. $\dfrac{a_{n+1}}{a_n} = \dfrac{(n+1)!}{3^{n+1}}\cdot\dfrac{3^n}{n!} = \dfrac{n+1}{3} > 1$ for $n \geq 3$, so eventually strictly increasing.

25. **(a)** Yes: a monotone sequence is increasing or decreasing; if it is increasing, then it is increasing and bounded above, so by Theorem 10.3.3 it converges; if decreasing, then use Theorem 10.3.4. The limit lies in the interval $[1, 2]$.

(b) Such a sequence may converge, in which case, by the argument in Part (a), its limit is ≤ 2. But convergence may not happen: for example, the sequence $\{-n\}_{n=1}^{+\infty}$ diverges.

27. **(a)** $\sqrt{2}$, $\sqrt{2+\sqrt{2}}$, $\sqrt{2+\sqrt{2+\sqrt{2}}}$

(b) $a_1 = \sqrt{2} < 2$ so $a_2 = \sqrt{2+a_1} < \sqrt{2+2} = 2$, $a_3 = \sqrt{2+a_2} < \sqrt{2+2} = 2$, and so on indefinitely.

(c) $a_{n+1}^2 - a_n^2 = (2+a_n) - a_n^2 = 2 + a_n - a_n^2 = (2-a_n)(1+a_n)$

(d) $a_n > 0$ and, from Part (b), $a_n < 2$ so $2 - a_n > 0$ and $1 + a_n > 0$ thus, from Part (c), $a_{n+1}^2 - a_n^2 > 0$, $a_{n+1} - a_n > 0$, $a_{n+1} > a_n$; $\{a_n\}$ is a strictly increasing sequence.

(e) The sequence is increasing and has 2 as an upper bound so it must converge to a limit L, $\lim_{n\to+\infty} a_{n+1} = \lim_{n\to+\infty} \sqrt{2+a_n}$, $L = \sqrt{2+L}$, $L^2 - L - 2 = 0$, $(L-2)(L+1) = 0$ thus $\lim_{n\to+\infty} a_n = 2$.

29. **(a)** The altitudes of the rectangles are $\ln k$ for $k = 2$ to n, and their bases all have length 1 so the sum of their areas is $\ln 2 + \ln 3 + \cdots + \ln n = \ln(2\cdot 3\cdots n) = \ln n!$. The area under the curve $y = \ln x$ for x in the interval $[1, n]$ is $\int_1^n \ln x\,dx$, and $\int_1^{n+1} \ln x\,dx$ is the area for x in the interval $[1, n+1]$ so, from the figure, $\int_1^n \ln x\,dx < \ln n! < \int_1^{n+1} \ln x\,dx$.

(b) $\int_1^n \ln x\,dx = (x\ln x - x)\Big]_1^n = n\ln n - n + 1$ and $\int_1^{n+1} \ln x\,dx = (n+1)\ln(n+1) - n$ so from Part (a), $n\ln n - n + 1 < \ln n! < (n+1)\ln(n+1) - n$, $e^{n\ln n - n + 1} < n! < e^{(n+1)\ln(n+1)-n}$, $e^{n\ln n}e^{1-n} < n! < e^{(n+1)\ln(n+1)}e^{-n}$, $\dfrac{n^n}{e^{n-1}} < n! < \dfrac{(n+1)^{n+1}}{e^n}$

(c) From Part (b), $\left[\dfrac{n^n}{e^{n-1}}\right]^{1/n} < \sqrt[n]{n!} < \left[\dfrac{(n+1)^{n+1}}{e^n}\right]^{1/n}$,

$\dfrac{n}{e^{1-1/n}} < \sqrt[n]{n!} < \dfrac{(n+1)^{1+1/n}}{e}$, $\dfrac{1}{e^{1-1/n}} < \dfrac{\sqrt[n]{n!}}{n} < \dfrac{(1+1/n)(n+1)^{1/n}}{e}$,

but $\dfrac{1}{e^{1-1/n}} \to \dfrac{1}{e}$ and $\dfrac{(1+1/n)(n+1)^{1/n}}{e} \to \dfrac{1}{e}$ as $n \to +\infty$ (why?), so $\lim_{n\to+\infty} \dfrac{\sqrt[n]{n!}}{n} = \dfrac{1}{e}$.

EXERCISE SET 10.4

1. **(a)** $s_1 = 2$, $s_2 = 12/5$, $s_3 = \dfrac{62}{25}$, $s_4 = \dfrac{312}{125}$ $s_n = \dfrac{2 - 2(1/5)^n}{1 - 1/5} = \dfrac{5}{2} - \dfrac{5}{2}(1/5)^n$,

$\lim_{n\to+\infty} s_n = \dfrac{5}{2}$, converges

(b) $s_1 = \dfrac{1}{4}$, $s_2 = \dfrac{3}{4}$, $s_3 = \dfrac{7}{4}$, $s_4 = \dfrac{15}{4}$ $s_n = \dfrac{(1/4) - (1/4)2^n}{1-2} = -\dfrac{1}{4} + \dfrac{1}{4}(2^n)$,

$\lim_{n\to+\infty} s_n = +\infty$, diverges

(c) $\dfrac{1}{(k+1)(k+2)} = \dfrac{1}{k+1} - \dfrac{1}{k+2}$, $s_1 = \dfrac{1}{6}$, $s_2 = \dfrac{1}{4}$, $s_3 = \dfrac{3}{10}$, $s_4 = \dfrac{1}{3}$;

$s_n = \dfrac{1}{2} - \dfrac{1}{n+2}$, $\lim\limits_{n\to+\infty} s_n = \dfrac{1}{2}$, converges

3. geometric, $a = 1$, $r = -3/4$, sum $= \dfrac{1}{1-(-3/4)} = 4/7$

5. geometric, $a = 7$, $r = -1/6$, sum $= \dfrac{7}{1+1/6} = 6$

7. $s_n = \displaystyle\sum_{k=1}^{n}\left(\frac{1}{k+2} - \frac{1}{k+3}\right) = \frac{1}{3} - \frac{1}{n+3}$, $\lim\limits_{n\to+\infty} s_n = 1/3$

9. $s_n = \displaystyle\sum_{k=1}^{n}\left(\frac{1/3}{3k-1} - \frac{1/3}{3k+2}\right) = \frac{1}{6} - \frac{1/3}{3n+2}$, $\lim\limits_{n\to+\infty} s_n = 1/6$

11. $\displaystyle\sum_{k=3}^{\infty}\frac{1}{k-2} = \sum_{k=1}^{\infty} 1/k$, the harmonic series, so the series diverges.

13. $\displaystyle\sum_{k=1}^{\infty}\frac{4^{k+2}}{7^{k-1}} = \sum_{k=1}^{\infty} 64\left(\frac{4}{7}\right)^{k-1}$; geometric, $a = 64$, $r = 4/7$, sum $= \dfrac{64}{1-4/7} = 448/3$

15. $0.4444\cdots = 0.4 + 0.04 + 0.004 + \cdots = \dfrac{0.4}{1-0.1} = 4/9$

17. $5.373737\cdots = 5 + 0.37 + 0.0037 + 0.000037 + \cdots = 5 + \dfrac{0.37}{1-0.01} = 5 + 37/99 = 532/99$

19. $0.782178217821\cdots = 0.7821 + 0.00007821 + 0.000000007821 + \cdots = \dfrac{0.7821}{1-0.0001} = \dfrac{7821}{9999} = \dfrac{79}{101}$

21. $d = 10 + 2\cdot\dfrac{3}{4}\cdot 10 + 2\cdot\dfrac{3}{4}\cdot\dfrac{3}{4}\cdot 10 + 2\cdot\dfrac{3}{4}\cdot\dfrac{3}{4}\cdot\dfrac{3}{4}\cdot 10 + \cdots$

$= 10 + 20\left(\dfrac{3}{4}\right) + 20\left(\dfrac{3}{4}\right)^2 + 20\left(\dfrac{3}{4}\right)^3 + \cdots = 10 + \dfrac{20(3/4)}{1-3/4} = 10 + 60 = 70$ meters

23. **(a)** $s_n = \ln\dfrac{1}{2} + \ln\dfrac{2}{3} + \ln\dfrac{3}{4} + \cdots + \ln\dfrac{n}{n+1} = \ln\left(\dfrac{1}{2}\cdot\dfrac{2}{3}\cdot\dfrac{3}{4}\cdots\dfrac{n}{n+1}\right) = \ln\dfrac{1}{n+1} = -\ln(n+1)$,

$\lim\limits_{n\to+\infty} s_n = -\infty$, series diverges.

(b) $\ln(1-1/k^2) = \ln\dfrac{k^2-1}{k^2} = \ln\dfrac{(k-1)(k+1)}{k^2} = \ln\dfrac{k-1}{k} + \ln\dfrac{k+1}{k} = \ln\dfrac{k-1}{k} - \ln\dfrac{k}{k+1}$,

$$s_n = \sum_{k=2}^{n+1}\left[\ln\frac{k-1}{k} - \ln\frac{k}{k+1}\right]$$

$$= \left(\ln\frac{1}{2} - \ln\frac{2}{3}\right) + \left(\ln\frac{2}{3} - \ln\frac{3}{4}\right) + \left(\ln\frac{3}{4} - \ln\frac{4}{5}\right) + \cdots + \left(\ln\frac{n}{n+1} - \ln\frac{n+1}{n+2}\right)$$

$= \ln\dfrac{1}{2} - \ln\dfrac{n+1}{n+2}$, $\lim\limits_{n\to+\infty} s_n = \ln\dfrac{1}{2} = -\ln 2$

25. (a) Geometric series, $a = x$, $r = -x^2$. Converges for $|-x^2| < 1$, $|x| < 1$;
$$S = \frac{x}{1-(-x^2)} = \frac{x}{1+x^2}.$$

(b) Geometric series, $a = 1/x^2$, $r = 2/x$. Converges for $|2/x| < 1$, $|x| > 2$;
$$S = \frac{1/x^2}{1-2/x} = \frac{1}{x^2-2x}.$$

(c) Geometric series, $a = e^{-x}$, $r = e^{-x}$. Converges for $|e^{-x}| < 1$, $e^{-x} < 1$, $e^x > 1$, $x > 0$;
$$S = \frac{e^{-x}}{1-e^{-x}} = \frac{1}{e^x - 1}.$$

27. $$s_n = (1-1/3) + (1/2-1/4) + (1/3-1/5) + (1/4-1/6) + \cdots + [1/n - 1/(n+2)]$$
$$= (1+1/2+1/3+\cdots+1/n) - (1/3+1/4+1/5+\cdots+1/(n+2))$$
$$= 3/2 - 1/(n+1) - 1/(n+2), \lim_{n\to+\infty} s_n = 3/2$$

29. $$s_n = \sum_{k=1}^{n} \frac{1}{(2k-1)(2k+1)} = \sum_{k=1}^{n}\left[\frac{1/2}{2k-1} - \frac{1/2}{2k+1}\right] = \frac{1}{2}\left[\sum_{k=1}^{n}\frac{1}{2k-1} - \sum_{k=1}^{n}\frac{1}{2k+1}\right]$$
$$= \frac{1}{2}\left[\sum_{k=1}^{n}\frac{1}{2k-1} - \sum_{k=2}^{n+1}\frac{1}{2k-1}\right] = \frac{1}{2}\left[1 - \frac{1}{2n+1}\right]; \quad \lim_{n\to+\infty} s_n = \frac{1}{2}$$

31. $$a_2 = \frac{1}{2}a_1 + \frac{1}{2},\ a_3 = \frac{1}{2}a_2 + \frac{1}{2} = \frac{1}{2^2}a_1 + \frac{1}{2^2} + \frac{1}{2},\ a_4 = \frac{1}{2}a_3 + \frac{1}{2} = \frac{1}{2^3}a_1 + \frac{1}{2^3} + \frac{1}{2^2} + \frac{1}{2},$$
$$a_5 = \frac{1}{2}a_4 + \frac{1}{2} = \frac{1}{2^4}a_1 + \frac{1}{2^4} + \frac{1}{2^3} + \frac{1}{2^2} + \frac{1}{2}, \ldots, a_n = \frac{1}{2^{n-1}}a_1 + \frac{1}{2^{n-1}} + \frac{1}{2^{n-2}} + \cdots + \frac{1}{2},$$
$$\lim_{n\to+\infty} a_n = \lim_{n\to+\infty}\frac{a_1}{2^{n-1}} + \sum_{n=1}^{\infty}\left(\frac{1}{2}\right)^n = 0 + \frac{1/2}{1-1/2} = 1$$

33. The series converges to $1/(1-x)$ only if $-1 < x < 1$.

35. By inspection, $\dfrac{\theta}{2} - \dfrac{\theta}{4} + \dfrac{\theta}{8} - \dfrac{\theta}{16} + \cdots = \dfrac{\theta/2}{1-(-1/2)} = \theta/3$

37. (b) $$\frac{2^k A}{3^k - 2^k} + \frac{2^k B}{3^{k+1} - 2^{k+1}} = \frac{2^k\left(3^{k+1} - 2^{k+1}\right)A + 2^k\left(3^k - 2^k\right)B}{\left(3^k - 2^k\right)\left(3^{k+1} - 2^{k+1}\right)}$$
$$= \frac{\left(3\cdot 6^k - 2\cdot 2^{2k}\right)A + \left(6^k - 2^{2k}\right)B}{\left(3^k - 2^k\right)\left(3^{k+1} - 2^{k+1}\right)} = \frac{(3A+B)6^k - (2A+B)2^{2k}}{\left(3^k - 2^k\right)\left(3^{k+1} - 2^{k+1}\right)}$$
so $3A + B = 1$ and $2A + B = 0$, $A = 1$ and $B = -2$.

(c) $$s_n = \sum_{k=1}^{n}\left[\frac{2^k}{3^k - 2^k} - \frac{2^{k+1}}{3^{k+1} - 2^{k+1}}\right] = \sum_{k=1}^{n}(a_k - a_{k+1}) \text{ where } a_k = \frac{2^k}{3^k - 2^k}.$$
But $s_n = (a_1 - a_2) + (a_2 - a_3) + (a_3 - a_4) + \cdots + (a_n - a_{n+1})$ which is a telescoping sum,
$$s_n = a_1 - a_{n+1} = 2 - \frac{2^{n+1}}{3^{n+1} - 2^{n+1}},\ \lim_{n\to+\infty} s_n = \lim_{n\to+\infty}\left[2 - \frac{(2/3)^{n+1}}{1-(2/3)^{n+1}}\right] = 2.$$

EXERCISE SET 10.5

1. **(a)** $\displaystyle\sum_{k=1}^{\infty}\frac{1}{2^k}=\frac{1/2}{1-1/2}=1$; $\displaystyle\sum_{k=1}^{\infty}\frac{1}{4^k}=\frac{1/4}{1-1/4}=1/3$; $\displaystyle\sum_{k=1}^{\infty}\left(\frac{1}{2^k}+\frac{1}{4^k}\right)=1+1/3=4/3$

(b) $\displaystyle\sum_{k=1}^{\infty}\frac{1}{5^k}=\frac{1/5}{1-1/5}=1/4$; $\displaystyle\sum_{k=1}^{\infty}\frac{1}{k(k+1)}=1$ (Example 5, Section 10 .4);

$$\sum_{k=1}^{\infty}\left[\frac{1}{5^k}-\frac{1}{k(k+1)}\right]=1/4-1=-3/4$$

3. **(a)** $p=3$, converges **(b)** $p=1/2$, diverges **(c)** $p=1$, diverges **(d)** $p=2/3$, diverges

5. **(a)** $\displaystyle\lim_{k\to+\infty}\frac{k^2+k+3}{2k^2+1}=\frac{1}{2}$; the series diverges. **(b)** $\displaystyle\lim_{k\to+\infty}\left(1+\frac{1}{k}\right)^k=e$; the series diverges.

(c) $\displaystyle\lim_{k\to+\infty}\cos k\pi$ does not exist; the series diverges. **(d)** $\displaystyle\lim_{k\to+\infty}\frac{1}{k!}=0$; no information

7. **(a)** $\displaystyle\int_1^{+\infty}\frac{1}{5x+2}=\lim_{\ell\to+\infty}\frac{1}{5}\ln(5x+2)\Big]_1^{\ell}=+\infty$, the series diverges by the Integral Test.

(b) $\displaystyle\int_1^{+\infty}\frac{1}{1+9x^2}dx=\lim_{\ell\to+\infty}\frac{1}{3}\tan^{-1}3x\Big]_1^{\ell}=\frac{1}{3}\left(\pi/2-\tan^{-1}3\right)$,
the series converges by the Integral Test.

9. $\displaystyle\sum_{k=1}^{\infty}\frac{1}{k+6}=\sum_{k=7}^{\infty}\frac{1}{k}$, diverges because the harmonic series diverges.

11. $\displaystyle\sum_{k=1}^{\infty}\frac{1}{\sqrt{k+5}}=\sum_{k=6}^{\infty}\frac{1}{\sqrt{k}}$, diverges because the p-series with $p=1/2\le 1$ diverges.

13. $\displaystyle\int_1^{+\infty}(2x-1)^{-1/3}dx=\lim_{\ell\to+\infty}\frac{3}{4}(2x-1)^{2/3}\Big]_1^{\ell}=+\infty$, the series diverges by the Integral Test.

15. $\displaystyle\lim_{k\to+\infty}\frac{k}{\ln(k+1)}=\lim_{k\to+\infty}\frac{1}{1/(k+1)}=+\infty$, the series diverges because $\displaystyle\lim_{k\to+\infty}u_k\neq 0$.

17. $\displaystyle\lim_{k\to+\infty}(1+1/k)^{-k}=1/e\neq 0$, the series diverges.

19. $\displaystyle\int_1^{+\infty}\frac{\tan^{-1}x}{1+x^2}dx=\lim_{\ell\to+\infty}\frac{1}{2}\left(\tan^{-1}x\right)^2\Big]_1^{\ell}=3\pi^2/32$, the series converges by the Integral Test, since

$$\frac{d}{dx}\frac{\tan^{-1}x}{1+x^2}=\frac{1-2x\tan^{-1}x}{(1+x^2)^2}<0 \text{ for } x\ge 1.$$

21. $\displaystyle\lim_{k\to+\infty}k^2\sin^2(1/k)=1\neq 0$, the series diverges.

23. $\displaystyle 7\sum_{k=5}^{\infty}k^{-1.01}$, p-series with $p>1$, converges

25. $\dfrac{1}{x(\ln x)^p}$ is decreasing for $x \geq e^p$, so use the Integral Test with $\displaystyle\int_{e^p}^{+\infty} \frac{dx}{x(\ln x)^p}$ to get

$$\lim_{\ell \to +\infty} \ln(\ln x)\Big]_{e^p}^{\ell} = +\infty \text{ if } p = 1, \qquad \lim_{\ell \to +\infty} \frac{(\ln x)^{1-p}}{1-p}\Big]_{e^p}^{\ell} = \begin{cases} +\infty & \text{if } p < 1 \\ \dfrac{p^{1-p}}{p-1} & \text{if } p > 1 \end{cases}$$

Thus the series converges for $p > 1$.

27. **(a)** $3\displaystyle\sum_{k=1}^{\infty} \frac{1}{k^2} - \sum_{k=1}^{\infty} \frac{1}{k^4} = \pi^2/2 - \pi^4/90$ **(b)** $\displaystyle\sum_{k=1}^{\infty} \frac{1}{k^2} - 1 - \frac{1}{2^2} = \pi^2/6 - 5/4$

(c) $\displaystyle\sum_{k=2}^{\infty} \frac{1}{(k-1)^4} = \sum_{k=1}^{\infty} \frac{1}{k^4} = \pi^4/90$

29. **(a)** diverges because $\displaystyle\sum_{k=1}^{\infty} (2/3)^{k-1}$ converges and $\displaystyle\sum_{k=1}^{\infty} 1/k$ diverges.

(b) diverges because $\displaystyle\sum_{k=1}^{\infty} 1/(3k+2)$ diverges and $\displaystyle\sum_{k=1}^{\infty} 1/k^{3/2}$ converges.

(c) converges because both $\displaystyle\sum_{k=2}^{\infty} \frac{1}{k(\ln k)^2}$ (Exercise 25) and $\displaystyle\sum_{k=2}^{\infty} 1/k^2$ converge.

31. **(a)** In Exercise 30 above let $f(x) = \dfrac{1}{x^2}$. Then $\displaystyle\int_n^{+\infty} f(x)\,dx = -\frac{1}{x}\Big]_n^{+\infty} = \frac{1}{n}$; use this result and the same result with $n+1$ replacing n to obtain the desired result.

(b) $s_3 = 1 + 1/4 + 1/9 = 49/36$; $58/36 = s_3 + \dfrac{1}{4} < \dfrac{1}{6}\pi^2 < s_3 + \dfrac{1}{3} = 61/36$

(d) $1/11 < \dfrac{1}{6}\pi^2 - s_{10} < 1/10$

33. Apply Exercise 30 in each case:

(a) $f(x) = \dfrac{1}{(2x+1)^2}$, $\displaystyle\int_n^{+\infty} f(x)\,dx = \frac{1}{2(2n+1)}$, so $\dfrac{1}{46} < \displaystyle\sum_{k=1}^{\infty} \frac{1}{(2k+1)^2} - s_{10} < \frac{1}{42}$

(b) $f(x) = \dfrac{1}{k^2+1}$, $\displaystyle\int_n^{+\infty} f(x)\,dx = \frac{\pi}{2} - \tan^{-1}(n)$, so

$$\pi/2 - \tan^{-1}(11) < \sum_{k=1}^{\infty} \frac{1}{k^2+1} - s_{10} < \pi/2 - \tan^{-1}(10)$$

(c) $f(x) = \dfrac{x}{e^x}$, $\displaystyle\int_n^{+\infty} f(x)\,dx = (n+1)e^{-n}$, so $12e^{-11} < \displaystyle\sum_{k=1}^{\infty} \frac{k}{e^k} - s_{10} < 11e^{-10}$

35. **(a)** $\displaystyle\int_n^{+\infty} \frac{1}{x^4}\,dx = \frac{1}{3n^3}$; choose n so that $\dfrac{1}{3n^3} - \dfrac{1}{3(n+1)^3} < 0.005$, $n = 4$; $S \approx 1.08$

37. p-series with $p = \ln a$; convergence for $p > 1, a > e$

39. (a) $f(x) = 1/(x^3+1)$ is decreasing and continuous on the interval $[1, +\infty]$, so the Integral Test applies.

(c)

n	10	20	30	40	50
s_n	0.681980	0.685314	0.685966	0.686199	0.686307

n	60	70	80	90	100
s_n	0.686367	0.686403	0.686426	0.686442	0.686454

(e) Set $g(n) = \displaystyle\int_n^{+\infty} \frac{1}{x^3+1}\,dx = \frac{\sqrt{3}}{6}\pi + \frac{1}{6}\ln\frac{n^3+1}{(n+1)^3} - \frac{\sqrt{3}}{3}\tan^{-1}\left(\frac{2n-1}{\sqrt{3}}\right)$; for $n \ge 13$, $g(n) - g(n+1) \le 0.0005$; $s_{13} + (g(13) + g(14))/2 \approx 0.6865$, so the sum ≈ 0.6865 to three decimal places.

EXERCISE SET 10.6

1. (a) $\dfrac{1}{5k^2-k} \le \dfrac{1}{5k^2-k^2} = \dfrac{1}{4k^2}$, $\displaystyle\sum_{k=1}^{\infty} \frac{1}{4k^2}$ converges

(b) $\dfrac{3}{k-1/4} > \dfrac{3}{k}$, $\displaystyle\sum_{k=1}^{\infty} 3/k$ diverges

3. (a) $\dfrac{1}{3^k+5} < \dfrac{1}{3^k}$, $\displaystyle\sum_{k=1}^{\infty} \frac{1}{3^k}$ converges

(b) $\dfrac{5\sin^2 k}{k!} < \dfrac{5}{k!}$, $\displaystyle\sum_{k=1}^{\infty} \frac{5}{k!}$ converges

5. compare with the convergent series $\displaystyle\sum_{k=1}^{\infty} 1/k^5$, $\rho = \lim_{k\to+\infty} \frac{4k^7-2k^6+6k^5}{8k^7+k-8} = 1/2$, converges

7. compare with the convergent series $\displaystyle\sum_{k=1}^{\infty} 5/3^k$, $\rho = \lim_{k\to+\infty} \frac{3^k}{3^k+1} = 1$, converges

9. compare with the divergent series $\displaystyle\sum_{k=1}^{\infty} \frac{1}{k^{2/3}}$,

$$\rho = \lim_{k\to+\infty} \frac{k^{2/3}}{(8k^2-3k)^{1/3}} = \lim_{k\to+\infty} \frac{1}{(8-3/k)^{1/3}} = 1/2, \text{ diverges}$$

11. $\rho = \displaystyle\lim_{k\to+\infty} \frac{3^{k+1}/(k+1)!}{3^k/k!} = \lim_{k\to+\infty} \frac{3}{k+1} = 0$, the series converges

13. $\rho = \displaystyle\lim_{k\to+\infty} \frac{k}{k+1} = 1$, the result is inconclusive

15. $\rho = \displaystyle\lim_{k\to+\infty} \frac{(k+1)!/(k+1)^3}{k!/k^3} = \lim_{k\to+\infty} \frac{k^3}{(k+1)^2} = +\infty$, the series diverges

17. $\rho = \displaystyle\lim_{k\to+\infty} \frac{3k+2}{2k-1} = 3/2$, the series diverges

19. $\rho = \displaystyle\lim_{k\to+\infty} \frac{k^{1/k}}{5} = 1/5$, the series converges

21. Ratio Test, $\rho = \lim_{k\to+\infty} 7/(k+1) = 0$, converges

23. Ratio Test, $\rho = \lim_{k\to+\infty} \dfrac{(k+1)^2}{5k^2} = 1/5$, converges

25. Ratio Test, $\rho = \lim_{k\to+\infty} e^{-1}(k+1)^{50}/k^{50} = e^{-1} < 1$, converges

27. Limit Comparison Test, compare with the convergent series $\sum_{k=1}^{\infty} 1/k^{5/2}$, $\rho = \lim_{k\to+\infty} \dfrac{k^3}{k^3+1} = 1$, converges

29. Limit Comparison Test, compare with the divergent series $\sum_{k=1}^{\infty} 1/k$, $\rho = \lim_{k\to+\infty} \dfrac{k}{\sqrt{k^2+k}} = 1$, diverges

31. Limit Comparison Test, compare with the convergent series $\sum_{k=1}^{\infty} 1/k^{5/2}$,

$\rho = \lim_{k\to+\infty} \dfrac{k^3+2k^{5/2}}{k^3+3k^2+3k} = 1$, converges

33. Limit Comparison Test, compare with the divergent series $\sum_{k=1}^{\infty} 1/\sqrt{k}$

35. Ratio Test, $\rho = \lim_{k\to+\infty} \dfrac{\ln(k+1)}{e\ln k} = \lim_{k\to+\infty} \dfrac{k}{e(k+1)} = 1/e < 1$, converges

37. Ratio Test, $\rho = \lim_{k\to+\infty} \dfrac{k+5}{4(k+1)} = 1/4$, converges

39. diverges because $\lim_{k\to+\infty} \dfrac{1}{4+2^{-k}} = 1/4 \neq 0$

41. $\dfrac{\tan^{-1} k}{k^2} < \dfrac{\pi/2}{k^2}$, $\sum_{k=1}^{\infty} \dfrac{\pi/2}{k^2}$ converges so $\sum_{k=1}^{\infty} \dfrac{\tan^{-1} k}{k^2}$ converges

43. Ratio Test, $\rho = \lim_{k\to+\infty} \dfrac{(k+1)^2}{(2k+2)(2k+1)} = 1/4$, converges

45. $u_k = \dfrac{k!}{1\cdot 3\cdot 5\cdots(2k-1)}$, by the Ratio Test $\rho = \lim_{k\to+\infty} \dfrac{k+1}{2k+1} = 1/2$; converges

47. Root Test: $\rho = \lim_{k\to+\infty} \dfrac{1}{3}(\ln k)^{1/k} = 1/3$, converges

49. **(b)** $\rho = \lim_{k\to+\infty} \dfrac{\sin(\pi/k)}{\pi/k} = 1$ and $\sum_{k=1}^{\infty} \pi/k$ diverges

51. Set $g(x) = \sqrt{x} - \ln x$; $\dfrac{d}{dx}g(x) = \dfrac{1}{2\sqrt{x}} - \dfrac{1}{x} = 0$ when $x = 4$. Since $\lim_{x\to 0+} g(x) = \lim_{x\to+\infty} g(x) = +\infty$ it follows that $g(x)$ has its minimum at $x = 4$, $g(4) = \sqrt{4} - \ln 4 > 0$, and thus $\sqrt{x} - \ln x > 0$ for $x > 0$.

(a) $\dfrac{\ln k}{k^2} < \dfrac{\sqrt{k}}{k^2} = \dfrac{1}{k^{3/2}}, \displaystyle\sum_{k=1}^{\infty} \frac{1}{k^{3/2}}$ converges so $\displaystyle\sum_{k=1}^{\infty} \frac{\ln k}{k^2}$ converges.

(b) $\dfrac{1}{(\ln k)^2} > \dfrac{1}{k}, \displaystyle\sum_{k=2}^{\infty} \frac{1}{k}$ diverges so $\displaystyle\sum_{k=2}^{\infty} \frac{1}{(\ln k)^2}$ diverges.

53. (a) If $\sum b_k$ converges, then set $M = \sum b_k$. Then $a_1 + a_2 + \cdots + a_n \le b_1 + b_2 + \cdots + b_n \le M$; apply Theorem 10.5.6 to get convergence of $\sum a_k$.

(b) Assume the contrary, that $\sum b_k$ converges; then use Part (a) of the Theorem to show that $\sum a_k$ converges, a contradiction.

EXERCISE SET 10.7

1. $a_{k+1} < a_k, \displaystyle\lim_{k\to+\infty} a_k = 0, a_k > 0$

3. diverges because $\displaystyle\lim_{k\to+\infty} a_k = \lim_{k\to+\infty} \frac{k+1}{3k+1} = 1/3 \neq 0$

5. $\{e^{-k}\}$ is decreasing and $\displaystyle\lim_{k\to+\infty} e^{-k} = 0$, converges

7. $\rho = \displaystyle\lim_{k\to+\infty} \frac{(3/5)^{k+1}}{(3/5)^k} = 3/5$, converges absolutely

9. $\rho = \displaystyle\lim_{k\to+\infty} \frac{3k^2}{(k+1)^2} = 3$, diverges

11. $\rho = \displaystyle\lim_{k\to+\infty} \frac{(k+1)^3}{ek^3} = 1/e$, converges absolutely

13. conditionally convergent, $\displaystyle\sum_{k=1}^{\infty} \frac{(-1)^{k+1}}{3k}$ converges by the Alternating Series Test but $\displaystyle\sum_{k=1}^{\infty} \frac{1}{3k}$ diverges

15. divergent, $\displaystyle\lim_{k\to+\infty} a_k \neq 0$

17. $\displaystyle\sum_{k=1}^{\infty} \frac{\cos k\pi}{k} = \sum_{k=1}^{\infty} \frac{(-1)^k}{k}$ is conditionally convergent, $\displaystyle\sum_{k=1}^{\infty} \frac{(-1)^k}{k}$ converges by the Alternating Series Test but $\displaystyle\sum_{k=1}^{\infty} 1/k$ diverges.

19. conditionally convergent, $\displaystyle\sum_{k=1}^{\infty} (-1)^{k+1} \frac{k+2}{k(k+3)}$ converges by the Alternating Series Test but $\displaystyle\sum_{k=1}^{\infty} \frac{k+2}{k(k+3)}$ diverges (Limit Comparison Test with $\sum 1/k$)

21. $\displaystyle\sum_{k=1}^{\infty} \sin(k\pi/2) = 1 + 0 - 1 + 0 + 1 + 0 - 1 + 0 + \cdots$, divergent ($\displaystyle\lim_{k\to+\infty} \sin(k\pi/2)$ does not exist)

23. conditionally convergent, $\sum_{k=2}^{\infty} \frac{(-1)^k}{k \ln k}$ converges by the Alternating Series Test but $\sum_{k=2}^{\infty} \frac{1}{k \ln k}$ diverges (Integral Test)

25. absolutely convergent, $\sum_{k=2}^{\infty} (1/\ln k)^k$ converges by the Root Test

27. conditionally convergent, let $f(x) = \frac{x^2+1}{x^3+2}$ then $f'(x) = \frac{x(4-3x-x^3)}{(x^3+2)^2} \le 0$ for $x \ge 1$ so $\{a_k\}_{k=2}^{+\infty} = \left\{\frac{k^2+1}{k^3+2}\right\}_{k=2}^{+\infty}$ is decreasing, $\lim_{k\to+\infty} a_k = 0$; the series converges by the Alternating Series Test but $\sum_{k=2}^{\infty} \frac{k^2+1}{k^3+2}$ diverges (Limit Comparison Test with $\sum 1/k$)

29. absolutely convergent by the Ratio Test, $\rho = \lim_{k\to+\infty} \frac{k+1}{(2k+1)(2k)} = 0$

31. $|\text{error}| < a_8 = 1/8 = 0.125$

33. $|\text{error}| < a_{100} = 1/\sqrt{100} = 0.1$

35. $|\text{error}| < 0.0001$ if $a_{n+1} \le 0.0001$, $1/(n+1) \le 0.0001$, $n+1 \ge 10,000$, $n \ge 9,999$, $n = 9,999$

37. $|\text{error}| < 0.005$ if $a_{n+1} \le 0.005$, $1/\sqrt{n+1} \le 0.005$, $\sqrt{n+1} \ge 200$, $n+1 \ge 40,000$, $n \ge 39,999$, $n = 39,999$

39. $a_k = \frac{3}{2^{k+1}}$, $|\text{error}| < a_{11} = \frac{3}{2^{12}} < 0.00074$; $s_{10} \approx 0.4995$; $S = \frac{3/4}{1-(-1/2)} = 0.5$

41. $a_k = \frac{1}{(2k-1)!}$, $a_{n+1} = \frac{1}{(2n+1)!} \le 0.005$, $(2n+1)! \ge 200$, $2n+1 \ge 6$, $n \ge 2.5$; $n = 3$, $s_3 = 1 - 1/6 + 1/120 \approx 0.84$

43. $a_k = \frac{1}{k2^k}$, $a_{n+1} = \frac{1}{(n+1)2^{n+1}} \le 0.005$, $(n+1)2^{n+1} \ge 200$, $n+1 \ge 6$, $n \ge 5$; $n = 5$, $s_5 \approx 0.41$

45. **(c)** $a_k = \frac{1}{2k-1}$, $a_{n+1} = \frac{1}{2n+1} \le 10^{-2}$, $2n+1 \ge 100$, $n \ge 49.5$; $n = 50$

47.
$$1 + \frac{1}{3^2} + \frac{1}{5^2} + \cdots = \left[1 + \frac{1}{2^2} + \frac{1}{3^2} + \cdots\right] - \left[\frac{1}{2^2} + \frac{1}{4^2} + \frac{1}{6^2} + \cdots\right]$$
$$= \frac{\pi^2}{6} - \frac{1}{2^2}\left[1 + \frac{1}{2^2} + \frac{1}{3^2} + \cdots\right] = \frac{\pi^2}{6} - \frac{1}{4}\frac{\pi^2}{6} = \frac{\pi^2}{8}$$

49. Every positive integer can be written in exactly one of the three forms $2k-1$ or $4k-2$ or $4k$, so a rearrangement is
$$\left(1 - \frac{1}{2} - \frac{1}{4}\right) + \left(\frac{1}{3} - \frac{1}{6} - \frac{1}{8}\right) + \left(\frac{1}{5} - \frac{1}{10} - \frac{1}{12}\right) + \cdots + \left(\frac{1}{2k-1} - \frac{1}{4k-2} - \frac{1}{4k}\right) + \cdots$$
$$= \left(\frac{1}{2} - \frac{1}{4}\right) + \left(\frac{1}{6} - \frac{1}{8}\right) + \left(\frac{1}{10} - \frac{1}{12}\right) + \cdots + \left(\frac{1}{4k-2} - \frac{1}{4k}\right) + \cdots = \frac{1}{2}\ln 2$$

51. **(a)** The distance d from the starting point is

$$d = 180 - \frac{180}{2} + \frac{180}{3} - \cdots - \frac{180}{1000} = 180\left[1 - \frac{1}{2} + \frac{1}{3} - \cdots - \frac{1}{1000}\right].$$

From Theorem 10.7.2, $1 - \frac{1}{2} + \frac{1}{3} - \cdots - \frac{1}{1000}$ differs from $\ln 2$ by less than $1/1001$ so $180(\ln 2 - 1/1001) < d < 180 \ln 2$, $124.58 < d < 124.77$.

(b) The total distance traveled is $s = 180 + \frac{180}{2} + \frac{180}{3} + \cdots + \frac{180}{1000}$, and from inequality (2) in Section 10.5,

$$\int_1^{1001} \frac{180}{x}dx < s < 180 + \int_1^{1000} \frac{180}{x}dx$$

$$180 \ln 1001 < s < 180(1 + \ln 1000)$$

$$1243 < s < 1424$$

EXERCISE SET 10.8

1. $f^{(k)}(x) = (-1)^k e^{-x}$, $f^{(k)}(0) = (-1)^k$; $\sum_{k=0}^{\infty} \frac{(-1)^k}{k!} x^k$

3. $f^{(k)}(0) = 0$ if k is odd, $f^{(k)}(0)$ is alternately π^k and $-\pi^k$ if k is even; $\sum_{k=0}^{\infty} \frac{(-1)^k \pi^{2k}}{(2k)!} x^{2k}$

5. $f^{(0)}(0) = 0$; for $k \geq 1$, $f^{(k)}(x) = \frac{(-1)^{k+1}(k-1)!}{(1+x)^k}$, $f^{(k)}(0) = (-1)^{k+1}(k-1)!$; $\sum_{k=1}^{\infty} \frac{(-1)^{k+1}}{k} x^k$

7. $f^{(k)}(0) = 0$ if k is odd, $f^{(k)}(0) = 1$ if k is even; $\sum_{k=0}^{\infty} \frac{1}{(2k)!} x^{2k}$

9. $f^{(k)}(x) = \begin{cases} (-1)^{k/2}(x \sin x - k \cos x) & k \text{ even} \\ (-1)^{(k-1)/2}(x \cos x + k \sin x) & k \text{ odd} \end{cases}$, $\quad f^{(k)}(0) = \begin{cases} (-1)^{1+k/2} k & k \text{ even} \\ 0 & k \text{ odd} \end{cases}$

$\sum_{k=0}^{\infty} \frac{(-1)^k}{(2k+1)!} x^{2k+2}$

11. $f^{(k)}(x_0) = e$; $\sum_{k=0}^{\infty} \frac{e}{k!}(x-1)^k$

13. $f^{(k)}(x) = \frac{(-1)^k k!}{x^{k+1}}$, $f^{(k)}(-1) = -k!$; $\sum_{k=0}^{\infty} (-1)(x+1)^k$

15. $f^{(k)}(1/2) = 0$ if k is odd, $f^{(k)}(1/2)$ is alternately π^k and $-\pi^k$ if k is even;

$\sum_{k=0}^{\infty} \frac{(-1)^k \pi^{2k}}{(2k)!}(x - 1/2)^{2k}$

17. $f(1)=0$, for $k\ge 1$, $f^{(k)}(x)=\dfrac{(-1)^{k-1}(k-1)!}{x^k}$; $f^{(k)}(1)=(-1)^{k-1}(k-1)!$;

$\sum_{k=1}^{\infty}\dfrac{(-1)^{k-1}}{k}(x-1)^k$

19. geometric series, $\rho=\lim_{k\to+\infty}\left|\dfrac{u_{k+1}}{u_k}\right|=|x|$, so the interval of convergence is $-1<x<1$, converges there to $\dfrac{1}{1+x}$ (the series diverges for $x=\pm 1$)

21. geometric series, $\rho=\lim_{k\to+\infty}\left|\dfrac{u_{k+1}}{u_k}\right|=|x-2|$, so the interval of convergence is $1<x<3$, converges there to $\dfrac{1}{1-(x-2)}=\dfrac{1}{3-x}$ (the series diverges for $x=1,3$)

23. **(a)** geometric series, $\rho=\lim_{k\to+\infty}\left|\dfrac{u_{k+1}}{u_k}\right|=|x/2|$, so the interval of convergence is $-2<x<2$, converges there to $\dfrac{1}{1+x/2}=\dfrac{2}{2+x}$; (the series diverges for $x=-2,2$)

(b) $f(0)=1$; $f(1)=2/3$

25. $\rho=\lim_{k\to+\infty}\dfrac{k+1}{k+2}|x|=|x|$, the series converges if $|x|<1$ and diverges if $|x|>1$. If $x=-1$, $\sum_{k=0}^{\infty}\dfrac{(-1)^k}{k+1}$ converges by the Alternating Series Test; if $x=1$, $\sum_{k=0}^{\infty}\dfrac{1}{k+1}$ diverges. The radius of convergence is 1, the interval of convergence is $[-1,1)$.

27. $\rho=\lim_{k\to+\infty}\dfrac{|x|}{k+1}=0$, the radius of convergence is $+\infty$, the interval is $(-\infty,+\infty)$.

29. $\rho=\lim_{k\to+\infty}\dfrac{5k^2|x|}{(k+1)^2}=5|x|$, converges if $|x|<1/5$ and diverges if $|x|>1/5$. If $x=-1/5$, $\sum_{k=1}^{\infty}\dfrac{(-1)^k}{k^2}$ converges; if $x=1/5$, $\sum_{k=1}^{\infty}1/k^2$ converges. Radius of convergence is $1/5$, interval of convergence is $[-1/5,1/5]$.

31. $\rho=\lim_{k\to+\infty}\dfrac{k|x|}{k+2}=|x|$, converges if $|x|<1$, diverges if $|x|>1$. If $x=-1$, $\sum_{k=1}^{\infty}\dfrac{(-1)^k}{k(k+1)}$ converges; if $x=1$, $\sum_{k=1}^{\infty}\dfrac{1}{k(k+1)}$ converges. Radius of convergence is 1, interval of convergence is $[-1,1]$.

33. $\rho=\lim_{k\to+\infty}\dfrac{\sqrt{k}}{\sqrt{k+1}}|x|=|x|$, converges if $|x|<1$, diverges if $|x|>1$. If $x=-1$, $\sum_{k=1}^{\infty}\dfrac{-1}{\sqrt{k}}$ diverges; if $x=1$, $\sum_{k=1}^{\infty}\dfrac{(-1)^{k-1}}{\sqrt{k}}$ converges. Radius of convergence is 1, interval of convergence is $(-1,1]$.

35. $\rho = \lim_{k\to+\infty} \dfrac{|x|^2}{(2k+3)(2k+2)} = 0$, radius of convergence is $+\infty$, interval of convergence is $(-\infty, +\infty)$.

37. $\rho = \lim_{k\to+\infty} \dfrac{3|x|}{k+1} = 0$, radius of convergence is $+\infty$, interval of convergence is $(-\infty, +\infty)$.

39. $\rho = \lim_{k\to+\infty} \dfrac{1+k^2}{1+(k+1)^2}|x| = |x|$, converges if $|x| < 1$, diverges if $|x| > 1$. If $x = -1$, $\sum_{k=0}^{\infty} \dfrac{(-1)^k}{1+k^2}$ converges; if $x = 1$, $\sum_{k=0}^{\infty} \dfrac{1}{1+k^2}$ converges. Radius of convergence is 1, interval of convergence is $[-1, 1]$.

41. $\rho = \lim_{k\to+\infty} \dfrac{k|x+1|}{k+1} = |x+1|$, converges if $|x+1| < 1$, diverges if $|x+1| > 1$. If $x = -2$, $\sum_{k=1}^{\infty} \dfrac{-1}{k}$ diverges; if $x = 0$, $\sum_{k=1}^{\infty} \dfrac{(-1)^{k+1}}{k}$ converges. Radius of convergence is 1, interval of convergence is $(-2, 0]$.

43. $\rho = \lim_{k\to+\infty} (3/4)|x+5| = \dfrac{3}{4}|x+5|$, converges if $|x+5| < 4/3$, diverges if $|x+5| > 4/3$. If $x = -19/3$, $\sum_{k=0}^{\infty} (-1)^k$ diverges; if $x = -11/3$, $\sum_{k=0}^{\infty} 1$ diverges. Radius of convergence is $4/3$, interval of convergence is $(-19/3, -11/3)$.

45. $\rho = \lim_{k\to+\infty} \dfrac{k^2+4}{(k+1)^2+4}|x+1|^2 = |x+1|^2$, converges if $|x+1| < 1$, diverges if $|x+1| > 1$. If $x = -2$, $\sum_{k=1}^{\infty} \dfrac{(-1)^{3k+1}}{k^2+4}$ converges; if $x = 0$, $\sum_{k=1}^{\infty} \dfrac{(-1)^k}{k^2+4}$ converges. Radius of convergence is 1, interval of convergence is $[-2, 0]$.

47. $\rho = \lim_{k\to+\infty} \dfrac{\pi|x-1|^2}{(2k+3)(2k+2)} = 0$, radius of convergence $+\infty$, interval of convergence $(-\infty, +\infty)$.

49. $\rho = \lim_{k\to+\infty} \sqrt[k]{|u_k|} = \lim_{k\to+\infty} \dfrac{|x|}{\ln k} = 0$, the series converges absolutely for all x so the interval of convergence is $(-\infty, +\infty)$.

51. (a)

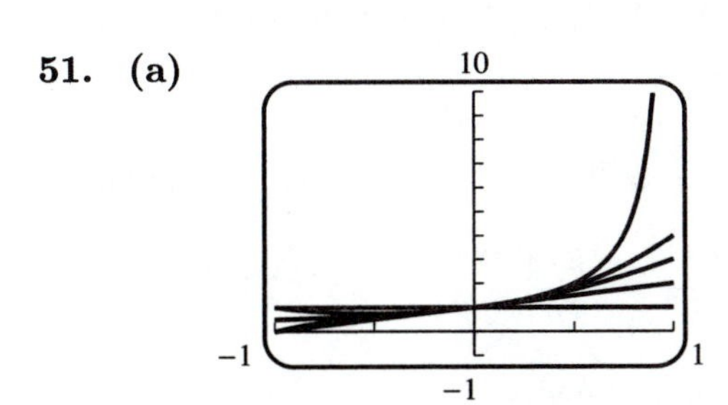

53. By the Ratio Test for absolute convergence,

$$\rho = \lim_{k\to+\infty} \frac{(pk+p)!(k!)^p}{(pk)![(k+1)!]^p}|x| = \lim_{k\to+\infty} \frac{(pk+p)(pk+p-1)(pk+p-2)\cdots(pk+p-[p-1])}{(k+1)^p}|x|$$

$$= \lim_{k\to+\infty} p\left(p-\frac{1}{k+1}\right)\left(p-\frac{2}{k+1}\right)\cdots\left(p-\frac{p-1}{k+1}\right)|x| = p^p|x|,$$

converges if $|x| < 1/p^p$, diverges if $|x| > 1/p^p$. Radius of convergence is $1/p^p$.

55. **(a)** By Theorem 10.5.3(b) both series converge or diverge together, so they have the same radius of convergence.

(b) By Theorem 10.5.3(a) the series $\sum(c_k+d_k)(x-x_0)^k$ converges if $|x-x_0| < R$; if $|x-x_0| > R$ then $\sum(c_k+d_k)(x-x_0)^k$ cannot converge, as otherwise $\sum c_k(x-x_0)^k$ would converge by the same Theorem. Hence the radius of convergence of $\sum(c_k+d_k)(x-x_0)^k$ is R.

(c) Let r be the radius of convergence of $\sum(c_k+d_k)(x-x_0)^k$. If $|x-x_0| < \min(R_1, R_2)$ then $\sum c_k(x-x_0)^k$ and $\sum d_k(x-x_0)^k$ converge, so $\sum(c_k+d_k)(x-x_0)^k$ converges. Hence $r \geq \min(R_1, R_2)$ (to see that $r > \min(R_1, R_2)$ is possible consider the case $c_k = -d_k = 1$). If in addition $R_1 \neq R_2$, and $R_1 < |x-x_0| < R_2$ (or $R_2 < |x-x_0| < R_1$) then $\sum(c_k+d_k)(x-x_0)^k$ cannot converge, as otherwise all three series would converge. Thus in this case $r = \min(R_1, R_2)$.

57. By assumption $\sum_{k=0}^{\infty} c_k x^k$ converges if $|x| < R$ so $\sum_{k=0}^{\infty} c_k x^{2k} = \sum_{k=0}^{\infty} c_k (x^2)^k$ converges if $|x^2| < R$, $|x| < \sqrt{R}$. Moreover, $\sum_{k=0}^{\infty} c_k x^{2k} = \sum_{k=0}^{\infty} c_k (x^2)^k$ diverges if $|x^2| > R$, $|x| > \sqrt{R}$. Thus $\sum_{k=0}^{\infty} c_k x^{2k}$ has radius of convergence $\sqrt{R}$.

EXERCISE SET 10.9

1. $\sin 4^\circ = \sin\left(\frac{\pi}{45}\right) = \frac{\pi}{45} - \frac{(\pi/45)^3}{3!} + \frac{(\pi/45)^5}{5!} - \cdots$

(a) Method 1: $|R_n(\pi/45)| \leq \frac{(\pi/45)^{n+1}}{(n+1)!} < 0.000005$ for $n+1 = 4, n = 3$;

$\sin 4^\circ \approx \frac{\pi}{45} - \frac{(\pi/45)^3}{3!} \approx 0.069756$

(b) Method 2: The first term in the alternating series that is less than 0.000005 is $\frac{(\pi/45)^5}{5!}$, so the result is the same as in Part (a).

3. $|R_n(0.1)| \leq \frac{(0.1)^{n+1}}{(n+1)!} \leq 0.000005$ for $n = 3$; $\cos 0.1 \approx 1 - (0.1)^2/2 = 0.99500$, calculator value $0.995004\ldots$

5. Expand about $\pi/2$ to get $\sin x = 1 - \frac{1}{2!}(x-\pi/2)^2 + \frac{1}{4!}(x-\pi/2)^4 - \cdots$, $85^\circ = 17\pi/36$ radians,

$|R_n(x)| \leq \frac{|x-\pi/2|^{n+1}}{(n+1)!}$, $|R_n(17\pi/36)| \leq \frac{|17\pi/36 - \pi/2|^{n+1}}{(n+1)!} = \frac{(\pi/36)^{n+1}}{(n+1)!} < 0.5 \times 10^{-4}$

if $n = 3$, $\sin 85^\circ \approx 1 - \frac{1}{2}(-\pi/36)^2 \approx 0.99619$, calculator value $0.99619\ldots$

7. $f^{(k)}(x) = \cosh x$ or $\sinh x$, $|f^{(k)}(x)| \le \cosh x \le \cosh 0.5 = \frac{1}{2}\left(e^{0.5}+e^{-0.5}\right) < \frac{1}{2}(2+1) = 1.5$ so $|R_n(x)| < \frac{1.5(0.5)^{n+1}}{(n+1)!} \le 0.5 \times 10^{-3}$ if $n = 4$, $\sinh 0.5 \approx 0.5 + \frac{(0.5)^3}{3!} \approx 0.5208$, calculator value $0.52109\ldots$

9. $f(x) = \sin x$, $f^{(n+1)}(x) = \pm\sin x$ or $\pm\cos x$, $|f^{(n+1)}(x)| \le 1$, $|R_n(x)| \le \frac{|x-\pi/4|^{n+1}}{(n+1)!}$, $\lim_{n\to+\infty}\frac{|x-\pi/4|^{n+1}}{(n+1)!} = 0$; by the Squeezing Theorem, $\lim_{n\to+\infty}|R_n(x)| = 0$ so $\lim_{n\to+\infty}R_n(x) = 0$ for all x.

11. **(a)** Let $x = 1/9$ in series (13).

(b) $\ln 1.25 \approx 2\left(1/9 + \frac{(1/9)^3}{3}\right) = 2(1/9 + 1/3^7) \approx 0.223$, which agrees with the calculator value $0.22314\ldots$ to three decimal places.

13. **(a)** $(1/2)^9/9 < 0.5 \times 10^{-3}$ and $(1/3)^7/7 < 0.5 \times 10^{-3}$ so

$$\tan^{-1}(1/2) \approx 1/2 - \frac{(1/2)^3}{3} + \frac{(1/2)^5}{5} - \frac{(1/2)^7}{7} \approx 0.4635$$

$$\tan^{-1}(1/3) \approx 1/3 - \frac{(1/3)^3}{3} + \frac{(1/3)^5}{5} \approx 0.3218$$

(b) From Formula (17), $\pi \approx 4(0.4635 + 0.3218) = 3.1412$

(c) Let $a = \tan^{-1}\frac{1}{2}, b = \tan^{-1}\frac{1}{3}$; then $|a - 0.4635| < 0.0005$ and $|b - 0.3218| < 0.0005$, so $|4(a+b) - 3.1412| \le 4|a - 0.4635| + 4|b - 0.3218| < 0.004$, so two decimal-place accuracy is guaranteed, but not three.

15. **(a)** $\cos x = 1 - \frac{x^2}{2!} + \frac{x^4}{4!} + (0)x^5 + R_5(x)$,

$|R_5(x)| \le \frac{|x|^6}{6!} \le \frac{(0.2)^6}{6!} < 9 \times 10^{-8}$

(b)

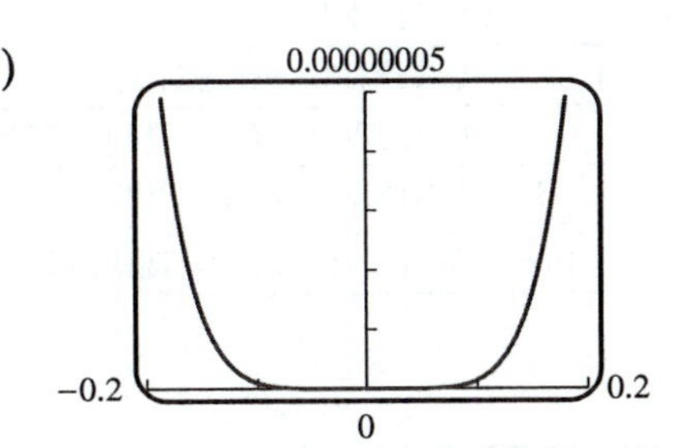

17. **(a)** $(1+x)^{-1} = 1 - x + \frac{-1(-2)}{2!}x^2 + \frac{-1(-2)(-3)}{3!}x^3 + \cdots + \frac{-1(-2)(-3)\cdots(-k)}{k!}x^k + \cdots$

$$= \sum_{k=0}^{\infty}(-1)^k x^k$$

(b) $(1+x)^{1/3} = 1 + (1/3)x + \frac{(1/3)(-2/3)}{2!}x^2 + \frac{(1/3)(-2/3)(-5/3)}{3!}x^3 + \cdots$

$$+ \frac{(1/3)(-2/3)\cdots(4-3k)/3}{k!}x^k + \cdots = 1 + x/3 + \sum_{k=2}^{\infty}(-1)^{k-1}\frac{2\cdot 5\cdots(3k-4)}{3^k k!}x^k$$

(c) $(1+x)^{-3} = 1 - 3x + \dfrac{(-3)(-4)}{2!}x^2 + \dfrac{(-3)(-4)(-5)}{3!}x^3 + \cdots + \dfrac{(-3)(-4)\cdots(-2-k)}{k!}x^k + \cdots$

$$= \sum_{k=0}^{\infty}(-1)^k\frac{(k+2)!}{2\cdot k!}x^k = \sum_{k=0}^{\infty}(-1)^k\frac{(k+2)(k+1)}{2}x^k$$

19. **(a)** $\dfrac{d}{dx}\ln(1+x) = \dfrac{1}{1+x}, \dfrac{d^k}{dx^k}\ln(1+x) = (-1)^{k-1}\dfrac{(k-1)!}{(1+x)^k}$; similarly $\dfrac{d}{dx}\ln(1-x) = -\dfrac{(k-1)!}{(1-x)^k}$,

so $f^{(n+1)}(x) = n!\left[\dfrac{(-1)^n}{(1+x)^{n+1}} + \dfrac{1}{(1-x)^{n+1}}\right]$.

(b) $\left|f^{(n+1)}(x)\right| \le n!\left|\dfrac{(-1)^n}{(1+x)^{n+1}}\right| + n!\left|\dfrac{1}{(1-x)^{n+1}}\right| = n!\left[\dfrac{1}{(1+x)^{n+1}} + \dfrac{1}{(1-x)^{n+1}}\right]$

(c) If $\left|f^{(n+1)}(x)\right| \le M$ on the interval $[0, 1/3]$ then $|R_n(1/3)| \le \dfrac{M}{(n+1)!}\left(\dfrac{1}{3}\right)^{n+1}$.

(d) If $0 \le x \le 1/3$ then $1+x \ge 1, 1-x \ge 2/3$, $\left|f^{(n+1)}(x)\right| \le M = n!\left[1 + \dfrac{1}{(2/3)^{n+1}}\right]$.

(e) $0.000005 \ge \dfrac{M}{(n+1)!}\left(\dfrac{1}{3}\right)^{n+1} = \dfrac{1}{n+1}\left[\left(\dfrac{1}{3}\right)^{n+1} + \dfrac{(1/3)^{n+1}}{(2/3)^{n+1}}\right] = \dfrac{1}{n+1}\left[\left(\dfrac{1}{3}\right)^{n+1} + \left(\dfrac{1}{2}\right)^{n+1}\right]$

21. $f(x) = \cos x, f^{(n+1)}(x) = \pm\sin x$ or $\pm\cos x, |f^{(n+1)}(x)| \le 1$, set $M = 1$,

$|R_n(x)| \le \dfrac{1}{(n+1)!}|x-a|^{n+1}$, $\lim_{n\to+\infty}\dfrac{|x-a|^{n+1}}{(n+1)!} = 0$ so $\lim_{n\to+\infty}R_n(x) = 0$ for all x.

23. **(a)** From Machin's formula and a CAS, $\dfrac{\pi}{4} \approx 0.7853981633974483096156608$, accurate to the 25th decimal place.

(b)

n	s_n
0	0.3183098 78...
1	0.3183098 861837906 067...
2	0.3183098 861837906 7153776 695...
3	0.3183098 861837906 7153776 752674502 34...
$1/\pi$	0.3183098 861837906 7153776 752674502 87...

EXERCISE SET 10.10

1. **(a)** Replace x with $-x$: $\dfrac{1}{1+x} = 1 - x + x^2 - \cdots + (-1)^k x^k + \cdots$; $R = 1$.

(b) Replace x with x^2 : $\dfrac{1}{1-x^2} = 1 + x^2 + x^4 + \cdots + x^{2k} + \cdots$; $R = 1$.

(c) Replace x with $2x$: $\dfrac{1}{1-2x} = 1 + 2x + 4x^2 + \cdots + 2^k x^k + \cdots$; $R = 1/2$.

(d) $\dfrac{1}{2-x} = \dfrac{1/2}{1-x/2}$; replace x with $x/2$: $\dfrac{1}{2-x} = \dfrac{1}{2} + \dfrac{1}{2^2}x + \dfrac{1}{2^3}x^2 + \cdots + \dfrac{1}{2^{k+1}}x^k + \cdots$; $R = 2$.

3. **(a)** From Section 10.9, Example 5(b), $\dfrac{1}{\sqrt{1+x}} = 1 - \dfrac{1}{2}x + \dfrac{1\cdot 3}{2^2\cdot 2!}x^2 - \dfrac{1\cdot 3\cdot 5}{2^3\cdot 3!}x^3 + \cdots$, so

$$(2+x)^{-1/2} = \frac{1}{\sqrt{2}\sqrt{1+x/2}} = \frac{1}{2^{1/2}} - \frac{1}{2^{5/2}}x + \frac{1\cdot 3}{2^{9/2}\cdot 2!}x^2 - \frac{1\cdot 3\cdot 5}{2^{13/2}\cdot 3!}x^3 + \cdots$$

(b) Example 5(a): $\dfrac{1}{(1+x)^2} = 1 - 2x + 3x^2 - 4x^3 + \cdots$, so $\dfrac{1}{(1-x^2)^2} = 1 + 2x^2 + 3x^4 + 4x^6 + \cdots$

5. **(a)** $2x - \dfrac{2^3}{3!}x^3 + \dfrac{2^5}{5!}x^5 - \dfrac{2^7}{7!}x^7 + \cdots;\ R = +\infty$

(b) $1 - 2x + 2x^2 - \dfrac{4}{3}x^3 + \cdots;\ R = +\infty$

(c) $1 + x^2 + \dfrac{1}{2!}x^4 + \dfrac{1}{3!}x^6 + \cdots;\ R = +\infty$

(d) $x^2 - \dfrac{\pi^2}{2}x^4 + \dfrac{\pi^4}{4!}x^6 - \dfrac{\pi^6}{6!}x^8 + \cdots;\ R = +\infty$

7. **(a)** $x^2\left(1 - 3x + 9x^2 - 27x^3 + \cdots\right) = x^2 - 3x^3 + 9x^4 - 27x^5 + \cdots;\ R = 1/3$

(b) $x\left(2x + \dfrac{2^3}{3!}x^3 + \dfrac{2^5}{5!}x^5 + \dfrac{2^7}{7!}x^7 + \cdots\right) = 2x^2 + \dfrac{2^3}{3!}x^4 + \dfrac{2^5}{5!}x^6 + \dfrac{2^7}{7!}x^8 + \cdots;\ R = +\infty$

(c) Substitute 3/2 for m and $-x^2$ for x in Equation (18) of Section 10.9, then multiply by x:
$x - \dfrac{3}{2}x^3 + \dfrac{3}{8}x^5 + \dfrac{1}{16}x^7 + \cdots;\ R = 1$

9. **(a)** $\sin^2 x = \dfrac{1}{2}(1-\cos 2x) = \dfrac{1}{2}\left[1 - \left(1 - \dfrac{2^2}{2!}x^2 + \dfrac{2^4}{4!}x^4 - \dfrac{2^6}{6!}x^6 + \cdots\right)\right]$

$$= x^2 - \frac{2^3}{4!}x^4 + \frac{2^5}{6!}x^6 - \frac{2^7}{8!}x^8 + \cdots$$

(b) $\ln\left[(1+x^3)^{12}\right] = 12\ln(1+x^3) = 12x^3 - 6x^6 + 4x^9 - 3x^{12} + \cdots$

11. **(a)** $\dfrac{1}{x} = \dfrac{1}{1-(1-x)} = 1 + (1-x) + (1-x)^2 + \cdots + (1-x)^k + \cdots$

$$= 1 - (x-1) + (x-1)^2 - \cdots + (-1)^k(x-1)^k + \cdots$$

(b) $(0, 2)$

13. **(a)** $(1 + x + x^2/2 + x^3/3! + x^4/4! + \cdots)(x - x^3/3! + x^5/5! - \cdots) = x + x^2 + x^3/3 - x^5/30 + \cdots$

(b) $(1 + x/2 - x^2/8 + x^3/16 - (5/128)x^4 + \cdots)(x - x^2/2 + x^3/3 - x^4/4 + x^5/5 - \cdots)$
$= x - x^3/24 + x^4/24 - (71/1920)x^5 + \cdots$

15. **(a)** $\dfrac{1}{\cos x} = 1\Big/\left(1 - \dfrac{1}{2!}x^2 + \dfrac{1}{4!}x^4 - \dfrac{1}{6!}x^6 + \cdots\right) = 1 + \dfrac{1}{2}x^2 + \dfrac{5}{24}x^4 + \dfrac{61}{720}x^6 + \cdots$

(b) $\dfrac{\sin x}{e^x} = \left(x - \dfrac{x^3}{3!} + \dfrac{x^5}{5!} - \cdots\right)\Big/\left(1 + x + \dfrac{x^2}{2!} + \dfrac{x^3}{3!} + \dfrac{x^4}{4!} + \cdots\right) = x - x^2 + \dfrac{1}{3}x^3 - \dfrac{1}{30}x^5 + \cdots$

17. $e^x = 1 + x + x^2/2 + x^3/3! + \cdots + x^k/k! + \cdots$, $e^{-x} = 1 - x + x^2/2 - x^3/3! + \cdots + (-1)^k x^k/k! + \cdots$;

$\sinh x = \dfrac{1}{2}\left(e^x - e^{-x}\right) = x + x^3/3! + x^5/5! + \cdots + x^{2k+1}/(2k+1)! + \cdots, R = +\infty$

$\cosh x = \dfrac{1}{2}\left(e^x + e^{-x}\right) = 1 + x^2/2 + x^4/4! + \cdots + x^{2k}/(2k)! + \cdots, R = +\infty$

19. $\dfrac{4x-2}{x^2-1} = \dfrac{-1}{1-x} + \dfrac{3}{1+x} = -\left(1+x+x^2+x^3+x^4+\cdots\right) + 3\left(1-x+x^2-x^3+x^4+\cdots\right)$

$= 2 - 4x + 2x^2 - 4x^3 + 2x^4 + \cdots$

21. **(a)** $\dfrac{d}{dx}\left(1 - x^2/2! + x^4/4! - x^6/6! + \cdots\right) = -x + x^3/3! - x^5/5! + \cdots = -\sin x$

(b) $\dfrac{d}{dx}\left(x - x^2/2 + x^3/3 - \cdots\right) = 1 - x + x^2 - \cdots = 1/(1+x)$

23. **(a)** $\displaystyle\int \left(1 + x + x^2/2! + \cdots\right) dx = \left(x + x^2/2! + x^3/3! + \cdots\right) + C_1$

$= \left(1 + x + x^2/2! + x^3/3! + \cdots\right) + C_1 - 1 = e^x + C$

(b) $\displaystyle\int \left(x + x^3/3! + x^5/5! + \cdots\right) = x^2/2! + x^4/4! + \cdots + C_1$

$= 1 + x^2/2! + x^4/4! + \cdots + C_1 - 1 = \cosh x + C$

25. **(a)** Substitute x^2 for x in the Maclaurin Series for $1/(1-x)$ (Table 10.9.1)

and then multiply by x: $\displaystyle\frac{x}{1-x^2} = x\sum_{k=0}^{\infty}(x^2)^k = \sum_{k=0}^{\infty} x^{2k+1}$

(b) $f^{(5)}(0) = 5!c_5 = 5!$, $f^{(6)}(0) = 6!c_6 = 0$ **(c)** $f^{(n)}(0) = n!c_n = \begin{cases} n! & \text{if } n \text{ odd} \\ 0 & \text{if } n \text{ even} \end{cases}$

27. **(a)** $\displaystyle\lim_{x\to 0}\frac{\sin x}{x} = \lim_{x\to 0}\left(1 - x^2/3! + x^4/5! - \cdots\right) = 1$

(b) $\displaystyle\lim_{x\to 0}\frac{\tan^{-1}x - x}{x^3} = \lim_{x\to 0}\frac{\left(x - x^3/3 + x^5/5 - x^7/7 + \cdots\right) - x}{x^3} = -1/3$

29. $\displaystyle\int_0^1 \sin\left(x^2\right) dx = \int_0^1 \left(x^2 - \frac{1}{3!}x^6 + \frac{1}{5!}x^{10} - \frac{1}{7!}x^{14} + \cdots\right) dx$

$\displaystyle = \frac{1}{3}x^3 - \frac{1}{7\cdot 3!}x^7 + \frac{1}{11\cdot 5!}x^{11} - \frac{1}{15\cdot 7!}x^{15} + \cdots\Big]_0^1$

$\displaystyle = \frac{1}{3} - \frac{1}{7\cdot 3!} + \frac{1}{11\cdot 5!} - \frac{1}{15\cdot 7!} + \cdots,$

but $\displaystyle\frac{1}{15\cdot 7!} < 0.5\times 10^{-3}$ so $\displaystyle\int_0^1 \sin(x^2)dx \approx \frac{1}{3} - \frac{1}{7\cdot 3!} + \frac{1}{11\cdot 5!} \approx 0.3103$

31. $\displaystyle\int_0^{0.2} \left(1 + x^4\right)^{1/3} dx = \int_0^{0.2}\left(1 + \frac{1}{3}x^4 - \frac{1}{9}x^8 + \cdots\right) dx$

$\displaystyle = x + \frac{1}{15}x^5 - \frac{1}{81}x^9 + \cdots\Big]_0^{0.2} = 0.2 + \frac{1}{15}(0.2)^5 - \frac{1}{81}(0.2)^9 + \cdots,$

but $\displaystyle\frac{1}{15}(0.2)^5 < 0.5\times 10^{-3}$ so $\displaystyle\int_0^{0.2}(1 + x^4)^{1/3}dx \approx 0.200$

33. **(a)** $\dfrac{x}{(1-x)^2} = x\dfrac{d}{dx}\left[\dfrac{1}{1-x}\right] = x\dfrac{d}{dx}\left[\sum_{k=0}^{\infty} x^k\right] = x\left[\sum_{k=1}^{\infty} kx^{k-1}\right] = \sum_{k=1}^{\infty} kx^k$

(b) $-\ln(1-x) = \int \dfrac{1}{1-x}dx - C = \int \left[\sum_{k=0}^{\infty} x^k\right] dx - C$

$$= \sum_{k=0}^{\infty} \frac{x^{k+1}}{k+1} - C = \sum_{k=1}^{\infty} \frac{x^k}{k} - C, -\ln(1-0) = 0 \text{ so } C = 0.$$

(c) Replace x with $-x$ in Part (b): $\ln(1+x) = -\sum_{k=1}^{+\infty} \dfrac{(-1)^k}{k}x^k = \sum_{k=1}^{+\infty} \dfrac{(-1)^{k+1}}{k}x^k$

(d) $\sum_{k=1}^{+\infty} \dfrac{(-1)^{k+1}}{k}$ converges by the Alternating Series Test.

(e) By Parts (c) and (d) and the remark, $\sum_{k=1}^{+\infty} \dfrac{(-1)^{k+1}}{k}x^k$ converges to $\ln(1+x)$ for $-1 < x \le 1$.

35. **(a)** $\sinh^{-1} x = \int (1+x^2)^{-1/2} dx - C = \int \left(1 - \dfrac{1}{2}x^2 + \dfrac{3}{8}x^4 - \dfrac{5}{16}x^6 + \cdots\right) dx - C$

$$= \left(x - \frac{1}{6}x^3 + \frac{3}{40}x^5 - \frac{5}{112}x^7 + \cdots\right) - C; \ \sinh^{-1} 0 = 0 \text{ so } C = 0.$$

(b) $(1+x^2)^{-1/2} = 1 + \sum_{k=1}^{\infty} \dfrac{(-1/2)(-3/2)(-5/2)\cdots(-1/2-k+1)}{k!}(x^2)^k$

$$= 1 + \sum_{k=1}^{\infty} (-1)^k \frac{1\cdot 3\cdot 5\cdots(2k-1)}{2^k k!}x^{2k},$$

$$\sinh^{-1} x = x + \sum_{k=1}^{\infty} (-1)^k \frac{1\cdot 3\cdot 5\cdots(2k-1)}{2^k k!(2k+1)}x^{2k+1}$$

(c) $R = 1$

37. **(a)** $y(t) = y_0 \sum_{k=0}^{\infty} \dfrac{(-1)^k (0.000121)^k t^k}{k!}$

(b) $y(1) \approx y_0(1 - 0.000121t)\Big]_{t=1} = 0.999879y_0$

(c) $y_0 e^{-0.000121} \approx 0.9998790073y_0$

39. $\theta_0 = 5° = \pi/36$ rad, $k = \sin(\pi/72)$

(a) $T \approx 2\pi\sqrt{\dfrac{L}{g}} = 2\pi\sqrt{1/9.8} \approx 2.00709$

(b) $T \approx 2\pi\sqrt{\dfrac{L}{g}}\left(1 + \dfrac{k^2}{4}\right) \approx 2.008044621$

(c) 2.008045644

41. **(a)** $F = \dfrac{mgR^2}{(R+h)^2} = \dfrac{mg}{(1+h/R)^2} = mg\left(1 - 2h/R + 3h^2/R^2 - 4h^3/R^3 + \cdots\right)$

(b) If $h = 0$, then the binomial series converges to 1 and $F = mg$.

(c) Sum the series to the linear term, $F \approx mg - 2mgh/R$.

(d) $\dfrac{mg - 2mgh/R}{mg} = 1 - \dfrac{2h}{R} = 1 - \dfrac{2 \cdot 29{,}028}{4000 \cdot 5280} \approx 0.9973$, so about 0.27% less.

43. Let $f(x) = \sum\limits_{k=0}^{\infty} a_k x^k = \sum\limits_{k=0}^{\infty} b_k x^k$ for $-r < x < r$. Then $a_k = f^{(k)}(0)/k! = b_k$ for all k.

CHAPTER 10 SUPPLEMENTARY EXERCISES

9. **(a)** always true by Theorem 10.5.2

(b) sometimes false, for example the harmonic series diverges but $\sum(1/k^2)$ converges

(c) sometimes false, for example $f(x) = \sin \pi x, a_k = 0, L = 0$

(d) always true by the comments which follow Example 3(d) of Section 10.2

(e) sometimes false, for example $a_n = \dfrac{1}{2} + (-1)^n \dfrac{1}{4}$

(f) sometimes false, for example $u_k = 1/2$

(g) always false by Theorem 10.5.3

(h) sometimes false, for example $u_k = 1/k, v_k = 2/k$

(i) always true by the Comparison Test

(j) always true by the Comparison Test

(k) sometimes false, for example $\sum(-1)^k/k$

(l) sometimes false, for example $\sum(-1)^k/k$

11. **(a)** geometric, $r = 1/5$, converges

(b) $1/(5^k + 1) < 1/5^k$, converges

(c) $\dfrac{9}{\sqrt{k}+1} \geq \dfrac{9}{\sqrt{k}+\sqrt{k}} = \dfrac{9}{2\sqrt{k}}$, $\sum\limits_{k=1}^{\infty} \dfrac{9}{2\sqrt{k}}$ diverges

13. **(a)** $\dfrac{1}{k^3 + 2k + 1} < \dfrac{1}{k^3}$, $\sum\limits_{k=1}^{\infty} 1/k^3$ converges, so $\sum\limits_{k=1}^{\infty} \dfrac{1}{k^3 + 2k + 1}$ converges by the Comparison Test

(b) Limit Comparison Test, compare with the divergent series $\sum\limits_{k=1}^{\infty} \dfrac{1}{k^{2/5}}$, diverges

(c) $\left|\dfrac{\cos(1/k)}{k^2}\right| < \dfrac{1}{k^2}$, $\sum\limits_{k=1}^{\infty} \dfrac{1}{k^2}$ converges, so $\sum\limits_{k=1}^{\infty} \dfrac{\cos(1/k)}{k^2}$ converges absolutely

15. $\sum\limits_{k=0}^{\infty} \dfrac{1}{5^k} - \sum\limits_{k=0}^{99} \dfrac{1}{5^k} = \sum\limits_{k=100}^{\infty} \dfrac{1}{5^k} = \dfrac{1}{5^{100}} \sum\limits_{k=0}^{\infty} \dfrac{1}{5^k} = \dfrac{1}{4 \cdot 5^{99}}$

17. **(a)** $p_0(x) = 1, p_1(x) = 1 - 7x, p_2(x) = 1 - 7x + 5x^2, p_3(x) = 1 - 7x + 5x^2 + 4x^3$, $p_4(x) = 1 - 7x + 5x^2 + 4x^3$

(b) If $f(x)$ is a polynomial of degree n and $k \geq n$ then the Maclaurin polynomial of degree k is the polynomial itself; if $k < n$ then it is the truncated polynomial.

19. $\sin x = x - x^3/3! + x^5/5! - x^7/7! + \cdots$ is an alternating series, so
$|\sin x - x + x^3/3! - x^5/5!| \le x^7/7! \le \pi^7/(4^7 7!) \le 0.00005$

21. **(a)** $\rho = \lim_{k\to+\infty} \left(\dfrac{2^k}{k!}\right)^{1/k} = \lim_{k\to+\infty} \dfrac{2}{\sqrt[k]{k!}} = 0$, converges

(b) $\rho = \lim_{k\to+\infty} u_k^{1/k} = \lim_{k\to+\infty} \dfrac{k}{\sqrt[k]{k!}} = e$, diverges

23. **(a)** $u_{100} = \sum_{k=1}^{100} u_k - \sum_{k=1}^{99} u_k = \left(2 - \dfrac{1}{100}\right) - \left(2 - \dfrac{1}{99}\right) = \dfrac{1}{9900}$

(b) $u_1 = 1$; for $k \ge 2, u_k = \left(2 - \dfrac{1}{k}\right) - \left(2 - \dfrac{1}{k-1}\right) = \dfrac{1}{k(k-1)}, \lim_{k\to+\infty} u_k = 0$

(c) $\sum_{k=1}^{\infty} u_k = \lim_{n\to+\infty} \sum_{k=1}^{n} u_k = \lim_{n\to+\infty} \left(2 - \dfrac{1}{n}\right) = 2$

25. **(a)** $e^2 - 1$ **(b)** $\sin\pi = 0$ **(c)** $\cos e$ **(d)** $e^{-\ln 3} = 1/3$

27. $e^{-x} = 1 - x + x^2/2! + \cdots$. Replace x with $-(\frac{x-100}{16})^2/2$ to obtain

$e^{-(\frac{x-100}{16})^2/2} = 1 - \dfrac{(x-100)^2}{2\cdot 16^2} + \dfrac{(x-100)^4}{8\cdot 16^4} + \cdots$, thus

$p \approx \dfrac{1}{16\sqrt{2\pi}} \displaystyle\int_{100}^{110} \left[1 - \dfrac{(x-100)^2}{2\cdot 16^2} + \dfrac{(x-100)^4}{8\cdot 16^4}\right] dx \approx 0.23406$ or 23.406%.

29. Let $A = 1 - \dfrac{1}{2^2} + \dfrac{1}{3^2} - \dfrac{1}{4^2} + \cdots$; since the series all converge absolutely,

$\dfrac{\pi^2}{6} - A = 2\dfrac{1}{2^2} + 2\dfrac{1}{4^2} + 2\dfrac{1}{6^2} + \cdots = \dfrac{1}{2}\left(1 + \dfrac{1}{2^2} + \dfrac{1}{3^2} + \cdots\right) = \dfrac{1}{2}\dfrac{\pi^2}{6}$, so $A = \dfrac{1}{2}\dfrac{\pi^2}{6} = \dfrac{\pi^2}{12}$.

31. **(a)** $x + \dfrac{1}{2}x^2 + \dfrac{3}{14}x^3 + \dfrac{3}{35}x^4 + \cdots$; $\rho = \lim_{k\to+\infty} \dfrac{k+1}{3k+1}|x| = \dfrac{1}{3}|x|$,

converges if $\dfrac{1}{3}|x| < 1$, $|x| < 3$ so $R = 3$.

(b) $-x^3 + \dfrac{2}{3}x^5 - \dfrac{2}{5}x^7 + \dfrac{8}{35}x^9 - \cdots$; $\rho = \lim_{k\to+\infty} \dfrac{k+1}{2k+1}|x|^2 = \dfrac{1}{2}|x|^2$,

converges if $\dfrac{1}{2}|x|^2 < 1$, $|x|^2 < 2$, $|x| < \sqrt{2}$ so $R = \sqrt{2}$.

33. If $x \ge 0$, then $\cos\sqrt{x} = 1 - \dfrac{(\sqrt{x})^2}{2!} + \dfrac{(\sqrt{x})^4}{4!} - \dfrac{(\sqrt{x})^6}{6!} + \cdots = 1 - \dfrac{x}{2!} + \dfrac{x^2}{4!} - \dfrac{x^3}{6!} + \cdots$; if $x \le 0$, then

$\cosh(\sqrt{-x}) = 1 + \dfrac{(\sqrt{-x})^2}{2!} + \dfrac{(\sqrt{-x})^4}{4!} + \dfrac{(\sqrt{-x})^6}{6!} + \cdots = 1 - \dfrac{x}{2!} + \dfrac{x^2}{4!} - \dfrac{x^3}{6!} + \cdots$.

35. $\left(1 - \dfrac{v^2}{c^2}\right)^{-1/2} \approx 1 + \dfrac{v^2}{2c^2}$, so $K = m_0c^2\left[\dfrac{1}{\sqrt{1 - v^2/c^2}} - 1\right] \approx m_0c^2(v^2/(2c^2)) = m_0v^2/2$

CHAPTER 11
Analytic Geometry in Calculus

EXERCISE SET 11.1

1.

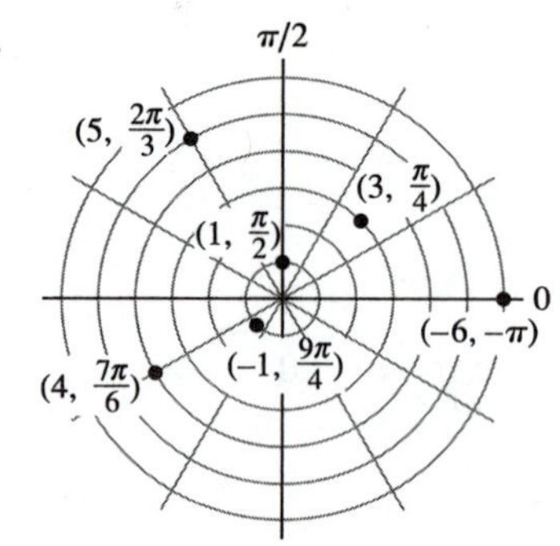

3. **(a)** $(3\sqrt{3}, 3)$ **(b)** $(-7/2, 7\sqrt{3}/2)$ **(c)** $(3\sqrt{3}, 3)$
(d) $(0, 0)$ **(e)** $(-7\sqrt{3}/2, 7/2)$ **(f)** $(-5, 0)$

5. **(a)** both $(5, \pi)$ **(b)** $(4, 11\pi/6), (4, -\pi/6)$ **(c)** $(2, 3\pi/2), (2, -\pi/2)$
(d) $(8\sqrt{2}, 5\pi/4), (8\sqrt{2}, -3\pi/4)$ **(e)** both $(6, 2\pi/3)$ **(f)** both $(\sqrt{2}, \pi/4)$

7. **(a)** $(5, 0.6435)$ **(b)** $(\sqrt{29}, 5.0929)$ **(c)** $(1.2716, 0.6658)$

9. **(a)** $r^2 = x^2 + y^2 = 4$; circle **(b)** $y = 4$; horizontal line
(c) $r^2 = 3r\cos\theta$, $x^2 + y^2 = 3x$, $(x - 3/2)^2 + y^2 = 9/4$; circle
(d) $3r\cos\theta + 2r\sin\theta = 6$, $3x + 2y = 6$; line

11. **(a)** $r\cos\theta = 7$ **(b)** $r = 3$
(c) $r^2 - 6r\sin\theta = 0$, $r = 6\sin\theta$
(d) $4(r\cos\theta)(r\sin\theta) = 9$, $4r^2\sin\theta\cos\theta = 9$, $r^2\sin 2\theta = 9/2$

13.

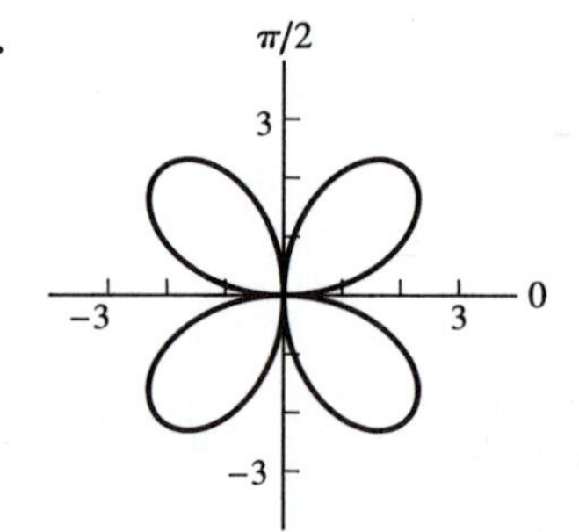

$r = 3\sin 2\theta$

15.

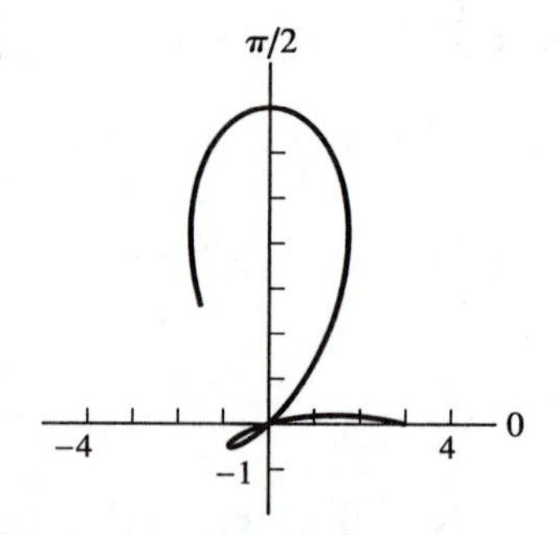

$r = 3 - 4\sin 3\theta$

17. **(a)** $r = 5$
(b) $(x - 3)^2 + y^2 = 9$, $r = 6\cos\theta$
(c) Example 6, $r = 1 - \cos\theta$

19. **(a)** Figure 11.1.18, $a = 3, n = 2, r = 3\sin 2\theta$

(b) From (8-9), symmetry about the y-axis and Theorem 11.1.1(b), the equation is of the form $r = a \pm b\sin\theta$. The cartesian points $(3, 0)$ and $(0, 5)$ give $a = 3$ and $5 = a + b$, so $b = 2$ and $r = 3 + 2\sin\theta$.

(c) Example 8, $r^2 = 9\cos 2\theta$

21.

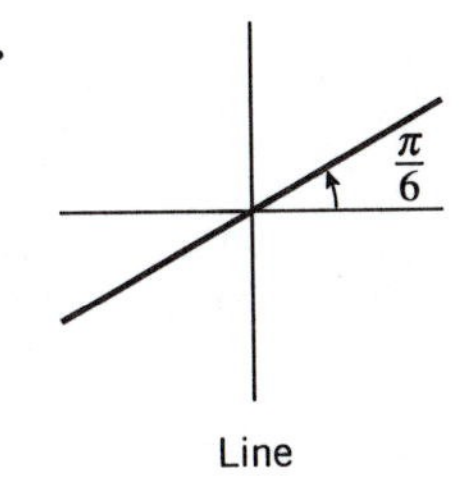

Line

23.

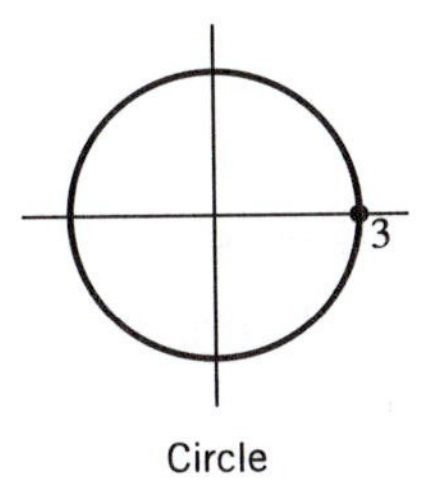

Circle

25.

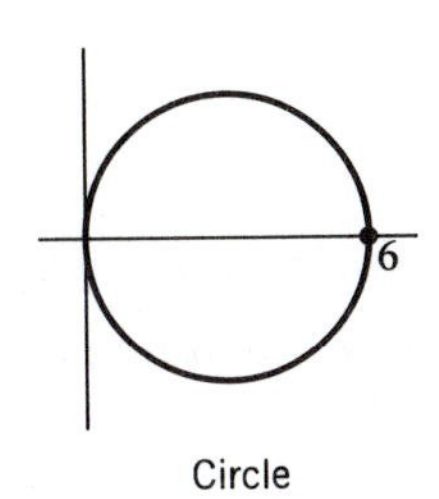

Circle

27.

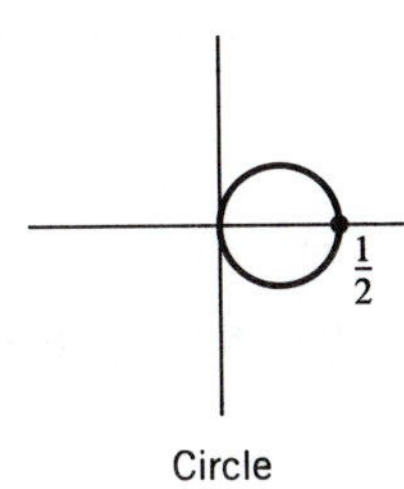

Circle

29.

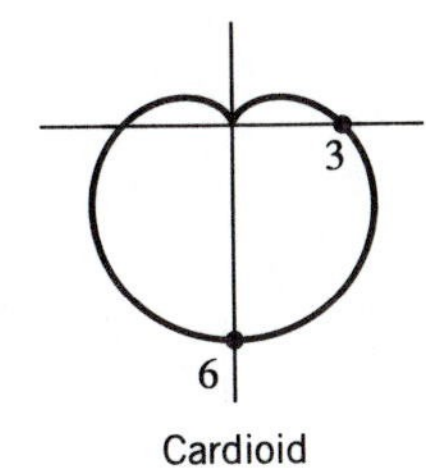

Cardioid

31.

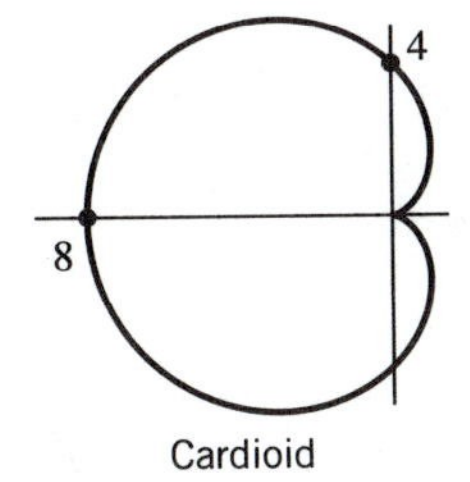

Cardioid

33.

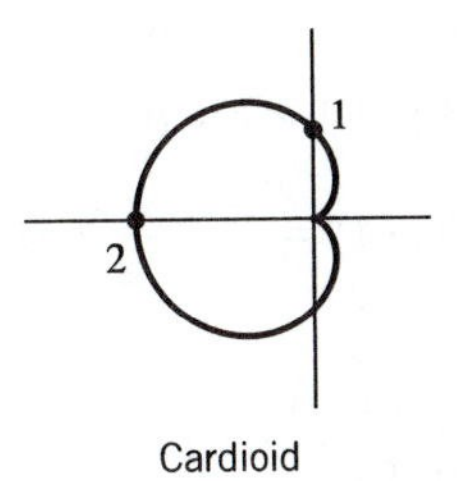

Cardioid

35.

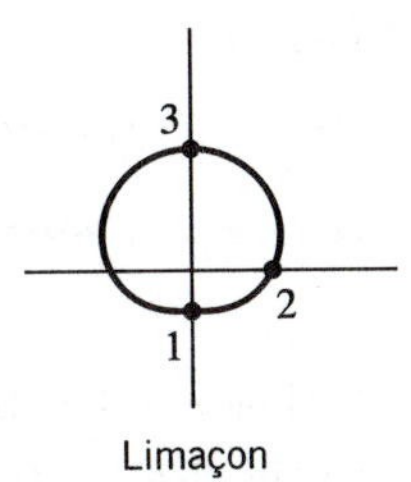

Limaçon

37.

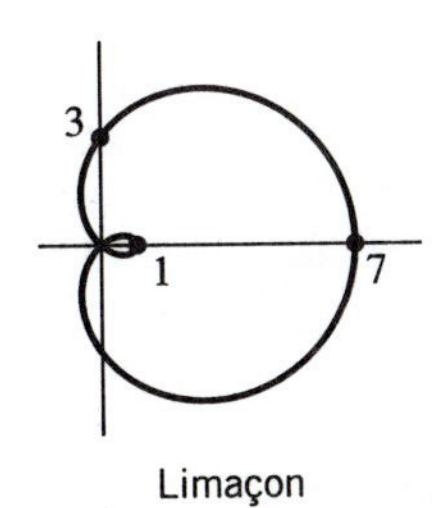

Limaçon

39.

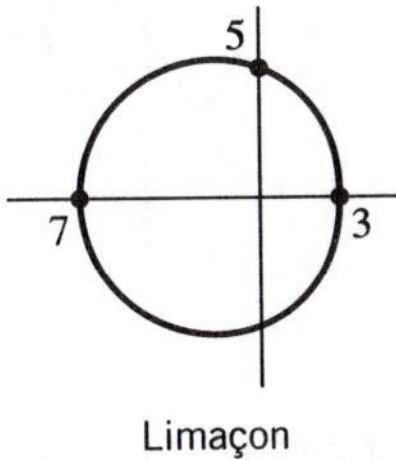

Limaçon

41.

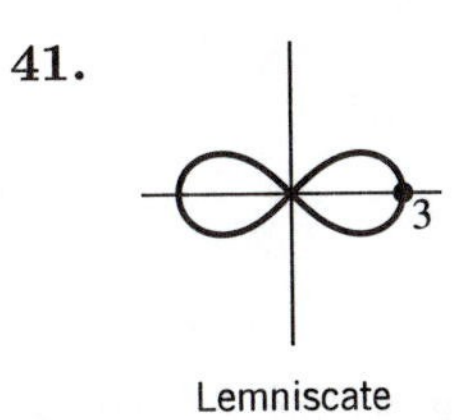

Lemniscate

43.

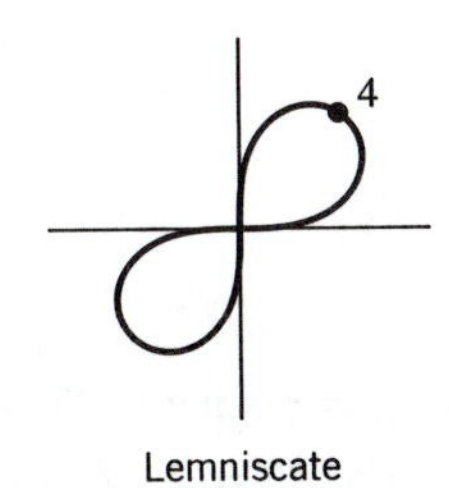

Lemniscate

45.

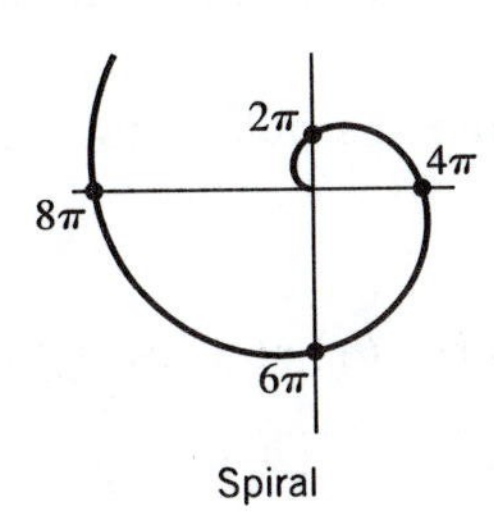

Spiral

47.

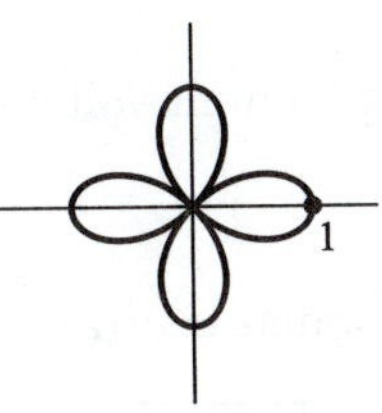

Four-petal rose

49.

Eight-petal rose

51.

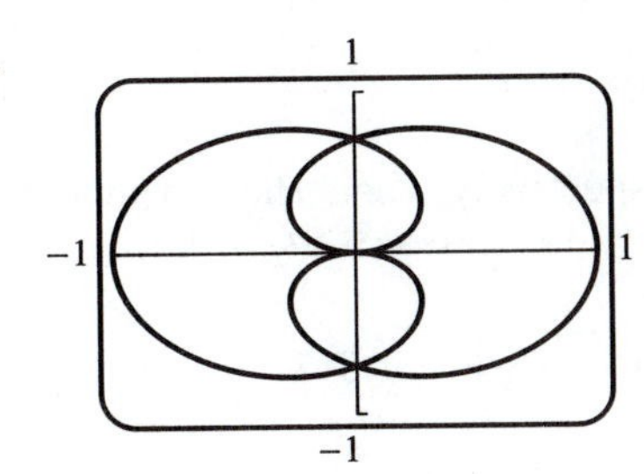

53.

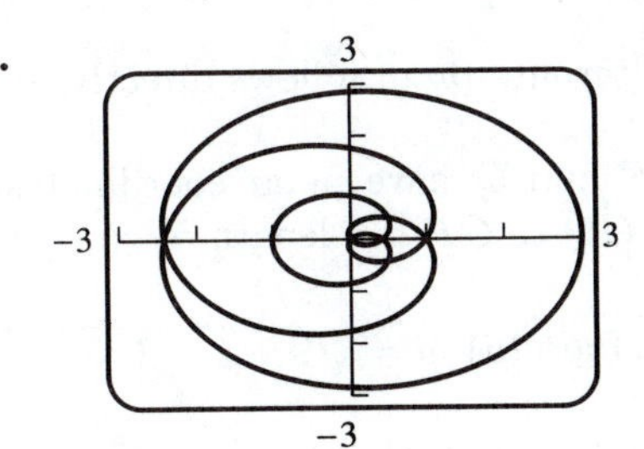

55.

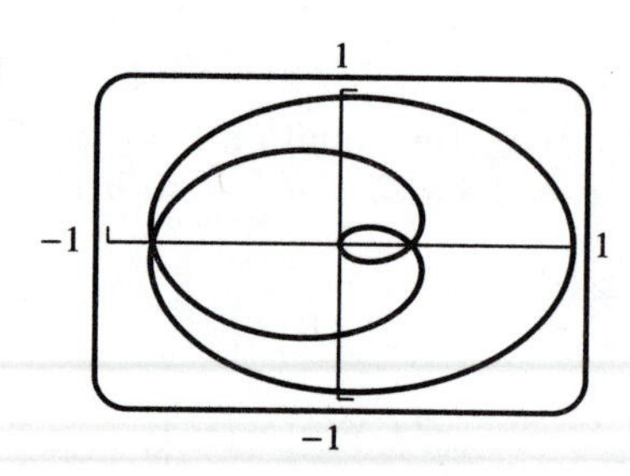

57. **(a)** $-4\pi < \theta < 4\pi$

59. **(a)** $r = a/\cos\theta, x = r\cos\theta = a$, a family of vertical lines

(b) $r = b/\sin\theta, y = r\sin\theta = b$, a family of horizontal lines

61. **(a)** $r = 1 + \cos(\theta - \pi/4) = 1 + \frac{\sqrt{2}}{2}(\cos\theta + \sin\theta)$

(b) $r = 1 + \cos(\theta - \pi/2) = 1 + \sin\theta$

(c) $r = 1 + \cos(\theta - \pi) = 1 - \cos\theta$

(d) $r = 1 + \cos(\theta - 5\pi/4) = 1 - \frac{\sqrt{2}}{2}(\cos\theta + \sin\theta)$

63. Either $r - 1 = 0$ or $\theta - 1 = 0$, so the graph consists of the circle $r = 1$ and the line $\theta = 1$.

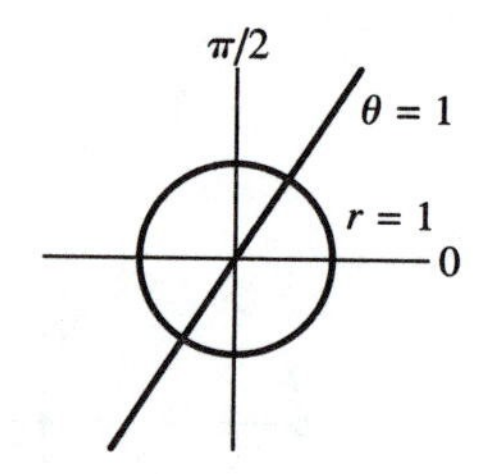

65. $y = r\sin\theta = (1 + \cos\theta)\sin\theta = \sin\theta + \sin\theta\cos\theta$,
$dy/d\theta = \cos\theta - \sin^2\theta + \cos^2\theta = 2\cos^2\theta + \cos\theta - 1 = (2\cos\theta - 1)(\cos\theta + 1)$;
$dy/d\theta = 0$ if $\cos\theta = 1/2$ or if $\cos\theta = -1$; $\theta = \pi/3$ or π (or $\theta = -\pi/3$, which leads to the minimum point).
If $\theta = \pi/3, \pi$, then $y = 3\sqrt{3}/4, 0$ so the maximum value of y is $3\sqrt{3}/4$ and the polar coordinates of the highest point are $(3/2, \pi/3)$.

67. **(a)** Let (x_1, y_1) and (x_2, y_2) be the rectangular coordinates of the points (r_1, θ_1) and (r_2, θ_2) then

$$d = \sqrt{(x_2 - x_1)^2 + (y_2 - y_1)^2} = \sqrt{(r_2\cos\theta_2 - r_1\cos\theta_1)^2 + (r_2\sin\theta_2 - r_1\sin\theta_1)^2}$$
$$= \sqrt{r_1^2 + r_2^2 - 2r_1r_2(\cos\theta_1\cos\theta_2 + \sin\theta_1\sin\theta_2)} = \sqrt{r_1^2 + r_2^2 - 2r_1r_2\cos(\theta_1 - \theta_2)}.$$

An alternate proof follows directly from the Law of Cosines.

(b) Let P and Q have polar coordinates $(r_1, \theta_1), (r_2, \theta_2)$, respectively, then the perpendicular from OQ to OP has length $h = r_2\sin(\theta_2 - \theta_1)$ and $A = \frac{1}{2}hr_1 = \frac{1}{2}r_1r_2\sin(\theta_2 - \theta_1)$.

(c) From Part (a), $d = \sqrt{9 + 4 - 2\cdot3\cdot2\cos(\pi/6 - \pi/3)} = \sqrt{13 - 6\sqrt{3}} \approx 1.615$

(d) $A = \frac{1}{2}2\sin(5\pi/6 - \pi/3) = 1$

69. $\lim_{\theta\to 0^+} y = \lim_{\theta\to 0^+} r\sin\theta = \lim_{\theta\to 0^+} \frac{\sin\theta}{\theta} = 1$, and $\lim_{\theta\to 0^+} x = \lim_{\theta\to 0^+} r\cos\theta = \lim_{\theta\to 0^+} \frac{\cos\theta}{\theta} = +\infty$.

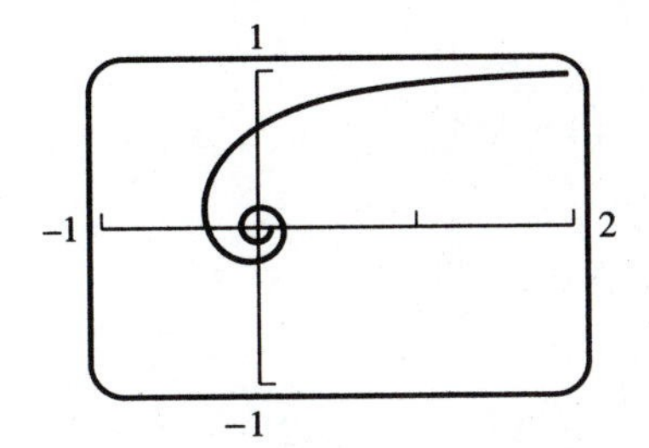

71. Note that $r \to \pm\infty$ as θ approaches odd multiples of $\pi/2$;
$x = r\cos\theta = 4\tan\theta\cos\theta = 4\sin\theta$,
$y = r\sin\theta = 4\tan\theta\sin\theta$
so $x \to \pm 4$ and $y \to \pm\infty$ as θ approaches odd multiples of $\pi/2$.

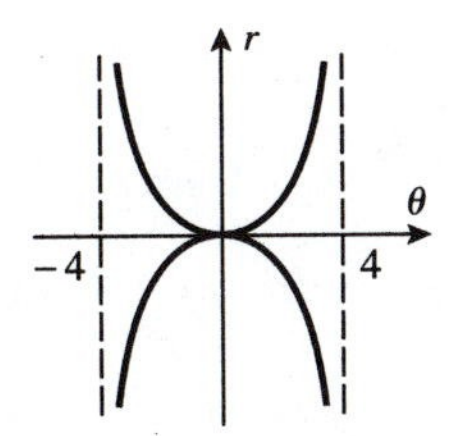

73. Let $r = a\sin n\theta$ (the proof for $r = a\cos n\theta$ is similar). If θ starts at 0, then θ would have to increase by some positive integer multiple of π radians in order to reach the starting point and begin to retrace the curve. Let (r, θ) be the coordinates of a point P on the curve for $0 \le \theta < 2\pi$. Now $a\sin n(\theta + 2\pi) = a\sin(n\theta + 2\pi n) = a\sin n\theta = r$ so P is reached again with coordinates $(r, \theta + 2\pi)$ thus the curve is traced out either exactly once or exactly twice for $0 \le \theta < 2\pi$. If for $0 \le \theta < \pi$, $P(r, \theta)$ is reached again with coordinates $(-r, \theta + \pi)$ then the curve is traced out exactly once for $0 \le \theta < \pi$, otherwise exactly once for $0 \le \theta < 2\pi$. But

$$a\sin n(\theta + \pi) = a\sin(n\theta + n\pi) = \begin{cases} a\sin n\theta, & n \text{ even} \\ -a\sin n\theta, & n \text{ odd} \end{cases}$$

so the curve is traced out exactly once for $0 \le \theta < 2\pi$ if n is even, and exactly once for $0 \le \theta < \pi$ if n is odd.

EXERCISE SET 11.2

1. **(a)** $dy/dx = \frac{1/2}{2t} = 1/(4t)$; $dy/dx\big|_{t=-1} = -1/4$; $dy/dx\big|_{t=1} = 1/4$

(b) $x = (2y)^2 + 1, dx/dy = 8y, dy/dx\big|_{y=\pm(1/2)} = \pm 1/4$

3. $\frac{d^2y}{dx^2} = \frac{d}{dx}\frac{dy}{dx} = \frac{d}{dt}\left(\frac{dy}{dx}\right)\frac{dt}{dx} = -\frac{1}{4t^2}(1/2t) = -1/(8t^3)$; positive when $t = -1$, negative when $t = 1$

5. $dy/dx = \frac{2}{1/(2\sqrt{t})} = 4\sqrt{t}$, $d^2y/dx^2 = \frac{2/\sqrt{t}}{1/(2\sqrt{t})} = 4$, $dy/dx\big|_{t=1} = 4$, $d^2y/dx^2\big|_{t=1} = 4$

7. $dy/dx = \frac{\sec^2 t}{\sec t\tan t} = \csc t$, $d^2y/dx^2 = \frac{-\csc t\cot t}{\sec t\tan t} = -\cot^3 t$,

$dy/dx\big|_{t=\pi/3} = 2/\sqrt{3}$, $d^2y/dx^2\big|_{t=\pi/3} = -1/(3\sqrt{3})$

9. $\dfrac{dy}{dx} = \dfrac{dy/d\theta}{dx/d\theta} = \dfrac{-\cos\theta}{2-\sin\theta}$; $\dfrac{d^2y}{dx^2} = \dfrac{d}{d\theta}\left(\dfrac{dy}{dx}\right)\Big/\dfrac{dx}{d\theta} = \dfrac{1}{(2-\sin\theta)^2}\dfrac{1}{2-\sin\theta} = \dfrac{1}{(2-\sin\theta)^3}$;

$\left.\dfrac{dy}{dx}\right|_{\theta=\pi/3} = \dfrac{-1/2}{2-\sqrt{3}/2} = \dfrac{-1}{4-\sqrt{3}}$; $\left.\dfrac{d^2y}{dx^2}\right|_{\theta=\pi/3} = \dfrac{1}{(2-\sqrt{3}/2)^3} = \dfrac{8}{(4-\sqrt{3})^3}$

11. **(a)** $dy/dx = \dfrac{-e^{-t}}{e^t} = -e^{-2t}$; for $t=1$, $dy/dx = -e^{-2}$, $(x,y) = (e, e^{-1})$; $y - e^{-1} = -e^{-2}(x-e)$, $y = -e^{-2}x + 2e^{-1}$

(b) $y = 1/x, dy/dx = -1/x^2, m = -1/e^2, y - e^{-1} = -\dfrac{1}{e^2}(x-e), y = -\dfrac{1}{e^2}x + \dfrac{2}{e}$

13. $dy/dx = \dfrac{4\cos t}{-2\sin t} = -2\cot t$

(a) $dy/dx = 0$ if $\cot t = 0$, $t = \pi/2 + n\pi$ for $n = 0, \pm 1, \cdots$

(b) $dx/dy = -\dfrac{1}{2}\tan t = 0$ if $\tan t = 0$, $t = n\pi$ for $n = 0, \pm 1, \cdots$

15. $x = y = 0$ when $t = 0, \pi$; $\dfrac{dy}{dx} = \dfrac{2\cos 2t}{\cos t}$; $\left.\dfrac{dy}{dx}\right|_{t=0} = 2$, $\left.\dfrac{dy}{dx}\right|_{t=\pi} = -2$, the equations of the tangent lines are $y = -2x, y = 2x$.

17. If $y = 4$ then $t^2 = 4$, $t = \pm 2$, $x = 0$ for $t = \pm 2$ so $(0, 4)$ is reached when $t = \pm 2$. $dy/dx = 2t/(3t^2 - 4)$. For $t = 2$, $dy/dx = 1/2$ and for $t = -2$, $dy/dx = -1/2$. The tangent lines are $y = \pm x/2 + 4$.

19. **(a)**

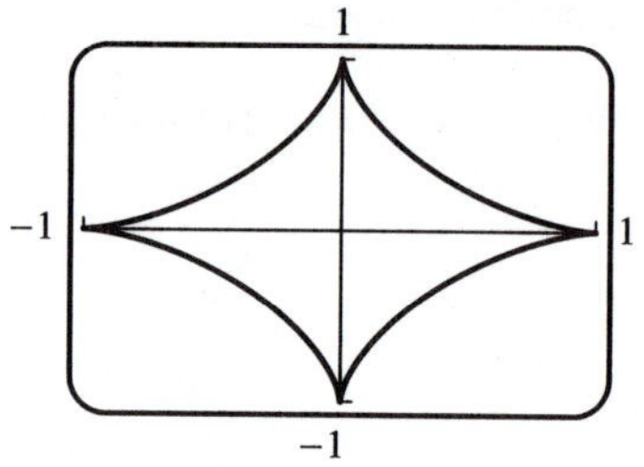

(b) $\dfrac{dx}{dt} = -3\cos^2 t \sin t$ and $\dfrac{dy}{dt} = 3\sin^2 t \cos t$ are both zero when $t = 0, \pi/2, \pi, 3\pi/2, 2\pi$, so singular points occur at these values of t.

21. Substitute $\theta = \pi/3$, $r = 1$, and $dr/d\theta = -\sqrt{3}$ in equation (7) gives slope $m = 1/\sqrt{3}$.

23. As in Exercise 21, $\theta = 2$, $dr/d\theta = -1/4$, $r = 1/2$, $m = \dfrac{\tan 2 - 2}{2\tan 2 + 1}$

25. As in Exercise 21, $\theta = 3\pi/4$, $dr/d\theta = -3\sqrt{2}/2$, $r = \sqrt{2}/2$, $m = -2$

27. $m = \dfrac{dy}{dx} = \dfrac{r\cos\theta + (\sin\theta)(dr/d\theta)}{-r\sin\theta + (\cos\theta)(dr/d\theta)} = \dfrac{\cos\theta + 2\sin\theta\cos\theta}{-\sin\theta + \cos^2\theta - \sin^2\theta}$; if $\theta = 0, \pi/2, \pi$, then $m = 1, 0, -1$.

29. $dx/d\theta = -a\sin\theta(1+2\cos\theta)$, $dy/d\theta = a(2\cos\theta - 1)(\cos\theta + 1)$

(a) horizontal if $dy/d\theta = 0$ and $dx/d\theta \neq 0$. $dy/d\theta = 0$ when $\cos\theta = 1/2$ or $\cos\theta = -1$ so $\theta = \pi/3$, $5\pi/3$, or π; $dx/d\theta \neq 0$ for $\theta = \pi/3$ and $5\pi/3$. For the singular point $\theta = \pi$ we find that $\lim\limits_{\theta\to\pi} dy/dx = 0$. There is a horizontal tangent line at $(3a/2, \pi/3)$, $(0, \pi)$, and $(3a/2, 5\pi/3)$.

(b) vertical if $dy/d\theta \neq 0$ and $dx/d\theta = 0$. $dx/d\theta = 0$ when $\sin\theta = 0$ or $\cos\theta = -1/2$ so $\theta = 0$, π, $2\pi/3$, or $4\pi/3$; $dy/d\theta \neq 0$ for $\theta = 0$, $2\pi/3$, and $4\pi/3$. The singular point $\theta = \pi$ was discussed in Part (a). There is a vertical tangent line at $(2a, 0)$, $(a/2, 2\pi/3)$, and $(a/2, 4\pi/3)$.

31. $dy/d\theta = (d/d\theta)(\sin^2\theta\cos^2\theta) = (\sin 4\theta)/2 = 0$ at $\theta = 0, \pi/4, \pi/2, 3\pi/4, \pi$; at the same points, $dx/d\theta = (d/d\theta)(\sin\theta\cos^3\theta) = \cos^2\theta(4\cos^2\theta - 3)$. Next, $\dfrac{dx}{d\theta} = 0$ at $\theta = \pi/2$, a singular point; and $\theta = 0, \pi$ both give the same point, so there are just three points with a horizontal tangent.

 **33.**

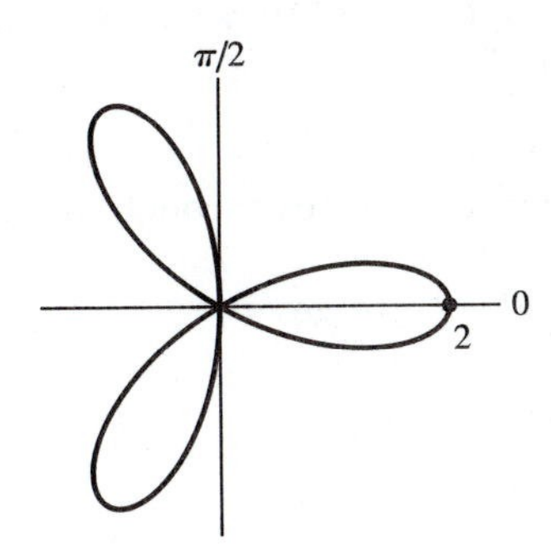

$\theta_0 = \pi/6, \pi/2, 5\pi/6$

35.

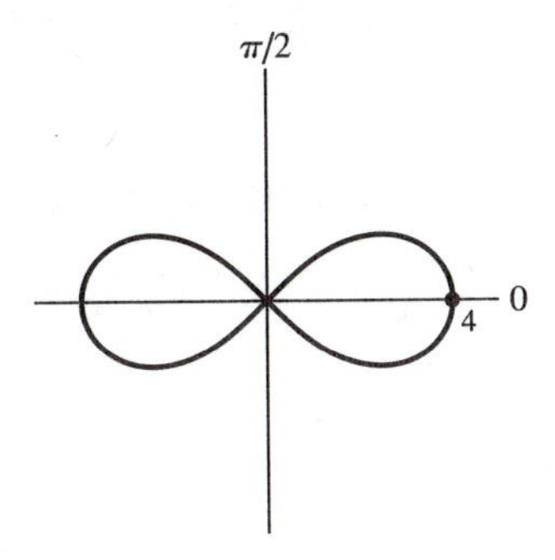

$\theta_0 = \pm\pi/4$

37.

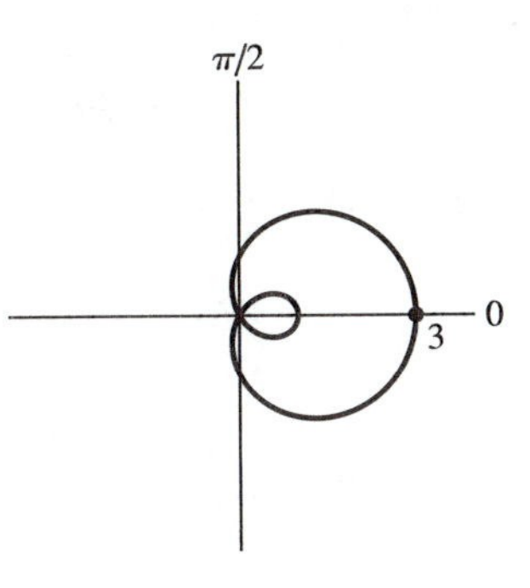

$\theta_0 = 2\pi/3, 4\pi/3$

39. $r^2 + (dr/d\theta)^2 = a^2 + 0^2 = a^2$, $L = \displaystyle\int_0^{2\pi} a\,d\theta = 2\pi a$

41. $r^2 + (dr/d\theta)^2 = [a(1-\cos\theta)]^2 + [a\sin\theta]^2 = 4a^2\sin^2(\theta/2)$, $L = 2\displaystyle\int_0^{\pi} 2a\sin(\theta/2)d\theta = 8a$

43. $r^2 + (dr/d\theta)^2 = (e^{3\theta})^2 + (3e^{3\theta})^2 = 10e^{6\theta}$, $L = \displaystyle\int_0^2 \sqrt{10}e^{3\theta}d\theta = \sqrt{10}(e^6 - 1)/3$

45. **(a)** From (3), $\dfrac{dy}{dx} = \dfrac{3\sin t}{1 - 3\cos t}$

(b) At $t = 10$, $\dfrac{dy}{dx} = \dfrac{3\sin 10}{1 - 3\cos 10} \approx -0.46402$, $\theta \approx \tan^{-1}(-0.46402) = -0.4345$

47. **(a)** $r^2 + (dr/d\theta)^2 = (\cos n\theta)^2 + (-n\sin n\theta)^2 = \cos^2 n\theta + n^2\sin^2 n\theta$
$= (1 - \sin^2 n\theta) + n^2\sin^2 n\theta = 1 + (n^2 - 1)\sin^2 n\theta$,

$$L = 2\int_0^{\pi/(2n)} \sqrt{1 + (n^2 - 1)\sin^2 n\theta}\,d\theta$$

(b) $L = 2\displaystyle\int_0^{\pi/4} \sqrt{1 + 3\sin^2 2\theta}\,d\theta \approx 2.42$

(c)

n	2	3	4	5	6	7	8	9	10	11
L	2.42211	2.22748	2.14461	2.10100	2.07501	2.05816	2.04656	2.03821	2.03199	2.02721

n	12	13	14	15	16	17	18	19	20
L	2.02346	2.02046	2.01802	2.01600	2.01431	2.01288	2.01167	2.01062	2.00971

49. $x' = 2t$, $y' = 2$, $(x')^2 + (y')^2 = 4t^2 + 4$

$$S = 2\pi \int_0^4 (2t)\sqrt{4t^2+4}dt = 8\pi \int_0^4 t\sqrt{t^2+1}dt = \frac{8\pi}{3}(t^2+1)^{3/2}\Big]_0^4 = \frac{8\pi}{3}(17\sqrt{17}-1)$$

51. $x' = -2\sin t\cos t$, $y' = 2\sin t\cos t$, $(x')^2 + (y')^2 = 8\sin^2 t\cos^2 t$

$$S = 2\pi \int_0^{\pi/2} \cos^2 t\sqrt{8\sin^2 t\cos^2 t}\, dt = 4\sqrt{2}\pi \int_0^{\pi/2} \cos^3 t\sin t\, dt = -\sqrt{2}\pi\cos^4 t\Big]_0^{\pi/2} = \sqrt{2}\pi$$

53. $x' = -r\sin t$, $y' = r\cos t$, $(x')^2 + (y')^2 = r^2$, $S = 2\pi \int_0^{\pi} r\sin t\sqrt{r^2}\, dt = 2\pi r^2 \int_0^{\pi} \sin t\, dt = 4\pi r^2$

55. **(a)** $\frac{dr}{dt} = 2$ and $\frac{d\theta}{dt} = 1$ so $\frac{dr}{d\theta} = \frac{dr/dt}{d\theta/dt} = \frac{2}{1} = 2$, $r = 2\theta + C$, $r = 10$ when $\theta = 0$ so $10 = C, r = 2\theta + 10$.

(b) $r^2 + (dr/d\theta)^2 = (2\theta+10)^2 + 4$, during the first 5 seconds the rod rotates through an angle of $(1)(5) = 5$ radians so $L = \int_0^5 \sqrt{(2\theta+10)^2+4}d\theta$, let $u = 2\theta + 10$ to get

$$L = \frac{1}{2}\int_{10}^{20} \sqrt{u^2+4}du = \frac{1}{2}\left[\frac{u}{2}\sqrt{u^2+4} + 2\ln|u+\sqrt{u^2+4}|\right]_{10}^{20}$$

$$= \frac{1}{2}\left[10\sqrt{404} - 5\sqrt{104} + 2\ln\frac{20+\sqrt{404}}{10+\sqrt{104}}\right] \approx 75.7 \text{ mm}$$

EXERCISE SET 11.3

1. **(a)** $\int_{\pi/2}^{\pi} \frac{1}{2}(1-\cos\theta)^2\, d\theta$ **(b)** $\int_0^{\pi/2} \frac{1}{2}4\cos^2\theta\, d\theta$ **(c)** $\int_0^{\pi/2} \frac{1}{2}\sin^2 2\theta\, d\theta$

(d) $\int_0^{2\pi} \frac{1}{2}\theta^2\, d\theta$ **(e)** $\int_{-\pi/2}^{\pi/2} \frac{1}{2}(1-\sin\theta)^2\, d\theta$ **(f)** $2\int_0^{\pi/4} \frac{1}{2}\cos^2 2\theta\, d\theta$

3. **(a)** $A = \int_0^{2\pi} \frac{1}{2}a^2\, d\theta = \pi a^2$ **(b)** $A = \int_0^{\pi} \frac{1}{2}4a^2\sin^2\theta\, d\theta = \pi a^2$

(c) $A = \int_{-\pi/2}^{\pi/2} \frac{1}{2}4a^2\cos^2\theta\, d\theta = \pi a^2$

5. $A = 2\int_0^{\pi} \frac{1}{2}(2+2\cos\theta)^2 d\theta = 6\pi$

7. $A = 6\int_0^{\pi/6} \frac{1}{2}(16\cos^2 3\theta)d\theta = 4\pi$

9. $A = 2\int_{2\pi/3}^{\pi} \frac{1}{2}(1+2\cos\theta)^2 d\theta = \pi - 3\sqrt{3}/2$

11. area $= A_1 - A_2 = \int_0^{\pi/2} \frac{1}{2}4\cos^2\theta\, d\theta - \int_0^{\pi/4} \frac{1}{2}\cos 2\theta\, d\theta = \pi/2 - \frac{1}{4}$

13. The circles intersect when $\cos t = \sqrt{3}\sin t, \tan t = 1/\sqrt{3}, t = \pi/6$, so

$$A = A_1 + A_2 = \int_0^{\pi/6} \frac{1}{2}(4\sqrt{3}\sin t)^2\,dt + \int_{\pi/6}^{\pi/2} \frac{1}{2}(4\cos t)^2\,dt = 2\pi - 3\sqrt{3} + 4\pi/3 - \sqrt{3} = 10\pi/3 - 4\sqrt{3}.$$

15. $A = 2\displaystyle\int_{\pi/6}^{\pi/2} \frac{1}{2}[25\sin^2\theta - (2+\sin\theta)^2]d\theta = 8\pi/3 + \sqrt{3}$

17. $A = 2\displaystyle\int_0^{\pi/3} \frac{1}{2}[(2+2\cos\theta)^2 - 9]d\theta = 9\sqrt{3}/2 - \pi$

19. $A = 2\left[\displaystyle\int_0^{2\pi/3} \frac{1}{2}(1/2+\cos\theta)^2 d\theta - \int_{2\pi/3}^{\pi} \frac{1}{2}(1/2+\cos\theta)^2 d\theta\right] = (\pi + 3\sqrt{3})/4$

21. $A = 2\displaystyle\int_0^{\cos^{-1}(3/5)} \frac{1}{2}(100 - 36\sec^2\theta)d\theta = 100\cos^{-1}(3/5) - 48$

23. **(a)** r is not real for $\pi/4 < \theta < 3\pi/4$ and $5\pi/4 < \theta < 7\pi/4$

(b) $A = 4\displaystyle\int_0^{\pi/4} \frac{1}{2}a^2\cos 2\theta\,d\theta = a^2$

(c) $A = 4\displaystyle\int_0^{\pi/6} \frac{1}{2}\left[4\cos 2\theta - 2\right]d\theta = 2\sqrt{3} - \frac{2\pi}{3}$

25. $A = \displaystyle\int_{2\pi}^{4\pi} \frac{1}{2}a^2\theta^2\,d\theta - \int_0^{2\pi} \frac{1}{2}a^2\theta^2\,d\theta = 8\pi^3 a^2$

27. $r^2 + \left(\dfrac{dr}{d\theta}\right)^2 = \cos^2\theta + \sin^2\theta = 1,$

so $S = \displaystyle\int_{-\pi/2}^{\pi/2} 2\pi\cos^2\theta\,d\theta = \pi^2.$

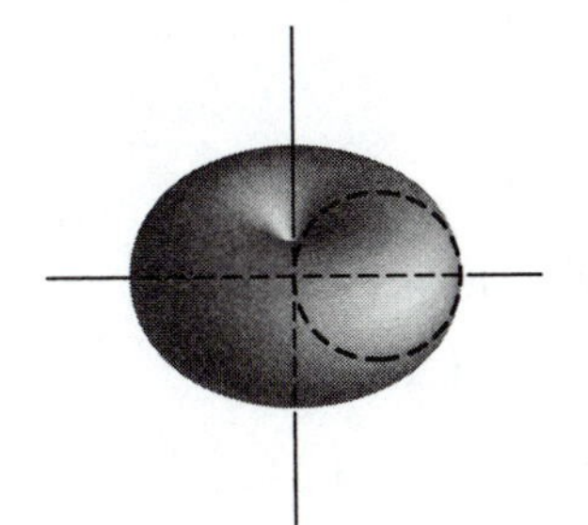

29. $S = \displaystyle\int_0^{\pi} 2\pi(1-\cos\theta)\sin\theta\sqrt{1 - 2\cos\theta + \cos^2\theta + \sin^2\theta}\,d\theta$

$= 2\sqrt{2}\pi\displaystyle\int_0^{\pi} \sin\theta(1-\cos\theta)^{3/2}\,d\theta = \frac{2}{5}2\sqrt{2}\pi(1-\cos\theta)^{5/2}\Big|_0^{\pi} = 32\pi/5$

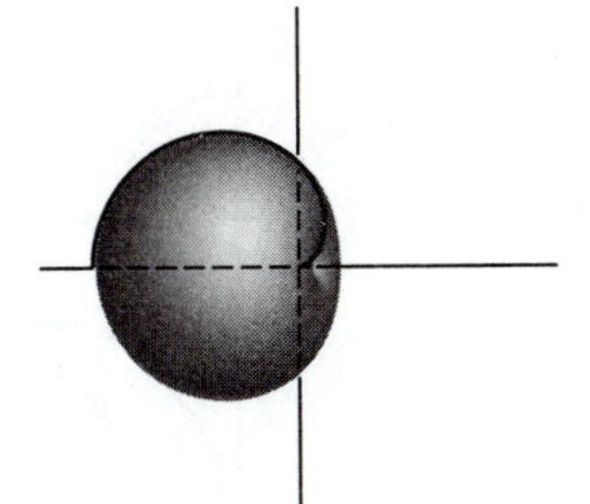

31. **(a)** $r^3\cos^3\theta - 3r^2\cos\theta\sin\theta + r^3\sin^3\theta = 0,\ r = \dfrac{3\cos\theta\sin\theta}{\cos^3\theta + \sin^3\theta}$

33. If the upper right corner of the square is the point (a, a) then the large circle has equation $r = \sqrt{2}a$ and the small circle has equation $(x - a)^2 + y^2 = a^2$, $r = 2a\cos\theta$, so

$$\text{area of crescent} = 2\int_0^{\pi/4} \frac{1}{2}\left[(2a\cos\theta)^2 - (\sqrt{2}a)^2\right] d\theta = a^2 = \text{area of square.}$$

EXERCISE SET 11.4

1. **(a)** $4px = y^2$, point $(1, 1), 4p = 1, x = y^2$

(b) $-4py = x^2$, point $(3, -3), 12p = 9, -3y = x^2$

(c) $a = 3, b = 2, \dfrac{x^2}{9} + \dfrac{y^2}{4} = 1$

(d) $a = 3, b = 2, \dfrac{x^2}{4} + \dfrac{y^2}{9} = 1$

(e) asymptotes: $y = \pm x$, so $a = b$; point $(0, 1)$, so $y^2 - x^2 = 1$

(f) asymptotes: $y = \pm x$, so $b = a$; point $(2, 0)$, so $\dfrac{x^2}{4} - \dfrac{y^2}{4} = 1$

3. **(a)**

(b)

5. **(a)**

(b)

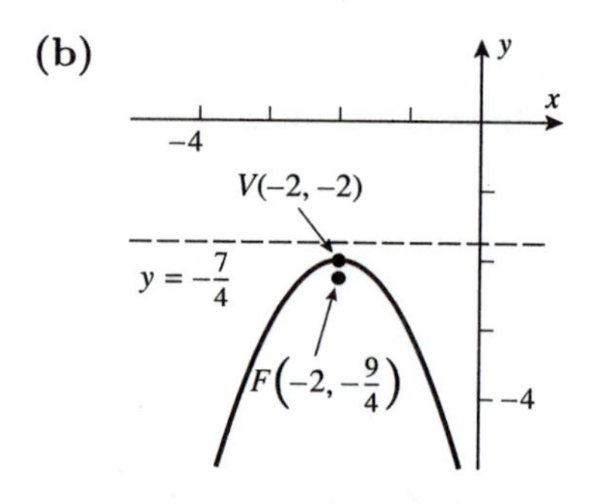

7. **(a)**

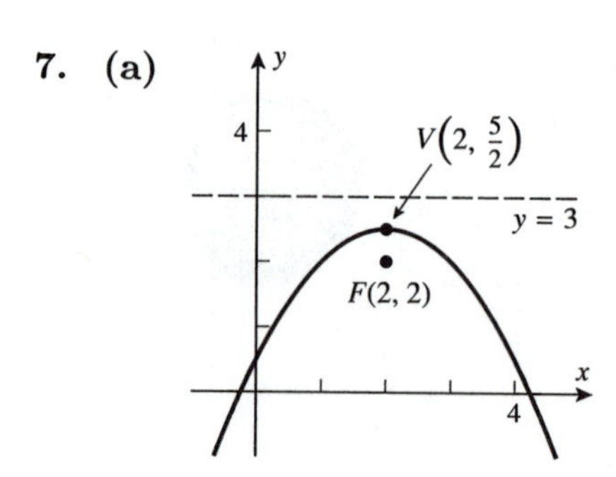

(b)

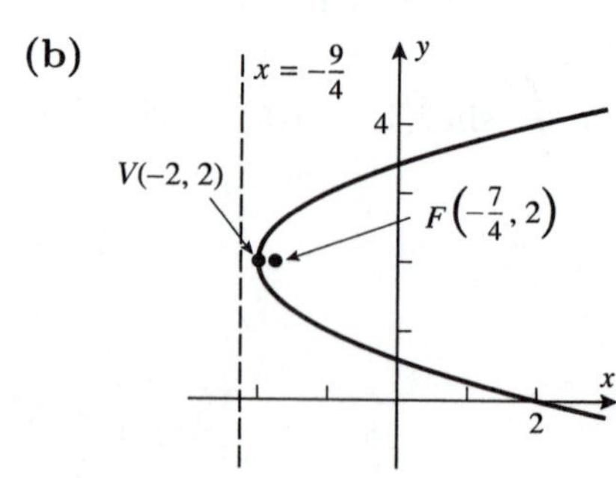

9. **(a)** $c^2 = 16 - 9 = 7$, $c = \sqrt{7}$

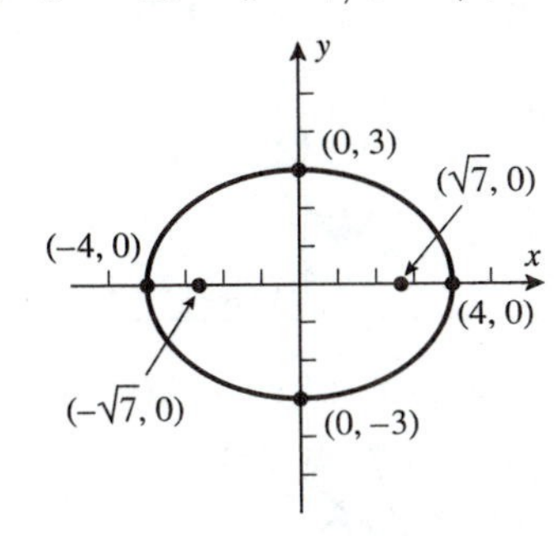

(b) $\frac{x^2}{1} + \frac{y^2}{9} = 1$

$c^2 = 9 - 1 = 8, c = \sqrt{8}$

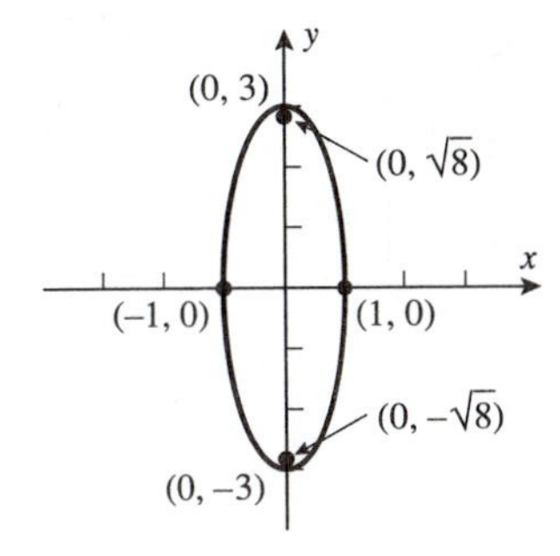

11. **(a)** $\frac{(x-1)^2}{16} + \frac{(y-3)^2}{9} = 1$

$c^2 = 16 - 9 = 7, c = \sqrt{7}$

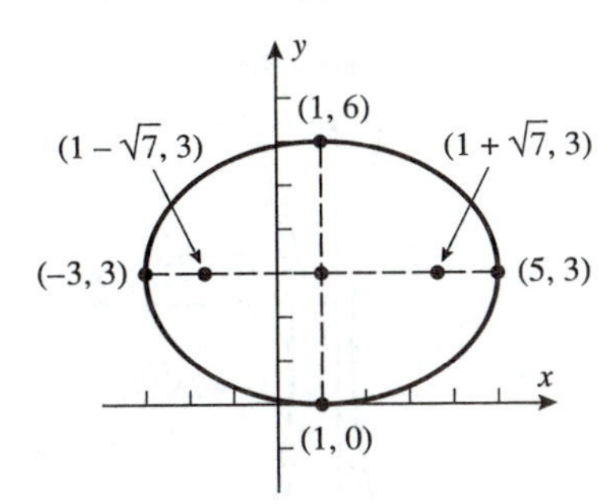

(b) $\frac{(x+2)^2}{4} + \frac{(y+1)^2}{3} = 1$

$c^2 = 4 - 3 = 1, c = 1$

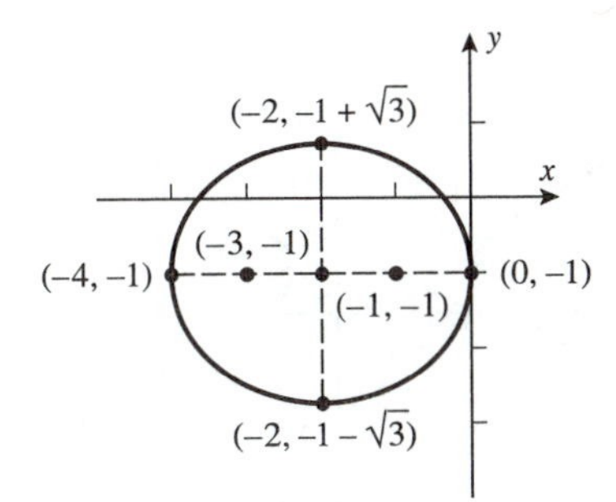

13. **(a)** $\frac{(x+1)^2}{9} + \frac{(y-1)^2}{1} = 1$

$c^2 = 9 - 1 = 8, c = \sqrt{8}$

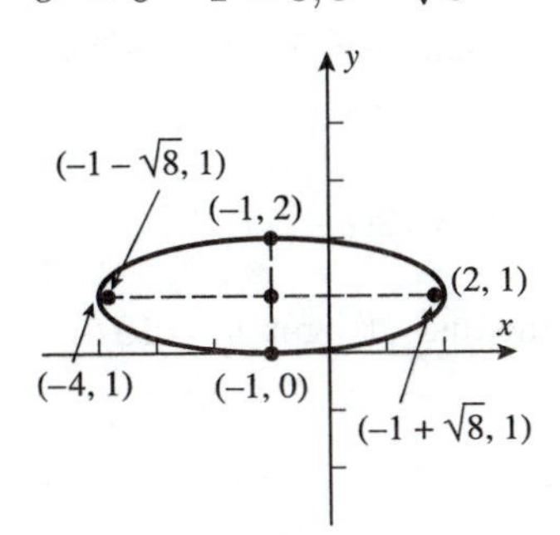

(b) $\frac{(x+1)^2}{4} + \frac{(y-5)^2}{16} = 1$

$c^2 = 16 - 4 = 12, c = 2\sqrt{3}$

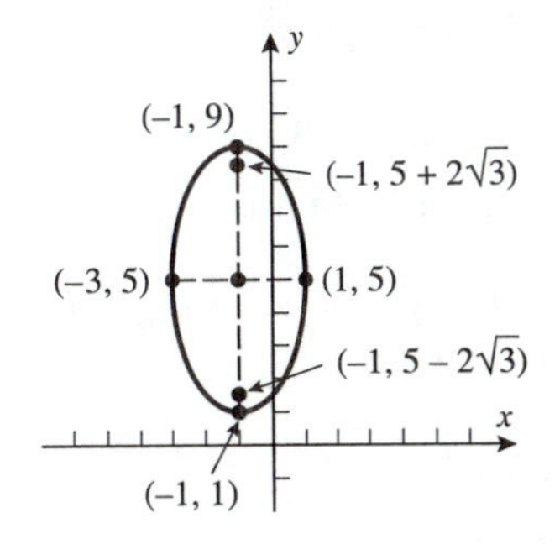

15. **(a)** $c^2 = a^2 + b^2 = 16 + 4 = 20, c = 2\sqrt{5}$

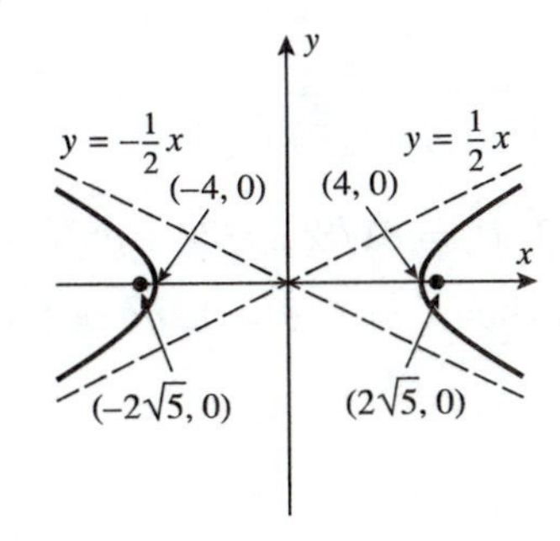

(b) $y^2/4 - x^2/9 = 1$

$c^2 = 4 + 9 = 13, c = \sqrt{13}$

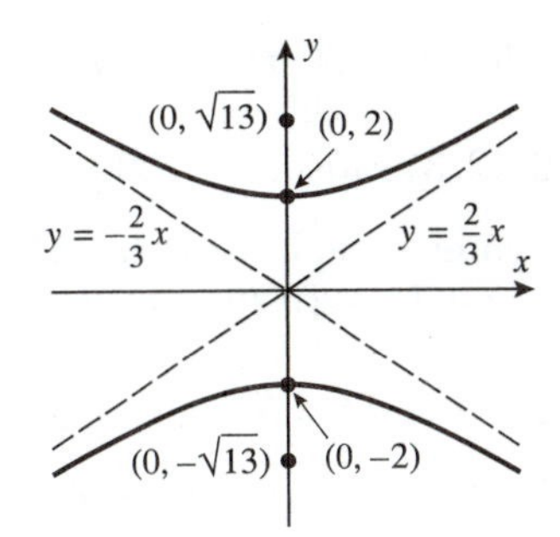

17. **(a)** $c^2 = 9 + 4 = 13, c = \sqrt{13}$

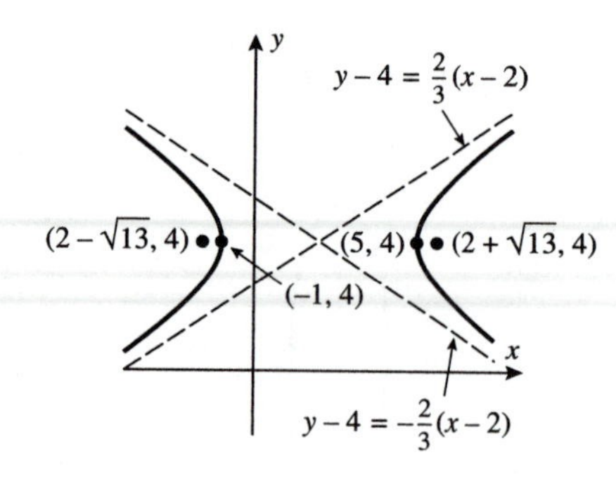

(b) $(y+3)^2/36 - (x+2)^2/4 = 1$
$c^2 = 36 + 4 = 40, c = 2\sqrt{10}$

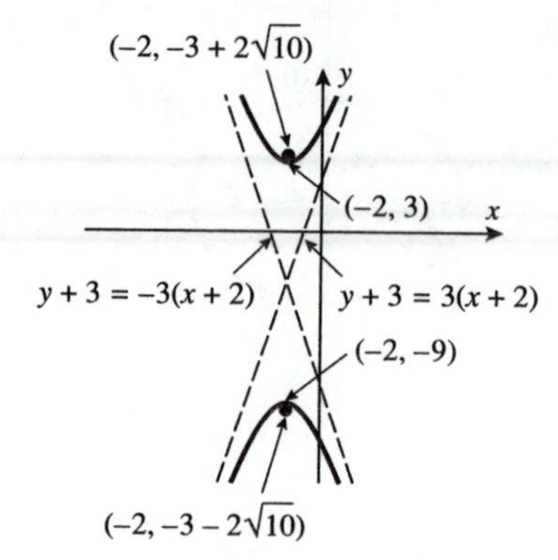

19. **(a)** $(x+1)^2/4 - (y-1)^2/1 = 1$
$c^2 = 4 + 1 = 5, c = \sqrt{5}$

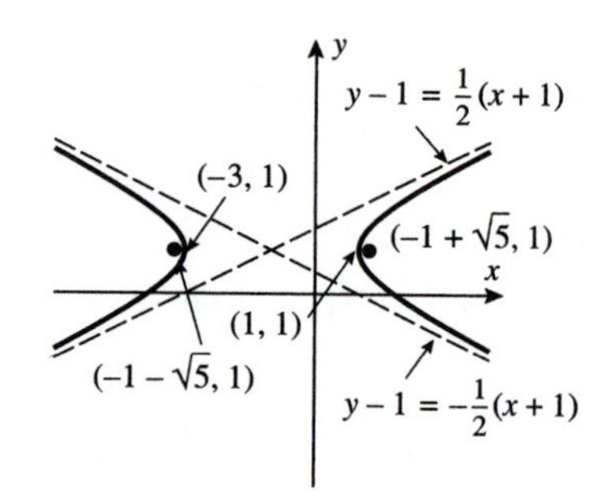

(b) $(x-1)^2/4 - (y+3)^2/64 = 1$
$c^2 = 4 + 64 = 68, c = 2\sqrt{17}$

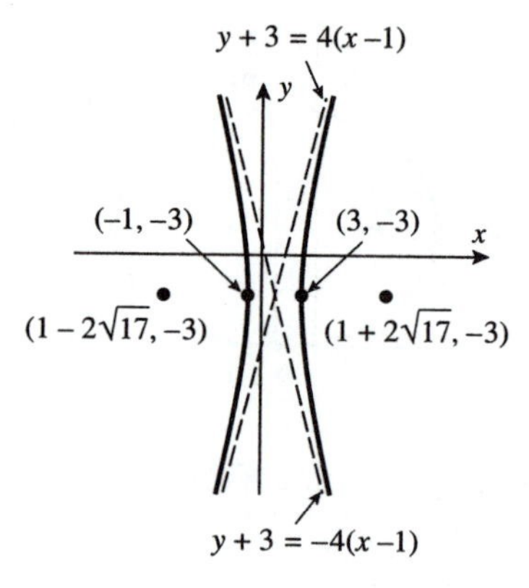

21. **(a)** $y^2 = 4px$, $p = 3$, $y^2 = 12x$ **(b)** $y^2 = -4px$, $p = 7$, $y^2 = -28x$

23. **(a)** $x^2 = -4py$, $p = 3$, $x^2 = -12y$

(b) The vertex is 3 units above the directrix so $p = 3$, $(x-1)^2 = 12(y-1)$.

25. $y^2 = a(x-h)$, $4 = a(3-h)$ and $9 = a(2-h)$, solve simultaneously to get $h = 19/5$, $a = -5$ so $y^2 = -5(x - 19/5)$

27. **(a)** $x^2/9 + y^2/4 = 1$

(b) $a = 26/2 = 13$, $c = 5$, $b^2 = a^2 - c^2 = 169 - 25 = 144$; $x^2/169 + y^2/144 = 1$

29. **(a)** $c = 1$, $a^2 = b^2 + c^2 = 2 + 1 = 3$; $x^2/3 + y^2/2 = 1$

(b) $b^2 = 16 - 12 = 4$; $x^2/16 + y^2/4 = 1$ and $x^2/4 + y^2/16 = 1$

31. **(a)** $a = 6$, $(2, 3)$ satisfies $x^2/36 + y^2/b^2 = 1$ so $4/36 + 9/b^2 = 1$, $b^2 = 81/8$; $x^2/36 + y^2/(81/8) = 1$

(b) The center is midway between the foci so it is at $(1, 3)$, thus $c = 1$, $b = 1$, $a^2 = 1 + 1 = 2$; $(x-1)^2 + (y-3)^2/2 = 1$

33. **(a)** $a = 2$, $c = 3$, $b^2 = 9 - 4 = 5$; $x^2/4 - y^2/5 = 1$

(b) $a = 1$, $b/a = 2$, $b = 2$; $x^2 - y^2/4 = 1$

35. **(a)** vertices along x-axis: $b/a = 3/2$ so $a = 8/3$; $x^2/(64/9) - y^2/16 = 1$
vertices along y-axis: $a/b = 3/2$ so $a = 6$; $y^2/36 - x^2/16 = 1$

(b) $c = 5$, $a/b = 2$ and $a^2 + b^2 = 25$, solve to get $a^2 = 20$, $b^2 = 5$; $y^2/20 - x^2/5 = 1$

37. **(a)** the center is at $(6, 4)$, $a = 4$, $c = 5$, $b^2 = 25 - 16 = 9$; $(x - 6)^2/16 - (y - 4)^2/9 = 1$

(b) The asymptotes intersect at $(1/2, 2)$ which is the center, $(y - 2)^2/a^2 - (x - 1/2)^2/b^2 = 1$ is the form of the equation because $(0, 0)$ is below both asymptotes, $4/a^2 - (1/4)/b^2 = 1$ and $a/b = 2$ which yields $a^2 = 3$, $b^2 = 3/4$; $(y - 2)^2/3 - (x - 1/2)^2/(3/4) = 1$.

39. **(a)** $y = ax^2 + b$, $(20, 0)$ and $(10, 12)$ are on the curve so $400a + b = 0$ and $100a + b = 12$. Solve for b to get $b = 16$ ft $=$ height of arch.

(b) $\dfrac{x^2}{a^2} + \dfrac{y^2}{b^2} = 1$, $400 = a^2, a = 20$; $\dfrac{100}{400} + \dfrac{144}{b^2} = 1$,
$b = 8\sqrt{3}$ ft $=$ height of arch.

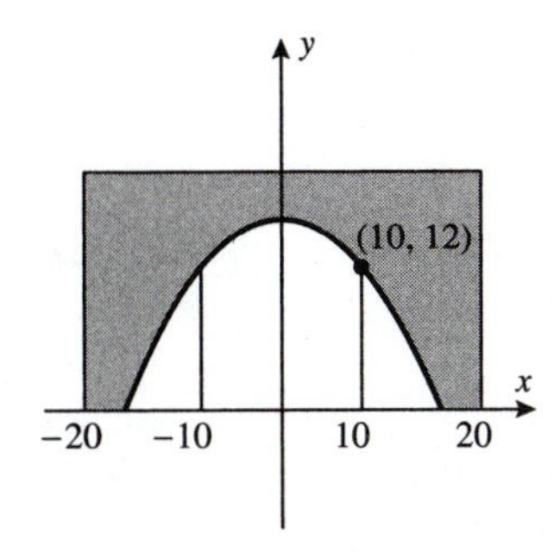

41. We may assume that the vertex is $(0, 0)$ and the parabola opens to the right. Let $P(x_0, y_0)$ be a point on the parabola $y^2 = 4px$, then by the definition of a parabola, $PF =$ distance from P to directrix $x = -p$, so $PF = x_0 + p$ where $x_0 \geq 0$ and PF is a minimum when $x_0 = 0$ (the vertex).

43. Use an xy-coordinate system so that $y^2 = 4px$ is an equation of the parabola, then $(1, 1/2)$ is a point on the curve so $(1/2)^2 = 4p(1)$, $p = 1/16$. The light source should be placed at the focus which is $1/16$ ft. from the vertex.

45. **(a)** $P: (b\cos t, b\sin t)$; $Q: (a\cos t, a\sin t)$; $R: (a\cos t, b\sin t)$

(b) For a circle, t measures the angle between the positive x-axis and the line segment joining the origin to the point. For an ellipse, t measures the angle between the x-axis and OPQ, not OR.

47. **(a)** For any point (x, y), the equation $y = b\tan t$ has a unique solution t where $-\pi/2 < t < \pi/2$. On the hyperbola, $\dfrac{x^2}{a^2} = 1 + \dfrac{y^2}{b^2} = 1 + \tan^2 t = \sec^2 t$, so $x = \pm a\sec t$.

(b)

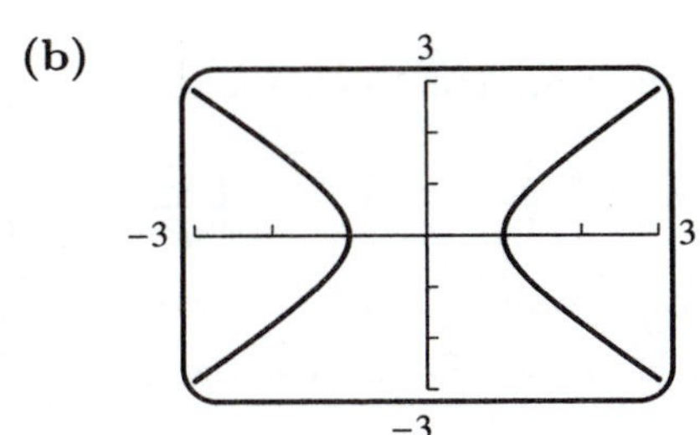

49. $(4, 1)$ and $(4, 5)$ are the foci so the center is at $(4, 3)$ thus $c = 2$, $a = 12/2 = 6$, $b^2 = 36 - 4 = 32$; $(x - 4)^2/32 + (y - 3)^2/36 = 1$

51. Let the ellipse have equation $\frac{4}{81}x^2 + \frac{y^2}{4} = 1$, then $A(x) = (2y)^2 = 16\left(1 - \frac{4x^2}{81}\right)$,
$V = 2\int_0^{9/2} 16\left(1 - \frac{4x^2}{81}\right) dx = 96$

53. Assume $\frac{x^2}{a^2} + \frac{y^2}{b^2} = 1, A = 4\int_0^a b\sqrt{1 - x^2/a^2}\, dx = \pi ab$

55. Assume $\frac{x^2}{a^2} + \frac{y^2}{b^2} = 1, \frac{dy}{dx} = -\frac{bx}{a\sqrt{a^2 - x^2}}, 1 + \left(\frac{dy}{dx}\right)^2 = \frac{a^4 - (a^2 - b^2)x^2}{a^2(a^2 - x^2)}$,

$$S = 2\int_0^a \frac{2\pi b}{a}\sqrt{1 - x^2/a^2}\sqrt{\frac{a^4 - (a^2 - b^2)x^2}{a^2 - x^2}}\, dx = 2\pi ab\left(\frac{b}{a} + \frac{a}{c}\sin^{-1}\frac{c}{a}\right), c = \sqrt{a^2 - b^2}$$

57. Open the compass to the length of half the major axis, place the point of the compass at an end of the minor axis and draw arcs that cross the major axis to both sides of the center of the ellipse. Place the tacks where the arcs intersect the major axis.

59. Let P denote the pencil tip, and let k be the difference between the length of the ruler and that of the string. Then $QP + PF_2 + k = QF_1$, and hence $PF_2 + k = PF_1, PF_1 - PF_2 = k$. But this is the definition of a hyperbola according to Definition 11.4.3.

61. $L = 2a = \sqrt{D^2 + p^2D^2} = D\sqrt{1 + p^2}$ (see figure), so $a = \frac{1}{2}D\sqrt{1 + p^2}$, but $b = \frac{1}{2}D$,
$T = c = \sqrt{a^2 - b^2} = \sqrt{\frac{1}{4}D^2(1 + p^2) - \frac{1}{4}D^2} = \frac{1}{2}pD.$

pD

D

63. By implicit differentiation, $\left.\frac{dy}{dx}\right|_{(x_0,y_0)} = -\frac{b^2}{a^2}\frac{x_0}{y_0}$ if $y_0 \neq 0$, the tangent line is

$y - y_0 = -\frac{b^2}{a^2}\frac{x_0}{y_0}(x - x_0)$, $a^2y_0y - a^2y_0^2 = -b^2x_0x + b^2x_0^2$, $b^2x_0x + a^2y_0y = b^2x_0^2 + a^2y_0^2$,
but (x_0, y_0) is on the ellipse so $b^2x_0^2 + a^2y_0^2 = a^2b^2$; thus the tangent line is $b^2x_0x + a^2y_0y = a^2b^2$, $x_0x/a^2 + y_0y/b^2 = 1$. If $y_0 = 0$ then $x_0 = \pm a$ and the tangent lines are $x = \pm a$ which also follows from $x_0x/a^2 + y_0y/b^2 = 1$.

65. Use $\frac{x^2}{a^2} + \frac{y^2}{b^2} = 1$ and $\frac{x^2}{A^2} - \frac{y^2}{B^2} = 1$ as the equations of the ellipse and hyperbola. If (x_0, y_0) is a point of intersection then $\frac{x_0^2}{a^2} + \frac{y_0^2}{b^2} = 1 = \frac{x_0^2}{A^2} - \frac{y_0^2}{B^2}$, so $x_0^2\left(\frac{1}{A^2} - \frac{1}{a^2}\right) = y_0^2\left(\frac{1}{B^2} + \frac{1}{b^2}\right)$ and $a^2A^2y_0^2(b^2 + B^2) = b^2B^2x_0^2(a^2 - A^2)$. Since the conics have the same foci, $a^2 - b^2 = c^2 = A^2 + B^2$, so $a^2 - A^2 = b^2 + B^2$. Hence $a^2A^2y_0^2 = b^2B^2x_0^2$. From Exercises 63 and 64, the slopes of the tangent lines are $-\frac{b^2x_0}{a^2y_0}$ and $\frac{B^2x_0}{A^2y_0}$, whose product is $-\frac{b^2B^2x_0^2}{a^2A^2y_0^2} = -1$. Hence the tangent lines are perpendicular.

67. Let (x_0, y_0) be such a point. The foci are at $(-\sqrt{5}, 0)$ and $(\sqrt{5}, 0)$, the lines are perpendicular if the product of their slopes is -1 so $\dfrac{y_0}{x_0+\sqrt{5}} \cdot \dfrac{y_0}{x_0-\sqrt{5}} = -1, y_0^2 = 5 - x_0^2$ and $4x_0^2 - y_0^2 = 4$. Solve to get $x_0 = \pm 3/\sqrt{5}, y_0 = \pm 4/\sqrt{5}$. The coordinates are $(\pm 3/\sqrt{5}, 4/\sqrt{5}), (\pm 3/\sqrt{5}, -4/\sqrt{5})$.

69. Let d_1 and d_2 be the distances of the first and second observers, respectively, from the point of the explosion. Then $t =$ (time for sound to reach the second observer) $-$ (time for sound to reach the first observer) $= d_2/v - d_1/v$ so $d_2 - d_1 = vt$. For constant v and t the difference of distances, d_2 and d_1 is constant so the explosion occurred somewhere on a branch of a hyperbola whose foci are where the observers are. Since $d_2 - d_1 = 2a, a = \dfrac{vt}{2}, b^2 = c^2 - \dfrac{v^2t^2}{4}$, and $\dfrac{x^2}{v^2t^2/4} - \dfrac{y^2}{c^2 - (v^2t^2/4)} = 1$.

71. **(a)** Use $\dfrac{x^2}{9} + \dfrac{y^2}{4} = 1$, $x = \dfrac{3}{2}\sqrt{4-y^2}$,

$$V = \int_{-2}^{-2+h} (2)(3/2)\sqrt{4-y^2}(18)dy = 54\int_{-2}^{-2+h} \sqrt{4-y^2}\,dy$$

$$= 54\left[\frac{y}{2}\sqrt{4-y^2} + 2\sin^{-1}\frac{y}{2}\right]_{-2}^{-2+h} = 27\left[4\sin^{-1}\frac{h-2}{2} + (h-2)\sqrt{4h-h^2} + 2\pi\right] \text{ ft}^3$$

(b) When $h = 4$ ft, $V_{\text{full}} = 108\sin^{-1}1 + 54\pi = 108\pi$ ft^3, so solve for h when $V = (k/4)V_{\text{full}}$, $k = 1, 2, 3$, to get $h = 1.19205, 2, 2.80795$ ft or $14.30465, 24, 33.69535$ in.

73. **(a)** $(x-1)^2 - 5(y+1)^2 = 5$, hyperbola

(b) $x^2 - 3(y+1)^2 = 0, x = \pm\sqrt{3}(y+1)$, two lines

(c) $4(x+2)^2 + 8(y+1)^2 = 4$, ellipse

(d) $3(x+2)^2 + (y+1)^2 = 0$, the point $(-2, -1)$ (degenerate case)

(e) $(x+4)^2 + 2y = 2$, parabola

(f) $5(x+4)^2 + 2y = -14$, parabola

75. distance from the point (x, y) to the focus $(0, -c)$ plus distance to the focus $(0, c) =$ const $= 2a$, $\sqrt{x^2+(y+c)^2} + \sqrt{x^2+(y-c)^2} = 2a, x^2 + (y+c)^2 = 4a^2 + x^2 + (y-c)^2 - 4a\sqrt{x^2+(y-c)^2}$, $\sqrt{x^2+(y-c)^2} = a - \dfrac{c}{a}y$, and since $a^2 - c^2 = b^2$, $\dfrac{x^2}{b^2} + \dfrac{y^2}{a^2} = 1$

77. Assume the equation of the parabola is $x^2 = 4py$. The tangent line at $P(x_0, y_0)$ (see figure) is given by $(y - y_0)/(x - x_0) = m = x_0/2p$. To find the y-intercept set $x = 0$ and obtain $y = -y_0$. Thus $Q : (0, -y_0)$. The focus is $(0, p) = (0, x_0^2/4y_0)$, the distance from P to the focus is $\sqrt{x_0^2 + (y_0-p)^2} = \sqrt{4py_0 + (y_0-p)^2} = \sqrt{(y_0+p)^2} = y_0 + p$, and the distance from the focus to the y-intercept is $p + y_0$, so triangle FPQ is isosceles, and angles FPQ and FQP are equal.

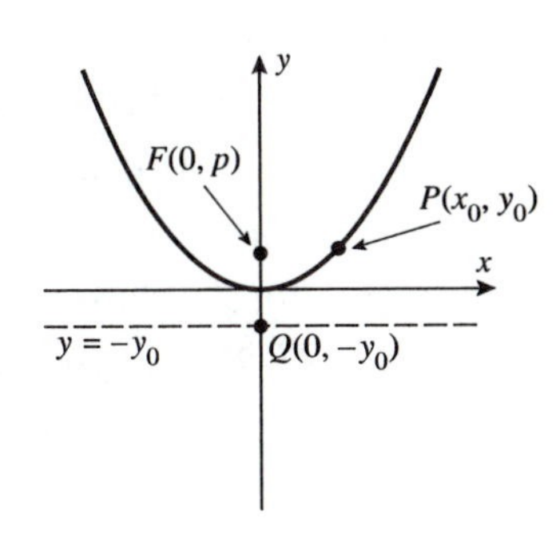

EXERCISE SET 11.5

1. **(a)** $\sin\theta = \sqrt{3}/2$, $\cos\theta = 1/2$

$x' = (-2)(1/2) + (6)(\sqrt{3}/2) = -1 + 3\sqrt{3}$, $y' = -(-2)(\sqrt{3}/2) + 6(1/2) = 3 + \sqrt{3}$

(b) $x = \frac{1}{2}x' - \frac{\sqrt{3}}{2}y' = \frac{1}{2}(x' - \sqrt{3}y')$, $y = \frac{\sqrt{3}}{2}x' + \frac{1}{2}y' = \frac{1}{2}(\sqrt{3}x' + y')$

$$\sqrt{3}\left[\frac{1}{2}(x' - \sqrt{3}y')\right]\left[\frac{1}{2}(\sqrt{3}x' + y')\right] + \left[\frac{1}{2}(\sqrt{3}x' + y')\right]^2 = 6$$

$$\frac{\sqrt{3}}{4}(\sqrt{3}x'^2 - 2x'y' - \sqrt{3}y'^2) + \frac{1}{4}(3x'^2 + 2\sqrt{3}x'y' + y'^2) = 6$$

$$\frac{3}{2}x'^2 - \frac{1}{2}y'^2 = 6,\ 3x'^2 - y'^2 = 12$$

(c)

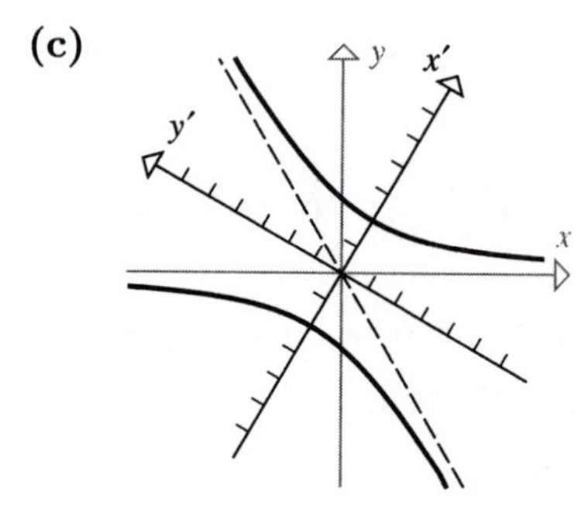

3. $\cot 2\theta = (0 - 0)/1 = 0$, $2\theta = 90°$, $\theta = 45°$

$x = (\sqrt{2}/2)(x' - y')$, $y = (\sqrt{2}/2)(x' + y')$

$y'^2/18 - x'^2/18 = 1$, hyperbola

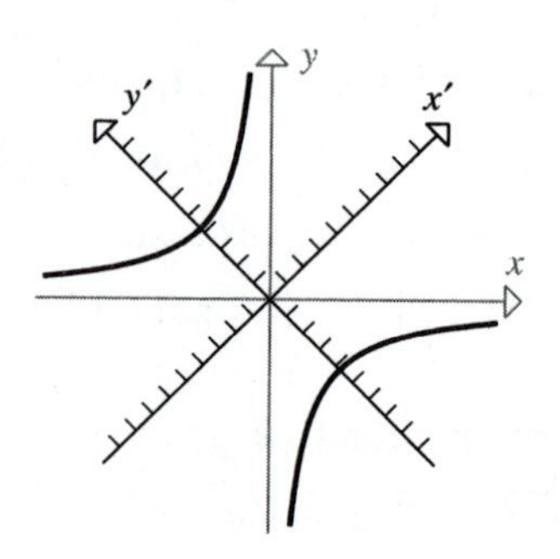

5. $\cot 2\theta = [1 - (-2)]/4 = 3/4$

$\cos 2\theta = 3/5$

$\sin\theta = \sqrt{(1 - 3/5)/2} = 1/\sqrt{5}$

$\cos\theta = \sqrt{(1 + 3/5)/2} = 2/\sqrt{5}$

$x = (1/\sqrt{5})(2x' - y')$

$y = (1/\sqrt{5})(x' + 2y')$

$x'^2/3 - y'^2/2 = 1$, hyperbola

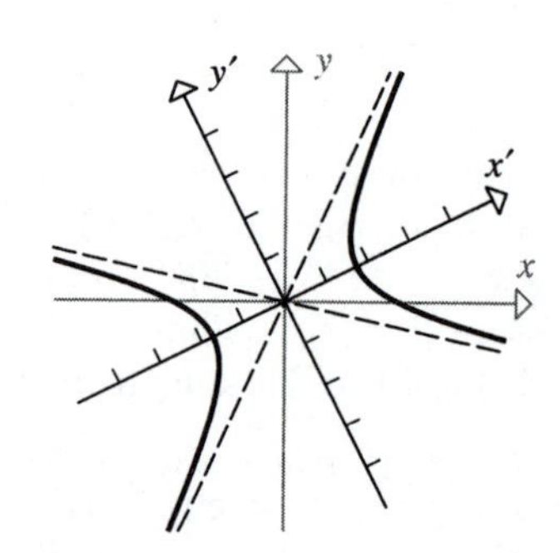

7. $\cot 2\theta = (1-3)/(2\sqrt{3}) = -1/\sqrt{3}$,
$2\theta = 120°, \theta = 60°$
$x = (1/2)(x' - \sqrt{3}y')$
$y = (1/2)(\sqrt{3}x' + y')$
$y' = x'^2$, parabola

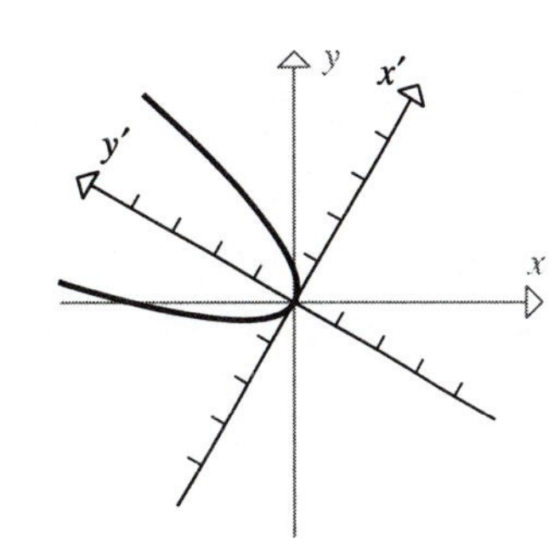

9. $\cot 2\theta = (9-16)/(-24) = 7/24$
$\cos 2\theta = 7/25$,
$\sin\theta = 3/5, \qquad \cos\theta = 4/5$
$x = (1/5)(4x' - 3y')$,
$y = (1/5)(3x' + 4y')$
$y'^2 = 4(x' - 1)$, parabola

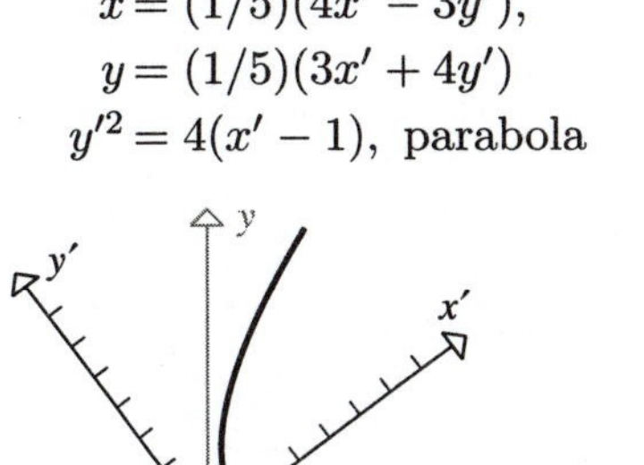

11. $\cot 2\theta = (52-73)/(-72) = 7/24$
$\cos 2\theta = 7/25, \qquad \sin\theta = 3/5$,
$\cos\theta = 4/5$
$x = (1/5)(4x' - 3y')$,
$y = (1/5)(3x' + 4y')$
$(x'+1)^2/4 + y'^2 = 1$, ellipse

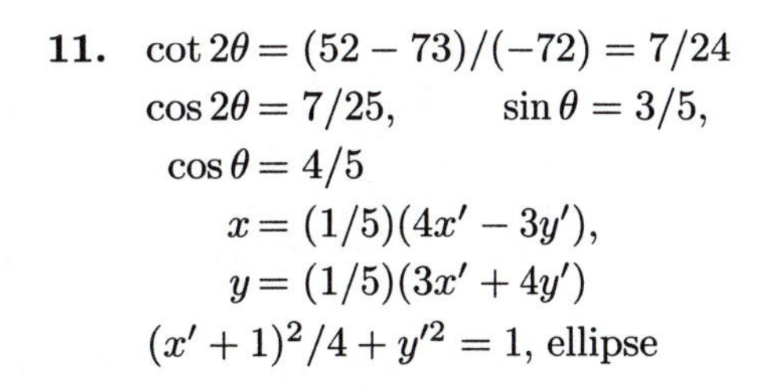

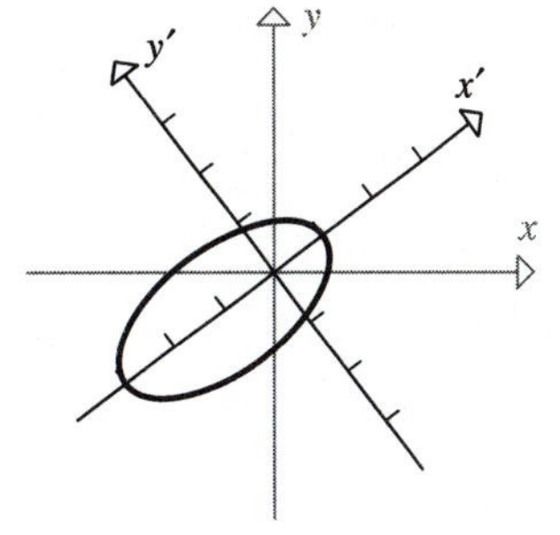

13. Let $x = x'\cos\theta - y'\sin\theta$, $y = x'\sin\theta + y'\cos\theta$ then $x^2 + y^2 = r^2$ becomes $(\sin^2\theta + \cos^2\theta)x'^2 + (\sin^2\theta + \cos^2\theta)y'^2 = r^2$, $x'^2 + y'^2 = r^2$. Under a rotation transformation the center of the circle stays at the origin of both coordinate systems.

15. $x' = (\sqrt{2}/2)(x+y)$, $y' = (\sqrt{2}/2)(-x+y)$ which when substituted into $3x'^2 + y'^2 = 6$ yields $x^2 + xy + y^2 = 3$.

17. $\sqrt{x} + \sqrt{y} = 1$, $\sqrt{x} = 1 - \sqrt{y}$, $x = 1 - 2\sqrt{y} + y$, $2\sqrt{y} = 1 - x + y$, $4y = 1 + x^2 + y^2 - 2x + 2y - 2xy$, $x^2 - 2xy + y^2 - 2x - 2y + 1 = 0$. $\cot 2\theta = \dfrac{1-1}{-2} = 0$, $2\theta = \pi/2$, $\theta = \pi/4$. Let $x = x'/\sqrt{2} - y'/\sqrt{2}$, $y = x'/\sqrt{2} + y'/\sqrt{2}$ to get $2y'^2 - 2\sqrt{2}x' + 1 = 0$, which is a parabola. From $\sqrt{x} + \sqrt{y} = 1$ we see that $0 \le x \le 1$ and $0 \le y \le 1$, so the graph is just a portion of a parabola.

19. Use (9) to express $B' - 4A'C'$ in terms of A, B, C, and θ, then simplify.

21. $\cot 2\theta = (A - C)/B = 0$ if $A = C$ so $2\theta = 90°$, $\theta = 45°$.

23. $B^2 - 4AC = (-1)^2 - 4(1)(1) = -3 < 0$; ellipse, point, or no graph. By inspection $(0, \pm\sqrt{2})$ lie on the curve, so it's an ellipse.

25. $B^2 - 4AC = (2\sqrt{3})^2 - 4(1)(3) = 0$; parabola, line, pair of parallel lines, or no graph. By inspection $(-\sqrt{3}, 3), (-\sqrt{3}, -1/3), (0, 0), (-2\sqrt{3}, 0), (0, 2/3)$ lie on the graph; since no three of these points are collinear, it's a parabola.

27. $B^2 - 4AC = (-24)^2 - 4(34)(41) = -5000 < 0$; ellipse, point, or no graph. By inspection $x = \pm 5/\sqrt{34}, y = 0$ satisfy the equation, so it's an ellipse.

29. Part (b): from (15), $A'C' < 0$ so A' and C' have opposite signs. By multiplying (14) through by -1, if necessary, assume that $A' < 0$ and $C' > 0$ so $(x' - h)^2/C' - (y' - k)^2/|A'| = K$. If $K \neq 0$ then the graph is a hyperbola (divide both sides by K), if $K = 0$ then we get the pair of intersecting lines $(x' - h)/\sqrt{C'} = \pm(y' - k)/\sqrt{|A'|}$.

Part (c): from (15), $A'C' = 0$ so either $A' = 0$ or $C' = 0$ but not both (this would imply that $A = B = C = 0$ which results in (14) being linear). Suppose $A' \neq 0$ and $C' = 0$ then complete the square to get $(x' - h)^2 = -E'y'/A' + K$. If $E' \neq 0$ the graph is a parabola, if $E' = 0$ and $K = 0$ the graph is the line $x' = h$, if $E' = 0$ and $K > 0$ the graph is the pair of parallel lines $x' = h \pm \sqrt{K}$, if $E' = 0$ and $K < 0$ there is no graph.

31. (a) $B^2 - 4AC = (9)^2 - 4(2)(1) > 0$ so the conic is a hyperbola (it contains the points $(2, -1), (2, -3/2)$).

(b) $y = -\frac{9}{2}x - \frac{1}{2} - \frac{1}{2}\sqrt{73x^2 + 42x + 17}$ or $y = -\frac{9}{2}x - \frac{1}{2} + \frac{1}{2}\sqrt{73x^2 + 42x + 17}$

(c)

EXERCISE SET 11.6

1. (a) $r = \dfrac{3/2}{1 - \cos\theta}, e = 1, d = 3/2$

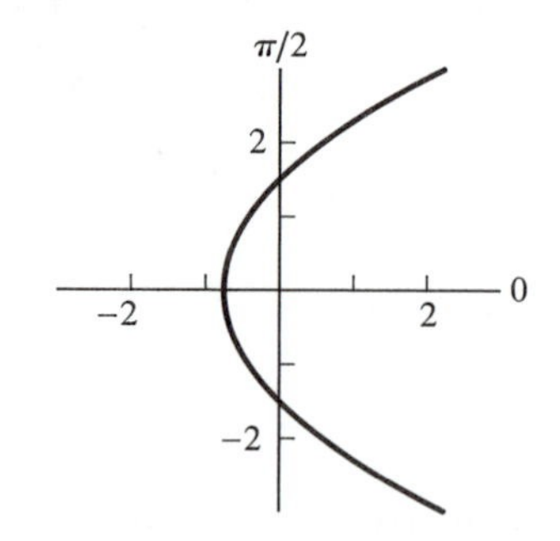

(b) $r = \dfrac{3/2}{1 + \frac{1}{2}\sin\theta}, e = 1/2, d = 3$

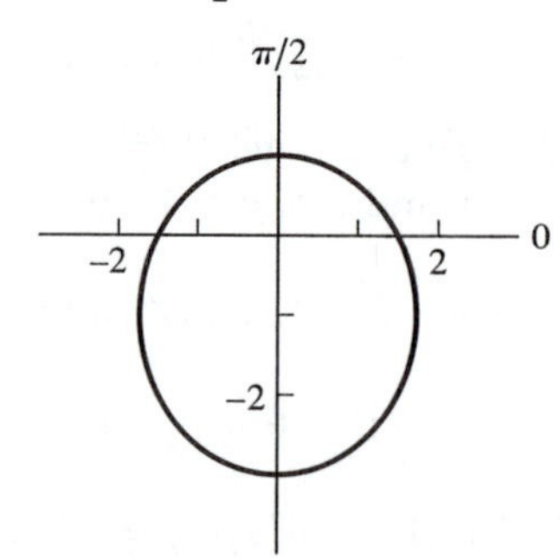

(c) $r = \dfrac{2}{1 + \frac{3}{2}\cos\theta}, e = 3/2, d = 4/3$

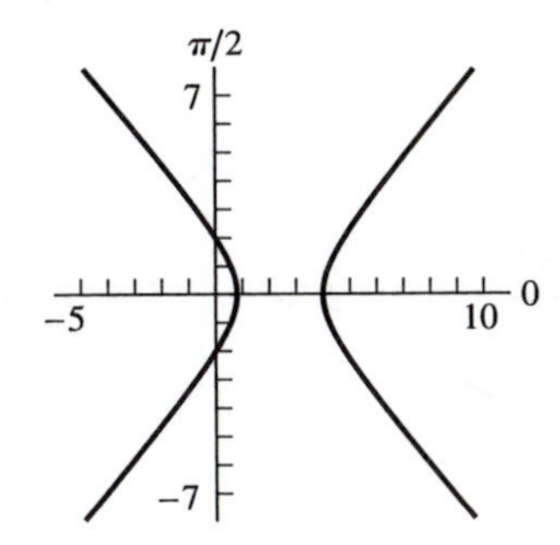

(d) $r = \dfrac{5/3}{1 + \sin\theta}, e = 1, d = 5/3$

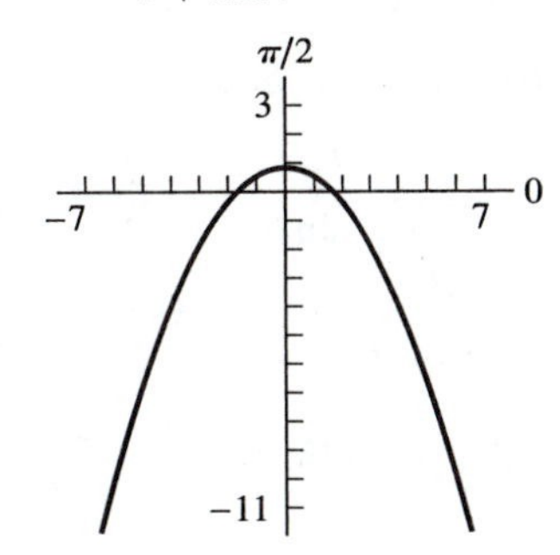

3. (a) $e = 1, d = 8$, parabola, opens up

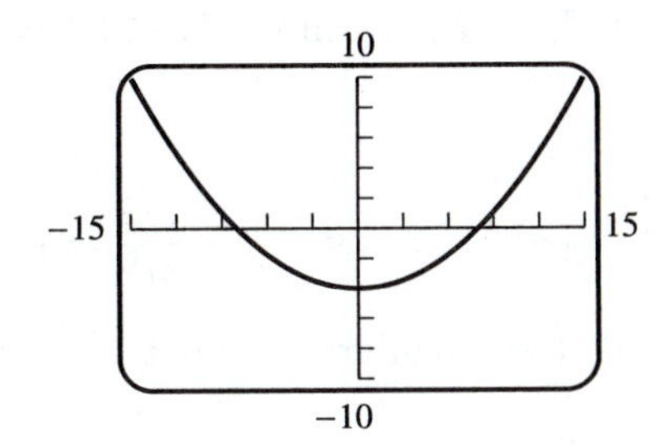

(b) $r = \dfrac{4}{1 + \frac{3}{4}\sin\theta}, e = 3/4, d = 16/3,$ ellipse, directrix 16/3 units above the pole

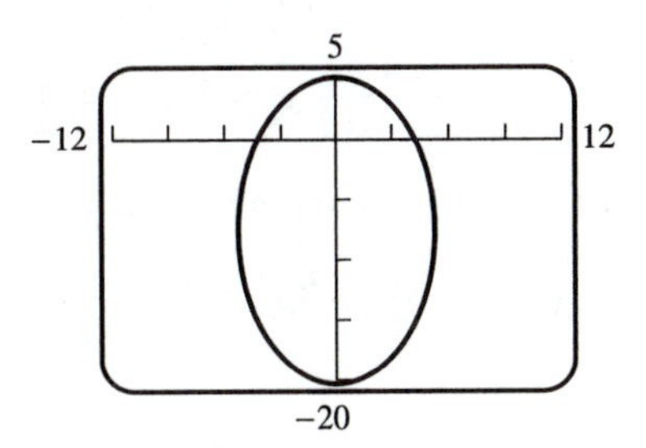

(c) $r = \dfrac{2}{1 - \frac{3}{2}\sin\theta}, e = 3/2, d = 4/3,$ hyperbola, directrix 4/3 units below the pole

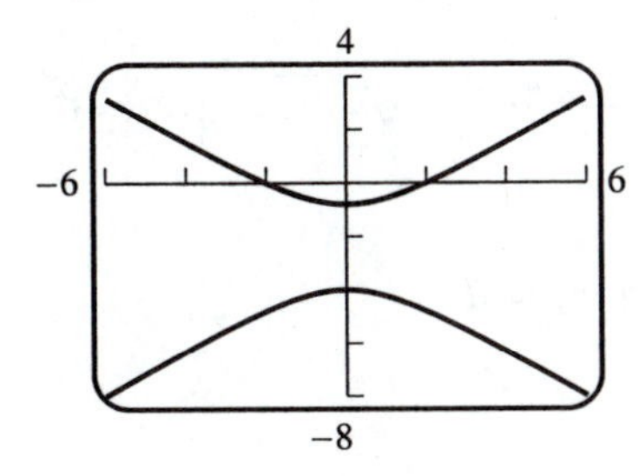

(d) $r = \dfrac{3}{1 + \frac{1}{4}\cos\theta}, e = 1/4, d = 12,$ ellipse, directrix 12 units to the right of the pole

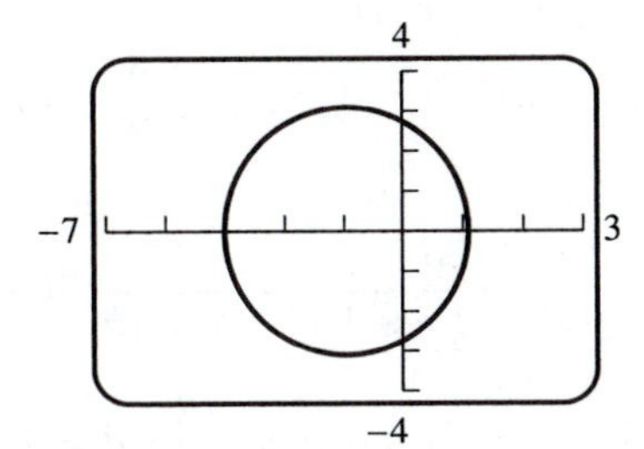

5. (a) $d = 1, r = \dfrac{ed}{1 + e\cos\theta} = \dfrac{2/3}{1 + \frac{2}{3}\cos\theta} = \dfrac{2}{3 + 2\cos\theta}$

(b) $e = 1, d = 1, r = \dfrac{ed}{1 - e\cos\theta} = \dfrac{1}{1 - \cos\theta}$

(c) $e = 3/2, d = 1, r = \dfrac{ed}{1 + e\sin\theta} = \dfrac{3/2}{1 + \frac{3}{2}\sin\theta} = \dfrac{3}{2 + 3\sin\theta}$

7. **(a)** $r = \dfrac{ed}{1 \pm e\cos\theta}, \theta = 0 : 6 = \dfrac{ed}{1 \pm e}, \theta = \pi : 4 = \dfrac{ed}{1 \mp e}, 6 \pm 6e = 4 \mp 4e, 2 = \mp 10e$, use bottom sign to get $e = 1/5, d = 24,\ \ r = \dfrac{24/5}{1 - \cos\theta} = \dfrac{24}{5 - 5\cos\theta}$

(b) $e = 1, r = \dfrac{d}{1 - \sin\theta}, 1 = \dfrac{d}{2}, d = 2, r = \dfrac{2}{1 - \sin\theta}$

(c) $r = \dfrac{ed}{1 \pm e\sin\theta}, \theta = \pi/2 : 3 = \dfrac{ed}{1 \pm e}, \theta = 3\pi/2 : -7 = \dfrac{ed}{1 \mp e}, ed = 3 \pm 3e = -7 \pm 7e, 10 = \pm 4e,$ $e = 5/2, d = 21/5, r = \dfrac{21/2}{1 + (5/2)\sin\theta} = \dfrac{21}{2 + 5\sin\theta}$

9. **(a)** $r = \dfrac{3}{1 + \frac{1}{2}\sin\theta}, e = 1/2, d = 6$, directrix 6 units above pole; if $\theta = \pi/2 : r_0 = 2$; if $\theta = 3\pi/2 : r_1 = 6, a = (r_0 + r_1)/2 = 4, b = \sqrt{r_0 r_1} = 2\sqrt{3}$, center $(0, -2)$ (rectangular coordinates), $\dfrac{x^2}{12} + \dfrac{(y+2)^2}{16} = 1$

(b) $r = \dfrac{1/2}{1 - \frac{1}{2}\cos\theta}, e = 1/2, d = 1$, directrix 1 unit left of pole; if $\theta = \pi : r_0 = \dfrac{1/2}{3/2} = 1/3$; if $\theta = 0 : r_1 = 1, a = 2/3, b = 1/\sqrt{3}$, center $= (1/3, 0)$ (rectangular coordinates), $\dfrac{9}{4}(x - 1/3)^2 + 3y^2 = 1$

11. **(a)** $r = \dfrac{2}{1 + 3\sin\theta}, e = 3, d = 2/3$, hyperbola, directrix 2/3 units above pole, if $\theta = \pi/2 :$ $r_0 = 1/2; \theta = 3\pi/2 : r_1 = 1$, center $(0, 3/4), a = 1/4, b = 1/\sqrt{2}, -2x^2 + 16\left(y - \dfrac{3}{4}\right)^2 = 1$

(b) $r = \dfrac{5/3}{1 - \frac{3}{2}\cos\theta}, e = 3/2, d = 10/9$, hyperbola, directrix 10/9 units left of pole, if $\theta = \pi :$ $r_0 = 2/3; \theta = 0 : r_1 = \dfrac{5/3}{1/2} = 10/3$, center $(-2, 0)$, $\ a = 4/3, b = \sqrt{20/9}, \dfrac{9}{16}(x+2)^2 - \dfrac{9}{20}y^2 = 1$

13. **(a)** $r = \dfrac{\frac{1}{2}d}{1 + \frac{1}{2}\cos\theta} = \dfrac{d}{2 + \cos\theta}$, if $\theta = 0 : r_0 = d/3; \theta = \pi, r_1 = d,$ $8 = a = \dfrac{1}{2}(r_1 + r_0) = \dfrac{2}{3}d, d = 12,\ \ r = \dfrac{12}{2 + \cos\theta}$

(b) $r = \dfrac{\frac{3}{5}d}{1 - \frac{3}{5}\sin\theta} = \dfrac{3d}{5 - 3\sin\theta}$, if $\theta = 3\pi/2 : r_0 = \dfrac{3}{8}d; \theta = \pi/2, r_1 = \dfrac{3}{2}d,$ $4 = a = \dfrac{1}{2}(r_1 + r_0) = \dfrac{15}{16}d, d = \dfrac{64}{15}, r = \dfrac{3(64/15)}{5 - 3\sin\theta} = \dfrac{64}{25 - 15\sin\theta}$

(c) $r = \dfrac{\frac{3}{5}d}{1 - \frac{3}{5}\cos\theta} = \dfrac{3d}{5 - 3\cos\theta}$, if $\theta = \pi : r_0 = \dfrac{3}{8}d; \theta = 0, r_1 = \dfrac{3}{2}d, 4 = b = \dfrac{3}{4}d,$ $d = 16/3,\ \ r = \dfrac{16}{5 - 3\cos\theta}$

(d) $r = \dfrac{\frac{1}{5}d}{1 + \frac{1}{5}\sin\theta} = \dfrac{d}{5 + \sin\theta}$, if $\theta = \pi/2 : r_0 = d/6; \theta = 3\pi/2, r_1 = d/4$,

$$5 = c = \frac{1}{2}d\left(\frac{1}{4} - \frac{1}{6}\right) = \frac{1}{24}d, d = 120, \quad r = \frac{120}{5 + \sin\theta}$$

15. **(a)** $e = c/a = \dfrac{\frac{1}{2}(r_1 - r_0)}{\frac{1}{2}(r_1 + r_0)} = \dfrac{r_1 - r_0}{r_1 + r_0}$

(b) $e = \dfrac{r_1/r_0 - 1}{r_1/r_0 + 1}, e(r_1/r_0 + 1) = r_1/r_0 - 1, \dfrac{r_1}{r_0} = \dfrac{1+e}{1-e}$

17. **(a)** $T = a^{3/2} = 39.5^{1.5} \approx 248$ yr

(b) $r_0 = a(1 - e) = 39.5(1 - 0.249) = 29.6645$ AU $\approx$ 4,449,675,000 km
$r_1 = a(1 + e) = 39.5(1 + 0.249) = 49.3355$ AU $\approx$ 7,400,325,000 km

(c) $r = \dfrac{a(1 - e^2)}{1 + e\cos\theta} \approx \dfrac{39.5(1 - (0.249)^2)}{1 + 0.249\cos\theta} \approx \dfrac{37.05}{1 + 0.249\cos\theta}$ AU

(d)

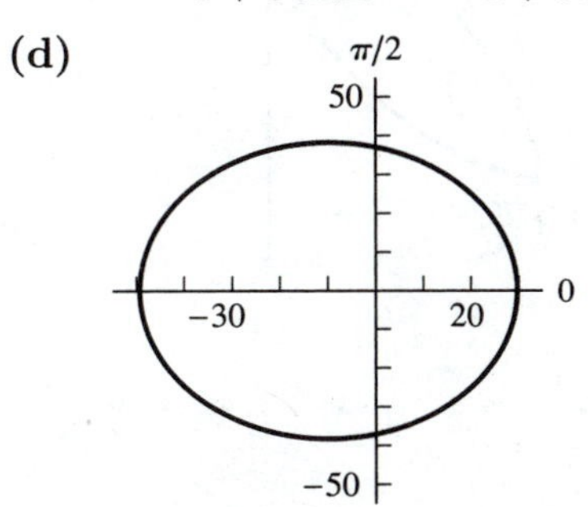

19. **(a)** $a = T^{2/3} = 2380^{2/3} \approx 178.26$ AU

(b) $r_0 = a(1 - e) \approx 0.8735$ AU, $r_1 = a(1 + e) \approx 355.64$ AU

(c) $r = \dfrac{a(1 - e^2)}{1 + e\cos\theta} \approx \dfrac{1.74}{1 + 0.9951\cos\theta}$ AU

(d)

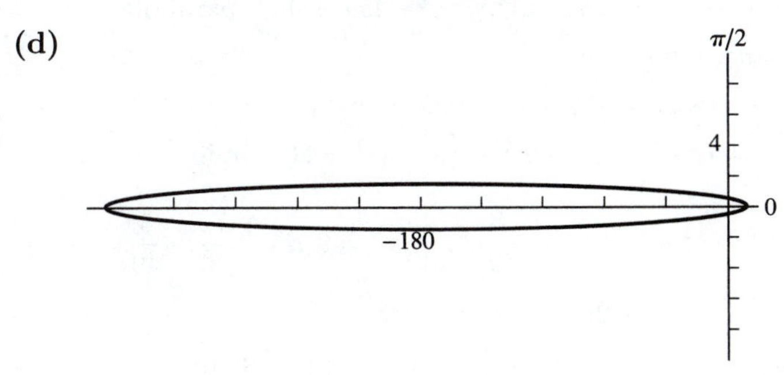

21. $r_0 = a(1 - e) \approx 7003$ km, $h_{\text{min}} \approx 7003 - 6440 = 563$ km,
$r_1 = a(1 + e) \approx 10{,}726$ km, $h_{\text{max}} \approx 10{,}726 - 6440 = 4286$ km

23. Since the foci are fixed, c is constant; since $e \to 0$, the distance $\dfrac{a}{e} = \dfrac{c}{e^2} \to +\infty$.

CHAPTER 11 SUPPLEMENTARY EXERCISES

3. **(a)** circle **(b)** rose **(c)** line **(d)** limaçon
(e) limaçon **(f)** none **(g)** none **(h)** spiral

5. 

(a)

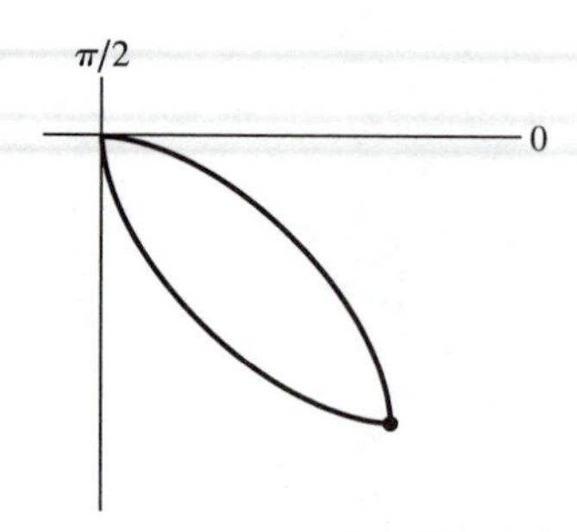

(b)

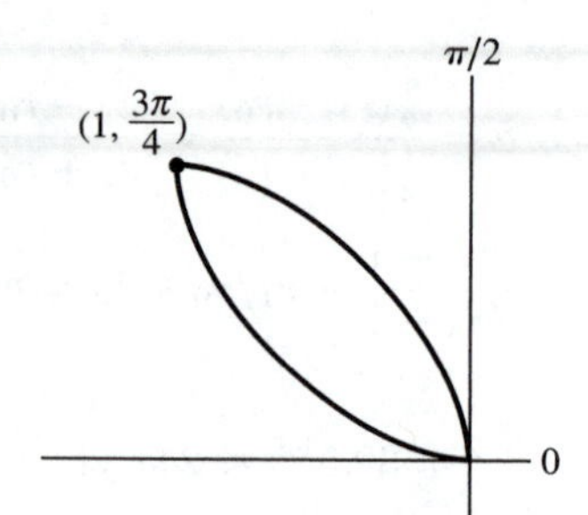

(c)

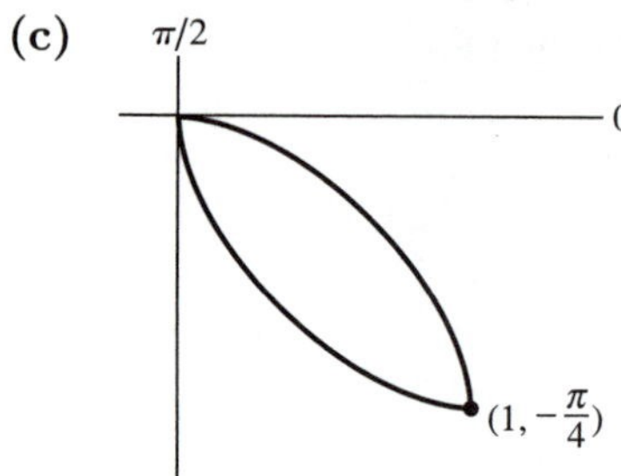

(d)

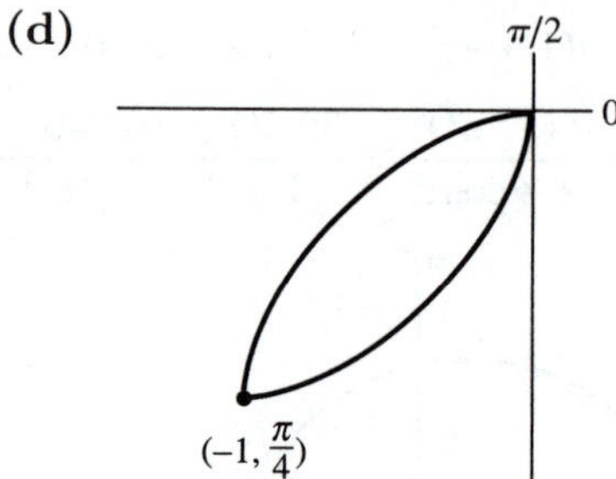

(e)

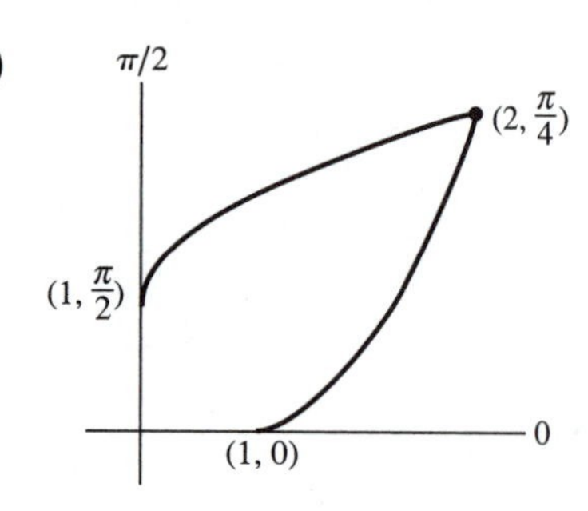

7. **(a)** $r = 2a/(1 + \cos\theta), r + x = 2a, x^2 + y^2 = (2a - x)^2, y^2 = -4ax + 4a^2$, parabola

(b) $r^2(\cos^2\theta - \sin^2\theta) = x^2 - y^2 = a^2$, hyperbola

(c) $r\sin(\theta - \pi/4) = (\sqrt{2}/2)r(\sin\theta - \cos\theta) = 4, y - x = 4\sqrt{2}$, line

(d) $r^2 = 4r\cos\theta + 8r\sin\theta, x^2 + y^2 = 4x + 8y, (x - 2)^2 + (y - 4)^2 = 20$, circle

9. **(a)** $\dfrac{c}{a} = e = \dfrac{2}{7}$ and $2b = 6, b = 3, a^2 = b^2 + c^2 = 9 + \dfrac{4}{49}a^2, \dfrac{45}{49}a^2 = 9, a = \dfrac{7}{\sqrt{5}}, \dfrac{5}{49}x^2 + \dfrac{1}{9}y^2 = 1$

(b) $x^2 = -4py$, directrix $y = 4$, focus $(-4, 0), 2p = 8, x^2 = -16y$

(c) For the ellipse, $a = 4, b = \sqrt{3}, c^2 = a^2 - b^2 = 16 - 3 = 13$, foci $(\pm\sqrt{13}, 0)$;
for the hyperbola, $c = \sqrt{13}, b/a = 2/3, b = 2a/3, 13 = c^2 = a^2 + b^2 = a^2 + \dfrac{4}{9}a^2 = \dfrac{13}{9}a^2$,
$a = 3, b = 2, \dfrac{x^2}{9} - \dfrac{y^2}{4} = 1$

11. (a)

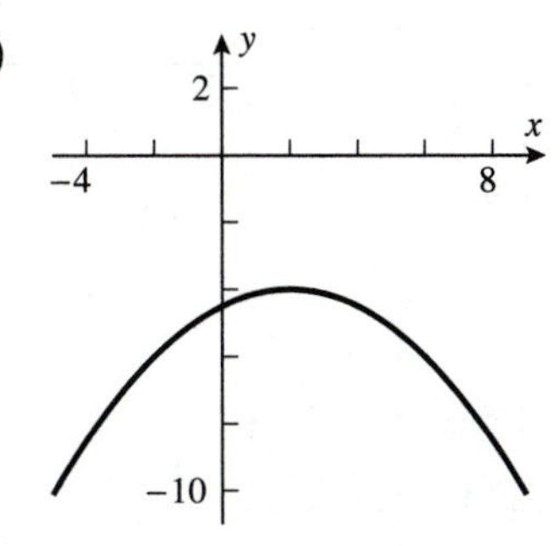

(b)

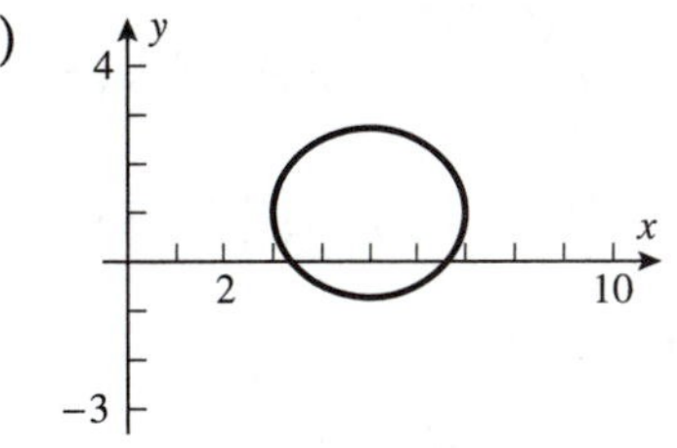

(c)

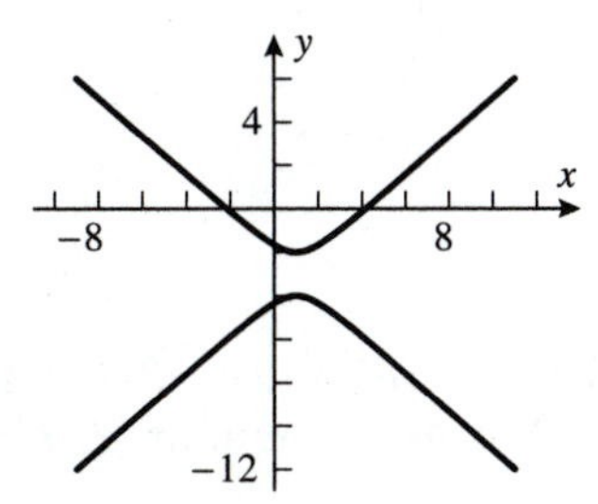

(d)

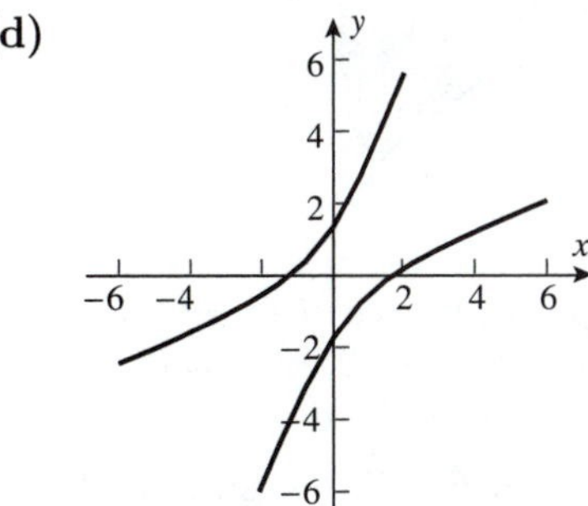

13. (a) The equation of the parabola is $y = ax^2$ and it passes through $(2100, 470)$, thus $a = \dfrac{470}{2100^2}$, $y = \dfrac{470}{2100^2}x^2$.

(b) $$L = 2\int_0^{2100} \sqrt{1 + \left(2\frac{470}{2100^2}x\right)^2}\,dx$$
$$= \frac{x}{220500}\sqrt{48620250000 + 2209x^2} + \frac{220500}{47}\sinh^{-1}\left(\frac{47}{220500}x\right) \approx 4336.3 \text{ ft}$$

15. $$= \int_0^{\pi/6} 4\sin^2\theta\,d\theta + \int_{\pi/6}^{\pi/4} 1\,d\theta = \int_0^{\pi/6} 2(1-\cos 2\theta)\,d\theta + \frac{\pi}{12} = (2\theta - \sin 2\theta)\Big]_0^{\pi/6} + \frac{\pi}{12}$$
$$= \frac{\pi}{3} - \frac{\sqrt{3}}{2} + \frac{\pi}{12} = \frac{5\pi}{12} - \frac{\sqrt{3}}{2}$$

17. (a) $r = 1/\theta, dr/d\theta = -1/\theta^2, r^2 + (dr/d\theta)^2 = 1/\theta^2 + 1/\theta^4$, $L = \displaystyle\int_{\pi/4}^{\pi/2} \frac{1}{\theta^2}\sqrt{1+\theta^2}\,d\theta \approx 0.9457$ by Endpaper Table Formula 93.

(b) The integral $\displaystyle\int_1^{+\infty} \frac{1}{\theta^2}\sqrt{1+\theta^2}\,d\theta$ diverges by the comparison test (with $1/\theta$), and thus the arc length is infinite.

19. (a) $V = \int_a^{\sqrt{a^2+b^2}} \pi\left(b^2x^2/a^2 - b^2\right)\,dx$

$= \dfrac{\pi b^2}{3a^2}(b^2 - 2a^2)\sqrt{a^2+b^2} + \dfrac{2}{3}ab^2\pi$

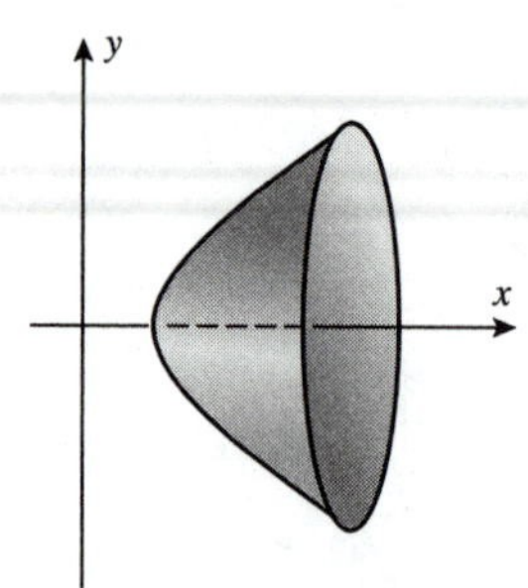

(b) $V = 2\pi \int_a^{\sqrt{a^2+b^2}} x\sqrt{b^2x^2/a^2 - b^2}\,dx$

$= (2b^4/3a)\pi$

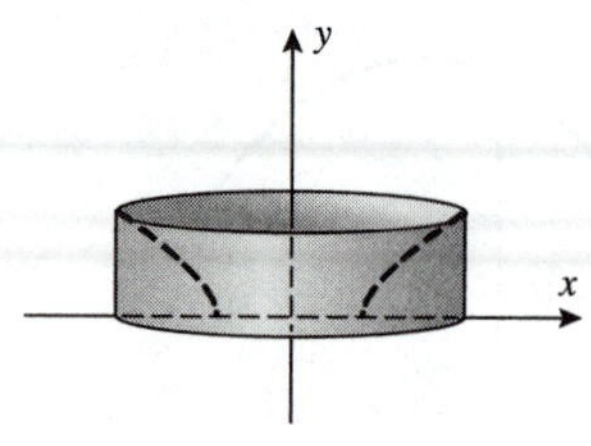

23. (a) $x = r\cos\theta = \cos\theta + \cos^2\theta, dx/d\theta = -\sin\theta - 2\sin\theta\cos\theta = -\sin\theta(1 + 2\cos\theta) = 0$ if $\sin\theta = 0$ or $\cos\theta = -1/2$, so $\theta = 0, \pi, 2\pi/3, 4\pi/3$; maximum $x = 2$ at $\theta = 0$, minimum $x = -1/4$ at $\theta = 2\pi/3, 4\pi/3$; $\theta = \pi$ is a local maximum for x

(b) $y = r\sin\theta = \sin\theta + \sin\theta\cos\theta, dy/d\theta = 2\cos^2\theta + \cos\theta - 1 = 0$ at $\cos\theta = 1/2, -1$, so $\theta = \pi/3, 5\pi/3, \pi$; maximum $y = 3\sqrt{3}/4$ at $\theta = \pi/3$, minimum $y = -3\sqrt{3}/4$ at $\theta = 5\pi/3$

25. The width is twice the maximum value of y for $0 \le \theta \le \pi/4$:
$y = r\sin\theta = \sin\theta\cos 2\theta = \sin\theta - 2\sin^3\theta, dy/d\theta = \cos\theta - 6\sin^2\theta\cos\theta = 0$ when $\cos\theta = 0$ or $\sin\theta = 1/\sqrt{6}, y = 1/\sqrt{6} - 2/(6\sqrt{6}) = \sqrt{6}/9$, so the width of the petal is $2\sqrt{6}/9$.

27. (a) The end of the inner arm traces out the circle $x_1 = \cos t, y_1 = \sin t$. Relative to the end of the inner arm, the outer arm traces out the circle $x_2 = \cos 2t, y_2 = -\sin 2t$. Add to get the motion of the center of the rider cage relative to the center of the inner arm:
$x = \cos t + \cos 2t, y = \sin t - \sin 2t$.

(b) Same as Part (a), except $x_2 = \cos 2t, y_2 = \sin 2t$, so $x = \cos t + \cos 2t, y = \sin t + \sin 2t$

(c) $L_1 = \int_0^{2\pi}\left[\left(\dfrac{dx}{dt}\right)^2 + \left(\dfrac{dy}{dt}\right)^2\right]^{1/2} dt = \int_0^{2\pi}\sqrt{5 - 4\cos 3t}\,dt \approx 13.36489321$,

$L_2 = \int_0^{2\pi}\sqrt{5 + 4\cos t}\,dt \approx 13.36489322$; L_1 and L_2 appear to be equal, and indeed, with the substitution $u = 3t - \pi$ and the periodicity of $\cos u$,

$L_1 = \dfrac{1}{3}\int_{-\pi}^{5\pi}\sqrt{5 - 4\cos(u+\pi)}\,du = \int_0^{2\pi}\sqrt{5 + 4\cos u}\,du = L_2$.

29. $C = 4\int_0^{\pi/2}\left[\left(\dfrac{dx}{dt}\right)^2 + \left(\dfrac{dy}{dt}\right)^2\right]^{1/2} dt = 4\int_0^{\pi/2}(a^2\sin^2 t + b^2\cos^2 t)^{1/2}\,dt$

$= 4\int_0^{\pi/2}(a^2\sin^2 t + (a^2 - c^2)\cos^2 t)^{1/2}\,dt = 4a\int_0^{\pi/2}(1 - e^2\cos^2 t)^{1/2}\,dt$

Set $u = \dfrac{\pi}{2} - t$, $C = 4a\int_0^{\pi/2}(1 - e^2\sin^2 t)^{1/2}\,dt$

31. (a) $\dfrac{r_0}{r_1} = \dfrac{59}{61} = \dfrac{1-e}{1+e}, e = \dfrac{1}{60}$

(b) $a = 93 \times 10^6, r_0 = a(1-e) = \dfrac{59}{60}(93 \times 10^6) = 91{,}450{,}000$ mi

(c) $C = 4 \times 93 \times 10^6 \displaystyle\int_0^{\pi/2} \left[1 - \left(\frac{\cos\theta}{60}\right)^2\right]^{1/2} d\theta \approx 584{,}295{,}652.5$ mi

33. $\alpha = \pi/4, y_0 = 3, x = v_0 t/\sqrt{2}, y = 3 + v_0 t/\sqrt{2} - 16t^2$

(a) Assume the ball passes through $x = 391, y = 50$, then $391 = v_0 t/\sqrt{2}, 50 = 3 + 391 - 16t^2$, $16t^2 = 344,\ t = \sqrt{21.5},\ v_0 = \sqrt{2}x/t \approx 119.2538820$ ft/s

(b) $\dfrac{dy}{dt} = \dfrac{v_0}{\sqrt{2}} - 32t = 0$ at $t = \dfrac{v_0}{32\sqrt{2}},\ y_{\max} = 3 + \dfrac{v_0}{\sqrt{2}}\dfrac{v_0}{32\sqrt{2}} - 16\dfrac{v_0^2}{2^{11}} = 3 + \dfrac{v_0^2}{128} \approx 114.1053779$ ft

(c) $y = 0$ when $t = \dfrac{-v_0/\sqrt{2} \pm \sqrt{v_0^2/2 + 192}}{-32}$, $t \approx -0.035339577$ (discard) and 5.305666365, dist $= 447.4015292$ ft

35. $\tan\psi = \tan(\phi - \theta) = \dfrac{\tan\phi - \tan\theta}{1 + \tan\phi\tan\theta} = \dfrac{\dfrac{dy}{dx} - \dfrac{y}{x}}{1 + \dfrac{y}{x}\dfrac{dy}{dx}}$

$$= \frac{\dfrac{r\cos\theta + (dr/d\theta)\sin\theta}{-r\sin\theta + (dr/d\theta)\cos\theta} - \dfrac{\sin\theta}{\cos\theta}}{1 + \left(\dfrac{r\cos\theta + (dr/d\theta)\sin\theta)}{-r\sin\theta + (dr/d\theta)\cos\theta)}\right)\left(\dfrac{\sin\theta}{\cos\theta}\right)} = \frac{r}{dr/d\theta}$$

37. $\tan\psi = \dfrac{r}{dr/d\theta} = \dfrac{ae^{b\theta}}{abe^{b\theta}} = \dfrac{1}{b}$ is constant, so ψ is constant.

CHAPTER 12
Three-Dimensional Space; Vectors

EXERCISE SET 12.1

1. **(a)** $(0,0,0), (3,0,0), (3,5,0), (0,5,0), (0,0,4), (3,0,4), (3,5,4), (0,5,4)$
(b) $(0,1,0), (4,1,0), (4,6,0), (0,6,0), (0,1,-2), (4,1,-2), (4,6,-2), (0,6,-2)$

3. corners: $(4,2,-2)$, (4,2,1), (4,1,1), $(4,1,-2)$,
$(-6,1,1)$, $(-6,2,1)$, $(-6,2,-2)$, $(-6,1,-2)$

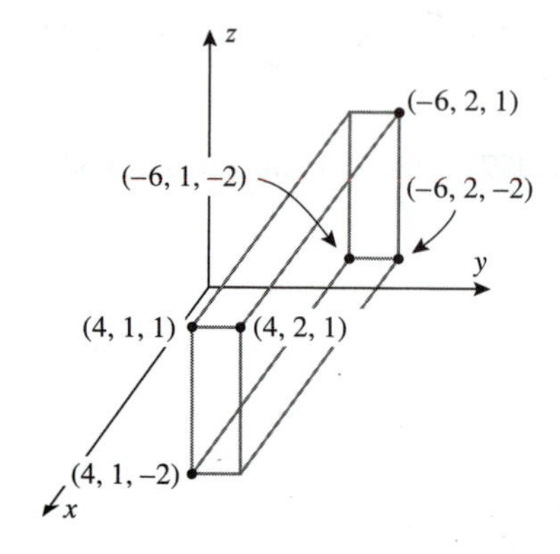

5. The diameter is $d = \sqrt{(1-3)^2 + (-2-4)^2 + (4+12)^2} = \sqrt{296}$, so the radius is $\sqrt{296}/2 = \sqrt{74}$. The midpoint $(2,1,-4)$ of the endpoints of the diameter is the center of the sphere.

7. **(a)** The sides have lengths 7, 14, and $7\sqrt{5}$; it is a right triangle because the sides satisfy the Pythagorean theorem, $(7\sqrt{5})^2 = 7^2 + 14^2$.
(b) (2,1,6) is the vertex of the $90°$ angle because it is opposite the longest side (the hypotenuse).
(c) area $= (1/2)(\text{altitude})(\text{base}) = (1/2)(7)(14) = 49$

9. **(a)** $(x-1)^2 + y^2 + (z+1)^2 = 16$
(b) $r = \sqrt{(-1-0)^2 + (3-0)^2 + (2-0)^2} = \sqrt{14}$, $(x+1)^2 + (y-3)^2 + (z-2)^2 = 14$
(c) $r = \dfrac{1}{2}\sqrt{(-1-0)^2 + (2-2)^2 + (1-3)^2} = \dfrac{1}{2}\sqrt{5}$, center $(-1/2, 2, 2)$,
$(x+1/2)^2 + (y-2)^2 + (z-2)^2 = 5/4$

11. $(x-2)^2 + (y+1)^2 + (z+3)^2 = r^2$,
(a) $r^2 = 3^2 = 9$ **(b)** $r^2 = 1^2 = 1$ **(c)** $r^2 = 2^2 = 4$

13. $(x+5)^2 + (y+2)^2 + (z+1)^2 = 49$; sphere, $C(-5,-2,-1)$, $r = 7$

15. $(x-1/2)^2 + (y-3/4)^2 + (z+5/4)^2 = 54/16$; sphere, $C(1/2, 3/4, -5/4)$, $r = 3\sqrt{6}/4$

17. $(x-3/2)^2 + (y+2)^2 + (z-4)^2 = -11/4$; no graph

19. **(a)**

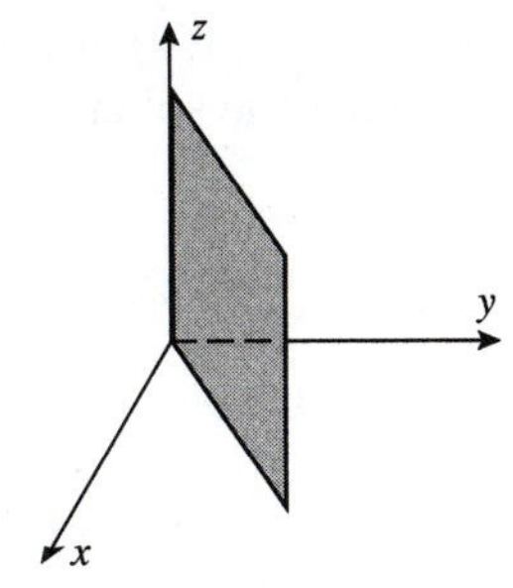

(b)

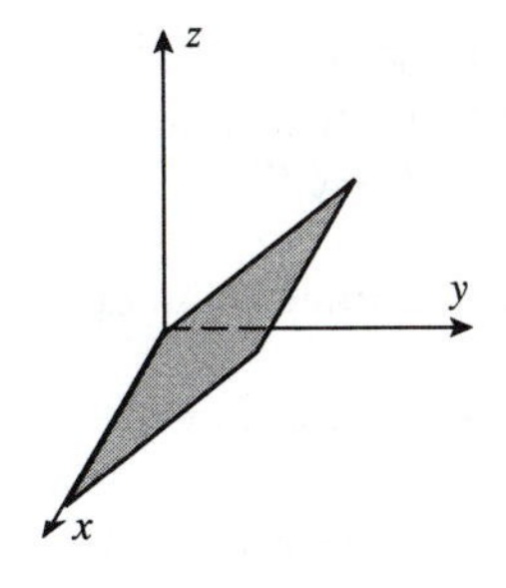

(c)

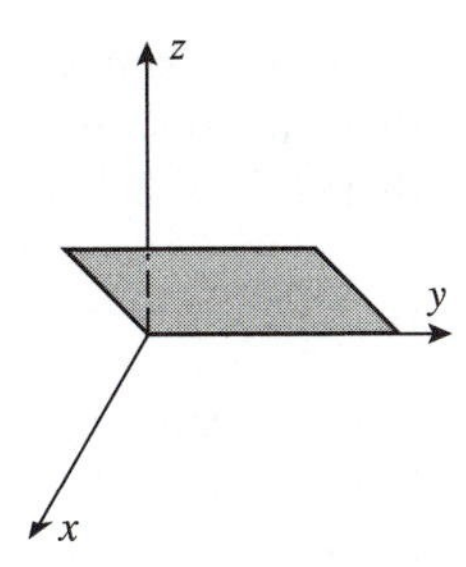

21. **(a)**

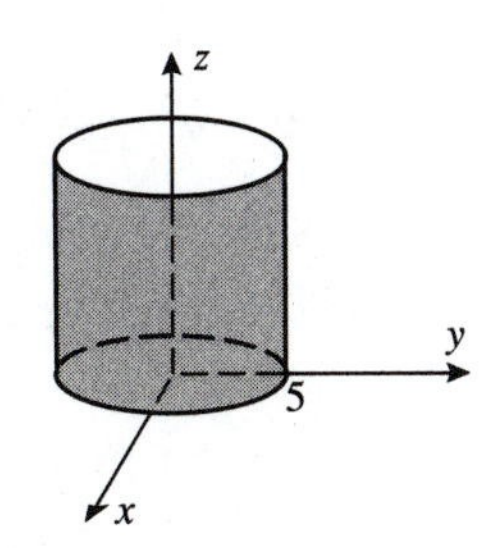

(b)

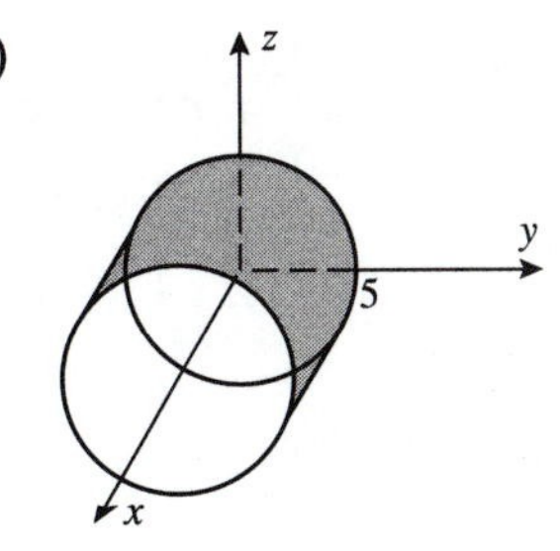

(c)

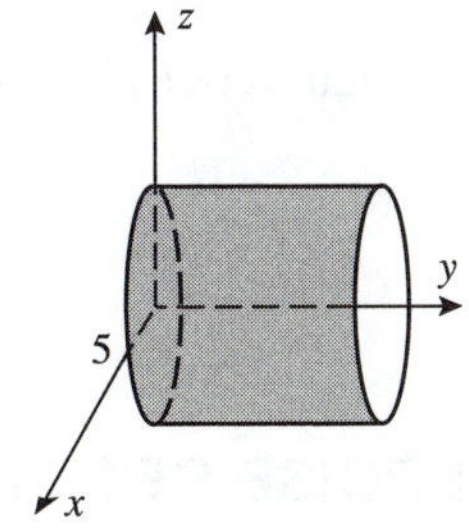

23. **(a)** $-2y + z = 0$ **(b)** $-2x + z = 0$

(c) $(x-1)^2 + (y-1)^2 = 1$ **(d)** $(x-1)^2 + (z-1)^2 = 1$

25.

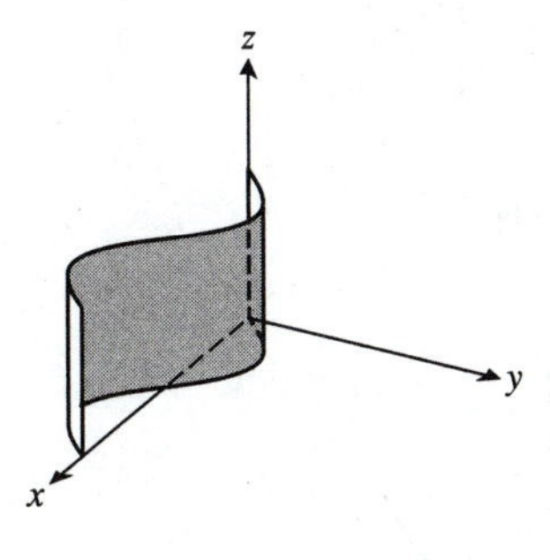

27.

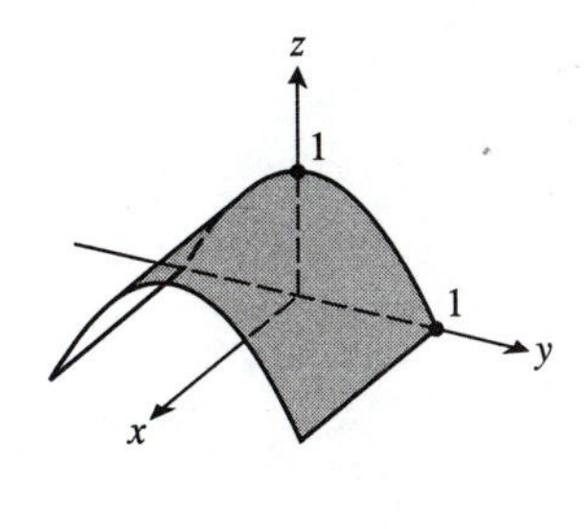

29.

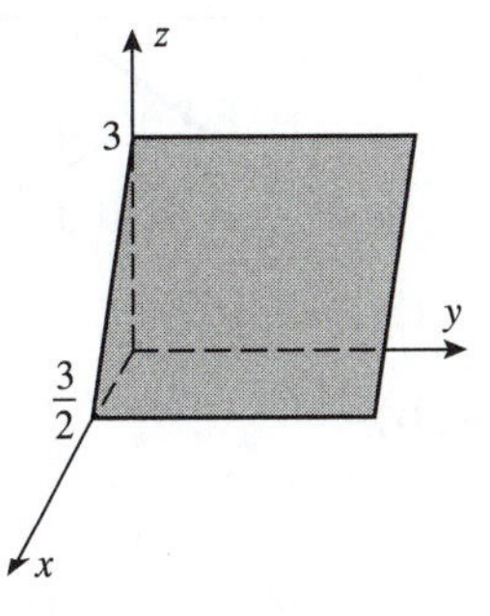

31.

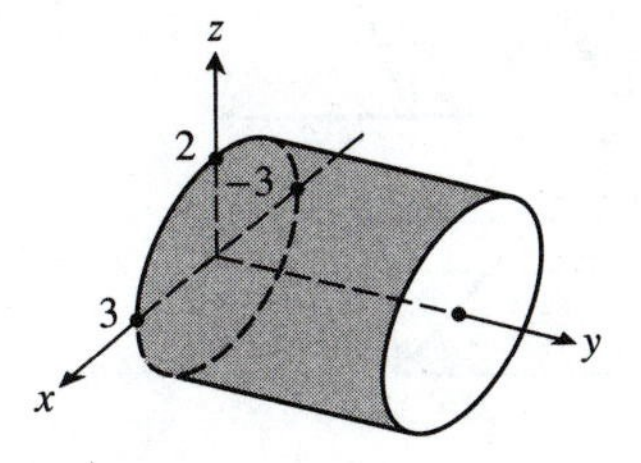

33.

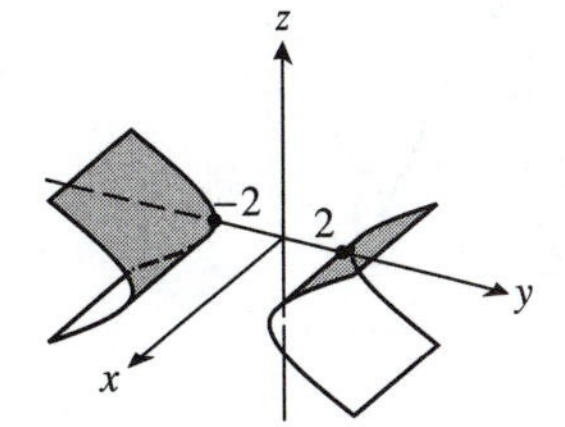

35. **(a)**

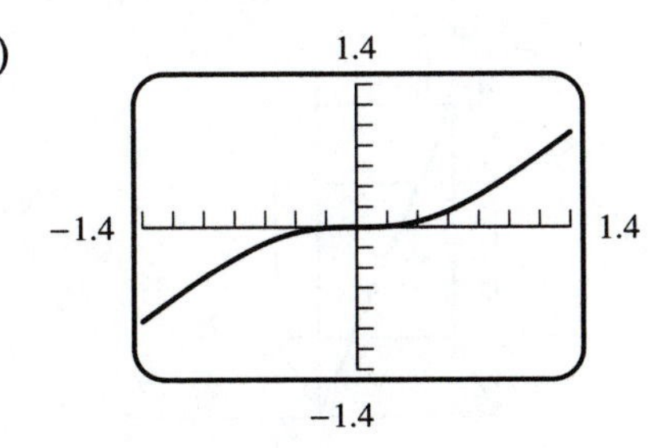

(b)

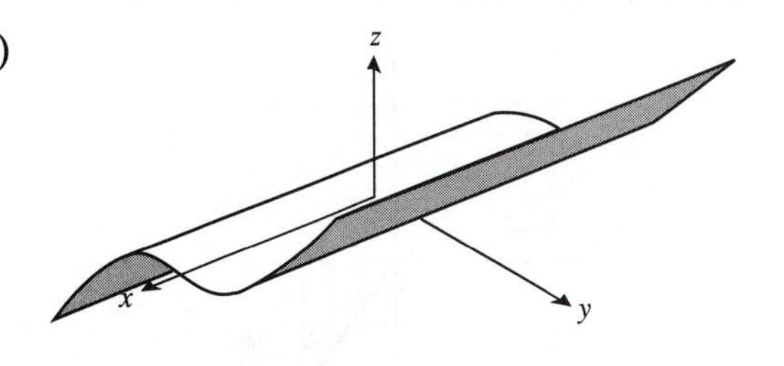

37. Complete the square to get $(x+1)^2+(y-1)^2+(z-2)^2=9$; center $(-1,1,2)$, radius 3. The distance between the origin and the center is $\sqrt{6}<3$ so the origin is inside the sphere. The largest distance is $3+\sqrt{6}$, the smallest is $3-\sqrt{6}$.

39. $(y+3)^2+(z-2)^2>16$; all points outside the circular cylinder $(y+3)^2+(z-2)^2=16$.

41. Let r be the radius of a styrofoam sphere. The distance from the origin to the center of the bowling ball is equal to the sum of the distance from the origin to the center of the styrofoam sphere nearest the origin and the distance between the center of this sphere and the center of the bowling ball so $\sqrt{3}R=\sqrt{3}r+r+R$, $(\sqrt{3}+1)r=(\sqrt{3}-1)R$, $r=\dfrac{\sqrt{3}-1}{\sqrt{3}+1}R=(2-\sqrt{3})R$.

43.
$$\begin{aligned}(a\sin\phi\cos\theta)^2+(a\sin\phi\sin\theta)^2+(a\cos\phi)^2 &= a^2\sin^2\phi\cos^2\theta+a^2\sin^2\phi\sin^2\theta+a^2\cos^2\phi\\ &= a^2\sin^2\phi(\cos^2\theta+\sin^2\theta)+a^2\cos^2\phi\\ &= a^2\sin^2\phi+a^2\cos^2\phi=a^2(\sin^2\phi+\cos^2\phi)=a^2\end{aligned}$$

EXERCISE SET 12.2

1. (a–c)

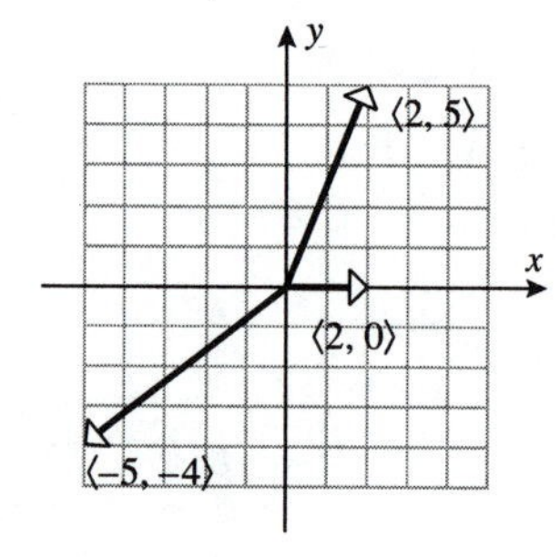

(d–f)

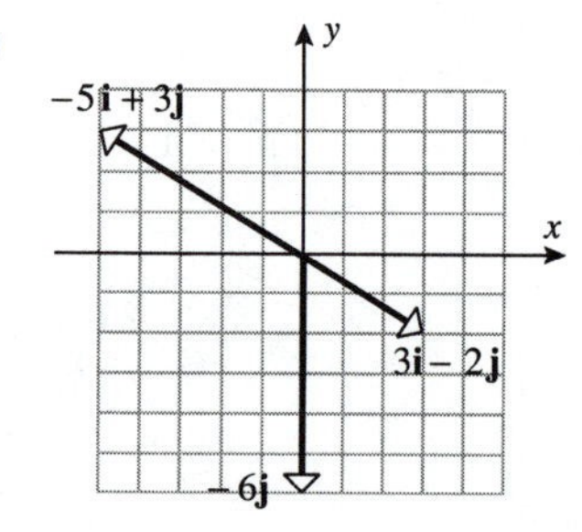

3. (a–b)

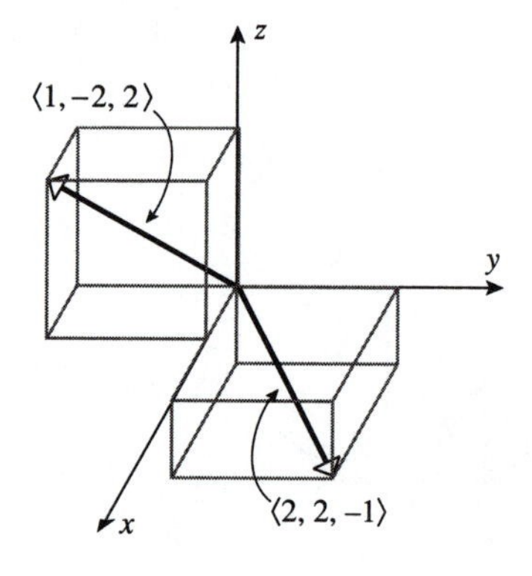

(c–d)

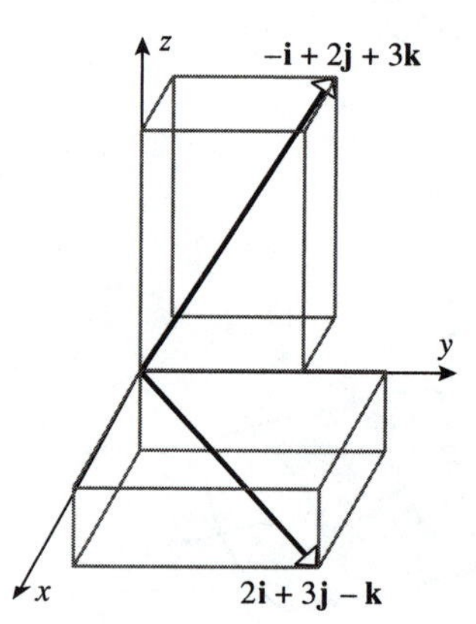

5. (a) $\langle 4-1, 1-5\rangle=\langle 3,-4\rangle$

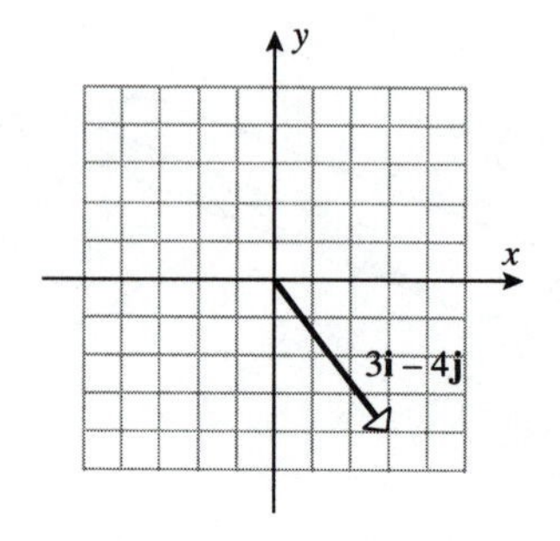

(b) $\langle 0-2, 0-3, 4-0\rangle=\langle -2,-3,4\rangle$

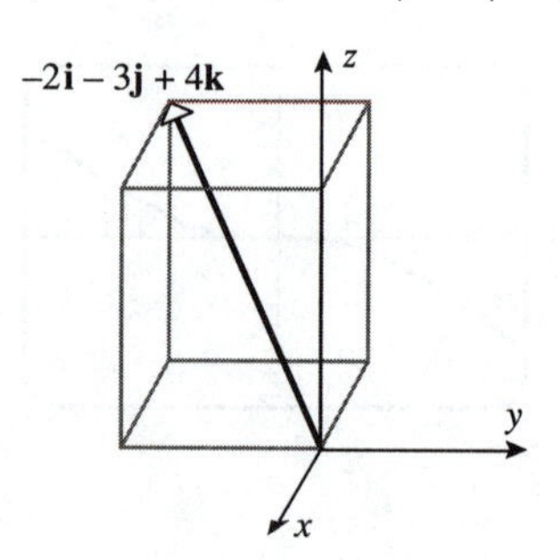

7. **(a)** $\langle 2-3, 8-5\rangle = \langle -1, 3\rangle$
(b) $\langle 0-7, 0-(-2)\rangle = \langle -7, 2\rangle$
(c) $\langle -3, 6, 1\rangle$

9. **(a)** Let (x, y) be the terminal point, then $x - 1 = 3$, $x = 4$ and $y - (-2) = -2$, $y = -4$. The terminal point is $(4, -4)$.
(b) Let (x, y, z) be the initial point, then $5 - x = -3$, $-y = 1$, and $-1 - z = 2$ so $x = 8$, $y = -1$, and $z = -3$. The initial point is $(8, -1, -3)$.

11. **(a)** $-\mathbf{i} + 4\mathbf{j} - 2\mathbf{k}$ **(b)** $18\mathbf{i} + 12\mathbf{j} - 6\mathbf{k}$ **(c)** $-\mathbf{i} - 5\mathbf{j} - 2\mathbf{k}$
(d) $40\mathbf{i} - 4\mathbf{j} - 4\mathbf{k}$ **(e)** $-2\mathbf{i} - 16\mathbf{j} - 18\mathbf{k}$ **(f)** $-\mathbf{i} + 13\mathbf{j} - 2\mathbf{k}$

13. **(a)** $\|\mathbf{v}\| = \sqrt{1+1} = \sqrt{2}$ **(b)** $\|\mathbf{v}\| = \sqrt{1+49} = 5\sqrt{2}$
(c) $\|\mathbf{v}\| = \sqrt{21}$ **(d)** $\|\mathbf{v}\| = \sqrt{14}$

15. **(a)** $\|\mathbf{u}+\mathbf{v}\| = \|2\mathbf{i} - 2\mathbf{j} + 2\mathbf{k}\| = 2\sqrt{3}$ **(b)** $\|\mathbf{u}\| + \|\mathbf{v}\| = \sqrt{14} + \sqrt{2}$
(c) $\|-2\mathbf{u}\| + 2\|\mathbf{v}\| = 2\sqrt{14} + 2\sqrt{2}$ **(d)** $\|3\mathbf{u} - 5\mathbf{v} + \mathbf{w}\| = \|-12\mathbf{j} + 2\mathbf{k}\| = 2\sqrt{37}$
(e) $(1/\sqrt{6})\mathbf{i} + (1/\sqrt{6})\mathbf{j} - (2/\sqrt{6})\mathbf{k}$ **(f)** 1

17. **(a)** $\|-\mathbf{i} + 4\mathbf{j}\| = \sqrt{17}$ so the required vector is $\left(-1/\sqrt{17}\right)\mathbf{i} + \left(4/\sqrt{17}\right)\mathbf{j}$
(b) $\|6\mathbf{i} - 4\mathbf{j} + 2\mathbf{k}\| = 2\sqrt{14}$ so the required vector is $(-3\mathbf{i} + 2\mathbf{j} - \mathbf{k})/\sqrt{14}$
(c) $\overrightarrow{AB} = 4\mathbf{i} + \mathbf{j} - \mathbf{k}$, $\|\overrightarrow{AB}\| = 3\sqrt{2}$ so the required vector is $(4\mathbf{i} + \mathbf{j} - \mathbf{k})/\left(3\sqrt{2}\right)$

19. **(a)** $-\frac{1}{2}\mathbf{v} = \langle -3/2, 2\rangle$ **(b)** $\|\mathbf{v}\| = \sqrt{85}$, so $\frac{\sqrt{17}}{\sqrt{85}}\mathbf{v} = \frac{1}{\sqrt{5}}\langle 7, 0, -6\rangle$ has length $\sqrt{17}$

21. **(a)** $\mathbf{v} = \|\mathbf{v}\|\langle\cos(\pi/4), \sin(\pi/4)\rangle = \langle 3\sqrt{2}/2, 3\sqrt{2}/2\rangle$
(b) $\mathbf{v} = \|\mathbf{v}\|\langle\cos 90^\circ, \sin 90^\circ\rangle = \langle 0, 2\rangle$
(c) $\mathbf{v} = \|\mathbf{v}\|\langle\cos 120^\circ, \sin 120^\circ\rangle = \langle -5/2, 5\sqrt{3}/2\rangle$
(d) $\mathbf{v} = \|\mathbf{v}\|\langle\cos\pi, \sin\pi\rangle = \langle -1, 0\rangle$

23. From (12), $\mathbf{v} = \langle\cos 30^\circ, \sin 30^\circ\rangle = \langle\sqrt{3}/2, 1/2\rangle$ and $\mathbf{w} = \langle\cos 135^\circ, \sin 135^\circ\rangle = \langle -\sqrt{2}/2, \sqrt{2}/2\rangle$, so $\mathbf{v} + \mathbf{w} = ((\sqrt{3} - \sqrt{2})/2, (1 + \sqrt{2})/2)$

25. **(a)** The initial point of $\mathbf{u} + \mathbf{v} + \mathbf{w}$ is the origin and the endpoint is $(-2, 5)$, so $\mathbf{u} + \mathbf{v} + \mathbf{w} = \langle -2, 5\rangle$.

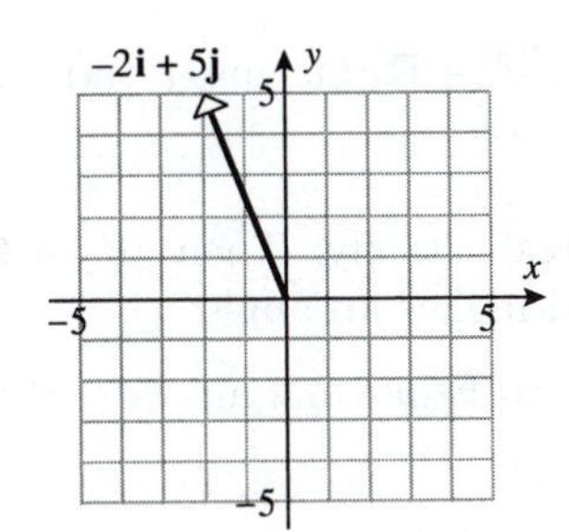

(b) The initial point of $\mathbf{u} + \mathbf{v} + \mathbf{w}$ is $(-5, 4)$ and the endpoint is $(-2, -4)$, so $\mathbf{u} + \mathbf{v} + \mathbf{w} = \langle 3, -8\rangle$.

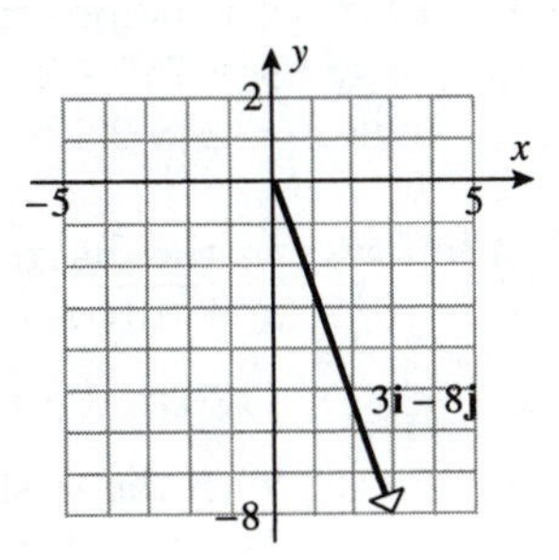

27. $6\mathbf{x} = 2\mathbf{u} - \mathbf{v} - \mathbf{w} = \langle -4, 6\rangle$, $\mathbf{x} = \langle -2/3, 1\rangle$

29. $\mathbf{u} = \frac{5}{7}\mathbf{i} + \frac{2}{7}\mathbf{j} + \frac{1}{7}\mathbf{k}$, $\mathbf{v} = \frac{8}{7}\mathbf{i} - \frac{1}{7}\mathbf{j} - \frac{4}{7}\mathbf{k}$

31. $\|(\mathbf{i}+\mathbf{j}) + (\mathbf{i}-2\mathbf{j})\| = \|2\mathbf{i}-\mathbf{j}\| = \sqrt{5}$, $\|(\mathbf{i}+\mathbf{j} - (\mathbf{i}-2\mathbf{j})\| = \|3\mathbf{j}\| = 3$

33. (a) $5 = \|k\mathbf{v}\| = |k|\|\mathbf{v}\| = \pm 3k$, so $\mathrm{k} = \pm 5/3$

(b) $6 = \|k\mathbf{v}\| = |k|\|\mathbf{v}\| = 2\|\mathbf{v}\|$, so $\|\mathbf{v}\| = 3$

35. (a) Choose two points on the line, for example $P_1(0,2)$ and $P_2(1,5)$; then $\overrightarrow{P_1P_2} = \langle 1,3\rangle$ is parallel to the line, $\|\langle 1,3\rangle\| = \sqrt{10}$, so $\langle 1/\sqrt{10}, 3/\sqrt{10}\rangle$ and $\langle -1/\sqrt{10}, -3/\sqrt{10}\rangle$ are unit vectors parallel to the line.

(b) Choose two points on the line, for example $P_1(0,4)$ and $P_2(1,3)$; then $\overrightarrow{P_1P_2} = \langle 1,-1\rangle$ is parallel to the line, $\|\langle 1,-1\rangle\| = \sqrt{2}$ so $\langle 1/\sqrt{2}, -1/\sqrt{2}\rangle$ and $\langle -1/\sqrt{2}, 1/\sqrt{2}\rangle$ are unit vectors parallel to the line.

(c) Pick any line that is perpendicular to the line $y = -5x+1$, for example $y = x/5$; then $P_1(0,0)$ and $P_2(5,1)$ are on the line, so $\overrightarrow{P_1P_2} = \langle 5,1\rangle$ is perpendicular to the line, so $\pm\frac{1}{\sqrt{26}}\langle 5,1\rangle$ are unit vectors perpendicular to the line.

37. (a) the circle of radius 1 about the origin

(b) the closed disk of radius 1 about the origin

(c) all points outside the closed disk of radius 1 about the origin

39. (a) the (hollow) sphere of radius 1 about the origin

(b) the closed ball of radius 1 about the origin

(c) all points outside the closed ball of radius 1 about the origin

41. Since $\phi = \pi/2$, from (14) we get $\|\mathbf{F}_1 + \mathbf{F}_2\|^2 = \|\mathbf{F}_1\|^2 + \|\mathbf{F}_2\|^2 = 3600 + 900$,

so $\|\mathbf{F}_1 + \mathbf{F}_2\| = 30\sqrt{5}$ lb, and $\sin\alpha = \dfrac{\|\mathbf{F}_2\|}{\|\mathbf{F}_1 + \mathbf{F}_2\|}\sin\phi = \dfrac{30}{30\sqrt{5}}, \alpha \approx 26.57°, \theta = \alpha \approx 26.57°$.

43. $\|\mathbf{F}_1 + \mathbf{F}_2\|^2 = \|\mathbf{F}_1\|^2 + \|\mathbf{F}_2\|^2 + 2\|\mathbf{F}_1\|\|\mathbf{F}_2\|\cos\phi = 160{,}000 + 160{,}000 - 2(400)(400)\dfrac{\sqrt{3}}{2}$,

so $\|\mathbf{F}_1 + \mathbf{F}_2\| \approx 207.06$ N, and $\sin\alpha = \dfrac{\|\mathbf{F}_2\|}{\|\mathbf{F}_1 + \mathbf{F}_2\|}\sin\phi \approx \dfrac{400}{207.06}\left(\dfrac{1}{2}\right), \alpha = 75.00°$,

$\theta = \alpha - 30° = 45.00°$.

45. Let $\mathbf{F}_1, \mathbf{F}_2, \mathbf{F}_3$ be the forces in the diagram with magnitudes $40, 50, 75$ respectively. Then $\mathbf{F}_1 + \mathbf{F}_2 + \mathbf{F}_3 = (\mathbf{F}_1 + \mathbf{F}_2) + \mathbf{F}_3$. Following the examples, $\mathbf{F}_1 + \mathbf{F}_2$ has magnitude 45.83 N and makes an angle 79.11° with the positive x-axis. Then $\|(\mathbf{F}_1+\mathbf{F}_2)+\mathbf{F}_3\|^2 \approx 45.83^2 + 75^2 + 2(45.83)(75)\cos 79.11°$, so $\mathbf{F}_1+\mathbf{F}_2+\mathbf{F}_3$ has magnitude ≈ 94.995 N and makes an angle $\theta = \alpha \approx 28.28°$ with the positive x-axis.

47. Let $\mathbf{F}_1, \mathbf{F}_2$ be the forces in the diagram with magnitudes $8, 10$ respectively. Then $\|\mathbf{F}_1 + \mathbf{F}_2\|$ has magnitude $\sqrt{8^2 + 10^2 + 2\cdot 8\cdot 10\cos 120°} = 2\sqrt{21} \approx 9.165$ lb, and makes an angle $60° + \sin^{-1}\dfrac{\|\mathbf{F}_1\|}{\|\mathbf{F}_1 + \mathbf{F}_2\|}\sin 120 \approx 109.11°$ with the positive x-axis, so $\mathbf{F}$ has magnitude 9.165 lb and makes an angle $-70.89°$ with the positive x-axis.

49. $\mathbf{F}_1 + \mathbf{F}_2 + \mathbf{F} = \mathbf{0}$, where $\mathbf{F}$ has magnitude 250 and makes an angle $-90°$ with the positive x-axis. Thus $\|\mathbf{F}_1 + \mathbf{F}_2\|^2 = \|\mathbf{F}_1\|^2 + \|\mathbf{F}_2\|^2 + 2\|\mathbf{F}_1\|\|\mathbf{F}_2\|\cos 105° = 250^2$ and

$$45° = \alpha = \sin^{-1}\left(\frac{\|\mathbf{F}_2\|}{250}\sin 105°\right), \text{ so } \frac{\sqrt{2}}{2} \approx \frac{\|\mathbf{F}_2\|}{250}0.9659, \|\mathbf{F}_2\| \approx 183.02 \text{ lb},$$

$$\|\mathbf{F}_1\|^2 + 2(183.02)(-0.2588)\|\mathbf{F}_1\| + (183.02)^2 = 62{,}500, \|\mathbf{F}_1\| = 224.13 \text{ lb}.$$

51. **(a)** $c_1\mathbf{v}_1 + c_2\mathbf{v}_2 = (2c_1 + 4c_2)\,\mathbf{i} + (-c_1 + 2c_2)\,\mathbf{j} = 4\mathbf{j}$, so $2c_1 + 4c_2 = 0$ and $-c_1 + 2c_2 = 4$ which gives $c_1 = -2$, $c_2 = 1$.

(b) $c_1\mathbf{v}_1 + c_2\mathbf{v}_2 = \langle c_1 - 2c_2, -3c_1 + 6c_2\rangle = \langle 3, 5\rangle$, so $c_1 - 2c_2 = 3$ and $-3c_1 + 6c_2 = 5$ which has no solution.

53. Place $\mathbf{u}$ and $\mathbf{v}$ tip to tail so that $\mathbf{u} + \mathbf{v}$ is the vector from the initial point of $\mathbf{u}$ to the terminal point of $\mathbf{v}$. The shortest distance between two points is along the line joining these points so $\|\mathbf{u} + \mathbf{v}\| \leq \|\mathbf{u}\| + \|\mathbf{v}\|$.

55. (d): $\mathbf{u} + (-\mathbf{u}) = (u_1\mathbf{i} + u_2\,\mathbf{j}) + (-u_1\mathbf{i} - u_2\,\mathbf{j}) = (u_1 - u_1)\mathbf{i} + (u_1 - u_1)\,\mathbf{j} = \mathbf{0}$
(g): $(k + l)\mathbf{u} = (k + l)(u_1\mathbf{i} + u_2\,\mathbf{j}) = ku_1\mathbf{i} + ku_2\,\mathbf{j} + lu_1\mathbf{i} + lu_2\,\mathbf{j} = k\mathbf{u} + l\mathbf{u}$
(h): $1\mathbf{u} = 1(u_1\mathbf{i} + u_2\,\mathbf{j}) = 1u_1\mathbf{i} + 1u_2\,\mathbf{j} = u_1\mathbf{i} + u_2\,\mathbf{j} = \mathbf{u}$

57. Let $\mathbf{a}$, $\mathbf{b}$, $\mathbf{c}$ be vectors along the sides of the triangle and A,B the midpoints of $\mathbf{a}$ and $\mathbf{b}$, then $\mathbf{u} = \frac{1}{2}\mathbf{a} - \frac{1}{2}\mathbf{b} = \frac{1}{2}(\mathbf{a} - \mathbf{b}) = \frac{1}{2}\mathbf{c}$ so $\mathbf{u}$ is parallel to $\mathbf{c}$ and half as long.

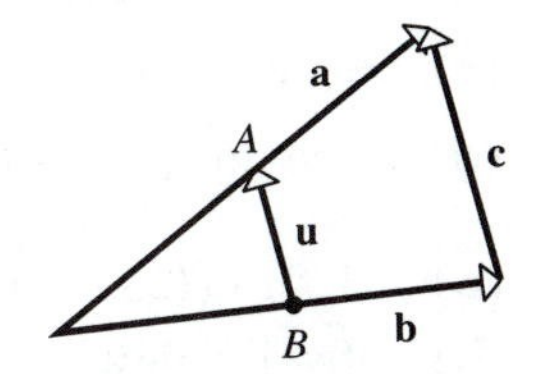

EXERCISE SET 12.3

1. **(a)** $(1)(6) + (2)(-8) = -10$; $\cos\theta = (-10)/[(\sqrt{5})(10)] = -1/\sqrt{5}$
(b) $(-7)(0) + (-3)(1) = -3$; $\cos\theta = (-3)/[(\sqrt{58})(1)] = -3/\sqrt{58}$
(c) $(1)(8) + (-3)(-2) + (7)(-2) = 0$; $\cos\theta = 0$
(d) $(-3)(4) + (1)(2) + (2)(-5) = -20$; $\cos\theta = (-20)/[(\sqrt{14})(\sqrt{45})] = -20/(3\sqrt{70})$

3. **(a)** $\mathbf{u}\cdot\mathbf{v} = -34 < 0$, obtuse **(b)** $\mathbf{u}\cdot\mathbf{v} = 6 > 0$, acute
(c) $\mathbf{u}\cdot\mathbf{v} = -1 < 0$, obtuse **(d)** $\mathbf{u}\cdot\mathbf{v} = 0$, orthogonal

5. Since $\mathbf{v}_1\cdot\mathbf{v}_i = \cos\phi_i$, the answers are, in order, $\sqrt{2}/2, 0, -\sqrt{2}/2, -1, -\sqrt{2}/2, 0, \sqrt{2}/2$

7. **(a)** $\overrightarrow{AB} = \langle 1, 3, -2\rangle, \overrightarrow{BC} = \langle 4, -2, -1\rangle, \overrightarrow{AB}\cdot\overrightarrow{BC} = 0$ so $\overrightarrow{AB}$ and $\overrightarrow{BC}$ are orthogonal; it is a right triangle with the right angle at vertex B.

(b) Let A, B, and C be the vertices $(-1,0)$, $(2,-1)$, and $(1,4)$ with corresponding interior angles α, β, and γ, then

$$\cos\alpha = \frac{\overrightarrow{AB}\cdot\overrightarrow{AC}}{\|\overrightarrow{AB}\|\,\|\overrightarrow{AC}\|} = \frac{\langle 3,-1\rangle\cdot\langle 2,4\rangle}{\sqrt{10}\sqrt{20}} = 1/(5\sqrt{2}),\ \alpha\approx 82^\circ$$

$$\cos\beta = \frac{\overrightarrow{BA}\cdot\overrightarrow{BC}}{\|\overrightarrow{BA}\|\,\|\overrightarrow{BC}\|} = \frac{\langle -3,1\rangle\cdot\langle -1,5\rangle}{\sqrt{10}\sqrt{26}} = 4/\sqrt{65},\ \beta\approx 60^\circ$$

$$\cos\gamma = \frac{\overrightarrow{CA}\cdot\overrightarrow{CB}}{\|\overrightarrow{CA}\|\,\|\overrightarrow{CB}\|} = \frac{\langle -2,-4\rangle\cdot\langle 1,-5\rangle}{\sqrt{20}\sqrt{26}} = 9/\sqrt{130},\ \gamma\approx 38^\circ$$

9. **(a)** $\mathbf{v}\cdot\mathbf{v}_1 = -ab+ba = 0;\ \ \mathbf{v}\cdot\mathbf{v}_2 = ab+b(-a) = 0$

(b) Let $\mathbf{v}_1 = 2\mathbf{i}+3\mathbf{j},\ \ \mathbf{v}_2 = -2\mathbf{i}-3\mathbf{j}$;

take $\mathbf{u}_1 = \dfrac{\mathbf{v}_1}{\|\mathbf{v}_1\|} = \dfrac{2}{\sqrt{13}}\mathbf{i}+\dfrac{3}{\sqrt{13}}\mathbf{j},\ \mathbf{u}_2 = -\mathbf{u}_1.$

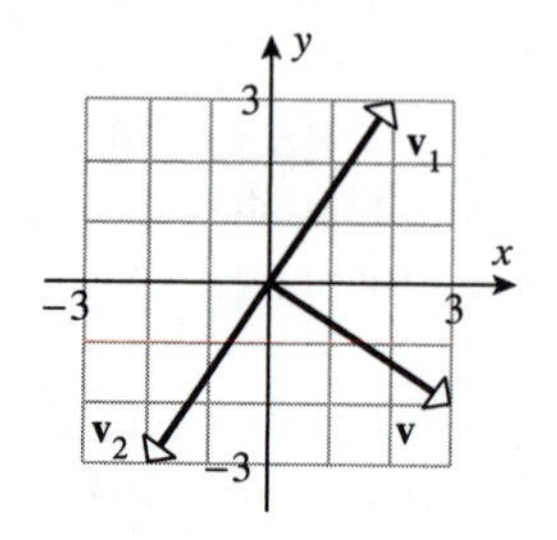

11. **(a)** The dot product of a vector $\mathbf{u}$ and a scalar $\mathbf{v}\cdot\mathbf{w}$ is not defined.

(b) The sum of a scalar $\mathbf{u}\cdot\mathbf{v}$ and a vector $\mathbf{w}$ is not defined.

(c) $\mathbf{u}\cdot\mathbf{v}$ is not a vector.

(d) The dot product of a scalar k and a vector $\mathbf{u}+\mathbf{v}$ is not defined.

13. **(a)** $\langle 1,2\rangle\cdot(\langle 28,-14\rangle+\langle 6,0\rangle) = \langle 1,2\rangle\cdot\langle 34,-14\rangle = 6$

(b) $\|6\mathbf{w}\| = 6\|\mathbf{w}\| = 36$ **(c)** $24\sqrt{5}$ **(d)** $24\sqrt{5}$

15. **(a)** $\|\mathbf{v}\| = \sqrt{3}$ so $\cos\alpha = \cos\beta = 1/\sqrt{3},\ \cos\gamma = -1/\sqrt{3},\ \alpha=\beta\approx 55^\circ,\ \gamma\approx 125^\circ$

(b) $\|\mathbf{v}\| = 3$ so $\cos\alpha = 2/3,\ \cos\beta = -2/3,\ \cos\gamma = 1/3,\ \alpha\approx 48^\circ,\ \beta\approx 132^\circ,\ \gamma\approx 71^\circ$

17. $\cos^2\alpha+\cos^2\beta+\cos^2\gamma = \dfrac{v_1^2}{\|\mathbf{v}\|^2}+\dfrac{v_2^2}{\|\mathbf{v}\|^2}+\dfrac{v_3^2}{\|\mathbf{v}\|^2} = \left(v_1^2+v_2^2+v_3^2\right)/\|\mathbf{v}\|^2 = \|\mathbf{v}\|^2/\|\mathbf{v}\|^2 = 1$

19. $\cos\alpha = \dfrac{\sqrt{3}}{2}\dfrac{1}{2} = \dfrac{\sqrt{3}}{4},\ \cos\beta = \dfrac{\sqrt{3}}{2}\dfrac{\sqrt{3}}{2} = \dfrac{3}{4},\ \cos\gamma = \dfrac{1}{2};\ \alpha\approx 64^\circ, \beta\approx 41^\circ, \gamma = 60^\circ$

21. **(a)** $\dfrac{\mathbf{b}}{\|\mathbf{b}\|} = \langle 3/5, 4/5\rangle$, so $\text{proj}_{\mathbf{b}}\mathbf{v} = \langle 6/25, 8/25\rangle$

and $\mathbf{v}-\text{proj}_{\mathbf{b}}\mathbf{v} = \langle 44/25, -33/25\rangle$

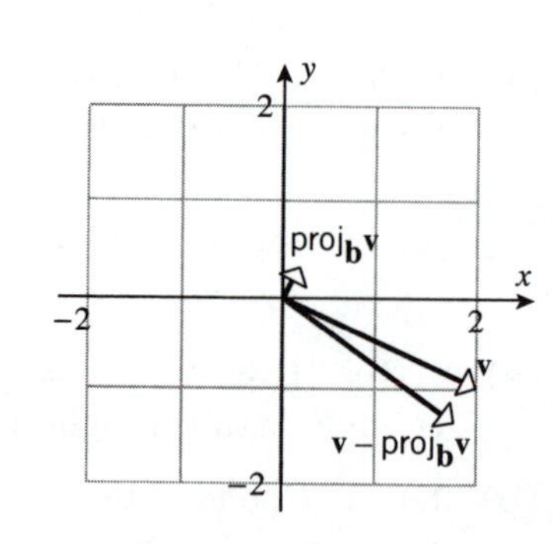

(b) $\dfrac{\mathbf{b}}{\|\mathbf{b}\|} = \langle 1/\sqrt{5}, -2/\sqrt{5}\rangle$, so $\text{proj}_{\mathbf{b}}\mathbf{v} = \langle -6/5, 12/5\rangle$
and $\mathbf{v} - \text{proj}_{\mathbf{b}}\mathbf{v} = \langle 26/5, 13/5\rangle$

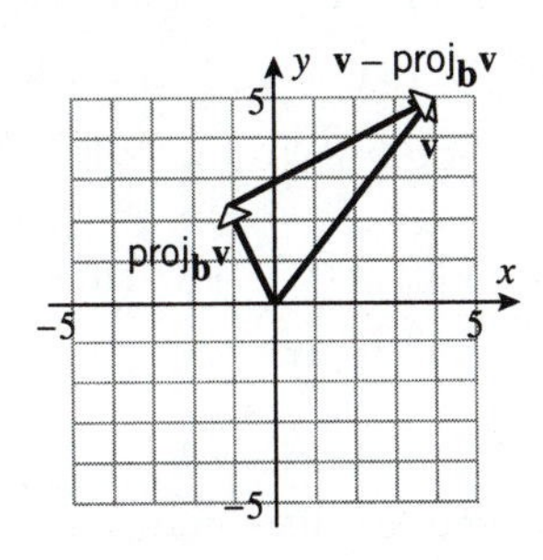

(c) $\dfrac{\mathbf{b}}{\|\mathbf{b}\|} = \langle 2/\sqrt{5}, 1/\sqrt{5}\rangle$, so $\text{proj}_{\mathbf{b}}\mathbf{v} = \langle -16/5, -8/5\rangle$
and $\mathbf{v} - \text{proj}_{\mathbf{b}}\mathbf{v} = \langle 1/5, -2/5\rangle$

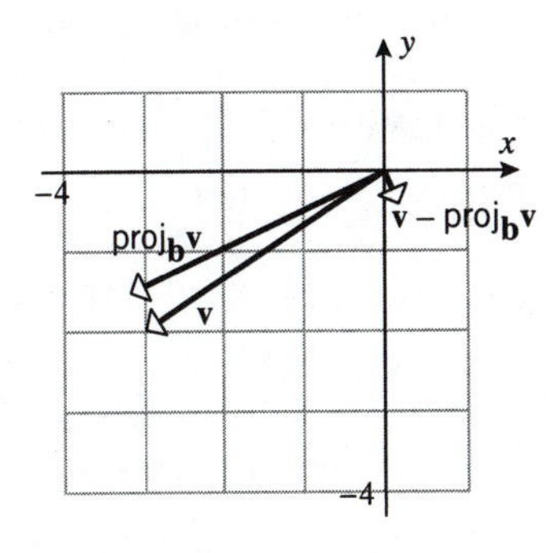

23. (a) $\text{proj}_{\mathbf{b}}\mathbf{v} = \langle -1, -1\rangle$, so $\mathbf{v} = \langle -1, -1\rangle + \langle 3, -3\rangle$
(b) $\text{proj}_{\mathbf{b}}\mathbf{v} = \langle 16/5, 0, -8/5\rangle$, so $\mathbf{v} = \langle 16/5, 0, -8/5\rangle + \langle -1/5, 1, -2/5\rangle$

25. $\overrightarrow{AP} = -\mathbf{i} + 3\mathbf{j}$, $\overrightarrow{AB} = 3\mathbf{i} + 4\mathbf{j}$, $\|\text{proj}_{\overrightarrow{AB}}\overrightarrow{AP}\| = |\overrightarrow{AP}\cdot\overrightarrow{AB}|/\|\overrightarrow{AB}\| = 9/5$
$\|\overrightarrow{AP}\| = \sqrt{10}$, $\sqrt{10 - 81/25} = 13/5$

27. Let $\mathbf{F}$ be the downward force of gravity on the block, then $\|\mathbf{F}\| = 10(9.8) = 98$ N, and if $\mathbf{F}_1$ and $\mathbf{F}_2$ are the forces parallel to and perpendicular to the ramp, then $\|\mathbf{F}_1\| = \|\mathbf{F}_2\| = 49\sqrt{2}$ N. Thus the block exerts a force of $49\sqrt{2}$ N against the ramp and it requires a force of $49\sqrt{2}$ N to prevent the block from sliding down the ramp.

29. Three forces act on the block: its weight $-300\mathbf{j}$; the tension in cable A, which has the form $a(-\mathbf{i}+\mathbf{j})$; and the tension in cable B, which has the form $b(\sqrt{3}\mathbf{i} - \mathbf{j})$, where a, b are positive constants. The sum of these forces is zero, which yields $a = 450 + 150\sqrt{3}, b = 150 + 150\sqrt{3}$. Thus the forces along cables A and B are, respectively,
$\|150(3+\sqrt{3})(\mathbf{i} - \mathbf{j})\| = 450\sqrt{2} + 150\sqrt{6}$ lb, and $\|150(\sqrt{3}+1)(\sqrt{3}\mathbf{i} - \mathbf{j})\| = 300 + 300\sqrt{3}$ lb.

31. Let P and Q be the points (1,3) and (4,7) then $\overrightarrow{PQ} = 3\mathbf{i} + 4\mathbf{j}$ so $W = \mathbf{F}\cdot\overrightarrow{PQ} = -12$ ft$\cdot$lb.

33. $W = \mathbf{F}\cdot 15\mathbf{i} = 15\cdot 50\cos 60^\circ = 375$ ft$\cdot$lb.

35. With the cube as shown in the diagram, and a the length of each edge,
$\mathbf{d}_1 = a\mathbf{i} + a\mathbf{j} + a\mathbf{k}, \mathbf{d}_2 = a\mathbf{i} + a\mathbf{j} - a\mathbf{k}$,
$\cos\theta = (\mathbf{d}_1\cdot\mathbf{d}_2)/(\|\mathbf{d}_1\|\,\|\mathbf{d}_2\|) = 1/3, \theta \approx 71^\circ$

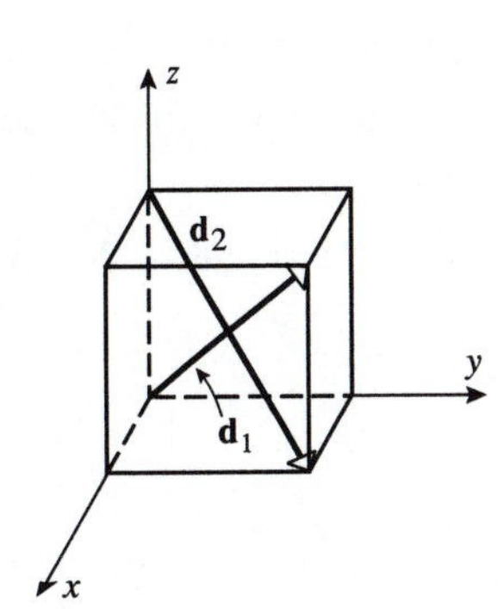

37. $\mathbf{u}+\mathbf{v}$ and $\mathbf{u}-\mathbf{v}$ are vectors along the diagonals,
$(\mathbf{u}+\mathbf{v})\cdot(\mathbf{u}-\mathbf{v}) = \mathbf{u}\cdot\mathbf{u} - \mathbf{u}\cdot\mathbf{v} + \mathbf{v}\cdot\mathbf{u} - \mathbf{v}\cdot\mathbf{v} = \|\mathbf{u}\|^2 - \|\mathbf{v}\|^2$ so $(\mathbf{u}+\mathbf{v})\cdot(\mathbf{u}-\mathbf{v}) = 0$
if and only if $\|\mathbf{u}\| = \|\mathbf{v}\|$.

39. $\|\mathbf{u}+\mathbf{v}\|^2 = (\mathbf{u}+\mathbf{v})\cdot(\mathbf{u}+\mathbf{v}) = \|\mathbf{u}\|^2 + 2\mathbf{u}\cdot\mathbf{v} + \|\mathbf{v}\|^2$ and
$\|\mathbf{u}-\mathbf{v}\|^2 = (\mathbf{u}-\mathbf{v})\cdot(\mathbf{u}-\mathbf{v}) = \|\mathbf{u}\|^2 - 2\mathbf{u}\cdot\mathbf{v} + \|\mathbf{v}\|^2$, add to get
$\|\mathbf{u}+\mathbf{v}\|^2 + \|\mathbf{u}-\mathbf{v}\|^2 = 2\|\mathbf{u}\|^2 + 2\|\mathbf{v}\|^2$

The sum of the squares of the lengths of the diagonals of a parallelogram is equal to twice the sum of the squares of the lengths of the sides.

41. $\mathbf{v} = c_1\mathbf{v}_1 + c_2\mathbf{v}_2 + c_3\mathbf{v}_3$ so $\mathbf{v}\cdot\mathbf{v}_i = c_i\mathbf{v}_i\cdot\mathbf{v}_i$ because $\mathbf{v}_i\cdot\mathbf{v}_j = 0$ if $i \neq j$,
thus $\mathbf{v}\cdot\mathbf{v}_i = c_i\|\mathbf{v}_i\|^2, c_i = \mathbf{v}\cdot\mathbf{v}_i/\|\mathbf{v}_i\|^2$ for $i = 1, 2, 3$.

43. (a) $\mathbf{u} = x\mathbf{i} + (x^2+1)\mathbf{j}, \mathbf{v} = x\mathbf{i} - (x+1)\mathbf{j}, \theta = \cos^{-1}[(\mathbf{u}\cdot\mathbf{v})/(\|\mathbf{u}\|\|\mathbf{v}\|)]$.
Use a CAS to solve $d\theta/dx = 0$ to find that the minimum value of θ occurs when $x \approx -3.136742$ so the minimum angle is about $40°$. NB: Since $\cos^{-1} u$ is a decreasing function of u, it suffices to maximize $(\mathbf{u}\cdot\mathbf{v})/(\|\mathbf{u}\|\|\mathbf{v}\|)$, or, what is easier, its square.

(b) Solve $\mathbf{u}\cdot\mathbf{v} = 0$ for x to get $x \approx -0.682328$.

45. Let $\mathbf{u} = \langle u_1, u_2, u_3\rangle, \mathbf{v} = \langle v_1, v_2, v_3\rangle, \mathbf{w} = \langle w_1, w_2, w_3\rangle$. Then
$$\mathbf{u}\cdot(\mathbf{v}+\mathbf{w}) = \langle u_1(v_1+w_1), u_2(v_2+w_2), u_3(v_3+w_3)\rangle = \langle u_1v_1+u_1w_1, u_2v_2+u_2w_2, u_3v_3+u_3w_3\rangle$$
$$= \langle u_1v_1, u_2v_2, u_3v_3\rangle + \langle u_1w_1, u_2w_2, u_3w_3\rangle = \mathbf{u}\cdot\mathbf{v} + \mathbf{u}\cdot\mathbf{w}$$
$\mathbf{0}\cdot\mathbf{v} = 0\cdot v_1 + 0\cdot v_2 + 0\cdot v_3 = 0$

EXERCISE SET 12.4

1. (a) $\mathbf{i}\times(\mathbf{i}+\mathbf{j}+\mathbf{k}) = \begin{vmatrix} \mathbf{i} & \mathbf{j} & \mathbf{k} \\ 1 & 0 & 0 \\ 1 & 1 & 1 \end{vmatrix} = -\mathbf{j}+\mathbf{k}$

(b) $\mathbf{i}\times(\mathbf{i}+\mathbf{j}+\mathbf{k}) = (\mathbf{i}\times\mathbf{i}) + (\mathbf{i}\times\mathbf{j}) + (\mathbf{i}\times\mathbf{k}) = -\mathbf{j}+\mathbf{k}$

3. $\langle 7, 10, 9\rangle$

5. $\langle -4, -6, -3\rangle$

7. (a) $\mathbf{v}\times\mathbf{w} = \langle -23, 7, -1\rangle, \mathbf{u}\times(\mathbf{v}\times\mathbf{w}) = \langle -20, -67, -9\rangle$

(b) $\mathbf{u}\times\mathbf{v} = \langle -10, -14, 2\rangle, (\mathbf{u}\times\mathbf{v})\times\mathbf{w} = \langle -78, 52, -26\rangle$

(c) $(\mathbf{u}\times\mathbf{v})\times(\mathbf{v}\times\mathbf{w}) = \langle -10, -14, 2\rangle \times \langle -23, 7, -1\rangle = \langle 0, -56, -392\rangle$

(d) $(\mathbf{v}\times\mathbf{w})\times(\mathbf{u}\times\mathbf{v}) = \langle 0, 56, 392\rangle$

9. $\mathbf{u}\times\mathbf{v} = (\mathbf{i}+\mathbf{j})\times(\mathbf{i}+\mathbf{j}+\mathbf{k}) = \mathbf{k}-\mathbf{j}-\mathbf{k}+\mathbf{i} = \mathbf{i}-\mathbf{j}$, the direction cosines are $\dfrac{1}{\sqrt{2}}, -\dfrac{1}{\sqrt{2}}, 0$

11. $\mathbf{n} = \overrightarrow{AB}\times\overrightarrow{AC} = \langle 1, 1, -3\rangle \times \langle -1, 3, -1\rangle = \langle 8, 4, 4\rangle$, unit vectors are $\pm\dfrac{1}{\sqrt{6}}\langle 2, 1, 1\rangle$

13. $A = \|\mathbf{u}\times\mathbf{v}\| = \|-7\mathbf{i}-\mathbf{j}+3\mathbf{k}\| = \sqrt{59}$

15. $A = \frac{1}{2}\|\overrightarrow{PQ} \times \overrightarrow{PR}\| = \frac{1}{2}\|\langle -1,-5,2\rangle \times \langle 2,0,3\rangle\| = \frac{1}{2}\|\langle -15,7,10\rangle\| = \sqrt{374}/2$

17. 80

19. -3

21. $V = |\mathbf{u}\cdot(\mathbf{v}\times\mathbf{w})| = |-16| = 16$

23. **(a)** $\mathbf{u}\cdot(\mathbf{v}\times\mathbf{w}) = 0$, yes **(b)** $\mathbf{u}\cdot(\mathbf{v}\times\mathbf{w}) = 0$, yes **(c)** $\mathbf{u}\cdot(\mathbf{v}\times\mathbf{w}) = 245$, no

25. **(a)** $V = |\mathbf{u}\cdot(\mathbf{v}\times\mathbf{w})| = |-9| = 9$ **(b)** $A = \|\mathbf{u}\times\mathbf{w}\| = \|3\mathbf{i} - 8\mathbf{j} + 7\mathbf{k}\| = \sqrt{122}$

(c) $\mathbf{v}\times\mathbf{w} = -3\mathbf{i} - \mathbf{j} + 2\mathbf{k}$ is perpendicular to the plane determined by $\mathbf{v}$ and $\mathbf{w}$; let θ be the angle between $\mathbf{u}$ and $\mathbf{v}\times\mathbf{w}$ then

$$\cos\theta = \frac{\mathbf{u}\cdot(\mathbf{v}\times\mathbf{w})}{\|\mathbf{u}\|\,\|\mathbf{v}\times\mathbf{w}\|} = \frac{-9}{\sqrt{14}\sqrt{14}} = -9/14$$

so the acute angle ϕ that $\mathbf{u}$ makes with the plane determined by $\mathbf{v}$ and $\mathbf{w}$ is $\phi = \theta - \pi/2 = \sin^{-1}(9/14)$.

27. **(a)** $\mathbf{u} = \overrightarrow{AP} = -4\mathbf{i} + 2\mathbf{k}$, $\mathbf{v} = \overrightarrow{AB} = -3\mathbf{i} + 2\mathbf{j} - 4\mathbf{k}$, $\mathbf{u}\times\mathbf{v} = -4\mathbf{i} - 22\mathbf{j} - 8\mathbf{k}$; distance $= \|\mathbf{u}\times\mathbf{v}\|/\|\mathbf{v}\| = 2\sqrt{141/29}$

(b) $\mathbf{u} = \overrightarrow{AP} = 2\mathbf{i} + 2\mathbf{j}$, $\mathbf{v} = \overrightarrow{AB} = -2\mathbf{i} + \mathbf{j}$, $\mathbf{u}\times\mathbf{v} = 6\mathbf{k}$; distance $= \|\mathbf{u}\times\mathbf{v}\|/\|\mathbf{v}\| = 6/\sqrt{5}$

29. $\overrightarrow{PQ} = \langle 3,-1,-3\rangle$, $\overrightarrow{PR} = \langle 2,-2,1\rangle$, $\overrightarrow{PS} = \langle 4,-4,3\rangle$,

$V = \frac{1}{6}|\overrightarrow{PQ}\cdot(\overrightarrow{PR}\times\overrightarrow{PS})| = \frac{1}{6}|-4| = 2/3$

31. From Theorems 12.3.3 and 12.4.5a it follows that $\sin\theta = \cos\theta$, so $\theta = \pi/4$.

33. **(a)** $\mathbf{F} = 10\mathbf{j}$ and $\overrightarrow{PQ} = \mathbf{i} + \mathbf{j} + \mathbf{k}$, so the vector moment of $\mathbf{F}$ about P is

$$\overrightarrow{PQ}\times\mathbf{F} = \begin{vmatrix} \mathbf{i} & \mathbf{j} & \mathbf{k} \\ 1 & 1 & 1 \\ 0 & 10 & 0 \end{vmatrix} = -10\mathbf{i} + 10\mathbf{k},$$ and the scalar moment is $10\sqrt{2}$ lb·ft.

The direction of rotation of the cube about P is counterclockwise looking along $\overrightarrow{PQ}\times\mathbf{F} = -10\mathbf{i} + 10\mathbf{k}$ toward its initial point.

(b) $\mathbf{F} = 10\mathbf{j}$ and $\overrightarrow{PQ} = \mathbf{j} + \mathbf{k}$, so the vector moment of $\mathbf{F}$ about P is

$$\overrightarrow{PQ}\times\mathbf{F} = \begin{vmatrix} \mathbf{i} & \mathbf{j} & \mathbf{k} \\ 0 & 1 & 1 \\ 0 & 10 & 0 \end{vmatrix} = -10\mathbf{i},$$ and the scalar moment is 10 lb·ft. The direction of rotation of the cube about P is counterclockwise looking along $-10\mathbf{i}$ toward its initial point.

(c) $\mathbf{F} = 10\mathbf{j}$ and $\overrightarrow{PQ} = \mathbf{j}$, so the vector moment of $\mathbf{F}$ about P is

$$\overrightarrow{PQ}\times\mathbf{F} = \begin{vmatrix} \mathbf{i} & \mathbf{j} & \mathbf{k} \\ 0 & 1 & 0 \\ 0 & 10 & 0 \end{vmatrix} = \mathbf{0},$$ and the scalar moment is 0 lb·ft. Since the force is parallel to the direction of motion, there is no rotation about P.

35. Take the center of the bolt as the origin of the plane. Then $\mathbf{F}$ makes an angle 72° with the positive x-axis, so $\mathbf{F} = 200\cos 72°\mathbf{i} + 200\sin 72°\mathbf{j}$ and $\vec{PQ} = 0.2\,\mathbf{i} + 0.03\,\mathbf{j}$. The scalar moment is given by

$$\left\| \begin{vmatrix} \mathbf{i} & \mathbf{j} & \mathbf{k} \\ 0.2 & 0.03 & 0 \\ 200\cos 72° & 200\sin 72° & 0 \end{vmatrix} \right\| = \left| 40\frac{1}{4}(\sqrt{5}-1) - 6\frac{1}{4}\sqrt{10+2\sqrt{5}} \right| \approx 36.1882 \text{ N·m}.$$

37. Let $\mathbf{u} = \langle u_1, u_2, u_3\rangle$ and $\mathbf{v} = \langle v_1, v_2, v_3\rangle$; show that $k(\mathbf{u}\times\mathbf{v})$, $(k\mathbf{u})\times\mathbf{v}$, and $\mathbf{u}\times(k\mathbf{v})$ are all the same; Part (e) is proved in a similar fashion.

39. $-8\mathbf{i} - 8\mathbf{k}, -8\mathbf{i} - 20\mathbf{j} + 2\mathbf{k}$

41. If $\mathbf{a}$, $\mathbf{b}$, $\mathbf{c}$, and $\mathbf{d}$ lie in the same plane then $\mathbf{a}\times\mathbf{b}$ and $\mathbf{c}\times\mathbf{d}$ are parallel so $(\mathbf{a}\times\mathbf{b})\times(\mathbf{c}\times\mathbf{d}) = \mathbf{0}$

43. $\overrightarrow{PQ'}\times\mathbf{F} = \overrightarrow{PQ}\times\mathbf{F} + \overrightarrow{QQ'}\times\mathbf{F} = \overrightarrow{PQ}\times\mathbf{F}$, since $\mathbf{F}$ and $\overrightarrow{QQ'}$ are parallel.

EXERCISE SET 12.5

In many of the Exercises in this section other answers are also possible.

1. **(a)** L_1: $P(1,0), \mathbf{v} = \mathbf{j}, x = 1, y = t$
L_2: $P(0,1), \mathbf{v} = \mathbf{i}, x = t, y = 1$
L_3: $P(0,0), \mathbf{v} = \mathbf{i}+\mathbf{j}, x = t, y = t$

(b) L_1: $P(1,1,0), \mathbf{v} = \mathbf{k}, x = 1, y = 1, z = t$
L_2: $P(0,1,1), \mathbf{v} = \mathbf{i}, x = t, y = 1, z = 1$
L_3: $P(1,0,1), \mathbf{v} = \mathbf{j}, x = 1, y = t, z = 1$
L_4: $P(0,0,0), \mathbf{v} = \mathbf{i}+\mathbf{j}+\mathbf{k}, x = t,$
$y = t, z = t$

3. **(a)** $\overrightarrow{P_1P_2} = \langle 2,3\rangle$ so $x = 3 + 2t$, $y = -2 + 3t$ for the line; for the line segment add the condition $0 \le t \le 1$.

(b) $\overrightarrow{P_1P_2} = \langle -3,6,1\rangle$ so $x = 5 - 3t$, $y = -2 + 6t$, $z = 1 + t$ for the line; for the line segment add the condition $0 \le t \le 1$.

5. **(a)** $x = 2 + t, y = -3 - 4t$

(b) $x = t, y = -t, z = 1 + t$

7. **(a)** $\mathbf{r}_0 = 2\mathbf{i} - \mathbf{j}$ so $P(2,-1)$ is on the line, and $\mathbf{v} = 4\mathbf{i} - \mathbf{j}$ is parallel to the line.
(b) At $t = 0$, $P(-1,2,4)$ is on the line, and $\mathbf{v} = 5\mathbf{i} + 7\mathbf{j} - 8\mathbf{k}$ is parallel to the line.

9. **(a)** $\langle x,y\rangle = \langle -3,4\rangle + t\langle 1,5\rangle; \mathbf{r} = -3\mathbf{i} + 4\mathbf{j} + t(\mathbf{i} + 5\mathbf{j})$
(b) $\langle x,y,z\rangle = \langle 2,-3,0\rangle + t\langle -1,5,1\rangle; \mathbf{r} = 2\mathbf{i} - 3\mathbf{j} + t(-\mathbf{i} + 5\mathbf{j} + \mathbf{k})$

11. $x = -5 + 2t$, $y = 2 - 3t$

13. $2x + 2yy' = 0$, $y' = -x/y = -(3)/(-4) = 3/4$, $\mathbf{v} = 4\mathbf{i} + 3\mathbf{j}$; $x = 3 + 4t$, $y = -4 + 3t$

15. $x = -1 + 3t$, $y = 2 - 4t$, $z = 4 + t$

17. The line is parallel to the vector $\langle 2,-1,2\rangle$ so $x = -2 + 2t$, $y = -t$, $z = 5 + 2t$.

19. **(a)** $y = 0, 2 - t = 0, t = 2, x = 7$

(b) $x = 0, 1 + 3t = 0, t = -1/3, y = 7/3$

(c) $y = x^2, 2 - t = (1 + 3t)^2, 9t^2 + 7t - 1 = 0, t = \dfrac{-7 \pm \sqrt{85}}{18}, x = \dfrac{-1 \pm \sqrt{85}}{6}, y = \dfrac{43 \mp \sqrt{85}}{18}$

21. **(a)** $z = 0$ when $t = 3$ so the point is $(-2, 10, 0)$
(b) $y = 0$ when $t = -2$ so the point is $(-2, 0, -5)$
(c) x is always -2 so the line does not intersect the yz-plane

23. $(1+t)^2 + (3-t)^2 = 16$, $t^2 - 2t - 3 = 0$, $(t+1)(t-3) = 0$; $t = -1, 3$. The points of intersection are $(0,4,-2)$ and (4,0,6).

25. The lines intersect if we can find values of t_1 and t_2 that satisfy the equations $2 + t_1 = 2 + t_2$, $2 + 3t_1 = 3 + 4t_2$, and $3 + t_1 = 4 + 2t_2$. Solutions of the first two of these equations are $t_1 = -1$, $t_2 = -1$ which also satisfy the third equation so the lines intersect at $(1,-1,2)$.

27. The lines are parallel, respectively, to the vectors $\langle 7,1,-3\rangle$ and $\langle -1,0,2\rangle$. These vectors are not parallel so the lines are not parallel. The system of equations $1 + 7t_1 = 4 - t_2$, $3 + t_1 = 6$, and $5 - 3t_1 = 7 + 2t_2$ has no solution so the lines do not intersect.

29. The lines are parallel, respectively, to the vectors $\mathbf{v}_1 = \langle -2,1,-1\rangle$ and $\mathbf{v}_2 = \langle -4,2,-2\rangle$; $\mathbf{v}_2 = 2\mathbf{v}_1$, $\mathbf{v}_1$ and $\mathbf{v}_2$ are parallel so the lines are parallel.

31. $\overrightarrow{P_1P_2} = \langle 3,-7,-7\rangle$, $\overrightarrow{P_2P_3} = \langle -9,-7,-3\rangle$; these vectors are not parallel so the points do not lie on the same line.

33. If t_2 gives the point $\langle -1+3t_2, 9-6t_2\rangle$ on the second line, then $t_1 = 4 - 3t_2$ yields the point $\langle 3-(4-3t_2), 1+2(4-3t_2)\rangle = \langle -1+3t_2, 9-6t_2\rangle$ on the first line, so each point of L_2 is a point of L_1; the converse is shown with $t_2 = (4-t_1)/3$.

35. The line segment joining the points (1,0) and $(-3,6)$.

37. $A(3,0,1)$ and $B(2,1,3)$ are on the line, and (method of Exercise 25) $\overrightarrow{AP} = -5\mathbf{i}+\mathbf{j}$, $\overrightarrow{AB} = -\mathbf{i}+\mathbf{j}+2\mathbf{k}$, $\|\text{proj}_{\overrightarrow{AB}}\overrightarrow{AP}\| = |\overrightarrow{AP}\cdot\overrightarrow{AB}|/\|\overrightarrow{AB}\| = \sqrt{6}$ and $\|\overrightarrow{AP}\| = \sqrt{26}$, so distance $= \sqrt{26-6} = 2\sqrt{5}$. Using the method of Exercise 26, distance $= \dfrac{\|\overrightarrow{AP}\times\overrightarrow{AB}\|}{\|\overrightarrow{AB}\|} = 2\sqrt{5}$.

39. The vectors $\mathbf{v}_1 = -\mathbf{i}+2\mathbf{j}+\mathbf{k}$ and $\mathbf{v}_2 = 2\mathbf{i}-4\mathbf{j}-2\mathbf{k}$ are parallel to the lines, $\mathbf{v}_2 = -2\mathbf{v}_1$ so $\mathbf{v}_1$ and $\mathbf{v}_2$ are parallel. Let $t = 0$ to get the points $P(2,0,1)$ and $Q(1,3,5)$ on the first and second lines, respectively. Let $\mathbf{u} = \overrightarrow{PQ} = -\mathbf{i}+3\mathbf{j}+4\mathbf{k}$, $\mathbf{v} = \frac{1}{2}\mathbf{v}_2 = \mathbf{i}-2\mathbf{j}-\mathbf{k}$; $\mathbf{u}\times\mathbf{v} = 5\mathbf{i}+3\mathbf{j}-\mathbf{k}$; by the method of Exercise 26 of Section 12.4, distance $= \|\mathbf{u}\times\mathbf{v}\|/\|\mathbf{v}\| = \sqrt{35/6}$.

41. **(a)** The line is parallel to the vector $\langle x_1-x_0, y_1-y_0, z_1-z_0\rangle$ so $x = x_0 + (x_1-x_0)t$, $y = y_0 + (y_1-y_0)t$, $z = z_0 + (z_1-z_0)t$

(b) The line is parallel to the vector $\langle a,b,c\rangle$ so $x = x_1 + at$, $y = y_1 + bt$, $z = z_1 + ct$

43. **(a)** It passes through the point $(1,-3,5)$ and is parallel to $\mathbf{v} = 2\mathbf{i}+4\mathbf{j}+\mathbf{k}$

(b) $\langle x,y,z\rangle = \langle 1+2t, -3+4t, 5+t\rangle$

45. **(a)** Let $t = 3$ and $t = -2$, respectively, in the equations for L_1 and L_2.

(b) $\mathbf{u} = 2\mathbf{i}-\mathbf{j}-2\mathbf{k}$ and $\mathbf{v} = \mathbf{i}+3\mathbf{j}-\mathbf{k}$ are parallel to L_1 and L_2, $\cos\theta = \mathbf{u}\cdot\mathbf{v}/(\|\mathbf{u}\|\,\|\mathbf{v}\|) = 1/(3\sqrt{11})$, $\theta \approx 84°$.

(c) $\mathbf{u}\times\mathbf{v} = 7\mathbf{i}+7\mathbf{k}$ is perpendicular to both L_1 and L_2, and hence so is $\mathbf{i}+\mathbf{k}$, thus $x = 7+t$, $y = -1$, $z = -2+t$.

47. (0,1,2) is on the given line ($t = 0$) so $\mathbf{u} = \mathbf{j}-\mathbf{k}$ is a vector from this point to the point (0,2,1), $\mathbf{v} = 2\mathbf{i}-\mathbf{j}+\mathbf{k}$ is parallel to the given line. $\mathbf{u}\times\mathbf{v} = -2\mathbf{j}-2\mathbf{k}$, and hence $\mathbf{w} = \mathbf{j}+\mathbf{k}$, is perpendicular to both lines so $\mathbf{v}\times\mathbf{w} = -2\mathbf{i}-2\mathbf{j}+2\mathbf{k}$, and hence $\mathbf{i}+\mathbf{j}-\mathbf{k}$, is parallel to the line we seek. Thus $x = t$, $y = 2+t$, $z = 1-t$ are parametric equations of the line.

49. **(a)** When $t = 0$ the bugs are at $(4, 1, 2)$ and (0,1,1) so the distance between them is $\sqrt{4^2 + 0^2 + 1^2} = \sqrt{17}$ cm.

(b)

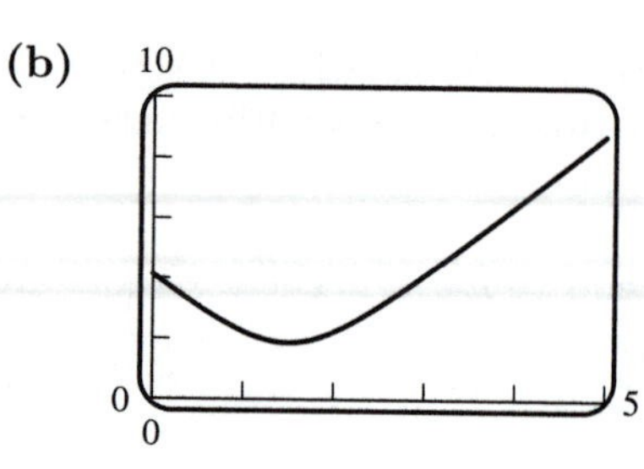

(c) The distance has a minimum value.

(d) Minimize D^2 instead of D (the distance between the bugs).
$D^2 = [t - (4 - t)]^2 + [(1 + t) - (1 + 2t)]^2 + [(1 + 2t) - (2 + t)]^2 = 6t^2 - 18t + 17$,
$d(D^2)/dt = 12t - 18 = 0$ when $t = 3/2$; the minimum
distance is $\sqrt{6(3/2)^2 - 18(3/2) + 17} = \sqrt{14}/2$ cm.

EXERCISE SET 12.6

1. $x = 3, y = 4, z = 5$

3. $(x - 2) + 4(y - 6) + 2(z - 1) = 0,\ x + 4y + 2z = 28$

5. $z = 0$

7. $\mathbf{n} = \mathbf{i} - \mathbf{j}, x - y = 0$

9. $\mathbf{n} = \mathbf{j} + \mathbf{k}, P(0, 1, 0), (y - 1) + z = 0, y + z = 1$

11. $\overrightarrow{P_1P_2} \times \overrightarrow{P_1P_3} = \langle 2, 1, 2\rangle \times \langle 3, -1, -2\rangle = \langle 0, 10, -5\rangle$,for convenience choose $\langle 0, 2, -1\rangle$ which is also normal to the plane. Use any of the given points to get $2y - z = 1$

13. **(a)** parallel, because $\langle 2, -8, -6\rangle$ and $\langle -1, 4, 3\rangle$ are parallel
(b) perpendicular, because $\langle 3, -2, 1\rangle$ and $\langle 4, 5, -2\rangle$ are orthogonal
(c) neither, because $\langle 1, -1, 3\rangle$ and $\langle 2, 0, 1\rangle$ are neither parallel nor orthogonal

15. **(a)** parallel, because $\langle 2, -1, -4\rangle$ and $\langle 3, 2, 1\rangle$ are orthogonal
(b) neither, because $\langle 1, 2, 3\rangle$ and $\langle 1, -1, 2\rangle$ are neither parallel nor orthogonal
(c) perpendicular, because $\langle 2, 1, -1\rangle$ and $\langle 4, 2, -2\rangle$ are parallel

17. **(a)** $3t - 2t + t - 5 = 0$, $t = 5/2$ so $x = y = z = 5/2$, the point of intersection is $(5/2, 5/2, 5/2)$
(b) $2(2 - t) + (3 + t) + t = 1$ has no solution so the line and plane do not intersect

19. $\mathbf{n}_1 = \langle 1, 0, 0\rangle, \mathbf{n}_2 = \langle 2, -1, 1\rangle, \mathbf{n}_1 \cdot \mathbf{n}_2 = 2$ so
$$\cos\theta = \frac{\mathbf{n}_1 \cdot \mathbf{n}_2}{\|\mathbf{n}_1\|\ \|\mathbf{n}_2\|} = \frac{2}{\sqrt{1}\sqrt{6}} = 2/\sqrt{6}, \theta = \cos^{-1}(2/\sqrt{6}) \approx 35^\circ$$

21. $\langle 4, -2, 7\rangle$ is normal to the desired plane and (0,0,0) is a point on it; $4x - 2y + 7z = 0$

23. Find two points P_1 and P_2 on the line of intersection of the given planes and then find an equation of the plane that contains P_1, P_2, and the given point $P_0(-1,4,2)$. Let (x_0, y_0, z_0) be on the line of intersection of the given planes; then $4x_0 - y_0 + z_0 - 2 = 0$ and $2x_0 + y_0 - 2z_0 - 3 = 0$, eliminate y_0 by addition of the equations to get $6x_0 - z_0 - 5 = 0$; if $x_0 = 0$ then $z_0 = -5$, if $x_0 = 1$ then $z_0 = 1$. Substitution of these values of x_0 and z_0 into either of the equations of the planes gives the corresponding values $y_0 = -7$ and $y_0 = 3$ so $P_1(0,-7,-5)$ and $P_2(1,3,1)$ are on the line of intersection of the planes. $\overrightarrow{P_0P_1} \times \overrightarrow{P_0P_2} = \langle 4, -13, 21\rangle$ is normal to the desired plane whose equation is $4x - 13y + 21z = -14$.

25. $\mathbf{n}_1 = \langle 2,1,1\rangle$ and $\mathbf{n}_2 = \langle 1,2,1\rangle$ are normals to the given planes, $\mathbf{n}_1 \times \mathbf{n}_2 = \langle -1,-1,3\rangle$ so $\langle 1,1,-3\rangle$ is normal to the desired plane whose equation is $x + y - 3z = 6$.

27. $\mathbf{n}_1 = \langle 2,-1,1\rangle$ and $\mathbf{n}_2 = \langle 1,1,-2\rangle$ are normals to the given planes,
$\mathbf{n}_1 \times \mathbf{n}_2 = \langle 1,5,3\rangle$ is normal to the desired plane whose equation is $x + 5y + 3z = -6$.

29. The plane is the perpendicular bisector of the line segment that joins $P_1(2,-1,1)$ and $P_2(3,1,5)$. The midpoint of the line segment is $(5/2, 0, 3)$ and $\overrightarrow{P_1P_2} = \langle 1,2,4\rangle$ is normal to the plane so an equation is $x + 2y + 4z = 29/2$.

31. The line is parallel to the line of intersection of the planes if it is parallel to both planes. Normals to the given planes are $\mathbf{n}_1 = \langle 1,-4,2\rangle$ and $\mathbf{n}_2 = \langle 2,3,-1\rangle$ so $\mathbf{n}_1 \times \mathbf{n}_2 = \langle -2,5,11\rangle$ is parallel to the line of intersection of the planes and hence parallel to the desired line whose equations are $x = 5 - 2t$, $y = 5t$, $z = -2 + 11t$.

33. $\mathbf{v} = \langle 0,1,1\rangle$ is parallel to the line.

(a) For any t, $6\cdot 0 + 4t - 4t = 0$, so $(0,t,t)$ is in the plane.

(b) $\mathbf{n} = \langle 5,-3,3\rangle$ is normal to the plane, $\mathbf{v}\cdot\mathbf{n} = 0$ so the line is parallel to the plane. (0,0,0) is on the line, $(0,0,1/3)$ is on the plane. The line is below the plane because (0,0,0) is below $(0,0,1/3)$.

(c) $\mathbf{n} = \langle 6,2,-2\rangle$, $\mathbf{v}\cdot\mathbf{n} = 0$ so the line is parallel to the plane. (0,0,0) is on the line, $(0,0,-3/2)$ is on the plane. The line is above the plane because (0,0,0) is above $(0,0,-3/2)$.

35. $\mathbf{v}_1 = \langle 1,2,-1\rangle$ and $\mathbf{v}_2 = \langle -1,-2,1\rangle$ are parallel, respectively, to the given lines and to each other so the lines are parallel. Let $t = 0$ to find the points $P_1(-2,3,4)$ and $P_2(3,4,0)$ that lie, respectively, on the given lines. $\mathbf{v}_1 \times \overrightarrow{P_1P_2} = \langle -7,-1,-9\rangle$ so $\langle 7,1,9\rangle$ is normal to the desired plane whose equation is $7x + y + 9z = 25$.

37. $\mathbf{n}_1 = \langle -2,3,7\rangle$ and $\mathbf{n}_2 = \langle 1,2,-3\rangle$ are normals to the planes, $\mathbf{n}_1 \times \mathbf{n}_2 = \langle -23,1,-7\rangle$ is parallel to the line of intersection. Let $z = 0$ in both equations and solve for x and y to get $x = -11/7$, $y = -12/7$ so $(-11/7, -12/7, 0)$ is on the line, a parametrization of which is
$x = -11/7 - 23t$, $y = -12/7 + t$, $z = -7t$.

39. $D = |2(1) - 2(-2) + (3) - 4|/\sqrt{4+4+1} = 5/3$

41. (0,0,0) is on the first plane so $D = |6(0) - 3(0) - 3(0) - 5|/\sqrt{36+9+9} = 5/\sqrt{54}$.

43. (1,3,5) and (4,6,7) are on L_1 and L_2, respectively. $\mathbf{v}_1 = \langle 7,1,-3\rangle$ and $\mathbf{v}_2 = \langle -1,0,2\rangle$ are, respectively, parallel to L_1 and L_2, $\mathbf{v}_1 \times \mathbf{v}_2 = \langle 2,-11,1\rangle$ so the plane $2x - 11y + z + 51 = 0$ contains L_2 and is parallel to L_1, $D = |2(1) - 11(3) + (5) + 51|/\sqrt{4+121+1} = 25/\sqrt{126}$.

45. The distance between $(2, 1, -3)$ and the plane is $|2-3(1)+2(-3)-4|/\sqrt{1+9+4} = 11/\sqrt{14}$ which is the radius of the sphere; an equation is $(x-2)^2 + (y-1)^2 + (z+3)^2 = 121/14$.

47. $\mathbf{v} = \langle 1, 2, -1 \rangle$ is parallel to the line, $\mathbf{n} = \langle 2, -2, -2 \rangle$ is normal to the plane, $\mathbf{v} \cdot \mathbf{n} = 0$ so $\mathbf{v}$ is parallel to the plane because $\mathbf{v}$ and $\mathbf{n}$ are perpendicular. $(-1, 3, 0)$ is on the line so
$D = |2(-1) - 2(3) - 2(0) + 3|/\sqrt{4+4+4} = 5/\sqrt{12}$

49. $D = |2(-3) + (5) - 1|/\sqrt{4+1} = 2/\sqrt{5}$

EXERCISE SET 12.7

1. (a) elliptic paraboloid, $a = 2, b = 3$
(b) hyperbolic paraboloid, $a = 1, b = 5$
(c) hyperboloid of one sheet, $a = b = c = 4$
(d) circular cone, $a = b = 1$
(e) elliptic paraboloid, $a = 2, b = 1$
(f) hyperboloid of two sheets, $a = b = c = 1$

3. (a) $-z = x^2 + y^2$, circular paraboloid opening down the negative z-axis

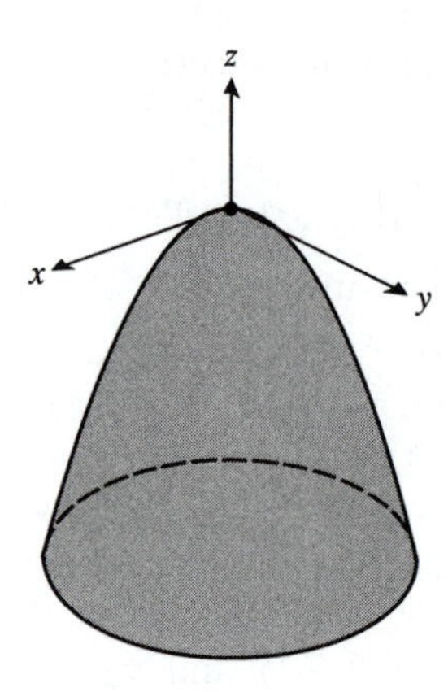

(b) $z = x^2 + y^2$, circular paraboloid, no change

(c) $z = x^2 + y^2$, circular paraboloid, no change
(d) $z = x^2 + y^2$, circular paraboloid, no change

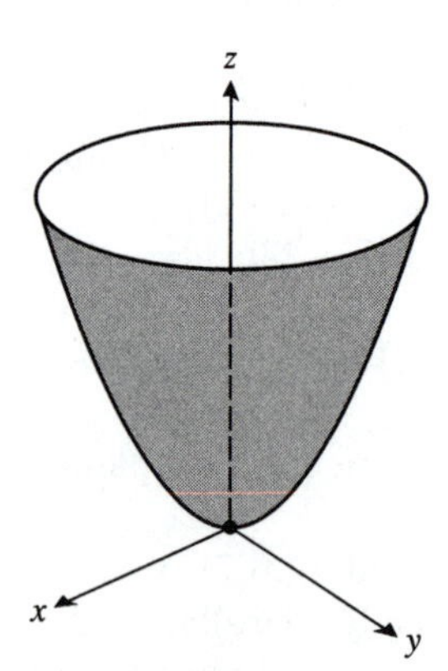

(e) $x = y^2 + z^2$, circular paraboloid opening along the positive x-axis

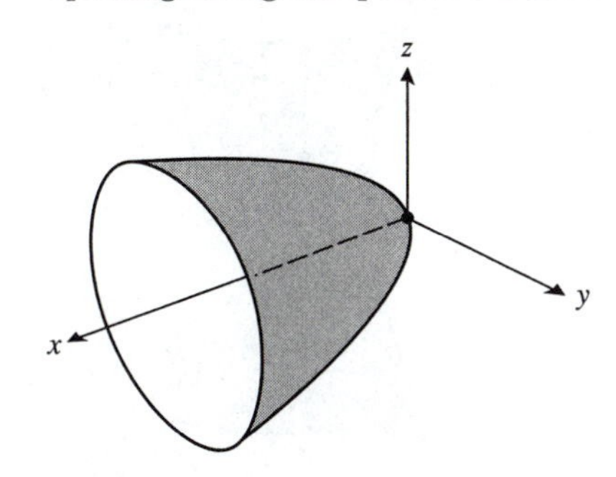

(f) $y = x^2 + z^2$, circular paraboloid opening along the positive y-axis

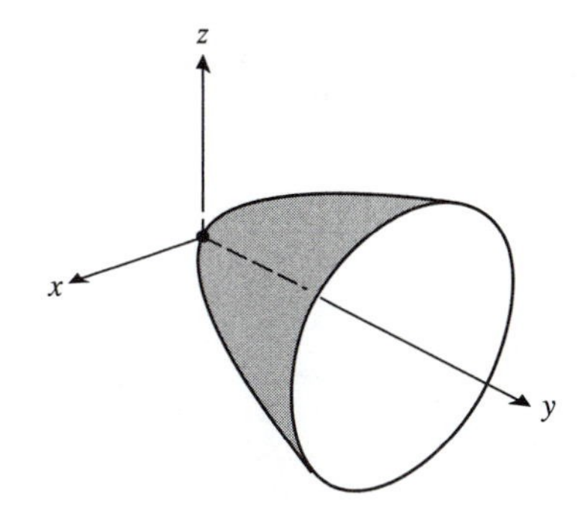

5. (a) hyperboloid of one sheet, axis is y-axis

(b) hyperboloid of two sheets separated by yz-plane

(c) elliptic paraboloid opening along the positive x-axis

(d) elliptic cone with x-axis as axis

(e) hyperbolic paraboloid straddling the z-axis

(f) paraboloid opening along the negative y-axis

7. (a) $x = 0 : \dfrac{y^2}{25} + \dfrac{z^2}{4} = 1; y = 0 : \dfrac{x^2}{9} + \dfrac{z^2}{4} = 1;$

$z = 0 : \dfrac{x^2}{9} + \dfrac{y^2}{25} = 1$

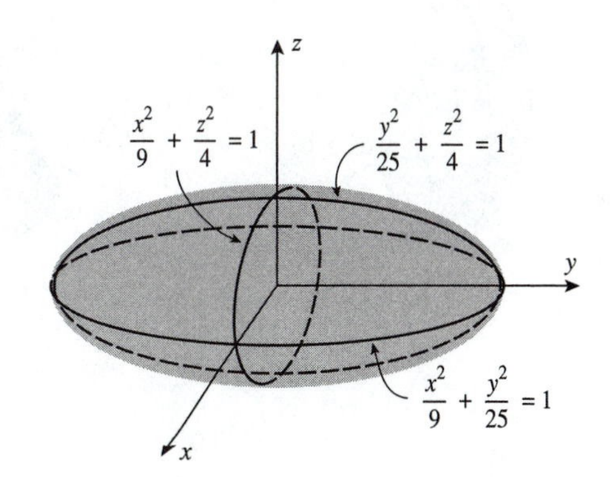

(b) $x = 0 : z = 4y^2; y = 0 : z = x^2;$

$z = 0 : x = y = 0$

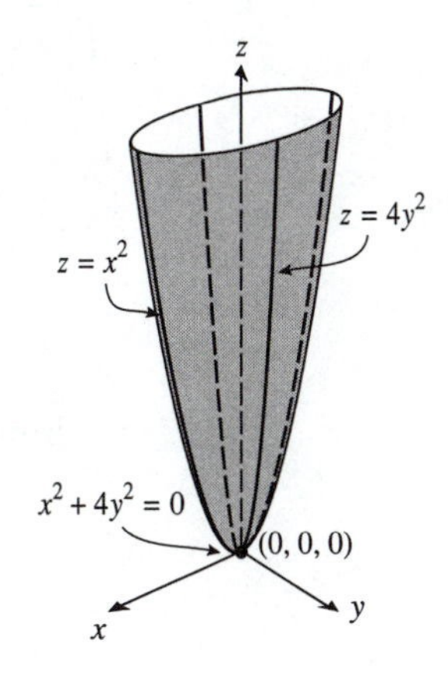

(c) $x = 0: \dfrac{y^2}{16} - \dfrac{z^2}{4} = 1; y = 0: \dfrac{x^2}{9} - \dfrac{z^2}{4} = 1;$

$z = 0: \dfrac{x^2}{9} + \dfrac{y^2}{16} = 1$

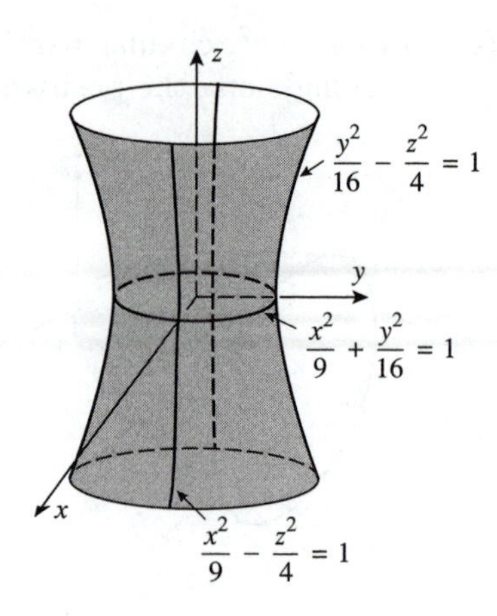

9. **(a)** $4x^2 + z^2 = 3$; ellipse **(b)** $y^2 + z^2 = 3$; circle **(c)** $y^2 + z^2 = 20$; circle
(d) $9x^2 - y^2 = 20$; hyperbola **(e)** $z = 9x^2 + 16$; parabola **(f)** $9x^2 + 4y^2 = 4$; ellipse

11.

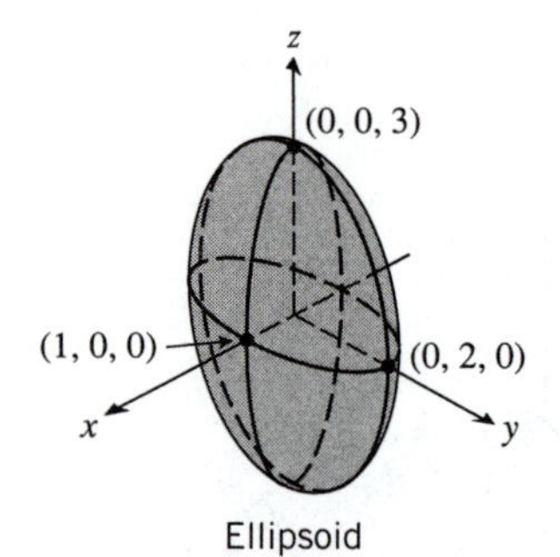

Ellipsoid

13.

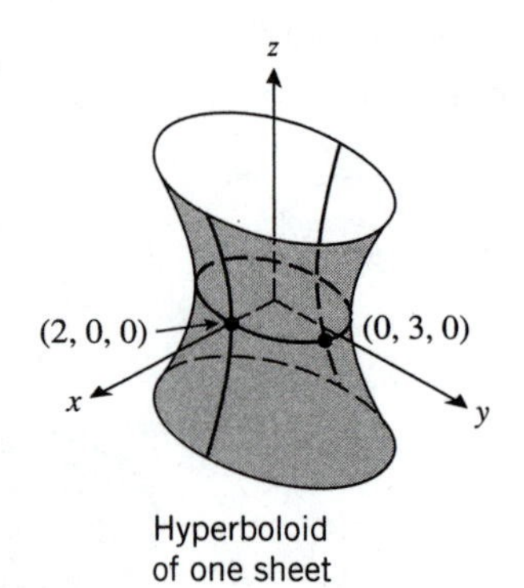

Hyperboloid of one sheet

15.

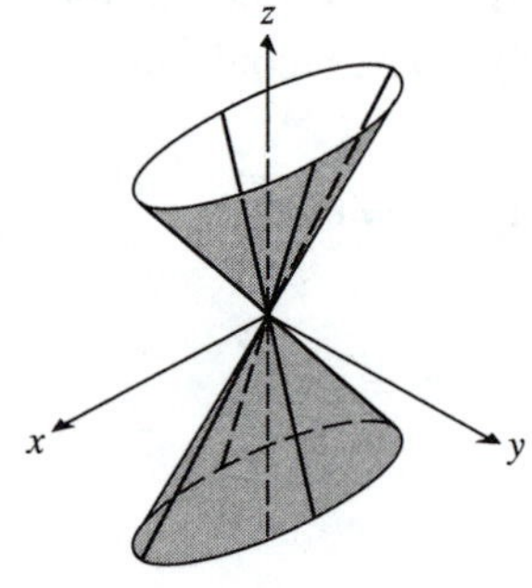

Elliptic cone

17.

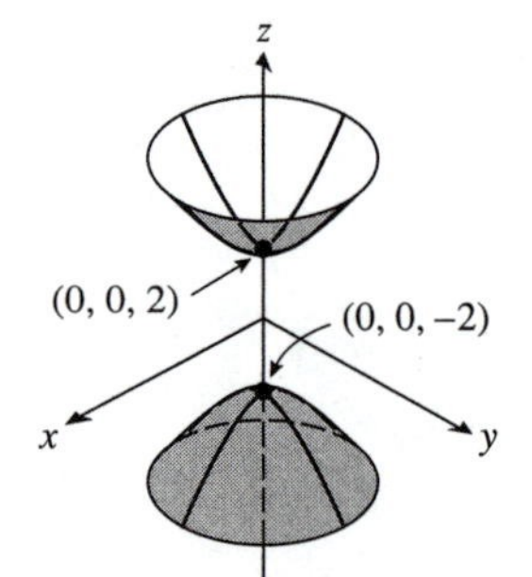

Hyperboloid of two sheets

19.

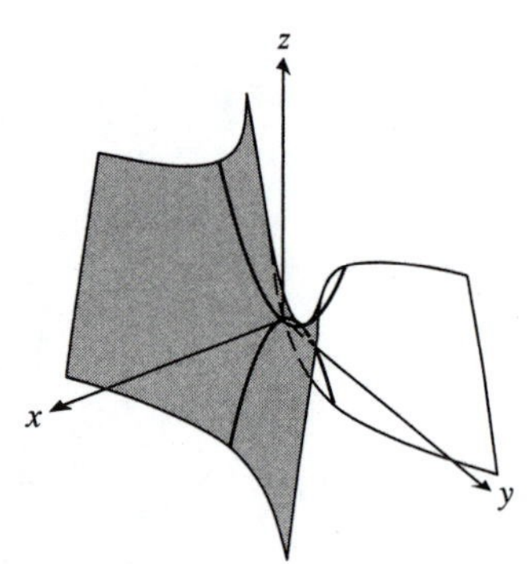

Hyperbolic paraboloid

21.

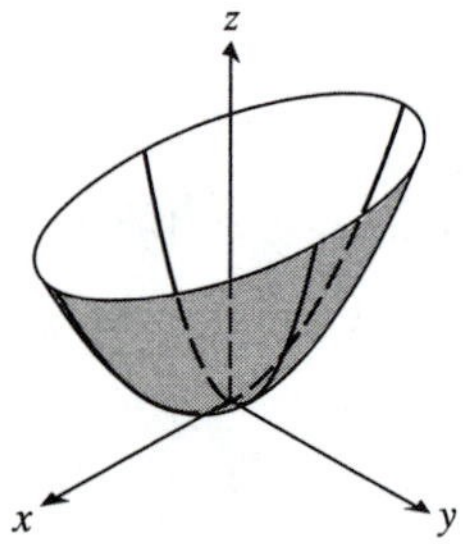

Elliptic paraboloid

23.

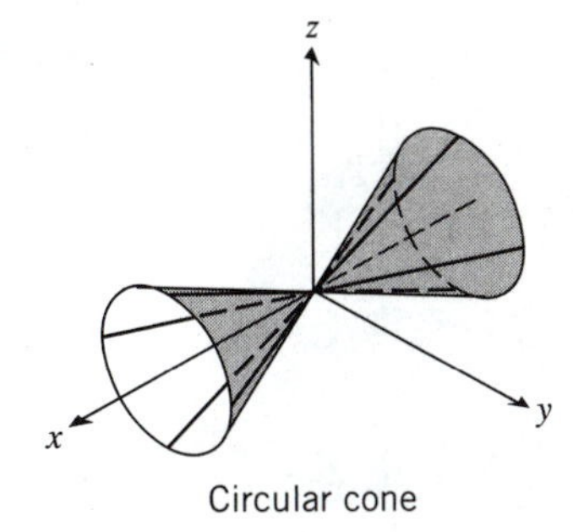

Circular cone

25.

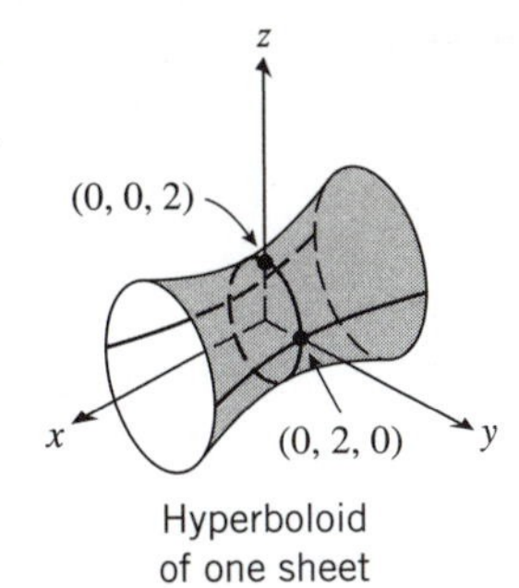

Hyperboloid of one sheet

27.

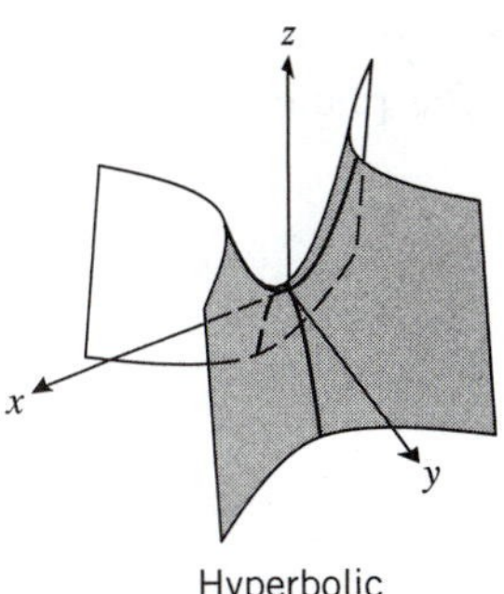

Hyperbolic paraboloid

29.

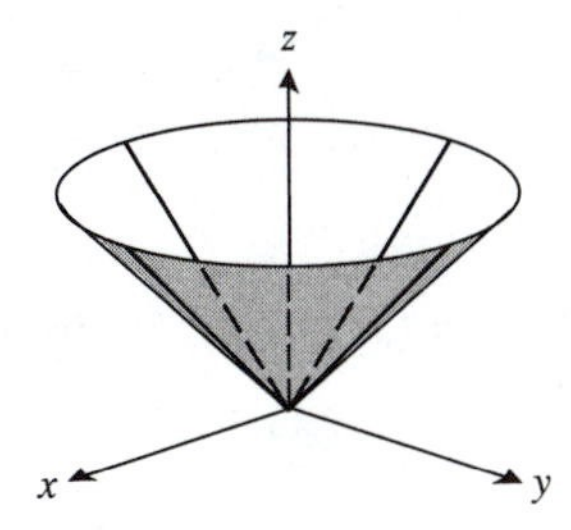

31.

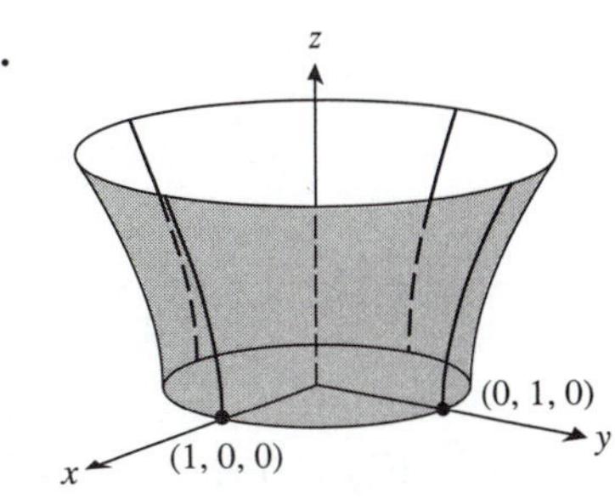

33.

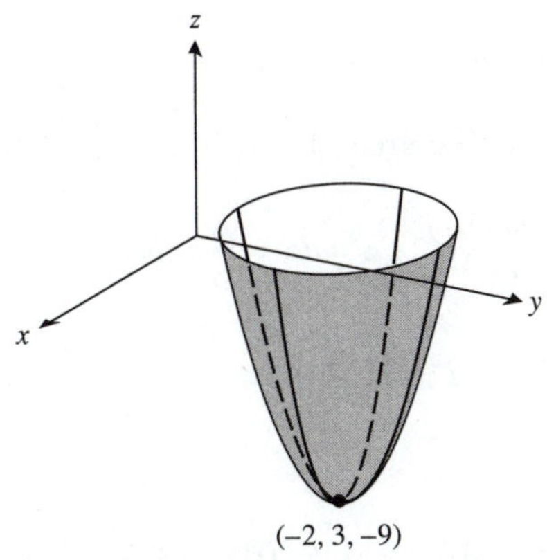

Circular paraboloid

35.

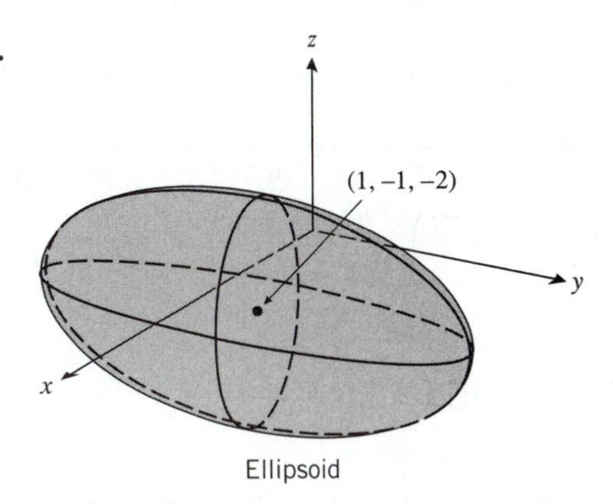

Ellipsoid

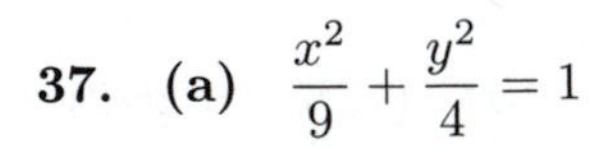

37. **(a)** $\dfrac{x^2}{9} + \dfrac{y^2}{4} = 1$ **(b)** $6, 4$ **(c)** $(\pm\sqrt{5}, 0, \sqrt{2})$

(d) The focal axis is parallel to the x-axis.

39. **(a)** $\dfrac{y^2}{4} - \dfrac{x^2}{4} = 1$ **(b)** $(0, \pm 2, 4)$ **(c)** $(0, \pm 2\sqrt{2}, 4)$

(d) The focal axis is parallel to the y-axis.

41. **(a)** $z + 4 = y^2$ **(b)** $(2, 0, -4)$ **(c)** $(2, 0, -15/4)$

(d) The focal axis is parallel to the z-axis.

43. $x^2+y^2=4-x^2-y^2, x^2+y^2=2$; circle of radius $\sqrt{2}$ in the plane $z=2$, centered at $(0,0,2)$

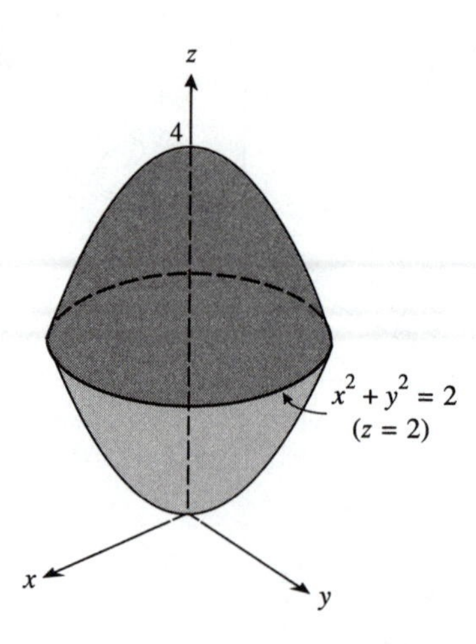

45. $y=4(x^2+z^2)$

47. $|z-(-1)|=\sqrt{x^2+y^2+(z-1)^2}$, $z^2+2z+1=x^2+y^2+z^2-2z+1$, $z=(x^2+y^2)/4$; circular paraboloid

49. If $z=0$, $\dfrac{x^2}{a^2}+\dfrac{y^2}{a^2}=1$; if $y=0$ then $\dfrac{x^2}{a^2}+\dfrac{z^2}{c^2}=1$; since $c<a$ the major axis has length $2a$, the minor axis length $2c$.

51. Each slice perpendicular to the z-axis for $|z|<c$ is an ellipse whose equation is $\dfrac{x^2}{a^2}+\dfrac{y^2}{b^2}=\dfrac{c^2-z^2}{c^2}$, or $\dfrac{x^2}{(a^2/c^2)(c^2-z^2)}+\dfrac{y^2}{(b^2/c^2)(c^2-z^2)}=1$, the area of which is $\pi\left(\dfrac{a}{c}\sqrt{c^2-z^2}\right)\left(\dfrac{b}{c}\sqrt{c^2-z^2}\right)=\pi\dfrac{ab}{c^2}\left(c^2-z^2\right)$ so $V=2\displaystyle\int_0^c \pi\frac{ab}{c^2}\left(c^2-z^2\right)dz=\frac{4}{3}\pi abc$.

EXERCISE SET 12.8

1. **(a)** $(8,\pi/6,-4)$ **(b)** $(5\sqrt{2},3\pi/4,6)$ **(c)** $(2,\pi/2,0)$ **(d)** $(8,5\pi/3,6)$

3. **(a)** $(2\sqrt{3},2,3)$ **(b)** $(-4\sqrt{2},4\sqrt{2},-2)$ **(c)** $(5,0,4)$ **(d)** $(-7,0,-9)$

5. **(a)** $(2\sqrt{2},\pi/3,3\pi/4)$ **(b)** $(2,7\pi/4,\pi/4)$ **(c)** $(6,\pi/2,\pi/3)$ **(d)** $(10,5\pi/6,\pi/2)$

7. **(a)** $(5\sqrt{6}/4,5\sqrt{2}/4,5\sqrt{2}/2)$ **(b)** $(7,0,0)$
(c) $(0,0,1)$ **(d)** $(0,-2,0)$

9. **(a)** $(2\sqrt{3},\pi/6,\pi/6)$ **(b)** $(\sqrt{2},\pi/4,3\pi/4)$
(c) $(2,3\pi/4,\pi/2)$ **(d)** $(4\sqrt{3},1,2\pi/3)$

11. **(a)** $(5\sqrt{3}/2,\pi/4,-5/2)$ **(b)** $(0,7\pi/6,-1)$
(c) $(0,0,3)$ **(d)** $(4,\pi/6,0)$

15.

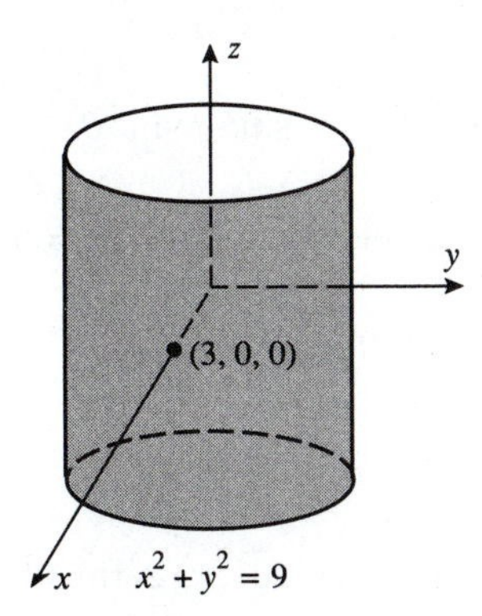

17.

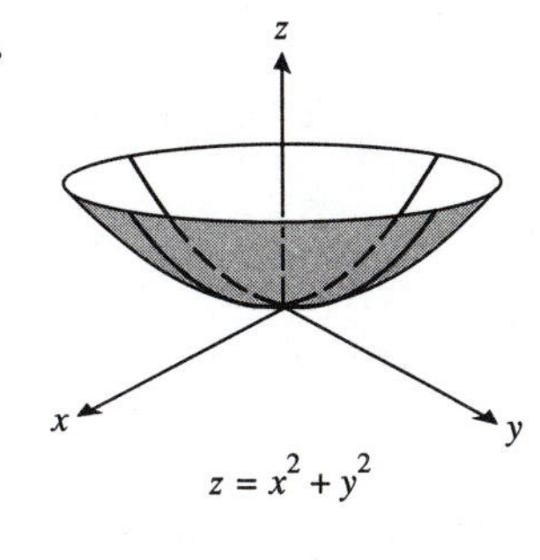

19.

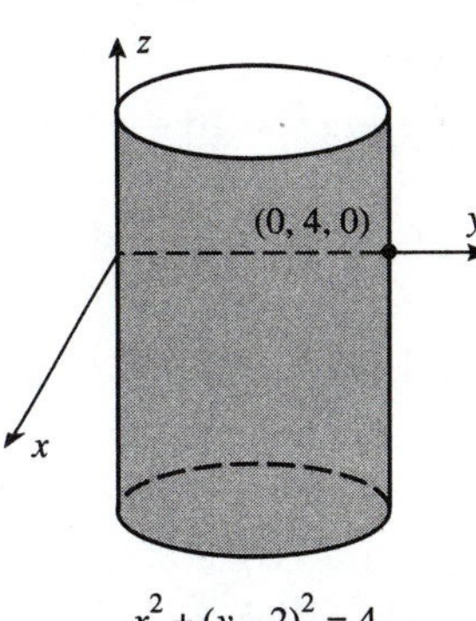

21.

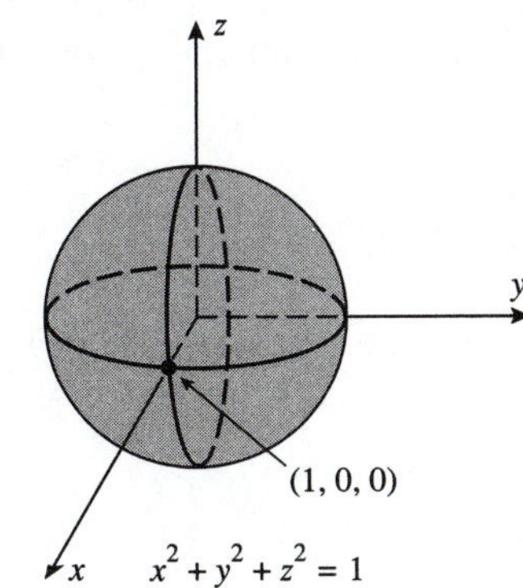

23.

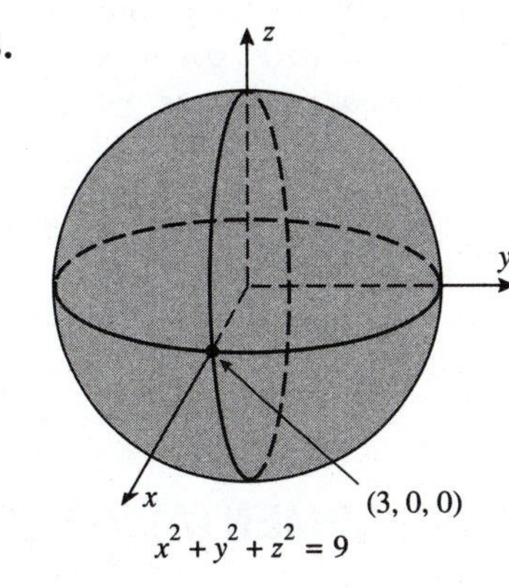

25.

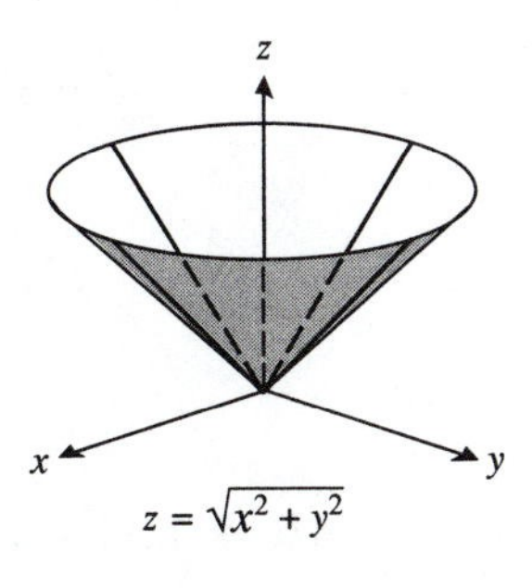

27.

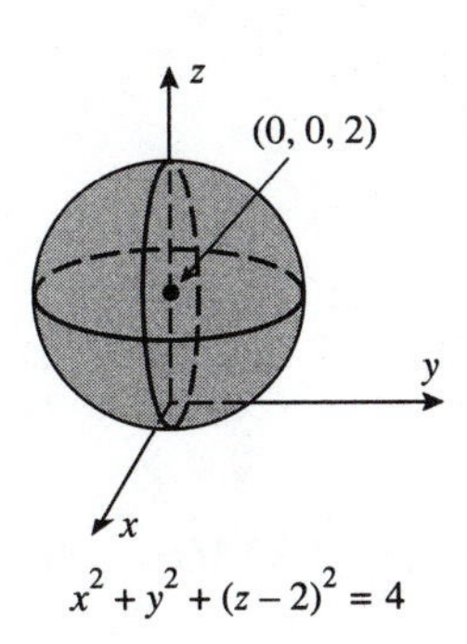

29.

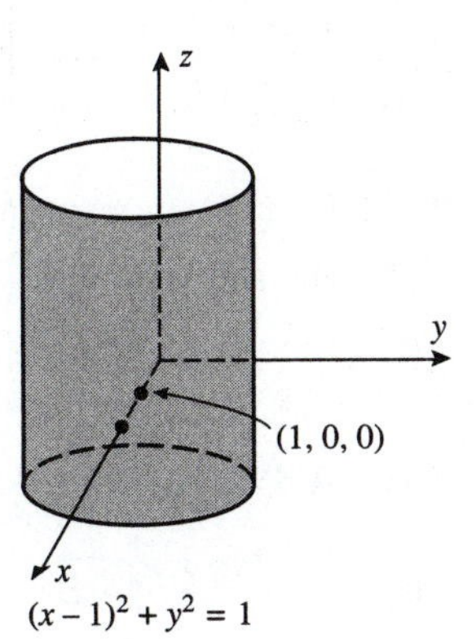

31. **(a)** $z = 3$ **(b)** $\rho\cos\phi = 3, \rho = 3\sec\phi$

33. **(a)** $z = 3r^2$ **(b)** $\rho\cos\phi = 3\rho^2\sin^2\phi, \rho = \frac{1}{3}\csc\phi\cot\phi$

35. **(a)** $r = 2$ **(b)** $\rho\sin\phi = 2, \rho = 2\csc\phi$

37. **(a)** $r^2 + z^2 = 9$ **(b)** $\rho = 3$

39. **(a)** $2r\cos\theta + 3r\sin\theta + 4z = 1$
(b) $2\rho\sin\phi\cos\theta + 3\rho\sin\phi\sin\theta + 4\rho\cos\phi = 1$

41. **(a)** $r^2\cos^2\theta = 16 - z^2$

(b) $x^2 = 16 - z^2$, $x^2 + y^2 + z^2 = 16 + y^2$, $\rho^2 = 16 + \rho^2\sin^2\phi\sin^2\theta$, $\rho^2\left(1 - \sin^2\phi\sin^2\theta\right) = 16$

43. all points on or above the paraboloid $z = x^2 + y^2$, that are also on or below the plane $z = 4$

45. all points on or between concentric spheres of radii 1 and 3 centered at the origin

47. $\theta = \pi/6$, $\phi = \pi/6$, spherical $(4000, \pi/6, \pi/6)$, rectangular $\left(1000\sqrt{3}, 1000, 2000\sqrt{3}\right)$

49. **(a)** $(10, \pi/2, 1)$ **(b)** $(0, 10, 1)$ **(c)** $(\sqrt{101}, \pi/2, \tan^{-1} 10)$

51. Using spherical coordinates: for point A, $\theta_A = 360° - 60° = 300°, \phi_A = 90° - 40° = 50°$; for point B, $\theta_B = 360° - 40° = 320°, \phi_B = 90° - 20° = 70°$. Unit vectors directed from the origin to the points A and B, respectively, are

$$\mathbf{u}_A = \sin 50° \cos 300°\mathbf{i} + \sin 50° \sin 300°\mathbf{j} + \cos 50°\mathbf{k},$$
$$\mathbf{u}_B = \sin 70° \cos 320°\mathbf{i} + \sin 70° \sin 320°\mathbf{j} + \cos 70°\mathbf{k}$$

The angle α between $\mathbf{u}_A$ and $\mathbf{u}_B$ is $\alpha = \cos^{-1}(\mathbf{u}_A \cdot \mathbf{u}_B) \approx 0.459486$ so the shortest distance is $6370\alpha \approx 2927$ km.

CHAPTER 12 SUPPLEMENTARY EXERCISES

3. **(b)** $x = \cos 120° = -1/2, y = \pm\sin 120° = \pm\sqrt{3}/2$

(d) true: $\|\mathbf{u} \times \mathbf{v}\| = \|\mathbf{u}\|\|\mathbf{v}\||\sin(\theta)| = 1$

5. **(b)** $(y, x, z), (x, z, y), (z, y, x)$

(c) the set of points $\{(5, \theta, 1)\}, 0 \le \theta \le 2\pi$

(d) the set of points $\{(\rho, \pi/4, 0)\}, 0 \le \rho < +\infty$

7. **(a)** $\overrightarrow{AB} = -\mathbf{i} + 2\mathbf{j} + 2\mathbf{k}$, $\overrightarrow{AC} = \mathbf{i} + \mathbf{j} - \mathbf{k}$, $\overrightarrow{AB} \times \overrightarrow{AC} = -4\mathbf{i} + \mathbf{j} - 3\mathbf{k}$, area $= \dfrac{1}{2}\|\overrightarrow{AB} \times \overrightarrow{AC}\| = \sqrt{26}/2$

(b) area $= \dfrac{1}{2}h\|\overrightarrow{AB}\| = \dfrac{3}{2}h = \dfrac{1}{2}\sqrt{26}$, $h = \sqrt{26}/3$

9. **(a)** $\mathbf{a} \cdot \mathbf{b} = 0$, $4c + 3 = 0$, $c = -3/4$

(b) Use $\mathbf{a} \cdot \mathbf{b} = \|\mathbf{a}\|\;\|\mathbf{b}\|\cos\theta$ to get $4c + 3 = \sqrt{c^2 + 1}(5)\cos(\pi/4)$, $4c + 3 = 5\sqrt{c^2 + 1}/\sqrt{2}$ Square both sides and rearrange to get $7c^2 + 48c - 7 = 0$, $(7c - 1)(c + 7) = 0$ so $c = -7$ (invalid) or $c = 1/7$.

(c) Proceed as in (b) with $\theta = \pi/6$ to get $11c^2 - 96c + 39 = 0$ and use the quadratic formula to get $c = \left(48 \pm 25\sqrt{3}\right)/11$.

(d) $\mathbf{a}$ must be a scalar multiple of $\mathbf{b}$, so $c\mathbf{i} + \mathbf{j} = k(4\mathbf{i} + 3\mathbf{j}), k = 1/3, c = 4/3$.

11. **(a)** the plane through the origin which is perpendicular to $\mathbf{r}_0$

(b) the plane through the tip of $\mathbf{r}_0$ which is perpendicular to $\mathbf{r}_0$

13. Since $\overrightarrow{AC} \cdot (\overrightarrow{AB} \times \overrightarrow{AD}) = \overrightarrow{AC} \cdot (\overrightarrow{AB} \times \overrightarrow{CD}) + \overrightarrow{AC} \cdot (\overrightarrow{AB} \times \overrightarrow{AC}) = \mathbf{0} + \mathbf{0} = \mathbf{0}$, the volume of the parallelopiped determined by $\overrightarrow{AB}, \overrightarrow{AC}$, and $\overrightarrow{AD}$ is zero, thus A, B, C, and D are coplanar (lie in the same plane). Since $\overrightarrow{AB} \times \overrightarrow{CD} \neq \mathbf{0}$, the lines are not parallel. Hence they must intersect.

15. **(a)** false, for example $\mathbf{i} \cdot \mathbf{j} = 0$ **(b)** false, for example $\mathbf{i} \times \mathbf{i} = \mathbf{0}$
(c) true; $0 = \|\mathbf{u}\| \cdot \|\mathbf{v}\| \cos\theta = \|\mathbf{u}\| \cdot \|\mathbf{v}\| \sin\theta$, so either $\mathbf{u} = \mathbf{0}$ or $\mathbf{v} = \mathbf{0}$ since $\cos\theta = \sin\theta = 0$ is impossible.

17. $\|\mathbf{u} - \mathbf{v}\|^2 = (\mathbf{u} - \mathbf{v}) \cdot (\mathbf{u} - \mathbf{v}) = \|\mathbf{u}\|^2 + \|\mathbf{v}\|^2 - 2\|\mathbf{u}\|\|\mathbf{v}\| \cos\theta = 2(1 - \cos\theta) = 4\sin^2(\theta/2)$, so $\|\mathbf{u} - \mathbf{v}\| = 2\sin(\theta/2)$

19. **(a)** $\langle 2, 1, -1\rangle \times \langle 1, 2, 1\rangle = \langle 3, -3, 3\rangle$, so the line is parallel to $\mathbf{i} - \mathbf{j} + \mathbf{k}$. By inspection, $(0, 2, -1)$ lies on both planes, so the line has an equation $\mathbf{r} = 2\mathbf{j} - \mathbf{k} + t(\mathbf{i} - \mathbf{j} + \mathbf{k})$, that is, $x = t, y = 2 - t, z = -1 + t$.
(b) $\cos\theta = \dfrac{\langle 2, 1, -1\rangle \cdot \langle 1, 2, 1\rangle}{\|\langle 2, 1, -1\rangle\|\|\langle 1, 2, 1\rangle\|} = 1/2$, so $\theta = \pi/3$

21. $5\langle \cos 60°, \cos 120°, \cos 135°\rangle = \langle 5/2, -5/2, -5\sqrt{2}/2\rangle$

23. **(a)** $(x - 3)^2 + 4(y + 1)^2 - (z - 2)^2 = 9$, hyperboloid of one sheet
(b) $(x + 3)^2 + (y - 2)^2 + (z + 6)^2 = 49$, sphere
(c) $(x - 1)^2 + (y + 2)^2 - z^2 = 0$, circular cone

25. **(a)** $r^2 = z; \rho^2 \sin^2\phi = \rho\cos\phi, \rho = \cot\phi \csc\phi$
(b) $r^2(\cos^2\theta - \sin^2\theta) - z^2 = 0, z^2 = r^2\cos 2\theta$;
$\rho^2 \sin^2\phi \cos^2\theta - \rho^2 \sin^2\phi \sin^2\theta - \rho^2\cos^2\phi = 0, \cos 2\theta = \cot^2\phi$
$\sin^2\phi(\cos^2\theta - \sin^2\theta) - \cos^2\phi = 0$

27. **(a)**

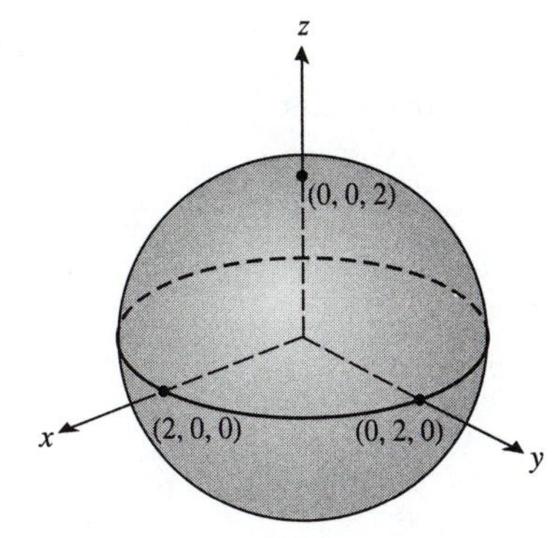

(b)

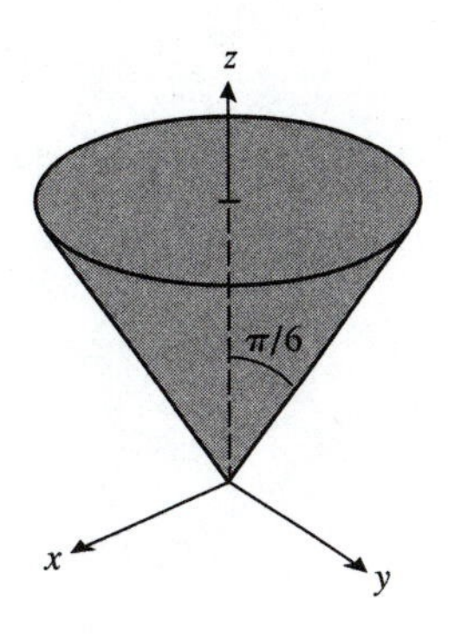

(c)

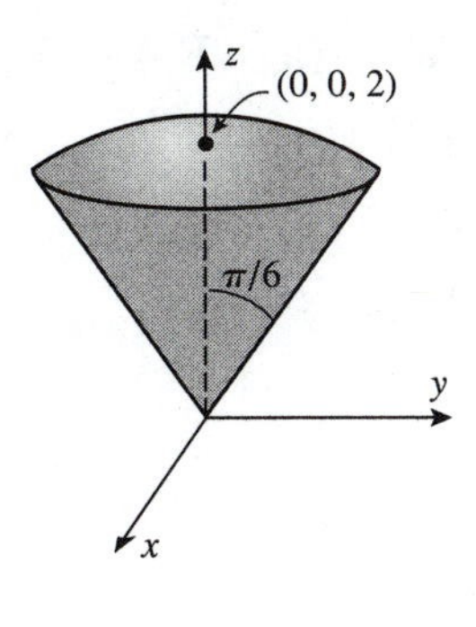

29. **(a)**

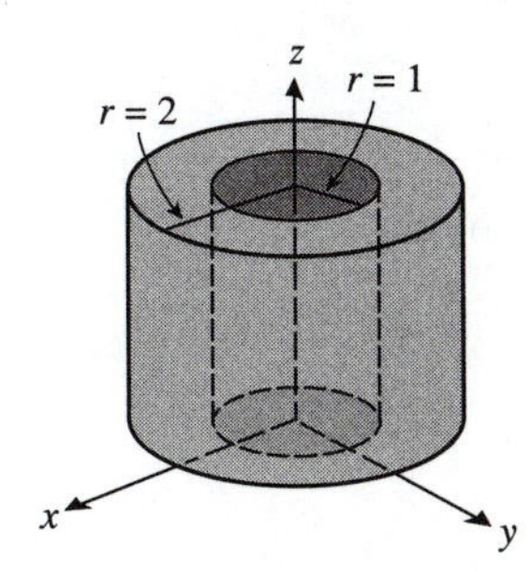

(b)

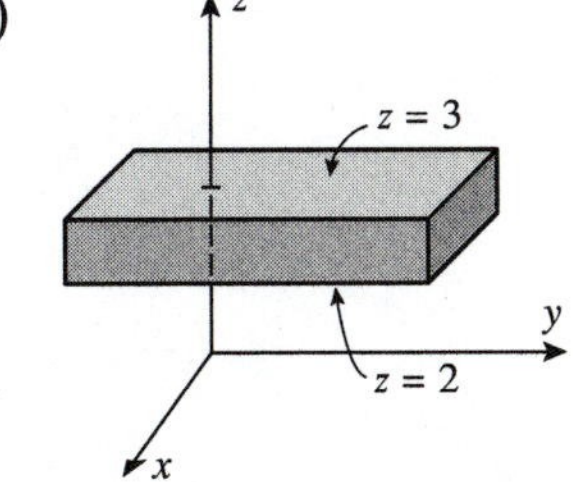

(c)

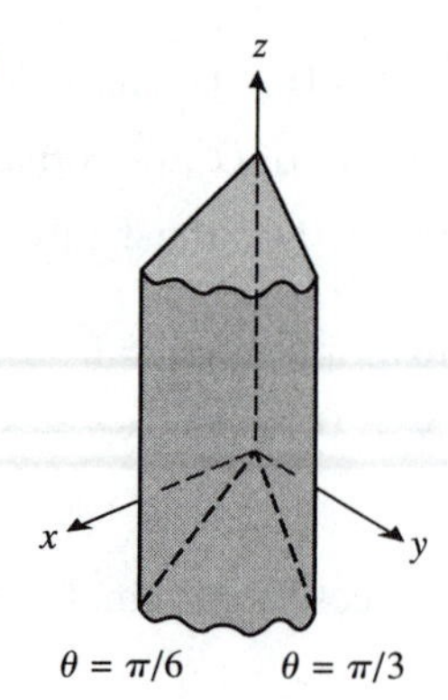

(d)

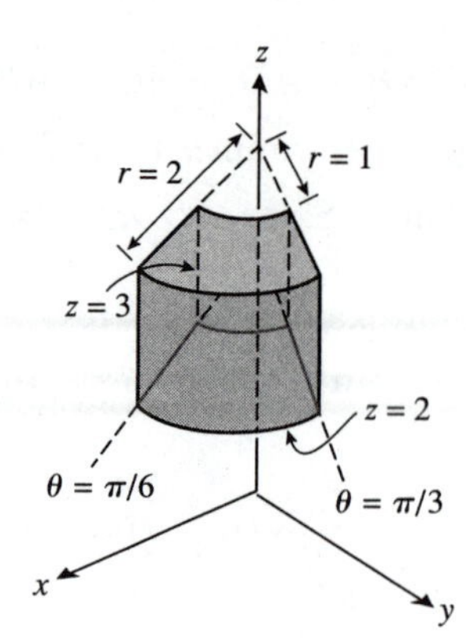

31. **(a)** At $x = c$ the trace of the surface is the circle $y^2 + z^2 = [f(c)]^2$, so the surface is given by $y^2 + z^2 = [f(x)]^2$

(b) $y^2 + z^2 = e^{2x}$

(c) $y^2 + z^2 = 4 - \frac{3}{4}x^2$, so let $f(x) = \sqrt{4 - \frac{3}{4}x^2}$

33.

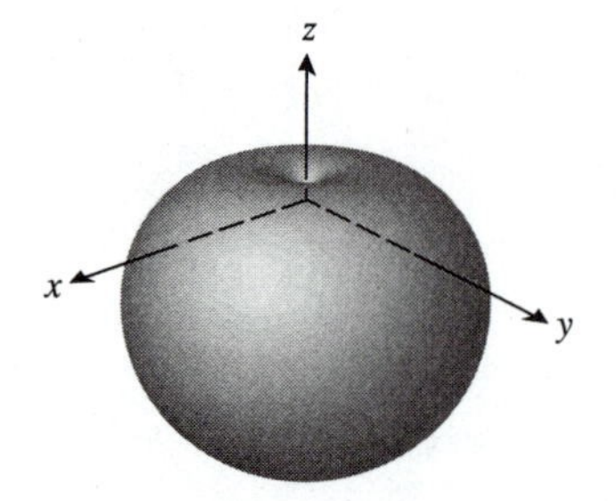

35. $\mathbf{F} = \mathbf{F}_1 + \mathbf{F}_2 = 2\mathbf{i} - \mathbf{j} + 3\mathbf{k}, \overrightarrow{PQ} = \mathbf{i} + 4\mathbf{j} - 3\mathbf{k}, W = \mathbf{F} \cdot \overrightarrow{PQ} = -11$ N·m $= -11$ J

37. **(a)** $\mathbf{F} = -6\mathbf{i} + 3\mathbf{j} - 6\mathbf{k}$

(b) $\overrightarrow{OA} = \langle 5, 0, 2\rangle$, so the vector moment is $\overrightarrow{OA} \times \mathbf{F} = -6\mathbf{i} + 18\mathbf{j} + 15\mathbf{k}$

CHAPTER 13
Vector-Valued Functions

EXERCISE SET 13.1

1. $(-\infty, +\infty)$; $\mathbf{r}(\pi) = -\mathbf{i} - 3\pi\mathbf{j}$

3. $[2, +\infty)$; $\mathbf{r}(3) = -\mathbf{i} - \ln 3\mathbf{j} + \mathbf{k}$

5. $\mathbf{r} = 3\cos t\mathbf{i} + (t + \sin t)\mathbf{j}$

7. $\mathbf{r} = 2t\mathbf{i} + 2\sin 3t\mathbf{j} + 5\cos 3t\mathbf{k}$

9. $x = 3t^2$, $y = -2$

11. $x = 2t - 1$, $y = -3\sqrt{t}$, $z = \sin 3t$

13. the line in 2-space through the point $(2, 0)$ and parallel to the vector $-3\mathbf{i} - 4\mathbf{j}$

15. the line in 3-space through the point $(0, -3, 1)$ and parallel to the vector $2\mathbf{i} + 3\mathbf{k}$

17. an ellipse in the plane $z = -1$, center at $(0, 0, -1)$, major axis of length 6 parallel to x-axis, minor axis of length 4 parallel to y-axis

19. **(a)** The line is parallel to the vector $-2\mathbf{i} + 3\mathbf{j}$; the slope is $-3/2$.

(b) $y = 0$ in the xz-plane so $1 - 2t = 0$, $t = 1/2$ thus $x = 2 + 1/2 = 5/2$ and $z = 3(1/2) = 3/2$; the coordinates are $(5/2, 0, 3/2)$.

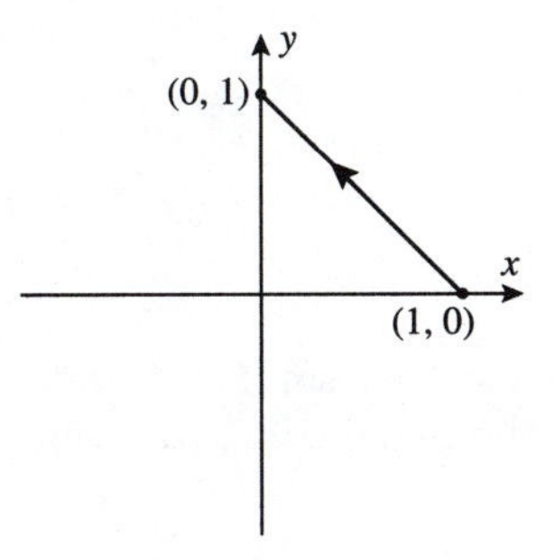

(b)

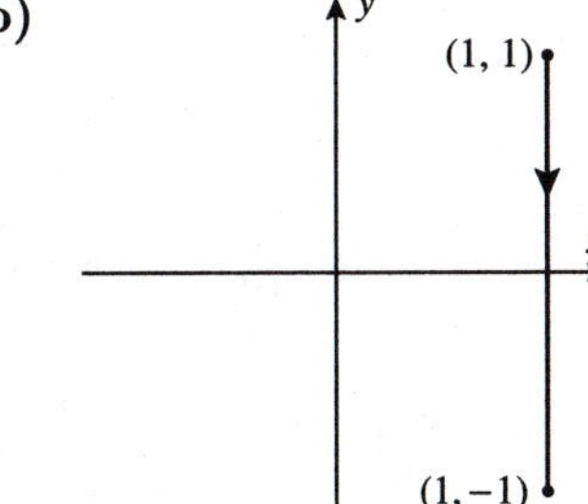

23. $\mathbf{r} = (1 - t)(3\mathbf{i} + 4\mathbf{j}), 0 \le t \le 1$

25. $x = 2$

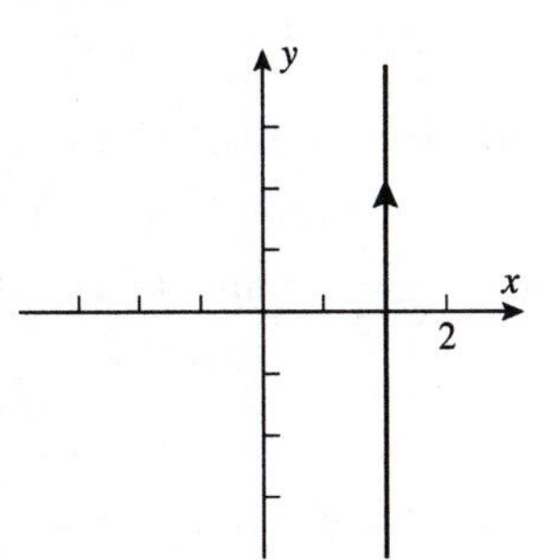

27. $(x - 1)^2 + (y - 3)^2 = 1$

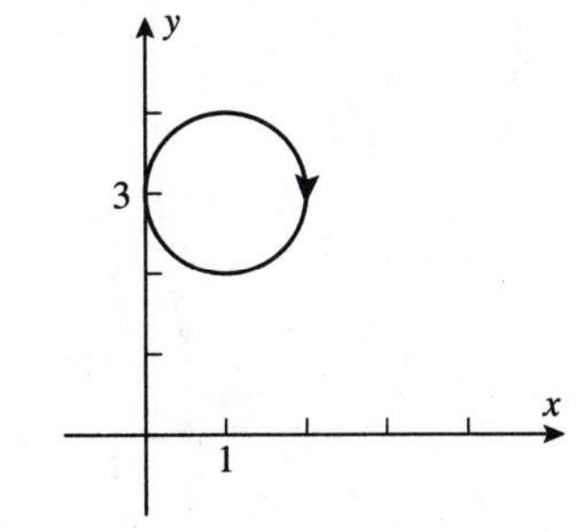

29. $x^2 - y^2 = 1$, $x \ge 1$

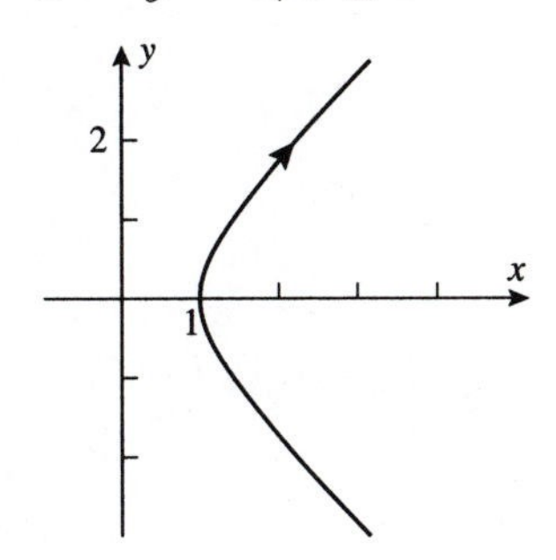

31.

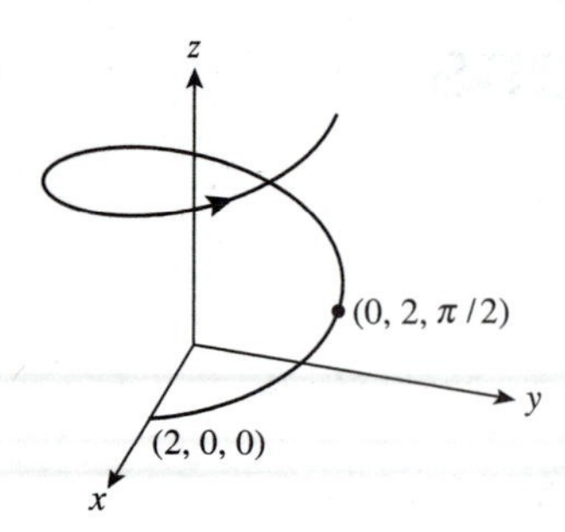

33.

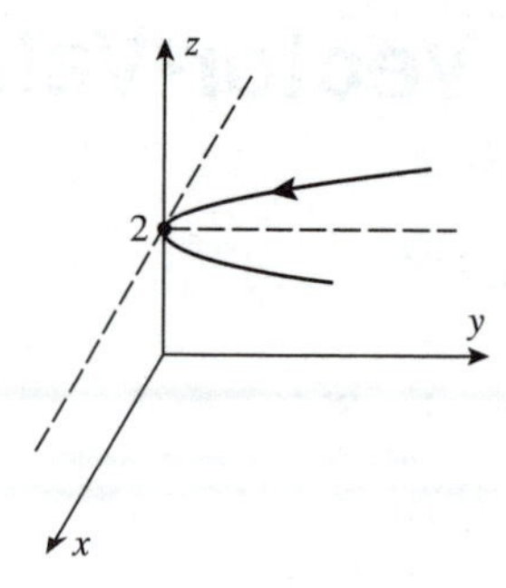

35. $x = t, y = t, z = 2t^2$

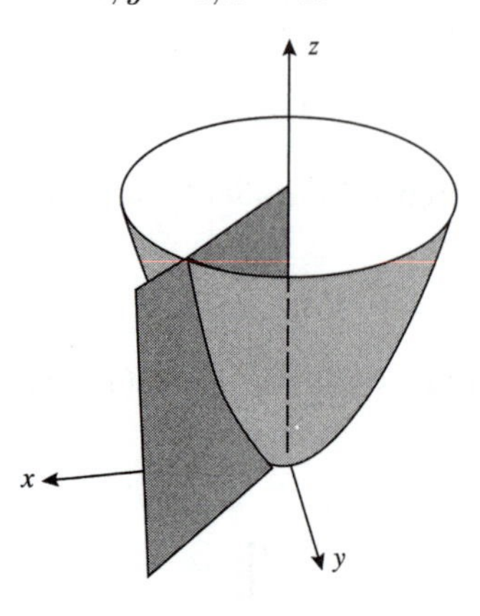

37. $\mathbf{r} = t\mathbf{i} + t^2\mathbf{j} \pm \frac{1}{3}\sqrt{81 - 9t^2 - t^4}\mathbf{k}$

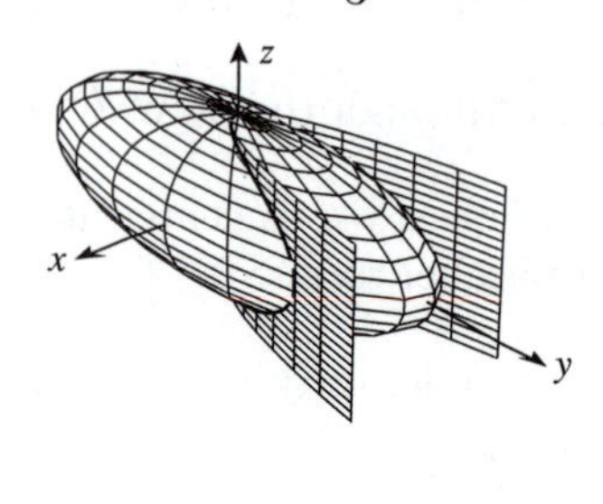

39. $x^2 + y^2 = (t \sin t)^2 + (t \cos t)^2 = t^2(\sin^2 t + \cos^2 t) = t^2 = z$

41. $x = \sin t$, $y = 2\cos t$, $z = \sqrt{3}\sin t$ so $x^2 + y^2 + z^2 = \sin^2 t + 4\cos^2 t + 3\sin^2 t = 4$ and $z = \sqrt{3}x$; it is the curve of intersection of the sphere $x^2 + y^2 + z^2 = 4$ and the plane $z = \sqrt{3}x$, which is a circle with center at $(0, 0, 0)$ and radius 2.

43. The helix makes one turn as t varies from 0 to 2π so $z = c(2\pi) = 3$, $c = 3/(2\pi)$.

45. $x^2 + y^2 = t^2 \cos^2 t + t^2 \sin^2 t = t^2$, $\sqrt{x^2 + y^2} = t = z$; a conical helix.

47. **(a)** III, since the curve is a subset of the plane $y = -x$

(b) IV, since only x is periodic in t, and y, z increase without bound

(c) II, since all three components are periodic in t

(d) I, since the projection onto the yz-plane is a circle and the curve increases without bound in the x-direction

49. **(a)** Let $x = 3\cos t$ and $y = 3\sin t$, then $z = 9\cos^2 t$.

(b)

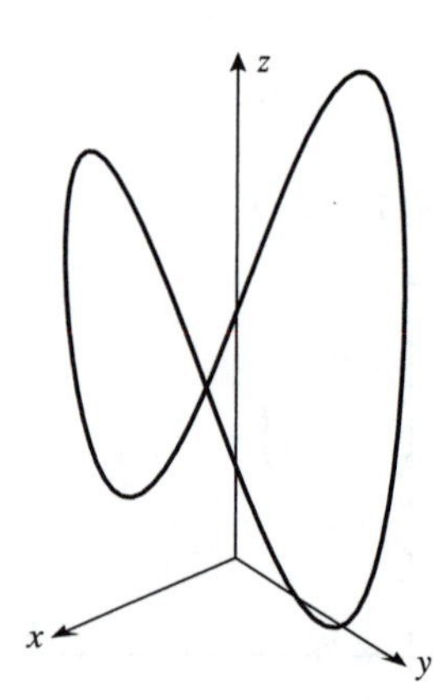

51. **(a)**

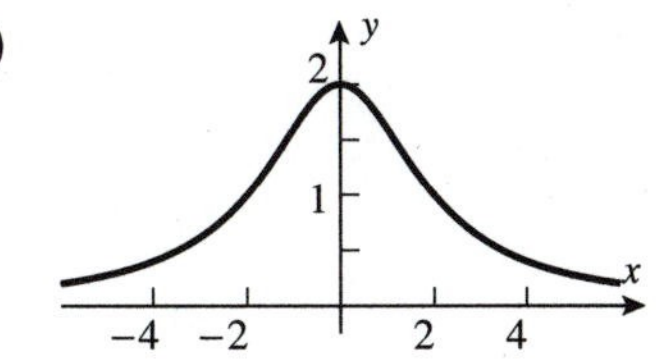

(b) In Part (a) set $x = 2t$;
then $y = 2/(1+(x/2)^2) = 8/(4+x^2)$

EXERCISE SET 13.2

1. $9\mathbf{i} + 6\mathbf{j}$

3. $\langle 1/3, 0 \rangle$

5. $2\mathbf{i} - 3\mathbf{j} + 4\mathbf{k}$

7. **(a)** continuous, $\lim_{t\to 0} \mathbf{r}(t) = \mathbf{0} = \mathbf{r}(0)$

(b) not continuous, $\lim_{t\to 0} \mathbf{r}(t)$ does not exist

9.

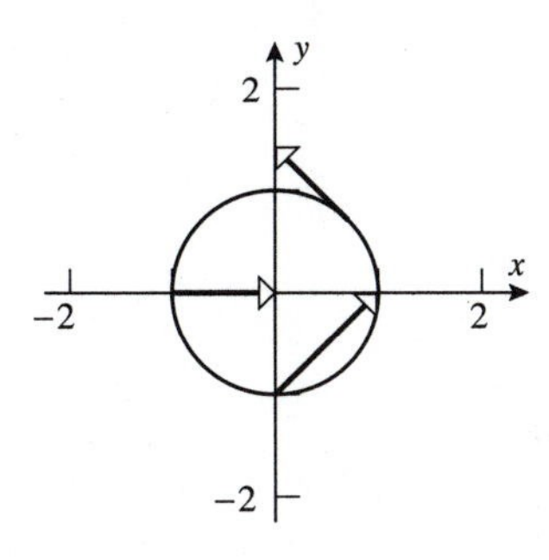

11. $\mathbf{r}'(t) = 5\mathbf{i} + (1-2t)\mathbf{j}$

13. $\mathbf{r}'(t) = -\dfrac{1}{t^2}\mathbf{i} + \sec^2 t\mathbf{j} + 2e^{2t}\mathbf{k}$

15. $\mathbf{r}'(t) = \langle 1, 2t \rangle$,
$\mathbf{r}'(2) = \langle 1, 4 \rangle$,
$\mathbf{r}(2) = \langle 2, 4 \rangle$

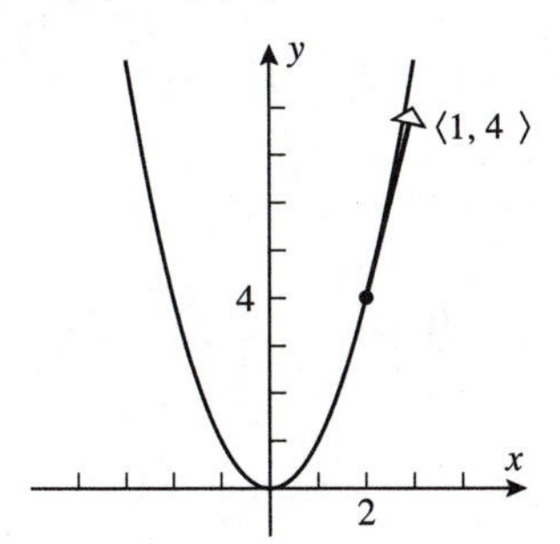

17. $\mathbf{r}'(t) = \sec t \tan t\mathbf{i} + \sec^2 t\mathbf{j}$,
$\mathbf{r}'(0) = \mathbf{j}$
$\mathbf{r}(0) = \mathbf{i}$

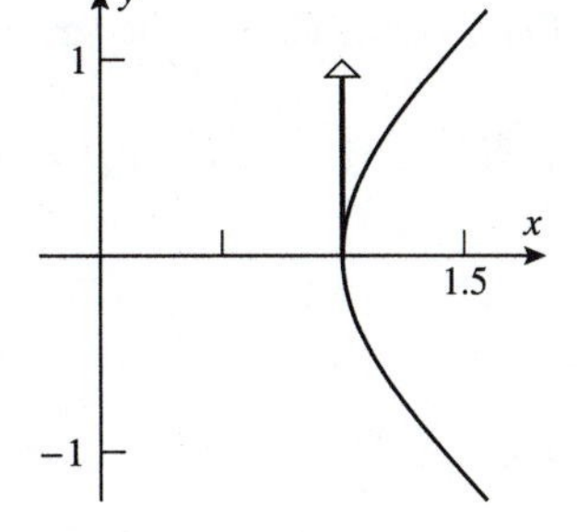

19. $\mathbf{r}'(t) = 2\cos t\mathbf{i} - 2\sin t\mathbf{k}$,
$\mathbf{r}'(\pi/2) = -2\mathbf{k}$,
$\mathbf{r}(\pi/2) = 2\mathbf{i} + \mathbf{j}$

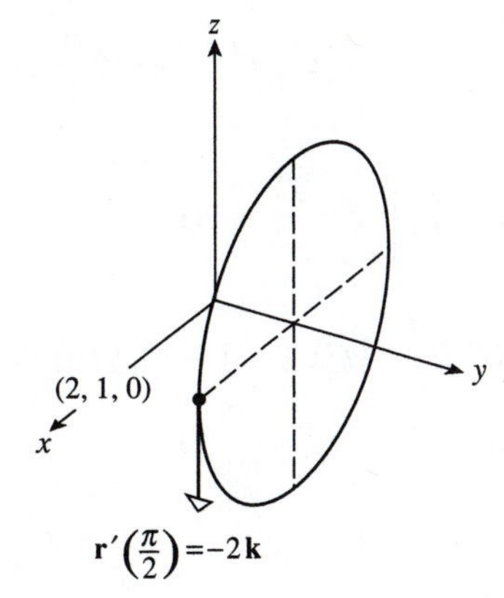

21.

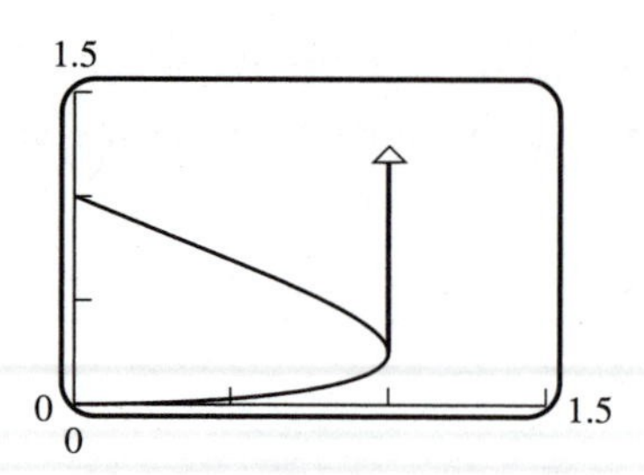

23. $\mathbf{r}'(t) = 2t\mathbf{i} - \dfrac{1}{t}\mathbf{j}$, $\mathbf{r}'(1) = 2\mathbf{i} - \mathbf{j}$, $\mathbf{r}(1) = \mathbf{i} + 2\mathbf{j}$; $x = 1 + 2t$, $y = 2 - t$

25. $\mathbf{r}'(t) = -2\pi \sin \pi t\mathbf{i} + 2\pi \cos \pi t\mathbf{j} + 3\mathbf{k}$, $\mathbf{r}'(1/3) = -\sqrt{3}\,\pi\mathbf{i} + \pi\mathbf{j} + 3\mathbf{k}$,
$\mathbf{r}(1/3) = \mathbf{i} + \sqrt{3}\,\mathbf{j} + \mathbf{k}$; $x = 1 - \sqrt{3}\,\pi t$, $y = \sqrt{3} + \pi t$, $z = 1 + 3t$

27. $\mathbf{r}'(t) = 2\mathbf{i} + \dfrac{3}{2\sqrt{3t+4}}\mathbf{j}$, $t = 0$ at P_0 so $\mathbf{r}'(0) = 2\mathbf{i} + \dfrac{3}{4}\mathbf{j}$,

$\mathbf{r}(0) = -\mathbf{i} + 2\mathbf{j}$; $\mathbf{r} = (-\mathbf{i} + 2\mathbf{j}) + t\left(2\mathbf{i} + \dfrac{3}{4}\mathbf{j}\right)$

29. $\mathbf{r}'(t) = 2t\mathbf{i} + \dfrac{1}{(t+1)^2}\mathbf{j} - 2t\mathbf{k}$, $t = -2$ at P_0 so $\mathbf{r}'(-2) = -4\mathbf{i} + \mathbf{j} + 4\mathbf{k}$,
$\mathbf{r}(-2) = 4\mathbf{i} + \mathbf{j}$; $\mathbf{r} = (4\mathbf{i} + \mathbf{j}) + t(-4\mathbf{i} + \mathbf{j} + 4\mathbf{k})$

31. **(a)** $\lim_{t\to 0}(\mathbf{r}(t) - \mathbf{r}'(t)) = \mathbf{i} - \mathbf{j} + \mathbf{k}$

(b) $\lim_{t\to 0}(\mathbf{r}(t) \times \mathbf{r}'(t)) = \lim_{t\to 0}(-\cos t\mathbf{i} - \sin t\mathbf{j} + k) = -\mathbf{i} + \mathbf{k}$

(c) $\lim_{t\to 0}(\mathbf{r}(t) \cdot \mathbf{r}'(t)) = 0$

33. **(a)** $\mathbf{r}_1' = 2\mathbf{i} + 6t\mathbf{j} + 3t^2\mathbf{k}$, $\mathbf{r}_2' = 4t^3\mathbf{k}$, $\mathbf{r}_1 \cdot \mathbf{r}_2 = t^7$; $\dfrac{d}{dt}(\mathbf{r}_1 \cdot \mathbf{r}_2) = 7t^6 = \mathbf{r}_1 \cdot \mathbf{r}_2' + \mathbf{r}_1' \cdot \mathbf{r}_2$

(b) $\mathbf{r}_1 \times \mathbf{r}_2 = 3t^6\mathbf{i} - 2t^5\mathbf{j}$, $\dfrac{d}{dt}(\mathbf{r}_1 \times \mathbf{r}_2) = 18t^5\mathbf{i} - 10t^4\mathbf{j} = \mathbf{r}_1 \times \mathbf{r}_2' + \mathbf{r}_1' \times \mathbf{r}_2$

35. $3t\mathbf{i} + 2t^2\mathbf{j} + \mathbf{C}$

37. $(-t\cos t + \sin t)\mathbf{i} + t\mathbf{j} + \mathbf{C}$

39. $(t^3/3)\mathbf{i} - t^2\mathbf{j} + \ln|t|\mathbf{k} + \mathbf{C}$

41. $\left\langle \dfrac{1}{3}\sin 3t, \dfrac{1}{3}\cos 3t \right\rangle\Big]_0^{\pi/3} = \langle 0, -2/3 \rangle$

43. $\displaystyle\int_0^2 \sqrt{t^2 + t^4}dt = \int_0^2 t(1+t^2)^{1/2}dt = \frac{1}{3}\left(1+t^2\right)^{3/2}\Big]_0^2 = (5\sqrt{5} - 1)/3$

45. $\left(\dfrac{2}{3}t^{3/2}\mathbf{i} + 2t^{1/2}\mathbf{j}\right)\Big]_1^9 = \dfrac{52}{3}\mathbf{i} + 4\mathbf{j}$

47. $\mathbf{y}(t) = \displaystyle\int \mathbf{y}'(t)dt = \tfrac{1}{3}t^3\mathbf{i} + t^2\mathbf{j} + \mathbf{C}, \mathbf{y}(0) = \mathbf{C} = \mathbf{i} + \mathbf{j}, \mathbf{y}(t) = (\tfrac{1}{3}t^3 + 1)\mathbf{i} + (t^2 + 1)\mathbf{j}$

49. $\mathbf{y}'(t) = \int \mathbf{y}''(t)dt = t\mathbf{i} + e^t\mathbf{j} + \mathbf{C}_1, \mathbf{y}'(0) = \mathbf{j} + \mathbf{C}_1 = \mathbf{j}$ so $\mathbf{C}_1 = \mathbf{0}$ and $\mathbf{y}'(t) = t\mathbf{i} + e^t\mathbf{j}$.

$\mathbf{y}(t) = \int \mathbf{y}'(t)dt = \frac{1}{2}t^2\mathbf{i} + e^t\mathbf{j} + \mathbf{C}_2$, $\mathbf{y}(0) = \mathbf{j} + \mathbf{C}_2 = 2\mathbf{i}$ so $\mathbf{C}_2 = 2\mathbf{i} - \mathbf{j}$ and

$\mathbf{y}(t) = \left(\frac{1}{2}t^2 + 2\right)\mathbf{i} + (e^t - 1)\mathbf{j}$

51. $\mathbf{r}'(t) = -4\sin t\mathbf{i} + 3\cos t\mathbf{j}$, $\mathbf{r}(t)\cdot\mathbf{r}'(t) = -7\cos t\sin t$, so $\mathbf{r}$ and $\mathbf{r}'$ are perpendicular for $t = 0, \pi/2, \pi, 3\pi/2, 2\pi$. Since

$\|\mathbf{r}(t)\| = \sqrt{16\cos^2 t + 9\sin^2 t}, \|\mathbf{r}'(t)\| = \sqrt{16\sin^2 t + 9\cos^2 t},$

$\|\mathbf{r}\|\|\mathbf{r}'\| = \sqrt{144 + 337\sin^2 t\cos^2 t}, \ \ \theta = \cos^{-1}\left[\frac{-7\sin t\cos t}{\sqrt{144 + 337\sin^2 t\cos^2 t}}\right]$, with the graph

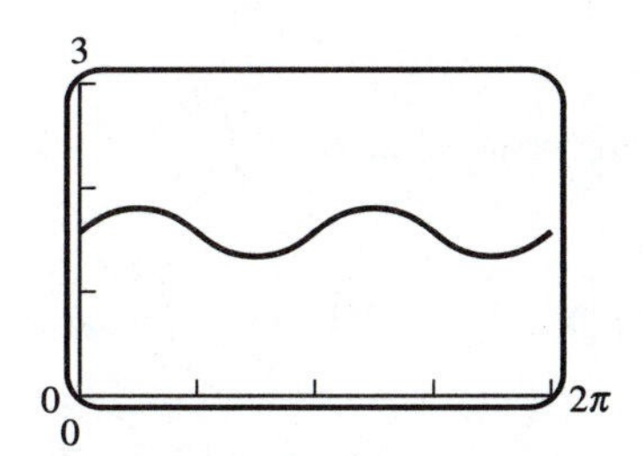

From the graph it appears that θ is bounded away from 0 and π, meaning that $\mathbf{r}$ and $\mathbf{r}'$ are never parallel. We can check this by considering them as vectors in 3-space, and then $\mathbf{r}\times\mathbf{r}' = 12\,\mathbf{k} \neq \mathbf{0}$, so they are never parallel.

53. (a) $2t - t^2 - 3t = -2$, $t^2 + t - 2 = 0$, $(t + 2)(t - 1) = 0$ so $t = -2, 1$. The points of intersection are $(-2, 4, 6)$ and $(1, 1, -3)$.

(b) $\mathbf{r}' = \mathbf{i} + 2t\mathbf{j} - 3\mathbf{k}$; $\mathbf{r}'(-2) = \mathbf{i} - 4\mathbf{j} - 3\mathbf{k}$, $\mathbf{r}'(1) = \mathbf{i} + 2\mathbf{j} - 3\mathbf{k}$, and $\mathbf{n} = 2\mathbf{i} - \mathbf{j} + \mathbf{k}$ is normal to the plane. Let θ be the acute angle, then
for $t = -2$: $\cos\theta = |\mathbf{n}\cdot\mathbf{r}'|/(\|\mathbf{n}\|\ \|\mathbf{r}'\|) = 3/\sqrt{156}$, $\theta \approx 76°$;
for $t = 1$: $\cos\theta = |\mathbf{n}\cdot\mathbf{r}'|/(\|\mathbf{n}\|\ \|\mathbf{r}'\|) = 3/\sqrt{84}$, $\theta \approx 71°$.

55. $\mathbf{r}_1(1) = \mathbf{r}_2(2) = \mathbf{i} + \mathbf{j} + 3\mathbf{k}$ so the graphs intersect at P; $\mathbf{r}_1'(t) = 2t\mathbf{i} + \mathbf{j} + 9t^2\mathbf{k}$ and

$\mathbf{r}_2'(t) = \mathbf{i} + \frac{1}{2}t\mathbf{j} - \mathbf{k}$ so $\mathbf{r}_1'(1) = 2\mathbf{i} + \mathbf{j} + 9\mathbf{k}$ and $\mathbf{r}_2'(2) = \mathbf{i} + \mathbf{j} - \mathbf{k}$ are tangent to the graphs at P,

thus $\cos\theta = \dfrac{\mathbf{r}_1'(1)\cdot\mathbf{r}_2'(2)}{\|\mathbf{r}_1'(1)\|\ \|\mathbf{r}_2'(2)\|} = -\dfrac{6}{\sqrt{86}\sqrt{3}}$, $\theta = \cos^{-1}(6/\sqrt{258}) \approx 68°$.

57. $\frac{d}{dt}[\mathbf{r}(t)\times\mathbf{r}'(t)] = \mathbf{r}(t)\times\mathbf{r}''(t) + \mathbf{r}'(t)\times\mathbf{r}'(t) = \mathbf{r}(t)\times\mathbf{r}''(t) + \mathbf{0} = \mathbf{r}(t)\times\mathbf{r}''(t)$

59. In Exercise 58, write each scalar triple product as a determinant.

61. Let $\mathbf{r}_1(t) = x_1(t)\mathbf{i} + y_1(t)\mathbf{j} + z_1(t)\mathbf{k}$ and $\mathbf{r}_2(t) = x_2(t)\mathbf{i} + y_2(t)\mathbf{j} + z_2(t)\mathbf{k}$, in both (6) and (7); show that the left and right members of the equalities are the same.

EXERCISE SET 13.3

1. **(a)** The tangent vector reverses direction at the four cusps.

(b) $\mathbf{r}'(t) = -3\cos^2 t \sin t\mathbf{i} + 3\sin^2 t \cos t\mathbf{j} = \mathbf{0}$ when $t = 0, \pi/2, \pi, 3\pi/2, 2\pi$.

3. $\mathbf{r}'(t) = 3t^2\mathbf{i} + (6t-2)\mathbf{j} + 2t\mathbf{k}$; smooth

5. $\mathbf{r}'(t) = (1-t)e^{-t}\mathbf{i} + (2t-2)\mathbf{j} - \pi\sin(\pi t)\mathbf{k}$; not smooth, $\mathbf{r}'(1) = \mathbf{0}$

7. $(dx/dt)^2 + (dy/dt)^2 + (dz/dt)^2 = (-3\cos^2 t \sin t)^2 + (3\sin^2 t\cos t)^2 + 0^2 = 9\sin^2 t\cos^2 t$,

$$L = \int_0^{\pi/2} 3\sin t\cos t\, dt = 3/2$$

9. $\mathbf{r}'(t) = \langle e^t, -e^{-t}, \sqrt{2}\rangle$, $\|\mathbf{r}'(t)\| = e^t + e^{-t}$, $L = \int_0^1 (e^t + e^{-t})dt = e - e^{-1}$

11. $\mathbf{r}'(t) = 3t^2\mathbf{i} + \mathbf{j} + \sqrt{6}\,t\mathbf{k}$, $\|\mathbf{r}'(t)\| = 3t^2 + 1$, $L = \int_1^3 (3t^2+1)dt = 28$

13. $\mathbf{r}'(t) = -3\sin t\mathbf{i} + 3\cos t\mathbf{j} + \mathbf{k}$, $\|\mathbf{r}'(t)\| = \sqrt{10}$, $L = \int_0^{2\pi} \sqrt{10}\, dt = 2\pi\sqrt{10}$

15. $(d\mathbf{r}/dt)(dt/d\tau) = (\mathbf{i} + 2t\mathbf{j})(4) = 4\mathbf{i} + 8t\mathbf{j} = 4\mathbf{i} + 8(4\tau+1)\mathbf{j}$;
$\mathbf{r}(\tau) = (4\tau+1)\mathbf{i} + (4\tau+1)^2\mathbf{j}$, $\mathbf{r}'(\tau) = 4\mathbf{i} + 2(4)(4\tau+1)\mathbf{j}$

17. $(d\mathbf{r}/dt)(dt/d\tau) = (e^t\mathbf{i} - 4e^{-t}\mathbf{j})(2\tau) = 2\tau e^{\tau^2}\mathbf{i} - 8\tau e^{-\tau^2}\mathbf{j}$;
$\mathbf{r}(\tau) = e^{\tau^2}\mathbf{i} + 4e^{-\tau^2}\mathbf{j}$, $\mathbf{r}'(\tau) = 2\tau e^{\tau^2}\mathbf{i} - 4(2)\tau e^{-\tau^2}\mathbf{j}$

19. **(a)** $\|\mathbf{r}'(t)\| = \sqrt{2}$, $s = \int_0^t \sqrt{2}\, dt = \sqrt{2}t$; $\mathbf{r} = \frac{s}{\sqrt{2}}\mathbf{i} + \frac{s}{\sqrt{2}}\mathbf{j}$, $x = \frac{s}{\sqrt{2}}$, $y = \frac{s}{\sqrt{2}}$

(b) Similar to Part (a), $x = y = z = \frac{s}{\sqrt{3}}$

21. **(a)** $\mathbf{r}(t) = \langle 1, 3, 4\rangle$ when $t = 0$,

so $s = \int_0^t \sqrt{1+4+4}\, du = 3t$, $x = 1 + s/3$, $y = 3 - 2s/3$, $z = 4 + 2s/3$

(b) $\mathbf{r}\Big]_{s=25} = \langle 28/3, -41/3, 62/3\rangle$

23. $x = 3 + \cos t$, $y = 2 + \sin t$, $(dx/dt)^2 + (dy/dt)^2 = 1$,

$s = \int_0^t du = t$ so $t = s$, $x = 3 + \cos s$, $y = 2 + \sin s$ for $0 \le s \le 2\pi$.

25. $x = t^3/3$, $y = t^2/2$, $(dx/dt)^2 + (dy/dt)^2 = t^2(t^2+1)$,

$s = \int_0^t u(u^2+1)^{1/2}du = \frac{1}{3}[(t^2+1)^{3/2} - 1]$ so $t = [(3s+1)^{2/3} - 1]^{1/2}$,

$x = \frac{1}{3}[(3s+1)^{2/3} - 1]^{3/2}$, $y = \frac{1}{2}[(3s+1)^{2/3} - 1]$ for $s \ge 0$

27. $x = e^t \cos t,\ y = e^t \sin t,\ (dx/dt)^2 + (dy/dt)^2 = 2e^{2t},\ s = \int_0^t \sqrt{2}\, e^u\, du = \sqrt{2}(e^t - 1)$ so
$t = \ln(s/\sqrt{2} + 1),\ x = (s/\sqrt{2} + 1)\cos[\ln(s/\sqrt{2} + 1)],\ y = (s/\sqrt{2} + 1)\sin[\ln(s/\sqrt{2} + 1)]$
for $0 \le s \le \sqrt{2}(e^{\pi/2} - 1)$

29. $dx/dt = -a\sin t,\ dy/dt = a\cos t,\ dz/dt = c,$
$s(t_0) = L = \int_0^{t_0} \sqrt{a^2 \sin^2 t + a^2 \cos^2 t + c^2}\, dt = \int_0^{t_0} \sqrt{a^2 + c^2}\, dt = t_0\sqrt{a^2 + c^2}$

31. $x = at - a\sin t,\ y = a - a\cos t,\ (dx/dt)^2 + (dy/dt)^2 = 4a^2 \sin^2(t/2),$
$s = \int_0^t 2a\sin(u/2)du = 4a[1 - \cos(t/2)]$ so $\cos(t/2) = 1 - s/(4a),\ t = 2\cos^{-1}[1 - s/(4a)],$
$\cos t = 2\cos^2(t/2) - 1 = 2[1 - s/(4a)]^2 - 1,$
$\sin t = 2\sin(t/2)\cos(t/2) = 2(1 - [1 - s/(4a)]^2)^{1/2}(2[1 - s/(4a)]^2 - 1),$
$x = 2a\cos^{-1}[1 - s/(4a)] - 2a(1 - [1 - s/(4a)]^2)^{1/2}(2[1 - s/(4a)]^2 - 1),$
$y = \dfrac{s(8a - s)}{8a}$ for $0 \le s \le 8a$

33. **(a)** $(dr/dt)^2 + r^2(d\theta/dt)^2 + (dz/dt)^2 = 9e^{4t},\ L = \int_0^{\ln 2} 3e^{2t}dt = \frac{3}{2}e^{2t}\Big]_0^{\ln 2} = 9/2$

(b) $(dr/dt)^2 + r^2(d\theta/dt)^2 + (dz/dt)^2 = 5t^2 + t^4 = t^2(5 + t^2),$
$L = \int_1^2 t(5 + t^2)^{1/2}dt = 9 - 2\sqrt{6}$

35. **(a)** $(d\rho/dt)^2 + \rho^2\sin^2\phi(d\theta/dt)^2 + \rho^2(d\phi/dt)^2 = 3e^{-2t},\ L = \int_0^2 \sqrt{3}e^{-t}dt = \sqrt{3}(1 - e^{-2})$

(b) $(d\rho/dt)^2 + \rho^2\sin^2\phi(d\theta/dt)^2 + \rho^2(d\phi/dt)^2 = 5,\ L = \int_1^5 \sqrt{5}dt = 4\sqrt{5}$

37. **(a)** $g(\tau) = \pi\tau$ **(b)** $g(\tau) = \pi(1 - \tau)$

39. Represent the helix by $x = a\cos t,\ y = a\sin t,\ z = ct$ with $a = 6.25$ and $c = 10/\pi$, so that the radius of the helix is the distance from the axis of the cylinder to the center of the copper cable, and the helix makes one turn in a distance of 20 in. ($t = 2\pi$). From Exercise 29 the length of the helix is $2\pi\sqrt{6.25^2 + (10/\pi)^2} \approx 44$ in.

41. $\mathbf{r}'(t) = (1/t)\mathbf{i} + 2\mathbf{j} + 2t\mathbf{k}$

(a) $\|\mathbf{r}'(t)\| = \sqrt{1/t^2 + 4 + 4t^2} = \sqrt{(2t + 1/t)^2} = 2t + 1/t$

(b) $\dfrac{ds}{dt} = 2t + 1/t$ **(c)** $\int_1^3 (2t + 1/t)dt = 8 + \ln 3$

43. Let $\mathbf{r}(t) = x(t)\mathbf{i} + y(t)\mathbf{j}$ and use the chain rule.

EXERCISE SET 13.4

1. (a)

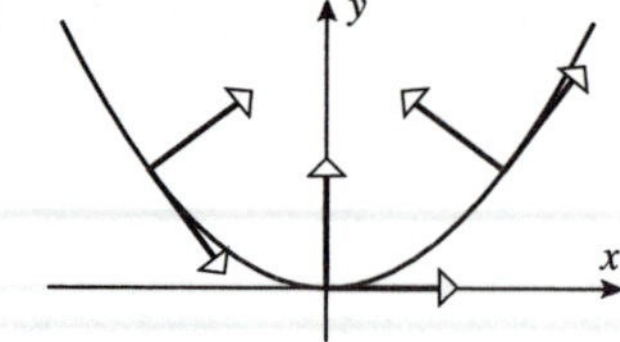

(b)

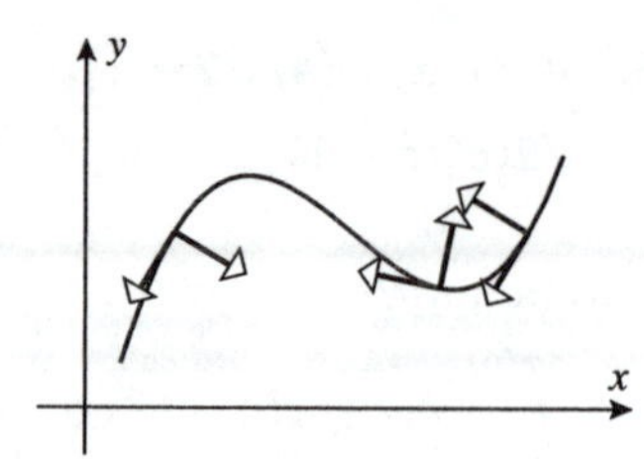

3. $\mathbf{r}'(t) = 2t\mathbf{i} + \mathbf{j}$, $\|\mathbf{r}'(t)\| = \sqrt{4t^2+1}$, $\mathbf{T}(t) = (4t^2+1)^{-1/2}(2t\mathbf{i}+\mathbf{j})$,
$\mathbf{T}'(t) = (4t^2+1)^{-1/2}(2\mathbf{i}) - 4t(4t^2+1)^{-3/2}(2t\mathbf{i}+\mathbf{j})$;
$\mathbf{T}(1) = \dfrac{2}{\sqrt{5}}\mathbf{i} + \dfrac{1}{\sqrt{5}}\mathbf{j}$, $\mathbf{T}'(1) = \dfrac{2}{5\sqrt{5}}(\mathbf{i} - 2\mathbf{j})$, $\mathbf{N}(1) = \dfrac{1}{\sqrt{5}}\mathbf{i} - \dfrac{2}{\sqrt{5}}\mathbf{j}$.

5. $\mathbf{r}'(t) = -5\sin t\mathbf{i} + 5\cos t\mathbf{j}$, $\|\mathbf{r}'(t)\| = 5$, $\mathbf{T}(t) = -\sin t\mathbf{i} + \cos t\mathbf{j}$, $\mathbf{T}'(t) = -\cos t\mathbf{i} - \sin t\mathbf{j}$;
$\mathbf{T}(\pi/3) = -\dfrac{\sqrt{3}}{2}\mathbf{i} + \dfrac{1}{2}\mathbf{j}$, $\mathbf{T}'(\pi/3) = -\dfrac{1}{2}\mathbf{i} - \dfrac{\sqrt{3}}{2}\mathbf{j}$, $\mathbf{N}(\pi/3) = -\dfrac{1}{2}\mathbf{i} - \dfrac{\sqrt{3}}{2}\mathbf{j}$

7. $\mathbf{r}'(t) = -4\sin t\mathbf{i} + 4\cos t\mathbf{j} + \mathbf{k}$, $\mathbf{T}(t) = \dfrac{1}{\sqrt{17}}(-4\sin t\mathbf{i} + 4\cos t\mathbf{j} + \mathbf{k})$,
$\mathbf{T}'(t) = \dfrac{1}{\sqrt{17}}(-4\cos t\mathbf{i} - 4\sin t\mathbf{j})$, $\mathbf{T}(\pi/2) = -\dfrac{4}{\sqrt{17}}\mathbf{i} + \dfrac{1}{\sqrt{17}}\mathbf{k}$
$\mathbf{T}'(\pi/2) = -\dfrac{4}{\sqrt{17}}\mathbf{j}$, $\mathbf{N}(\pi/2) = -\mathbf{j}$

9. $\mathbf{r}'(t) = e^t[(\cos t - \sin t)\mathbf{i} + (\cos t + \sin t)\mathbf{j} + \mathbf{k}]$, $\mathbf{T}(t) = \dfrac{1}{\sqrt{3}}[(\cos t - \sin t)\mathbf{i} + (\cos t + \sin t)\mathbf{j} + \mathbf{k}]$,
$\mathbf{T}'(t) = \dfrac{1}{\sqrt{3}}[(-\sin t - \cos t)\mathbf{i} + (-\sin t + \cos t)\mathbf{j}]$,
$\mathbf{T}(0) = \dfrac{1}{\sqrt{3}}\mathbf{i} + \dfrac{1}{\sqrt{3}}\mathbf{j} + \dfrac{1}{\sqrt{3}}\mathbf{k}$, $\mathbf{T}'(0) = \dfrac{1}{\sqrt{3}}(-\mathbf{i} + \mathbf{j})$, $\mathbf{N}(0) = -\dfrac{1}{\sqrt{2}}\mathbf{i} + \dfrac{1}{\sqrt{2}}\mathbf{j}$

11. From the remark, the line is parametrized by normalizing $\mathbf{v}$, but $\mathbf{T}(t_0) = \mathbf{v}/\|\mathbf{v}\|$, so $\mathbf{r} = \mathbf{r}(t_0) + t\mathbf{v}$ becomes $\mathbf{r} = \mathbf{r}(t_0) + s\mathbf{T}(t_0)$.

13. $\mathbf{r}'(t) = \cos t\mathbf{i} - \sin t\mathbf{j} + t\mathbf{k}$, $\mathbf{r}'(0) = \mathbf{i}$, $\mathbf{r}(0) = \mathbf{j}$, $\mathbf{T}(0) = \mathbf{i}$, so the tangent line has the parametrization $x = s, y = 1$.

15. $\mathbf{T} = \dfrac{3}{5}\cos t\,\mathbf{i} - \dfrac{3}{5}\sin t\,\mathbf{j} + \dfrac{4}{5}\mathbf{k}$, $\mathbf{N} = -\sin t\,\mathbf{i} - \cos t\,\mathbf{j}$, $\mathbf{B} = \mathbf{T}\times\mathbf{N} = \dfrac{4}{5}\cos t\,\mathbf{i} - \dfrac{4}{5}\sin t\,\mathbf{j} - \dfrac{3}{5}\mathbf{k}$. Check:
$\mathbf{r}' = 3\cos t\,\mathbf{i} - 3\sin t\,\mathbf{j} + 4\,\mathbf{k}$, $\mathbf{r}'' = -3\sin t\,\mathbf{i} - 3\cos t\,\mathbf{j}$, $\mathbf{r}'\times\mathbf{r}'' = 12\cos t\,\mathbf{i} - 12\sin t\,\mathbf{j} - 9\,\mathbf{k}$,
$\|\mathbf{r}'\times\mathbf{r}''\| = 15$, $(\mathbf{r}'\times\mathbf{r}'')/\|\mathbf{r}'\times\mathbf{r}''\| = \frac{4}{5}\cos t\,\mathbf{i} - \frac{4}{5}\sin t\,\mathbf{j} - \frac{3}{5}\,\mathbf{k} = \mathbf{B}$.

17. $\mathbf{r}'(t) = t\sin t\mathbf{i} + t\cos t\mathbf{j}$, $\|\mathbf{r}'\| = t$, $\mathbf{T} = \sin t\mathbf{i} + \cos t\mathbf{j}$, $\mathbf{N} = \cos t\mathbf{i} - \sin t\mathbf{j}$, $\mathbf{B} = \mathbf{T}\times\mathbf{N} = -\mathbf{k}$. Check:
$\mathbf{r}' = t\sin t\,\mathbf{i} + t\cos t\,\mathbf{j}$, $\mathbf{r}'' = (\sin t + t\cos t)\,\mathbf{i} + (\cos t - t\sin t)\,\mathbf{j}$, $\mathbf{r}'\times\mathbf{r}'' = -2e^{2t}\,\mathbf{k}$,
$\|\mathbf{r}'\times\mathbf{r}''\| = 2e^{2t}$, $(\mathbf{r}'\times\mathbf{r}'')/\|\mathbf{r}'\times\mathbf{r}''\| = -\mathbf{k} = \mathbf{B}$.

19. $\mathbf{r}(\pi/4) = \dfrac{\sqrt{2}}{2}\mathbf{i} + \dfrac{\sqrt{2}}{2}\mathbf{j} + \mathbf{k}$, $\mathbf{T} = -\sin t\mathbf{i} + \cos t\mathbf{j} = \dfrac{\sqrt{2}}{2}(-\mathbf{i} + \mathbf{j})$, $\mathbf{N} = -(\cos t\mathbf{i} + \sin t\mathbf{j}) = -\dfrac{\sqrt{2}}{2}(\mathbf{i} + \mathbf{j})$,
$\mathbf{B} = \mathbf{k}$; the rectifying, osculating, and normal planes are given (respectively) by $x + y = \sqrt{2}$, $z = 1$, $-x + y = 0$.

21. **(a)** By formulae (1) and (11), $\mathbf{N}(t) = \mathbf{B}(t) \times \mathbf{T}(t) = \dfrac{\mathbf{r}'(t) \times \mathbf{r}''(t)}{\|\mathbf{r}'(t) \times \mathbf{r}''(t)\|} \times \dfrac{\mathbf{r}'(t)}{\|\mathbf{r}'(t)\|}$.

(b) Since $\mathbf{r}'$ is perpendicular to $\mathbf{r}' \times \mathbf{r}''$ it follows from Lagrange's Identity (Exercise 32 of Section 12.4) that $\|(\mathbf{r}'(t) \times \mathbf{r}''(t)) \times \mathbf{r}'(t)\| = \|\mathbf{r}'(t) \times \mathbf{r}''(t)\| \|\mathbf{r}'(t)\|$, and the result follows.

(c) From Exercise 39 of Section 12.4,
$(\mathbf{r}'(t) \times \mathbf{r}''(t)) \times \mathbf{r}'(t) = \|\mathbf{r}'(t)\|^2\mathbf{r}''(t) - (\mathbf{r}'(t) \cdot \mathbf{r}''(t))\mathbf{r}'(t) = \mathbf{u}(t)$, so $\mathbf{N}(t) = \mathbf{u}(t)/\|\mathbf{u}(t)\|$

23. $\mathbf{r}'(t) = \cos t\mathbf{i} - \sin t\mathbf{j} + \mathbf{k}, \mathbf{r}''(t) = -\sin t\mathbf{i} - \cos t\mathbf{j}, \mathbf{u} = -2(\sin t\mathbf{i} + \cos t\mathbf{j}), \|\mathbf{u}\| = 2, \mathbf{N} = -\sin t\mathbf{i} - \cos t\mathbf{j}$

EXERCISE SET 13.5

1. $\kappa \approx \dfrac{1}{0.5} = 2$

3. $\mathbf{r}'(t) = 2t\mathbf{i} + 3t^2\mathbf{j}$, $\mathbf{r}''(t) = 2\mathbf{i} + 6t\mathbf{j}$, $\kappa = \|\mathbf{r}'(t) \times \mathbf{r}''(t)\|/\|\mathbf{r}'(t)\|^3 = \dfrac{6}{t(4+9t^2)^{3/2}}$

5. $\mathbf{r}'(t) = 3e^{3t}\mathbf{i} - e^{-t}\mathbf{j}$, $\mathbf{r}''(t) = 9e^{3t}\mathbf{i} + e^{-t}\mathbf{j}$, $\kappa = \|\mathbf{r}'(t) \times \mathbf{r}''(t)\|/\|\mathbf{r}'(t)\|^3 = \dfrac{12e^{2t}}{(9e^{6t} + e^{-2t})^{3/2}}$

7. $\mathbf{r}'(t) = -4\sin t\mathbf{i} + 4\cos t\mathbf{j} + \mathbf{k}$, $\mathbf{r}''(t) = -4\cos t\mathbf{i} - 4\sin t\mathbf{j}$,
$\kappa = \|\mathbf{r}'(t) \times \mathbf{r}''(t)\|/\|\mathbf{r}'(t)\|^3 = 4/17$

9. $\mathbf{r}'(t) = \sinh t\mathbf{i} + \cosh t\mathbf{j} + \mathbf{k}$, $\mathbf{r}''(t) = \cosh t\mathbf{i} + \sinh t\mathbf{j}$, $\kappa = \|\mathbf{r}'(t) \times \mathbf{r}''(t)\|/\|\mathbf{r}'(t)\|^3 = \dfrac{1}{2\cosh^2 t}$

11. $\mathbf{r}'(t) = -3\sin t\mathbf{i} + 4\cos t\mathbf{j} + \mathbf{k}$, $\mathbf{r}''(t) = -3\cos t\mathbf{i} - 4\sin t\mathbf{j}$,
$\mathbf{r}'(\pi/2) = -3\mathbf{i} + \mathbf{k}$, $\mathbf{r}''(\pi/2) = -4\mathbf{j}$; $\kappa = \|4\mathbf{i} + 12\mathbf{k}\|/\| - 3\mathbf{i} + \mathbf{k}\|^3 = 2/5, \rho = 5/2$

13. $\mathbf{r}'(t) = e^t(\cos t - \sin t)\mathbf{i} + e^t(\cos t + \sin t)\mathbf{j} + e^t\mathbf{k}$,
$\mathbf{r}''(t) = -2e^t \sin t\mathbf{i} + 2e^t \cos t\mathbf{j} + e^t\mathbf{k}$, $\mathbf{r}'(0) = \mathbf{i} + \mathbf{j} + \mathbf{k}$,
$\mathbf{r}''(0) = 2\mathbf{j} + \mathbf{k}$; $\kappa = \| - \mathbf{i} - \mathbf{j} + 2\mathbf{k}\|/\|\mathbf{i} + \mathbf{j} + \mathbf{k}\|^3 = \sqrt{2}/3, \rho = 3\sqrt{2}/2$

15. $\mathbf{r}'(s) = \dfrac{1}{2}\cos\left(1 + \dfrac{s}{2}\right)\mathbf{i} - \dfrac{1}{2}\sin\left(1 + \dfrac{s}{2}\right)\mathbf{j} + \dfrac{\sqrt{3}}{2}\mathbf{k}$, $\|\mathbf{r}'(s)\| = 1$, so
$\dfrac{d\mathbf{T}}{ds} = -\dfrac{1}{4}\sin\left(1 + \dfrac{s}{2}\right)\mathbf{i} - \dfrac{1}{4}\cos\left(1 + \dfrac{s}{2}\right)\mathbf{j}$, $\kappa = \left\|\dfrac{d\mathbf{T}}{ds}\right\| = \dfrac{1}{4}$

17. **(a)** $\mathbf{r}' = x'\mathbf{i} + y'\mathbf{j}, \mathbf{r}'' = x''\mathbf{i} + y''\mathbf{j}$, $\|\mathbf{r}' \times \mathbf{r}''\| = |x'y'' - x''y'|$, $\kappa = \dfrac{|x'y'' - y'x''|}{(x'^2 + y'^2)^{3/2}}$

(b) Set $x = t$, $y = f(x) = f(t)$, $x' = 1$, $x'' = 0$, $y' = \dfrac{dy}{dx}$, $y'' = \dfrac{d^2y}{dx^2}, \kappa = \dfrac{|d^2y/dx^2|}{(1 + (dy/dx)^2)^{3/2}}$

19. $\kappa(x) = \dfrac{|\sin x|}{(1 + \cos^2 x)^{3/2}}$, $\kappa(\pi/2) = 1$

21. $\kappa(x) = \dfrac{2|x|^3}{(x^4 + 1)^{3/2}}$, $\kappa(1) = 1/\sqrt{2}$

23. $\kappa(x) = \dfrac{2\sec^2 x|\tan x|}{(1+\sec^4 x)^{3/2}}$, $\kappa(\pi/4) = 4/(5\sqrt{5})$

25. $x'(t) = 2t$, $y'(t) = 3t^2$, $x''(t) = 2$, $y''(t) = 6t$,
$x'(1/2) = 1$, $y'(1/2) = 3/4$, $x''(1/2) = 2$, $y''(1/2) = 3$; $\kappa = 96/125$

27. $x'(t) = 3e^{3t}$, $y'(t) = -e^{-t}$, $x''(t) = 9e^{3t}$, $y''(t) = e^{-t}$,
$x'(0) = 3$, $y'(0) = -1$, $x''(0) = 9$, $y''(0) = 1$; $\kappa = 6/(5\sqrt{10})$

29. $x'(t) = 1, y'(t) = -1/t^2, x''(t) = 0, y''(t) = 2/t^3$
$x'(1) = 1, y'(1) = -1, x''(1) = 0, y''(1) = 2; \kappa = 1/\sqrt{2}$

31. **(a)** $\kappa(x) = \dfrac{|\cos x|}{(1+\sin^2 x)^{3/2}}$,

$\rho(x) = \dfrac{(1+\sin^2 x)^{3/2}}{|\cos x|}$

$\rho(0) = \rho(\pi) = 1.$

(b) $\kappa(t) = \dfrac{2}{(4\sin^2 t + \cos^2 t)^{3/2}}$,

$\rho(t) = \dfrac{1}{2}(4\sin^2 t + \cos^2 t)^{3/2}$,

$\rho(0) = 1/2$, $\rho(\pi/2) = 4$

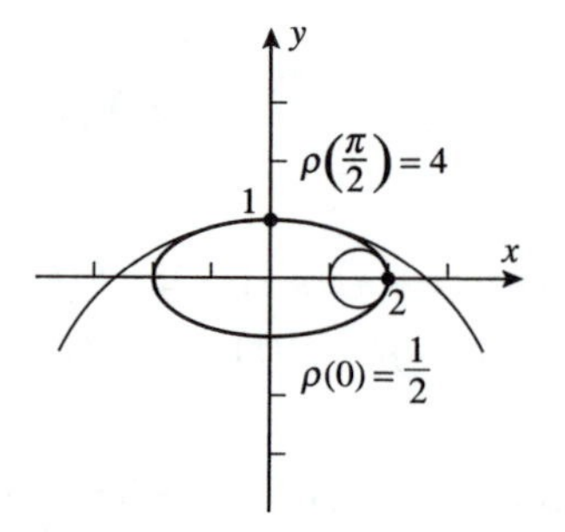

33. **(a)** At $x = 0$ the curvature of I has a large value, yet the value of II there is zero, so II is not the curvature of I; hence I is the curvature of II.

(b) I has points of inflection where the curvature is zero, but II is not zero there, and hence is not the curvature of I; so I is the curvature of II.

35. **(a)**

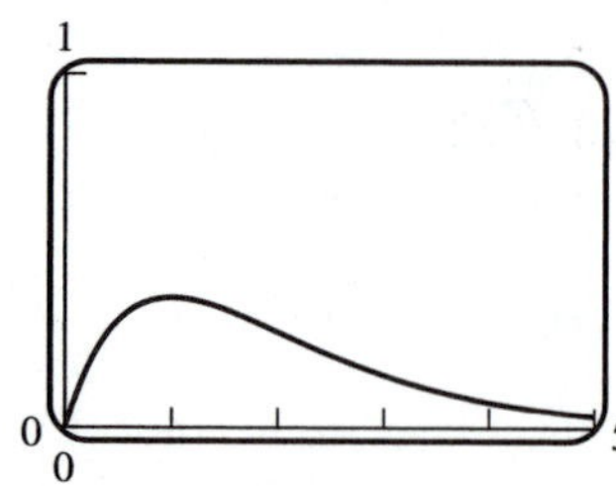

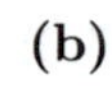

(b)

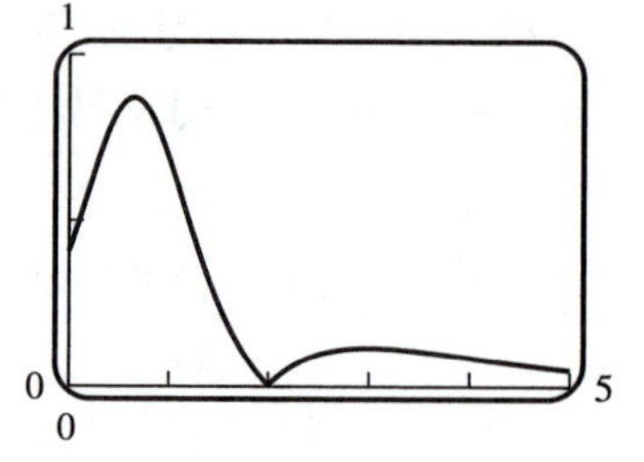

37. **(a)** $\kappa = \dfrac{|12x^2 - 4|}{(1+(4x^3-4x)^2)^{3/2}}$

(b)

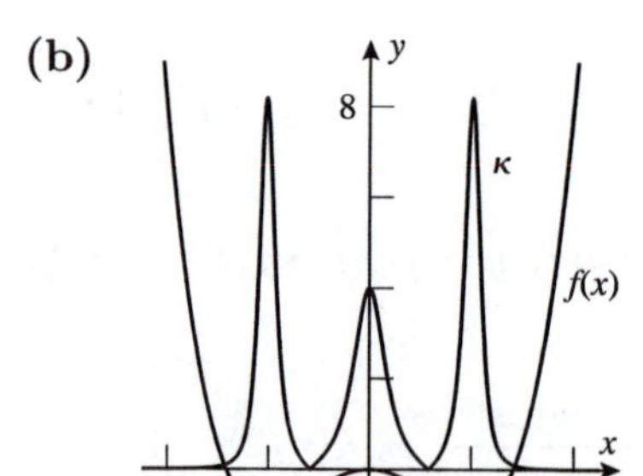

(c) $f'(x) = 4x^3 - 4x = 0$ at $x = 0, \pm 1, f''(x) = 12x^2 - 4$, so extrema at $x = 0, \pm 1$, and $\rho = 1/4$ for $x = 0$ and $\rho = 1/8$ when $x = \pm 1$.

39. $\mathbf{r}'(\theta) = \left(-r\sin\theta + \cos\theta\dfrac{dr}{d\theta}\right)\mathbf{i} + \left(r\cos\theta + \sin\theta\dfrac{dr}{d\theta}\right)\mathbf{j}$;

$$\mathbf{r}''(\theta) = \left(-r\cos\theta - 2\sin\theta\frac{dr}{d\theta} + \cos\theta\frac{d^2r}{d\theta^2}\right)\mathbf{i} + \left(-r\sin\theta + 2\cos\theta\frac{dr}{d\theta} + \sin\theta\frac{d^2r}{d\theta^2}\right)\mathbf{j};$$

$$\kappa = \frac{\left|r^2 + 2\left(\dfrac{dr}{d\theta}\right)^2 - r\dfrac{d^2r}{d\theta^2}\right|}{\left[r^2 + \left(\dfrac{dr}{d\theta}\right)^2\right]^{3/2}}.$$

41. $\kappa(\theta) = \dfrac{3}{2\sqrt{2}(1+\cos\theta)^{1/2}}$, $\kappa(\pi/2) = \dfrac{3}{2\sqrt{2}}$

43. $\kappa(\theta) = \dfrac{10 + 8\cos^2 3\theta}{(1 + 8\cos^2\theta)^{3/2}}$, $\kappa(0) = \dfrac{2}{3}$

45. The radius of curvature is zero when $\theta = \pi$, so there is a cusp there.

47. Let $y = t$, then $x = \dfrac{t^2}{4p}$ and $\kappa(t) = \dfrac{1/|2p|}{[t^2/(4p^2) + 1]^{3/2}}$;

$t = 0$ when $(x, y) = (0, 0)$ so $\kappa(0) = 1/|2p|$, $\rho = 2|p|$.

49. Let $x = 3\cos t$, $y = 2\sin t$ for $0 \le t < 2\pi$, $\kappa(t) = \dfrac{6}{(9\sin^2 t + 4\cos^2 t)^{3/2}}$ so

$\rho(t) = \dfrac{1}{6}(9\sin^2 t + 4\cos^2 t)^{3/2} = \dfrac{1}{6}(5\sin^2 t + 4)^{3/2}$ which, by inspection, is minimum when $t = 0$ or π. The radius of curvature is minimum at $(3, 0)$ and $(-3, 0)$.

51. $\mathbf{r}'(t) = -\sin t\mathbf{i} + \cos t\mathbf{j} - \sin t\mathbf{k}$, $\mathbf{r}''(t) = -\cos t\mathbf{i} - \sin t\mathbf{j} - \cos t\mathbf{k}$, $\|\mathbf{r}'(t) \times \mathbf{r}''(t)\| = \|-\mathbf{i} + \mathbf{k}\| = \sqrt{2}$, $\|\mathbf{r}'(t)\| = (1 + \sin^2 t)^{1/2}$; $\kappa(t) = \sqrt{2}/(1 + \sin^2 t)^{3/2}$, $\rho(t) = (1 + \sin^2 t)^{3/2}/\sqrt{2}$. The minimum value of ρ is $1/\sqrt{2}$; the maximum value is 2.

53. From Exercise 39: $dr/d\theta = ae^{a\theta} = ar$, $d^2r/d\theta^2 = a^2e^{a\theta} = a^2r$; $\kappa = 1/[\sqrt{1 + a^2}\, r]$.

55. (a) $d^2y/dx^2 = 2$, $\kappa(\phi) = |2\cos^3\phi|$

(b) $dy/dx = \tan\phi = 1, \phi = \pi/4$, $\kappa(\pi/4) = |2\cos^3(\pi/4)| = 1/\sqrt{2}$, $\rho = \sqrt{2}$

(c)

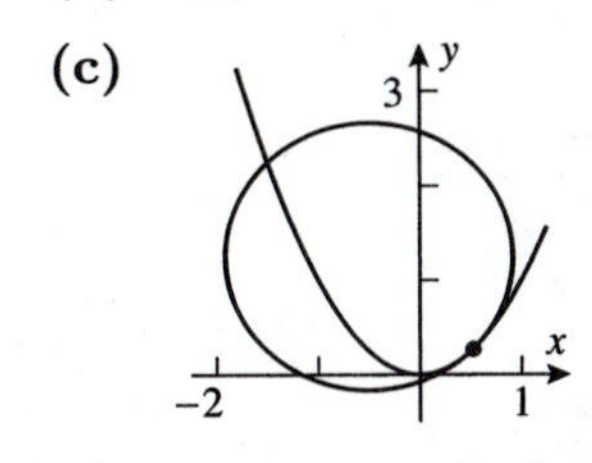

57. $\kappa = 0$ along $y = 0$; along $y = x^2$, $\kappa(x) = 2/(1 + 4x^2)^{3/2}$, $\kappa(0) = 2$. Along $y = x^3$, $\kappa(x) = 6|x|/(1 + 9x^4)^{3/2}$, $\kappa(0) = 0$.

59. $\kappa = 1/r$ along the circle; along $y = ax^2$, $\kappa(x) = 2a/(1+4a^2x^2)^{3/2}$, $\kappa(0) = 2a$ so $2a = 1/r$, $a = 1/(2r)$.

61. The result follows from the definitions $\mathbf{N} = \dfrac{\mathbf{T}'(s)}{\|\mathbf{T}'(s)\|}$ and $\kappa = \|\mathbf{T}'(s)\|$.

63. $\dfrac{d\mathbf{N}}{ds} = \mathbf{B} \times \dfrac{d\mathbf{T}}{ds} + \dfrac{d\mathbf{B}}{ds} \times \mathbf{T} = \mathbf{B} \times (\kappa\mathbf{N}) + (-\tau\mathbf{N}) \times \mathbf{T} = \kappa\mathbf{B} \times \mathbf{N} - \tau\mathbf{N} \times \mathbf{T}$, but $\mathbf{B} \times \mathbf{N} = -\mathbf{T}$ and $\mathbf{N} \times \mathbf{T} = -\mathbf{B}$ so $\dfrac{d\mathbf{N}}{ds} = -\kappa\mathbf{T} + \tau\mathbf{B}$

65. $\mathbf{r} = a\cos(s/w)\mathbf{i} + a\sin(s/w)\mathbf{j} + (cs/w)\mathbf{k}$, $\mathbf{r}' = -(a/w)\sin(s/w)\mathbf{i} + (a/w)\cos(s/w)\mathbf{j} + (c/w)\mathbf{k}$, $\mathbf{r}'' = -(a/w^2)\cos(s/w)\mathbf{i} - (a/w^2)\sin(s/w)\mathbf{j}$, $\mathbf{r}''' = (a/w^3)\sin(s/w)\mathbf{i} - (a/w^3)\cos(s/w)\mathbf{j}$, $\mathbf{r}' \times \mathbf{r}'' = (ac/w^3)\sin(s/w)\mathbf{i} - (ac/w^3)\cos(s/w)\mathbf{j} + (a^2/w^3)\mathbf{k}$, $(\mathbf{r}' \times \mathbf{r}'') \cdot \mathbf{r}''' = a^2c/w^6$, $\|\mathbf{r}''(s)\| = a/w^2$, so $\tau = c/w^2$ and $\mathbf{B} = (c/w)\sin(s/w)\mathbf{i} - (c/w)\cos(s/w)\mathbf{j} + (a/w)\mathbf{k}$

67. $\mathbf{r}' = 2\mathbf{i} + 2t\mathbf{j} + t^2\mathbf{k}$, $\mathbf{r}'' = 2\mathbf{j} + 2t\mathbf{k}$, $\mathbf{r}''' = 2\mathbf{k}$, $\mathbf{r}' \times \mathbf{r}'' = 2t^2\mathbf{i} - 4t\mathbf{j} + 4\mathbf{k}$, $\|\mathbf{r}' \times \mathbf{r}''\| = 2(t^2+2)$, $\tau = 8/[2(t^2+2)]^2 = 2/(t^2+2)^2$

69. $\mathbf{r}' = e^t\mathbf{i} - e^{-t}\mathbf{j} + \sqrt{2}\mathbf{k}$, $\mathbf{r}'' = e^t\mathbf{i} + e^{-t}\mathbf{j}$, $\mathbf{r}''' = e^t\mathbf{i} - e^{-t}\mathbf{j}$, $\mathbf{r}' \times \mathbf{r}'' = -\sqrt{2}e^{-t}\mathbf{i} + \sqrt{2}e^t\mathbf{j} + 2\mathbf{k}$, $\|\mathbf{r}' \times \mathbf{r}''\| = \sqrt{2}(e^t + e^{-t})$, $\tau = (-2\sqrt{2})/[2(e^t+e^{-t})^2] = -\sqrt{2}/(e^t+e^{-t})^2$

EXERCISE SET 13.6

1. $\mathbf{v}(t) = -3\sin t\mathbf{i} + 3\cos t\mathbf{j}$
$\mathbf{a}(t) = -3\cos t\mathbf{i} - 3\sin t\mathbf{j}$
$\|\mathbf{v}(t)\| = \sqrt{9\sin^2 t + 9\cos^2 t} = 3$
$\mathbf{r}(\pi/3) = (3/2)\mathbf{i} + (3\sqrt{3}/2)\mathbf{j}$
$\mathbf{v}(\pi/3) = -(3\sqrt{3}/2)\mathbf{i} + (3/2)\mathbf{j}$
$\mathbf{a}(\pi/3) = -(3/2)\mathbf{i} - (3\sqrt{3}/2)\mathbf{j}$

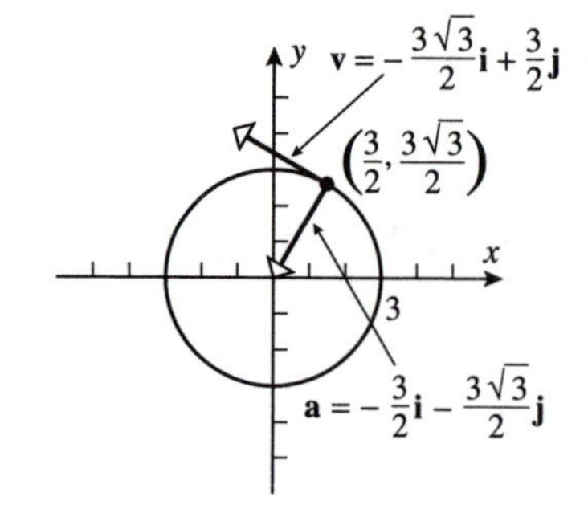

3. $\mathbf{v}(t) = e^t\mathbf{i} - e^{-t}\mathbf{j}$
$\mathbf{a}(t) = e^t\mathbf{i} + e^{-t}\mathbf{j}$
$\|\mathbf{v}(t)\| = \sqrt{e^{2t} + e^{-2t}}$
$\mathbf{r}(0) = \mathbf{i} + \mathbf{j}$
$\mathbf{v}(0) = \mathbf{i} - \mathbf{j}$
$\mathbf{a}(0) = \mathbf{i} + \mathbf{j}$

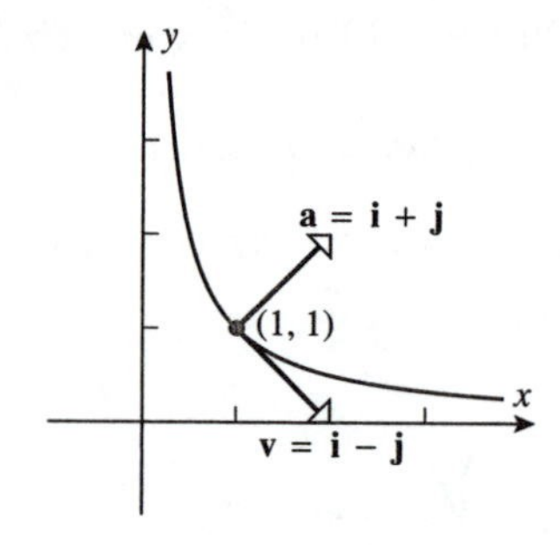

5. $\mathbf{v} = \mathbf{i} + t\mathbf{j} + t^2\mathbf{k}$, $\mathbf{a} = \mathbf{j} + 2t\mathbf{k}$; at $t = 1$, $\mathbf{v} = \mathbf{i} + \mathbf{j} + \mathbf{k}$, $\|\mathbf{v}\| = \sqrt{3}$, $\mathbf{a} = \mathbf{j} + 2\mathbf{k}$

7. $\mathbf{v} = -2\sin t\mathbf{i} + 2\cos t\mathbf{j} + \mathbf{k}$, $\mathbf{a} = -2\cos t\mathbf{i} - 2\sin t\mathbf{j}$;
at $t = \pi/4$, $\mathbf{v} = -\sqrt{2}\mathbf{i} + \sqrt{2}\mathbf{j} + \mathbf{k}$, $\|\mathbf{v}\| = \sqrt{5}$, $\mathbf{a} = -\sqrt{2}\mathbf{i} - \sqrt{2}\mathbf{j}$

9. (a) $\mathbf{v} = -a\omega \sin \omega t\mathbf{i} + b\omega \cos \omega t\mathbf{j}$, $\mathbf{a} = -a\omega^2 \cos \omega t\mathbf{i} - b\omega^2 \sin \omega t\mathbf{j} = -\omega^2\mathbf{r}$

(b) From Part (a), $\|\mathbf{a}\| = \omega^2\|\mathbf{r}\|$

11. $\mathbf{v} = (6/\sqrt{t})\mathbf{i} + (3/2)t^{1/2}\mathbf{j}$, $\|\mathbf{v}\| = \sqrt{36/t + 9t/4}$, $d\|\mathbf{v}\|/dt = (-36/t^2 + 9/4)/(2\sqrt{36/t + 9t/4}) = 0$ if $t = 4$ which yields a minimum by the first derivative test. The minimum speed is $3\sqrt{2}$ when $\mathbf{r} = 24\mathbf{i} + 8\mathbf{j}$.

13. (a)

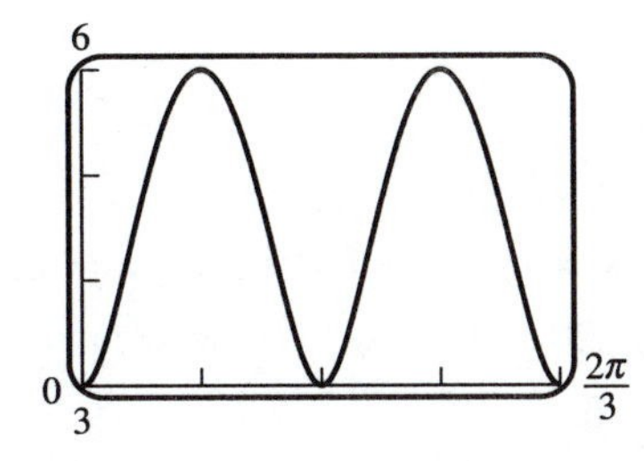

(b) $\mathbf{v} = 3\cos 3t\mathbf{i} + 6\sin 3t\mathbf{j}$, $\|\mathbf{v}\| = \sqrt{9\cos^2 3t + 36\sin^2 3t} = 3\sqrt{1 + 3\sin^2 3t}$; by inspection, maximum speed is 6 and minimum speed is 3

(d) $\dfrac{d}{dt}\|\mathbf{v}\| = \dfrac{27\sin 6t}{2\sqrt{1 + 3\sin^2 3t}} = 0$ when $t = 0, \pi/6, \pi/3, \pi/2, 2\pi/3$; the maximum speed is 6 which occurs first when $\sin 3t = 1, t = \pi/6$.

15. $\mathbf{v}(t) = -\sin t\mathbf{i} + \cos t\mathbf{j} + \mathbf{C}_1$, $\mathbf{v}(0) = \mathbf{j} + \mathbf{C}_1 = \mathbf{i}$, $\mathbf{C}_1 = \mathbf{i} - \mathbf{j}$, $\mathbf{v}(t) = (1 - \sin t)\mathbf{i} + (\cos t - 1)\mathbf{j}$;
$\mathbf{r}(t) = (t + \cos t)\mathbf{i} + (\sin t - t)\mathbf{j} + \mathbf{C}_2$, $\mathbf{r}(0) = \mathbf{i} + \mathbf{C}_2 = \mathbf{j}$,
$\mathbf{C}_2 = -\mathbf{i} + \mathbf{j}$ so $\mathbf{r}(t) = (t + \cos t - 1)\mathbf{i} + (\sin t - t + 1)\mathbf{j}$

17. $\mathbf{v}(t) = -\cos t\mathbf{i} + \sin t\mathbf{j} + e^t\mathbf{k} + \mathbf{C}_1$, $\mathbf{v}(0) = -\mathbf{i} + \mathbf{k} + \mathbf{C}_1 = \mathbf{k}$ so
$\mathbf{C}_1 = \mathbf{i}$, $\mathbf{v}(t) = (1 - \cos t)\mathbf{i} + \sin t\mathbf{j} + e^t\mathbf{k}$; $\mathbf{r}(t) = (t - \sin t)\mathbf{i} - \cos t\mathbf{j} + e^t\mathbf{k} + \mathbf{C}_2$,
$\mathbf{r}(0) = -\mathbf{j} + \mathbf{k} + \mathbf{C}_2 = -\mathbf{i} + \mathbf{k}$ so $\mathbf{C}_2 = -\mathbf{i} + \mathbf{j}$, $\mathbf{r}(t) = (t - \sin t - 1)\mathbf{i} + (1 - \cos t)\mathbf{j} + e^t\mathbf{k}$.

19. If $\mathbf{a} = \mathbf{0}$ then $x''(t) = y''(t) = z''(t) = 0$, so $x(t) = x_1 t + x_0, y(t) = y_1 t + y_0, z(t) = z_1 t + z_0$, the motion is along a straight line and has constant speed.

21. $\mathbf{v} = 3t^2\mathbf{i} + 2t\mathbf{j}$, $\mathbf{a} = 6t\mathbf{i} + 2\mathbf{j}$; $\mathbf{v} = 3\mathbf{i} + 2\mathbf{j}$ and $\mathbf{a} = 6\mathbf{i} + 2\mathbf{j}$ when $t = 1$ so
$\cos\theta = (\mathbf{v} \cdot \mathbf{a})/(\|\mathbf{v}\|\,\|\mathbf{a}\|) = 11/\sqrt{130}$, $\theta \approx 15°$.

23. (a) displacement $= \mathbf{r}_1 - \mathbf{r}_0 = 0.7\mathbf{i} + 2.7\mathbf{j} - 3.4\mathbf{k}$

(b) $\Delta\mathbf{r} = \mathbf{r}_1 - \mathbf{r}_0$, so $\mathbf{r}_0 = \mathbf{r}_1 - \Delta\mathbf{r} = -0.7\mathbf{i} - 2.9\mathbf{j} + 4.8\mathbf{k}$.

25. $\Delta\mathbf{r} = \mathbf{r}(3) - \mathbf{r}(1) = 8\mathbf{i} + (26/3)\mathbf{j}$; $\mathbf{v} = 2t\mathbf{i} + t^2\mathbf{j}$, $s = \displaystyle\int_1^3 t\sqrt{4 + t^2}\,dt = (13\sqrt{13} - 5\sqrt{5})/3$.

27. $\Delta\mathbf{r} = \mathbf{r}(\ln 3) - \mathbf{r}(0) = 2\mathbf{i} - (2/3)\mathbf{j} + \sqrt{2}(\ln 3)\mathbf{k}$; $\mathbf{v} = e^t\mathbf{i} - e^{-t}\mathbf{j} + \sqrt{2}\,\mathbf{k}$, $s = \displaystyle\int_0^{\ln 3} (e^t + e^{-t})dt = 8/3$.

29. In both cases, the equation of the path in rectangular coordinates is $x^2 + y^2 = 4$, the particles move counterclockwise around this circle; $\mathbf{v}_1 = -6\sin 3t\mathbf{i} + 6\cos 3t\mathbf{j}$ and
$\mathbf{v}_2 = -4t\sin(t^2)\mathbf{i} + 4t\cos(t^2)\mathbf{j}$ so $\|\mathbf{v}_1\| = 6$ and $\|\mathbf{v}_2\| = 4t$.

31. (a) $\mathbf{v} = -e^{-t}\mathbf{i} + e^t\mathbf{j}$, $\mathbf{a} = e^{-t}\mathbf{i} + e^t\mathbf{j}$; when $t = 0$, $\mathbf{v} = -\mathbf{i} + \mathbf{j}$, $\mathbf{a} = \mathbf{i} + \mathbf{j}$, $\|\mathbf{v}\| = \sqrt{2}$, $\mathbf{v} \cdot \mathbf{a} = 0$, $\mathbf{v} \times \mathbf{a} = -2\mathbf{k}$ so $a_T = 0$, $a_N = \sqrt{2}$.

(b) $a_T\mathbf{T} = \mathbf{0}, a_N\mathbf{N} = \mathbf{a} - a_T\mathbf{T} = \mathbf{i} + \mathbf{j}$ (c) $\kappa = 1/\sqrt{2}$

33. (a) $\mathbf{v} = (3t^2 - 2)\mathbf{i} + 2t\mathbf{j}$, $\mathbf{a} = 6t\mathbf{i} + 2\mathbf{j}$; when $t = 1$, $\mathbf{v} = \mathbf{i} + 2\mathbf{j}$, $\mathbf{a} = 6\mathbf{i} + 2\mathbf{j}$, $\|\mathbf{v}\| = \sqrt{5}$, $\mathbf{v} \cdot \mathbf{a} = 10$, $\mathbf{v} \times \mathbf{a} = -10\mathbf{k}$ so $a_T = 2\sqrt{5}$, $a_N = 2\sqrt{5}$

(b) $a_T\mathbf{T} = \dfrac{2\sqrt{5}}{\sqrt{5}}(\mathbf{i} + 2\mathbf{j}) = 2\mathbf{i} + 4\mathbf{j}$, $a_N\mathbf{N} = \mathbf{a} - a_T\mathbf{T} = 4\mathbf{i} - 2\mathbf{j}$

(c) $\kappa = 2/\sqrt{5}$

35. (a) $\mathbf{v} = (-1/t^2)\mathbf{i} + 2t\mathbf{j} + 3t^2\mathbf{k}$, $\mathbf{a} = (2/t^3)\,\mathbf{i} + 2\mathbf{j} + 6t\mathbf{k}$; when $t = 1$, $\mathbf{v} = -\mathbf{i} + 2\mathbf{j} + 3\mathbf{k}$, $\mathbf{a} = 2\mathbf{i} + 2\mathbf{j} + 6\mathbf{k}$, $\|\mathbf{v}\| = \sqrt{14}$, $\mathbf{v} \cdot \mathbf{a} = 20$, $\mathbf{v} \times \mathbf{a} = 6\mathbf{i} + 12\mathbf{j} - 6\mathbf{k}$ so $a_T = 20/\sqrt{14}$, $a_N = 6\sqrt{3}/\sqrt{7}$

(b) $a_T\mathbf{T} = -\dfrac{10}{7}\mathbf{i} + \dfrac{20}{7}\mathbf{j} + \dfrac{30}{7}\mathbf{k}$, $a_N\mathbf{N} = \mathbf{a} - a_T\mathbf{T} = \dfrac{24}{7}\mathbf{i} - \dfrac{6}{7}\mathbf{j} + \dfrac{12}{7}\mathbf{k}$

(c) $\kappa = \dfrac{6\sqrt{6}}{14^{3/2}} = \left(\dfrac{3}{7}\right)^{3/2}$

37. (a) $\mathbf{v} = 3\cos t\mathbf{i} - 2\sin t\mathbf{j} - 2\cos 2t\mathbf{k}$, $\mathbf{a} = -3\sin t\mathbf{i} - 2\cos t\mathbf{j} + 4\sin 2t\mathbf{k}$; when $t = \pi/2$, $\mathbf{v} = -2\mathbf{j} + 2\mathbf{k}$, $\mathbf{a} = -3\mathbf{i}$, $\|\mathbf{v}\| = 2\sqrt{2}$, $\mathbf{v} \cdot \mathbf{a} = 0$, $\mathbf{v} \times \mathbf{a} = -6\mathbf{j} - 6\mathbf{k}$ so $a_T = 0$, $a_N = 3$

(b) $a_T\mathbf{T} = \mathbf{0}, a_N\mathbf{N} = \mathbf{a} = -3\mathbf{i}$

(c) $\kappa = \dfrac{3}{8}$

39. $\|\mathbf{v}\| = 4$, $\mathbf{v} \cdot \mathbf{a} = -12$, $\mathbf{v} \times \mathbf{a} = 8\mathbf{k}$ so $a_T = -3$, $a_N = 2$, $\mathbf{T} = -\mathbf{j}$, $\mathbf{N} = (\mathbf{a} - a_T\mathbf{T})/a_N = \mathbf{i}$

41. $\|\mathbf{v}\| = 3$, $\mathbf{v} \cdot \mathbf{a} = 4$, $\mathbf{v} \times \mathbf{a} = 4\mathbf{i} - 3\mathbf{j} - 2\mathbf{k}$ so $a_T = 4/3$, $a_N = \sqrt{29}/3$, $\mathbf{T} = (1/3)(2\mathbf{i} + 2\mathbf{j} + \mathbf{k})$, $\mathbf{N} = (\mathbf{a} - a_T\mathbf{T})/a_N = (\mathbf{i} - 8\mathbf{j} + 14\mathbf{k})/(3\sqrt{29})$

43. $a_T = \dfrac{d^2s}{dt^2} = \dfrac{d}{dt}\sqrt{3t^2 + 4} = 3t/\sqrt{3t^2 + 4}$ so when $t = 2$, $a_T = 3/2$.

45. $a_T = \dfrac{d^2s}{dt^2} = \dfrac{d}{dt}\sqrt{(4t - 1)^2 + \cos^2 \pi t} = [4(4t - 1) - \pi \cos \pi t \sin \pi t]/\sqrt{(4t - 1)^2 + \cos^2 \pi t}$ so when $t = 1/4$, $a_T = -\pi/\sqrt{2}$.

47. $a_N = \kappa(ds/dt)^2 = (1/\rho)(ds/dt)^2 = (1/1)(2.9 \times 10^5)^2 = 8.41 \times 10^{10}$ km/s^2

49. $a_N = \kappa(ds/dt)^2 = [2/(1 + 4x^2)^{3/2}](3)^2 = 18/(1 + 4x^2)^{3/2}$

51. $\mathbf{a} = a_T\mathbf{T} + a_N\mathbf{N}$; by the Pythagorean Theorem $a_N = \sqrt{\|\mathbf{a}\|^2 - a_T^2} = \sqrt{9 - 9} = 0$

53. Let $c = ds/dt, a_N = \kappa\left(\dfrac{ds}{dt}\right)^2, a_N = \dfrac{1}{1000}c^2$, so $c^2 = 1000a_N$, $c \le 10\sqrt{10}\sqrt{1.5} \approx 38.73$ m/s.

55. (a) $v_0 = 320$, $\alpha = 60°$, $s_0 = 0$ so $x = 160t$, $y = 160\sqrt{3}t - 16t^2$.

(b) $dy/dt = 160\sqrt{3} - 32t$, $dy/dt = 0$ when $t = 5\sqrt{3}$ so $y_{\max} = 160\sqrt{3}(5\sqrt{3}) - 16(5\sqrt{3})^2 = 1200$ ft.

(c) $y = 16t(10\sqrt{3} - t)$, $y = 0$ when $t = 0$ or $10\sqrt{3}$ so $x_{\max} = 160(10\sqrt{3}) = 1600\sqrt{3}$ ft.

(d) $\mathbf{v}(t) = 160\mathbf{i} + (160\sqrt{3} - 32t)\mathbf{j}$, $\mathbf{v}(10\sqrt{3}) = 160(\mathbf{i} - \sqrt{3}\mathbf{j})$, $\|\mathbf{v}(10\sqrt{3})\| = 320$ ft/s.

57. $v_0 = 80$, $\alpha = -60°$, $s_0 = 168$ so $x = 40t$, $y = 168 - 40\sqrt{3}\,t - 16t^2$; $y = 0$ when $t = -7\sqrt{3}/2$ (invalid) or $t = \sqrt{3}$ so $x(\sqrt{3}) = 40\sqrt{3}$ ft.

59. $\alpha = 30°$, $s_0 = 0$ so $x = \sqrt{3}v_0t/2$, $y = v_0t/2 - 16t^2$; $dy/dt = v_0/2 - 32t$, $dy/dt = 0$ when $t = v_0/64$ so $y_{\max} = v_0^2/256 = 2500$, $v_0 = 800$ ft/s.

61. $v_0 = 800$, $s_0 = 0$ so $x = (800\cos\alpha)t$, $y = (800\sin\alpha)t - 16t^2 = 16t(50\sin\alpha - t)$; $y = 0$ when $t = 0$ or $50\sin\alpha$ so $x_{\max} = 40,000\sin\alpha\cos\alpha = 20,000\sin 2\alpha = 10,000$, $2\alpha = 30°$ or $150°$, $\alpha = 15°$ or $75°$.

63. (a) Let $\mathbf{r}(t) = x(t)\mathbf{i} + y(t)\mathbf{j}$ with $\mathbf{j}$ pointing up. Then $\mathbf{a} = -32\mathbf{j} = x''(t)\mathbf{i} + y''(t)\mathbf{j}$, so $x(t) = At + B, y(t) = -16t^2 + Ct + D$. Next, $x(0) = 0, y(0) = 4$ so $x(t) = At, y(t) = -16t^2 + Ct + 4$; $y'(0)/x'(0) = \tan 60° = \sqrt{3}$, so $C = \sqrt{3}A$; and $40 = v_0 = \sqrt{x'(0)^2 + y'(0)^2} = \sqrt{A^2 + 3A^2}, A = 20$, thus $\mathbf{r}(t) = 20t\,\mathbf{i} + (-16t^2 + 20\sqrt{3}t + 4)\,\mathbf{j}$. When $x = 15$, $t = \dfrac{3}{4}$, and $y = 4 + 20\sqrt{3}\dfrac{3}{4} - 16\left(\dfrac{3}{4}\right)^2 \approx 20.98$ ft, so the water clears the corner point A with 0.98 ft to spare.

(b) $y = 20$ when $-16t^2 + 20\sqrt{3}t - 16 = 0, t = 0.668$ (reject) or $1.497, x(1.497) \approx 29.942$ ft, so the water hits the roof.

(c) about $29.942 - 15 = 14.942$ ft

65. (a) $x = (35\sqrt{2}/2)t$, $y = (35\sqrt{2}/2)t - 4.9t^2$, from Exercise 17a in Section 13.5

$$\kappa = \frac{|x'y'' - x''y'|}{[(x')^2 + (y')^2]^{3/2}},\ \kappa(0) = \frac{9.8}{35^2\sqrt{2}} = 0.004\sqrt{2} \approx 0.00565685; \rho = 1/\kappa \approx 176.78 \text{ m}$$

(b) $y'(t) = 0$ when $t = \dfrac{25}{14}\sqrt{2}, y = \dfrac{125}{4}$ m

67. $s_0 = 0$ so $x = (v_0\cos\alpha)t$, $y = (v_0\sin\alpha)t - gt^2/2$

(a) $dy/dt = v_0\sin\alpha - gt$ so $dy/dt = 0$ when $t = (v_0\sin\alpha)/g$, $y_{\max} = (v_0\sin\alpha)^2/(2g)$

(b) $y = 0$ when $t = 0$ or $(2v_0\sin\alpha)/g$, so $x = R = (2v_0^2\sin\alpha\cos\alpha)/g = (v_0^2\sin 2\alpha)/g$ when $t = (2v_0\sin\alpha)/g$; R is maximum when $2\alpha = 90°$, $\alpha = 45°$, and the maximum value of R is v_0^2/g.

69. $v_0 = 80$, $\alpha = 30°$, $s_0 = 5$ so $x = 40\sqrt{3}t$, $y = 5 + 40t - 16t^2$

(a) $y = 0$ when $t = (-40 \pm \sqrt{(40)^2 - 4(-16)(5)})/(-32) = (5 \pm \sqrt{30})/4$, reject $(5 - \sqrt{30})/4$ to get $t = (5 + \sqrt{30})/4 \approx 2.62$ s.

(b) $x \approx 40\sqrt{3}(2.62) \approx 181.5$ ft.

71. (a) $v_0(\cos\alpha)(2.9) = 259\cos 23°$ so $v_0\cos\alpha \approx 82.21061$, $v_0(\sin\alpha)(2.9) - 16(2.9)^2 = -259\sin 23°$ so $v_0\sin\alpha \approx 11.50367$; divide $v_0\sin\alpha$ by $v_0\cos\alpha$ to get $\tan\alpha \approx 0.139929$, thus $\alpha \approx 8°$ and $v_0 \approx 82.21061/\cos 8° \approx 83$ ft/s.

(b) From Part (a), $x \approx 82.21061t$ and $y \approx 11.50367t - 16t^2$ for $0 \le t \le 2.9$; the distance traveled is $\int_0^{2.9} \sqrt{(dx/dt)^2 + (dy/dt)^2}dt \approx 268.76$ ft.

EXERCISE SET 13.7

1. The results follow from formulae (1) and (7) of Section 11.6.

3. (a) From (15) and (6), at $t = 0$,
$\mathbf{C} = \mathbf{v}_0 \times \mathbf{b}_0 - GM\mathbf{u} = v_0\mathbf{j} \times r_0v_0\mathbf{k} - GM\mathbf{u} = r_0v_0^2\mathbf{i} - GM\mathbf{i} = (r_0v_0^2 - GM)\mathbf{i}$

(b) From (22), $r_0v_0^2 - GM = GMe$, so from (7) and (17), $\mathbf{v} \times \mathbf{b} = GM(\cos\theta\mathbf{i} + \sin\theta\mathbf{j}) + GMe\mathbf{i}$, and the result follows.

(c) From (10) it follows that $\mathbf{b}$ is perpendicular to $\mathbf{v}$, and the result follows.

(d) From Part (c) and (10), $\|\mathbf{v} \times \mathbf{b}\| = \|\mathbf{v}\|\|\mathbf{b}\| = vr_0v_0$. From Part (b),
$\|\mathbf{v} \times \mathbf{b}\| = GM\sqrt{(e + \cos\theta)^2 + \sin^2\theta} = GM\sqrt{e^2 + 2e\cos\theta + 1}$. By (10) and Part (c), $\|\mathbf{v} \times \mathbf{b}\| = \|\mathbf{v}\|\|\mathbf{b}\| = v(r_0v_0)$ thus $v = \dfrac{GM}{r_0v_0}\sqrt{e^2 + 2e\cos\theta + 1}$. From (22), $r_0v_0^2/(GM) = 1 + e$, $GM/(r_0v_0) = v_0/(1 + e)$ so $v = \dfrac{v_0}{1+e}\sqrt{e^2 + 2e\cos\theta + 1}$.

5. $v_{\max}$ occurs when $\theta = 0$ so $v_{\max} = v_0$; $v_{\min}$ occurs when $\theta = \pi$ so
$v_{\min} = \dfrac{v_0}{1+e}\sqrt{e^2 - 2e + 1} = v_{\max}\dfrac{1-e}{1+e}$, thus $v_{\max} = v_{\min}\dfrac{1+e}{1-e}$.

7. $r_0 = 6440 + 200 = 6640$ km so $v = \sqrt{3.99 \times 10^5/6640} \approx 7.75$ km/s.

9. From (23) with $r_0 = 6440 + 300 = 6740$ km, $v_{\text{esc}} = \sqrt{\dfrac{2(3.99) \times 10^5}{6740}} \approx 10.88$ km/s.

11. (a) At perigee, $r = r_{\min} = a(1 - e) = 238{,}900\,(1 - 0.055) \approx 225{,}760$ mi; at apogee, $r = r_{\max} = a(1 + e) = 238{,}900(1 + 0.055) \approx 252{,}040$ mi. Subtract the sum of the radius of the Moon and the radius of the Earth to get
minimum distance $= 225{,}760 - 5080 = 220{,}680$ mi,
and maximum distance $= 252{,}040 - 5080 = 246{,}960$ mi.

(b) $T = 2\pi\sqrt{a^3/(GM)} = 2\pi\sqrt{(238{,}900)^3/(1.24 \times 10^{12})} \approx 659$ hr ≈ 27.5 days.

13. (a) $r_0 = 4000 + 180 = 4180$ mi, $v = \sqrt{\dfrac{GM}{r_0}} = \sqrt{1.24 \times 10^{12}/4180} \approx 17{,}224$ mi/h

(b) $r_0 = 4180$ mi, $v_0 = \sqrt{\dfrac{GM}{r_0}} + 600$; $e = \dfrac{r_0v_0^2}{GM} - 1 = 1200\sqrt{\dfrac{r_0}{GM}} + (600)^2\dfrac{r_0}{GM} \approx 0.071$;
$r_{\max} = 4180(1 + 0.071)/(1 - 0.071) \approx 4819$ mi; the apogee altitude is $4819 - 4000 = 819$ mi.

CHAPTER 13 SUPPLEMENTARY EXERCISES

7. **(a)** $\mathbf{r}(t) = \int_0^t \cos\left(\frac{\pi u^2}{2}\right) du\,\mathbf{i} + \int_0^t \sin\left(\frac{\pi u^2}{2}\right) du\,\mathbf{j}$;

$$\left\|\frac{d\mathbf{r}}{dt}\right\|^2 = x'(t)^2 + y'(t)^2 = \cos^2\left(\frac{\pi t^2}{2}\right) + \sin^2\left(\frac{\pi t^2}{2}\right) = 1 \text{ and } \mathbf{r}(0) = \mathbf{0}$$

(b) $\mathbf{r}'(s) = \cos\left(\frac{\pi s^2}{2}\right)\mathbf{i} + \sin\left(\frac{\pi s^2}{2}\right)\mathbf{j}, \mathbf{r}''(s) = -\pi s \sin\left(\frac{\pi s^2}{2}\right)\mathbf{i} + \pi s \cos\left(\frac{\pi s^2}{2}\right)\mathbf{j}$,

$\kappa = \|\mathbf{r}''(s)\| = \pi|s|$

(c) $\kappa(s) \to +\infty$, so the spiral winds ever tighter.

9. **(a)** $\|\mathbf{r}(t)\| = 1$, so by Theorem 13.2.9, $\mathbf{r}'(t)$ is always perpendicular to the vector $\mathbf{r}(t)$. Then $\mathbf{v}(t) = R\omega(-\sin\omega t\mathbf{i} + \cos\omega t\mathbf{j}), v = \|\mathbf{v}(t)\| = R\omega$

(b) $\mathbf{a} = -R\omega^2(\cos\omega t\mathbf{i} + \sin\omega t\mathbf{j}), a = \|\mathbf{a}\| = R\omega^2$, and $\mathbf{a} = -\omega^2\mathbf{r}$ is directed toward the origin.

(c) The smallest value of t for which $\mathbf{r}(t) = \mathbf{r}(0)$ satisfies $\omega t = 2\pi$, so $T = t = \frac{2\pi}{\omega}$.

11. **(a)** Let $\mathbf{r} = x\mathbf{i} + y\mathbf{j} + z\mathbf{k}$, then $x^2 + z^2 = t^2(\sin^2\pi t + \cos^2\pi t) = t^2 = y^2$

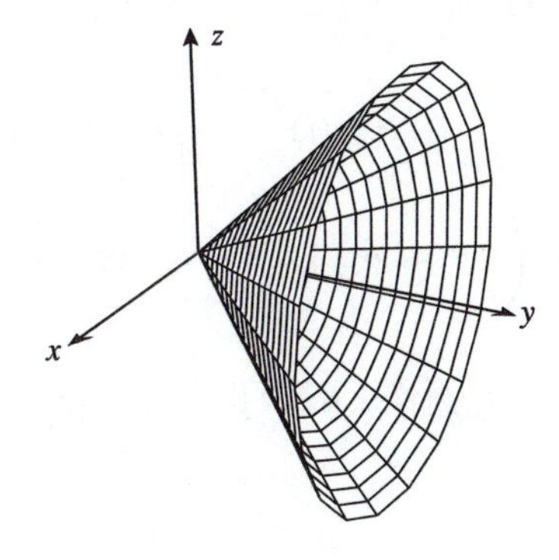

(b) Let $x = t$, then $y = t^2, z = \pm\sqrt{4 - t^2/3 - t^4/6}$

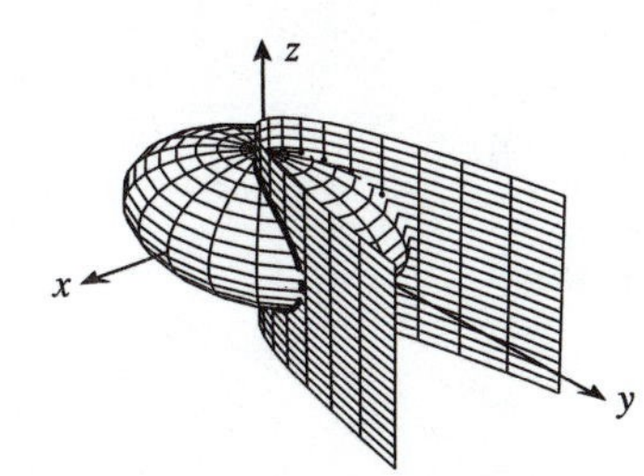

13. **(a)** $\|\mathbf{e}_r(t)\|^2 = \cos^2\theta + \sin^2\theta = 1$, so $\mathbf{e}_r(t)$ is a unit vector; $\mathbf{r}(t) = r(t)\mathbf{e}(t)$, so they have the same direction if $r(t) > 0$, opposite if $r(t) < 0$. $\mathbf{e}_\theta(t)$ is perpendicular to $\mathbf{e}_r(t)$ since $\mathbf{e}_r(t) \cdot \mathbf{e}_\theta(t) = 0$, and it will result from a counterclockwise rotation of $\mathbf{e}_r(t)$ provided $\mathbf{e}(t) \times \mathbf{e}_\theta(t) = \mathbf{k}$, which is true.

(b) $\frac{d}{dt}\mathbf{e}_r(t) = \frac{d\theta}{dt}(-\sin\theta\mathbf{i} + \cos\theta\mathbf{j}) = \frac{d\theta}{dt}\mathbf{e}_\theta(t)$ and $\frac{d}{dt}\mathbf{e}_\theta(t) = -\frac{d\theta}{dt}(\cos\theta\mathbf{i} + \sin\theta\mathbf{j}) = -\frac{d\theta}{dt}\mathbf{e}_r(t)$, so

$$\mathbf{v}(t) = \frac{d}{dt}\mathbf{r}(t) = \frac{d}{dt}(r(t)\mathbf{e}_r(t)) = r'(t)\mathbf{e}_r(t) + r(t)\frac{d\theta}{dt}\mathbf{e}_\theta(t)$$

(c) From Part (b), $\mathbf{a} = \frac{d}{dt}\mathbf{v}(t)$

$$= r''(t)\mathbf{e}_r(t) + r'(t)\frac{d\theta}{dt}\mathbf{e}_\theta(t) + r'(t)\frac{d\theta}{dt}\mathbf{e}_\theta(t) + r(t)\frac{d^2\theta}{dt^2}\mathbf{e}_\theta(t) - r(t)\left(\frac{d\theta}{dt}\right)^2\mathbf{e}_r(t)$$

$$= \left[\frac{d^2r}{dt^2} - r\left(\frac{d\theta}{dt}\right)^2\right]\mathbf{e}_r(t) + \left[r\frac{d^2\theta}{dt^2} + 2\frac{dr}{dt}\frac{d\theta}{dt}\right]\mathbf{e}_\theta(t)$$

15. $\mathbf{r} = \mathbf{r}_0 + t\overrightarrow{PQ} = (t-1)\mathbf{i} + (4-2t)\mathbf{j} + (3+2t)\mathbf{k}$; $\left\|\frac{d\mathbf{r}}{dt}\right\| = 3$, $\mathbf{r}(s) = \frac{s-3}{3}\mathbf{i} + \frac{12-2s}{3}\mathbf{j} + \frac{9+2s}{3}\mathbf{k}$

17. $\mathbf{r}'(1) = 3\mathbf{i} + 10\mathbf{j} + 10\mathbf{k}$, so if $\mathbf{r}'(t) = 3t^2\mathbf{i} + 10\mathbf{j} + 10t\mathbf{k}$ is perpendicular to $\mathbf{r}'(1)$, then $9t^2 + 100 + 100t = 0, t = -10, -10/9$,
so $\mathbf{r} = -1000\mathbf{i} - 100\mathbf{j} + 500\mathbf{k}, -(1000/729)\mathbf{i} - (100/9)\mathbf{j} + (500/81)\mathbf{k}$.

19. (a) $\frac{d\mathbf{v}}{dt} = 2t^2\mathbf{i} + \mathbf{j} + \cos 2t\mathbf{k}$, $\mathbf{v}_0 = \mathbf{i} + 2\mathbf{j} - \mathbf{k}$, so $x'(t) = \frac{2}{3}t^3 + 1, y'(t) = t + 2, z'(t) = \frac{1}{2}\sin 2t - 1$,

$x(t) = \frac{1}{6}t^4 + t, y(t) = \frac{1}{2}t^2 + 2t, z(t) = -\frac{1}{4}\cos 2t - t + \frac{1}{4}$, since $\mathbf{r}(0) = \mathbf{0}$. Hence

$$\mathbf{r}(t) = \left(\frac{1}{6}t^4 + t\right)\mathbf{i} + \left(\frac{1}{2}t^2 + 2t\right)\mathbf{j} - \left(\frac{1}{4}\cos 2t + t - \frac{1}{4}\right)\mathbf{k}$$

(b) $\left.\frac{ds}{dt}\right]_{t=1} = \|\mathbf{r}'(t)\|\Big]_{t=1}\sqrt{(5/3)^2 + 9 + (1 - (\sin 2)/2)^2} \approx 3.475$

CHAPTER 14
Partial Derivatives

EXERCISE SET 14.1

1. **(a)** $f(2,1) = (2)^2(1) + 1 = 5$ **(b)** $f(1,2) = (1)^2(2) + 1 = 3$
(c) $f(0,0) = (0)^2(0) + 1 = 1$ **(d)** $f(1,-3) = (1)^2(-3) + 1 = -2$
(e) $f(3a,a) = (3a)^2(a) + 1 = 9a^3 + 1$ **(f)** $f(ab,\, a-b) = (ab)^2(a-b) + 1 = a^3b^2 - a^2b^3 + 1$

3. **(a)** $f(x+y, x-y) = (x+y)(x-y) + 3 = x^2 - y^2 + 3$
(b) $f\left(xy, 3x^2y^3\right) = (xy)\left(3x^2y^3\right) + 3 = 3x^3y^4 + 3$

5. $F(g(x), h(y)) = F\left(x^3, 3y+1\right) = x^3e^{x^3(3y+1)}$

7. **(a)** $t^2 + 3t^{10}$ **(b)** 0 **(c)** 3076

9. **(a)** At $T = 25$ there is a drop in temperature of 12 degrees when v changes from 5 to 10, thus $WCI \approx (2/5)(-12) + 22 = 22 - 24/5 = 17.2°$ F.
(b) At $v = 5$ there is an increase in temperature of 5 degrees as T changes from 25 to 30 degrees, thus $WCI \approx (3/5)5 + 22 = 25°$ F.

11. **(a)** The depression is $20 - 16 = 4$, so the relative humidity is 66%.
(b) The relative humidity $\approx 77 - (1/2)7 = 73.5\%$.
(c) The relative humidity $\approx 59 + (2/5)4 = 60.6\%$.

13. **(a)** 19 **(b)** -9 **(c)** 3
(d) $a^6 + 3$ **(e)** $-t^8 + 3$ **(f)** $(a+b)(a-b)^2b^3 + 3$

15. $F\left(x^2, y+1, z^2\right) = (y+1)e^{x^2(y+1)z^2}$

17. **(a)** $f(\sqrt{5}, 2, \pi, -3\pi) = 80\sqrt{\pi}$ **(b)** $f(1,1,\ldots,1) = \sum_{k=1}^{n} k = n(n+1)/2$

19.

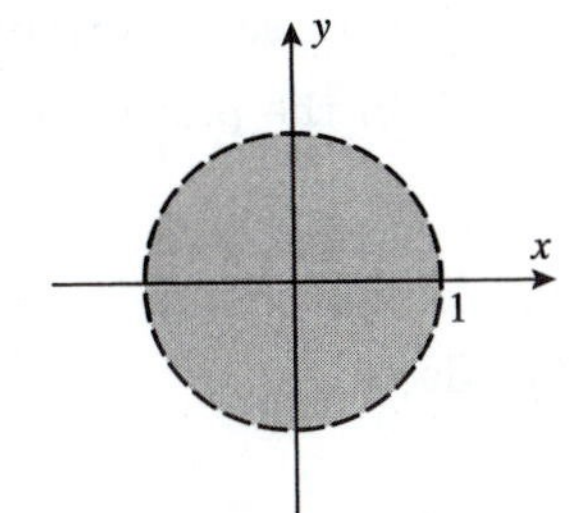

21.

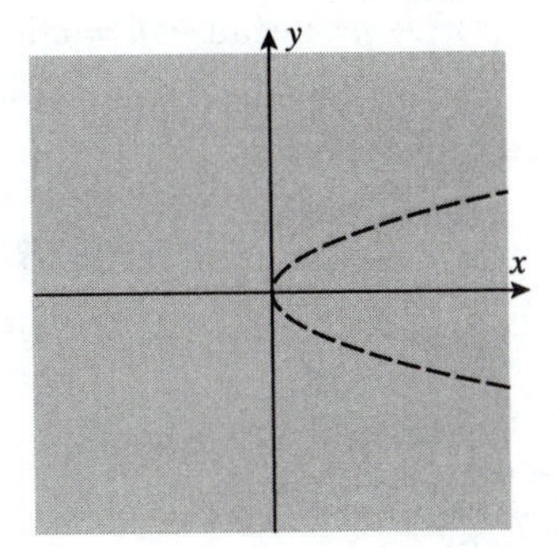

23. **(a)** all points above or on the line $y = -2$
(b) all points on or within the sphere $x^2 + y^2 + z^2 = 25$
(c) all points in 3-space

25.

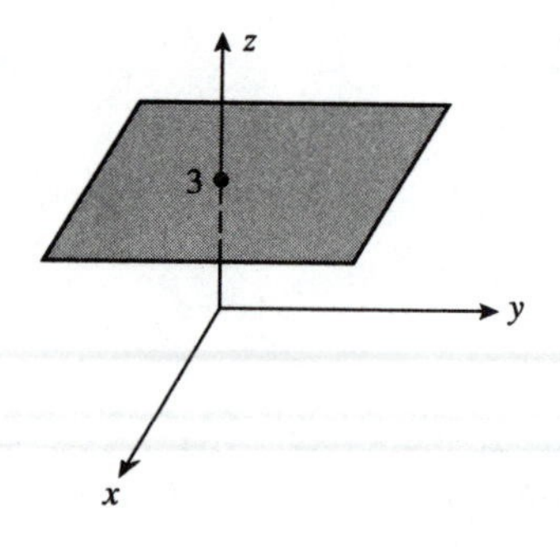

27.

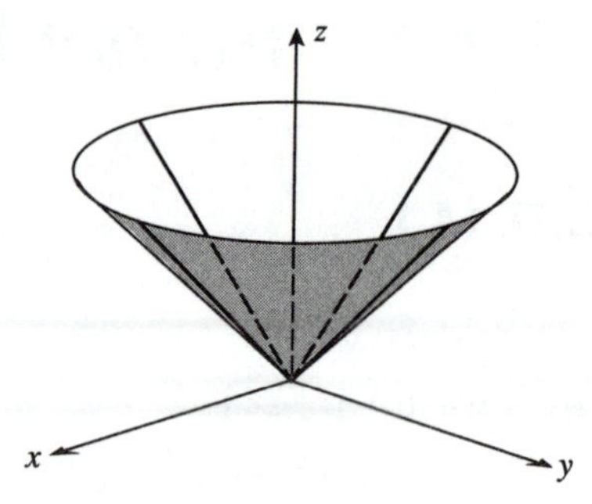

29.

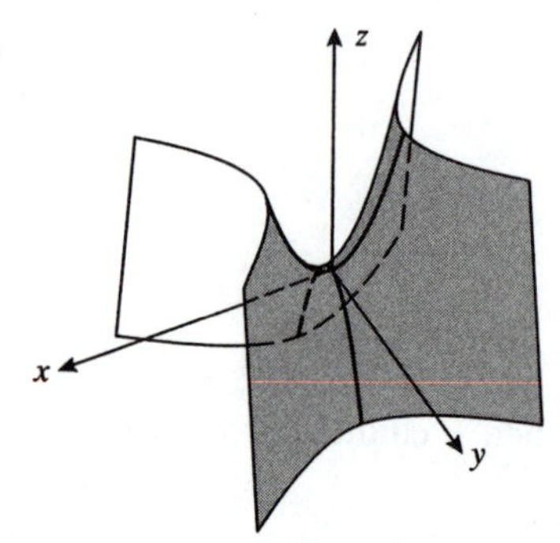

31.

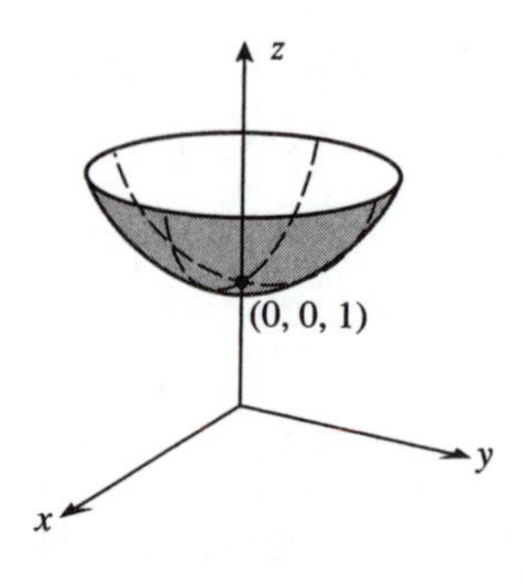

33.

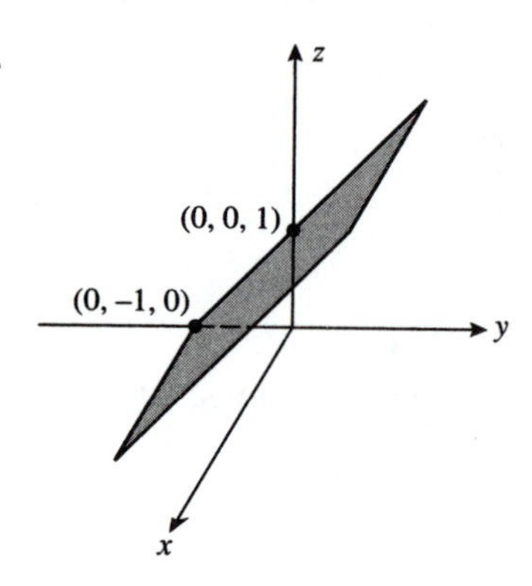

35. **(a)** $f(x,y) = 1 - x^2 - y^2$, because $f = c$ is a circle of radius $\sqrt{1-c}$ (provided $c \leq 1$), and the radii in (a) decrease as c increases.

(b) $f(x,y) = \sqrt{x^2+y^2}$ because $f = c$ is a circle of radius c, and the radii increase uniformly.

(c) $f(x,y) = x^2 + y^2$ because $f = c$ is a circle of radius $\sqrt{c}$ and the radii in the plot grow like the square root function.

37. **(a)** A **(b)** B **(c)** increase

(d) decrease **(e)** increase **(f)** decrease

39.

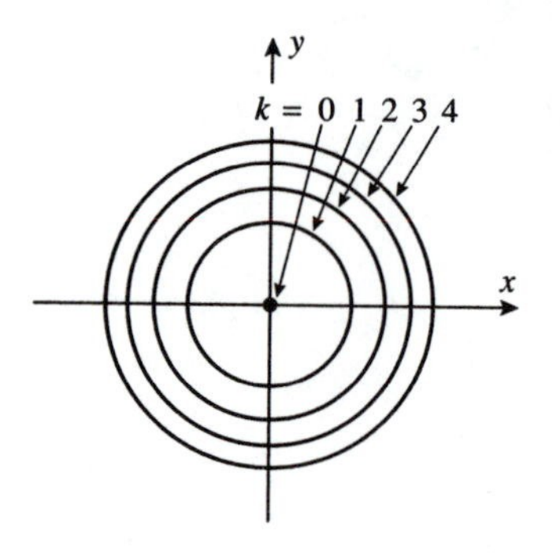

41.

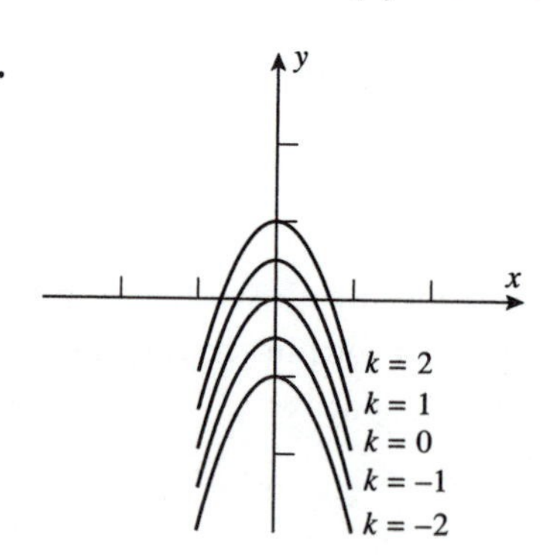

43.

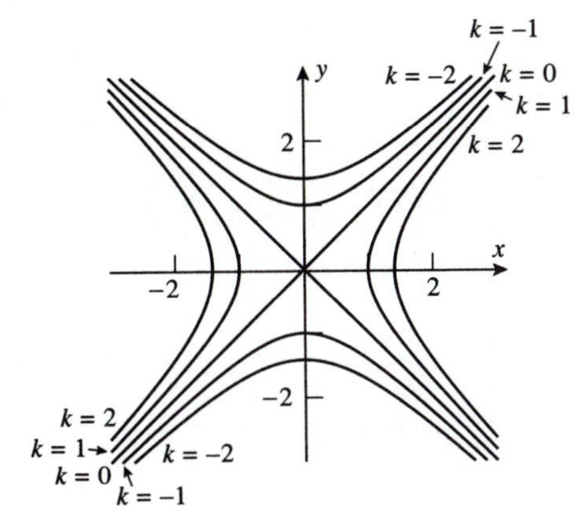

45.

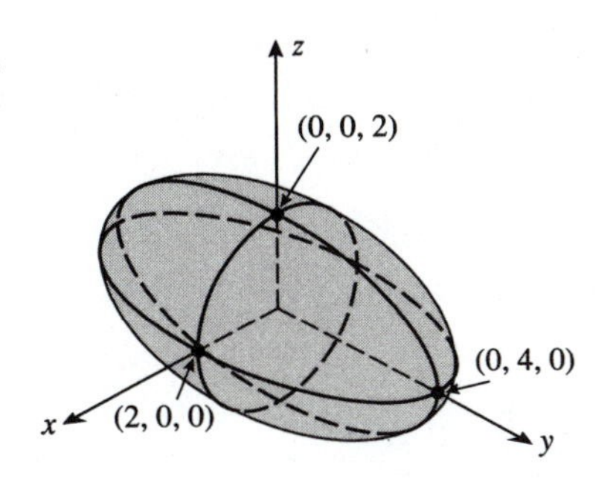

47.

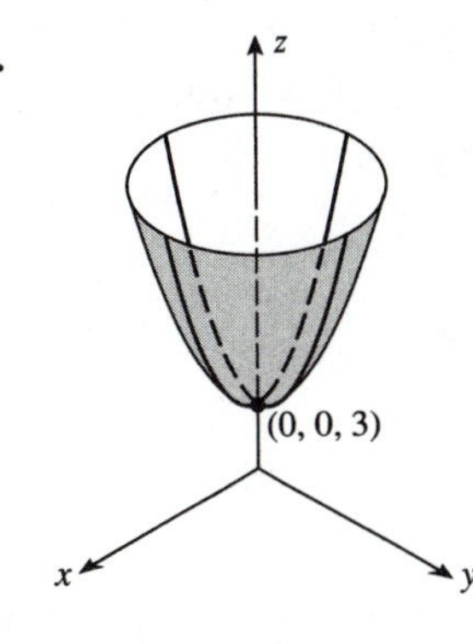

49. concentric spheres, common center at (2,0,0)

51. concentric cylinders, common axis the y-axis

53. **(a)** $f(-1,1) = 0$; $x^2 - 2x^3 + 3xy = 0$ **(b)** $f(0,0) = 0$; $x^2 - 2x^3 + 3xy = 0$
(c) $f(2,-1) = -18$; $x^2 - 2x^3 + 3xy = -18$

55. **(a)** $f(1,-2,0) = 5$; $x^2 + y^2 - z = 5$ **(b)** $f(1,0,3) = -2$; $x^2 + y^2 - z = -2$
(c) $f(0,0,0) = 0$; $x^2 + y^2 - z = 0$

57. **(a)**

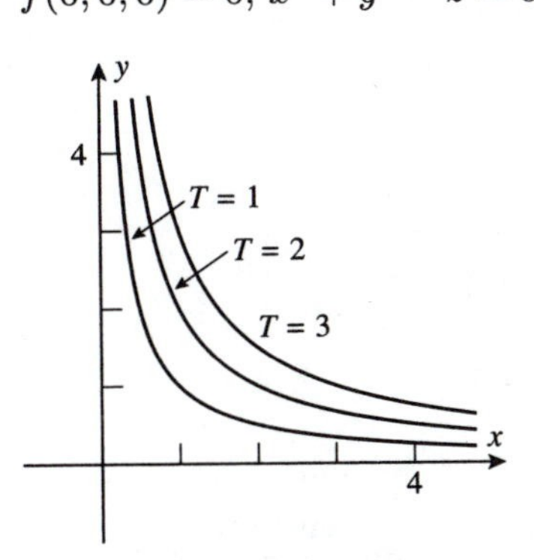

(b) At $(1,4)$ the temperature is $T(1,4) = 4$ so the temperature will remain constant along the path $xy = 4$.

59. **(a)**

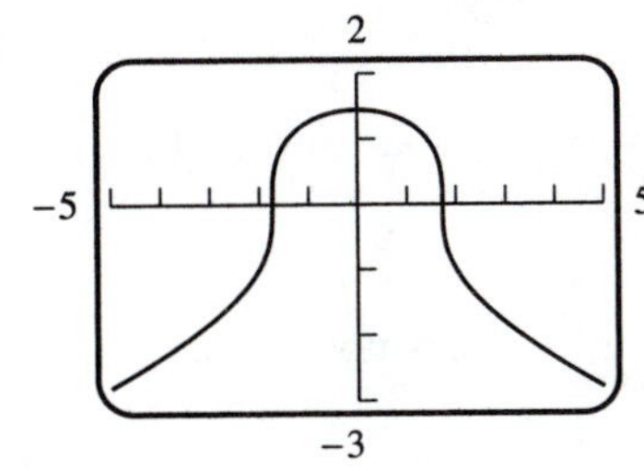

(b)

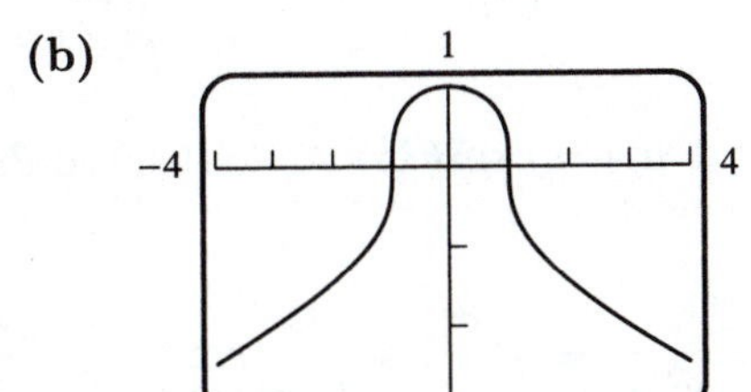

61. (a)

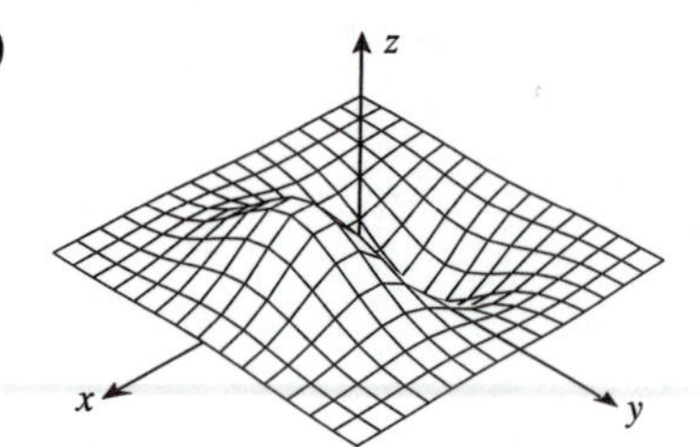

(b)

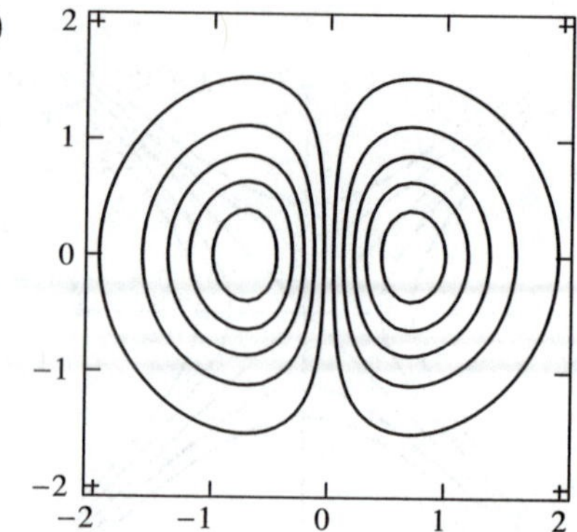

63. (a) The graph of g is the graph of f shifted one unit in the positive x-direction.
(b) The graph of g is the graph of f shifted one unit up the z-axis.
(c) The graph of g is the graph of f shifted one unit down the y-axis and then inverted with respect to the plane $z = 0$.

EXERCISE SET 14.2

1. 35 **3.** −8 **5.** 0

7. (a) Along $x = 0$ $\lim\limits_{(x,y)\to(0,0)} \dfrac{3}{x^2 + 2y^2} = \lim\limits_{y\to 0} \dfrac{3}{2y^2}$ does not exist.

(b) Along $x = 0$, $\lim\limits_{(x,y)\to(0,0)} \dfrac{x + y}{x + y^2} = \lim\limits_{y\to 0} \dfrac{1}{y}$ does not exist.

9. Let $z = x^2 + y^2$, then $\lim\limits_{(x,y)\to(0,0)} \dfrac{\sin(x^2 + y^2)}{x^2 + y^2} = \lim\limits_{z\to 0+} \dfrac{\sin z}{z} = 1$

11. Let $z = x^2 + y^2$, then $\lim\limits_{(x,y)\to(0,0)} e^{-1/(x^2+y^2)} = \lim\limits_{z\to 0+} e^{-1/z} = 0$

13. $\lim\limits_{(x,y)\to(0,0)} \dfrac{(x^2 + y^2)(x^2 - y^2)}{x^2 + y^2} = \lim\limits_{(x,y)\to(0,0)} (x^2 - y^2) = 0$

15. along $y = 0$: $\lim\limits_{x\to 0} \dfrac{0}{3x^2} = \lim\limits_{x\to 0} 0 = 0$; along $y = x$: $\lim\limits_{x\to 0} \dfrac{x^2}{5x^2} = \lim\limits_{x\to 0} 1/5 = 1/5$
so the limit does not exist.

17. 8/3

19. Let $t = \sqrt{x^2 + y^2 + z^2}$, then $\lim\limits_{(x,y,z)\to(0,0,0)} \dfrac{\sin(x^2 + y^2 + z^2)}{\sqrt{x^2 + y^2 + z^2}} = \lim\limits_{t\to 0+} \dfrac{\sin(t^2)}{t} = 0$

21. $y\ln(x^2 + y^2) = r\sin\theta\ln r^2 = 2r(\ln r)\sin\theta$, so $\lim\limits_{(x,y)\to(0,0)} y\ln(x^2 + y^2) = \lim\limits_{r\to 0+} 2r(\ln r)\sin\theta = 0$

23. $\dfrac{e^{\sqrt{x^2+y^2+z^2}}}{\sqrt{x^2 + y^2 + z^2}} = \dfrac{e^{\rho}}{\rho}$, so $\lim\limits_{(x,y,z)\to(0,0,0)} \dfrac{e^{\sqrt{x^2+y^2+z^2}}}{\sqrt{x^2 + y^2 + z^2}} = \lim\limits_{\rho\to 0+} \dfrac{e^{\rho}}{\rho}$ does not exist.

25. **(a)** No, since there seem to be points near $(0,0)$ with $z=0$ and other points near $(0,0)$ with $z \approx 1/2$.

(b) $\lim\limits_{x\to 0} \dfrac{mx^3}{x^4+m^2x^2} = \lim\limits_{x\to 0} \dfrac{mx}{x^2+m^2} = 0$ **(c)** $\lim\limits_{x\to 0} \dfrac{x^4}{2x^4} = \lim\limits_{x\to 0} 1/2 = 1/2$

(d) A limit must be unique if it exists, so $f(x,y)$ cannot have a limit as $(x,y)\to(0,0)$.

27. **(a)** $\lim\limits_{t\to 0} \dfrac{abct^3}{a^2t^2+b^4t^4+c^4t^4} = \lim\limits_{t\to 0} \dfrac{abct}{a^2+b^4t^2+c^4t^2} = 0$

(b) $\lim\limits_{t\to 0} \dfrac{t^4}{t^4+t^4+t^4} = \lim\limits_{t\to 0} 1/3 = 1/3$

29. $-\pi/2$ because $\dfrac{x^2-1}{x^2+(y-1)^2} \to -\infty$ as $(x,y)\to(0,1)$

31. No, because $\lim\limits_{(x,y)\to(0,0)} \dfrac{x^2}{x^2+y^2}$ does not exist.

Along $x=0$: $\lim\limits_{y\to 0}(0/y^2) = \lim\limits_{y\to 0} 0 = 0$; along $y=0$: $\lim\limits_{x\to 0}(x^2/x^2) = \lim\limits_{x\to 0} 1 = 1$.

33.

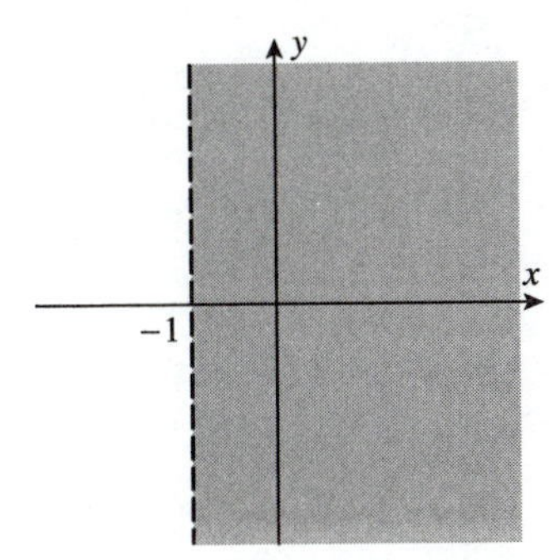

35.

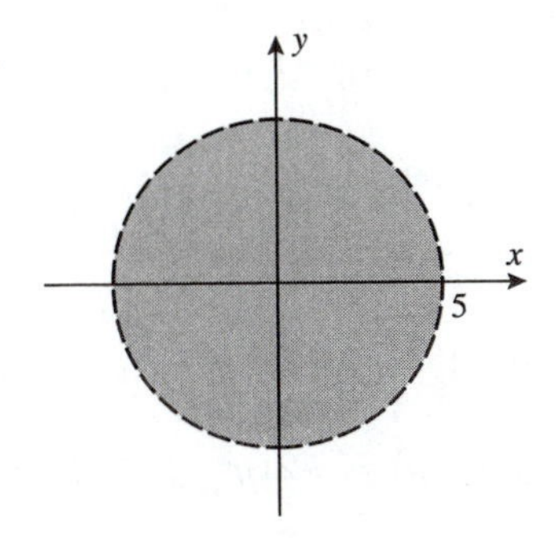

37.

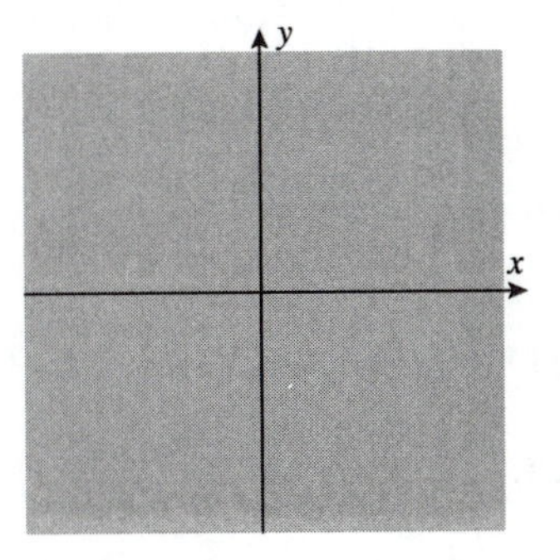

39.

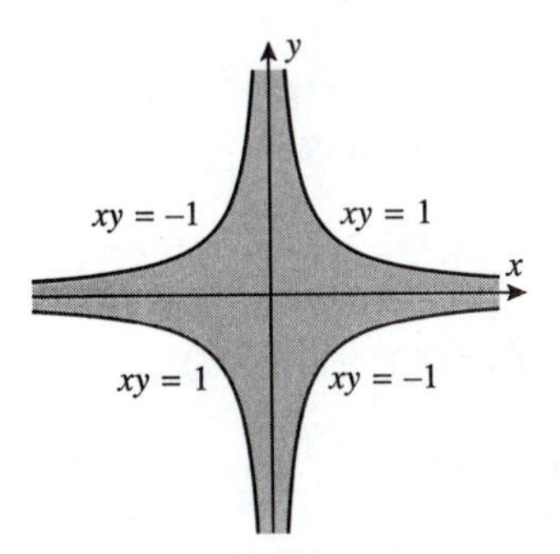

41. all of 3-space

43. all points not on the cylinder $x^2+z^2=1$

EXERCISE SET 14.3

1. **(a)** $9x^2y^2$ **(b)** $6x^3y$ **(c)** $9y^2$ **(d)** $9x^2$
(e) $6y$ **(f)** $6x^3$ **(g)** 36 **(h)** 12

3. **(a)** $\dfrac{\partial z}{\partial x} = \dfrac{3}{2\sqrt{3x+2y}}$; slope $= \dfrac{3}{8}$ **(b)** $\dfrac{\partial z}{\partial y} = \dfrac{1}{\sqrt{3x+2y}}$; slope $= \dfrac{1}{4}$

5. **(a)** $\dfrac{\partial z}{\partial x} = -4\cos(y^2 - 4x)$; rate of change $= -4\cos 7$

(b) $\dfrac{\partial z}{\partial y} = 2y\cos(y^2 - 4x)$; rate of change $= 2\cos 7$

7. $\partial z/\partial x =$ slope of line parallel to xz-plane $= -4$; $\partial z/\partial y =$ slope of line parallel to yz-plane $= 1/2$

9. **(a)** The right-hand estimate is $\partial r/\partial v \approx (222 - 197)/(85 - 80) = 5$; the left-hand estimate is $\partial r/\partial v \approx (197 - 173)/(80 - 75) = 4.8$; the average is $\partial r/\partial v \approx 4.9$.

(b) The right-hand estimate is $\partial r/\partial \theta \approx (200 - 197)/(45 - 40) = 0.6$; the left-hand estimate is $\partial r/\partial \theta \approx (197 - 188)/(40 - 35) = 1.8$; the average is $\partial r/\partial \theta \approx 1.2$.

11. $\partial z/\partial x = 8xy^3e^{x^2y^3}$, $\partial z/\partial y = 12x^2y^2e^{x^2y^3}$

13. $\partial z/\partial x = x^3/(y^{3/5} + x) + 3x^2\ln(1 + xy^{-3/5})$, $\partial z/\partial y = -(3/5)x^4/(y^{8/5} + xy)$

15. $\dfrac{\partial z}{\partial x} = -\dfrac{y(x^2 - y^2)}{(x^2 + y^2)^2}, \ \dfrac{\partial z}{\partial y} = \dfrac{x(x^2 - y^2)}{(x^2 + y^2)^2}$

17. $f_x(x, y) = (3/2)x^2y\left(5x^2 - 7\right)\left(3x^5y - 7x^3y\right)^{-1/2}$
$f_y(x, y) = (1/2)x^3\left(3x^2 - 7\right)\left(3x^5y - 7x^3y\right)^{-1/2}$

19. $f_x(x, y) = \dfrac{y^{-1/2}}{y^2 + x^2}, \ f_y(x, y) = -\dfrac{xy^{-3/2}}{y^2 + x^2} - \dfrac{3}{2}y^{-5/2}\tan^{-1}(x/y)$

21. $f_x(x, y) = -(4/3)y^2\sec^2 x\left(y^2\tan x\right)^{-7/3}, \ f_y(x, y) = -(8/3)y\tan x\left(y^2\tan x\right)^{-7/3}$

23. $f_x(x, y) = -2x, \ f_x(3, 1) = -6; \ f_y(x, y) = -21y^2, \ f_y(3, 1) = -21$

25. $\partial z/\partial x = x(x^2{+}4y^2)^{-1/2}, \ \partial z/\partial x\,|_{(1,2)} = 1/\sqrt{17}\,; \ \partial z/\partial y = 4y(x^2{+}4y^2)^{-1/2}, \ \partial z/\partial y\,|_{(1,2)} = 8/\sqrt{17}$

27. **(a)** $2xy^4z^3 + y$ **(b)** $4x^2y^3z^3 + x$ **(c)** $3x^2y^4z^2 + 2z$
(d) $2y^4z^3 + y$ **(e)** $32z^3 + 1$ **(f)** 438

29. $f_x = 2z/x, \ f_y = z/y, \ f_z = \ln(x^2y\cos z) - z\tan z$

31. $f_x = -y^2z^3/\left(1 + x^2y^4z^6\right), \ f_y = -2xyz^3/\left(1 + x^2y^4z^6\right), \ f_z = -3xy^2z^2/\left(1 + x^2y^4z^6\right)$

33. $\partial w/\partial x = yze^z\cos xz, \ \partial w/\partial y = e^z\sin xz, \ \partial w/\partial z = ye^z(\sin xz + x\cos xz)$

35. $\partial w/\partial x = x/\sqrt{x^2 + y^2 + z^2}, \ \partial w/\partial y = y/\sqrt{x^2 + y^2 + z^2}, \ \partial w/\partial z = z/\sqrt{x^2 + y^2 + z^2}$

37. **(a)** e **(b)** $2e$ **(c)** e

39. **(a)**

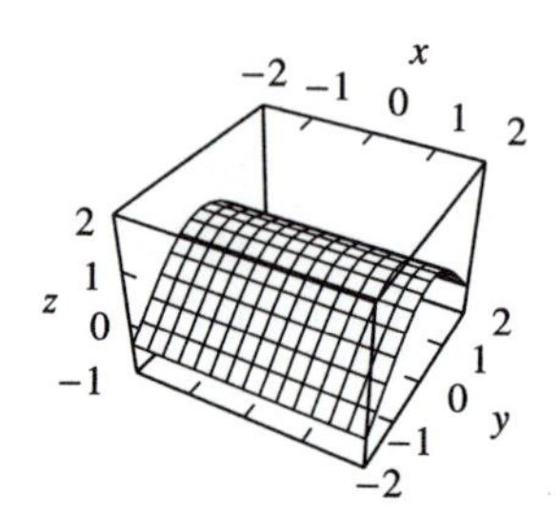

(b)

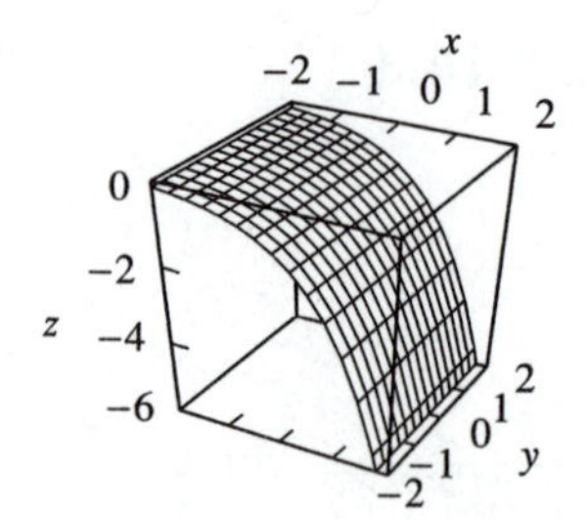

41. $\partial z/\partial x = 2x + 6y(\partial y/\partial x) = 2x,\ \partial z/\partial x]_{(2,1)} = 4$

43. $\partial z/\partial x = -x\left(29 - x^2 - y^2\right)^{-1/2},\ \partial z/\partial x]_{(4,3)} = -2$

45. **(a)** $\partial V/\partial r = 2\pi rh$ **(b)** $\partial V/\partial h = \pi r^2$

(c) $\partial V/\partial r]_{r=6,\ h=4} = 48\pi$ **(d)** $\partial V/\partial h]_{r=8,\ h=10} = 64\pi$

47. **(a)** $P = 10T/V,\ \partial P/\partial T = 10/V,\ \partial P/\partial T]_{T=80,\ V=50} = 1/5\ \text{lb}/(\text{in}^2\text{K})$

(b) $V = 10T/P, \partial V/\partial P = -10T/P^2$, if $V = 50$ and $T = 80$ then $P = 10(80)/(50) = 16,\ \partial V/\partial P]_{T=80,\ P=16} = -25/8(\text{in}^5/\text{lb})$

49. **(a)** $V = lwh, \partial V/\partial l = wh = 6$

(b) $\partial V/\partial w = lh = 15$

(c) $\partial V/\partial h = lw = 10$

51. $\partial V/\partial r = \dfrac{2}{3}\pi rh = \dfrac{2}{r}(\dfrac{1}{3}\pi r^2 h) = 2V/r$

53. **(a)** $2x - 2z(\partial z/\partial x) = 0,\ \partial z/\partial x = x/z = \pm 3/(2\sqrt{6}) = \pm\sqrt{6}/4$

(b) $z = \pm\sqrt{x^2+y^2-1},\ \partial z/\partial x = \pm x/\sqrt{x^2+y^2-1} = \pm\sqrt{6}/4$

55. $\dfrac{3}{2}\left(x^2+y^2+z^2\right)^{1/2}\left(2x + 2z\dfrac{\partial z}{\partial x}\right) = 0,\ \partial z/\partial x = -x/z$; similarly, $\partial z/\partial y = -y/z$

57. $2x + z\left(xy\dfrac{\partial z}{\partial x} + yz\right)\cos xyz + \dfrac{\partial z}{\partial x}\sin xyz = 0,\ \dfrac{\partial z}{\partial x} = -\dfrac{2x + yz^2\cos xyz}{xyz\cos xyz + \sin xyz}$;

$z\left(xy\dfrac{\partial z}{\partial y} + xz\right)\cos xyz + \dfrac{\partial z}{\partial y}\sin xyz = 0,\ \dfrac{\partial z}{\partial y} = -\dfrac{xz^2\cos xyz}{xyz\cos xyz + \sin xyz}$

59. $(3/2)\left(x^2+y^2+z^2+w^2\right)^{1/2}\left(2x + 2w\dfrac{\partial w}{\partial x}\right) = 0,\ \partial w/\partial x = -x/w$; similarly, $\partial w/\partial y = -y/w$ and $\partial w/\partial z = -z/w$

61. $\dfrac{\partial w}{\partial x} = -\dfrac{yzw\cos xyz}{2w + \sin xyz},\ \dfrac{\partial w}{\partial y} = -\dfrac{xzw\cos xyz}{2w + \sin xyz},\ \dfrac{\partial w}{\partial z} = -\dfrac{xyw\cos xyz}{2w + \sin xyz}$

63. $f_x = e^{x^2},\ f_y = -e^{y^2}$

65. **(a)** $-\dfrac{1}{4x^{3/2}}\cos y$ **(b)** $-\sqrt{x}\cos y$ **(c)** $-\dfrac{\sin y}{2\sqrt{x}}$ **(d)** $-\dfrac{\sin y}{2\sqrt{x}}$

67. $f_x = 8x - 8y^4,\ f_y = -32xy^3 + 35y^4,\ f_{xy} = f_{yx} = -32y^3$

69. $f_x = e^x\cos y,\ f_y = -e^x\sin y,\ f_{xy} = f_{yx} = -e^x\sin y$

71. $f_x = 4/(4x-5y),\ f_y = -5/(4x-5y),\ f_{xy} = f_{yx} = 20/(4x-5y)^2$

73. $f_x = 2y/(x+y)^2, f_y = -2x/(x+y)^2, f_{xy} = f_{yx} = 2(x-y)/(x+y)^3$

75. III is a plane, and its partial derivatives are constants, so III cannot be $f(x,y)$. If I is the graph of $z=f(x,y)$ then (by inspection) f_y is constant as y varies, but neither II nor III is constant as y varies. Hence $z=f(x,y)$ has II as its graph, and as II seems to be an odd function of x and an even function of y, f_x has I as its graph and f_y has III as its graph.

77. (a) $\dfrac{\partial^3 f}{\partial x^3}$ (b) $\dfrac{\partial^3 f}{\partial y^2 \partial x}$ (c) $\dfrac{\partial^4 f}{\partial x^2 \partial y^2}$ (d) $\dfrac{\partial^4 f}{\partial y^3 \partial x}$

79. (a) $30xy^4-4$ (b) $60x^2y^3$ (c) $60x^3y^2$

81. (a) $f_{xyy}(0,1)=-30$ (b) $f_{xxx}(0,1)=-125$ (c) $f_{yyxx}(0,1)=150$

83. (a) $f_{xy}=15x^2y^4z^7+2y$ (b) $f_{yz}=35x^3y^4z^6+3y^2$
(c) $f_{xz}=21x^2y^5z^6$ (d) $f_{zz}=42x^3y^5z^5$
(e) $f_{zyy}=140x^3y^3z^6+6y$ (f) $f_{xxy}=30xy^4z^7$
(g) $f_{zyx}=105x^2y^4z^6$ (h) $f_{xxyz}=210xy^4z^6$

85. (a) $f_x=2x+2y,\ f_{xx}=2,\ f_y=-2y+2x,\ f_{yy}=-2;\ f_{xx}+f_{yy}=2-2=0$

(b) $z_x=e^x\sin y-e^y\sin x,\ z_{xx}=e^x\sin y-e^y\cos x,\ z_y=e^x\cos y+e^y\cos x,$
$z_{yy}=-e^x\sin y+e^y\cos x;\ z_{xx}+z_{yy}=e^x\sin y-e^y\cos x-e^x\sin y+e^y\cos x=0$

(c) $$z_x=\frac{2x}{x^2+y^2}-2\frac{y}{x^2}\frac{1}{1+(y/x)^2}=\frac{2x-2y}{x^2+y^2},\ z_{xx}=-2\frac{x^2-y^2-2xy}{(x^2+y^2)^2},$$
$$z_y=\frac{2y}{x^2+y^2}+2\frac{1}{x}\frac{1}{1+(y/x)^2}=\frac{2y+2x}{x^2+y^2},\ z_{yy}=-2\frac{y^2-x^2+2xy}{(x^2+y^2)^2};$$
$$z_{xx}+z_{yy}=-2\frac{x^2-y^2-2xy}{(x^2+y^2)^2}-2\frac{y^2-x^2+2xy}{(x^2+y^2)^2}=0$$

87. $u_x=\omega\sin c\omega t\cos\omega x,\ u_{xx}=-\omega^2\sin c\omega t\sin\omega x,\ u_t=c\omega\cos c\omega t\sin\omega x,\ u_{tt}=-c^2\omega^2\sin c\omega t\sin\omega x;$
$u_{xx}-\dfrac{1}{c^2}u_{tt}=-\omega^2\sin c\omega t\sin\omega x-\dfrac{1}{c^2}(-c^2)\omega^2\sin c\omega t\sin\omega x=0$

89. $\partial u/\partial x=\partial v/\partial y$ and $\partial u/\partial y=-\partial v/\partial x$ so $\partial^2u/\partial x^2=\partial^2v/\partial x\partial y$, and $\partial^2u/\partial y^2=-\partial^2v/\partial y\partial x$, $\partial^2u/\partial x^2+\partial^2u/\partial y^2=\partial^2v/\partial x\partial y-\partial^2v/\partial y\partial x$, if $\partial^2v/\partial x\partial y=\partial^2v/\partial y\partial x$ then $\partial^2u/\partial x^2+\partial^2u/\partial y^2=0$; thus u satisfies Laplace's equation. The proof that v satisfies Laplace's equation is similar. Adding Laplace's equations for u and v gives Laplaces' equation for $u+v$.

91. $\partial f/\partial v=8vw^3x^4y^5,\ \partial f/\partial w=12v^2w^2x^4y^5,\ \partial f/\partial x=16v^2w^3x^3y^5,\ \partial f/\partial y=20v^2w^3x^4y^4$

93. $\partial f/\partial v_1=2v_1/\left(v_3^2+v_4^2\right),\ \partial f/\partial v_2=-2v_2/\left(v_3^2+v_4^2\right),\ \partial f/\partial v_3=-2v_3\left(v_1^2-v_2^2\right)/\left(v_3^2+v_4^2\right)^2,$
$\partial f/\partial v_4=-2v_4\left(v_1^2-v_2^2\right)/\left(v_3^2+v_4^2\right)^2$

95. (a) 0 (b) 0 (c) 0 (d) 0
(e) $2(1+yw)e^{yw}\sin z\cos z$ (f) $2xw(2+yw)e^{yw}\sin z\cos z$

97. $\partial w/\partial x_i=-i\sin(x_1+2x_2+\ldots+nx_n)$

99. (a) xy-plane, $f_x=12x^2y+6xy,\ f_y=4x^3+3x^2,\ f_{xy}=f_{yx}=12x^2+6x$
(b) $y\neq 0, f_x=3x^2/y,\ f_y=-x^3/y^2,\ f_{xy}=f_{yx}=-3x^2/y^2$

101. $f_x(2,-1) = \lim_{x\to 2} \dfrac{f(x,-1)-f(2,-1)}{x-2} = \lim_{x\to 2} \dfrac{2x^2+3x+1-15}{x-2} = \lim_{x\to 2}(2x+7) = 11$ and

$f_y(2,-1) = \lim_{y\to -1} \dfrac{f(2,y)-f(2,-1)}{y+1} = \lim_{y\to -1} \dfrac{8-6y+y^2-15}{y+1} = \lim_{y\to -1} y-7 = -8$

103. (a) $f_y(0,0) = \dfrac{d}{dy}[f(0,y)]\Big]_{y=0} = \dfrac{d}{dy}[y]\Big]_{y=0} = 1$

(b) If $(x,y) \neq (0,0)$, then $f_y(x,y) = \dfrac{1}{3}(x^3+y^3)^{-2/3}(3y^2) = \dfrac{y^2}{(x^3+y^3)^{2/3}}$;
$f_y(x,y)$ does not exist when $y \neq 0$ and $y = -x$

EXERCISE SET 14.4

1. (a) Let $f(x,y) = e^x \sin y$; $f(0,0) = 0$, $f_x(0,0) = 0$, $f_y(0,0) = 1$, so $e^x \sin y \approx y$

(b) Let $f(x,y) = \dfrac{2x+1}{y+1}$; $f(0,0) = 1$, $f_x(0,0) = 2$, $f_y(0,0) = -1$, so $\dfrac{2x+1}{y+1} \approx 1+2x-y$

3. (a) Let $f(x,y,z) = xyz+2$, then $f_x = f_y = f_z = 1$ at $x = y = z = 1$, and
$L(x,y,z) = f(1,1,1)+f_x(x-1)+f_y(y-1)+f_z(z-1) = 3+x-1+y-1+z-1 = x+y+z$

(b) Let $f(x,y,z) = \dfrac{4x}{y+z}$, then $f_x = 2$, $f_y = -1$, $f_z = -1$ at $x = y = z = 1$, and

$$\begin{aligned} L(x,y,z) &= f(1,1,1)+f_x(x-1)+f_y(y-1)+f_z(z-1) \\ &= 2+2(x-1)-(y-1)-(z-1) = 2x-y-z+2 \end{aligned}$$

5. $f(x,y) \approx f(3,4)+f_x(x-3)+f_y(y-4) = 5+2(x-3)-(y-4)$ and
$f(3.01, 3.98) \approx 5+2(0.01)-(-0.02) = 5.04$

7. $L(x,y) = f(1,1)+f_x(1,1)(x-1)+f_y(1,1)(y-1)$ and
$L(1.1, 0.9) = 3.15 = 3+2(0.1)+f_y(1,1)(-0.1)$ so $f_y(1,1) = -0.05/(-0.1) = 0.5$

9. $L(x,y,z) = f(1,2,3)+(x-1)+2(y-2)+3(z-3)$,
$f(1.01, 2.02, 3.03) \approx 4+0.01+2(0.02)+3(0.03) = 4.14$

11. $x-y+2z-2 = L(x,y,z) = f(3,2,1)+f_x(3,2,1)(x-3)+f_y(3,2,1)(y-2)+f_z(3,2,1)(z-1)$, so
$f_x(3,2,1) = 1$, $f_y(3,2,1) = -1$, $f_z(3,2,1) = 2$ and $f(3,2,1) = L(3,2,1) = 1$

13. $L(x,y) = f(x_0,y_0)+f_x(x_0,y_0)(x-x_0)+f_y(x_0,y_0)(y-y_0)$,
$2y-2x-2 = x_0^2+y_0^2+2x_0(x-x_0)+2y_0(y-y_0)$, from which it follows that $x_0 = -1$, $y_0 = 1$.

15. $L(x,y,z) = f(x_0,y_0,z_0)+f_x(x_0,y_0,z_0)(x-x_0)+f_y(x_0,y_0,z_0)(y-y_0)+f_z(x_0,y_0,z_0)(z-z_0)$,
$y+2z-1 = x_0y_0+z_0^2+y_0(x-x_0)+x_0(y-y_0)+2z_0(z-z_0)$, so that $x_0 = 1$, $y_0 = 0$, $z_0 = 1$.

17. (a) $f(P) = 1/5$, $f_x(P) = -x/(x^2+y^2)^{-3/2}\Big]_{(x,y)=(4,3)} = -4/125$,

$f_y(P) = -y/(x^2+y^2)^{-3/2}\Big]_{(x,y)=(4,3)} = -3/125$, $L(x,y) = \dfrac{1}{5} - \dfrac{4}{125}(x-4) - \dfrac{3}{125}(y-3)$

(b) $L(Q) - f(Q) = \frac{1}{5} - \frac{4}{125}(-0.08) - \frac{3}{125}(0.01) - 0.2023342382 \approx -0.0000142382$,

$|PQ| = \sqrt{0.08^2 + 0.01^2} \approx 0.0008062257748, |L(Q) - f(Q)|/|PQ| \approx 0.000176603$

19. **(a)** $f(P) = 0, f_x(P) = 0, f_y(P) = 0, L(x, y) = 0$

(b) $L(Q) - f(Q) = -0.003\sin(0.004) \approx -0.000012, |PQ| = \sqrt{0.003^2 + 0.004^2} = 0.005$,

$|L(Q) - f(Q)|/|PQ| \approx 0.0024$

21. **(a)** $f(P) = 6, f_x(P) = 6, f_y(P) = 3, f_z(P) = 2, L(x, y) = 6 + 6(x - 1) + 3(y - 2) + 2(z - 3)$

(b) $L(Q) - f(Q) = 6 + 6(0.001) + 3(0.002) + 2(0.003) - 6.018018006 = -.000018006$,

$|PQ| = \sqrt{0.001^2 + 0.002^2 + 0.003^2} \approx .0003741657387; \quad |L(Q) - f(Q)|/|PQ| \approx -0.000481$

23. **(a)** $f(P) = e, f_x(P) = e, f_y(P) = -e, f_z(P) = -e, L(x, y) = e + e(x - 1) - e(y + 1) - e(z + 1)$

(b) $L(Q) - f(Q) = e - 0.01e + 0.01e - 0.01e - 0.99e^{0.9999} = 0.99(e - e^{0.9999})$,

$|PQ| = \sqrt{0.01^2 + 0.01^2 + 0.01^2} \approx 0.01732, |L(Q) - f(Q)|/|PQ| \approx 0.01554$

25. $dz = 7dx - 2dy$

27. $dz = 3x^2y^2dx + 2x^3ydy$

29. $dz = \left[y/\left(1 + x^2y^2\right)\right]dx + \left[x/\left(1 + x^2y^2\right)\right]dy$

31. $dw = 8dx - 3dy + 4dz$

33. $dw = 3x^2y^2zdx + 2x^3yzdy + x^3y^2dz$

35. $dw = \frac{yz}{1 + x^2y^2z^2}dx + \frac{xz}{1 + x^2y^2z^2}dy + \frac{xy}{1 + x^2y^2z^2}dz$

37. $df = (2x + 2y - 4)dx + 2xdy$; $x = 1$, $y = 2$, $dx = 0.01$, $dy = 0.04$ so

$df = 0.10$ and $\Delta f = 0.1009$

39. $df = -x^{-2}dx - y^{-2}dy$; $x = -1$, $y = -2$, $dx = -0.02$, $dy = -0.04$ so

$df = 0.03$ and $\Delta f \approx 0.029412$

41. $df = 2y^2z^3dx + 4xyz^3dy + 6xy^2z^2dz, x = 1, y = -1, z = 2, dx = -0.01, dy = -0.02, dz = 0.02$ so
$df = 0.96$ and $\Delta f \approx 0.97929$

43. Label the four smaller rectangles A, B, C, D starting with the lower left and going clockwise. Then the increase in the area of the rectangle is represented by B, C and D; and the portions B and D represent the approximation of the increase in area given by the total differential.

45. $A = xy$, $dA = ydx + xdy$, $dA/A = dx/x + dy/y$, $|dx/x| \le 0.03$ and $|dy/y| \le 0.05$,
$|dA/A| \le |dx/x| + |dy/y| \le 0.08 = 8\%$

47. $z = \sqrt{x^2 + y^2}$, $dz = \frac{x}{\sqrt{x^2 + y^2}}dx + \frac{y}{\sqrt{x^2 + y^2}}dy$,

$$\frac{dz}{z} = \frac{x}{x^2 + y^2}dx + \frac{y}{x^2 + y^2}dy = \frac{x^2}{x^2 + y^2}\left(\frac{dx}{x}\right) + \frac{y^2}{x^2 + y^2}\left(\frac{dy}{y}\right),$$

$\left|\frac{dz}{z}\right| \le \frac{x^2}{x^2 + y^2}\left|\frac{dx}{x}\right| + \frac{y^2}{x^2 + y^2}\left|\frac{dy}{y}\right|$, if $\left|\frac{dx}{x}\right| \le r/100$ and $\left|\frac{dy}{y}\right| \le r/100$ then

$\left|\frac{dz}{z}\right| \le \frac{x^2}{x^2 + y^2}(r/100) + \frac{y^2}{x^2 + y^2}(r/100) = \frac{r}{100}$ so the percentage error in z is at most about $r\%$.

49. $dT = \dfrac{\pi}{g\sqrt{L/g}}dL - \dfrac{\pi L}{g^2\sqrt{L/g}}dg$, $\dfrac{dT}{T} = \dfrac{1}{2}\dfrac{dL}{L} - \dfrac{1}{2}\dfrac{dg}{g}$; $|dL/L| \le 0.005$ and $|dg/g| \le 0.001$ so $|dT/T| \le (1/2)(0.005) + (1/2)(0.001) = 0.003 = 0.3\%$

51. (a) $\left|\dfrac{d(xy)}{xy}\right| = \left|\dfrac{y\,dx + x\,dy}{xy}\right| = \left|\dfrac{dx}{x} + \dfrac{dy}{y}\right| \le \left|\dfrac{dx}{x}\right| + \left|\dfrac{dy}{y}\right| \le \dfrac{r}{100} + \dfrac{s}{100}$; $(r+s)\%$

(b) $\left|\dfrac{d(x/y)}{x/y}\right| = \left|\dfrac{y\,dx - x\,dy}{xy}\right| = \left|\dfrac{dx}{x} - \dfrac{dy}{y}\right| \le \left|\dfrac{dx}{x}\right| + \left|\dfrac{dy}{y}\right| \le \dfrac{r}{100} + \dfrac{s}{100}$; $(r+s)\%$

(c) $$\left|\frac{d(x^2y^3)}{x^2y^3}\right| = \left|\frac{2xy^3\,dx + 3x^2y^2\,dy}{x^2y^3}\right| = \left|2\frac{dx}{x} + 3\frac{dy}{y}\right| \le 2\left|\frac{dx}{x}\right| + 3\left|\frac{dy}{y}\right|$$
$$\le 2\frac{r}{100} + 3\frac{s}{100};\ (2r+3s)\%$$

(d) $$\left|\frac{d(x^3y^{1/2})}{x^3y^{1/2}}\right| = \left|\frac{3x^2y^{1/2}\,dx + (1/2)x^3y^{-1/2}\,dy}{x^3y^{1/2}}\right| = \left|3\frac{dx}{x} + \frac{1}{2}\frac{dy}{y}\right| \le 3\left|\frac{dx}{x}\right| + \frac{1}{2}\left|\frac{dy}{y}\right|$$
$$\le 3\frac{r}{100} + \frac{1}{2}\frac{s}{100};\ (3r + \tfrac{1}{2}s)\%$$

53. $dA = \dfrac{1}{2}b\sin\theta da + \dfrac{1}{2}a\sin\theta db + \dfrac{1}{2}ab\cos\theta d\theta$,

$$|dA| \le \frac{1}{2}b\sin\theta|da| + \frac{1}{2}a\sin\theta|db| + \frac{1}{2}ab\cos\theta|d\theta|$$
$$\le \frac{1}{2}(50)(1/2)(1/2) + \frac{1}{2}(40)(1/2)(1/4) + \frac{1}{2}(40)(50)\left(\sqrt{3}/2\right)(\pi/90)$$
$$= 35/4 + 50\pi\sqrt{3}/9 \approx 39 \text{ ft}^2$$

55. If $f(x,y) = f(x_0,y_0)$ for all (x,y) then $L(x,y) = f(x_0,y_0)$ since the first partial derivatives of f are zero. Thus the error E is zero and f is differentiable. The proof for three variables is analogous.

57. $f_x = 2x\sin y$, $f_y = x^2\cos y$ are both continuous everywhere, so f is differentiable everywhere.

59. $f_x = 2x$, $f_y = 2y$, $f_z = 2z$ so $L(x,y,z) = 0$, $E = f - L = x^2 + y^2 + z^2$, and $\displaystyle\lim_{(x,y,z)\to(0,0,0)} \frac{E(x,y,z)}{\sqrt{x^2+y^2+z^2}} = \lim_{(x,y,z)\to(0,0,0)} \sqrt{x^2+y^2+z^2} = 0$, so f is differentiable at $(0,0,0)$.

61. Let $\epsilon > 0$. Then $\displaystyle\lim_{x\to x_0} \frac{f(x)}{g(x)} = 0$ if and only if there exists $\delta > 0$ such that $\left|\dfrac{f(x)}{g(x)}\right| < \epsilon$ whenever $|x - x_0| < \delta$. But this condition is equivalent to $\left|\dfrac{f(x)}{|g(x)|}\right| < \epsilon$, and thus the two limits both exist or neither exists.

63. If f is differentiable at (x_0,y_0) then $L(x,y)$ exists and is a linear function and thus differentiable, and thus the difference $E = f - L$ is also differentiable.

65. Let $x > 0$. Then $\dfrac{f(x,y) - f(x,0)}{y - 0}$ can be $-1/y$ or 0 depending on whether $y > 0$ or $y < 0$. Thus the partial derivative $f_y(x,0)$ cannot exist. A similar argument works for $f_x(0,y)$ if $y > 0$.

EXERCISE SET 14.5

1. $42t^{13}$

3. $3t^{-2}\sin(1/t)$

5. $-\dfrac{10}{3}t^{7/3}e^{1-t^{10/3}}$

7. $165t^{32}$

9. $-2t\cos\left(t^2\right)$

11. 3264

13. $\partial z/\partial u = 24u^2v^2 - 16uv^3 - 2v + 3$, $\partial z/\partial v = 16u^3v - 24u^2v^2 - 2u - 3$

15. $\partial z/\partial u = -\dfrac{2\sin u}{3\sin v}$, $\partial z/\partial v = -\dfrac{2\cos u\cos v}{3\sin^2 v}$

17. $\partial z/\partial u = e^u$, $\partial z/\partial v = 0$

19. $\partial T/\partial r = 3r^2\sin\theta\cos^2\theta - 4r^3\sin^3\theta\cos\theta$
$\partial T/\partial\theta = -2r^3\sin^2\theta\cos\theta + r^4\sin^4\theta + r^3\cos^3\theta - 3r^4\sin^2\theta\cos^2\theta$

21. $\partial t/\partial x = \left(x^2+y^2\right)/\left(4x^2y^3\right)$, $\partial t/\partial y = \left(y^2-3x^2\right)/\left(4xy^4\right)$

23. $\partial z/\partial r = (dz/dx)(\partial x/\partial r) = 2r\cos^2\theta/\left(r^2\cos^2\theta+1\right)$,
$\partial z/\partial\theta = (dz/dx)(\partial x/\partial\theta) = -2r^2\sin\theta\cos\theta/\left(r^2\cos^2\theta+1\right)$

25. $\partial w/\partial\rho = 2\rho\left(4\sin^2\phi+\cos^2\phi\right)$, $\partial w/\partial\phi = 6\rho^2\sin\phi\cos\phi$, $\partial w/\partial\theta = 0$

27. $-\pi$

29. $\sqrt{3}e^{\sqrt{3}}$, $\left(2-4\sqrt{3}\right)e^{\sqrt{3}}$

31. $F(x,y) = x^2y^3 + \cos y$, $\dfrac{dy}{dx} = -\dfrac{2xy^3}{3x^2y^2 - \sin y}$

33. $F(x,y) = e^{xy} + ye^y - 1$, $\dfrac{dy}{dx} = -\dfrac{ye^{xy}}{xe^{xy} + ye^y + e^y}$

35. $\dfrac{\partial F}{\partial x} + \dfrac{\partial F}{\partial z}\dfrac{\partial z}{\partial x} = 0$ so $\dfrac{\partial z}{\partial x} = -\dfrac{\partial F/\partial x}{\partial F/\partial z}$.

37. $\dfrac{\partial z}{\partial x} = \dfrac{2x+yz}{6yz-xy}$, $\dfrac{\partial z}{\partial y} = \dfrac{xz-3z^2}{6yz-xy}$

39. $ye^x - 5\sin 3z - 3z = 0$; $\dfrac{\partial z}{\partial x} = -\dfrac{ye^x}{-15\cos 3z - 3} = \dfrac{ye^x}{15\cos 3z + 3}$, $\dfrac{\partial z}{\partial y} = \dfrac{e^x}{15\cos 3z + 3}$

41. $D = \left(x^2+y^2\right)^{1/2}$ where x and y are the distances of cars A and B, respectively, from the intersection and D is the distance between them.
$dD/dt = \left[x/\left(x^2+y^2\right)^{1/2}\right](dx/dt) + \left[y/\left(x^2+y^2\right)^{1/2}\right](dy/dt)$, $dx/dt = -25$ and $dy/dt = -30$ when $x = 0.3$ and $y = 0.4$ so $dD/dt = (0.3/0.5)(-25) + (0.4/0.5)(-30) = -39$ mph.

43. $A = \dfrac{1}{2}ab\sin\theta$ but $\theta = \pi/6$ when $a = 4$ and $b = 3$ so $A = \dfrac{1}{2}(4)(3)\sin(\pi/6) = 3$.

Solve $\dfrac{1}{2}ab\sin\theta = 3$ for θ to get $\theta = \sin^{-1}\left(\dfrac{6}{ab}\right)$, $0 \le \theta \le \pi/2$.

$$\frac{d\theta}{dt} = \frac{\partial\theta}{\partial a}\frac{da}{dt} + \frac{\partial\theta}{\partial b}\frac{db}{dt} = \frac{1}{\sqrt{1-\dfrac{36}{a^2b^2}}}\left(-\frac{6}{a^2b}\right)\frac{da}{dt} + \frac{1}{\sqrt{1-\dfrac{36}{a^2b^2}}}\left(-\frac{6}{ab^2}\right)\frac{db}{dt}$$

$$= -\frac{6}{\sqrt{a^2b^2-36}}\left(\frac{1}{a}\frac{da}{dt} + \frac{1}{b}\frac{db}{dt}\right), \ \frac{da}{dt} = 1 \text{ and } \frac{db}{dt} = 1$$

when $a = 4$ and $b = 3$ so $\dfrac{d\theta}{dt} = -\dfrac{6}{\sqrt{144-36}}\left(\dfrac{1}{4}+\dfrac{1}{3}\right) = -\dfrac{7}{12\sqrt{3}} = -\dfrac{7}{36}\sqrt{3}$ radians/s

45. $V = (\pi/4)D^2h$ where D is the diameter and h is the height, both measured in inches, $dV/dt = (\pi/2)Dh(dD/dt) + (\pi/4)D^2(dh/dt)$, $dD/dt = 3$ and $dh/dt = 24$ when $D = 30$ and $h = 240$, so $dV/dt = (\pi/2)(30)(240)(3) + (\pi/4)(30)^2(24) = 16{,}200\pi$ in^3/year.

47. **(a)** $V = \ell wh$, $\dfrac{dV}{dt} = \dfrac{\partial V}{\partial \ell}\dfrac{d\ell}{dt} + \dfrac{\partial V}{\partial w}\dfrac{dw}{dt} + \dfrac{\partial V}{\partial h}\dfrac{dh}{dt} = wh\dfrac{d\ell}{dt} + \ell h\dfrac{dw}{dt} + \ell w\dfrac{dh}{dt}$

$= (3)(6)(1) + (2)(6)(2) + (2)(3)(3) = 60$ in^3/s

(b) $D = \sqrt{\ell^2+w^2+h^2}$; $dD/dt = (\ell/D)d\ell/dt + (w/D)dw/dt + (h/D)dh/dt$

$= (2/7)(1) + (3/7)(2) + (6/7)(3) = 26/7$ in/s

49. **(a)** $f(tx,ty) = 3t^2x^2 + t^2y^2 = t^2f(x,y)$; $n = 2$

(b) $f(tx,ty) = \sqrt{t^2x^2+t^2y^2} = tf(x,y)$; $n = 1$

(c) $f(tx,ty) = t^3x^2y - 2t^3y^3 = t^3f(x,y)$; $n = 3$

(d) $f(tx,ty) = 5/\left(t^2x^2+2t^2y^2\right)^2 = t^{-4}f(x,y)$; $n = -4$

51. **(a)** $\dfrac{\partial z}{\partial x} = \dfrac{dz}{du}\dfrac{\partial u}{\partial x}$, $\dfrac{\partial z}{\partial y} = \dfrac{dz}{du}\dfrac{\partial u}{\partial y}$

(b) $\dfrac{\partial^2 z}{\partial x^2} = \dfrac{dz}{du}\dfrac{\partial^2 u}{\partial x^2} + \dfrac{\partial}{\partial x}\left(\dfrac{dz}{du}\right)\dfrac{\partial u}{\partial x} = \dfrac{dz}{du}\dfrac{\partial^2 u}{\partial x^2} + \dfrac{d^2z}{du^2}\left(\dfrac{\partial u}{\partial x}\right)^2$;

$$\frac{\partial^2 z}{\partial y\partial x} = \frac{dz}{du}\frac{\partial^2 u}{\partial y\partial x} + \frac{\partial}{\partial y}\left(\frac{dz}{du}\right)\frac{\partial u}{\partial x} = \frac{dz}{du}\frac{\partial^2 u}{\partial y\partial x} + \frac{d^2z}{du^2}\frac{\partial u}{\partial x}\frac{\partial u}{\partial y}$$

$$\frac{\partial^2 z}{\partial y^2} = \frac{dz}{du}\frac{\partial^2 u}{\partial y^2} + \frac{\partial}{\partial y}\left(\frac{dz}{du}\right)\frac{\partial u}{\partial y} = \frac{dz}{du}\frac{\partial^2 u}{\partial y^2} + \frac{d^2z}{du^2}\left(\frac{\partial u}{\partial y}\right)^2$$

53. Let $z = f(u)$ where $u = x + 2y$; then $\partial z/\partial x = (dz/du)(\partial u/\partial x) = dz/du$, $\partial z/\partial y = (dz/du)(\partial u/\partial y) = 2dz/du$ so $2\partial z/\partial x - \partial z/\partial y = 2dz/du - 2dz/du = 0$

55. $\dfrac{\partial w}{\partial x} = \dfrac{dw}{du}\dfrac{\partial u}{\partial x} = \dfrac{dw}{du}$, $\dfrac{\partial w}{\partial y} = \dfrac{dw}{du}\dfrac{\partial u}{\partial y} = 2\dfrac{dw}{du}$, $\dfrac{\partial w}{\partial z} = \dfrac{dw}{du}\dfrac{\partial u}{\partial z} = 3\dfrac{dw}{du}$, so $\dfrac{\partial w}{\partial x} + \dfrac{\partial w}{\partial y} + \dfrac{\partial w}{\partial z} = 6\dfrac{dw}{du}$

57. $z = f(u,v)$ where $u = x - y$ and $v = y - x$,

$\dfrac{\partial z}{\partial x} = \dfrac{\partial z}{\partial u}\dfrac{\partial u}{\partial x} + \dfrac{\partial z}{\partial v}\dfrac{\partial v}{\partial x} = \dfrac{\partial z}{\partial u} - \dfrac{\partial z}{\partial v}$ and $\dfrac{\partial z}{\partial y} = \dfrac{\partial z}{\partial u}\dfrac{\partial u}{\partial y} + \dfrac{\partial z}{\partial v}\dfrac{\partial v}{\partial y} = -\dfrac{\partial z}{\partial u} + \dfrac{\partial z}{\partial v}$ so $\dfrac{\partial z}{\partial x} + \dfrac{\partial z}{\partial y} = 0$

59. (a) $1 = -r\sin\theta\dfrac{\partial\theta}{\partial x} + \cos\theta\dfrac{\partial r}{\partial x}$ and $0 = r\cos\theta\dfrac{\partial\theta}{\partial x} + \sin\theta\dfrac{\partial r}{\partial x}$; solve for $\partial r/\partial x$ and $\partial\theta/\partial x$.

(b) $0 = -r\sin\theta\dfrac{\partial\theta}{\partial y} + \cos\theta\dfrac{\partial r}{\partial y}$ and $1 = r\cos\theta\dfrac{\partial\theta}{\partial y} + \sin\theta\dfrac{\partial r}{\partial y}$; solve for $\partial r/\partial y$ and $\partial\theta/\partial y$.

(c) $\dfrac{\partial z}{\partial x} = \dfrac{\partial z}{\partial r}\dfrac{\partial r}{\partial x} + \dfrac{\partial z}{\partial\theta}\dfrac{\partial\theta}{\partial x} = \dfrac{\partial z}{\partial r}\cos\theta - \dfrac{1}{r}\dfrac{\partial z}{\partial\theta}\sin\theta.$

$\dfrac{\partial z}{\partial y} = \dfrac{\partial z}{\partial r}\dfrac{\partial r}{\partial y} + \dfrac{\partial z}{\partial\theta}\dfrac{\partial\theta}{\partial y} = \dfrac{\partial z}{\partial r}\sin\theta + \dfrac{1}{r}\dfrac{\partial z}{\partial\theta}\cos\theta.$

(d) Square and add the results of Parts (a) and (b).

(e) From Part (c),

$$\frac{\partial^2 z}{\partial x^2} = \frac{\partial}{\partial r}\left(\frac{\partial z}{\partial r}\cos\theta - \frac{1}{r}\frac{\partial z}{\partial\theta}\sin\theta\right)\frac{\partial r}{\partial x} + \frac{\partial}{\partial\theta}\left(\frac{\partial z}{\partial r}\cos\theta - \frac{1}{r}\frac{\partial z}{\partial\theta}\sin\theta\right)\frac{\partial\theta}{\partial x}$$

$$= \left(\frac{\partial^2 z}{\partial r^2}\cos\theta + \frac{1}{r^2}\frac{\partial z}{\partial\theta}\sin\theta - \frac{1}{r}\frac{\partial^2 z}{\partial r\partial\theta}\sin\theta\right)\cos\theta$$

$$+\left(\frac{\partial^2 z}{\partial\theta\partial r}\cos\theta - \frac{\partial z}{\partial r}\sin\theta - \frac{1}{r}\frac{\partial^2 z}{\partial\theta^2}\sin\theta - \frac{1}{r}\frac{\partial z}{\partial\theta}\cos\theta\right)\left(-\frac{\sin\theta}{r}\right)$$

$$= \frac{\partial^2 z}{\partial r^2}\cos^2\theta + \frac{2}{r^2}\frac{\partial z}{\partial\theta}\sin\theta\cos\theta - \frac{2}{r}\frac{\partial^2 z}{\partial\theta\partial r}\sin\theta\cos\theta + \frac{1}{r^2}\frac{\partial^2 z}{\partial\theta^2}\sin^2\theta + \frac{1}{r}\frac{\partial z}{\partial r}\sin^2\theta.$$

Similarly, from Part (c),

$$\frac{\partial^2 z}{\partial y^2} = \frac{\partial^2 z}{\partial r^2}\sin^2\theta - \frac{2}{r^2}\frac{\partial z}{\partial\theta}\sin\theta\cos\theta + \frac{2}{r}\frac{\partial^2 z}{\partial\theta\partial r}\sin\theta\cos\theta + \frac{1}{r^2}\frac{\partial^2 z}{\partial\theta^2}\cos^2\theta + \frac{1}{r}\frac{\partial z}{\partial r}\cos^2\theta.$$

Add to get $\dfrac{\partial^2 z}{\partial x^2} + \dfrac{\partial^2 z}{\partial y^2} = \dfrac{\partial^2 z}{\partial r^2} + \dfrac{1}{r^2}\dfrac{\partial^2 z}{\partial\theta^2} + \dfrac{1}{r}\dfrac{\partial z}{\partial r}.$

61. (a) By the chain rule, $\dfrac{\partial u}{\partial r} = \dfrac{\partial u}{\partial x}\cos\theta + \dfrac{\partial u}{\partial y}\sin\theta$ and $\dfrac{\partial v}{\partial\theta} = -\dfrac{\partial v}{\partial x}r\sin\theta + \dfrac{\partial v}{\partial y}r\cos\theta$, use the Cauchy-Riemann conditions $\dfrac{\partial u}{\partial x} = \dfrac{\partial v}{\partial y}$ and $\dfrac{\partial u}{\partial y} = -\dfrac{\partial v}{\partial x}$ in the equation for $\dfrac{\partial u}{\partial r}$ to get $\dfrac{\partial u}{\partial r} = \dfrac{\partial v}{\partial y}\cos\theta - \dfrac{\partial v}{\partial x}\sin\theta$ and compare to $\dfrac{\partial v}{\partial\theta}$ to see that $\dfrac{\partial u}{\partial r} = \dfrac{1}{r}\dfrac{\partial v}{\partial\theta}$. The result $\dfrac{\partial v}{\partial r} = -\dfrac{1}{r}\dfrac{\partial u}{\partial\theta}$ can be obtained by considering $\dfrac{\partial v}{\partial r}$ and $\dfrac{\partial u}{\partial\theta}$.

(b) $u_x = \dfrac{2x}{x^2+y^2}$, $v_y = 2\dfrac{1}{x}\dfrac{1}{1+(y/x)^2} = \dfrac{2x}{x^2+y^2} = u_x$;

$u_y = \dfrac{2y}{x^2+y^2}$, $v_x = -2\dfrac{y}{x^2}\dfrac{1}{1+(y/x)^2} = -\dfrac{2y}{x^2+y^2} = -u_y$;

$u = \ln r^2, v = 2\theta, u_r = 2/r, v_\theta = 2$, so $u_r = \dfrac{1}{r}v_\theta$, $u_\theta = 0, v_r = 0$, so $v_r = -\dfrac{1}{r}u_\theta$

63. $\partial w/\partial\rho = (\sin\phi\cos\theta)\partial w/\partial x + (\sin\phi\sin\theta)\partial w/\partial y + (\cos\phi)\,\partial w/\partial z$

$\partial w/\partial\phi = (\rho\cos\phi\cos\theta)\partial w/\partial x + (\rho\cos\phi\sin\theta)\partial w/\partial y - (\rho\sin\phi)\partial w/\partial z$

$\partial w/\partial\theta = -(\rho\sin\phi\sin\theta)\partial w/\partial x + (\rho\sin\phi\cos\theta)\partial w/\partial y$

65. $w_r = e^r / (e^r + e^s + e^t + e^u)$, $w_{rs} = -e^r e^s / (e^r + e^s + e^t + e^u)^2$,
$w_{rst} = 2e^r e^s e^t / (e^r + e^s + e^t + e^u)^3$,
$w_{rstu} = -6e^r e^s e^t e^u / (e^r + e^s + e^t + e^u)^4 = -6e^{r+s+t+u}/e^{4w} = -6e^{r+s+t+u-4w}$

67. **(a)** $dw/dt = \sum_{i=1}^{4} (\partial w/\partial x_i)(dx_i/dt)$

(b) $\partial w/\partial v_j = \sum_{i=1}^{4} (\partial w/\partial x_i)(\partial x_i/\partial v_j)$ for $j = 1, 2, 3$

69. $dF/dx = (\partial F/\partial u)(du/dx) + (\partial F/\partial v)(dv/dx)$
$= f(u)g'(x) - f(v)h'(x) = f(g(x))g'(x) - f(h(x))h'(x)$

71. Let (a, b) be any point in the region, if (x, y) is in the region then by the result of Exercise 70 $f(x, y) - f(a, b) = f_x(x^*, y^*)(x - a) + f_y(x^*, y^*)(y - b)$ where (x^*, y^*) is on the line segment joining (a, b) and (x, y). If $f_x(x, y) = f_y(x, y) = 0$ throughout the region then
$f(x, y) - f(a, b) = (0)(x - a) + (0)(y - b) = 0$, $f(x, y) = f(a, b)$ so $f(x, y)$ is constant on the region.

EXERCISE SET 14.6

1. $\nabla f(x, y) = (3y/2)(1 + xy)^{1/2}\mathbf{i} + (3x/2)(1 + xy)^{1/2}\mathbf{j}$, $\nabla f(3, 1) = 3\mathbf{i} + 9\mathbf{j}$,
$D_{\mathbf{u}}f = \nabla f \cdot \mathbf{u} = 12/\sqrt{2} = 6\sqrt{2}$

3. $\nabla f(x, y) = \left[2x/\left(1 + x^2 + y\right)\right]\mathbf{i} + \left[1/\left(1 + x^2 + y\right)\right]\mathbf{j}$, $\nabla f(0, 0) = \mathbf{j}$, $D_{\mathbf{u}}f = -3/\sqrt{10}$

5. $\nabla f(x, y, z) = 20x^4y^2z^3\mathbf{i} + 8x^5yz^3\mathbf{j} + 12x^5y^2z^2\mathbf{k}$, $\nabla f(2, -1, 1) = 320\mathbf{i} - 256\mathbf{j} + 384\mathbf{k}$, $D_{\mathbf{u}}f = -320$

7. $\nabla f(x, y, z) = \dfrac{2x}{x^2 + 2y^2 + 3z^2}\mathbf{i} + \dfrac{4y}{x^2 + 2y^2 + 3z^2}\mathbf{j} + \dfrac{6z}{x^2 + 2y^2 + 3z^2}\mathbf{k}$,
$\nabla f(-1, 2, 4) = (-2/57)\mathbf{i} + (8/57)\mathbf{j} + (24/57)\mathbf{k}$, $D_{\mathbf{u}}f = -314/741$

9. $\nabla f(x, y) = 12x^2y^2\mathbf{i} + 8x^3y\mathbf{j}$, $\nabla f(2, 1) = 48\mathbf{i} + 64\mathbf{j}$, $\mathbf{u} = (4/5)\mathbf{i} - (3/5)\mathbf{j}$, $D_{\mathbf{u}}f = \nabla f \cdot \mathbf{u} = 0$

11. $\nabla f(x, y) = \left(y^2/x\right)\mathbf{i} + 2y\ln x\mathbf{j}$, $\nabla f(1, 4) = 16\mathbf{i}$, $\mathbf{u} = (-\mathbf{i} + \mathbf{j})/\sqrt{2}$, $D_{\mathbf{u}}f = -8\sqrt{2}$

13. $\nabla f(x, y) = -\left[y/\left(x^2 + y^2\right)\right]\mathbf{i} + \left[x/\left(x^2 + y^2\right)\right]\mathbf{j}$,
$\nabla f(-2, 2) = -(\mathbf{i} + \mathbf{j})/4$, $\mathbf{u} = -(\mathbf{i} + \mathbf{j})/\sqrt{2}$, $D_{\mathbf{u}}f = \sqrt{2}/4$

15. $\nabla f(x, y, z) = \left(3x^2z - 2xy\right)\mathbf{i} - x^2\mathbf{j} + \left(x^3 + 2z\right)\mathbf{k}$, $\nabla f(2, -1, 1) = 16\mathbf{i} - 4\mathbf{j} + 10\mathbf{k}$,
$\mathbf{u} = (3\mathbf{i} - \mathbf{j} + 2\mathbf{k})/\sqrt{14}$, $D_{\mathbf{u}}f = 72/\sqrt{14}$

17. $\nabla f(x, y, z) = -\dfrac{1}{z + y}\mathbf{i} - \dfrac{z - x}{(z + y)^2}\mathbf{j} + \dfrac{y + x}{(z + y)^2}\mathbf{k}$, $\nabla f(1, 0, -3) = (1/3)\mathbf{i} + (4/9)\mathbf{j} + (1/9)\mathbf{k}$,
$\mathbf{u} = (-6\mathbf{i} + 3\mathbf{j} - 2\mathbf{k})/7$, $D_{\mathbf{u}}f = -8/63$

19. $\nabla f(x, y) = (y/2)(xy)^{-1/2}\mathbf{i} + (x/2)(xy)^{-1/2}\mathbf{j}$, $\nabla f(1, 4) = \mathbf{i} + (1/4)\mathbf{j}$,
$\mathbf{u} = \cos\theta\mathbf{i} + \sin\theta\mathbf{j} = (1/2)\mathbf{i} + \left(\sqrt{3}/2\right)\mathbf{j}$, $D_{\mathbf{u}}f = 1/2 + \sqrt{3}/8$

21. $\nabla f(x,y) = 2\sec^2(2x+y)\mathbf{i} + \sec^2(2x+y)\mathbf{j}$, $\nabla f(\pi/6,\pi/3) = 8\mathbf{i}+4\mathbf{j}$, $\mathbf{u} = (\mathbf{i}-\mathbf{j})/\sqrt{2}$, $D_{\mathbf{u}}f = 2\sqrt{2}$

23. $\nabla f(x,y) = y(x+y)^{-2}\mathbf{i} - x(x+y)^{-2}\mathbf{j}$, $\nabla f(1,0) = -\mathbf{j}$, $\overrightarrow{PQ} = -2\mathbf{i}-\mathbf{j}$, $\mathbf{u} = (-2\mathbf{i}-\mathbf{j})/\sqrt{5}$, $D_{\mathbf{u}}f = 1/\sqrt{5}$

25. $\nabla f(x,y) = \dfrac{ye^y}{2\sqrt{xy}}\mathbf{i} + \left(\sqrt{xy}e^y + \dfrac{xe^y}{2\sqrt{xy}}\right)\mathbf{j}$, $\nabla f(1,1) = (e/2)(\mathbf{i}+3\mathbf{j})$, $\mathbf{u} = -\mathbf{j}$, $D_{\mathbf{u}}f = -3e/2$

27. $\nabla f(2,1,-1) = -\mathbf{i}+\mathbf{j}-\mathbf{k}$. $\overrightarrow{PQ} = -3\mathbf{i}+\mathbf{j}+\mathbf{k}$, $\mathbf{u} = (-3\mathbf{i}+\mathbf{j}+\mathbf{k})/\sqrt{11}$, $D_{\mathbf{u}}f = 3/\sqrt{11}$

29. Solve the system $(3/5)f_x(1,2) - (4/5)f_y(1,2) = -5$, $(4/5)f_x(1,2) + (3/5)f_y(1,2) = 10$ for

(a) $f_x(1,2) = 5$ **(b)** $f_y(1,2) = 10$

(c) $\nabla f(1,2) = 5\mathbf{i}+10\mathbf{j}$, $\mathbf{u} = (-\mathbf{i}-2\mathbf{j})/\sqrt{5}$, $D_{\mathbf{u}}f = -5\sqrt{5}$.

31. f increases the most in the direction of III.

33. $\nabla z = 4\mathbf{i} - 8\mathbf{j}$

35. $\nabla w = \dfrac{x}{x^2+y^2+z^2}\mathbf{i} + \dfrac{y}{x^2+y^2+z^2}\mathbf{j} + \dfrac{z}{x^2+y^2+z^2}\mathbf{k}$

37. $\nabla f(x,y) = 3(2x+y)\left(x^2+xy\right)^2\mathbf{i} + 3x\left(x^2+xy\right)^2\mathbf{j}$, $\nabla f(-1,-1) = -36\mathbf{i} - 12\mathbf{j}$

39. $\nabla f(x,y,z) = [y/(x+y+z)]\mathbf{i} + [y/(x+y+z) + \ln(x+y+z)]\mathbf{j} + [y/(x+y+z)]\mathbf{k}$, $\nabla f(-3,4,0) = 4\mathbf{i}+4\mathbf{j}+4\mathbf{k}$

41. $f(1,2) = 3$,
level curve $4x - 2y + 3 = 3$,
$2x - y = 0$;
$\nabla f(x,y) = 4\mathbf{i} - 2\mathbf{j}$
$\nabla f(1,2) = 4\mathbf{i} - 2\mathbf{j}$

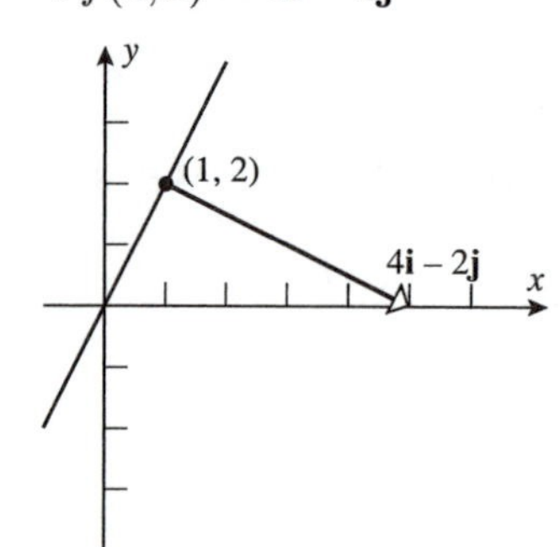

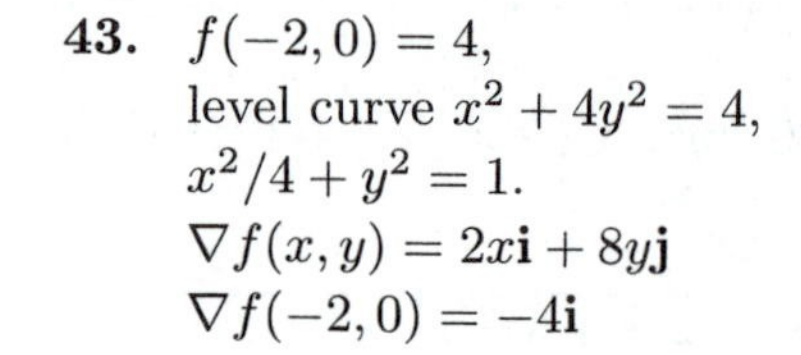

43. $f(-2,0) = 4$,
level curve $x^2 + 4y^2 = 4$,
$x^2/4 + y^2 = 1$.
$\nabla f(x,y) = 2x\mathbf{i} + 8y\mathbf{j}$
$\nabla f(-2,0) = -4\mathbf{i}$

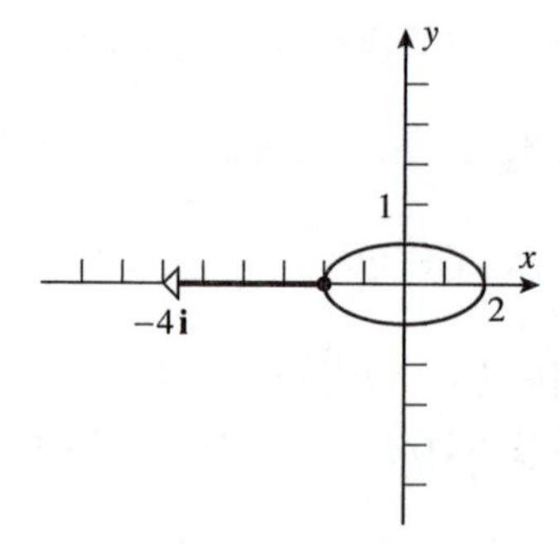

45. $\nabla f(x,y) = 8xy\mathbf{i} + 4x^2\mathbf{j}$, $\nabla f(1,-2) = -16\mathbf{i} + 4\mathbf{j}$ is normal to the level curve through P so $\mathbf{u} = \pm(-4\mathbf{i}+\mathbf{j})/\sqrt{17}$.

47. $\nabla f(x,y) = 12x^2y^2\mathbf{i} + 8x^3y\mathbf{j}$, $\nabla f(-1,1) = 12\mathbf{i} - 8\mathbf{j}$, $\mathbf{u} = (3\mathbf{i}-2\mathbf{j})/\sqrt{13}$, $\|\nabla f(-1,1)\| = 4\sqrt{13}$

49. $\nabla f(x,y) = x\left(x^2+y^2\right)^{-1/2}\mathbf{i} + y\left(x^2+y^2\right)^{-1/2}\mathbf{j}$,
$\nabla f(4,-3) = (4\mathbf{i}-3\mathbf{j})/5$, $\mathbf{u} = (4\mathbf{i}-3\mathbf{j})/5$, $\|\nabla f(4,-3)\| = 1$

51. $\nabla f(1,1,-1) = 3\mathbf{i} - 3\mathbf{j}$, $\mathbf{u} = (\mathbf{i}-\mathbf{j})/\sqrt{2}$, $\|\nabla f(1,1,-1)\| = 3\sqrt{2}$

53. $\nabla f(1,2,-2) = (-\mathbf{i}+\mathbf{j})/2$, $\mathbf{u} = (-\mathbf{i}+\mathbf{j})/\sqrt{2}$, $\|\nabla f(1,2,-2)\| = 1/\sqrt{2}$

55. $\nabla f(x,y) = -2x\mathbf{i} - 2y\mathbf{j}$, $\nabla f(-1,-3) = 2\mathbf{i} + 6\mathbf{j}$, $\mathbf{u} = -(\mathbf{i}+3\mathbf{j})/\sqrt{10}$, $-\|\nabla f(-1,-3)\| = -2\sqrt{10}$

57. $\nabla f(x,y) = -3\sin(3x-y)\mathbf{i} + \sin(3x-y)\mathbf{j}$,
$\nabla f(\pi/6,\pi/4) = (-3\mathbf{i}+\mathbf{j})/\sqrt{2}$, $\mathbf{u} = (3\mathbf{i}-\mathbf{j})/\sqrt{10}$, $-\|\nabla f(\pi/6,\pi/4)\| = -\sqrt{5}$

59. $\nabla f(5,7,6) = -\mathbf{i} + 11\mathbf{j} - 12\mathbf{k}$, $\mathbf{u} = (\mathbf{i} - 11\mathbf{j} + 12\mathbf{k})/\sqrt{266}$, $-\|\nabla f(5,7,6)\| = -\sqrt{266}$

61. $\nabla f(4,-5) = 2\mathbf{i} - \mathbf{j}$, $\mathbf{u} = (5\mathbf{i}+2\mathbf{j})/\sqrt{29}$, $D_{\mathbf{u}}f = 8/\sqrt{29}$

63. **(a)** At $(1,2)$ the steepest ascent seems to be in the direction $\mathbf{i}+\mathbf{j}$ and the slope in that direction seems to be $0.5/(\sqrt{2}/2) = 1/\sqrt{2}$, so $\nabla f \approx \frac{1}{2}\mathbf{i} + \frac{1}{2}\mathbf{j}$, which has the required direction and magnitude.

(b) The direction of $-\nabla f(4,4)$ appears to be $-\mathbf{i}-\mathbf{j}$ and its magnitude appears to be $1/0.8 = 5/4$.

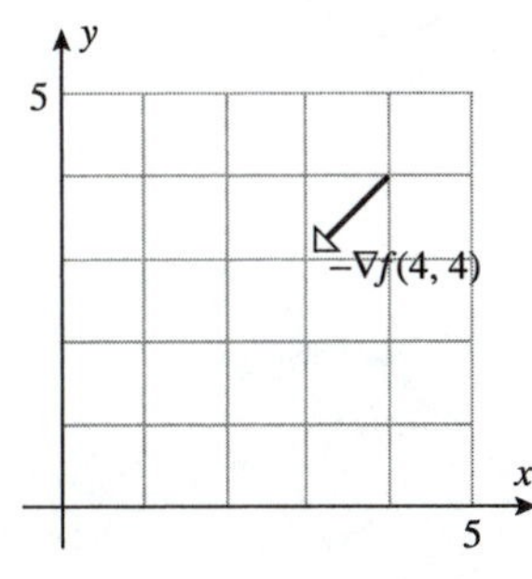

65. $\nabla z = 6x\mathbf{i} - 2y\mathbf{j}$, $\|\nabla z\| = \sqrt{36x^2 + 4y^2} = 6$ if $36x^2 + 4y^2 = 36$; all points on the ellipse $9x^2 + y^2 = 9$.

67. $\mathbf{r} = t\mathbf{i} - t^2\mathbf{j}$, $d\mathbf{r}/dt = \mathbf{i} - 2t\mathbf{j} = \mathbf{i} - 4\mathbf{j}$ at the point $(2,-4)$, $\mathbf{u} = (\mathbf{i}-4\mathbf{j})/\sqrt{17}$;
$\nabla z = 2x\mathbf{i} + 2y\mathbf{j} = 4\mathbf{i} - 8\mathbf{j}$ at $(2,-4)$, hence $dz/ds = D_{\mathbf{u}}z = \nabla z \cdot \mathbf{u} = 36/\sqrt{17}$.

69. **(a)** $\nabla V(x,y) = -2e^{-2x}\cos 2y\mathbf{i} - 2e^{-2x}\sin 2y\mathbf{j}$, $\mathbf{E} = -\nabla V(\pi/4, 0) = 2e^{-\pi/2}\mathbf{i}$

(b) $V(x,y)$ decreases most rapidly in the direction of $-\nabla V(x,y)$ which is $\mathbf{E}$.

71. Let $\mathbf{u}$ be the unit vector in the direction of $\mathbf{a}$, then
$D_{\mathbf{u}}f(3,-2,1) = \nabla f(3,-2,1) \cdot \mathbf{u} = \|\nabla f(3,-2,1)\|\cos\theta = 5\cos\theta = -5$, $\cos\theta = -1$, $\theta = \pi$ so $\nabla f(3,-2,1)$ is oppositely directed to $\mathbf{u}$; $\nabla f(3,-2,1) = -5\mathbf{u} = -10/3\mathbf{i} + 5/3\mathbf{j} + 10/3\mathbf{k}$.

73. **(a)** $\nabla r = \dfrac{x}{\sqrt{x^2+y^2}}\mathbf{i} + \dfrac{y}{\sqrt{x^2+y^2}}\mathbf{j} = \mathbf{r}/r$

(b) $\nabla f(r) = \dfrac{\partial f(r)}{\partial x}\mathbf{i} + \dfrac{\partial f(r)}{\partial y}\mathbf{j} = f'(r)\dfrac{\partial r}{\partial x}\mathbf{i} + f'(r)\dfrac{\partial r}{\partial y}\mathbf{j} = f'(r)\nabla r$

75. $\mathbf{u}_r = \cos\theta\mathbf{i} + \sin\theta\mathbf{j}$, $\mathbf{u}_\theta = -\sin\theta\mathbf{i} + \cos\theta\mathbf{j}$,

$$\nabla z = \frac{\partial z}{\partial x}\mathbf{i} + \frac{\partial z}{\partial y}\mathbf{j} = \left(\frac{\partial z}{\partial r}\cos\theta - \frac{1}{r}\frac{\partial z}{\partial \theta}\sin\theta\right)\mathbf{i} + \left(\frac{\partial z}{\partial r}\sin\theta + \frac{1}{r}\frac{\partial z}{\partial \theta}\cos\theta\right)\mathbf{j}$$

$$= \frac{\partial z}{\partial r}(\cos\theta\mathbf{i} + \sin\theta\mathbf{j}) + \frac{1}{r}\frac{\partial z}{\partial \theta}(-\sin\theta\mathbf{i} + \cos\theta\mathbf{j}) = \frac{\partial z}{\partial r}\mathbf{u}_r + \frac{1}{r}\frac{\partial z}{\partial \theta}\mathbf{u}_\theta$$

77. $\mathbf{r}'(t) = \mathbf{v}(t) = k(x, y)\nabla\mathbf{T} = -8k(x, y)x\mathbf{i} - 2k(x, y)y\mathbf{j}$; $\dfrac{dx}{dt} = -8kx$, $\dfrac{dy}{dt} = -2ky$. Divide and solve to get $y^4 = 256x$; one parametrization is $x(t) = e^{-8t}$, $y(t) = 4e^{-2t}$.

79.

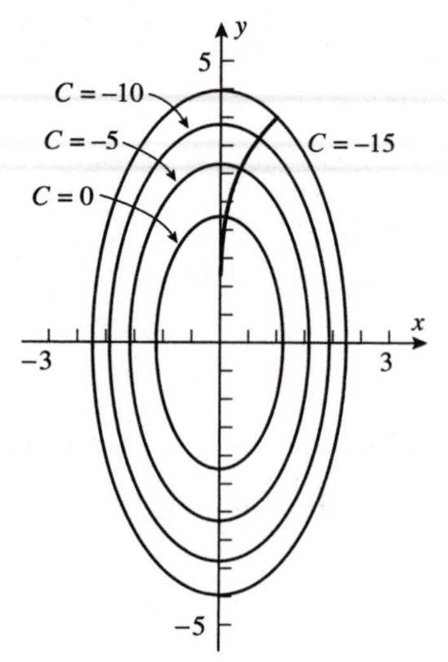

81. **(a)**

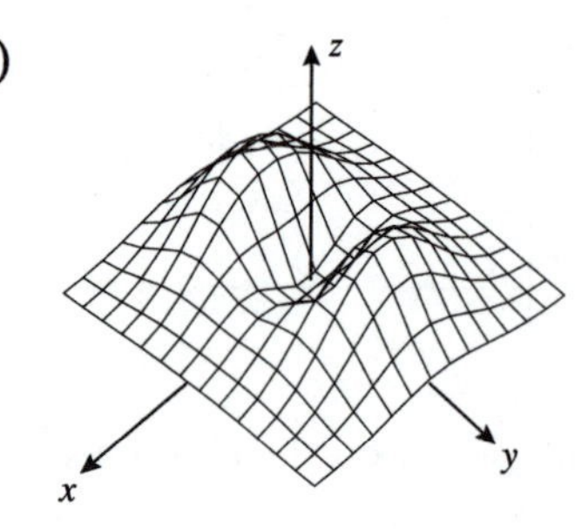

(c) $\nabla f = [2x - 2x(x^2 + 3y^2)]e^{-(x^2+y^2)}\mathbf{i} + [6y - 2y(x^2 + 3y^2)]e^{-(x^2+y^2)}\mathbf{j}$

(d) $\nabla f = \mathbf{0}$ if $x = y = 0$ or $x = 0, y = \pm 1$ or $x = \pm 1, y = 0$.

83. $\nabla f(x, y) = f_x(x, y)\mathbf{i} + f_y(x, y)\mathbf{j}$, if $\nabla f(x, y) = 0$ throughout the region then $f_x(x, y) = f_y(x, y) = 0$ throughout the region, the result follows from Exercise 71, Section 14.5.

85.
$$\nabla f(u, v, w) = \frac{\partial f}{\partial x}\mathbf{i} + \frac{\partial f}{\partial y}\mathbf{j} + \frac{\partial f}{\partial z}\mathbf{k}$$
$$= \left(\frac{\partial f}{\partial u}\frac{\partial u}{\partial x} + \frac{\partial f}{\partial v}\frac{\partial v}{\partial x} + \frac{\partial f}{\partial w}\frac{\partial w}{\partial x}\right)\mathbf{i} + \left(\frac{\partial f}{\partial u}\frac{\partial u}{\partial y} + \frac{\partial f}{\partial v}\frac{\partial v}{\partial y} + \frac{\partial f}{\partial w}\frac{\partial w}{\partial y}\right)\mathbf{j}$$
$$+ \left(\frac{\partial f}{\partial u}\frac{\partial u}{\partial z} + \frac{\partial f}{\partial v}\frac{\partial v}{\partial z} + \frac{\partial f}{\partial w}\frac{\partial w}{\partial z}\right)\mathbf{k} = \frac{\partial f}{\partial u}\nabla u + \frac{\partial f}{\partial v}\nabla v + \frac{\partial f}{\partial w}\nabla w$$

87. **(a)** $\dfrac{d}{ds}f(x_0+su_1, y_0+su_2)$ at $s = 0$ is by definition equal to $\displaystyle\lim_{s\to 0}\frac{f(x_0 + su_1, y_0 + su_2) - f(x_0, y_0)}{s}$, and from Exercise 86(a) this value is equal to $f_x(x_0, y_0)u_1 + f_y(x_0, y_0)u_2$.

(b) For any number $\epsilon > 0$ a number $\delta > 0$ exists such that whenever $0 < |s| < \delta$ then
$$\left|\frac{f(x_0 + su_1, y_0 + su_2) - f(x_0, y_0) - f_x(x_0, y_0)su_1 - f_y(x_0, y_0)su_2}{s}\right| < \epsilon.$$

(c) For any number $\epsilon > 0$ there exists a number $\delta > 0$ such that $\dfrac{|E(x,y)|}{\sqrt{(x-x_0)^2+(y-y_0)^2}} < \epsilon$ whenever $0 < \sqrt{(x-x_0)^2+(y-y_0)^2} < \delta$.

(d) For any number $\epsilon > 0$ there exists a number $\delta > 0$ such that

$$\left|\frac{f(x_0+su_1, y_0+su_2) - f(x_0,y_0) - f_x(x_0,y_0)su_1 - f_y(x_0,y_0)su_2|}{s}\right| < \epsilon \text{ when } 0 < |s| < \delta.$$

(e) Since f is differentiable at (x_0, y_0), by Part (c) the Equation (5) of Definition 14.2.1 holds. By Part (d), for any $\epsilon > 0$ there exists $\delta > 0$ such that

$$\left|\frac{f(x_0+su_1, y_0+su_2) - f(x_0,y_0) - f_x(x_0,y_0)su_1 - f_y(x_0,y_0)su_2}{s}\right| < \epsilon \text{ when } 0 < |s| < \delta.$$

By Part (a) it follows that the limit in Part (a) holds, and thus that

$$\frac{d}{ds}f(x_0+su_1, y_0+su_2)\Big]_{s=0} = f_x(x_0,y_0)u_1 + f_y(x_0,y_0)u_2,$$

which proves Equation (4) of Theorem 14.6.3.

EXERCISE SET 14.7

1. At P, $\partial z/\partial x = 48$ and $\partial z/\partial y = -14$, tangent plane $48x - 14y - z = 64$, normal line $x = 1 + 48t$, $y = -2 - 14t$, $z = 12 - t$.

3. At P, $\partial z/\partial x = 1$ and $\partial z/\partial y = -1$, tangent plane $x - y - z = 0$, normal line $x = 1 + t$, $y = -t$, $z = 1 - t$.

5. At P, $\partial z/\partial x = 0$ and $\partial z/\partial y = 3$, tangent plane $3y - z = -1$, normal line $x = \pi/6$, $y = 3t$, $z = 1 - t$.

7. By implicit differentiation $\partial z/\partial x = -x/z$, $\partial z/\partial y = -y/z$ so at P, $\partial z/\partial x = 3/4$ and $\partial z/\partial y = 0$, tangent plane $3x - 4z = -25$, normal line $x = -3 + 3t/4$, $y = 0$, $z = 4 - t$.

9. The tangent plane is horizontal if the normal $\partial z/\partial x\mathbf{i} + \partial z/\partial y\mathbf{j} - \mathbf{k}$ is parallel to $\mathbf{k}$ which occurs when $\partial z/\partial x = \partial z/\partial y = 0$.

(a) $\partial z/\partial x = 3x^2y^2$, $\partial z/\partial y = 2x^3y$; $3x^2y^2 = 0$ and $2x^3y = 0$ for all (x, y) on the x-axis or y-axis, and $z = 0$ for these points, the tangent plane is horizontal at all points on the x-axis or y-axis.

(b) $\partial z/\partial x = 2x - y - 2$, $\partial z/\partial y = -x + 2y + 4$; solve the system $2x - y - 2 = 0$, $-x + 2y + 4 = 0$, to get $x = 0$, $y = -2$. $z = -4$ at $(0, -2)$, the tangent plane is horizontal at $(0, -2, -4)$.

11. $\partial z/\partial x = -6x$, $\partial z/\partial y = -4y$ so $-6x_0\mathbf{i} - 4y_0\mathbf{j} - \mathbf{k}$ is normal to the surface at a point (x_0, y_0, z_0) on the surface. This normal must be parallel to the given line and hence to the vector $-3\mathbf{i} + 8\mathbf{j} - \mathbf{k}$ which is parallel to the line so $-6x_0 = -3$, $x_0 = 1/2$ and $-4y_0 = 8$, $y_0 = -2$. $z = -3/4$ at $(1/2, -2)$. The point on the surface is $(1/2, -2, -3/4)$.

13. **(a)** $2t + 7 = (-1 + t)^2 + (2 + t)^2$, $t^2 = 1$, $t = \pm 1$ so the points of intersection are $(-2, 1, 5)$ and $(0, 3, 9)$.

(b) $\partial z/\partial x = 2x$, $\partial z/\partial y = 2y$ so at $(-2, 1, 5)$ the vector $\mathbf{n} = -4\mathbf{i} + 2\mathbf{j} - \mathbf{k}$ is normal to the surface. $\mathbf{v} = \mathbf{i} + \mathbf{j} + 2\mathbf{k}$ is parallel to the line; $\mathbf{n} \cdot \mathbf{v} = -4$ so the cosine of the acute angle is $[\mathbf{n} \cdot (-\mathbf{v})]/(\|\mathbf{n}\|\,\| - \mathbf{v}\|) = 4/\left(\sqrt{21}\sqrt{6}\right) = 4/\left(3\sqrt{14}\right)$. Similarly, at (0,3,9) the vector $\mathbf{n} = 6\mathbf{j} - \mathbf{k}$ is normal to the surface, $\mathbf{n} \cdot \mathbf{v} = 4$ so the cosine of the acute angle is $4/\left(\sqrt{37}\sqrt{6}\right) = 4/\sqrt{222}$.

15. (a) $f(x,y,z)=x^2+y^2+4z^2, \nabla f=2x\mathbf{i}+2y\mathbf{j}+8z\mathbf{k}, \nabla f(2,2,1)=4\mathbf{i}+4\mathbf{j}+8\mathbf{k}$,
$\mathbf{n}=\mathbf{i}+\mathbf{j}+2\mathbf{k},\ x+y+2z=6$

(b) $\mathbf{r}(t)=2\mathbf{i}+2\mathbf{j}+\mathbf{k}+t(\mathbf{i}+\mathbf{j}+2\mathbf{k}), x(t)=2+t, y(t)=2+t, z(t)=1+2t$

(c) $\cos\theta=\dfrac{\mathbf{n}\cdot\mathbf{k}}{\|\mathbf{n}\|}=\dfrac{\sqrt{2}}{\sqrt{3}}, \theta\approx 35.26^\circ$

17. Set $f(x,y)=z+x-z^4(y-1)$, then $f(x,y,z)=0, \mathbf{n}=\pm\nabla f(3,5,1)=\pm(\mathbf{i}-\mathbf{j}-19\mathbf{k})$,
unit vectors $\pm\dfrac{1}{\sqrt{363}}(\mathbf{i}-\mathbf{j}-19\mathbf{k})$

19. $f(x,y,z)=x^2+y^2+z^2$, if (x_0,y_0,z_0) is on the sphere then $\nabla f(x_0,y_0,z_0)=2(x_0\mathbf{i}+y_0\mathbf{j}+z_0\mathbf{k})$ is normal to the sphere at (x_0,y_0,z_0), the normal line is $x=x_0+x_0t,\ y=y_0+y_0t,\ z=z_0+z_0t$ which passes through the origin when $t=-1$.

21. $f(x,y,z)=x^2+y^2-z^2$, if (x_0,y_0,z_0) is on the surface then $\nabla f(x_0,y_0,z_0)=2(x_0\mathbf{i}+y_0\mathbf{j}-z_0\mathbf{k})$ is normal there and hence so is $\mathbf{n}_1=x_0\mathbf{i}+y_0\mathbf{j}-z_0\mathbf{k}$; $\mathbf{n}_1$ must be parallel to $\overrightarrow{PQ}=3\mathbf{i}+2\mathbf{j}-2\mathbf{k}$ so $\mathbf{n}_1=c\,\overrightarrow{PQ}$ for some constant c. Equate components to get $x_0=3c,\ y_0=2c$ and $z_0=2c$ which when substituted into the equation of the surface yields $9c^2+4c^2-4c^2=1,\ c^2=1/9,\ c=\pm1/3$ so the points are $(1,2/3,2/3)$ and $(-1,-2/3,-2/3)$.

23. $\mathbf{n}_1=2\mathbf{i}-2\mathbf{j}-\mathbf{k}, \mathbf{n}_2=2\mathbf{i}-8\mathbf{j}+4\mathbf{k}, \mathbf{n}_1\times\mathbf{n}_2=-16\mathbf{i}-10\mathbf{j}-12\mathbf{k}$ is tangent to the line, so $x(t)=1+8t, y(t)=-1+5t, z(t)=2+6t$

25. $f(x,y,z)=x^2+z^2-25,\ g(x,y,z)=y^2+z^2-25, \mathbf{n}_1=\nabla f(3,-3,4)=6\mathbf{i}+8\mathbf{k}$,
$\mathbf{n}_2=\nabla g(3,-3,4)=-6\mathbf{j}+8\mathbf{k}, \mathbf{n}_1\times\mathbf{n}_2=48\mathbf{i}-48\mathbf{j}-36\mathbf{k}$ is tangent to the line,
$x(t)=3+4t, y(t)=-3-4t, z(t)=4-3t$

27. Use implicit differentiation to get $\partial z/\partial x=-c^2x/(a^2z),\ \partial z/\partial y=-c^2y/(b^2z)$. At (x_0,y_0,z_0), $z_0\neq 0$, a normal to the surface is $-\left[c^2x_0/(a^2z_0)\right]\mathbf{i}-\left[c^2y_0/(b^2z_0)\right]\mathbf{j}-\mathbf{k}$ so the tangent plane is

$$-\frac{c^2x_0}{a^2z_0}x-\frac{c^2y_0}{b^2z_0}y-z=-\frac{c^2x_0^2}{a^2z_0}-\frac{c^2y_0^2}{b^2z_0}-z_0,\ \frac{x_0x}{a^2}+\frac{y_0y}{b^2}+\frac{z_0z}{c^2}=\frac{x_0^2}{a^2}+\frac{y_0^2}{b^2}+\frac{z_0^2}{c^2}=1$$

29. $\mathbf{n}_1=f_x(x_0,y_0)\mathbf{i}+f_y(x_0,y_0)\mathbf{j}-\mathbf{k}$ and $\mathbf{n}_2=g_x(x_0,y_0)\mathbf{i}+g_y(x_0,y_0)\mathbf{j}-\mathbf{k}$ are normal, respectively, to $z=f(x,y)$ and $z=g(x,y)$ at P; $\mathbf{n}_1$ and $\mathbf{n}_2$ are perpendicular if and only if $\mathbf{n}_1\cdot\mathbf{n}_2=0$,
$f_x(x_0,y_0)g_x(x_0,y_0)+f_y(x_0,y_0)g_y(x_0,y_0)+1=0$,
$f_x(x_0,y_0)g_x(x_0,y_0)+f_y(x_0,y_0)g_y(x_0,y_0)=-1$.

31. $\nabla f=f_x\mathbf{i}+f_y\mathbf{j}+f_z\mathbf{k}$ and $\nabla g=g_x\mathbf{i}+g_y\mathbf{j}+g_z\mathbf{k}$ evaluated at (x_0,y_0,z_0) are normal, respectively, to the surfaces $f(x,y,z)=0$ and $g(x,y,z)=0$ at (x_0,y_0,z_0). The surfaces are orthogonal at (x_0,y_0,z_0) if and only if $\nabla f\cdot\nabla g=0$ so $f_xg_x+f_yg_y+f_zg_z=0$.

33. $z=\dfrac{k}{xy}$; at a point $\left(a,b,\dfrac{k}{ab}\right)$ on the surface, $\left\langle-\dfrac{k}{a^2b},-\dfrac{k}{ab^2},-1\right\rangle$ and hence $\langle bk,ak,a^2b^2\rangle$ is normal to the surface so the tangent plane is $bkx+aky+a^2b^2z=3abk$. The plane cuts the x, y, and z-axes at the points 3a, 3b, and $\dfrac{3k}{ab}$, respectively, so the volume of the tetrahedron that is formed is $V=\dfrac{1}{3}\left(\dfrac{3k}{ab}\right)\left[\dfrac{1}{2}(3a)(3b)\right]=\dfrac{9}{2}k$, which does not depend on a and b.

EXERCISE SET 14.8

1. **(a)** minimum at $(2, -1)$, no maxima **(b)** maximum at $(0, 0)$, no minima

(c) no maxima or minima

3. $f(x, y) = (x - 3)^2 + (y + 2)^2$, minimum at $(3, -2)$, no maxima

5. $f_x = 6x + 2y = 0$, $f_y = 2x + 2y = 0$; critical point (0,0); $D = 8 > 0$ and $f_{xx} = 6 > 0$ at (0,0), relative minimum.

7. $f_x = 2x - 2xy = 0$, $f_y = 4y - x^2 = 0$; critical points (0,0) and $(\pm 2, 1)$; $D = 8 > 0$ and $f_{xx} = 2 > 0$ at (0,0), relative minimum; $D = -16 < 0$ at $(\pm 2, 1)$, saddle points.

9. $f_x = y + 2 = 0$, $f_y = 2y + x + 3 = 0$; critical point $(1, -2)$; $D = -1 < 0$ at $(1, -2)$, saddle point.

11. $f_x = 2x + y - 3 = 0$, $f_y = x + 2y = 0$; critical point $(2, -1)$; $D = 3 > 0$ and $f_{xx} = 2 > 0$ at $(2, -1)$, relative minimum.

13. $f_x = 2x - 2/\left(x^2 y\right) = 0$, $f_y = 2y - 2/\left(xy^2\right) = 0$; critical points $(-1, -1)$ and $(1, 1)$; $D = 32 > 0$ and $f_{xx} = 6 > 0$ at $(-1, -1)$ and $(1, 1)$, relative minima.

15. $f_x = 2x = 0$, $f_y = 1 - e^y = 0$; critical point $(0, 0)$; $D = -2 < 0$ at $(0, 0)$, saddle point.

17. $f_x = e^x \sin y = 0$, $f_y = e^x \cos y = 0$, $\sin y = \cos y = 0$ is impossible, no critical points.

19. $f_x = -2(x + 1)e^{-(x^2+y^2+2x)} = 0$, $f_y = -2ye^{-(x^2+y^2+2x)} = 0$; critical point $(-1, 0)$; $D = 4e^2 > 0$ and $f_{xx} = -2e < 0$ at $(-1, 0)$, relative maximum.

21.

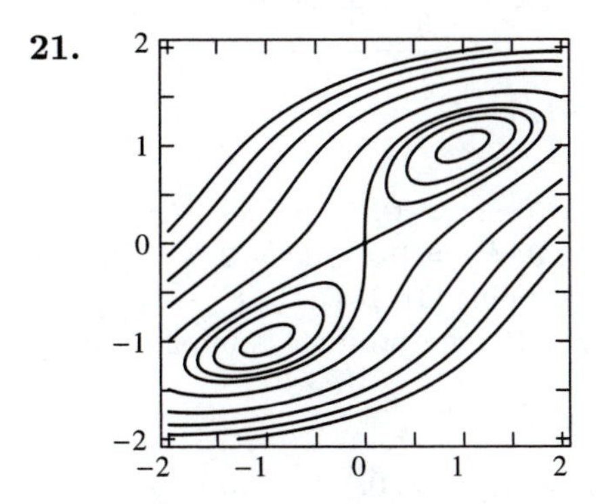

$\nabla f = (4x - 4y)\mathbf{i} - (4x - 4y^3)\mathbf{j} = \mathbf{0}$ when $x = y, x = y^3$, so $x = y = 0$ or $x = y = \pm 1$. At $(0, 0), D = -16$, a saddle point; at $(1, 1)$ and $(-1, -1), D = 32 > 0, f_{xx} = 4$, a relative minimum.

23. **(a)** critical point (0,0); $D = 0$

(b) $f(0, 0) = 0$, $x^4 + y^4 \geq 0$ so $f(x, y) \geq f(0, 0)$, relative minimum.

25. **(a)** $f_x = 3e^y - 3x^2 = 3\left(e^y - x^2\right) = 0$, $f_y = 3xe^y - 3e^{3y} = 3e^y\left(x - e^{2y}\right) = 0$, $e^y = x^2$ and $e^{2y} = x$, $x^4 = x$, $x\left(x^3 - 1\right) = 0$ so $x = 0, 1$; critical point $(1, 0)$; $D = 27 > 0$ and $f_{xx} = -6 < 0$ at $(1, 0)$, relative maximum.

(b) $\lim\limits_{x \to -\infty} f(x, 0) = \lim\limits_{x \to -\infty} \left(3x - x^3 - 1\right) = +\infty$ so no absolute maximum.

27. $f_x = y - 1 = 0,\ f_y = x - 3 = 0$; critical point (3,1).

$$\begin{aligned} \text{Along } y = 0:&\quad u(x) = -x;\ \text{no critical points,} \\ \text{along } x = 0:&\quad v(y) = -3y;\ \text{no critical points,} \\ \text{along } y = -\frac{4}{5}x + 4:&\quad w(x) = -\frac{4}{5}x^2 + \frac{27}{5}x - 12;\ \text{critical point } (27/8, 13/10). \end{aligned}$$

(x,y)	$(3,1)$	$(0,0)$	$(5,0)$	$(0,4)$	$(27/8,13/10)$
$f(x,y)$	-3	0	-5	-12	$-231/80$

Absolute maximum value is 0, absolute minimum value is -12.

29. $f_x = 2x - 2 = 0,\ f_y = -6y + 6 = 0$; critical point (1,1).

$$\begin{aligned} \text{Along } y = 0:&\quad u_1(x) = x^2 - 2x;\ \text{critical point } (1,0), \\ \text{along } y = 2:&\quad u_2(x) = x^2 - 2x;\ \text{critical point } (1,2) \\ \text{along } x = 0:&\quad v_1(y) = -3y^2 + 6y;\ \text{critical point } (0,1), \\ \text{along } x = 2:&\quad v_2(y) = -3y^2 + 6y;\ \text{critical point } (2,1) \end{aligned}$$

(x,y)	$(1,1)$	$(1,0)$	$(1,2)$	$(0,1)$	$(2,1)$	$(0,0)$	$(0,2)$	$(2,0)$	$(2,2)$
$f(x,y)$	2	-1	-1	3	3	0	0	0	0

Absolute maximum value is 3, absolute minimum value is -1.

31. $f_x = 2x - 1 = 0,\ f_y = 4y = 0$; critical point $(1/2, 0)$.
Along $x^2 + y^2 = 4 : y^2 = 4 - x^2,\ u(x) = 8 - x - x^2$ for $-2 \le x \le 2$; critical points $(-1/2, \pm\sqrt{15}/2)$.

(x,y)	$(1/2,0)$	$(-1/2,\sqrt{15}/2)$	$(-1/2,-\sqrt{15}/2)$	$(-2,0)$	$(2,0)$
$f(x,y)$	$-1/4$	$33/4$	$33/4$	6	2

Absolute maximum value is 33/4, absolute minimum value is $-1/4$.

33. Maximize $P = xyz$ subject to $x + y + z = 48,\ x > 0,\ y > 0,\ z > 0$. $z = 48 - x - y$ so $P = xy(48 - x - y) = 48xy - x^2y - xy^2$, $P_x = 48y - 2xy - y^2 = 0$, $P_y = 48x - x^2 - 2xy = 0$. But $x \neq 0$ and $y \neq 0$ so $48 - 2x - y = 0$ and $48 - x - 2y = 0$; critical point (16,16). $P_{xx}P_{yy} - P_{xy}^2 > 0$ and $P_{xx} < 0$ at (16,16), relative maximum. $z = 16$ when $x = y = 16$, the product is maximum for the numbers 16,16,16.

35. Maximize $w = xy^2z^2$ subject to $x + y + z = 5,\ x > 0,\ y > 0,\ z > 0$. $x = 5 - y - z$ so $w = (5 - y - z)y^2z^2 = 5y^2z^2 - y^3z^2 - y^2z^3$, $w_y = 10yz^2 - 3y^2z^2 - 2yz^3 = yz^2(10 - 3y - 2z) = 0$, $w_z = 10y^2z - 2y^3z - 3y^2z^2 = y^2z(10 - 2y - 3z) = 0$, $10 - 3y - 2z = 0$ and $10 - 2y - 3z = 0$; critical point when $y = z = 2$; $w_{yy}w_{zz} - w_{yz}^2 = 320 > 0$ and $w_{yy} = -24 < 0$ when $y = z = 2$, relative maximum. $x = 1$ when $y = z = 2$, xy^2z^2 is maximum at (1,2,2).

37. The diagonal of the box must equal the diameter of the sphere, thus we maximize $V = xyz$ or, for convenience, $w = V^2 = x^2y^2z^2$ subject to $x^2 + y^2 + z^2 = 4a^2,\ x > 0,\ y > 0,\ z > 0$; $z^2 = 4a^2 - x^2 - y^2$ hence $w = 4a^2x^2y^2 - x^4y^2 - x^2y^4$, $w_x = 2xy^2(4a^2 - 2x^2 - y^2) = 0$, $w_y = 2x^2y\left(4a^2 - x^2 - 2y^2\right) = 0$, $4a^2 - 2x^2 - y^2 = 0$ and $4a^2 - x^2 - 2y^2 = 0$; critical point $\left(2a/\sqrt{3}, 2a/\sqrt{3}\right)$;
$w_{xx}w_{yy} - w_{xy}^2 = \dfrac{4096}{27}a^8 > 0$ and $w_{xx} = -\dfrac{128}{9}a^4 < 0$ at $\left(2a/\sqrt{3}, 2a/\sqrt{3}\right)$, relative maximum.
$z = 2a/\sqrt{3}$ when $x = y = 2a/\sqrt{3}$, the dimensions of the box of maximum volume are $2a/\sqrt{3}, 2a/\sqrt{3}, 2a/\sqrt{3}$.

39. Let x, y, and z be, respectively, the length, width, and height of the box. Minimize $C = 10(2xy) + 5(2xz + 2yz) = 10(2xy + xz + yz)$ subject to $xyz = 16$. $z = 16/(xy)$ so $C = 20(xy + 8/y + 8/x)$, $C_x = 20(y - 8/x^2) = 0$, $C_y = 20(x - 8/y^2) = 0$; critical point (2,2); $C_{xx}C_{yy} - C_{xy}^2 = 1200 > 0$ and $C_{xx} = 40 > 0$ at (2,2), relative minimum. $z = 4$ when $x = y = 2$. The cost of materials is minimum if the length and width are 2 ft and the height is 4 ft.

41. (a) $x = 0 : f(0,y) = -3y^2$, minimum -3, maximum 0;

$x = 1, f(1,y) = 4 - 3y^2 + 2y, \dfrac{\partial f}{\partial y}(1,y) = -6y + 2 = 0$ at $y = 1/3$, minimum 3, maximum $13/3$;

$y = 0, f(x,0) = 4x^2$, minimum 0, maximum 4;

$y = 1, f(x,1) = 4x^2 + 2x - 3, \dfrac{\partial f}{\partial x}(x,1) = 8x + 2 \neq 0$ for $0 < x < 1$, minimum -3, maximum 3

(b) $f(x,x) = 3x^2$, minimum 0, maximum 3; $f(x,1-x) = -x^2+8x-3, \dfrac{d}{dx}f(x,1-x) = -2x+8 \neq 0$ for $0 < x < 1$, maximum 4, minimum -3

(c) $f_x(x,y) = 8x + 2y = 0, f_y(x,y) = -6y + 2x = 0$, solution is $(0,0)$, which is not an interior point of the square, so check the sides: minimum -3, maximum $13/3$.

43. Minimize $S = xy + 2xz + 2yz$ subject to $xyz = V$, $x > 0$, $y > 0$, $z > 0$ where x, y, and z are, respectively, the length, width, and height of the box. $z = V/(xy)$ so $S = xy + 2V/y + 2V/x$, $S_x = y - 2V/x^2 = 0$, $S_y = x - 2V/y^2 = 0$; critical point $(\sqrt[3]{2V}, \sqrt[3]{2V})$; $S_{xx}S_{yy} - S_{xy}^2 = 3 > 0$ and $S_{xx} = 2 > 0$ at this point so there is a relative minimum there. The length and width are each $\sqrt[3]{2V}$, the height is $z = \sqrt[3]{2V}/2$.

45. (a) $$\frac{\partial g}{\partial m} = \sum_{i=1}^{n} 2(mx_i + b - y_i)x_i = 2\left(m\sum_{i=1}^{n} x_i^2 + b\sum_{i=1}^{n} x_i - \sum_{i=1}^{n} x_i y_i\right) = 0 \text{ if}$$

$$\left(\sum_{i=1}^{n} x_i^2\right)m + \left(\sum_{i=1}^{n} x_i\right)b = \sum_{i=1}^{n} x_i y_i,$$

$$\frac{\partial g}{\partial b} = \sum_{i=1}^{n} 2(mx_i + b - y_i) = 2\left(m\sum_{i=1}^{n} x_i + bn - \sum_{i=1}^{n} y_i\right) = 0 \text{ if } \left(\sum_{i=1}^{n} x_i\right)m + nb = \sum_{i=1}^{n} y_i$$

(b) $$\sum_{i=1}^{n}(x_i - \bar{x})^2 = \sum_{i=1}^{n}(x_i^2 - 2\bar{x}x_i + \bar{x}^2) = \sum_{i=1}^{n} x_i^2 - 2\bar{x}\sum_{i=1}^{n} x_i + n\bar{x}^2$$

$$= \sum_{i=1}^{n} x_i^2 - \frac{2}{n}\left(\sum_{i=1}^{n} x_i\right)^2 + \frac{1}{n}\left(\sum_{i=1}^{n} x_i\right)^2$$

$$= \sum_{i=1}^{n} x_i^2 - \frac{1}{n}\left(\sum_{i=1}^{n} x_i\right)^2 \geq 0 \text{ so } n\sum_{i=1}^{n} x_i^2 - \left(\sum_{i=1}^{n} x_i\right)^2 \geq 0$$

This is an equality if and only if $\sum_{i=1}^{n}(x_i - \bar{x})^2 = 0$, which means $x_i = \bar{x}$ for each i.

(c) The system of equations $Am + Bb = C, Dm + Eb = F$ in the unknowns m and b has a unique solution provided $AE \neq BD$, and if so the solution is $m = \dfrac{CE - BF}{AE - BD}, b = \dfrac{F - Dm}{E}$, which after the appropriate substitution yields the desired result.

47. $n = 3, \sum_{i=1}^{3} x_i = 3, \sum_{i=1}^{3} y_i = 7, \sum_{i=1}^{3} x_i y_i = 13, \sum_{i=1}^{3} x_i^2 = 11, y = \frac{3}{4}x + \frac{19}{12}$

49. $\sum_{i=1}^{4} x_i = 10, \sum_{i=1}^{4} y_i = 8.2, \sum_{i=1}^{4} x_i^2 = 30, \sum_{i=1}^{4} x_i y_i = 23, n = 4;\ m = 0.5,\ b = 0.8,\ y = 0.5x + 0.8.$

51. **(a)** $y = \frac{8843}{140} + \frac{57}{200}t \approx 63.1643 + 0.285t$ **(b)**

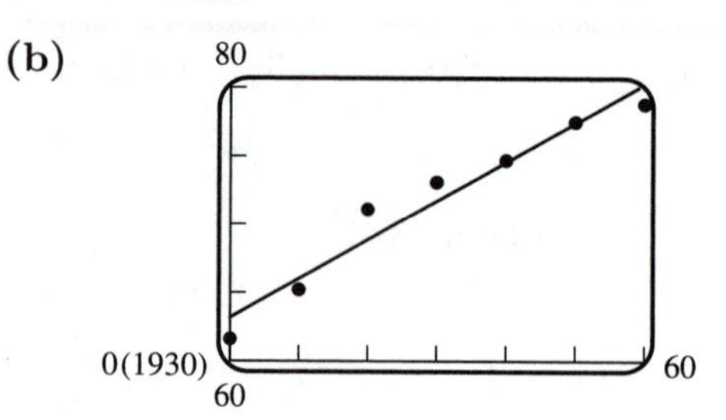

(c) $y = \frac{2909}{35} \approx 83.1143$

53. **(a)** $P = \frac{2798}{21} + \frac{171}{350}T \approx 133.2381 + 0.4886T$ **(b)**

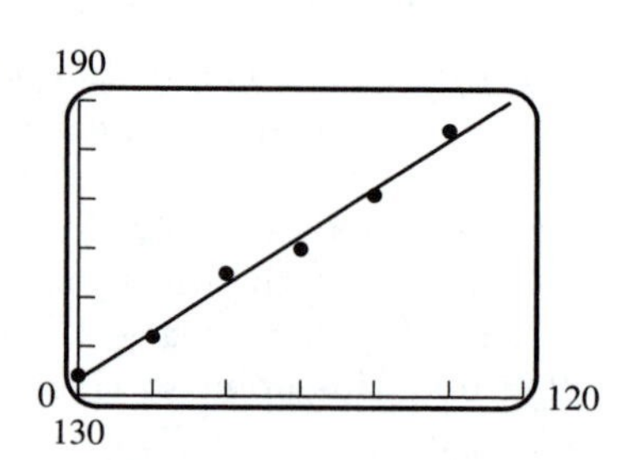

(c) $T \approx -\frac{139{,}900}{513} \approx -272.7096°$ C

55. $f(x_0, y_0) \geq f(x, y)$ for all (x, y) inside a circle centered at (x_0, y_0) by virtue of Definition 14.8.1. If r is the radius of the circle, then in particular $f(x_0, y_0) \geq f(x, y_0)$ for all x satisfying $|x - x_0| < r$ so $f(x, y_0)$ has a relative maximum at x_0. The proof is similar for the function $f(x_0, y)$.

EXERCISE SET 14.9

1. **(a)** $xy = 4$ is tangent to the line, so the maximum value of f is 4.

(b) $xy = 2$ intersects the curve and so gives a smaller value of f.

(c) Maximize $f(x, y) = xy$ subject to the constraint $g(x, y) = x + y - 4 = 0, \nabla f = \lambda \nabla g$, $y\mathbf{i} + x\mathbf{j} = \lambda(\mathbf{i} + \mathbf{j})$, so solve the equations $y = \lambda, x = \lambda$ with solution $x = y = \lambda$, but $x + y = 4$, so $x = y = 2$, and the maximum value of f is $f = xy = 4$.

3. **(a)**

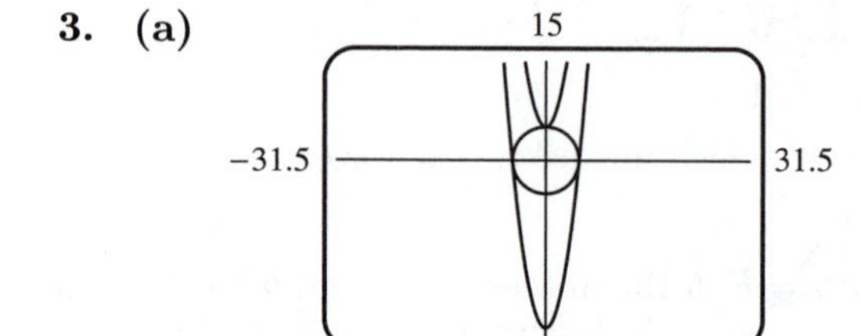

(b) one extremum at $(0, 5)$ and one at approximately $(\pm 5, 0)$, so minimum value -5, maximum value ≈ 25

(c) Find the minimum and maximum values of $f(x, y) = x^2 - y$ subject to the constraint $g(x, y) = x^2 + y^2 - 25 = 0, \nabla f = \lambda \nabla g, 2x\mathbf{i} - \mathbf{j} = 2\lambda x\mathbf{i} + 2\lambda y\mathbf{j}$, so solve $2x = 2\lambda x, -1 = 2\lambda y, x^2 + y^2 - 25 = 0$. If $x = 0$ then $y = \pm 5, f = \mp 5$, and if $x \neq 0$ then $\lambda = 1, y = -1/2, x^2 = 25 - 1/4 = 99/4, f = 99/4 + 1/2 = 101/4$, so the maximum value of f is $101/4$ at $(\pm 3\sqrt{11}/2, -1/2)$ and the minimum value of f is -5 at $(0, 5)$.

5. $y = 8x\lambda$, $x = 16y\lambda$; $y/(8x) = x/(16y)$, $x^2 = 2y^2$ so $4\left(2y^2\right) + 8y^2 = 16$, $y^2 = 1$, $y = \pm 1$. Test $\left(\pm\sqrt{2}, -1\right)$ and $(\pm\sqrt{2}, 1)$. $f\left(-\sqrt{2}, -1\right) = f\left(\sqrt{2}, 1\right) = \sqrt{2}$, $f\left(-\sqrt{2}, 1\right) = f\left(\sqrt{2}, -1\right) = -\sqrt{2}$. Maximum $\sqrt{2}$ at $\left(-\sqrt{2}, -1\right)$ and $\left(\sqrt{2}, 1\right)$, minimum $-\sqrt{2}$ at $\left(-\sqrt{2}, 1\right)$ and $\left(\sqrt{2}, -1\right)$.

7. $12x^2 = 4x\lambda$, $2y = 2y\lambda$. If $y \neq 0$ then $\lambda = 1$ and $12x^2 = 4x$, $12x(x - 1/3) = 0$, $x = 0$ or $x = 1/3$ so from $2x^2 + y^2 = 1$ we find that $y = \pm 1$ when $x = 0$, $y = \pm\sqrt{7}/3$ when $x = 1/3$. If $y = 0$ then $2x^2 + (0)^2 = 1$, $x = \pm 1/\sqrt{2}$. Test $(0, \pm 1)$, $(1/3, \pm\sqrt{7}/3)$, and $(\pm 1/\sqrt{2}, 0)$. $f(0, \pm 1) = 1$, $f\left(1/3, \pm\sqrt{7}/3\right) = 25/27$, $f\left(1/\sqrt{2}, 0\right) = \sqrt{2}$, $f\left(-1/\sqrt{2}, 0\right) = -\sqrt{2}$. Maximum $\sqrt{2}$ at $\left(1/\sqrt{2}, 0\right)$, minimum $-\sqrt{2}$ at $\left(-1/\sqrt{2}, 0\right)$.

9. $2 = 2x\lambda, 1 = 2y\lambda, -2 = 2z\lambda$; $1/x = 1/(2y) = -1/z$ thus $x = 2y$, $z = -2y$ so $(2y)^2 + y^2 + (-2y)^2 = 4$, $y^2 = 4/9$, $y = \pm 2/3$. Test $(-4/3, -2/3, 4/3)$ and $(4/3, 2/3, -4/3)$. $f(-4/3, -2/3, 4/3) = -6$, $f(4/3, 2/3, -4/3) = 6$. Maximum 6 at $(4/3, 2/3, -4/3)$, minimum -6 at $(-4/3, -2/3, 4/3)$.

11. $yz = 2x\lambda$, $xz = 2y\lambda$, $xy = 2z\lambda$; $yz/(2x) = xz/(2y) = xy/(2z)$ thus $y^2 = x^2$, $z^2 = x^2$ so $x^2 + x^2 + x^2 = 1$, $x = \pm 1/\sqrt{3}$. Test the eight possibilities with $x = \pm 1/\sqrt{3}$, $y = \pm 1/\sqrt{3}$, and $z = \pm 1/\sqrt{3}$ to find the maximum is $1/\left(3\sqrt{3}\right)$ at $\left(1/\sqrt{3}, 1/\sqrt{3}, 1/\sqrt{3}\right)$, $\left(1/\sqrt{3}, -1/\sqrt{3}, -1/\sqrt{3}\right)$, $\left(-1/\sqrt{3}, 1/\sqrt{3}, -1/\sqrt{3}\right)$, and $\left(-1/\sqrt{3}, -1/\sqrt{3}, 1/\sqrt{3}\right)$; the minimum is $-1/\left(3\sqrt{3}\right)$ at $\left(1/\sqrt{3}, 1/\sqrt{3}, -1/\sqrt{3}\right)$, $\left(1/\sqrt{3}, -1/\sqrt{3}, 1/\sqrt{3}\right)$, $\left(-1/\sqrt{3}, 1/\sqrt{3}, 1/\sqrt{3}\right)$, and $\left(-1/\sqrt{3}, -1/\sqrt{3}, -1/\sqrt{3}\right)$.

13. $f(x, y) = x^2 + y^2$; $2x = 2\lambda$, $2y = -4\lambda$; $y = -2x$ so $2x - 4(-2x) = 3$, $x = 3/10$. The point is $(3/10, -3/5)$.

15. $f(x, y, z) = x^2 + y^2 + z^2$; $2x = \lambda$, $2y = 2\lambda$, $2z = \lambda$; $y = 2x$, $z = x$ so $x + 2(2x) + x = 1$, $x = 1/6$. The point is $(1/6, 1/3, 1/6)$.

17. $f(x, y) = (x - 1)^2 + (y - 2)^2$; $2(x - 1) = 2x\lambda$, $2(y - 2) = 2y\lambda$; $(x - 1)/x = (y - 2)/y$, $y = 2x$ so $x^2 + (2x)^2 = 45$, $x = \pm 3$. $f(-3, -6) = 80$ and $f(3, 6) = 20$ so (3,6) is closest and $(-3, -6)$ is farthest.

19. $f(x, y, z) = x + y + z$, $x^2 + y^2 + z^2 = 25$ where x, y, and z are the components of the vector; $1 = 2x\lambda$, $1 = 2y\lambda$, $1 = 2z\lambda$; $1/(2x) = 1/(2y) = 1/(2z)$; $y = x$, $z = x$ so $x^2 + x^2 + x^2 = 25$, $x = \pm 5/\sqrt{3}$. $f\left(-5/\sqrt{3}, -5/\sqrt{3}, -5/\sqrt{3}\right) = -5\sqrt{3}$ and $f\left(5/\sqrt{3}, 5/\sqrt{3}, 5/\sqrt{3}\right) = 5\sqrt{3}$ so the vector is $5(\mathbf{i} + \mathbf{j} + \mathbf{k})/\sqrt{3}$.

21. Minimize $f = x^2 + y^2 + z^2$ subject to $g(x, y, z) = x + y + z - 27 = 0$. $\nabla f = \lambda \nabla g$, $2x\mathbf{i} + 2y\mathbf{j} + 2z\mathbf{k} = \lambda\mathbf{i} + \lambda\mathbf{j} + \lambda\mathbf{k}$, solution $x = y = z = 9$, minimum value 243

23. Minimize $f = x^2 + y^2 + z^2$ subject to $x^2 - yz = 5, \nabla f = \lambda \nabla g, 2x = 2x\lambda, 2y = -z\lambda, 2z = -y\lambda$. If $\lambda \neq \pm 2$, then $y = z = 0, x = \pm\sqrt{5}, f = 5$; if $\lambda = \pm 2$ then $x = 0$, and since $-yz = 5$, $y = -z = \pm\sqrt{5}, f = 10$, thus the minimum value is 5 at $(\pm\sqrt{5}, 0, 0)$.

25. Let x, y, and z be, respectively, the length, width, and height of the box. Minimize $f(x, y, z) = 10(2xy) + 5(2xz + 2yz) = 10(2xy + xz + yz)$ subject to $g(x, y, z) = xyz - 16 = 0$, $\nabla f = \lambda \nabla g, 20y + 10z = \lambda yz, 20x + 10z = \lambda xz, 10x + 10y = \lambda xy$. Since $V = xyz = 16, x, y, z \neq 0$, thus $\lambda z = 20 + 10(z/y) = 20 + 10(z/x)$, so $x = y$. From this and $10x + 10y = \lambda xy$ it follows that $20 = \lambda x$, so $10z = 20x, z = 2x = 2y, V = 2x^3 = 16$ and thus $x = y = 2$ ft, $z = 4$ ft, $f(2, 2, 4) = 240$ cents.

27. Maximize $A(a, b, \alpha) = ab \sin\alpha$ subject to $g(a, b, \alpha) = 2a + 2b - \ell = 0$, $\nabla_{(a,b,\alpha)} f = \lambda \nabla_{(a,b,\alpha)} g$, $b \sin\alpha = 2\lambda$, $a \sin\alpha = 2\lambda$, $ab \cos\alpha = 0$ with solution $a = b\ (= \ell/4)$, $\alpha = \pi/2$ maximum value if parallelogram is a square.

29. **(a)** Maximize $f(\alpha, \beta, \gamma) = \cos\alpha \cos\beta \cos\gamma$ subject to $g(\alpha, \beta, \gamma) = \alpha + \beta + \gamma - \pi = 0$, $\nabla f = \lambda \nabla g$, $-\sin\alpha \cos\beta \cos\gamma = \lambda$, $-\cos\alpha \sin\beta \cos\gamma = \lambda$, $-\cos\alpha \cos\beta \sin\gamma = \lambda$ with solution $\alpha = \beta = \gamma = \pi/3$, maximum value $1/8$

(b) for example, $f(\alpha, \beta) = \cos\alpha \cos\beta \cos(\pi - \alpha - \beta)$

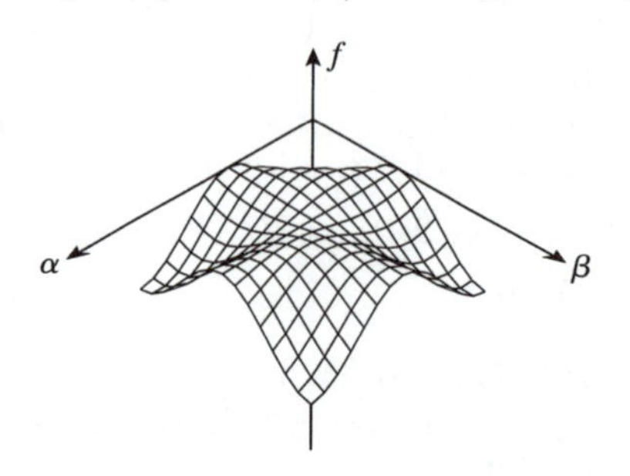

CHAPTER 14 SUPPLEMENTARY EXERCISES

1. **(a)** They approximate the profit per unit of any additional sales of the standard or high-resolution monitors, respectively.

(b) The rates of change with respect to the two directions x and y, and with respect to time.

3. $z = \sqrt{x^2 + y^2} = c$ implies $x^2 + y^2 = c^2$, which is the equation of a circle; $x^2 + y^2 = c$ is also the equation of a circle (for $c > 0$).

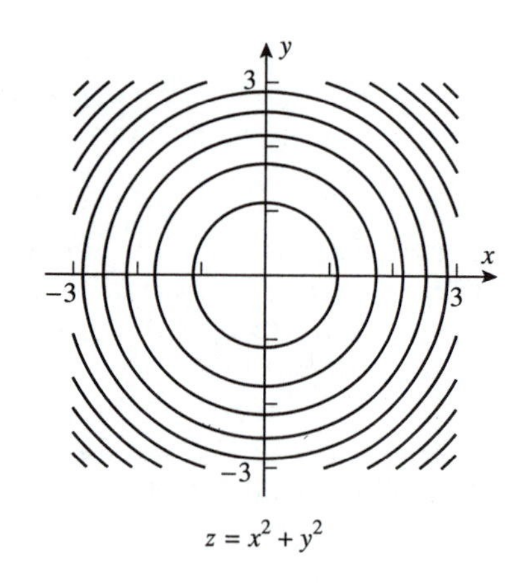

$z = x^2 + y^2$

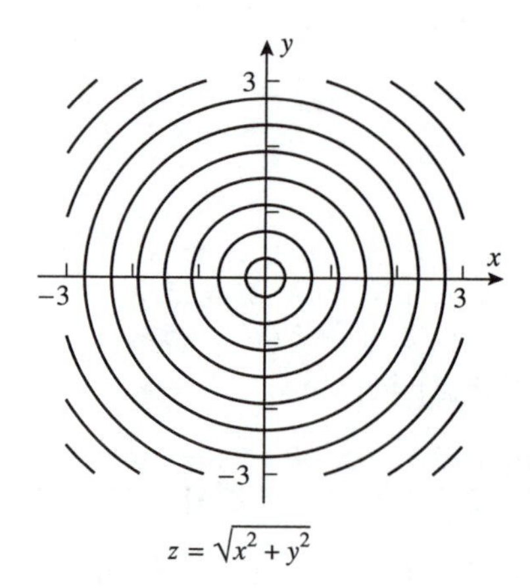

$z = \sqrt{x^2 + y^2}$

5. **(b)** $f(x, y, z) = z - x^2 - y^2$

7. **(a)** $f(\ln y, e^x) = e^{\ln y} \ln e^x = xy$ **(b)** $e^{r+s} \ln(rs)$

9. $w_x = 2x \sec^2(x^2 + y^2) + \sqrt{y}$, $w_{xy} = 8xy \sec^2(x^2 + y^2) \tan(x^2 + y^2) + \dfrac{1}{2} y^{-1/2}$,

$w_y = 2y \sec^2(x^2 + y^2) + \dfrac{1}{2} x y^{-1/2}$, $w_{yx} = 8xy \sec^2(x^2 + y^2) \tan(x^2 + y^2) + \dfrac{1}{2} y^{-1/2}$

11. $F_x = -6xz, F_{xx} = -6z, F_y = -6yz, F_{yy} = -6z, F_z = 6z^2 - 3x^2 - 3y^2,$
$F_{zz} = 12z, F_{xx} + F_{yy} + F_{zz} = -6z - 6z + 12z = 0$

13. (a) $P = \dfrac{10T}{V},$

$$\frac{dP}{dt} = \frac{\partial P}{\partial T}\frac{dT}{dt} + \frac{\partial P}{\partial V}\frac{dV}{dt} = \frac{10}{V}\cdot 3 - \frac{10T}{V^2}\cdot 0 = \frac{30}{V} = \frac{30}{2.5} = 12\ \text{N/(m}^2\text{min)} = 12\ \text{Pa/min}$$

(b) $$\frac{dP}{dt} = \frac{\partial P}{\partial T}\frac{dT}{dt} + \frac{\partial P}{\partial V}\frac{dV}{dt} = \frac{10}{V}\cdot 0 - \frac{10T}{V^2}\cdot(-3) = \frac{30T}{V^2} = \frac{30\cdot 50}{(2.5)^2} = 240\ \text{Pa/min}$$

15. $x^4 - x + y - x^3y = (x^3-1)(x-y)$, limit $= -1$, not defined on the line $y = x$ so not continuous at $(0,0)$

17. Use the unit vectors $\mathbf{u} = \langle \frac{1}{\sqrt{2}}, \frac{1}{\sqrt{2}}\rangle, \mathbf{v} = \langle 0, -1\rangle, \mathbf{w} = \langle -\frac{1}{\sqrt{5}}, -\frac{2}{\sqrt{5}}\rangle = -\frac{\sqrt{2}}{\sqrt{5}}\mathbf{u} + \frac{1}{\sqrt{5}}\mathbf{v}$, so that

$$D_{\mathbf{w}}f = -\frac{\sqrt{2}}{\sqrt{5}}D_{\mathbf{u}}f + \frac{1}{\sqrt{5}}D_{\mathbf{v}}f = -\frac{\sqrt{2}}{\sqrt{5}}2\sqrt{2} + \frac{1}{\sqrt{5}}(-3) = -\frac{7}{\sqrt{5}}$$

19. The origin is not such a point, so assume that the normal line at $(x_0, y_0, z_0) \neq (0,0,0)$ passes through the origin, then $\mathbf{n} = z_x\mathbf{i} + z_y\mathbf{j} - \mathbf{k} = -y_0\mathbf{i} - x_0\mathbf{j} - \mathbf{k}$; the line passes through the origin and is normal to the surface if it has the form $\mathbf{r}(t) = -y_0t\mathbf{i} - x_0t\mathbf{j} - t\mathbf{k}$ and $(x_0, y_0, z_0) = (x_0, y_0, 2 - x_0y_0)$ lies on the line if $-y_0t = x_0, -x_0t = y_0, -t = 2 - x_0y_0$, with solutions $x_0 = y_0 = -1$, $x_0 = y_0 = 1, x_0 = y_0 = 0$; thus the points are $(0,0,2), (1,1,1), (-1,-1,1)$.

21. A tangent to the line is $6\mathbf{i} + 4\mathbf{j} + \mathbf{k}$, a normal to the surface is $\mathbf{n} = 18x\mathbf{i} + 8y\mathbf{j} - \mathbf{k}$, so solve $18x = 6k, 8y = 4k, -1 = k;\ k = -1, x = -1/3, y = -1/2, z = 2$

23. $dV = \frac{2}{3}xh\,dx + \frac{1}{3}x^2dh = \frac{2}{3}2(-0.1) + \frac{1}{3}(0.2) = -0.06667\ \text{m}^3; \Delta V = -0.07267\ \text{m}^3$

25. $\nabla f = (2xy - 6x)\mathbf{i} + (x^2 - 12y)\mathbf{j} = \mathbf{0}$ if $2xy - 6x = 0, x^2 - 12y = 0$; if $x = 0$ then $y = 0$, and if $x \neq 0$ then $y = 3, x = \pm 6$, thus the gradient vanishes at $(0,0), (-6,3), (6,3)$; $f_{xx} = 0$ at all three points, $f_{yy} = -12 < 0, D = -4x^2$, so $(\pm 6, 3)$ are saddle points, and near the origin we write $f(x,y) = (y-3)x^2 - 6y^2$; since $y - 3 < 0$ when $|y| < 3$, f has a local maximum by inspection.

27. (a)

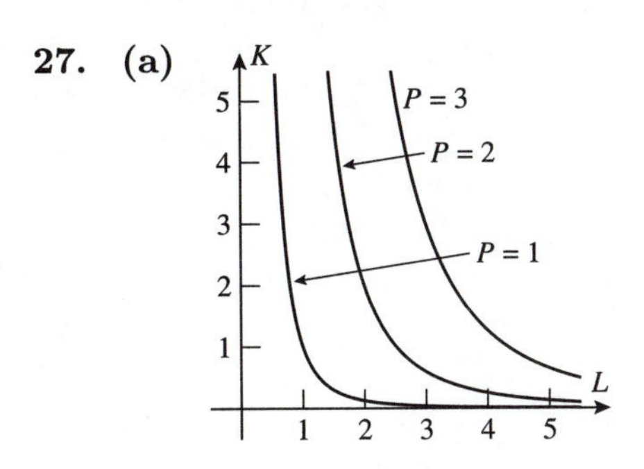

(b)

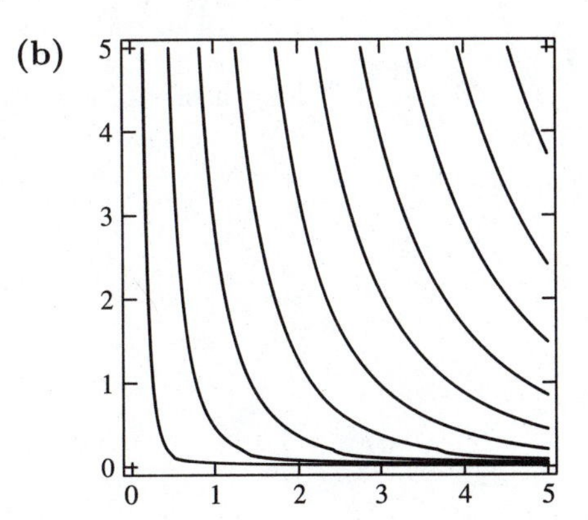

29. (a) Maximize $P = 1000L^{0.6}(200{,}000 - L)^{0.4}$ subject to $50L + 100K = 200{,}000$ or $L = 2K = 4000$.
$600\left(\frac{K}{L}\right)^{0.4} = \lambda,\ 400\left(\frac{L}{K}\right)^{0.6} = 2\lambda,\ L + 2K = 4000$; so $\frac{2}{3}\left(\frac{L}{K}\right) = 2$, thus $L = 3K$, $L = 2400, K = 800, P(2400, 800) = 1000\cdot 2400^{0.6}\cdot 800^{0.4} = 1000\cdot 3^{0.6}\cdot 800 = 800{,}000\cdot 3^{0.6} \approx$ \$1,546,545.64

(b) The value of labor is $50L = 120{,}000$ and the value of capital is $100K = 80{,}000$.

31. Let the first octant corner of the box be (x, y, z), so that $(x/a)^2 + (y/b)^2 + (z/c)^2 = 1$. Maximize $V = 8xyz$ subject to $g(x, y, z) = (x/a)^2 + (y/b)^2 + (z/c)^2 = 1$, solve $\nabla V = \lambda\nabla g$, or $8(yz\mathbf{i} + xz\mathbf{j} + xy\mathbf{k}) = (2\lambda x/a^2)\mathbf{i} + (2\lambda y/b^2)\mathbf{j} + (2\lambda z/c^2)\mathbf{k}, 8a^2yz = 2\lambda x, 8b^2xz = 2\lambda y, 8c^2xy = 2\lambda z$. For the maximum volume, $x, y, z \neq 0$; divide the first equation by the second to obtain $a^2y^2 = b^2x^2$; the first by the third to obtain $a^2z^2 = c^2x^2$, and finally $b^2z^2 = c^2y^2$. From $g = 1$ get $3(x/a)^2 = 1, x = \pm a/\sqrt{3}$, and then $y = \pm b/\sqrt{3}, z = \pm c/\sqrt{3}$. The dimensions of the box are $\frac{2a}{\sqrt{3}} \times \frac{2b}{\sqrt{3}} \times \frac{2c}{\sqrt{3}}$, and the maximum volume is $8abc/(3\sqrt{3})$.

33. $\frac{dy}{dx} = -\frac{f_x}{f_y}, \frac{d^2y}{dx^2} = -\frac{f_y(d/dx)f_x - f_x(d/dx)f_y}{f_y^2} = -\frac{f_y(f_{xx} + f_{xy}(dy/dx)) - f_x(f_{xy} + f_{yy}(dy/dx))}{f_y^2}$

$$= -\frac{f_y(f_{xx} + f_{xy}(-f_x/f_y)) - f_x(f_{xy} + f_{yy}(-f_x/f_y))}{f_y^2} = \frac{-f_y^2 f_{xx} + 2f_x f_y f_{xy} - f_x^2 f_{yy}}{f_y^3}$$

35. Solve $(t-1)^2/4 + 16e^{-2t} + (2-\sqrt{t})^2 = 1$ for t to get $t = 1.833223, 2.839844$; the particle strikes the surface at the points $P_1(0.83322, 0.639589, 0.646034)$, $P_2(1.83984, 0.233739, 0.314816)$. The velocity vectors are given by $\mathbf{v} = \frac{dx}{dt}\mathbf{i} + \frac{dy}{dt}\mathbf{j} + \frac{dz}{dt}\mathbf{k} = \mathbf{i} - 4e^{-t}\mathbf{j} - 1/(2\sqrt{t})\mathbf{k}$, and a normal to the surface is $\mathbf{n} = \nabla(x^2/4 + y^2 + z^2) = x/2\mathbf{i} + 2y\mathbf{j} + 2z\mathbf{k}$. At the points P_i these are
$\mathbf{v}_1 = \mathbf{i} - 0.639589\mathbf{j} - 0.369286\mathbf{k}, \mathbf{v}_2 = \mathbf{i} - 0.233739\mathbf{j} + 0.296704\mathbf{k}$;
$\mathbf{n}_1 = 0.41661\mathbf{i} + 1.27918\mathbf{j} + 1.29207\mathbf{k}$ and $\mathbf{n}_2 = 0.91992\mathbf{i} + 0.46748\mathbf{j} + 0.62963\mathbf{k}$ so
$\cos^{-1}[(\mathbf{v}_i \cdot \mathbf{n}_i)/(\|\mathbf{v}_i\|\,\|\mathbf{n}_i\|)] = 112.3°, 61.1°$; the acute angles are $67.7°, 61.1°$.

37. Let x, y, z be the lengths of the sides opposite angles α, β, γ, located at A, B, C respectively. Then $x^2 = y^2 + z^2 - 2yz\cos\alpha$ and $x^2 = 100 + 400 - 2(10)(20)/2 = 300, x = 10\sqrt{3}$ and

$$2x\frac{dx}{dt} = 2y\frac{dy}{dt} + 2z\frac{dz}{dt} - 2\left(y\frac{dz}{dt}\cos\alpha + z\frac{dy}{dt}\cos\alpha - yz(\sin\alpha)\frac{d\alpha}{dt}\right)$$

$$= 2(10)(4) + 2(20)(2) - 2\left(10(2)\frac{1}{2} + 20(4)\frac{1}{2} - 10(20)\frac{\sqrt{3}}{2}\frac{\pi}{60}\right) = 60 + \frac{10\pi}{\sqrt{3}}$$

so $\frac{dx}{dt} = \sqrt{3} + \frac{\pi}{6}$, the length of BC is increasing.

CHAPTER 15
Multiple Integrals

EXERCISE SET 15.1

1. $\int_0^1 \int_0^2 (x+3)dy\,dx = \int_0^1 (2x+6)dx = 7$

3. $\int_2^4 \int_0^1 x^2 y\,dx\,dy = \int_2^4 \frac{1}{3} y\,dy = 2$

5. $\int_0^{\ln 3} \int_0^{\ln 2} e^{x+y} dy\,dx = \int_0^{\ln 3} e^x dx = 2$

7. $\int_{-1}^0 \int_2^5 dx\,dy = \int_{-1}^0 3\,dy = 3$

9. $\int_0^1 \int_0^1 \frac{x}{(xy+1)^2} dy\,dx = \int_0^1 \left(1 - \frac{1}{x+1}\right) dx = 1 - \ln 2$

11. $\int_0^{\ln 2} \int_0^1 xy\,e^{y^2 x} dy\,dx = \int_0^{\ln 2} \frac{1}{2}(e^x - 1)dx = (1 - \ln 2)/2$

13. $\int_{-1}^1 \int_{-2}^2 4xy^3 dy\,dx = \int_{-1}^1 0\,dx = 0$

15. $\int_0^1 \int_2^3 x\sqrt{1-x^2}\,dy\,dx = \int_0^1 x(1-x^2)^{1/2} dx = 1/3$

17. (a) $x_k^* = k/2 - 1/4, k = 1,2,3,4; y_l^* = l/2 - 1/4, l = 1,2,3,4,$

$$\int\int_R f(x,y)\,dxdy \approx \sum_{k=1}^4 \sum_{l=1}^4 f(x_k^*, y_l^*)\Delta A_{kl} = \sum_{k=1}^4 \sum_{l=1}^4 [(k/2-1/4)^2 + (l/2-1/4)](1/2)^2 = 37/4$$

(b) $\int_0^2 \int_0^2 (x^2+y)\,dxdy = 28/3$; the error is $|37/4 - 28/3| = 1/12$

19. $V = \int_3^5 \int_1^2 (2x+y)dy\,dx = \int_3^5 (2x+3/2)dx = 19$

21. $V = \int_0^2 \int_0^3 x^2 dy\,dx = \int_0^2 3x^2 dx = 8$

23. (a)

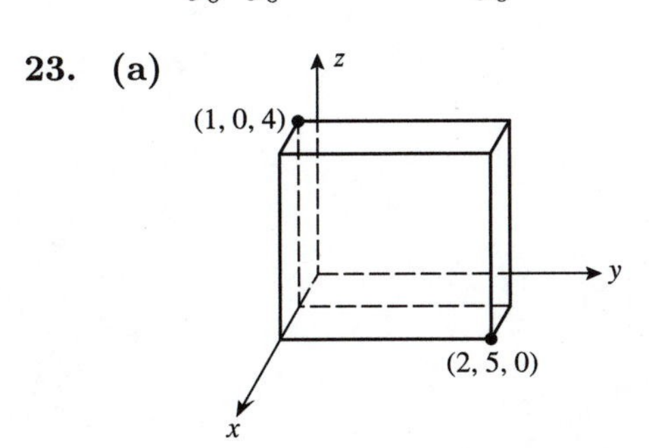

(b)

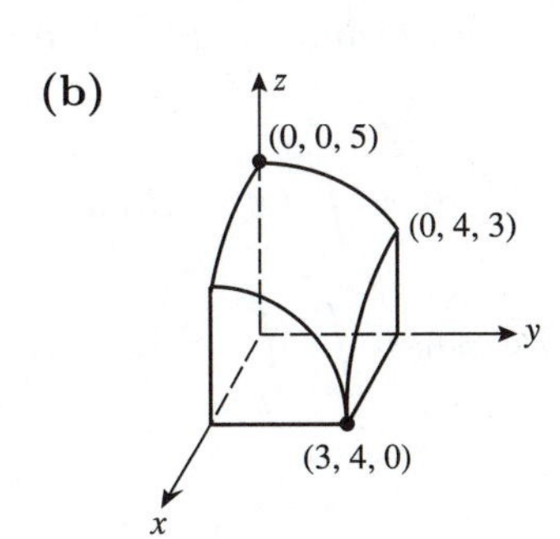

25. $$\int_0^{1/2}\int_0^{\pi} x\cos(xy)\cos^2\pi x\,dy\,dx = \int_0^{1/2}\cos^2\pi x\sin(xy)\Big]_0^{\pi}dx$$
$$= \int_0^{1/2}\cos^2\pi x\sin\pi x\,dx = -\frac{1}{3\pi}\cos^3\pi x\Big]_0^{1/2} = \frac{1}{3\pi}$$

27. $$f_{\text{ave}} = \frac{2}{\pi}\int_0^{\pi/2}\int_0^1 y\sin xy\,dx\,dy = \frac{2}{\pi}\int_0^{\pi/2}\left(-\cos xy\Big]_{x=0}^{x=1}\right)dy = \frac{2}{\pi}\int_0^{\pi/2}(1-\cos y)\,dy = 1-\frac{2}{\pi}$$

29. $$T_{\text{ave}} = \frac{1}{2}\int_0^1\int_0^2(10-8x^2-2y^2)\,dy\,dx = \frac{1}{2}\int_0^1\left(\frac{44}{3}-16x^2\right)dx = \left(\frac{14}{3}\right)^{\circ}$$

31. 1.381737122

33. $$\iint_R f(x,y)dA = \int_a^b\left[\int_c^d g(x)h(y)dy\right]dx = \int_a^b g(x)\left[\int_c^d h(y)dy\right]dx$$
$$= \left[\int_a^b g(x)dx\right]\left[\int_c^d h(y)dy\right]$$

35. The first integral equals 1/2, the second equals −1/2. No, because the integrand is not continuous.

EXERCISE SET 15.2

1. $$\int_0^1\int_{x^2}^{x} xy^2dy\,dx = \int_0^1\frac{1}{3}(x^4-x^7)dx = 1/40$$

3. $$\int_0^3\int_0^{\sqrt{9-y^2}} y\,dx\,dy = \int_0^3 y\sqrt{9-y^2}\,dy = 9$$

5. $$\int_{\sqrt{\pi}}^{\sqrt{2\pi}}\int_0^{x^3}\sin(y/x)dy\,dx = \int_{\sqrt{\pi}}^{\sqrt{2\pi}}[-x\cos(x^2)+x]dx = \pi/2$$

7. $$\int_{\pi/2}^{\pi}\int_0^{x^2}\frac{1}{x}\cos(y/x)dy\,dx = \int_{\pi/2}^{\pi}\sin x\,dx = 1$$

9. $$\int_0^1\int_0^x y\sqrt{x^2-y^2}\,dy\,dx = \int_0^1\frac{1}{3}x^3dx = 1/12$$

11. (a) $$\int_0^2\int_0^{x^2} xy\,dy\,dx = \int_0^2\frac{1}{2}x^5dx = \frac{16}{3}$$

(b) $$\int_1^3\int_{-(y-5)/2}^{(y+7)/2} xy\,dx\,dy = \int_1^3(3y^2+3y)dy = 38$$

13. (a) $$\int_4^8\int_{16/x}^{x} x^2dy\,dx = \int_4^8(x^3-16x)dx = 576$$

(b) $$\int_2^4\int_{16/y}^8 x^2\,dxdy + \int_4^8\int_y^8 x^2\,dx\,dy = \int_4^8\left[\frac{512}{3}-\frac{4096}{3y^3}\right]dy + \int_4^8\frac{512-y^3}{3}dy$$
$$= \frac{640}{3}+\frac{1088}{3} = 576$$

15. (a) $\displaystyle\int_{-1}^{1}\int_{-\sqrt{1-x^2}}^{\sqrt{1-x^2}}(3x-2y)dy\,dx = \int_{-1}^{1} 6x\sqrt{1-x^2}\,dx = 0$

(b) $\displaystyle\int_{-1}^{1}\int_{-\sqrt{1-y^2}}^{\sqrt{1-y^2}}(3x-2y)\,dxdy = \int_{-1}^{1} -4y\sqrt{1-y^2}\,dy = 0$

17. $\displaystyle\int_0^4\int_0^{\sqrt{y}} x(1+y^2)^{-1/2}dx\,dy = \int_0^4 \frac{1}{2}y(1+y^2)^{-1/2}dy = (\sqrt{17}-1)/2$

19. $\displaystyle\int_0^2\int_{y^2}^{6-y} xy\,dx\,dy = \int_0^2 \frac{1}{2}(36y-12y^2+y^3-y^5)dy = 50/3$

21. $\displaystyle\int_0^1\int_{x^3}^{x}(x-1)dy\,dx = \int_0^1(-x^4+x^3+x^2-x)dx = -7/60$

23. (a)

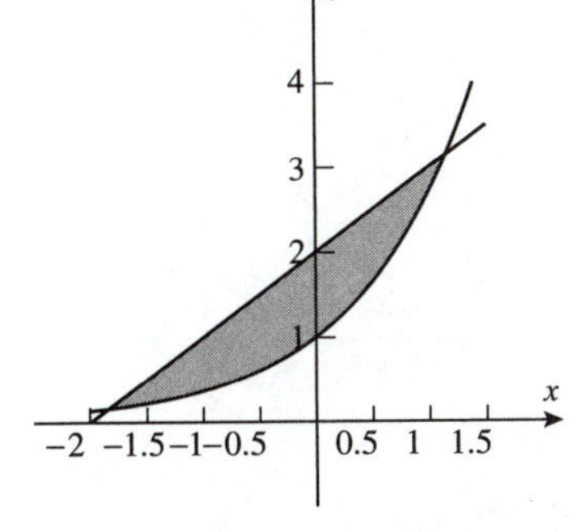

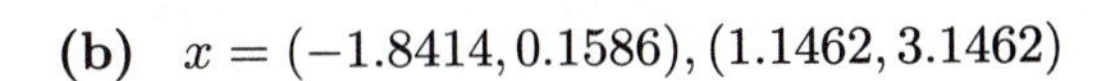
(b) $x = (-1.8414, 0.1586), (1.1462, 3.1462)$

(c) $\displaystyle\iint_R x\,dA \approx \int_{-1.8414}^{1.1462}\int_{e^x}^{x+2} x\,dydx = \int_{-1.8414}^{1.1462} x(x+2-e^x)\,dx \approx -0.4044$

(d) $\displaystyle\iint_R x\,dA \approx \int_{0.1586}^{3.1462}\int_{y-2}^{\ln y} x\,dxdy = \int_{0.1586}^{3.1462}\left[\frac{\ln^2 y}{2} - \frac{(y-2)^2}{2}\right]dy \approx -0.4044$

25. $\displaystyle A = \int_0^{\pi/4}\int_{\sin x}^{\cos x} dy\,dx = \int_0^{\pi/4}(\cos x - \sin x)dx = \sqrt{2}-1$

27. $\displaystyle A = \int_{-3}^{3}\int_{1-y^2/9}^{9-y^2} dx\,dy = \int_{-3}^{3} 8(1-y^2/9)dy = 32$

29. $\displaystyle\int_0^4\int_0^{6-3x/2}(3-3x/4-y/2)\,dy\,dx = \int_0^4[(3-3x/4)(6-3x/2)-(6-3x/2)^2/4]\,dx = 12$

31. $\displaystyle V = \int_{-3}^{3}\int_{-\sqrt{9-x^2}}^{\sqrt{9-x^2}}(3-x)dy\,dx = \int_{-3}^{3}(6\sqrt{9-x^2}-2x\sqrt{9-x^2})dx = 27\pi$

33. $\displaystyle V = \int_0^3\int_0^2(9x^2+y^2)dy\,dx = \int_0^3(18x^2+8/3)dx = 170$

35. $\displaystyle V = \int_{-3/2}^{3/2}\int_{-\sqrt{9-4x^2}}^{\sqrt{9-4x^2}}(y+3)dy\,dx = \int_{-3/2}^{3/2} 6\sqrt{9-4x^2}\,dx = 27\pi/2$

37. $V = 8\int_0^5 \int_0^{\sqrt{25-x^2}} \sqrt{25-x^2}dy\,dx = 8\int_0^5 (25-x^2)dx = 2000/3$

39. $V = 4\int_0^1 \int_0^{\sqrt{1-x^2}} (1-x^2-y^2)dy\,dx = \frac{8}{3}\int_0^1 (1-x^2)^{3/2}dx = \pi/2$

41. $\int_0^{\sqrt{2}} \int_{y^2}^{2} f(x,y)dx\,dy$

43. $\int_1^{e^2} \int_{\ln x}^{2} f(x,y)dy\,dx$

45. $\int_0^{\pi/2} \int_0^{\sin x} f(x,y)dy\,dx$

47. $\int_0^4 \int_0^{y/4} e^{-y^2}dx\,dy = \int_0^4 \frac{1}{4}ye^{-y^2}dy = (1-e^{-16})/8$

49. $\int_0^2 \int_0^{x^2} e^{x^3}dy\,dx = \int_0^2 x^2e^{x^3}dx = (e^8-1)/3$

51. $\int_0^2 \int_0^{y^2} \sin(y^3)dx\,dy = \int_0^2 y^2\sin(y^3)dy = (1-\cos 8)/3$

53. (a) $\int_0^4 \int_{\sqrt{x}}^2 \sin \pi y^3\, dy\, dx$; the inner integral is non-elementary.

$$\int_0^2 \int_0^{y^2} \sin\left(\pi y^3\right)\, dx\, dy = \int_0^2 y^2 \sin\left(\pi y^3\right)\, dy = -\frac{1}{3\pi}\cos\left(\pi y^3\right)\Big]_0^2 = 0$$

(b) $\int_0^1 \int_{\sin^{-1} y}^{\pi/2} \sec^2(\cos x)dx\,dy$; the inner integral is non-elementary.

$$\int_0^{\pi/2} \int_0^{\sin x} \sec^2(\cos x)dy\,dx = \int_0^{\pi/2} \sec^2(\cos x)\sin x\,dx = \tan 1$$

55. The region is symmetric with respect to the y-axis, and the integrand is an odd function of x, hence the answer is zero.

57. Area of triangle is 1/2, so $\bar{f} = 2\int_0^1 \int_x^1 \frac{1}{1+x^2}dy\,dx = 2\int_0^1 \left[\frac{1}{1+x^2} - \frac{x}{1+x^2}\right]dx = \frac{\pi}{2} - \ln 2$

59. $T_{\text{ave}} = \frac{1}{A(R)}\iint_R (5xy + x^2)\,dA$. The diamond has corners $(\pm 2, 0), (0, \pm 4)$ and thus has area $A(R) = 4\frac{1}{2}2(4) = 16\text{m}^2$. Since $5xy$ is an odd function of x (as well as y), $\iint_R 5xy\,dA = 0$. Since x^2 is an even function of both x and y,

$$T_{\text{ave}} = \frac{4}{16}\iint_{\substack{R\\x,y>0}} x^2\,dA = \frac{1}{4}\int_0^2 \int_0^{4-2x} x^2\,dydx = \frac{1}{4}\int_0^2 (4-2x)x^2\,dx = \frac{1}{4}\left(\frac{4}{3}x^3 - \frac{1}{2}x^4\right)\Big]_0^2 = \frac{2}{3}{}^\circ\text{C}$$

61. $y = \sin x$ and $y = x/2$ intersect at $x = 0$ and $x = a = 1.895494$, so

$$V = \int_0^a \int_{x/2}^{\sin x} \sqrt{1 + x + y}\, dy\, dx = 0.676089$$

EXERCISE SET 15.3

1. $\displaystyle \int_0^{\pi/2} \int_0^{\sin\theta} r\cos\theta dr\, d\theta = \int_0^{\pi/2} \frac{1}{2}\sin^2\theta\cos\theta\, d\theta = 1/6$

3. $\displaystyle \int_0^{\pi/2} \int_0^{a\sin\theta} r^2 dr\, d\theta = \int_0^{\pi/2} \frac{a^3}{3}\sin^3\theta\, d\theta = \frac{2}{9}a^3$

5. $\displaystyle \int_0^{\pi} \int_0^{1-\sin\theta} r^2\cos\theta\, dr\, d\theta = \int_0^{\pi} \frac{1}{3}(1-\sin\theta)^3\cos\theta\, d\theta = 0$

7. $\displaystyle A = \int_0^{2\pi} \int_0^{1-\cos\theta} r\, dr\, d\theta = \int_0^{2\pi} \frac{1}{2}(1-\cos\theta)^2 d\theta = 3\pi/2$

9. $\displaystyle A = \int_{\pi/4}^{\pi/2} \int_{\sin 2\theta}^{1} r\, dr\, d\theta = \int_{\pi/4}^{\pi/2} \frac{1}{2}(1-\sin^2 2\theta)d\theta = \pi/16$

11. $\displaystyle A = 2\int_{\pi/6}^{\pi/2} \int_2^{4\sin\theta} r\, dr\, d\theta = \int_{\pi/6}^{\pi/2} (16\sin^2\theta - 4)d\theta = 4\pi/3 + 2\sqrt{3}$

13. $\displaystyle V = 8\int_0^{\pi/2} \int_1^3 r\sqrt{9-r^2}\, dr\, d\theta = \frac{128}{3}\sqrt{2}\int_0^{\pi/2} d\theta = \frac{64}{3}\sqrt{2}\pi$

15. $\displaystyle V = 2\int_0^{\pi/2} \int_0^{\cos\theta} (1-r^2)r\, dr\, d\theta = \frac{1}{2}\int_0^{\pi/2} (2\cos^2\theta - \cos^4\theta)d\theta = 5\pi/32$

17. $\displaystyle V = \int_0^{\pi/2} \int_0^{3\sin\theta} r^2\sin\theta\, dr\, d\theta = 9\int_0^{\pi/2} \sin^4\theta\, d\theta = 27\pi/16$

19. $\displaystyle \int_0^{2\pi} \int_0^1 e^{-r^2} r\, dr\, d\theta = \frac{1}{2}(1-e^{-1})\int_0^{2\pi} d\theta = (1-e^{-1})\pi$

21. $\displaystyle \int_0^{\pi/4} \int_0^2 \frac{1}{1+r^2} r\, dr\, d\theta = \frac{1}{2}\ln 5 \int_0^{\pi/4} d\theta = \frac{\pi}{8}\ln 5$

23. $\displaystyle \int_0^{\pi/2} \int_0^1 r^3 dr\, d\theta = \frac{1}{4}\int_0^{\pi/2} d\theta = \pi/8$

25. $\displaystyle \int_0^{\pi/2} \int_0^{2\cos\theta} r^2 dr\, d\theta = \frac{8}{3}\int_0^{\pi/2} \cos^3\theta\, d\theta = 16/9$

27. $\displaystyle \int_0^{\pi/2} \int_0^a \frac{r}{(1+r^2)^{3/2}} dr\, d\theta = \frac{\pi}{2}\left(1 - 1/\sqrt{1+a^2}\right)$

29. $\displaystyle\int_0^{\pi/4}\int_0^2 \frac{r}{\sqrt{1+r^2}}dr\,d\theta = \frac{\pi}{4}(\sqrt{5}-1)$

31. $\displaystyle V = \int_0^{2\pi}\int_0^a hr\,dr\,d\theta = \int_0^{2\pi} h\frac{a^2}{2}\,d\theta = \pi a^2 h$

33. $\displaystyle V = 2\int_0^{\pi/2}\int_0^{a\sin\theta} \frac{c}{a}(a^2-r^2)^{1/2}r\,dr\,d\theta = \frac{2}{3}a^2c\int_0^{\pi/2}(1-\cos^3\theta)d\theta = (3\pi-4)a^2c/9$

35. $\displaystyle A = \int_{\pi/6}^{\pi/4}\int_{\sqrt{8\cos 2\theta}}^{4\sin\theta} r\,dr\,d\theta + \int_{\pi/4}^{\pi/2}\int_0^{4\sin\theta} r\,dr\,d\theta$

$\displaystyle = \int_{\pi/6}^{\pi/4}(8\sin^2\theta - 4\cos 2\theta)d\theta + \int_{\pi/4}^{\pi/2} 8\sin^2\theta\,d\theta = 4\pi/3 + 2\sqrt{3} - 2$

37. **(a)** $\displaystyle I^2 = \left[\int_0^{+\infty} e^{-x^2}dx\right]\left[\int_0^{+\infty} e^{-y^2}dy\right] = \int_0^{+\infty}\left[\int_0^{+\infty} e^{-x^2}dx\right]e^{-y^2}dy$

$\displaystyle = \int_0^{+\infty}\int_0^{+\infty} e^{-x^2}e^{-y^2}dx\,dy = \int_0^{+\infty}\int_0^{+\infty} e^{-(x^2+y^2)}dx\,dy$

(b) $\displaystyle I^2 = \int_0^{\pi/2}\int_0^{+\infty} e^{-r^2}r\,dr\,d\theta = \frac{1}{2}\int_0^{\pi/2} d\theta = \pi/4$ **(c)** $I = \sqrt{\pi}/2$

39. $\displaystyle V = \int_0^{2\pi}\int_0^R D(r)r\,dr\,d\theta = \int_0^{2\pi}\int_0^R ke^{-r}r\,dr\,d\theta = -2\pi k(1+r)e^{-r}\Big]_0^R = 2\pi k[1-(R+1)e^{-R}]$

EXERCISE SET 15.4

1. **(a)**

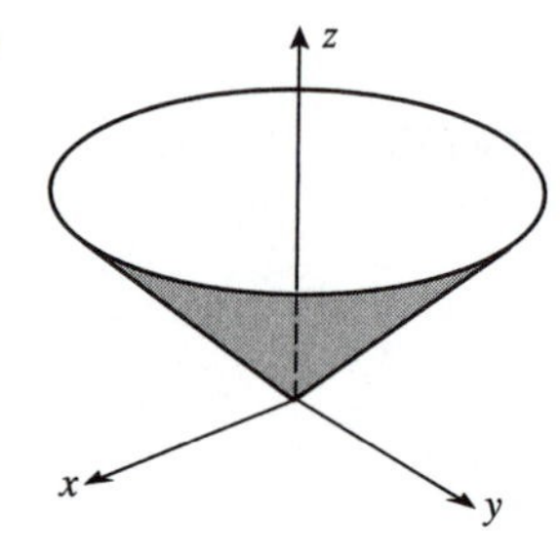

(b)

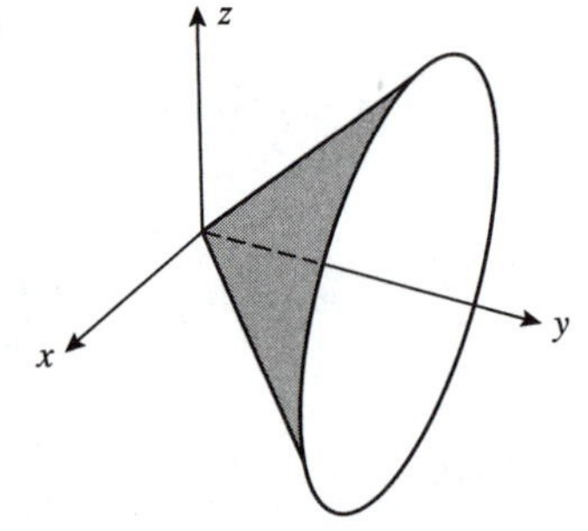

(c)

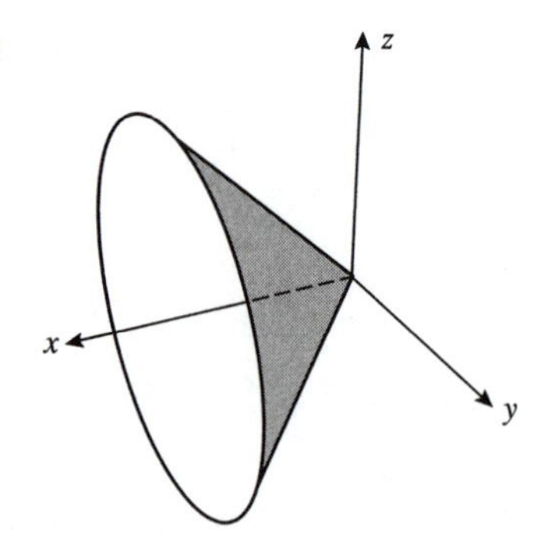

3. **(a)** $x = u, y = v, z = \dfrac{5}{2} + \dfrac{3}{2}u - 2v$ **(b)** $x = u, y = v, z = u^2$

5. (a) $x = 5\cos u, y = 5\sin u, z = v; 0 \le u \le 2\pi, 0 \le v \le 1$

(b) $x = 2\cos u, y = v, z = 2\sin u; 0 \le u \le 2\pi, 1 \le v \le 3$

7. $x = u, y = \sin u \cos v, z = \sin u \sin v$

9. $x = r\cos\theta, y = r\sin\theta, z = \dfrac{1}{1+r^2}$

11. $x = r\cos\theta, y = r\sin\theta, z = 2r^2\cos\theta\sin\theta$

13. $x = r\cos\theta, y = r\sin\theta, z = \sqrt{9-r^2}; r \le \sqrt{5}$

15. $x = \dfrac{1}{2}\rho\cos\theta, y = \dfrac{1}{2}\rho\sin\theta, z = \dfrac{\sqrt{3}}{2}\rho$

17. $z = x - 2y$; a plane

19. $(x/3)^2 + (y/2)^2 = 1; 2 \le z \le 4$; part of an elliptic cylinder

21. $(x/3)^2 + (y/4)^2 = z^2; 0 \le z \le 1$; part of an elliptic cone

23. (a) $x = r\cos\theta, y = r\sin\theta, z = r, 0 \le r \le 2; x = u, y = v, z = \sqrt{u^2+v^2}; 0 \le u^2 + v^2 \le 4$

25. (a) $0 \le u \le 3, 0 \le v \le \pi$

(b) $0 \le u \le 4, -\pi/2 \le v \le \pi/2$

27. (a) $0 \le \phi \le \pi/2, 0 \le \theta \le 2\pi$

(b) $0 \le \phi \le \pi, 0 \le \theta \le \pi$

29. $u = 1, v = 2, \mathbf{r}_u \times \mathbf{r}_v = -2\mathbf{i} - 4\mathbf{j} + \mathbf{k}; 2x + 4y - z = 5$

31. $u = 0, v = 1, \mathbf{r}_u \times \mathbf{r}_v = 6\mathbf{k}; z = 0$

33. $\mathbf{r}_u \times \mathbf{r}_v = (\sqrt{2}/2)\mathbf{i} - (\sqrt{2}/2)\mathbf{j} + (1/2)\mathbf{k}; x - y + \dfrac{\sqrt{2}}{2}z = \dfrac{\pi\sqrt{2}}{8}$

35. $z = \sqrt{9-y^2},\ z_x = 0,\ z_y = -y/\sqrt{9-y^2},\ z_x^2 + z_y^2 + 1 = 9/(9-y^2),$

$$S = \int_0^2 \int_{-3}^3 \frac{3}{\sqrt{9-y^2}}\,dy\,dx = \int_0^2 3\pi\,dx = 6\pi$$

37. $z^2 = 4x^2 + 4y^2,\ 2zz_x = 8x$ so $z_x = 4x/z$, similarly $z_y = 4y/z$ thus

$$z_x^2 + z_y^2 + 1 = (16x^2 + 16y^2)/z^2 + 1 = 5,\ S = \int_0^1 \int_{x^2}^x \sqrt{5}\,dy\,dx = \sqrt{5}\int_0^1 (x - x^2)dx = \sqrt{5}/6$$

39. $z_x = -2x,\ z_y = -2y,\ z_x^2 + z_y^2 + 1 = 4x^2 + 4y^2 + 1,$

$$S = \iint_R \sqrt{4x^2+4y^2+1}\,dA = \int_0^{2\pi}\int_0^1 r\sqrt{4r^2+1}\,dr\,d\theta$$

$$= \frac{1}{12}(5\sqrt{5}-1)\int_0^{2\pi} d\theta = (5\sqrt{5}-1)\pi/6$$

41. $\partial\mathbf{r}/\partial u = \cos v\mathbf{i} + \sin v\mathbf{j} + 2u\mathbf{k},\ \partial\mathbf{r}/\partial v = -u\sin v\mathbf{i} + u\cos v\mathbf{j},$

$$\|\partial\mathbf{r}/\partial u \times \partial\mathbf{r}/\partial v\| = u\sqrt{4u^2+1};\ S = \int_0^{2\pi}\int_1^2 u\sqrt{4u^2+1}\,du\,dv = (17\sqrt{17} - 5\sqrt{5})\pi/6$$

43. $z_x = y$, $z_y = x$, $z_x^2 + z_y^2 + 1 = x^2 + y^2 + 1$,

$$S = \iint_R \sqrt{x^2 + y^2 + 1}\, dA = \int_0^{\pi/6} \int_0^3 r\sqrt{r^2 + 1}\, dr\, d\theta = \frac{1}{3}(10\sqrt{10} - 1) \int_0^{\pi/6} d\theta = (10\sqrt{10} - 1)\pi/18$$

45. On the sphere, $z_x = -x/z$ and $z_y = -y/z$ so $z_x^2 + z_y^2 + 1 = (x^2 + y^2 + z^2)/z^2 = 16/(16 - x^2 - y^2)$; the planes $z = 1$ and $z = 2$ intersect the sphere along the circles $x^2 + y^2 = 15$ and $x^2 + y^2 = 12$;

$$S = \iint_R \frac{4}{\sqrt{16 - x^2 - y^2}} dA = \int_0^{2\pi} \int_{\sqrt{12}}^{\sqrt{15}} \frac{4r}{\sqrt{16 - r^2}} dr\, d\theta = 4 \int_0^{2\pi} d\theta = 8\pi$$

47. $\mathbf{r}(u, v) = a \cos u \sin v\mathbf{i} + a \sin u \sin v\mathbf{j} + a \cos v\mathbf{k}$, $\|\mathbf{r}_u \times \mathbf{r}_v\| = a^2 \sin v$,

$$S = \int_0^{\pi} \int_0^{2\pi} a^2 \sin v\, du\, dv = 2\pi a^2 \int_0^{\pi} \sin v\, dv = 4\pi a^2$$

49. $z_x = \dfrac{h}{a} \dfrac{x}{\sqrt{x^2 + y^2}}$, $z_y = \dfrac{h}{a} \dfrac{y}{\sqrt{x^2 + y^2}}$, $z_x^2 + z_y^2 + 1 = \dfrac{h^2x^2 + h^2y^2}{a^2(x^2 + y^2)} + 1 = (a^2 + h^2)/a^2$,

$$S = \int_0^{2\pi} \int_0^a \frac{\sqrt{a^2 + h^2}}{a} r\, dr\, d\theta = \frac{1}{2} a\sqrt{a^2 + h^2} \int_0^{2\pi} d\theta = \pi a\sqrt{a^2 + h^2}$$

51. $\partial\mathbf{r}/\partial u = -(a + b \cos v) \sin u\mathbf{i} + (a + b \cos v) \cos u\mathbf{j}$,
$\partial\mathbf{r}/\partial v = -b \sin v \cos u\mathbf{i} - b \sin v \sin u\mathbf{j} + b \cos v\mathbf{k}$, $\|\partial\mathbf{r}/\partial u \times \partial\mathbf{r}/\partial v\| = b(a + b \cos v)$;

$$S = \int_0^{2\pi} \int_0^{2\pi} b(a + b \cos v) du\ dv = 4\pi^2 ab$$

53. $z = -1$ when $v \approx 0.27955$, $z = 1$ when $v \approx 2.86204$, $\|\mathbf{r}_u \times \mathbf{r}_v\| = |\cos v|$;

$$S = \int_0^{2\pi} \int_{0.27955}^{2.86204} |\cos v|\, dv\, du \approx 9.099$$

55. $(x/a)^2 + (y/b)^2 + (z/c)^2 = \cos^2 v(\cos^2 u + \sin^2 u) + \sin^2 v = 1$, ellipsoid

57. $(x/a)^2 + (y/b)^2 - (z/c)^2 = \sinh^2 v + \cosh^2 v(\sinh^2 u - \cosh^2 u) = -1$, hyperboloid of two sheets

EXERCISE SET 15.5

1. $$\int_{-1}^1 \int_0^2 \int_0^1 (x^2 + y^2 + z^2) dx\, dy\, dz = \int_{-1}^1 \int_0^2 (1/3 + y^2 + z^2) dy\, dz = \int_{-1}^1 (10/3 + 2z^2) dz = 8$$

3. $$\int_0^2 \int_{-1}^{y^2} \int_{-1}^z yz\, dx\, dz\, dy = \int_0^2 \int_{-1}^{y^2} (yz^2 + yz) dz\, dy = \int_0^2 \left(\frac{1}{3}y^7 + \frac{1}{2}y^5 - \frac{1}{6}y\right) dy = \frac{47}{3}$$

5. $$\int_0^3 \int_0^{\sqrt{9-z^2}} \int_0^x xy\, dy\, dx\, dz = \int_0^3 \int_0^{\sqrt{9-z^2}} \frac{1}{2}x^3 dx\, dz = \int_0^3 \frac{1}{8}(81 - 18z^2 + z^4) dz = 81/5$$

7. $$\int_0^2 \int_0^{\sqrt{4-x^2}} \int_{-5+x^2+y^2}^{3-x^2-y^2} x\, dz\, dy\, dx = \int_0^2 \int_0^{\sqrt{4-x^2}} [2x(4 - x^2) - 2xy^2] dy\, dx$$

$$= \int_0^2 \frac{4}{3}x(4 - x^2)^{3/2} dx = 128/15$$

9. $\displaystyle\int_0^{\pi}\int_0^1\int_0^{\pi/6} xy\sin yz\,dz\,dy\,dx = \int_0^{\pi}\int_0^1 x[1-\cos(\pi y/6)]dy\,dx = \int_0^{\pi}(1-3/\pi)x\,dx = \pi(\pi-3)/2$

11. $\displaystyle\int_0^{\sqrt{2}}\int_0^{x}\int_0^{2-x^2} xyz\,dz\,dy\,dx = \int_0^{\sqrt{2}}\int_0^{x}\frac{1}{2}xy(2-x^2)^2dy\,dx = \int_0^{\sqrt{2}}\frac{1}{4}x^3(2-x^2)^2dx = 1/6$

13. $\displaystyle\int_0^3\int_1^2\int_{-2}^1 \frac{\sqrt{x+z^2}}{y}\,dzdydx \approx 9.425$

15. $\displaystyle V = \int_0^4\int_0^{(4-x)/2}\int_0^{(12-3x-6y)/4} dz\,dy\,dx = \int_0^4\int_0^{(4-x)/2}\frac{1}{4}(12-3x-6y)dy\,dx$

$\displaystyle = \int_0^4 \frac{3}{16}(4-x)^2dx = 4$

17. $\displaystyle V = 2\int_0^2\int_{x^2}^4\int_0^{4-y} dz\,dy\,dx = 2\int_0^2\int_{x^2}^4(4-y)dy\,dx = 2\int_0^2\left(8-4x^2+\frac{1}{2}x^4\right)dx = 256/15$

19. The projection of the curve of intersection onto the xy-plane is $x^2+y^2=1$,

$\displaystyle V = 4\int_0^1\int_0^{\sqrt{1-x^2}}\int_{4x^2+y^2}^{4-3y^2} dz\,dy\,dx$

21. $\displaystyle V = 2\int_{-3}^3\int_0^{\sqrt{9-x^2}/3}\int_0^{x+3} dz\,dy\,dx$

23. (a)

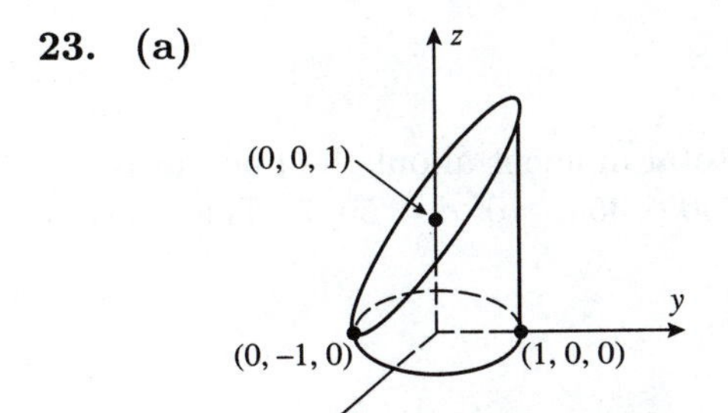

(b)

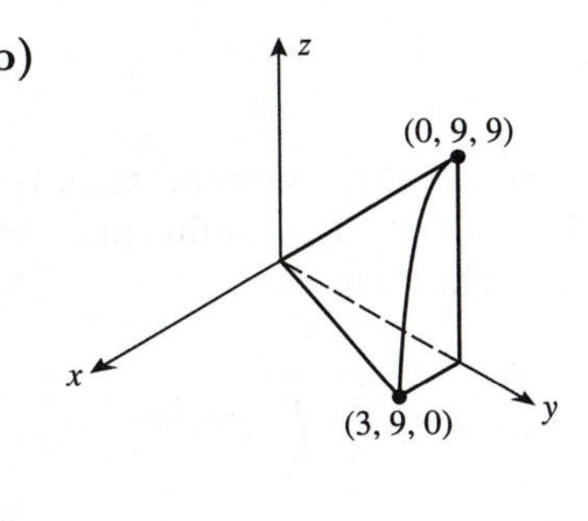

(c)

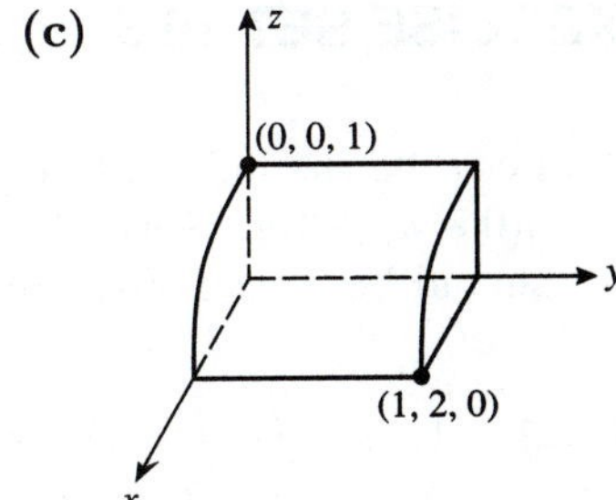

25. $\displaystyle V = \int_0^1\int_0^{1-x}\int_0^{1-x-y} dz\,dy\,dx = 1/6,\ f_{\text{ave}} = 6\int_0^1\int_0^{1-x}\int_0^{1-x-y}(x+y+z)\,dz\,dy\,dx = \frac{3}{4}$

27. The volume $V = \dfrac{3\pi}{\sqrt{2}}$, and thus

$\displaystyle r_{\text{ave}} = \frac{\sqrt{2}}{3\pi}\iiint\limits_G \sqrt{x^2+y^2+z^2}\,dV = \frac{\sqrt{2}}{3\pi}\int_{-1/\sqrt{2}}^{1/\sqrt{2}}\int_{-\sqrt{1-2x^2}}^{\sqrt{1-2x^2}}\int_{5x^2+5y^2}^{6-7x^2-y^2}\sqrt{x^2+y^2+z^2}dzdydx \approx 3.291$

29. (a) $\displaystyle\int_0^a\int_0^{b(1-x/a)}\int_0^{c(1-x/a-y/b)} dz\,dy\,dx,\ \int_0^b\int_0^{a(1-y/b)}\int_0^{c(1-x/a-y/b)} dz\,dx\,dy,$

$\displaystyle\int_0^c\int_0^{a(1-z/c)}\int_0^{b(1-x/a-z/c)} dy\,dx\,dz,\ \int_0^a\int_0^{c(1-x/a)}\int_0^{b(1-x/a-z/c)} dy\,dz\,dx,$

$\displaystyle\int_0^c\int_0^{b(1-z/c)}\int_0^{a(1-y/b-z/c)} dx\,dy\,dz,\ \int_0^b\int_0^{c(1-y/b)}\int_0^{a(1-y/b-z/c)} dx\,dz\,dy$

(b) Use the first integral in Part (a) to get

$$\int_0^a \int_0^{b(1-x/a)} c\left(1 - \frac{x}{a} - \frac{y}{b}\right) dy\, dx = \int_0^a \frac{1}{2}bc\left(1 - \frac{x}{a}\right)^2 dx = \frac{1}{6}abc$$

31. **(a)** $\displaystyle\int_0^2 \int_0^{\sqrt{4-x^2}} \int_0^5 f(x,y,z)\, dzdydx$

(b) $\displaystyle\int_0^9 \int_0^{3-\sqrt{x}} \int_y^{3-\sqrt{x}} f(x,y,z)\, dzdydx$ **(c)** $\displaystyle\int_0^2 \int_0^{4-x^2} \int_y^{8-y} f(x,y,z)\, dzdydx$

33. **(a)** At any point outside the closed sphere $\{x^2 + y^2 + z^2 \leq 1\}$ the integrand is negative, so to maximize the integral it suffices to include all points inside the sphere; hence the maximum value is taken on the region $G = \{x^2 + y^2 + z^2 \leq 1\}$.

(b) 4.934802202

(c) $\displaystyle\int_0^{2\pi} \int_0^{\pi} \int_0^1 (1 - \rho^2)\rho\, d\rho\, d\phi\, d\theta = \frac{\pi^2}{2}$

35. **(a)** $\displaystyle\left[\int_{-1}^1 x\, dx\right]\left[\int_0^1 y^2 dy\right]\left[\int_0^{\pi/2} \sin z\, dz\right] = (0)(1/3)(1) = 0$

(b) $\displaystyle\left[\int_0^1 e^{2x} dx\right]\left[\int_0^{\ln 3} e^y dy\right]\left[\int_0^{\ln 2} e^{-z} dz\right] = [(e^2 - 1)/2](2)(1/2) = (e^2 - 1)/2$

EXERCISE SET 15.6

1. Let a be the unknown coordinate of the fulcrum; then the total moment about the fulcrum is $5(0 - a) + 10(5 - a) + 20(10 - a) = 0$ for equilibrium, so $250 - 35a = 0$, $a = 50/7$. The fulcrum should be placed 50/7 units to the right of m_1.

3. $\displaystyle A = 1,\ \overline{x} = \int_0^1 \int_0^1 x\, dy\, dx = \frac{1}{2},\ \overline{y} = \int_0^1 \int_0^1 y\, dy\, dx = \frac{1}{2}$

5. $\displaystyle A = 1/2,\ \iint_R x\, dA = \int_0^1 \int_0^x x\, dy\, dx = 1/3,\ \iint_R y\, dA = \int_0^1 \int_0^x y\, dy\, dx = 1/6;$

centroid $(2/3, 1/3)$

7. $\displaystyle A = \int_0^1 \int_x^{2-x^2} dy\, dx = 7/6,\ \iint_R x\, dA = \int_0^1 \int_x^{2-x^2} x\, dy\, dx = 5/12,$

$\displaystyle\iint_R y\, dA = \int_0^1 \int_x^{2-x^2} y\, dy\, dx = 19/15$; centroid $(5/14, 38/35)$

9. $\overline{x} = 0$ from the symmetry of the region,

$$A = \frac{1}{2}\pi(b^2 - a^2),\ \iint\limits_R y\,dA = \int_0^{\pi}\int_a^b r^2 \sin\theta\,dr\,d\theta = \frac{2}{3}(b^3 - a^3);\ \text{centroid } \overline{x} = 0,\ \overline{y} = \frac{4(b^3 - a^3)}{3\pi(b^2 - a^2)}.$$

11. $$M = \iint\limits_R \delta(x,y)dA = \int_0^1\int_0^1 |x + y - 1|\,dxdy$$

$$= \int_0^1 \left[\int_0^{1-x} (1 - x - y)\,dy + \int_{1-x}^1 (x + y - 1)\,dy\right] dx = \frac{1}{3}$$

$$\overline{x} = 3\int_0^1\int_0^1 x\delta(x,y)\,dy\,dx = 3\int_0^1 \left[\int_0^{1-x} x(1 - x - y)\,dy + \int_{1-x}^1 x(x + y - 1)\,dy\right] dx = \frac{1}{2}$$

By symmetry, $\overline{y} = \frac{1}{2}$ as well; center of gravity $(1/2, 1/2)$

13. $M = \int_0^1\int_0^{\sqrt{x}} (x + y)dy\,dx = 13/20$, $M_x = \int_0^1\int_0^{\sqrt{x}} (x + y)y\,dy\,dx = 3/10$,

$M_y = \int_0^1\int_0^{\sqrt{x}} (x + y)x\,dy\,dx = 19/42$, $\overline{x} = M_y/M = 190/273$, $\overline{y} = M_x/M = 6/13$;

the mass is $13/20$ and the center of gravity is at $(190/273, 6/13)$.

15. $M = \int_0^{\pi/2}\int_0^a r^3 \sin\theta\cos\theta\,dr\,d\theta = a^4/8$, $\overline{x} = \overline{y}$ from the symmetry of the density and the region, $M_y = \int_0^{\pi/2}\int_0^a r^4 \sin\theta\cos^2\theta\,dr\,d\theta = a^5/15$, $\overline{x} = 8a/15$; mass $a^4/8$, center of gravity $(8a/15, 8a/15)$.

17. $V = 1, \overline{x} = \int_0^1\int_0^1\int_0^1 x\,dz\,dy\,dx = \frac{1}{2}$, similarly $\overline{y} = \overline{z} = \frac{1}{2}$; centroid $\left(\frac{1}{2}, \frac{1}{2}, \frac{1}{2}\right)$

19. $\overline{x} = \overline{y} = \overline{z}$ from the symmetry of the region, $V = 1/6$,

$$\overline{x} = \frac{1}{V}\int_0^1\int_0^{1-x}\int_0^{1-x-y} x\,dz\,dy\,dx = (6)(1/24) = 1/4;\ \text{centroid } (1/4, 1/4, 1/4)$$

21. $\overline{x} = 1/2$ and $\overline{y} = 0$ from the symmetry of the region,

$$V = \int_0^1\int_{-1}^1\int_{y^2}^1 dz\,dy\,dx = 4/3,\ \overline{z} = \frac{1}{V}\iiint\limits_G z\,dV = (3/4)(4/5) = 3/5;\ \text{centroid } (1/2, 0, 3/5)$$

23. $\overline{x} = \overline{y} = \overline{z}$ from the symmetry of the region, $V = \pi a^3/6$,

$$\overline{x} = \frac{1}{V}\int_0^a\int_0^{\sqrt{a^2-x^2}}\int_0^{\sqrt{a^2-x^2-y^2}} x\,dz\,dy\,dx = \frac{1}{V}\int_0^a\int_0^{\sqrt{a^2-x^2}} x\sqrt{a^2 - x^2 - y^2}\,dy\,dx$$

$$= \frac{1}{V}\int_0^{\pi/2}\int_0^a r^2\sqrt{a^2 - r^2}\cos\theta\,dr\,d\theta = \frac{6}{\pi a^3}(\pi a^4/16) = 3a/8;\ \text{centroid } (3a/8, 3a/8, 3a/8)$$

25. $M = \int_0^a \int_0^a \int_0^a (a-x)dz\,dy\,dx = a^4/2$, $\overline{y} = \overline{z} = a/2$ from the symmetry of density and region, $\overline{x} = \frac{1}{M}\int_0^a \int_0^a \int_0^a x(a-x)dz\,dy\,dx = (2/a^4)(a^5/6) = a/3$;
mass $a^4/2$, center of gravity $(a/3, a/2, a/2)$

27. $M = \int_{-1}^1 \int_0^1 \int_0^{1-y^2} yz\,dz\,dy\,dx = 1/6$, $\overline{x} = 0$ by the symmetry of density and region,

$\overline{y} = \frac{1}{M}\iiint\limits_G y^2 z\,dV = (6)(8/105) = 16/35$, $\overline{z} = \frac{1}{M}\iiint\limits_G yz^2 dV = (6)(1/12) = 1/2$;

mass $1/6$, center of gravity $(0, 16/35, 1/2)$

29. **(a)** $M = \int_0^1 \int_0^1 k(x^2+y^2)dy\,dx = 2k/3$, $\overline{x} = \overline{y}$ from the symmetry of density and region,
$\overline{x} = \frac{1}{M}\iint\limits_R kx(x^2+y^2)dA = \frac{3}{2k}(5k/12) = 5/8$; center of gravity $(5/8, 5/8)$

(b) $\overline{y} = 1/2$ from the symmetry of density and region,
$M = \int_0^1 \int_0^1 kx\,dy\,dx = k/2$, $\overline{x} = \frac{1}{M}\iint\limits_R kx^2 dA = (2/k)(k/3) = 2/3$,
center of gravity $(2/3, 1/2)$

31. $V = \iiint\limits_G dV = \int_0^{\pi} \int_0^{\sin x} \int_0^{1/(1+x^2+y^2)} dz\ dy\ dx = 0.666633$,

$\overline{x} = \frac{1}{V}\iiint\limits_G x dV = 1.177406$, $\overline{y} = \frac{1}{V}\iiint\limits_G y dV = 0.353554$, $\overline{z} = \frac{1}{V}\iiint\limits_G z dV = 0.231557$

33. Let $x = r\cos\theta$, $y = r\sin\theta$, and $dA = r\,dr\,d\theta$ in formulas (11) and (12).

35. $\overline{x} = \overline{y}$ from the symmetry of the region, $A = \int_0^{\pi/2} \int_0^{\sin 2\theta} r\,dr\,d\theta = \pi/8$,

$\overline{x} = \frac{1}{A}\int_0^{\pi/2} \int_0^{\sin 2\theta} r^2 \cos\theta\,dr\,d\theta = (8/\pi)(16/105) = \frac{128}{105\pi}$; centroid $\left(\frac{128}{105\pi}, \frac{128}{105\pi}\right)$

37. $\overline{x} = 0$ from the symmetry of the region, $\pi a^2/2$ is the area of the semicircle, $2\pi\overline{y}$ is the distance traveled by the centroid to generate the sphere so $4\pi a^3/3 = (\pi a^2/2)(2\pi\overline{y})$, $\overline{y} = 4a/(3\pi)$

39. $\overline{x} = k$ so $V = (\pi ab)(2\pi k) = 2\pi^2 abk$

41. The region generates a cone of volume $\frac{1}{3}\pi ab^2$ when it is revolved about the x-axis, the area of the region is $\frac{1}{2}ab$ so $\frac{1}{3}\pi ab^2 = \left(\frac{1}{2}ab\right)(2\pi\overline{y})$, $\overline{y} = b/3$. A cone of volume $\frac{1}{3}\pi a^2 b$ is generated when the region is revolved about the y-axis so $\frac{1}{3}\pi a^2 b = \left(\frac{1}{2}ab\right)(2\pi\overline{x})$, $\overline{x} = a/3$. The centroid is $(a/3, b/3)$.

43. $I_x = \int_0^{2\pi}\int_0^a r^3\sin^2\theta\,\delta\,dr\,d\theta = \delta\pi a^4/4$; $I_y = \int_0^{2\pi}\int_0^a r^3\cos^2\theta\,\delta\,dr\,d\theta = \delta\pi a^4/4 = I_x$;
$I_z = I_x + I_y = \delta\pi a^4/2$

EXERCISE SET 15.7

1. $\int_0^{2\pi}\int_0^1\int_0^{\sqrt{1-r^2}} zr\,dz\,dr\,d\theta = \int_0^{2\pi}\int_0^1 \frac{1}{2}(1-r^2)r\,dr\,d\theta = \int_0^{2\pi}\frac{1}{8}d\theta = \pi/4$

3. $\int_0^{\pi/2}\int_0^{\pi/2}\int_0^1 \rho^3\sin\phi\cos\phi\,d\rho\,d\phi\,d\theta = \int_0^{\pi/2}\int_0^{\pi/2}\frac{1}{4}\sin\phi\cos\phi\,d\phi\,d\theta = \int_0^{\pi/2}\frac{1}{8}d\theta = \pi/16$

5. $V = \int_0^{2\pi}\int_0^3\int_{r^2}^9 r\,dz\,dr\,d\theta = \int_0^{2\pi}\int_0^3 r(9-r^2)dr\,d\theta = \int_0^{2\pi}\frac{81}{4}d\theta = 81\pi/2$

7. $r^2+z^2=20$ intersects $z=r^2$ in a circle of radius 2; the volume consists of two portions, one inside the cylinder $r=\sqrt{20}$ and one outside that cylinder:

$$V = \int_0^{2\pi}\int_0^2\int_{-\sqrt{20-r^2}}^{r^2} r\,dz\,dr\,d\theta + \int_0^{2\pi}\int_2^{\sqrt{20}}\int_{-\sqrt{20-r^2}}^{\sqrt{20-r^2}} r\,dz\,dr\,d\theta$$
$$= \int_0^{2\pi}\int_0^2 r\left(r^2+\sqrt{20-r^2}\right)dr\,d\theta + \int_0^{2\pi}\int_2^{\sqrt{20}} 2r\sqrt{20-r^2}\,dr\,d\theta$$
$$= \frac{4}{3}(10\sqrt{5}-13)\int_0^{2\pi}d\theta + \frac{128}{3}\int_0^{2\pi}d\theta = \frac{152}{3}\pi + \frac{80}{3}\pi\sqrt{5}$$

9. $V = \int_0^{2\pi}\int_0^{\pi/3}\int_0^4 \rho^2\sin\phi\,d\rho\,d\phi\,d\theta = \int_0^{2\pi}\int_0^{\pi/3}\frac{64}{3}\sin\phi\,d\phi\,d\theta = \frac{32}{3}\int_0^{2\pi}d\theta = 64\pi/3$

11. In spherical coordinates the sphere and the plane $z=a$ are $\rho=2a$ and $\rho=a\sec\phi$, respectively. They intersect at $\phi=\pi/3$,

$$V = \int_0^{2\pi}\int_0^{\pi/3}\int_0^{a\sec\phi}\rho^2\sin\phi\,d\rho\,d\phi\,d\theta + \int_0^{2\pi}\int_{\pi/3}^{\pi/2}\int_0^{2a}\rho^2\sin\phi\,d\rho\,d\phi\,d\theta$$
$$= \int_0^{2\pi}\int_0^{\pi/3}\frac{1}{3}a^3\sec^3\phi\sin\phi\,d\phi\,d\theta + \int_0^{2\pi}\int_{\pi/3}^{\pi/2}\frac{8}{3}a^3\sin\phi\,d\phi\,d\theta$$
$$= \frac{1}{2}a^3\int_0^{2\pi}d\theta + \frac{4}{3}a^3\int_0^{2\pi}d\theta = 11\pi a^3/3$$

13. $\int_0^{\pi/2}\int_0^a\int_0^{a^2-r^2} r^3\cos^2\theta\,dz\,dr\,d\theta = \int_0^{\pi/2}\int_0^a (a^2r^3-r^5)\cos^2\theta\,dr\,d\theta$
$$= \frac{1}{12}a^6\int_0^{\pi/2}\cos^2\theta\,d\theta = \pi a^6/48$$

15. $\int_0^{\pi/2}\int_0^{\pi/4}\int_0^{\sqrt{8}} \rho^4\cos^2\phi\sin\phi\,d\rho\,d\phi\,d\theta = 32(2\sqrt{2}-1)\pi/15$

17. (a) $\displaystyle\int_{\pi/3}^{\pi/2}\int_{1}^{4}\int_{-2}^{2}\frac{r\tan^3\theta}{\sqrt{1+z^2}}\,dz\,dr\,d\theta = \left(\int_{\pi/6}^{\pi/3}\tan^3\theta\,d\theta\right)\left(\int_1^4 r\,dr\right)\left(\int_{-2}^{2}\frac{1}{\sqrt{1+z^2}}\,dz\right)$

$$= \left(\frac{4}{3}-\frac{1}{2}\ln 3\right)\frac{15}{2}\left(-2\ln(\sqrt{5}-2)\right) = \frac{5}{2}(-8+3\ln 3)\ln(\sqrt{5}-2)$$

(b) $\displaystyle\int_{\pi/3}^{\pi/2}\int_{1}^{4}\int_{-2}^{2}\frac{y\tan^3 z}{\sqrt{1+x^2}}\,dx\,dy\,dz$; the region is a rectangular solid with sides $\pi/6$, 3, 4.

19. (a) $\displaystyle V = 2\int_0^{2\pi}\int_0^a\int_0^{\sqrt{a^2-r^2}} r\,dz\,dr\,d\theta = 4\pi a^3/3$

(b) $\displaystyle V = \int_0^{2\pi}\int_0^{\pi}\int_0^a \rho^2\sin\phi\,d\rho\,d\phi\,d\theta = 4\pi a^3/3$

21. $\displaystyle M = \int_0^{2\pi}\int_0^3\int_r^3 (3-z)r\,dz\,dr\,d\theta = \int_0^{2\pi}\int_0^3 \frac{1}{2}r(3-r)^2 dr\,d\theta = \frac{27}{8}\int_0^{2\pi} d\theta = 27\pi/4$

23. $\displaystyle M = \int_0^{2\pi}\int_0^{\pi}\int_0^a k\rho^3\sin\phi\,d\rho\,d\phi\,d\theta = \int_0^{2\pi}\int_0^{\pi}\frac{1}{4}ka^4\sin\phi\,d\phi\,d\theta = \frac{1}{2}ka^4\int_0^{2\pi}d\theta = \pi ka^4$

25. $\bar{x}=\bar{y}=0$ from the symmetry of the region,

$$V = \int_0^{2\pi}\int_0^1\int_{r^2}^{\sqrt{2-r^2}} r\,dz\,dr\,d\theta = \int_0^{2\pi}\int_0^1 (r\sqrt{2-r^2}-r^3)dr\,d\theta = (8\sqrt{2}-7)\pi/6,$$

$$\bar{z} = \frac{1}{V}\int_0^{2\pi}\int_0^1\int_{r^2}^{\sqrt{2-r^2}} zr\,dz\,dr\,d\theta = \frac{6}{(8\sqrt{2}-7)\pi}(7\pi/12) = 7/(16\sqrt{2}-14);$$

centroid $\left(0, 0, \dfrac{7}{16\sqrt{2}-14}\right)$

27. $\bar{x}=\bar{y}=\bar{z}$ from the symmetry of the region, $V=\pi a^3/6$,

$$\bar{z} = \frac{1}{V}\int_0^{\pi/2}\int_0^{\pi/2}\int_0^a \rho^3\cos\phi\sin\phi\,d\rho\,d\phi\,d\theta = \frac{6}{\pi a^3}(\pi a^4/16) = 3a/8;$$

centroid $(3a/8, 3a/8, 3a/8)$

29. $\bar{y}=0$ from the symmetry of the region, $\displaystyle V = 2\int_0^{\pi/2}\int_0^{2\cos\theta}\int_0^{r^2} r\,dz\,dr\,d\theta = 3\pi/2,$

$$\bar{x} = \frac{2}{V}\int_0^{\pi/2}\int_0^{2\cos\theta}\int_0^{r^2} r^2\cos\theta\,dz\,dr\,d\theta = \frac{4}{3\pi}(\pi) = 4/3,$$

$$\bar{z} = \frac{2}{V}\int_0^{\pi/2}\int_0^{2\cos\theta}\int_0^{r^2} rz\,dz\,dr\,d\theta = \frac{4}{3\pi}(5\pi/6) = 10/9;\ \text{centroid } (4/3, 0, 10/9)$$

31. $\displaystyle V = \int_0^{\pi/2}\int_{\pi/6}^{\pi/3}\int_0^2 \rho^2\sin\phi\,d\rho\,d\phi\,d\theta = \int_0^{\pi/2}\int_{\pi/6}^{\pi/3}\frac{8}{3}\sin\phi\,d\phi\,d\theta = \frac{4}{3}(\sqrt{3}-1)\int_0^{\pi/2}d\theta$

$= 2(\sqrt{3}-1)\pi/3$

33. $\bar{x} = \bar{y} = 0$ from the symmetry of density and region,

$$M = \int_0^{2\pi}\int_0^1\int_0^{1-r^2} (r^2 + z^2)r\, dz\, dr\, d\theta = \pi/4,$$

$$\bar{z} = \frac{1}{M}\int_0^{2\pi}\int_0^1\int_0^{1-r^2} z(r^2+z^2)r\, dz\, dr\, d\theta = (4/\pi)(11\pi/120) = 11/30;$$ center of gravity $(0, 0, 11/30)$

35. $\bar{x} = \bar{y} = 0$ from the symmetry of density and region,

$$M = \int_0^{2\pi}\int_0^{\pi/2}\int_0^a k\rho^3 \sin\phi\, d\rho\, d\phi\, d\theta = \pi k a^4/2,$$

$$\bar{z} = \frac{1}{M}\int_0^{2\pi}\int_0^{\pi/2}\int_0^a k\rho^4 \sin\phi\cos\phi\, d\rho\, d\phi\, d\theta = \frac{2}{\pi k a^4}(\pi k a^5/5) = 2a/5;$$ center of gravity $(0, 0, 2a/5)$

37. $M = \int_0^{2\pi}\int_0^{\pi}\int_0^R \delta_0 e^{-(\rho/R)^3}\rho^2 \sin\phi\, d\rho\, d\phi\, d\theta = \int_0^{2\pi}\int_0^{\pi} \frac{1}{3}(1 - e^{-1})R^3\delta_0 \sin\phi\, d\phi\, d\theta$

$= \frac{4}{3}\pi(1 - e^{-1})\delta_0 R^3$

39. $I_z = \int_0^{2\pi}\int_0^a\int_0^h r^2\delta\, r\, dz\, dr\, d\theta = \delta\int_0^{2\pi}\int_0^a\int_0^h r^3 dz\, dr\, d\theta = \frac{1}{2}\delta\pi a^4 h$

41. $I_z = \int_0^{2\pi}\int_{a_1}^{a_2}\int_0^h r^2\delta\, r\, dz\, dr\, d\theta = \delta\int_0^{2\pi}\int_{a_1}^{a_2}\int_0^h r^3 dz\, dr\, d\theta = \frac{1}{2}\delta\pi h(a_2^4 - a_1^4)$

EXERCISE SET 15.8

1. $\frac{\partial(x, y)}{\partial(u, v)} = \begin{vmatrix} 1 & 4 \\ 3 & -5 \end{vmatrix} = -17$

3. $\frac{\partial(x, y)}{\partial(u, v)} = \begin{vmatrix} \cos u & -\sin v \\ \sin u & \cos v \end{vmatrix} = \cos u \cos v + \sin u \sin v = \cos(u - v)$

5. $x = \frac{2}{9}u + \frac{5}{9}v,\ y = -\frac{1}{9}u + \frac{2}{9}v;\ \frac{\partial(x, y)}{\partial(u, v)} = \begin{vmatrix} 2/9 & 5/9 \\ -1/9 & 2/9 \end{vmatrix} = \frac{1}{9}$

7. $x = \sqrt{u+v}/\sqrt{2},\ y = \sqrt{v-u}/\sqrt{2};\ \frac{\partial(x, y)}{\partial(u, v)} = \begin{vmatrix} \frac{1}{2\sqrt{2}\sqrt{u+v}} & \frac{1}{2\sqrt{2}\sqrt{u+v}} \\ -\frac{1}{2\sqrt{2}\sqrt{v-u}} & \frac{1}{2\sqrt{2}\sqrt{v-u}} \end{vmatrix} = \frac{1}{4\sqrt{v^2 - u^2}}$

9. $\frac{\partial(x, y, z)}{\partial(u, v, w)} = \begin{vmatrix} 3 & 1 & 0 \\ 1 & 0 & -2 \\ 0 & 1 & 1 \end{vmatrix} = 5$

11. $y = v, x = u/y = u/v, z = w - x = w - u/v; \dfrac{\partial(x,y,z)}{\partial(u,v,w)} = \begin{vmatrix} 1/v & -u/v^2 & 0 \\ 0 & 1 & 0 \\ -1/v & u/v^2 & 1 \end{vmatrix} = 1/v$

13.

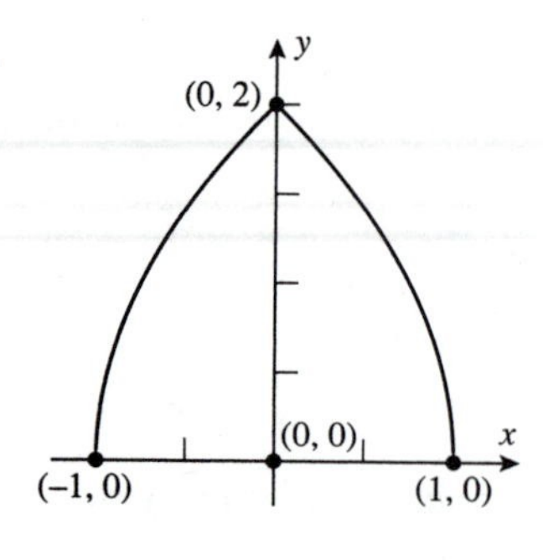

15.

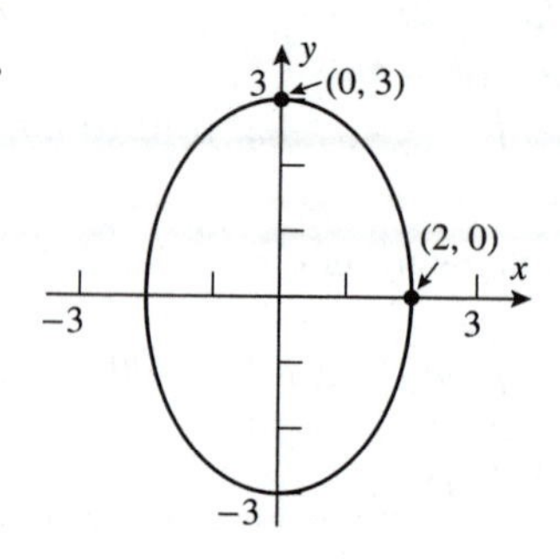

17. $x = \dfrac{1}{5}u + \dfrac{2}{5}v,\ y = -\dfrac{2}{5}u + \dfrac{1}{5}v,\ \dfrac{\partial(x,y)}{\partial(u,v)} = \dfrac{1}{5};\ \dfrac{1}{5}\iint\limits_S \dfrac{u}{v}dA_{uv} = \dfrac{1}{5}\int_1^3\int_1^4 \dfrac{u}{v}\,du\,dv = \dfrac{3}{2}\ln 3$

19. $x = u + v,\ y = u - v,\ \dfrac{\partial(x,y)}{\partial(u,v)} = -2$; the boundary curves of the region S in the uv-plane are $v = 0, v = u$, and $u = 1$ so $2\iint\limits_S \sin u \cos v dA_{uv} = 2\int_0^1\int_0^u \sin u \cos v\, dv\, du = 1 - \dfrac{1}{2}\sin 2$

21. $x = 3u, y = 4v, \dfrac{\partial(x,y)}{\partial(u,v)} = 12$; S is the region in the uv-plane enclosed by the circle $u^2 + v^2 = 1$. Use polar coordinates to obtain $\iint\limits_S 12\sqrt{u^2+v^2}(12)\,dA_{uv} = 144\int_0^{2\pi}\int_0^1 r^2 dr\,d\theta = 96\pi$

23. Let S be the region in the uv-plane bounded by $u^2 + v^2 = 1$, so $u = 2x, v = 3y$, $x = u/2, y = v/3, \dfrac{\partial(x,y)}{\partial(u,v)} = \begin{vmatrix} 1/2 & 0 \\ 0 & 1/3 \end{vmatrix} = 1/6$, use polar coordinates to get

$$\frac{1}{6}\iint\limits_S \sin(u^2+v^2)du\,dv = \frac{1}{6}\int_0^{\pi/2}\int_0^1 r\sin r^2\,dr\,d\theta = \frac{\pi}{24}(-\cos r^2)\Big]_0^1 = \frac{\pi}{24}(1-\cos 1)$$

25. $x = u/3, y = v/2, z = w, \dfrac{\partial(x,y,z)}{\partial(u,v,w)} = 1/6$; S is the region in uvw-space enclosed by the sphere $u^2 + v^2 + w^2 = 36$ so

$$\iiint\limits_S \frac{u^2}{9}\frac{1}{6}dV_{uvw} = \frac{1}{54}\int_0^{2\pi}\int_0^{\pi}\int_0^6 (\rho\sin\phi\cos\theta)^2\rho^2\sin\phi\,d\rho\,d\phi\,d\theta$$
$$= \frac{1}{54}\int_0^{2\pi}\int_0^{\pi}\int_0^6 \rho^4\sin^3\phi\cos^2\theta d\rho\,d\phi\,d\theta = \frac{192}{5}\pi$$

27. $u = \theta = \cot^{-1}(x/y), v = r = \sqrt{x^2+y^2}$

29. $u = \frac{3}{7}x - \frac{2}{7}y, v = -\frac{1}{7}x + \frac{3}{7}y$

31. Let $u = y - 4x, v = y + 4x$, then $x = \frac{1}{8}(v - u), y = \frac{1}{2}(v + u)$ so $\frac{\partial(x,y)}{\partial(u,v)} = -\frac{1}{8}$;

$$\frac{1}{8}\iint_S \frac{u}{v} dA_{uv} = \frac{1}{8}\int_2^5 \int_0^2 \frac{u}{v} du\, dv = \frac{1}{4}\ln\frac{5}{2}$$

33. Let $u = x - y, v = x + y$, then $x = \frac{1}{2}(v + u), y = \frac{1}{2}(v - u)$ so $\frac{\partial(x,y)}{\partial(u,v)} = \frac{1}{2}$; the boundary curves of the region S in the uv-plane are $u = 0, v = u$, and $v = \pi/4$; thus

$$\frac{1}{2}\iint_S \frac{\sin u}{\cos v} dA_{uv} = \frac{1}{2}\int_0^{\pi/4}\int_0^v \frac{\sin u}{\cos v} du\, dv = \frac{1}{2}[\ln(\sqrt{2} + 1) - \pi/4]$$

35. Let $u = y/x, v = x/y^2$, then $x = 1/(u^2 v), y = 1/(uv)$ so $\frac{\partial(x,y)}{\partial(u,v)} = \frac{1}{u^4 v^3}$;

$$\iint_S \frac{1}{u^4 v^3} dA_{uv} = \int_1^4 \int_1^2 \frac{1}{u^4 v^3} du\, dv = 35/256$$

37. $x = u, y = w/u, z = v + w/u, \frac{\partial(x,y,z)}{\partial(u,v,w)} = -\frac{1}{u}$;

$$\iiint_S \frac{v^2 w}{u} dV_{uvw} = \int_2^4 \int_0^1 \int_1^3 \frac{v^2 w}{u} du\, dv\, dw = 2\ln 3$$

39. **(b)** If $x = x(u,v), y = y(u,v)$ where $u = u(x,y), v = v(x,y)$, then by the chain rule

$$\frac{\partial x}{\partial u}\frac{\partial u}{\partial x} + \frac{\partial x}{\partial v}\frac{\partial v}{\partial x} = \frac{\partial x}{\partial x} = 1, \quad \frac{\partial x}{\partial u}\frac{\partial u}{\partial y} + \frac{\partial x}{\partial v}\frac{\partial v}{\partial y} = \frac{\partial x}{\partial y} = 0$$

$$\frac{\partial y}{\partial u}\frac{\partial u}{\partial x} + \frac{\partial y}{\partial v}\frac{\partial v}{\partial x} = \frac{\partial y}{\partial x} = 0, \quad \frac{\partial y}{\partial u}\frac{\partial u}{\partial y} + \frac{\partial y}{\partial v}\frac{\partial v}{\partial y} = \frac{\partial y}{\partial y} = 1$$

41. $\frac{\partial(u,v)}{\partial(x,y)} = 3xy^4 = 3v$ so $\frac{\partial(x,y)}{\partial(u,v)} = \frac{1}{3v}$; $\frac{1}{3}\iint_S \frac{\sin u}{v} dA_{uv} = \frac{1}{3}\int_1^2 \int_\pi^{2\pi} \frac{\sin u}{v} du\, dv = -\frac{2}{3}\ln 2$

43. $\frac{\partial(u,v)}{\partial(x,y)} = -2(x^2 + y^2)$ so $\frac{\partial(x,y)}{\partial(u,v)} = -\frac{1}{2(x^2 + y^2)}$;

$$(x^4 - y^4)e^{xy}\left|\frac{\partial(x,y)}{\partial(u,v)}\right| = \frac{x^4 - y^4}{2(x^2 + y^2)}e^{xy} = \frac{1}{2}(x^2 - y^2)e^{xy} = \frac{1}{2}ve^u \text{ so}$$

$$\frac{1}{2}\iint_S ve^u dA_{uv} = \frac{1}{2}\int_3^4 \int_1^3 ve^u du\, dv = \frac{7}{4}(e^3 - e)$$

45. **(a)** Let $u = x + y, v = y$, then the triangle R with vertices $(0,0), (1,0)$ and $(0,1)$ becomes the triangle in the uv-plane with vertices $(0,0), (1,0), (1,1)$, and

$$\iint_R f(x+y)dA = \int_0^1 \int_0^u f(u)\frac{\partial(x,y)}{\partial(u,v)} dv\, du = \int_0^1 uf(u)\, du$$

(b) $\displaystyle\int_0^1 ue^u\, du = (u-1)e^u\Big]_0^1 = 1$

CHAPTER 15 SUPPLEMENTARY EXERCISES

3. **(a)** $\displaystyle\iint_R dA$ **(b)** $\displaystyle\iiint_G dV$ **(c)** $\displaystyle\iint_R \sqrt{1 + \left(\frac{\partial z}{\partial x}\right)^2 + \left(\frac{\partial z}{\partial y}\right)^2}\, dA$

7. $\displaystyle\int_0^1 \int_{1-\sqrt{1-y^2}}^{1+\sqrt{1-y^2}} f(x,y)\, dx\, dy$

9. **(a)** $(1,2) = (b,d), (2,1) = (a,c)$, so $a = 2, b = 1, c = 1, d = 2$

(b) $\displaystyle\iint_R dA = \int_0^1 \int_0^1 \frac{\partial(x,y)}{\partial(u,v)} du\, dv = \int_0^1 \int_0^1 3du\, dv = 3$

11. $\displaystyle\int_{1/2}^1 2x\cos(\pi x^2)\, dx = \frac{1}{\pi}\sin(\pi x^2)\Big]_{1/2}^1 = -1/(\sqrt{2}\pi)$

13. $\displaystyle\int_0^1 \int_{2y}^2 e^x e^y\, dx\, dy$

15.

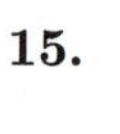

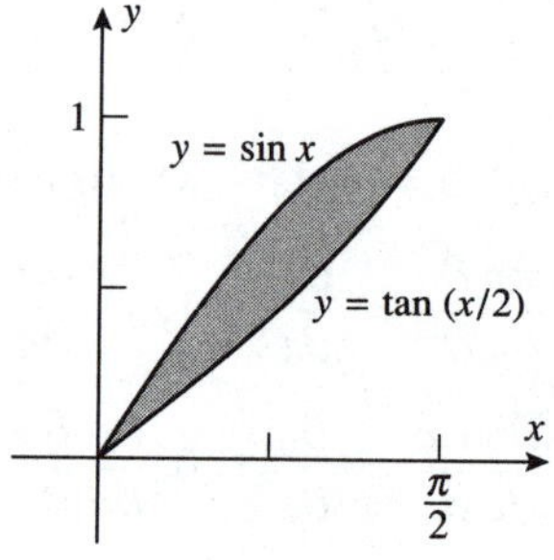

17. $\displaystyle 2\int_0^8 \int_0^{y^{1/3}} x^2 \sin y^2 dx\, dy = \frac{2}{3}\int_0^8 y\sin y^2\, dy = -\frac{1}{3}\cos y^2\Big]_0^8 = \frac{1}{3}(1 - \cos 64) \approx 0.20271$

19. $\sin 2\theta = 2\sin\theta\cos\theta = \dfrac{2xy}{x^2+y^2}$, and $r = 2a\sin\theta$ is the circle $x^2 + (y-a)^2 = a^2$, so

$$\int_0^a \int_{a-\sqrt{a^2-x^2}}^{a+\sqrt{a^2-x^2}} \frac{2xy}{x^2+y^2} dy\, dx = \int_0^a x\left[\ln\left(a + \sqrt{a^2-x^2}\right) - \ln\left(a - \sqrt{a^2-x^2}\right)\right] dx = a^2$$

21. $\displaystyle\int_0^{2\pi} \int_0^2 \int_{r^4}^{16} r^2\cos^2\theta\, r\, dz\, dr\, d\theta = \int_0^{2\pi} \cos^2\theta\, d\theta \int_0^2 r^3(16 - r^4)\, dr = 32\pi$

23. **(a)** $\displaystyle\int_0^{2\pi} \int_0^{\pi/3} \int_0^a (\rho^2\sin^2\phi)\rho^2\sin\phi\, d\rho\, d\phi\, d\theta = \int_0^{2\pi} \int_0^{\pi/3} \int_0^a \rho^4\sin^3\phi\, d\rho\, d\phi\, d\theta$

(b) $\displaystyle\int_0^{2\pi}\int_0^{\sqrt{3}a/2}\int_{r/\sqrt{3}}^{\sqrt{a^2-r^2}} r^2\,dz\,r dr\,d\theta = \int_0^{2\pi}\int_0^{\sqrt{3}a/2}\int_{r/\sqrt{3}}^{\sqrt{a^2-r^2}} r^3\,dz\,dr\,d\theta$

(c) $\displaystyle\int_{-\sqrt{3}a/2}^{\sqrt{3}a/2}\int_{-\sqrt{(3a^2/4)-x^2}}^{\sqrt{(3a^2/4)-x^2}}\int_{\sqrt{x^2+y^2}/\sqrt{3}}^{\sqrt{a^2-x^2-y^2}} (x^2+y^2)\,dz\,dy\,dx$

25. $\displaystyle\int_0^2\int_{(y/2)^{1/3}}^{2-y/2} dx\,dy = \int_0^2\left(2-\frac{y}{2}-\left(\frac{y}{2}\right)^{1/3}\right)dy = \left(2y-\frac{y^2}{4}-\frac{3}{2}\left(\frac{y}{2}\right)^{4/3}\right)\Big]_0^2 = \frac{3}{2}$

27. $\displaystyle V = \int_0^{2\pi}\int_0^{a/\sqrt{3}}\int_{\sqrt{3}r}^{a} r\,dz\,dr\,d\theta = 2\pi\int_0^{a/\sqrt{3}} r(a-\sqrt{3}r)\,dr = \frac{\pi a^3}{9}$

29. $\|\mathbf{r}_u\times\mathbf{r}_v\| = \sqrt{2u^2+2v^2+4}$,

$$S = \iint_{u^2+v^2\le 4}\sqrt{2u^2+2v^2+4}\,dA = \int_0^{2\pi}\int_0^2 \sqrt{2}\sqrt{r^2+2}\,r\,dr\,d\theta = \frac{8\pi}{3}(3\sqrt{3}-1)$$

31. $(\mathbf{r}_u\times\mathbf{r}_v)\Big]_{\substack{u=1\\v=2}} = \langle -2,-4,1\rangle$, tangent plane $2x+4y-z=5$

33. $\displaystyle A = \int_{-4}^{4}\int_{y^2/4}^{2+y^2/8} dx\,dy = \int_{-4}^{4}\left(2-\frac{y^2}{8}\right)dy = \frac{32}{3}; \bar{y}=0$ by symmetry;

$$\int_{-4}^{4}\int_{y^2/4}^{2+y^2/8} x\,dx\,dy = \int_{-4}^{4}\left(2+\frac{1}{4}y^2-\frac{3}{128}y^4\right)dy = \frac{256}{15},\ \bar{x} = \frac{3}{32}\frac{256}{15} = \frac{8}{5};\ \text{centroid}\ \left(\frac{8}{5},0\right)$$

35. $V = \frac{1}{3}\pi a^2 h, \bar{x}=\bar{y}=0$ by symmetry,

$$\int_0^{2\pi}\int_0^a\int_0^{h-rh/a} rz\,dz\,dr\,d\theta = \pi\int_0^a rh^2\left(1-\frac{r}{a}\right)^2 dr = \pi a^2h^2/12,\ \text{centroid}\ (0,0,h/4)$$

37. The two quarter-circles with center at the origin and of radius A and $\sqrt{2}A$ lie inside and outside of the square with corners $(0,0),(A,0),(A,A),(0,A)$, so the following inequalities hold:

$$\int_0^{\pi/2}\int_0^A \frac{1}{(1+r^2)^2} r dr\,d\theta \le \int_0^A\int_0^A \frac{1}{(1+x^2+y^2)^2}dx\,dy \le \int_0^{\pi/2}\int_0^{\sqrt{2}A}\frac{1}{(1+r^2)^2} r dr\,d\theta$$

The integral on the left can be evaluated as $\dfrac{\pi A^2}{4(1+A^2)}$ and the integral on the right equals $\dfrac{2\pi A^2}{4(1+2A^2)}$. Since both of these quantities tend to $\dfrac{\pi}{4}$ as $A\to+\infty$, it follows by sandwiching that

$$\int_0^{+\infty}\int_0^{+\infty}\frac{1}{(1+x^2+y^2)^2}dx\,dy = \frac{\pi}{4}.$$

39. **(a)** Let S_1 be the set of points (x,y,z) which satisfy the equation $x^{2/3}+y^{2/3}+z^{2/3}=a^{2/3}$, and let S_2 be the set of points (x,y,z) where $x=a(\sin\phi\cos\theta)^3, y=a(\sin\phi\sin\theta)^3, z=a\cos^3\phi$, $0\le\phi\le\pi, 0\le\theta<2\pi$.

If (x,y,z) is a point of S_2 then

$x^{2/3}+y^{2/3}+z^{2/3} = a^{2/3}[(\sin\phi\cos\theta)^3+(\sin\phi\sin\theta)^3+\cos^3\phi] = a^{2/3}$

so (x,y,z) belongs to S_1.

If (x, y, z) is a point of S_1 then $x^{2/3} + y^{2/3} + z^{2/3} = a^{2/3}$. Let
$x_1 = x^{1/3}, y_1 = y^{1/3}, z_1 = z^{1/3}, a_1 = a^{1/3}$. Then $x_1^2 + y_1^2 + z_1^2 = a_1^2$, so in spherical coordinates $x_1 = a_1 \sin\phi\cos\theta, y_1 = a_1 \sin\phi\sin\theta, z_1 = a_1\cos\phi$, with
$\theta = \tan^{-1}\left(\dfrac{y_1}{x_1}\right) = \tan^{-1}\left(\dfrac{y}{x}\right)^{1/3}, \phi = \cos^{-1}\dfrac{z_1}{a_1} = \cos^{-1}\left(\dfrac{z}{a}\right)^{1/3}$. Then
$x = x_1^3 = a_1^3(\sin\phi\cos\theta)^3 = a(\sin\phi\cos\theta)^3$, similarly $y = a(\sin\phi\sin\theta)^3, z = a\cos\phi$ so (x, y, z) belongs to S_2. Thus $S_1 = S_2$

(b) Let $a = 1$ and $\mathbf{r} = (\cos\theta\sin\phi)^3\mathbf{i} + (\sin\theta\sin\phi)^3\mathbf{j} + \cos^3\phi\mathbf{k}$, then

$$S = 8\int_0^{\pi/2}\int_0^{\pi/2} \|\mathbf{r}_\theta \times \mathbf{r}_\phi\| d\phi\, d\theta$$

$$= 72\int_0^{\pi/2}\int_0^{\pi/2} \sin\theta\cos\theta\sin^4\phi\cos\phi\sqrt{\cos^2\phi + \sin^2\phi\sin^2\theta\cos^2\theta}\, d\theta\, d\phi \approx 4.4506$$

(c)
$$\frac{\partial(x,y,z)}{\partial(\rho,\theta,\phi)} = \left\|\begin{array}{ccc} \sin^3\phi\cos^3\theta & 3\rho\sin^2\phi\cos\phi\cos^3\theta & -3\rho\sin^3\phi\cos^2\theta\sin\theta \\ \sin^3\phi\sin^3\theta & 3\rho\sin^2\phi\cos\phi\sin^3\theta & 3\rho\sin^3\phi\sin^2\theta\cos\theta \\ \cos^3\phi & -3\rho\cos^2\phi\sin\phi & 0 \end{array}\right\|$$

$$= 9\rho^2\cos^2\theta\sin^2\theta\cos^2\phi\sin^5\phi,$$

$$V = 9\int_0^{2\pi}\int_0^{\pi}\int_0^{a} \rho^2\cos^2\theta\sin^2\theta\cos^2\phi\sin^5\phi\, d\rho\, d\phi\, d\theta = \frac{4}{35}\pi a^3$$

41. (a) $(x/a)^2 + (y/b)^2 + (z/c)^2 = \sin^2\phi\cos^2\theta + \sin^2\phi\sin^2\theta + \cos^2\phi = \sin^2\phi + \cos^2\phi = 1$, an ellipsoid

(b) $\mathbf{r}(\phi,\theta) = \langle 2\sin\phi\cos\theta, 3\sin\phi\sin\theta, 4\cos\phi\rangle$; $\mathbf{r}_\phi \times \mathbf{r}_\theta = 2\langle 6\sin^2\phi\cos\theta, 4\sin^2\phi\sin\theta, 3\cos\phi\sin\phi\rangle$,

$\|\mathbf{r}_\phi \times \mathbf{r}_\theta\| = 2\sqrt{16\sin^4\phi + 20\sin^4\phi\cos^2\theta + 9\sin^2\phi\cos^2\phi}$,

$$S = \int_0^{2\pi}\int_0^{\pi} 2\sqrt{16\sin^4\phi + 20\sin^4\phi\cos^2\theta + 9\sin^2\phi\cos^2\phi}\, d\phi\, d\theta \approx 111.5457699$$

CHAPTER 16
Topics in Vector Calculus

EXERCISE SET 16.1

1. **(a)** III because the vector field is independent of y and the direction is that of the negative x-axis for negative x, and positive for positive

(b) IV, because the y-component is constant, and the x-component varies priodically with x

3. **(a)** true **(b)** true **(c)** true

5.

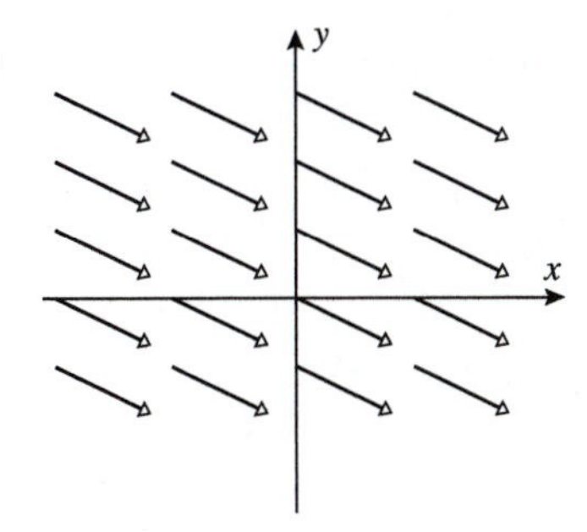

7.

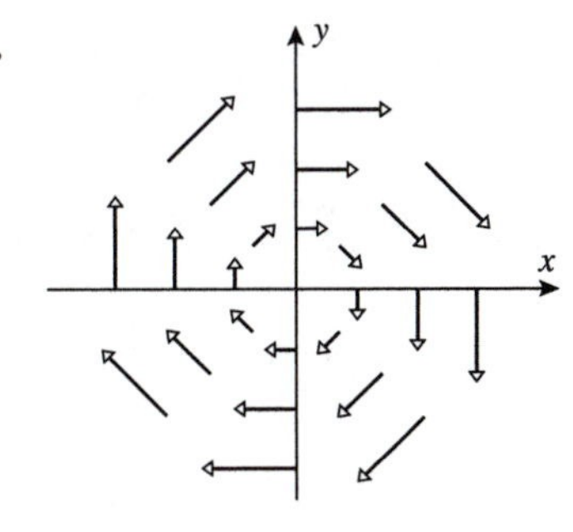

9.

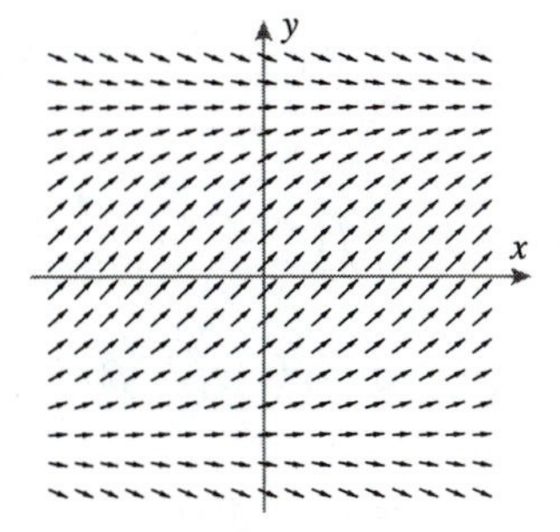

11. **(a)** $\nabla\phi = \phi_x\mathbf{i} + \phi_y\mathbf{j} = \dfrac{y}{1+x^2y^2}\mathbf{i} + \dfrac{x}{1+x^2y^2}\mathbf{j} = \mathbf{F}$, so $\mathbf{F}$ is conservative for all x, y

(b) $\nabla\phi = \phi_x\mathbf{i} + \phi_y\mathbf{j} = 2x\mathbf{i} - 6y\mathbf{j} + 8z\mathbf{k} = \mathbf{F}$ so $\mathbf{F}$ is conservative for all x, y

13. div $\mathbf{F} = 2x + y$, curl $\mathbf{F} = z\mathbf{i}$

15. div $\mathbf{F} = 0$, curl $\mathbf{F} = (40x^2z^4 - 12xy^3)\mathbf{i} + (14y^3z + 3y^4)\mathbf{j} - (16xz^5 + 21y^2z^2)\mathbf{k}$

17. div $\mathbf{F} = \dfrac{2}{\sqrt{x^2+y^2+z^2}}$, curl $\mathbf{F} = \mathbf{0}$

19. $\nabla\cdot(\mathbf{F}\times\mathbf{G}) = \nabla\cdot(-(z+4y^2)\mathbf{i} + (4xy+2xz)\mathbf{j} + (2xy-x)\mathbf{k}) = 4x$

21. $\nabla\cdot(\nabla\times\mathbf{F}) = \nabla\cdot(-\sin(x-y)\mathbf{k}) = 0$

23. $\nabla\times(\nabla\times\mathbf{F}) = \nabla\times(xz\mathbf{i} - yz\mathbf{j} + y\mathbf{k}) = (1+y)\mathbf{i} + x\mathbf{j}$

27. Let $\mathbf{F} = f\mathbf{i} + g\mathbf{j} + h\mathbf{k}$; div $(k\mathbf{F}) = k\dfrac{\partial f}{\partial x} + k\dfrac{\partial g}{\partial y} + k\dfrac{\partial h}{\partial z} = k$ div $\mathbf{F}$

29. Let $\mathbf{F} = f(x,y,z)\mathbf{i} + g(x,y,z)\mathbf{j} + h(x,y,z)\mathbf{k}$ and $\mathbf{G} = P(x,y,z)\mathbf{i} + Q(x,y,z)\mathbf{j} + R(x,y,z)\mathbf{k}$, then

$$\text{div }(\mathbf{F}+\mathbf{G}) = \left(\frac{\partial f}{\partial x} + \frac{\partial P}{\partial x}\right) + \left(\frac{\partial g}{\partial y} + \frac{\partial Q}{\partial y}\right) + \left(\frac{\partial h}{\partial z} + \frac{\partial R}{\partial z}\right)$$

$$= \left(\frac{\partial f}{\partial x} + \frac{\partial g}{\partial y} + \frac{\partial h}{\partial z}\right) + \left(\frac{\partial P}{\partial x} + \frac{\partial Q}{\partial y} + \frac{\partial R}{\partial z}\right) = \text{div }\mathbf{F} + \text{div }\mathbf{G}$$

31. Let $\mathbf{F} = f\mathbf{i} + g\mathbf{j} + h\mathbf{k}$;

$$\text{div }(\phi\mathbf{F}) = \left(\phi\frac{\partial f}{\partial x} + \frac{\partial \phi}{\partial x}f\right) + \left(\phi\frac{\partial g}{\partial y} + \frac{\partial \phi}{\partial y}g\right) + \left(\phi\frac{\partial h}{\partial z} + \frac{\partial \phi}{\partial z}h\right)$$

$$= \phi\left(\frac{\partial f}{\partial x} + \frac{\partial g}{\partial y} + \frac{\partial h}{\partial z}\right) + \left(\frac{\partial \phi}{\partial x}f + \frac{\partial \phi}{\partial y}g + \frac{\partial \phi}{\partial z}h\right)$$

$$= \phi \text{ div }\mathbf{F} + \nabla\phi\cdot\mathbf{F}$$

33. Let $\mathbf{F} = f\mathbf{i} + g\mathbf{j} + h\mathbf{k}$;

$$\text{div}(\text{curl } \mathbf{F}) = \frac{\partial}{\partial x}\left(\frac{\partial h}{\partial y} - \frac{\partial g}{\partial z}\right) + \frac{\partial}{\partial y}\left(\frac{\partial f}{\partial z} - \frac{\partial h}{\partial x}\right) + \frac{\partial}{\partial z}\left(\frac{\partial g}{\partial x} - \frac{\partial f}{\partial y}\right)$$

$$= \frac{\partial^2 h}{\partial x \partial y} - \frac{\partial^2 g}{\partial x \partial z} + \frac{\partial^2 f}{\partial y \partial z} - \frac{\partial^2 h}{\partial y \partial x} + \frac{\partial^2 g}{\partial z \partial x} - \frac{\partial^2 f}{\partial z \partial y} = 0,$$

assuming equality of mixed second partial derivatives

35. $\nabla \cdot (k\mathbf{F}) = k\nabla \cdot \mathbf{F}$, $\nabla \cdot (\mathbf{F} + \mathbf{G}) = \nabla \cdot \mathbf{F} + \nabla \cdot \mathbf{G}$, $\nabla \cdot (\phi \mathbf{F}) = \phi \nabla \cdot \mathbf{F} + \nabla \phi \cdot \mathbf{F}$, $\nabla \cdot (\nabla \times \mathbf{F}) = 0$

37. **(a)** $\text{curl } \mathbf{r} = 0\mathbf{i} + 0\mathbf{j} + 0\mathbf{k} = \mathbf{0}$

(b) $\nabla \|\mathbf{r}\| = \nabla \sqrt{x^2 + y^2 + z^2} = \dfrac{x}{\sqrt{x^2+y^2+z^2}}\mathbf{i} + \dfrac{y}{\sqrt{x^2+y^2+z^2}}\mathbf{j} + \dfrac{z}{\sqrt{x^2+y^2+z^2}}\mathbf{k} = \dfrac{\mathbf{r}}{\|\mathbf{r}\|}$

39. **(a)** $\nabla f(r) = f'(r)\dfrac{\partial r}{\partial x}\mathbf{i} + f'(r)\dfrac{\partial r}{\partial y}\mathbf{j} + f'(r)\dfrac{\partial r}{\partial z}\mathbf{k} = f'(r)\nabla r = \dfrac{f'(r)}{r}\mathbf{r}$

(b) $\text{div}[f(r)\mathbf{r}] = f(r)\text{div } \mathbf{r} + \nabla f(r) \cdot \mathbf{r} = 3f(r) + \dfrac{f'(r)}{r}\mathbf{r} \cdot \mathbf{r} = 3f(r) + rf'(r)$

41. $f(r) = 1/r^3, f'(r) = -3/r^4$, $\text{div}(\mathbf{r}/r^3) = 3(1/r^3) + r(-3/r^4) = 0$

43. **(a)** At the point (x, y) the slope of the line along which the vector $-y\mathbf{i} + x\mathbf{j}$ lies is $-x/y$; the slope of the tangent line to C at (x, y) is dy/dx, so $dy/dx = -x/y$.

(b) $ydy = -xdx$, $y^2/2 = -x^2/2 + K_1$, $x^2 + y^2 = K$

45. $dy/dx = 1/x, y = \ln x + K$

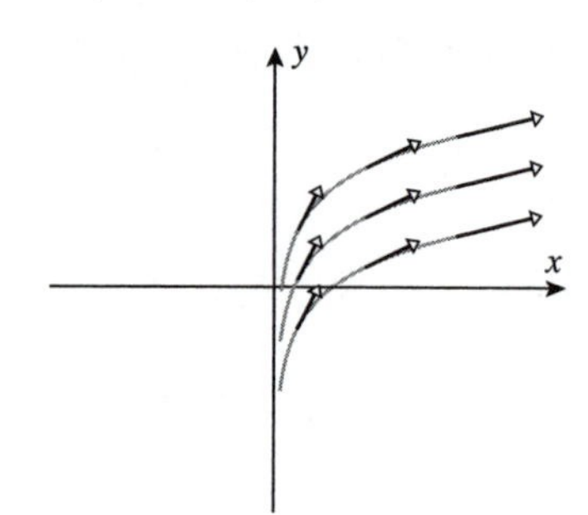

EXERCISE SET 16.2

1. **(a)** $\displaystyle\int_0^1 dy = 1$ because $s = y$ is arclength measured from $(0, 0)$

(b) 0, because $\sin xy = 0$ along C

3. **(a)** $ds = \sqrt{\left(\dfrac{dx}{dt}\right)^2 + \left(\dfrac{dy}{dt}\right)^2}\,dt$, so $\displaystyle\int_0^1 (2t - 3t^2)\sqrt{4 + 36t^2}\,dt = -\frac{11}{108}\sqrt{10} - \frac{1}{36}\ln(\sqrt{10} - 3) - \frac{4}{27}$

(b) $\displaystyle\int_0^1 (2t - 3t^2)2\,dt = 0$

(c) $\displaystyle\int_0^1 (2t - 3t^2)6t\,dt = -\frac{1}{2}$

5. **(a)** $C: x = t,\ y = t,\ 0 \le t \le 1;\ \int_0^1 6t\,dt = 3$

(b) $C: x = t,\ y = t^2,\ 0 \le t \le 1;\ \int_0^1 (3t + 6t^2 - 2t^3)dt = 3$

(c) $C: x = t,\ y = \sin(\pi t/2),\ 0 \le t \le 1;$

$$\int_0^1 [3t + 2\sin(\pi t/2) + \pi t\cos(\pi t/2) - (\pi/2)\sin(\pi t/2)\cos(\pi t/2)]dt = 3$$

(d) $C: x = t^3,\ y = t,\ 0 \le t \le 1;\ \int_0^1 (9t^5 + 8t^3 - t)dt = 3$

7. $\int_0^3 \frac{\sqrt{1+t}}{1+t}\,dt = \int_0^3 (1+t)^{-1/2}\,dt = 2$

9. $\int_0^1 3(t^2)(t^2)(2t^3/3)(1+2t^2)\,dt = 2\int_0^1 t^7(1+2t^2)\,dt = 13/20$

11. $\int_0^{\pi/4} (8\cos^2 t - 16\sin^2 t - 20\sin t\cos t)dt = 1 - \pi$

13. $C: x = (3-t)^2/3,\ y = 3 - t,\ 0 \le t \le 3;\ \int_0^3 \frac{1}{3}(3-t)^2 dt = 3$

15. $C: x = \cos t,\ y = \sin t,\ 0 \le t \le \pi/2;\ \int_0^{\pi/2} (-\sin t - \cos^2 t)dt = -1 - \pi/4$

17. $\int_0^1 (-3)e^{3t}dt = 1 - e^3$

19. **(a)** $\int_0^{\ln 2} \left(e^{3t} + e^{-3t}\right)\sqrt{e^{2t} + e^{-2t}}\,dt$

$$= \frac{63}{64}\sqrt{17} + \frac{1}{4}\ln(4 + \sqrt{17}) - \frac{1}{8}\ln\frac{\sqrt{17}+1}{\sqrt{17}-1} - \frac{1}{4}\ln(\sqrt{2}+1) + \frac{1}{8}\ln\frac{\sqrt{2}+1}{\sqrt{2}-1}$$

(b) $\int_0^{\pi/2} [\sin t\cos t\,dt - \sin^2 t\,dt] = \frac{1}{2} - \frac{\pi}{4}$

21. **(a)** $C_1: (0,0)$ to $(1,0); x = t, y = 0, 0 \le t \le 1$
$C_2: (1,0)$ to $(0,1); x = 1 - t, y = t, 0 \le t \le 1$
$C_3: (0,1)$ to $(0,0); x = 0, y = 1 - t, 0 \le t \le 1$

$$\int_0^1 (0)dt + \int_0^1 (-1)dt + \int_0^1 (0)dt = -1$$

(b) $C_1: (0,0)$ to $(1,0); x = t, y = 0, 0 \le t \le 1$
$C_2: (1,0)$ to $(1,1); x = 1, y = t, 0 \le t \le 1$
$C_3: (1,1)$ to $(0,1); x = 1 - t, y = 1, 0 \le t \le 1$
$C_4: (0,1)$ to $(0,0); x = 0, y = 1 - t, 0 \le t \le 1$

$$\int_0^1 (0)dt + \int_0^1 (-1)dt + \int_0^1 (-1)dt + \int_0^1 (0)dt = -2$$

23. $C_1 : x = t, y = z = 0, 0 \le t \le 1, \int_0^1 0\, dt = 0$; $C_2 : x = 1, y = t, z = 0, 0 \le t \le 1, \int_0^1 (-t)\, dt = -\frac{1}{2}$

$C_3 : x = 1, y = 1, z = t, 0 \le t \le 1, \int_0^1 3\, dt = 3$; $\int_C x^2 z\, dx - yx^2\, dy + 3\, dz = 0 - \frac{1}{2} + 3 = \frac{5}{2}$

25. $\int_0^\pi (0) dt = 0$

27. $\int_0^1 e^{-t} dt = 1 - e^{-1}$

29. Represent the circular arc by $x = 3\cos t, y = 3\sin t, 0 \le t \le \pi/2$.

$\int_C x\sqrt{y} ds = 9\sqrt{3} \int_0^{\pi/2} \sqrt{\sin\ t} \cos t\ dt = 6\sqrt{3}$

31. $\int_C \frac{kx}{1+y^2} ds = 15k \int_0^{\pi/2} \frac{\cos t}{1 + 9\sin^2 t} dt = 5k \tan^{-1} 3$

33. $C : x = t^2,\ y = t,\ 0 \le t \le 1;\ W = \int_0^1 3t^4 dt = 3/5$

35. $W = \int_0^1 (t^3 + 5t^6) dt = 27/28$

37. Since $\mathbf{F}$ and $\mathbf{r}$ are parallel, $\mathbf{F} \cdot \mathbf{r} = \|\mathbf{F}\|\|\mathbf{r}\|$, and since $\mathbf{F}$ is constant,

$\int \mathbf{F} \cdot d\mathbf{r} = \int_C d(\mathbf{F} \cdot \mathbf{r}) = \int_C d(\|\mathbf{F}\|\|\mathbf{r}\|) = \sqrt{2} \int_{-4}^4 \sqrt{2} dt = 16$

39. $C : x = 4\cos t, y = 4\sin t, 0 \le t \le \pi/2$

$\int_0^{\pi/2} \left(-\frac{1}{4}\sin t + \cos t\right) dt = 3/4$

41. Represent the parabola by $x = t, y = t^2, 0 \le t \le 2$.

$\int_C 3x ds = \int_0^2 3t\sqrt{1 + 4t^2}\ dt = (17\sqrt{17} - 1)/4$

43. **(a)** $2\pi rh = 2\pi(1)2 = 4\pi$

(b) $S = \int_C z(t)\, dt$

(c) $C : x = \cos t, y = \sin t, 0 \le t \le 2\pi; S = \int_0^{2\pi} (2 + (1/2)\sin 3t)\, dt = 4\pi$

45. $W = \int_C \mathbf{F} \cdot d\mathbf{r} = \int_0^1 (\lambda t^2(1-t), t - \lambda t(1-t)) \cdot (1, \lambda - 2\lambda t)\, dt = -\lambda/12$, $W = 1$ when $\lambda = -12$

EXERCISE SET 16.3

1. $\partial x/\partial y = 0 = \partial y/\partial x$, conservative so $\partial\phi/\partial x = x$ and $\partial\phi/\partial y = y$, $\phi = x^2/2 + k(y)$, $k'(y) = y$, $k(y) = y^2/2 + K$, $\phi = x^2/2 + y^2/2 + K$

3. $\partial(x^2 y)/\partial y = x^2$ and $\partial(5xy^2)/\partial x = 5y^2$, not conservative

5. $\partial(\cos y + y\cos x)/\partial y = -\sin y + \cos x = \partial(\sin x - x\sin y)/\partial x$, conservative so $\partial\phi/\partial x = \cos y + y\cos x$ and $\partial\phi/\partial y = \sin x - x\sin y$, $\phi = x\cos y + y\sin x + k(y)$, $-x\sin y + \sin x + k'(y) = \sin x - x\sin y$, $k'(y) = 0$, $k(y) = K$, $\phi = x\cos y + y\sin x + K$

7. **(a)** $\partial(y^2)/\partial y = 2y = \partial(2xy)/\partial x$, independent of path

(b) $C : x = -1 + 2t,\ y = 2 + t,\ 0 \le t \le 1;\ \int_0^1 (4 + 14t + 6t^2)dt = 13$

(c) $\partial\phi/\partial x = y^2$ and $\partial\phi/\partial y = 2xy$, $\phi = xy^2 + k(y)$, $2xy + k'(y) = 2xy$, $k'(y) = 0$, $k(y) = K$, $\phi = xy^2 + K$. Let $K = 0$ to get $\phi(1,3) - \phi(-1,2) = 9 - (-4) = 13$

9. $\partial(3y)/\partial y = 3 = \partial(3x)/\partial x$, $\phi = 3xy$, $\phi(4,0) - \phi(1,2) = -6$

11. $\partial(2xe^y)/\partial y = 2xe^y = \partial(x^2e^y)/\partial x$, $\phi = x^2e^y$, $\phi(3,2) - \phi(0,0) = 9e^2$

13. $\partial(2xy^3)/\partial y = 6xy^2 = \partial(3x^2y^2)/\partial x$, $\phi = x^2y^3$, $\phi(-1,0) - \phi(2,-2) = 32$

15. $\phi = x^2y^2/2$, $W = \phi(0,0) - \phi(1,1) = -1/2$

17. $\phi = e^{xy}$, $W = \phi(2,0) - \phi(-1,1) = 1 - e^{-1}$

19. $\partial(e^y + ye^x)/\partial y = e^y + e^x = \partial(xe^y + e^x)/\partial x$ so $\mathbf{F}$ is conservative, $\phi(x,y) = xe^y + ye^x$ so $\int_C \mathbf{F}\cdot d\mathbf{r} = \phi(0, \ln 2) - \phi(1,0) = \ln 2 - 1$

21.
$$\mathbf{F}\cdot d\mathbf{r} = [(e^y + ye^x)\mathbf{i} + (xe^y + e^x)\mathbf{j}]\cdot[(\pi/2)\cos(\pi t/2)\mathbf{i} + (1/t)\mathbf{j}]dt$$
$$= \left(\frac{\pi}{2}\cos(\pi t/2)(e^y + ye^x) + (xe^y + e^x)/t\right)dt,$$
so
$$\int_C \mathbf{F}\cdot d\mathbf{r} = \int_1^2 \left(\frac{\pi}{2}\cos(\pi t/2)\left(t + (\ln t)e^{\sin(\pi t/2)}\right) + \left(\sin(\pi t/2) + \frac{1}{t}e^{\sin(\pi t/2)}\right)\right)dt = \ln 2 - 1$$

23. No; a closed loop can be found whose tangent everywhere makes an angle $< \pi$ with the vector field, so the line integral $\int_C \mathbf{F}\cdot d\mathbf{r} > 0$, and by Theorem 16.3.2 the vector field is not conservative.

25. If $\mathbf{F}$ is conservative, then $\mathbf{F} = \nabla\phi = \dfrac{\partial\phi}{\partial x}\mathbf{i} + \dfrac{\partial\phi}{\partial y}\mathbf{j} + \dfrac{\partial\phi}{\partial z}\mathbf{k}$ and hence $f = \dfrac{\partial\phi}{\partial x}$, $g = \dfrac{\partial\phi}{\partial y}$, and $h = \dfrac{\partial\phi}{\partial z}$.

Thus $\dfrac{\partial f}{\partial y} = \dfrac{\partial^2\phi}{\partial y\partial x}$ and $\dfrac{\partial g}{\partial x} = \dfrac{\partial^2\phi}{\partial x\partial y}$, $\dfrac{\partial f}{\partial z} = \dfrac{\partial^2\phi}{\partial z\partial x}$ and $\dfrac{\partial h}{\partial x} = \dfrac{\partial^2\phi}{\partial x\partial z}$, $\dfrac{\partial g}{\partial z} = \dfrac{\partial^2\phi}{\partial z\partial y}$ and $\dfrac{\partial h}{\partial y} = \dfrac{\partial^2\phi}{\partial y\partial z}$.

The result follows from the equality of mixed second partial derivatives.

27.
$$\frac{\partial}{\partial y}(h(x)[x\sin y + y\cos y]) = h(x)[x\cos y - y\sin y + \cos y]$$
$$\frac{\partial}{\partial x}(h(x)[x\cos y - y\sin y]) = h(x)\cos y + h'(x)[x\cos y - y\sin y],$$

equate these two partial derivatives to get $(x\cos y - y\sin y)(h'(x) - h(x)) = 0$ which holds for all x and y if $h'(x) = h(x)$, $h(x) = Ce^x$ where C is an arbitrary constant.

29. **(a)** See Exercise 28, $c = 1$; $W = \int_P^Q \mathbf{F} \cdot d\mathbf{r} = \phi(3,2,1) - \phi(1,1,2) = -\frac{1}{\sqrt{14}} + \frac{1}{\sqrt{6}}$

(b) C begins at $P(1,1,2)$ and ends at $Q(3,2,1)$ so the answer is again $W = -\frac{1}{\sqrt{14}} + \frac{1}{\sqrt{6}}$.

(c) The circle is not specified, but cannot pass through $(0,0,0)$, so Φ is continuous and differentiable on the circle. Start at any point P on the circle and return to P, so the work is $\Phi(P) - \Phi(P) = 0$.
C begins at, say, $(3,0)$ and ends at the same point so $W = 0$.

31. If C is composed of smooth curves $C_1, C_2, \ldots, C_n$ and curve C_i extends from (x_{i-1}, y_{i-1}) to (x_i, y_i) then $\int_C \mathbf{F} \cdot d\mathbf{r} = \sum_{i=1}^{n} \int_{C_i} \mathbf{F} \cdot d\mathbf{r} = \sum_{i=1}^{n} [\phi(x_i, y_i) - \phi(x_{i-1}, y_{i-1})] = \phi(x_n, y_n) - \phi(x_0, y_0)$
where (x_0, y_0) and (x_n, y_n) are the endpoints of C.

33. Let C_1 be an arbitrary piecewise smooth curve from (a,b) to a point (x, y_1) in the disk, and C_2 the vertical line segment from (x, y_1) to (x, y). Then

$$\phi(x,y) = \int_{C_1} \mathbf{F} \cdot d\mathbf{r} + \int_{C_2} \mathbf{F} \cdot d\mathbf{r} = \int_{(a,b)}^{(x,y_1)} \mathbf{F} \cdot d\mathbf{r} + \int_{C_2} \mathbf{F} \cdot d\mathbf{r}.$$

The first term does not depend on y;

hence $\frac{\partial \phi}{\partial y} = \frac{\partial}{\partial y} \int_{C_2} \mathbf{F} \cdot d\mathbf{r} = \frac{\partial}{\partial y} \int_{C_2} f(x,y)dx + g(x,y)dy$.

However, the line integral with respect to x is zero along C_2, so $\frac{\partial \phi}{\partial y} = \frac{\partial}{\partial y} \int_{C_2} g(x,y)\, dy$.

Express C_2 as $x = x, y = t$ where t varies from y_1 to y, then $\frac{\partial \phi}{\partial y} = \frac{\partial}{\partial y} \int_{y_1}^{y} g(x,t)\, dt = g(x,y)$.

EXERCISE SET 16.4

1. $\iint_R (2x - 2y)dA = \int_0^1 \int_0^1 (2x - 2y)dy\, dx = 0$; for the line integral, on $x = 0, y^2\, dx = 0, x^2\, dy = 0$; on $y = 0, y^2\, dx = x^2\, dy = 0$; on $x = 1, y^2\, dx + x^2\, dy = dy$; and on $y = 1, y^2\, dx + x^2\, dy = dx$, hence $\oint_C y^2\, dx + x^2\, dy = \int_0^1 dy + \int_1^0 dx = 1 - 1 = 0$

3. $\int_{-2}^{4} \int_1^2 (2y - 3x)dy\, dx = 0$

5. $\int_0^{\pi/2} \int_0^{\pi/2} (-y \cos x + x \sin y)dy\, dx = 0$

7. $\iint_R [1 - (-1)]dA = 2 \iint_R dA = 8\pi$

9. $\displaystyle\iint_R \left(-\frac{y}{1+y} - \frac{1}{1+y}\right) dA = -\iint_R dA = -4$

11. $\displaystyle\iint_R \left(-\frac{y^2}{1+y^2} - \frac{1}{1+y^2}\right) dA = -\iint_R dA = -1$

13. $\displaystyle\int_0^1 \int_{x^2}^{\sqrt{x}} (y^2 - x^2) dy\, dx = 0$

15. **(a)** $C: x = \cos t, y = \sin t, 0 \le t \le 2\pi;$

$$\oint_C = \int_0^{2\pi} \left(e^{\sin t}(-\sin t) + \sin t \cos t e^{\cos t}\right) dt \approx -3.550999378;$$

$$\iint_R \left[\frac{\partial}{\partial x}(ye^x) - \frac{\partial}{\partial y}e^y\right] dA = \iint_R \left[ye^x - e^y\right] dA$$

$$= \int_0^{2\pi} \int_0^1 \left[r \sin\theta e^{r\cos\theta} - e^{r\sin\theta}\right] r\, dr\, d\theta \approx -3.550999378$$

(b) $C_1: x = t, y = t^2, 0 \le t \le 1; \displaystyle\int_{C_1} [e^y\, dx + ye^x\, dy] = \int_0^1 \left[e^{t^2} + 2t^3 e^t\right] dt \approx 2.589524432$

$C_2: x = t^2, y = t, 0 \le t \le 1; \displaystyle\int_{C_2} [e^y\, dx + ye^x\, dy] = \int_0^1 \left[2te^t + te^{t^2}\right] dt = \frac{e+3}{2} \approx 2.859140914$

$$\int_{C_1} - \int_{C_2} \approx -0.269616482; \quad \iint_R = \int_0^1 \int_{x^2}^{\sqrt{x}} [ye^x - e^y]\, dy\, dx \approx -0.269616482$$

17. $\displaystyle A = \frac{1}{2}\oint_C -y\, dx + x\, dy = \frac{1}{2}\int_0^{2\pi} (3a^2 \sin^4\phi \cos^2\phi + 3a^2 \cos^4\phi \sin^2\phi) d\phi$

$$= \frac{3}{2}a^2 \int_0^{2\pi} \sin^2\phi \cos^2\phi\, d\phi = \frac{3}{8}a^2 \int_0^{2\pi} \sin^2 2\phi\, d\phi = 3\pi a^2/8$$

19. $C_1: (0,0)$ to $(a,0)$; $x = at,\ y = 0,\ 0 \le t \le 1$

$C_2: (a,0)$ to $(a\cos t_0, b\sin t_0)$; $x = a\cos t,\ y = b\sin t,\ 0 \le t \le t_0$

$C_3: (a\cos t_0, b\sin t_0)$ to $(0,0)$; $x = -a(\cos t_0)t,\ y = -b(\sin t_0)t,\ -1 \le t \le 0$

$$A = \frac{1}{2}\oint_C -y\, dx + x\, dy = \frac{1}{2}\int_0^1 (0)\, dt + \frac{1}{2}\int_0^{t_0} ab\, dt + \frac{1}{2}\int_{-1}^0 (0)\, dt = \frac{1}{2}ab\, t_0$$

21. $\displaystyle W = \iint_R y\, dA = \int_0^{\pi} \int_0^5 r^2 \sin\theta\, dr\, d\theta = 250/3$

23. $\displaystyle\oint_C y\, dx - x\, dy = \iint_R (-2) dA = -2\int_0^{2\pi} \int_0^{a(1+\cos\theta)} r\, dr\, d\theta = -3\pi a^2$

25. $A = \int_0^1 \int_{x^3}^{x} dy\,dx = \frac{1}{4}$; $C_1 : x = t, y = t^3, 0 \le t \le 1$, $\int_{C_1} x^2\,dy = \int_0^1 t^2(3t^2)\,dt = \frac{3}{5}$

$C_2 : x = t, y = t, 0 \le t \le 1$; $\int_{C_2} x^2\,dy = \int_0^1 t^2\,dt = \frac{1}{3}$, $\oint_C x^2\,dy = \int_{C_1} - \int_{C_2} = \frac{3}{5} - \frac{1}{3} = \frac{4}{15}$, $\bar{x} = \frac{8}{15}$

$\int_C y^2\,dx = \int_0^1 t^6\,dt - \int_0^1 t^2\,dt = \frac{1}{7} - \frac{1}{3} = -\frac{4}{21}$, $\bar{y} = \frac{8}{21}$, centroid $\left(\frac{8}{15}, \frac{8}{21}\right)$

27. $\bar{x} = 0$ from the symmetry of the region,
$C_1 : (a, 0)$ to $(-a, 0)$ along $y = \sqrt{a^2 - x^2}$; $x = a\cos t$, $y = a\sin t$, $0 \le t \le \pi$
$C_2 : (-a, 0)$ to $(a, 0)$; $x = t$, $y = 0$, $-a \le t \le a$

$$A = \pi a^2/2, \quad \bar{y} = -\frac{1}{2A}\left[\int_0^\pi -a^3 \sin^3 t\,dt + \int_{-a}^{a} (0)dt\right]$$

$$= -\frac{1}{\pi a^2}\left(-\frac{4a^3}{3}\right) = \frac{4a}{3\pi}; \text{ centroid } \left(0, \frac{4a}{3\pi}\right)$$

29. From Green's Theorem, the given integral equals $\iint_R (1 - x^2 - y^2)dA$ where R is the region enclosed by C. The value of this integral is maximum if the integration extends over the largest region for which the integrand $1 - x^2 - y^2$ is nonnegative so we want $1 - x^2 - y^2 \ge 0$, $x^2 + y^2 \le 1$. The largest region is that bounded by the circle $x^2 + y^2 = 1$ which is the desired curve C.

31. $\int_C \mathbf{F} \cdot d\mathbf{r} = \int_C (x^2 + y)\,dx + (4x - \cos y)\,dy = 3\iint_R dA = 3(25 - 2) = 69$

EXERCISE SET 16.5

1. R is the annular region between $x^2 + y^2 = 1$ and $x^2 + y^2 = 4$;

$$\iint_\sigma z^2 dS = \iint_R (x^2 + y^2)\sqrt{\frac{x^2}{x^2 + y^2} + \frac{y^2}{x^2 + y^2} + 1}\,dA$$

$$= \sqrt{2}\iint_R (x^2 + y^2)dA = \sqrt{2}\int_0^{2\pi}\int_1^2 r^3 dr\,d\theta = \frac{15}{2}\pi\sqrt{2}.$$

3. Let $\mathbf{r}(u, v) = \cos u\mathbf{i} + v\mathbf{j} + \sin u\mathbf{k}, 0 \le u \le \pi, 0 \le v \le 1$. Then $\mathbf{r}_u = -\sin u\mathbf{i} + \cos u\mathbf{k}, \mathbf{r}_v = \mathbf{j}$,

$\mathbf{r}_u \times \mathbf{r}_v = -\cos u\mathbf{i} - \sin u\mathbf{k}, \|\mathbf{r}_u \times \mathbf{r}_v\| = 1$, $\iint_\sigma x^2 y\,dS = \int_0^1 \int_0^\pi v\cos^2 u\,du\,dv = \pi/4$

5. If we use the projection of σ onto the xz-plane then $y = 1 - x$ and R is the rectangular region in the xz-plane enclosed by $x = 0$, $x = 1$, $z = 0$ and $z = 1$;

$$\iint_\sigma (x - y - z)dS = \iint_R (2x - 1 - z)\sqrt{2}dA = \sqrt{2}\int_0^1\int_0^1 (2x - 1 - z)dz\,dx = -\sqrt{2}/2$$

7. There are six surfaces, parametrized by projecting onto planes:
$\sigma_1 : z = 0;\ 0 \le x \le 1,\ 0 \le y \le 1$ (onto xy-plane), $\sigma_2 : x = 0;\ 0 \le y \le 1,\ 0 \le z \le 1$ (onto yz-plane), $\sigma_3 : y = 0;\ 0 \le x \le 1,\ 0 \le z \le 1$ (onto xz-plane), $\sigma_4 : z = 1;\ 0 \le x \le 1,\ 0 \le y \le 1$ (onto xy-plane), $\sigma_5 : x = 1;\ 0 \le y \le 1,\ 0 \le z \le 1$ (onto yz-plane), $\sigma_6 : y = 1;\ 0 \le x \le 1,\ 0 \le z \le 1$ (onto xz-plane).

By symmetry the integrals over σ_1, σ_2 and σ_3 are equal, as are those over σ_4, σ_5 and σ_6, and

$$\iint_{\sigma_1} (x+y+z)dS = \int_0^1\int_0^1 (x+y)dx\,dy = 1; \quad \iint_{\sigma_4} (x+y+z)dS = \int_0^1\int_0^1 (x+y+1)dx\,dy = 2,$$

thus, $\displaystyle\iint_{\sigma} (x+y+z)dS = 3\cdot 1 + 3\cdot 2 = 9.$

9. R is the circular region enclosed by $x^2+y^2=1$;

$$\iint_{\sigma} \sqrt{x^2+y^2+z^2}\,dS = \iint_R \sqrt{2(x^2+y^2)}\sqrt{\frac{x^2}{x^2+y^2}+\frac{y^2}{x^2+y^2}+1}\,dA$$
$$= \lim_{r_0\to 0^+} 2\iint_{R'} \sqrt{x^2+y^2}\,dA$$

where R' is the annular region enclosed by $x^2+y^2=1$ and $x^2+y^2=r_0^2$ with r_0 slightly larger than 0 because $\sqrt{\dfrac{x^2}{x^2+y^2}+\dfrac{y^2}{x^2+y^2}+1}$ is not defined for $x^2+y^2=0$, so

$$\iint_{\sigma} \sqrt{x^2+y^2+z^2}\,dS = \lim_{r_0\to 0^+} 2\int_0^{2\pi}\int_{r_0}^1 r^2 dr\,d\theta = \lim_{r_0\to 0^+} \frac{4\pi}{3}(1-r_0^3) = \frac{4\pi}{3}.$$

11. (a) $\displaystyle\frac{\sqrt{29}}{16}\int_0^6\int_0^{(12-2x)/3} xy(12-2x-3y)dy\,dx$

(b) $\displaystyle\frac{\sqrt{29}}{4}\int_0^3\int_0^{(12-4z)/3} yz(12-3y-4z)dy\,dz$

(c) $\displaystyle\frac{\sqrt{29}}{9}\int_0^3\int_0^{6-2z} xz(12-2x-4z)dx\,dz$

13. $18\sqrt{29}/5$

15. $\displaystyle\int_0^4\int_1^2 y^3 z\sqrt{4y^2+1}\,dy\,dz;\ \frac{1}{2}\int_0^4\int_1^4 xz\sqrt{1+4x}\,dx\,dz$

17. $391\sqrt{17}/15 - 5\sqrt{5}/3$

19. $z=\sqrt{4-x^2},\ \dfrac{\partial z}{\partial x} = -\dfrac{x}{\sqrt{4-x^2}},\ \dfrac{\partial z}{\partial y} = 0;$

$$\iint_{\sigma} \delta_0 dS = \delta_0\iint_R \sqrt{\frac{x^2}{4-x^2}+1}\,dA = 2\delta_0\int_0^4\int_0^1 \frac{1}{\sqrt{4-x^2}}dx\,dy = \frac{4}{3}\pi\delta_0.$$

21. $z = 4 - y^2$, R is the rectangular region enclosed by $x = 0$, $x = 3$, $y = 0$ and $y = 3$;

$$\iint_\sigma y\,dS = \iint_R y\sqrt{4y^2+1}\,dA = \int_0^3\int_0^3 y\sqrt{4y^2+1}\,dy\,dx = \frac{1}{4}(37\sqrt{37}-1).$$

23. $M = \iint_\sigma \delta(x,y,z)dS = \iint_\sigma \delta_0 dS = \delta_0 \iint_\sigma dS = \delta_0 S$

25. By symmetry $\bar{x} = \bar{y} = 0$.

$$\iint_\sigma dS = \iint_R \sqrt{x^2+y^2+1}\,dA = \int_0^{2\pi}\int_0^{\sqrt{8}} \sqrt{r^2+1}\,r\,dr\,d\theta = \frac{52\pi}{3},$$

$$\iint_\sigma z\,dS = \iint_R z\sqrt{x^2+y^2+1}\,dA = \frac{1}{2}\iint_R (x^2+y^2)\sqrt{x^2+y^2+1}\,dA$$

$$= \frac{1}{2}\int_0^{2\pi}\int_0^{\sqrt{8}} r^3\sqrt{r^2+1}\,dr\,d\theta = \frac{596\pi}{15}$$

so $\bar{z} = \dfrac{596\pi/15}{52\pi/3} = \dfrac{149}{65}$. The centroid is $(\bar{x},\bar{y},\bar{z}) = (0, 0, 149/65)$.

27. $\partial\mathbf{r}/\partial u = \cos v\mathbf{i} + \sin v\mathbf{j} + 3\mathbf{k}$, $\partial\mathbf{r}/\partial v = -u\sin v\mathbf{i} + u\cos v\mathbf{j}$, $\|\partial\mathbf{r}/\partial u \times \partial\mathbf{r}/\partial v\| = \sqrt{10}u$;

$$3\sqrt{10}\iint_R u^4 \sin v\cos v\,dA = 3\sqrt{10}\int_0^{\pi/2}\int_1^2 u^4\sin v\cos v\,du\,dv = 93/\sqrt{10}$$

29. $\partial\mathbf{r}/\partial u = \cos v\mathbf{i} + \sin v\mathbf{j} + 2u\mathbf{k}$, $\partial\mathbf{r}/\partial v = -u\sin v\mathbf{i} + u\cos v\mathbf{j}$, $\|\partial\mathbf{r}/\partial u \times \partial\mathbf{r}/\partial v\| = u\sqrt{4u^2+1}$;

$$\iint_R u\,dA = \int_0^\pi\int_0^{\sin v} u\,du\,dv = \pi/4$$

31. $\partial z/\partial x = -2xe^{-x^2-y^2}$, $\partial z/\partial y = -2ye^{-x^2-y^2}$,

$(\partial z/\partial x)^2 + (\partial z/\partial y)^2 + 1 = 4(x^2+y^2)e^{-2(x^2+y^2)} + 1$; use polar coordinates to get

$$M = \int_0^{2\pi}\int_0^3 r^2\sqrt{4r^2e^{-2r^2}+1}\,dr\,d\theta \approx 57.895751$$

EXERCISE SET 16.6

1. (a) zero (b) zero (c) positive
(d) negative (e) zero (f) zero

3. (a) positive (b) zero (c) positive
(d) zero (e) positive (f) zero

5. (a) $\mathbf{n} = -\cos v\mathbf{i} - \sin v\mathbf{j}$ (b) inward, by inspection

7. $\mathbf{n} = -z_x\mathbf{i} - z_y\mathbf{j} + \mathbf{k}$, $\displaystyle\iint_R \mathbf{F}\cdot\mathbf{n}\,dS = \iint_R (2x^2 + 2y^2 + 2(1 - x^2 - y^2))\,dS = \int_0^{2\pi}\int_0^1 2r\,dr\,d\theta = 2\pi$

9. R is the annular region enclosed by $x^2 + y^2 = 1$ and $x^2 + y^2 = 4$;

$$\iint_\sigma \mathbf{F}\cdot\mathbf{n}\,dS = \iint_R \left(-\frac{x^2}{\sqrt{x^2+y^2}} - \frac{y^2}{\sqrt{x^2+y^2}} + 2z\right) dA$$

$$= \iint_R \sqrt{x^2+y^2}\,dA = \int_0^{2\pi}\int_1^2 r^2 dr\,d\theta = \frac{14\pi}{3}.$$

11. R is the circular region enclosed by $x^2 + y^2 - y = 0$; $\displaystyle\iint_\sigma \mathbf{F}\cdot\mathbf{n}\,dS = \iint_R (-x)dA = 0$ since the region R is symmetric across the y-axis.

13. $\partial\mathbf{r}/\partial u = \cos v\mathbf{i} + \sin v\mathbf{j} - 2u\mathbf{k}$, $\partial\mathbf{r}/\partial v = -u\sin v\mathbf{i} + u\cos v\mathbf{j}$,
$\partial\mathbf{r}/\partial u \times \partial\mathbf{r}/\partial v = 2u^2\cos v\mathbf{i} + 2u^2\sin v\mathbf{j} + u\mathbf{k}$;

$$\iint_R (2u^3 + u)\,dA = \int_0^{2\pi}\int_1^2 (2u^3 + u)du\,dv = 18\pi$$

15. $\partial\mathbf{r}/\partial u = \cos v\mathbf{i} + \sin v\mathbf{j} + 2\mathbf{k}$, $\partial\mathbf{r}/\partial v = -u\sin v\mathbf{i} + u\cos v\mathbf{j}$,
$\partial\mathbf{r}/\partial u \times \partial\mathbf{r}/\partial v = -2u\cos v\mathbf{i} - 2u\sin v\mathbf{j} + u\mathbf{k}$;

$$\iint_R u^2\,dA = \int_0^{\pi}\int_0^{\sin v} u^2 du\,dv = 4/9$$

17. In each part, divide σ into the six surfaces
$\sigma_1 : x = -1$ with $|y| \le 1$, $|z| \le 1$, and $\mathbf{n} = -\mathbf{i}$, $\sigma_2 : x = 1$ with $|y| \le 1$, $|z| \le 1$, and $\mathbf{n} = \mathbf{i}$,
$\sigma_3 : y = -1$ with $|x| \le 1$, $|z| \le 1$, and $\mathbf{n} = -\mathbf{j}$, $\sigma_4 : y = 1$ with $|x| \le 1$, $|z| \le 1$, and $\mathbf{n} = \mathbf{j}$,
$\sigma_5 : z = -1$ with $|x| \le 1$, $|y| \le 1$, and $\mathbf{n} = -\mathbf{k}$, $\sigma_6 : z = 1$ with $|x| \le 1$, $|y| \le 1$, and $\mathbf{n} = \mathbf{k}$,

(a) $\displaystyle\iint_{\sigma_1} \mathbf{F}\cdot\mathbf{n}\,dS = \iint_{\sigma_1} dS = 4$, $\displaystyle\iint_{\sigma_2} \mathbf{F}\cdot\mathbf{n}\,dS = \iint_{\sigma_2} dS = 4$, and $\displaystyle\iint_{\sigma_i} \mathbf{F}\cdot\mathbf{n}\,dS = 0$ for $i = 3, 4, 5, 6$ so $\displaystyle\iint_\sigma \mathbf{F}\cdot\mathbf{n}\,dS = 4 + 4 + 0 + 0 + 0 + 0 = 8$.

(b) $\displaystyle\iint_{\sigma_1} \mathbf{F}\cdot\mathbf{n}\,dS = \iint_{\sigma_1} dS = 4$, similarly $\displaystyle\iint_{\sigma_i} \mathbf{F}\cdot\mathbf{n}\,dS = 4$ for $i = 2, 3, 4, 5, 6$ so

$$\iint_\sigma \mathbf{F}\cdot\mathbf{n}\,dS = 4 + 4 + 4 + 4 + 4 + 4 = 24.$$

(c) $\displaystyle\iint_{\sigma_1} \mathbf{F}\cdot\mathbf{n}\,dS = -\iint_{\sigma_1} dS = -4$, $\displaystyle\iint_{\sigma_2} \mathbf{F}\cdot\mathbf{n}\,dS = 4$, similarly $\displaystyle\iint_{\sigma_i} \mathbf{F}\cdot\mathbf{n}\,dS = -4$ for $i = 3, 5$ and $\displaystyle\iint_{\sigma_i} \mathbf{F}\cdot\mathbf{n}\,dS = 4$ for $i = 4, 6$ so $\displaystyle\iint_{\sigma} \mathbf{F}\cdot\mathbf{n}\,dS = -4 + 4 - 4 + 4 - 4 + 4 = 0$.

19. R is the circular region enclosed by $x^2 + y^2 = 1$; $x = r\cos\theta, y = r\sin\theta, z = r$, $\mathbf{n} = \cos\theta\mathbf{i} + \sin\theta\mathbf{j} - \mathbf{k}$;

$$\iint_{\sigma} \mathbf{F}\cdot\mathbf{n}\,dS = \iint_{R} (\cos\theta + \sin\theta - 1)\,dA = \int_0^{2\pi}\int_0^1 (\cos\theta + \sin\theta - 1)\,r\,dr\,d\theta = -\pi.$$

21. (a) $\mathbf{n} = \dfrac{1}{\sqrt{3}}[\mathbf{i} + \mathbf{j} + \mathbf{k}]$,

$$V = \int_{\sigma} \mathbf{F}\cdot\mathbf{n}\,dS = \int_0^1\int_0^{1-x} (2x - 3y + 1 - x - y)\,dy\,dx = 0 \text{ m}^3/\text{s}$$

(b) $m = 0 \cdot 806 = 0$ kg/s

23. (a) $G(x, y, z) = x - g(y, z)$, $\nabla G = \mathbf{i} - \dfrac{\partial g}{\partial y}\mathbf{j} - \dfrac{\partial g}{\partial z}\mathbf{k}$, apply Theorem 16.6.3:

$\displaystyle\iint_{\sigma} \mathbf{F}\cdot\mathbf{n}dS = \iint_{R} \mathbf{F}\cdot\left(\mathbf{i} - \frac{\partial x}{\partial y}\mathbf{j} - \frac{\partial x}{\partial z}\mathbf{k}\right) dA$, if σ is oriented by front normals, and

$\displaystyle\iint_{\sigma} \mathbf{F}\cdot\mathbf{n}dS = \iint_{R} \mathbf{F}\cdot\left(-\mathbf{i} + \frac{\partial x}{\partial y}\mathbf{j} + \frac{\partial x}{\partial z}\mathbf{k}\right) dA$, if σ is oriented by back normals,

where R is the projection of σ onto the yz-plane.

(b) R is the semicircular region in the yz-plane enclosed by $z = \sqrt{1 - y^2}$ and $z = 0$;

$$\iint_{\sigma} \mathbf{F}\cdot\mathbf{n}\,dS = \iint_{R} (-y - 2yz + 16z)dA = \int_{-1}^{1}\int_0^{\sqrt{1-y^2}} (-y - 2yz + 16z)dz\,dy = \frac{32}{3}.$$

25. (a) On the sphere, $\|\mathbf{r}\| = a$ so $\mathbf{F} = a^k\mathbf{r}$ and $\mathbf{F}\cdot\mathbf{n} = a^k\mathbf{r}\cdot(\mathbf{r}/a) = a^{k-1}\|\mathbf{r}\|^2 = a^{k-1}a^2 = a^{k+1}$, hence $\displaystyle\iint_{\sigma} \mathbf{F}\cdot\mathbf{n}\,dS = a^{k+1}\iint_{\sigma} dS = a^{k+1}(4\pi a^2) = 4\pi a^{k+3}$.

(b) If $k = -3$, then $\displaystyle\iint_{\sigma} \mathbf{F}\cdot\mathbf{n}\,dS = 4\pi$.

EXERCISE SET 16.7

1. $\sigma_1 : x = 0, \mathbf{F}\cdot\mathbf{n} = -x = 0, \displaystyle\iint_{\sigma_1} (0)dA = 0$ $\qquad$ $\sigma_2 : x = 1, \mathbf{F}\cdot\mathbf{n} = x = 1, \displaystyle\iint_{\sigma_2} (1)dA = 1$

$\sigma_3 : y = 0, \mathbf{F}\cdot\mathbf{n} = -y = 0, \displaystyle\iint_{\sigma_3} (0)dA = 0$ $\qquad$ $\sigma_4 : y = 1, \mathbf{F}\cdot\mathbf{n} = y = 1, \displaystyle\iint_{\sigma_4} (1)dA = 1$

$\sigma_5 : z = 0, \mathbf{F}\cdot\mathbf{n} = -z = 0, \iint\limits_{\sigma_5} (0)dA = 0$ $\qquad$ $\sigma_6 : z = 1, \mathbf{F}\cdot\mathbf{n} = z = 1, \iint\limits_{\sigma_6} (1)dA = 1$

$$\iint\limits_{\sigma} \mathbf{F}\cdot\mathbf{n} = 3; \iiint\limits_{G} \text{div } \mathbf{F}dV = \iiint\limits_{G} 3dV = 3$$

3. $\sigma_1 : z = 1, \mathbf{n} = \mathbf{k}, \mathbf{F}\cdot\mathbf{n} = z^2 = 1, \iint\limits_{\sigma_1} (1)dS = \pi,$

$\sigma_2 : \mathbf{n} = 2x\mathbf{i} + 2y\mathbf{j} - \mathbf{k}, \mathbf{F}\cdot\mathbf{n} = 4x^2 - 4x^2y^2 - x^4 - 3y^4,$

$$\iint\limits_{\sigma_2} \mathbf{F}\cdot\mathbf{n}\,dS = \int_0^{2\pi}\int_0^1 \left[4r^2\cos^2\theta - 4r^4\cos^2\theta\sin^2\theta - r^4\cos^4\theta - 3r^4\sin^4\theta\right] r\,dr\,d\theta = \frac{\pi}{3};$$

$$\iint\limits_{\sigma} = \frac{4\pi}{3}$$

$$\iiint\limits_{G} \text{div } \mathbf{F}dV = \iiint\limits_{G} (2+z)dV = \int_0^{2\pi}\int_0^1\int_{r^2}^1 (2+z)dz\,r\,dr\,d\theta = 4\pi/3$$

5. G is the rectangular solid; $\iiint\limits_{G} \text{div } \mathbf{F}\,dV = \int_0^2\int_0^1\int_0^3 (2x-1)\,dx\,dy\,dz = 12.$

7. G is the cylindrical solid;

$$\iiint\limits_{G} \text{div } \mathbf{F}\,dV = 3\iiint\limits_{G} dV = (3)(\text{volume of cylinder}) = (3)[\pi a^2(1)] = 3\pi a^2.$$

9. G is the cylindrical solid;

$$\iiint\limits_{G} \text{div } \mathbf{F}\,dV = 3\iiint\limits_{G} (x^2+y^2+z^2)dV = 3\int_0^{2\pi}\int_0^2\int_0^3 (r^2+z^2)r\,dz\,dr\,d\theta = 180\pi.$$

11. G is the hemispherical solid bounded by $z = \sqrt{4-x^2-y^2}$ and the xy-plane;

$$\iiint\limits_{G} \text{div } \mathbf{F}\,dV = 3\iiint\limits_{G} (x^2+y^2+z^2)dV = 3\int_0^{2\pi}\int_0^{\pi/2}\int_0^2 \rho^4\sin\phi\,d\rho\,d\phi\,d\theta = \frac{192\pi}{5}.$$

13. G is the conical solid;

$$\iiint\limits_{G} \text{div } \mathbf{F}\,dV = 2\iiint\limits_{G} (x+y+z)dV = 2\int_0^{2\pi}\int_0^1\int_r^1 (r\cos\theta + r\sin\theta + z)r\,dz\,dr\,d\theta = \frac{\pi}{2}.$$

15. G is the solid bounded by $z = 4 - x^2$, $y + z = 5$, and the coordinate planes;

$$\iiint\limits_{G} \text{div } \mathbf{F}\,dV = 4\iiint\limits_{G} x^2dV = 4\int_{-2}^2\int_0^{4-x^2}\int_0^{5-z} x^2dy\,dz\,dx = \frac{4608}{35}.$$

17. $\iint\limits_{\sigma} \mathbf{r}\cdot\mathbf{n}\,dS = \iiint\limits_{G} \text{div}\,\mathbf{r}\,dV = 3\iiint\limits_{G} dV = 3\text{vol}(G)$

19. $\iint_\sigma \text{curl } \mathbf{F} \cdot \mathbf{n}\, dS = \iiint_G \text{div}(\text{curl } \mathbf{F}) dV = \iiint_G (0) dV = 0$

21. $\iint_\sigma (f\nabla g) \cdot \mathbf{n} = \iiint_G \text{div } (f\nabla g) dV = \iiint_G (f\nabla^2 g + \nabla f \cdot \nabla g) dV$ by Exercise 31, Section 16.1.

23. Since $\mathbf{v}$ is constant, $\nabla \cdot \mathbf{v} = \mathbf{0}$. Let $\mathbf{F} = f\mathbf{v}$; then $\text{div}\mathbf{F} = (\nabla f)\mathbf{v}$ and by the Divergence Theorem

$$\iint_\sigma f\mathbf{v} \cdot \mathbf{n}\, dS = \iint_\sigma \mathbf{F} \cdot \mathbf{n}\, dS = \iiint_G \text{div}\mathbf{F}\, dV = \iiint_G (\nabla f) \cdot \mathbf{v}\, dV$$

25. **(a)** The flux through any cylinder whose axis is the z-axis is positive by inspection; by the Divergence Theorem, this says that the divergence cannot be negative at the origin, else the flux through a small enough cylinder would also be negative (impossible), hence the divergence at the origin must be ≥ 0.

(b) Similar to Part (a), ≤ 0.

27. $\text{div } \mathbf{F} = 0$; no sources or sinks.

29. $\text{div } \mathbf{F} = 3x^2 + 3y^2 + 3z^2$; sources at all points except the origin, no sinks.

31. Let σ_1 be the portion of the paraboloid $z = 1 - x^2 - y^2$ for $z \geq 0$, and σ_2 the portion of the plane $z = 0$ for $x^2 + y^2 \leq 1$. Then

$$\iint_{\sigma_1} \mathbf{F} \cdot \mathbf{n}\, dS = \iint_R \mathbf{F} \cdot (2x\mathbf{i} + 2y\mathbf{j} + \mathbf{k})\, dA$$

$$= \int_{-1}^{1} \int_{-\sqrt{1-x^2}}^{\sqrt{1-x^2}} (2x[x^2 y - (1 - x^2 - y^2)^2] + 2y(y^3 - x) + (2x + 2 - 3x^2 - 3y^2))\, dy\, dx$$

$$= 3\pi/4;$$

$z = 0$ and $\mathbf{n} = -\mathbf{k}$ on σ_2 so $\mathbf{F} \cdot \mathbf{n} = 1 - 2x$, $\iint_{\sigma_2} \mathbf{F} \cdot \mathbf{n}\, dS = \iint_{\sigma_2} (1 - 2x) dS = \pi$. Thus

$\iint_\sigma \mathbf{F} \cdot \mathbf{n}\, dS = 3\pi/4 + \pi = 7\pi/4$. But $\text{div } \mathbf{F} = 2xy + 3y^2 + 3$ so

$$\iiint_G \text{div } \mathbf{F}\, dV = \int_{-1}^{1} \int_{-\sqrt{1-x^2}}^{\sqrt{1-x^2}} \int_0^{1-x^2-y^2} (2xy + 3y^2 + 3)\, dz\, dy\, dx = 7\pi/4.$$

EXERCISE SET 16.8

1. **(a)** The flow is independent of z and has no component in the direction of $\mathbf{k}$, and so by inspection the only nonzero component of the curl is in the direction of $\mathbf{k}$. However both sides of (9) are zero, as the flow is orthogonal to the curve C_a. Thus the curl is zero.

(b) Since the flow appears to be tangential to the curve C_a, it seems that the right hand side of (9) is nonzero, and thus the curl is nonzero, and points in the positive z-direction.

3. If σ is oriented with upward normals then C consists of three parts parametrized as $C_1 : \mathbf{r}(t) = (1-t)\mathbf{i} + t\mathbf{j}$ for $0 \le t \le 1$, $C_2 : \mathbf{r}(t) = (1-t)\mathbf{j} + t\mathbf{k}$ for $0 \le t \le 1$, $C_3 : \mathbf{r}(t) = t\mathbf{i} + (1-t)\mathbf{k}$ for $0 \le t \le 1$.

$$\int_{C_1} \mathbf{F} \cdot d\mathbf{r} = \int_{C_2} \mathbf{F} \cdot d\mathbf{r} = \int_{C_3} \mathbf{F} \cdot d\mathbf{r} = \int_0^1 (3t-1)dt = \frac{1}{2} \text{ so}$$

$\oint_C \mathbf{F} \cdot d\mathbf{r} = \frac{1}{2} + \frac{1}{2} + \frac{1}{2} = \frac{3}{2}$. curl $\mathbf{F} = \mathbf{i} + \mathbf{j} + \mathbf{k}$, $z = 1 - x - y$, R is the triangular region in the xy-plane enclosed by $x + y = 1$, $x = 0$, and $y = 0$;

$$\iint_\sigma (\text{curl } \mathbf{F}) \cdot \mathbf{n}\, dS = 3 \iint_R dA = (3)(\text{area of } R) = (3)\left[\frac{1}{2}(1)(1)\right] = \frac{3}{2}.$$

5. If σ is oriented with upward normals then C can be parametrized as $\mathbf{r}(t) = a\cos t\mathbf{i} + a\sin t\mathbf{j}$ for $0 \le t \le 2\pi$.

$$\oint_C \mathbf{F} \cdot d\mathbf{r} = \int_0^{2\pi} 0\, dt = 0; \text{ curl } \mathbf{F} = \mathbf{0} \text{ so } \iint_\sigma (\text{curl } \mathbf{F}) \cdot \mathbf{n}\, dS = \iint_\sigma 0\, dS = 0.$$

7. Take σ as the part of the plane $z = 0$ for $x^2 + y^2 \le 1$ with $\mathbf{n} = \mathbf{k}$; curl $\mathbf{F} = -3y^2\mathbf{i} + 2z\mathbf{j} + 2\mathbf{k}$,

$$\iint_\sigma (\text{curl } \mathbf{F}) \cdot \mathbf{n}\, dS = 2 \iint_\sigma dS = (2)(\text{area of circle}) = (2)[\pi(1)^2] = 2\pi.$$

9. C is the boundary of R and curl $\mathbf{F} = 2\mathbf{i} + 3\mathbf{j} + 4\mathbf{k}$, so

$$\oint \mathbf{F} \cdot \mathbf{r} = \iint_R \text{curl}\, \mathbf{F} \cdot \mathbf{n}\, dS = \iint_R 4\, dA = 4(\text{area of } R) = 16\pi$$

11. curl $\mathbf{F} = x\mathbf{k}$, take σ as part of the plane $z = y$ oriented with upward normals, R is the circular region in the xy-plane enclosed by $x^2 + y^2 - y = 0$;

$$\iint_\sigma (\text{curl } \mathbf{F}) \cdot \mathbf{n}\, dS = \iint_R x\, dA = \int_0^\pi \int_0^{\sin\theta} r^2 \cos\theta\, dr\, d\theta = 0.$$

13. curl $\mathbf{F} = \mathbf{i} + \mathbf{j} + \mathbf{k}$, take σ as the part of the plane $z = 0$ with $x^2 + y^2 \le a^2$ and $\mathbf{n} = \mathbf{k}$;

$$\iint_\sigma (\text{curl } \mathbf{F}) \cdot \mathbf{n}\, dS = \iint_\sigma dS = \text{ area of circle } = \pi a^2.$$

15. **(a)** Take σ as the part of the plane $2x + y + 2z = 2$ in the first octant, oriented with downward normals; curl $\mathbf{F} = -x\mathbf{i} + (y-1)\mathbf{j} - \mathbf{k}$,

$$\oint_C \mathbf{F} \cdot \mathbf{T}\, ds = \iint_\sigma (\text{curl } \mathbf{F}) \cdot \mathbf{n}\, dS$$

$$= \iint_R \left(x - \frac{1}{2}y + \frac{3}{2}\right) dA = \int_0^1 \int_0^{2-2x} \left(x - \frac{1}{2}y + \frac{3}{2}\right) dy\, dx = \frac{3}{2}.$$

(b) At the origin curl $\mathbf{F} = -\mathbf{j} - \mathbf{k}$ and with $\mathbf{n} = \mathbf{k}$, curl $\mathbf{F}(0,0,0) \cdot \mathbf{n} = (-\mathbf{j} - \mathbf{k}) \cdot \mathbf{k} = -1$.

(c) The rotation of $\mathbf{F}$ has its maximum value at the origin about the unit vector in the same direction as curl $\mathbf{F}(0,0,0)$ so $\mathbf{n} = -\frac{1}{\sqrt{2}}\mathbf{j} - \frac{1}{\sqrt{2}}\mathbf{k}$.

17. Since $\oint_C \mathbf{E} \cdot \mathbf{r}d\mathbf{r} = \iint_\sigma \text{curl } \mathbf{E} \cdot \mathbf{n}\,dS$, it follows that $\iint_\sigma \text{curl } \mathbf{E} \cdot \mathbf{n}dS = -\iint_\sigma \frac{\partial \mathbf{B}}{\partial t} \cdot \mathbf{n}dS$. This relationship holds for any surface σ, hence curl $\mathbf{E} = -\frac{\partial \mathbf{B}}{\partial t}$.

CHAPTER 16 SUPPLEMENTARY EXERCISES

3. (a) $\int_a^b \left[f(x(t), y(t))\frac{dx}{dt} + g(x(t), y(t))\frac{dy}{dt} \right] dt$

(b) $\int_a^b f(x(t), y(t))\sqrt{x'(t)^2 + y'(t)^2}\, dt$

11. $\iint_\sigma f(x,y,z)dS = \iint_R f(x(u,v), y(u,v), z(u,v))\|\mathbf{r}_u \times \mathbf{r}_v\|\, du\, dv$

13. Let O be the origin, P the point with polar coordinates $\theta = \alpha, r = f(\alpha)$, and Q the point with polar coordinates $\theta = \beta, r = f(\beta)$. Let

$C_1 : O$ to P; $x = t\cos\alpha,\ y = t\sin\alpha,\ 0 \le t \le f(\alpha), -y\frac{dx}{dt} + x\frac{dy}{dt} = 0$

$C_2 : P$ to Q; $x = f(t)\cos t,\ y = f(t)\sin t,\ \alpha \le \theta \le \beta, -y\frac{dx}{dt} + x\frac{dy}{dt} = f(t)^2$

$C_3 : Q$ to O; $x = -t\cos\beta,\ y = -t\sin\beta,\ -f(\beta) \le t \le 0, -y\frac{dx}{dt} + x\frac{dy}{dt} = 0$

$A = \frac{1}{2}\oint_C -y\,dx + x\,dy = \frac{1}{2}\int_\alpha^\beta f(t)^2\,dt$; set $t = \theta$ and $r = f(\theta) = f(t)$, $A = \frac{1}{2}\int_\alpha^\beta r^2\,d\theta$.

15. (a) Assume the mass M is located at the origin and the mass m at (x,y,z), then

$\mathbf{r} = x\mathbf{i} + y\mathbf{j} + z\mathbf{k}, \mathbf{F}(x,y,z) = -\frac{GmM}{(x^2+y^2+z^2)^{3/2}}\mathbf{r}$,

$$W = -\int_{t_1}^{t_2} \frac{GmM}{(x^2+y^2+z^2)^{3/2}}\left(x\frac{dx}{dt} + y\frac{dy}{dt} + z\frac{dz}{dt}\right) dt$$
$$= GmM(x^2+y^2+z^2)^{-1/2}\Big]_{t_1}^{t_2} = GmM\left(\frac{1}{r_2} - \frac{1}{r_1}\right)$$

(b) $W = 3.99 \times 10^5 \times 10^3 \left[\frac{1}{7170} - \frac{1}{6970}\right] \approx -1596.801594 \text{ km}^2\text{kg/s}^2 \approx -1.597 \times 10^9 \text{ J}$

17. $\bar{x} = 0$ by symmetry; by Exercise 16, $\bar{y} = -\frac{1}{2A}\int_C y^2\,dx$; $C_1 : y = 0, -a \le x \le a, y^2\,dx = 0$;

$C_2 : x = a\cos\theta, y = a\sin\theta, 0 \le \theta \le \pi$, so

$$\bar{y} = -\frac{1}{2(\pi a^2/2)}\int_0^\pi a^2\sin^2\theta(-a\sin\theta)\,d\theta = \frac{4a}{3\pi}$$

19. $\bar{y} = 0$ by symmetry; $\bar{x} = \dfrac{1}{2A}\displaystyle\int_C x^2\,dy$; $A = \alpha a^2$; $C_1 : x = t\cos\alpha, y = -t\sin\alpha, 0 \le t \le a$;

$C_2 : x = a\cos\theta, y = a\sin\theta, -\alpha \le \theta \le \alpha$;

$C_3 : x = t\cos\alpha, y = t\sin\alpha, 0 \le t \le a$ (reverse orientation);

$$2A\bar{x} = -\int_0^a t^2\cos^2\alpha\sin\alpha\,dt + \int_{-\alpha}^{\alpha} a^3\cos^3\theta\,d\theta - \int_0^a t^2\cos^2\alpha\sin\alpha\,dt,$$

$$= -\frac{2a^3}{3}\cos^2\alpha\sin\alpha + 2a^3\int_0^{\alpha}\cos^3\theta\,d\theta = -\frac{2a^3}{3}\cos^2\alpha\sin\alpha + 2a^3\left[\sin\alpha - \frac{1}{3}\sin^3\alpha\right]$$

$$= \frac{4}{3}a^3\sin\alpha;\ \text{since} A = \alpha a^2, \bar{x} = \frac{2a}{3}\frac{\sin\alpha}{\alpha}$$

21. **(a)** $\displaystyle\int_C f(x)\,dx + g(y)\,dy = \iint_R \left(\frac{\partial}{\partial x}g(y) - \frac{\partial}{\partial y}f(x)\right)dA = 0$

(b) $W = \displaystyle\int_C \mathbf{F}\cdot d\mathbf{r} = \int_C f(x)\,dx + g(y)\,dy = 0$, so the work done by the vector field around any simple closed curve is zero. The field is conservative.

23. Yes; by imagining a normal vector sliding around the surface it is evident that the surface has two sides.

25. By Exercise 24, $\displaystyle\iint_\sigma D_{\mathbf{n}}f\,dS = -\iiint_G [f_{xx} + f_{yy} + f_{zz}]\,dV = -6\iiint_G dV = -6\text{vol}(G) = -8\pi$

27. **(a)** If $h(x)\mathbf{F}$ is conservative, then $\dfrac{\partial}{\partial y}(yh(x)) = \dfrac{\partial}{\partial x}(-2xh(x))$, or $h(x) = -2h(x) - 2xh'(x)$ which has the general solution $x^3h(x)^2 = C_1, h(x) = Cx^{-3/2}$, so $C\dfrac{y}{x^{3/2}}\mathbf{i} - C\dfrac{2}{x^{1/2}}\mathbf{j}$ is conservative, with potential function $\phi = -2Cy/\sqrt{x}$.

(b) If $g(y)\mathbf{F}(x,y)$ is conservative then $\dfrac{\partial}{\partial y}(yg(y)) = \dfrac{\partial}{\partial x}(-2xg(y))$, or $g(y) + yg'(y) = -2g(y)$, with general solution $g(y) = C/y^3$, so $\mathbf{F} = C\dfrac{1}{y^2}\mathbf{i} - C\dfrac{2x}{y^3}\mathbf{j}$ is conservative, with potential function Cx/y^2.

29. **(a)** conservative, $\phi(x,y,z) = xz^2 - e^{-y}$ **(b)** not conservative, $f_y \ne g_x$

APPENDIX A
Real Numbers, Intervals, and Inequalities

EXERCISE SET A

1. **(a)** rational **(b)** integer, rational **(c)** integer, rational
(d) rational **(e)** integer, rational **(f)** irrational
(g) rational **(h)** 467 integer, rational

3. **(a)** $x = 0.123123123\ldots$, $1000x = 123 + x$, $x = 123/999 = 41/333$
(b) $x = 12.7777\ldots$, $10(x - 12) = 7 + (x - 12)$, $9x = 115$, $x = 115/9$
(c) $x = 38.07818181\ldots$, $100x = 3807.81818181\ldots$, $99x = 100x - x = 3769.74$,
$x = \dfrac{3769.74}{99} = \dfrac{376974}{9900} = \dfrac{20943}{550}$
(d) $\dfrac{4296}{10000} = \dfrac{537}{1250}$

5. **(a)** If r is the radius, then $D = 2r$ so $\left(\dfrac{8}{9}D\right)^2 = \left(\dfrac{16}{9}r\right)^2 = \dfrac{256}{81}r^2$. The area of a circle of radius r is πr^2 so $256/81$ was the approximation used for π.
(b) $22/7 \approx 3.1429$ is better than $256/81 \approx 3.1605$.

7.

Line	2	3	4	5	6	7
Blocks	3, 4	1, 2	3, 4	2, 4, 5	1, 2	3, 4

9. **(a)** always correct (add -3 to both sides of $a \leq b$)
(b) not always correct (correct only if $a = b$)
(c) not always correct (correct only if $a = b$)
(d) always correct (multiply both sides of $a \leq b$ by 6)
(e) not always correct (correct only if $a \geq 0$ or $a = b$)
(f) always correct (multiply both sides of $a \leq b$ by the nonnegative quantity a^2)

11. **(a)** all values because $a = a$ is always valid **(b)** none

13. **(a)** yes, because $a \leq b$ means $a < b$ or $a = b$, thus $a < b$ certainly means $a \leq b$
(b) no, because $a < b$ is false if $a = b$ is true

15. **(a)** $\{x : x \text{ is a positive odd integer}\}$ **(b)** $\{x : x \text{ is an even integer}\}$
(c) $\{x : x \text{ is irrational}\}$ **(d)** $\{x : x \text{ is an integer and } 7 \leq x \leq 10\}$

17. **(a)** false, there are points inside the triangle that are not inside the circle
(b) true, all points inside the triangle are also inside the square
(c) true **(d)** false **(e)** true
(f) true, a is inside the circle **(g)** true

19. **(a)** 4 **(b)** −3
(c) −1, 7 **(d)** −3, 3
(e) −3, 3 **(f)** −3, 3

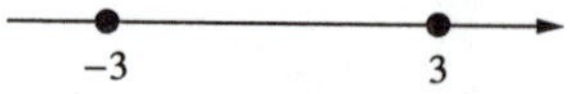

21. **(a)** $[-2, 2]$ **(b)** $(-\infty, -2) \cup (2, +\infty)$

23. $3x < 10$; $(-\infty, 10/3)$

25. $2x \le -11$; $(-\infty, -11/2]$

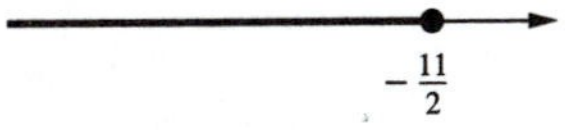

27. $2x \le 1$ and $2x > -3$; $(-3/2, 1/2]$

29. $\dfrac{x}{x-3} - 4 < 0,\ \dfrac{12-3x}{x-3} < 0,\ \dfrac{4-x}{x-3} < 0;$
$(-\infty, 3) \cup (4, +\infty)$

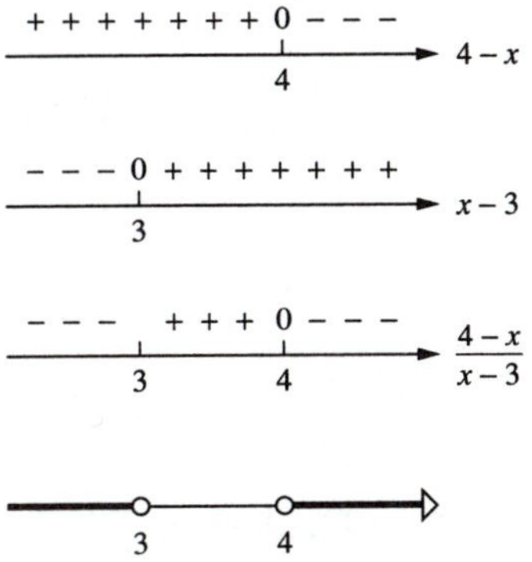

31. $\dfrac{3x+1}{x-2} - 1 = \dfrac{2x+3}{x-2} < 0,\ \dfrac{x+3/2}{x-2} < 0;$
$(-\frac{3}{2}, 2)$

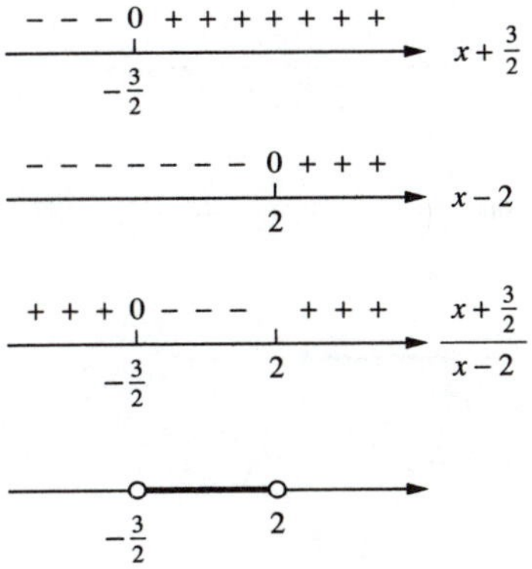

33. $\dfrac{4}{2-x} - 1 = \dfrac{x+2}{2-x} \le 0; (-\infty, -2] \cup (2, +\infty)$

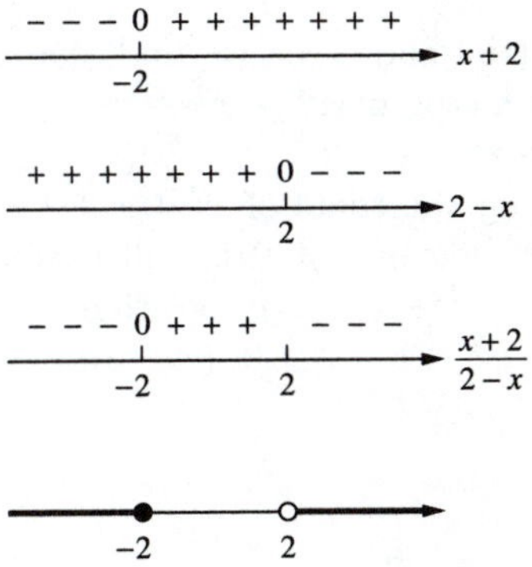

35. $x^2 - 9 = (x+3)(x-3) > 0$; $(-\infty, -3) \cup (3, +\infty)$

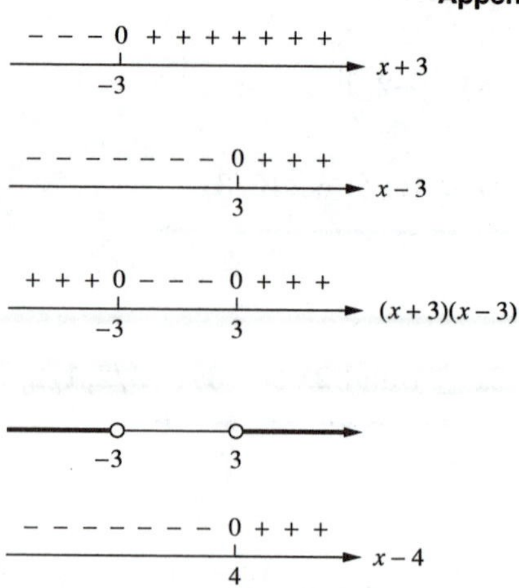

37. $(x-4)(x+2) > 0$; $(-\infty, -2) \cup (4, +\infty)$

39. $(x-4)(x-5) \le 0$; $[4, 5]$

41. $\dfrac{3}{x-4} - \dfrac{2}{x} = \dfrac{x+8}{x(x-4)} > 0$; $(-8, 0) \cup (4, +\infty)$

43. By trial-and-error we find that $x = 2$ is a root of the equation $x^3 - x^2 - x - 2 = 0$ so $x - 2$ is a factor of $x^3 - x^2 - x - 2$. By long division we find that $x^2 + x + 1$ is another factor so $x^3 - x^2 - x - 2 = (x-2)(x^2 + x + 1)$. The linear factors of $x^2 + x + 1$ can be determined by first finding the roots of $x^2 + x + 1 = 0$ by the quadratic formula. These roots are complex numbers so $x^2 + x + 1 \neq 0$ for all real x; thus $x^2 + x + 1$ must be always positive or always negative. Since $x^2 + x + 1$ is positive when $x = 0$, it follows that $x^2 + x + 1 > 0$ for all real x. Hence $x^3 - x^2 - x - 2 > 0$, $(x-2)(x^2 + x + 1) > 0$, $x - 2 > 0$, $x > 2$, so $S = (2, +\infty)$.

45. $\sqrt{x^2+x-6}$ is real if $x^2+x-6 \geq 0$. Factor to get $(x+3)(x-2) \geq 0$ which has as its solution $x \leq -3$ or $x \geq 2$.

47. $25 \leq \frac{5}{9}(F-32) \leq 40$, $45 \leq F-32 \leq 72$, $77 \leq F \leq 104$

49. **(a)** Assume m and n are rational, then $m = \frac{p}{q}$ and $n = \frac{r}{s}$ where p, q, r, and s are integers so $m+n = \frac{p}{q} + \frac{r}{s} = \frac{ps+rq}{qs}$ which is rational because $ps+rq$ and qs are integers.

(b) (proof by contradiction) Assume m is rational and n is irrational, then $m = \frac{p}{q}$ where p and q are integers. Suppose that $m+n$ is rational, then $m+n = \frac{r}{s}$ where r and s are integers so $n = \frac{r}{s} - m = \frac{r}{s} - \frac{p}{q} = \frac{rq-ps}{sq}$. But $rq-ps$ and sq are integers, so n is rational which contradicts the assumption that n is irrational.

51. $a = \sqrt{2}, b = \sqrt{3}, c = \sqrt{6}, d = -\sqrt{2}$ are irrational, and $a+d = 0$, a rational; $a+a = 2\sqrt{2}$, an irrational; $ad = -2$, a rational; and $ab = c$, an irrational.

53. The average of a and b is $\frac{1}{2}(a+b)$; if a and b are rational then so is the average, by Exercise 49(a) and Exercise 50(a). On the other hand if $a = b = \sqrt{2}$ then the average of a and b is irrational, but the average of a and $-b$ is rational.

55. $8x^3 - 4x^2 - 2x + 1$ can be factored by grouping terms: $(8x^3-4x^2)-(2x-1) = 4x^2(2x-1)-(2x-1) = (2x-1)(4x^2-1) = (2x-1)^2(2x+1)$. The problem, then, is to solve $(2x-1)^2(2x+1) < 0$. By inspection, $x = 1/2$ is not a solution. If $x \neq 1/2$, then $(2x-1)^2 > 0$ and it follows that $2x+1 < 0$, $2x < -1$, $x < -1/2$, so $S = (-\infty, -1/2)$.

57. If $a < b$, then $ac < bc$ because c is positive; if $c < d$, then $bc < bd$ because b is positive, so $ac < bd$ (Theorem A.1(a)). (Note that the result is still true if one of a, b, c, d is allowed to be negative, that is $a < 0$ or $c < 0$.)

APPENDIX B
Absolute Value

EXERCISE SET B

1. **(a)** 7 **(b)** $\sqrt{2}$ **(c)** k^2 **(d)** k^2

3. $|x-3| = |3-x| = 3-x$ if $3-x \geq 0$, which is true if $x \leq 3$

5. All real values of x because $x^2+9>0$.

7. $|3x^2+2x| = |x(3x+2)| = |x||3x+2|$. If $|x||3x+2| = x|3x+2|$, then $|x||3x+2| - x|3x+2| = 0$, $(|x|-x)|3x+2| = 0$, so either $|x|-x=0$ or $|3x+2|=0$. If $|x|-x=0$, then $|x|=x$, which is true for $x \geq 0$. If $|3x+2|=0$, then $x=-2/3$. The statement is true for $x \geq 0$ or $x=-2/3$.

9. $\sqrt{(x+5)^2} = |x+5| = x+5$ if $x+5 \geq 0$, which is true if $x \geq -5$.

13. **(a)** $|7-9| = |-2| = 2$
(b) $|3-2| = |1| = 1$
(c) $|6-(-8)| = |14| = 14$
(d) $|-3-\sqrt{2}| = |-(3+\sqrt{2})| = 3+\sqrt{2}$
(e) $|-4-(-11)| = |7| = 7$
(f) $|-5-0| = |-5| = 5$

15. **(a)** B is 6 units to the left of A; $b = a-6 = -3-6 = -9$.
(b) B is 9 units to the right of A; $b = a+9 = -2+9 = 7$.
(c) B is 7 units from A; either $b = a+7 = 5+7 = 12$ or $b = a-7 = 5-7 = -2$. Since it is given that $b > 0$, it follows that $b = 12$.

17. $|6x-2| = 7$

Case 1:
$$6x-2 = 7$$
$$6x = 9$$
$$x = 3/2$$

Case 2:
$$6x-2 = -7$$
$$6x = -5$$
$$x = -5/6$$

19. $|6x-7| = |3+2x|$

Case 1:
$$6x-7 = 3+2x$$
$$4x = 10$$
$$x = 5/2$$

Case 2:
$$6x-7 = -(3+2x)$$
$$8x = 4$$
$$x = 1/2$$

21. $|9x|-11 = x$

Case 1:
$$9x-11 = x$$
$$8x = 11$$
$$x = 11/8$$

Case 2:
$$-9x-11 = x$$
$$-10x = 11$$
$$x = -11/10$$

23. $\left|\dfrac{x+5}{2-x}\right| = 6$

Case 1:
$$\frac{x+5}{2-x} = 6$$
$$x+5 = 12-6x$$
$$7x = 7$$
$$x = 1$$

Case 2:
$$\frac{x+5}{2-x} = -6$$
$$x+5 = -12+6x$$
$$-5x = -17$$
$$x = 17/5$$

25.
$$|x+6| < 3$$
$$-3 < x+6 < 3$$
$$-9 < x < -3$$
$$S = (-9,-3)$$

27.
$$|2x-3| \leq 6$$
$$-6 \leq 2x-3 \leq 6$$
$$-3 \leq 2x \leq 9$$
$$-3/2 \leq x \leq 9/2$$
$$S = [-3/2, 9/2]$$

29. $|x+2| > 1$

Case 1: $x+2 > 1$, $x > -1$

Case 2: $x+2 < -1$, $x < -3$

$S = (-\infty, -3) \cup (-1, +\infty)$

31. $|5-2x| \geq 4$

Case 1: $5-2x \geq 4$, $-2x \geq -1$, $x \leq 1/2$

Case 2: $5-2x \leq -4$, $-2x \leq -9$, $x \geq 9/2$

$S = (-\infty, 1/2] \cup [9/2, +\infty)$

33. $\dfrac{1}{|x-1|} < 2, x \neq 1$

$|x-1| > 1/2$

Case 1: $x-1 > 1/2$, $x > 3/2$

Case 2: $x-1 < -1/2$, $x < 1/2$

$S = (-\infty, 1/2) \cup (3/2, +\infty)$

35.

$$\frac{3}{|2x-1|} \geq 4, x \neq 1/2$$

$$\frac{|2x-1|}{3} \leq \frac{1}{4}$$

$$|2x-1| \leq 3/4$$

$$-3/4 \leq 2x-1 \leq 3/4$$

$$1/4 \leq 2x \leq 7/4$$

$$1/8 \leq x \leq 7/8$$

$$S = [1/8, 1/2) \cup (1/2, 7/8]$$

37. $\sqrt{(x^2-5x+6)^2} = x^2-5x+6$ if $x^2-5x+6 \geq 0$ or, equivalently, if $(x-2)(x-3) \geq 0$; $x \in (-\infty, 2] \cup [3, +\infty)$.

39. If $u = |x-3|$ then $u^2 - 4u = 12$, $u^2 - 4u - 12 = 0$, $(u-6)(u+2) = 0$, so $u = 6$ or $u = -2$. If $u = 6$ then $|x-3| = 6$, so $x = 9$ or $x = -3$. If $u = -2$ then $|x-3| = -2$ which is impossible. The solutions are -3 and 9.

41.

$$\begin{aligned} |a-b| &= |a+(-b)| \\ &\leq |a| + |-b| \text{ (triangle inequality)} \\ &= |a| + |b|. \end{aligned}$$

43. From Exercise 42
(i) $|a| - |b| \leq |a-b|$; but $|b| - |a| \leq |b-a| = |a-b|$, so (ii) $|a| - |b| \geq -|a-b|$.
Combining (i) and (ii): $-|a-b| \leq |a| - |b| \leq |a-b|$, so $||a| - |b|| \leq |a-b|$.

APPENDIX C
Coordinate Planes and Lines

EXERCISE SET C

1.

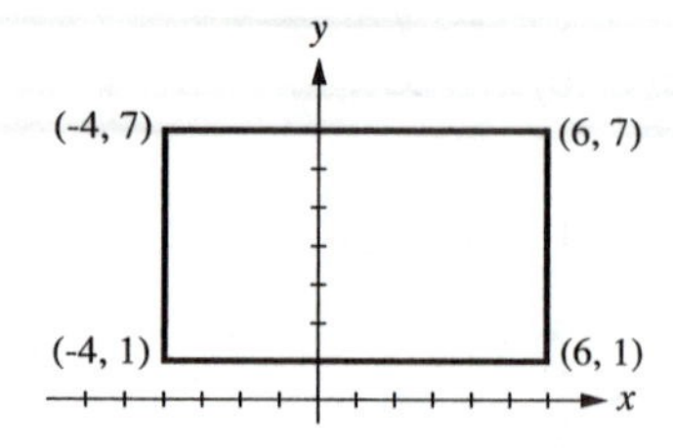

3. **(a)** $x = 2$

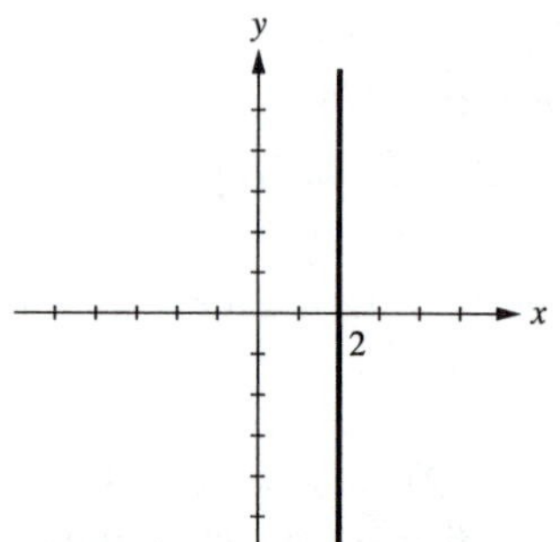
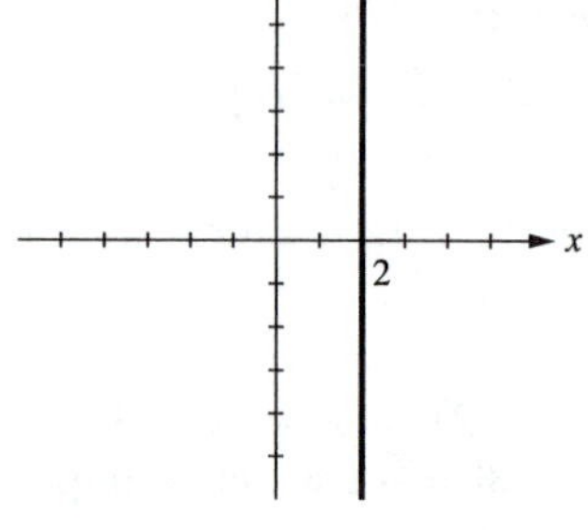

(b) $y = -3$

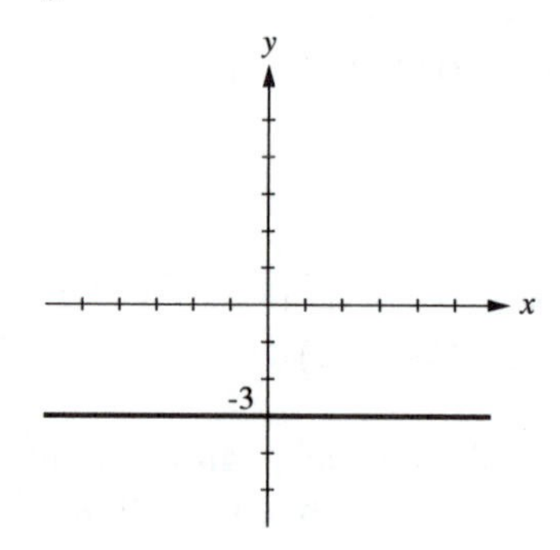

(c) $x \geq 0$

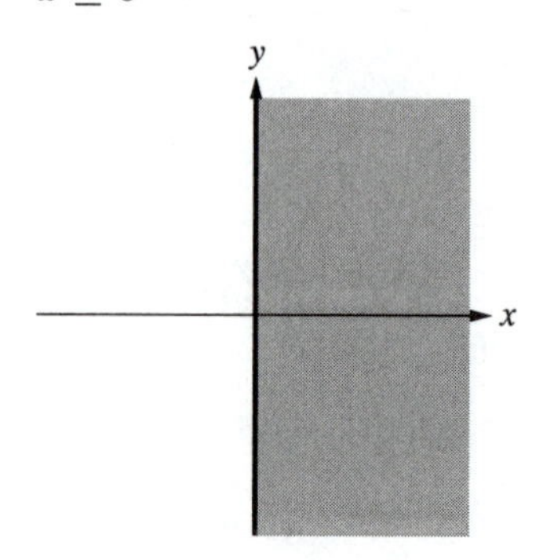

(d) $y = x$

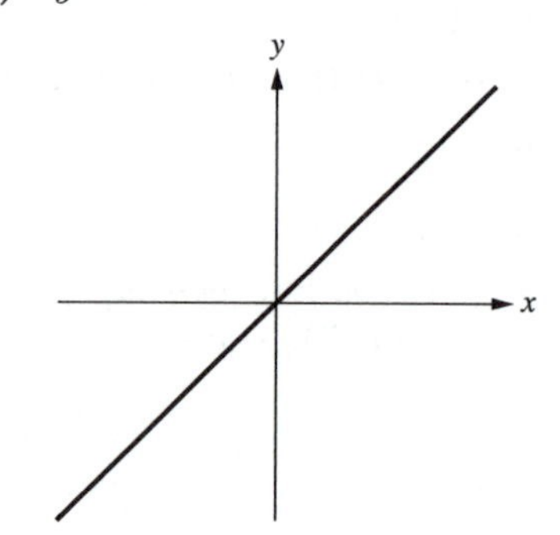

(e) $y \geq x$

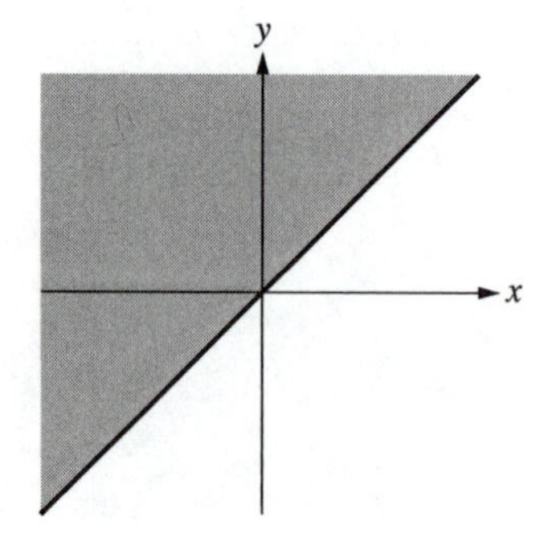

(f) $|x| \geq 1$

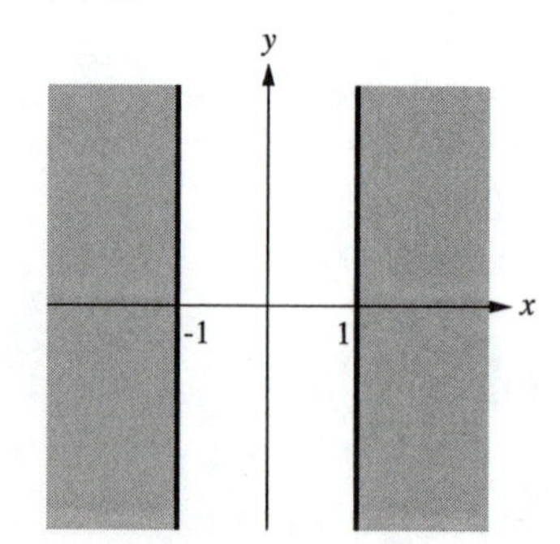

5. $y = 4 - x^2$

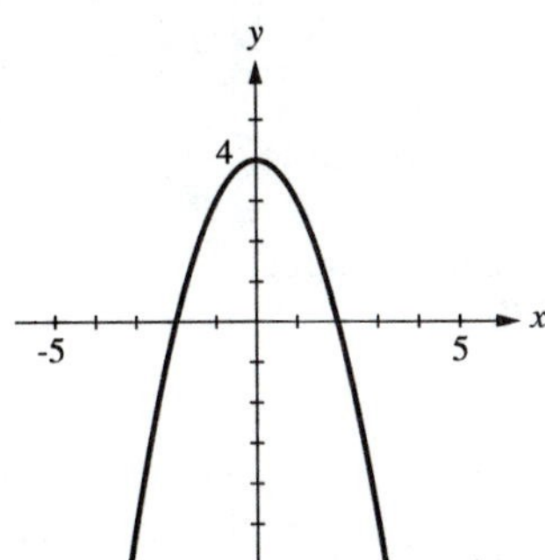

7. $y = \sqrt{x-4}$

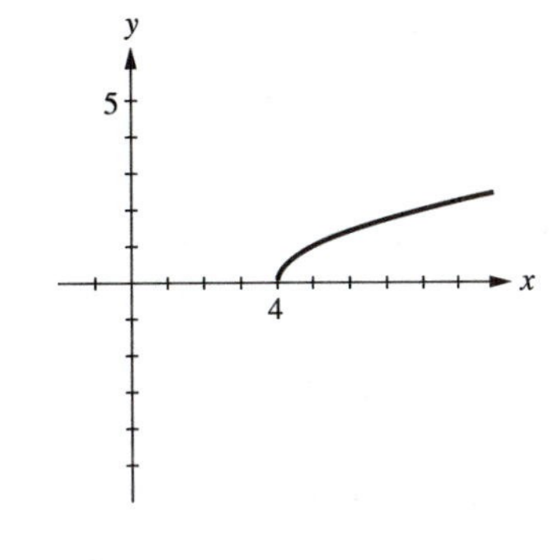

9. $x^2 - x + y = 0$

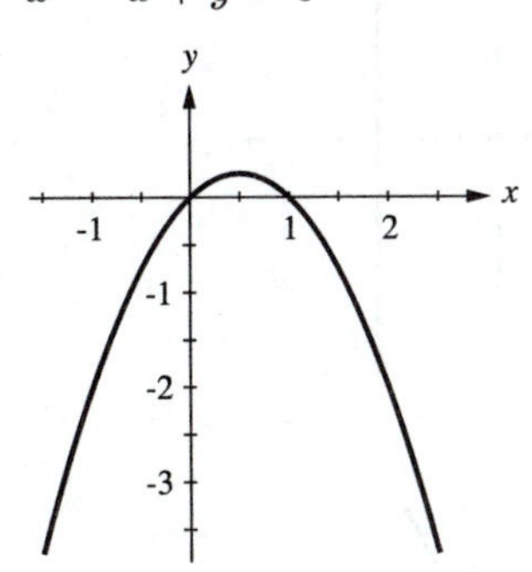

11. $x^2 y = 2$

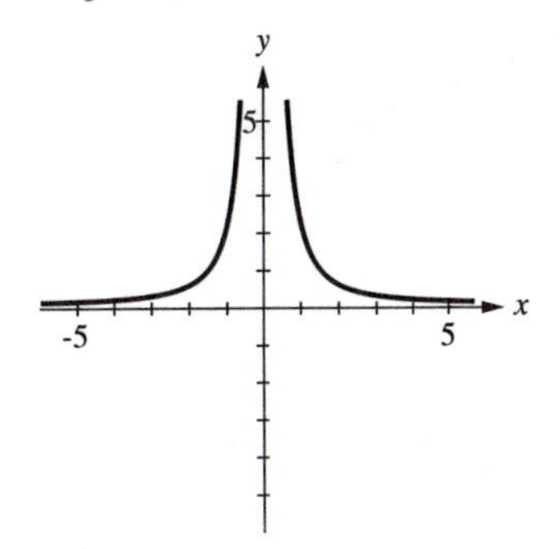

13. **(a)** $m = \dfrac{4-2}{3-(-1)} = \dfrac{1}{2}$ **(b)** $m = \dfrac{1-3}{7-5} = -1$

(c) $m = \dfrac{\sqrt{2}-\sqrt{2}}{-3-4} = 0$ **(d)** $m = \dfrac{12-(-6)}{-2-(-2)} = \dfrac{18}{0}$, not defined

15. **(a)** The line through $(1,1)$ and $(-2,-5)$ has slope $m_1 = \dfrac{-5-1}{-2-1} = 2$, the line through $(1,1)$ and $(0,-1)$ has slope $m_2 = \dfrac{-1-1}{0-1} = 2$. The given points lie on a line because $m_1 = m_2$.

(b) The line through $(-2,4)$ and $(0,2)$ has slope $m_1 = \dfrac{2-4}{0+2} = -1$, the line through $(-2,4)$ and $(1,5)$ has slope $m_2 = \dfrac{5-4}{1+2} = \dfrac{1}{3}$. The given points do not lie on a line because $m_1 \neq m_2$.

17.

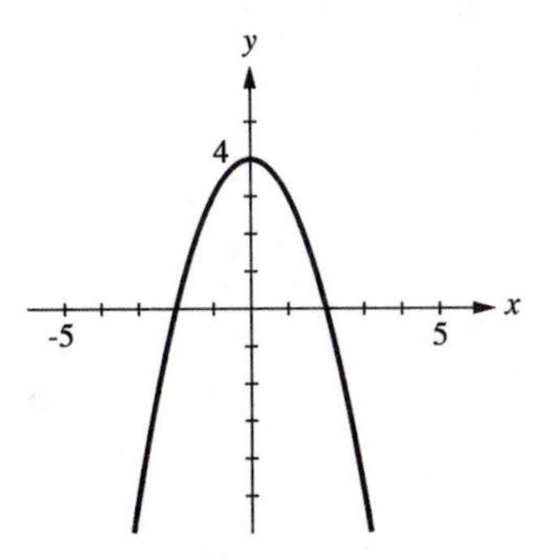

19. III < II < IV < I

21. Use the points $(1,2)$ and (x,y) to calculate the slope: $(y-2)/(x-1) = 3$

(a) if $x = 5$, then $(y-2)/(5-1) = 3$, $y - 2 = 12$, $y = 14$

(b) if $y = -2$, then $(-2-2)/(x-1) = 3$, $x - 1 = -4/3$, $x = -1/3$

23. Using $(3, k)$ and $(-2, 4)$ to calculate the slope, we find $\dfrac{k-4}{3-(-2)} = 5$, $k - 4 = 25$, $k = 29$.

25. $\dfrac{0-2}{x-1} = -\dfrac{0-5}{x-4}$, $-2x + 8 = 5x - 5$, $7x = 13$, $x = 13/7$

27. Show that opposite sides are parallel by showing that they have the same slope: using $(3, -1)$ and $(6, 4)$, $m_1 = 5/3$; using $(6, 4)$ and $(-3, 2)$, $m_2 = 2/9$; using $(-3, 2)$ and $(-6, -3)$, $m_3 = 5/3$; using $(-6, -3)$ and $(3, -1)$, $m_4 = 2/9$. Opposite sides are parallel because $m_1 = m_3$ and $m_2 = m_4$.

29. **(a)**

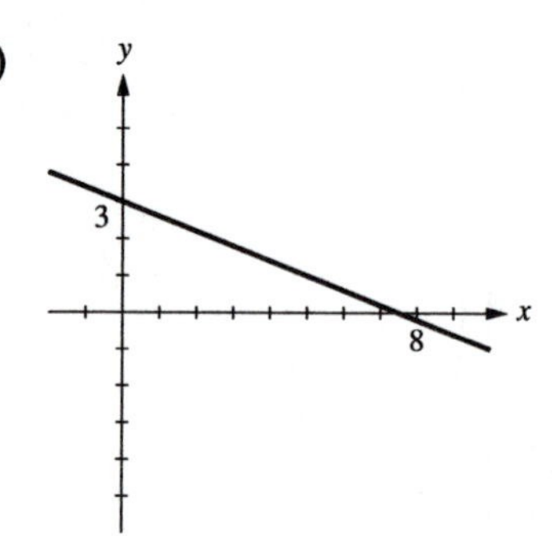

(b)

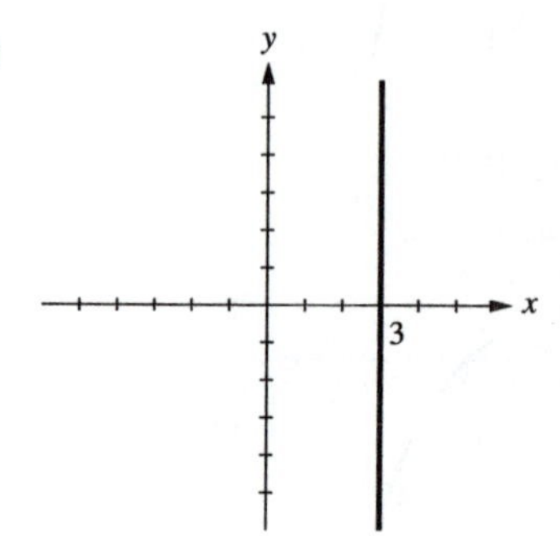

(c)

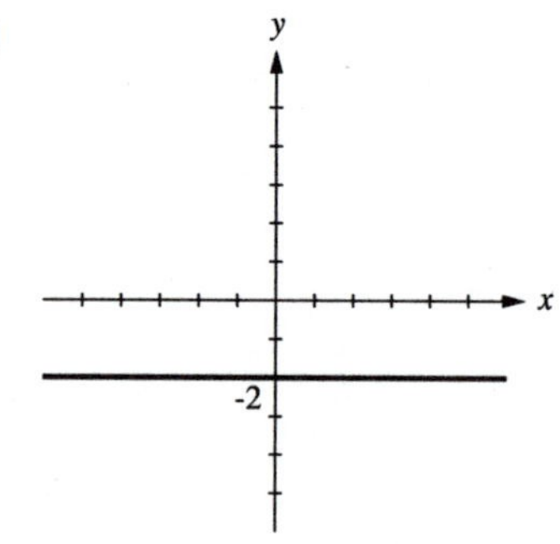

(d)

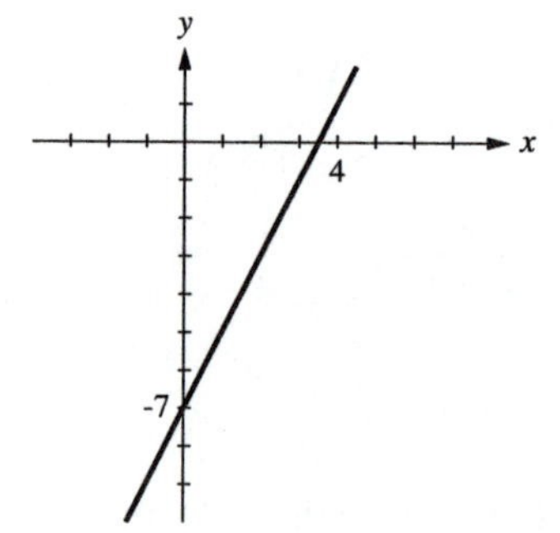

31. **(a)**

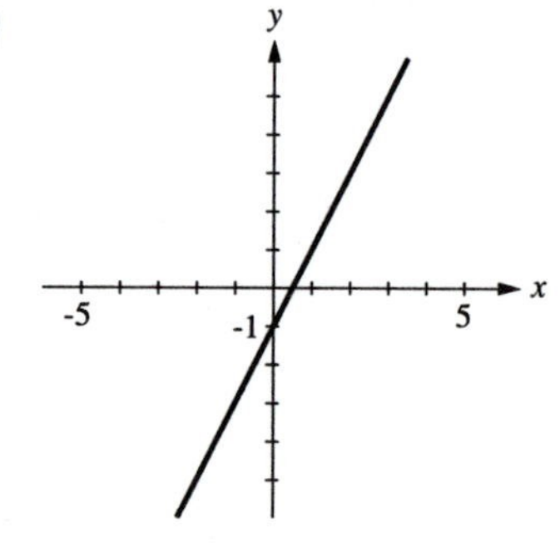

(b)

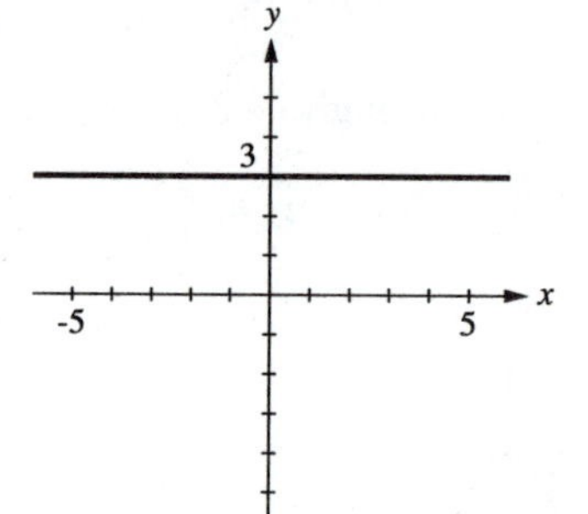

(c)

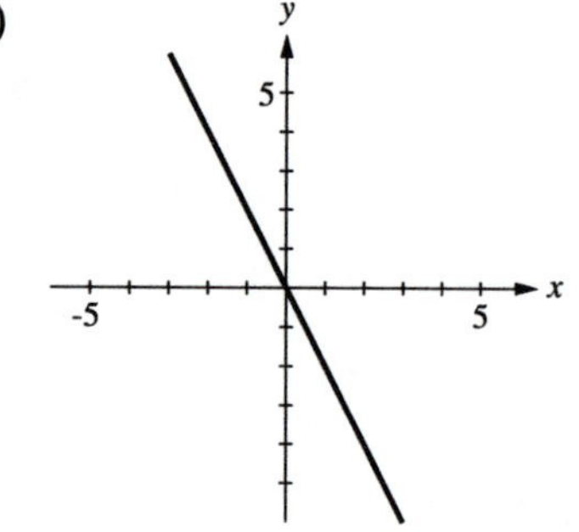

33. **(a)** $m = 3$, $b = 2$ **(b)** $m = -\frac{1}{4}$, $b = 3$

(c) $y = -\frac{3}{5}x + \frac{8}{5}$ so $m = -\frac{3}{5}$, $b = \frac{8}{5}$ **(d)** $m = 0$, $b = 1$

(e) $y = -\frac{b}{a}x + b$ so $m = -\frac{b}{a}$, y-intercept b

35. **(a)** $m = (0 - (-3))/(2 - 0)) = 3/2$ so $y = 3x/2 - 3$
(b) $m = (-3 - 0)/(4 - 0) = -3/4$ so $y = -3x/4$

37. $y = -2x + 4$

39. The slope m of the line must equal the slope of $y = 4x - 2$, thus $m = 4$ so the equation is $y = 4x + 7$.

41. The slope m of the line must be the negative reciprocal of the slope of $y = 5x + 9$, thus $m = -1/5$ and the equation is $y = -x/5 + 6$.

43. $y - 4 = \frac{-7-4}{1-2}(x - 2) = 11(x - 2)$, $y = 11x - 18$.

45. The line passes through $(0, 2)$ and $(-4, 0)$, thus $m = \frac{0-2}{-4-0} = \frac{1}{2}$ so $y = \frac{1}{2}x + 2$.

47. $y = 1$

49. **(a)** $m_1 = 4$, $m_2 = 4$; parallel because $m_1 = m_2$
(b) $m_1 = 2$, $m_2 = -1/2$; perpendicular because $m_1 m_2 = -1$
(c) $m_1 = 5/3$, $m_2 = 5/3$; parallel because $m_1 = m_2$
(d) If $A \neq 0$ and $B \neq 0$, then $m_1 = -A/B$, $m_2 = B/A$ and the lines are perpendicular because $m_1 m_2 = -1$. If either A or B (but not both) is zero, then the lines are perpendicular because one is horizontal and the other is vertical.
(e) $m_1 = 4$, $m_2 = 1/4$; neither

51. $y = (-3/k)x + 4/k$, $k \neq 0$
(a) $-3/k = 2$, $k = -3/2$
(b) $4/k = 5$, $k = 4/5$
(c) $3(-2) + k(4) = 4$, $k = 5/2$
(d) The slope of $2x - 5y = 1$ is $2/5$ so $-3/k = 2/5$, $k = -15/2$.
(e) The slope of $4x + 3y = 2$ is $-4/3$ so the slope of a line perpendicular to it is $3/4$; $-3/k = 3/4$, $k = -4$.

53. $(x - y)(x + y) = 0$: the union of the graphs of $x - y = 0$ and $x + y = 0$

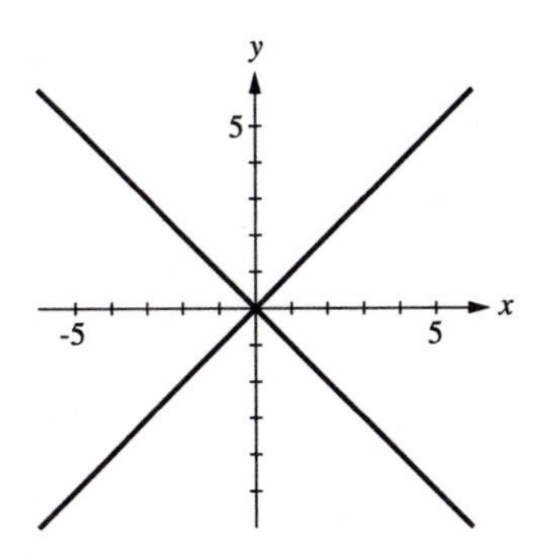

55. $u = 3v^2$

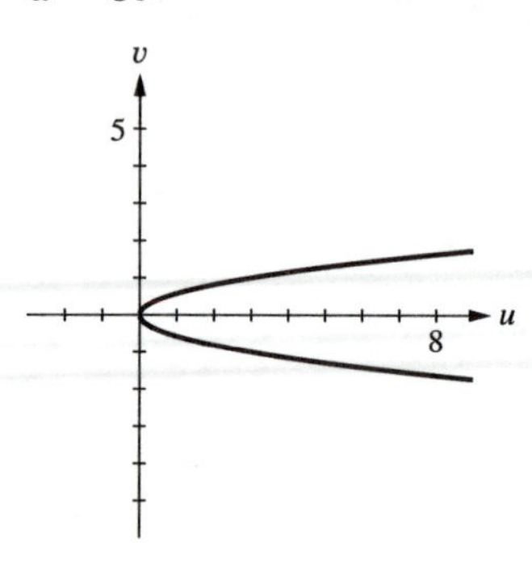

57. Solve $x = 5t + 2$ for t to get $t = \frac{1}{5}x - \frac{2}{5}$, so $y = \left(\frac{1}{5}x - \frac{2}{5}\right) - 3 = \frac{1}{5}x - \frac{17}{5}$, which is a line.

59. An equation of the line through $(1, 4)$ and $(2, 1)$ is $y = -3x + 7$. It crosses the y-axis at $y = 7$, and the x-axis at $x = 7/3$, so the area of the triangle is $\frac{1}{2}(7)(7/3) = 49/6$.

61. **(a)** yes **(b)** yes **(c)** no **(d)** yes
(e) yes **(f)** yes **(g)** no

APPENDIX D

Distances, Circles, and Quadratic Equations

EXERCISE SET D

1. in the proof of Theorem D.1

3. **(a)** $d = \sqrt{(1-7)^2 + (9-1)^2} = \sqrt{36+64} = \sqrt{100} = 10$

(b) $\left(\dfrac{7+1}{2}, \dfrac{1+9}{2}\right) = (4,5)$

5. **(a)** $d = \sqrt{[-7-(-2)]^2 + [-4-(-6)]^2} = \sqrt{25+4} = \sqrt{29}$

(b) $\left(\dfrac{-2+(-7)}{2}, \dfrac{-6+(-4)}{2}\right) = (-9/2, -5)$

7. Let $A(5,-2)$, $B(6,5)$, and $C(2,2)$ be the given vertices and a, b, and c the lengths of the sides opposite these vertices; then

$a = \sqrt{(2-6)^2 + (2-5)^2} = \sqrt{25} = 5$ and $b = \sqrt{(2-5)^2 + (2+2)^2} = \sqrt{25} = 5$.

Triangle ABC is isosceles because it has two equal sides ($a = b$).

9. $P_1(0,-2)$, $P_2(-4,8)$, and $P_3(3,1)$ all lie on a circle whose center is $C(-2,3)$ if the points P_1, P_2 and P_3 are equidistant from C. Denoting the distances between P_1, P_2, P_3 and C by d_1, d_2 and d_3 we find that $d_1 = \sqrt{(0+2)^2 + (-2-3)^2} = \sqrt{29}$, $d_2 = \sqrt{(-4+2)^2 + (8-3)^2} = \sqrt{29}$, and $d_3 = \sqrt{(3+2)^2 + (1-3)^2} = \sqrt{29}$, so P_1, P_2 and P_3 lie on a circle whose center is $C(-2,3)$ because $d_1 = d_2 = d_3$.

11. If $(2,k)$ is equidistant from (3,7) and (9,1), then

$\sqrt{(2-3)^2 + (k-7)^2} = \sqrt{(2-9)^2 + (k-1)^2}$, $1 + (k-7)^2 = 49 + (k-1)^2$,

$1 + k^2 - 14k + 49 = 49 + k^2 - 2k + 1$, $-12k = 0$, $k = 0$.

13. The slope of the line segment joining $(2,8)$ and $(-4,6)$ is $\dfrac{6-8}{-4-2} = \dfrac{1}{3}$ so the slope of the perpendicular bisector is -3. The midpoint of the line segment is $(-1,7)$ so an equation of the bisector is $y - 7 = -3(x+1)$; $y = -3x + 4$.

15. Method (see figure): Find an equation of the perpendicular bisector of the line segment joining $A(3,3)$ and $B(7,-3)$. All points on this perpendicular bisector are equidistant from A and B, thus find where it intersects the given line.

The midpoint of AB is $(5,0)$, the slope of AB is $-3/2$ thus the slope of the perpendicular bisector is $2/3$ so an equation is

$$y - 0 = \frac{2}{3}(x-5)$$
$$3y = 2x - 10$$
$$2x - 3y - 10 = 0.$$

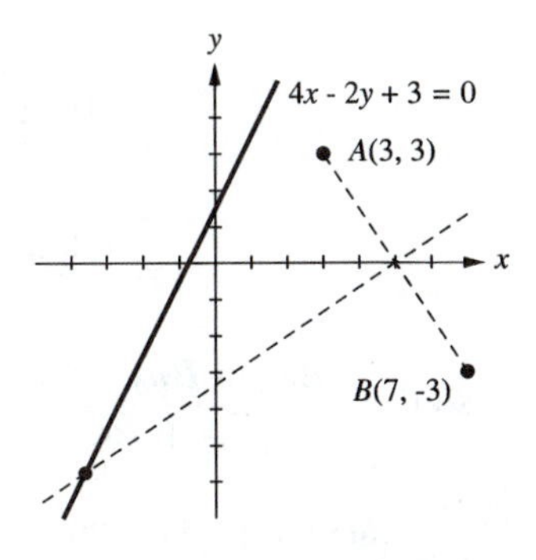

The solution of the system

$$\begin{cases} 4x - 2y + 3 = 0 \\ 2x - 3y - 10 = 0 \end{cases}$$

gives the point $(-29/8, -23/4)$.

17. Method (see figure): write an equation of the line that goes through the given point and that is perpendicular to the given line; find the point P where this line intersects the given line; find the distance between P and the given point.

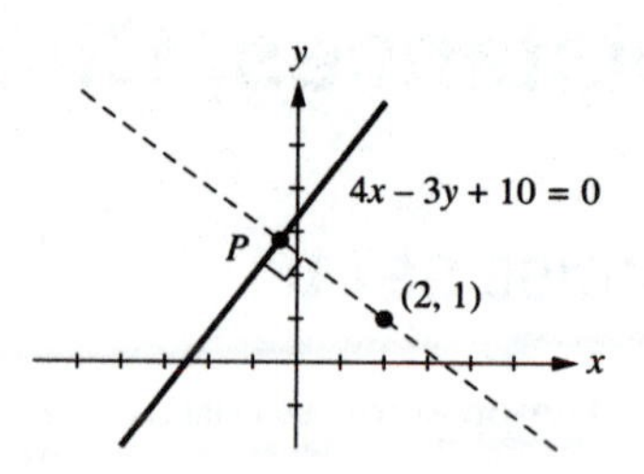

The slope of the given line is $4/3$, so the slope of a line perpendicular to it is $-3/4$.

The line through $(2,1)$ having a slope of $-3/4$ is $y-1=-\dfrac{3}{4}(x-2)$ or, after simplification, $3x+4y=10$ which when solved simultaneously with $4x-3y+10=0$ yields $(-2/5,14/5)$ as the point of intersection. The distance d between $(-2/5,14/5)$ and $(2,1)$ is $d=\sqrt{(2+2/5)^2+(1-14/5)^2}=3$.

19. If $B=0$, then the line $Ax+C=0$ is vertical and $x=-C/A$ for each point on the line. The line through (x_0,y_0) and perpendicular to the given line is horizontal and intersects the given line at the point $(-C/A,\ y_0)$. The distance d between $(-C/A,\ y_0)$ and (x_0,y_0) is

$$d=\sqrt{(x_0+C/A)^2+(y_0-y_0)^2}=\sqrt{\frac{(Ax_0+C)^2}{A^2}}=\frac{|Ax_0+C|}{\sqrt{A^2}}$$

which is the value of $\dfrac{|Ax_0+By_0+C|}{\sqrt{A^2+B^2}}$ for $B=0$.

If $B\neq 0$, then the slope of the given line is $-A/B$ and the line through (x_0,y_0) and perpendicular to the given line is

$y-y_0=\dfrac{B}{A}(x-x_0)$, $Ay-Ay_0=Bx-Bx_0$, $Bx-Ay=Bx_0-Ay_0$.

The point of intersection of this line and the given line is obtained by solving

$Ax+By=-C$ and $Bx-Ay=Bx_0-Ay_0$.

Multiply the first equation through by A and the second by B and add the results to get

$(A^2+B^2)x=B^2x_0-ABy_0-AC$ so $x=\dfrac{B^2x_0-ABy_0-AC}{A^2+B^2}$

Similarly, by multiplying by B and $-A$, we get $y=\dfrac{-ABx_0+A^2y_0-BC}{A^2+B^2}$.

The square of the distance d between (x,y) and (x_0,y_0) is

$$\begin{aligned}
d^2&=\left[x_0-\frac{B^2x_0-ABy_0-AC}{A^2+B^2}\right]^2+\left[y_0-\frac{-ABx_0+A^2y_0-BC}{A^2+B^2}\right]^2\\
&=\frac{(A^2x_0+ABy_0+AC)^2}{(A^2+B^2)^2}+\frac{(ABx_0+B^2y_0+BC)^2}{(A^2+B^2)^2}\\
&=\frac{A^2(Ax_0+By_0+C)^2+B^2(Ax_0+By_0+C)^2}{(A^2+B^2)^2}\\
&=\frac{(Ax_0+By_0+C)^2(A^2+B^2)}{(A^2+B^2)^2}=\frac{(Ax_0+By_0+C)^2}{A^2+B^2}
\end{aligned}$$

so $d=\dfrac{|Ax_0+By_0+C|}{\sqrt{A^2+B^2}}$.

21. $d=\dfrac{|5(8)+12(4)-36|}{\sqrt{5^2+12^2}}=\dfrac{|52|}{\sqrt{169}}=\dfrac{52}{13}=4.$

23. **(a)** center (0,0), radius 5 **(b)** center (1,4), radius 4
(c) center $(-1, -3)$, radius $\sqrt{5}$ **(d)** center $(0, -2)$, radius 1

25. $(x-3)^2 + (y-(-2))^2 = 4^2$, $(x-3)^2 + (y+2)^2 = 16$

27. $r = 8$ because the circle is tangent to the x-axis, so $(x+4)^2 + (y-8)^2 = 64$.

29. $(0,0)$ is on the circle, so $r = \sqrt{(-3-0)^2 + (-4-0)^2} = 5$; $(x+3)^2 + (y+4)^2 = 25$.

31. The center is the midpoint of the line segment joining $(2,0)$ and $(0,2)$ so the center is at $(1,1)$. The radius is $r = \sqrt{(2-1)^2 + (0-1)^2} = \sqrt{2}$, so $(x-1)^2 + (y-1)^2 = 2$.

33. $(x^2 - 2x) + (y^2 - 4y) = 11$, $(x^2 - 2x + 1) + (y^2 - 4y + 4) = 11 + 1 + 4$, $(x-1)^2 + (y-2)^2 = 16$; center (1,2) and radius 4

35. $2(x^2 + 2x) + 2(y^2 - 2y) = 0$, $2(x^2 + 2x + 1) + 2(y^2 - 2y + 1) = 2 + 2$, $(x+1)^2 + (y-1)^2 = 2$; center $(-1, 1)$ and radius $\sqrt{2}$

37. $(x^2 + 2x) + (y^2 + 2y) = -2$, $(x^2 + 2x + 1) + (y^2 + 2y + 1) = -2 + 1 + 1$, $(x+1)^2 + (y+1)^2 = 0$; the point $(-1, -1)$

39. $x^2 + y^2 = 1/9$; center $(0,0)$ and radius $1/3$

41. $x^2 + (y^2 + 10y) = -26$, $x^2 + (y^2 + 10y + 25) = -26 + 25$, $x^2 + (y+5)^2 = -1$; no graph

43. $16\left(x^2 + \frac{5}{2}x\right) + 16(y^2 + y) = 7$, $16\left(x^2 + \frac{5}{2}x + \frac{25}{16}\right) + 16\left(y^2 + y + \frac{1}{4}\right) = 7 + 25 + 4$,
$(x + 5/4)^2 + (y + 1/2)^2 = 9/4$; center $(-5/4, -1/2)$ and radius $3/2$

45. **(a)** $y^2 = 16 - x^2$, so $y = \pm\sqrt{16 - x^2}$. The bottom half is $y = -\sqrt{16 - x^2}$.
(b) Complete the square in y to get $(y-2)^2 = 3 - 2x - x^2$, so $y - 2 = \pm\sqrt{3 - 2x - x^2}$, or $y = 2 \pm \sqrt{3 - 2x - x^2}$. The top half is $y = 2 + \sqrt{3 - 2x - x^2}$.

47. **(a)**

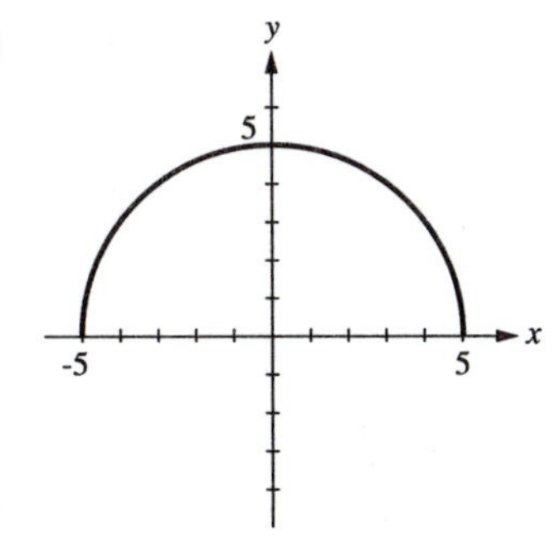

(b)
$$\begin{aligned} y &= \sqrt{5 + 4x - x^2} \\ &= \sqrt{5 - (x^2 - 4x)} \\ &= \sqrt{5 + 4 - (x^2 - 4x + 4)} \\ &= \sqrt{9 - (x-2)^2} \end{aligned}$$

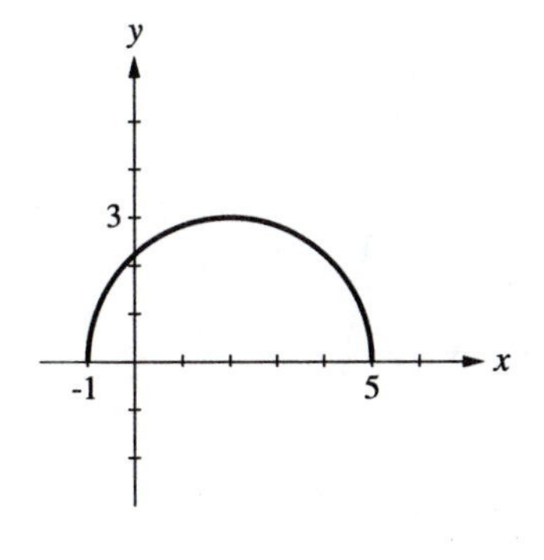

49. The tangent line is perpendicular to the radius at the point. The slope of the radius is 4/3, so the slope of the perpendicular is $-3/4$. An equation of the tangent line is $y - 4 = -\frac{3}{4}(x-3)$, or $y = -\frac{3}{4}x + \frac{25}{4}$.

51. **(a)** The center of the circle is at (0,0) and its radius is $\sqrt{20} = 2\sqrt{5}$. The distance between P and the center is $\sqrt{(-1)^2 + (2)^2} = \sqrt{5}$ which is less than $2\sqrt{5}$, so P is inside the circle.

(b) Draw the diameter of the circle that passes through P, then the shorter segment of the diameter is the shortest line that can be drawn from P to the circle, and the longer segment is the longest line that can be drawn from P to the circle (can you prove it?). Thus, the smallest distance is $2\sqrt{5} - \sqrt{5} = \sqrt{5}$, and the largest is $2\sqrt{5} + \sqrt{5} = 3\sqrt{5}$.

53. Let (a, b) be the coordinates of T (or T'). The radius from $(0, 0)$ to T (or T') will be perpendicular to L (or L') so, using slopes, $b/a = -(a-3)/b$, $a^2+b^2 = 3a$. But (a, b) is on the circle so $a^2+b^2 = 1$, thus $3a = 1$, $a = 1/3$. Let $a = 1/3$ in $a^2 + b^2 = 1$ to get $b^2 = 8/9$, $b = \pm\sqrt{8}/3$. The coordinates of T and T' are $(1/3, \sqrt{8}/3)$ and $(1/3, -\sqrt{8}/3)$.

55. **(a)** $[(x-4)^2 + (y-1)^2] + [(x-2)^2 + (y+5)^2] = 45$
$x^2 - 8x + 16 + y^2 - 2y + 1 + x^2 - 4x + 4 + y^2 + 10y + 25 = 45$
$2x^2 + 2y^2 - 12x + 8y + 1 = 0$, which is a circle.

(b) $2(x^2 - 6x) + 2(y^2 + 4y) = -1$, $2(x^2 - 6x + 9) + 2(y^2 + 4y + 4) = -1 + 18 + 8$,
$(x-3)^2 + (y+2)^2 = 25/2$; center $(3, -2)$, radius $5/\sqrt{2}$.

57. $y = x^2 + 2$

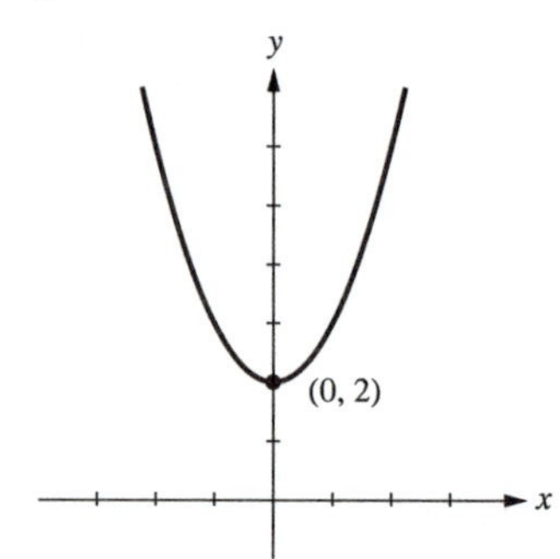

59. $y = x^2 + 2x - 3$

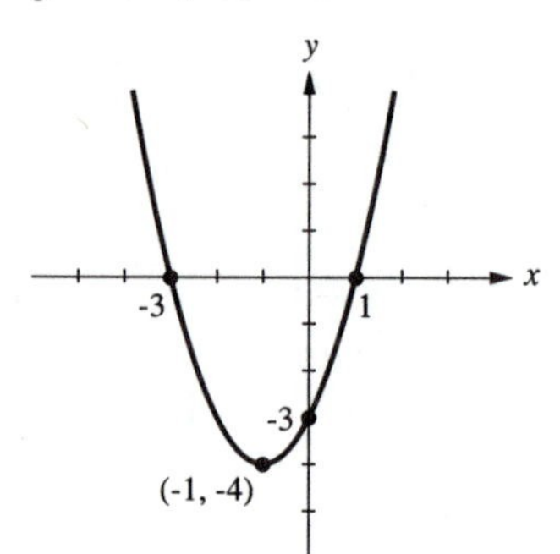

61. $y = -x^2 + 4x + 5$

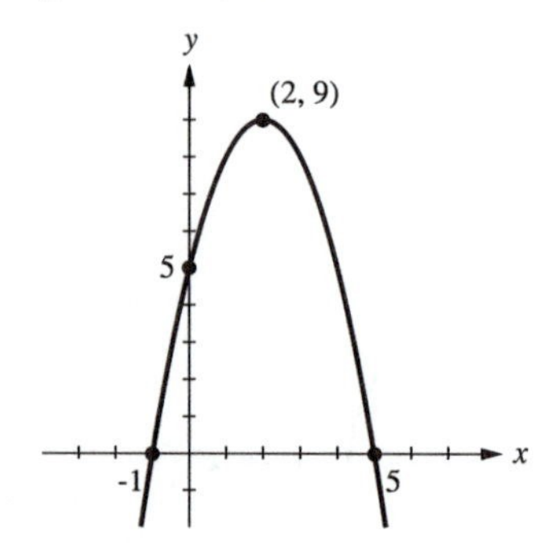

63. $y = (x-2)^2$

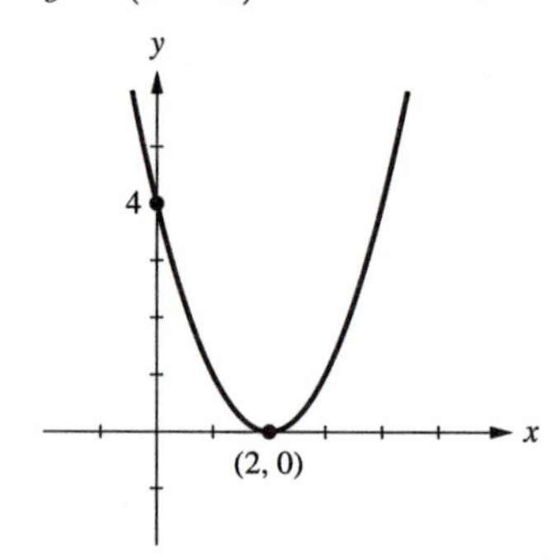

65. $x^2 - 2x + y = 0$

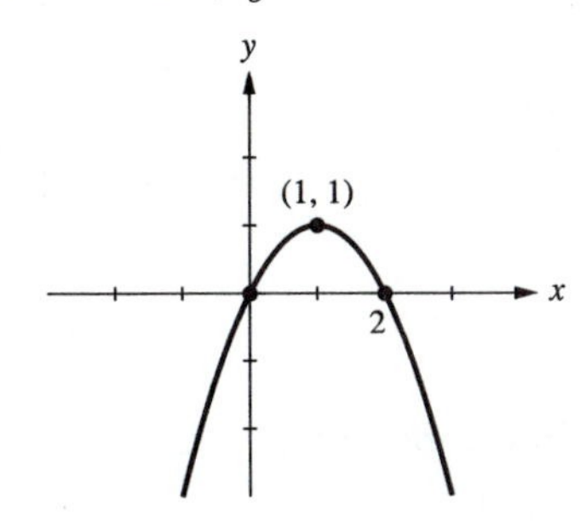

67. $y = 3x^2 - 2x + 1$

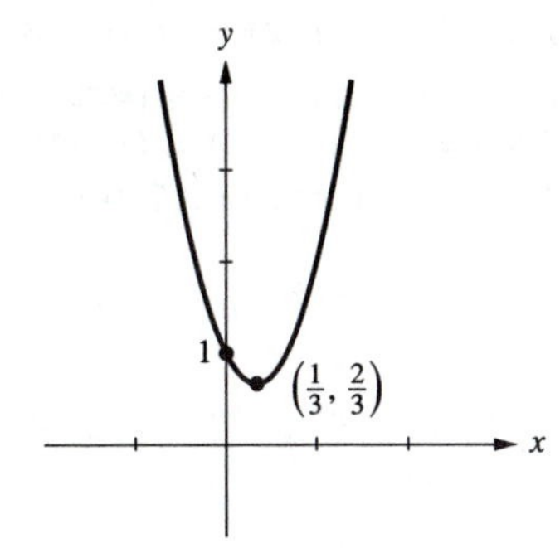

69. $x = -y^2 + 2y + 2$

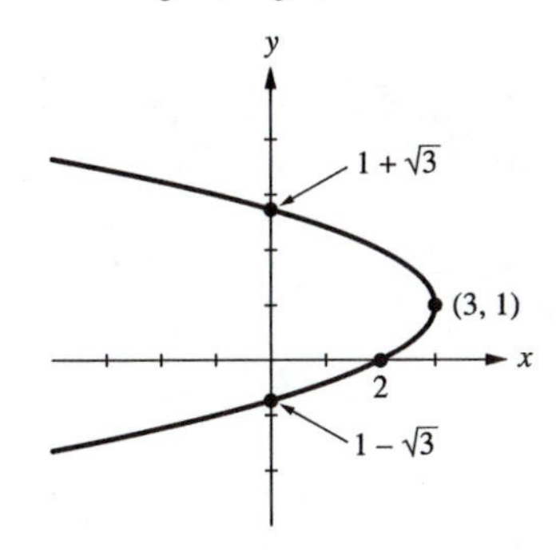

71. **(a)** $x^2 = 3 - y$, $x = \pm\sqrt{3-y}$. The right half is $x = \sqrt{3-y}$.
(b) Complete the square in x to get $(x-1)^2 = y+1$, $x = 1 \pm \sqrt{y+1}$. The left half is $x = 1 - \sqrt{y+1}$.

73. **(a)**

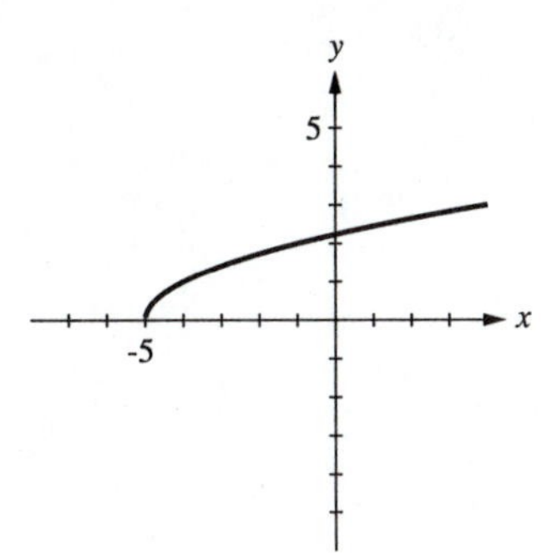

(b)

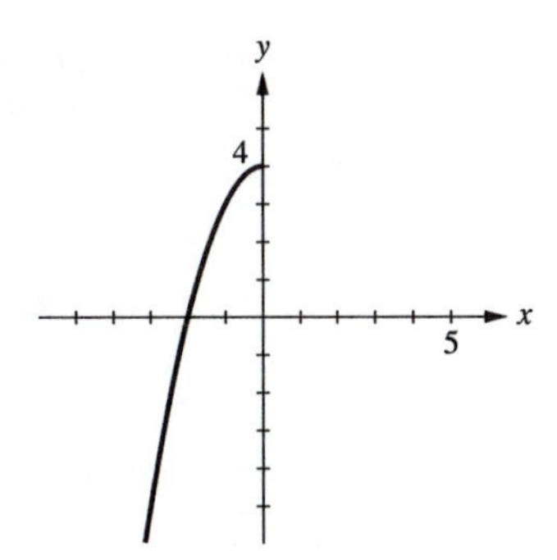

75. **(a)**

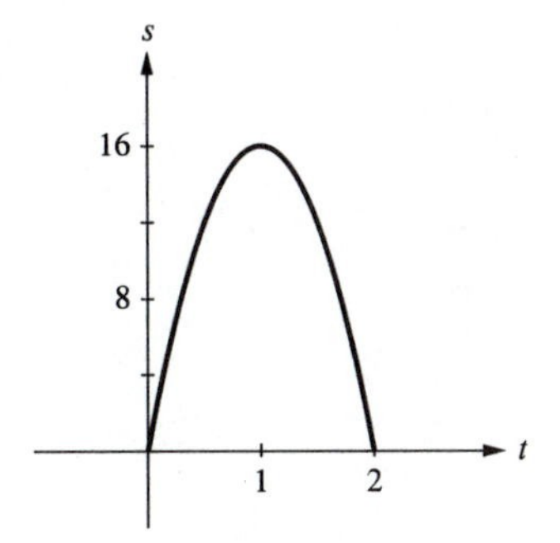

(b) The ball will be at its highest point when $t = 1$ sec; it will rise 16 ft.

77. **(a)** $(3)(2x) + (2)(2y) = 600$, $6x + 4y = 600$, $y = 150 - 3x/2$

(b) $A = xy = x(150 - 3x/2) = 150x - 3x^2/2$

(c) The graph of A versus x is a parabola with its vertex (high point) at $x = -b/(2a) = -150/(-3) = 50$, so the maximum value of A is $A = 150(50) - 3(50)^2/2 = 3{,}750$ ft^2.

79. **(a)** The parabola $y = 2x^2 + 5x - 1$ opens upward and has x-intercepts of $x = (-5 \pm \sqrt{33})/4$, so $2x^2 + 5x - 1 < 0$ if $(-5 - \sqrt{33})/4 < x < (-5 + \sqrt{33})/4$.

(b) The parabola $y = x^2 - 2x + 3$ opens upward and has no x-intercepts, so $x^2 - 2x + 3 > 0$ if $-\infty < x < +\infty$.

81. **(a)** The t-coordinate of the vertex is $t = -40/[(2)(-16)] = 5/4$, so the maximum height is $s = 5 + 40(5/4) - 16(5/4)^2 = 30$ ft.

(b) $s = 5 + 40t - 16t^2 = 0$ if $t \approx 2.6$ s

(c) $s = 5 + 40t - 16t^2 > 12$ if $16t^2 - 40t + 7 < 0$, which is true if $(5 - 3\sqrt{2})/4 < t < (5 + 3\sqrt{2})/4$. The length of this interval is $(5 + 3\sqrt{2})/4 - (5 - 3\sqrt{2})/4 = 3\sqrt{2}/2 \approx 2.1$ s.

APPENDIX E
Trigonometry Review

EXERCISE SET E

1. (a) $5\pi/12$ (b) $13\pi/6$ (c) $\pi/9$ (d) $23\pi/30$

3. (a) $12°$ (b) $(270/\pi)°$ (c) $288°$ (d) $540°$

5.

	$\sin\theta$	$\cos\theta$	$\tan\theta$	$\csc\theta$	$\sec\theta$	$\cot\theta$
(a)	$\sqrt{21}/5$	$2/5$	$\sqrt{21}/2$	$5/\sqrt{21}$	$5/2$	$2/\sqrt{21}$
(b)	$3/4$	$\sqrt{7}/4$	$3/\sqrt{7}$	$4/3$	$4/\sqrt{7}$	$\sqrt{7}/3$
(c)	$3/\sqrt{10}$	$1/\sqrt{10}$	3	$\sqrt{10}/3$	$\sqrt{10}$	$1/3$

7. $\sin\theta = 3/\sqrt{10}$, $\cos\theta = 1/\sqrt{10}$

9. $\tan\theta = \sqrt{21}/2$, $\csc\theta = 5/\sqrt{21}$

11. Let x be the length of the side adjacent to θ, then $\cos\theta = x/6 = 0.3$, $x = 1.8$.

13.

	θ	$\sin\theta$	$\cos\theta$	$\tan\theta$	$\csc\theta$	$\sec\theta$	$\cot\theta$
(a)	$225°$	$-1/\sqrt{2}$	$-1/\sqrt{2}$	1	$-\sqrt{2}$	$-\sqrt{2}$	1
(b)	$-210°$	$1/2$	$-\sqrt{3}/2$	$-1/\sqrt{3}$	2	$-2/\sqrt{3}$	$-\sqrt{3}$
(c)	$5\pi/3$	$-\sqrt{3}/2$	$1/2$	$-\sqrt{3}$	$-2/\sqrt{3}$	2	$-1/\sqrt{3}$
(d)	$-3\pi/2$	1	0	—	1	—	0

15.

	$\sin\theta$	$\cos\theta$	$\tan\theta$	$\csc\theta$	$\sec\theta$	$\cot\theta$
(a)	$4/5$	$3/5$	$4/3$	$5/4$	$5/3$	$3/4$
(b)	$-4/5$	$3/5$	$-4/3$	$-5/4$	$5/3$	$-3/4$
(c)	$1/2$	$-\sqrt{3}/2$	$-1/\sqrt{3}$	2	$-2\sqrt{3}$	$-\sqrt{3}$
(d)	$-1/2$	$\sqrt{3}/2$	$-1/\sqrt{3}$	-2	$2/\sqrt{3}$	$-\sqrt{3}$
(e)	$1/\sqrt{2}$	$1/\sqrt{2}$	1	$\sqrt{2}$	$\sqrt{2}$	1
(f)	$1/\sqrt{2}$	$-1/\sqrt{2}$	-1	$\sqrt{2}$	$-\sqrt{2}$	-1

17. (a) $x = 3\sin 25° \approx 1.2679$ (b) $x = 3/\tan(2\pi/9) \approx 3.5753$

19.

	$\sin\theta$	$\cos\theta$	$\tan\theta$	$\csc\theta$	$\sec\theta$	$\cot\theta$
(a)	$a/3$	$\sqrt{9-a^2}/3$	$a/\sqrt{9-a^2}$	$3/a$	$3/\sqrt{9-a^2}$	$\sqrt{9-a^2}/a$
(b)	$a/\sqrt{a^2+25}$	$5/\sqrt{a^2+25}$	$a/5$	$\sqrt{a^2+25}/a$	$\sqrt{a^2+25}/5$	$5/a$
(c)	$\sqrt{a^2-1}/a$	$1/a$	$\sqrt{a^2-1}$	$a/\sqrt{a^2-1}$	a	$1/\sqrt{a^2-1}$

21. (a) $\theta = 3\pi/4 \pm n\pi$, $n = 0, 1, 2, \ldots$
(b) $\theta = \pi/3 \pm 2n\pi$ and $\theta = 5\pi/3 \pm 2n\pi$, $n = 0, 1, 2, \ldots$

23. (a) $\theta = \pi/6 \pm n\pi,\ n = 0, 1, 2, \ldots$
(b) $\theta = 4\pi/3 \pm 2n\pi$ and $\theta = 5\pi/3 \pm 2n\pi,\ n = 0, 1, 2, \ldots$

25. (a) $\theta = 3\pi/4 \pm n\pi,\ n = 0, 1, 2, \ldots$ (b) $\theta = \pi/6 \pm n\pi,\ n = 0, 1, 2, \ldots$

27. (a) $\theta = \pi/3 \pm 2n\pi$ and $\theta = 2\pi/3 \pm 2n\pi,\ n = 0, 1, 2, \ldots$
(b) $\theta = \pi/6 \pm 2n\pi$ and $\theta = 11\pi/6 \pm 2n\pi,\ n = 0, 1, 2, \ldots$

29. $\sin\theta = 2/5$, $\cos\theta = -\sqrt{21}/5$, $\tan\theta = -2/\sqrt{21}$, $\csc\theta = 5/2$, $\sec\theta = -5/\sqrt{21}$, $\cot\theta = -\sqrt{21}/2$

31. (a) $\theta = \pm n\pi,\ n = 0, 1, 2, \ldots$ (b) $\theta = \pi/2 \pm n\pi,\ n = 0, 1, 2, \ldots$
(c) $\theta = \pm n\pi,\ n = 0, 1, 2, \ldots$ (d) $\theta = \pm n\pi,\ n = 0, 1, 2, \ldots$
(e) $\theta = \pi/2 \pm n\pi,\ n = 0, 1, 2, \ldots$ (f) $\theta = \pm n\pi,\ n = 0, 1, 2, \ldots$

33. (a) $s = r\theta = 4(\pi/6) = 2\pi/3$ cm (b) $s = r\theta = 4(5\pi/6) = 10\pi/3$ cm

35. $\theta = s/r = 2/5$

37. (a) $2\pi r = R(2\pi - \theta)$, $r = \dfrac{2\pi - \theta}{2\pi}R$

(b) $h = \sqrt{R^2 - r^2} = \sqrt{R^2 - (2\pi - \theta)^2 R^2/(4\pi^2)} = \dfrac{\sqrt{4\pi\theta - \theta^2}}{2\pi}R$

39. Let h be the altitude as shown in the figure, then
$h = 3\sin 60° = 3\sqrt{3}/2$ so $A = \dfrac{1}{2}(3\sqrt{3}/2)(7) = 21\sqrt{3}/4$.

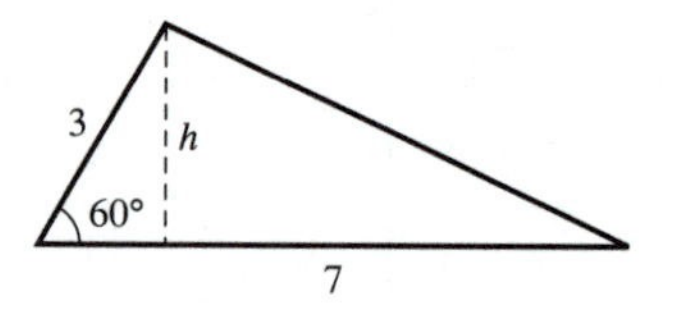

41. Let x be the distance above the ground, then $x = 10\sin 67° \approx 9.2$ ft.

43. From the figure, $h = x - y$ but $x = d\tan\beta$,
$y = d\tan\alpha$ so $h = d(\tan\beta - \tan\alpha)$.

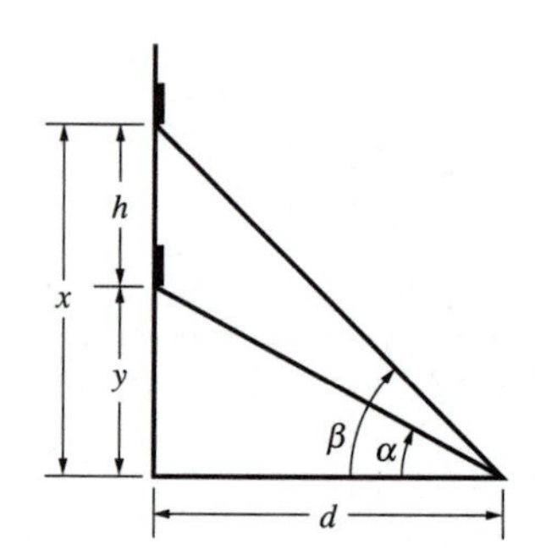

45. (a) $\sin 2\theta = 2\sin\theta\cos\theta = 2(\sqrt{5}/3)(2/3) = 4\sqrt{5}/9$
(b) $\cos 2\theta = 2\cos^2\theta - 1 = 2(2/3)^2 - 1 = -1/9$

47. $\sin 3\theta = \sin(2\theta + \theta) = \sin 2\theta\cos\theta + \cos 2\theta\sin\theta = (2\sin\theta\cos\theta)\cos\theta + (\cos^2\theta - \sin^2\theta)\sin\theta$
$= 2\sin\theta\cos^2\theta + \sin\theta\cos^2\theta - \sin^3\theta = 3\sin\theta\cos^2\theta - \sin^3\theta$; similarly, $\cos 3\theta = \cos^3\theta - 3\sin^2\theta\cos\theta$

49. $\dfrac{\cos\theta\tan\theta + \sin\theta}{\tan\theta} = \dfrac{\cos\theta(\sin\theta/\cos\theta) + \sin\theta}{\sin\theta/\cos\theta} = 2\cos\theta$

51. $\tan\theta + \cot\theta = \dfrac{\sin\theta}{\cos\theta} + \dfrac{\cos\theta}{\sin\theta} = \dfrac{\sin^2\theta + \cos^2\theta}{\sin\theta\cos\theta} = \dfrac{1}{\sin\theta\cos\theta} = \dfrac{2}{2\sin\theta\cos\theta} = \dfrac{2}{\sin 2\theta} = 2\csc 2\theta$

53. $\dfrac{\sin\theta+\cos 2\theta-1}{\cos\theta-\sin 2\theta}=\dfrac{\sin\theta+(1-2\sin^2\theta)-1}{\cos\theta-2\sin\theta\cos\theta}=\dfrac{\sin\theta(1-2\sin\theta)}{\cos\theta(1-2\sin\theta)}=\tan\theta$

55. Using (47), $2\cos 2\theta\sin\theta=2(1/2)[\sin(-\theta)+\sin 3\theta]=\sin 3\theta-\sin\theta$

57. $\tan(\theta/2)=\dfrac{\sin(\theta/2)}{\cos(\theta/2)}=\dfrac{2\sin(\theta/2)\cos(\theta/2)}{2\cos^2(\theta/2)}=\dfrac{\sin\theta}{1+\cos\theta}$

59. From the figures, area $=\dfrac{1}{2}hc$ but $h=b\sin A$ so area $=\dfrac{1}{2}bc\sin A$. The formulas area $=\dfrac{1}{2}ac\sin B$ and area $=\dfrac{1}{2}ab\sin C$ follow by drawing altitudes from vertices B and C, respectively.

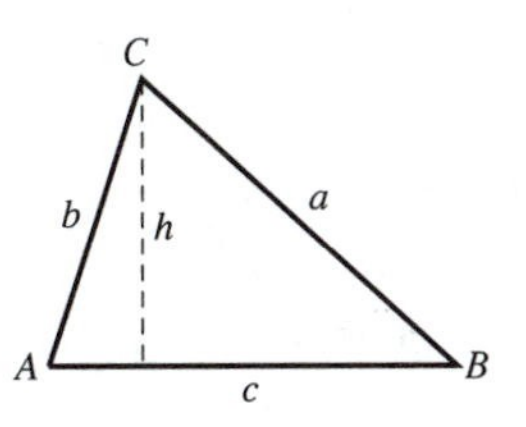

61. **(a)** $\sin(\pi/2+\theta)=\sin(\pi/2)\cos\theta+\cos(\pi/2)\sin\theta=(1)\cos\theta+(0)\sin\theta=\cos\theta$
(b) $\cos(\pi/2+\theta)=\cos(\pi/2)\cos\theta-\sin(\pi/2)\sin\theta=(0)\cos\theta-(1)\sin\theta=-\sin\theta$
(c) $\sin(3\pi/2-\theta)=\sin(3\pi/2)\cos\theta-\cos(3\pi/2)\sin\theta=(-1)\cos\theta-(0)\sin\theta=-\cos\theta$
(d) $\cos(3\pi/2+\theta)=\cos(3\pi/2)\cos\theta-\sin(3\pi/2)\sin\theta=(0)\cos\theta-(-1)\sin\theta=\sin\theta$

63. **(a)** Add (34) and (36) to get $\sin(\alpha-\beta)+\sin(\alpha+\beta)=2\sin\alpha\cos\beta$ so $\sin\alpha\cos\beta=(1/2)[\sin(\alpha-\beta)+\sin(\alpha+\beta)]$.
(b) Subtract (35) from (37). **(c)** Add (35) and (37).

65. $\sin\alpha+\sin(-\beta)=2\sin\dfrac{\alpha-\beta}{2}\cos\dfrac{\alpha+\beta}{2}$, but $\sin(-\beta)=-\sin\beta$ so
$\sin\alpha-\sin\beta=2\cos\dfrac{\alpha+\beta}{2}\sin\dfrac{\alpha-\beta}{2}$.

67. Consider the triangle having a, b, and d as sides. The angle formed by sides a and b is $\pi-\theta$ so from the law of cosines, $d^2=a^2+b^2-2ab\cos(\pi-\theta)=a^2+b^2+2ab\cos\theta$, $d=\sqrt{a^2+b^2+2ab\cos\theta}$.

APPENDIX F
Solving Polynomial Equations

EXERCISE SET F

1. **(a)** $q(x) = x^2 + 4x + 2, r(x) = -11x + 6$
(b) $q(x) = 2x^2 + 4, r(x) = 9$
(c) $q(x) = x^3 - x^2 + 2x - 2, r(x) = 2x + 1$

3. **(a)** $q(x) = 3x^2 + 6x + 8, r(x) = 15$
(b) $q(x) = x^3 - 5x^2 + 20x - 100, r(x) = 504$
(c) $q(x) = x^4 + x^3 + x^2 + x + 1, r(x) = 0$

5.

x	0	1	-3	7
$p(x)$	-4	-3	101	5001

7. **(a)** $q(x) = x^2 + 6x + 13, r = 20$ **(b)** $q(x) = x^2 + 3x - 2, r = -4$

9. Assume $r = a/b$ a and b integers with $a > 0$:
(a) b divides 1, $b = \pm 1$; a divides 24, $a = 1, 2, 3, 4, 6, 8, 12, 24$;
the possible candidates are $\{\pm 1, \pm 2, \pm 3, \pm 4, \pm 6, \pm 8, \pm 12, \pm 24\}$
(b) b divides 3 so $b = \pm 1, \pm 3$; a divides -10 so $a = 1, 2, 5, 10$;
the possible candidates are $\{\pm 1, \pm 2, \pm 5, \pm 10, \pm 1/3, \pm 2/3, \pm 5/3, \pm 10/3\}$
(c) b divides 1 so $b = \pm 1$; a divides 17 so $a = 1, 17$;
the possible candidates are $\{\pm 1, \pm 17\}$

11. $(x + 1)(x - 1)(x - 2)$

13. $(x + 3)^3(x + 1)$

15. $(x + 3)(x + 2)(x + 1)^2(x - 3)$

17. -3 is the only real root.

19. $x = -2, -2/3, -1 \pm \sqrt{3}$ are the real roots.

21. $-2, 2, 3$ are the only real roots.

23. If $x - 1$ is a factor then $p(1) = 0$, so $k^2 - 7k + 10 = 0$, $k^2 - 7k + 10 = (k - 2)(k - 5)$, so $k = 2, 5$.

25. If the side of the cube is x then $x^2(x - 3) = 196$; the only real root of this equation is $x = 7$ cm.

27. Use the Factor Theorem with x as the variable and y as the constant c.
(a) For any positive integer n the polynomial $x^n - y^n$ has $x = y$ as a root.
(b) For any positive even integer n the polynomial $x^n - y^n$ has $x = -y$ as a root.
(c) For any positive odd integer n the polynomial $x^n + y^n$ has $x = -y$ as a root.

CHAPTER 1
Sample Exams

SECTION 1.1

1. Answer true or false. Given the equation $y = x^2 - 5x + 4$, the values of x for which $y = 0$ are -4 and -1.

2. Answer true or false. Given the equation $y = x^2 - 5x + 6$, $y \geq 0$ for all $x \geq 0$.

3. Answer true or false. Given the equation $y = 1 - \sqrt{x}$, $y = 1$, when $x = 0$.

4. Answer true or false. Given the equation $y = -x^2 + 4$, it can be determined that y has a minimum value.

5. Answer true or false. Referring to the graph of $y = \sqrt[3]{x}$, y can be determined to have a maximum value.

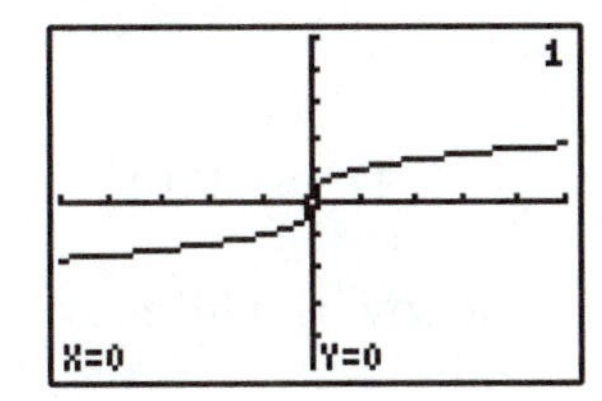

6. Assume the function $y = x^3 - 4x^2$ is used to describe the profit/loss of a company over the first 10 years after having started business. Initially, the company expects to lose money. After how many years will the company start to make a profit?

A. 1 year B. 2 years C. 4 years D. 8 years

7. Use the equation $y = x^2 - 6x + 8$. For what values of x is $y \geq 0$?

A. $\{x : 2 \leq x \leq 4\}$ B. $\{x : x \leq 2 \text{ or } x \geq 4\}$
C. $\{x : -4 \leq x \leq -2\}$ D. $\{x : x \leq -4 \text{ or } x \leq -2\}$

8. From the graph of $y = x^2 - 5x$ determine the value(s) of x where $y = 0$.

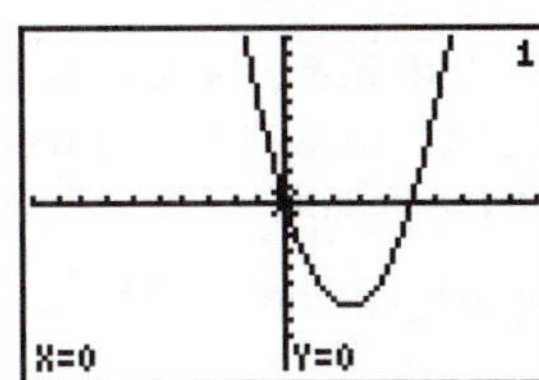

A. 0 B. 5 C. 0, 5 D. $-5, 0$

9. From the graph of $y = 2x^2 + 4x - 7$ determine the minimum value of y.

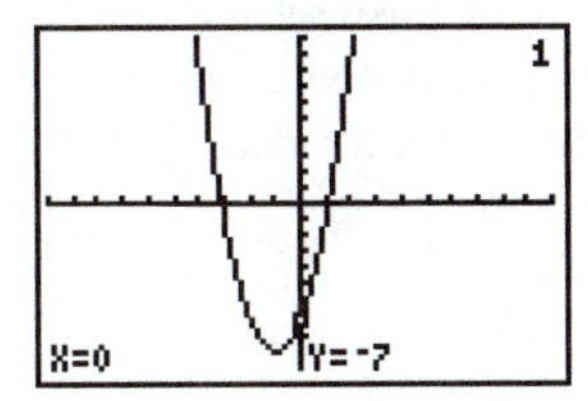

A. -1 B. 1 C. -9 D. 9

10. From the graph of $y = 2x^2 + 6x - 8$ determine at what x the graph has a minimum.

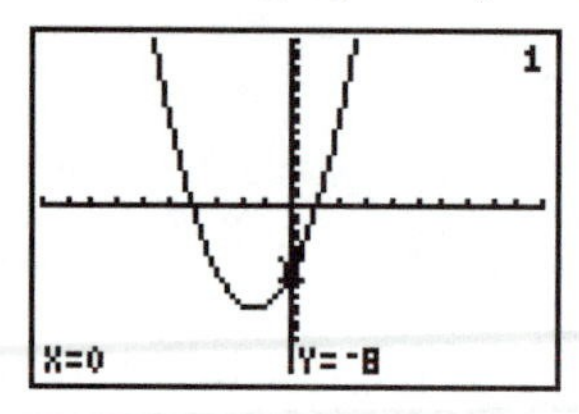

A. $\frac{-3}{2}$ B. $\frac{3}{2}$ C. 0 D. -1

11. From the graph of $y = x^4 - 3x^2 - 4$ determine for what x values the graph appears to be below the x-axis.

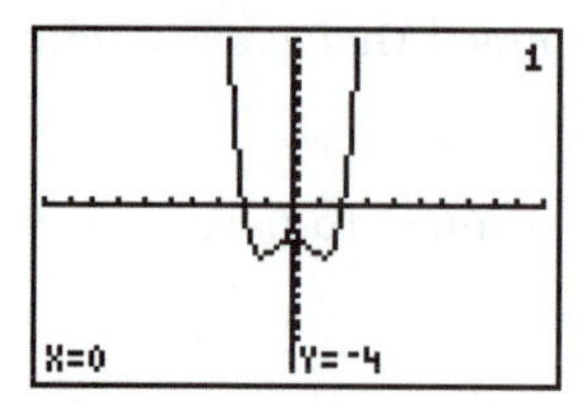

A. $(-2, 2)$ B. $(0, 2)$ C. $(-5, 0)$ D. $(-6, 0)$

12. Assume it is possible to measure the observations given below. Which, if any, would most likely generate a broken graph?

A. the temperature over a 6 hour period
B. the number of hamburgers served at a fast food restaurant
C. the speed of an automobile for the first 10 minutes after leaving a red light
D. none of the above

13. If a company recorded its profit over a year graphically with time represented on the horizontal axis, the graph would be

A. continuous (unbroken)
B. broken because money is earned in lump sums at sales
C. broken because no graph is continuous (unbroken)
D. broken because the printer cannot really handle continuous (unbroken) graphs

14. A manufacturer makes boxes taking rectangular sheets of cardboard that initially are twice as long as they are wide by cutting squares from the corners and turning up the sides. A customer requests that the such a box have a minimum volume. Why is this not practical?

A. If the length of the square that is cut out has half the original width of the cardboard, the volume would be zero, hence there would be no box.
B. Tools can cut only a certain size square
C. The squares cut from the corners would be very small.
D. The manufacturer would have to cut rectangles, not squares, from the corners.

15. A person makes a cube, then finds the surface area to volume ratio is improved by beating the points down with a hammer. As the object distorts, more and more points form, each of which is beat down with the hammer to further improve the surface area to volume ratio. What three dimensional object would this suggest has the best surface area to volume ratio?

A. multi-sided polygon B. cone
C. pyramid D. sphere

SECTION 1.2

1. Answer true or false. If $f(x) = 4x^2 - 1$, then $f(0) = 0$.

2. Answer true or false. If $f(x) = \frac{1}{x}$, then $f(0) = 0$.

3. $f(x) = \frac{1}{x^2 - 1}$. The natural domain of the function is

A. all real numbers
B. all real numbers except 1
C. all real numbers except -1 and 1
D. all real numbers except -1, 0, and 1

4. Use a graphing utility to determine the natural domain of $h(x) = \frac{1}{|x| - 1}$.

A. all real numbers
B. all real numbers except 1
C. all real numbers except -1 and 1
D. all real numbers except -1, 0, and 1

5. Use a graphing utility to determine the natural domain of $g(x) = \sqrt{9 - x^2}$.

A. $\{x : -3 \le x \le 3\}$
B. $\{x : -9 \le x \le 3\}$
C. $\{x : x \ge -9\}$
D. $\{x : x \ge -3\}$

6. Answer true or false. $f(x) = x - |x + 1|$ can be represented in the piecewise form by
$$f(x) = \begin{cases} 2x + 1, & \text{if } x \le -1 \\ -1, & \text{if } x > -1 \end{cases}.$$

7. Find the x-coordinate of any hole(s) in the graph of $f(x) = \frac{(x^2 - 9)(x + 4)}{(x + 3)(x + 4)}$.

A. 3
B. -3 and -4
C. 3 and 4
D. -12

8. Answer true or false. $f(x) = \frac{x^2 - 4}{x + 2}$ and $g(x) = x - 2$ are identical except $f(x)$ has a hole at $x = -2$.

9. Use a graphing device to plot the function, then find the indicated value from observing the graph. A flying object attains a height $h(t)$ over time according to the equation $h(t) = 3 \sin t - \cos t$. If $t = \frac{\pi}{4}$, $h\left(\frac{\pi}{4}\right)$ is approximately

A. 0
B. 0.707
C. 1.414
D. 3

10. A box is made from a piece of sheet metal by cutting a square whose sides measure x from each corner. If the sheet of metal initially measured 20 cm by 10 cm, the volume of the box is given by

A. $V(x) = (20 - 2x)(10)x$ cm^3
B. $V(x) = (20 - 2x)(10 - x)x$ cm^3
C. $V(x) = (20 - 2x)(10 - 2x)x$ cm^3
D. $V(x) = (20 - 2x)(10 - 2x)(2x)$ cm^3

11. Answer true or false. At a given location on a certain day the temperature T in °F changed according to the equation $T(t) = t \sin\left(\frac{t\pi}{12}\right) + 4$ where t represents time in hours starting at midnight. The temperature at 6 P.M. was approximately 58°F.

12. The speed of a truck in miles/hour for the first 10 seconds after leaving a red light is given by $f(x) = \frac{x^2}{2}$. Find the speed of the truck 6 seconds after leaving the red light.

A. 18 miles/hour
B. 36 miles/hour
C. 3 miles/hour
D. 6 miles/hour

13. Find $f(2)$ if $f(x) = \left\{\frac{4}{x}, \text{ if } x < 2 \text{ and } 3x \text{ if } x \geq 2\right\}$.

A. 2
B. 4
C. 6
D. it cannot be determined.

14. Determine all x-values where there are holes in the graph of $f(x) = \frac{x^2 - 4}{(x+2)^2(x-2)}$.

A. $-2, 2$
B. -2
C. 2
D. none

15. Assume the hourly temperature in °F starting at midnight is given by $T(x) = 36 - (x - 15)^2 + x$. Find the temperature at 3 P.M.

A. 36°F
B. 51°F
C. 39°F
D. 27°F

SECTION 1.3

1. Answer true or false. $f(x) = x^3 - 2x^2 - x + 5$ has one localized maximum and one localized minimum. The window on a graphing utility with $-10 \leq x \leq 10$ and $-10 \leq y \leq 10$ will show enough detail of the graph of $f(x)$ to include both of these.

2. Answer true or false. $f(x) = x^4 - 8$ has a minimum value. This can be shown on a graphing utility with a window defined by $-5 \leq x \leq 5$ and $-5 \leq y \leq 5$.

3. The smallest domain that is needed to show the entire graph of $f(x) = \sqrt{25 - x^2}$ on a graphing utility is

A. $-10 \leq x \leq 10$
B. $-5 \leq x \leq 5$
C. $0 \leq x \leq 10$
D. $0 \leq x \leq 5$

4. The smallest range that is needed to show the entire graph of $f(x) = \sqrt{36 - x^2}$ on a graphing utility is

A. $0 \leq y \leq 6$
B. $-6 \leq y \leq 6$
C. $0 \leq y \leq 10$
D. $-10 \leq y \leq 10$

5. If xScl is changed from 1 to 2 on a graphing utility which of these statements describes what happens?

A. The domain becomes twice as large as the original domain.
B. The domain becomes half as large as the original domain.
C. The domain becomes four times as large as the original domain.
D. The domain remains the same.

6. Using a graphing utility, the graph of $y = \frac{x}{x^2 - 4}$ will generate how many false line segments on a $-10 \leq x \leq 10$ domain?

A. 3
B. 0
C. 1
D. 2

7. How many functions are needed to graph the circle $x^2 + y^2 = 10$ on a graphing utility?

A. 1 B. 2 C. 3 D. 4

8. A student tries to graph the circle $x^2 + y^2 = 9$ on a graphing utility, but the graph appears to be elliptical. To view this as a circle the student could

A. increase the range of x
B. increase the range of y
C. increase xScl
D. increase yScl

9. Answer true or false. A student wishes to graph $f(x) = \begin{cases} x^3, & \text{if } x \le 2 \\ x-4, & \text{if } 2 < x \le 4 \\ x^2, & \text{if } x > 4 \end{cases}$. This can be accomplished by graphing three functions, then sketching the graph from the information obtained.

10. The graph of $f(x) = |x-1| + |x+2|$ touches the x-axis

A. nowhere B. at 1 point C. at 2 points D. at 4 points

11. Answer true or false. The graph of $y = x\cos x$ has its greatest difference between consecutive localized maxima and minima when x is near 0.

12. If a graphing utility is not generating the complete graph of $f(x) = x^{2/3}$, the graph of what function should remedy this?

A. $g(x) = |x|^{2/3}$
B. $g(x) = \left(\frac{|x|}{x}\right)|x|^{2/3}$
C. $g(x) = \frac{x}{x^{1/3}}$
D. $g(x) = \begin{cases} -x^{2/3}, & \text{if } x < 0 \\ x^{2/3}, & \text{if } x \ge 0 \end{cases}$

13. Which of these functions generates a graph that goes negative?

A. $f(x) = |\sin x|$
B. $G(x) = |\sin |x||$
C. $h(x) = \sin |x|$
D. $F(x) = |\sin x| + |\cos x|$

14. The window that best shows $f(x) = \sqrt{5x+10}$ should include what restriction on the x-values?

A. $x \le 2$ B. $x \le -2$ C. $x \ge 2$ D. $x \ge -2$

15. The window that best shows $f(x) = \sqrt{5x+10} - 5$ should include what restriction on the y-values?

A. $y \le 5$ B. $y \le -5$ C. $y \ge 5$ D. $y \ge -5$

SECTION 1.4

1.

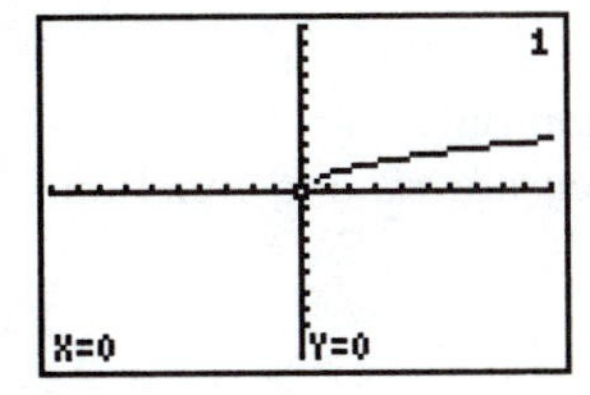

The graph on the left is the graph of $f(x) = \sqrt{x}$. The graph on the right is the graph of

A. $y = f(x+4)$ B. $y = f(x-4)$ C. $y = f(x) - 4$ D. $y = f(x) + 4$

2. The graph of $y = 1 + (x+2)^2$ is obtained from the graph of $y = x^2$ by

A. translating horizontally 2 units to the right, then translating vertically 1 unit up
B. translating horizontally 2 units to the left, then translating vertically 1 unit up
C. translating horizontally 2 units to the right, then translating vertically 1 unit down
D. translating horizontally 2 units to the left, then translating vertically 1 unit down

3. The graph of $y = \sqrt{x}$ and $y = \sqrt{-x}$ are related. The graph of $y = \sqrt{-x}$ is obtained by

A. reflecting the graph of $y = \sqrt{x}$ about the x-axis
B. reflecting the graph of $y = \sqrt{x}$ about the y-axis
C. reflecting the graph of $y = \sqrt{x}$ about the origin
D. The equations are not both defined.

4. The graphs of $y = x^3$ and $y = 4 - 2(x-1)^3$ are related. Of reflection, stretching, vertical translation, and horizontal translation, which should be done first?

A. reflection
B. stretching
C. vertical translation
D. horizontal translation

5. Answer true or false. $f(x) = x^2$ and $g(x) = x - 2$. Then $(f - g)(x) = x^2 - x - 2$.

6. Answer true or false. $f(x) = x^2 - 9$ and $g(x) = x^3 + 8$. f/g has the same domain as g/f.

7. $f(x) = \sqrt{x^2+1}$ and $g(x) = x^2 - 2$. $f \circ g(x) =$

A. $\sqrt{x^4 - 1}$
B. $\sqrt{x^2 - 1}$
C. $\sqrt{x^4 - 4x^2 + 5}$
D. $\sqrt{x^2 + 2x - 1}$

8. $f(x) = |(x+3)^3|$ is the composition of

A. $f(x) = x + 3$; $g(x) = |x^3|$
B. $f(x) = (x+3)^3$; $g(x) = |x^3|$
C. $f(x) = \sqrt[3]{x+3}$; $g(x) = |x^3|$
D. $f(x) = x^3$; $g(x) = |x+3|$

9. $f(x) = \sqrt[3]{x}$. Find $f(3x)$.

A. $3\sqrt[3]{x}$
B. $\sqrt[3]{3x}$
C. $\dfrac{\sqrt[3]{x}}{3}$
D. $\sqrt[3]{\dfrac{x}{3}}$

10. $f(x) = x^2 + 1$. Find $f(f(x))$.

A. $x^4 + 2$
B. $2x^2 + 2$
C. $x^4 + 2x^2 + 2$
D. $x^4 + 2x^2$

11. $f(x) = |x+2|$ is

A. an even function only
B. an odd function only
C. both an even and an odd function
D. neither an even nor an odd function

12. $f(x) = 0$ is

A. an even function only
B. an odd function only
C. both an even and an odd function
D. neither an even nor an odd function

13.

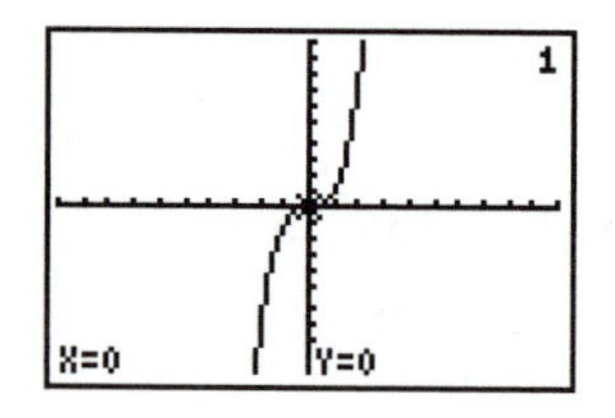

The function graphed at the bottom of page 388 is

A. so even function only B. an odd function only
C. both an even and an odd function D. neither an even nor an odd function

14. Answer true or false. $f(x) = |x| + \cos x$ is an even function.

15. $f(x) = 3|x| + 2\cos x$ is symmetric about

A. the x-axis B. the y-axis C. the origin D. nothing

16. $f(x) = 5x^3 - 2x$ is symmetric about

A. the x-axis B. the y-axis C. the origin D. nothing

SECTION 1.5

1. Answer true or false. The points $(1, 1)$, $(2, 3)$, and $(4, 7)$ lie on the same line.

2. A particle, initially at $(4, 3)$, moves along a line of slope $m = 3$ to a new position (x, y). Find y if $x = 6$.

A. 8 B. 9 C. 16 D. 5

3. Find the angle of inclination of the line $3x + 2y = 5$ to the nearest degree.

A. $56°$ B. $-56°$ C. $21°$ D. $-21°$

4. The slope-intercept form of a line having a slope of 4 and a y-intercept of 5 is

A. $x = 4y + 5$ B. $y = 4x + 5$ C. $y = -4x - 5$ D. $x = -4y - 5$

5. Answer true or false. The lines $y = 5x + 4$ and $y = -5x + 4$ are parallel.

6. Answer true or false. The lines $y = 2x - 1$ and $x + 2y = 6$ are perpendicular.

7.

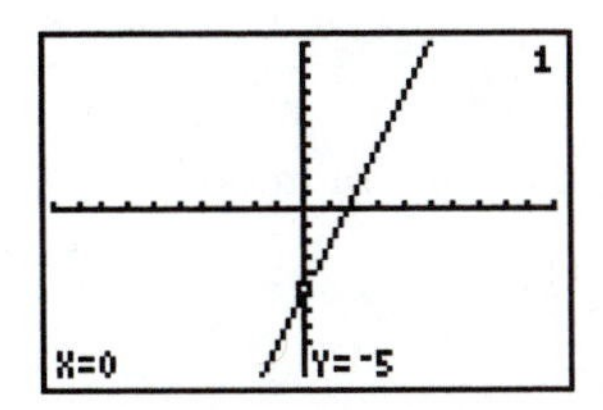

The slope-intercept form of the equation of the graphed line is

A. $y = 3x - 5$ B. $y = 3x + 5$ C. $y = -3x + 5$ D. $y = -3x - 5$

8. A particle moving along an x-axis with a constant velocity is at the point $x = 3$ when $t = 1$ and $x = 8$ when $t = 2$. The velocity of the particle if x is in meters and t is in seconds

A. 5 m/s B. $\frac{1}{2}$ m/s C. $\frac{8}{3}$ m/s D. $\frac{3}{8}$ m/s

9. Answer true or false. A particle moving along an x-axis with constant acceleration has velocity $v = 5$ m/s at time $t = 1$ s and velocity $v = 9$ m/s at time $t = 2$ s. The acceleration of the particle is 4 m/s^2.

10. A family travels north along a highway at 60 mi/hr, then turns back and travels south at 65 mi/hr until returning to the starting point. Their average velocity is

A. 62.5 mi/hr B. 125 mi/hr C. 5 mi/hr D. 0 mi/hr

11. Answer true or false. An arrow is shot upward at 100 ft/s. If the effect of gravity is to cause velocity as a function of time to be $v = 100 - 32t$, the arrow will be moving downward when $t = 4$s.

12. A spring with a natural length of 4.00 m is stretched to the length of 4.05 m when an object weighing 5.00 N is suspended from it. If a 50-N object is later suspended from it, it will stretch to

A. 4.50 m B. 45.0 m C. 6.50 m D. 4.55 m

13. A company makes a certain object for \$6 each, and sells each such object for \$10. If the company has a monthly overhead expense of \$10,000, how many of these objects must the company make and sell each month not to lose money?

A. 25 B. 25,000 C. 2,500 D. 250

14. Answer true or false. A particle has a velocity with respect to time given by the function $v = t^2 - 3t + 6$. At time $t = 0$ the particle is not moving.

15. A circuit has a 20 volt battery, and a variable resistor. The current I in amperes (A) is given as a function of resistance, R, and is given in ohms (Ω) by $I = \dfrac{20}{R}$. What is the current when $R = 4\Omega$?

A. 5 A B. 80 A C. $\dfrac{1}{5}$ A D. $\dfrac{1}{80}$ A

SECTION 1.6

1. What do all members of the family of lines of the form $y = 5x + b$ have in common?

A. Their slope is 5.
B. Their slope is -5.
C. They go through the origin.
D. They cross the x-axis at the point $(5, 0)$.

2. What points do all graphs of equations of the form $y = x^n$, n is odd, have in common?

A. $(0,0)$ only B. $(0,0)$ and $(1,1)$
C. $(-1,-1)$, $(0,0)$, and $(1,1)$ D. none

3.

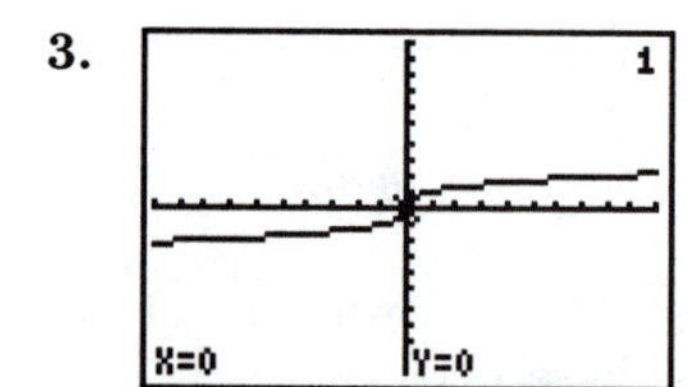

The equation whose graph is given is

A. $y = \sqrt{x}$ B. $y = \sqrt[3]{x}$ C. $y = \dfrac{1}{x^2}$ D. $y = \dfrac{1}{x^3}$

4. Answer true or false. The graph of $y = -3(x-4)^3$ can be obtained by making transformations to the graph of $y = x^3$.

5. Answer true or false. The graph of $y = x^2 + 6x + 9$ can be obtained by transforming the graph of $y = x^2$ to the left three units.

6. Answer true or false. There is no difference in the graphs of $y = \sqrt{|x|}$ and $y = |\sqrt{x}|$.

7. Determine the vertical asymptote(s) of $y = \dfrac{x+3}{x^2+2x-8}$.

A. $x = -4$, $x = 2$ B. $x = 4$ C. $x = -2$, $x = 4$ D. $x = 8$

8. Find the vertical asymptote(s) of $y = \dfrac{x^6}{3x^6 - 3}$.

A. $x = 0$ B. $x = -1$, $x = 1$ C. $x = 3$ D. $x = \dfrac{1}{3}$

9. For which of the given angles, if any, is all of the trigonometric functions negative?

A. $\dfrac{\pi}{3}$ B. $\dfrac{2\pi}{3}$ C. $\dfrac{4\pi}{3}$ D. No such angle exists

10. Use the trigonometric function of a calculating utility set to the radian mode to evaluate $\sin\left(\dfrac{\pi}{7}\right)$.

A. 0.0078 B. 0.4339 C. 0.1424 D. 0.1433

11. A rolling wheel of radius 2.00 m turns through an angle of 180°. How far does the wheel travel as it turns?

A. 6.28 m B. 200 m C. 2.00 m D. 4.00 m

12. Answer true or false. The amplitude of $5\cos(3x - \pi)$ is 3.

13. Answer true or false. The phase shift of $2\sin\left(3x - \dfrac{\pi}{3}\right)$ is $\dfrac{\pi}{3}$.

14. Use a graphing utility to graph $y_1 = \sin\left(x - \dfrac{\pi}{3}\right)$ and $y_2 = \sin\left(2\left(x - \dfrac{\pi}{3}\right)\right)$.

A. y_1 has the greatest phase shift.
B. y_2 has the greatest phase shift.
C. y_1 and y_2 have the same phase shift.
D. Neither y_1 nor y_2 have a phase shift.

15. Answer true or false. The period of $y = \cos\left(5x - \dfrac{\pi}{3}\right)$ is 10π.

16. Answer true or false. A force acting on an object, $F = \dfrac{k}{x^2}$, that is inversely proportional to the square of the distance from the object to the source of the force is found to be 25 N when $x = 1$ m. The force will be 100 N if x becomes 2 m.

SECTION 1.7

1. Answer true or false: An object's position from the origin is given by the accompanying table. Using the quadratic regression feature of a calculator, the position equation is best given by $s(t) = 4.1t^2 - 2.5t - 51.0$.

Time (sec)	Position (cm)
0	−51.00
1	−51.21
2	−51.34
3	−51.38
4	−51.34
5	−51.23
6	−51.02

2. Answer true or false: An object's position from the origin is given by the accompanying table. Using the quadratic regression feature of a calculator, the position equation is best given by $s(t) = 7.1t^2 - 1.4t - 6.0$.

Time (sec)	Position (cm)
0	−6.0
1	−0.3
2	19.6
3	53.7
4	102.0
5	164.5
6	241.2

3. Answer true or false: An object's position from the origin is given by the accompanying table. Using the linear regression feature of a calculator, the position equation is best given by $s(t) = 9.4t - 6.1$.

Time (sec)	Position (cm)
0	−6.1
1	−5.16
2	−4.22
3	−3.28
4	−2.34
5	−1.40
6	−0.46

4. A metal rod's length varies with its temperature according to the table. What is the least squares line for this collection of data points?

Temperature (°C)	length (m)
72	7.00
78	7.12
84	7.26
90	7.35
96	7.47

A. $y = 0.180x + 7.000$
B. $y = 0.195x + 7.000$
C. $y = 0.180x + 5.602$
D. $y = 0.195x + 5.602$

5. A metal rod's length varies with its temperature according to the table. What is the correlation coefficient for this collection of data points?

Temperature (°C)	length (m)
72	7.00
78	7.12
84	7.26
90	7.35
96	7.47

A. 0.995
B. 0.996
C. 0.998
D. 0.999

6. The accompanying table gives the number of centimeters a spring is stretched by various attached weights. Use a linear regression to express the amount of stretch of the spring as a function of the weight attached.

Weight (N)	Stretch (cm)
0.0	0.0
0.5	2.1
1.0	4.3
1.5	6.1
2.0	8.5
2.5	10.4

A. $S = 4.114Wt - 0.057$
B. $S = 4.114Wt + 0.057$
C. $S = 4.114Wt + 2.1$
D. $S = 4.114Wt$

7. The accompanying table gives the number of centimeters a spring is stretched by various attached weights. Use a linear regression to determine the correlation coefficient.

Weight (N)	Stretch (cm)
0.0	0.0
0.5	2.1
1.0	4.3
1.5	6.1
2.0	8.5
2.5	10.2

A. 0.996 B. 0.997 C. 0.999 D. 1.00

8. A particles has a position function t seconds after an experiment begins given by the accompanying table. The position function is close to

Time (sec)	position (cm)
0	2.0
1	8.0
2	16.1
3	26.2
4	38.4
5	52.1
6	68.0

A. quadratic B. linear C. sine D. none of these

9. A particles has a position function t seconds after an experiment begins given by the accompanying table. The position function is close to

Time (sec)	Position (cm)
0	3.5
1	11.7
2	19.9
3	28.1
4	36.3
5	44.5
6	52.7

A. quadratic B. linear C. sine D. none of these

10. A particles has a position function t seconds after an experiment begins given by t he accompanying table. The position function is close to

Time (sec)	Position (cm)
0	0.00
1	0.84
2	0.91
3	0.14
4	−0.76
5	−0.96
6	−0.28

A. quadratic B. linear C. sine D. none of these

SECTION 1.8

1. If $x = 3\cos^2(\pi t)$ and $y = \sin\left(\frac{\pi}{2}t\right)$ $(0 \leq t \leq 4)$, where t is time in seconds, describe the motion of particle, then the x- and y-coordinates of the position of the particle at time $t = 25$ are

A. $(3, 0)$ B. $(0, 1)$ C. $(-3, 0)$ D. $(0, -1)$

2. Answer true or false. Given the parametric equations $x = 5t$ and $y = t - 2$, eliminating the parameter t gives $x = 5(y + 2)$.

3. Use a graphing utility to graph $x = 4\cos t$ and $y = 2\sin t (0 \leq t \leq 2\pi)$. The resulting graph is

A. A circle B. A hyperbola C. An ellipse D. A parabola

4. Identify the equation in rectangular coordinates that is a representation of $x = 2\cos t$, $y = 3\sin t (0 \leq t \leq 2\pi)$.

A. $2x^2 + 3y^2 = 1$ B. $4x^2 + 3y^2 = 1$ C. $\frac{x^2}{2} + \frac{y^2}{3} = 1$ D. $\frac{x^2}{4} + \frac{y^2}{9} = 1$

5. Answer true or false. The graph in the rectangular coordinate system of $x = \cot t$, $y = \tan t (0 \leq t \leq \pi/2)$ is a hyperbola.

6. Answer true or false. The parametric representation of $4x^2 + 4y^2 = 1$ is $x = 2\sin t$, $y = 2\cos t (0 \leq t \leq 2\pi)$.

7. The circle represented by $x = 2 + 3\cos t$, $y = 4 + 3\sin t (0 \leq t \leq 2\pi)$ is centered at

A. $(0, 0)$ B. $(2, 4)$ C. $(-2, -4)$ D. $\left(\frac{2}{3}, \frac{4}{3}\right)$

8. Answer true or false. The trajectory of a particle is given by $x = t^3$, $y = t^2 - 4t + 6$ over the interval $-12 \leq t \leq 12$. The time at which the particle crosses the y-axis is 6.

9. Answer true or false. $x = t - 4$, $y = t (1 \leq t \leq 3)$ is the parametric representation of the line segment from P to Q, where P is the point $(-3, 1)$ and Q is the point $(-1, 3)$.

10. $x = a$, $y = t$, where a is a constant, is the parametric representation of a

A. horizontal line B. vertical line C. line with slope $+1$ D. line with slope -1

11. Use a graphing utility to graph $x = 4y^2 - 2y + 6$. The resulting graph is a parabola that opens

A. upward B. downward C. left D. right

12. Answer true or false. $x = 3 + 2t$, $y = 3 + 2t$ represents a line passing through the point $(3, 5)$.

13. The parametric form of a horizontal line passing through $(0, 2)$ is

A. $x = 2,\ y = t$ B. $x = t,\ y = 2$ C. $x = -2,\ y = t$ D. $x = t,\ y = -2$

14. Answer true or false. The curve represented by the piecewise parametric equation $x = 3t$, $y = t (0 \leq 2)$; $x = \frac{3t^2}{4}$, $y = t^2 (2 < t \leq 4)$ can be graphed as a continuous curve over the interval $(0 \leq t \leq 4)$.

15. Answer true or false. A ball is thrown at an angle of 30° above the horizontal with an initial speed $v_0 = 10$ m/s. The ball will rise 1.28 m (rounded to the nearest hundredth of a meter) in the absence of air resistance.

CHAPTER 1 TEST

1. Answer true or false. For the equation $y = x^2 + 10x + 21$, the values of x that cause y to be zero are 3 and 7.

2. Answer true or false. The graph of $y = x^2 - 4x + 9$ has a minimum value.

3. A company has a profit/loss given by $P(x) = 0.1x^2 - 2x - 10{,}000$, where x is time in years, good for the first 20 years. After how many years will the graph of the profit/loss equation first begin to rise?

A. 5 years B. 10 years C. 15 years D. 0 years

4. Use a graphing utility to determine the natural domain of $g(x) = \frac{1}{|x - 2|}$.

A. all real numbers
B. all real numbers except 2
C. all real numbers except -2
D. all real numbers except -2 and 2

5. Answer true or false. If $f(x) = \frac{2x}{x^2 + 1}$, then $f(1) = 1$.

6. Find the hole(s) in the graph of $f(x) = \frac{x - 4}{x^2 - 5x + 4}$.

A. $x = 4$ B. $x = 1$ C. $x = 1, 4$ D. $x = -4$

7. The cumulative number of electrons passing through an experiment over time, given in seconds, is given by $n(t) = t^2 + 4t + 6$. How many electrons pass through the experiment in the first 5 seconds?

A. 13 B. 39 C. 15 D. 51

8. Use a graphing utility to determine the entire domain of $f(x) = \sqrt{49 - x^2}$.

A. all real numbers B. $0 \leq x \leq 7$ C. $-7 \leq x \leq 7$ D. $0 \leq x \leq 49$

9. Answer true or false. The graph of $f(x) = |x + 5|$ touches the x-axis.

10. Answer true or false. The graph of $y = \dfrac{x^2 - 8}{x - 9}$ would produce a false line segment on a graphing utility on the domain of $-10 \le x \le 10$.

11. The graph of $y = (x - 2)^3$ is obtained from the graph of $y = x^3$ by

A. translating vertically 2 units upward
B. translating vertically 2 units downward
C. translating horizontally 2 units to the left
D. translating horizontally 2 units to the right

12. If $f(x) = x^3 + 2$ and $g(x) = x^2$, then $g \circ f(x) =$

A. $x^5 + 2$ B. $(x^3 + 2)^2$ C. $x^3 + x^2 + 2$ D. $x^6 + 2$

13. Answer true or false. $f(x) = |x| + \sin x$ is an odd function.

14. A particle initially at $(1, 3)$ moves along a line of slope $m = 4$ to a new position (x, y). Find y if $x = 4$.

A. 12 B. 15 C. 16 D. 19

15. The slope-intercept form of a line having a slope of 5 and a y-intercept of 3 is

A. $x = 5y + 3$ B. $x = 5y - 3$ C. $y = 5x + 3$ D. $y = 5x - 3$

16. A spring is initially 3 m long. When 2 kg are suspended from the spring it stretches 4 cm. How long will the spring be if 10 kg are suspended from it?

A. 3.20 m B. 3.02 m C. 3.40 m D. 3.04 m

17. The equation whose graph is given is

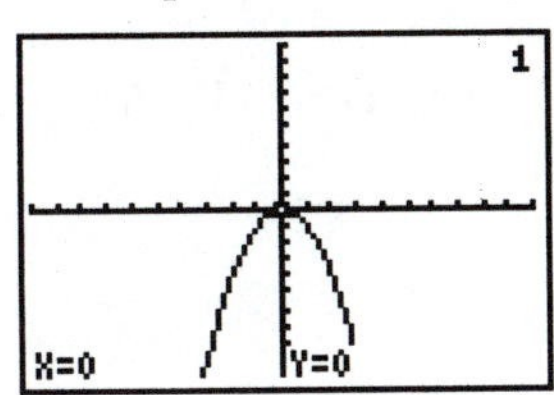

A. $y = x^2$ B. $y = (-x)^2$ C. $y = -x^2$ D. $y = x^{-2}$

18. Answer true or false. The only asymptote of $y = \dfrac{x + 5}{x^2 + 6x + 5}$ is $y = -1$.

19. A rolling wheel of radius 2 cm turns through an angle of 330°. How far does the wheel travel while rolling through this angle? (Round to the nearest hundredth of a centimeter.)

A. 11.52 cm B. 11.50 cm C. 11.48 cm D. 11.46 cm

20. If $x = 4\cos(\pi t)$ and $y = \sin(2\pi t)(0 \le t \le 2)$, where t is time in seconds, describe the motion of a particle, the x- and y-coordinates of the position of the particle at $t = 0.5$ s are

A. $(0, 0)$ B. $(-1, 1)$ C. $(0, 1)$ D. $(1, 1)$

21. The ellipse represented by $x = 6\cos t$, $y = 2\sin t(0 \le t \le 2\pi)$ is centered at

A. $(6, 2)$ B. $(-6, -2)$ C. $(\sqrt{6}, \sqrt{2})$ D. $(0, 0)$

22. The graph of $x = 5 + 2\cos t$, $y = 1 + 2\sin t(0 \le t \le 2\pi)$ is

A. a circle B. a hyperbola C. an ellipse D. a parabola

ANSWERS TO SAMPLE TESTS

SECTION 1.1:

1. F 2. F 3. T 4. F 5. F 6. C 7. B 8. C 9. C 10. A 11. A 12. B
13. B 14. A 15. D

SECTION 1.2:

1. F 2. F 3. C 4. C 5. A 6. T 7. B 8. T 9. C 10. C 11. F 12. A
13. C 14. C 15. B

SECTION 1.3:

1. T 2. F 3. B 4. A 5. D 6. D 7. B 8. B 9. T 10. A 11. F 12. A
13. C 14. D 15. D

SECTION 1.4:

1. B 2. B 3. B 4. D 5. F 6. F 7. C 8. D 9. B 10. C 11. D 12. C
13. B 14. T 15. B 16. C

SECTION 1.5:

1. T 2. B 3. B 4. B 5. F 6. T 7. A 8. A 9. T 10. D 11. T 12. A
13. C 14. F 15. A

SECTION 1.6:

1. A 2. C 3. B 4. T 5. T 6. F 7. A 8. B 9. D 10. B 11. A 12. F
13. F 14. A 15. F 16. F

SECTION 1.7:

1. F 2. T 3. T 4. D 5. C 6. C 7. C 8. A 9. B 10. C

SECTION 1.8:

1. B 2. T 3. C 4. D 5. T 6. F 7. B 8. F 9. T 10. B 11. D 12. F
13. B 14. F 15. F

CHAPTER 1 TEST:

1. F 2. T 3. B 4. B 5. T 6. A 7. D 8. C 9. T 10. T 11. D 12. B
13. F 14. B 15. C 16. A 17. C 18. F 19. A 20. A 21. D 22. A

CHAPTER 2
Sample Exams

SECTION 2.1

1. 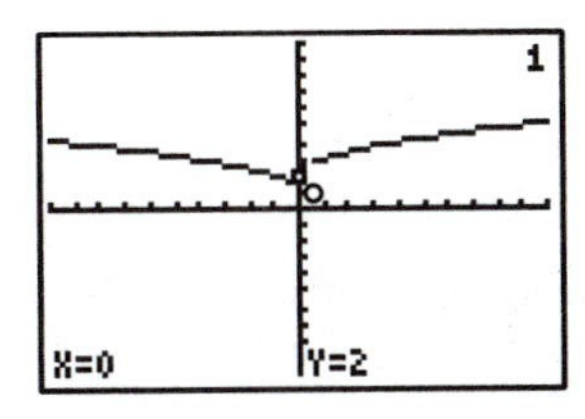

The function $f(x)$ is graphed. $\lim_{x \to 0^-} f(x) =$

A. 1 B. 2 C. $\frac{3}{2}$ D. undefined

2. 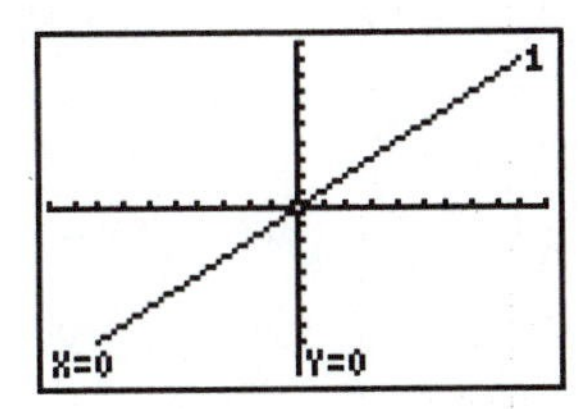

Answer true or false. For the function graphed $\lim_{x \to 2} f(x)$ is undefined.

3. Approximate the $\lim_{x \to 2^-} \frac{x^2 - 4}{x - 2}$ by evaluating $f(x) = \frac{x^2 - 4}{x - 2}$ at $x = 3$, 2.5, 2.1, 2.01, 2.001, 1, 1.5, 1.9, 1.99, and 1.999.

A. 2 B. -2 C. 0 D. 4

4. Answer true or false. If $\lim_{x \to 0^+} f(x) = 4$ and $\lim_{x \to 0^-} f(x) = 4$, then $\lim_{x \to 0} f(x) = 4$

5. Approximate the $\lim_{x \to 2^-} \frac{x}{x - 2}$ by evaluating $f(x) = \frac{x}{x - 2}$ at appropriate values of x.

A. 1 B. 0 C. ∞ D. $-\infty$

6. Approximate the limit by evaluating $f(x) = \frac{\sin x}{5x}$ at appropriate values of x. $\lim_{x \to 0^-} \frac{\sin x}{5x} =$

A. 1 B. 5 C. $\frac{1}{5}$ D. ∞

7. Approximate the limit by evaluating $f(x) = \frac{x}{\sin x}$ at appropriate values of x. $\lim_{x \to 0} \frac{x}{\sin x} =$

A. 1 B. -1 C. 0 D. ∞

8. Approximate the limit by evaluating $f(x) = \frac{\sqrt{x + 4} - 2}{x}$ at appropriate values of x.

$\lim_{x \to 0^-} \frac{\sqrt{x + 4} - 2}{x} =$

A. $\frac{1}{4}$ B. 0 C. ∞ D. $-\infty$

9. Use a graphing utility to approximate the y-coordinates of any horizontal asymptote of $y = f(x) = \dfrac{4x-9}{x+3}$.

A. 4 B. 1 C. None exist D. -3

10. Use a graphing utility to approximate the y-coordinate of any horizontal asymptote of $y = f(x) = \dfrac{\cos x}{x}$.

A. 0 B. 1 C. -1 and 1 D. -1

11. Use a graphing utility to approximate the y-coordinate of any horizontal asymptote of $y = f(x) = \dfrac{x^2+4}{x-2}$

A. 0 B. None exist. C. 1 D. -1 and 1

12. Answer true or false. A graphing utility can be used to show $f(x) = \left(1 + \dfrac{2}{x}\right)^x$ has a horizontal asymptote.

13. Answer true or false. A graphing utility can be used to show $f(x) = \left(6 + \dfrac{1}{x}\right)^x$ has a horizontal asymptote.

14. Answer true or false. $\lim\limits_{x \to +\infty} \dfrac{2+x}{1-x}$ is equivalent to $\lim\limits_{x \to 0^+} \left(\dfrac{\frac{2}{x}+1}{\frac{1}{x}-1}\right)$.

15. Answer true or false. $\lim\limits_{x \to +\infty} \dfrac{\sin(2\pi x)}{x-2}$ is equivalent to $\lim\limits_{x \to 0^+} \sin(2\pi x)$.

16. Answer true or false $f(x) = \dfrac{x}{x^2-4}$ has no horizontal asymptote.

SECTION 2.2

1. Given that $\lim\limits_{x \to a} f(x) = 3$ and $\lim\limits_{x \to a} g(x) = 5$, find, if it exists, $\lim\limits_{x \to 0}[f(x) - g(x)]^2$.

A. -4 B. 4 C. 22 D. It does not exist.

2. $\lim\limits_{x \to 3} 4 =$

A. 4 B. 3 C. 7 D. 12

3. Answer true or false. $\lim\limits_{x \to 2} 4x = 8$.

4. $\lim\limits_{x \to -3} \dfrac{x^2-9}{x+3} =$

A. $-\infty$ B. -6 C. 6 D. 1

5. $\lim_{x \to 5} \frac{4}{x-5} =$

A. $+\infty$ B. $-\infty$ C. 0 D. It does not exist.

6. $\lim_{x \to 1} \frac{4x}{x^2 - 6x + 5} =$

A. $+\infty$ B. $-\infty$ C. 0 D. It does not exist.

7. $\lim_{x \to 1} \frac{x-4}{\sqrt{x} - 1} =$

A. $+\infty$ B. $-\infty$ C. 1 D. It does not exist

8. Let $f(x) = \begin{cases} x+4, & x \leq 2 \\ x^2, & x > 2 \end{cases}$. $\lim_{x \to 2} f(x) =$

A. 6 B. 4 C. 3 D. It does not exist.

9. Let $g(x) = \begin{cases} x^2+4, & x \leq 1 \\ x^3, & x > 1 \end{cases}$. $\lim_{x \to 1} g(x) =$

A. 5 B. 1 C. 3 D. It does not exist.

10. Answer true or false. $\lim_{x \to 0} \frac{\sqrt{x^2+25} - 5}{x} = \frac{1}{10}$.

SECTION 2.3

1. $\lim_{x \to +\infty} \frac{10x}{x^2 - 5x + 3} =$

A. 0 B. 2 C. 5 D. It does not exist.

2. $\lim_{x \to -\infty} \frac{2x^2 - x}{x^2} =$

A. 2 B. $-\infty$ C. ∞ D. It does not exist.

3. $\lim_{x \to -\infty} \sqrt[4]{\frac{32x^8 - 6x^5 + 2}{2x^8 - 3x^3 + 1}} =$

A. $+\infty$ B. $-\infty$ C. 2 D. It does not exist.

4. $\lim_{x \to +\infty} (x^4 - 500x^3)$

A. $+\infty$ B. $-\infty$ C. -500 D. It does not exist.

5. Answer true or false. $\lim_{x \to +\infty} \frac{\sqrt{x^2+9} - 3}{x}$ does not exist.

SECTION 2.4

1. Find a least number δ such that $|f(x) - L| < \epsilon$ if $0 < |x - a| < \delta$. $\lim_{x\to 5} 4x = 20$; $\epsilon = 0.1$

A. 0.1 B. 0.25 C. 0.5 D. 0.025

2. Find a least number δ such that $|f(x) - L| < \epsilon$ if $0 < |x - a| < \delta$. $\lim_{x\to 2} 3x - 4 = 2$; $\epsilon = 0.1$

A. 0.033 B. 0.33 C. 3.0 D. 0.3

3. Answer true or false. A least number δ such that $|f(x) - L| < \epsilon$ if $0 < |x - a| < \delta$. $\lim_{x\to 3} x^3 = 27$; $\epsilon = 0.05$ is $\delta\sqrt[3]{27.05} - 3$.

4. Find a least number δ such that $|f(x) - L| < \epsilon$ if $0 < |x - a| < \delta$. $\lim_{x\to 5} \dfrac{x^2 - 25}{x - 5} = 10$; $\epsilon = 0.001$

A. 0.001 B. 0.000001 C. 0.005 D. 0.025

5. Find a least positive integer N such that $|f(x) - L| < \epsilon$ if $x > N$. $\lim_{x\to +\infty} \dfrac{12}{x^3} = 0$; $\epsilon = 0.1$

A. $N = 100$ B. $N = 1{,}000$ C. $N = 4$ D. $N = 5$

6. Find a greatest negative integer N such that $|f(x) - L| < \epsilon$ if $x < N$. $\lim_{x\to -\infty} \dfrac{1}{x^5} = 0$; $\epsilon = 0.1$

A. $N = -100{,}000$ B. $N = -10{,}000$ C. $N = -1$ D. $N = -2$

7. Answer true or false. It is possible to prove that $\lim_{x\to +\infty} \dfrac{1}{x^2 + 1} = 0$.

8. Answer true or false. It is possible to prove that $\lim_{x\to -\infty} \dfrac{1}{x + 6} = 0$.

9. Answer true or false. It is possible to prove that $\lim_{x\to +\infty} \dfrac{x}{x - 5} = 0$.

10. Answer true or false. It is possible to prove that $\lim_{x\to 5} \dfrac{1}{x - 5} = +\infty$.

11. To prove that $\lim_{x\to 5}(x + 2) = 7$ a reasonable relationship between δ and ϵ would be

A. $\delta = 5\epsilon$ B. $\delta = \epsilon$ C. $\delta = \sqrt{\epsilon}$ D. $\delta = \dfrac{1}{\epsilon}$

12. Answer true or false. To use a δ-ϵ approach to show that $\lim_{x\to 0^+} \dfrac{1}{x^3} = +\infty$, a reasonable first step would be to change the limit to $\lim_{x\to +\infty} x^3 = 0$.

13. Answer true or false. It is possible to show that $\lim_{x\to 4^-} \dfrac{1}{x - 4} = -\infty$.

14. To prove that $\lim_{x\to 3} f(x) = 6$ where $f(x) = \begin{cases} 2x, & x < 3 \\ x + 3, & x \geq 3 \end{cases}$ a reasonable relationship between δ and ϵ would be

A. $d4 = 2\epsilon$ B. $\delta = \epsilon$ C. $\delta = \epsilon + 3$ D. $\delta = 2\epsilon + 3$

15. Answer true or false. It is possible to show that $\lim_{x\to +\infty} \dfrac{x}{5} = 5$.

SECTION 2.5

1.

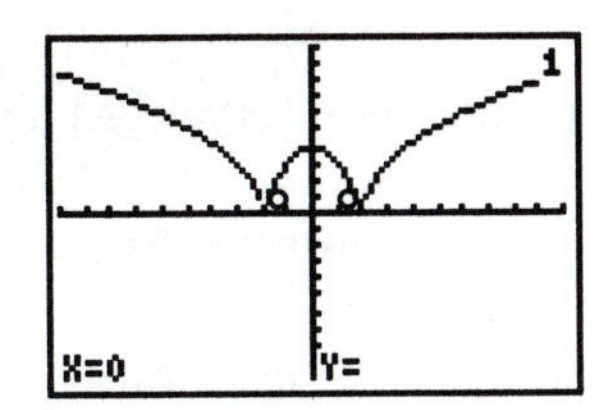

On the interval of $[-10, 10]$, where is f not continuous?

A. $-2, 2$ B. 2 C. -2 D. nowhere

2.

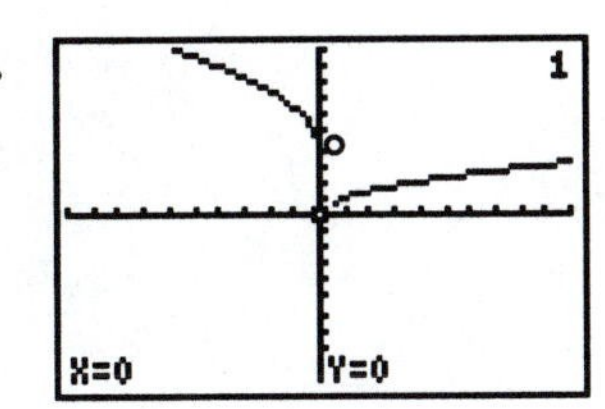

On the interval of $[-10, 10]$, where is f not continuous?

A. 3 B. $0, 3$ C. 0 D. nowhere

3. Answer true or false. $f(x) = x^6 - 4x^4 + 2x^2 - 8$ has no point of discontinuity.

4. Answer true or false. $f(x) = |x + 4|$ has a point of discontinuity at $x = -4$.

5. Find the x-coordinates for all points of discontinuity for $f(x) = \dfrac{x+6}{x^2 + 8x + 12}$.

A. $-6, -2$ B. -2 C. $2, 6$ D. 2

6. Find the x-coordinates for all points of discontinuity for $f(x) = \dfrac{|x+5|}{x^2 + 5x}$.

A. 0 B. -5 C. $-5, 0$ D. $-5, 0, 5$

7. Find the x-coordinates for all points of discontinuity for $f(x) = \begin{cases} x^2 - 5, & x \le 1 \\ -4, & x > 1 \end{cases}$.

A. 1 B. $-\sqrt{5}, \sqrt{5}$ C. $-4, -\sqrt{5}, \sqrt{5}$ D. Non exists.

8. Find the value of k, if possible, that will make the function continuous. $f(x) = \begin{cases} x - 2, & x \le 1 \\ kx^3, & x > 1 \end{cases}$

A. 1 B. -1 C. 2 D. None exists.

9. Answer true or false. The function $f(x) = \dfrac{x^3 - 1}{x - 1}$ has a removable discontinuity at $x = 1$.

10. Answer true or false. The function $f(x) = \begin{cases} x^2, & x \le 2 \\ x - 4, & x > 2 \end{cases}$ is continuous everywhere.

11. Answer true or false. If f and g are each continuous at c, $f + g$ may be discontinuous at c.

12. Answer true or false. The Intermediate-Value Theorem can be used to approximate the locations of all discontinuities for $f(x) = \frac{x}{x^3 - 5x + 13}$. [Hint: It may be applied to the denominator only.]

13. Answer true or false. $f(x) = x^5 - 6x^2 + 2 = 0$ has at least one solution on the interval $[-1, 0]$.

14. Answer true or false. $f(x) = x^3 + 2x + 9 = 0$ has at least one solution on the interval $[0, 1]$.

15. Use the fact that $\sqrt{8}$ is a solution of $x^2 - 8 = 0$ to approximate $\sqrt{8}$ with an error of at most 0.005.

A. 2.82 B. 2.81 C. 2.83 D. 2.84

SECTION 2.6

1. Answer true or false. $f(x) = \cos(x^2 + 5)$ has no point of discontinuity.

2. A point of discontinuity of $f(x) = \frac{1}{|-1 + 2\cos x|}$ is at

A. $\frac{\pi}{2}$ B. $\frac{\pi}{3}$ C. $\frac{\pi}{4}$ D. $\frac{\pi}{6}$

3. Find the limit. $\lim_{x \to +\infty} \cos\left(\frac{4}{x}\right) =$

A. 0 B. 1 C. -1 D. $+\infty$

4. Find the limit. $\lim_{x \to 0^-} \frac{\sin x}{x^3} =$

A. $+\infty$ B. 0 C. 1 D. $-\infty$

5. Find the limit. $\lim_{x \to 0} \frac{\sin(3x)}{\sin(8x)} =$

A. $+\infty$ B. 0 C. $\frac{3}{8}$ D. 1

6. Find the limit. $\lim_{x \to 0} \frac{4}{1 - \sin x} =$

A. 1 B. -1 C. 4 D. 0

7. Find the limit. $\lim_{x \to 0} \frac{1 + \cos x}{1 - \cos x} =$

A. 0 B. $+\infty$ C. $-\infty$ D. 2

8. Find the limit. $\lim_{x \to 0} \frac{\tan x}{\cos x} =$

A. 0 B. 1 C. $+\infty$ D. $-\infty$

9. Find the limit. $\lim_{x \to 0^-} \cos\frac{1}{x} =$

A. 1 B. -1 C. $-\infty$ D. does not exist

10. Find the limit. $\lim_{x\to 0^+} \dfrac{-1.3x+\cos x}{x} =$

A. 0 B. 1 C. −1 D. $+\infty$

11. Answer true or false. The value of k that makes f continuous for $f(x) = \begin{cases} \dfrac{\cos x - 1}{x}, & x \le 0 \\ \cos x + k, & x > 0 \end{cases}$ is 0.

12. Answer true or false. The fact that $\lim_{x\to 0} \dfrac{1-\cos x}{x} = 0$ and that $\lim_{x\to 0} x = 0$ guarantees that $\lim_{x\to 0} \dfrac{(1-\cos x)^2}{x} = 0$ by the Squeeze Theorem.

13. Answer true or false. The Squeeze Theorem can be used to show $\lim_{x\to 0} \dfrac{\sin(4x)}{6x} = 1$ utilizing $\lim_{x\to 0} \dfrac{\sin(4x)}{4x} = 1$ and $\lim_{x\to 0} \dfrac{\sin(6x)}{6x} = 1$.

14. Answer true or false. The Intermediate-Value Theorem can be used to show that the equation $y^3 = \sin^2 x$ has at least one solution on the interval $[-5\pi/6, 5\pi/6]$.

15. $\lim_{x\to 0} \left(\dfrac{\sin x}{2x} + \dfrac{x}{2\sin x} \right) =$

A. 1 B. 2 C. $\dfrac{1}{2}$ D. 0

CHAPTER 2 TEST

1.

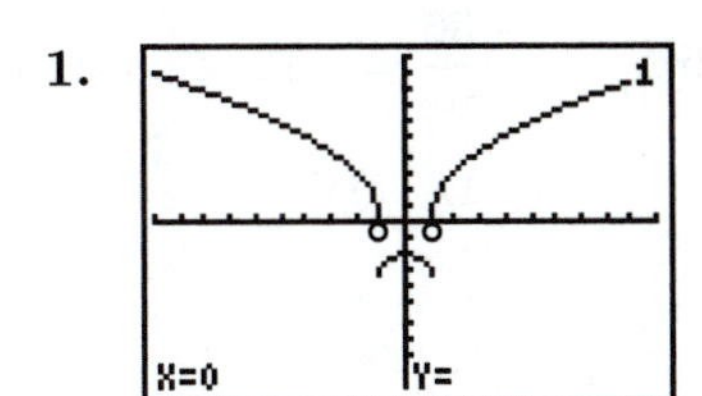

The function f is graphed. $\lim_{x\to -1} f(x) =$

A. 2 B. −2 C. 0 D. undefined

2. Approximate $\lim_{x\to -5} \dfrac{x^2-25}{x+5}$ by evaluating $f(x) = \dfrac{x^2-25}{x+5}$ at $x = -4, -4.5, -4.9, -4.99, -4.999, -6, -5.5, -5.1, -5.01$, and -5.001.

A. 5 B. −5 C. 10 D. −10

3. Use a graphing utility to approximate the y-coordinate of the horizontal asymptote of $y = f(x) = \dfrac{9x+3}{4x+2}$

A. $\dfrac{9}{4}$ B. $\dfrac{4}{9}$ C. $\dfrac{3}{2}$ D. $\dfrac{2}{3}$

4. Answer true or false. A graphing utility can be used to show that $f(x) = \left(8 + \frac{2}{2x}\right)^{2x}$ has a horizontal asymptote.

5. Answer true or false. $\lim_{x \to +\infty} \frac{5}{x^2}$ is equivalent to $\lim_{x \to 0+} 5x^2$.

6. Given that $\lim_{x \to a} f(x) = 5$ and $\lim_{x \to a} g(x) = -5$, find $\lim_{x \to a} [2f(x) = 4g(x)]$.

A. 0 B. −5 C. −10 D. 30

7. $\lim_{x \to 6} 4 =$

A. 6 B. −6 C. 4 D. does not exist

8. $\lim_{x \to 1} \frac{x^8 - 1}{x - 1} =$

A. 0 B. $+\infty$ C. 4 D. 8

9. $\lim_{x \to 6} \frac{1}{x - 6} =$

A. $\frac{1}{12}$ B. 0 C. $+\infty$ D. does not exist

10. Let $f(x) = \begin{cases} x^2, & x \le 1 \\ x, & x > 1 \end{cases}$. $\lim_{x \to 1} f(x) =$

A. 1 B. −1 C. 0 D. does not exist

11. Find a least number δ such that $|f(x) - L| < \epsilon$ if $0 < |x - a| < \delta$. $\lim_{x \to 5} 4x = 20$; $\epsilon < 0.01$

A. 0.01 B. 0.025 C. 0.05 D. 0.0025

12. Find a least number δ such that $|f(x) - L| < \epsilon$ if $0 < |x - a| < \delta$. $\lim_{x \to -6} \frac{x^2 - 36}{x + 6} = -12$; $\epsilon < 0.001$

A. 0.001 B. 0.000001 C. 0.006 D. 0.03

13. Answer true or false. It is possible to prove that $\lim_{x \to -\infty} \frac{1}{x^7} = 0$.

14. To prove $\lim_{x \to 2} (3x + 1) = 7$, a reasonable relationship between δ and ϵ would be

A. $\delta = \frac{\epsilon}{3}$ B. $\delta = 3\epsilon$ C. $\delta = \epsilon$ D. $\delta = \epsilon - 1$

15. Answer true or false. It is possible to show that $\lim_{x \to +\infty} (x - 3) = -3$.

16. On the interval $[-10, 10]$ where is f not continuous. f is the function graphed to the left.

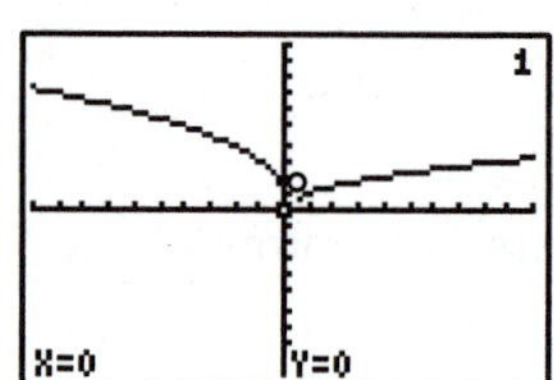

A. 0 B. −2 C. 2 D. nowhere

17. Find the x-coordinate of each point of discontinuity of $f(x) = \dfrac{x+2}{x^2-3x-10}$.

A. 5 B. $-2, 5$ C. $2, 5$ D. $-5, 2$

18. Find the value of k, if possible, that will make the function $f(x) = \begin{cases} kx+3, & x \le 3 \\ x^2, & x > 3 \end{cases}$ continuous.

A. 3 B. 0 C. 2 D. None exists.

19. Answer true or false. $f(x) = \dfrac{1}{x-6}$ has a removable discontinuity at $x = 6$.

20. Answer true or false. $f(x) = x^5 + 6 = 0$ has at least one solution on the interval $[-2, -1]$.

21. Find $\lim_{x \to 0} \dfrac{\sin(-3x)}{\sin(2x)}$.

A. 0 B. $\dfrac{-3}{2}$ C. $\dfrac{3}{2}$ D. not defined

22. Find $\lim_{x \to 0} \dfrac{\sin x}{\tan x}$.

A. 0 B. -1 C. 1 D. undefined

23. Answer true or false. $\lim_{x \to 0} \dfrac{\sin x}{1 - \cos x} = 0$.

ANSWERS TO SAMPLE TESTS

SECTION 2.1:

1. A 2. F 3. D 4. T 5. D 6. C 7. A 8. A 9. A 10. A 11. B 12. T
13. F 14. T 15. F 16. F

SECTION 2.2:

1. B 2. A 3. T 4. B 5. D 6. D 7. D 8. A 9. D 10. F

SECTION 2.3:

1. A 2. A 3. C 4. A 5. F

SECTION 2.4:

1. D 2. A 3. F 4. A 5. C 6. C 7. T 8. T 9. F 10. F 11. B 12. T
13. T 14. A 15. F

SECTION 2.5:

1. A 2. C 3. T 4. F 5. A 6. C 7. D 8. B 9. T 10. F 11. F 12. T
13. T 14. F 15. C

SECTION 2.6:

1. T 2. B 3. B 4. A 5. C 6. C 7. B 8. A 9. D 10. D 11. F 12. T
13. F 14. F 15. A

CHAPTER 2 TEST:

1. D 2. D 3. A 4. F 5. T 6. C 7. C 8. D 9. D 10. A 11. D 12. A
13. T 14. A 15. F 16. A 17. B 18. C 19. F 20. T 21. B 22. C 23. F

CHAPTER 3
Sample Exams

SECTION 3.1

1. Find the average rate of change of y with respect to x over the interval $[1, 5]$. $y = f(x) = \dfrac{1}{x^2}$.

A. 0.24 B. -0.24 C. 0.48 D. -0.48

2. Find the average rate of change of y with respect to x over the interval $[1, 4]$. $y = f(x) = x^3$.

A. 21 B. -21 C. 31.5 D. -31.5

3. Find the instantaneous rate of change of $y = x^4$ with respect to x at $x_0 = 3$.

A. 108 B. 27 C. 54 D. 13.5

4. Find the instantaneous rate of change of $y = \dfrac{1}{x}$ with respect to x at $x_0 = -2$.

A. 0.25 B. 0.5 C. -0.25 D. -0.5

5. Find the instantaneous rate of change of $y = 2x^3$ with respect to x at a general point x_0.

A. $6x_0^2$ B. $4x_0^2$ C. $\dfrac{2x_0^2}{3}$ D. x_0^4

6. Find the instantaneous rate of change of $y = \dfrac{2}{x}$ with respect to x at a general point x_0.

A. $-\dfrac{3}{x_0}$ B. $-\dfrac{3x_0}{2}$ C. $-\dfrac{3}{x_0^2}$ D. $-\dfrac{2}{x_0^2}$

7. Find the slope of the tangent to the graph of $f(x) = x^3 - 2$ at a general point x_0.

A. $2x_0 - 2$ B. $2x_0^2 - 2$ C. $2x_0^2$ D. $2x_0$

8. Answer true or false. The slope of the tangent line to the graph of $f(x) = x^2 - 4$ at $x_0 = 3$ is 2.

9. Answer true or false. Use a graphing utility to graph $y = x^3$ on $[0, 5]$. If this graph represents a position versus time curve for a particle, the instantaneous velocity of the particle is increasing over the graphed domain.

10. Use a graphing utility to graph $y = x^2 - 6x + 4$ on $[0, 10]$. If this graph represents a position versus time curve for a particle, the instantaneous velocity of the particle is zero at what time? Assume time is in seconds.

A. 0 s B. 3 s C. 5 s D. 10 s

11. A rock is dropped from a height of 64 feet and falls toward earth in a straight line. What is the instantaneous velocity downward when it hits the ground?

A. 64 ft/s B. 32 ft/s C. 2 ft/s D. 16 ft/s

12. Answer true or false. The magnitude of the instantaneous velocity is never less than the magnitude of the average velocity.

13. Answer true or false. If a rock is thrown straight upward from the ground, when it returns to earth its average velocity will be zero.

14. Answer true or false. If an object is thrown straight upward with a positive instantaneous velocity, its instantaneous velocity when it returns to the ground will be negative.

15. An object moves in a straight line so that after t s its distance in mm from its original position is given by $s = t^3 + t$. Its instantaneous velocity at $t = 5$ s is

A. 128 mm B. 28 mm C. 27 mm D. 76 mm

SECTION 3.2

1. Find the equation of the tangent line to $y = f(x) = 4x^2$ at $x = 2$.

A. $y = 8x$ B. $y = 8x + 16$ C. $y = 8x - 16$ D. $y = 16x + 16$

2. Find the equation of the tangent line to $y = f(x) = \sqrt{x+2}$ at $x = 7$.

A. $y = \dfrac{x}{6} + \dfrac{11}{6}$ B. $y = \dfrac{x}{6} - 4$ C. $y = \dfrac{x}{6}$ D. $y = \dfrac{1}{3}$

3. $y = x^2$. $dy/dx =$

A. 2 B. $2x^2$ C. $2x$ D. $\dfrac{x}{2}$

4. $y = \sqrt{x}$. $dy/dx =$

A. $\dfrac{\sqrt{x}}{2x}$ B. $\dfrac{\sqrt{x}}{x}$ C. $\dfrac{\sqrt{x}}{2}$ D. $\dfrac{2\sqrt{x}}{x}$

5.

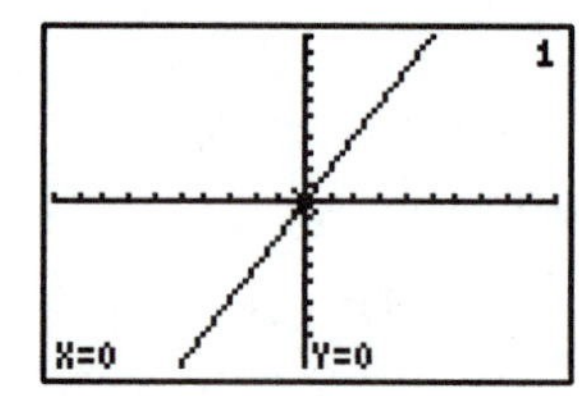

Answer true or false. The derivative of the function graphed on the left is graphed on the right.

6. Answer true or false. Use a graphing utility to help in obtaining the graph of $y = f(x) = |x|$. The derivative $f'(x)$ is not defined at $x = 0$.

7. Find $f'(t)$ if $f(t) = 4x^3 - 2$.

A. $12t^3 - 2$ B. $3t^2$ C. $4t^2$ D. $12t^2$

8. $\displaystyle\lim_{h \to 0} \frac{(4+h)^3 - 64}{h}$ represents the derivative of $f(x) = x^3$ at $x = a$. Find a.

A. 4 B. -4 C. 64 D. -64

9. $\displaystyle\lim_{h \to 0} \frac{\sqrt[3]{(27+h)} - 3}{h}$ represents the derivative of $f(x) = \sqrt[3]{x}$ at $x = a$. Find a.

A. 27 B. 3 C. -3 D. -27

10. Find an equation for the tangent line to the curve $y = x^3 - x^2 + x + 1$ at $(1, 2)$.

A. $y = -2$ B. $y = 2$ C. $y = x - 2$ D. $y = 2x$

11. Let $f(x) = \cos x$. Estimate $f'\left(\frac{\pi}{4}\right)$ by using a graphing utility.

A. $\frac{1}{4}$ B. $-\frac{\sqrt{2}}{2}$ C. $\frac{1}{2}$ D. $\frac{\pi}{4}$

12. An air source constantly increases the air supply rate of a balloon. The volume V in cubic feet is given by $V(t) = t^2$ $[0 \le t \le 5]$, where t is time in seconds. How fast is the balloon increasing at $t = 3$ s?

A. 6 ft^3/s B. 9 ft^3/s C. 18 ft^3/s D. 3 ft^3/s

13. Answer true or false. Using a graphing utility it can be shown that $f(x) = \sqrt[5]{x}$ is differentiable everywhere on $[-10, 10]$.

14. Answer true or false. A graphing utility can be used to determine that $f(x) = \begin{cases} x^3, & x \le 1 \\ x^2, & x > 1 \end{cases}$ is differentiable at $x = 1$.

15. Answer true or false. A graphing utility can be used to determine that $f(x) = \begin{cases} x^3, & x \le 0 \\ x^5, & x > 0 \end{cases}$ is differentiable at $x = 0$.

SECTION 3.3

1. Find dy/dx if $y = 7x^9$.

A. $16x^9$ B. $63x^9$ C. $16x^8$ D. $63x^8$

2. Find dy/dx if $y = e^3$.

A. $3e^2$ B. $2e^2$ C. $2e^3$ D. 0

3. Find dy/dx is $y = 3(x^2 - 2x + 4)$.

A. $6x - 6$ B. $3x^2 - 6x + 4$
C. $9x^3 - 12x^2 + 4x$ D. $2x - 2$

4. Answer true or false. If $f(x) = \sqrt{x} + x^2$, $f'(x) = \frac{\sqrt{x}}{2} + 2x$.

5. Answer true or false. If $y = \frac{1}{4x + 2}$, $y'(x) = \frac{1}{4}$.

6. If $y = \frac{2x}{x - 4}$, $dy/dx|_1 =$

A. $-\frac{8}{3}$ B. $\frac{8}{3}$ C. $-\frac{8}{9}$ D. $\frac{8}{9}$

7. $y = \frac{4}{x+5}$, $y'(0) =$

A. 0 B. $\frac{4}{5}$ C. $-\frac{4}{25}$ D. $-\frac{4}{5}$

8. $g(x) = x^2 f(x)$. Find $g'(2)$, given that $f(2) = 4$ and $f'(2) = 8$.

A. 48 B. -16 C. 16 D. 32

9. $y = 9x^2 + 4x + 6$. Find d^2y/dx^2.

A. 19 B. $18x + 4$ C. 18 D. 9

10. $y = x^{-6} + x^3$. Find y'''.

A. $-120x^{-4} + 6$ B. $-336x^{-9} + 6$ C. $120x^{-4} + 6$ D. $336x^{-9} + 6$

11. Answer true or false. $y = y''' + 12y' - 48x^3$ is satisfied by $y = x^4 + 4x^3 + 2$.

12. Use a graphing utility to locate all horizontal tangent lines to the curve $y = x^3 + 6x^2 + 2$.

A. $x = 0, 4$ B. $x = -4, 0$ C. $x = 0$ D. $x = -4$

13. Find the x-coordinate of the point on the graph of $y = x^4$ where the tangent line is parallel to the secant line that cuts the curve at $x = 0$ and at $x = 1$.

A. $\frac{1}{\sqrt[3]{4}}$ B. 4 C. 1 D. $\frac{1}{2}$

14. The position of a moving particle is given by $s(t) = 3t^2 + 2t + 1$ where t is time in seconds. The velocity in m/s is given by ds/dt. Find the velocity at $t = 2$.

A. 6 m/s B. 8 m/s C. 16 m/s D. 14 m/s

15. Answer true or false. If f, g, and h are differentiable functions, and $h \neq 0$ anywhere on its domain, then $\left(\frac{fg}{h}\right)' = \frac{hf'g' - fgh'}{y^2t}$.

SECTION 3.4

1. Find $f'(x)$ if $f(x) = x^4 \sin x$.

A. $4x^3 \cos x$ B. $-4x^3 \cos x$
C. $4x^3 \sin x + x^4 \cos x$ D. $4x^3 \sin x - x^4 \cos x$

2. Find $f'(x)$ if $f(x) = \sin x \tan x$.

A. $\cos x$ B. $(\sin x)(1 + \sec^2 x)$
C. $(\sin x)(1 - \sec^2 x)$ D. $-\cos x$

3. Find $f'(x)$ if $f(x) = \sin^2 x + \cos x$.

A. $\sin x$ B. $2 \sin x \cos x - \sin x$
C. $2 \sin x - \cos x$ D. $3 \sin x$

4. Find d^2y/dx^2 if $y = x \sin x$.

A. $-x \sin x + 2 \cos x$ B. 0
C. $-\cos x$ D. $\cos x$

5. Answer true or false. If $y = \sec x$, $d^2y/dx^2 = \sec x$.

6. Find the equation of the line tangent to the graph of $y = \sin x$ at the point where $x = 0$.

A. $y = -1$ B. $y = -x$ C. $y = x$ D. $y = 1$

7. Find the x-coordinates of all points in the interval $[-2\pi, 2\pi]$ at which the graph of $f(x) = \csc x$ has a horizontal tangent line.

A. $-3\pi/2$, $-\pi/2$, $\pi/2$, $3\pi/3$ B. $-\pi$, π
C. $-\pi$, 0, π D. $-3\pi/2$, 0, $3\pi/2$

8. Find $d^{93} \cos x/dx^{93}$.

A. $\cos x$ B. $y = -\cos x$ C. $\sin x$ D. $-\sin x$

9. Find all x-values on $(0, 2\pi)$ where $f(x)$ is not differentiable. $f(x) = \tan x \cos x$

A. $\pi/2$, $3\pi/2$ B. π C. $\pi/2$, π, $3\pi/2$ D. None

10. Answer true or false. If x is given in radians, the derivative formula for $y = \tan x$ in degrees is $y' = \dfrac{\pi}{180} \sec^2 x$.

11. A rock at an elevation angle of θ is falling in a straight line. If at a given instant it has an angle of elevation of $\theta = \pi/4$ and has a horizontal distance s from an observer, find the rate at which the rock is falling with respect to θ.

A. $\sec\left(\frac{\pi}{4}\right)$ B. $s \sec^2\left(\frac{\pi}{4}\right)$ C. $\dfrac{\sec\left(\frac{\pi}{4}\right)}{s}$ D. $\sec\left(\frac{s\pi}{4}\right)$

12. Answer true or false. If $f(x) = \tan x \cos x - \cot x \sin x$, $f'(x) = \cos x - \sin x$.

13. Answer true or false. If $f(x) = \dfrac{1}{\tan x}$, $f'(x) = -\csc^2 x$.

14. Answer true or false. $f(x) = \dfrac{\sin x}{1 - \cos x}$ is differentiable everywhere.

15. If $y = x^3 \sin x$, find d^2y/dx^2.

A. $6x \sin x$ B. $6x \sin x + 6x^2 \cos x + x^3 \sin x$
C. $6x \sin x + 6x^2 \cos x - x^3 \sin x$ D. $6x \sin x - x^3 \sin x$

SECTION 3.5

1. $f(x) = \sqrt{x^2 - 4x + 3}$. $f'(x) =$

A. $\dfrac{x-2}{\sqrt{x^2 - 4x + 3}}$ B. $\dfrac{x-4}{\sqrt{x^2 - 4x + 3}}$ C. $\dfrac{1}{2\sqrt{x^2 - 4x + 3}}$ D. $2x - 4$

2. $f(x) = (x^5 - 2)^{20}$. $f'(x) =$

A. $20(x^5 - 2)^{19}$ B. $100x^4(x^5 - 2)^{19}$ C. $100x^5(x - 2)^{20}$ D. $100x^4$

3. $f(x) = \sin(3x)$. $f'(x) =$

A. $\cos(3x)$ B. $3\cos(3x)$ C. $-\cos(3x)$ D. $-3\cos(3x)$

4. Answer true or false. If $f(x) = \sqrt{\sin^2 x + 3}$, $f'(x) = \dfrac{\cos x}{\sqrt{\sin^2 x + 3}}$.

5. $f(x) = x^3\sqrt{x^2 - 2}$. $f'(x) =$

A. $\dfrac{x^3}{2\sqrt{x^2 - 2}} + 3x^2\sqrt{x^2 - 2}$ B. $3x^2\sqrt{x^2 - 2}$

C. $6x^4$ D. $\dfrac{x^4}{\sqrt{x^2 - 2}} + 3x^2\sqrt{x^2 - 2}$

6. $y = \sin(\cos x)$. Find dy/dx.

A. $-\cos(\cos x)\sin x$ B. $\cos(\cos x)\sin x$

C. $\sin(\cos x)\cos x$ D. $-\sin(\cos x)\cos x$

7. $y = x^4\tan(6x)$. Find dy/dx

A. $24x^3\sec(6x)$ B. $4x^3\tan(6x)$

C. $6x^4\sec^2(6x) + 4x^2\tan(6x)$ D. $6x^4\sec^2 x + 4x^3\tan x$

8. $y = \left(\dfrac{1 + \sin^2 x}{\cos x}\right)$. $dy/dx =$

A. $\dfrac{2\sin x + \sin x\cos^2 x}{\cos^2 x}$ B. $\dfrac{\sin x + 2\sin x\cos^2 x - \sin^3 x}{\cos^2 x}$

C. $\dfrac{2\cos x}{\sin x}$ D. $-\dfrac{2\cos x}{\sin x}$

9. Answer true or false. If $y = \cos(5x^3)$, $d^2y/dx^2 = -\cos(5x^3)$.

10. Answer true or false. $y = \sin x^3 - \cos x^2$. $d^2y/dx^2 = 6x\cos x^3 - 9x^4\sin x^3 + 2\sin x^2 + 4x^2\cos x^2$.

11. Find an equation for the tangent line to the graph of $y = x\tan x$ at $x = \pi/4$.

A. $y - \frac{\pi}{4} = \left(\frac{\pi}{2} + 1\right)\left(x - \frac{\pi}{4}\right)$ B. $y - 1 = x - \frac{\pi}{4}$

C. $y - \frac{\pi}{4} = x - 1$ D. $y - \frac{\pi}{4} = \frac{\sqrt{2}}{2}\left(x - \frac{\pi}{4}\right)$

12. $y = \sin^3(\pi - 2\theta)$. Find $dy/d\theta$.

A. $-6\sin^2(\pi - 2\theta)\cos(\pi - 2\theta)$ B. $-6\sin^2(\pi - 2\theta)$

C. $-6\sin^3(\pi - 2\theta)$ D. $3\sin^2(\pi - 2\theta)\cos(\pi - 2\theta)$

13. Use a graphing utility to obtain the graph of $f(x) = (x + 2)^3\sqrt{x}$. Determine the slope of the tangent line to the graph at $x = 1$.

A. 1 B. 54 C. 27 D. 40.5

14. Find the value of the constant A so that $y = A\cos 3t$ satisfies $d^2y/dt^2 + 3y = \cos 3t$.

A. $-\dfrac{1}{12}$ B. $\dfrac{1}{6}$ C. $-\dfrac{1}{6}$ D. $-\dfrac{9}{2}$

15. Answer true or false. Given $f'(x) = \sqrt{x+2}$ and $g(x) = x^3$, then $F'(x) = \sqrt{x^3+2}(3x^2)$ if $F(x) = f(g(x))$.

SECTION 3.6

1. Answer true or false. If $y = \sqrt[5]{4x+2}$, $\dfrac{dy}{dx} = \dfrac{4}{5\sqrt{4x+2}}$.

2. Answer true or false. If $y^3 = x$, $\dfrac{dy}{dx} = \dfrac{1}{3y^2}$.

3. Find dy/dx if $\sqrt[3]{y} - \cos x = 2$.

A. $dy/dx = -3y^{2/3}\sin x$ B. $dy/dx = 3y^{2/3}\sin x$
C. $dy/dx = -6y^{2/3}\sin x$ D. $dy/dx = 6y^{2/3}\sin x$

4. Find dy/dx if $x^2 + y^2 = 25$.

A. $\dfrac{25x}{y}$ B. $\dfrac{x}{y}$ C. $-\dfrac{x}{y}$ D. $-\dfrac{25x}{y}$

5. Answer true or false. If $y^2 + 2xy = 5x$, $\dfrac{dy}{dx} = \dfrac{5}{2y+2x}$.

6. $2x^2 + 3y^2 = 9$. Find d^2y/dx^2.

A. $\dfrac{d^2y}{dx^2} = \dfrac{16x^2+4y}{36y^3}$ B. $\dfrac{d^2y}{dx^2} = \dfrac{16x^2+4y}{6y^3}$
C. $\dfrac{d^2y}{dx^2} = \dfrac{16x^2-4y}{36y^3}$ D. $\dfrac{d^2y}{dx^2} = \dfrac{-18y-12x^2}{27y^3}$

7. Find the slope of the tangent line to $x^2 + y^2 = 5$ at $(1, 2)$.

A. $\dfrac{1}{2}$ B. $-\dfrac{1}{2}$ C. $\dfrac{5}{2}$ D. $-\dfrac{5}{2}$

8. Find the slope of the tangent line to $xy^2 = 4$ at $(1, 2)$.

A. 4 B. -4 C. 1 D. -1

9. Find dy/dx if $xy^4 = x^3$.

A. $\dfrac{3x}{4y^4}$ B. $\dfrac{3x^2-y^4}{4xy^3}$ C. $-\dfrac{3x^2}{4y^4}$ D. $\dfrac{3x^2+y^4}{4xy^3}$

10. Find dy/dx if $x = \cos(xy)$.

A. $-\dfrac{1}{\sin(xy)}$ B. $\dfrac{1}{\sin(xy)}$ C. $\dfrac{1+y\sin(xy)}{x\sin(xy)}$ D. $-\dfrac{1+y\sin(xy)}{x\sin(xy)}$

11. Answer true or false. If $\cos x = \sin y$, $dy/dx = \tan x$.

12. Answer true or false. If $\tan(xy) = 4$, $\frac{dy}{dx} = -\frac{y\sec^2(xy)}{x}$.

13. $xy^2 = x^3 - 2x$ has a tangent line parallel to the x-axis at

A. $(-1, 1)$ B. $(0, 0)$ C. $(2, 2)$ D. $(1, 2)$

14. $x^2 + y^2 = 16$ has tangent lines parallel to the y-axis at

A. $(0, -16)$ and $(0, 16)$
B. $(0, -4)$ and $(0, 4)$
C. $(-4, 0)$ and $(4, 0)$
D. $(-16, 0)$ and $(16, 0)$

15. Find dy/dx if $h^2 + t^2 y = 2$ and $dt/dx = 3$.

A. $\frac{-3x^2 - 6ty}{2yt + t^2 - 1}$ B. $-3y^2$ C. $\frac{-3y^2 - y}{2y + 1}$ D. $\frac{-3y^2 - 6ty}{2yt + t^2}$

SECTION 3.7

1. The volume of a sphere is given by $V = \frac{4}{3}\pi r^3$. Find $\frac{dV}{dt}$ in terms of $\frac{dr}{dt}$.

A. $\frac{dV}{dr} = 4\pi r^2 \frac{dr}{dt}$ B. $\frac{dV}{dr} = \frac{4}{3}\pi r^3 \frac{dr}{dt}$ C. $\frac{dV}{dr} = \frac{4}{3}\pi r^2 \frac{dr}{dt}$ D. $\frac{dV}{dr} = 3r^2 \frac{dr}{dt}$

2. A cylinder of length 2 m and radius 1 m is expanding such that $dl/dt = 0.01$ m and $dr/dt = 0.02$ m. Find dV/dt.

A. 0.0013 m B. 0.2827 m C. 0.015 m D. 0.03 m

3. A 10-ft ladder rests against a wall. If it were to slip so that when the bottom of the ladder is 6 feet from the wall it will be moving at 0.02 ft/s, how fast would the ladder be moving down the wall?

A. 0.02 ft/s B. 0.0025 ft/s C. 0.015 ft/s D. 0.12 ft/s

4. A plane is approaching an observer with a horizontal speed of 200 ft/s and is currently 10,000 ft from being directly overhead at an altitude of 20,000 ft. Find the rate at which the angle of elevation, θ, is changing with respect to time, $d\theta/dt$?

A. 0.020 rad/s B. 0.010 rad/s C. 0.08 rad/s D. 0.009 rad/s

5. Answer true or false. Suppose $z = y^5 x^3$. $dz/dt = (dy/dt)^5 + (dx/dt)^3$.

6. Suppose $z = \sqrt{x^2 + y^2}$, $dz/dt =$

A. $2x\,dx/dt + 2y\,dy/dt$
B. $dx/dt + dy/dt$
C. $\sqrt{(dx/dt)^2 + (dy/dt)^2}$
D. $\frac{2x\,dx/dt + 2y\,dy/dt}{2\sqrt{x^2 + y^2}}$

7. The power in watts for a circuit is given by $P = I^2 R$. How fast is the power changing if the resistance, R, of the circuit is 1,000 Ω, the current, I, is 2 A, and the current is decreasing with respect to time at a rate of 0.03 A/s.

A. -0.09 w/s B. -60 w/s C. -120 w/s D. -1.8 w/s

8. Gravitational force is inversely proportional to the distance between two objects squared. If $F = \frac{5}{d^2}$ N at a distance $d = 3$ m, how fast is the force diminishing if the objects are moving away from each other at 2 m/s?

A. 2 N/s B. −0.74 N/s C. −6.7 N/s D. −1.1 N/s

9. A point P is moving along a curve whose equation is $y = \sqrt{x^4 + 9}$. When $P = (2, 5)$, y is increasing at a rate of 2 units/s. How fast is x changing?

A. 2 units/s B. $\frac{5}{8}$ units/s C. 64 units/s D. $\frac{5}{16}$ units/s

10. Water is running out of an inverted conical rank at a rate of 3 ft^3/s. How fast is the height of the water in the tank changing if the height is currently 5 ft and the radius is 5 ft.

A. 0.344 ft/s B. 0.377 ft/s C. 1.131 ft/s D. 9.425 ft/s

11. Answer true or false. If $z = x \ln y$, $\frac{dz}{dt} = \left(\frac{dx}{dt}\right)\left(\frac{dy}{dt}\right)$.

12. Answer true or false. If $z = e^x \ln y$, $\frac{dz}{dt} = \frac{e^x}{y}\frac{dy}{dt} + e^x \ln y \frac{dx}{dt}$.

13. Answer true or false. If $\sin\theta = 3xy$, $\frac{d\theta}{dt} = 3x\frac{dy}{dt} = 3y\frac{dx}{dt}$.

14. Answer true or false. If $V = 3x^2y$, $\frac{dy}{dt} = \frac{dV}{dt} - 6x\frac{dx}{dt}$.

15. Answer true or false. If $V = 10x^3$, $\frac{dx}{dt} = \frac{1}{30x^2}\frac{dV}{dt}$.

SECTION 3.8

1. If $y = \sqrt[3]{x}$, find the formula for Δy.

A. $\Delta y = \sqrt[3]{x + \Delta x} - \sqrt[3]{x}$

B. $\Delta y = \sqrt[3]{x + \Delta x}$

C. $\Delta y = \frac{1}{3\sqrt[3]{(x+\Delta x)^2}} - \frac{1}{3\sqrt[3]{x^2}}$

D. $\Delta y = \frac{1}{3\sqrt[3]{(x+\Delta x)^2}}$

2. If $y = x^4$, find the formula for Δy.

A. $\Delta y = (x + \Delta x)^4$

B. $\Delta y = 4x^3 \Delta x$

C. $\Delta y = 4(x - \Delta x)^3$

D. $\Delta y = (x + \Delta x)^4 - x^4$

3. If $y = \cos x$, find the formula for Δy.

A. $\Delta y = \cos(x + \Delta x) - \cos x$

B. $\Delta y = \cos(x + \Delta x)$

C. $\Delta y = -\sin x \Delta x$

D. $\Delta y = \Delta x + \cos x$

4. Answer true or false. The formula for dy is obtained from the formula for Δy by replacing Δx with dx.

5. $y = x^5$. Find the formula for dy.

A. $dy = (x + dx)^5$
B. $dy = (x + dx)^5 - x^5$
C. $dy = x^5 + (dx)^5$
D. $dy = 5x^4dx$

6. $y = \tan x$. Find the formula for dy.

A. $dy = \sec^2 xdx$
B. $dy = \tan xdx$
C. $dy = \tan(x + dx) - \tan x$
D. $dy = \tan(x + dx)$

7. $y = x^3 \sin x$. Find the formula for dy.

A. $dy = (3x^2 \sin x + x^3 \cos x)dx$
B. $dy = 3(x + dx)^3 \sin(x + dx)$
C. $dy = 3(x + dx)^3 \sin(x + dx) - 3x^3 \sin x$
D. $dy = (x + dx)^3 \sin(x + dx) - x^3 \sin x$

8. Let $y = \frac{1}{x^2}$. Find dy at $x = 1$ if $dx = 0.01$.

A. 0.02 B. −0.02 C. −0.001 D. 0.001

9. Let $y = x^5$. Find dy at $x = 1$ if $dx = -0.01$.

A. 0.00000005 B. −0.00000005 C. 0.05 D. −0.000000005

10. Let $y = \sqrt{x}$. Find Δy at $x = 3$ if $\Delta x = 1$.

A. −0.268 B. 0.268 C. 0.289 D. 0.250

11. Use dy to approximate $\sqrt{3.96}$ starting at $x = 4$.

A. 2.01 B. 1.99 C. 4.01 D. 3.99

12. Answer true or false. A circular spill is spreading so that when its radius r is 2 m, $dr = 0.05$ m. The corresponding change in the area covered by the spill, A, is, to the nearest hundredth of a square meter, 0.63 m^2.

13. A small suspended droplet of radius 10 microns is evaporating. If $dr = -0.001$ micron find the change in the volume, dV, to the nearest thousandth of a cubic micron.

A. −420.237 B. −1.257 C. −0.419 D. −4.189

14. Answer true or false. A cube is expanding as temperature increases. If the length of the cube is changing at a rate of $\Delta x = 2$ mm when x is 1 m, the volume is experiencing a corresponding change of 0.006 mm^3, correct to the nearest thousandth.

15. Answer true or false. The radius of the base of a cylinder is 2 mm with a possible error of ± 0.01 mm. The height of the cylinder is exactly 4m. Using differentials to estimate the maximum error of the volume, it is found to be 502.65 mm^3, correct to the nearest hundredth.

CHAPTER 3 TEST

1. Find the average rate of change of y with respect to x over the interval $[1, 2]$. $y = f(x) = 3x^2$.

A. 9 B. -9 C. 11 D. -11

2. Find the instantaneous rate of change of $y = 2x^4$ with respect to x at $x_0 = 3$.

A. 216 B. 54 C. 108 D. 27

3. An object moves in a straight line so that after t s its distance from its original position is given by $s = t^3 - 2t$. Its instantaneous velocity at $t = 4$ s is

A. 56 B. 46 C. 14 D. 11.5

4. Find the equation of the tangent line to $y = f(x) = 3x^3$ at $x = 3$.

A. $y = 27x - 60$ B. $y = 27x + 60$ C. $y = 81x + 162$ D. $y = 81x - 162$

5. If $y = x^8$, $dy/dx =$

A. $8x^7$ B. $8x^8$ C. $7x^7$ D. $7x^8$

6.

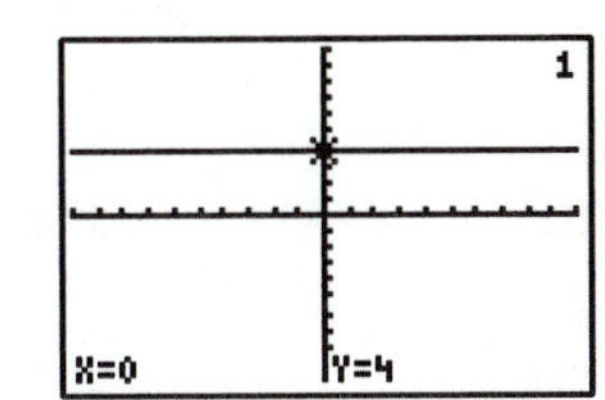

Answer true or false. The derivative of the function graphed on the left is graphed on the right.

7. $\lim\limits_{h \to 0} \dfrac{(6 - 2h)^2 - (2h)^2}{h}$ represents the derivative of $f(x) = (2x)^2$ at $x =$

A. 6 B. 3 C. -6 D. -3

8. Let $f(x) = \tan x$. Estimate $f'(5\pi/4)$ by using a graphing utility

A. 1 B. -1 C. 0 D. 2

9. Find dy/dx if $y = \pi^5$.

A. $5\pi^4$ B. π^5 C. 0 D. $4\pi^5$

10. Answer true or false. If $f(x) = \sqrt{x^3} + x^3$, $f'(x) = \dfrac{1}{2\sqrt{x^3}} + 3x^2$.

11. If $y = \dfrac{5x}{x - 2}$, $\left.\dfrac{dy}{dx}\right|_1 =$

A. 10 B. -10 C. 8 D. -8

12. $f(x) = \sqrt{x}\, f(x)$. Find $g'(4)$ given that $f(4) = 4$ and $f'(4) = 6$.

A. 24 B. 1.5 C. 13 D. 14

13. Find $f'(x)$ if $f(x) = x^3 \tan x$.

A. $3x^2 \sec^2 x$
B. $-3x^2 \sec^2 x$
C. $3x^2 \tan x + x^3 \sec^2 x$
D. $3x^2 \tan x - x^3 \sec^2 x$

14. Find d^2y/dx^2 if $y = -2(\sin x)(\cos x)$

A. $8(\cos x)(\sin x)$ B. $-8(\cos x)(\sin x)$ C. $2(\cos x)(\sin x)$ D. $-2(\cos x)(\sin x)$

15. Find the x-coordinates of all the points over the interval $(0, \pi)$ where the graph of $f(x) = \cot x$ has a horizontal tangent line.

A. $\pi/4$, $\pi/2$, $3\pi/4$ B. $\pi/4$, $3\pi/4$ C. $\pi/2$ D. None exist.

16. Answer true or false. $\dfrac{d^{105}}{dx^{105}} \sin x = \cos x$.

17. Answer true or false. If $f(x) = \sqrt{x^3 - 2x^2}$, $f'(x) = \dfrac{3x^2 - 4x}{\sqrt{x^3 - 2x^2}}$.

18. If $f(x) = \sin(6x)$, $f'(x) =$

A. $6\cos(6x)$ B. $-6\cos(6x)$ C. $\cos(6x)$ D. $-\cos(6x)$

19. If $y = \sqrt[9]{x}$, find the formula for Δy.

A. $\Delta y = \sqrt[9]{x - \Delta x} + \sqrt[9]{x}$
B. $\Delta y = \sqrt[9]{x + \Delta x} - \sqrt[9]{x}$
C. $\Delta y = \dfrac{\Delta x}{9\sqrt[9]{x^8}}$
D. $\Delta y = \sqrt[9]{x + \Delta x} + \sqrt[9]{x}$

20. If $y = x^2 \tan x$, find the formula for dy.

A. $dy = 2x \tan x dx$
B. $dy = x^2 \sec^2 x dx$
C. $dy = (2x \tan x + x^2 \sec^2 x)dx$
D. $dy = (2x \tan x - x^2 \sec^2 x)dx$

21. Answer true or false. If $y = \dfrac{1}{x^5}$, dy at $x = 2$ is $-\dfrac{5}{16}dx$.

22. Answer true or false. A spherical balloon is deflating. The rate the volume is changing at $r = 2$ m is given by $dV = -16\pi dr$.

ANSWERS TO SAMPLE TESTS

SECTION 3.1:

1. B 2. A 3. A 4. C 5. A 6. D 7. D 8. F 9. T 10. B 11. A 12. F
13. T 14. T 15. D

SECTION 3.2:

1. B 2. A 3. C 4. A 5. T 6. T 7. D 8. A 9. B 10. D 11. B 12. A
13. F 14. F 15. T

SECTION 3.3:

1. D 2. D 3. A 4. F 5. F 6. C 7. C 8. A 9. C 10. B 11. F 12. B
13. A 14. D 15. F

SECTION 3.4:

1. C 2. B 3. B 4. A 5. F 6. C 7. A 8. D 9. A 10. F 11. B 12. F
13. T 14. F 15. C

SECTION 3.5:

1. A 2. B 3. B 4. F 5. D 6. A 7. C 8. A 9. F 10. T 11. A 12. A
13. D 14. C 15. T

SECTION 3.6:

1. F 2. T 3. A 4. C 5. F 6. D 7. B 8. D 9. B 10. D 11. F 12. F
13. A 14. B 15. A

SECTION 3.7:

1. A 2. B 3. C 4. C 5. F 6. D 7. C 8. B 9. B 10. A 11. F 12. T
13. F 14. F 15. T

SECTION 3.8:

1. A 2. D 3. A 4. F 5. D 6. A 7. A 8. B 9. D 10. B 11. B 12. T
13. B 14. T 15. T

CHAPTER 3 TEST:

1. A 2. A 3. B 4. D 5. A 6. T 7. B 8. D 9. C 10. F 11. B 12. C
13. C 14. A 15. C 16. T 17. F 18. A 19. B 20. C 21. F 22. T

CHAPTER 4
Sample Exams

SECTION 4.1

1. Answer true or false. The functions $f(x) = \sqrt[3]{x+3}$ and $g(x) = x^3 - 3$ are inverses of each other.

2. Answer true or false. The functions $f(x) = \sqrt[6]{x}$ and $g(x) = x^6$ are inverses of each other.

3. Answer true or false. $\sec x$ is a one-to-one function.

4. Find $f^{-1}(x)$ if $f(x) = x^7$.

A. $\sqrt[7]{x}$ B. $\dfrac{1}{x^7}$ C. $-\sqrt[7]{x}$ D. $-\dfrac{1}{x^7}$

5. Find $f^{-1}(x)$ if $f(x) = 3x - 4$.

A. $\dfrac{1}{3x-4}$ B. $\dfrac{x+4}{3}$ C. $\dfrac{x}{3} + 4$ D. $\dfrac{1}{3x} - 4$

6. Find $f^{-1}(x)$ if $f(x) = \sqrt[9]{x+3}$.

A. $x^9 - 3$ B. $(x+3)^9$ C. $x^9 + 3$ D. $\dfrac{1}{\sqrt[9]{x+3}}$

7. Find $f^{-1}(x)$, if it exists, for the function $f(x) = \begin{cases} -x^2, & x < 0 \\ x^2, & x \geq 0 \end{cases}$.

A. $\begin{cases} -\sqrt{x}, & x < 0 \\ \sqrt{x}, & x \geq 0 \end{cases}$ B. $\begin{cases} -\dfrac{1}{x}, & x < 0 \\ \dfrac{1}{x}, & x \geq 0 \end{cases}$ C. $\sqrt{|x|}$ D. It does not exist.

8. Answer true or false. If f has a domain of $0 \leq x \leq 10$, then f^{-1} has a domain of $0 \leq x \leq 10$.

9. The graphs of f and f^{-1} are reflections of each other about the

A. x-axis B. y-axis C. line $x = y$ D. origin.

10. Answer true or false. A 200 foot fence is used as a perimeter about a rectangular plot. The formula for the length of the fence is the inverse of the formula for the width of the fence.

11. Find the domain of $f^{-1}(x)$ if $f(x) = (x+3)^2$, $x \geq -3$.

A. $x \geq -3$ B. $x \geq 3$ C. $x \geq 0$ D. $x \leq 0$

12. Find the domain of $f^{-1}(x)$ if $f(x) = -\sqrt{x-5}$.

A. $x \leq 0$ B. $x \geq 0$ C. $x \leq 5$ D. $x \geq 5$

13. Let $f(x) = x^2 + 6x + 9$. Find the smallest value of k such that $f(x)$ is a one-to-one function on the interval $[k, \infty)$.

A. 0 B. -3 C. 3 D. -9

14. Answer true or false. $f(x) = -x$ is its own inverse.

15. Answer true or false. to have an inverse a trigonometric function must have its domain restricted to $[0, 2\pi]$.

SECTION 4.2

1. $2^{-5} =$ A. $\frac{1}{10}$ B. $-\frac{1}{10}$ C. $\frac{1}{32}$ D. $-\frac{1}{32}$

2. Use a calculating utility to approximate $\sqrt[6]{29}$. Round to four decimal places.

   A. 1.7528 B. 5.3852 C. 1.7530 D. 5.3854

3. Use a calculating utility to approximate $\log 28.4$. Round to four decimal places.

   A. 3.3464 B. 3.3462 C. 1.4533 D. 1.4535

4. Find the exact value of $\log_3 81$. A. 12 B. $\frac{3}{4}$ C. $\frac{1}{4}$ D. 4

5. Use a calculating utility to approximate $\ln 39.1$ to four decimal places.

   A. 1.5920 B. 3.6661 C. 1.5922 D. 3.6663

6. Answer true or false. $\ln \frac{a^3 b}{c^2} = 3\ln a + \ln b - 2\ln c$.

7. Answer true or false. $\log(5x\sqrt{x-2}) = (\log 5)(\log x)(\log^{1/2}(x-2))$.

8. Rewrite the expression as a single logarithm. $5\log 2 - \log 12 + \log 24$

   A. $\log 64$ B. $\log 22$ C. $\frac{\log 24}{5}$ D. $\log 44$

9. Solve $\log_{10}(x+5) = 1$ for x. A. 5 B. -5 C. 0 D. no solution

10. Solve for x. $\log_{10} x^{5/2} - \log_{10} x^{3/2} = 4$.

    A. 4 B. 40 C. 1,000 D. 10,000

11. Solve $4^{-2x} = 6$ for x to four decimal places.

    A. 0.6462 B. -0.6462 C. 1.2925 D. -1.2925

12. Solve for x. $3e^x - xe^x = 0$ A. 3 B. -3 C. $\frac{1}{3}$ D. $-\frac{1}{3}$

13. 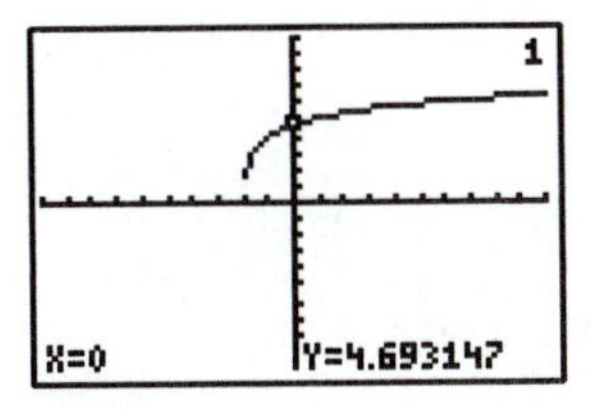

    This is a graph of

    A. $4 - \ln(2+x)$ B. $4 + \ln(2+x)$ C. $4 - \log(x-2)$ D. $4 + \log(x-2)$.

14. Use a calculating utility and change of base formula to find $\log_3 4$.

    A. 0.2007 B. 0.7925 C. 1.2619 D. 0.4621

15. The equation $Q = 6e^{-0.052t}$ gives the mass Q in grams of a certain radioactive substance remaining after t hours. How much remains after 6 hours?

A. 4.3919 g B. 4.3920 g C. 4.3921 g D. 2.3922 g

SECTION 4.3

1. If $y = \ln 6x$ find dy/dx. A. $\frac{1}{6x}$ B. $\frac{6}{x}$ C. $\frac{1}{x}$ D. $\frac{6\ln 6x}{x}$

2. If $y = \ln(\cos x)$ find dy/dx. A. $\tan x$ B. $-\tan x$ C. $\frac{1}{\cos x}$ D. $-\frac{1}{\cos x}$

3. If $y = \sqrt{3 + \ln^2 x^2}$, $dy/dx =$

A. $\frac{2}{x\sqrt{3+\ln^2 x^2}}$ B. $\frac{2}{\sqrt{3+\ln^2 x^2}}$ C. $\frac{1}{\sqrt{3+\ln^2 x^2}}$ D. $\frac{2\ln x^2}{x\sqrt{3+\ln^2 x^2}}$

4. Answer true or false. If $y = x^8 e^{7x}$, $dy/dx = 56x^7 e^{7x}$.

5. Answer true or false. If $y = \ln(x^3)$, $\frac{dy}{dx} = \frac{3}{x}$.

6. If $y = (\ln x)e^{2x}$, $dy/dx =$

A. $2(\ln x)e^{2x} + \frac{e^{2x}}{x}$ B. $2(\ln x)e^{2x}$ C. $2(\ln x)e^{2x}\frac{e^{2x-1}}{x}$ D. $\frac{e^{2x}}{x}$

7. Answer true or false. If $x + e^{xy} = 2$, $\frac{dy}{dx} = \frac{-ye^{xy} - 1}{xe^{xy}}$.

8. $y = \ln\left[\frac{1}{\sin x}\right]$. Find $\frac{dy}{dx}$.

A. $\cos x \sin x$ B. $\cot x$ C. $\tan x$ D. $-\cot x$

9. If $y = \sqrt[8]{\frac{x+2}{x+3}}$, find $\frac{dy}{dx}$ by logarithmic differentiation.

A. $\frac{1}{8}\left(\frac{x+2}{x+3}\right)^{-7/8}$ B. $\frac{1}{8(x+3)^2}$

C. $\frac{1}{8}\left(\frac{1}{x+2} - \frac{1}{x+3}\right)\sqrt[8]{\frac{x+2}{x+3}}$ D. $\sqrt[7]{\frac{x+2}{x+3}}$

10. $f(x) = 5^x$. Find $df(x)/dx$.

A. $5^x \ln 5$ B. 5^{x-1} C. $x \ln 5^x$ D. $5^x \ln x$

11. Answer true or false. If $f(x) = \pi^{\sin x + \cos x}$, $dy/dx = (\sin x + \cos x)\pi^{\sin x + \cos x - 1}$

12. $y = \ln(kx)$. $d^n y/dx^n =$ A. $\frac{1}{k^n x^n}$ B. $\frac{(-1)^n}{k^n x^n}$ C. $\frac{(-1)^{n+1}}{x^n}(k-1)!$ D. $\frac{1}{x}$

13. Answer true or false. If $y = x^{\cos x}$, $\frac{dy}{dx} = \left(\frac{\cos x}{x} - \sin x \ln x\right) x^{\cos x}$.

14. $y'x = -yx$ is satisfied by $y =$ A. e^x B. $\cos x$ C. $\sin x$ D. e^{-x}

15. $\lim\limits_{h\to 0} \dfrac{3^h - 1}{2h} =$ A. 1 B. 0 C. $+\infty$ D. $\dfrac{\ln 3}{2}$

SECTION 4.4

1. Find the exact value of $\sin^{-1}(1)$. A. 0 B. $\pi/2$ C. π D. $3\pi/2$

2. Find the exact value of $\cos^{-1}(\cos(3\pi/4))$.

A. $3\pi/4$ B. $\pi/4$ C. $-\pi/4$ D. $5\pi/4$

3. Use a calculating utility to approximate x if $\tan x = 3.1$, $-\pi/2 < x < \pi/2$.

A. 1.2582 B. 1.2588 C. 1.2593 D. 1.2595

4. Use a calculating utility to approximate x if $\sin x = 0.15$, $\pi/2 < x < 3\pi/2$.

A. 0.1506 B. 2.9910 C. 3.2932 D. no solution

5. Answer true or false. $\cos^{-1} x = \dfrac{1}{\cos x}$ for all x.

6. $y = \sin^{-1}(2x)$. Find dy/dx.

A. $\dfrac{2}{\sqrt{1-2x^2}}$ B. $\dfrac{2}{\sqrt{1-x^2}}$ C. $\dfrac{1}{\sqrt{1-4x^2}}$ D. $\dfrac{2}{\sqrt{1-4x^2}}$

7. $y = \tan^{-1}\sqrt{x}$. Find dy/dx.

A. $\dfrac{\sqrt{x}}{2x(1+x)}$ B. $\dfrac{x}{1+x}$ C. $\dfrac{x}{2(1+x^2)}$ D. $\sqrt{\dfrac{1}{1+x^2}}$

8. $y = x^{\sin^{-1} x}$. $dy/dx =$

A. $\dfrac{e^{\sin^{-1} x}}{\sqrt{1-x^2}}$ B. $-\sin^{-2} x e^{\sin^{-1} x}$ C. $\dfrac{1}{\cos x e^{\sin x}}$ D. $-\dfrac{\cos^{-1} x e^{\sin^{-1} x}}{\sqrt{1-x^2}}$

9. $y = \ln(x \cos^{-1} x)$. Find dy/dx.

A. $\dfrac{1}{x\cos^{-1} x}$

B. $\dfrac{\sqrt{1-x^2}}{-x + \cos^{-1} x\sqrt{1-x^2}}$

C. $\dfrac{-2\sqrt{1-x^2}\sin^{-1}(x) + \pi\sqrt{1-x^2} - 2x}{2x\sqrt{1-x^2}\cos^{-1}(x)}$

D. $\dfrac{1}{x}$

10. $y = \sqrt{\sin^{-1} x}$. Find dy/dx.

A. $y = \dfrac{1}{\sqrt[4]{1-x^2}}$

B. $y = \dfrac{1}{2(\sqrt{\sin^{-1} x})(\sqrt{1-x^2})}$

C. $y = \dfrac{1}{2\sqrt[4]{1-x^2}}$

D. $y = -\dfrac{1}{2(\sqrt{\sin^{-1} x})(\sqrt[4]{1-x^2})}$

11. $x^2 - \sin^{-1} y = \ln x$. Find dy/dx.

A. $\left(\dfrac{1}{x} - 2x\right)\sqrt{1-y^2}$

B. $\left(-\dfrac{1}{x} + 2x\right)\sqrt{1-y^2}$

C. $\dfrac{\sin^{-2} y - 2x^2}{x\sin^{-2} y}$

D. $\dfrac{-\sin^{-2} y + 2x^2}{x\sin^{-2} y}$

12. Approximate $\sin(\sin^{-1} 0.3)$.

A. 3.0000 B. 0.3096 C. 0.3000 D. 0

13. A ball is thrown at 5 m/s and travels 245 m horizontally before coming back to its original height. Given that the acceleration due to gravity is 9.8 m/s^2, and air resistance is negligible, the range formula is $R = \frac{v^2}{9.8}\sin 2\theta$, where θ is the angle above the horizontal at which the ball is thrown. Find all possible positive angles in radians above the horizontal at which the ball can be thrown.

A. 1.5708 B. 1.5708 and 3.1416 C. 0.7854 and 1.5708 D. 0.7854

14. Answer true or false. $\sin^{-1} x$ is an odd function.

15. Answer true or false. $\tan^{-1}(1) + \tan^{-1}(2) = \tan^{-1}(-3)$.

SECTION 4.5

1. $\lim_{x\to 0} \frac{\sin 7x}{\sin x} =$ A. 7 B. $\frac{1}{7}$ C. -7 D. $-\frac{1}{7}$

2. $\lim_{x\to +\infty} \frac{x^2-9}{x^2-3x} =$ A. 1 B. $+\infty$ C. $-\infty$ D. 0

3. $\lim_{x\to 0} \frac{\tan^2 x}{x} =$ A. 1 B. $+\infty$ C. $-\infty$ D. 0

4. $\lim_{x\to 0+} \frac{\ln(x+1)}{e^x - 1} =$ A. 0 B. 1 C. $+\infty$ D. $-\infty$

5. $\lim_{x\to +\infty} \frac{e^x}{x^4} =$ A. 1 B. 0 C. $+\infty$ D. $-\infty$

6. $\lim_{x\to +\infty} \ln xe^{-x} =$ A. 0 B. 1 C. $+\infty$ D. $-\infty$

7. $\lim_{x\to 0} (1+3x)^{1/x} =$ A. 0 B. $+\infty$ C. $-\infty$ D. 3

8. $\lim_{x\to 0+} \frac{\sin x}{\ln(x+1)} =$ A. 10 B. 1 C. $+\infty$ D. $-\infty$

9. Answer true or false. $\lim_{x\to 0} \frac{\cos(\frac{1}{x})}{\cos(\frac{2}{x})} = \frac{1}{2}$

10. $\lim_{x\to \infty} \left(e^{-x} - \frac{1}{x}\right) =$ A. $+\infty$ B. $-\infty$ C. 1 D. 0

11. $\lim_{x\to 0+} (1 - \ln x)^x =$ A. 0 B. 1 C. $+\infty$ D. $-\infty$

12. Answer true or false. $\lim_{x\to 0} \frac{\sin 3x}{1-\cos x} = 1$

13. Answer true or false. $\lim_{x\to +\infty} \frac{2x^3 - 2x^2 + x - 3}{x^3 + 2x^2 - x + 1} = 2$

14. Answer true or false. $\lim_{x\to+\infty}(\sqrt{x^2-2x}-x)=0$

15. $\lim_{x\to 0^+}\left(\frac{\sin x}{x}-\frac{1}{x}\right)=$ A. 0   B. 1   C. $+\infty$ D. $-\infty$

CHAPTER 4 TEST

1. Answer true or false. The functions $f(x)=\sqrt[3]{x-3}$ and $g(x)=x^3+3$ are inverses of each other.

2. If $f(x)=\frac{1}{x^3+2}$, find $f^{-1}(x)$.

A. $\sqrt[3]{x-2}$ B. $\sqrt[3]{\frac{1-2x}{x}}$ C. $\sqrt[3]{1+2x}$ D. $\sqrt[3]{\frac{1+2x}{x}}$

3. Find the domain of $f^{-1}(x)$ if $f(x)=\sqrt{x-6}$.

A. $x\geq 0$ B. $x\leq 0$ C. $x\geq 6$ D. $x\geq -6$

4. Use a calculating utility to approximate $\log 41.3$.

A. 1.610   B. 1.613   C. 1.616   D. 1.618

5. Answer true or false. $\log\frac{ab^3}{\sqrt{c}}=\log a+3\log b-\frac{1}{2}\log c$.

6. Solve for x. $5^{2x}=8$.   A. 1.292   B. 0.646   C. 0.204   D. 0.102

7. If $y=\ln(5x^2)$ find dy/dx.   A. $\frac{2}{x}$ B. $\frac{2}{x^2}$ C. $\frac{2}{5x^2}$ D. $\frac{1}{x^2}$

8. Answer true or false. If $y=3\ln xe^{3x}$, $\frac{dy}{dx}=\frac{3e^{3x}}{x}+9\ln xe^{3x}$

9. If $f(x)=8^x$ find $df(x)/dx$.   A. $8^x\ln x$ B. $x\ln 8^x$ C. 8^{x-1} D. $8^x\ln 8$

10. Use a calculating utility to approximate x if $\sin x=0.42$, $3\pi/2<x<5\pi/2$.

A. 6.715   B. 6.717   C. 6.719   D. 6.723

11. $y=\tan^{-1}\sqrt[4]{x}$. Find dy/dx.

A. $\frac{1}{4(\sqrt[4]{x^3}+\sqrt{x})}$ B. $\frac{\sqrt{x}}{1+\sqrt{x}}$ C. $\frac{\sqrt{x}}{2(1+x)}$ D. $\sqrt[4]{\frac{1}{1+x}}$

12. Answer true or false. If $y=\sqrt{\cos^{-1}x+1}$, $\frac{dy}{dx}=\frac{-1}{2(\sqrt{\cos^{-1}x+1})(\sqrt[4]{s^2-1})}$.

13. $\lim_{x\to 0}\frac{\sin 8x}{\sin 9x}=$ A. 1   B. $+\infty$ C. $-\infty$ D. $\frac{8}{9}$

14. $\lim_{x\to 0}\frac{\sin 2x}{2x}=$ A. 0   B. 1   C. $\frac{1}{2}$ D. $+\infty$

15. Answer true or false. $\lim_{x\to 0}\left(8+\frac{1}{x}\right)^x=e^8$.

ANSWERS TO SAMPLE TESTS

SECTION 4.1

1. T 2. F 3. F 4. A 5. B 6. A 7. A 8. F 9. C 10. T 11. C 12. A
13. B 14. T 15. F

SECTION 4.2

1. C 2. A 3. C 4. D 5. B 6. T 7. F 8. A 9. A 10. D 11. B 12. A
13. B 14. C 15. A

SECTION 4.3

1. C 2. B 3. D 4. F 5. T 6. A 7. T 8. D 9. C 10. A 11. F 12. C
13. T 14. D 15. D

SECTION 4.4

1. B 2. B 3. B 4. B 5. F 6. D 7. A 8. A 9. C 10. B 11. B 12. C
13. D 14. T 15. F

SECTION 4.5

1. A 2. A 3. D 4. B 5. C 6. A 7. B 8. B 9. F 10. D 11. B 12. F
13. T 14. F 15. D

CHAPTER 4 TEST

1. T 2. B 3. A 4. C 5. T 6. B 7. A 8. T 9. D 10. B 11. A 12. F
13. D 14. B 15. F

CHAPTER 5
Sample Exams

SECTION 5.1

1. Answer true or false. If $f'(x) > 0$ for all x on the interval I, then $f(x)$ is concave up on the interval I.

2. Answer true or false. A point of inflection always has an x-coordinate where $f''(x) = 0$.

3. The largest interval over which f is increasing for $f(x) = (x-5)^4$ is

A. $[5, \infty)$ B. $[-5, \infty)$ C. $(-\infty, 5]$ D. $(-\infty, -5]$.

4. The largest interval over which f is decreasing for $f(x) = x^3 - 12x + 7$ is

A. $(-\infty, -2]$ B. $[2, \infty)$ C. $[-2, 2]$ D. $[-2, \infty)$.

5. The largest interval over which f is increasing for $f(x) = \sqrt[3]{x-2}$ is

A. $[2, \infty)$ B. $(-\infty, 2]$ C. $(-\infty, \infty)$ D. nowhere.

6. The largest open interval over which f is concave up for $f(x) = \sqrt[3]{x-5}$ is

A. $(-\infty, 5)$ B. $(5, \infty)$ C. $(-\infty, \infty)$ D. nowhere.

7. The largest open interval over which f is concave up for $f(x) = e^{x^4}$ is

A. $(-\infty, 0)$ B. $(0, \infty)$ C. $(-\infty, \infty)$ D. nowhere.

8. The function $f(x) = x^{4/5}$ has a point of inflection with an x-coordinate of

A. 0 B. $\frac{4}{5}$ C. $-\frac{4}{5}$ D. None exist.

9. The function $f(x) = e^{x^6}$ has a point of inflection with an x-coordinate of

A. $-e$ B. e C. 0 D. None exist.

10. Use a graphing utility to determine where $f(x) = \sin x$ is decreasing on $[0, 2\pi]$.

A. $[0, \pi]$ B. $[\pi, 2\pi]$ C. $[\pi/2, 3\pi/2]$ D. $[0, 2\pi]$

11. Answer true or false. $\cot x$ has a point of inflection on $(0, \pi)$.

12. Answer true or false. All functions of the form $f(x) = ax^n$ have an inflection point.

13. $f(x) = x^4 - 16x^2 + 6$ is concave up on the interval(s) $I =$

A. $(-\infty, 0) \cup (0, \infty)$ B. $[-2, \infty)$ C. $(-\infty, -2]$ D. $[-2, 2]$.

14. Answer true or false. If $f''(-2) = -3$ and $f''(2) = 3$, then there must be a point of inflection on $(-2, 2)$.

15. The function $f(x) = \dfrac{x^2}{x^2 - 4}$ has

A. points of inflection at $x = -4$ and $x = 4$
B. points of inflection at $x = -2$ and $x = 2$
C. a point of inflection at $x = 0$
D. no point of inflection.

SECTION 5.2

1. Determine the x-coordinate of each stationary point of $f(x) = 2x^3 - 3x^2 - 72x + 6$.

A. $x = -4$ and $x = 3$ B. $x = -3$ and $x = 4$ C. $x = -6$ D. None exists

2. Determine the x-coordinate of each critical point of $f(x) = \sqrt[5]{x-3}$.

A. 0 B. 3 C. -3 D. None exist.

3. Answer true or false. $f(x) = x^{2/7}$ has a critical point.

4. Answer true or false. A function has a relative extrema at every critical point.

5. $f(x) = x^2 + 6x + 8$ has a

A. relative maximum at $x = -3$
B. relative minimum at $x = -3$
C. relative maximum at $x = 3$
D. relative minimum at $x = 3$.

6. $f(x) = \cos^2 x$ on $0 < x < 2\pi$ has

A. a relative maximum at $x = \pi$; relative minima at $x = \pi/2$ and $x = 3\pi/2$
B. relative maxima at $x = \pi/2$ and $x = 3\pi/2$; a relative minimum at $x = \pi$
C. a relative maximum at $x = \pi$; no relative minimum
D. no relative maximum; a relative minimum at $x = \pi$.

7. $f(x) = x^4 - 4x^3$ has

A. a relative maximum at $x = 0$; no relative minimum
B. no relative maximum; a relative minimum at $x = 3$
C. a relative maximum at $x = 0$; a relative minimum at $x = 3$
D. a relative maximum at $x = 0$; relative minima at $x = -3$ and $x = 3$.

8. Answer true or false. $f(x) = |\tan x|$ has no relative extrema on $(-\pi/2, \pi/2)$.

9. $f(x) = e^{2x}$ has

A. a relative maximum at $x = 0$
B. a relative minimum at $x = 0$
C. relative minimum at $x = 2$
D. no relative extrema.

10. $f(x) = |x^2 - 9|$ has

A. no relative maximum; a relative minimum at $x = 3$
B. a relative maximum at $x = 3$; no relative minimum
C. relative minima at $x = -3$ and $x = 3$; a relative maximum at $x = 0$
D. relative minima at $x = -9$ and $x = 9$; a relative maximum at $x = 0$.

11. $f(x) = \ln(x^2 + 2)$ has

A. a relative maximum only
B. a relative minimum only
C. both a relative maximum and a relative minimum
D. no relative extrema.

12. On the interval $(0, 2\pi)$, $f(x) = \sin x \cos(2x)$ has

A. a relative maximum only
B. a relative minimum only
C. both a relative maximum and a relative minimum
D. no relative extrema.

13. Answer true or false. $f(x) = e^x \ln x$ has a relative minimum on $(0, \infty)$.

14. Answer true or false. A graphing utility can be used to show $f(x) = |x|$ has a relative minimum.

15. Answer true or false. A graphing utility can be used to show $f(x) = x^4 - 3x^2 + 3$ has two relative minima on $[-10, 10]$.

SECTION 5.3

1. Answer true or false. If $f''(-2) = -1$ and $f''(2) = 1$, then there must be an inflection point on $(-2, 2)$.

2. The polynomial function $x^2 - 6x + 8$ has

A. one stationary point that is at $x = 3$
B. two stationary points, one at $x = 0$ and one at $x = 3$
C. one stationary point that is at $x = -3$
D. one stationary point that is at $x = 0$.

3. The rational function $\dfrac{3x+6}{x^2-1}$ has

A. a horizontal asymptote at $y = 0$
B. a horizontal asymptote at $y = -2$
C. horizontal asymptotes at $x = -1$ and $x = 1$
D. no horizontal asymptote.

4. The rational function $\dfrac{3x+6}{x^2-1}$ has

A. a stationary point at $x = -2$
B. a stationary point at $x = 2$
C. two stationary points, one at $x = -1$ and one at $x = 1$
D. three stationary points, one at $x = -2$, one at $x = -1$, and one at $x = 1$.

5. Answer true or false. The rational function $x^3 - \dfrac{1}{x^2}$ has no vertical asymptote.

6. On a $[-10, 10]$ by $[-10, 10]$ window on a graphing utility the rational function $f(x) = \dfrac{x^3+8}{x^3-8}$ can be determined to have

A. one horizontal asymptote, no vertical asymptote
B. no horizontal asymptote; one vertical asymptote
C. one horizontal asymptote; one vertical asymptote
D. one horizontal asymptote; three vertical asymptotes.

7. Use a graphing utility to graph $f(x) = x^{1/7}$. How many points of inflection does the function have?

A. 0 B. 1 C. 2 D. 3

8. Use a graphing utility to graph $f(x) = x^{-1/7}$. How many points of inflection are there?

A. 0 B. 1 C. 2 D. 3

9. Determine which function is graphed.

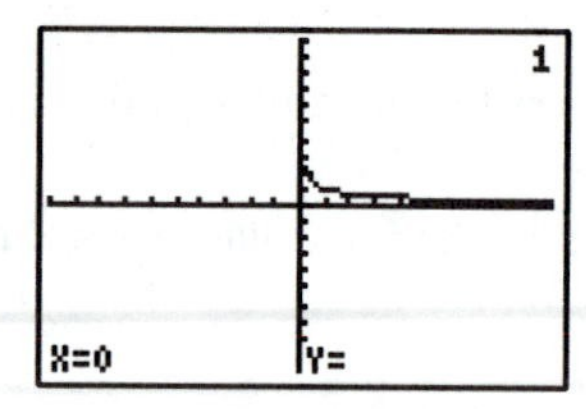

A. $f(x) = x^{1/2}$ B. $f(x) = x^{-1/3}$ C. $f(x) = x^{-1/2}$ D. $f(x) = x^{1/3}$

10. Use a graphing utility to generate the graph of $f(x) = x^2e^{3x}$, then determine the x-coordinate of all relative extrema on $(-10, 10)$ and identify them as a relative maximum or a relative minimum.

A. There is a relative maximum at $x = 0$.
B. There is a relative minimum at $x = 0$.
C. There is a relative minimum at $x = 0$ and relative maxima at $x = -1$ and $x = 1$.
D. There is no relative extremum.

11. Answer true or false. Using a graphing utility it can be shown that $f(x) = x^2 \tan^2 x$ has a maximum on $0 < x < 2\pi$.

12. Answer true or false. $\lim_{x\to 0^+} \sqrt{x} \ln x = 0$.

13. $\lim_{x\to +\infty} x^{3/2} \ln x =$ A. 0 B. 1 C. $+\infty$ D. It does not exist.

14. Answer true or false. A fence is to be used to enclose a rectangular plot of land. If there are 160 feet of fencing, it can be shown that a 40 ft by 40 ft square is the rectangle that can be enclosed with the greatest area. (A square is considered a rectangle.)

15. Answer true or false. $f(x) = \dfrac{x^2 + 3x + 4}{x - 1}$ has an oblique asymptote.

SECTION 5.4

1.

X=5 Y=2.236068

The graph represents a position function. Determine what is happening to the velocity.

A. It is speeding up. B. It is slowing down.
C. It is constant. D. There is insufficient information to tell.

2.

X=5 Y=2.236068

The graph represents a position function. Determine what is happening to the acceleration.

A. It is positive.
B. It is negative.
C. It is zero.
D. There is insufficient information to tell.

3. 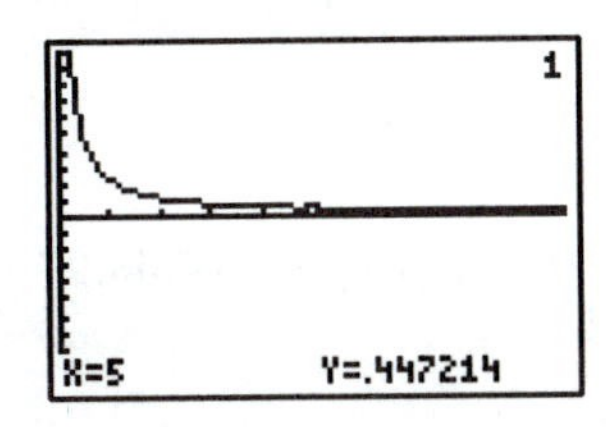

The graph represents a velocity function. The acceleration is

A. positive
B. negative
C. zero
D. There is insufficient information to tell.

4. 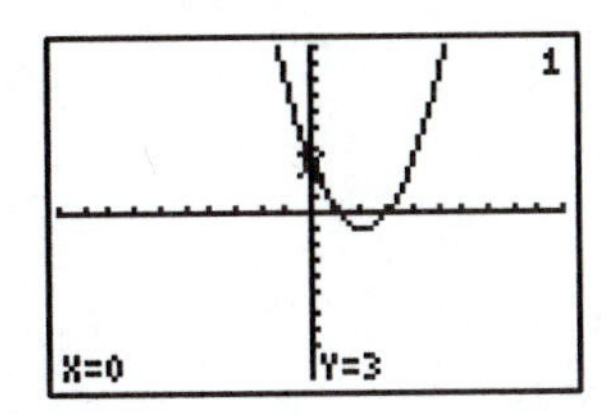

Answer true or false. This is the graph of a particle that is moving to the right at $t = 0$.

5. 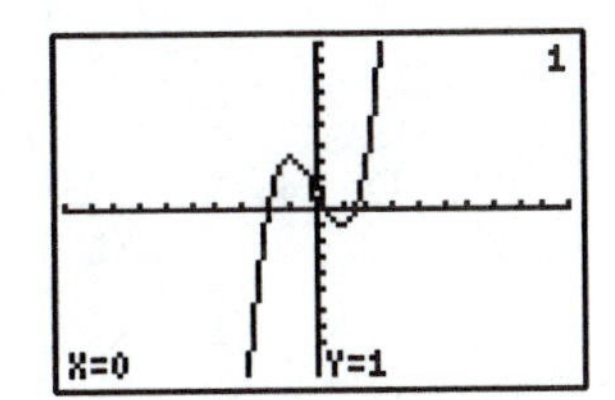

Answer true or false. For the position function graphed, the acceleration at $t = 1$ is positive.

6. 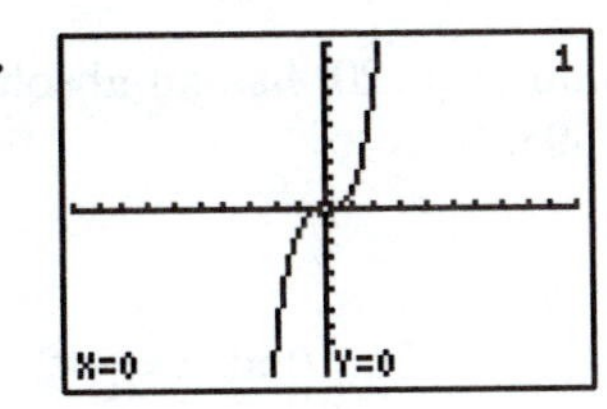

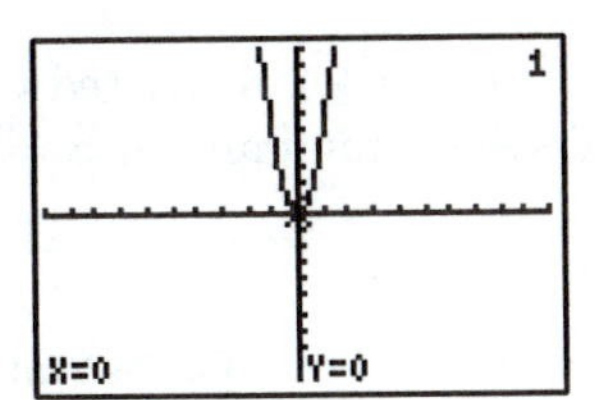

Answer true or false. If the graph on the left is a position function, the graph on the right represents the corresponding velocity function.

7. Let $s(t) = \cos t$ be a position function of a particle. At $t = \frac{\pi}{2}$ the particle's velocity is

A. positive
B. negative
C. zero.

8. Let $s(t) = t^3 - t$ be a position function of a particle. At $t = 0$ the particle's acceleration is

A. positive
B. negative
C. zero.

9. $s(t) = t^4 - 2t$, $t \geq 0$. The velocity function is

A. $t^3 - 2$
B. $4t^3 - 2$
C. $12t^2$
D. $12t^2 - 2$.

10. $s(t) = t^4 - 2t$, $t \geq 0$. The acceleration function is

A. $t^3 - 2$ B. $4t^3 - 2$ C. $12t^2$ D. $12t^2 - 2$.

11. A projectile is thrown upward at 100 m/s. About how long does it take the projectile to reach its highest point?

A. 1,020 s B. 510 s C. 5 s D. 10 s

12. Answer true or false. If a particle is dropped a distance of 100 m. It has a speed of 44.27 m/s (rounded to the nearest hundredth of a m/s) when it hits the ground.

13. $s(t) = t^4 - 4t^2$. Find t when $a = 0$.

A. 12 B. -12 C. $\sqrt{\frac{2}{3}}$ D. $-\frac{2}{3}$

14. $s(t) = t^3 - 3t$, $t \geq 0$. Find s when $a = 0$.

A. 1 B. 2 C. $0, 2$ D. -1

15. Let $s(t) = \sqrt{3t^2 - 2}$ be a position function. Find v when $t = 1$.

A. 3 B. 6 C. 1 D. 0

SECTION 5.5

1. $f(x) = 3x^2 - x + 2$ has an absolute maximum on $[-2, 2]$ of

A. 16 B. 2 C. 12 D. 4.

2. $f(x) = |5 - 2x|$ has an absolute minimum of

A. 0 B. 3 C. 1 D. 5.

3. Answer true or false. $f(x) = x^3 - x^2 + 2$ has an absolute maximum and an absolute minimum.

4. Answer true or false. $f(x) = x^3 - 18x^2 + 20x + 2$ restricted to a domain of $[0, 20]$ has an absolute maximum at $x = 2$ of -22, and an absolute minimum at $x = 10$ of -598.

5. $f(x) = \sqrt{x - 2}$ has an absolute minimum of

A. 0 at $x = 2$ B. 0 at $x = 0$ C. -2 at $x = 0$ D. 0 at $x = -2$.

6. $f(x) = \sqrt{x^2 + 5}$ has an absolute maximum, if one exists, at

A. $x = -5$ B. $x = 5$ C. $x = 0$ D. None exist.

7. Find the location of the absolute maximum of $\tan x$ on $[0, \pi]$, if it exists.

A. 0 B. π C. $\frac{\pi}{2}$ D. None exist.

8. $f(x) = x^2 - 3x + 2$ on $(-\infty, \infty)$ has

A. only a absolute maximum
B. only an absolute minimum
C. both an absolute maximum and an absolute minimum
D. neither an absolute maximum nor an absolute minimum.

9. $f(x) = \dfrac{1}{x^2}$ on $[1,3]$ has

A. an absolute maximum at $x = 1$ and an absolute minimum at $x = 3$
B. an absolute minimum at $x = 1$ and an absolute maximum at $x = 3$
C. no absolute extrema
D. an absolute minimum at $x = 2$ and absolute maxima at $x = 1$ and $x = 3$.

10. Answer true or false. $f(x) = \sin x \cos x$ on $[0, \pi]$ has an absolute maximum at $x = \dfrac{\pi}{2}$.

11. Use a graphing utility to assist in determining the location of the absolute maximum of $f(x) = -(x^2 - 3)^2$ on $(-\infty, \infty)$, if it exists.

A. $x = \sqrt{3}$ and $x = -\sqrt{3}$
B. $x = \sqrt{3}$ only
C. $x = 0$
D. None exist

12. Answer true or false. If $f(x)$ has an absolute minimum at $x = 2$, $-f(x)$ also has an absolute minimum at $x = 2$.

13. Answer true or false. Every function has an absolute maximum and an absolute minimum if its domain is restricted to where f is defined on an interval $[-a, a]$, where a is finite.

14. Use a graphing utility to locate the value of x where $f(x) = x^4 - 3x + 2$ has an absolute minimum, if it exists.

A. 1
B. $\sqrt[3]{\dfrac{3}{4}}$
C. 0
D. None exist

15. Use a graphing utility to estimate the absolute maximum of $f(x) = (x - 5)^2$ on $[0, 6]$, if it exists.

A. 25
B. 0
C. 1
D. None exist

SECTION 5.6

1. Express the number 20 as the sum of two nonnegative numbers whose product is as large as possible.

A. 5, 15
B. 1, 19
C. 10, 10
D. 0, 20

2. A right triangle has a perimeter of 16. What are the lengths of each side if the area contained within the triangle is to be maximized?

A. $\dfrac{16}{3}, \dfrac{16}{3}, \dfrac{16}{3}$
B. 5, 5, 6
C. $16 - 8\sqrt{2}$, $16 - 8\sqrt{2}$, $-16 + 16\sqrt{2}$
D. 4, 5, 7

3. A rectangular sheet of cardboard 2 m by 1 m is used to make an open box by cutting squares of equal size from the four corners and folding up the sides. What size squares should be cut to obtain the largest possible volume?

A. $\dfrac{3 + \sqrt{3}}{6}$
B. $\dfrac{3 - \sqrt{3}}{6}$
C. $\dfrac{1}{2}$
D. $\dfrac{1}{4}$

4. Suppose that the number of bacteria present in a culture bacteria at time t is given by $N = 10{,}000e^{-t/10}$. Find the smallest number of bacteria in the culture during the time interval $0 \le t \le 50$.

A. 67
B. 10,000
C., 3,679
D. 73,891

5. An object moves a distance s away from the origin according to the equation $s(t) = 4t^4 - 2t + 1$, where $0 \le t \le 10$. At what time is the object farthest from the origin?

A. 0 B. 2 C. 10 D. $\frac{1}{8}$

6. An electrical generator produces a current in amperes starting at $t = 0$ s and running until $t = 6\pi$ s that is given by $\sin(2t)$. Find the maximum current produced.

A. 1 A B. 0 A C. 2 A D. $\frac{1}{2}$ A

7. A storm is passing with the wind speed in mph changing over time according to $v(t) = -x^2 + 10x + 55$, for $0 \le t \le 10$. Find the highest wind speed that occurs.

A. 55 mph B. 80 mph C. 110 mph D. 30 mph

8. A company has a cost of operation function given by $C(t) = 0.01t^2 - 6t + 1{,}000$ for $0 \le t \le 500$. Find the minimum cost of operation.

A. 1,000 B. 100 C. 500 D. 0

9. Find the point on the curve $x^2 + y^2 - 4$ closest to $(0, 3)$.

A. $(0, 4)$ B. $(0, 2)$ C. $(2, 0)$ D. $(4, 0)$

10. Answer true or false. The point on the parabola $y = x^2$ closest to $(0, 0.9)$ is $(0, 0)$.

11. For a triangle with sides 3 m, 4 m, and 5 m, the smallest circle that contains the triangle has a diameter of

A. 3 m B. 4 m C. 5 m D. 10 m.

12. Answer true or false. If $f(t) = 3e^{4t}$ represents a growth function over the time interval $[a, b]$, the absolute maximum must occur at $t = b$.

13. Answer true or false. The rectangle with the largest area that can be inscribed inside a circle is a square.

14. Answer true or false. The rectangle with the largest area that can be inscribed in a semi-circle is a square.

15. Answer true or false. An object that is thrown upward and reaches a height of $s(t) = 50 + 120t - 32t^2$ for $0 \le t \le 2$. The object is highest at $t = 2$.

SECTION 5.7

1. Approximate $\sqrt{3}$ by applying Newton's Method to the equation $x^2 - 3 = 0$.

A. 1.73205080757 B. 1.73205079216 C. 1.73205084126 D. 1.73205094712

2. Approximate $\sqrt[3]{9}$ by applying Newton's Method to the equation $x^3 - 9 = 0$.

A. 2.08008381347 B. 2.08008382305 C. 2.08008397111 D. 2.08008382176

3. Use Newton's Method to approximate the solutions of $x^3 + 2x^2 + 5x + 10 = 0$.

A. -2.000, 2.236 B. -5, 0, 5 C. -3.1623, 0, 3.1623 D. -3.1623, 3.1623

4. Use Newton's Method to find the largest positive solution of $x^3 - x^2 + 2x - 4 = 0$.

A. 1.478 B. 1.000 C. 2.828 D. 3.721

5. Use Newton's Method to find the largest positive solution of $x^3 - x^2 + 3x - 3 = 0$.

A. 1.7325 B. 1.000 C. 1.7319 D. 1.7316

6. Use Newton's Method to find the largest positive solution of $x^4 + 6x^3 - x^2 - 6 = 0$.

A. 6.000 B. 1.000 C. 1.732 D. 1.412

7. Use Newton's Method to find the largest positive solution of $x^4 + x^3 - 4x - 4 = 0$.

A. 4.000 B. 1.000 C. 0.500 D. 1.587

8. Use Newton's Method to find the largest positive solution of $x^5 - 2x^3 - 14x^2 + 28 = 0$.

A. 3.742 B. 2.410 C. 1.414 D. 1.260

9. Use Newton's Method to find the largest positive solution of $x^5 + 2x^3 - 2x^2 - 4 = 0$.

A. 1.260 B. 1.414 C. 1.587 D. 2.000

10. Use Newton's Method to find the largest positive solution of $x^4 - 13x^2 + 30 = 0$.

A. 3.162 B. 2.340 C. 5.477 D. 1.732

11. Use Newton's Method to find the largest positive solution of $x^5 + x^4 + x^3 - 5x^2 - 5x - 5 = 0$.

A. 2.236 B. 1.380 C. 1.710 D. 1.621

12. Use Newton's Method to find the x-coordinate of the intersection of $y = 2x^3 - 2x^2$ and $y = -x^5 + 4$.

A. 3.742 B. 2.410 C. 1.414 D. 1.260

13. Use Newton's Method to approximate the greatest x-coordinate of the intersection of $y = 2x^3 - 2x^2 + 1$ and $y = -x^5 + 5$.

A. 3.742 B. 1.414 C. 2.410 D. 1.260

14. Use Newton's Method to approximate the x-coordinate intersection of $y = x^5 + x^3 - 5$ and $y = -x^4 + 5x^2 + 5x$.

A. 2.236 B. 1.380 C. 1.710 D. 1.627

15. Use Newton's Method to find the greatest x-coordinate of the intersection of $y = x^4 - 7x^2$ and $y = 6x^2 - 30$.

A. 3.162 B. 2.340 C. 5.477 D. 1.732

SECTION 5.8

1. Answer true or false. $f(x) = \dfrac{1}{x}$ on $[-1, 1]$ satisfies the hypotheses of Rolle's Theorem.

2. Find the value c such that the conclusion of Rolle's Theorem are satisfied for $f(x) = x^2 - 4$ on $[-2, 2]$.

A. 0 B. -1 C. 1 D. 0.5

3. Answer true or false. The Mean-Value Theorem is used to find the average of a function.

4. Answer true or false. The Mean-Value Theorem can be used on $f(x) = |x|$ on $[-2, 1]$.

5. Answer true or false. The Mean-Value Theorem guarantees there is at least one c on $[0, 1]$ such that $f'(x) = 1$ when $f(x) = \sqrt{x}$.

6. If $f(x) = \sqrt[3]{x}$ on $[0, 1]$, find the value c that satisfies the Mean-Value Theorem.

A. 1 B. $\frac{1}{3}$ C. $\left(\frac{1}{3}\right)^{3/2}$ D. $\frac{1}{9}$

7. Answer true or false. The hypotheses of the Mean-Value Theorem are satisfied for $f(x) = \sqrt[3]{|x|}$ on $[-1, 1]$.

8. Answer true or false. The hypotheses of the Mean-Value Theorem are satisfied for $f(x) = \sin x$ and $[0, 4\pi]$.

9. Answer true or false. The hypotheses of the Mean-Value Theorem are satisfied for $f(x) = \frac{1}{\sin x}$ on $[0, 4\pi]$.

10. Find the value for which $f(x) = x^2 + 3$ on $[1, 3]$ satisfies the Mean-Value Theorem.

A. 2 B. $\frac{9}{4}$ C. $\frac{7}{3}$ D. $\frac{11}{4}$

11. Find the value for which $f(x) = x^3$ on $[2, 3]$ satisfies the Mean-Value Theorem.

A. 1.291 B. 2.5000 C. 2.2500 D. 2.1250

12. Answer true or false. A graphing utility can be used to show that Rolle's Theorem can be applied to show that $f(x) = (x - 2)^2$ has a point where $f'(x) = 0$.

13. Answer true or false. According to Rolle's Theorem if a function does not cross the x-axis its derivative cannot be zero anywhere.

14. Find the value c that satisfies Rolle's Theorem for $f(x) = \sin x$ on $[0, \pi]$.

A. $\frac{\pi}{4}$ B. $\frac{\pi}{2}$ C. $\frac{3\pi}{4}$ D. $\frac{\pi}{3}$

15. Find the value c that satisfies the Mean-Value Theorem for $f(x) = x^3 + 3x$ on $[0, 1]$.

A. $\frac{\sqrt{3}}{3}$ B. $\frac{\sqrt{3}}{2}$ C. $\frac{\sqrt{2}}{2}$ D. $\frac{\sqrt{2}}{3}$

CHAPTER 5 TEST

1. The largest interval on which $f(x) = x^2 - 4x + 7$ is increasing is

A. $[0, \infty)$ B. $(-\infty, 0]$ C. $[2, \infty)$ D. $(-\infty, 2]$.

2. Answer true or false. The function $f(x) = \sqrt{x - 2}$ is concave down on its entire domain, except at $x = 2$.

3. The function $f(x) = x^3 - 8$ is concave down on

A. $(-\infty, 2)$ B. $(2, \infty)$ C. $(-\infty, 0)$ D. $(0, \infty)$.

4. Answer true or false. $f(x) = x^4 - 2x + 3$ has a point of inflection.

5. $f(x) = [x^2 - 4]$ is concave up on

A. $(-\infty, -2) \cup (2, \infty)$ B. $(-\infty, -4) \cup (4, \infty)$ C. $(-2, 2)$ D. $(-4, 4)$.

6. The largest open interval on which $f(x) = e^{2x^6}$ is concave up is

A. $(-\infty, 0)$ B. $(0, \infty)$ C. $(-\infty, \infty)$ D. $(-\infty, e)$.

7. Use a graphing utility to determine where $f(x) = \sin x$ is increasing on $[0, 2\pi]$.

A. $(0, \pi)$ B. $(\pi, 2\pi)$
C. $(\pi/2, 3\pi/2)$ D. $(0, \pi/2) \cup (3\pi/2, 2\pi)$

8. Answer true or false. $f(x) = x^4 - 2x^2 + 10$ has a point of inflection.

9. $f(x) = -x^4 - 6x^3 + 2x^2$ is concave down on

A. $(-\infty, \infty)$ B. $(-\infty, -81)$ C. $(-\infty, -9)$ D. nowhere.

10. Answer true or false. If $f''(-1) = 4$ and $f''(1) = -4$, and if f is continuous on $[-1, 1]$, then there is a point of inflection on $(-1, 1)$.

11. Determine the x-coordinate of each stationary point of $f(x) = 4x^4 - 16$.

A. -1 B. 0 C. 16 D. 1

12. Answer true or false. $f(x) = x^{4/9}$ has a critical point at $x = 0$.

13. $f(x) = x^2 - 8x + 9$ has

A. a relative maximum at $x = 4$ B. a relative minimum at $x = 4$
C. a relative maximum at $x = -4$ D. a relative minimum at $x = -4$.

14. $f(x) = -e^{7x}$ has

A. a relative maximum at $x = 0$ B. a relative minimum at $x = 0$
C. a relative maximum at $x = 7$ D. no relative extremum.

15. $f(x) = |16x^2|$ has

A. no relative maximum; a relative minimum at $x = 4$
B. a relative maximum at $x = 4$; no relative minimum
C. a relative maximum at $x = 0$; relative minima at $x = -4$ and $x = 4$
D. no relative maximum; a relative minimum at $x = 0$.

16. Answer true or false. $f(x) = -e^{2x}\ln(2x)$ has a relative minimum on $(0, \infty)$.

17. The rational function $\dfrac{4x + 20}{x^2 - 25}$ has

A. a horizontal asymptote at $y = 0$ B. a horizontal asymptote at $y = 5$
C. a horizontal asymptote at $y = 4$ D. no horizontal asymptote.

18. Answer true or false $f(x) = \dfrac{3}{x - 5}$ has a vertical asymptote.

19. 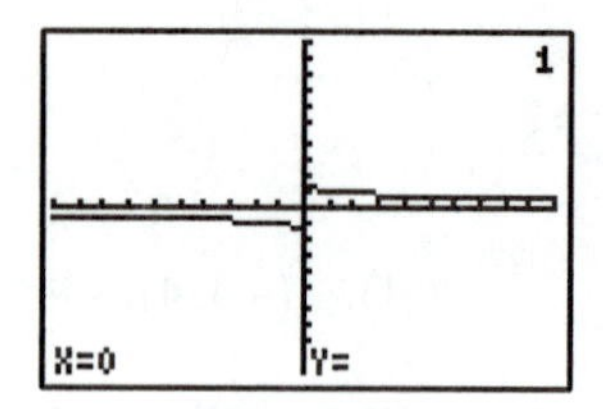

This is the graph that would appear on a graphing utility if the function that is graphed is

A. $f(x) = x^{1/5}$ B. $f(x) = x^{1/6}$ C. $f(x) = x^{-1/5}$ D. $f(x) = x^{-1/6}$.

20. Answer true or false. $\lim_{x \to 0^+} \sqrt[3]{x} \ln x = 0$

21. A weekly profit function for a company is $P(x) = -0.01x^2 + 3x - 2{,}000$, where x is the number of the company's only product that is made and sold. How many individual items of the product must the company make and sell weekly to maximize the profit?

A. 300 B. 150 C. 600 D. 60

22. $f(x) = 6x^2 - 2$ has an absolute minimum on $[-3, 3]$ of

A. 2 B. -2 C. 52 D. -52.

23. $f(x) = x^3 + 3$ has an absolute maximum on $[-2, 2]$ of

A. 0 B. 6 C. 11 D. 8.

24. $f(x) = 3\sin(x + 2)$ has an absolute minimum of

A. -5 B. -3 C. $-\frac{1}{3}$ D. $-\frac{2}{3}$.

25. $f(x) = \frac{1}{x^5}$ has an absolute maximum on $[1, 3]$ of

A. 1 B. $\frac{1}{243}$ C. 243 D. None exist.

26. Answer true or false. $f(x) = \frac{1}{x^7}$ has an absolute maximum of 1 on $[1, 1]$.

27. Express the number 40 as the sum of two nonnegative numbers whose product is as large as possible.

A. 5, 35 B. 10, 30 C. 20, 20 D. 1, 39

28. An object moves a distance s away from the origin as given by $s(t) = t^4 - 2$, $0 \leq t \leq 10$. At what time is the object farthest from the origin?

A. 0 B. 2 C. 8 D. 10

29. Find the point on; the curve $x^2 + y^2 = 16$ closest to $(0, 5)$.

A. $(0, 4)$ B. $(4, 0)$ C. $(-4, 0)$ D. $(0, -4)$

30. Answer true or false. A growth function $f(x) = 4e^{0.02t}$, $0, \leq t \leq 10$ has an absolute maximum at $t = 10$.

31. 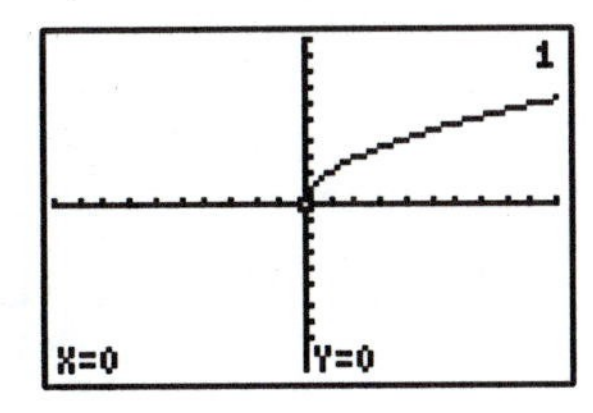

The graph represents a position function. Determine what is happening to the velocity.

A. It is increasing. B. It is decreasing.
C. It is constant. D. More information is needed.

32. 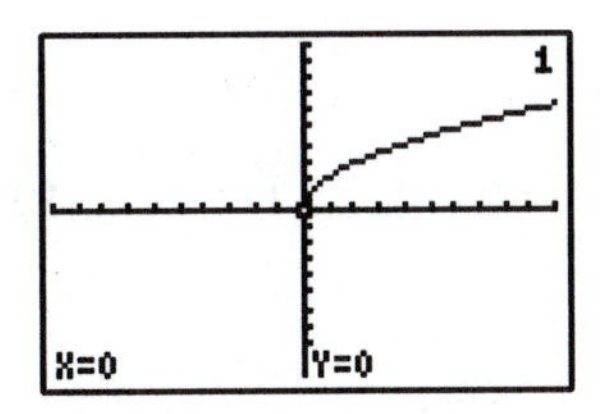

The graph represents a position function. Determine what is happening to the acceleration.

A. It is positive. B. It is negative.
C. It is zero. D. More information is needed.

33. Let $s(t) = t^4 - 2$ be a position function particle. The particle's acceleration for $t > 0$ is

A. positive B. negative
C. zero D. More information is needed.

34. Let $s(t) = 4 - t^2$ be a position function. The particle's velocity for $t > 0$ is

A. positive B. negative
C. zero D. More information is needed.

35. $s(t) = 4t^2 - 8$. $a = 0$ when $t =$

A. 0 B. 8 C. 2 D. nowhere.

36. Approximate $\sqrt{11}$ using Newton's Method.

A. 3.31662479036 B. 3.31662478727 C. 3.31662479002 D. 3.31662478841

37. Use Newton's Method to approximate the greatest positive solution of $x^3 + 4x^2 - 5x - 20 = 0$.

A. 4.000 B. 2.236 C. 5.292 D. 3.037

38. Use Newton's Method to approximate the greatest x-coordinate where the graphs of $y = x^3 + 2x^2$ and $y = -2x^2 + 5x + 20$ cross.

A. 4.000 B. 2.236 C. 5.292 D. 3.037

39. Answer true or false. The hypotheses of Rolle's Theorem are satisfied for $f(x) = \dfrac{1}{x^6} - 1$ on $[-1, 1]$.

40. Answer true or false. Given $f(x) = x^2 - 16$ on $[-4, 4]$, the value c that satisfies Rolle's Theorem is 0.

41. Answer true or false. $f(x) = x^5$ on $[-1, 1]$. The value c that satisfies the Mean-Value Theorem is 0.

ANSWERS TO SAMPLE TESTS

SECTION 5.1:

1. F 2. F 3. A 4. C 5. C 6. A 7. C 8. D 9. D 10. C 11. T 12. F
13. A 14. F. 15. D

SECTION 5.2:

1. B 2. B 3. T 4. F 5. B 6. A 7. B 8. F 9. D 10. C 11. B 12. C
13. F 14. T 15. F

SECTION 5.3:

1. F 2. A 3. A 4. A 5. F 6. C 7. B 8. A 9. C 10. B 11. F 12. T
13. A 14. T 15. T

SECTION 5.4:

1. B 2. B 3. B 4. F 5. T 6. T 7. C 8. C 9. B 10. C 11. D 12. T
13. A 14. C 15. A

SECTION 5.5:

1. A 2. A 3. F 4. F 5. A 6. D 7. D 8. B 9. A 10. F 11. A 12. F
13. F 14. B 15. A

SECTION 5.6:

1. C 2. C 3. B 4. A 5. C 6. A 7. B 8. B 9. B 10. F 11. C 12. T
13. T 14. F 15. F

SECTION 5.7:

1. A 2. B 3. A 4. A 5. B 6. B 7. D 8. B 9. A 10. A 11. C 12. D
13. B 14. C 15. A

SECTION 5.8:

1. F 2. A 3. F 4. F 5. T 6. C 7. F 8. T 9. F 10. A 11. A 12. F
13. F 14. B 15. A

CHAPTER 5 TEST:

1. C 2. T 3. C 4. F 5. A 6. C 7. D 8. F 9. A 10. T 11. B 12. T
13. B 14. D 15. D 16. F 17. A 18. T 19. C 20. T 21. B 22. B 23. C
24. B 25. A 26. F 27. C 28. D 29. A 30. T 31. A 32. B 33. A 34. B
35. D 36. A 37. B 38. B 39. F 40. T 41. F

CHAPTER 6
Sample Exams

SECTION 6.1

1. $f(x) = 2x$; $[0, 1]$. Use the rectangle method to approximate the area using 4 rectangles.

A. 1 B. 0.5 C. 0.75 D. 0.875

2. $f(x) = 5 + x$; $[0, 2]$. Use the rectangle method to approximate the area using 4 rectangles.

A. 5.625 B. 5.375 C. 6.000 D. 7.000

3. $f(x) = \sqrt{1 + x}$; $[0, 1]$. Use the rectangle method to approximate the area using 4 rectangles.

A. 1.219 B. 1.250 C. 1.500 D. 1.141

SECTION 6.2

1. $\int x^6 dx =$ A. $\frac{x^5}{5} + C$ B. $\frac{x^7}{7} + C$ C. $\frac{x^7}{6} + C$ D. $\frac{x^6}{7} + C$

2. $\int x^{2/5} dx =$ A. $\frac{5}{2x^{3/5}} + C$ B. $\frac{5}{7}x^{7/5} + C$ C. $-\frac{5}{7}x^{7/5} + c$ D. $-\frac{5}{2x^{3/5}} + C$

3. $\int \sqrt[3]{x}\, dx =$ A. $\frac{3}{2x^{2/3}} + C$ B. $\frac{3}{4}x^{4/3} + C$ C. $-\frac{3}{4}x^{4/3} + C$ D. $-\frac{3}{x^{2/3}} + C$

4. $\int x^{-2} dx =$ A. $-\frac{3}{x^3} + C$ B. $-\frac{1}{x} + C$ C. $\frac{1}{x^3} + C$ D. $\frac{3}{x^3} + C$

5. $\int \sin x\, dx =$ A. $\sin^2 x + C$ B. $\cos x + C$ C. $-\cos x + C$ D. $-\sin^2 x + C$

6. $\int 3e^x dx =$ A. $3e^x + C$ B. $\frac{e^x}{3} + C$ C. $-3e^x + C$ D. $-\frac{e^x}{3} + C$

7. $\int \frac{\cos x}{\sin^2 x} dx =$ A. $-\frac{1}{\sin^3 x} + C$ B. $-\frac{1}{\sin x} + C$ C. $\frac{1}{\sin x} + C$ D. $\frac{1}{\sin^3 x} + C$

8. $\int \frac{5}{x}\, dx =$ A. $\frac{5}{2x^2} + C$ B. $-\frac{5}{2x^2} + C$ C. $-5 \ln x + C$ D. $5 \ln x + C$

9. Answer true or false. $\int \frac{1}{x} + 3e^x dx = \ln x + 3e^x + C$

10. Answer true or false. $\int \sin x \cos x\, dx = \sin x \cos x + C$

11. Answer true or false. $\int x + \frac{1}{\sin x} dx = \frac{x^2}{2} + \ln|\sin x| + C$

12. Answer true or false. $\int \sqrt{x} + \sqrt[3]{x} + \sqrt[4]{x}\, dx = x^{3/2} + x^{4/3} + x^{5/4} + C$

13. Answer true or false. $\int \sin x + \cos x\, dx = -\cos x + \sin x + C$

14. Find $y(x)$. $\frac{dy}{dx} = x^2$, $y(0) = 1$

A. $\frac{x^3}{3} + 1$ B. $\frac{x^3}{3}$ C. $\frac{x^3}{3} - 1$ D. $\frac{x^3 - 1}{3}$

15. Find $y(x)$. $\frac{dy}{dx} = e^x$, $y(0) = 2$

A. $e^x + 1$ B. $e^x + 2$ C. e^x D. $e^x - 2$

SECTION 6.3

1. $\int 2x(x^2 - 5)^{30} dx =$

A. $\frac{(x^2 - 5)^{31}}{31} + C$ B. $\frac{(x^2 - 5)^{29}}{29} + C$ C. $31(x^2 - 5)^{31} + C$ D. $29(x^2 - 5)^{29} + C$

2. $\int \sin^2 x \cos x\, dx =$

A. $\frac{\sin^3 x}{3} + C$ B. $\frac{\cos^3 x}{3} + C$ C. $\cos^3 x \sin^2 x + C$ D. $3 \sin x + C$

3. $\int 2xe^{x^2} dx =$

A. $\frac{e^{x^2}}{2x} + C$ B. $2e^{x^2} + C$ C. $x^2 + e^{x^2} + C$ D. $e^{x^2} + C$

4. $\int \frac{\ln x}{x} dx =$

A. $\ln x + C$ B. $(\ln x)^2 + C$ C. $\frac{(\ln x)^2}{2} + C$ D. $2(\ln x)^2 + C$

5. $\int e^{-8} dx =$

A. $-8e^{-8x} + C$ B. $-\frac{e^{-8x}}{8} + C$ C. $8e^{-8x} + C$ D. $\frac{e^{-8x}}{8} + C$

6. $\int (x + 9)^5 dx =$

A. $5(x + 9) + C$ B. $\left(\frac{x^2}{2} + 9\right)^5 + C$ C. $\frac{(x + 9)^6}{6} + C$ D. $\frac{(x + 9)^4}{4} + C$

7. $\int \frac{1}{5x} dx =$

A. $\ln 5x + C$ B. $\frac{\ln x}{5} + C$ C. $5 \ln x + C$ D. $\ln x + C$

8. Answer true or false. $\int x\sqrt{x + 2}\, dx = \frac{2}{5}(x + 2)^{5/2} - \frac{4}{3}(x + 2)^{3/2} + C$

9. Answer true or false. For $\int x \sin x^2 dx$ a good choice for u is x^2.

10. Answer true or false. For $\int \frac{e^x}{e^x - 2} dx$ a good choice for u is $e^x - 2$.

11. Answer true or false. $\int x\sqrt{3x^3 - 2}\, dx$ can be easily solved by letting $u = x^3$.

12. Answer true or false. $\int x^2(x^3 - 4)^{2/5} dx$ can be easily solved by letting $u = x^3 - 4$.

13. Answer true or false. $\int e^{x^2} dx$ can be easily solved by letting $u = x^2$.

14. Answer true or false. $\int \sin^3 x\, dx$ can be easily solved by letting $u = \sin x$.

15. Answer true or false. $\int e^{9x} dx$ can be easily solved by letting $u = 9x$.

SECTION 6.4

1. $\sum_{k=1}^{5} k^2 =$ A. 15 B. 55 C. 26 D. 6

2. $\sum_{j=2}^{4} 3^j =$ A. 90 B. 18 C. 120 D. 117

3. $\sum_{k=1}^{4} \sin(k\pi) =$ A. 4 B. 0 C. 2 D. 3

4. Answer true or false. $\sum_{i=1}^{4}(i+1) = 14$

5. Express in sigma notation, but do not evaluate. $1 + 2 + 3 + 4$

A. $\sum_{i=0}^{3} i$ B. $\sum_{i=1}^{4} i$ C. $\sum_{i=1}^{4} i^2$ D. $\sum_{i=0}^{4} i + 1$

6. Express in sigma notation, but do not evaluate. $1 + 4 + 9 + 16 + 25$

A. $\sum_{i=1}^{5} i$ B. $\sum_{i=0}^{5} i + 1$ C. $\sum_{i=1}^{5} i^2$ D. $\sum_{i=2}^{5} i^2$

7. Express in sigma notation, but do not evaluate. $2 + 8 + 18 + 50$

A. $\sum_{i=1}^{4} i^2$ B. $\sum_{i=1}^{4} 2i^2$ C. $\sum_{i=1}^{4} 2i + 6$ D. $\sum_{i=1}^{4} 2(i+6)$

8. Answer true or false. $4 + 9 + 16 + 25 + 36$ can be expressed in sigma notation as $\sum_{i=2}^{6} i^2$.

9. Answer true or false. $8 + 27 + 64$ can be expressed in sigma notation as $\sum_{i=1}^{3} i^3$.

10. $\sum_{i=5}^{100} =$ A. 5,040 B. 5,050 C. 5,035 D. 5,000

11. $\lim_{n \to +\infty} \sum_{k=1}^{n} \left(\frac{1}{5^k}\right) =$ A. 0 B. $\frac{4}{5}$ C. $\frac{1}{5}$ D. $\frac{1}{4}$

12. $\lim_{n \to +\infty} \sum_{k=1}^{n} \left(\frac{4}{5}\right)^k =$ A. $\frac{4}{5}$ B. 4 C. 20 D. $\frac{5}{4}$

13. Answer true or false. $\sum_{i=1}^{n} x_i^3 = \left(\sum_{i=1}^{n} x_i\right)^3$

14. Answer true or false. $\sum_{i=1}^{n}(a_i + b_i) = \sum_{i=1}^{n} a_i + \sum_{i=1}^{n} b_i$

15. Answer true or false. $\sum_{i=1}^{n} 2a_i = 2\sum_{i=1}^{n} a_i$

SECTION 6.5

1. $\int_1^5 x\,dx =$ A. 12 B. 6 C. 24 D. 4

2. $\int_0^4 3\,dx =$ A. 3 B. 12 C. 6 D. 24

3. $\int_{-5}^{5} |x - 10| dx =$ A. 100 B. -100 C. 0 D. 10

4. $\int_{-2}^{2} x\sqrt{4 - x^2}\,dx =$ A. $\frac{8}{3}$ B. $\frac{16}{3}$ C. $-\frac{16}{3}$ D. 0

5. $\int_0^1 \frac{x}{2 + x} dx =$ A. 0.095 B. 0.189 C. 0 D. 1

6. $\int_{-\pi/4}^{\pi/4} \sin x\,dx =$ A. 0 B. 0.134 C. 0.268 D. 0.293

7. Answer true or false. $\int_1^3 [f(x) + 3g(x)]dx = 5$ if $\int_1^3 f(x)dx = -1$ and $\int_1^3 g(x)dx = 2$

8. $\int_1^4 x + x^3 dx =$ A. 71.25 B. 162.75 C. 3 D. 4.5

9. Answer true or false. $\int_0^5 \frac{x^2}{1+x}dx$ is positive.

10. Answer true or false. $\int_{-2}^{0} |x+2|dx$ is negative.

11. Answer true or false. $\int_{-2}^{-1} \frac{1}{x^2}dx$ is negative.

12. Answer true or false. $\int 3x\,dx = \lim_{\max \Delta x \to 0} \sum_{i=1}^{k} 3i\Delta x_i$

13. $\int_{-1}^{1} x^3 dx =$ A. 0 B. 3 C. 13.5 D. 9

14. $\int_0^1 x - 1\,dx =$ A. 0.5 B. −0.5 C. 1 D. −1

15. $\int_{-2}^{2} x\sqrt{x^2+5}\,dx =$ A. 0 B. 5.27 C. 10.55 D. 4

SECTION 6.6

1. Answer true or false. $\int_2^4 x\,dx = \left.\frac{x^2}{2}\right]_2^4$

2. Answer true or false. $\int_0^{\pi} \sin x\,dx = \cos x\Big]_0^{\pi}$

3. $\int_{-2}^{2} x^4 dx =$ A. 0 B. 12.8 C. 6.4 D. 4

4. $\int_1^e \frac{1}{x}\,dx =$ A. 1 B. e C. $\frac{1}{e}$ D. 0

5. Find the area under the curve $y = x^2 - 2$ on $[2, 3]$.

A. 1 B. 2.17 C. 4.33 D. 8.66

6. Find the area under the curve $y = -(x-3)(x+2)$ and above the x-axis.

A. 20.83 B. 41.67 C. 5 D. 0

7. Find the area under the curve $y = e^x$ and above the x-axis on $[0, 1]$.

A. 0 B. 0.63 C. 1.72 D. 2.7

8. Use the Fundamental Theorem of Calculus. $\int_1^2 x^{-3/4}dx =$

A. 0.76 B. −0.76 C. 1 D. 0

9. $\int_{-\pi/4}^{\pi/4} \tan x\, dx =$ A. 0 B. $\frac{\pi}{2}$ C. 0.70 D. 0.35

10. Answer true or false. $\int_{-5}^{5} x^5 dx = 0$

11. Answer true or false. $\int_{-1}^{1} |x| dx = \int_{-1}^{0} -x\, dx + \int_{0}^{1} x\, dx$

12. Answer true or false. $\int_{1}^{2} x^2 dx = \int_{3}^{4} x^2 dx$

13. Answer true or false. $\int_{-8}^{8} x^2 dx = (x*)^2(8-(-8))$ is satisfied when $x* = 0$.

14. Answer true or false. $\frac{d}{dx}\int_{0}^{x} t^3 dt = x^3$

15. Answer true or false. $\frac{d}{dx}\int_{0}^{x} \sin t\, dt = \cos x$

SECTION 6.7

1. Find the displacement of a particle if $v(t) = \cos t$; $[0, 2\pi]$.

   A. 0 B. 1 C. 2 D. 2π

2. Find the displacement of a particle if $v(t) = \sin t$; $[0, \pi/2]$.

   A. 1 B. 0 C. 2 D. 2π

3. Find the displacement of a particle if $v(t) = t^5$; $[0, 1]$.

   A. 0 B. −0.17 C. 0.17 D. 1

4. Find the displacement of a particle if $v(t) = t^2 - 2$; $[0, 3]$.

   A. 3 B. 9 C. 1.5 D. 6

5. Find the displacement of a particle if $v(t) = e^t + 2$; $[0, 2]$.

   A. 1.72 B. 16.39 C. 4.81 D. 3.15

6. Find the area between the curve and the x-axis on the given interval. $y = x^2 - 4$; $[0, 4]$

   A. 5.33 B. 10.67 C. 16 D. 8

7. Answer true or false. The area between the curve $y = x^3 - 2$ and the x-axis on $[0, 2]$ is given by $\int_{0}^{2} x^3 - 2\, dx$.

8. Answer true or false. If a velocity $v(t) = t^5$ on $[-2, 2]$, the displacement is given by $\int_{-2}^{0} -t^5 dt + \int_{0}^{2} t^5 dt$.

9. Find the total area between $y = e^x$ and the x-axis on $[0, 2]$.

A. 6.39 B. 7.39 C. 6.45 D. 8.39

10. Find the total area between $y = \frac{1}{x}$ and the x-axis on $[0.5, 1]$.

A. 0 B. 1 C. 0.41 D. 0.69

11. Answer true or false. The area between $y = \frac{1}{x}$ and the x-axis on $[2, 3]$ is $-\int_2^3 \frac{1}{x}\,dx$.

12. Answer true or false. The area between $y = x^4 + \sin x$ and the x-axis on $[0, 7]$ is $\int_0^7 x^{-4} + \sin x\,dx$.

13. Answer true or false. The area between $y = \frac{1}{x^3}$ and the x-axis on $[-2, -1]$ is $-\int_{-2}^{-1} \frac{1}{x^3}dx$.

14. Answer true or false. The area between $y = x^4 - x^3$ and the x-axis on $[1, 2]$ is $\int_1^2 x^4 dx + \int_1^2 x^3 dx$.

15. If the velocity of a particle is given by $v(t) = 4$; $[0, 2]$ the displacement is

A. 0 B. 4 C. 8 D. 2

SECTION 6.8

1. $\int_0^2 (x + 5)^8 dx =$

A. 4,268,720 B. 8 C. 3,842,780 D. 57

2. $\int_0^1 \frac{1}{5x + 3}\,dx =$

A. 0.197 B. 0.039 C. 0.981 D. 0.392

3. Answer true or false. $\int \tan^3 x \sec^2 x\,dx = \int u^3 du$ if $u = \tan x$

4. Answer true or false. $\int_0^1 (x + 4)(x - 2)^{15}\,dx = \int_{02}^{-1} u^{15} du$ if $u = x - 2$

5. $\int_0^1 (5x + 2)^3 dx =$

A. 0.33 B. 119.25 C. 1.67 D. 3.67

6. Answer true or false. For $\int_0^1 e^x(1 + 5e^x)^2 dx$ a good choice for u is e^x.

7. Answer true or false. For $\int_0^1 e^x(7 + 8e^x)\,dx$ a good choice for u is $7 + 8e^x$.

8. Answer true or false. For $\int \frac{1}{\sqrt{x}(2+\sqrt{x})}\,dx$ a good choice for u is $2+\sqrt{x}$.

9. Answer true or false. For $\int_0^2 e^{5x}dx$ a good choice for u is $5x$.

10. $\int_0^1 e^{-3x}dx =$ A. 0.216 B. 0.148 C. 0.317 D. 0.519

11. Answer true or false. $\int_{-2}^{2} \sin^2 x\,dx = 2\int_0^2 \sin^2 x\,dx$

12. Answer true or false. $\int_{-4}^{4} x^4 dx = 2\int_0^4 x^4 dx$

13. $\int_0^{\pi/2} 2\sin 4x\,dx =$ A. 2.359 B. -2.359 C. 0 D. 1

14. $\int_1^2 x\sqrt{x+4}\,dx =$ A. 3.53 B. 1.08 C. 0 D. 7.06

SECTION 6.9

1. Simplify. $e^{3\ln x} =$

A. x^3 B. $3x$ C. $\frac{x}{3}$ D. e^3

2. Simplify. $\ln(e^{-6x}) =$

A. x^6 B. x^{-6} C. -6 D. 6

3. Simplify. $\ln(xe^{4x}) =$

A. 4 B. $4+\ln x$ C. $4x+\ln x$ D. $4\ln x$

4. Approximate $\ln 7/2$ to 3 decimal places.

A. 1.250 B. 1.253 C. 1.256 D. 1.259

5. Approximate $\ln 7$ to 3 decimal places.

A. 1.946 B. 1.948 C. 1.950 D. 1.952

6. Approximate $\ln 6.1$ to 3 decimal places.

A. 1.800 B. 1.803 C. 1.805 D. 1.808

7. Let $f(x) = e^{-3x}$, the simplest exact value of $f(\ln 2) =$

A. 6^{-6} B. $\frac{1}{8}$ C. -8 D. $-\frac{1}{8}$

8. Answer true or false. If $F(x) = \int_1^{x^2} \frac{2}{t}\,dt$, $F'(x) = \frac{2}{x}$

9. Answer true or false. If $F(x) = \int_1^{x^3} \frac{2}{t}\,dt$, $F'(x) = \frac{4}{x}$

10. Answer true or false. $\lim_{x\to\infty} \left(1 + \frac{1}{x}\right)^{3x} = 0$

11. Answer true or false. $\lim_{x\to 0} (1 + 5x)^{1/(5x)} = e$

12. Answer true or false. $\lim_{x\to 0} (1 + 2x)^{1/(2x)} = 0$

13. Approximate $\ln 9.1$ to 3 decimal places.

A. 2.011 B. 2.104 C. 2.208 D. 2.211

14. Approximate $\ln 4.1$ to 3 decimal places.

A. 1.411 B. 2.014 C. 2.116 D. 2.120

15. Approximate $\ln 5.2$ to 3 decimal places.

A. 1.641 B. 1.649 C. 1.654 D. 1.695

CHAPTER 6 TEST

1. $f(x) = x$; $[0, 4]$. Use the rectangle method to approximate the area using 4 rectangles. Use the left side of the rectangles.

A. 8.0 B. 2.0 C. 1.25 D. 4.0

2. Use the antiderivative method to find the area under $y = x^2$ on $[0, 1]$.

A. 0.40 B. 0.33 C. 0.50 D. 0.67

3. Answer true or false. $\int x^9 dx = 9x^8 + C$

4. $\int 3\cos x\,dx =$

A. $3\sin x + C$ B. $-3\sin x + C$ C. $3\cos x + C$ D. $-3\cos x + C$

5. $\int 5e^x dx =$

A. $5e^x + C$ B. $\frac{e^x}{5} + C$ C. $-5e^x + C$ D. $-\frac{e^x}{5} + C$

6. Answer true or false. $\int x^2 + e^x dx = x^3 + e^x + C$

7. $\sum_{i=3}^{6} i^2 =$ A. 50 B. 4 C. 86 D. 100

8. Answer true or false. $3 + 6 + 9 + 12 + 15 = \sum_{i=1}^{5} 3i$

9. Answer true or false. $\int 2x(x^2+2)^5 dx = \dfrac{(x^2+2)^6}{6} + C$

10. Answer true or false. For $\int e^{3x} dx$, a good choice for u is $3x$.

11. Answer true or false. For $\int x\sqrt{x+2}\, dx$, a good choice for u is $x+2$.

12. Answer true or false. $\sum_{i=1}^{4} 3i = 3\sum_{i=1}^{4} i$

13. $\lim_{n\to+\infty} \sum_{k=1}^{n} \left(\frac{2}{3}\right)^k =$

A. $\frac{2}{3}$ B. $\frac{1}{2}$ C. 2 D. $\frac{3}{2}$

14. $\int_6^{10} |x+5| dx =$

A. 60 B. 52 C. 5 D. 20

15. Answer true or false. $\int_5^6 \frac{x}{2+x} dx$ is positive.

16. $\int_{-3}^{3} x^5 - x^3 + 3x\, dx =$

A. 0 B. 114.75 C. −114.75 D. 229.5

17. $\int_1^e \frac{3}{x}\, dx =$ A. 3.00 B. 6.48 C. 5.14 D. 3.30

18. Find the displacement of a particle if $v(t) = t^5$; $[0, 2]$.

A. 4 B. 8.33 C. 10.67 D. 2

19. Answer true or false. $\int_1^2 \frac{1}{3x+1} dx = \frac{\ln 7 - \ln 1}{3}$

20. Approximate ln 12.4 to 3 decimal places.

A. 2.487 B. 2.501 C. 2.518 D. 2.531

21. Answer true or false. $\lim_{x\to+\infty} \left(1 + \frac{1}{5x}\right)^{5x} = e$

ANSWERS TO SAMPLE TESTS

SECTION 6.1

1. B 2. D 3. A

SECTION 6.2

1. B 2. B 3. B 4. B 5. C 6. A 7. B 8. D 9. T 10. F 11. F 12. F 13. T 14. A 15. A

SECTION 6.3

1. A 2. A 3. D 4. C 5. B 6. C 7. B 8. T 9. T 10. T 11. F 12. T 13. F 14. F 15. T

SECTION 6.4

1. B 2. D 3. B 4. T 5. B 6. C 7. B 8. T 9. F 10. A 11. D 12. B 13. F 14. T 15. T

SECTION 6.5

1. A 2. B 3. A 4. D 5. B 6. A 7. T 8. A 9. T 10. F 11. F 12. T 13. A 14. B 15. A

SECTION 6.6

1. T 2. F 3. B 4. A 5. C 6. A 7. C 8. A 9. A 10. T 11. T 12. F 13. T 14. T 15. F

SECTION 6.7

1. A 2. A 3. C 4. A 5. B 6. C 7. F 8. F 9. A 10. D 11. F 12. F 13. T 14. F 15. C

SECTION 6.8

1. A 2. A 3. T 4. F 5. B 6. F 7. T 8. T 9. T 10. C 11. T 12. T 13. B 14. A

SECTION 6.9

1. A. 2. C 3. B 4. B 5. A 6. D 7. B 8. F 9. T 10. F 11. T 12. F 13. C 14. A 15. B

CHAPTER 6 TEST

1. A 2. B 3. F 4. A 5. A 6. F 7. C 8. T 9. T 10. T 11. T 12. T 13. C 14. B 15. T 16. A 17. A 18. C 19. T 20. C 21. T

CHAPTER 7
Sample Exams

SECTION 7.1

1. Find the area of the region enclosed by the curves $y = x^2$ and $y = -x$ by integrating with respect to x.

A. $\frac{1}{6}$ B. 1 C. $\frac{1}{4}$ D. $\frac{1}{16}$

2. Answer true or false. $\int_0^2 8x - x^3\,dx = \int_0^2 y - \sqrt[3]{y}\,dy$

3. Find the area enclosed by the curves $y = -x^5$, $y = -\sqrt[3]{x}$, $x = 0$, $x = 1/2$.

A. 0.295 B. 0.315 C. 0.273 D. 0.279

4. Find the area enclosed by the curves $y = \sin 3x$, $y = 2x$, $x = 0$, $x = \pi$.

A. 4.27 B. 2.38 C. 9.32 D. 10.68

5. Find the area between the curves $y = |x - 2|$, $y = \frac{x}{2} + 2$.

A. 3.0240 B. 12.000 C. 3.0251 D. 3.0262

6. Find the area between the curves $x = 2|y|$, $x = -2y + 4$, $y = 0$.

A. 1 B. 2 C. 0.5 D. 0.3

7. Use a graphing utility to find the area of the region enclosed by the curves $y = x^3 - 2x^2 + 5x + 2$, $y = 0$, $x = 0$, $x = 2$.

A. 11.67 B. 12.33 C. 12.67 D. 13.33

8. Use a graphing utility to find the area enclosed by the curves $y = -x^5$, $y = x^2$, $x = 0$, $x = 3$.

A. 130.5 B. 120.75 C. 140.5 D. 125.25

9. Use a graphing utility to find the region enclosed by the curves $x = 2y^4$, $x = \sqrt{4y}$.

A. 2 B. 1 C. 0.76 D. 0.93

10. Answer true or false. The curves $y = x^2 + 5$ and $y = 6x$ intersect at $x = 1$ and $x = 2$.

11. Answer true or false. The curves $x = y^2 + 3$ and $x = 11y$ intersect at $y = 3$ and $y = 6$.

12. Answer true or false. The curves $y = \cos x - 1$, $y = x^2$ intersect at $x = 0$ and $x = \pi$.

13. Answer true or false. The curves $y = 2\sin(\pi x/2)$, $y = 2x^3$ intersect at $x = 0$ and $x = 1$.

14. Find a vertical line $x = k$ that divides the area enclosed by $y = -\sqrt{x}$, $y = 0$ and $x = 4$ into equal parts.

A. $k = 4$ B. $k = 4^{2/3}$ C. $k = 4^{3/2}$ D. $k = 2$

15. Approximate the area of the region that lies between $y = 3\cos x$ and above $y = 0.3x$, where $0 \le x \le \pi$.

A. 9.998 B. 2.67 C. 2.70 D. 4.43

SECTION 7.2

1. Use the method of disks to find the volume of the solid that results by revolving the region enclosed by the curves $y = -x^3$, $x = -4$, $x = 0$, $y = 0$ about the x-axis.

A. 7,353.122 B. 3,676.561 C. 14,706.244 D. 46,201.028

2. Use the method of disks to find the volume of the solid that results by revolving the region enclosed by the curves $y = \sqrt{\cos x}$, $x = 0$, $x = \pi/2$, $y = 0$ about the x-axis.

A. $\pi/4$ B. 2π C. $\pi/2$ D. π

3. Use the method of disks to find the volume of the solid that results by revolving the region enclosed by the curves $y = 2x^2 - 4$, $y = 0$, $x = 0$, $y = 2$ about the x-axis.

A. 12.57 B. 157.91 C. 50.27 D. 39.18

4. Use the method of disks to find the volume of the solid that results by revolving the region enclosed by the curves $y = x + e^x$, $y = 0$, $x = 0$, $x = 1$ about the x-axis.

A. 5.53 B. 11.05 C. 17.37 D. 34.73

5. Answer true or false. The volume of the solid that results when the region enclosed by the curves $y = x^{10}$, $y = 0$, $x = 0$, $x = 2$ is revolved about the x-axis is given by $\int_0^2 \pi x^{20}\,dx$.

6. Answer true or false. The volume of the solid that results when the region enclosed by the curves $y = \sqrt[3]{x}$, $y = 0$, $x = 0$, $x = 3$ is revolved about the x-axis is given by $\left(\int_0^3 \pi\sqrt[3]{x}\,dx\right)^2$.

7. Use the method of disks to find the volume of the solid that results by revolving the region enclosed by the curves $y = -x^5$, $x = 0$, $y = -1$ about the y-axis.

A. 0.83 B. 2.62 C. 8.23 D. 2.24

8. Use the method of disks to find the volume of the solid that results by revolving the region enclosed by the curves $x = \sqrt{4y + 16}$, $x = 0$, $y = 1$, $y = 0$ about the y-axis.

A. 28.3 B. 98.8 C. 13.3 D. 41.9

9. Use the method of disks to find the volume of the solid that results by revolving the region enclosed by the curves $x = y^2$, $x = -y + 6$ about the y-axis.

A. 20.83 B. 65.45 C. 523.60 D. 1,028.08

10. Answer true or false. The volume of the solid that results when the region enclosed by the curves $x = y^6$, $x = y^8$ is revolved about the y-axis is given by $\int_0^1 \left(y^6 - y^8\right)^2 dy$.

11. Find the volume of the solid whose base is enclosed by the circle $(x-2)^2 + (y+3)^2 = 9$ and whose cross sections taken perpendicular to the base are semicircles.

A. 113.10 B. 355.31 C. 56.5 D. 117.65

12. Answer true or false. A right-circular cylinder of radius 8 cm contains a hollow sphere of radius 4 cm. If the cylinder is filled to a height h with water and the sphere floats so that its highest point is 1 cm above the water level, there is $16\pi h - 8\pi/3$ cm^3 of water in the cylinder.

13. Use the method of disks to find the volume of the solid that results by revolving the region enclosed by the curves $y = \cos^6 x$, $x = 2\pi$, $x = 5\pi/2$ about the x-axis.

A. 0.35 B. 1.11 C. 0.49 D. 0.76

14. Use the method of disks to find the volume of the solid that results by revolving the region enclosed by the curves $y = -e^{2x}$, $x = 2$, $y = -1$ about the x-axis.

A. 574,698.32 B. 2,334.17 C. 693.39 D. 2,178.36

15. Answer true or false. The volume of the solid that results when the region enclosed by the curves $y = x^2$ and $x = y$ is revolved about $x = 1$, correct to three decimals, is 0.133.

SECTION 7.3

1. Use cylindrical shells to find the volume of the solid when the region enclosed by $y = x^2$, $x = -2$, $x = -1$, $y = 0$ is revolved about the y-axis.

A. $\frac{15\pi^2}{4}$ B. $\frac{15\pi}{4}$ C. $\frac{15\pi}{8}$ D. $\frac{15\pi}{2}$

2. Use cylindrical shells to find the volume of the solid when the region enclosed by $y = \sqrt{-x}$, $x = 0$, $x = -1$, $y = 0$ is revolved about the y-axis.

A. 0.4π B. 0.8π C. 0.2π D. $0.2\pi^2$

3. Use cylindrical shells to find the volume of the solid when the region enclosed by $y = \frac{3}{x^2}$, $x = 1$, $x = 2$, $y = 0$ is revolved about the y-axis.

A. 2.08π B. 4.16π C. $2.08\pi^2$ D. $4.16\pi^2$

4. Use cylindrical shells to find the volume of the solid when the region enclosed by $y = -\frac{1}{x^3}$, $x = 1$, $x = 2$, $y = 0$ is revolved about the y-axis.

A. π^2 B. 0.5π C. π D. $0.05\pi^2$

5. Use cylindrical shells to find the volume of the solid when the region enclosed by $y = -\sin(-x^2)$, $y = 0$, $x = 0$, $x = 1$ is revolved about the y-axis.

A. 0.560π B. 0.520π C. 0.460π D. 0.500π

6. Use cylindrical shells to find the volume of the solid when the region enclosed by $y = 2e^{x^2}$, $x = 1$, $x = 2$, $y = 0$ is revolved about the y-axis.

A. 103.8π B. 12.970π C. 6.485π D. 25.940π

7. Use cylindrical shells to find the volume of the solid when the region enclosed by $y = 4 - x$, $y = x$, $x = 4$, $y = 0$ is revolved about the y-axis.

A. $\frac{80\pi}{3}$ B. $\frac{32\pi}{3}$ C. $\frac{16\pi}{3}$ D. $\frac{8\pi}{3}$

8. Use cylindrical shells to find the volume of the solid when the region enclosed by $y = 2x^2 - 6x$, $y = 0$ is revolved about the y-axis.

A. $\frac{27\pi}{2}$ B. 27π C. $\frac{27\pi}{8}$ D. $\frac{27\pi}{4}$

9. Use cylindrical shells to find the volume of the solid when the region enclosed by $x = -y^2$, $x = 0$, $y = -2$ is revolved about the x-axis.

A. 4 B. 4π C. 8 D. 8π

10. Use cylindrical shells to find the volume of the solid when the region enclosed by $y = \sqrt[3]{8x}$, $x = 1$, $y = 0$ is revolved about the x-axis.

A. $\dfrac{\pi}{10}$ B. $\dfrac{2\pi}{5}$ C. $\dfrac{\pi}{20}$ D. $\dfrac{48\sqrt[3]{2}\,\pi}{7}$

11. Use cylindrical shells to find the volume of the solid when the region enclosed by $xy = 7$, $x+y = -6$ is revolved about the x-axis.

A. $\dfrac{\pi}{3}$ B. 16.4π C. 196.1π D. 65.6π

12. Use cylindrical shells to find the volume of the solid when the region enclosed by $xy = 7$, $x+y = -6$ is revolved about the y-axis.

A. $\dfrac{\pi}{3}$ B. 16.4π C. 196.1π D. 65.6π

13. Use cylindrical shells to find the volume of the solid when the region enclosed by $y = -x^2$, $x = 1$, $x = 2$, $y = 0$ is revolved about the line $x = 1$.

A. 68.27π B. 3.333π C. 0.833π D. $4.958\pi^2$

14. Use cylindrical shells to find the volume of the solid when the region enclosed by $y = x^2$, $x = -1$, $x = -2$, $y = 0$ is revolved about the line $x = 1$.

A. 1.521π B. $\dfrac{453\pi}{2}$ C. 6.083π D. 24.333π

15. Answer true or false. The region enclosed by revolving the semicircle $y = \sqrt{16 - x^2}$ about the x-axis is $\dfrac{32\pi}{3}$.

SECTION 7.4

1. Find the arc length of the curve $y = -2x^{3/2}$ from $x = 0$ to $x = 3$.

A. 10.9 B. 10.9π C. 6.8 D. 6.8π

2. Find the arc length of the curve $y = \dfrac{1}{2}(x^2 + 2)^{3/2} + 4$ from $x = 0$ to $x = 2$.

A. 6.49 B. 12.99 C. 25.98 D. 51.96

3. Answer true or false. The arc length of the curve $y = (x - 2)^{5/2}$ from $x = 0$ to $x = 5$ is given by $\displaystyle\int_0^5 \sqrt{1 + (x-2)^5}\, dx$.

4. Answer true or false. The arc length of the curve $y = e^x + e^{2x}$ from $x = 0$ to $x = 4$ is given by $\displaystyle\int_0^4 \sqrt{1 + (e^{3x})^2}\, dx$.

5. The arc length of the curve $x = \dfrac{1}{6}(y^2 + 4)^{3/2}$ from $y = -1$ to $y = 0$ is

A. 2.333 B. 1.667 C. 4.667 D. 9.333.

6. Find the arc length of the parametric curve $x = \frac{3}{2}t^2$, $y = t^3$, $0 \le t \le 2$.

A. 3.328 B. 3.324 C. 10.180 D. 3.348

7. Find the arc length of the parametric curve $x = -\cos t$, $y = \sin t$, $0 \le t \le \pi/2$.

A. $\frac{\pi}{2}$ B. $\frac{\pi^2}{4}$ C. $\sqrt{\pi}$ D. π

8. Answer true or false. The arc length of the parametric curve $x = 3e^t$, $y = e^t$, $0 \le t \le 3$ is given by $\int_0^3 \sqrt{4e^t}\, dt$.

9. The arc length of the parametric curve $x = -\cos 2t$, $y = \sin 2t$, $0 \le t \le 1$ is

A. 2 B. $\sqrt{2}$ C. π D. 2π.

10. Answer true or false. The arc length of the parametric curve $x = e^{3t}$, $y = e^{3t}$, $0 \le t \le 2$ is given by $\int_0^2 \sqrt{3}\, e^t\, dt$.

11. Use a CAS or a calculator with integration capabilities to approximate the arc length $y = \sin(-x)$ from $x = 0$ to $x = \pi/2$.

A. 1.43 B. 1.74 C. 1.86 D. 1.91

12. Use a CAS or a calculator with integration capabilities to approximate the arc length $x = \sin(-3y)$ from $y = 0$ to $y = \pi$.

A. 2.042 B. 6.987 C. 2.051 D. 2.916

13. Use a CAS or a calculator with integration capabilities to approximate the arc length $y = \sin(-3x)$ from $x = 0$ to $x = \pi$.

A. 2.042 B. 6.987 C. 2.051 D. 2.916

14. Use a CAS or a calculator with integration capabilities to approximate the arc length $y = -xe^x$ from $x = 0$ to $x = 2$.

A. 21.02 B. 4.17 C. 15.04 D. 19.71

15. Answer true or false. The arc length of $y = x \cos x$ from $x = 0$ to $x = \pi$ can be approximated by a CAS or a calculator with integration capabilities to be 4.698.

SECTION 7.5

1. Find the area of the surface generated by revolving $y = -2x$, $0 \le x \le 1$ about the x-axis.

A. 4.47 B. 14.05 C. 28.10 D. 88.28

2. Find the area of the surface generated by revolving $y = \sqrt{1+x}$, $-1 \le x \le 0$ about the x-axis.

A. 4.47 B. 28.07 C. 5.33 D. 7.02

3. Find the area of the surface generated by revolving $x = -2y$, $0 \le y \le 1$ about the y-axis.

A. 4.47 B. 14.05 C. 28.10 D. 88.28

4. Find the area of the surface generated by revolving $x = \sqrt{y+1}$, $0 \le y \le 1$ about the y-axis.

A. 67.88 B. 3.44 C. 8.28 D. 21.60

5. Answer true or false. The area of the surface generated by revolving $x = \sqrt{3y}$, $1 \le y \le 5$ about the y-axis is given by $\int_1^5 2\pi y \left(1 + \frac{3}{4\sqrt{3x}}\right) dy$.

6. Answer true or false. The area of the surface generated by revolving $x = e^{y+2}$, $0 \le y \le 1$ about the y-axis is given by $\int_0^1 2\pi y\sqrt{1 + e^{2y+4}}\, dy$.

7. Answer true or false. The area of the surface generated by revolving $x = \sin y$, $0 \le y \le \pi$ about the y-axis is given by $\int_0^\pi 2\pi y\sqrt{1 - \cos^2 x}\, dx$.

8. Use a CAS or a scientific calculator with numerical integration capabilities to approximate the area of the surface generated by revolving the curve $y = e^{x+1}$, $-1 \le x \le -0.5$ about the x-axis.

A. 18.54 B. 9.27 C. 1.48 D. 1.36

9. Use a CAS or a scientific calculator with numerical integration capabilities to approximate the area of the surface generated by revolving the curve $x = e^{y-1}$, $1 \le y \le 1.5$ about the y-axis.

A. 18.54 B. 9.27 C. 1.48 D. 1.63

10. Answer true or false. A CAS or a calculator with numerical integration capabilities can be used to approximate the area of the surface generated by revolving the curve $y = \cos x$, $0 \le x \le \pi/2$ about the x-axis to be 1.

11. Answer true or false. A CAS or a calculator with numerical integration capabilities can be used to approximate the area of the surface generated by revolving the curve $y = \sin x$, $0 \le x \le \pi/2$ about the x-axis to be 1.

12. Answer true or false. A CAS or a calculator with numerical integration capabilities can be used to approximate the area of the surface generated by revolving the curve $x = \cos y$, $0 \le y \le \pi/2$ about the y-axis to be 8.08.

13. Answer true or false. The area of the surface generated by revolving the curve $x = t^2$, $y = e^t$, $0 \le t \le 1$ about the x-axis is given by $2\pi \int_0^1 t\sqrt{e^{2t} + 4t^2}\, dt$.

14. Answer true or false. The area of the surface generated by revolving the curve $x = t^2$, $y = e^t$, $0 \le t \le 1$ about the y-axis is given by $2\pi \int_0^1 t^2\sqrt{e^{2t} + 4t^2}\, dt$.

15. The area of the surface generated by revolving the curve $x = 4\sin t$, $y = 4\cos t$, $0 \le t \le \pi$ about the x-axis is

A. $\frac{16\pi}{3}$ B. $\frac{8\pi}{3}$ C. $\frac{4\pi}{3}$ D. 64π

SECTION 7.6

1. Find the work done when a constant force of 20 lb in the positive x direction moves an object from $x = 3$ to $x = 4$ ft.

   A. 20 ft-lb B. 140 ft-lb C. 40 ft-lb D. 100 ft-lb

2. A spring whose natural length is 35 cm is stretched to a length of 40 cm by a 2-N force. Find the work done in stretching the spring.

   A. 0.05 J B. 0.4 J C. 0.7 J D. 0.3 J

3. Assume 20 J of work stretch a spring from 60 cm to 64 cm. Find the spring constant in J/cm.

   A. 80 B. 5 C. 1.25 D. 1.6

4. Answer true or false. Assume a spring is stretched from 100 cm to 140 cm by a force of 500 N. The work needed to do this is 200 J.

5. A cylindrical tank of radius 5 m and height 10 m is filled with a liquid whose density is 1.84 kg/m^3. How much work is needed to lift the liquid out of the tank?

   A. 7,225.7 J B. 7,125.8 J C. 7,334.1 J D. 7,310.2 J

6. Answer true or false. The amount of work needed to lift a liquid of density 0.95 kg/m^3 from a spherical tank of radius 4 m is $\int_0^8 0.95(8 - x)\pi x^2 \, dx$.

7. An object in deep space is initially considered to be stationary. If a force of 250 N acts on the object over a distance of 400 m, how much work is done on the object?

   A. 0 J B. 100,000 J C. 50,000 J D. 25,000 J

8. Find the work done when a variable force of $F(x) = \dfrac{3}{x^2}$ N in the positive x-direction moves an object from $x = 2$ m to $x = 8$ m.

   A. 5.64 J B. 4.50 J C. 2.25 J D. 4.16 J

9. Find the work done when a variable force of $F(x) = \dfrac{1}{x^2}$ N in the positive x direction moves an object from $x = -5$ m to $x = -4$ m.

   A. 0.113 J B. 1.00 J C. 0.15 J D. 0.25 J

10. Find the work done when a variable force of $F(x) = 30x$ N in the positive x direction moves an object from $x = -4$ m to $x = 0$ m.

    A. 240 J B. 320 J C. 80 J D. 160 J

11. If the Coulomb force is proportional to x^{-2}, the work it does is proportional to

    A. x^{-1} B. x^{-3} C. x D. x^{-2}.

12. Answer true or false. It takes the same amount of work to move an object from 100,000 km above the earth to 200,000 km above the earth as it does to move the object from 200,000 km above the earth to 300,000 km above the earth.

13. Answer true or false. It takes twice as much work to elevate an object to 120 m above the earth as it does to elevate the same object to 60 m above the earth.

14. Answer true or false. It takes twice as much work to stretch a spring 100 cm as it does to stretch the same spring 50 cm.

15. A 1-kg object is moving at 10.0 m/s. If a force in the direction of the motion does 40.0 J of work on the object, what is the object's final speed?

A. 13.4 m/s B. 5.5 m/s C. 5.0 m/s D. 11 m/s

SECTION 7.7

1. A flat rectangular plate is submerged horizontally in water to a depth of 6.0 ft. If the top surface of the plate has an area of 50 ft^2, and the liquid in which it is submerged is water, find the force on the top of the plate. Neglect the effect of the atmosphere above the liquid. (The density of water is 62.4 lb/ft^3.)

A. 300 lb B. 33.4 lb C. 2,080 lb D. 18,720 lb

2. Find the force (in N) on the top of a submerged object it its surface is 10.0 m^2 and the pressure acting on it is 3.2×10^5 Pa. Neglect the effect of the atmosphere above the liquid.

A. 37,000 N B. 3,200,000 N C. 6,400,000 N D. 1,700,000 N

3. Find the force on a 100-ft wide by 5-ft deep wall of a swimming pool filled with water. Neglect the effect of the atmosphere above the liquid. (The density of water is 62.4 lb/ft^3.)

A. 78,000 lb B. 124,800 lb C. 62,400 lb D. 624 lb

4. Answer true or false. If a completely full pyramid-shaped tank is inverted, the force on the side wall perpendicular to it will be the same as before inversion for all sides.

5. Answer true or false. The force a liquid of density ρ exerts on an equilateral triangle with edges h in length submerged point down is given by $\int_0^h \frac{\rho}{3} x^2\, dx$.

6. A right triangle is submerged vertically with one side at the surface in a liquid of density ρ. The triangle has a leg that is 20 m long located at the surface and a leg 10 m long straight down. Find the force exerted on the triangular surface, in terms of the density. Neglect the effect of the atmosphere above the liquid.

A. 667 ρ N B. 500 ρ N C. 600 ρ N D. 200 ρ N

7. Answer true or false. A semicircular wall 10 ft across the top forms one vertical end of a tank. The total force exerted on this wall by water, if the tank is full of a liquid is 260 lb. Neglect the effect of the atmosphere above the liquid. (The density of the liquid is 124.8 lb/ft^3.)

8. Answer true or false. A glass circular window on the side of a submarine has the same force acting on the top half as on the bottom half.

9. Find the force on a 30 ft^2 horizontal surface 20 ft deep in water. Neglect the effect of the atmosphere above the liquid. (The density of water is 62.4 lb/ft^3.)

A. 600 lb B. 37,440 lb C. 1,200 lb D. 30,000 lb

10. Answer true or false. A flat sheet of material is submerged vertically in water. The force acting on each side must be the same.

11. Answer true or false. If a submerged horizontal object is elevated to half its original depth, the force exerted on the top of the object will be half the force originally exerted on the object. Assume there is a vacuum above the liquid surface.

12. Answer true or false. If a square, flat surface is suspended vertically in water and its center is 20 m deep, the force on the object will double if the object is relocated to a depth of 40 m. Neglect the effect of the atmosphere above the liquid.

13. Answer true or false. The force on a semicircular, vertical wall with top d is given by $\int_0^{1/2} 2\rho x\sqrt{\frac{d^2}{4} - x^2}\,dx$. Neglect the effect of the atmosphere above the liquid.

14. Answer true or false. The force exerted by water on a surface of a square, vertical plate with edges of 3 m if it is suspended with its top 2 m below the surface is 18 lb. (The density of water is 62.4 lb/ft^3.)

15. Answer true or false. If a submerged rectangle is rotated 90° about an axis through its center and perpendicular to its surface, the force exerted on one side of it will be the same, provided the entire rectangle remains submerged.

SECTION 7.8

1. Evaluate sinh (7).

A. Not defined B. 551.1614 C. 548.3161 D. 549.4283

2. 2. Evaluate $\cosh^{-1}(2)$.

A. 1.3170 B. 1.3165 C. 1.3152 D. 1.3174

3. Find dy/dx if $y = \sinh(5x + 1)$.

A. $(5x + 1)\cosh(5x + 1)$ B. $5\cosh(5x + 1)$
C. $-(5x + 1)\cosh(5x + 1)$ D. $-5\cosh(5x + 1)$

4. Find dy/dx if $y = \sinh(3x^2)$.

A. $6x\cosh(3x^2)$ B. $-6x\cosh(3x^2)$ C. $2x^2\cosh(3x^2)$ D. $-2x^2\cosh(3x^2)$

5. Find dy/dx if $y = 2\sqrt{\text{sech}\,(x + 5) - x^3}$.

A. $-\sinh(x + 5)\tanh(x + 5) - 3x^2$ B. $\dfrac{(x + 5)\cosh(x + 5) - 3x^2}{\sqrt{\sinh(x + 5) - x^3}}$

C. $\dfrac{-\cosh(x + 5) + 3x^2}{\sqrt{\sinh(x + 5) - x^3}}$ D. $\dfrac{-(x + 5)\cosh(x + 5) + 3x^2}{\sqrt{\sinh(x + 5) - x^3}}$

6. $\int \sinh(3x+6)\,dx =$

A. $3\cosh(3x+6)+C$

B. $\dfrac{\cosh(3x+6)}{3}+C$

C. $-3\cosh(3x+6)+C$

D. $\dfrac{-\cosh(3x+6)}{3}+C$

7. $\int \cosh^7 x \sinh x\,dx =$

A. $\dfrac{\cosh^8 x}{8}+C$

B. $8\cosh^8 x+C$

C. $7\cosh^6 x+C$

D. $\dfrac{\cosh^6 x}{6}+C$

8. $\int \cosh^9 x \sinh x\,dx =$

A. $\dfrac{\cosh^{10} x}{10}+C$

B. $10\cosh^{10} x+C$

C. $9\cosh^8 x+C$

D. $\dfrac{\cosh^8 x}{8}+C$

9. Find dy/dx if $y=\sinh^{-1}\left(\dfrac{x}{6}\right)$.

A. $\dfrac{1}{\sqrt{36+x^2}}$

B. $\dfrac{1}{6\sqrt{36+x^2}}$

C. $\dfrac{1}{\sqrt{36-x^2}}$

D. $\dfrac{1}{6\sqrt{36-x^2}}$

10. Answer true or false. If $y=-\coth^{-1}(x+3)$ when $|x|>0$, $dy/dx=-\dfrac{1}{x^2+6x+8}$.

11. $\int \dfrac{dx}{\sqrt{1+16x^2}} =$

A. $\dfrac{\sinh^{-1}(4x)}{4}+C$

B. $\dfrac{\coth^{-1}(4x)}{4}+C$

C. $\dfrac{\cosh^{-1}(4x)}{4}+C$

D. $\dfrac{\tanh^{-1}(4x)}{4}+C$

12. Answer true or false. $\int \dfrac{4\,dx}{1+e^{2x}} = 4\sinh^{-1} e^x + C$

13. Answer true or false. $\int \dfrac{e^x}{\sqrt{1+e^{2x}}}\,dx = \sinh^{-1}(e^{2x})+C$

14. Answer true or false. $\lim_{x\to+\infty}(\cosh x)^2 = 0$

15. Answer true or false. $\lim_{x\to-\infty}(\coth x)^2 = 1$

SECTION 7 TEST

1. Find the area of the region enclosed by $y=x^2$ and $y=x$ by integrating with respect to x.

A. $\dfrac{1}{6}$

B. 1

C. $\dfrac{1}{4}$

D. $\dfrac{1}{16}$

2. Find the region of the area enclosed by $y=\cos(x-\pi/2)$, $y=-x$, $x=0$, $x=\pi/2$.

A. 1.1169

B. 2.2337

C. 4.4674

D. 1

3. Find the volume of the solid that results when the region enclosed by the curves $y = \sqrt{-\sin(-x)}$, $y = 0$, $x = \pi/4$ is revolved about the x-axis.

A. 0.143 B. 0.920 C. 1.408 D. 2.816

4. Find the volume of the solid that results when the region enclosed by the curves $x = -e^y$, $x = 1$, $y = 1$ is revolved about the y-axis.

A. 10.036 B. 3.195 C. 10.205 D. 32.060

5. Answer true or false. Cylindrical shells can be used to find the volume of the solid when the region enclosed by $y = \sqrt[3]{x}$, $x = -3$, $x = 0$, $y = 0$ is revolved about the y-axis is 5.563π.

6. Answer true or false. Cylidrical shells can be used to find the volume of the solid when the region enclosed by $x = y^2$, $x = 0$, $y = -2$ is revolved about the x-axis is 4π.

7. Answer true or false. The arc length of $y = \cos(-x)$ from 0 to $\pi/2$ is 1.

8. Answer true or false. The arc length of $x = \sin t$, $y = -\cos t$, $0 \leq t \leq \pi/2$ is $\pi/2$.

9. Answer true or false. The surface area of the curve $y = \sin(x + \pi)$, $-\pi \leq y \leq 0$ revolved about the x-axis is given by $\int_0^{\pi} 2\pi x\sqrt{1 + \sin^2(x + \pi)}\, dx$.

10. Use a CAS to find the surface area that results when the curve $y = -e^x$, $0 \leq x \leq 0.5$ is revolved about the x-axis.

A. 18.54 B. 9.27 C. 1.48 D. 5.03

11. Assume a spring whose natural length is 2.0 m is stretched 0.8 m by a 150 N force. How much work is done in stretching the spring?

A. 60 J B. 6,120 J C. 6,000 J D. 240 J

12. Find the work done when a constant force of $F(x) = 15$ N in the positive x direction moves an object from $x = 4$ m to $x = 10$ m.

A. 45 J B. 90 J C. 180 J D. 150 J

13. Find the work done when a variable force of $F(x) = \dfrac{4}{x^2}$ N in the positive x direction moves an object from $x = 1$ m to $x = 3$ m.

A. 0 J B. 2.7 J C. 0.6 J D. 1.7 J

14. Answer true or false. A semicircular wall 20 ft across at the top forms one end of a tank. The total force exerted on this wall by a liquid that fills the tank is 24,800 lb. Ignore the force of the air above the liquid. (The density of liquid is 124.8 lb/ft^3.)

15. A horizontal table top is submerged in 10 ft of water. If the dimensions of the table are 6 ft by 1 ft, find the force on the top of the table that exceeds the force that would be exerted by the atmosphere if the table were at the surface of the water. (The density of water is 62.4 lb/ft^3.)

A. 3,744 lb B. 1,872 lb C. 4,000 lb D. 60 lb

16. Find dy/dx if $y = \tanh(x^5)$.

A. $5x^4 \operatorname{sech}^2(x^5)$ B. $-5x^4 \operatorname{sech}^2(x^5)$ C. $5x^4 \tanh(x^5)$ D. $\operatorname{sech}^2(5x^4)$

17. $\int \tanh^5 x \operatorname{sech}^2 x \, dx =$

A. $4\tanh^4 x + C$ B. $5\tanh^6 x + C$ C. $6\tanh^6 x + C$ D. $\dfrac{\tanh^6 x}{6} + C$

18. Answer true or false. $\displaystyle\int \frac{4\,dx}{\sqrt{e^{2x}-1}} = 4\cosh^{-1}(e^x)$

19. Answer true or false. $\displaystyle\lim_{x\to\infty} (\coth x)^2 = 1$

20. Evaluate $\cosh(1)$.

A. 1.543 B. 1.551 C. 1.562 D. 1.580

ANSWERS TO SAMPLE TESTS

SECTION 7.1

1. A 2. F 3. A 4. C 5. B 6. B 7. C 8. A 9. D 10. F 11. F 12. F
13. T 14. B 15. A

SECTION 7.2

1. A 2. D 3. D 4 C 5. T 6. F 7. D 8. A 9. C 10. F 11. C 12. F
13. B 14. B 15. F

SECTION 7.3

1. D 2. B 3. B 4. C 5. C 6. A 7. A 8. B 9. D 10. D 11. C 12. C
13. A 14. B 15. F

SECTION 7.4

1. A 2. A 3. F 4. F 5. B 6. C 7. A 8. T 9. A 10. F 11. D 12. B
13. B 14. C 15. F

SECTION 7.5

1. B 2. C 3. B 4. C 5. F 6. T 7. F 8. D 9. D 10. F 11. F 12. F
13. T 14. F 15. D

SECTION 7.6

1. A 2. A 3. D 4. T 5. A 6. F 7. B 8. D 9. C 10. A 11. A 12. F
13. F 14. F 15. A

SECTION 7.7

1. D 2. A 3. A 4. F 5. F 6. A 7. F 8. F 9. B 10. T 11. T 12. T
13. F 14. F 15. T

SECTION 7.8

1. C 2. A 3. B 4. A 5. A 6. B 7. A 8. A 9. A 10. F 11. A 12. F
13. F 14. F 15. F

CHAPTER 7 TEST

1. A 2. B 3. B 4. A 5. F 6. F 7. F 8. T 9. F 10. D 11. A 12. B
13. B 14. F 15. A 16. A 17. D 18. F 19. T 20. A

CHAPTER 8
Sample Exams

SECTION 8.1

1. Evaluate $\int (8-2x)^5 dx$.

A. $\frac{(8-2x)^6}{6}+C$ B. $\frac{-(8-2x)^6}{6}+C$ C. $\frac{-(8-2x)^6}{12}+C$ D. $\frac{-(8-2x)^6}{3}+C$

2. Evaluate $\int \sqrt{4x+9}\, dx$.

A. $\frac{1}{8\sqrt{4x+9}}+C$ B. $\frac{2}{\sqrt{4x+9}}+C$ C. $\frac{(4x+9)^{3/2}}{6}+C$ D. $\frac{(4x+9)^{3/2}}{2}+C$

3. $\int 3x\sin(x^2) =$

A. $3\cos(x^2)+C$ B. $6\cos(x^2)+C$ C. $-6\cos(x^2)+C$ D. $\frac{-3\cos(x^2)}{2}+C$

4. $\int xe^{x^2}\, dx =$

A. $\frac{e^{x^2}}{2}+C$ B. $2e^{x^2}+C$ C. $e^{x^2}+C$ D. $\frac{x^2e^{x^2}}{2}+C$

5. $\int \sin x e^{\cos x} dx =$

A. $\cos x e^{\cos x}+C$ B. $-e^{\cos x}+C$ C. $-\sin x e^{\cos x}+C$ D. $-xe^{\cos x}+C$

6. $\int \cos^8 x \sin x\, dx =$

A. $\frac{\cos^9 x}{9}+C$ B. $\frac{-\cos^9 x}{9}+C$ C. $9\cos^9 x+C$ D. $-9\cos^9 x+C$

7. $\int \frac{e^{\sqrt{x}}}{2\sqrt{x}}\, dx =$

A. $\frac{e^{\sqrt{x}}}{2x^{3/2}}+C$ B. $\frac{3e^{\sqrt{x}}}{4x^{3/2}}+C$ C. $\frac{e^{\sqrt{x}}}{4}+C$ D. $e^{\sqrt{x}}+C$

8. $\int \frac{e^x}{5+e^x}\, dx =$

A. $\frac{5e^x}{5+e^x}+C$ B. $5\ln|5+e^x|+C$ C. $\ln|5+e^x|+C$ D. $\frac{e^x}{5+e^x}+C$

9. $\int \frac{4x\, dx}{4-x^2} =$

A. $4\ln|4-x^4|+C$ B. $-2\ln|4-x^2|+C$ C. $2\ln|4-x^2|+C$ D. $4\ln|2-x^2|+C$

10. $\int 4\sinh^2 x\cosh x\, dx =$

A. $\frac{4\sinh^3 x}{3}+C$ B. $12\sinh^3 x+C$ C. $4\sinh^2 x+C$ D. $2\sinh^2 x+C$

11. Answer true or false: In evaluating $\int x7^{x^2}\,dx$ a good choice for u would be x^2.

12. Answer true or false: In evaluating $\int \cos^6 x \sin x\,dx$ a good choice for u would be $\sin x$.

13. Answer true or false: In evaluating $\int e^x(e^x+8)\,dx$ a good choice for u would be e^x+8.

14. Answer true or false: In evaluating $\int \frac{4\sin x}{\cos x}\,dx$ a good choice for u would be $\sin x$.

15. Answer true or false: In evaluating $\int x^3\cos(x^4)\,dx$ a good choice for u would be x^4.

SECTION 8.2

1. $\int xe^{6x}\,dx =$

A. $\frac{e^{6x}}{6}(6x-1)+C$ B. $\frac{e^{6x}}{36}(6x-1)+C$ C. $e^{6x}+C$ D. $\frac{e^{6x}}{6}+C$

2. $\int 9x^2e^{3x}\,dx =$

A. $\frac{e^{3x}}{3}(9x^2-6x+2)+C$ B. $xe^{3x}+C$ C. $27e^{3x}+C$ D. $3e^{3x}+C$

3. $\int x\cos 9x\,dx =$

A. $\frac{x\sin 9x}{9}+C$ B. $\frac{\sin 9x}{9}+C$
C. $\frac{\cos 9x}{81}+\frac{x\sin 9x}{9}+C$ D. $\frac{\cos 9x}{81}+\frac{x\cos 9x}{9}+C$

4. $\int 2x\sin x\,dx =$

A. $2\sin x-2x\cos x+C$ B. $2\sin x-2\cos x+C$
C. $2\cos x-2x\cos x+C$ D. $2\cos x+2x\cos x+C$

5. $\int 2x^2\sin 2x\,dx =$

A. $x\sin 2x+\frac{\cos 2x}{2}-\frac{x^2\cos^2 2x}{4}+C$ B. $x\sin 2x+\frac{\cos 2x}{4}-\frac{x^2\cos^2 2x}{2}+C$
C. $x\sin 2x+\frac{\cos 2x}{2}-x^2\cos^2 2x+C$ D. $2x\sin 2x+\cos 2x-\frac{x^2\cos^2 2x}{2}+C$

6. $\int 5x\ln(2x)\,dx =$

A. $\frac{x^2\ln(2x)}{4}-\frac{x^2}{4}+C$ B. $\frac{5x^2\ln(2x)}{2}-\frac{5x^2}{4}+C$
C. $5x^2\ln(2x)-5x^2+C$ D. $x^2\ln(2x)-x^2+C$

7. $\int \sin^{-1}(6x)\,dx =$

A. $x\sin^{-1}(6x) + \dfrac{\sqrt{1-36x^2}}{6} + C$

B. $x\cos^{-1}(6x) + \dfrac{\sqrt{1-36x^2}}{6} + C$

C. $6\cos^{-1}(6x) + C$

D. $\dfrac{\cos^{-1}(6x)}{6} + C$

8. $\int 5e^{4x}\sin 3x\,dx =$

A. $5e^{4x}(3\sin 3x + \cos 3x) + C$

B. $5e^{4x}(3\sin 3x - \cos 3x) + C$

C. $\dfrac{5e^{4x}}{7}(3\sin 3x - 3\cos 3x) + C$

D. $\dfrac{e^{4x}}{5}(4\sin 3x - 3\cos 3x) + C$

9. $\int_0^1 4xe^{6x}\,dx =$

A. 223.9 B. 224.2 C. 224.7 D. 225.1

10. $\int_1^3 x^2 \ln x\,dx =$

A. 7.00 B. 7.03 C. 6.96 D. 6.92

11. $\int_1^3 \cos(\ln x)\,dx =$

A. 1.57 B. 1.52 C. 1.48 D. 1.42

12. Answer true or false: $\int_0^{\pi/4} x\sin 2x\,dx = 1/4$

13. Answer true or false: $\int_{\pi/4}^{3\pi/4} x\tan x\,dx = 1$

14. Answer true or false: $\int_0^1 \ln(x^2+20)\,dx = 1$

15. Answer true or false: $\int_0^1 4xe^{-3x}\,dx = 0$

SECTION 8.3

1. $\int \cos^{13}x \sin x\,dx =$

A. $\dfrac{\cos^{14}x}{14} + C$ B. $\dfrac{-\cos^{14}x}{14} + C$ C. $\dfrac{\cos^{14}x}{13} + C$ D. $\dfrac{-\cos^{14}x}{13} + C$

2. $\int 2\sin^2(5x)\,dx =$

A. $x - \dfrac{\sin(10x)}{10} + C$

B. $2x - 2\sin(5x) + C$

C. $x - \dfrac{\sin(5x)}{5} + C$

D. $x - \dfrac{\sin(5x)}{10} + C$

3. $\int -\sin^3(2x)\,dx =$

A. $\dfrac{\cos(2x)}{2} + \dfrac{\sin^3(2x)}{6} + C$

B. $\dfrac{\cos(2x)}{2} + \dfrac{\cos^3(2x)}{6} + C$

C. $\dfrac{-\cos(2x)}{2} + \dfrac{\cos^3(2x)}{6} + C$

D. $\dfrac{-\cos(2x)}{2} - \dfrac{\sin^3(2x)}{6} + C$

4. $\int 4\cos^4 x\,dx =$

A. $\dfrac{3x}{2} + 3\,\dfrac{\sin(2x)}{4} + \cos^3 x \sin x + C$

B. $\dfrac{3x}{2} - 3\,\dfrac{\sin(2x)}{4} + \dfrac{\cos^3 x \sin x}{8} + C$

C. $\dfrac{3\sin(2x)}{4} + \sin^3 x \cos x + C$

D. $\dfrac{3x}{2} + 3\,\dfrac{\cos(2x)}{4} + \sin^3 x \cos x + C$

5. $\int \sin^2(4x)\cos^2(4x)\,dx =$

A. $\dfrac{x}{8} - \dfrac{\sin(16x)}{128} + C$

B. $\dfrac{x}{8} - \dfrac{\cos(16x)}{128} + C$

C. $\dfrac{x}{8} - \dfrac{\sin(8x)}{128} + C$

D. $\dfrac{x}{8} - \dfrac{\cos(8x)}{128} + C$

6. $\int \tan(9x)\,dx =$

A. $\dfrac{\ln|\cos(9x)|}{9} + C$

B. $\dfrac{-\ln|\cos(9x)|}{9} + C$

C. $\dfrac{\tan^2(9x)}{18} + C$

D. $\dfrac{-\tan^2(9x)}{18} + C$

7. Answer true or false: $\int \sin(6x)\cos(4x)\,dx = \dfrac{-\cos(2x)}{4} - \dfrac{\cos(10x)}{20} + C$

8. $\int \sec^2(5x+8)\,dx =$

A. $\dfrac{\tan(5x+8)}{5} + C$

B. $\dfrac{\tan(5x+8)}{5x+8} + C$

C. $\dfrac{-\tan(5x+8)}{5} + C$

D. $\dfrac{-\tan(5x+8)}{5x+8} + C$

9. $\int \csc(6x)\,dx =$

A. $\dfrac{\ln|\tan(3x)|}{6} + C$

B. $\dfrac{-\ln|\tan(3x)|}{6} + C$

C. $\dfrac{\ln|\tan(6x)|}{6} + C$

D. $\dfrac{\ln|\tan(6x)|}{12} + C$

10. Answer true or false: $\int \tan^{11} x \sec^2 x\,dx = \dfrac{\tan^{12} x}{12} + C$

11. $\int \tan x \sec^5 x\,dx =$

A. $\sec^5 x + C$ B. $\dfrac{\sec^6 x}{6} + C$ C. $\dfrac{\sec^5 x}{5} + C$ D. $\sec^6 x + C$

12. Answer true or false: $\int \cot^5(4x)\csc^2(4x)\,dx = \dfrac{-\cot^6(4x)}{24} + C$

13. Answer true or false: $\int_0^{\pi/4} \tan^2(6x)\,dx = 1.00$

14. Answer true or false: $\int_0^{\pi/4} \sec^2 x\,dx = 0$

15. Answer true or false: $\int_{-\pi/3}^{\pi/3} \tan(2x)\,dx = 0$

SECTION 8.4

1. $\int \sqrt{9-x^2}\,dx =$

A. $\dfrac{x\sqrt{9-x^2}}{6} + 4\sin^{-1}\left(\dfrac{x}{9}\right) + C$

B. $\dfrac{x\sqrt{9-x^2}}{2} + \dfrac{9}{2}\sin^{-1}\left(\dfrac{x}{3}\right) + C$

C. $\dfrac{x\sqrt{9-x^2}}{2} + 2\sin^{-1}\left(\dfrac{x}{3}\right) + C$

D. $\dfrac{x\sqrt{9-x^2}}{2} + 3\sin^{-1}\left(\dfrac{x}{3}\right) + C$

2. $\int \dfrac{dx}{\sqrt{5-x^2}} =$

A. $\dfrac{\sin^{-1}(\sqrt{5}x)}{5} + C$

B. $\dfrac{\sin^{-1}(5x)}{5} + C$

C. $\sin^{-1}\left(\dfrac{\sqrt{5}x}{5}\right) + C$

D. $\dfrac{\sqrt{5}}{5}\sin^{-1}\left(\dfrac{\sqrt{5}\,x}{5}\right) + C$

3. $\int_{-1}^{0} 3e^x\sqrt{4-2e^{2x}}\,dx =$

A. 1.381 B. 1.388 C. 3.271 D. 1.398

4. $\int_2^3 \dfrac{dx}{(-x)^2\sqrt{(-x)^2-1}} =$

A. 14.941 B. 0.077 C. 0.093 D. 17.01

5. $\int_1^2 \dfrac{4\,dx}{x\sqrt[4]{x^2+5}} =$

A. 2.003 B. 1.684 C. 1.692 D. 1.698

6. $\int_0^1 \dfrac{x^3dx}{(9+x^2)^{5/2}} =$

A. 0.00086 B. 0.00031 C. 0.00041 D. 0.00082

7. $\displaystyle\int_1^4 \frac{\sqrt{3x^2-2}\,dx}{x} =$

A. 4.72 B. 4.79 C. 3.12 D. 3.17

8. $\displaystyle\int_\pi^{2\pi} \frac{\cos\theta\,d\theta}{\sqrt{4-\sin^2\theta}} =$

A. 1 B. 0 C. -1 D. 4

9. $\displaystyle\int \frac{dx}{x^2+3x+9} =$

A. $2\sqrt{3}\tan^{-1}\left(\dfrac{2\sqrt{3}x+3\sqrt{3}}{9}\right)+C$ B. $\dfrac{2}{3}\tan^{-1}\left(\dfrac{2x+3}{3}\right)+C$

C. $\dfrac{2}{3}\tan^{-1}(2x+3)+C$ D. $\dfrac{2}{11}\tan^{-1}\left(\dfrac{2x+3}{11}\right)+C$

10. Answer true or false: $\displaystyle\int \frac{dx}{\sqrt{x^2-4x+2}} = \int \frac{dx}{\sqrt{(x-2)^2-2}}$

11. Answer true or false: $\displaystyle\int \frac{dx}{x^2-7x+3} = \int \frac{dx}{x^2} - 7\int \frac{dx}{x} + \int \frac{dx}{3}$

12. $\displaystyle\int_0^1 \frac{dx}{2\sqrt{5x-x^2}} =$

A. 0.46 B. 0.47 C. 1 D. 0

13. $\displaystyle\int_1^2 \frac{dx}{3x^2+9x+4} =$

A. 0.021 B. 0.034 C. 0.043 D. 0.053

14. Answer true or false: $\displaystyle\int_0^2 4^x\sqrt{16^x-1}\,dx = 90.9$

15. Answer true or false: $\displaystyle\int_0^\pi \cos x\sin x\sqrt{1-\sin^2 x}\,dx = 0$

SECTION 8.5

1. Write out the partial fraction decomposition of $\dfrac{5x+10}{(x-2)(x-5)}$.

A. $\dfrac{1}{x+3}+\dfrac{4}{x-2}$ B. $\dfrac{2}{x+3}+\dfrac{3}{x-2}$ C. $\dfrac{4}{x+3}+\dfrac{1}{x-2}$ D. $\dfrac{3}{x+3}+\dfrac{2}{x-2}$

2. Write out the partial fraction decomposition of $\dfrac{5x-2}{x^2-x}$.

A. $\dfrac{1}{x-1}+\dfrac{2}{x}$ B. $\dfrac{2}{x-1}+\dfrac{1}{x}$ C. $\dfrac{3}{x-1}+\dfrac{2}{x}$ D. $\dfrac{2}{x-1}+\dfrac{3}{x}$

3. Write out the partial fraction decomposition of $\dfrac{3x^2-2x+2}{(x^2+2)(x-1)}$.

A. $\dfrac{2}{x^2+2}+\dfrac{1}{x-1}$ B. $\dfrac{1}{x^2+2}+\dfrac{2}{x-1}$ C. $\dfrac{2x}{x^2+2}+\dfrac{1}{x-1}$ D. $\dfrac{x}{x^2+2}+\dfrac{2}{x-1}$

4. $\displaystyle\int \frac{4x+10}{x^2+5x-5}dx =$

A. $2\ln|x^2+5x-6|+C$ B. $2\ln|x^2+5x-5|+C$
C. $2\ln|x+6|+\ln|x-1|+C$ D. $2\ln|x+2|+\ln|x+3|+C$

5. $\displaystyle\int \frac{x^2+2x+5}{x^3+2x^2+x+2}dx =$

A. $\ln|4x^2+1|+\ln|2x+4|+C$ B. $2\ln|x^2+1|+\ln|x+2|+C$
C. $2\tan^{-1}x+\ln|x+2|+C$ D. $\tan^{-1}x+\ln|x+2|+C$

6. Answer true or false: $\displaystyle\int \frac{x^2+2x+1}{(x+2)(x+4)}dx = \frac{x^3}{3}+x^2+x+\ln|x+2|+\ln|x+4|+C$

7. Answer true or false: $\displaystyle\int \frac{dx}{(x+3)(x-2)} = \ln|x+3|+\ln|x-2|$

8. Answer true or false: $\displaystyle\int \frac{2x^3+4x^2+2x+2}{(x^2+4)(x^2+2)}dx = \tan^{-1}x+\ln|x^2+2|+C$

9. $\displaystyle\int \frac{x^2+4}{(x-1)^3} =$

A. $\ln|x-1|-\dfrac{2}{x-1}-\dfrac{5}{2(x-1)^2}+C$ B. $\ln|x-1|+C$
C. $\dfrac{x^3}{3}+2x+\ln^3|x-1|+C$ D. $\dfrac{x^3}{3}+2x-\dfrac{1}{2(x-1)^2}+C$

10. Answer true or false: $\displaystyle\int \frac{x^3+x+3}{x(x+5)}dx = \ln|x|+\ln|x+5|+C$

11. Answer true or false: $\displaystyle\int \frac{1}{(x-6)^3}dx = \ln^3|x-6|+C$

12. Answer true or false: $\displaystyle\int \frac{2x+1}{(x^2+2)(x+2)}dx = \ln|x^2+2|+\ln|x+2|+C$

13. Answer true or false: $\displaystyle\int \frac{x^2-3x-17}{(x+7)(x^2+4)}dx = \ln|x+7|-\frac{3}{2}\tan^{-1}\left(\frac{x}{2}\right)+C$

14. Answer true or false: $\displaystyle\int \frac{1}{(x^2+1)(x^2+4)}dx = \tan^{-1}x+\frac{1}{2}\tan^{-1}\left(\frac{x}{2}\right)+C$

15. Answer true or false: $\displaystyle\int \frac{x}{(x+3)^2}dx = \ln|x+3|+C$

SECTION 8.6

1. $\int \frac{6x}{4x+3} dx =$

A. $\frac{3}{2} + \frac{3}{8} \ln|4x+3| + C$

B. $\frac{3}{2} - \frac{3}{8} \ln|4x+3| + C$

C. $6\ln|4x+3| + \frac{x^2}{2} + C$

D. $\frac{3x}{2} - \frac{9}{8} \ln|4x+3| + C$

2. $\int \frac{16x}{(4-5x)^2} dx =$

A. $\frac{2}{5} \ln|4-5x| + C$

B. $\frac{16(5x-4)\ln|5x-4|}{25(5x-4)} + C$

C. $\frac{4}{25(4-5x)} - \frac{1}{25} \ln|4-5x| + C$

D. $\frac{2}{25} \ln|4-5x| + C$

3. $\int \sin 3x \sin 7x \, dx =$

A. $\frac{\sin 4x}{8} - \frac{\sin 10x}{20} + C$

B. $\frac{\sin 7x}{8} - \frac{\sin 3x}{20} + C$

C. $\frac{\cos 7x}{7} - \frac{\sin 3x}{3} + C$

D. $\frac{-\cos 7x}{7} + \frac{\sin 3x}{3} + C$

4. $\int x^5 \ln 3x \, dx =$

A. $\frac{x^6}{6} + \frac{1}{x} + C$

B. $x^6 \left(\frac{\ln 3x}{6} - \frac{1}{36} \right) + C$

C. $\frac{x^6 \ln 3x}{6} - \frac{1}{36} + C$

D. $\frac{x^6}{6} - \frac{1}{x} + C$

5. $\int 4\sqrt{x} \ln x \, dx =$

A. $\frac{8x^{3/2}}{3} \ln x - \frac{16}{9} x^{3/2} + C$

B. $4x^{3/2} \ln x - \frac{8}{3} x^{3/2} + C$

C. $\frac{8x^{3/2}}{3} \ln x - \frac{16}{9} + C$

D. $\frac{8x^{3/2}}{3} \ln x + C$

6. $\int e^{4x} \cos 2x \, dx =$

A. $e^{4x} \left(\frac{\cos 2x}{3} + \frac{\sin 2x}{2} \right) + C$

B. $\frac{e^4 x}{20} (\cos 2x - \sin 2x) + C$

C. $\frac{e^4 x}{20} (4\cos 2x + 2\sin 2x) + C$

D. $\frac{e^4 x}{6} (3\cos 2x + 2\sin 2x) + C$

7. $\int e^{-3x} \sin 4x \, dx =$

A. $\frac{e^{-3x}}{25}(-3\sin 4x - 4\sin 4x) + C$

B. $\frac{e^{-3x}}{5}(-3\cos 4x + 4\sin 4x) + C$

C. $\frac{e^{-3x}}{25}(3\cos 4x + 4\sin 4x) + C$

D. $\frac{e^{-3x}}{5}(3\cos 4x + 4\sin 4x) + C$

8. $\int \frac{1}{x^2\sqrt{2x^2+6}} dx =$

A. $\frac{-6x}{\sqrt{2x^2+6}} + C$

B. $\frac{6x}{\sqrt{2x^2+6}} + C$

C. $\frac{\sqrt{2x^2+6}}{6x} + C$

D. $\frac{-\sqrt{2x^2+6}}{6x} + C$

9. $\int \ln(4x+2)\, dx =$

A. $\frac{4\ln^2(4x+2)}{2} + C$

B. $\frac{x\ln(4x+2) - x}{4} + C$

C. $4x\ln(4x+2) - 4x + C$

D. $\frac{\ln^2(4x+2)}{2} + C$

10. Answer true or false: For $\int x\ln(6-2x^2)\, dx$ a good choice for u is $6 - 2x^2$

11. Answer true or false: $\int \cos\sqrt{x}\, dx = 2\sin\sqrt{x}$

12. Answer true of false: $\int e^{\sqrt{x}} dx = e^{\sqrt{x}}\left(\sqrt{x}+1\right) + C$

13. $\int x\sqrt{x+7}\, dx =$

A. $\frac{2(x-7)^{5/2}}{5} - \frac{14(x-7)^{3/2}}{3} + C$

B. $(x-7)^{3/2} + C$

C. $\frac{2(x-7)^{3/2}}{3} + x + C$

D. $\frac{2(x-7)^{5/2}}{3} - \frac{5(x-7)^{3/2}}{2} + C$

14. Answer true or false: The area enclosed by $y = \sqrt{16-x^2}, y = 0, x = 0, x = 4$ is $\frac{128\pi}{3}$.

15. Answer true or false: $\int_3^x \frac{1}{t\sqrt{3t-5}} dt = \frac{1}{3} + x$

SECTION 8.7

1. Use $n = 10$ to approximate the integral by the midpoint rule. $\int_0^1 2x^{3/2}dx$

A. 0.804 B. 0.798 C. 4.98 D. 8.04

2. Use $n = 10$ to approximate the integral by the midpoint rule. $\int_0^1 x^3 + 1dx$

A. 3.49 B. 1.249 C. 1.257 D. 3.51

3. Use $n = 10$ to approximate the integral by the midpoint rule. $\int_1^2 x^3 + 2\,dx$

A. 39.56 B. 5.756 C. 5.746 D. 39.46

4. Use $n = 10$ to approximate the integral by the midpoint rule. $\int_1^2 x^7 - 1\,dx$

A. 0.317 B. 316 C. 30.7 D. 2.17

5. Use $n = 10$ to approximate the integral by the midpoint rule. $\int_0^1 \sin x + 2\,dx$

A. 2.4599 B. 2.4601 C. 2.4603 D. 2.4605

6. Use the trapezoid rule with $n = 10$ to approximate the integral. $\int_0^1 x^{3/2} + 1\,dx$

A. 1.126 B. 1.151 C. 1.821 D. 1.113

7. Use the trapezoid rule with $n = 10$ to approximate the integral. $\int_0^1 x^3 + 4\,dx$

A. 4.2525 B. 4.2528 C. 4.2521 D. 4.2517

8. Use the trapezoid rule with $n = 10$ to approximate the integral. $\int_1^2 x^3 - 1\,dx$

A. 2.754 B. 2.752 C. 2.760 D. 2.758

9. Use the trapezoid rule with $n = 10$ to approximate the integral. $\int_0^1 \sin x + 4\,dx$

A. 4.461 B. 4.510 C. 4.621 D. 4.713

10. Use the trapezoid rule with $n = 10$ to approximate the integral. $\int_0^1 \cos x + 1\,dx$

A. 1.841 B. 1.837 C. 1.834 D. 1.830

11. Use Simpson's Rule with $n = 10$ to approximate the integral. $\int_0^1 x^{3/2} + 3\,dx$

A. 3.362 B. 3.364 C. 3.369 D. 3.512

12. Use Simpson's Rule with $n = 10$ to approximate the integral. $\int_0^1 x^3 + 5\,dx$

A. 5.15 B. 5.20 C. 5.25 D. 5.30

13. Use Simpson's Rule with $n = 10$ to approximate the integral. $\int_1^2 x^3 - 1\,dx$

A. 2.75 B. 2.475 C. 2.485 D. 2.495

14. Use Simpson's Rule with $n = 10$ to approximate the integral. $\int_0^1 \sin x + 8\,dx$

A. 8.437 B. 8.433 C. 8.926 D. 9.102

15. Use Simpson's Rule with $n = 10$ to approximate the integral. $\int_0^1 \cos x - 0.5\,dx$

A. 2.951 B. 2.942 C. 2.713 D. 2.614

SECTION 8.8

1. Answer true or false: $\int_0^8 \frac{dx}{x-4}$ is an improper integral.

2. Answer true or false: $\int_0^4 \frac{dx}{x-1}$ is an improper integral.

3. Answer true or false: $\int_{-\infty}^2 e^{5x}\,dx$ is an improper integral.

4. $\int_1^\infty \frac{dx}{x^4} =$

A. $\frac{1}{3}$ B. $\frac{1}{6}$ C. $\frac{1}{2}$ D. Diverges

5. $\int_6^\infty \frac{dx}{\sqrt{x}} =$

A. $\frac{1}{2}$ B. $\frac{1}{6}$ C. 2 D. Diverges

6. $\int_{-\infty}^0 e^{6x}\,dx =$

A. $-\frac{1}{6}$ B. $\frac{1}{6}$ C. 6 D. Diverges

7. $\int_1^0 \frac{dx}{\sqrt{1-x^2}} =$

A. 1.5 B. -1.5 C. 0 D. $-\frac{\pi}{2}$

8. $\displaystyle\int_0^{\pi/2} \cot x\, dx =$

A. 0 B. 1 C. -30.08 D. Diverges

9. Answer true or false: $\displaystyle\int_0^3 \frac{1}{x^2}\, dx$ diverges.

10. Answer true or false: $\displaystyle\int_1^\infty \frac{1}{x^2} dx$ diverges.

11. Answer true or false: $\displaystyle\int_0^1 \ln 2x\, dx$ diverges.

12. Answer true or false: $\displaystyle\int_0^\infty \cos x\, dx$ diverges.

13. Answer true or false: $\displaystyle\int_0^\infty e^{-6x} dx$ diverges.

14. Answer true or false: $\displaystyle\int_{-\infty}^0 e^{-5x}\, dx$ diverges.

15. Answer true or false: $\displaystyle\int_{-\infty}^4 \frac{1}{x^3} dx = 1 - \lim_{b \to -\infty} \frac{-2}{b^2} = 1$

CHAPTER 8 TEST

1. $\displaystyle\int \sinh^{12} x \cosh x\, dx =$

A. $\dfrac{\sinh^{13} x}{13} + C$ B. $13 \sinh^{13} x + C$ C. $\dfrac{\sinh^{11} x}{11} + C$ D. $11 \sinh^{11} x + C$

2. $\displaystyle\int \frac{2x\, dx}{\sqrt{9 - x^4}} =$

A. $\sin^{-1}\left(\frac{x}{3}\right) + C$ B. $\sin^{-1}\left(\frac{x^2}{3}\right) + C$ C. $\cos^{-1}\left(\frac{x}{3}\right) + C$ D. $\cos^{-1}\left(\frac{x^2}{3}\right) + C$

3. Answer true or false: In evaluating $\displaystyle\int e^x (4e^x + 5)\, dx$ a good choice for u is $4e^x + 5$.

4. $\displaystyle\int 3x \cos x\, dx =$

A. $3 \cos x + 3x \sin x + C$
B. $3 \sin x + 3x \sin x + C$
C. $3 \cos x - 3x \sin x + C$
D. $3 \cos x + 3x \cos x + C$

5. $\displaystyle\int e^{3x} \sin 6x\, dx =$

A. $e^{3x}(3 \sin 6x + 6 \cos 6x) + C$
B. $e^{3x}(3 \sin 6x - 6 \cos 6x) + C$
C. $\dfrac{-e^{3x}}{3}(3 \sin 6x - 6 \sin 6x) + C$
D. $\dfrac{e^{3x}}{45}(3 \sin 6x - 6 \cos 6x) + C$

6. Answer true or false: $\int_{5\pi/4}^{7\pi/4} x \cot x \, dx = 0$

7. $\int \tan 6x \, dx \ =$

A. $\frac{1}{6} \ln|\cos 6x| + C$

B. $-\frac{1}{6} \ln|\cos 6x| + C$

C. $\frac{1}{12} \tan^2 6x + C$

D. $-\frac{1}{12} \tan^2 6x + C$

8. Answer true or false: $\int \sin 8x \cos 6x \, dx = -\frac{1}{4} \cos 2x - \frac{\cos 14x}{28} + C$

9. $\int \sqrt{9 - x^2} \, dx =$

A. $\frac{1}{2} \, x\sqrt{9 - x^2} + 9 \sin^{-1}\left(\frac{x}{9}\right) + C$

B. $\frac{1}{2} \, x\sqrt{9 - x^2} + \frac{9}{2} \sin^{-1}\left(\frac{x}{3}\right) + C$

C. $\frac{1}{2} \, x\sqrt{9 - x^2} - 9 \sin^{-1}\left(\frac{x}{9}\right) + C$

D. $\frac{1}{2} \, x\sqrt{9 - x^2} - \frac{9}{2} \sin^{-1}\left(\frac{x}{3}\right) + C$

10. $\int_0^1 \frac{2 \, dx}{\sqrt{6x - x^2}} =$

A. 1.68 B. 1.66 C. 1.72 D. 1.84

11. Answer true or false: $\frac{2}{x+4} + \frac{1}{x-8}$ is the partial fraction decomposition of $\frac{3x - 12}{(x+4)(x-8)}$.

12. $\int \frac{2x^2 + 9x + 20}{x^3 + 2x^2 + x + 2} dx =$

A. $\ln|9x^2 + 1| + \ln|2x + 4| + C$

B. $9 \ln|x^2 + 1| + 2 \ln|x + 2| + C$

C. $9 \tan^{-1} x + 2 \ln|x + 2| + C$

D. $3 \tan^{-1} x + 2 \ln|x + 2| + C$

13. $\int x^8 \ln^2 x \, dx =$

A. $\frac{x^9(81 \ln^2(x) - 18 \ln(x) + 2)}{729} + C$

B. $\frac{x^9 \ln^2 x}{9} - \frac{1}{81} + C$

C. $\frac{x^9 \ln^2 x}{9} + C$

D. $\frac{x^9 \ln^2 x}{9} - \frac{1}{9} + C$

14. Answer true or false: $\int x \sin 4x \, dx = \frac{x^2}{4} - \frac{x \sin 14x}{14} - \frac{\cos 14x}{8}$

15. Use $n = 10$ subdivisions to approximate the integral by the midpoint rule. $\int_0^1 \cos x + 1 \, dx =$

A. 1.8424 B. 1.8422 C. 1.8420 D. 1.8418

16. Use $n = 10$ subdivisions to approximate the integral by the trapezoid rule. $\int_1^2 x^7 - 1 \, dx =$

A. 31.26 B. 31.32 C. 31.24 D. 31.20

17. Use $n = 10$ subdivisions to approximate the integral by Simpson's Rule. $\int_0^1 x^5 + 4\,dx =$

A. 4.1667 B. 4.1892 C. 4.1995 D. 4.2001

18. $\int_0^\infty e^{-5x}\,dx =$

A. $\frac{1}{5}$ B. $-\frac{1}{5}$ C. 5 D. Diverges

19. $\int_0^3 \frac{dx}{\sqrt{9-x^2}} =$

A. 1.5 B. −1.5 C. 0 D. Diverges

20. Answer true or false: $\int_4^\infty e^{2x}\,dx$ diverges.

ANSWERS TO SAMPLE TESTS

SECTION 8.1

1. C 2. C 3. D 4. A 5. B 6. B 7. D 8. C 9. B 10. A 11. T 12. F
13. T 14. F 15. T

SECTION 8.2

1. B 2. A 3. C 4. A 5. C 6. B 7. A 8. D 9. B 10. A 11. B 12. T
13. F 14. F 15. F

SECTION 8.3

1. B 2. A 3. B 4. A 5. A 6. B 7. T 8. A 9. A 10. T 11. C 12. T
13. F 14. F 15. T

SECTION 8.4

1. B 2. C 3. C 4. B 5. D 6. A 7. A 8. B 9. A 10. T 11. F 12. A
13. C 14. T 15. T

SECTION 8.5

1. A 2. C 3. C 4. B 5. C 6. F 7. F 8. F 9. A 10. F 11. F 12. F
13. T 14. F 15. F

SECTION 8.6

1. D 2. B 3. A 4. B 5. A 6. C 7. A 8. D 9. B 10. T 11. F 12. F
13. A 14. F 15. F

SECTION 8.7

1. B 2. B 3. C 4. C 5. A 6. A 7.A 8. D 9. A 10. A 11. C 12. C
13. A 14. C 15. A

SECTION 8.8

1. T 2. T 3. T 4. A 5. D 6. B 7. D 8. D 9. T 10. F 11. F 12. T
13. F 14. T 15. F

CHAPTER 8 TEST

1. A 2. B 3. T 4. A 5. D 6. F 7. B 8. T 9. B 10. A 11. T 12. C
13. A 14. F 15. D 16. C 17. A 18. A 19. D 20. T

CHAPTER 9
Sample Exams

SECTION 9.1

1. State the order of the differential equation $3y'' + 7y = 0$.

A. 0 B. 1 C. 2 D. 3

2. State the order of the differential equation $y' - 5y = 0$.

A. 0 B. 1 C. 2 D. 3

3. Answer true or false: The differential equation $y' - 5y = 0$ is solved by $y = Ce^{-5t}$.

4. Answer true or false: The differential equation $(3 + x)\dfrac{dy}{dx} = 1$ is solved by $\ln|3 + x| + C$, when $x \geq 0$.

5. Solve the differential equation $\dfrac{dy}{dx} - 2y = 0$.

A. $y = Ce^{2x}$ B. $y = Ce^{-x}$ C. $y = e^{Cx}$ D. $y = e^{-Cx}$

6. Solve the differential equation $\dfrac{dy}{dx} - 3y = -2e^{t}$.

A. $y = Ce^{-t}$ B. $y = Ce^{t}$ C. $y = e^{t} + Ce^{3t}$ D. $y = -\ln|t| + C$

7. Solve the differential equation $\dfrac{dy}{dt} + y^3 = 0$.

A. $e^{3t} + C$ B. $e^{t} + C$ C. $\dfrac{\sqrt[3]{t}}{3} + C$ D. $\dfrac{\sqrt{2t}}{2t} + C$

8. Solve the differential equation $y' = 4y$; $y(1) = e^4$.

A. $y = e^{4t}$ B. $y = 4e^{4t}$ C. e^{-4t} D. $y = -\dfrac{1}{t}$

9. Solve the differential equation $\dfrac{dy}{dt} = y^2$; $y(1) = -1$.

A. $y^3 - 4$ B. $y^3 + 4$ C. $\dfrac{\sqrt[3]{t}}{3} + 2$ D. $-\dfrac{1}{t}$

10. Solve the differential equation $\dfrac{2x}{y} = y'$.

A. $y = \sqrt{2}x + C$ B. $y = Cx$ C. $y = Ce^{2x}$ D. $y = e^{2Cx}$

11. Solve the differential equation $\dfrac{x}{y} = y'$; $y(1) = 3$.

A. $y = x + 2$ B. $y = 2x$ C. $y = 2e^{x}$ D. $y = e^{2x}$

12. Solve the differential equation $\frac{dy}{dt} = t^3$; $y(0) = 4$.

A. $y = \frac{t^4}{4} + 4$ B. $y = 4t^4 + 4$ C. $y = 0$ D. $y = \frac{t^2}{2} + 4$

13. Solve the differential equation $\frac{dy}{dt} = \sqrt[5]{t}$.

A. $t^{6/5} + C$ B. $y = 2t^{6/5} + C$ C. $y = \frac{6t^{6/5}}{5} + C$ D. $y = \frac{5t^{6/5}}{6} + C$

14. Solve the differential equation $\frac{dy}{dt} = \frac{1}{y^4}$.

A. $\sqrt[5]{5t} + C$ B. $625t^5 + C$ C. $t + C$ D. $5t + C$

SECTION 9.2

1. If $y' = -x + 5y$, find the direction field at (1,2).

A. 9 B. -9 C. $\frac{1}{9}$ D. $-\frac{1}{9}$

2. If $y' = \cos(xy)$, find the direction field at (0,3).

A. 1 B. π C. -1 D. 0

3. If $y' = \cos(3xy)$, find the direction field at (5,0).

A. 1 B. π C. -1 D. 0

4. If $y' = x\cos y$, find the direction field at (7,0).

A. 7 B. -7 C. 0 D. 1

5. If $y' = ye^x$, find the direction field at (6,0).

A. $6e^6$ B. e^6 C. 0 D. 6

6. If $y' = 4x - 4y$, find the direction field at (1,1).

A. 4 B. -8 C. 8 D. 0

7. If $y' = \frac{x}{3y}$, find the direction field at (5,2).

A. $\frac{1}{6}$ B. 0 C. -1 D. $\frac{5}{6}$

8. If $y' = y\cosh x$, find the direction field at (5,0).

A. 5 B. -5 C. 1 D. 0

9. If $y' = (\sin x)(\cos x)$, find the direction field at $\left(\frac{\pi}{2}, \frac{\pi}{2}\right)$.

A. 1 B. -1 C. 0 D. $\frac{\sqrt{2}}{2}$

10. If $y' = 3\ln x - 2\ln y$, find the direction field at (1,1).

A. 0 B. 2 C. 1 D. $2e$

11. Answer true or fasle: If Euler's method is used to approximate $y' = 4x - y$, $y(1) = 2$; $y(4)$ will be approximately 0. Choose the step size to be approximately 0.1.

12. Answer true or fasle: If Euler's method is used to approximate $y' = \sin(x - y)$, $y(1) = 1$; $y(3)$ will be approximately 0. Choose the step size to be approximately 0.1.

13. Answer true or fasle: If Euler's method is used to approximate $y' = xe^y$, $y(2) = 0$; $y(1)$ will be approximately 1. Choose the step size to be approximately 0.1.

14. Answer true or fasle: If Euler's method is used to approximate $y' = \ln y$, $y(1) = 1$; $y(2)$ will be approximately 2. Choose the step size to be approximately 0.1.

15. Answer true or fasle: If Euler's method is used to approximate $y' = 3\ln(xy)$, $y(1) = 1$; $y(2)$ will be approximately 3. Choose the step size to be approximately 0.1.

SECTION 9.3

1. Answer true or false: Suppose that a quantity $y = y(t)$ changes in such a way that $dy/dx = k\sqrt[5]{y}$, where $k > 0$. It can be said that y increases at a rate that is proportional to the fifth root of the amount present.

2. Answer true or false: Suppose that a quantity $y = y(t)$ changes in such a way that $dy/dx = k\sqrt{y}$, where $k > 0$. It can be said that y increases at a rate that is proportional to the square root of the time.

3. Suppose that an initial population of 5,000 bacteria grows exponentially at a rate of 3% per hour, the number $y = y(t)$ of bacteria present t hours later is

A. 5,000 t B. $5{,}000\,e^{0.03t}$ C. $5{,}000\,e^{-0.03t}$ D. $5{,}000(1.03)^t$.

4. Suppose a radioactive substance decays with a half-life of 122 years. Find a formula that relates the amount present to time, if there are 500 g of the substance present initially.

A. $y(t) = 500e^{-0.00568t}$ B. $y(t) = 500e^{242t}$
C. $y(t) = 500e^{2t}$ D. $y(t) = e^{-0.5t}$

5. Suppose a radioactive substance decays with a half-life of 151 years. Find a formula that relates the amount present to time, if there are 500 g of the substance present initially.

A. $y(t) = 500e^{-0.00459t}$ B. $y(t) = 500e^{151t}$
C. $y(t) = 500e^{2t}$ D. $y(t) = e^{-0.5t}$

6. If 400 g of a radioactive substance decay to 60 g in 12 years find the half-life of the substance.

A. 0.15 years B. 0.11 years C. 0.22 years D. 4.38 years

7. If 800 g of a radioactive substance decay to 120 g in 12 years find the half-life of the substance.

A. 0.15 years B. 0.11 years C. 0.22 years D. 4.38 years

8. Answer true or false: The differential equation that is used to find the position function $y(t)$ of mass 8 kg suspended by a vertical spring that has a spring constant 4 N/m is given by $y''(t) = -2y(t)$.

9. Answer true or false: $y''(t) = 25y(t)$ is solved by $C_1\cos(5t) + C_2\sin(5t)$.

10. If $y = y_0e^{kt}$, $k > 0$, where $y_0 > 0$, the situation modeled is

A. Increasing
B. Decreasing
C. Remaining constant
D. More information is needed.

11. Find the exponential growth model $y = y_0e^{kt}$, that satisfies $y_0 = 7$ if the doubling time is $T = 10$.

A. $y = 7e^{0.0693t}$
B. $y = 7e^{2t}$
C. $y = 7e^{0.5t}$
D. $y = e^{0.841t}$

12. Answer true or false: Every exponential growth model $y = y_0e^{kt}$ used to find the time for a substance to double must use 0.5 for k.

13. Answer true or false: If $y(0) = 20$ and the substance represented increases at a rate of 8%, then $y = 20(0.08)^t$.

14. Answer true or false: If $y(0) = 40$ and the substance represented decreases at a rate of 16%, then $y = 40(0.16)^t$.

15. Answer true or false: If $y(0) = 40$ and the substance represented decreases at a rate of 8%, then $y = 40(-0.92)^t$.

SECTION 9.4

1. Solve the differential equation $\dfrac{d^2y}{dt^2} = 9y$.

A. $C_1e^{3t} + C_2$
B. $C_1e^{C_2t}$
C. $C_1e^{3t} + C_2e^{-3t}$
D. $C_1 \sin C_2t$

2. Answer true or false: $y'' + 2y'x = 4$ is a second order differential equation.

3. Answer true or false: $2y'' + \sin y = 0$ is a second order differential equation.

4. Answer true or false: $4y'' = \sin(2x)$ is solved by $y = C_1 \sin(2x) + C_2 \cos(2x)$

5. Solve $y'' + y = 0$.

A. $y = C_1e^x + C_2e^{-x}$
B. $y = C_1(e^x + e^{-2x})$
C. $y = C_1 + C_2e^x$
D. $y = C_1 \sin x + C_2$

6. Solve $y'' = 4$.

A. $y = 2x^2 + C_1x + C_2$
B. $y = 2x^2 + 4C_1x + C_2$
C. $y = C_1 \sin x + C_2 \cos x$
D. $y = C_1e^x + C_2e^{-x}$

7. Solve $y'' = e^x$.

A. $y = C_1e^x + C_2e^{-x}$
B. $y = C_1 \sin x + C_2 \cos x$
C. $y = e^x + C_1x + C_2$
D. $y = e^x + C$

8. Solve $y'' + 16y = 0$.

A. $y = C_1e^{4x} + C_2e^{-4x}$
B. $y = x^2 + C_1x + C_2$
C. $y = x^2 + C$
D. $y = \sin C_1x + \cos C_2x$

9. Solve $\frac{y''}{x} = 3$.

A. $y = C_1e^x + C_2$

B. $y = e^x + C$

C. $y = \frac{3x^2}{2} + 3C_1x + C_2$

D. $y = x^2 + C_1x + C_2$

10. Solve $\frac{y''}{25y} = -1$.

A. $y = C_1e^{5x} + C_2e^{-5x}$

B. $y = \sinh x + \cosh x + C$

C. $y = C_1\sin(2x) + C_2\cos(2x)$

D. $y = C_1e^x + C_2e^{2x}$

CHAPTER 9 TEST

1. State the order of the differential equation $4xy'' = 5y$.

A. 0 B. 1 C. 2 D. 3

2. Answer true or false: The differential equation $y'' - 5y = 0$ is solved by $y = Ce^t$.

3. Solve the differential equation $\frac{dy}{dx} - 9y = 0$.

A. $y = Ce^{-t}$

B. $y = Ce^{9t}$

C. $y = e^t + 9e^{9t} + C$

D. $y = -\ln|t| + C$

4. Solve the differential equation $75y'' - 25y = 0$.

A. $\sin t + C$

B. $C_1\cos 5t + C_2\sin 5t$

C. Ce^t

D. Ce^{-t}

5. Answer true or false: $y = \cos 3t + C$ solves $y' + y = 0$.

6. Answer true or false: $y = Ce^t - 4$ solves $y' - y = 4$.

7. If $y' = x\cos y$, find the direction field at (5,0).

A. 0 B. -5 C. 5 D. 1

8. If $y' = 4x + 6y$, the direction field at (1,2) is

A. 0 B. 16 C. -16 D. 1.

9. If $y' = e^{xy}$, the field direction at (8,0) is

A. 0 B. 1 C. 8 D. e^8.

10. $y' = e^{3xy}$, the direction field at (7,0) is

A. 0 B. 1 C. 7 D. 21.

11. Answer true or false: Suppose that a quantity $y = y(t)$ changes in such a way that $dy/dt = k\sqrt[6]{t}$, where $k > 0$. It can be said that y increases at a rate that is proportional to the sixth root of time.

12. Answer true or false: Suppose that a quantity $y = y(t)$ changes in such a way that $dy/dt = k\sqrt{t}$, where $k > 0$. It can be said that y increases at a rate that is proportional to the square root of time.

13. Suppose that an initial population of 20,000 bacteria grows exponentially at a rate of 4% per hour, and that $y = y(t)$ is the number of bacteria present after t hours. Write an expression for y.

A. $20,000t$ B. $20,000e^{0.04t}$ C. $20,000e^{-0.04t}$ D. $20,000(1.04)^t$

14. Suppose a radioactive substance decays with a half-life of 70 days. Find a formula that relates the amount present to t, if initially 60 g of the substance are present.

A. $y(t) = 60e^{-0.0099t}$ B. $y(t) = 60e^{-70t}$
C. $y(t) = 60e^{-2t}$ D. $y(t) = 60e^{-0.5t}$

15. If 480 g of a radioactive substance decay to 72 g in 12 years, the half-life if the substance is

A. 0.15 years B. 0.11 years C. 0.22 years D. 4.38 years.

16. Answer true or false: $y''(t) = 25t$ is solved by $y(t) = C_1 \cos 25t + C_2 \sin 25t$.

17. If $y = y_0 e^{kt}$, $k < 0$, the function modeled is

A. Increasing B. Decreasing
C. Remaining constant D. More information is needed.

18. Answer true or false: An exponential decay model $y = y_0 e^{kt}$ used to find the half-life of a substance always uses -0.2 for k.

19. Answer true or false: If $y(0) = 20$ and a substance grows at a rate of 13%, a model for this situation is $y = 20(0.13)^t$.

20. Answer true or false: If $y(0) = 20$ and a substance decreases at a rate of 8%, a model for this situation is $y = 20(0.08)^t$.

ANSWERS TO SAMPLE TESTS

SECTION 9.1

1. C 2. B 3. F 4. T 5. A 6. C 7. D 8. A 9. D 10. A 11. A 12. A 13. D 14. A

SECTION 9.2

1. A 2. A 3. A 4. A 5. C 6. D 7. D 8. D 9. C 10. A 11. F 12. F 13. F 14. F 15. F

SECTION 9.3

1. T 2. F 3. B 4. A 5. A 6. D 7. D 8. T 9. T 10. A 11. A 12. F 13. F 14. F 15. F

SECTION 9.4

1. C 2. T 3. T 4. F 5. A 6. B 7. C 8. A 9. C 10. A

CHAPTER 9 TEST

1. C 2. F 3. B 4. B 5. F 6. T 7. C 8. B 9. B 10. B 11. T 12. T 13. B 14. A 15. D 16. F 17. B 18. F 19. F 20. F

CHAPTER 10
Sample Exams

SECTION 10.1

1. Find the Maclaurin polynomial of order 2 for e^{3x}.

A. $1 + 3x + \dfrac{9x^2}{2}$ B. $1 - 3x + 3x^2$ C. $1 + x + x^2$ D. $1 + 3x + 9x^2$

2. Find the Maclaurin polynomial of order 2 for $\sin \dfrac{x}{2}$.

A. $1 + \dfrac{1}{8}x^2$ B. $1 - \dfrac{x}{2}$ C. $1 + \dfrac{1}{2}x + \dfrac{1}{8}x^2$ D. $1 + \dfrac{1}{2}x - \dfrac{1}{8}x^2$

3. Find the Maclaurin polynomial of order 2 for $\cos x$.

A. $1 + \dfrac{x^2}{2}$ B. $1 + x^2$ C. $1 - x^2$ D. $1 - \dfrac{x^2}{2}$

4. Find the Maclaurin polynomial of order 2 for e^{7x}.

A. $1 + 7x + \dfrac{49x^2}{2}$ B. $1 - 7x + \dfrac{49x^2}{2}$ C. $1 + 7x + 49x^2$ D. $1 - 7x + 49x^2$

5. Find the Maclaurin polynomial of order 2 for $2e^{-6x}$.

A. $2 - 12x + 36x^2$ B. $2 + 12x + 36x^2$ C. $2 - 12x + 72x^2$ D. $2 + 12x + 72x^2$

6. Find a Taylor polynomial for $f(x) = e^x$ of order 2 about $x = 3$.

A. $e^3 - e^3(x-3) + e^3(x-3)^2$ B. $e^3 - e^3(x-3) + \dfrac{e^3(x-3)^2}{2}$

C. $e^3 + e^3(x-3) + \dfrac{e^3(x-3)^2}{2}$ D. $e^3 + e^3(x-3) + e^3(x-3)^2$

7. Find a Taylor polynomial for $f(x) = e^{-3x}$ of order 2 about $x = 3$.

A. $e^{-9} + 3e^{-9}(x-3) + \dfrac{9e^{-9}(x-3)^2}{2}$ B. $e^{-9} - 3e^{-9}(x-3) + \dfrac{9e^{-9}(x-3)^2}{2}$

C. $e^{-9} - 3e^{-9}(x-3) + 9e^{-9}(x-3)^2$ D. $e^{-9} + 3e^{-9}(x-3) + 9e^{-9}(x-3)^2$

8. Find a Taylor polynomial for $f(x) = 3\ln x$ of order 2 about $x = 2$.

A. $3\ln 2 + \dfrac{3}{2}(x-2) - \dfrac{3}{4}(x-2)^2$ B. $3\ln 2 + \dfrac{3}{2}(x-2) - \dfrac{3}{8}(x-2)^2$

C. $3\ln 2 + \dfrac{3}{2}(x-2) + \dfrac{3}{4}(x-2)^2$ D. $3\ln 2 + \dfrac{3}{2}(x-2) + \dfrac{3}{8}(x-2)^2$

9. Find a Taylor polynomial for $f(x) = 4\sin x$ of order 2 about $x = \dfrac{\pi}{2}$.

A. $4 - 4x^2$ B. $4 + 4x^2$ C. $\dfrac{4 - (2x - \pi)^2}{2}$ D. $\dfrac{4 + (2x - \pi)^2}{2}$

10. Answer true or false: The Maclaurin polynomial of order 3 for e^{5x} is $1 + 5x + \dfrac{25x^2}{2} + \dfrac{32x^3}{3}$.

11. Answer true or false: The Maclaurin polynomial of order 3 for $\ln(3 + x)$ is
$\ln 3 + \ln 3\, x + \dfrac{\ln 3}{2}x^2 + \dfrac{\ln 3}{6}x^3$.

12. Answer true or false: The Maclaurin polynomial of order 3 for $\cosh x^2$ is $\cosh x^2 + 2x^2 \sinh x^2 + 2x^4 \cosh x^2 + 2x^6 \sinh x^2$.

13. Answer true or false: The Taylor polynomial for e^x of order 3 about $x = 3$ is $e^3 + e^3(x-3) + \dfrac{e^3(x-3)^2}{2} + \dfrac{e^3(x-3)^3}{6}$.

14. Answer true or false: The Taylor polynomial for $\cos xe^{-x}$ of order 2 about $x = \dfrac{\pi}{2}$ is $\left(x - \dfrac{\pi}{2}\right) - \left(x - \dfrac{\pi}{2}\right)^3$.

15. Answer true or false: The Taylor polynomial for $\ln x$ of order 3 about $x = 4$ is $\ln 4 + \ln 4(x-4) + \ln 4\dfrac{(x-4)^2}{2} + \ln 4\dfrac{(x-4)^3}{6}$.

SECTION 10.2

1. The general term for the sequence 3, 3/8, 1/9, 3/64, ... is

A. $\dfrac{3}{n^3}$ B. $\dfrac{3}{n^2}$ C. $\dfrac{1}{n}$ D. $\sqrt[3]{n}$

2. Write out the first five terms of $\left\{\dfrac{2n}{n+5}\right\}_{n=1}^{+\infty}$.

A. 2, 1, 2/3, 1/2, 2/5
B. 1/3, 4/7, 3/4, 8/9, 1
C. 2/5, 1/3, 2/7, 1/4, 2/9
D. 2/5, 2/3, 6/7, 1, 10/9

3. Write out the first five terms of $\{3 \sin n\pi\}_{n=1}^{+\infty}$.

A. 3π, 6π, 9π, 12π, 15π
B. −3, 3, −3, 3, −3
C. 0, 0, 0, 0, 0
D. 3, 0, −3, 0, 3

4. Write out the first five terms of $\{(-1)^{n-1}n^2\}_{n=1}^{+\infty}$.

A. 1, −4, 9, −16, 25
B. 3, 6, 9, 12, 15
C. −1, 4, −9, 16, −25
D. 1, 4, 9, 16, 25

5. Write out the first five terms of $\{(-1)^{n-1}3n\}_{n=1}^{+\infty}$.

A. −3, 6, −9, 12, −15
B. 3, 6, 9, 12, 15
C. −3, −6, −9, −12, −15
D. 3, −6, 9, −12, 15

6. Write out the first five terms of $\left\{2 + \dfrac{3}{n}\right\}_{n=1}^{+\infty}$.

A. 5, 7/2, 3, 11/4, 13/5
B. 4, 5/2, 2, 7/4, 8/5
C. 5, 6, 7, 8, 9
D. 4, 5, 6, 7, 8

7. Answer true or false: $\left\{\dfrac{3n^3}{n+1}\right\}_{n=1}^{+\infty}$ converges.

8. Answer true or false: $\left\{\dfrac{7n+8}{2n+5}\right\}_{n=1}^{+\infty}$ converges.

9. Answer true or false: $\left\{\frac{2}{n^3}+6\right\}_{n=1}^{+\infty}$ converges.

10. Answer true or false: $\{\cos 2n\pi\}_{n=1}^{+\infty}$ converges.

11. Answer true or false: $\left\{\sin\left(n\pi+\frac{\pi}{2}\right)\right\}_{n=1}^{+\infty}$ converges.

12. If the sequence converges, find its limit. If it does not converge, answer diverges. $\{(-1)e^{2n}\}_{n=1}^{+\infty}$

A. 0 B. $\frac{1}{e}$ C. $-e$ D. Diverges

13. If the sequence converges, find its limit. If it does not converge, answer diverges. $\left\{\frac{3n}{8^n}\right\}_{n=1}^{+\infty}$

A. 3/4 B. 3 C. 0 D. Diverges

14. If the sequence converges, find its limit. If it does not converge, answer diverges. $\frac{1}{6^2}, \frac{1}{6^3}, \frac{1}{6^4}, \frac{1}{6^5}, \frac{1}{6^6}, \ldots$

A. 1/6 B. 1/36 C. 0 D. Diverges

15. If the sequence converges, find its limit. If it does not converge, answer diverges. 1, 3, 5, 7, 9, ...

A. 1 B. 2 C. 1/2 D. Diverges

SECTION 10.3

1. Determine which answer best describes the sequence $\left\{\frac{4}{3n^2}\right\}_{n=1}^{+\infty}$.

A. Strictly increasing B. Strictly decreasing
C. Increasing, but not strictly increasing D. Decreasing, but not strictly decreasing

2. Determine which answer best describes the sequence $\{2e^n\}_{n=1}^{+\infty}$.

A. Strictly increasing B. Strictly decreasing
C. Increasing, but not strictly increasing D. Decreasing, but not strictly decreasing

3. Determine which answer best describes the sequence $\{n-5\}_{n=1}^{+\infty}$.

A. Strictly increasing B. Strictly decreasing
C. Increasing, but not strictly increasing D. Decreasing, but not strictly decreasing

4. Determine which answer best describes the sequence $\left\{\frac{(n-1)^2-n}{2}\right\}_{n=1}^{+\infty}$.

A. Strictly increasing B. Strictly decreasing
C. Increasing, but not strictly increasing D. Decreasing, but not strictly decreasing

5. Determine which answer best describes the sequence $\{e^{-3n}\}_{n=1}^{+\infty}$.

A. Strictly increasing B. Strictly decreasing
C. Increasing, but not strictly increasing D. Decreasing, but not strictly decreasing

6. Determine which answer best describes the sequence $\left\{\sin\left(\frac{2\pi}{n}\right)n^{3n}\right\}_{n=1}^{+\infty}$.

A. Strictly increasing
B. Strictly decreasing
C. Increasing, but not strictly increasing
D. Decreasing, but not strictly decreasing

7. Determine which answer best describes the sequence $\left\{3\cos\left(\frac{\pi}{2n}\right)\right\}_{n=1}^{+\infty}$.

A. Strictly increasing
B. Strictly decreasing
C. Increasing, but not strictly increasing
D. Decreasing, but not strictly decreasing

8. Determine which answer best describes the sequence $\{2n - 2n^2\}_{n=1}^{+\infty}$.

A. Strictly increasing
B. Strictly decreasing
C. Increasing, but not strictly increasing
D. Decreasing, but not strictly decreasing

9. Determine which answer best describes the sequence $\left\{\frac{6}{n}\right\}_{n=1}^{+\infty}$.

A. Strictly increasing
B. Strictly decreasing
C. Increasing, but not strictly increasing
D. Decreasing, but not strictly decreasing

10. Determine which answer best describes the sequence $\left\{8 - \frac{3}{n^3}\right\}_{n=1}^{+\infty}$.

A. Strictly increasing
B. Strictly decreasing
C. Increasing, but not strictly increasing
D. Decreasing, but not strictly decreasing

11. Determine which answer best describes the sequence $\left\{7 + \frac{2}{n^4}\right\}_{n=1}^{+\infty}$.

A. Strictly increasing
B. Strictly decreasing
C. Increasing, but not strictly increasing
D. Decreasing, but not strictly decreasing

12. Determine which answer best describes the sequence $\{4n^4 - 3n^2\}_{n=1}^{+\infty}$.

A. Strictly increasing
B. Strictly decreasing
C. Increasing, but not strictly increasing
D. Decreasing, but not strictly decreasing

13. Determine which answer best describes the sequence $\{((n-1)! - 1)e^n\}_{n=1}^{+\infty}$.

A. Strictly increasing
B. Strictly decreasing
C. Increasing, but not strictly increasing
D. Decreasing, but not strictly decreasing

14. Determine which answer best describes the sequence $\{((n-1)! - 1)(-e^{4n})\}_{n=1}^{+\infty}$.

A. Strictly increasing
B. Strictly decreasing
C. Increasing, but not strictly increasing
D. Decreasing, but not strictly decreasing

15. Determine which answer best describes the sequence $\{-n\cos(2n\pi + \pi)\}_{n=1}^{+\infty}$.

A. Strictly increasing
B. Strictly decreasing
C. Increasing, but not strictly increasing
D. Decreasing, but not strictly decreasing

SECTION 10.4

1. Answer true or false: The series $5 + 5/2 + 5/3 + \cdots + 5/n$ converges.

2. Answer true or false: The series $7/2 + 7/4 + 7/8 + 7/16 + \cdots + 7\left(\frac{1}{2}\right)^n$ converges.

3. Answer true or false: The series $\sum_{k=1}^{\infty} 8\left(\frac{1}{5}\right)^k$ converges.

4. Answer true or false: The series $\sum_{k=1}^{\infty} \left(\frac{7}{3}\right)^k$ converges.

5. Answer true or false: The series $\sum_{k=1}^{\infty} \frac{1}{(k+7)(k+6)}$ converges.

6. Answer true or false: The series $\sum_{k=1}^{\infty} \frac{6}{k}$ converges.

7. Determine whether the series $\sum_{k=1}^{\infty} \left(\frac{1}{5}\right)^k$ converges, and if so, find its sum.

A. 1/4 B. 6 C. 4 D. Diverges

8. Determine whether the series $\sum_{k=1}^{\infty} \left(\frac{4}{(k+9)(k+10)} + \frac{4}{k+9}\right)$ converges, and if so, find its sum.

A. 0 B. 1 C. 2/45 D. Diverges

9. Determine whether the series $\sum_{k=1}^{\infty} 10\left(\frac{4}{3}\right)^k$ converges, and if so, find its sum.

A. 16/3 B. 10/3 C. 2000/3 D. Diverges

10. Determine whether the series $\sum_{k=1}^{\infty} 5^{k-3}9^k$ converges, and if so, find its sum.

A. 0 B. 1/45 C. 45 D. Diverges

11. Determine whether the series $\sum_{k=1}^{\infty} (-1)^{k-1}\frac{18}{5^k}$ converges, and if so, find its sum.

A. 18/5 B. 3 C. 9/2 D. Diverges

12. Determine whether the series $\sum_{k=1}^{\infty} \left(\frac{3}{4}\right)^k$ converges, and if so, find its sum.

A. 3/4 B. 3 C. 5/4 D. Diverges

13. Write 0.3939... as a fraction.

A. 39/100 B. 39/99 C. 393/1000 D. 3939/10000

14. Write 0.1717... as a fraction.

A. 17/99 B. 171/999 C. 17/100 D. 171/1000

15. Write 4.34343... as a fraction.

A. 430/99 B. 43/99 C. 434/999 D. 217/50

SECTION 10.5

1. The series $\sum_{k=1}^{\infty} \frac{1}{k^9}$

A. Converges B. Diverges

2. The series $\sum_{k=1}^{\infty} \frac{1}{k^5}$

A. Converges B. Diverges

3. The series $\sum_{k=1}^{\infty} \frac{1}{k+7}$

A. Converges B. Diverges

4. The series $\sum_{k=1}^{\infty} \frac{4k^2+7k+2}{3k^2-6}$

A. Converges B. Diverges

5. The series $\sum_{k=1}^{\infty} 12\cos k\pi$

A. Converges B. Diverges

6. The series $\sum_{k=1}^{\infty} 7k^{-5/2}$

A. Converges B. Diverges

7. The series $\sum_{k=1}^{\infty} 3k^{-3/4}$

A. Converges B. Diverges

8. The series $\sum_{k=1}^{\infty} \frac{1}{5k+2}$

A. Converges B. Diverges

9. The series $\sum_{k=1}^{\infty} \frac{1}{k+2}$

A. Converges B. Diverges

10. The series $\sum_{k=1}^{\infty} \frac{k^2+8}{k^2+7}$

A. Converges B. Diverges

11. The series $\sum_{k=1}^{\infty} \frac{k+4}{k^3}$

A. Converges B. Diverges

12. The series $\sum_{k=1}^{\infty} \frac{1}{\sqrt{k+5}}$

A. Converges B. Diverges

13. The series $\sum_{k=1}^{\infty} \frac{1}{\sqrt[3]{2k+7}}$

A. Converges B. Diverges

14. The series $\sum_{k=1}^{\infty} \frac{3^k}{2}$

A. Converges B. Diverges

15. The series $\sum_{k=1}^{\infty} \frac{4}{k^4} + \frac{2}{k^3}$

A. Converges B. Diverges

SECTION 10.6

1. The series $\sum_{k=1}^{\infty} \frac{1}{9k^2 - 2k}$

A. Converges B. Diverges
C. Convergence cannot be determined

2. The series $\sum_{k=1}^{\infty} \frac{1}{7k^3 + 3k}$

A. Converges B. Diverges
C. Convergence cannot be determined

3. The series $\sum_{k=1}^{\infty} \frac{1}{k - 5}$

A. Converges B. Diverges
C. Convergence cannot be determined

4. The series $\sum_{k=1}^{\infty} \frac{4\sin^2 k}{k!}$

A. Converges B. Diverges
C. Convergence cannot be determined

5. The series $\sum_{k=1}^{\infty} \frac{k!}{k^8}$

A. Converges B. Diverges
C. Convergence cannot be determined

6. The series $\sum_{k=1}^{\infty} \frac{k^{3k}}{k!}$

A. Converges B. Diverges
C. Convergence cannot be determined

7. The series $\sum_{k=1}^{\infty} \frac{k!}{k^{5k}}$

A. Converges B. Diverges
C. Convergence cannot be determined

8. The series $\sum_{k=1}^{\infty} \frac{1}{k^3}$

A. Converges B. Diverges
C. Convergence cannot be determined

9. The series $\sum_{k=1}^{\infty} \frac{(6k)!}{k^k}$

A. Converges B. Diverges
C. Convergence cannot be determined

10. The series $\sum_{k=1}^{\infty} \frac{k}{7^k}$

A. Converges B. Diverges
C. Convergence cannot be determined

11. The series $\sum_{k=1}^{\infty} \frac{9k + 1}{2k - 3}$

A. Converges B. Diverges
C. Convergence cannot be determined

12. The series $\sum_{k=1}^{\infty} \frac{6}{e^k}$

A. Converges B. Diverges
C. Convergence cannot be determined

13. The series $\sum_{k=1}^{\infty} \frac{1}{(7\ln(k+2))^k}$

A. Converges B. Diverges
C. Convergence cannot be determined

14. The series $\sum_{k=1}^{\infty} \frac{4}{(k+1)^k}$

A. Converges B. Diverges
C. Convergence cannot be determined

15. The series $\sum_{k=1}^{\infty} \frac{|\sin kx|}{3^k}$

A. Converges B. Diverges
C. Convergence cannot be determined

SECTION 10.7

1. $\sum_{k=1}^{\infty} \frac{(-1)^k}{6k+1}$

A. Converges absolutely
B. Converges conditionally
C. Diverges

2. $\sum_{k=1}^{\infty} \frac{(-1)^k}{5k}$

A. Converges absolutely
B. Converges conditionally
C. Diverges

3. $\sum_{k=1}^{\infty} (-1)^k \left(\frac{3}{4}\right)^k$

A. Converges absolutely
B. Converges conditionally
C. Diverges

4. $\sum_{k=1}^{\infty} \frac{(-1)^k (2k)!}{2k^{2k}}$

A. Converges absolutely
B. Converges conditionally
C. Diverges

5. $\sum_{k=1}^{\infty} \left(-\frac{9}{2}\right)^k$

A. Converges absolutely
B. Converges conditionally
C. Diverges

6. $\sum_{k=1}^{\infty} \left(-\frac{1}{k^4}\right)^k$

A. Converges absolutely
B. Converges conditionally
C. Diverges

7. $\sum_{k=1}^{\infty} \frac{k}{(-3)^k}$

A. Converges absolutely
B. Converges conditionally
C. Diverges

8. $\sum_{k=1}^{\infty} \frac{(-1)^k}{k+6}$

A. Converges absolutely
B. Converges conditionally
C. Diverges

9. $\sum_{k=1}^{\infty} \frac{\cos \pi k}{4^k}$

A. Converges absolutely
B. Converges conditionally
C. Diverges

10. $\sum_{k=1}^{\infty} \frac{\cos(\pi k + 2)}{k}$

A. Converges absolutely
B. Converges conditionally
C. Diverges

11. $\sum_{k=1}^{\infty} \frac{4k}{\cos \pi k}$

A. Converges absolutely
B. Converges conditionally
C. Diverges

12. $\sum_{k=1}^{\infty} \frac{(-1)^k}{k!}$. Find the fifth partial sum.

A. -1.266 B. -1.191
C. -1.306 D. -1.318

13. $\sum_{k=1}^{\infty} 4\left(-\frac{1}{2}\right)^k$. Find the fifth partial sum.

A. 1.375 B. -1.375
C. 1.396 D. -1.396

14. $\sum_{k=1}^{\infty} \frac{(-1)^{k+1}4}{2^k}$. Find the fifth partial sum.

A. -4.886 B. 4.886
C. 1.396 D. -1.396

15. Answer true or false: For $\sum_{k=1}^{\infty} \frac{(-1)^{k+1}3}{\sqrt{k}+1}$ the fourth partial sum is 0.5475.

SECTION 10.8

1. Find the radius of convergence for $\sum_{k=1}^{\infty} \frac{2x^k}{k+2}$.

A. 2 B. 1 C. $\frac{1}{2}$ D. ∞

2. Find the radius of convergence for $\sum_{k=1}^{\infty} 8^k x^k$.

A. 8 B. 1 C. $\frac{1}{8}$ D. ∞

3. Find the radius of convergence for $\sum_{k=1}^{\infty} \frac{x^k}{k+7}$.

A. $\frac{1}{7}$ B. 1 C. 7 D. ∞

4. Find the radius of convergence for $\sum_{k=1}^{\infty} \frac{3x^k}{\ln k}$.

A. 3 B. 1 C. $\frac{1}{3}$ D. ∞

5. Find the radius of convergence for $\sum_{k=1}^{\infty} (-1)^k \frac{x^k}{\sqrt{x}}$.

A. 1 B. 2 C. $\frac{1}{2}$ D. ∞

6. Find the interval of convergence for $\sum_{k=1}^{\infty} (-1)^k \frac{x^k}{4}$.

A. $(-1,1)$ B. $(-4,4)$ C. $\left(-\frac{1}{4},\frac{1}{4}\right)$ D. $(-\infty,\infty)$

7. Find the interval of convergence for $\sum_{k=1}^{\infty} (-1)^k \frac{(x-3)^k}{3}$.

A. $(-4,-2)$ B. (2,4) C. $(-1,1)$ D. $(-\infty,\infty)$

8. Find the interval of convergence for $\sum_{k=1}^{\infty} \frac{2(3x-5)^k}{5^k}$.

A. $\left(-\frac{10}{3},0\right)$ B. $(-5,5)$ C. $\left(0,\frac{10}{3}\right)$ D. $(-\infty,\infty)$

9. Answer true or false: The interval of convergence $\sum_{k=1}^{\infty} \frac{6^k x^k}{k!}$ is $(-1,1)$.

10. Answer true or false: The interval of convergence for $\sum_{k=1}^{\infty} \frac{6^k x^{k+2}}{(3k!)}$ is $(-\infty,\infty)$.

11. Answer true or false: The interval of convergence for $\sum_{k=1}^{\infty} \frac{3^k(x-3)^k}{k!}$ is $(-\infty, \infty)$.

12. Answer true or false: The interval of convergence for $\sum_{k=1}^{\infty} (x-8)^k$ is (7,9).

13. Answer true or false: The interval of convergence for $\sum_{k=1}^{\infty} (4x-1)^k$ is (0,1).

14. Answer true or false: The interval of convergence for $\sum_{k=1}^{\infty} \frac{6^k x^k}{k!}$ is $\left(-\frac{2}{3}, 0\right)$.

15. Answer true or false: The interval of convergence for $\sum_{k=1}^{\infty} \frac{4(x+4)^k}{3^k}$ is $(-7, -1)$.

SECTION 10.9

1. Estimate $\sin 5°$ to 5 decimal-place accuracy.

   A. 0.08721 B. 0.08716 C. 0.00720 D. 0.00730

2. Estimate $\tan 9°$ to 5 decimal-place accuracy.

   A. 0.15827 B. 0.15832 C. 0.15838 D. 0.15849

3. Estimate $\sin^{-1}(0.3)$ to 5 decimal-place accuracy.

   A. 0.30458 B. 0.30462 C. 0.30469 D. 0.30475

4. Estimate $\sinh(0.3)$ to 5 decimal-place accuracy.

   A. 0.30452 B. 0.30456 C. 0.30459 D. 0.30462

5. Estimate $\cosh(0.2)$ to 5 decimal-place accuracy.

   A. 1.01989 B. 1.02007 C. 1.02019 D. 1.02040

6. Estimate $\sqrt[5]{e}$ to 5 decimal-place accuracy.

   A. 1.22134 B. 1.22140 C. 1.22146 D. 1.22153

7. Estimate $\frac{1}{e^3}$ to 5 decimal-place accuracy.

   A. 0.04959 B. 0.04965 C. 0.04972 D. 0.04979

8. Estimate $\sin(0.6)$ to 5 decimal-place accuracy.

   A. 0.56453 B. 0.56458 C. 0.56464 D. 0.56472

9. Answer true or false: $\cos(0.8)$ can be approximated to 4 decimal places to be 0.6967.

10. Answer true or false: $\ln 4$ can be approximated to 3 decimal places to be 1.388.

11. Answer true or false: e^7 can be approximated to 3 decimal places to be 1,096.610.

12. Answer true or false: $\cosh 0.7$ can be approximated to 3 decimal places to be 1.255.

13. Answer true or false: $\tanh^{-1} 0.14$ can be approximated to 3 decimal places to be 0.141.

14. Answer true or false: $\sinh^{-1} 0.17$ can be approximated to 3 decimal places to be 0.162.

15. Answer true or false: $\cosh^{-1} 1.19$ can be approximated to 3 decimal places to be 1.421.

SECTION 10.10

1. Answer true or false: The Maclaurin series for $e^{2x} - e^{-x}$ can be obtained by subtracting the Maclaurin series for e^{-x} from the Maclaurin series for e^{2x}.

2. Answer true or false: The Maclaurin series for $x^2 \sin x$ can be obtained by multiplying the Maclaurin series for $\sin x$ by x^2.

3. Answer true or false: The Maclaurin series for $\sin^2 x$ can be obtained by multiplying the Maclaurin series for $\sin x$ by itself.

4. Answer true or false: The Maclaurin series for $\sin 2x$ can be obtained by multiplying the Maclaurin series for $\sin x$ by itself.

5. Answer true or false: The Maclaurin series for $2\cosh x$ can be obtained by multiplying the Maclaurin series for $\cosh x$ by 2.

6. Answer true or false: The Maclaurin series for $\tan x$ can be obtained by dividing the Maclaurin series for $\sin x$ by the Maclaurin series for $\cos x$.

7. Answer true or false: The Maclaurin series for $e^x \sinh x$ can be obtained by multiplying the Maclaurin series for e^x by the Maclaurin series for $\sinh x$.

8. The Maclaurin series for $\dfrac{\ln(5+x)}{1+x}$ can be obtained by dividing the Maclaurin series for $\ln(2+x)$ by $1+x$.

9. Answer true or false: The Maclaurin series for $\ln(3+x)$ can be differentiated term by term to determine that the derivative of $\ln(3+x)$ is $\dfrac{1}{3+x}$.

10. Answer true or false: The Maclaurin series for $\ln(6x+2)$ can be differentiated term by term to determine that the derivative of $\ln(6x+2)$ is $\dfrac{1}{x}$.

11. Answer true or false: The Maclaurin series for $\sin 3x$ can be differentiated term by term to determine that the derivative of $\sin 3x$ is $\cos x$.

12. Answer true or false: The Maclaurin series for $\sinh 7x$ can be differentiated term by term to determine that the derivative of $\sinh 7x$ is $\cosh 7x$.

13. Answer true or false: The Maclaurin series for e^x can be integrated term by term to determine that the integral of e^x is $e^x + C$.

14. Answer true or false: The Maclaurin series for $\cos(4x)$ can be integrated term by term to determine that the integral of $\cos(4x)$ is $-4\sin(4x) + C$.

15. Answer true or false: The Maclaurin series for $\dfrac{1}{7+x}$ can be integrated term by term to determine that the integral of $\dfrac{1}{7+x}$ is $\ln(7+x) + C$.

CHAPTER 10 TEST

1. The general term for the sequence $2,\ 2\sqrt{2},\ 2\sqrt{3},\ 4,\ 2\sqrt{5},\ldots$ is

A. $2\sqrt{n}$ B. $2\sqrt{n+1}$ C. $2\sqrt{n-1}$ D. $2n\sqrt{n}$.

2. Write out the first five terms of the sequence $\left\{\dfrac{-n-5}{n+8}\right\}_{n=1}^{\infty}$.

A. $-5/8,\ -2/3,\ -7/10,\ -8/11,\ -3/4$
B. $-2/3,\ -7/10,\ -8/11,\ -3/4,\ -10/13$
C. $-5/9,\ -3/5,\ -7/11,\ -2/3,\ -9/13$
D. $-5/8,\ -5/8,\ -5/8,\ -5/8,\ -5/8$

3. If the sequence $\left\{\dfrac{n^2+5}{n^5-8}\right\}_{n=1}^{\infty}$ converges, find its limit. If not, answer diverges.

A. 0 B. $-5/7$ C. 1 D. Diverges

4. Determine which answer best describes the sequence $\left\{\dfrac{n^3}{n^4+5}\right\}_{n=1}^{\infty}$.

A. Strictly increasing
B. Strictly decreasing
C. Increasing, but not strictly increasing
D. Decreasing, but not strictly decreasing

5. Determine which answer best describes the sequence $\{(n-1)!n^7\}_{n=1}^{\infty}$.

A. Strictly increasing
B. Strictly decreasing
C. Increasing, but not strictly increasing
D. Decreasing, but not strictly decreasing

6. Answer true or false: The series $\dfrac{9}{2}+\dfrac{9}{4}+\dfrac{9}{8}+\dfrac{9}{16}+\cdots+9\left(\dfrac{1}{2}\right)^n$ converges.

7. Answer true or false: The series $\displaystyle\sum_{k=1}^{\infty} 6\left(\frac{4}{5}\right)^k$ converges to 24.

8. Write $2.1919\ldots$ as a fraction.

A. 73/33 B. 219/100 C. 217/99 D. 219/500

9. Answer true or false: The series $\displaystyle\sum_{k=1}^{\infty}\frac{1}{(k+5)^8}$ converges.

10. Answer true or false: The series $\displaystyle\sum_{k=1}^{\infty}\frac{k^3+2}{5k^3+1}$ converges.

11. Answer true or false: The Maclaurin polynomial of order 3 for e^{7x} is $1+7x+49x^2+243x^3$.

12. Answer true or false: The Maclaurin polynomial of order 3 for $\ln(x+5)$ is
$\ln 5 + x\ln 5 + \dfrac{x^2\ln 5}{2} + \dfrac{x^3 \ln 5}{6}$.

13. Answer true or false: The Maclaurin polynomial of order 3 for $\cosh x^3$ is
$\cosh x^3 + 3x^2\sinh x^3 + 9x^4\cosh x^3 + 27x^6\sinh x^3$.

14. Answer true or false: The Taylor polynomial for e^x of order 3 about $x = 5$ is $e^5 + e^5(x-5) + \dfrac{e^5(x-5)^2}{2} + \dfrac{e^5(x-5)^3}{6}$.

15. The series $\displaystyle\sum_{k=1}^{\infty} \frac{1}{k^6 - 7}$

A. Converges B. Diverges
C. Convergence cannot be determined

16. The series $\displaystyle\sum_{k=1}^{\infty} \frac{1}{5k^4 - k^2}$

A. Converges B. Diverges
C. Convergence cannot be determined

17. The series $\displaystyle\sum_{k=1}^{\infty} \frac{(k+1)!}{k^{9k}}$

A. Converges B. Diverges
C. Convergence cannot be determined

18. $\displaystyle\sum_{k=1}^{\infty} \frac{(-1)^k}{7k+2}$

A. Converges absolutely
B. Converges conditionally
C. Diverges

19. Find the radius of convergence for $\displaystyle\sum_{k=0}^{\infty} 6^k x^k$.

A. 6 B. 1 C. $\dfrac{1}{6}$ D. ∞

20. Find the interval of convergence for $\displaystyle\sum_{k=0}^{\infty} (-1)^k \frac{x^k}{9}$.

A. $(-1, 1)$ B. $(-9, 9)$ C. $\left(-\dfrac{1}{9}, \dfrac{1}{9}\right)$ D. $(-\infty, \infty)$

21. Estimate $\sin 11°$ to 5 decimal-place accuracy.

A. 0.19081 B. 0.19089 C. 0.19094 D. 0.19100

22. Answer true or false: The Maclaurin series for $e^{-x} + \ln x$ can be obtained by adding the Maclaurin series for e^{-x} and $\ln x$.

ANSWERS TO SAMPLE TESTS

SECTION 10.1

1. A 2. B 3. D 4. A 5. A 6. C 7. B 8. D 9. C 10. F 11. F 12. F 13. T 14. F 15. F

SECTION 10.2

1. A 2. B 3. C 4. A 5. D 6. A 7. F 8. T 9. T 10. T 11. F 12. D 13. C 14. C 15. D

SECTION 10.3

1. B 2. A 3. A 4. C 5. B 6. C 7. A 8. B 9. B 10. A 11. B 12. A 13. C 14. D 15. A

SECTION 10.4

1. F 2. T 3. T 4. F 5. T 6. F 7. A 8. D 9. D 10. D 11. C 12. B 13. B 14. A 15. A

SECTION 10.5

1. A 2. A 3. B 4. B 5. B 6. A 7. B 8. B 9. B 10. B 11. A 12. B 13. B 14. B 15. A

SECTION 10.6

1. A 2. A 3. B 4. A 5. B 6. B 7. A 8. A 9. A 10. A 11. B 12. A 13. A 14. A 15. A

SECTION 10.7

1. B 2. B 3. A 4. A 5. C 6. A 7. A 8. B 9. A 10. B 11. C 12. A 13. B 14. A 15. T

SECTION 10.8

1. B 2. C 3. B 4. B 5. A 6. A 7. B 8. C 9. F 10. T 11. T 12. T 13. F 14. F 15. T

SECTION 10.9

1. B 2. C 3. C 4. A 5. B 6. B 7. D 8. C 9. T 10. F 11. F 12. T 13. T 14. F 15. F

SECTION 10.10

1. T 2. T 3. T 4. F 5. T 6. F 7. T 8. T 9. T 10. F 11. F 12. F 13. T 14. F 15. T

CHAPTER 10 TEST

1. A 2. B 3. A 4. B 5. A 6. T 7. T 8. C 9. T 10. F 11. F 12. F 13. F 14. T 15. A 16. A 17. A 18. B 19. C 20. A 21. A 22. T

CHAPTER 11
Sample Exams

SECTION 11.1

1. Answer true or false: To plot $(6, \pi/4)$ in polar coordinates go out 6 units from the pole to the right, then rotate $\pi/4$ radians clockwise.

2. Find the rectangular coordinates of $(1, \pi/4)$.

A. $(\sqrt{2}, \sqrt{2})$ B. $(\sqrt{2}/2, \sqrt{2}/2)$ C. $(2\sqrt{2}, 2\sqrt{2})$ D. $(4\sqrt{2}, 4\sqrt{2})$

3. Find the rectangular coordinates of $(5, -\pi/2)$.

A. (5,0) B. (0,5) C. $(-5, 0)$ D. $(0, -5)$

4. Find the rectangular coordinates of $(-4, -\pi/4)$.

A. $(-2\sqrt{2}, -2\sqrt{2})$ B. $(2\sqrt{2}, 2\sqrt{2})$ C. $(-2\sqrt{2}, 2\sqrt{2})$ D. $(2\sqrt{2}, -2\sqrt{2})$

5. Use a calculating utility to approximate the polar coordinates of (7,3).

A. (58,1.1903) B. (7.6158,0.4049) C. (58,1.1659) D. (7.6158,1.1659)

6. Use a calculating utility to approximate the polar coordinates of $(-2, -5)$.

A. (29,4.3319) B. (5.3852,4.3319) C. (58,1.19031) D. (5.3319,1.1903)

7. Describe the curve $\theta = \pi/2$.

A. A vertical line B. A horizontal line C. A circle D. A semicircle

8. Describe the curve $r = 4\cos\theta$.

A. A circle left of the origin
B. A circle above the origin
C. A circle right of the origin
D. A circle below the origin

9. Describe the curve $r = 8\sin\theta$.

A. A circle left of the origin
B. A circle above the origin
C. A circle right of the origin
D. A circle below the origin

10. What is the radius of the circle $r = 8\sin\theta$?

A. 8 B. 16 C. 4 D. 1

11. How many petals does the rose $r = 7\sin 3\theta$ have?

A. 1 B. 4 C. 3 D. 6

12. Describe the curve $r = 10 + 10\sin\theta$.

A. Limacon with inner loop
B. Cardioid
C. Dimpled limacon
D. Convex limacon

13. Describe the curve $r = 4 + 6\cos\theta$.

A. Limacon with inner loop
B. Cardioid
C. Dimpled limacon
D. Convex limacon

14. Describe the curve $r = 2 + 8\cos\theta$.

A. Limacon with inner loop
B. Cardioid
C. Dimpled limacon
D. Convex limacon

15. Answer true or false: $r = 7\theta$ graphs as an Archimedean spiral.

SECTION 11.2

1. $x = 3t^2$, $y = 9t$. Find dy/dx.

A. $\frac{3}{2t}$ B. $\frac{2t}{3}$ C. $6t$ D. $\frac{3t}{2}$

2. $x = \sin t$, $y = 2\cos t$. Find dy/dx.

A. $2\tan t$ B. $2\cot t$ C. $-2\tan t$ D. $-2\cot t$

3. $x = e^{2t}$, $y = t$. Find dy/dx.

A. $\frac{e^{-2t}}{2}$ B. $2e^t$ C. $\frac{t}{2^{2t}}$ D. $2te^{2t}$

4. Answer true or false: If $x = t^3$ and $y = t^2 - 2$, $d^2y/dx^2 = -\frac{1}{3t}$.

5. Answer true or false: If $x = \sin t$ and $y = \cos t$, $d^2y/dx^2 = \cot t$.

6. Find the value of t for which the tangent to $x = t^4 - 2$, $y = 3t^2 - 2t + 6$ is horizontal.

A. 1/3 B. 0 C. 2/3 D. 8

7. Find the value(s) of t for which the tangent to $x = 3\sin t + 8$, $y = 5t^2 + 7$ is/are horizontal.

A. $\pi/2, 3\pi/2$ B. 0 C. $-3/5$ D. $0, \pi/2, 3\pi/2$

8. Find the value(s) of t for which the tangent to $x = 2e^t - 5$, $y = 7t^2 + 3t + 1$ is/are horizontal.

A. 1 B. $-3/14$ C. $3/2$ D. $1, -3/2$

9. Find the value(s) of t for which the tangent to $x = t^3 - t^2 - 5t$, $y = t^4 + 2$ is/are horizontal.

A. 0 B. 5/2 C. 5 D. 0,5/2

10. Find the value(s) of t for which the tangent to $x = 6t^{3/2}$, $y = 4\sin t$ is/are horizontal.

A. 0 B. $0, \pi/2$ C. $\pi/2, 3\pi/2$ D. $0, \pi/2, 3\pi/2$

11. Answer true or false: If $r = 7\sin\theta$, the tangent to the curve at the origin is the line $\theta = 0$.

12. Find the arc length of the spiral $r = e^{5\theta}$ between $\theta = 0$ and $\theta = 1$.

A. $\frac{\sqrt{26}}{5}e - \frac{\sqrt{26}}{5}$ B. $\frac{2}{5}e - \frac{2}{5}$ C. $\frac{\sqrt{26}}{5}e^5 - \frac{\sqrt{26}}{5}$ D. $\frac{2}{5}e^5 - \frac{2}{5}$

13. Find the arc length of the spiral $r = \sin\theta$ between $\theta = 0$ and $\theta = 2\pi$.

A. 2 B. 2π C. $2\sqrt{2}\,\pi$ D. $2\sqrt{2}$

14. Answer true or false: The arc length of the curve $r = \sin 2\theta$ between $\theta = 0$ and π is π.

15. Answer true or false: The arc length of the curve $r = 8\theta$ between $\theta = 0$ and π is 8π.

SECTION 11.3

1. Find the area of the region enclosed by $r = 4 - 4\cos\theta$.

A. 75.40 B. 56.55 C. 18.85 D. 9.42

2. Find the area of the region enclosed by $r = 4 - 4\sin\theta$.

A. 75.40 B. 56.55 C. 18.85 D. 9.42

3. Find the area of the region enclosed by $r = 2 - 6\cos\theta$ from $\theta = 0$ to $\theta = \pi/2$.

A. 5.28 B. 58.56 C. 183.96 D. 23.00

4. Find the area of the region enclosed by $r = 2 - 6\sin\theta$ from $\theta = 0$ to $\theta = \pi/2$.

A. 183.96 B. 58.56 C. 1.53 D. 23.00

5. Answer true or false: The area of the region bounded by the curve $r = 3\cos 2\theta$ from $\theta = 3\pi/2$ to 2π is 3.53.

6. Answer true or false: The area between the circle $r = 5$ and the curve $r = 2 + 2\cos\theta$ is $\pi/2$.

7. Answer true or false: The area of one petal of $\sin 2\theta$ is given by $\int_0^{\pi/2} 0.5\sin 2\theta\, d\theta$.

8. Answer true or false: The area in one petal of $2\sin 6\theta$ is given by $\int_0^{\pi/3} 2(\sin 6\theta)^2\, d\theta$.

9. Answer true or false: The area in all of the petals of $\cos 6\theta$ is given by $\int_0^{\pi/3} 3(\sin 6\theta)^2\, d\theta$.

10. Find the region bounded by $r = 6\theta$ from 0 to π.

A. 3.1 B. 2.6 C. 186.04 D. 10.3

11. Find the region bounded by $r = 10\theta$ from 0 to 2π.

A. 41,034 B. 2,067 C. 8,268 D. 4,134

12. Answer true or false: The region between $r = 2\cos\theta$ and $r = 2\sin\theta$ is given by $\int_0^{\pi/4} 2(\sin\theta - \cos\theta)^2\, d\theta$.

13. Find the area bounded by $r = 6 - 4\cos\theta$ from π to $3\pi/2$.

A. 12.6 B. 29.3 C. 117.1 D. 58.6

14. Find the area bounded by $r = 6 - 4\cos\theta$ from $\pi/2$ to π.

A. 12.6 B. 29.3 C. 117.1 D. 58.6

15. Find the area bounded by $r = 6 - 3\sin\theta$ from π to $3\pi/2$.

A. 12.1 B. 29.3 C. 117.1 D. 49.81

SECTION 11.4

1. The vertex of the parabola $y^2 = 9x$ is

A. (1,9) B. (0,0) C. (9,1) D. (1,1)

2. The vertex of the parabola $(y-6)^2 = 3(x-2)$ is

A. $(-2,-6)$ B. (2,6) C. $(-3,-6)$ D. (3,3)

3. A parabola has a vertex at (4,5) and a directrix $x = 0$. Find the focus.

A. (4,0) B. (4,10) C. (8,5) D. (8,10)

4. The graph of the parabola $x = 9y^2 + 2$ opens

A. Right B. Left C. Up D. Down

5. What are the ends of the minor axis for the ellipse $\dfrac{x^2}{36} + \dfrac{y^2}{4} = 1$?

A. $(6,0),(-6,0)$ B. $(36,0),(-36,0)$ C. $(0,6),(0,-6)$ D. $(0,2),(0,-2)$

6. Answer true or false: The foci of $\dfrac{x^2}{49} + \dfrac{y^2}{81} = 1$ are $(7,0)$ and $(-7,0)$.

7. The foci of the ellipse $\dfrac{x^2}{81} + \dfrac{y^2}{49} = 1$ are

A. $(-4\sqrt{2}, 4\sqrt{2}),(7,0)$ B. $(0,-7),(0,7)$ C. $(-9,0),(9,0)$ D. $(0,-9),(0,9)$

8. Answer true or false: The foci of the ellipse $\dfrac{x^2}{64} + \dfrac{y^2}{36} = 1$ are (8,0) and $(-8,0)$.

9. Answer true or false: The foci of the hyperbola $\dfrac{x^2}{81} - \dfrac{y^2}{16} = 1$ are $(0,-\sqrt{97})$ and $(0,\sqrt{97})$.

10. Answer true or false: The foci of the hyperbola $\dfrac{x^2}{64} - \dfrac{y^2}{36} = 1$ are (10,0) and $(-10,0)$.

11. Answer true or false: The hyperbola $\dfrac{y^2}{12} - \dfrac{x^2}{7} = 1$ opens up and down.

12. Answer true or false: $x^2 + \dfrac{y^2}{7} = 1$ has a vertical major axis.

13. Answer true or false: $y = 2x^2 + 8$ has a vertex (0,0).

14. Answer true or false: $y = x^2 - 9$ has a vertex $(0,-9)$.

15. Answer true or false: $x = 5y^2 + 12$ has a vertex (0,0).

SECTION 11.5

1. Name the conic section that is the graph of $xy = 12$.

A. parabola B. circle C. ellipse D. hyperbola

2. Name the conic section that is the graph of $x^2 + 6xy - 3y^2 = 0$.

A. parabola B. circle C. hyperbola D. ellipse

3. Name the conic section that is the graph of $x^2 + 5xy - 7y^2 + 25 = 0$.

A. parabola B. circle C. ellipse D. hyperbola

4. Name the conic section that is the graph of $3xy = 5$.

A. parabola B. circle C. ellipse D. hyperbola

5. Name the conic section or degenerate conic section that is the graph of $4x^2 - 4y^2 = 0$.

A. hyperbola B. circle
C. line D. pair of intersecting lines

6. Answer true or false: $x^2 + 7xy + 2y^2 + 7x + 5y - 2 = 0$ represents an ellipse.

7. Answer true or false: $x^2 - 5xy + 6y^2 + 5x - 3y + 4 = 0$ represents a pair of intersecting lines.

8. Answer true or false: $6x^2 - 6y^2 + 8 = 0$ represents a pair of intersecting lines.

9. Answer true or false: $x^2 + 3xy + 2y^2 = 0$ represents a point.

10. Answer true or false: $x^2 + 3xy + 2y^2 - 6 = 0$ represents a circle.

SECTION 11.6

1. The eccentricity of $r = \dfrac{8}{1 + 2\sin\theta}$ is

A. 4 B. 1 C. 2 D. 6.

2. The eccentricity of $r = \dfrac{8}{4 + 8\cos\theta}$ is

A. 8 B. 4 C. 2 D. 1.

3. The eccentricity of $r = \dfrac{3}{5 + 15\sin\theta}$ is

A. 12 B. 4 C. 2 D. 3.

4. Answer true or false: $r = \dfrac{5}{1 - 3\cos\theta}$ has its directrix left of the pole.

5. Write the equation of the ellipse that has $e = 3$ and directrix $x = 1$.

A. $r = \dfrac{3}{1 + 3\cos\theta}$ B. $r = \dfrac{3}{1 - 3\cos\theta}$

C. $r = \dfrac{3}{1 + 3\sin\theta}$ D. $r = \dfrac{3}{1 - 3\sin\theta}$

6. The graph of $r = \dfrac{6}{5 - 2\sin\theta}$ is

A. A parabola B. An ellipse C. A circle D. A hyperbola.

7. The graph of $r = \dfrac{10}{2 + 4\cos\theta}$ is

A. A parabola B. An ellipse C. A circle D. A hyperbola.

8. The graph of $r = \dfrac{7}{6 + 4\cos\theta}$ is

A. A parabola B. An ellipse C. A circle D. A hyperbola.

9. Answer true or false: The graph of $r = \dfrac{1}{5 - \cos\theta}$ orients horizontally.

10. Answer true or false: $r = \dfrac{3}{1 - 6\sin\theta}$ is a hyperbola that opens left and right.

11. Answer true or false: $r = \dfrac{6}{1 - \sin\theta}$ is a parabola that opens to the left.

12. Answer true or false: $r = \dfrac{4}{1 + \cos\theta}$ is a parabola that opens to the left.

13. Answer true or false: $r = \dfrac{1}{6 - 3\sin\theta}$ is a parabola that opens up.

14. Answer true or false: $r = \dfrac{10}{6 - 6\sin\theta}$ is a hyperbola oriented up and down.

15. A small planet is found 8 times as far from the sun as the earth. What is its period?

A. 22.6 years B. 64 years C. 32 years D. 4 years

CHAPTER 11 TEST

1. Find the rectangular coordinates of $(-4, \pi/4)$.

A. $(-\sqrt{2}, -\sqrt{2})$ B. $\left(\dfrac{-\sqrt{2}}{2}, \dfrac{\sqrt{2}}{2}\right)$ C. $(-2\sqrt{2}, -2\sqrt{2})$ D. $(-4\sqrt{2}, -4\sqrt{2})$

2. Use a calculating utility to approximate the polar coordinates of the point (6,8).

A. (10,0.9273) B. (100,0.9273) C. (10,0.6435) D. (100,0.6435)

3. Describe the curve $r = 8\cos\theta$.

A. A circle left of the origin B. A circle above the origin
C. A circle right of the origin D. A circle below the origin

4. What is the radius of the circle $r = 24\cos\theta$?

A. 24 B. 48 C. 12 D. 1

5. How many petals does the rose $r = 6\sin 3\theta$ have?

A. 1 B. 4 C. 3 D. 6

6. Describe the curve $r = 10 + 5\cos\theta$.

A. Limacon with inner loop B. Cardioid
C. Dimpled limacon D. Convex limacon

7. Answer true or false: $r = 7/\theta$ graphs as a hyperbolic spiral.

8. $x = -2\cos t$, $y = -2\sin t$. Find dy/dt.

A. $\tan t$ B. $\cot t$ C. $-\tan t$ D. $-\cot t$

9. Answer true or false: If $x = \sqrt{\sin t}$ and $y = \sqrt{\cos t}$, $\dfrac{d^2y}{dx^2} = \cot t$.

10. Find the value(s) of t for which the tangent to $x = \sin t$, $y = 3t^2 + 10$ is/are horizontal.

A. $\pi/2, 3\pi/2$ B. 0 C. $-3/5$ D. $0, \pi/2, 3\pi/2$

11. Answer true or false: If $r = 12\sin\theta$, the tangent to the curve at the origin is the line $\theta = 0$.

12. Answer true or false: The arc length of the curve $r = 2\cos 3\theta$ between $\theta = 0$ and π is 4π.

13. Find the area of the region enclosed by $r = 4 - 4\sin\theta$.

A. 24π B. 18π C. 6π D. 3π

14. Answer true or false: The area of the region bounded by the curve $r = 3\sin 2\theta$ from 0 to π is 3.14.

15. Answer true or false: The area between the circle $r = 12$ and the curve $r = 5 + 5\ \cos\theta$ is π.

16. Answer true or false: The area in one petal of $r = \sin 4\theta$ is given by $\displaystyle\int_0^{2\pi} 0.5(\sin 4\theta)\, d\theta$.

17. Find the area bounded by $r = 6 - 4\cos\theta$.

A. 12.6 B. 29.3 C. 117.1 D. 138.23

18. The vertex of the parabola $x^2 = -3y$ is

A. $(0,0)$ B. $(3,1)$ C. $(1,3)$ D. $(1,1)$.

19. The eccentricity of $r = \dfrac{6}{1 + 3\cos\theta}$ is

A. 6 B. 1 C. 3 D. 12.

20. Answer true or false: $r = \dfrac{8}{1 - 2\cos\theta}$ has its directrix left of the pole.

ANSWERS TO SAMPLE TESTS

SECTION 11.1

1. F 2. B 3. D 4. C 5. B 6. B 7. A 8. C 9. B 10. C 11. C 12. B
13. A 14. A 15. T

SECTION 11.2

1. A 2. C 3. C 4. F 5. F 6. A 7. B 8. B 9. A 10. C 11. T 12. C
13. B 14. F 15. F

SECTION 11.3

1. A 2. A 3. A 4. C 5. T 6. F 7. F 8. F 9. F 10. C 11. D 12. F
13. D 14. D 15. D

SECTION 11.4

1. B 2. B 3. C 4. A 5. D 6. F 7. A 8. F 9. F 10. T 11. T 12. T
13. F 14. T 15. F

SECTION 11.5

1. D 2. C 3. D 4. D 5. D 6. F 7. F 8. T 9. F 10. F

SECTION 11.6

1. C 2. C 3. D 4. T 5. A 6. B 7. D 8. B 9. T 10. F 11. F 12. T
13. F 14. F 15. A

CHAPTER 11 TEST

1. C 2. A 3. A 4. C 5. C 6. D 7. T 8. D 9. F 10. B 11. T 12. F
13. A 14. F 15. F 16. F 17. D 18. A 19. C 20. T

CHAPTER 12
Sample Exams

SECTION 12.1

1. Answer true or false: A box has a corner at the origin and corners at $(5,0,0)$, $(0,2,0)$, and $(0,2,1)$. If three of the edges of the box lie on the axes, the point $(5,2,1)$ is a corner point of the box.

2. Answer true or false: $(5,7,1)$, $(-1,11,3)$, and $(1,5,7)$ are vertices of an equilateral triangle.

3. Find the distance from $(1,3,2)$ to the xy-plane.

A. 1 B. 2 C. 3 D. $\sqrt{14}$

4. Find the distance from $(1,-2,-3)$ to the origin.

A. 1 B. 2 C. 3 D. $\sqrt{14}$

5. The surface described by $x^2+y^2+z^2=15$ is a(n)

A. sphere B. cylinder C. cone D. ellipsoid.

6. The spherical surface $(x-2)^2+(y-3)^2+(z+4)^2=5$ is centered at

A. $(4,9,-16)$ B. $(-4,-9,16)$ C. $(2,3,-4)$ D. $(-2,-3,4)$.

7. Answer true or false: The sphere $x^2+(y-2)^2+z^2=16$ has a radius of 4.

8. The graph of $x^2+z^2=5$ is an infinitely long cylinder whose central axis is the

A. x-axis B. y-axis C. z-axis D. line $x=y$.

9. $z=\sin y$ describes a surface. In what direction would it be possible to travess the surface in a straight line?

A. parallel to the x-axis
B. parallel to the y-axis
C. parallel to the z-axis
D. parallel to the line $y=z$

10. The equation for a cylinder with radius 5 oriented symmetrically about the z-axis is

A. $x^2+y^2=5$ B. $x^2+y^2=25$ C. $z^2=5$ D. $z^2=25$.

11. Answer true or false: $x^2+2x+y^2+2y+z^2+2z=16$ describes a sphere of radius 4.

12. $x^2+y^2+z^2=-1$ graphs as

A. a sphere
B. a point
C. Nothing, there is no such graph
D. a cylinder.

13. Answer true or false: $x^2+7x+y^2+y+z^2+z=1$ describes a sphere centered at the origin.

14. Find the distance the surface $x^2+y^2+z^2=1$ is from the point $(0,0,-3)$.

A. 1 B. 2 C. $\sqrt{2}$ D. 0

15. Answer true or false: $(x-3)^2+(y-6)^2+(z+2)^2=25$ represents a sphere centered at $(-3,-6,2)$.

SECTION 12.2

1. The vector with initial point $P_1(1,2)$, and terminal point $P_2(3,10)$ is

A. $\langle 2,8\rangle$ B. $\langle -2,-8\rangle$ C. $\langle 6,10\rangle$ D. $\langle -6,-10\rangle$.

2. The vector with initial point $P_1(0,1,2)$, and terminal point $P_2(2,3,1)$ is

A. $\langle 4,6,5\rangle$ B. $\langle -4,-6,-5\rangle$ C. $\langle 2,2,-1\rangle$ D. $\langle -2,-2,1\rangle$.

3. Find the terminal point of $\mathbf{v} = \mathbf{i} + 4\mathbf{j} + \mathbf{k}$, if the initial point is $(0,0,1)$

A. $(1,2,0)$ B. $(1,3,1)$ C. $(1,4,2)$ D. $(1,1,1)$.

4. Let $\mathbf{v} = \langle 3,-2\rangle$. Find the norm of $\mathbf{v}$.

A. $-\sqrt{5}$ B. $\sqrt{5}$ C. $\sqrt{13}$ D. $-\sqrt{13}$

5. Answer true or false: $\mathbf{u} = 3\mathbf{i} + 4\mathbf{j} + 2\mathbf{k}$. The norm of $\mathbf{u}$ is $\sqrt{29}$.

6. Answer true or false: Let $\mathbf{v} = 5\mathbf{i} - 5\mathbf{j}$. The norm of $\mathbf{v}$ is 0.

7. Add $\mathbf{u} = 2\mathbf{i} + 2\mathbf{j} + 3\mathbf{k}$ to $\mathbf{v} = \mathbf{i} + 2\mathbf{j} + 2\mathbf{k}$.

A. $3\mathbf{i} + 4\mathbf{j} + 5\mathbf{k}$ B. $\mathbf{i} + 2\mathbf{j} + 3\mathbf{k}$ C. $2\mathbf{i} + 3\mathbf{j} + 4\mathbf{k}$ D. $5\mathbf{i} + 6\mathbf{j} + 7\mathbf{k}$

8. If $\mathbf{u} = 3\mathbf{i} + 5\mathbf{j} + \mathbf{k}, 6\mathbf{u} =$

A. $18\mathbf{i} + 6\mathbf{j} + \mathbf{k}$ B. $8\mathbf{i} + 10\mathbf{j} + 6\mathbf{k}$ C. $8\mathbf{i} + 6\mathbf{j} + \mathbf{k}$ D. $18\mathbf{i} + 30\mathbf{j} + 6\mathbf{k}$.

9. Let $\mathbf{u} = \langle 2,1\rangle$, $\mathbf{v} = \langle 1,3\rangle$, and $\mathbf{w} = \langle 5,1\rangle$. Find the vector $\mathbf{x}$ that satisfies $2\mathbf{u} + \mathbf{v} - \mathbf{x} = \mathbf{w} + \mathbf{x}$.

A. $\langle 0,0\rangle$ B. $\langle 0,-2\rangle$ C. $\langle 5,5\rangle$ D. $\langle 0,2\rangle$

10. Given that $\|\mathbf{v}\| = 2$, find all values of k such that $\|k\mathbf{v}\| = 4$.

A. $-2, 2$ B. 2 C. $-4, 4$ D. 4

11. Answer true or false: If $\|\mathbf{v}\| = 5$ and ϕ, the angle the vector makes with the positive x-axis, is $\pi/3$, then $\mathbf{v} = \langle 5/2, 5\sqrt{2}/2\rangle$.

12. Answer true or false: If $\|\mathbf{v}\| = 16$ and ϕ, the angle the vector makes with the positive x-axis, is $45°$, then $\mathbf{v} = \langle \sqrt{2}/2, \sqrt{2}/2\rangle$.

13. Answer true or false: Two forces, one 60 N and the other 80 N, act at right angles. The resultant force has a magnitude of 100 N.

14. A particle is said to be in a static equilibrium if the resultant of all forces applied to it is zero. Find the force F that must be applied to a particle to produce static equilibrium if there are two forces, each of 20 N, applied so that one acts 60° above the positive x-axis and the other acts 60° below the positive x-axis. Give the magnitude of the resultant acting in the negative x direction.

A. 20 N B. 40 N C. $40\sqrt{2}$ N D. $80\sqrt{2}$ N

15. Let $\mathbf{u} = \langle 1,1,0\rangle$, $\mathbf{v} = \langle 0,1,0\rangle$, and $\mathbf{w} = \langle 0,0,2\rangle$. Find C_1, C_2, and C_3 such that $\langle 6,6,4\rangle = C_1\mathbf{u} + C_2\mathbf{v} + C_3\mathbf{w}$.

A. $6, 0, 2$ B. $6, 6, 4$
C. $6, 6, 2$ D. No such constraints exist.

SECTION 12.3

1. Find the dot product $\langle 3, 1\rangle \cdot \langle 5, 2\rangle$.

A. 28 B. 17 C. 11 D. $\sqrt{17}$

2. Find the dot product $\langle 0, 1, 1\rangle \cdot \langle 2, -1, 1\rangle$.

A. 0 B. 2 C. -2 D. 4

3. Find the dot product $\langle 5, 3, 2\rangle \cdot \langle 2, 7, 3\rangle$.

A. 70 B. 48 C. 37 D. 111

4. Answer true or false: $\mathbf{0} \cdot \mathbf{v} = 0$.

5. Find the dot product $\mathbf{u} \cdot \mathbf{v}$ where $\mathbf{u} = 4\mathbf{i} + 3\mathbf{j}$ and $\mathbf{v} = 2\mathbf{i} - 2\mathbf{j}$.

A. 8 B. -8 C. -2 D. 2

6. Answer true or false: If $\mathbf{u} = 2\mathbf{i} + 3\mathbf{j} - \mathbf{k}$ and $\mathbf{v} = 5\mathbf{i} - 2\mathbf{j} + 4\mathbf{k}$, $\mathbf{u} \cdot \mathbf{v} = 0$.

7. Find the angle between $\mathbf{u} = \langle 4, 2\rangle$ and $\mathbf{v} = \langle 10, 4\rangle$.

A. 4.21° B. 4.76° C. 5.12° D. 8.13°

8. Find the angle between $\mathbf{u} = 4\mathbf{i} + 2\mathbf{k}$ and $\mathbf{v} = 5\mathbf{i} + 2\mathbf{k}$.

A. 4.21° B. 4.76° C. 5.12° D. 8.13°

9. Let $\mathbf{u} = \langle 4, 1\rangle$, $\mathbf{v} = \langle 2, 8\rangle$, and $\mathbf{w} = \langle 10, 3\rangle$. Find $\mathbf{u} \cdot (\mathbf{w} - 2\mathbf{v})$.

A. -37 B. 11 C. -3 D. -19

10. Answer true or false: Let $\mathbf{u} = \mathbf{i} + \mathbf{j} + 3\mathbf{k}$, then the direction cosines are $\cos\alpha = \frac{\sqrt{11}}{11}$, $\cos\beta = \frac{\sqrt{11}}{11}$, and $\cos\gamma = \frac{3\sqrt{11}}{11}$.

11. Answer true or false: $\mathbf{u} \cdot \mathbf{u} = 0$

12. Answer true or false. Let $\mathbf{u} = 3\mathbf{i} + 5\mathbf{j} + \mathbf{k}$. The direction cosines are $\cos\alpha = \frac{1}{3}$, $\cos\beta = \frac{5}{12}$, and $\cos\gamma = \frac{1}{4}$.

13. Answer true or false: $\mathbf{u} \cdot \mathbf{v} = 0$ does not necessarily mean $\mathbf{u} = 0$.

14. If $\mathbf{v}$ is a three-space vector that has direction cosines α and β each equal to $\frac{1}{3}$, then $\cos\gamma$ must be 0.

15. A box is pulled across a frictionless surface by applying a 5-N force. The force is applied by pulling on a rope at an angle of 60° above the horizontal. If the box is moved by the force a total of 60 m, how much work is done?

A. 300 N·m B. 150 N·m C. $150\sqrt{2}$ N·m D. $150\sqrt{3}$ N·m

SECTION 12.4

1. Find $(\mathbf{i}+\mathbf{j}+\mathbf{k})\times -\mathbf{j}$.

A. $\mathbf{i}-\mathbf{k}$ B. $-\mathbf{j}$ C. $-\mathbf{i}-\mathbf{j}$ D. $\mathbf{j}$

2. If $\mathbf{u}=-2\mathbf{i}-3\mathbf{j}-4\mathbf{k}$ and $\mathbf{v}=-\mathbf{i}+3\mathbf{j}-2\mathbf{k}$, $\mathbf{u}\times\mathbf{v}=$

A. $18\mathbf{i}-9\mathbf{k}$ B. $18\mathbf{i}+4\mathbf{j}-9\mathbf{k}$ C. $18\mathbf{i}-4\mathbf{j}-9\mathbf{k}$ D. $12\mathbf{i}+8\mathbf{j}-3\mathbf{k}$.

3. If $\mathbf{u}=\langle 0,2,1\rangle$ and $\mathbf{v}=\langle -1,-3,0\rangle$, $\mathbf{u}\times\mathbf{v}=$

A. $\langle 3,-1,2\rangle$ B. $\langle -3,1,-2\rangle$ C. $\langle -3,-1,-2\rangle$ D. $\langle -3,-1,2\rangle$.

4. If $\mathbf{a}=4\mathbf{i}+\mathbf{k}$ and $\mathbf{b}=-2\mathbf{j}$, find $\mathbf{a}\times\mathbf{b}$.

A. $\mathbf{0}$ B. $2\mathbf{i}+8\mathbf{k}$ C. $-2\mathbf{i}+8\mathbf{k}$ D. $2\mathbf{i}-8\mathbf{k}$

5. A parallelogram has $\mathbf{u}=-2\mathbf{j}-\mathbf{k}$ and $\mathbf{v}=\mathbf{i}+3\mathbf{j}$ as adjacent sides. The area of the parallelogram is

A. $\sqrt{14}$ B. $\sqrt{5}$ C. $\frac{\sqrt{14}}{2}$ D. $\frac{\sqrt{5}}{2}$.

6. If $\mathbf{u}=\mathbf{i}+2\mathbf{j}+\mathbf{k}$, $\mathbf{v}=3\mathbf{i}-2\mathbf{j}+\mathbf{k}$, and $\mathbf{w}=-3\mathbf{i}-4\mathbf{j}+\mathbf{k}$, find $\mathbf{u}\cdot(\mathbf{w}\times\mathbf{v})$.

A. 28 B. -28 C. 4 D. 0

7. If $\mathbf{u}=\langle 1,2,3\rangle$, $\mathbf{v}=\langle -1,-7,-2\rangle$, and $\mathbf{w}=\langle 4,1,2\rangle$, find $\mathbf{u}\cdot(\mathbf{w}\times\mathbf{v})$.

A. 57 B. -57 C. -81 D. 81

8. Answer true or false: If $\mathbf{u}\cdot(\mathbf{v}\times\mathbf{w})=8$, $\mathbf{u}\cdot(\mathbf{w}\times\mathbf{v})=-6$.

9. If $\mathbf{u}=\mathbf{i}+\mathbf{j}+\mathbf{k}$ and $\mathbf{v}=-4\mathbf{i}-3\mathbf{j}$, the area of the parallelogram that has $\mathbf{u}$ and $\mathbf{v}$ as adjacent sides is

A. 6 B. 74 C. $\sqrt{26}$ D. 14.

10. Let $\mathbf{u}=-\mathbf{i}-\mathbf{j}+\mathbf{k}$ and $\mathbf{v}=-4\mathbf{i}+3\mathbf{j}$, the area of the parallelogram that has $\mathbf{u}$ and $\mathbf{v}$ as adjacent sides is

A. 6 B. 74 C. $\sqrt{74}$ D. 14.

11. Calculate the triple scalar product of $\mathbf{u}=2\mathbf{i}+3\mathbf{j}+4\mathbf{k}$, $\mathbf{v}=\mathbf{i}-\mathbf{j}+\mathbf{k}$, and $\mathbf{w}=3\mathbf{i}-\mathbf{j}+5\mathbf{k}$.

A. 6 B. 28 C. -28 D. -6

12. Answer true or false: The volume of the parallelpiped that has $\mathbf{u}$, $\mathbf{v}$, and $\mathbf{w}$ as adjacent edges, where $\mathbf{u}=-2\mathbf{i}-\mathbf{j}+\mathbf{k}$, $\mathbf{v}=\mathbf{i}-4\mathbf{j}+3\mathbf{k}$, and $\mathbf{w}=2\mathbf{i}+2\mathbf{j}+2\mathbf{k}$, is 42.

13. Answer true or false: $\mathbf{u}=\langle 1,1,1\rangle$, $\mathbf{v}=\langle 1,0,2\rangle$, and $\mathbf{w}=\langle 3,4,9\rangle$. $\mathbf{u}$, $\mathbf{v}$, and $\mathbf{w}$ lie in the same plane.

14. Answer true or false: $\mathbf{u}=\langle 1,0,0\rangle$, $\mathbf{v}=\langle 0,3,4\rangle$, and $\mathbf{w}=\langle 1,6,8\rangle$ lie in the same plane.

15. Answer true or false: A force of 50 N acts in the positive z-direction at a point $(2,1,3)$. If the object is free to rotate about the point $(0,0,0)$, the scalar moment about $(0,0,0)$ is $50\sqrt{5}$ N·m.

SECTION 12.5

1. Answer true or false: The parametric equations for the line joining $P_1(3,5)$ and $P_2(1,1)$ are $x = 3 - 2t,\ y = 5 - 4t$.

2. Answer true or false: The parametric equations of the line passing through $(-2,1,6)$ and parallel to $\mathbf{v} = \langle 2,3,5\rangle$ are $x = -2 + 2t,\ y = 1 + 3t,\ z = 6 + 5t$.

3. Let $L_1 : x = 5 - 2t,\ y = 2 + t,\ z = 3 + t$; $L_2 : x = -7 + 2t,\ y = 8 - t,\ z = 9 - t$. These lines intersect at

A. $(5,2,3)$ B. $(-1,5,3)$ C. $(2,2,4)$ D. $(0,0,0)$.

4. Find the parametric equations for the line whose vector is given by $\langle x,y\rangle = \langle 7,0\rangle + t\langle 2,-3\rangle$.

A. $x = 7,\ y = 2t - 3$ B. $x = 7 + 2t,\ y = 3$
C. $x = 7 + 2t,\ y = -3t$ D. $x = \frac{2t}{7},\ y = 0$

5. Find the parametric equations for the line whose vector is given by $\langle x,y,z\rangle = \langle -2,1,3\rangle + t\langle 3,2,5\rangle$.

A. $x = -2 + 3t,\ y = 1 + 2t,\ z = 3 + 5t$ B. $x = 2 + 3t,\ y = -1 + 2t,\ z = -3 + 5t$
C. $x = -2 - 3t,\ y = 1 - 2t,\ z = 3 - 5t$ D. $x = 2 - 3t,\ y = -1 - 2t,\ z = -3 - 5t$

6. Express $x = 5 + 3t,\ y = -3 + 4t$ in bracket notation.

A. $\langle x,y\rangle = \langle 5,3\rangle + t\langle 3,4\rangle$ B. $\langle x,y\rangle = t\langle 7,1\rangle$
C. $\langle x,y\rangle = t\langle -3,4\rangle$ D. $\langle x,y\rangle = \langle 5,3\rangle + t\langle -3,4\rangle$

7. Lines $x = -3 + 5t,\ y = 6 + 3t,\ z = 8 - 2t$ and $x = -5 + 5t,\ y = 2 + 3t,\ z = 7 - 2t$ are

A. intersecting at one single point B. skew
C. parallel D. perpendicular.

8. Lines $x = -2t,\ y = 9 + 2t,\ z = 5 - 6t$ and $x = 2t,\ y = 9 + 10t,\ z = 5 - 14t$ are

A. intersecting at a single point B. parallel
C. skew D. perpendicular.

9. The lines $x = -2t,\ y = -2t,\ z = -2t$ and $x = 4t,\ y = 4t,\ z = 4t$ are

A. parallel B. perpendicular C. the same line D. skew.

10. The lines $x = 5 - t,\ y = 1 + 2t$ and $x = 12 + t,\ y = -3 - 2t$ are

A. parallel B. skew C. the same line D. perpendicular.

11. Where does the line $x = 4 - 2t,\ y = 6 + 3t,\ z = 4 - 2t$ intersect the yz-plane?

A. $(0,12,0)$ B. $(2,9,2)$ C. $(4,6,0)$ D. $(4,6,4)$

12. Where does the line $x = 6 - 6t,\ y = 5 + 2t,\ z = 2 - 8t$ intersect the yz-plane?

A. $(21,0,22)$ B. $(0,7,-6)$ C. $(6,5,2)$ D. $(-3,1,-4)$

13. Where does the line $x = 5 - 4t,\ y = 7 + 3t,\ z = 2 + t$ intersect the plane parallel to the xy plane that includes the point $(0,0,1)$?

A. $(9,4,1)$ B. $(6,8,3)$ C. $(5,7,2)$ D. $(-3,4,2)$

14. Where does the line $x = 6 - 6t$, $y = 3 - 3t$ intersect $x = 0$, $y = 1 - t$?

A. $(0, 0)$ B. $(4, 11)$ C. $(3, 14)$ D. Does not exist.

15. How far are the vector $\langle x, y, z\rangle = t\langle 1, 2, 4\rangle$ and $\langle x, y, z\rangle = \langle 0, 3, 4\rangle + t\langle 1, 2, 4\rangle$ apart?

A. 0 B. 7 C. 5 D. 25

SECTION 12.6

1. The equation of the plane that passes through $P(8, 1, 3)$ and has $\mathbf{n} = \langle 1, 5, -2\rangle$ as a normal vector is

A. $(x + 8) + 5(y + 1) - 2(z + 3) = 0$ B. $(x - 8) + 5(y - 1) - 2(z - 3) = 0$
C. $(x + 8) + (5y + 1) - (2z + 3) = 0$ D. $(x - 8) + (5y - 1) - (2z - 3) = 0$.

2. The equation of the plane that passes through $P(-5, -3, -1)$ and has $\mathbf{n} = \langle 3, 1, 2\rangle$ as a normal vector is

A. $3(x - 5) + (y - 3) + 2(x - 1) = 0$ B. $(3x - 5) + (y - 3) + (2x - 1) = 0$
C. $(3x + 5) + (y + 3) + (2x + 1) = 0$ D. $3(x + 5) + (y + 3) + 2(x + 1) = 0$.

3. Find an equation of the plane that passes through $P_1(2, 7, 1)$, $P_2(-1, -1, -3)$, and $P_3(5, 2, 7)$.

A. $66(x - 2) - 6(y - 7) + 9(z - 1) = 0$ B. $-66(x - 2) + 6(y - 2) - 9(z - 1) = 0$
C. $-66(x + 2) - 6(y + 7) + 9(z + 1) = 0$ D. $66(x + 2) + 6(y + 7) - 9(z + 1) = 0$

4. Answer true or false: The planes $x - 2y + z = 5$ and $4x - 8y + 4z = 5$ are parallel.

5. Answer true or false: The planes $x - y + 3z = 6$ and $-4x + 4y + 3z = 7$ are parallel.

6. Answer true or false: The planes $x + 2y + z = 5$ and $\frac{7}{4}x + \frac{7}{2}y + \frac{7}{4}z = 1$ are parallel.

7. Answer true or false: The line $x = 9 + t$, $y = 7 - t$, $z = 3 - 3t$ is parallel to the plane $x - 2y + z = 5$.

8. Answer true or false: The line $x = 6 - t$, $y = 1 + 3t$, $z = 9 + 5t$ is parallel to the plane $x + y + z = 8$.

9. Find the distance between the point $(1, 2, 2)$ and $2x + y + 2z + 1 = 0$.

A. -3 B. -9 C. 3 D. 9

10. Find the distance between the point (0,3,4) and $2x + 3y + 6z + 10 = 0$.

A. 1 B. $\frac{1}{7}$ C. 7 D. $\frac{31}{7}$

11. Determine whether the planes $2x + 4y - 2z = 4$ and $5x + 10y - 5z = 2$ are parallel, perpendicular, or neither.

A. parallel B. perpendicular C. neither

12. Determine whether the planes $3x + 3y - 3z = 2$ and $x + y - 2z = 3$ are parallel, perpendicular, or neither.

A. parallel B. perpendicular C. neither

13. Find the acute angle of intersection of $2x + 3y - z = 5$ and $x + 3y + 4z = 2$. (Round answer to nearest degree.)

A. $60°$ B. $63°$ C. $68°$ D. $71°$

14. Answer true or false: The equation of the plane passing through the point $(1, 2, 7)$ and perpendicular to $\mathbf{n} = \langle 5, 2, 8 \rangle$ is $\langle 5, 2, 8 \rangle \cdot \langle x - 1, y - 2, z - 7 \rangle = 0$.

15. Answer true or false: The equation of the plane passing through the point $(5, 2, 3)$ and perpendicular to $\mathbf{n} = \langle 3, 3, 4 \rangle$ is $\langle 3, 3, 4 \rangle \cdot \langle x + 5, y + 2, z + 3 \rangle = 0$.

SECTION 12.7

1. Identify the quadratic surface defined by $x = \dfrac{y^2}{3} + \dfrac{z^2}{7}$.

A. Ellipsoid B. Elliptic cone
C. Elliptic paraboloid D. Hyperbolic paraboloid

2. Identify the quadratic surface defined by $4x^2 + 4y^2 - z^2 = 1$.

A. Sphere B. Ellipsoid
C. Hyperboloid of one sheet D. Hyperboloid of two sheets

3. Identify the quadratic surface defined by $z^2 - 2x^2 - 5y^2 = 0$.

A. Ellipsoid B. Hyperboloid of one sheet
C. Hyperboloid of two sheets D. Elliptic cone

4. Identify the quadratic surface defined by $x^2 + 3y^2 + 8z^2 = 1$.

A. Ellipsoid B. Hyperboloid of one sheet
C. Hyperboloid of two sheets D. Elliptic cone

5. Identify the quadratic surface defined by $z^2 - 2x^2 - 5y^2 = 0$.

A. Ellipsoid B. Hyperboloid of one sheet
C. Elliptic cone D. Elliptic paraboloid

6. Identify the trace of the surface $2x^2 + 5y^2 - z^2 = 5$ where $x = 1$.

A. Circle B. Ellipse C. Parabola D. Hyperbola

7. Identify the trace of the surface $3x^2 + 5y^2 + 5z^2 = 10$ where $x = 0$.

A. Circle B. Ellipse C. Parabola D. Hyperbola

8. Identify the trace of the surface $z = x^2 - 3y^2$ where $x = 5$.

A. Circle B. Ellipse C. Parabola D. Hyperbola

9. Identify the trace of the surface $z = 3x^2 + 2y^2$ where $y = 2$.

A. Circle B. Ellipse C. Parabola D. Hyperbola

10. Identify the trace of the surface $y = x^2 - 2z^2$ where $y = 7$.

A. Circle B. Ellipse C. Parabola D. Hyperbola

11. Identify the trace of the surface $y = 4x^2 - z^2$ where $z = 1$.

A. Circle B. Ellipse C. Parabola D. Hyperbola

12. Identify the trace of the surface $4x^2 + 9y^2 - 7z^2 = 1$ where $z = 0$.

A. Circle B. Ellipse C. Parabola D. Hyperbola

13. Identify the trace of the surface $2x^2 + 3y - z^2 = 5$ where $x = 1$.

A. Circle B. Ellipse C. Parabola D. Hyperbola

14. Identify the trace of the surface $6x^2 + 6y^2 + 3z^2 = 25$ where $z = 0$.

A. Circle B. Ellipse C. Parabola D. Hyperbola

15. Identify the trace of the surface $4x^2 - 4y^2 - 4z^2 = 0$ where $x = 2$.

A. Circle B. Ellipse C. Parabola D. Hyperbola

SECTION 12.8

1. Convert $(6, 8, 9)$ from rectangular coordinates to cylindrical coordinates.

A. $(10, 0.927, 9)$ B. $(10, 0.644, 9)$ C. $(100, 0.927, 9)$ D. $(100, 0.644, 9)$

2. Convert $(6, 3, 6)$ from rectangular coordinates to spherical coordinates.

A. $(81, 0.464, 0.841)$ B. $(9, 0.464, 0.841)$ C. $(81, 1.107, 0.730)$ D. $(9, 1.107, 0.730)$

3. Convert $(5, \pi/2, \pi/2)$ from spherical coordinates to rectangular coordinates.

A. $(0, 0, 5)$ B. $(5, 0, 0)$ C. $(0, 5, 0)$ D. $(5, 5, 5)$

4. Convert $(4, \pi/4, \pi/6)$ from spherical coordinates to rectangular coordinates.

A. $(\sqrt{2}, \sqrt{2}, 2\sqrt{3})$ B. $(4\sqrt{2}, 4\sqrt{2}, 8\sqrt{3})$
C. $(\sqrt{10}, \sqrt{10}, 2\sqrt{15})$ D. $(\sqrt{10}, \sqrt{10}, 2\sqrt{10})$

5. Answer true or false: There is no way to convert from spherical coordinates directly to cylindrical coordinates.

6. Convert the equation $\rho = 5$ from spherical coordinates to cylindrical coordinates.

A. $z^2 = 25 - r^2$ B. $5z^2 = 1 - 5r^2$ C. $z^2 = 5 - 5r^2$ D. $5z^2 = 1 - r^2$

7. Convert the equation $\rho = 3$ from spherical coordinates to rectangular coordinates.

A. $x^2 + y^2 + z^2 = 9$ B. $x^2 + y^2 + z^2 = 3$
C. $9x^2 + 9y^2 + 9z^2 = 1$ D. $9x^2 + 9y^2 + 9z^2 = 9$

8. Convert the equation $9z = x^2 + y^2$ from rectangular coordinates to cylindrical coordinates.

A. $9z = r^2$ B. $z = 9r^2$ C. $3z = r$ D. $z = 3r$

9. Answer true or false: (3,0,0) in rectangular coordinates and $(3, 0, 0)$ in cylindrical coordinates identify the same point.

10. Answer true or false: $(0, 2, 0)$ in rectangular coordinates and $(0, 2, 0)$ in cylindrical coordinates identify the same point.

11. Answer true or false: The equation in rectangular coordinates, $z = 5x^2 + 5y^2$, converts to the equation $z = 5r^2$ in cylindrical coordinates.

12. Answer true or false: The equation $z = 8\rho\cos\phi$ in spherical coordinates converts to $\cos\phi = \dfrac{8z}{\sqrt{x^2 + y^2 + z^2}}$ in rectangular coordinates.

13. Answer true or false: The equation $z = 12$ in cylindrical coordinates converts to $z = 12$ in rectangular coordinates.

14. Answer true or false: $z = \sqrt{6x^2 + 6y^2}$ in rectangular coordinates converts to $z = \sqrt{6}r$ in cylindrical coordinates.

15. Answer true or false: The equation $\rho = 3$ in spherical coordinates converts to $3z = 3\sqrt{x^2 + y^2}$ in rectangular coordinates.

CHAPTER 12 TEST

1. Find the distance from $(5, 3, -6)$ to the xy-plane.

A. 3 B. 5 C. 6 D. $2\sqrt{15}$

2. The surface described by $x^2 + y^2 + z^2 = 10$ is a(n)

A. sphere B. cylinder C. cone D. ellipsoid.

3. Answer true or false: $(x + 3)^2 + (y - 2)^2 + (z - 3)^2 = 6$ describes a sphere centered at $(-3, 2, 3)$ with radius $\sqrt{6}$.

4. Find the norm of $\mathbf{u} = 2\mathbf{i} + 4\mathbf{j} + 3\mathbf{k}$.

A. 29 B. $\sqrt{29}$ C. 9 D. 3

5. If $\mathbf{u} = 4\mathbf{i} + 5\mathbf{j} - 2\mathbf{k}$ and $\mathbf{v} = \mathbf{i} - \mathbf{j} + 2\mathbf{k}$, $\mathbf{u} - \mathbf{v} =$

A. $3\mathbf{i} + 5\mathbf{j} - 2\mathbf{k}$ B. $3\mathbf{i} + 4\mathbf{j}$ C. $3\mathbf{i} + 6\mathbf{j} - 4\mathbf{k}$ D. $3\mathbf{i} + 6\mathbf{j}$.

6. Let $\mathbf{u} = \langle 1, 3 \rangle$ and $\mathbf{v} = \langle 5, 9 \rangle$. Find $\mathbf{x}$ that satisfies $2\mathbf{u} + 2\mathbf{x} = 4\mathbf{v} - \mathbf{x}$.

A. $\langle 6, 10 \rangle$ B. $\langle 12, 20 \rangle$ C. $\langle 18, 30 \rangle$ D. $\langle 3, 5 \rangle$

7. Answer true or false: If $\|\mathbf{v}\| = 8$ and ϕ, the angle the vector makes with the positive x-axis is $45°$, then $\mathbf{v} = \langle 4\sqrt{2}, 4\sqrt{2} \rangle$.

8. Find the dot product $\langle 3, 8 \rangle \cdot \langle 2, 5 \rangle$.

A. 18 B. 46 C. 31 D. 77

9. Find the dot product $\langle 2, 4 \rangle \cdot \langle 3, 5 \rangle$.

A. 26 B. 32 C. 2 D. 48

10. Let $\mathbf{u} = \langle 5, 2 \rangle$, $\mathbf{v} = \langle 1, 3 \rangle$, and $\mathbf{w} = \langle 1, 2 \rangle$. Find $(\mathbf{u} \cdot \mathbf{v}) + (\mathbf{u} \cdot \mathbf{w})$.

A. 26 B. 22 C. 20 D. 48

11. Answer true or false: Let $\mathbf{u} = 5\mathbf{i} + 3\mathbf{j} + 4\mathbf{k}$. The direction cosines are $\cos\alpha = \frac{5}{12}$, $\cos\beta = \frac{1}{4}$, and $\cos\gamma = \frac{1}{3}$.

12. If $\mathbf{u} = 2\mathbf{i} + 3\mathbf{j} + 4\mathbf{k}$ and $\mathbf{v} = \mathbf{i} - 3\mathbf{j} + 2\mathbf{k}$, $\mathbf{u} \times \mathbf{v} =$

A. $18\mathbf{i} - 9\mathbf{k}$ B. $18\mathbf{i} + 4\mathbf{j} - 9\mathbf{k}$ C. $18\mathbf{i} - 4\mathbf{j} - 9\mathbf{k}$ D. $12\mathbf{i} + 8\mathbf{j} - 3\mathbf{k}$.

13. If $\mathbf{a} = -2\mathbf{j}$ and $\mathbf{b} = -4\mathbf{i} - \mathbf{k}$, find $\mathbf{a} \times \mathbf{b}$.

A. $\mathbf{0}$ B. $2\mathbf{i} + 8\mathbf{k}$ C. $-2\mathbf{i} - 8\mathbf{k}$ D. $2\mathbf{i} - 8\mathbf{k}$

14. A parallelogram has $\mathbf{u} = 2\mathbf{j} + \mathbf{k}$ and $\mathbf{v} = \mathbf{i} + 3\mathbf{j}$ as adjacent sides. The area of the parallelogram is

A. $\sqrt{14}$ B. $\sqrt{5}$ C. $\frac{\sqrt{14}}{2}$ D. $\frac{\sqrt{5}}{2}$.

15. If $\mathbf{u} = \langle 2, 3, 2 \rangle$, $\mathbf{v} = \langle 3, -2, 1 \rangle$, and $\mathbf{w} = \langle 3, 4, -1 \rangle$, find $\mathbf{u} \cdot (\mathbf{v} \times \mathbf{w})$.

A. $\langle 50 \rangle$ B. $\langle 14 \rangle$ C. $\langle 24 \rangle$ D. $\langle 12 \rangle$

16. Answer true or false: The volume of the parallelpiped that has $\mathbf{u}$, $\mathbf{v}$, and $\mathbf{w}$ as adjacent edges, where $\mathbf{u} = -\mathbf{i} - \mathbf{j} - \mathbf{k}$, $\mathbf{v} = 2\mathbf{i} + \mathbf{j} - \mathbf{k}$, and $\mathbf{w} = \mathbf{i} - 4\mathbf{j} + 3\mathbf{k}$ is 21.

17. Answer true or false: The parametric equations to the line passing through $(-1, 3, 6)$ and parallel to $\mathbf{v} = \langle 2, 4, 8 \rangle$ are $x = 2 - t$, $y = 4 + 3t$, $z = 8 + 6t$.

18. Answer true or false: The parametric equations for the line whose vector are given by $\langle x, y, z \rangle = \langle 2, 1, 3 \rangle + t\langle 1, 4, 7 \rangle$ is $x = 1 + 2t$, $y = 4 + t$, $z = 7 + 3t$.

19. The lines $\langle x, y, z \rangle = \langle 1, 4, 7 \rangle + t\langle 1, 1, 1 \rangle$ and $\langle x, y, z \rangle = \langle 3, 8, 1 \rangle + t\langle 1, 1, 1 \rangle$ are

A. skew B. perpendicular C. parallel D. The same line.

20. The equation for the plane that passes through $P(2, 1, 4)$ and has $\mathbf{n} = \langle 4, 1, 5 \rangle$ as a normal vector is

A. $(4x - 2) + (y - 1) + (5z - 4) = 0$ B. $(4x + 2) + (y + 1) + (5z + 4) = 0$
C. $4(x - 2) + (y - 1) + 5(z - 4) = 0$ D. $4(x + 2) + (y + 2) + 5(z + 4) = 0$.

21. Find the equation of the plane that passes through $P_1(0, 0, 0)$, $P_2(-2, -1, -3)$, and $P_3(-5, -2, -4)$.

A. $2x - 7y + z = 0$ B. $(x - 2) + (y + 7) + (z - 1) = 0$
C. $-2x + 7y - z = 0$ D. $(x + 2) + (y - 7) + (z + 1) = 0$

22. Answer true or false: The planes $x + 3y - 2z = 6$ and $-4x - 12y - 8z = 1$ are parallel.

23. Identify the quadratic surface defined by $x^2 - 2y^2 + 4z^2 = 1$.

A. Sphere B. Ellipsoid
C. Hyperboloid of one sheet D. Hyperboloid of two sheets

24. Identify the trace of the surface $x^2 - 2y^2 + 3z^2 = 1$ where $y = 1$.

A. Circle B. Ellipse C. Parabola D. Hyperbola

25. Convert $(4, 8, 8)$ from rectangular coordinates to spherical coordinates.

A. $(144, 0.469, 0.841)$ B. $(12, 1.107, 0.841)$ C. $(144, 1.107, 0.730)$ D. $(12, 0.464, 0.730)$

26. Convert the equation $\sqrt{x^2 + y^2 + z^2} = 25$ from rectangular coordinates to spherical coordinates.

A. $\rho = 50$ B. $\rho = 5$ C. $\rho = 10$ D. $\rho = 25$

ANSWERS TO SAMPLE TESTS

SECTION 12.1

1. T 2. T 3. B 4. D 5. A 6. C 7. T 8. B 9. A 10. B 11. F 12. C
13. F 14. B 15. F

SECTION 12.2

1. A 2. C 3. C 4. C 5. T 6. F 7. A 8. D 9. D 10. A 11. T 12. F
13. T 14. A 15. A

SECTION 12.3

1. B 2. A 3. C 4. T 5. D 6. T 7. B 8. B 9. B 10. T 11. F 12. F
13. T 14. F 15. B

SECTION 12.4

1. A 2. A 3. A 4. D 5. A 6. A 7. B 8. T 9. C 10. C 11. D 12. F
13. F 14. T 15. T

SECTION 12.5

1. T 2. T 3. B 4. C 5. A 6. D 7. C 8. A 9. C 10. A 11. A 12. B
13. A 14. A 15. C

SECTION 12.6

1. B 2. D 3. B 4. T 5. F 6. T 7. T 8. F 9. C 10. C 11. A 12. C
13. C 14. T 15. F

SECTION 12.7

1. C 2. C 3. D 4. A 5. C 6. D 7. A 8. C 9. C 10. D 11. C 12. B
13. C 14. A 15. A

SECTION 12.8

1. A 2. B 3. C 4. A 5. F 6. A 7. A 8. A 9. T 10. F 11. T 12. F
13. T 14. T 15. F

CHAPTER 12 TEST

1. C 2. A 3. T 4. B 5. C 6. A 7. T 8. B 9. A 10. C 11. F 12. A
13. D 14. A 15. A 16. F 17. F 18. F 19. C 20. C 21. A 22. T 23. C
24. C 25. B 26. D

CHAPTER 13
Sample Exams

SECTION 13.1

1. Find the domain of $\mathbf{r}(t) = (5 + \cos t)\mathbf{i} - 2t\mathbf{j}$; $t_0 = 0$.

A. $0 \le t < \infty$ B. $-\infty < t < \infty$ C. $0 \le t \le 2\pi$ D. $-\pi \le t \le \pi$

2. Find the domain of $\mathbf{r}(t) = \sqrt{t-4}\mathbf{i} + t^2\mathbf{j} - 3t\mathbf{k}$; $t_0 = 5$.

A. $4 \le t < \infty$ B. $0 \le t < \infty$ C. $5 \le t < \infty$ D. $-4 \le t < \infty$

3. Find the domain of $\mathbf{r}(t) = \langle t^2, t-2, \sqrt{t+1}\,\rangle$; $t_0 = 5$.

A. $0 \le t < \infty$ B. $1 \le t < \infty$ C. $-1 \le t < \infty$ D. $-\infty < t < \infty$

4. Answer true or false: $\mathbf{r}(t) = t^2\mathbf{i} + t^3\mathbf{j}$ can be expressed as a parametric equation by $x = \sin t$, $y = \cos t$.

5. Answer true or false: $\mathbf{r}(t) = \cos t\mathbf{i} - \sin t\mathbf{j}$ can be expressed as a parametric equation by $x^2 - y^2 = t$.

6. Answer true or false: The parametric equation $x = t^2$, $y = t$ can be expressed by the single vector equation $\mathbf{r}(t) = t^3\mathbf{i} + t^3\mathbf{j}$.

7. Answer true or false: The parametric equation $x = \sin t$, $y = 8t$, $z = t$ can be expressed by the single vector equation $\mathbf{r}(t) = \sin t\mathbf{i} + 8t\mathbf{j} + t\mathbf{k}$.

8. Describe the graph of $\mathbf{r}(t) = 6t\mathbf{i} + 2t\mathbf{j} + t\mathbf{k}$.

A. Twisted cubic B. Straight line C. Spiral D. Parabola

9. Describe the graph of $\mathbf{r}(t) = 4\mathbf{i} + 2\cos t\mathbf{j} + 2\sin t\mathbf{k}$.

A. Straight line B. Spiral C. Parabola D. Circle

10. Describe the graph of $\mathbf{r}(t) = -2\mathbf{i} + \sin t\mathbf{j} - \cos t\mathbf{k}$.

A. Straight line B. Spiral C. Parabola D. Circle

11. Describe the graph of $\mathbf{r}(t) = 4t\mathbf{i} + 8\sin t\mathbf{j} + 8\cos t\mathbf{k}$.

A. Straight line B. Spiral C. Parabola D. Circle

12. Describe the graph of $\mathbf{r}(t) = t^3\mathbf{i} + t^2\mathbf{j} + t\mathbf{k}$.

A. Cubic B. Twisted cubic C. Spiral D. Parabola

13. As t increases, the graph of $\mathbf{r}(t) + \langle 4\sin t, \cos t, 3t\rangle$ sketches

A. Clockwise and up B. Counter-clockwise and up
C. Clockwise and down D. Counter-clockwise and down.

14. As t increases, the graph of $\mathbf{r}(t) = \langle \cos t, 5\sin t, -3t\rangle$ sketches

A. Clockwise and up B. Counter-clockwise and up
C. Clockwise and down D. Counter-clockwise and down.

15. As t increases, the graph of $\mathbf{r}(t) = \langle 2\cos t, 9\sin t, 4t\rangle$ sketches

A. Clockwise and up
B. Counter-clockwise and up
C. Clockwise and down
D. Counter-clockwise and down.

SECTION 13.2

1. If $\mathbf{r}(t) = (6 - 2t)\mathbf{i} + (t^2 - 3)\mathbf{j}$, find $\mathbf{r}'(t)$.

A. $t^2\mathbf{i} + \dfrac{t^3}{3}\mathbf{j}$
B. $(6t - t^2)\mathbf{i} + \left(\dfrac{t^3}{3} - 3\right)\mathbf{j}$
C. $-2\mathbf{i} + 2t\mathbf{j}$
D. $-3t$

2. If $\mathbf{r}(t) = 4t\,\mathbf{i} - 3t\mathbf{j} + \cos t\mathbf{k}$, find $\mathbf{r}'(t)$.

A. $4\mathbf{i} - 3\mathbf{j} - \sin t\mathbf{k}$
B. $4\mathbf{i} - 3\mathbf{j} + \sin t\mathbf{k}$
C. $-7\sin t$
D. $7\sin t$

3. Find $\mathbf{r}'(\pi/2)$ if $\mathbf{r}(t) = \cos t\,\mathbf{i} + \sin t\,\mathbf{j} + \mathbf{k}$.

A. $\mathbf{i}$ B. $-\mathbf{i}$ C. $-\mathbf{j}$ D. $\mathbf{j}$

4. Find $\mathbf{r}'(0)$ if $\mathbf{r}(t) = 7t^4\,\mathbf{i} + 2t^3\,\mathbf{j} + t\mathbf{k}$.

A. $15\mathbf{i} + 6\mathbf{j} + \mathbf{k}$ B. $\mathbf{0}$ C. $\mathbf{k}$ D. $5\mathbf{i} + 2\mathbf{j} + \mathbf{k}$

5. $\lim\limits_{t\to 2} 4t^2\mathbf{i} + 2t\,\mathbf{j} =$

A. 20 B. $16\mathbf{i} + 4\mathbf{j}$ C. Not defined D. 2

6. $\lim\limits_{t\to \pi} \langle 2\sin t,\ 3\cos t,\ t\rangle =$

A. $\langle 0, -3, \pi\rangle$ B. $\langle 0, -3, 0\rangle$ C. $-\pi$ D. π

7. Answer true or false: $\mathbf{r}(t) = \sin t\mathbf{i} + 5\cos t\mathbf{j}$ is continuous at $t = 0$.

8. Answer true or false: $\mathbf{r}(t) = 4\ln t\mathbf{i} + \cos t\mathbf{j} - 4\ln t\mathbf{k}$ is continuous at $t = 0$.

9. $\mathbf{r}(t) = t^3\mathbf{i} + 2t^2\mathbf{j} + t\mathbf{k}$. Find $\mathbf{r}''(t)$.

A. $3t\mathbf{i} - 4t\mathbf{j} + \mathbf{k}$ B. $6t\mathbf{i} - 4t\mathbf{j} + \mathbf{k}$ C. $6t\mathbf{i} + 4\mathbf{j}$ D. $6t\mathbf{i} + 4t\mathbf{j} + \mathbf{k}$

10. $\displaystyle\int (4t\mathbf{i} + 6\mathbf{j})\,dt =$

A. $2t^2\mathbf{i} + 6t\mathbf{j} + C$
B. $(2t^2 + C)\mathbf{i} + (6t + C)\mathbf{j}$
C. $(2t^2 + C_1)\mathbf{i} + (6t + C_2)\mathbf{j}$
D. $2t^2 + 6t + C$

11. $\displaystyle\int_0^{\pi/2} \langle \cos t, \sin t\rangle\, dt =$

A. $\langle -1, -1\rangle$ B. $\langle -1, 1\rangle$ C. $\langle 1, 1\rangle$ D. $\langle 1, -1\rangle$

12. $\int_0^3 \langle t, t^2, 2\rangle\, dt =$

A. $\langle 9/2, 9, 6\rangle$ B. $\langle 9/2, 9, 2\rangle$ C. $\langle 9, 9, 6\rangle$ D. $\langle 3, 9, 2\rangle$

13. Answer true or false: If $\mathbf{r}(t) = t^3\mathbf{i} + 2t\mathbf{j}$, the tangent line at $t_0 = 4$ is given by $\mathbf{r} = 3t^2\mathbf{i} + 2\mathbf{j}$.

14. If $y'(t) = 8t\mathbf{i} + 3t^2\mathbf{j}$, $y(0) = \mathbf{i} + 2\mathbf{j}$, find $y(t)$.

A. $(12t^2 + 1)\mathbf{i} + (12t^3 + 2)\mathbf{j}$
B. $(4t^2 + 1)\mathbf{i} + (t^3 + 2)\mathbf{j}$
C. $4t^2\mathbf{i} + 3t^3\mathbf{j}$
D. $12t^2\mathbf{i} + 12t^3\mathbf{j}$

15. If $y'(t) = 9t^2\mathbf{i} + 2t\mathbf{j}$, and $y(1) = \mathbf{i} + \mathbf{j}$, find $y(t)$.

A. $(3t^3 - 2)\mathbf{i} + t^2\mathbf{j}$
B. $(3t^3 + 2)\mathbf{i} + t^2\mathbf{j}$
C. $(3t^3 + 1)\mathbf{i} + (t^2 + 1)\mathbf{j}$
D. $(3t^3 + 4)\mathbf{i} + t^2\mathbf{j}$

SECTION 13.3

1. Answer true or false: $\mathbf{r}(t) = 3t^2\mathbf{i} + 2t^3\mathbf{j} + \sin t\mathbf{k}$ is a smooth function of the parameter t.

2. Answer true or false: $\mathbf{r}(t) = 4t^2\mathbf{i} - t^3 + \mathbf{j} + \sin(2t)\mathbf{k}$ is a smooth function of the parameter t.

3. Answer true or false: $\mathbf{r}(t) = \sqrt[5]{t}\,\mathbf{i} + 4t^2\mathbf{j} + 8t^3\mathbf{k}$ is a smooth function of the parameter t.

4. Find the arc length of the graph of $\mathbf{r}(t) = 2t\mathbf{i} + 4\mathbf{j} + 3\mathbf{k}$, $4 \le t \le 7$.

A. 6 B. -6 C. 21 D. -21

5. Find the arc length of the graph of $\mathbf{r}(t) = -\sin t\mathbf{i} - \cos t\mathbf{j} + 9\mathbf{k}$, $0 \le t \le \pi$.

A. 2 B. 2π C. π D. 0

6. Find the arc length of the graph of $\mathbf{r}(t) = 2e^t\mathbf{i} + 2e^t\mathbf{j} + e^t\mathbf{k}$, $0 \le t \le 1$.

A. $3e$
B. $5e - 5$
C. $3e - 3$
D. $(1 + 2\sqrt{2})e - 1 - 2\sqrt{2}$

7. Find the arc length of the parametric curve $x = 3e^t$, $y = 4e^t$, $z = 2$; $0 \le t \le 1$.

A. $3e$
B. $5e - 5$
C. $3e - 3$
D. $(1 + 2\sqrt{2})e - 1 - 2\sqrt{2}$

8. Find the arc length of the parametric curve $x = -\sin t$, $y = -7$, $z = -\cos t$; $0 \le t \le \pi$.

A. 2 B. 2π C. π D. 0

9. Find the arc length parametrization of the line $x = 6t + 2$, $y = 4t - 1$ that has the same orientation as the given line and uses $(2, -1)$ as a reference point.

A. $x = \frac{s}{2\sqrt{13}}$, $y = \frac{s}{\sqrt{13}}$
B. $x = \frac{3s}{\sqrt{13}}$, $y = \frac{2s}{\sqrt{13}}$
C. $x = \frac{s}{2\sqrt{13}} + 2$, $y = \frac{s}{\sqrt{13}} - 1$
D. $x = \frac{3s}{\sqrt{13}} + 2$, $y = \frac{2s}{\sqrt{13}} - 1$

10. Find the arc length parametrization of the line $x = 2\cos t + 3$, $y = 2\sin t - 4$ that has the same orientation as the given curve and uses $(6, -3)$ as a reference point.

A. $x = 2\cos\left(\frac{s}{2}\right) + 3$, $y = 2\sin\left(\frac{s}{2}\right) - 4$ B. $x = \cos s + 3$, $y = \sin s - 4$

C. $x = 2\cos\left(\frac{s}{2}\right) + \frac{3}{2}$, $y = 2\sin\left(\frac{s}{2}\right) - 2$ D. $x = \cos s + \frac{3}{2}$, $y = \sin s - 2$

11. Answer true or false: If $\mathbf{r} = 2t\mathbf{i} + (-4t + 3)\mathbf{j}$, the arc length paramentization of the curve relative to the reference point $(0, 3)$ involves the parameter $t = 2\sqrt{5}s$.

12. Answer true or false: If $\mathbf{r} = (2t + 2)\mathbf{i} + (6t + 2)\mathbf{j} + (4t - 1)\mathbf{k}$, the arc length paramentization of the curve relative to the reference point $(2, 2, -1)$ involves the parameter $t = \dfrac{s}{2\sqrt{19}}$.

13. Answer true or false: If $\mathbf{r} = 2\sin t\mathbf{i} - 2\cos t\mathbf{j} + t\mathbf{k}$, the arc length paramentization of the curve relative to the reference point $(1, 0, 0)$ involves the parameter $t = \dfrac{s}{\sqrt{5}}$.

14. Answer true or false: If $\mathbf{r} = (2t - 2)\mathbf{i} + (t - 3)\mathbf{j} + 4t\mathbf{k}$, the arc length paramentization of the curve relative to the reference point $(-2, -3, 0)$ involves the parameter $t = \dfrac{s}{\sqrt{21}}$.

15. Answer true or false: If $\mathbf{r} = (3t - 1)\mathbf{i} + (4t - 1)\mathbf{j} + (2t + 1)\mathbf{k}$, the arc length paramentization of the curve relative to the reference point $(-1, -1, 1)$ involves the parameter $t = \dfrac{s}{\sqrt{3}}$.

SECTION 13.4

1. $\mathbf{r}(t) = 4t^2\mathbf{i} + 8t\mathbf{j}$. Find $\mathbf{T}(t)$ for $t = 2$.

A. $\frac{2}{\sqrt{5}}\mathbf{i} + \frac{1}{\sqrt{5}}\mathbf{j}$ B. $\frac{1}{\sqrt{5}}\mathbf{k}$ C. $-\frac{1}{\sqrt{5}}\mathbf{k}$ D. $\mathbf{i}$

2. Answer true or false: $\mathbf{r}(t) = 3t^2\mathbf{i} + 6t\mathbf{j}$. $\mathbf{N}(t)$ for $t = 2$ is $0.49\mathbf{i} - 0.87\mathbf{j}$.

3. $\mathbf{r}(t) = 7t^2\mathbf{i} + 14t\mathbf{j}$. Find $\mathbf{B}(t)$ for $t = 2$.

A. $\mathbf{k}$ B. $-0.99\mathbf{k}$ C. $0.99\mathbf{k}$ D. $-0.55\mathbf{i}$

4. $\mathbf{r}(t) = 4(t^2 + 2)\mathbf{i} + 4e^t\mathbf{j} + 4e^t\mathbf{k}$. Find $\mathbf{T}(t)$ for $t = 0$.

A. $2\mathbf{i} + \frac{1}{\sqrt{2}}\mathbf{j} + \frac{1}{\sqrt{2}}\mathbf{k}$ B. $\frac{1}{\sqrt{2}}\mathbf{j} + \frac{1}{\sqrt{2}}\mathbf{k}$

C. $2\mathbf{i} + \frac{1}{\sqrt{6}}\mathbf{j} + \frac{1}{\sqrt{6}}\mathbf{k}$ D. $\frac{1}{\sqrt{3}}\mathbf{j} + \frac{1}{\sqrt{3}}\mathbf{k}$

5. $\mathbf{r}(t) = 4(t^2 + 2)\mathbf{i} + 4e^t\mathbf{j} + 4e^t\mathbf{k}$. Find $\mathbf{N}(t)$ for $t = 0$.

A. $\mathbf{0}$ B. $1.334\mathbf{i}$

C. $-2\mathbf{i}$ D. $1.334\mathbf{i} + 1.334\mathbf{j} + 1.334\mathbf{k}$

6. $\mathbf{r}(t) = 4(t^2 + 2)\mathbf{i} + 4e^t\mathbf{j} + 4e^t\mathbf{k}$. Find $\mathbf{B}(t)$ for $t = 0$.

A. $2\mathbf{i} + 0.707\mathbf{j} + 0.707\mathbf{k}$ B. $0.707\mathbf{j} - 0.707\mathbf{k}$

C. $2\mathbf{i} + 0.662\mathbf{j} + 0.662\mathbf{k}$ D. $0.662\mathbf{j} + 0.662\mathbf{k}$

7. $\mathbf{r}(t) = 6(t^2 + t)\mathbf{i} + 6t^2\mathbf{j} + 18t^2\mathbf{k}$. Find $\mathbf{T}(t)$ when $t = 0$.

A. $\mathbf{i}$ B. $\mathbf{j}$ C. $\mathbf{k}$ D. $\mathbf{0}$

8. $\mathbf{r}(t) = 6(t^2 + t)\mathbf{i} + 6t^2\mathbf{j} + 18t^2\mathbf{k}$. Find $\mathbf{N}(t)$ when $t = 0$.

A. $\mathbf{i}$ B. $\mathbf{i} - 0.022\mathbf{j} - 0.022\mathbf{k}$
C. $\mathbf{i} + 0.022\mathbf{j} + 0.022\mathbf{k}$ D. $\mathbf{0}$

9. $\mathbf{r}(t) = 6(t^2 + t)\mathbf{i} + 6t^2\mathbf{j} + 18t^2\mathbf{k}$. Find $\mathbf{B}(t)$ when $t = 0$.

A. $0.136\mathbf{i}$ B. $\mathbf{i} - 0.022\mathbf{j} - 0.022\mathbf{k}$
C. $\mathbf{i} + 0.95\mathbf{j} + 0.316\mathbf{k}$ D. $\mathbf{0}$

10. $\mathbf{r}(t) = 9e^t\mathbf{i} + 9e^{2t}\mathbf{j} + 9e^{3t}\mathbf{k}$. Find $\mathbf{T}(t)$ when $t = 0$.

A. $\frac{3}{7\sqrt{7}}\mathbf{i} - \frac{3}{7\sqrt{7}}\mathbf{j} + \frac{1}{7\sqrt{7}}\mathbf{k}$ B. $6\mathbf{i} - 6\mathbf{j} + 2\mathbf{k}$
C. $\frac{1}{\sqrt{14}}\mathbf{i} + \frac{2}{\sqrt{14}}\mathbf{j} + \frac{3}{\sqrt{14}}\mathbf{k}$ D. $\frac{1}{7\sqrt{2}}\mathbf{i} + \frac{4}{7\sqrt{2}}\mathbf{j} + \frac{9}{7\sqrt{2}}\mathbf{k}$

11. Answer true or false: $\mathbf{r}(t) = 9e^t\mathbf{i} + 9e^{2t}\mathbf{j} + 9e^{3t}\mathbf{k}$. When $t = 0$, $\mathbf{N}(t) = \frac{1}{7\sqrt{2}}\mathbf{i} + \frac{4}{7\sqrt{2}}\mathbf{j} + \frac{9}{7\sqrt{2}}\mathbf{k}$.

12. Answer true or false: $\mathbf{r}(t) = 9e^t\mathbf{i} + 9e^{2t}\mathbf{j} + 9e^{3t}\mathbf{k}$. When $t = 0$, $\mathbf{B}(t) = \frac{3}{7\sqrt{7}}\mathbf{i} - \frac{3}{7\sqrt{7}}\mathbf{j} + \frac{1}{7\sqrt{7}}\mathbf{k}$.

13. $\mathbf{r}(t) = 8t\mathbf{i} + 8t^2\mathbf{j} + 8t^3\mathbf{k}$. Find $\mathbf{T}(t)$ when $t = 0$.

A. $\mathbf{i}$ B. $\frac{1}{\sqrt{10}}\mathbf{j}$ C. $\frac{1}{\sqrt{10}}\mathbf{j} + \frac{3}{\sqrt{10}}\mathbf{k}$ D. $-\frac{1}{\sqrt{14}}\mathbf{k}$

14. Answer true or false: $\mathbf{r}(t) = 8t\mathbf{i} + 8t^2\mathbf{j} + 8t^3\mathbf{k}$. When $t = 0$, $\mathbf{N}(t)\ \frac{1}{\sqrt{10}}\mathbf{j}$.

15. Answer true or false: $\mathbf{r}(t) = 8t\mathbf{i} + 8t^2\mathbf{j} + 8t^3\mathbf{k}$. When $t = 0$, $\mathbf{B}(t) = -\frac{1}{2\sqrt{35}}\mathbf{k}$.

SECTION 13.5

1. Find the curviture $k(t)$ for $\mathbf{r}(t) = -\sin t\mathbf{i} - \cos t\mathbf{j}$.

A. 1 B. -1 C. 0 D. $\sin^2 t - \cos^2 t$

2. Find the curviture $k(t)$ for $\mathbf{r}(t) = -\cos t\mathbf{i} - \sin t\mathbf{j} - 6\mathbf{k}$.

A. $\sqrt{11}$ B. 1 C. $\sin^2 t - \cos^2 t$ D. $\sin^2 t$

3. Find the curviture $k(t)$ for $\mathbf{r}(t) = 2e^t\mathbf{i} + 2e^t\mathbf{j} + 6\mathbf{k}$.

A. 0 B. $\frac{2e^t}{\sqrt{2}}$ C. $\frac{3}{\sqrt{2}e^t}$ D. $\frac{1}{\sqrt{2}}$

4. Find the curviture $k(t)$ for $\mathbf{r}(t) = 6\mathbf{i} + 2t\mathbf{j} + 3t^2\mathbf{k}$ at $t = 0$.

A. $\frac{3}{40\sqrt{10}}$ B. $\frac{4}{3(1 + 9t)^{3/2}}$ C. 0 D. 1

5. Answer true or false: If $\mathbf{r}(t) = (t^3+2)\mathbf{i} + (t^4-5)\mathbf{j} + (t^5+1)\mathbf{k}$, the curvature $k(t)$ is $\sqrt{36t^2+144t^4+400t^6}$.

6. Answer true or false: If $\mathbf{r}(t) = t\mathbf{i} + t^3\mathbf{j} + t^4\mathbf{k}$, the curvature $k(t)$ at $t=1$ is $\dfrac{9}{13\sqrt{26}}$.

7. If $\mathbf{r}(t) = (2t^2-1)\mathbf{i} + (t-3)\mathbf{j} + (4t+8)\mathbf{k}$, find the curvature $k(t)$ at $t=1$.

A. $\dfrac{2\sqrt{17}}{\sqrt{33}}$ B. $\dfrac{4\sqrt{17}}{33\sqrt{33}}$ C. $\dfrac{34}{\sqrt{33}}$ D. $\dfrac{\sqrt{34}}{\sqrt{33}}$

8. If $\mathbf{r}(s) = 2\cos\left(\frac{s}{2}\right)\mathbf{i} + \left(8+2\cos\left(\frac{s}{2}\right)\right)\mathbf{j} + 3\mathbf{k}$, find $k(s)$.

A. $\dfrac{\sqrt{2}}{2}$ B. $\dfrac{1}{4}$ C. 2 D. 4

9. If $\mathbf{r}(s) = 6\mathbf{i} + 3\cos\left(\frac{s}{3}\right)\mathbf{j} + \left(2+3\cos\left(\frac{s}{3}\right)\right)\mathbf{k}$, find $k(s)$.

A. $\dfrac{1}{9}$ B. $\dfrac{\sqrt{2}}{3}$ C. 3 D. 9

10. If $y = -\cos x + 4$, find the curvature at $x = \dfrac{\pi}{2}$.

A. 0 B. 1 C. -1 D. $\dfrac{1}{2\sqrt{2}}$

11. If $y = -7 + \sin x$, find the curvature at $x = \dfrac{\pi}{2}$.

A. 0 B. 1 C. -1 D. $\dfrac{1}{2\sqrt{2}}$

12. If $x = t^3+8$, $y = t^2-1$, then $k(t)$ at $t=1$ is

A. $\dfrac{6}{13\sqrt{13}}$ B. $\dfrac{6}{\sqrt{13}}$ C. 0 D. $\dfrac{18}{13\sqrt{13}}$.

13. Answer true or false: The curve $y = 4x^3$ has a maximum curvature at $x=4$.

14. At what points does $12x^2 + 75y^2 = 300$ have maximum curvature?

A. $(0,-4)$, $(0,4)$ B. $(-4,0)$, $(4,0)$ C. $(-5,0)$, $(5,0)$ D. $(0,-5)$, $(0,5)$

15. At what points does $16x^2 + 100y^2 = 400$ have minimum curvature?

A. $(0,-4)$, $(0,4)$ B. $(-4,0)$, $(4,0)$ C. $(-5,0)$, $(5,0)$ D. $(0,-5)$, $(0,5)$

SECTION 13.6

1. $\mathbf{r}(t) = (4t^3+7)\mathbf{i} + (2t-3)\mathbf{j}$ is the position vector of a particle moving in a plane. Find the velocity.

A. $12t^2\mathbf{i} + 2\mathbf{j}$ B. $12\mathbf{i}$ C. $24t\mathbf{i} + 2\mathbf{j}$ D. $24t\mathbf{i}$

2. $\mathbf{r}(t) = (4t^3+8)\mathbf{i} + (2t-1)\mathbf{j}$ is the position vector of a particle moving in a plane. Find the acceleration.

A. $12t^2\mathbf{i} + 2\mathbf{j}$ B. $12\mathbf{i}$ C. $24t\mathbf{i} + 2\mathbf{j}$ D. $24t\mathbf{i}$

3. $\mathbf{r}(t) = (4t^3 + 5)\mathbf{i} + (2t - 2)\mathbf{j}$ is the position vector of a particle moving in a plane. Find the speed at $t = 1$.

A. $2\sqrt{37}$ B. 12 C. 24 D. 0

4. Find the velocity of a particle moving along the curve $\mathbf{r}(t) = (t^3 - 5)\mathbf{i} + (4t - 6)\mathbf{j} - (t^2 - 1)\mathbf{k}$ at $t = 1$.

A. $3\mathbf{i} + 4\mathbf{j} - 2\mathbf{k}$ B. $6\mathbf{i} - 2\mathbf{j}$ C. $3\mathbf{i} + 4\mathbf{j} + 2\mathbf{k}$ D. $\mathbf{0}$

5. Find the acceleration of a particle moving along the curve $\mathbf{r}(t) = (t^3 + 3t)\mathbf{i} + (4t - 2)\mathbf{j} - (t^2 + t - 2)\mathbf{k}$ at $t = 1$.

A. $3\mathbf{i} + 4\mathbf{j} - 2\mathbf{k}$ B. $6\mathbf{i} - 2\mathbf{k}$ C. $3\mathbf{i} + 4\mathbf{j} + 2\mathbf{k}$ D. $\mathbf{0}$

6. Find the speed of a particle moving along the curve $\mathbf{r}(t) = (t^3 + 6)\mathbf{i} + (4t - 6)\mathbf{j} - (t^2 - 4)\mathbf{k}$ at $t = 1$.

A. $\sqrt{29}$ B. $\sqrt{21}$ C. $4\sqrt{2}$ D. $2\sqrt{10}$

7. Answer true or false: If $\mathbf{a}(t) = 2\sin t\mathbf{i} + t\mathbf{j}$, the velocity vector is $-2\cos t\mathbf{i} + \dfrac{t^2}{2}\mathbf{j}$, if $\mathbf{v}(0) = -\mathbf{j}$.

8. Answer true or false: If $\mathbf{a}(t) = \sin t\mathbf{i} + t\mathbf{j}$, the position vector is $\mathbf{r}(t) = -\sin t\mathbf{i} + \left(\dfrac{t^3}{3} + 1\right)\mathbf{j}$ if $\mathbf{v}(0) = \mathbf{i}$ and $\mathbf{r}(0) = \mathbf{i}$.

9. If $\mathbf{v} = \mathbf{i}$ and $\mathbf{a} = \mathbf{i} - 3\mathbf{j}$, find $\mathbf{a_T}$.

A. 1 B. 2 C. $\dfrac{1}{2}$ D. 8

10. If $\mathbf{v} = 4\mathbf{i}$ and $\mathbf{a} = \mathbf{i} - 3\mathbf{j}$, find $\mathbf{a_N}$.

A. 6 B. -6 C. $\dfrac{3}{4}$ D. 3

11. If $\mathbf{v} = 2\mathbf{j}$ and $\mathbf{a} = \mathbf{j} - 3\mathbf{k}$, find k.

A. 6 B. -6 C. $\dfrac{3}{4}$ D. 3

12. $\mathbf{r}(t) = 2t^3\mathbf{i} - 4t\mathbf{j}$; $1 \le t \le 2$. Find the displacement.

A. $14\mathbf{i} - 4\mathbf{j}$ B. $18\mathbf{i} - 8\mathbf{j}$ C. $14\mathbf{i} + 4\mathbf{j}$ D. $18\mathbf{i} + 8\mathbf{j}$

13. $\mathbf{r}(t) = 2t^3\mathbf{i} - 4t\mathbf{j}$; $1 \le t \le 2$. Find the distance.

A. $2\sqrt{53}$ B. $6\sqrt{5}$ C. $2\sqrt{85}$ D. 18

14. $\mathbf{v}(t) = 2\mathbf{i} + 3\mathbf{j}$. Find $\mathbf{T}(t)$.

A. $\dfrac{2}{\sqrt{13}}\mathbf{i} + \dfrac{3}{\sqrt{13}}\mathbf{j}$ B. $\dfrac{2}{\sqrt{5}}\mathbf{i} + \dfrac{3}{\sqrt{5}}\mathbf{j}$

C. $\dfrac{1}{\sqrt{13}}\mathbf{i} + \dfrac{1}{\sqrt{13}}\mathbf{j}$ D. $\dfrac{1}{\sqrt{5}}\mathbf{i} + \dfrac{1}{\sqrt{5}}\mathbf{j}$

15. Find $\mathbf{a_N}$ if $\|\mathbf{a}\| = 4$ and $\theta = \pi/6$.

A. 2 B. $\sqrt{2}$ C. $\sqrt{3}$ D. 1

SECTION 13.7

1. Answer true or false: According to Keplers second law a planet moves fastest at a point on its semiminor axis.

2. If an object orbits the sun with $r_{\text{max}} = 120{,}000{,}000$ miles and $r_{\text{min}} = 110{,}000{,}000$ miles, the elliptical orbit has eccentricity

A. 23 B. $\frac{1}{23}$ C. 20 D. $\frac{1}{20}$

3. If an object orbits the sun with $r_{\text{max}} = 610{,}000{,}000$ miles and $r_{\text{min}} = 600{,}000{,}000$ miles, the elliptical orbit has eccentricity

A. 121 B. $\frac{1}{121}$ C. 120 D. $\frac{1}{120}$

4. Answer true or false: Object 1 has $r_{\text{max}} = 110{,}000{,}000$ miles and $r_{\text{min}} = 100{,}000{,}000$ miles. Object 2 has $r_{\text{max}} = 320{,}000{,}000$ miles and $r_{\text{min}} = 310{,}000{,}000$ miles. Both elliptical orbits have the same eccentricity.

5. Find the speed of a particle in a circular orbit with radius 10^{28} m around an object of mass 10^{25} kg. ($G = 6.67 \times 10^{-11}$ m/kg·s^2)

A. 1.50×10^{13} m/s B. 6.67×10^{-14} m/s
C. 3.87×10^{6} m/s D. 2.58×10^{-7} m/s

6. Find the speed of a particle in a circular orbit with radius 10^{30} m around an object of mass 10^{27} kg. ($G = 6.67 \times 10^{-11}$ m/kg·s^2)

A. 1.50×10^{13} m/s B. 6.67×10^{-14} m/s
C. 3.87×10^{6} m/s D. 2.58×10^{-7} m/s

7. An object in orbit has $r_{\text{max}} = 10^{25}$ km and $e = 0.58$. Find r_{min}.

A. 2.66×10^{23} km B. 2.70×10^{23} km
C. 2.66×10^{24} km D. 2.70×10^{24} km

8. An object in orbit has $r_{\text{min}} = 10^{26}$ km and $e = 0.58$. Find r_{max}.

A. 3.76×10^{26} km B. 3.80×10^{26} km
C. 3.80×10^{25} km D. 3.76×10^{24} km

9. An object in orbit has $r_{\text{max}} = 10^{26}$ km and $e = 0.52$. Find r_{min}.

A. 3.16×10^{25} km B. 3.21×10^{25} km
C. 3.24×10^{25} km D. 3.27×10^{25} km

10. An object in orbit has $r_{\text{min}} = 10^{26}$ km and $e = 0.52$. Find r_{max}.

A. 3.15×10^{26} km B. 3.17×10^{26} km
C. 3.19×10^{26} km D. 3.21×10^{26} km

11. If, for an elliptical orbit, $r_{\text{min}} = 10^{26}$ km and $e = 0.59$, find a, the semimajor axis.

A. 2.40×10^{26} km B. 2.44×10^{26} km
C. 2.47×10^{26} km D. 2.51×10^{26} km

12. If, for an elliptical orbit, $r_{\text{max}} = 10^{26}$ km and $e = 0.59$, find a, the semimajor axis.

A. 6.25×10^{25} km
B. 6.27×10^{25} km
C. 6.29×10^{25} km
D. 6.31×10^{25} km

13. If, for an elliptical orbit, $r_{\text{min}} = 10^{27}$ km and $e = 0.81$, find a, the semimajor axis.

A. 5.23×10^{27} km
B. 5.26×10^{27} km
C. 5.29×10^{27} km
D. 5.32×10^{27} km

14. If, for an elliptical orbit, $r_{\text{max}} = 10^{27}$ km and $e = 0.81$, find a, the semimajor axis.

A. 5.41×10^{27} km
B. 5.49×10^{27} km
C. 5.44×10^{27} km
D. 5.52×10^{26} km

15. Answer true or false: If $a = 1.50 \times 10^{12}$ km and $e = 0.10$, r_{max} of an elliptical orbit is 1.65×10^{12} km, where a denotes the semimajor axis.

CHAPTER 13 TEST

1. Find the domain of $\mathbf{r}(t) = \langle\sqrt{t-6},\, t^3,\, t-5\rangle$; $t_0 = 6$.

A. $0 \le t < \infty$
B. $6 \le t < \infty$
C. $-6 \le t < \infty$
D. $-\infty < t < \infty$

2. Answer true or false: The vector equation $\mathbf{r} = \cos t\mathbf{i} + \sin t\mathbf{k}$ can be expressed in parametric form by $x = \cos t$, $y = 0$, $z = \sin t$.

3. Describe the graph of $\mathbf{r}(t) = 15\mathbf{i} + \cos t\mathbf{j} + \sin t\mathbf{k}$.

A. Srtaight line
B. Spiral
C. Parabola
D. Circle

4. Describe the graph of $\mathbf{r}(t) = t\mathbf{i} + t^3\mathbf{j} + t^2\mathbf{k}$.

A. Cubic
B. Twisted cubic
C. Spiral
D. Parabola

5. If $\mathbf{r}(t) = 2\mathbf{i} + 5t^3\mathbf{j} + 3\cos t\mathbf{k}$, find $\mathbf{r}'(t)$.

A. $15t^2\mathbf{j} - 3\sin t\mathbf{k}$
B. $15t^2\mathbf{j} + 3\sin t\mathbf{k}$
C. $t\mathbf{i} + 15t^2\mathbf{j} - 3\sin t\mathbf{k}$
D. $t\mathbf{i} + 15t^2\mathbf{j} + 3\sin t\mathbf{k}$

6. Answer true or false: $\mathbf{r}(t) = (t^2+4)\mathbf{i} + 2t^4\mathbf{j} - (3t+7)\mathbf{k}$ is continuous at $t = 0$.

7. $\displaystyle\int (9t\mathbf{i} + 3\mathbf{j})dt =$

A. $\dfrac{9}{2}\,t^2\mathbf{i} + 3t\mathbf{j} + C$
B. $\left(\dfrac{9}{2}\,t^2 + C\right)\mathbf{i} + (3t + C)\mathbf{j}$
C. $\left(\dfrac{9}{2}\,t^2 + C_1\right)\mathbf{i} + (3t + C_2)\mathbf{j}$
D. $\dfrac{9}{2}\,t^2 + 3t + C$

8. $\displaystyle\int_0^{\pi/2} \langle\cos t, \sin t, 9\sin t\rangle dt =$

A. $\langle 1, 1, 9\rangle$
B. $\langle 1, -1, -9\rangle$
C. $\langle -1, 1, 9\rangle$
D. $\langle -1, -1, -9\rangle$

9. Answer true or false: If $\mathbf{r}(t) = \sin t\mathbf{i} + 3\cos t\mathbf{j}$, the tangent line at $t_0 = \pi$ is given by $\mathbf{r}(t) = -\mathbf{i}$.

10. Answer true or false: $\mathbf{r}(t) = 6t\mathbf{i} + 9\cos t\mathbf{j} + 3t^5\mathbf{k}$ is a smooth function of the parameter t.

11. Find the arc length of the graph of $\mathbf{r}(t) = -\sin t\mathbf{i} + 6\mathbf{j} - \cos t\mathbf{k}$, $0 \leq t \leq \pi$.

A. 2 B. 2π C. π D. 0

12. Find the arc length of the parametric curve $x = \sin t$, $y = 10$, $z = \cos t$, $0 \leq t \leq \pi$.

A. 2 B. 2π C. π D. 0

13. Answer true or false: If $\mathbf{r} = (t-3)\mathbf{i} + (4t+2)\mathbf{j} + (\sqrt{3} + 2t)\mathbf{k}$, the arc length paramentization of the curve relative to the reference point $(-3, 2, 0)$ involves the parameter $t = \dfrac{s}{\sqrt{19}}$.

14. $\mathbf{r}(t) = (t^2 - 1)\mathbf{i} + (2t - 5)\mathbf{j}$, $t = 2$. $\mathbf{T}(2) =$

A. $\dfrac{2}{\sqrt{5}}\mathbf{i} + \dfrac{1}{\sqrt{5}}\mathbf{j}$ B. $\dfrac{1}{\sqrt{5}}\mathbf{k}$ C. $-\dfrac{1}{\sqrt{5}}\mathbf{k}$ D. $\mathbf{i}$

15. Answer true or false: $\mathbf{r}(t) = (t^2 - 1)\mathbf{i} + (2t - 3)\mathbf{j}, t = 2$. $\mathbf{N}(t)$ for the given value of t is $\mathbf{i}$.

16. Answer true or false: $\mathbf{r}(t) = t^2\mathbf{i} + (2t-1)\mathbf{j}, t = 2$. $\mathbf{B}(2) = -\dfrac{1}{\sqrt{5}}$.

17. Find the curviture $k(t)$ for $\mathbf{r}(t) = 2t\mathbf{i} + 3t^2\mathbf{j} + 5\mathbf{k}$ at $t = 0$.

A. $\dfrac{3}{40\sqrt{10}}$ B. $\dfrac{4}{3}$ C. 0 D. 1

18. Answer true or false: If $\mathbf{r}(t) = t^5\mathbf{i} + t^4\mathbf{j} + t^3\mathbf{k}$ the curviture $k(t)$ is $20t^6 - 30t^5 + 336t^4$.

19. Answer true or false: If $\mathbf{r}(t) = (t+2)\mathbf{i} + (2t^2+1)\mathbf{j} + 4t\mathbf{k}$, the curviture $k(t)$ at $t = 1$ is $\dfrac{4\sqrt{17}}{33\sqrt{33}}$.

20. If an object orbits the sun with $r_{\max} = 290{,}000{,}000$ miles and $r_{\min} + 280{,}000{,}000$ miles, the elliptical orbit has eccentricity

A. 57 B. $\dfrac{1}{57}$ C. 56 D. $\dfrac{1}{56}$

21. If $y = 4 + \sin x$, find the curviture at $x = \dfrac{\pi}{2}$.

A. 0 B. 1 C. -1 D. $\dfrac{1}{2\sqrt{2}}$

22. If $x = t^3 + 2$, $y = t^2 - 1$, then $k(t)$ at $t = 1$ is

A. $\dfrac{6}{13\sqrt{13}}$ B. $\dfrac{6}{\sqrt{13}}$ C. 0 D. $\dfrac{180}{13\sqrt{13}}$

23. $\mathbf{r}(t) = (4t^3 + 7)\mathbf{i} + (2t - 3)\mathbf{j}$ is the position vector of a particle moving in a plane. Find the velocity.

A. $12t^2\mathbf{i} + 2\mathbf{j}$ B. $12\mathbf{i}$ C. $24t\mathbf{i} + 2\mathbf{j}$ D. $24t\mathbf{i}$

24. $\mathbf{r}(t) = (4t^3 + 5)\mathbf{i} + (6t - 9)\mathbf{j}$ is the position vector of a particle moving in a plane. Find the acceleration.

A. $12t^2\mathbf{i} + 2\mathbf{j}$ B. $12\mathbf{i}$ C. $24t\mathbf{i} + 2\mathbf{j}$ D. $24t\mathbf{i}$

25. $\mathbf{r}(t) = (4t^3 + 3)\mathbf{i} + (2t - 6)\mathbf{j}$ is the position vector of a particle moving in a plane. Find the speed at $t = 1$.

A. $2\sqrt{37}$ B. 12 C. 24 D. 0

26. Answer true or false: If $\mathbf{a}(t) = \sin t\mathbf{i} + 2t\mathbf{j} + \mathbf{k}$, the position vector is $-\sin t\mathbf{i}$ if $\mathbf{v}(0) = -\mathbf{i}$ and $\mathbf{r}(0) = \mathbf{0}$.

27. Answer true or false: Each planet moves in a circular orbit with the sun at the center of the circle.

ANSWERS TO SAMPLE TESTS

SECTION 13.1

1. B 2. A 3. C 4. F 5. F 6. F 7. T 8. B 9. D 10. D 11. B 12. B 13. A 14. D 15. B

SECTION 13.2

1. C 2. A 3. D 4. C 5. B 6. A 7. T 8. F 9. C 10. C 11. C 12. A 13. F 14. B 15. A

SECTION 13.3

1. T 2. T 3. F 4. A 5. C 6. C 7. B 8. C 9. D 10. A 11. F 12. T 13. T 14. T 15. F

SECTION 13.4

1. A 2. F 3. C 4. B 5. C 6. B 7. A 8. B 9. C 10. C 11. F 12. F 13. F 14. F 15. F

SECTION 13.5

1. A 2. B 3. A 4. B 5. F 6. T 7. B 8. A 9. B 10. A 11. C 12. A 13. F 14. C 15. C

SECTION 13.6

1. A 2. D 3. A 4. A 5. B 6. A 7. F 8. F 9. A 10. D 11. C 12. A 13. A 14. A 15. C

SECTION 13.7

1. T 2. B 3. B 4. F 5. D 6. D 7. C 8. A 9. A 10. B 11. B 12. C 13. B 14. D 15. T

CHAPTER 13 TEST

1. B 2. T 3. D 4. B 5. A 6. T 7. C 8. A 9. F 10. T 11. C 12. C 13. F 14. A 15. F 16. F 17. B 18. F 19. T 20. B 21. C 22. D 23. A 24. D 25. A 26. F 27. F

CHAPTER 14
Sample Exams

SECTION 14.1

1. $f(x, y, z) = x^2 - yz$. Find $f(1, 2, 3)$.

A. -4 B. -5 C. -7 D. 5

2. $f(x, y, z) = 3e^{xy} + z$. Find $f(1, 0, 6)$.

A. 9 B. $3e + 6$ C. $6e$ D. $9e$

3. $f(x, y, z) = \sqrt{x + y + z}$. Find $f(1, 2, 1)$.

A. 4 B. 0 C. 2 D. 1

4. Answer true or false: $f(x, y) = 7$ describes a plane parallel to the xy-plane 7 units above it.

5. Answer true or false: $f(x, y) = x^2 + y^2$ graphs in 3-space as a circle of radius 1 centered at $(0, 0)$ and confined to the xy-plane.

6. Answer true or false: $f(x, y) = 3\sqrt{x^2 + y^2}$ graphs as a semicircle.

7. Answer true or false: $f(x, y) = \sqrt{x^2 + y^2 + 2}$ graphs as a hemisphere.

8. The graph of $z = 4x^2 + 4y^2$ for $z = 0$ is

A. A circle of radius 4 B. A circle of radius 2
C. A circle of radius 16 D. A point.

9. The graph of $z = 4x^2 - 2y^2$ for $z = 0$ includes the point

A. $(0, 0, 0)$ B. $(1, 0, 0)$
C. $(0, 1, 1)$ D. None of the above.

10. Let $f(x, y, z) = 3x^2 + y^2 - z$. Find an equation of the level surface passing through $(1, 0, 1)$.

A. $3x^2 + y^2 - z = 4$ B. $3x^2 + y^2 - z = 0$
C. $3x^2 + y^2 - z = 2$ D. $3x^2 + y^2 - z = -2$

11. Let $f(x, y, z) = 2x^2 + y^2 - z^2$. Find an equation of a level surface passing through $(0, 0, 1)$.

A. $-z^2 = 1$ B. $z^2 = 1$
C. $2x^2 + y^2 - z^2 = 1$ D. $2x^2 + y^2 - z^2 = -1$

12. $f(x, y, z) = e^{xyz}$. Find an equation of the level surface that passes through $(0, 1, 2)$.

A. $e^{xyz} = 2$ B. $e^{xyz} = 3$ C. $e^{xyz} = 1$ D. $e^{xyz} = 0$

13. Answer true or false: If $V(x, y)$ is the voltage potential at a point (x, y) in the xy-plane, then the level curve for V, called the equipotential curve, is $V(x, y) = \dfrac{2}{\sqrt{x^2 + y^2}}$, and it passes through $(1, 0)$ when $V(x, y) = 1$.

14. Answer true or false: If $V(x, y)$ is the voltage potential at a point (x, y) in the xy-plane, then the level curve for V, called the equipotential curve, is $V(x, y) = \dfrac{2}{\sqrt{x^2 + y^2}}$, and it passes through $(0, 1)$ when $V(x, y) = 1$.

15. What is/are the domain restriction(s) for $f(x, y) = \ln(xy^2)$?

A. $x > 0, y \neq 0$ B. $x > 0, y > 0$
C. $x \neq 0, y \neq 0$ D. No restrictions exist

SECTION 14.2

1. $\displaystyle\lim_{(x,y)\to(3,4)} 2x + y =$

A. 10 B. 7 C. 14 D. Does not exist.

2. $\displaystyle\lim_{(x,y)\to(\pi,0)} (1 + y)\sin x =$

A. 0 B. 1 C. 2 D. Does not exist.

3. Answer true or false: $\displaystyle\lim_{(x,y)\to(0,0)} \frac{5}{x^2 + 4y^2}$ does not exist.

4. Answer true or false: $\displaystyle\lim_{(x,y)\to(0,0)} \frac{2}{3x^2 + y^2}$ does not exist.

5. Answer true or false: $\displaystyle\lim_{(x,y)\to(0,0)} \frac{y - 2x}{x^2 + y^2}$ does not exist.

6. Answer true or false: $\displaystyle\lim_{(x,y)\to(0,0)} x + y + 2$ does not exist.

7. $\displaystyle\lim_{(x,y)\to(1,1)} xy =$

A. 0 B. 1 C. 2 D. Does not exist.

8. $\displaystyle\lim_{(x,y)\to(0,0)} 8e^{2xy} =$

A. 0 B. 1 C. 8 D. Does not exist.

9. $\displaystyle\lim_{(x,y)\to(0,0)} \frac{4\sin(x^2 + y^2)}{\sqrt{x^2 + y^2 + 1}} =$

A. 4 B. 0 C. 1 D. Does not exist.

10. $\displaystyle\lim_{(x,y,z)\to(1,1,2)} x^2yz =$

A. 2 B. 4 C. 0 D. Does not exist.

11. $\displaystyle\lim_{(x,y)\to(0,0)} \frac{x+1}{y+2} =$

A. $\frac{1}{2}$ B. 0 C. 1 D. Does not exist.

12. Answer true or false: $f(x,y,z) = 4x^2y^2z$ is continuous everywhere.

13. Answer true or false: $f(x,y,z) = \cos(xyz)$ is continuous everywhere.

14. Answer true or false: $f(x,y,z) = \dfrac{2z}{\sin(xy)}$ is continuous everywhere.

15. Answer true or false: $f(x,y,z) = yz\ln|x|$ is continuous everywhere.

SECTION 14.3

1. $f(x,y) = 6x^4y^7$. Find $f_x(x,y)$.

A. $24x^3y^7$ B. $168x^3y^6$ C. $24x^3y^7 + 42x^4y^6$ D. $42x^2y^6$

2. $f(x,y) = \ln(xy)$. Find $f_x(2,3)$.

A. $\frac{3}{2}$ B. $\frac{1}{2}$ C. $\frac{5}{6}$ D. $\frac{1}{6}$

3. $z = e^{3xy}$. Find $\dfrac{\partial z}{\partial y}$.

A. $3xe^{3xy}$ B. $3e^{3xy}$ C. $3ye^{3xy}$ D. $3xye^{3xy}$

4. Answer true or false: If $f(x,y) = \sqrt{x^4+3y^2}$, $f_x(x,y) = \dfrac{4x^3}{2\sqrt{x^4+3y^2}}$.

5. $z = \sin(x^2y^4)$. Find $\dfrac{\partial z}{\partial y}$.

A. $(4x^2y^3 + 2xy^4)\cos(x^2y^4)$ B. $8xy^3\cos(x^2y^4)$
C. $4y^3\cos(x^2y^4)$ D. $4x^2y^3\cos(x^2y^4)$

6. $f(x,y) = x^4y^3$. $f_{xx} =$

A. $12x^2y^3$ B. $6x^4y$ C. $12x^3y^2$ D. $12x^2$

7. Answer true or false: If $(x^2+y^3+z^4)^{1/3} = 2$, $\dfrac{\partial f(x,y,z)}{\partial x} = \dfrac{6x}{(x^2+y^3+z^4)^{2/3}}$.

8. $f(x,y,z) = (x^2+y^2+z^2)^{1/4}$. $f_x(1,2,3) =$

A. $\dfrac{1}{2\sqrt[4]{14^3}}$ B. $\dfrac{1}{\sqrt[4]{14^3}}$ C. $\dfrac{1}{4\sqrt[4]{14^3}}$ D. $\dfrac{7}{\sqrt[4]{14^3}}$

9. $f(x,y,z) = xe^{yz}$. $f_{xz} =$

A. xye^{yz} B. ye^{yz} C. 0 D. y

10. $f(x,y,z) = y^2e^{xz}$. $f_{zz} =$

A. e^{xz} B. x^2e^{xz} C. y^2e^z D. $x^2y^2e^{xz}$

11. Answer true or false: $x \sin x$ solves the wave equation.

12. Answer true or false: If $z = \sin x \cos y$, $\dfrac{\partial z}{\partial x} = -\dfrac{\partial z}{\partial y}$.

13. Answer true or false: The tangent line to $z = x^2 y$ at $(1,1,1)$ in the y-direction has a slope of 2.

14. $f(x,y,z) = e^{4xyz}$. $f_{xyy} =$

A. $4xyze^{4xyz}$ B. $64x^2yz^3e^{4xyz}$ C. $64xy^2e^{4xyz}$ D. $4xy^2e^{4xyz}$

15. $f(x,y,z) = x\sin(yz)$. $f_{yz} =$

A. $xyz\sin(yz)$ B. $-xyz\sin(yz)$ C. $x\sin(yz)$ D. $-x\sin(yz)$

SECTION 14.4

1. Answer true or false: If $f(x,y) = x^3 + 2y^2$, then $df(x,y) = 3x^2dx + 4ydy$.

2. Answer true or false: If $f(x,y) = x^4y^7$, then $df = 4x^3dx + 7y^6dy$.

3. For the gas $PV = nRT$, where n and R are constants, estimate the change in nRT as P goes from 1 atm to 1.001 atm and V goes from 2 m^3 to 2.002 m^3. Answer in atm·m^3.

A. 0.004 B. 0.003 C. 0.002 D. 0.0015

4. Use the total differential to approximate the change in $f(x,y) = x^2 + 3y^3$ as (x,y) varies from $(3,4)$ to $(3.01, 3.98)$.

A. -0.01 B. -0.03 C. -2.79 D. -0.78

5. Use the total differential to approximate the change in $f(x,y) = xy$ as (x,y) varies from $(3,4)$ to $(3.01, 3.98)$.

A. -0.05 B. -1.10 C. -0.02 D. -0.10

SECTION 14.5

1. $w = r^2 - 3s$; $r = 2x$, $s = x + 4y$. Find $\left.\dfrac{\partial w}{\partial x}\right|_{x=1,y=3}$.

A. 5 B. 7 C. 4 D. 3

2. $w = 4x\sin y$; $x = t^2$, $y = 3t$. Find $\left.\dfrac{dw}{dt}\right|_{\pi}$.

A. $8\pi + 3$ B. 8π C. $-4\pi^2$ D. 5π

3. Let $f(x,y) = xy^8$. Find f_{xyx}.

A. $8y^7$ B. 0 C. 1 D. $8xy^7$

4. Let $f(x, y) = e^{2xy}$. Find f_{xyy}.

A. $2xy^2e^{2xy}$ B. $4x^2ye^{2xy}$ C. $2x^2ye^{2xy}$ D. $8x^2ye^{2xy}$

5. $z = 5x^2y^3$; $x = u + v$, $y = u - v$. Find $\frac{\partial z}{\partial u}$.

A. $15(u - v)^2 + 10(u + v)$ B. $25u^9$
C. $15(u + v)^2(u - v)^2 + 10(u + v)(u - v)^3$ D. $15u^4$

6. $z = e^{xy}$; $x = u^2$, $y = u - v$. Find $\frac{\partial z}{\partial u}$.

A. $2ue^{2u-v}$ B. $e^{u^3-u^2v}$ C. $(3u^2 - 2uv)e^{u^3-u^2v}$ D. $2e^{u^3-u^2v}$

7. $z = 4x - 2y$; $x = u^2$, $y = u - 3v$. Find $\frac{\partial z}{\partial u}$.

A. $8u + 4$ B. 6 C. $8u - 2 + 6v$ D. $8u - 2$

8. Answer true or false: If $z = f(v)$ and $v = g(x, y)$, then $\frac{\partial^2 z}{\partial x^2} = \frac{dz}{dv}\frac{\partial^2 v}{\partial x^2} + \frac{d^2 z}{dv^2}\frac{\partial^2 v}{\partial x^2}$.

9. Answer true or false: If $z = x^{1/3}y^3$, f_{xy} and f_{yx} differ on the xy-plane.

10. Answer true or false: If $z = x^7y^{1/5}$, f_{xy} and f_{yx} are equal where $y \neq 0$.

11. A right triangle initially has legs of 1 m. If they are increasing, one by 3 m/s and the other by 4 m/s, how fast is the hypotenuse increasing?

A. 5 m/s B. 7 m/s C. $7\sqrt{2}$ m/s D. $\frac{7\sqrt{2}}{2}$ m/s

SECTION 14.6

1. $z = 5x + 3y$. Find ∇z.

A. $5\mathbf{i} + 3\mathbf{j}$ B. $5x\mathbf{i} + 3y\mathbf{j}$ C. $x\mathbf{i} + y\mathbf{j}$ D. $-5\mathbf{i} - 3\mathbf{j}$

2. $z = 3x^2 + 4y^2$. Find ∇z.

A. $3x\mathbf{i} + 4y\mathbf{j}$ B. $6x\mathbf{i} + 8y\mathbf{j}$ C. $3\mathbf{i} + 4\mathbf{j}$ D. $x\mathbf{i} + y\mathbf{j}$

3. $f(x, y) = (x^2 + y)^{3/2}$. Find the gradient of f at $(1, 3)$.

A. $12\mathbf{i} + 2\mathbf{j}$ B. $3\mathbf{i} + \mathbf{j}$ C. $3\mathbf{i} + \frac{1}{2}\mathbf{j}$ D. $6\mathbf{i} + 3\mathbf{j}$

4. $f(x, y) = xy$. Find the gradient of f at $(2, 1)$.

A. $\mathbf{i} + 2\mathbf{j}$ B. $2\mathbf{i} + \mathbf{j}$ C. $\mathbf{i} + \mathbf{j}$ D. $3\mathbf{i} + 3\mathbf{j}$

5. $f(x, y) = e^{4xy}$; $P = (1, 2)$; $\mathbf{u} = \frac{2}{\sqrt{13}}\mathbf{i} + \frac{3}{\sqrt{13}}\mathbf{j}$. Find $D_{\mathbf{u}}f$ at P.

A. $\frac{28e^4}{\sqrt{13}}$ B. $\frac{28e^8}{\sqrt{13}}$ C. $\frac{28}{\sqrt{13}}$ D. e^8

6. $f(x, y) = ye^x$; $P = (0, 3)$; $u = \frac{2}{\sqrt{13}}\mathbf{i} + \frac{3}{\sqrt{13}}\mathbf{j}$. Find $D_u f$ at P.

A. $\frac{9}{\sqrt{13}}$ B. 0 C. $\frac{14}{\sqrt{13}}$ D. $\frac{5}{\sqrt{13}}$

7. Answer true or false: If $f(x, y) = 5e^{xy} + 3x$ and $\mathbf{a} = 4\mathbf{i} + 3\mathbf{j}$ is a vector, the direction derivative of f with respect to $\mathbf{a}$ at $(2, 3)$ is $\frac{4}{5}(15e^6 + 3)\mathbf{i} + 6e^6\mathbf{j}$.

8. Find the largest value among all possible directional derivatives of $f(x, y) = 3x^3 + 2y$.

A. $\sqrt{81x^4 + 4}$ B. $\sqrt{9x^6 + 4y^2}$ C. $81x^4 + 4$ D. $9x^2 + 2$

9. Find the smallest value among all possible directional derivatives of $f(x, y) = 4x + y$.

A. $-\sqrt{5}$ B. $\sqrt{5}$ C. $-\sqrt{17}$ D. $\sqrt{17}$

10. A particle is located at the point $(3, 5)$ on a metal surface whose temperature at a point (x, y) is $T(x, y) = 25 - 3x^2 - 2y^2$. Find the equation for the trajectory of a particle moving continuously in the direction of maximum temperature increase. $y =$

A. $x^{2/3}$ B. $\frac{5}{3^{2/3}}x^{2/3}$ C. $\frac{(3x)^{2/3}}{5}$ D. $\frac{5}{3}x^{2/3}$

11. A particle is located at the point $(3, 5)$ on a metal surface whose temperature at a point (x, y) is $T(x, y) = 16 - 2x^2 - 3y^2$. Find the equation for the trajectory of a particle moving continuously in the direction of maximum temperature increase. $y =$

A. $x^{2/3}$ B. $\frac{5}{3^{3/2}}x^{3/2}$ C. $\frac{(3x)^{2/3}}{5}$ D. $\frac{5}{3}x^{2/3}$

12. Answer true or false: $z = 4x^2 + y^2$. $\|\nabla z\| = 10$ at $(1, 1)$.

13. Answer true or false: The gradient of $f(x, y) = 7x - 3y^2$ at $(1, 2)$ is $7\mathbf{i} - 6\mathbf{j}$.

14. Answer true or false: The gradient of $f(x, y) = 5x^2 - 7y$ at $(2, 3)$ is $10\mathbf{i} - 7\mathbf{j}$.

15. The gradient of $f(x, y) = 5e^{xy}$ at $(1, 0)$ is

A. $5\mathbf{i}$ B. $5\mathbf{j}$ C. $\mathbf{i}$ D. $\mathbf{j}$

16. $f(x, y, z) = 5x + 3y^2 + z^3$. Find the gradient at $(1, 2, 1)$.

A. $9\mathbf{i} + 2\mathbf{j} + 3\mathbf{k}$ B. $9\mathbf{i} + 24\mathbf{j} + 3\mathbf{k}$ C. $\mathbf{i} + 2\mathbf{j} + \mathbf{k}$ D. $9\mathbf{i} + 2\mathbf{j} + \mathbf{k}$

17. $f(x, y, z) = x^2yz$. Find the gradient at $(2, 3, 4)$.

A. $\mathbf{i}$ B. $2\mathbf{i}$ C. $\mathbf{j}$ D. $\mathbf{k}$

18. $f(x, y, z) = \ln(xyz)$ and $\mathbf{u} = 2\mathbf{i} + \mathbf{j} - 4\mathbf{k}$. Find the directional derivative of f at $(1, 2, 3)$ in the direction of $\mathbf{u}$.

A. $2\mathbf{i} + \frac{1}{2}\mathbf{j} - \frac{4}{3}\mathbf{k}$ B. $2\mathbf{i} + \mathbf{j} - 4\mathbf{k}$

C. $\frac{1}{3}\mathbf{i} + \frac{1}{6}\mathbf{j} - \frac{2}{3}\mathbf{k}$ D. $2\ln 6\mathbf{i} + \ln 6\mathbf{j} - 4\ln 6\mathbf{k}$

19. $f(x, y, z) = e^{-4xyz}$ and $\mathbf{u} = 3\mathbf{i} + 4\mathbf{j} - \mathbf{k}$. Find the directional derivative of f at $(3, 2, 4)$ in the direction of $\mathbf{u}$.

A. $4\mathbf{i}$ B. $4\mathbf{k}$ C. $\frac{4}{e}\mathbf{j}$ D. $4\mathbf{j}$

20. Answer true or false: $f(x,y,z) = 4x^2 + 3y^2 + 7z^2$. The directional derivative of f at $(1,1,1)$ that has the largest value is in the direction $8\mathbf{i} + 6\mathbf{j} + 14\mathbf{k}$.

21. Answer true or false: $f(x,y,z) = 4x^2 + 3y^2 + 7z^2$. The directional derivative of f at $(1,1,1)$ that has the smallest value is in the direction $-8\mathbf{i} - 6\mathbf{j} - 14\mathbf{k}$.

22. Answer true or false: $f(x,y,z) = 4x^2 + 3y^2 + 7z^2$. The directional derivative of f at $(1,1,1)$ that has the largest value is $\sqrt{296}$.

23. Answer true or false: $f(x,y,z) = 4x^2 + 3y^2 + 7z^2$. The directional derivative of f at $(1,1,1)$ that has the smallest value is $-\sqrt{296}$.

24. The gradient of $\sin x + \cos x + z^2$ at $(0,0,2)$ is

A. $\mathbf{i} + 4\mathbf{k}$ B. $\mathbf{j} + 4\mathbf{k}$ C. $\mathbf{i} + \mathbf{j} + 4\mathbf{k}$ D. $-\mathbf{i} + 4\mathbf{k}$.

25. The gradient of $\sin x + y^3 - \cos z$ at $(\pi/2, 2, \pi)$ is

A. $12\mathbf{j}$ B. $\mathbf{i} + 12\mathbf{j} + \mathbf{k}$ C. $-\mathbf{i} + 12\mathbf{j} + \mathbf{k}$ D. $\mathbf{i} + 12\mathbf{j} - \mathbf{k}$.

SECTION 14.7

1. Find an equation for the tangent plane to $z = 5x^2y$ at $P = (1,2,0)$.

A. $10 + 20(x-1) + 10(y-2) = 0$ B. $20 + 40(x-1) + 20(y-2) = 0$
C. $10 + 20(x-1) + 10(y-2) = 1$ D. $20 + 40(x-1) + 20(y-2) = 1$

2. For $z = 5x^2y$, find the parametric normal lines to the surface at $P(1,2,4)$.

A. $x = 1 - 20t$, $y = 2 - 5t$, $z = 4 + t$ B. $x = 1 + 20t$, $y = 2 + 5t$, $z = 4 - t$
C. $x = 1 - 20t$, $y = 2 - 5t$, $z = 4 - t$ D. $x = 1 + 20t$, $y = 2 + 5t$, $z = 4 + t$

3. Find an equation for the tangent plane to $z = 4x^7y^2$ at $P = (1,2,0)$.

A. $8 + 56(x-1) + 16(y-2) = 1$ B. $8 + 56(x-1) + 16(y-2) = 0$
C. $16 + 112(x-1) + 64(y-2) = 1$ D. $16 + 112(x-1) + 64(y-2) = 0$

4. For $z = 4x^7y^2$, find the parametric normal lines to the surface at $P(1,2,5)$.

A. $x = 1 - 112t$, $y = 2 - 16t$, $z = 5 + t$ B. $x = 1 + 112t$, $y = 2 + 16t$, $z = 5 - t$
C. $x = 1 - 112t$, $y = 2 - 16t$, $z = 5 - t$ D. $x = 1 + 112t$, $y = 2 + 16t$, $z = 5 + t$

5. Find an equation for the tangent plane to $z = \sin(2x)\cos(3y)$ at $P = (\pi, \pi, 2)$.

A. $-2(x-\pi) - 3(y-\pi) - (z-2) = 0$ B. $-2(x-\pi) - 3(y-\pi) + (z-2) = 0$
C. $-2(x-\pi) - (z-2) = 0$ D. $2(x-\pi) + (z-2) = 0$

6. For $z = \sin(2x)\cos(3y)$ find the parametric normal lines to the surface at $P = (\pi, \pi, 2)$.

A. $x = \pi + 2t, y = \pi, z = 2 - t$ B. $x = \pi + 2t, y = \pi, z = 2 + t$
C. $x = \pi - 2t, y = \pi, z = 2 + t$ D. $x = \pi - 2t, y = \pi, z = 2 - t$

7. Find an equation for the tangent plane to $3x^2 + 4y^2 = 9$, at $P = (1,0,2)$.

A. $-6(x-1) - 8y - (z-2) = 0$ B. $6(x-1) + 8y - (z-2) = 0$
C. $-6(x-1) - 8y + (z-2) = 0$ D. $6(x-1) + 8y + (z-2) = 0$

8. For $3x^2 + 4y^2 + z^2 = 9$, find the parametric normal lines to the surface at $P = (1, 0, 2)$.

A. $x = 1 - 6t, y = -8t, z = 2 - t$
B. $x = 1 + 6t, y = +8t, z = 2 - t$
C. $x = 1 - 6t, y = -8t, z = 2 + t$
D. $x = 1 + 6t, y = +8t, z = 2 + t$

9. Find an equation for the tangent plane to $3x^2y = 9$, at $P = (1, -1, 2)$.

A. $-\frac{6}{\sqrt{18}}(x - 1) + \frac{3}{2\sqrt{18}}(y + 1) - (z - 2) = 0$

B. $\frac{6}{\sqrt{18}}(x - 1) + \frac{3}{2\sqrt{18}}(y + 1) + (z - 2) = 0$

C. $-\frac{1}{\sqrt{18}}(x + 1) + \frac{1}{2\sqrt{18}}(y - 1) - (z + 2) = 0$

D. $-\frac{1}{\sqrt{18}}(x + 1) + \frac{1}{2\sqrt{18}}(y - 1) + (z + 2) = 0$

10. For $3x^2y - z^3 = 9$, find the parametric normal lines to the surface at $P = (1, -1, 2)$.

A. $x = 1 - 6t, y = -1 - 3t, z = 2 - t$

B. $x = 1 - t, y = -1 - t, z = 2 + t$

C. $x = 1 + \frac{1}{\sqrt{18}}t, y = -1 + \frac{1}{2\sqrt{18}}t, z = 2 - t$

D. $x = 1 + \frac{1}{\sqrt{18}}t, y = -1 + \frac{1}{2\sqrt{18}}t, z = 2 + t$

SECTION 14.8

1. $f(x, y) = 2xy + 4x + 2y - 8$. There is a critical point at

A. $(-2, -1)$ B. $(-1, -2)$ C. $(2, 1)$ D. $(1, 2)$.

2. $f(x, y) = 5x^2 + 2y^2 - 9$. There is a critical point at

A. $(10, 4)$ B. $(5, 2)$ C. $(-10, -4)$ D. $(0, 0)$.

3. $f(x, y) = 6x^2 - 2y^2 + 11$. There is a critical point at

A. $(6, -2)$ B. $(-6, 2)$ C. $(-12, -4)$ D. $(0, 0)$.

4. $f(x, y) = e^{xy} + 4$. There is a critical point at

A. $(0, 0)$ B. $(4, 4)$ C. $(-4, -4)$ D. None exist.

5. Answer true or false: $f(x, y) = e^x + e^y - 3$. There is no critical point.

6. $f(x, y) = x^3 - 12x - 9y$. $(2, 0)$ is

A. A relative maximum
B. A relative minimum
C. A saddle point
D. Cannot be determined.

7. $f(x, y) = x^5 - 80x + 3y$. $(2, 0)$ is

A. A relative maximum
B. A relative minimum
C. A saddle point
D. Cannot be determined.

8. $f(x, y) = 4xy - 8x$. $(0, 2)$ is

A. A relative maximum
B. A relative minimum
C. A saddle point
D. Cannot be determined.

9. $f(x, y) = xy^2 + x^2y$. $(0, 0)$ is

A. A relative maximum
B. A relative minimum
C. A saddle point
D. Cannot be determined.

10. $f(x, y) = x^2 + 4xy + y^2 + 8x + 2y$. $\left(\frac{2}{3}, -\frac{7}{3}\right)$ is

A. A relative maximum
B. A relative minimum
C. A saddle point
D. Cannot be determined.

11. $f(x, y) = x^2y^2 - x^2$. $(0, 1)$ is

A. A relative maximum
B. A relative minimum
C. A saddle point
D. Cannot be determined.

12. $f(x, y) = x^2 + 2x + y^2$. $(-1, 0)$ is

A. A relative maximum
B. A relative minimum
C. A saddle point
D. Cannot be determined.

13. Answer true or false: If $f(x, y)$ has two critical points, one must be a relative maximum and the other must be a relative minimum.

14. Answer true or false: Every function $f(x, y)$ has a relative maximum.

15. $f(x, y) = e^{3xy}$. $(0, 0)$ is

A. A relative maximum
B. A relative minimum
C. A saddle point
D. Cannot be determined.

SECTION 14.9

1. $4xy$ subject to $2x + 2y = 20$ is maximized at

A. $(2, 2)$
B. $(4, 4)$
C. $(5, 5)$
D. $(0, 0)$.

2. xy subject to $4x + 2y = 8$ is maximized at

A. $(1, 2)$
B. $(2, 1)$
C. $(8, 8)$
D. $(2, 4)$.

3. Answer true or false: To maximize $3x^2y$ subject to $4x - 2xy = 10$, $\nabla f(x, y) = \lambda \nabla g(x, y)$ can be written as $6x\mathbf{i} + 3x^2\mathbf{j} = 4\lambda\mathbf{i} - 2\lambda\mathbf{j}$.

4. Answer true or false: There are no relative extrema of $f(x, y, z) = x^2 + (y + 3)^2 + (z - 3)^2$ subject to the constraint $x^2 + y^2 + z^2 = 1$.

5. Answer true or false: $3xyz$ subject to $x^2y + z^2 = 6$ has an extrema at $(0, 0, 0)$.

6. Answer true or false: x^2yz^2, subject to $x + y + z = 5$ has an extrema at $(0, 0, 0)$.

7. Answer true or false: x^2yz^2, subject to $x + y + z = 5$ has an extrema at $(2, 1, 2)$.

8. Answer true or false: x^2yz^2, subject to $x + y + z = 5$ has an extrema at $(1, 2, 1)$.

9. Answer true or false: To find an extrema subject to a constraint it is always necessary to find λ.

10. Answer true or false: To find an extrema for $f(x, y, z) = (x - 4)^2 + (y - 4)^2 + (z - 4)^2$ subject to $x^4 + y^4 + z^4 = 1$, $\nabla f(x, y, z) = \lambda \nabla g(x, y, z)$ gives $x - 4 = 2x^3\lambda, y - 4 = 2y^3\lambda, z - 4 = 2z^3\lambda$.

11. Answer true or false: To find an extrema for $f(x, y, z) = (x - 4)^2 + (y - 4)^2 + (z - 4)^2$ subject to $\frac{1}{x^2} + \frac{1}{y^2} - \frac{1}{z^2} = 0$, $\nabla f(x, y, z) = \lambda \nabla g(x, y, z)$ gives $x - 4 = -x\lambda, y - 4 = -y\lambda, z - 4 = z\lambda$.

12. Answer true or false: $f(x, y, z) = (x - 4)^2 + (y - 4)^2 + (z - 4)^2$ subject to $\frac{9}{x^2} + \frac{9}{y^2} + \frac{9}{z^2} = 1$ has an extrema at $(0, 0, 0)$.

13. Answer true or false: $x^3 + 2x^2y + y^3$ has an extrema when subjected to $x^3 + y^3 = 5$ at $(1, 1)$.

14. Answer true or false: $x^3 + 2x^2y + y^3$ has an extrema when subjected to $x^3 + y^3 = 5$ at $(-1, -1)$.

15. Answer true or false: $x^3 + 2x^2y + y^3$ has an extrema when subjected to $x^3 + y^3 = 2$ at $(1, 1)$.

CHAPTER 14 TEST

1. $f(x, y, z) = 2x^2 + yz$. Find $f(1, 2, 1)$.

   A. 4 B. 0 C. 1 D. 5

2. Answer true or false: $f(x, y) = \sqrt{x^2 + y^2 + 4}$ graphs as a hemisphere.

3. Let $f(x, y, z) = 5xyz$. Find an equation of the level surface passing through $(2, 3, 1)$.

   A. $5xyz = 6$ B. $5xyz = 30$ C. $5xyz = 0$ D. $5xyz = 5$

4. $\lim_{(x,y)\to(1,2)} (x - y) =$

   A. -1 B. 1 C. 3 D. Does not exist.

5. Answer true or false: $\lim_{(x,y)\to(0,0)} \frac{5}{x^2 + y^2}$ does not exist.

6. Answer true or false: $f(x, y, z) = yz \ln |x|$ is continuous everywhere.

7. $z = e^{5xy}$. Find $\frac{\partial z}{\partial y}$.

   A. $5xe^{5xy}$ B. $5e^{5xy}$ C. $5ye^{5xy}$ D. $5xye^{5xy}$

8. Answer true or false: If $f(x, y) = \sqrt{x^6 + 4y^5}$, $f_y(x, y) = \dfrac{10y^4}{\sqrt{x^6 + 4y^5}}$.

9. xy subject to $2x + 4y = 16$ is maximized at

A. $(4, 2)$ B. $(2, 1)$ C. $(8, 4)$ D. $(0, 0)$

10. $z = 4x - 2y$; $x = u^2$, $y = u - 3v$. Find $\dfrac{\partial z}{\partial v}$.

A. $-6uv$ B. -6 C. $6v$ D. 6

11. A right triangle initially has legs of 1 m. If they are increasing, one by 6 m/s and the other by 8 m/s, how fast is the hypotenuse increasing?

A. 10 m/s B. 14 m/s C. $14\sqrt{2}$ m/s D. $7\sqrt{2}$ m/s

12. Find an equation for the tangent plane to $z = 4x^7y^2$ at $P = (1, 2, 1)$.

A. $28x^6y^2(x - 1) + 8x^7(y - 2) - (z - 1) = 0$
B. $112(x - 1) + 16(y - 2) - (z - 1) = 0$
C. $28x^6y^2(x - 1) + 8x^7y(y - 2) + (z - 1) = 0$
D. $112(x - 1) + 16(y - 2) + (z - 1) = 0$

13. For $z = 4x^7y^2$, find the parametric normal lines to the surface at $P(1, 2, 1)$.

A. $x = 1 - 112t, y = 2 - 16t, z = 1 + t$
B. $x = 1 + 112t, y = 2 + 16t, z = 1 - t$
C. $x = 1 - 112t, y = 2 - 16t, z = 1 - t$
D. $x = 1 + 112t, y = 2 + 16t, z = 1 + t$

14. Answer true or false: If $f(x, y) = x^4y^7$, then $df = 4x^3dx + 7y^6dy$.

15. $z = 5x^2 + y^2$. Find ∇z.

A. $5\mathbf{i} + \mathbf{j}$ B. $10\mathbf{i} + \mathbf{j}$ C. $10x\mathbf{i} + 2y\mathbf{j}$ D. $5\mathbf{i} + 2\mathbf{j}$

16. $f(x, y) = xe^y$; $P = (3, 0)$; $u = \dfrac{2}{\sqrt{13}}\mathbf{i} + \dfrac{3}{\sqrt{13}}\mathbf{j}$. Find D_uf at P.

A. $\dfrac{11}{\sqrt{13}}$ B. 0 C. $\dfrac{4}{\sqrt{13}}$ D. $\dfrac{6}{\sqrt{13}}$

17. Answer true or false: If $f(x, y) = 5e^{xy} + 3x$ and $\mathbf{a} = 8\mathbf{i} + 6\mathbf{j}$ is a vector, the direction derivative of f with respect to $\mathbf{a}$ at $(2, 3)$ is $\dfrac{4}{5}(15e^6 + 3)\mathbf{i} + 6e^6\mathbf{j}$.

18. Answer true or false: $f(x, y, z) = |y| - e^{xyz}$ is differentiable everywhere.

19. $f(x, y, z) = 7x + 4y^2 + z^3$. Find the gradient at $(2, 1, 2)$.

A. $7\mathbf{i} + 8\mathbf{j} + 12\mathbf{k}$ B. $7\mathbf{i} + 8\mathbf{j} + 3\mathbf{k}$ C. $7\mathbf{i} + 4\mathbf{j} + \mathbf{k}$ D. $2\mathbf{i} + \mathbf{j} + 2\mathbf{k}$

20. $f(x, y) = 9x^2 - 3y^2 - 3$. There is a critical point at

A. $(9, -3)$ B. $(-18, 9)$ C. $(18, -9)$ D. $(0, 0)$

ANSWERS TO SAMPLE TESTS

SECTION 14.1

1. B 2. A 3. C 4. T 5. F 6. F 7. F 8. D 9. A 10. C 11. D 12. C 13. T 14. F 15. A

SECTION 14.2

1. A 2. A 3. T 4. T 5. T 6. F 7. B 8. C 9. B 10. A 11. A 12. T 13. T 14. F 15. F

SECTION 14.3

1. A 2. B 3. A 4. T 5. D 6. A 7. F 8. A 9. B 10. D 11. F 12. F 13. F 14. B 15. B

SECTON 14.4

1. T 2. F 3. A 4. C 5. C

SECTION 14.5

1. A 2. C 3. B 4. D 5. C 6. C 7. D 8. F 9. F 10. T 11. D

SECTION 14.6

1. A 2. B 3. D 4. A 5. B 6. A 7. F 8. A 9. C 10. B 11. B 12. F 13. F 14. F 15. B 16. B 17. C 18. A 19. D 20. T 21. T 22. T 23. T 24. A 25. A

SECTION 14.7

1. A 2. B 3. B 4. B 5. C 6. D 7. B 8. A 9. A 10. A

SECTION 14.8

1. B 2. D 3. D 4. A 5. T 6. D 7. D 8. C 9. D 10. C 11. D 12. D 13. F 14. F 15. D

SECTION 14.9

1. C 2. A 3. F 4. T 5. F 6. F 7. F 8. F 9. F 10. T 11. F 12. F 13. F 14. F 15. T

CHAPTER 14 TEST

1. A 2. F 3. B 4. A 5. T 6. F 7. A 8. T 9. A 10. D 11. D 12. B 13. B 14. F 15. C 16. A 17. T 18. F 19. A 20. D

CHAPTER 15
Sample Exams

SECTION 15.1

1. $\int_0^1 \int_0^3 (x+4)\,dx\,dy =$

A. $\frac{33}{2}$ B. $\frac{27}{2}$ C. $\frac{9}{2}$ D. $\frac{11}{2}$

2. $\int_0^1 \int_0^3 (x+4)\,dy\,dx =$

A. $\frac{33}{2}$ B. $\frac{21}{2}$ C. $\frac{9}{2}$ D. $\frac{11}{2}$

3. $\int_0^2 \int_0^3 dx\,dy =$

A. 5 B. 6 C. 0 D. 36

4. $\int_0^1 \int_0^2 e^x\,dx\,dy =$

A. e^2-1 B. e^2-2 C. $2e$ D. e^2

5. $\int_0^{\pi} \int_{\pi/2}^{\pi} \cos x\,dx\,dy =$

A. π B. $-\pi$ C. $\frac{\pi}{2}$ D. $-\frac{\pi}{2}$

6. Evaluate $\iint\limits_R 2xy^2 dA; R = \{(x,y) : -1 \le x \le 2, 1 \le y \le 2\}$.

A. 4 B. 2 C. $\frac{35}{3}$ D. 7

7. Answer true or false: $\iint\limits_R 5xy^3 dA; R = \{(x,y) : -2 \le x \le 4, -1 \le y \le 2$ is $\int_{-2}^{4} \int_{-1}^{2} 5xy^3 dx\,dy$.

8. Answer true or false: $\iint\limits_R x^2 \sin y\,dA; R = \{(x,y) : 0 \le x \le 2, 1 \le y \le 3\}$ is $\int_1^3 \int_0^2 x^2 \sin y\,dx\,dy$.

9. Find the volume of the solid bounded by $z = -2x - 2y$ and the rectangle $R = [0,1] \times [0,3]$.

A. 10 B. 12 C. 20 D. 25

10. Find the volume of the solid bounded by $z = 10 - 4x - 2y$ and the rectangle $R = [0,2] \times [0,1]$.

A. 1 B. 10 C. 9 D. 2

11. Answer true or false: The average value of the function $f(x,y) = \sin x \cos y$ over the rectangle $[0,\pi] \times [0,2\pi]$ is $\frac{1}{\pi}\int_0^{\pi}\int_0^{2\pi} \sin x \cos y\,dy\,dx$.

12. Answer true or false: The average value of the function $f(x, y) = x^2y^3$ over the rectangle $[0, 5] \times [0, 2]$ is $\frac{1}{10}\int_0^5\int_0^2 x^2y^3 dy\,dx$.

13. Answer true or false: The volume of the solid bounded by $z = x^3y$ and $R = \{(x, y) : -1 \le x \le 1, -1 \le y \le 2\}$ is $\int_{-1}^1\int_{-1}^2 x^3y\,dy\,dx$.

14. Answer true or false: The volume of the solid bounded by $z = \sin x \cos y$ and $R = \{(x, y) : -1 \le x \le 1, -1 \le y \le 2\}$ is $\int_{-1}^1\int_{-1}^2 \sin x \cos y\,dy\,dx$.

15. Answer true or false: The volume of the solid bounded by $z = e^{3xy}$ and $R = \{(x, y) : -1 \le x \le 1, -1 \le y \le 2\}$ is $\int_{-1}^1\int_{-1}^2 e^{3xy} dy\,dx$.

SECTION 15.2

1. $\int_0^2\int_0^x xy\,dy\,dx =$

A. 1 B. 2 C. 3 D. 4

2. $\int_0^{\pi/2}\int_0^{\cos x} dy\,dx =$

A. 1 B. 0 C. -1 D. π

3. $\int_0^1\int_0^x e^y dy\,dx =$

A. $e - 2$ B. $e + 1$ C. $-e$ D. e

4. $\int_0^1\int_0^x \sqrt{x^2 + 1}\,dy\,dx =$

A. $\frac{2}{3}(2^{3/2} + 1)$ B. $\frac{2^{3/2} + 1}{3}$ C. $\frac{2}{3}(2^{3/2} - 1)$ D. $\frac{2^{3/2} - 1}{3}$

5. Answer true or false: $\iint_R x^4 dA$, where R is the region bounded by $y = x + 4$, $y = 2x$, and $x = 16$ is $\int_{16}^{28}\int_{x+4}^{2x} x^4 dy\,dx$.

6. Answer true or false: $\iint_R xy\,dA$, where R is the region bounded by $y = x$, $y = 0$, and $x = 4$, is $\int_0^4\int_0^x xy\,dy\,dx$.

7. $\int_0^2 \int_x^{x^2} y\,dy\,dx =$

A. $\frac{2^5}{10}$ B. $\frac{2^6}{15}$ C. $\frac{28}{15}$ D. $\frac{2^5}{10} + \frac{2^3}{3}$

8. Find the area of the plane enclosed by $y = x$ and $y = x^2$, for $0 \le x \le 2$.

A. $\frac{1}{3}$ B. $\frac{7}{6}$ C. $\frac{2}{3}$ D. $\frac{1}{6}$

9. Find the area of the plane enclosed by $y = x$ and $y = x^2$, for $1 \le x \le 3$.

A. 4 B. $\frac{14}{3}$ C. $\frac{16}{3}$ D. 6

10. $\int_1^2 \int_x^{x^2} (xy - 4)\,dy\,dx =$

A. 0.412 B. 0.042 C. 0.419 D. 0.423

11. Answer true or false: $\int_0^1 \int_x^{x^2} dy\,dx = -\frac{2}{3}$.

12. $\int_{-1}^0 \int_0^x dy\,dx =$

A. -1 B. 1 C. $\frac{1}{2}$ D. $-\frac{1}{2}$

13. Answer true or false: $\int_0^2 \int_0^{x^2} f(x, y)\,dy\,dx = \int_0^4 \int_0^{\sqrt{y}} f(x, y)\,dx\,dy$.

14. Answer true or false: $\int_0^3 \int_0^{x^2} f(x, y)\,dy\,dx = \int_0^3 \int_0^{y^2} f(x, y)\,dx\,dy$.

15. Answer true or false: $\int_1^2 \int_1^{\ln x} f(x, y)\,dy\,dx = \int_1^2 \int_1^{e^y} f(x, y)\,dx\,dy$.

SECTION 15.3

1. $\int_0^{\pi/2} \int_{\sin\theta}^0 r\cos\theta\,dr\,d\theta =$

A. $-\frac{1}{6}$ B. $\frac{1}{6}$ C. $\frac{\pi}{2}$ D. $-\frac{\pi}{2}$

2. $\int_{-\pi/2}^{\pi/2} \int_0^{\sin\theta} r^3\,dr\,d\theta =$

A. -0.33 B. -0.29 C. 0.29 D. 0.33

3. $\int_0^{\pi} \int_0^{\cos\theta} r^2\,dr\,d\theta =$

A. 1 B. -1 C. 0 D. 2

4. $\displaystyle\int_0^{\pi}\int_0^{\cos 2\theta} dr\,d\theta =$

A. 1 B. -1 C. 0 D. 2

5. $\displaystyle\int_0^{\pi/6}\int_0^{\sin 3\theta} dr\,d\theta =$

A. 3 B. $\frac{1}{3}$ C. -3 D. $-\frac{1}{3}$

6. Answer true or false: $\displaystyle\int_0^1\int_0^x x^2+y^2 dy\,dx = \int_0^1\int_0^{\cos\theta} dr\,d\theta.$

7. Answer true or false: $\displaystyle\int_0^1\int_0^{\sqrt{1-x^2}} e^{x^2+y^2} dy\,dx = \int_0^{\pi}\int_0^1 re^{r^3} dr\,d\theta.$

8. Answer true or false: $\displaystyle\int_0^1\int_0^{\sqrt{1-x^2}} dy\,dx = \int_0^{\pi/2}\int_0^1 r\,dr\,d\theta.$

9. Answer true or false: $\displaystyle\int_{-1}^0\int_0^{\sqrt{4-x^2}} dy\,dx = \int_0^{\pi}\int_0^2 r\,dr\,d\theta.$

10. Find the volume of the solid formed by the left hemisphere $r^2 + z^2 = 4$.

A. $\frac{8\pi}{3}$ B. $\frac{16\pi}{3}$ C. $\frac{32\pi}{3}$ D. $\frac{2\pi}{3}$

11. Find the volume of the solid formed by the left hemisphere $r^2 + z^2 = 25$.

A. $\frac{25\pi}{3}$ B. $\frac{25\pi}{6}$ C. $\frac{250\pi}{3}$ D. $\frac{125\pi}{3}$

12. Find the volume of the solid formed by the left hemisphere $x^2 + y^2 + z^2 = 25$.

A. $\frac{25\pi}{3}$ B. $\frac{25\pi}{6}$ C. $\frac{250\pi}{3}$ D. $\frac{125\pi}{3}$

13. Find the area enclosed by the three-petaled rose $r = 2\sin 3\theta$.

A. $\frac{\pi}{4}$ B. $\frac{\pi}{2}$ C. $\frac{\pi}{8}$ D. π

14. Find the volume between $x^2 + y^2 = 4$ and $x^2 + z^2 = 4$ above the xy-plane.

A. $\frac{8\pi}{3}$ B. $\frac{16\pi}{3}$ C. 8π D. 4π

15. Find the region inside the circle $r = 25\sin\theta$.

A. 25 B. 5 C. 25π D. 5π

SECTION 15.4

1. The surface expressed parametrically by $x = r\cos\theta$, $y = r\sin\theta$, $z = 9 - r^2$ is

A. a sphere B. an ellipsoid C. a paraboloid D. a cone.

2. The surface expressed parametrically by $x = r\cos\theta$, $y = r\sin\theta$, $z = \sqrt{9 - r^2}$ is

A. a sphere B. an ellipsoid C. a paraboloid D. a cone.

3. Answer true or false: A parametric representation of the surface $z + x^2 + y^2 = 5$ in terms of the parameters $u = x$, and $v = y$ is $x = u$, $y = v$, $z = 5 - u^2 - v^2$.

4. Answer true or false: The parametric equations for $x^2 + y^2 = 16$ from the plane $z = 1$ to the plane $z = 2$ are $x = 4\cos v$, $y = 4\sin v$, $z = u$; $0 \leq v \leq 2\pi$, $1 \leq u \leq 2$.

5. Answer true or false: Parametric equations for $x^2 + z^2 = 25$ from $y = 0$ to $y = 1$ are $x = 5\cos u$, $y = v$, $z = 5\sin u$; $0 \leq u \leq 2\pi$, $0 \leq v \leq 1$.

6. The cylindrical paramentation of $z = xe^{x^2+y^2}$ is

A. $x = r\cos\theta$, $y = r\sin\theta$, $z = e^r$
B. $x = r\sin\theta$, $y = r\cos\theta$, $z = e^r$
C. $x = r\cos\theta$, $y = r\sin\theta$, $z = re^r\cos\theta$
D. $x = r\sin\theta$, $y = r\cos\theta$, $z = re^r\cos\theta$.

7. The equation of the tangent plane to $x = u$, $y = v$, $z = u + v^2$ where $u = 0$ and $v = 1$ is

A. $x + 2(y - 2) + z - 1 = 0$
B. $x + 2(y - 2) - z + 1 = 0$
C. $x + 2y - 2 + z + 1 = 0$
D. $x + 2y - 2 + z - 1 = 0$.

8. Answer true or false: To find the portion of the surface $z = x + y^2$ that lies above the rectangle $0 \leq x \leq 2$, $0 \leq y \leq 3$, evaluate $\int_0^2 \int_0^3 x + y^2 dy\, dx$.

9. Answer true or false: To find the portion of the surface $z = 3x^2 + 4y^2$ that lies above the rectangle $0 \leq x \leq 2$, $1 \leq y \leq 3$, evaluate $\int_1^3 \int_0^2 \sqrt{6x^2 + 8y^2}\, dx\, dy$.

10. Answer true or false: To find the portion of the surface $z = x^2 + 3y^2 + 4$ that lies above the rectangle $1 \leq x \leq 2$, $2 \leq y \leq 4$, evaluate $\int_2^4 \int_1^2 \sqrt{4x^2 + 36y^2 + 1}\, dx\, dy$.

11. Answer true or false: To find the portion of the surface $z = 5xy + 3$ that lies above the rectangle $1 \leq x \leq 3$, $2 \leq y \leq 5$, evaluate $\int_2^5 \int_1^3 \sqrt{25x^2 + 25y^2 + 1}\, dx\, dy$.

12. Answer true or false: To find the portion of the surface $z = x^2 - x + y^2$ that lies above the rectangle $0 \leq x \leq 4$, $0 \leq y \leq 3$, evaluate $\int_0^3 \int_0^4 \sqrt{4x^2 + 4y^2}\, dx\, dy$.

13. Answer true or false: To find the portion of the surface $z = x^2 - y$ that lies above the rectangle $0 \leq x \leq 2$, $0 \leq y \leq 4$, evaluate $\int_0^4 \int_0^2 2x\, dx\, dy$.

14. Answer true or false: To find the portion of the surface $z = x^2 - 2y$ that lies above the rectangle $1 \leq x \leq 2$, $0 \leq y \leq 1$, evaluate $\int_0^1 \int_1^2 \sqrt{4x^2 - 1}\, dx\, dy$.

15. Answer true or false: To find the portion of the surface $z = x^3 + y^3$ that lies above the rectangle $0 \leq x \leq 1$, $0 \leq y \leq 3$, evaluate $\int_0^3 \int_0^1 \sqrt{9x^4 + 9y^4 + 1}\, dx\, dy$.

SECTION 15.5

1. $\int_0^1 \int_0^2 \int_0^3 x^2 yz \, dx \, dy \, dz =$

A. 18 B. 9 C. 27 D. 6

2. $\int_0^2 \int_0^{\pi/2} \int_0^{\pi/2} \sin x \sin y \, dx \, dy \, dz =$

A. -2 B. 2 C. 1 D. -1

3. $\int_0^2 \int_0^{z^2} \int_0^y y \, dx \, dy \, dz =$

A. 4 B. $\frac{128}{21}$ C. 8 D. $\frac{8}{5}$

4. $\int_0^1 \int_0^x \int_0^{y^2} 3 \, dz \, dy \, dx =$

A. 1 B. $\frac{1}{12}$ C. $\frac{1}{3}$ D. $\frac{1}{4}$

5. $\int_{-1}^1 \int_0^z \int_0^y x^5 \, dx \, dy \, dz =$

A. $\frac{1}{7}$ B. $\frac{2}{7}$ C. $\frac{1}{5}$ D. 0

6. $\int_{-1}^1 \int_0^z \int_0^2 z^3 \, dx \, dy \, dz =$

A. $\frac{1}{3}$ B. $-\frac{1}{3}$ C. $\frac{2}{3}$ D. $\frac{4}{5}$

7. $\int_1^5 \int_0^{\pi/2} \int_0^{\cos y} \sin y \, dx \, dy \, dz =$

A. 2 B. 4 C. 0 D. 1

8. $\int_0^2 \int_0^z \int_0^y z^3 y \, dx \, dy \, dz =$

A. 2 B. 6.082 C. 6.095 D. 5

9. $\int_0^1 \int_0^z \int_0^{\sqrt{y^2+5}} yz \, dx \, dy \, dz =$

A. 0.29 B. 0 C. 2 D. 4

10. $\int_1^2 \int_0^{z^2} \int_0^3 dx \, dy \, dz =$

A. 8 B. 7 C. 6 D. 5

11. $\int_1^2 \int_{-z^2}^0 \int_0^{3\pi} \sin x \, dx \, dy \, dz =$

A. 8 B. $\frac{14}{3}$ C. 6 D. 5

12. $\int_0^4 \int_0^x \int_0^{x^2+y^2} dz \, dy \, dx =$

A. 64 B. $\frac{256}{3}$ C. $\frac{128}{3}$ D. 4

13. Answer true or false: $\int_0^5 \int_0^x \int_0^{x+y} dz \, dy \, dx = x^2$.

14. Answer true or false: $\int_0^5 \int_0^z \int_0^y dx \, dy \, dz = 25$.

15. Answer true or false: $\int_0^4 \int_0^{z^2} \int_0^{y^3} dx \, dy \, dz = 4^5$.

SECTION 15.6

1. A uniform beam 10 m in length is supported at its center by a fulcrum. A mass of 10 kg is placed at the left end, a mass of 4 kg is placed on the beam 4 m from the left end, and a third mass is placed 2 m from the right end. What mass should the third mass be to achieve equilibrium?

A. 14 kg B. 18 kg C. 8 kg D. 10 kg

2. A lamina with density $\delta(x, y) = xy$ is bounded by $x = 2$, $x = 0$, $y = x$, $y = 0$. Find its mass.

A. 2 B. 4 C. 1 D. 8

3. A lamina with density $\delta(x, y) = xy$ is bounded by $x = 2$, $x = 0$, $y = 0$, $y = x$. Find its center of mass.

A. $\left(\frac{8}{5}, \frac{16}{5}\right)$ B. $\left(\frac{8}{5}, \frac{8}{5}\right)$ C. $\left(\frac{16}{5}, \frac{16}{5}\right)$ D. $\left(\frac{16}{15}, \frac{16}{15}\right)$

4. A lamina with density $\delta(x, y) = x^2 + 2y^2$ is bounded by $x = y$, $x = 0$, $y = 0$, $y = 2$. Find its mass.

A. 4 B. $\frac{20}{3}$ C. $\frac{20}{5}$ D. 2

5. A lamina with density $\delta(x, y) = x^2 + 2y^2$ is bounded by $x = y$, $x = 0$, $y = 0$, $y = 2$. Find its center of mass.

A. $\left(\frac{1}{3}, \frac{1}{3}\right)$ B. $\left(\frac{34}{25}, \frac{88}{15}\right)$ C. $\left(\frac{25}{72}, \frac{25}{72}\right)$ D. $\left(\frac{1}{2}, 1\right)$

6. A lamina with density $\delta(x, y) = x^2 + 2y^2$ is bounded by $x = y$, $x = 0$, $y = 0$, $y = 2$. Find its moment of inertia about the x-axis.

A. $\frac{56}{9}$ B. 8 C. $\frac{25}{72}$ D. $\frac{56}{3}$

7. A lamina with density $\delta(x, y) = x^2 + 2y^2$ is bounded by $x = y$, $x = 0$, $y = 0$, $y = 2$. Find its moment of inertia about the y-axis.

A. $\frac{25}{72}$ B. 16 C. $\frac{32}{5}$ D. $\frac{56}{3}$

8. A lamina with density $\delta(x, y) = xy$ is bounded by $x = 2$, $x = 0$, $y = x$, $y = 0$. Find its moment of inertia about the x-axis.

A. $\frac{16}{5}$ B. $\frac{32}{5}$ C. $\frac{8}{3}$ D. $\frac{8}{5}$

9. A lamina with density $\delta(x, y) = xy$ is bounded by $x = 2$, $x = 0$, $y = x$, $y = 0$. Find its moment of inertia about the y-axis.

A. $\frac{16}{5}$ B. $\frac{32}{5}$ C. 2 D. $\frac{8}{5}$

10. Answer true or false: The moment of inertia about $x = a$, where a is the x-coordinate of the center of mass, is 0.

11. Answer true or false: The centroid given by $z = \sqrt{x^2 + y^2}$ is 0.

12. The centroid of a rectangular solid in the first octant with vertices $(0, 0, 0)$, $(0, 0, 1)$, and $(1, 1, 1)$ is

A. $\left(\frac{1}{2}, \frac{1}{2}, \frac{1}{2}\right)$ B. $(1, 1, 1)$ C. $\left(\frac{1}{3}, \frac{1}{3}, \frac{1}{3}\right)$ D. $\left(\frac{1}{4}, \frac{1}{4}, \frac{1}{4}\right)$.

13. The centroid of a rectangular solid in the first octant with vertices $(0, 0, 0)$, $(0, 0, 2)$, and $(2, 2, 2)$ is

A. $\left(\frac{1}{2}, \frac{1}{2}, \frac{1}{2}\right)$ B. $(0, 1, 2)$ C. $(1, 1, 1)$ D. $\left(\frac{2}{3}, \frac{2}{3}, \frac{2}{3}\right)$.

14. The centroid of the solid given by $(x - 2)^2 + y^2 + (z + 3)^2 = 9$ is

A. $(2, 0, 3)$ B. $(-2, 0, 3)$ C. $(0, 0, 0)$ D. $(2, 0, -3)$.

15. The centroid of the solid given by $\frac{(x-3)^2}{4} + \frac{(y-5)^2}{16} + \frac{(z+2)^2}{9} = 1$ is

A. $(-3, -5, 2)$ B. $(0, 0, 0)$ C. $(2, 4, 3)$ D. $(3, 5, -2)$.

SECTION 15.7

1. $\int_0^{\pi/2} \int_0^{\pi/2} \int_0^1 \rho^4 \sin\phi \cos\theta \, d\rho \, d\phi \, d\theta =$

A. 1 B. $\frac{1}{5}$ C. -1 D. $-\frac{1}{5}$

2. $\int_0^{\pi/2} \int_0^{\pi/2} \int_0^2 \rho^2 \sin\phi \cos\theta \, d\rho \, d\phi \, d\theta =$

A. $\frac{8}{3}$ B. $-\frac{8}{3}$ C. $2\pi^3$ D. π^3

3. Answer true or false: $\int_0^{2\pi} \int_0^{\pi} \int_0^{\sqrt{36-r^2}} 2r \, dz \, dr \, d\theta = \frac{16\pi\sqrt{2} - 24\pi}{3}$

4. $\int_0^{2\pi} \int_0^{\pi} \int_{-2}^{2} \sin\phi \cos\theta \, \rho \, d\rho \, d\theta \, d\phi =$

A. 0 B. 4 C. -4 D. 6

5. $\int_0^{\pi} \int_{-\pi/2}^{\pi/2} \int_1^3 \sin\phi \, d\rho \, d\theta \, d\phi =$

A. π B. 3π C. 4π D. $\frac{3\pi}{2}$

6. $\displaystyle\int_{-2}^{2}\int_{0}^{\pi/2}\int_{0}^{2\pi} \rho^3 \sin\phi \cos\theta \, d\phi \, d\theta \, d\rho =$

A. 16 B. 0 C. 8 D. 4

7. $\displaystyle\int_{0}^{2\pi}\int_{3}^{6}\int_{1}^{2} dz \, dr \, d\theta =$

A. 6π B. 2π C. 4π D. π

8. Find the center of gravity of the sphere $x^2 + y^2 + z^2 = 4$ where $\delta(x, y, z) = 6x^2y^2z^2$.

A. $(2, 2, 2)$ B. $(4, 4, 4)$ C. $(1, 1, 1)$ D. $(0, 0, 0)$

9. Answer true or false: The center of gravity of the solid enclosed by $z = \sqrt{x^2 + y^2}$ and $z = -\sqrt{x^2 + y^2}$, if the density is $\delta(x, y, z) = x^2 + y^2 + z^2$, is at the origin.

10. Answer true or false: The center of gravity of the solid enclosed by $x^2 + y^2 = 1$ and $2y^2 + z^2 = 1$ is at the origin if $\delta(x, y, z)$ is constant.

11. Answer true or false: The center of gravity of the solid enclosed by $x^2 + y^2 = 1$ and $3y^2 + z^2 = 1$ is at the origin if $\delta(x, y, z) = x^2 + 1$.

12. Answer true or false: $\displaystyle\int_{0}^{2\pi}\int_{0}^{1-\sin^2\theta}\int_{0}^{1} \sin\theta \, d\rho \, d\phi \, d\theta = 0$

13. $\displaystyle\int_{0}^{1}\int_{0}^{2\pi}\int_{0}^{3} \delta(r, \theta, z) \, dr \, d\theta \, dz$, where $\delta(r, \theta, z) = r$, is

A. 6π B. 9π C. 3 D. 3π.

14. $\displaystyle\int_{0}^{1}\int_{0}^{2\pi}\int_{0}^{3} \delta(r, \theta, z) \, dr \, d\theta \, dz$, where $\delta(r, \theta, z) = rz$, is

A. 3π B. $\dfrac{9\pi}{2}$ C. $\dfrac{3}{2}$ D. $\dfrac{3\pi}{2}$.

15. $\displaystyle\int_{0}^{1}\int_{0}^{2\pi}\int_{0}^{3} \delta(r, \theta, z) \, dr \, d\theta \, dz$, where $\delta(r, \theta, z) = z^2$, is

A. 6π B. $\dfrac{27\pi}{2}$ C. 3 D. 2π.

SECTION 15.8

1. Find $\dfrac{\partial(x, y)}{\partial(u, v)}$, if $x = u + 2v$ and $y = 3u + v$.

A. 5 B. -5 C. 7 D. -7

2. Find $\dfrac{\partial(x, y)}{\partial(u, v)}$, if $x = u^2$ and $y = u - v$.

A. $2u + 1$ B. $2u$ C. $-2u$ D. $-2u - 1$

3. Find the Jacobian if $x = e^u$ and $y = e^v$.

A. 0 B. e^{uv} C. e^{u-v} D. e^{u+v}

4. Find the Jacobian if $u = e^x$ and $v = ye^x$.

A. $\dfrac{\ln v - v}{u}$ B. $\dfrac{\ln v + v}{u}$ C. $\dfrac{1}{u^2}$ D. $-\dfrac{1}{u^2}$

5. Find the Jacobian if $x = 3u + w$, $y = vw$, and $z = u^2v$.

A. $3u^2v + 2uw$ B. $3u^2v + 2uw^2$ C. $6(u^2v + uw^2)$ D. $-6(u^2v + uw^2)$

6. Find the Jacobian if $u = x$, $v = \dfrac{y}{x}$, and $w = x + z$.

A. $-u$ B. $-uvw$ C. u D. uvw

7. Answer true or false: If $x = uv$, $y = u + 2v$, then $\dfrac{\partial(x,y)}{\partial(u,v)} = \begin{vmatrix} v & u \\ 1 & 2 \end{vmatrix}$.

8. Answer true or false: If $x = u^2$, $y = v^2$, then $\dfrac{\partial(x,y)}{\partial(u,v)} = \begin{vmatrix} 2u & 1 \\ 1 & 2v \end{vmatrix}$.

9. Answer true or false: If $x = u + 2v + 3w$, $y = 7 + uv + 4v$, $z = e^u + vw$, then $\dfrac{\partial(x,y,z)}{\partial(u,v,w)} = \begin{vmatrix} 1 & 2v & 3 \\ 7 & 4 & u \\ e^u & w & v \end{vmatrix}$.

10. Answer true or false: If $x = e^u$, $y = e^v$, $z = e^w$, then $\dfrac{\partial(x,y,z)}{\partial(u,v,w)} = \begin{vmatrix} e^u & 0 & 0 \\ 0 & e^v & 0 \\ 0 & 0 & e^w \end{vmatrix}$.

11. Answer true or false: If $u = x + 2y$ and $v = 3x + v$, $\displaystyle\int_0^1 \int_0^1 e^{x+2y} e^{3x+y}\, dx\, dy = \int_0^3 \int_0^4 e^u e^v\, du\, dv$.

12. Answer true or false: If $u = 4x + y$ and $v = x^2y$, $\displaystyle\int_1^2 \int_1^2 \frac{4x+y}{x^2y}\, dx\, dy = \int_5^{10} \int_1^4 \frac{u}{v}\, dv\, du$.

13. Answer true or false: If $u = x + y$ and $v = 2x - y$, $\displaystyle\int_1^2 \int_1^2 \frac{(x+y)^2}{2x-y}\, dy\, dx = \int_2^4 \int_1^2 -\frac{u^2}{v}\, dv\, du$.

14. Answer true or false: If $x = uvw$, $y = e^u$, and $z = 2u$, the Jacobian is 0.

15. Answer true or false: If $x = uvw$, $y = u - v^2w$, and $z = u$, the Jacobian has no dependence on v, nor on w.

CHAPTER 15 TEST

1. $\int_0^4 \int_0^2 dx\, dy =$

A. 6 B. 8 C. 0 D. 20

2. $\iint_R 2yx^2 dA;\ R = \{(x, y) : 0 \leq x \leq 1,\ 0 \leq y \leq 2\}.$

A. $\frac{4}{3}$ B. $\frac{8}{3}$ C. 0 D. $\frac{1}{3}$

3. Answer true or false: The volume of the solid bounded by $z = x^7y^3$ and $R = \{(x, y) : -1 \leq x \leq 1,$ $-1 \leq y \leq 2\}$ is $\int_{-1}^{1} \int_{-1}^{2} x^7y^3 dy\, dx.$

4. Answer true or false: The average value of the function $f(x, y) = x^3y^4$ over the rectangle $[0, 2] \times [0, 3]$ is $\frac{1}{6} \int_0^2 \int_0^3 x^3y^4 dy\, dx.$

5. $\int_0^{\pi/2} \int_0^{\cos x} dy\, dx =$

A. 1 B. 0 C. -1 D. π

6. Answer true or false: $\iint_R \sin x\, dA$, where R is the region bounded by $y = x + 4$, $y = x$, and $x = 16$ is $\int_{16}^{28} \int_{x+4}^{2x} \sin x\, dy\, dx.$

7. $\int_0^3 \int_0^{x^2} y\, dy\, dx =$

A. $\frac{3^5}{10}$ B. $\frac{3^5}{5}$ C. $\frac{3^5}{10} - 9$ D. $\frac{3^5}{10} + 9$

8. $\int_{-\pi/2}^{0} \int_0^{\sin\theta} r \cos\theta\, dr\, d\theta =$

A. $-\frac{1}{6}$ B. $\frac{1}{6}$ C. $\frac{\pi}{2}$ D. $-\frac{\pi}{2}$

9. Answer true or false: $\int_{-1}^{1} \int_0^{\sqrt{9-x^2}} dx\, dy = \int_0^{\pi} \int_0^3 r\, dr\, d\theta$

10. Find the volume of the solid left hemisphere if $r^2 + z^2 = 1$.

A. $\frac{4\pi}{3}$ B. $\frac{8\pi}{3}$ C. $\frac{2\pi}{3}$ D. $\frac{16\pi}{3}$

11. Find the area enclosed by one petal of the three-petaled rose $r = 2\sin 3\theta$.

A. $\frac{\pi}{12}$ B. $\frac{\pi}{6}$ C. $\frac{\pi}{24}$ D. $\frac{\pi}{3}$

12. The surface expressed parametrically by $x = r\cos\theta$, $y = r\sin\theta$, $z = \sqrt{16 - r^2}$ is

A. a sphere B. an ellipsoid C. a paraboloid D. a cone.

13. The cylindrical paramentation of $z = (x^2 + y^2)e^x$ is

A. $x = r\cos\theta,\ y = r\sin\theta,\ z = e^r$
B. $x = r\sin\theta,\ y = r\cos\theta,\ z = e^r$
C. $x = r\cos\theta,\ y = r\sin\theta,\ z = e^{r\cos\theta}$
D. $x = r\sin\theta,\ y = r\cos\theta,\ z = e^{r\cos\theta}$.

14. The equation of the tangent plane to $x = u,\ y = v,\ z = u + v^2$ where $u = 1$ and $v = 1$ is

A. $x - 1 + 2(y - 2) + z - 2 = 0$
B. $x - 1 - 2(y - 1) - z + 2 = 0$
C. $x - 1 + 2y - 2 + z + 2 = 0$
D. $x - 1 + 2y - 2 - z + 2 = 0$.

15. Answer true or false: To find the portion of the surface $z = 3x^2 + 4y^2$ that lies above the rectangle $0 \le x \le 2,\ 1 \le y \le 3$, evaluate $\int_1^3 \int_0^2 \sqrt{36x^2 + 64y^2 + 1}\, dx\, dy$.

16. $\int_1^3 \int_0^{\pi/2} \int_0^{\pi/2} \sin x \sin y\, dx\, dy\, dz =$

A. -2 B. 2 C. 1 D. -1

17. $\int_1^5 \int_0^{\pi/2} \int_0^{\sin y} \cos y\, dx\, dy\, dz =$

A. 2 B. 4 C. 0 D. 1

18. A lamina with density $\delta(x, y) = xy$ is bounded by $x = 0,\ x = y,\ y = 0,\ y = 2$. Find its mass.

A. 2 B. 4 C. 1 D. 8

19. A lamina with density $\delta(x, y) = xy$ is bounded by $x = 0,\ x = y,\ y = 0,\ y = 2$. Find its center of mass.

A. $\left(\frac{8}{5}, \frac{16}{5}\right)$ B. $\left(\frac{8}{5}, \frac{8}{5}\right)$ C. $\left(\frac{16}{5}, \frac{16}{5}\right)$ D. $\left(\frac{16}{15}, \frac{16}{15}\right)$

20. The centroid of a rectangular in the first octant with vertices $(0, 0, 0)$, $(0, 1, 1)$, and $(1, 0, 0)$ is

A. $(0, 1, 2)$ B. $\left(\frac{1}{2}, \frac{1}{2}, \frac{1}{2}\right)$ C. $(1, 1, 1)$ D. $\left(\frac{2}{3}, \frac{2}{3}, \frac{2}{3}\right)$

21. $\int_0^{2\pi} \int_0^{\pi/2} \int_0^2 \rho^2 \sin\phi\, d\rho\, d\phi\, d\theta =$

A. $\frac{16\pi}{3}$ B. $-\frac{16\pi}{3}$ C. $4\pi^3$ D. $-4\pi^3$

22. $\int_0^{\pi/2} \int_{-\pi}^{\pi} \int_4^8 \sin\phi\, d\rho\, d\theta\, d\phi =$

A. 4π B. 8π C. 0 D. 2π

23. Find the center of gravity of the sphere $x^2 + y^2 + z^2 = 6$ where $\delta(x, y, z) = 6x^2y^2z^2$.

A. $(3, 3, 3)$ B. $(6, 6, 6)$ C. $(1, 1, 1)$ D. $(0, 0, 0)$

24. Answer true or false: The center of gravity of the solid enclosed by $x^2 + y^2 = 1$ and $y^2 + z^2 = 1$ is at the origin if $\delta(x, y, z) = x + 1$.

25. Answer true or false: $\int_0^{2\pi} \int_0^{1-\cos^2\theta} \int_0^1 \cos\theta\, d\rho\, d\phi\, d\theta = 0$

26. Find $\dfrac{\partial(x,y)}{\partial(u,v)}$, if $x = 5u + 2v$ and $y = 7u + v$.

A. -9 B. 9 C. -19 D. 19

27. Find the Jacobian if $u = xy$ and $v = 2x$.

A. $-\dfrac{2u}{v^2}$ B. $-\dfrac{2u}{v^2} - \dfrac{1}{v}$ C. $-\dfrac{2u}{v^2} + \dfrac{1}{v}$ D. $\dfrac{1}{v}$

28. Find the Jacobian if $x = 4u + w$, $y = vw$, and $z = u^2v$.

A. $4u^2v - 2uvw$ B. $4u^2v + 2uw^2$ C. $8(u^2v + uw^2)$ D. $-8(u^2v + uw^2)$

29. Answer true or false: $\displaystyle\int_1^2 \int_1^2 \left(\frac{x+y}{2x-y}\right)^2 dx\,dy = \int_2^4 \int_1^2 -\frac{u^2}{v^2} dv\,du$, where $u = x+y$ and $v = 2x-y$.

ANSWERS TO SAMPLE TESTS

SECTION 15.1

1. A 2. B 3. B 4. A 5. B 6. D 7. F 8. T 9. B 10. B 11. F 12. T 13. T 14. T 15. T

SECTION 15.2

1. B 2. A 3. A 4. D 5. F 6. T 7. C 8. C 9. B 10. B 11. F 12. D 13. T 14. F 15. F

SECTION 15.3

1. A 2. C 3. C 4. C 5. B 6. F 7. F 8. T 9. F 10. B 11. C 12. C 13. B 14. B 15. C

SECTION 15.4

1. C 2. A 3. T 4. T 5. T 6. C 7. D 8. F 9. F 10. T 11. T 12. F 13. T 14. T 15. T

SECTION 15.5

1. B 2. B 3. B 4. D 5. D 6. D 7. A 8. C 9. A 10. B 11. B 12. A 13. F 14. F 15. F

SECTION 15.6

1. B 2. A 3. B 4. B 5. B 6. A 7. A 8. C 9. A 10. F 11. F 12. A 13. C 14. D 15. D

SECTION 15.7

1. B 2. A 3. F 4. A 5. C 6. B 7. A 8. D 9. T 10. F 11. F 12. T 13. B 14. B 15. D

SECTION 15.8

1. B 2. C 3. D 4. C 5. A 6. C 7. T 8. F 9. F 10. T 11. F 12. F 13. T 14. F 15. F

CHAPTER 15 TEST

1. B 2. A 3. T 4. T 5. A 6. F 7. A 8. B 9. T 10. C 11. B 12. A 13. C 14. B 15. T 16. B 17. A 18. A 19. D 20. B 21. A 22. B 23. D 24. F 25. F 26. A 27. D 28. A 29. F

CHAPTER 16

Sample Exams

SECTION 16.1

1. Answer true or false: $\phi(x, y) = x^2y$ is the potential function for $\mathbf{F}(x, y) = 2x\mathbf{i} + \mathbf{j}$.

2. Answer true or false: $\phi(x, y) = \sin x + \cos y$ is the potential function for $\mathbf{F}(x, y) = \cos x\mathbf{i} - \sin y\mathbf{j}$.

3. $\mathbf{F}(x, y, z) = x^3\mathbf{i} + 3\mathbf{j} + xz\mathbf{k}$. Find div$\mathbf{F}$.

A. $3x^2 + x$ B. $3x^2\mathbf{i} + x\mathbf{k}$ C. $3x^3$ D. $3x^2 - x$

4. $\mathbf{F}(x, y, z) = x^3\mathbf{i} + 3\mathbf{j} + xz\mathbf{k}$. Find curl$\mathbf{F}$.

A. $z\mathbf{j}$ B. $-z\mathbf{j}$ C. z D. $-z$

5. $\mathbf{F}(x, y, z) = xyz\mathbf{i} + y\mathbf{j} + x\mathbf{k}$. Find div$\mathbf{F}$.

A. $yz + 1$ B. $yz - 1$ C. $xy - 2 - xz$ D. $xy - 2 + xz$

6. $\mathbf{F}(x, y, z) = xyz\mathbf{i} + y\mathbf{j} + x\mathbf{k}$. Find curl$\mathbf{F}$.

A. $xz\mathbf{i} + \mathbf{j}$ B. $xz\mathbf{i} - x\mathbf{j}$ C. $(xy - 1)\mathbf{j} - xz\mathbf{k}$ D. $(xy - 1)\mathbf{j} + xz\mathbf{k}$

7. $\mathbf{F}(x, y, z) = e^x\mathbf{i} + \sqrt{x^2 + y^2}\mathbf{j} + ye^x\mathbf{k}$. Find div$\mathbf{F}$.

A. $e^x + \dfrac{2y}{\sqrt{x^2 + y^2}}$ B. $e^x + \dfrac{y}{\sqrt{x^2 + y^2}}$

C. $e^x + ye^x + \dfrac{2x}{\sqrt{x^2 + y^2}}$ D. $e^x + ye^x + \dfrac{x}{\sqrt{x^2 + y^2}}$

8. $\mathbf{F}(x, y, z) = e^x\mathbf{i} + \sqrt{x^2 + y^2}\mathbf{j} + ye^x\mathbf{k}$. Find curl$\mathbf{F}$.

A. $e^x\mathbf{i} + \dfrac{2y}{\sqrt{x^2 + y^2}}\mathbf{j}$ B. $e^x\mathbf{i} + \dfrac{y}{\sqrt{x^2 + y^2}}\mathbf{j}$

C. $e^x\mathbf{i} + ye^x\mathbf{j} + \dfrac{2x}{\sqrt{x^2 + y^2}}\mathbf{k}$ D. $e^x\mathbf{i} - ye^x\mathbf{j} + \dfrac{x}{\sqrt{x^2 + y^2}}\mathbf{k}$

9. Answer true or false: $\nabla^2\phi = 6xy^2 + 2x^3$, if $\phi = x^3y^2$

10. Answer true or false: $\nabla^2\phi = \cos x - \sin(xy)$, if $\phi = \sin x + \cos(xy)$

11. Answer true or false: $\nabla^2\phi = e^x + 2xye^{xy}$, if $\phi = e^{xy}$

12. Answer true or false: $\nabla^2\phi = (xy + yz + xz)^2e^{xyz}$, if $\phi = e^{xyz}$

13. Answer true or false: $\nabla^2\phi = 0$, if $\phi = 7x + 3y + 2z$

14. Answer true or false: $\nabla^2\phi = 6x$, if $\phi = x^3 + y + z$

15. Answer true or false: $\nabla^2\phi = -\sin x - \sin y - \sin z$, if $\phi = \sin x + \sin y + \sin z$

SECTION 16.2

1. $\int_C (1 + x^2y)\, ds$, where $x = 3t$ and $y = t$, $(0 \le t \le 1)$, is

A. 3.25 B. 0.3 C. 13 D. 2.

2. $\int_C (1 + x^2y)\, dx$, where $x = 3t$ and $y = t$, $(0 \le t \le 1)$, is

A. 6.5 B. 3.75 C. 6.90 D. 3.25.

3. $\int_C (1 + x^2y)\, dy$, where $x = 3t$ and $y = t$, $(0 \le t \le 1)$, is

A. 6.75 B. 3.75 C. 9.75 D. 2.3.

4. $\int_C xy + z\, ds$, where $x = t$, $y = 2t$, and $z = -2t$, $(0 \le t \le 1)$, is

A. -1 B. -2 C. 0 D. 0.05.

5. $\int_C xy + z\, dy$, where $x = t$, $y = 2t$, and $z = -2t$, $(0 \le t \le 1)$, is

A. -1 B. 1 C. $\frac{2}{3}$ D. $-\frac{2}{3}$.

6. $\int_C xy + z\, dz$, where $x = t$, $y = 2t$, and $z = -2t$, $(0 \le t \le 1)$, is

A. -1 B. 1 C. $\frac{2}{3}$ D. $-\frac{2}{3}$.

7. $\int_C 3xy^2 dx + 5xy^2 dy$ along the curve $x = y^2$ from $(0,0)$ to $(1,1)$ is

A. 2 B. 7 C. -2 D. -7.

8. $\int_C xy\, ds$, where $x = \cos t$, $y = \sin t$, $0 \le t \le \pi$, is

A. 0 B. 1 C. 2 D. 4.

9. Answer true or false: If $x = t$, $y = 5t$, $(0 \le t \le 1)$, $\int_C xy\, ds = \int_0^1 5t^2 dt$

10. Answer true or false: If $x = \cos t$, $y = \sin t$, $(0 \le t \le 2\pi)$, $\int_C x - 2y\, ds = \int_0^{2\pi} \cos t - 2\sin t\, dt$

11. Answer true or false: If $x = \cos t$, $y = -\sin t$, $z = 3t (0 \le t \le 2\pi)$, $\int_C xy - z\, ds = 2\int_0^{2\pi} -\sin t \cos t - 3t\, dt$

12. Answer true or false: If $x = e^t$, $y = 3e^t$, $z = e^t$ $(0 \le t \le 2\pi)$, $\int_C x + y + z\, ds = \int_0^1 5\sqrt{11} e^t dt$

13. Find the work done by $\mathbf{F}(x, y) = x\mathbf{i} + xy\mathbf{j}$ along the curve $y = x^2$ from $(0, 0)$ to $(1, 1)$ is

A. 0.75 B. 1.50 C. 1 D. 2.

14. Answer true or false: The work done by $\mathbf{F}(x, y) = x\mathbf{i} + ye^x\mathbf{j}$ along the curve $y = x^2$ from $(-1, -1)$ to $(0, 0)$ is the same as the work moving the same particle along the same curve from $(0, 0)$ to $(1, 1)$.

15. Answer true or false: The work done by $\mathbf{F}(x, y) = x\mathbf{i} + ye^x\mathbf{j}$ along the curve $y = x^3$ from $(0, 0)$ to $(1, 1)$ is the same as the work moving the same particle along the same curve from $(-1, -1)$ to $(0, 0)$.

SECTION 16.3

1. Answer true or false: $\mathbf{F}(x, y) = 2x\mathbf{i} + 3y\mathbf{j}$ is a conservative vector field.

2. Answer true or false: $\mathbf{F}(x, y) = xy\mathbf{i} + xy\mathbf{j}$ is a conservative vector field.

3. Answer true or false: $\mathbf{F}(x, y) = x^2y\mathbf{i} + \frac{x^3}{3}\mathbf{j}$ is a conservative vector field.

4. Answer true or false: $\mathbf{F}(x, y) = \sin x\mathbf{i} + \sin y\mathbf{j}$ is a conservative vector field.

5. Answer true or false: $\mathbf{F}(x, y) = \sin y\mathbf{i} + \sin x\mathbf{j}$ is a conservative vector field.

6. $\int_{(1,0)}^{(2,3)} 2x\,dx + 3y\,dy =$

A. $\frac{27}{2}$ B. $\frac{39}{2}$ C. $\frac{33}{2}$ D. 0

7. $\int_{(0,1)}^{(1,3)} x^2y\,dx + \frac{x^3}{3}\,dy =$

A. 0 B. $\frac{9}{2}$ C. 2 D. $\frac{7}{2}$

8. $\int_{(1,2)}^{(2,5)} 3x^2y\,dx + x^3dy =$

A. 76 B. 38 C. 80 D. 40

9. $\int_{(0,0)}^{(\pi,\pi)} \sin x\,dx + \sin y\,dy =$

A. 4 B. 2 C. −4 D. 0

10. $\int_{(0,0)}^{(\pi,\pi)} \sin x\,dx - \sin y\,dy =$

A. 4 B. 2 C. −4 D. 0

11. Answer true or false: If $\mathbf{F}(x, y) = 5y\mathbf{i} + 5x\mathbf{j}$, then $\phi = 5$.

12. For $\mathbf{F}(x,y) = x^2y^3\mathbf{i}+x^3y^2\mathbf{j}$ the work done by the force field on a particle moving along an arbitrary smooth curve from $P(1,1)$ to $Q(0,0)$ is

A. $\frac{1}{2}$ B. 1 C. $-\frac{1}{2}$ D. $\frac{2}{3}$.

13. For $\mathbf{F}(x,y) = x^2y\mathbf{i} + \frac{x^3}{3}\mathbf{j}$ the work done by the force field on a particle moving along an arbitrary smooth curve from $P(0,0)$ to $Q(1,2)$ is

A. $\frac{4}{3}$ B. $-\frac{5}{6}$ C. $\frac{1}{3}$ D. $-\frac{1}{3}$.

14. For $\mathbf{F}(x,y) = \mathbf{i}+\mathbf{j}$ the work done by the force field on a particle moving along an arbitrary smooth curve from $P(0,0)$ to $Q(2,3)$ is

A. 2 B. 3 C. 5 D. 0.

15. For $\mathbf{F}(x,y) = x^2\mathbf{i} + y\mathbf{j}$ the work done by the force field on a particle moving along an arbitrary smooth curve from $P(0,0)$ to $Q(3,2)$ is

A. 11 B. 7 C. -11 D. -7.

SECTION 16.4

1. Evaluate $\oint_C 5xy\,dx + 7xy\,dy$, where C is the rectangle $x = 1$, $x = 2$, $y = 1$, $y = 3$.

A. 13 B. $\frac{45}{95}$ C. $\frac{47}{97}$ D. $\frac{51}{101}$

2. The area enclosed in the ellipse $\frac{x^2}{9} + \frac{y^2}{4} = 1$ is

A. 36 B. 36π C. 6 D. 6π.

3. The area enclosed in the ellipse $\frac{(x-2)^2}{25} + \frac{(y+3)^2}{9} = 1$ is

A. 225 B. 225π C. 15 D. 15π.

4. The work done by the field $\mathbf{F}(x,y) = y^3\mathbf{i} + (y - x^3)\mathbf{j}$ on a particle that travels once around a unit circle $x^2 + y^2 = 1$ in a counterclockwise direction is

A. $\frac{3\pi}{4}$ B. $\frac{3\pi}{2}$ C. $\frac{\pi}{4}$ D. π.

5. The work done by the field $\mathbf{F}(x,y) = (x^2 + y^3)\mathbf{i} + (\sin y - x^3)\mathbf{j}$ on a particle that travels once around a unit circle $x^2 + y^2 = 1$ in a counterclockwise direction is

A. $\frac{3\pi}{4}$ B. $\frac{3\pi}{2}$ C. $\frac{\pi}{4}$ D. π.

6. The work done by the field $\mathbf{F}(x,y) = y^3\mathbf{i} + (y - x^3)\mathbf{j}$ on a particle that travels once around a circle $x^2 + y^2 = 4$ in a counterclockwise direction is

A. 96π B. $\frac{3\pi}{2}$ C. $\frac{\pi}{4}$ D. π.

7. The work done by the field $\mathbf{F}(x, y) = (x^2 + y^3)\mathbf{i} + (\sin y - x^3)\mathbf{j}$ on a particle that travels once around a circle $x^2 + y^2 = 4$ in a counterclockwise direction is

A. 96π B. $\frac{3\pi}{2}$ C. $\frac{\pi}{4}$ D. π.

8. $\oint_C (x^2 + y)\,dx + (y^2 + x)\,dy$, where C is $x^2 + y^2 = 9$, is

A. 6π B. 9π C. 18π D. 0.

9. $\oint_C (e^x + 2y)\,dx + (3y^4 + 2x)\,dy$, where C is $x^2 + y^2 = 1$, is

A. π B. 2π C. 4π D. 0.

10. $\oint_C (e^x + 2y)\,dx + (3y^4 + 2x)\,dy$, where C is $\frac{x^2}{9} + \frac{y^2}{4} = 1$, is

A. 6π B. 6 C. 0 D. 12.

11. $\oint_C (e^x + 2y)\,dx + (3y^4 + 2x)\,dy$, where C is $4x^2 + 9y^2 = 36$, is

A. 6π B. 6 C. 0 D. 12.

12. Use a line integral to find the area of the triangle with vertices $(0, 0)$, $(a, 0)$, and (a, b).

A. $\frac{1}{2}$ B. ab C. $\frac{ab}{2}$ D. $2ab$

13. Answer true or false: $\oint_C (e^y + 2x)\,dx + (3x^4 + 2y)\,dy$; $C = x^2 + y^2 = 1$, is 2π

14. Answer true or false: $\oint_C \sin x\,dx + \cos y\,dy$; $C = x^2 + y^2 = 1$, is π

15. Answer true or false: $\oint_C \sin y\,dx + \cos x\,dy$; $C = x^2 + y^2 = 1$, is 2π

SECTION 16.5

1. Evaluate $\iint_\sigma xz\,dS$, where σ is the part of the plane $x + y + 2z = 1$ in the first octant.

A. $\frac{\sqrt{6}}{24}$ B. $\frac{\sqrt{3}}{12}$ C. $\frac{5\sqrt{6}}{16}$ D. $\sqrt{3}$

2. Answer true or false: $\iint_\sigma xz\,dS$, where σ is the part of the plane $x + 2y + z = 1$ in the first octant is $\int_0^1 \int_0^{1/2-x} (x - x^2 - xy)\,dy\,dx$.

3. Find the surface area of the cone $z = \sqrt{x^2 + y^2}$ that lies below the plane $z = 4$.

A. 16π B. $16\pi\sqrt{2}$ C. 8π D. $8\pi\sqrt{2}$

4. Find the surface area of the cone $z = \sqrt{x^2 + y^2}$ that lies between the planes $z = 3$ and $z = 4$.

A. 7π B. $7\pi\sqrt{2}$ C. 8π D. $8\pi\sqrt{2}$

5. Find the surface of $x^2 + y^2 + (z - 2)^2 = 4$ that lies below $z = 2$.

A. 8π B. 16π C. 32π D. 64π

6. Find the surface of $x^2 + y^2 + (z - 2)^2 = 4$ that lies below $z = 4$.

A. 8π B. 32π C. 16π D. 64π

7. Answer true or false: If σ is the part of $x + y + z = 2$ that lies in the first octant $\iint_\sigma xz\,dS =$
$\sqrt{3}\int_0^1\int_0^{1-z}(1 - y - z)z\,dy\,dz$.

8. Answer true or false: If σ is the part of $\sin x + \cos y + z = 0$ that lies in the first octant $\iint_\sigma x^2 z\,dS =$
$\sqrt{2}\iint_R x^2(-\sin x - \cos y)\,dy\,dx$.

9. Answer true or false: If σ is the part of $x + y - 2z = 5$ that lies in the first octant $\iint_\sigma x\sin y\,dS =$
$\sqrt{6}\iint_R \sin y(5 - y + 2z)\,dA$.

10. Answer true or false: If σ is the part of $x + y + 3z = 8$ that lies in the first octant $\iint_\sigma xe^y\,dS =$
$\sqrt{11}\iint_R e^y(8 - 3z - y)\,dA$.

11. Answer true or false: If σ is the part of $z = x^2 + y^2 + 4$ that lies in the first octant $\iint_\sigma y^2 z^2\,dS =$
$\iint_R x^4 + y^2\sqrt{4x^2 + 4y^2}\,dA$.

12. Answer true or false: If σ is the part of $x + y + z = 6$ that lies in the first octant $\iint_\sigma e^x e^y\,dS =$
$\sqrt{3}\iint_R e^x e^{6-y-x}\,dA$.

13. Answer true or false: If σ is the part of $x + y + z = 6$ that lies in the first octant $\iint_\sigma e^x e^y\,dS =$
$\sqrt{3}\iint_R e^y e^{6-x-y}\,dA$.

14. Answer true or false: If σ is the part of $x + 2y + z = 5$ that lies in the first octant $\iint_\sigma zx^2y\,dS =$ $\sqrt{6}\iint_R x^2y(5 - x - 2y)\,dA$.

15. Answer true or false: If σ is the part of $x + 2y + z = 5$ that lies in the first octant $\iint_\sigma zx^2\,dS =$ $\sqrt{6}\iint_R zx^2\left(\frac{5 - x - z}{2}\right)\,dA$.

SECTION 16.6

1. Find the flux of the vector field $\mathbf{F}(x, y, z) = 2z\mathbf{k}$ across the sphere $x^2 + y^2 + z^2 = 9$ oriented outward.

   A. 72π B. 36π C. 0 D. 108π

2. Find the flux of the vector field $\mathbf{F}(x, y, z) = 5z\mathbf{k}$ across the sphere $x^2 + y^2 + z^2 = 1$ oriented outward.

   A. $\frac{20\pi}{3}$ B. $\frac{500\pi}{3}$ C. 0 D. $\frac{4\pi}{3}$

3. Answer true or false: If σ is the portion of the surface $z = 4 - x^2 - y^2$ that lies above the xy-plane, and σ is oriented up, the flux of the vector field $\mathbf{F}(x, y, z) = x\mathbf{i} + y\mathbf{j} + z\mathbf{k}$ across σ is $\Phi = \int_0^{2\pi}\int_0^1 (x^2 + y^2 + 4)\,dA$.

4. Answer true or false: If σ is the portion of the surface $z = 1 - x^2 - y^2$ that lies above the xy-plane, and σ is oriented up, the flux of the vector field $\mathbf{F}(x, y, z) = x^2\mathbf{i} + y\mathbf{j} + z\mathbf{k}$ across σ is $\Phi = \iint_R (x^3 + y^2 - x^2 + 2)\,dA$.

5. Let $\mathbf{F}(x, y, z) = 5x\mathbf{i}$. The flux outward between the planes $z = 0$ and $z = 2$ is

   A. 0 B. $\frac{25}{2}$ C. 25 D. 5.

6. Answer true or false: Let $\mathbf{F}(x, y, z) = x\mathbf{i} + y\mathbf{j} + 3\mathbf{k}$. The flux outward through the surface $x^2 + y^2 + z^2 = 1$ is $\int_0^{2\pi}\int_0^{\pi} (2\sin^2\phi\cos\theta\sin\theta + 3\sin\phi\cos\phi)\,d\phi\,d\theta$.

7. Let $\mathbf{F}(x, y, z) = x\mathbf{i} + y\mathbf{j} + z\mathbf{k}$ and σ be the portion of the surface $z = 5 - x^2 - y^2$ that lies above the xy-plane. Find the flux of the vector field across σ.

   A. $\frac{575\pi}{2}$ B. $\frac{5\pi}{2}$ C. $\frac{15\pi}{2}$ D. 0

8. Answer true or false: If $\mathbf{F}(x, y, z) = 2x\mathbf{i} + y\mathbf{k}$, the flux through the portion of the surface σ that lies above the xy-plane, where σ is defined by $z = 6 - x^2 - y^2$, is $\iint_R (x^2 + y^2 + 6)\,dA$.

9. Answer true or false: If $\mathbf{F}(x, y, z) = 2x\mathbf{i} + y\mathbf{k}$, the flux through the portion of the surface σ that lies above the xy-plane, where σ is defined by $z = 3x^2 + 3y^2 + 1$, is $\iint_R (3x^2 + 3y^2 + z)\, dA$.

10. If $\mathbf{F}(x, y, z) = x\mathbf{i} + y\mathbf{k}$, the magnitude of the flux through the portion of the surface σ that lies in front of the xz-plane, where σ is defined by $y = 1 - x^2 - z^2$, is

A. $\frac{5\pi}{2}$ B. $\frac{15\pi}{2}$ C. π D. 0.

11. If $\mathbf{F}(x, y, z) = y\mathbf{j} + z\mathbf{k}$, the magnitude of the flux through the portion of the surface σ that lies right of the yz-plane, where σ is defined by $x = 1 - y^2 - z^2$, is

A. $\frac{5\pi}{2}$ B. $\frac{15\pi}{2}$ C. π D. 0.

12. Answer true or false: The surface $z = -x^3 - y^2 + 5$ has a normal vector $\mathbf{n} = \frac{-3x^2\mathbf{i} - 2y\mathbf{j} + \mathbf{k}}{\sqrt{9x^2 + 4y^2 + 1}}$.

13. Answer true or false: The surface $y = -x^2 - z^2 + 8$ has a normal vector $\mathbf{n} = \frac{-2x\mathbf{i} - 2z\mathbf{j} + \mathbf{k}}{\sqrt{4x^2 + 4z^2 + 1}}$.

14. Answer true or false: The surface $z = -2x^3 - 2y^3 + 7$ has a normal vector $\mathbf{n} = \frac{-6x\mathbf{i} - 6y\mathbf{j} + \mathbf{k}}{\sqrt{36x^2 + 36y^2 + 1}}$.

15. Answer true or false: The surface $z = 4x + 2y$ has a normal vector $\mathbf{n} = \frac{-4\mathbf{i} - 2\mathbf{j} + \mathbf{k}}{\sqrt{21}}$.

SECTION 16.7

1. Find the outward flux of the vector field $\mathbf{F}(x, y, z) = x\mathbf{i}$ across the sphere $x^2 + y^2 + z^2 = 4$.

A. $\frac{32\pi}{3}$ B. $\frac{256\pi}{3}$ C. 0 D. $\frac{16\pi}{3}$

2. Find the outward flux of the vector field $\mathbf{F}(x, y, z) = \frac{x}{2}\mathbf{i}$ across the sphere $x^2 + y^2 + z^2 = 4$.

A. $\frac{32\pi}{3}$ B. $\frac{256\pi}{3}$ C. 0 D. $\frac{16\pi}{3}$

3. Find the outward flux of the vector field $\mathbf{F}(x, y, z) = 3z\mathbf{k}$ across the sphere $x^2 + y^2 + z^2 = 4$.

A. $\frac{32\pi}{3}$ B. 32π C. $\frac{256\pi}{3}$ D. 256π

4. Let $\mathbf{F}(x, y, z) = \frac{5x\mathbf{i} + 5y\mathbf{j} + 5z\mathbf{k}}{(x^2 + y^2 + z^2)^{3/2}}$ and let σ be a closed, orientable surface that surrounds the origin. Then $\mathbf{\Phi} =$

A. 5π B. 20π C. 100π D. 25π.

5. Let $\mathbf{F}(x, y, z) = \frac{8x\mathbf{i} + 8y\mathbf{j} + 8z\mathbf{k}}{(4x^2 + 4y^2 + 4z^2)^{3/2}}$ and let σ be a closed, orientable surface that surrounds the origin. Then $\mathbf{\Phi} =$

A. 4π B. 8π C. 2π D. 16π.

6. Answer true or false: Let $\mathbf{F}(x, y, z) = \dfrac{x^2\mathbf{i} + y^2\mathbf{j} + z^2\mathbf{k}}{(x^4 + y^4 + z^4)^{3/2}}$ and let σ be a closed, orientable surface that surrounds the origin. Then $\mathbf{\Phi} = \pi$.

7. Find the outward flux of $\mathbf{F}(x, y, z) = x\mathbf{i} + (2y + 3)\mathbf{j} + 6z^2\mathbf{k}$ across the unit cube in the first octant that has a vertex at the origin.

A. 1 B. 0 C. 9 D. 8

8. Find the outward flux of $\mathbf{F}(x, y, z) = x\mathbf{i} + y\mathbf{j} + (z - 2)\mathbf{k}$ across the rectangle with vertices $(0, 0, 0)$, $(0, 0, 1)$, $(2, 1, 0)$, and $(2, 1, 1)$.

A. 3 B. 6 C. 0 D. 1

9. Find the outward flux of $\mathbf{F}(x, y, z) = (x - 1)\mathbf{i} + (y - 3)\mathbf{j} + z\mathbf{k}$ across the rectangle with vertices $(0, 0, 0)$, $(0, 0, 1)$, $(2, 1, 0)$, and $(2, 1, 1)$.

A. 3 B. 6 C. 0 D. 1

10. Find the outward flux of $\mathbf{F}(x, y, z) = 2x^2\mathbf{i} + 3y\mathbf{j} + z\mathbf{k}$ across the rectangle with vertices $(0, 0, 0)$, $(0, 0, 1)$, $(2, 1, 0)$, and $(2, 1, 1)$.

A. 12 B. 20 C. 30 D. 7

11. Answer true or false: The outward flux of the vector field $\mathbf{F}(x, y, z) = \dfrac{x^3}{3}\mathbf{i} + \dfrac{y^3}{3}\mathbf{j} + \dfrac{z^3 - 1}{3}\mathbf{k}$ across the surface of the region that is enclosed by the hemisphere $z = \sqrt{16 - x^2 - y^2}$ and the plane $z = 0$ is $\dfrac{2\pi}{5}$.

12. Answer true or false: The outward flux of the vector field $\mathbf{F}(x, y, z) = x^2\mathbf{i} + 2y\mathbf{j} + z\mathbf{k}$ across the cube bounded by the axes and the planes $x = 2$, $y = 3$, and $z = 4$ is 120.

13. Answer true or false: The outward flux of the vector field $\mathbf{F}(x, y, z) = 2x^2\mathbf{i} + 2y^2\mathbf{j} + 2z^2\mathbf{k}$ across the cube bounded by the axes and the planes $x = 2$, $y = 3$, and $z = 4$ is 58.

14. Answer true or false: The outward flux of the vector field $\mathbf{F}(x, y, z) = (2x^2 + 3)\mathbf{i} + (y^2 + 2)\mathbf{j} + 2z^2\mathbf{k}$ across the cube bounded by the axes and the planes $x = 2$, $y = 3$, and $z = 4$ is 178.

15. Answer true or false: The outward flux of the vector field $\mathbf{F}(x, y, z) = x^3\mathbf{i} + y\mathbf{j} + z\mathbf{k}$ across the cube bounded by the axes and the planes $x = 2$, $y = 3$, and $z = 4$ is 192.

SECTION 16.8

1. Answer true or false: If σ is the surface $z = -x^2 - y^2 + 4$ and $\mathbf{F}(x, y, z) = 3x\mathbf{i} + 4y\mathbf{j} + z\mathbf{k}$, then $\displaystyle\iint_\sigma (\text{curl } \mathbf{F}) \cdot \mathbf{n}\, dS = \iint_R (6x + 8y + 1)\, dA$.

2. Answer true or false: If σ is the surface $z = -x^2 - y^2 + 4$ and $\mathbf{F}(x, y, z) = 5x\mathbf{i} + 2y\mathbf{j} + z^2\mathbf{k}$, then $\displaystyle\iint_\sigma (\text{curl } \mathbf{F}) \cdot \mathbf{n}\, dS = \iint_R (10x + 4y + 2z)\, dA$.

3. Answer true or false: If σ is the surface $z = -x^2 - y^2 + 4$ and $\mathbf{F}(x, y, z) = (4x + 2y)\mathbf{i} + (3x + 5y)\mathbf{j} + x\mathbf{k}$, then $\iint_\sigma (\text{curl } \mathbf{F}) \cdot \mathbf{n}\, dS = \iint_R (12x + 8y + z)\, dA$.

4. Answer true or false: If σ is the surface $z = -x^2 - y^2 + 4$ and $\mathbf{F}(x, y, z) = 6x\mathbf{i} + 2y\mathbf{j} + 3z\mathbf{k}$, then $\iint_\sigma (\text{curl } \mathbf{F}) \cdot \mathbf{n}\, dS = 0$.

5. Answer true or false: If σ is the surface $z = -x^2 - y^2 + 4$ and $\mathbf{F}(x, y, z) = xyz\mathbf{i} + xz\mathbf{j} + x^2yz\mathbf{k}$, then $\iint_\sigma (\text{curl } \mathbf{F}) \cdot \mathbf{n}\, dS = \iint_R ((2x^2(xz - 1) + 2y(xyz - 2xz) + z(1 - x))\, dA$.

6. Answer true or false: The amount of work needed to move a particle around the rectangle $(0, 0, 0)$, $(0, 2, 3)$, $(1, 2, 3)$, $(1, 0, 0)$, and back to $(0, 0, 0)$ by $\mathbf{F}(x, y, z) = xyz\mathbf{i} + yz\mathbf{j} + xy\mathbf{k}$ is $\int_0^2 \int_0^1 \frac{2(-y + xy)}{3} - x - 2y\, dy\, dx$.

7. The work done by $\mathbf{F}(x, y, z) = x\mathbf{i} + y\mathbf{j} + z\mathbf{k}$ to move a particle completely around the rectangle $(0, 0, 0)$, $(0, 0, 2)$, $(1, 0, 2)$, and $(1, 0, 0)$ is 0.

8. The work done by $\mathbf{F}(x, y, z) = x^2\mathbf{i} + y^3\mathbf{j} + z\mathbf{k}$ to move a particle completely around the quadrangle $(0, 0, 0)$, $(0, 1, 3)$, $(2, 1, 3)$, and $(2, 0, 0)$ is 0.

9. Answer true or false: If $\mathbf{F}(x, y, z) = x\mathbf{i} + 2y\mathbf{j} + z^2\mathbf{k}$ and σ is a surface such that $z = -x^2 - y^2 + 2$, where $z \geq 0$, $\iint_\sigma (\text{curl } \mathbf{F}) \cdot \mathbf{n}\, dS = \iint_R (-2x\mathbf{i} - 9y\mathbf{j} + 2z\mathbf{k})\, dA$.

10. Answer true or false: If $\mathbf{F}(x, y, z) = y\mathbf{i} + z\mathbf{j} + x\mathbf{k}$ and σ is a surface such that $z = -x^2 - y^2 + 2$, where $z \geq 0$, $\iint_\sigma (\text{curl } \mathbf{F}) \cdot \mathbf{n}\, dS = \iint_R (-2x + 2y + 1)\, dA$.

11. Answer true or false: If $\mathbf{F}(x, y, z) = y\mathbf{i} + z\mathbf{j} + x\mathbf{k}$ and σ is a surface such that $z = -\frac{x^2}{2} - \frac{y^2}{2} + 1$, where $z \geq 0$, $\iint_\sigma (\text{curl } \mathbf{F}) \cdot \mathbf{n}\, dS = \iint_R (-x + y + 1)\, dA$.

12. Answer true or false: If $\mathbf{F}(x, y, z) = y\mathbf{i} + z\mathbf{j} + x\mathbf{k}$ and σ is a surface such that $z = 2x + 3y + 4$, where $z \geq 0$, $\iint_\sigma (\text{curl } \mathbf{F}) \cdot \mathbf{n}\, dS = \iint_R (2x + 3y + z)\, dA$.

13. Answer true or false: $\mathbf{F}(x, y, z) = x^2\mathbf{i} + y^3\mathbf{j} + x^2z\mathbf{k}$. The work needed to move a particle around a triangle $(0, 0, 0)$, $(0, 2, 3)$, $(0, 0, 3)$ is 0.

14. Answer true or false: $\mathbf{F}(x, y, z) = \sin x\mathbf{i} + e^y\mathbf{j} + z^4\mathbf{k}$. The work needed to move a particle around a triangle $(0, 0, 0)$, $(0, 2, 3)$, $(0, 0, 3)$ is 0.

15. Answer true or false: $\mathbf{F}(x, y, z) = \sin y\mathbf{i} + e^x\mathbf{j} + z^4\mathbf{k}$. The work needed to move a particle around a triangle $(0, 0, 0)$, $(0, 2, 3)$, $(0, 0, 3)$ is 0.

CHAPTER 16 TEST

1. Answer true or false: $\phi(x, y) = xe^y$ is the potential function for $\mathbf{F}(x, y) = e^y\mathbf{i} + xe^y\mathbf{j}$.

2. $\mathbf{F}(x, y, z) = z\mathbf{i} + y\mathbf{j} + xyz\mathbf{k}$. Find div$\mathbf{F}$.

A. $1 + xy$ B. $1 - xy$ C. $xz + 1 - yz$ D. $xz + 1 - yz$

3. $\mathbf{F}(x, y, z) = x^3\mathbf{i} + 3\mathbf{j} + xz\mathbf{k}$. Find curl$\mathbf{F}$.

A. $\mathbf{j} + xy\mathbf{k}$ B. $\mathbf{j} - z\mathbf{j}$ C. $xz\mathbf{i} - z\mathbf{j}$ D. $xz\mathbf{i} - (1 - yz)\mathbf{j}$

4. Answer true or false: $\nabla^2\phi$, if $\phi = x^4y$, is $12x^2y + 4x^3$.

5. $\int_C (x^2y - 1)\,ds$, where $x = 3t$ and $y = t(0 \le t \le 1)$ is

A. 3.95 B. 1.75 C. 7 D. 1.

6. $\int_C 6xy^2 dx + 15xy^2\,dy$ along the curve $x = y^2$ from $(0, 0)$ to $(1, 1)$ is

A. 5 B. 21 C. -6 D. -2.

7. Answer true or false: If $x = \sin t$, $y = \cos t(0 \le t \le 2\pi)$, $\int_C (x - 2y)\,ds = \int_0^{2\pi} \sin t - 2\cos t\,dt$

8. Answer true or false: The work done by $\mathbf{F}(x, y) = x\mathbf{i} + ye^x\mathbf{j}$ along the curve $y = x^2$ from $(0, 0)$ to $(1, 1)$ is the same as the work done moving the same particle along the same curve from $(1, 1)$ to $(4, 2)$.

9. Answer true or false: $\mathbf{F}(x, y) = 6x\mathbf{i} + 2y\mathbf{j}$ is a conservative vector field.

10. Answer true or false: $\mathbf{F}(x, y) = 3x^2y\mathbf{i} + x^3\mathbf{j}$ is a conservative vector field.

11. $\int_{(0,1)}^{(2,4)} 6x\,dx + 2y\,dy =$

A. 12 B. 27 C. 28 D. 11

12. $\mathbf{F}(x, y) = 6x\mathbf{i} + y\mathbf{j}$ the work done by the force field on a particle moving along an arbitrary smooth curve from $P(0, 0)$ to $Q(1, 2)$ is

A. 7 B. $\frac{7}{2}$ C. 10 D. $-\frac{7}{2}$.

13. The area enclosed in the ellipse $\frac{x^2}{25} + \frac{y^2}{9} = 1$ is

A. 225 B. 225π C. 15 D. 15π.

14. The work done on a particle by the field $\mathbf{F}(x, y) = (\sin x + y^3)\mathbf{i} - (x^3 - \cos ye^y)\mathbf{j}$ on a particle that travels once around a unit circle $x^2 + y^2 = 1$ in a counterclockwise direction is

A. $\frac{3\pi}{4}$ B. $\frac{3\pi}{2}$ C. $\frac{\pi}{4}$ D. π.

15. $\oint_C (e^x + 2y)\,dx + (3y^4 + 2x)\,dy$, where C is $9x^2 + 4y^2 = 36$, is

A. π B. 12π C. 4π D. 0.

16. Evaluate $\iint_\sigma xz\,dS$ where σ is the part of the plane $x + y + \frac{z}{2} = 1$ in the first octant.

A. 0.204 B. 0.121 C. 0.131 D. 0.142

17. Answer true or false: If σ is the part of $x + y + z = 2$ that lies in the first octant, $\iint_\sigma xy\,dS = \sqrt{3}\int_0^1 \int_0^{1-x} x(1 - x - z)\,dz\,dx$.

18. Find the flux of the vector field $\mathbf{F}(x, y, z) = 4z\mathbf{k}$ across the sphere $x^2 + y^2 + z^2 = 9$ oriented outward.

A. $\frac{16\pi}{3}$ B. $\frac{64\pi}{3}$ C. 0 D. 144π

19. Answer true or false: If σ is the portion of the surface $z = 5 - x^2 - y^2$ that lies above the xy-plane, and σ is oriented up, the flux of the vector field $\mathbf{F}(x, y, z) = x\mathbf{i} + y\mathbf{j} + z\mathbf{k}$ across σ is $\Phi = \int_0^{2\pi} \int_0^5 (x^2 + y^2 + 5)\,dA$.

20. Answer true or false: The surface $z = 4x + 2y$ has a normal vector $\mathbf{n} = \frac{4\mathbf{i} + 2\mathbf{j} + \mathbf{k}}{\sqrt{21}}$.

21. Find the outward flux of the vector field $\mathbf{F}(x, y, z) = y\mathbf{j}$ across the sphere $x^2 + y^2 + z^2 = 4$.

A. $\frac{32\pi}{3}$ B. $\frac{256\pi}{3}$ C. 0 D. $\frac{16\pi}{3}$

22. Answer true or false: Let $\mathbf{F}(x, y, z) = \frac{3x\mathbf{i} + 3y\mathbf{j} + 3z\mathbf{k}}{(9x^2 + 9y^2 + 9z^2)^{3/2}}$ and σ be a closed orientable surface that surrounds the origin. Then $\Phi = 4\pi$.

23. Find the outward flux of $\mathbf{F}(x, y, z) = (x - 2)\mathbf{i} + (y - 3)\mathbf{j} + z\mathbf{k}$ across the rectangle with vertices $(2, 3, 0)$, $(2, 3, 1)$, $(4, 4, 0)$, and $(4, 4, 1)$.

A. 2 B. 4 C. 6 D. 8

24. Answer true or false: If σ is the surface $z = -x^2 - y^2 - 4$, and $\mathbf{F}(x, y, z) = 6x\mathbf{i} + 2y\mathbf{j} + z\mathbf{k}$, then $\iint_\sigma (\text{curl } \mathbf{F}) \cdot \mathbf{n}\,dS = \iint_R (12x + 4y + 1)\,dA$.

ANSWERS TO SAMPLE TESTS

SECTION 16.1

1. F 2. T 3. A 4. B 5. A 6. C 7. B 8. D 9. T 10. F 11. F 12. F
13. T 14. T 15. T

SECTION 16.2

1. B 2. C 3. D 4. A 5. D 6. C 7. A 8. A 9. F 10. T 11. F 12. F
13. A 14. F 15. F

SECTION 16.3

1. T 2. F 3. T 4. T 5. F 6. C 7. C 8. A 9. A 10. D 11. F 12. D
13. A 14. C 15. A

SECTION 16.4

1. A 2. D 3. D 4. B 5. B 6. A 7. A 8. D 9. D 10. C 11. C 12. C
13. F 14. T 15. F

SECTION 16.5

1. A 2. F 3. B 4. B 5. A 6. C 7. F 8. F 9. T 10. T 11. F 12. T
13. T 14. T 15. F

SECTION 16.6

1. A 2. A 3. F 4. F 5. A 6. F 7. A 8. F 9. F 10. C 11. C 12. T
13. T 14. F 15. T

SECTION 16.7

1. A 2. D 3. B 4. B 5. A 6. F 7. C 8. B 9. B 10. A 11. F 12. T
13. F 14. F 15. F

SECTION 16.8

1. F 2. F 3. F 4. T 5. T 6. F 7. T 8. F 9. F 10. F 11. F 12. F
13. F 14. T 15. F

CHAPTER 16 TEST

1. T 2. A 3. C 4. F 5. A 6. A 7. T 8. F 9. T 10. T 11. B 12. A
13. D 14. B 15. D 16. A 17. F 18. D 19. F 20. F 21. A 22. F 23. C
24. F